The Visualization Handbook

The Visualization Handbook

Edited by

Charles D. Hansen
Associate Director, Scientific Computing and Imaging Institute
Associate Professor, School of Computing
University of Utah
Salt Lake City, Utah

Chris R. Johnson
Director, Scientific Computing and Imaging Institute
Distinguished Professor, School of Computing
University of Utah
Salt Lake City, Utah

ELSEVIER
BUTTERWORTH
HEINEMANN

AMSTERDAM • BOSTON • HEIDELBERG • LONDON
NEW YORK • OXFORD • PARIS • SAN DIEGO
SAN FRANCISCO • SINGAPORE • SYDNEY • TOKYO

Elsevier Butterworth–Heinemann
30 Corporate Drive, Suite 400, Burlington, MA 01803, USA
Linacre House, Jordan Hill, Oxford OX2 8DP, UK

Library of Congress Cataloging-in-Publication Data
The visualization handbook / edited by Charles D. Hansen, Chris R. Johnson.
    p. cm.
  Includes bibliographical references and index.
  ISBN 0-12-387582-X
  1. Information visualization. 2. Computer graphics. I. Hansen, Charles D. II. Johnson, Chris R. III. Title.

TK7882.I6V59 2005
006.6—DC22

2004020457

British Library Cataloguing-in-Publication Data
A catalogue record for this book is available from the British Library.

ISBN: 0-12-387582-X

For information on all Elsevier Butterworth–Heinemann publications visit our Web site at www.books.elsevier.com

05 06 07 08 09 10 11 10 9 8 7 6 5 4 3 2 1

Printed in the United States of America

Contents

Contributors

James Ahrens (27, 36)
Advanced Computing Laboratory
Los Alamos National Laboratory
Los Alamos, New Mexico

Mihael Ankerst (43)
The Boeing Company
Seattle, Washington

Alan H. Barr (15)
Department of Computer Science
California Institute of Technology
Pasadena, California

Wes Bethel (29)
Lawrence Berkeley National Laboratory
Berkeley, California

David Borland (46)
Department of Computer Science
University of North Carolina at
    Chapel Hill
Chapel Hill, North Carolina

J. Dean Brederson (22)
Scientific Computing and Imaging Institute
University of Utah
Salt Lake City, Utah

Frederick P. Brooks, Jr. (46)
Department of Computer Science
University of North Carolina at Chapel Hill
Chapel Hill, North Carolina

Steve Bryson (21)
NASA Ames Research Center
Moffett Field, California

Kirsten Cater (41)
University of Bristol
Bristol, United Kingdom

Alan Chalmers (41)
University of Bristol
Bristol, United Kingdom

Jim X. Chen (24)
George Mason University
Fairfax, Virginia

Paolo Cignoni (4)
Istituto di Scienza e Tecnologie
    dell'Informazione
Consiglio Nazionale delle Ricerche
Pisa, Italy

Jonathan D. Cohen (20)
Johns Hopkins University
Baltimore, Maryland

Matthew D. Cooper (35)
Manchester Visualization Centre
The University of Manchester
Manchester, United Kingdom

Roger Crawfis (8)
The Ohio State University
Columbus, Ohio

David S. Ebert (39)
School of Electrical and Computer
    Engineering
Purdue University
West Lafayette, Indiana

Stephen G. Eick (42)
SSS Research
Warrenville, Illinois
National Center for Data Mining
University of Illinois
Chicago, Illinois

Gordon Erlebacher (12, 13)
Florida State University
Tallahassee, Florida

Thomas Ertl (10)
Visualization and Interactive Systems Group
University of Stuttgart
Stuttgart, Germany

Mike Falvo (46)
Curriculum on Applied and Materials Science
University of North Carolina at Chapel Hill
Chapel Hill, North Carolina

Jean M. Favre (33)
Swiss National Supercomputing Center
Manno, Switzerland

Berk Geveci (36)
Kitware, Inc.
Clifton Park, New York

Martin Guthold (46)
Department of Physics
Wake Forest University
Winston-Salem, North Carolina

Hans Hagen (19)
University of Kaiserslautern
Kaiserslautern, Germany

Charles D. Hansen (9)
Scientific Computing and Imaging Institute
University of Utah
Salt Lake City, Utah

Philip D. Heermann (25, 28)
Sandia National Laboratories
Albuquerque, New Mexico

Hans-Christian Hege (38)
Zuse Institute Berlin
Berlin, Germany

W. T. Hewitt (35)
Manchester Visualization Centre
The University of Manchester
Manchester, United Kingdom

Bill Hibbard (34)
University of Wisconsin
Madison, Wisconsin

Ingrid Hotz (19)
University of Kaiserslautern
Kaiserslautern, Germany

Tom Hudson (46)
Department of Computer Science
University of North Carolina at Wilmington
Wilmington, North Carolina

Milan Ikits (22)
Scientific Computing and Imaging Institute
University of Utah
Salt Lake City, Utah

Victoria Interrante (40)
Department of Computer Science and
    Engineering
University of Minnesota
Minneapolis, Minnesota

Takayuki Itoh (5)
IBM Japan Tokyo Research Laboratory
Tokyo, Japan

Kevin Jeffay (46)
Department of Computer Science
University of North Carolina at
    Chapel Hill
Chapel Hill, North Carolina

Ming Jiang (14)
Department of Computer and
    Information Science
The Ohio State University
Columbus, Ohio

Bruno Jobard (13)
Université de Pau
Pau, France

Nigel W. John (35)
Manchester Visualization Centre
The University of Manchester
Manchester, United Kingdom

Gail Jones (46)
School of Education
University of North Carolina at
    Chapel Hill
Chapel Hill, North Carolina

Greg M. Jones (31)
Scientific Computing and Imaging Institute
University of Utah
Salt Lake City, Utah

Arie Kaufman (7)
Center for Visual Computing
Stony Brook University
Stony Brook, New York

Daniel F. Keefe (45)
Department of Computer Science
Brown University
Providence, Rhode Island

Daniel A. Keim (43)
University of Konstanz
Konstanz, Germany

Gordon Kindlmann (9, 16)
Scientific Computing and Imaging Institute
University of Utah
Salt Lake City, Utah

Robert M. Kirby (45)
Scientific Computing and Imaging Institute
University of Utah
Salt Lake City, Utah

Joe Kniss (9)
Scientific Computing and Imaging Institute
University of Utah
Salt Lake City, Utah

Koji Koyamada (5)
Kyoto University Center for the Promotion
    of Excellence in Higher Education
Kyoto, Japan

Martin Kraus (10)
Visualization and Interactive Systems Group
University of Stuttgart
Stuttgart, Germany

K. Yien Kwok (35)
Manchester Visualization Centre
The University of Manchester
Manchester, United Kingdom

David H. Laidlaw (16, 45)
Department of Computer Science
Brown University
Providence, Rhode Island

Charles Law (36)
Kitware, Inc.
Clifton Park, New York

George W. Leaver (35)
Manchester Visualization Centre
The University of Manchester
Manchester, United Kingdom

Joanna M. Leng (35)
Manchester Visualization Centre
The University of Manchester
Manchester, United Kingdom

Paul G. Lever (35)
Manchester Visualization Centre
The University of Manchester
Manchester, United Kingdom

Yarden Livnat (2)
Scientific Computing and Imaging Institute
University of Utah
Salt Lake City, Utah

R. Bowen Loftin (24)
Old Dominion University
Norfolk, Virginia

Eric B. Lum (26)
University of California at Davis
Davis, California

Kwan-Liu Ma (26, 47)
University of California at Davis
Davis, California

Raghu Machiraju (14)
Department of Computer and
    Information Science
The Ohio State University
Columbus, Ohio

Dinesh Manocha (20)
University of North Carolina at
    Chapel Hill
Chapel Hill, North Carolina

David Marshburn (46)
Department of Computer Science
University of North Carolina at
    Chapel Hill
Chapel Hill, North Carolina

Kenneth M. Martin (1, 30)
Kitware, Inc.
Clifton Park, New York

Patrick McCormick (27)
Advanced Computing Laboratory
Los Alamos National Laboratory
Los Alamos, New Mexico

Mary J. McDerby (35)
Manchester Visualization Centre
The University of Manchester
Manchester, United Kingdom

Don Middleton (44)
National Center for Atmospheric
    Research
Boulder, Colorado

Claudio Montani (4)
Istituto di Scienza e Tecnologie
    dell'Informazione
Consiglio Nazionale delle Ricerche,
Pisa, Italy

Klaus Mueller (7)
Center for Visual Computing
Stony Brook University
Stony Brook, New York

Steven Parker (31)
Scientific Computing and
    Imaging Institute
University of Utah
Salt Lake City, Utah

Stergios J. Papadakis (46)
Department of Physics and Astronomy
University of North Carolina at
    Chapel Hill
Chapel Hill, North Carolina

Constantine Pavlakos (25, 28)
Sandia National Laboratories
Albuquerque, New Mexico

James S. Perrin (35)
Manchester Visualization Centre
The University of Manchester
Manchester, United Kingdom

Hanspeter Pfister (11)
Mitsubishi Electric Research Laboratories
Cambridge, Massachusetts

Enrico Puppo (4)
Dipartimento di Informatica e Scienze
    dell'Informazione
Universitá degli Studi di Genova
Genova, Italy

Lu-Chang Qin (46)
Department of Physics and Astronomy and
    Curriculum on Applied and Materials Science
University of North Carolina at Chapel Hill
Chapel Hill, North Carolina

William Ribarsky (23)
College of Computing
Georgia Institute of Technology
Atlanta, Georgia

Mark Riding (35)
Manchester Visualization Centre
The University of Manchester
Manchester, United Kingdom

Warren Robinett (46)
http://www.warrenrobinett.com

Larry Rosenblum (24)
U.S. Naval Research Laboratory
Washington, DC

Jarek Rossignac (18)
College of Computing and Graphics,
    Visualization, Usability Center
Georgia Institute of Technology
Atlanta, Georgia

I. Ari Sadarjoen (35)
Manchester Visualization Centre
The University of Manchester
Manchester, United Kingdom

Tim Scheitlin (44)
National Center for Atmospheric
    Research
Boulder, Colorado

Gerik Scheuermann (17)
University of Kaiserslautern
Kaiserslautern, Germany

Tobias M. Schiebeck (35)
Manchester Visualization Centre
The University of Manchester
Manchester, United Kingdom

William J. Schroeder (1, 30)
Kitware, Inc.
Clifton Park, New York

Greg Schussman (47)
University of California at Davis
Davis, California

Roberto Scopigno (4)
Istituto di Scienza e Tecnologie
    dell'Informazione
Consiglio Nazionale delle Ricerche
Pisa, Italy

Adam Seeger (46)
Department of Computer Science
University of North Carolina at
    Chapel Hill
Chapel Hill, North Carolina

John Shalf (29)
Lawrence Berkeley National Laboratory
Berkeley, California

Mike Sips (43)
University of Konstanz
Konstanz, Germany

Han-Wei Shen (3)
Department of Computer Science and
    Engineering
The Ohio State University
Columbus, Ohio

Jenny Simpson (31)
Scientific Computing and Imaging Institute
University of Utah
Salt Lake City, Utah

F. Donelson Smith (46)
Department of Computer Science
University of North Carolina at
    Chapel Hill
Chapel Hill, North Carolina

Dianne Sonnenwald (46)
School of Information and
    Library Science
University of North Carolina at
    Chapel Hill
Chapel Hill, North Carolina

Detlev Stalling (38)
Zuse Institute Berlin
Berlin, Germany

Richard Superfine (46)
Department of Physics and Astronomy and
    Curriculum on Applied and
    Materials Science
University of North Carolina at Chapel Hill
Chapel Hill, North Carolina

Russell M. Taylor II (46)
Departments of Computer Science and
    Physics and Astronomy and Curriculum
    on Applied and Materials Science
University of North Carolina at Chapel Hill
Chapel Hill, North Carolina

David Thompson (14)
Department of Aerospace Engineering
Mississippi State University
Mississippi State, Mississippi

Xavier Tricoche (17)
University of Kaiserslautern
Kaiserslautern, Germany

Mario Valle (33)
Swiss National Supercomputing Center
Manno, Switzerland

Colin C. Venters (35)
Manchester Visualization Centre
The University of Manchester
Manchester, United Kingdom

Leandra Vicci (46)
Department of Computer Science
University of North Carolina at Chapel Hill
Chapel Hill, North Carolina

Jeremy Walton (32)
The Numerical Algorithms Group, Ltd.
Oxford, United Kingdom

Sean Washburn (46)
Curriculum on Applied and Materials Science
    and Department of Physics and Astronomy
University of North Carolina at Chapel Hill
Chapel Hill, North Carolina

Chris Weigle (46)
Department of Computer Science
University of North Carolina at Chapel Hill
Chapel Hill, North Carolina

David M. Weinstein (31)
Scientific Computing and Imaging Institute
University of Utah
Salt Lake City, Utah

Daniel Weiskopf (12, 13)
University of Stuttgart
Stuttgart, Germany

Malte Westerhoff (38)
Zuse Institute Berlin
Berlin, Germany

Ross T. Whitaker (6)
School of Computing
University of Utah
Salt Lake City, Utah

Mary Whitton (46)
Department of Computer Science
University of North Carolina at
    Chapel Hill
Chapel Hill, North Carolina

Bob Wilhelmson (44)
National Center for Supercomputing
    Applications
University of Illinois
Champaign, Illinois

Phillip Williams (46)
NASA Langley Research Center
Hampton, Virginia

Brett Wilson (47)
University of California at Davis
Davis, California

Daqing Xue (8)
The Ohio State Unversity
Columbus, Ohio

Terry S. Yoo (37)
Office of High Performance Computing and
    Communications
The National Library of Medicine, National
    Institutes of Health
Bethesda, Maryland

Caixia Zhang (8)
The Ohio State University
Columbus, Ohio

Song Zhang (16)
Department of Computer Science
Brown University
Providence, Rhode Island

Leonid Zhukov (15)
Department of Computer Science
California Institute of Technology
Pasadena, California

Kurt Zimmerman (31)
Scientific Computing and Imaging Institute
University of Utah
Salt Lake City, Utah

Preface

The field of visualization is focused on creating images that convey salient information about underlying data and processes. In the past three decades, the field has seen unprecedented growth in computational and acquisition technologies, which has resulted in an increased ability both to sense the physical world with very detailed precision and to model and simulate complex physical phenomena. Given these capabilities, visualization plays a crucial enabling role in our ability to comprehend such large and complex data—data that, in two, three, or more dimensions, conveys insight into such diverse applications as medical processes, earth and space sciences, complex flow of fluids, and biological processes, among many other areas.

The field was aptly described in the 1987 National Science Foundation's Visualization in Scientific Computing Workshop report, which explained:

> Visualization is a method of computing. It transforms the symbolic into the geometric, enabling researchers to *observe* their simulations and computations. Visualization offers a method for seeing the unseen. It enriches the process of scientific discovery and fosters profound and unexpected insights. In many fields it is already revolutionizing the way scientists do science... The goal of visualization is to leverage existing scientific methods by providing new scientific insight through visual methods.

While visualization is a relatively young field, the goal of visualization—that is, the creation of a visual representation to help explain complex phenomena—is certainly not new. One has only to look at the Da Vinci notebooks to understand the great power of illustration to bring out salient details of complex processes. Another fine example, the drawing by Charles Minard (1781–1870) of the ill-fated Russian

campaign by Napoleon's troops, elegantly incorporates both spatial and temporal data in a comprehensive visualization created by drawing the sequence of events and the resulting effects on the troop size.

The discipline of visualization as it is currently understood was born with the advent of scientific computing and the use of computer graphics for depicting computational data. Simultaneously, devices capable of sensing the physical world, from medical scanners to geophysical sensing to satellite-borne sensing, and the need to interpret the vast amount of data either computed or acquired, have also driven the field. In addition to the rapid growth in visualization of scientific and medical data, data that typically lacks a spatial domain has caused the rise of the field of information visualization.

With this Handbook, we have tried to compile a thorough overview of our young field by presenting the basic concepts of visualization, providing a snapshot of current visualization software systems, and examining research topics that are advancing the field.

We have organized the book into parts to reflect a taxonomy we use in our teaching to explain scientific visualization: basic visualization algorithms, scalar data isosurface methods, scalar data volume rendering, vector data, tensor data, geometric modeling, virtual environments, large-scale data, visualization software and frameworks, perceptual issues, and selected application topics including information visualization. While, as we say, this taxonomy represents topics covered in a standard visualization course, this Handbook is not meant to serve as a textbook. Rather, it is meant to reach a broad audience, including not only the expert in visualization seeking advanced methods to solve a particular problem but also the novice looking for general background information on visualization topics.

I. Introduction

Part I looks at basic algorithms for scientific visualization. In practice, a typical algorithm may be thought of as a transformation from one data form into another. These operations may also change the dimensionality of the data. For example, generating a streamline from a specified starting point in an input 3D dataset produces a 1D curve. The input may be represented as a finite element mesh, while the output may be a represented as a polyline. Such operations are typical of scientific visualization systems that repeatedly transform data into different forms and ultimately into a representation that can be rendered by the computer system.

II. Scalar Field Visualization: Isosurfaces

The analysis of scalar data benefits from the extraction of lines (2D) or surfaces (3D) of constant value. As described in Part I, *marching cubes* is the most widely used method for the extraction of isosurfaces. In this section, methods for the acceleration of isosurface extraction are presented by the various contributors. Yarden Livnat introduces the span space, a representation of acceleration of isosurfaces. Based on this concept, methods that use the span space are described. Han-Wei Shen presents a method for exploiting temporal locality for isosurface extraction, in recognition of the fact that temporal information is becoming increasingly crucial to comprehension of time-dependent scalar fields. Roberto Scopigno, Paolo Cignoni, Claudio Montani, and Enrico Puppo present a method for optimally using the span space for isosurface extraction based on the interval tree. Koji Koyamada and Takayuki Itoh describe a method for isosurface extraction based on the extrema graph. To conclude this section, Ross Whitaker presents an overview of level-sets and their relation to isosurface extraction.

III. Scalar Field Visualization: Volume Rendering

Direction scalar field visualization is accomplished with volume rendering, which produces an image directly from the data without an intermediate geometrical representation. Arie Kaufman and Klaus Mueller provide an excellent survey of volume rendering algorithms. Roger Crawfis, Daqing Xue, and Caixia Zhang provide a more detailed look at the splatting method for volume rendering. Joe Kniss, Gordon Kindlmann, and Chuck Hansen describe how to exploit multidimensional transfer functions for extracting the material boundaries of objects. Martin Kraus and Thomas Ertl describe a method by which volume rendering can be accelerated through the precomputation of the volume integral. Finally, Hanspeter Pfister provides an overview of another approach to the acceleration of volume rendering, the use of hardware methods.

IV. Vector Field Visualization

Flow visualization is an important topic in scientific visualization and has been the subject of active research for many years. Typically, data originates from numerical simulations, such as those of computational fluid dynamics (CFD), and must be analyzed by means of visualization to provide an understanding of the flow. Daniel Weiskopf and Gordon Erlebacher present an overview of such methods, including a specific technique for the rapid visualization of flow data that exploits hardware available on most graphics cards. Gordon Erlebacher, Bruno Jobard, and Daniel Weiskopf describe their method for flow textures in the next chapter. Ming Jiang, Raghu Machiraju, and David Thompson then provide an overview and solution to the problem of gaining insight into flow fields through the localization of vortices.

V.   Tensor Field Visualization

Computational and sensed data can also represent tensor information. The visualization of such fields is the topic of this part. Leonid Zhukov and Alan Barr describe the reconstruction of oriented tensors in a method similar to streamlines for vector fields. Song Zhang, Gordon Kindlmann, and David Laidlaw describe the use of visualization methods for the analysis of Diffusion Tensor Magnetic Resonance Imaging (DT-MRI or DTI). Finally, Gerik Scheuermann and Xavier Tricoche describe a more abstract representation of tensor fields through the use of topological methods.

VI.   Geometric Modeling for Visualization

Geometric modeling plays an important role in visualization. For example, in the first chapter of this part, Jarek Rossignac describes techniques for the compression of 3D meshes, which can be enormous and which are commonly used to represent isosurfaces. Next, Hans Hagen and Ingrid Hotz present the basic principles of variational modeling techniques, already powerful tools for free-form modeling in CAD/CAM whose basic principles are now being imported for use in scientific visualization. To complete this part, Jonathan Cohen and Dinesh Manocha give an overview of model simplification, which is critical for interactive applications.

VII.   Virtual Environments for Visualization

Virtual environments provide a natural interface to 3D data. They are becoming more prevalent in the visualization field. Steve Bryson describes the use of direction manipulation as a modality of data interaction in the visualization process. Milan Ikits and Dean Brederson explore the use of haptics in visualization. Bill Ribarsky describes how geographic information systems can benefit from a virtual environment interface. And, in the last chapter in this section, Bowen Loftin provides an overview of virtual environments for visualization.

VIII.   Large-Scale Data Visualization

With the dramatic increase in computational capabilities in recent years, the problem of visualization of the massive datasets produced by computation is an active area of research. Philip Heermann and Constantine Pavlakos describe the problems involved in providing scientists with access to such enormous data. Kwan-Liu Ma and Eric Lum explore methods for time-varying scalar data. Patrick McCormick and James Ahrens present an analytical approach to large data visualization, describing their own method, which identifies four fundamental techniques for addressing the large-data problem. Constantine Pavlakos and Philip Heermann give an overview of the large-data problem from the DOE ASCI perspective. Finally, Wes Bethel and John Shalf present a GRID method for the visualization of large data across wide-area networks.

IX.   Visualization Software and Frameworks

There are many visualization packages available to assist scientists and developers in the analysis of data. Several of these are described in this part.

X.   Perceptual Issues in Visualization

Since the primary purpose of visualization is to convey information to users, it is important that visualizers understand and address issues involving perception. To open this section, David Ebert describes the importance of perception in visualization. Next, Victoria Interrante explores ways in which art and science have been combined since the Renaissance to produce inspirational results. In the last chapter of this section, Alan Chalmers and Kirsten Cater discuss the exploitation of human visual perception in visualization to produce more effective results.

XI. Selected Topics and Applications

The visualization of nonspatial data is becoming increasingly important. Methods for such visualization are known as information visualization techniques. This section presents two applications that employ information visualization: the visualization of networks and data mining. Stephen G. Eick defines the concept of visual scalability for the visualization of very large networks, illustrates it with three examples, and describes techniques to increase network visualization scalability. Information visualization and visual data mining can help with the exploration and analysis of the current flood of information facing modern society.

Daniel Keim, Mike Sips, and Mihael Ankerst provide an overview of information visualization and visual data-mining techniques, using examples to elucidate those techniques.

Weather and climate research is an area that has traditionally employed visualization techniques. Don Middleton, Bob Wilhelmson, and Tim Scheitlin describe an overview of this application area. Then Robert Kirby, Daniel Keefe, and David Laidlaw explore the relationship between art and visualization, building on their work in layering of information for visualization. The research group at the University of North Carolina at Chapel Hill describes the use of visualization to assist in providing users with the fine motor control required by modern microscopy instruments. In the last chapter of this section, Kwan-Liu Ma, Greg Schussman, and Brett Wilson present several novel techniques for using computational accelerator physics as an application area for visualization.

Acknowledgments

This book is the result of a multiyear effort of collecting material from the leaders in the field. It has been a pleasure working with the chapter authors, though as always the book has taken longer to bring to publication than we anticipated. We appreciate the contributors' patience. We would like to thank Donna Prisbrey and Piper Bessinger-West for their superb administration skills, without which this Handbook would not have seen the light of day. We also would like to thank all the students, staff, and faculty of the SCI Institute for making each and every day an exciting intellectual adventure.

PART I
Introduction

1   Overview of Visualization

WILLIAM J. SCHROEDER and KENNETH M. MARTIN
Kitware, Inc.

1.1   Introduction

In this chapter, we look at basic algorithms for scientific visualization. In practice, a typical algorithm can be thought of as a transformation from one data form into another. These operations may also change the dimensionality of the data. For example, generating a streamline from a specification of a starting point in an input 3D dataset produces a 1D curve. The input may be represented as a finite element mesh, while the output may be represented as a polyline. Such operations are typical of scientific visualization systems that repeatedly transform data into different forms and ultimately transform it into a representation that can be rendered by the computer system.

The algorithms that transform data are the heart of data visualization. To describe the various transformations available, we need to categorize algorithms according to the *structure* and *type* of transformation. By *structure*, we mean the effects that transformation has on the topology and geometry of the dataset. By *type*, we mean the type of dataset that the algorithm operates on.

Structural transformations can be classified in four ways, depending on how they affect the geometry, topology, and attributes of a dataset. Here, we consider the topology of the dataset as the relationship of discrete data samples (one to another) that are invariant with respect to geometric transformation. For example, a regular, axis-aligned sampling of data in three dimensions is referred to as a *volume*, and its topology is a rectangular (structured) lattice with clearly defined

neighborhood voxels and samples. On the other hand, the topology of a finite element mesh is represented by an (unstructured) list of elements, each defined by an ordered list of points. Geometry is a specification of the topology in space (typically 3D), including point coordinates and interpolation functions. Attributes are data associated with the topology and/or geometry of the dataset, such as temperature, pressure, or velocity. Attributes are typically categorized as being scalars (single value per sample), vectors (*n*-vector of values), tensor (matrix), surface normals, texture coordinates, or general field data. Given these terms, the following transformations are typical of scientific visualization systems:

- *Geometric transformations* alter input geometry but do not change the topology of the dataset. For example, if we translate, rotate, and/or scale the points of a polygonal dataset, the topology does not change, but the point coordinates, and therefore the geometry, do.

- *Topological transformations* alter input topology but do not change geometry and attribute data. Converting a dataset type from polygonal to unstructured grid, or from image to unstructured grid, changes the topology but not the geometry. More often, however, the geometry changes whenever the topology does, so topological transformation is uncommon.

- *Attribute transformations* convert data attributes from one form to another, or create new attributes from the input data. The structure of the dataset remains unaffected.

Text and images taken with permission from the book *The Visualization Toolkit: An Object-Oriented Approach to 3D Graphics*, 3rd ed., published by Kitware, Inc. http://www.kitware.com/products/vtktextbook.html.

Computing vector magnitude and creating scalars based on elevation are data attribute transformations.

- *Combined transformations* change both dataset structure and attribute data. For example, computing contour lines or surfaces is a combined transformation.

We also may classify algorithms according to the *type* of data they operate on. The meaning of the word "type" is often somewhat vague. Typically, "type" means the type of attribute data, such as scalars or vectors. These categories include the following:

- *Scalar algorithms* operate on scalar data. An example is the generation of contour lines of temperature on a weather map.
- *Vector algorithms* operate on vector data. Showing oriented arrows of airflow (direction and magnitude) is an example of vector visualization.
- *Tensor algorithms* operate on tensor matrices. One example of a tensor algorithm is to show the components of stress or strain in a material using oriented icons.
- *Modeling algorithms* generate dataset topology or geometry, or surface normals or texture data. "Modeling algorithms" tends to be the catch-all category for algorithms that do not fit neatly into any single category mentioned above. For example, generating glyphs oriented according to the vector direction and then scaled according to the scalar value is a combined scalar/vector algorithm. For convenience, we classify such an algorithm as a modeling algorithm because it does not fit squarely into any other category.

Note that an alternative classification scheme is to refer to the topological type of the input data (e.g., image, volume, or unstructured mesh) that a particular algorithm operates on. In the remainder of the chapter we will classify the type of the algorithm as the type of attribute data on which it operates. Be forewarned, though, that alternative classification schemes do exist and may be better suited to describing the true nature of the algorithm.

1.1.1 Generality Vs. Efficiency

Most algorithms can be implemented specifically for a particular data type or, more generally, for treating any data type. The advantage of a specific algorithm is that it is usually faster than a comparable general algorithm. An implementation of a specific algorithm may also be more memory-efficient, and it may better reflect the relationship between the algorithm and the dataset type it operates on.

One example of this is contour surface creation. Algorithms for extracting contour surfaces were originally developed for volume data, mainly for medical applications. The regularity of volumes lends itself to efficient algorithms. However, the specialization of volume-based algorithms precludes their use for more general datasets such as structured or unstructured grids. Although the contour algorithms can be adapted to these other dataset types, they are less efficient than those for volume datasets.

The presentation of algorithms in this chapter favors more general implementations. In some special cases, authors will describe performance-improving techniques for particular dataset types. Various other chapters in this book also include detailed descriptions of specialized algorithms.

1.1.2 Algorithms as Filters

In a typical visualization system, algorithms are implemented as filters that operate on data. This approach is due in some part to the success of early systems like the Application Visualization System [2] and Data Explorer [9] and the popularity of systems like SCIRun [37] and the Visualization Toolkit [36] that are built around the abstraction of data flow. This abstraction is natural because of the transformative nature of visualization. The basic idea is that two types of objects—data objects and process objects—are connected together into visualization pipelines.

The process objects, or filters, are the algorithms that operate on the data objects and in turn produce data objects as output. In this abstraction, filters that initiate the pipeline are referred to as *sources*; filters that terminate the pipeline are known as *sinks* (or *mappers*). Depending on their particular implementation, filters may have multiple inputs and/or may produce multiple outputs.

1.2 Scalar Algorithms

Scalars are single data values associated with each point and/or cell of a dataset. Because scalar data is commonly found in real-world applications, and because it is so easy to work with, there are many different algorithms to visualize it.

1.2.1 Color Mapping

Color mapping is a common scalar visualization technique that maps scalar data to colors and displays the colors using the standard coloring and shading facilities of the graphics library. The scalar mapping is implemented by indexing into a *color lookup table*. Scalar values serve as indices into the lookup table.

The mapping proceeds as follows. The lookup table holds an array of colors (e.g., red, green, blue, and alpha transparency components or other comparable representations). Associated with the table is a minimum and maximum *scalar range (min, max)* into which

the scalar values are mapped. Scalar values greater than the maximum range are clamped to the maximum color, and scalar values less than the minimum range are clamped to the minimum color value. For each scalar value s_i, the index i into the color table with n entries (and 0-offset) is given by Fig. 1.1.

A more general form of the lookup table is called a *transfer function*. A transfer function is any expression that maps scalar value into a color specification. For example, Fig. 1.2 maps scalar values into separate intensity values for the red, green, and blue color components. We can also use transfer functions to map scalar data into other information, such as local transparency. A lookup table is a discrete sampling of a transfer function. We can create a lookup table from any transfer function by sampling the transfer function at a set of discrete points.

Color mapping is a 1D visualization technique. It maps one piece of information (i.e., a scalar value) into a color specification. However, the display of color information is not limited to 1D displays. Often the colors are mapped onto 2D or 3D objects. This is a simple way to increase the information content of the visualizations.

The key to color mapping for scalar visualization is to choose the lookup table entries carefully. Fig. 1.3 shows four different lookup tables used to visualize gas density as fluid flows through a combustion chamber. The first lookup table is grey-scale. Grey-scale tables often provide better structural detail to the eye.

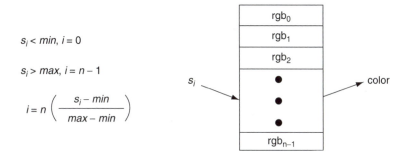

$$s_i < min, i = 0$$

$$s_i > max, i = n - 1$$

$$i = n \left(\frac{s_i - min}{max - min} \right)$$

Figure 1.1 Mapping scalars to colors via a lookup table.

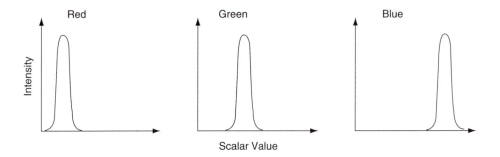

Figure 1.2   Transfer function for color components red, green, and blue as a function of scalar value.

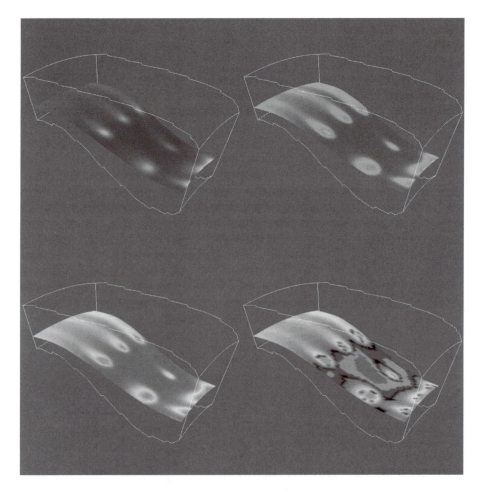

Figure 1.3   Flow density colored with different lookup tables. (Top left) Grey-scale; (top right) rainbow (blue to red); (lower left) rainbow (red to blue); (lower right) large contrast. (See also color insert.)

The other three images in Fig. 1.3 use different color lookup tables. The second uses rainbow hues from blue to red. The third uses rainbow hues arranged from red to blue. The last image uses a table designed to enhance contrast. Careful use of colors can often enhance important features of a dataset. However, any type of lookup table can exaggerate unimportant details or create visual artifacts because of unforeseen interactions among data, color choice, and human physiology.

Designing lookup tables is as much an art as it is a science. From a practical point of view, tables should accentuate important features while minimizing less important or extraneous details. It is also desirable to use palettes that inherently contain scaling information. For example, a color rainbow scale from blue to red is often used to represent temperature scale, since many people associate blue with cold temperatures and red with hot temperatures. However, even this scale is problematic: a physicist would say that blue is hotter than red, since hotter objects emit more blue (i.e., shorter-wavelength) light than red. Also, there is no need to limit ourselves to "linear" lookup tables. Even though the mapping of scalars into colors has been presented as a linear operation (Fig. 1.1), the table itself need not be linear; that is, tables can be designed to enhance small variations in scalar value using logarithmic or other schemes.

1.2.2 Contouring

One natural extension to color mapping is *contouring*. When we see a surface colored with data values, the eye often separates similarly colored areas into distinct regions. When we contour data, we are effectively constructing the boundary between these regions. A particular boundary can be expressed as the n-dimensional separating surfaces

$$F(\overline{x}) = c \qquad (1.1)$$

between the two regions $F(\overline{x}) < c$ and $F(\overline{x}) > c$, where c is the *contour value* and $\overline{x}$ is an n-dimen-

sional point in the dataset. These two regions are typically referred to as the *inside* or *outside* regions of the contour.

Examples of 2D contour displays include weather maps annotated with lines of constant temperature (isotherms) or topological maps drawn with lines of constant elevation. 3D contours are called *isosurfaces* and can be approximated by many polygonal primitives. Examples of isosurfaces include constant medical image intensity corresponding to body tissues such as skin, bone, or other organs. Other abstract isosurfaces, such as surfaces of constant pressure or temperature in fluid flow, may also be created.

Consider the 2D structured grid shown in Fig. 1.4. Scalar values are shown next to the points that define the grid. Contouring always begins when one specifies a contour value defining the contour line or surface to be generated. To generate the contours, some form of interpolation must be used. This is because we have scalar values at a discrete set of (sample) points in the dataset, and our contour value may lie between the point values. Since the most common interpolation technique is linear, we generate points on the contour surface by linear interpolation along the edges. If an edge has scalar values 10 and 0 at its two endpoints, for example, and if we are trying to generate a contour line of value 5, then edge interpolation computes that

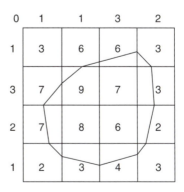

Figure 1.4   Contouring a 2D structured grid with contour line value = 5.

the contour passes through the midpoint of the edge.

Once the points on cell edges are generated, we can connect these points into contours using a few different approaches. One approach detects an edge intersection (i.e., the passing of a contour through an edge) and then "tracks" this contour as it moves across cell boundaries. We know that if a contour edge enters a cell, it must exit a cell as well. The contour is tracked until it closes back on itself or exits a dataset boundary. If it is known that only a single contour exists, then the process stops. Otherwise, every edge in the dataset must be checked to see whether other contour lines exist.

Another approach uses a divide-and-conquer technique, treating cells independently. This is called the *marching squares* algorithm in 2D and the *marching cubes* algorithm [23] in 3D. The basic assumption of these techniques is that a contour can pass through a cell in only a finite number of ways. A case table is constructed that enumerates all possible topological *states* of a cell, given combinations of scalar values at the cell points. The number of topological states depends on the number of cell vertices and the number of inside/outside relationships a vertex can have with respect to the contour value. A vertex is considered inside a contour if its scalar value is larger than the scalar value of the contour line. Vertices with scalar values less than the contour value are said to be outside the contour. For example, if a cell has four vertices

and each vertex can be either inside or outside the contour, there are $2^4 = 16$ possible ways that the contour passes through the cell. In the case table, we are not interested in where the contour passes through the cell (e.g., geometric intersection), just how it passes through the cell (i.e., topology of the contour in the cell).

Fig. 1.5 shows the 16 combinations for a square cell. An index into the case table can be computed by encoding the state of each vertex as a binary digit. For 2D data represented on a rectangular grid, we can represent the 16 cases with a 4-bit index. Once the proper case is selected, the location of the contour line/cell edge intersection can be calculated using interpolation. The algorithm processes a cell and then moves, or *marches*, to the next cell. After all the cells are visited, the contour will be completed. In summary, the marching algorithms proceed as follows:

1. Select a cell.

2. Calculate the inside/outside state of each vertex of the cell.

3. Create an index by storing the binary state of each vertex in a separate bit.

4. Use the index to look up the topological state of the cell in a case table.

5. Calculate the contour location (via interpolation) for each edge in the case table.

This procedure will construct independent geometric primitives in each cell. At the cell

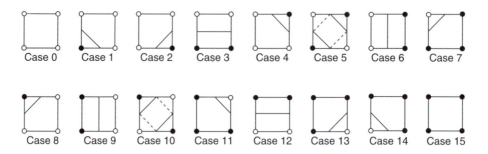

Figure 1.5 Sixteen different marching squares cases. Dark vertices indicate scalar value is above contour value. Cases 5 and 10 are ambiguous.

boundaries, duplicate vertices and edges may be created. These duplicates can be eliminated by use of a special coincident point-merging operation. Note that interpolation along each edge should be done in the same direction. If it is not, numerical round-off will likely cause points to be generated that are not precisely coincident and will thus not merge properly.

There are advantages and disadvantages to both the edge-tracking and the marching cubes approaches. The marching squares algorithm is easy to implement. This is particularly important when we extend the technique into three dimensions, where isosurface tracking becomes much more difficult. On the other hand, the algorithm creates disconnected line segments and points, and the required merging operation requires extra computation resources. The tracking algorithm can be implemented to generate a single polyline per contour line, avoiding the need to merge coincident points.

As mentioned previously, the 3D analogy of marching squares is marching cubes. Here, there are 256 different combinations of scalar value, given that there are eight points in a cubical cell (i.e., 2^8 combinations). Figure 1.6 shows these combinations reduced to 15 cases by arguments of symmetry. We use combinations of rotation and mirroring to produce topologically equivalent cases. (This is the so-called marching cubes case table.)

An important issue is *contouring ambiguity*. Careful observation of marching squares cases 5 and 10 and marching cubes cases 3, 6, 7, 10, 12, and 13 show that there are configurations where a cell can be contoured in more than one way. (This ambiguity also exists in an edge-tracking approach to contouring.) Contouring ambiguity arises on a 2D square or the face of a 3D cube when adjacent edge points are in different states but diagonal vertices are in the same state.

In two dimensions, contour ambiguity is simple to treat: for each ambiguous case, we implement one of the two possible cases. The choice for a particular case is independent of all other choices. Depending on the choice, the

contour may either extend or break the current contour, as illustrated in Fig. 1.8. Either choice is acceptable since the resulting contour lines will be continuous and closed (or will end at the dataset boundary).

In three dimensions the problem is more complex. We cannot simply choose an ambiguous case independent of all other ambiguous cases. For example, Fig. 1.9 shows what happens if we carelessly implement two cases independent of one another. In this figure we have used the usual case 3 but replaced case 6 with its *complementary* case. Complementary cases are formed by exchanging the "dark" vertices with "light" vertices. (This is equivalent to swapping vertex scalar value from above the isosurface value to below the isosurface value, and vice versa.) The result of pairing these two cases is that a hole is left in the isosurface.

Several different approaches have been taken to remedy this problem. One approach tessellates the cubes with tetrahedra and uses a *marching tetrahedra* technique. This works because the marching tetrahedra exhibit no ambiguous cases. Unfortunately, the marching tetrahedra algorithm generates isosurfaces consisting of more triangles, and the tessellation of a cube with tetrahedra requires one to make a choice regarding the orientation of the tetrahedra. This choice may result in artificial "bumps" in the isosurface because of interpolation along the face diagonals, as shown in Fig. 1.7. Another approach evaluates the asymptotic behavior of the surface and then chooses the cases to either join or break the contour. Nielson and Hamann [28] have developed a technique based on this approach that they call the *asymptotic decider*. It is based on an analysis of the variation of the scalar variable across an ambiguous face. The analysis determines how the edges of isosurface polygons should be connected.

A simple and effective solution extends the original 15 marching cubes cases by adding additional complementary cases. These cases are designed to be compatible with neighboring cases and prevent the creation of holes in the

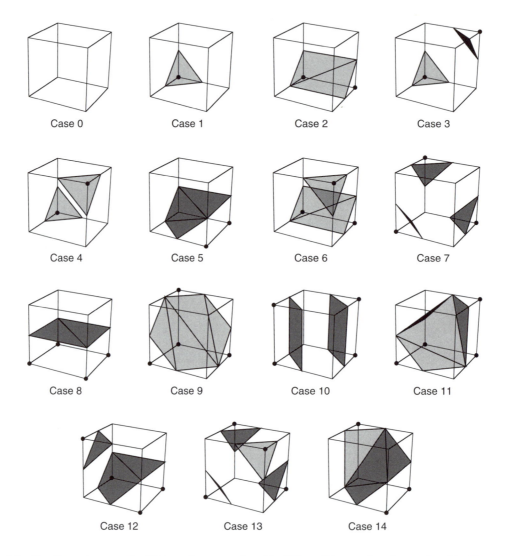

Figure 1.6 Marching cubes cases for 3D isosurface generation. The 256 possible cases have been reduced to 15 cases using symmetry. Vertices with a dot are greater than the selected isosurface value.

isosurface. There are six complementary cases required, corresponding to the marching cubes cases 3, 6, 7, 10, 12, and 13. The complementary marching cubes cases are shown in Fig. 1.10. In practice the simplest approach is to create a case table consisting of all 256 possible combinations and to design them in such a way as to prevent holes. A successful marching cubes case table will always produce manifold surfaces (i.e., interior edges are used by exactly two triangles; boundary edges are used by exactly one triangle).

We can extend the general approach of marching squares and marching cubes to other topological types such as triangles, tetrahedra, pyramids, and wedges. In addition, although we refer to regular types such as squares and cubes, marching cubes can be applied to any cell type

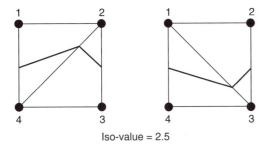

Iso-value = 2.5

Figure 1.7 Using marching triangles or marching tetrahedra to resolve ambiguous cases on rectangular lattice (only the face of the cube is shown). Choice of diagonal orientation can result in "bumps" in the contour surface. In two dimensions, diagonal orientation can be chosen arbitrarily, but in three dimensions the diagonal is constrained by the neighbor.

topologically equivalent to a cube (e.g., a hexahedron or noncubical voxel).

Fig. 1.11 shows four applications of contouring. In Fig. 1.11a we see 2D contour lines of CT density value corresponding to different tissue types. These lines were generated using marching squares. Figs 1.11b through 1.11d are isosurfaces created by marching cubes. Fig. 1.11b is a surface of constant image intensity from a computed tomography (CT) x-ray imaging system. (Fig. 1.11a is a 2D subset of these data.) The intensity level corresponds to human bone. Fig. 1.11c is an isosurface of constant flow density. Figure 1.11d is an isosurface of electron potential of an iron protein molecule. The image shown in Fig. 1.11b is immediately recognizable because of our fa-

miliarity with human anatomy. However, for those practitioners in the fields of computational fluid dynamics (CFD) and molecular biology, Figs. 1.11c and 1.11d are equally familiar. As these examples show, methods for contouring are powerful, yet general, techniques for visualizing data from a variety of fields.

1.2.3 Scalar Generation

The two visualization techniques presented thus far, color mapping and contouring, are simple, effective methods to display scalar information. It is natural to turn to these techniques first when visualizing data. However, often our data are not in a form convenient to these techniques. The data may not be single-valued (i.e., a scalar), or they may be a mathematical or other complex relationship. That is part of the fun and creative challenge of visualization: we must tap our creative resources to convert data into a form on which we can bring our existing tools to bear.

For example, consider terrain data. We assume that the data are x-y-z coordinates, where x and y represent the coordinates in the plane and z represents the elevation above sea level. Our desired visualization is to color the terrain according to elevation. This requires us to create a color map—possibly using white for high altitudes, blue for sea level and below, and various shades of green and brown for different elevations between sea level and high altitude. We also need scalars to index into the color

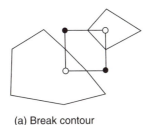

(a) Break contour

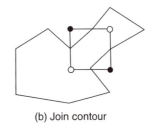

(b) Join contour

Figure 1.8 Choosing a particular contour case will (a) break or (b) join the current contour. The case shown is marching squares case 10.

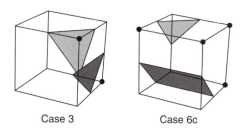

Figure 1.9 Arbitrarily choosing marching cubes cases leads to holes in the isosurface.

map. The obvious choice here is to extract the z coordinate. That is, scalars are simply the z-coordinate value.

This example can be made more interesting by generalizing the problem. Although we could easily create a filter to extract the z coordinate, we can create a filter that produces elevation scalar values where the elevation is measured along any axis. Given an oriented line starting at the (low) point p_l (e.g., sea level) and ending at the (high) point p_h (e.g., mountain top), we compute the elevation scalar s_i at point

$p_i = (x_i, y_i, z_i)$ using the dot product as shown in Fig. 1.12. The scalar is normalized using the magnitude of the oriented line and may be clamped between minimum and maximum scalar values (if necessary). The bottom half of this figure shows the results of applying this technique to a terrain model of Honolulu, Hawaii. A lookup table of 256 points ranging from deep blue (water) to yellow-white (mountain top) is used to color map this figure.

Scalar visualization techniques are deceptively powerful. Color mapping and isocontour generation are the predominant methods used in scientific visualization. Scalar visualization techniques are easily adapted to a variety of situations through creation of a relationship that transforms data at a point into a scalar value. Other examples of scalar mapping include an index value into a list of data, computing vector magnitude or matrix determinant, evaluating surface curvature, or determining distance between points. Scalar generation, when coupled with color mapping or contouring, is a simple yet effective technique for visualizing many types of data.

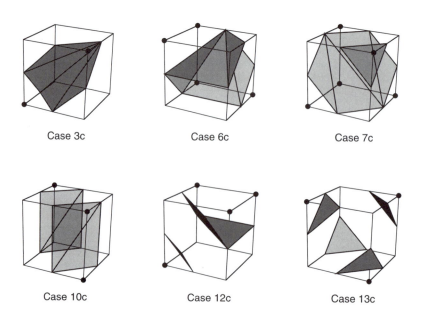

Figure 1.10   Marching cubes complementary cases.

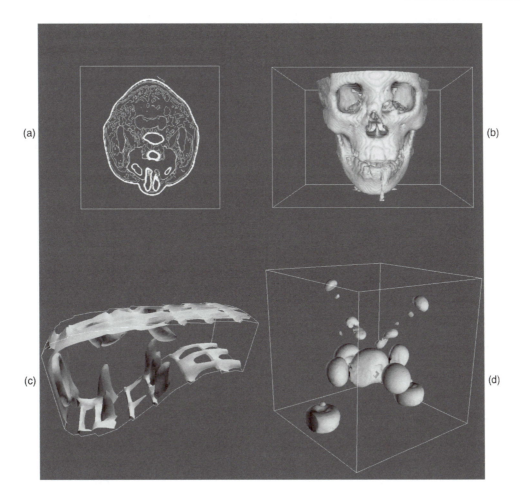

Figure 1.11 Contouring examples. (a) Marching squares used to generate contour lines; (b) marching cubes surface of human bone; (c) marching cubes surface of flow density; (d) marching cubes surface of iron–protein.

1.3 Vector Algorithms

Vector data is a 3D representation of direction and magnitude. Vector data often results from the study of fluid flow or data derivatives.

1.3.1 Hedgehogs and Oriented Glyphs

A natural vector visualization technique is to draw an oriented, scaled line for each vector in a dataset (Fig. 1.13a). The line begins at the point with which the vector is associated and is oriented in the direction of the vector components (v_x, v_y, v_z). Typically, the resulting line must be scaled up or down to control the size of its visual representation. This technique is often referred to as a *hedgehog* because of the bristly result.

There are many variations of this technique (Fig. 1.13b). Arrows may be added to indicate the direction of the line. The lines may be colored according to vector magnitude or some other scalar quantity (e.g., pressure or temperature). Also, instead of using a line, oriented

$$s_i = \frac{(p_i - p_l) \cdot (p_h - p_l)}{\left|p_h - p_l\right|^2}$$

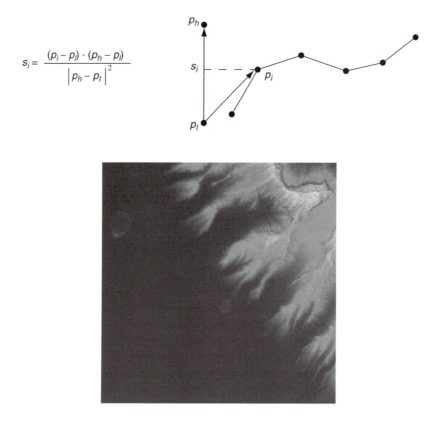

Figure 1.12   Computing scalars using normalized dot product. The bottom half of the figure illustrates a technique applied to terrain data from Honolulu, HI. (See also color insert.)

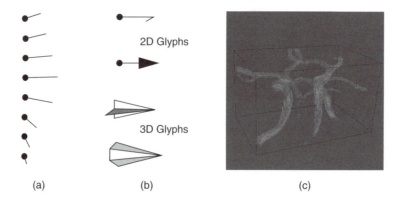

Figure 1.13   Vector visualization techniques. (a) Oriented lines; (b) oriented glyphs; (c) complex vector visualization. (See also color insert.)

"glyphs" can be used. By glyph we mean any 2D or 3D geometric representation, such as an oriented triangle or cone.

Care should be used in applying these techniques. In three dimensions it is often difficult to understand the position and orientation of a vector because of its projection into the 2D view plane. Also, using large numbers of vectors can clutter the display to the point where the visualization becomes meaningless. Figure 1.13c shows 167,000 3D vectors (using oriented and scaled lines) in the region of the human carotid artery. The larger vectors lie inside the arteries, and the smaller vectors lie outside the arteries and are randomly oriented (measurement error) but small in magnitude. Clearly, the details of the vector field are not discernible from this image.

Scaling glyphs also poses interesting problems. In what Tufte [39] has termed a "visualization lie," scaling a 2D or 3D glyph results in nonlinear differences in appearance. The surface area of an object increases with the square of its scale factor, so two vectors differing by a factor of two in magnitude may appear up to four times different based on surface area. Such scaling issues are common in data visualization, and great care must be taken to avoid misleading viewers.

1.3.2 Warping

Vector data is often associated with "motion." The motion is in the form of velocity or displacement. An effective technique for displaying such vector data is to "warp" or deform geometry according to the vector field. For example, imagine representing the displacement of a structure under load by deforming the structure. If we are visualizing the flow of fluid, we can create a flow profile by distorting a straight line inserted perpendicular to the flow.

Figure 1.14 shows two examples of vector warping. In the first example the motion of a vibrating beam is shown. The original undeformed outline is shown in wireframe. The second example shows warped planes in a structured grid dataset. The planes are warped according to flow momentum. The relative back and forward flows are clearly visible in the deformation of the planes.

Typically, we must scale the vector field to control geometric distortion. Too small a distortion might not be visible, while too large a distortion can cause the structure to turn inside out or self-intersect. In such a case, the viewer of the visualization is likely to lose context, and the visualization will become ineffective.

1.3.3 Displacement Plots

Vector displacement on the surface of an object can be visualized with displacement plots. A displacement plot shows the motion of an object in the direction perpendicular to its surface. The object motion is caused by an applied vector field. In a typical application the vector field is a displacement or strain field.

Vector displacement plots draw on the ideas in Section 1.2.3. Vectors are converted to scalars by computation of the dot product between the surface normal and vector at each point (Fig. 1.15a). If positive values result, the motion at the point is in the direction of the surface normal (i.e., positive displacement). Negative values indicate that the motion is opposite the surface normal (i.e., negative displacement).

A useful application of this technique is the study of vibration. In vibration analysis, we are interested in the eigenvalues (i.e., natural resonant frequencies) and eigenvectors (i.e., mode shapes) of a structure. To understand mode shapes, we can use displacement plots to indicate regions of motion. There are special regions in the structure where positive displacement changes to negative displacement. These are regions of zero displacement. When plotted on the surface of the structure, these regions appear as the so-called *modal* lines of vibration. The study of modal lines has long been an important visualization tool for understanding mode shapes.

Figure 1.15b shows modal lines for a vibrating rectangular beam. The vibration mode in this figure is the second torsional mode, clearly

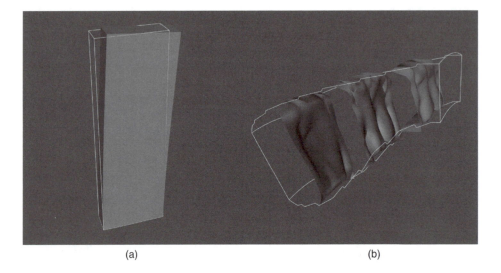

Figure 1.14   Warping geometry to show vector field. (a) Beam displacement; (b) flow momentum. (See also color insert.)

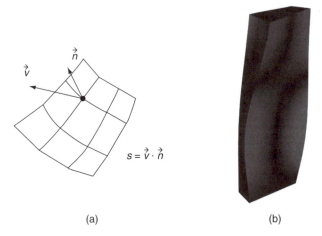

$$s = \vec{v} \cdot \vec{n}$$

(a)   (b)

Figure 1.15   Vector displacement plots. (a) Vector converted to scalar via dot product computation; (b) surface plot of vibrating plate. Dark areas show nodal lines and bright areas show maximum motion. (See also color insert.)

indicated by the crossing modal lines. (The aliasing in the figure is a result of the coarseness of the analysis mesh.) To create the figure we combined the procedure of Fig. 1.15a with a special lookup table. The lookup table was arranged with dark areas in the center (corresponding to zero dot products) and bright areas at the beginning and end of the table (corresponding to 1 or −1 dot products). As a result, regions of large normal displacement are bright and regions near the modal lines are dark.

1.3.4   Time Animation

Some of the techniques described so far can be thought of as moving a point or object over a small time-step. The hedgehog line is an approximation of a point's motion over a time

Figure 1.16 Time animation of a point C. Although the spacing between points varies, the time increment between each point is constant.

period whose duration is given by the scale factor. In other words, if the vector is considered to be a velocity $\vec{V} = dx/dt$, then the displacement of a point is

$$dx = \vec{V}dt \qquad (1.2)$$

This suggests an extension to our previous techniques: repeatedly displace points over many time-steps. Fig. 1.16 shows such an approach. Beginning with a sphere S centered about some point C, we move S repeatedly to generate the bubbles shown. The eye tends to trace out a path by connecting the bubbles, giving the observer a qualitative understanding of the vector field in that area. The bubbles may be displayed as an animation over time (giving the illusion of motion) or as a multiple-exposure sequence (giving the appearance of a path).

Such an approach can be misused. For one thing, the velocity at a point is instantaneous. Once we move away from the point, the velocity is likely to change. Using Equation 1.2 assumes that the velocity is constant over the entire step. By taking large steps, we are likely to jump over changes in the velocity. Using smaller steps, we will end in a different position. Thus, the choice of step size is a critical parameter in constructing accurate visualization of particle paths in a vector field.

To evaluate Equation 1.2, we can express it as an integral:

$$\vec{x}(t) = \int_t \vec{V}dt \qquad (1.3)$$

Although this form cannot be solved analytically for most real-world data, its solution can be approximated using numerical integration techniques. Accurate numerical integration is a

topic beyond the scope of this book, but it is known that the accuracy of the integration is a function of the step size dt. Because the path is an integration throughout the dataset, the accuracy of the cell interpolation functions and the accuracy of the original vector data play important roles in realizing accurate solutions. No definitive study that relates cell size or interpolation function characteristics to visualization error is yet available. But the lesson is clear: the result of numerical integration must be examined carefully, especially in regions with large vector field gradients. However, as with many other visualization algorithms, the insight gained by using vector-integration techniques is qualitatively beneficial, despite the unavoidable numerical errors.

The simplest form of numerical integration is Euler's method,

$$\vec{x}_{i+1} = \vec{x}_i + \vec{V}_i \Delta t \qquad (1.4)$$

where the position at time $\vec{x}_{i+1}$ is the vector sum of the previous position plus the instantaneous velocity times the incremental time step Δt.

Euler's method has error on the order of $O(\Delta t^2)$, which is not accurate enough for some applications. One such example is shown in Fig. 1.17. The velocity field describes perfect rotation about a central point. Using Euler's method, we find that we will always diverge and, instead of generating circles, will generate spirals.

In this chapter we will use the Runge-Kutta technique of order 2 [8]. This is given by the expression

$$\vec{x}_{i+1} = \vec{x}_i + \frac{\Delta t}{2}(\vec{V}_i + \vec{V}_{i+1}) \qquad (1.5)$$

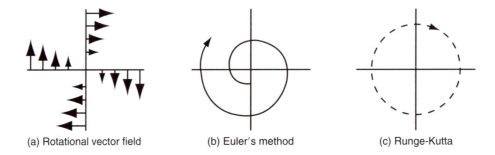

(a) Rotational vector field   (b) Euler's method   (c) Runge-Kutta

Figure 1.17   Euler's integration (b) and Runge-Kutta integration of order 2 (c) applied to a uniform rotational vector field (a). Euler's method will always diverge.

where the velocity $\vec{V}_{i+1}$ is computed using Euler's method. The error of this method is $O(\Delta t^3)$. Compared to Euler's method, the Runge-Kutta technique allows us to take a larger integration step at the expense of one additional function evaluation. Generally, this tradeoff is beneficial, but like any numerical technique, the best method to use depends on the particular nature of the data. Higher-order techniques are also available, but generally not necessary, because the higher accuracy is countered by error in interpolation function or inherent in the data values. If you are interested in other integration formulas, please check the references at the end of the chapter.

One final note about accuracy concerns: the error involved in either perception or computation of visualizations is an open research area. The discussion in the preceding paragraph is a good example of this: there, we characterized the error in streamline integration using conventional numerical integration arguments. But there is a problem with this argument. In visualization applications, we are integrating across cells whose function values are continuous but whose derivatives are not. As the streamline crosses the cell boundary, subtle effects may occur that are not treated by the standard numerical analysis. Thus, the standard arguments need to be extended for visualization applications.

Integration formulas require repeated transformation from global to local coordinates.

Consider moving a point through a dataset under the influence of a vector field. The first step is to identify the cell that contains the point. This operation is a search plus a conversion to local coordinates. Once the cell is found, then the next step is to compute the velocity at that point by interpolating the velocity from the cell points. The point is then incrementally repositioned (using the integration formula in Equation 1.5). The process is then repeated until the point exits the dataset or the distance or time traversed exceeds some specified value.

This process can be computationally demanding. There are two important steps we can take to improve performance:

1. *Improve search procedures.* There are two distinct types of searches. Initially, the starting location of the particle must be determined by a global search procedure. Once the initial location of the point is determined in the dataset, an incremental search procedure can be used. Incremental searching is efficient because the motion of the point is limited within a single cell, or, at most, across a cell boundary. Thus, the search space is greatly limited, and the incremental search is faster relative to the global search.

2. *Coordinate transformation.* The cost of a coordinate transformation from global to local coordinates can be reduced if either of the

following conditions is true: the local and global coordinate systems are identical to each other (or vary by *x*-*y*-*z* translation), or the vector field is transformed from global space to local coordinate space. The image data coordinate system is an example of local coordinates that are parallel to global coordinates, and thus a situation in which global-to-local coordinate transformation can be greatly accelerated. If the vector field is transformed into local coordinates (either as a preprocessing step or on a cell-by-cell basis), then the integration can proceed completely in local space. Once the integration path is computed, selected points along the path can be transformed into global space for the sake of visualization.

1.3.5 Streamlines

A natural extension of the previous time animation techniques is to connect the point position $\vec{x}(t)$ over many time-steps. The result is a numerical approximation to a particle trace represented as a line.

Borrowing terminology from the study of fluid flow, we can define three related line-representation schemes for vector fields.

- *Particle traces* are trajectories traced by fluid particles over time.
- *Streaklines* are the set of particle traces at a particular time t_i that have previously passed through a specified point x_i.
- *Streamlines* are integral curves along a curve *s* satisfying the equation

$$s = \int_t \vec{V} ds, \ \text{with} \ s = s(x, t) \tag{1.6}$$

for a particular time *t*.

Streamlines, streaklines, and particle traces are equivalent to one another if the flow is steady. In time-varying flow, a given streamline exists only at one moment in time. Visualization systems generally provide facilities to compute particle traces. However, if time is fixed, the same facility can be used to compute streamlines. In general, we will use the term *streamline* to refer to the method of tracing trajectories in a vector field. Please bear in mind the differences in these representations if the flow is time-varying.

Fig. 1.18 shows 40 streamlines in a small kitchen. The room has two windows, a door (with air leakage), and a cooking area with a

Figure 1.18 Flow velocity computed for a small kitchen (top and side view). Forty streamlines start along the rake positioned under the window. Some eventually travel over the hot stove and are convected upwards. (See also color insert.)

hot stove. The air leakage and temperature variation combine to produce air convection currents throughout the kitchen. The starting positions of the streamlines were defined by creating a *rake*, or curve (and its associated points). There, the rake was a straight line. These streamlines clearly show features of the flow field. By releasing many streamlines simultaneously, we obtain even more information, as the eye tends to assemble nearby streamlines into a "global" understanding of flow field features.

Many enhancements of streamline visualization exist. Lines can be colored according to velocity magnitude to indicate speed of flow. Other scalar quantities such as temperature or pressure also may be used to color the lines. We also may create constant-time dashed lines. Each dash represents a constant time increment. Thus, in areas of high velocity, the length of the dash will be greater relative to regions of lower velocity. These techniques are illustrated in Fig. 1.19 for air flow around a blunt fin. This example consists of a wall with half of a rounded fin projecting into the fluid flow. (Using arguments of symmetry, only half of the domain was modeled.) Twenty-five streamlines are released upstream of the fin. The boundary layer effects near the junction of the fin and wall are clearly evident from the stream-

lines. In this area, flow recirculation and the reduced flow speed are apparent.

1.4   Tensor Algorithms

Tensor visualization is an active area of research. However, there are a few simple techniques that we can use to visualize 3×3 real symmetric tensors. Such tensors are used to describe the state of displacement or stress in a 3D material. The stress and strain tensors for an elastic material are shown in Fig. 1.20.

In these tensors, the diagonal coefficients are the so-called normal stresses and strains, and the off-diagonal terms are the shear stresses and strains. Normal stresses and strains act perpendicularly to a specified surface, while shear stresses and strains act tangentially to the surface. Normal stress is either compression or tension, depending on the sign of the coefficient.

A 3×3 real symmetric matrix can be characterized by three vectors in 3D called the eigenvectors and three numbers called the eigenvalues of the matrix. The eigenvectors form a 3D coordinate system whose axes are mutually perpendicular. In some applications, particularly the study of materials, these axes are also referred to as the principal axes of the tensor and are physically significant. For example, if

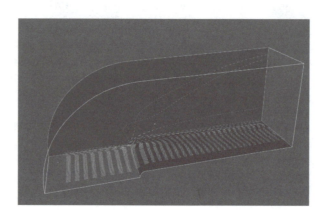

Figure 1.19   Dashed streamlines around a blunt fin. Each dash is a constant time increment. Fast-moving particles create longer dashes than slower-moving particles. The streamlines also are colored by flow density scalar.

$$\begin{bmatrix} \sigma_x & \tau_{xy} & \tau_{xz} \\ \tau_{yx} & \sigma_y & \tau_{yz} \\ \tau_{zx} & \tau_{zy} & \sigma_z \end{bmatrix}$$

$$\begin{bmatrix} \dfrac{\partial u}{\partial x} & \left(\dfrac{\partial u}{\partial y} + \dfrac{\partial v}{\partial z}\right) & \left(\dfrac{\partial u}{\partial z} + \dfrac{\partial w}{\partial x}\right) \\ \left(\dfrac{\partial u}{\partial y} + \dfrac{\partial v}{\partial z}\right) & \dfrac{\partial v}{\partial y} & \left(\dfrac{\partial v}{\partial z} + \dfrac{\partial w}{\partial y}\right) \\ \left(\dfrac{\partial u}{\partial z} + \dfrac{\partial w}{\partial x}\right) & \left(\dfrac{\partial v}{\partial z} + \dfrac{\partial w}{\partial y}\right) & \dfrac{\partial w}{\partial z} \end{bmatrix}$$

(a)   (b)

Figure 1.20   (a) Stress and (b) strain tensors. Normal stresses in the x-y-z coordinate directions are indicated as σ_x, σ_y, σ_z, and shear stresses are indicated as τ_{ij}. Material displacement is represented by u, v, w components.

the tensor is a stress tensor, then the principal axes are the directions of normal stress and no shear stress. Associated with each eigenvector is an eigenvalue. The eigenvalues are often physically significant as well. In the study of vibration, eigenvalues correspond to the resonant frequencies of a structure, and the eigenvectors are the associated mode shapes.

Mathematically we can represent eigenvalues and eigenvectors as follows. Given a matrix A, the eigenvector $\vec{x}$ and eigenvalue λ must satisfy the relation

$$A \cdot \vec{x} = \lambda \vec{x} \tag{1.7}$$

For Equation 1.7 to hold, the matrix determinate must satisfy

$$\det|A - \lambda I| = 0 \tag{1.8}$$

Expanding this equation yields an nth-degree polynomial in λ whose roots are the eigenvalues. Thus, there are always n eigenvalues, although they may not be distinct. In general, Equation 1.8 is not solved using polynomial root searching because of poor computational performance. (For matrices of order 3, root searching is acceptable because we can solve for the eigenvalues analytically.) Once we determine the eigenvalues, we can substitute each into Equation 1.8 to solve for the associated eigenvectors.

We can express the eigenvectors of the 3×3 system as

$$\vec{v}_i = \lambda_i \vec{e}_i, \ with \ i = 1, 2, 3 \tag{1.9}$$

with $\vec{e}_i$ a unit vector in the direction of the eigenvalue, and λ_i the eigenvalues of the system. If we order eigenvalues such that

$$\lambda_1 \geq \lambda_2 \geq \lambda_3 \tag{1.10}$$

then we refer to the corresponding eigenvectors $\vec{v}_1$, $\vec{v}_2$, and $\vec{v}_3$ as the *major, medium*, and *minor* eigenvectors.

1.4.1   Tensor Ellipsoids

This leads us to the tensor ellipsoid technique for the visualization of real, symmetric 3×3 matrices. The first step is to extract eigenvalues and eigenvectors as described in the previous section. Since eigenvectors are known to be orthogonal, the eigenvectors form a local coordinate system. These axes can be taken as the *minor, medium*, and *major* axes of an ellipsoid. Thus, the shape and orientation of the ellipsoid represent the relative size of the eigenvalues and the orientation of the eigenvectors.

To form the ellipsoid we begin by positioning a sphere at the tensor location. The sphere is then rotated around its origin using the eigenvectors, which in the form of Equation 1.9 are direction cosines. The eigenvalues are used to scale the sphere. Using 4×4 transformation matrices, we form the ellipsoid by transforming the sphere centered at the origin using the matrix T:

$$T = T_T \cdot T_R \cdot T_S \qquad (1.11)$$

where

$$T_T = \begin{bmatrix} 1 & 0 & 0 & t_x \\ 0 & 1 & 0 & t_y \\ 0 & 0 & 1 & t_z \\ 0 & 0 & 0 & 1 \end{bmatrix}$$

$$T_S = \begin{bmatrix} s_x & 0 & 0 & 0 \\ 0 & s_y & 0 & 0 \\ 0 & 0 & s_z & 0 \\ 0 & 0 & 0 & 1 \end{bmatrix} \qquad (1.12)$$

$$T_R = \begin{bmatrix} \cos\theta_{x'x} & \cos\theta_{x'y} & \cos\theta_{x'z} & 0 \\ \cos\theta_{y'x} & \cos\theta_{y'y} & \cos\theta_{y'z} & 0 \\ \cos\theta_{z'x} & \cos\theta_{z'y} & \cos\theta_{z'z} & 0 \\ 0 & 0 & 0 & 1 \end{bmatrix}$$

where T_T, T_S, and T_R are translation, scale, and rotation matrices. The eigenvectors can be directly plugged in to create the rotation matrix, while the point coordinates x-y-z and eigenvalues $\lambda_1 \geq \lambda_2 \geq \lambda_3$ are inserted into the translation and scaling matrices. A concatenation of these matrices in the correct order forms the final transformation matrix T.

Fig. 1.21a depicts the tensor ellipsoid technique. In Fig. 1.22b we show this technique to visualize material stress near a point load on the surface of a semi-infinite domain. (This is the so-called Boussinesq's problem.) From Saada [33] we have the analytic expression for the stress components in Cartesian coordinates shown in Fig. 1.21c. Note that the z direction is defined as the axis originating at the point of application of the force P. The variable ρ is the distance from the point of load application to a point x-y-z. The orientations of the x and y axes are in the plane perpendicular to the z axis. The rotation in the plane of these axes is unimportant since the solution is symmetric around the z axis. The parameter v is Poisson's ratio, which is a property of the material. Poisson's ratio relates the lateral contraction of a material to axial elongation under a uniaxial stress condition [33,35].

In Fig. 1.22 we visualize the analytical results of Boussinesq's problem from Saada. The left-hand portion of the figure shows the results by displaying the scaled and oriented principal axes of the stress tensor. (These are called *tensor axes*.) In the right-hand portion we use tensor ellipsoids to show the same result. Tensor ellipsoids and tensor axes are a form of *glyph* (see Section 1.5.4) specialized to tensor visualization.

A certain amount of care must be taken to visualize this result, because there is a stress singularity at the point of contact of the load. In a real application, loads are applied over a small area and not at a single point. Plastic behavior prevents stress levels from exceeding a certain point. The results of the visualization, as with any computer process, are only as good as the underlying model.

1.5 Modeling Algorithms

"Modeling algorithms" is the catch-all category for our taxonomy of visualization techniques. Modeling algorithms will typically transform the type of input dataset or use combinations of input data and parameters to affect their result.

1.5.1 Source Objects

Source objects begin the visualization pipeline. Often, source objects are used to create geometry such as spheres, cones, or cubes to support visualization context, or are used to read in data files. Source objects also may be used to create dataset attributes. Some examples of source objects and their use are as follows.

1.5.1.1 Modeling Simple Geometry

Spheres, cones, cubes, and other simple geometric objects can be used alone or in combination to model geometry. Often, we visualize real-world applications such as air flow in a room and need to show real-world objects such as furniture, windows, or doors. Real-world objects often can be represented using these simple geometric representations. These source

(a) Tensor ellipsoid

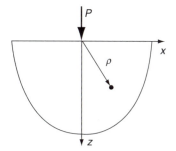

(b) Point load on semi-infinite domain

$$\sigma_x = -\frac{P}{2\pi\rho^2}\left(\frac{3zx^2}{\rho^3} - (1-2v)\left(\frac{z}{\rho} - \frac{\rho}{\rho+z} + \frac{x^2(2\rho+z)}{\rho(\rho+z)^2}\right)\right)$$

$$\sigma_y = -\frac{P}{2\pi\rho^2}\left(\frac{3zy^2}{\rho^3} - (1-2v)\left(\frac{z}{\rho} - \frac{\rho}{\rho+z} + \frac{y^2(2\rho+z)}{\rho(\rho+z)^2}\right)\right)$$

$$\sigma_y = -\frac{3Pz^3}{2\pi\rho^5}$$

$$\tau_{xy} = \tau_{yx} = -\frac{P}{2\pi\rho^2}\left(\frac{3xyz}{\rho^3} - (1-2v)\left(\frac{xy(2\rho+z)}{\rho(\rho+z)^2}\right)\right)$$

$$\tau_{xz} = \tau_{zx} = -\frac{3Pxz^2}{2\pi\rho^5}$$

$$\tau_{yz} = \tau_{zy} = -\frac{3Pyz^2}{2\pi\rho^5}$$

c) Analytic solution

Figure 1.21 Tensor ellipsoids. (a) Ellipsoid oriented along eigenvalues (i.e., principal axes) of tensor; (b) pictorial description of Boussinesq's problem; (c) analytic results according to Saada.

objects generate their data procedurally. Alternatively, we may use reader objects to access geometric data defined in data files. These data files may contain more complex geometry, such as that produced by a 3D Computer-Aided Design (CAD) system.

1.5.1.2 Supporting Geometry

During the visualization process, we may use source objects to create supporting geometry. This may be as simple as three lines to represent a coordinate axis or as complex as tubes wrapped around line segments to thicken and enhance their appearance. Another common use is as supplemental input to objects such as streamlines or probe filters. These filters take a second input that defines a set of points. For streamlines, the points determine the initial positions for generating the streamlines. The probe filter uses the points as the position to compute attribute values such as scalars, vectors, or tensors.

1.5.1.3 Data Attribute Creation

Source objects can be used as procedures to create data attributes. For example, we

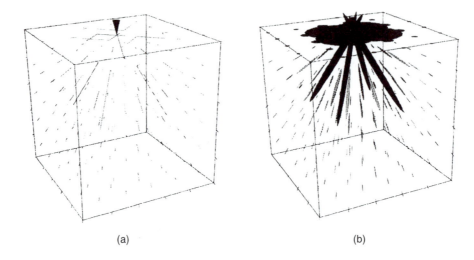

(a)   (b)

Figure 1.22   Tensor visualization techniques. (a) Tensor axes; (b) tensor ellipsoids.

can procedurally create textures and texture coordinates. Another use is to create scalar values over a uniform grid. If the scalar values are generated from a mathematical function, then we can use the visualization techniques described here to visualize the function. In fact, this leads us to a very important class of source objects: implicit functions.

1.5.2   Implicit Functions

Implicit functions are functions of the form

$$F(\bar{x}) = c \tag{1.13}$$

where c is an arbitrary constant. Implicit functions have three important properties:

- *Simple geometric description*. Implicit functions are convenient tools to describe common geometric shapes, including planes, spheres, cylinders, cones, ellipsoids, and quadrics.
- *Region separation*. Implicit functions separate 3D Euclidean space into three distinct regions. These regions are inside, on, and outside the implicit function. These regions are defined as $F(x, y, z) < 0$, $F(x, y, z) = 0$, and $F(x, y, z) > 0$, respectively.

- *Scalar generation*. Implicit functions convert a position in space into a scalar value. That is, given an implicit function, we can sample it at a point (x_i, y_i, z_i) to generate a scalar value c_i.

An example of an implicit function is the equation for a sphere of radius R

$$F(x, y, z) = x^2 + y^2 + z^2 - R^2 \tag{1.14}$$

This simple relationship defines the three regions $F(x, y, z) = 0$ (on the surface of the sphere), $F(x, y, z) < 0$ (inside the sphere), and $F(x, y, z) > 0$ (outside the sphere). Any point may be classified inside, on, or outside the sphere simply by evaluating Equation 1.14.

If you have been paying attention, you will note that Equation 1.14 is identical to the equation defining a contour (Equation 1.1). This should provide you with a clue as to the many ways in which implicit functions can be used. These include geometric modeling, selection of data, and visualization of complex mathematical descriptions.

1.5.2.1   Modeling Objects

Implicit functions can be used alone or in combination to model geometric objects.

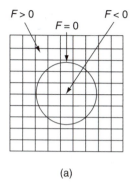 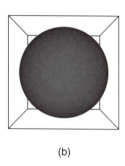 

(a)   (b)   (c)

Figure 1.23 Sampling functions. (a) 2D depiction of sphere sampling; (b) isosurface of sampled sphere; (c) Boolean combination of two spheres, a cone, and two planes. (One sphere intersects the other; the planes clip the cone.)

For example, to model a surface described by an implicit function, we sample F on a dataset and generate an isosurface at a contour value c_i. The result is a polygonal representation of the function. Fig. 1.23b shows an isosurface for a sphere of radius = 1 sampled on a volume. Note that we can choose nonzero contour values to generate a family of offset surfaces. This is useful for creating blending functions and other special effects.

Implicit functions can be combined to create complex objects using the Boolean operators union, intersection, and difference. The union operation $F \cup G$ between two functions $F(x, y, z)$ and $G(x, y, z)$ at a point (x_0, y_0, z_0) is the minimum value

$$F \cup G = \min(F(x_0, y_0, z_0), \\ G(x_0, y_0, z_0)) \tag{1.15}$$

The intersection between two implicit functions is given by

$$F \cap G = \max(F(x_0, y_0, z_0), \\ G(x_0, y_0, z_0)) \tag{1.16}$$

The difference of two implicit functions is given by

$$F - G = \max(F(x_0, y_0, z_0), \\ -G(x_0, y_0, z_0)) \tag{1.17}$$

Fig. 1.23c shows a combination of simple implicit functions to create an ice cream cone.

The cone is created by clipping the (infinite) cone function with two planes. The ice cream is constructed by performing a difference operation on a larger sphere with a smaller offset sphere to create the "bite." The resulting surface was extracted using surface contouring with isosurface value 0.0.

1.5.2.2 Selecting Data

We can take advantage of the properties of implicit functions to select and cut data. In particular, we will use the region separation property to select data. (We defer the discussion on cutting to Section 1.5.5.)

Selecting or extracting data with an implicit function means choosing cells and points (and associated attribute data) that lie within a particular region of the function. To determine whether a point x-y-z lies within a region, we simply evaluate the point and examine the sign of the result. A cell lies in a region if all its points lie in the region.

Fig. 1.24a shows a 2D implicit function, here an ellipse, used to select the data (i.e., points, cells, and data attributes) contained within it. Boolean combinations also can be used to create complex selection regions, as illustrated in Fig. 1.24b. Here, two ellipses are used in combination to select voxels within a volume dataset. Note that extracting data

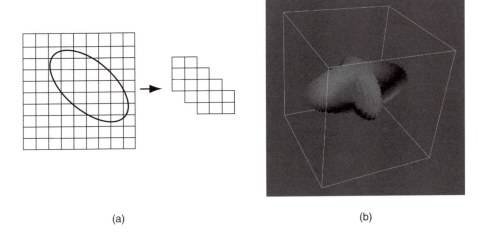

(a)   (b)

Figure 1.24   Implicit functions used to select data: (a) 2D cells lying in ellipse are selected; (b) two ellipsoids combined using the union operation used to select voxels from a volume. Voxels shrank 50%. (See also color insert.)

often changes the structure of the dataset. In Fig. 1.24 the input type is a volume dataset, while the output type is an unstructured grid dataset.

1.5.2.3   *Visualizing Mathematical Descriptions*

Some functions, often discrete or probabilistic in nature, cannot be cast into the form of Equation 1.13. However, by applying some creative thinking, we can often generate scalar values that can be visualized. An interesting example of this is the so-called strange attractor.

Strange attractors arise in the study of nonlinear dynamics and chaotic systems. In these systems, the usual types of dynamic motion—equilibrium, periodic motion, and quasi-periodic motion—are not present. Instead, the system exhibits chaotic motion. The resulting behavior of the system can change radically as a result of small perturbations in its initial conditions.

A classical strange attractor was developed by Lorenz [24] in 1963. Lorenz developed a simple model for thermally induced fluid convection in the atmosphere. Convection causes rings of rotating fluid and can be developed

from the general Navier-Stokes partial differential equations for fluid flow. The Lorenz equations can be expressed in nondimensional form as

$$\frac{dx}{dt} = \sigma(y - x)$$
$$\frac{dy}{dt} = \rho x - y - xz \qquad (1.18)$$
$$\frac{dz}{dt} = xy - \beta z$$

where x is proportional to the fluid velocity in the fluid ring, y and z measure the fluid temperature in the plane of the ring, the parameters σ and ρ are related to the Prandtl number and Raleigh number, respectively, and β is a geometric factor.

Certainly these equations are not in the implicit form of Equation 1.13, so how do we visualize them? Our solution is to treat the variables x, y, and z as the coordinates of a 3D space, and integrate Equation 1.18 to generate the system "trajectory," that is, the state of the system through time. The integration is carried out within a volume and scalars are created by counting the number of times each voxel is visited. By integrating long enough, we can create a volume representing the "surface" of the

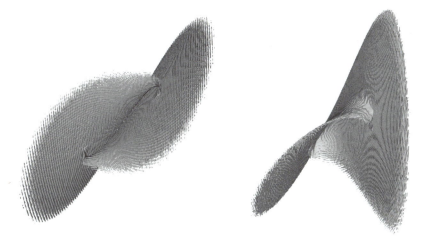

Figure 1.25 Visualizing a Lorenz strange attractor by integrating the Lorenz equations in a volume. The number of visits in each voxel is recorded as a scalar function. The surface is extracted via marching cubes using a visit value of 50. The number of integration steps is 10 million, in a volume of dimensions 200^3. The surface roughness is caused by the discrete nature of the evaluation function. (See also color insert.)

strange attractor, Fig. 1.25. The surface of the strange attractor is extracted by using marching cubes and a scalar value specifying the number of visits in a voxel.

1.5.3   Implicit Modeling

In the previous section, we saw how implicit functions, or Boolean combinations of implicit functions, could be used to model geometric objects. The basic approach is to evaluate these functions on a regular array of points, or volume, and then to generate scalar values at each point in the volume. Then either volume rendering or isosurface generation is used to display the model.

An extension of this approach, called *implicit modeling*, is similar to modeling with implicit functions. The difference lies in the fact that scalars are generated using a distance function instead of the usual implicit function. The distance function is computed as a Euclidean distance to a set of generating primitives such as points, lines, or polygons. For example, Fig. 1.26 shows the distance functions to a point, line, and triangle. Because distance functions

are well-behaved monotonic functions, we can define a series of offset surfaces by specifying different isocontour values, where the value is the distance to the generating primitive. The isocontours form approximations to the true offset surfaces, but using high-volume resolution we can achieve satisfactory results.

Used alone the generating primitives are limited in their ability to model complex geometry. By using Boolean combinations of the primitives, however, complex geometry can be easily modeled. The Boolean operations union, intersection, and difference (Equations 1.15, 1.16, and 1.17, respectively) are illustrated in Fig. 1.27. Fig. 1.28 shows the application of implicit modeling to "thicken" the line segments in the text symbol "HELLO." The isosurface is generated on a $110 \times 40 \times 20$ volume at a distance offset of 0.25 units. The generating primitives were combined using the Boolean union operator. Although Euclidean distance is always a nonnegative value, it is possible to use a signed distance function for objects that have an outside and an inside. A negative distance is the negated distance of a point inside the object to the surface of the object. Using a

Figure 1.26 Distance functions to a point, line, and triangle.

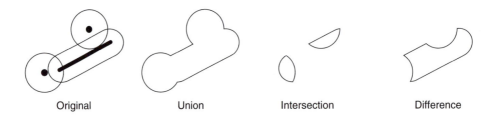

| Original | Union | Intersection | Difference |

Figure 1.27 Boolean operations using points and lines as generating primitives.

Figure 1.28 Implicit modeling used to thicken a stroked font. Original lines can be seen within the translucent implicit surface.

signed distance function allows us to create offset surfaces that are contained within the actual surface.

Another interesting feature of implicit modeling is that when isosurfaces are generated, more than one connected surface can result. These situations occur when the generating primitives form concave features. Fig. 1.29 illustrates this situation. If desired, multiple surfaces can be extracted by using a connectivity segmentation algorithm.

1.5.4 Glyphs

Glyphs, sometimes referred to as icons, are a versatile technique to visualize data of every type. A glyph is an "object" that is affected by its input data. This object may be geometry, a dataset, or a graphical image. The glyph may orient, scale, translate, deform, or somehow alter the appearance of the object in response to data. We have already seen a simple form of glyph: hedgehogs are lines that are oriented, translated, and scaled according to the position and vector value of a point. A variation of this is to use oriented cones or arrows (see Section 1.3.1).

More elaborate glyphs are possible. In one creative visualization technique, Chernoff [6] tied data values to an iconic representation of the human face. Eyebrows, nose, mouth, and other features were modified according to financial data values. This interesting technique built on the human capability to recognize

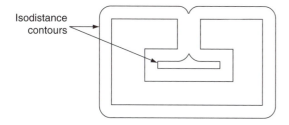

Isodistance contours

Figure 1.29 Concave features can result in multiple contour lines/surfaces.

facial expression. By tying appropriate data values to facial characteristics, rapid identification of important data points is possible.

In a sense, glyphs represent the fundamental result of the visualization process. Moreover, all the visualization techniques we present can be treated as concrete representations of an abstract glyph class. For example, while hedgehogs are an obvious manifestation of a vector glyph, isosurfaces can be considered a topologically 2D glyph for scalar data. Delmarcelle and Hesselink [11] have developed a unified framework for flow visualization based on types of glyphs. They classify glyphs according to one of three categories:

- *Elementary icons* represent their data across the extent of their spatial domain. For example, an oriented arrow can be used to represent a surface normal.

- *Local icons* represent elementary information plus a local distribution of the values around the spatial domain. A surface normal vector colored by local curvature is one example of a local icon, because local data beyond the elementary information is encoded.

- *Global icons* show the structure of the complete dataset. An isosurface is an example of a global icon.

This classification scheme can be extended to other visualization techniques such as vector and tensor data, or even to nonvisual forms

such as sound or tactile feedback. We have found this classification scheme to be helpful when designing visualizations or creating visualization techniques. Often, it gives insight into ways of representing data that can be overlooked.

Fig. 1.30 is an example of glyphing. Small 3D cones are oriented on a surface to indicate the direction of the surface normal. A similar approach could be used to show other surface properties such as curvature or anatomical key points.

1.5.5 Cutting

Often, we want to cut through a dataset with a surface and then display the interpolated data values on the surface. We refer to this technique as *data cutting* or simply *cutting*. The data cutting operation requires two pieces of information: a definition for the surface and a dataset to cut. We will assume that the cutting surface is defined by an implicit function. A typical application of cutting is to slice through a dataset with a plane, and color map the scalar data and/or warp the plane according to vector value.

A property of implicit functions is to convert a position into a scalar value (see Section 1.5.2).

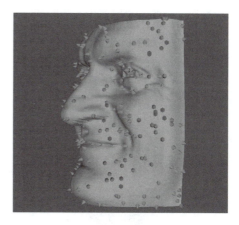

Figure 1.30 Glyphs indicate surface normals on a model of a human face. Glyph positions are randomly selected. (See also color insert.)

We can use this property in combination with a contouring algorithm (e.g., marching cubes) to generate cut surfaces. The basic idea is to generate scalars for each point of each cell of a dataset (using the implicit cut function) and then contour the surface value $F(x, y, z) = 0$.

The cutting algorithm proceeds as follows. For each cell, function values are generated by evaluating $F(x, y, z)$ for each cell point. If all the points evaluate positive or negative, then the surface does not cut the cell. However, if the points evaluate positive and negative, then the surface passes through the cell. We can use the cell contouring operation to generate the isosurface $F(x, y, z) = 0$. Data-attribute values can then be computed by interpolating along cut edges.

Fig. 1.31 illustrates a plane cut through a structured grid dataset. The plane passes through the center of the dataset with normal $(-0.287, 0, 0.9579)$. For comparison purposes, a portion of the grid geometry is also shown. The grid geometry is the grid surface $k = 9$ (shown in wireframe). One benefit of cut surfaces is that we can view data on (nearly) arbitrary surfaces. Thus, the structure of the dataset does not constrain how we view the data.

We can easily make multiple planar cuts through a structured grid dataset by specifying multiple iso-values for the cutting algorithm. Fig. 1.32 shows 100 cut planes generated perpendicular to the camera's view plane normal. Rendering the planes from back to front with an

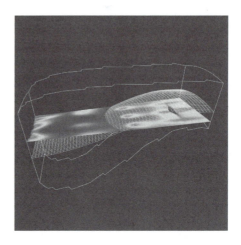

Figure 1.31   Cut through structured grid with plane. The cut plane is shown solid shaded. A computational plane of constant k value is shown in wireframe for comparison. The colors correspond to flow density. Cutting surfaces are not necessarily planes: implicit functions such as spheres, cylinders, and quadrics can also be used. (See also color insert.)

opacity of 0.05 produces a simulation of volume rendering.

This example illustrates that cutting the volumetric data in a structured grid dataset produces polygonal cells. Similarly, cutting polygonal data produces lines. Using a single plane equation, we can extract "contour lines" from a surface model defined with polygons. Fig. 1.33 shows contours extracted from a surface model of the skin. At each vertex in the surface model, we evaluate the equation of the plane $F(x, y, z) = c$ and store the value

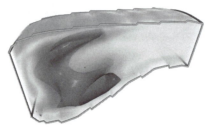

Figure 1.32   100 cut planes with opacity of 0.05, rendered back-to-front to simulate volume rendering. (See also color insert.)

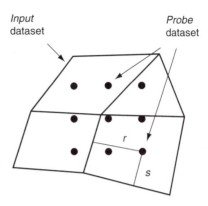

Figure 1.34 Probing data. The geometry of one dataset (*Probe*) is used to extract dataset attributes from another dataset (*Input*).

Figure 1.33 Cutting a surface model of the skin with a series of planes produces contour lines. Lines are wrapped with tubes for visual clarity. (See also color insert.)

of the function as a scalar value. Cutting the data with 46 iso-values from 1.5 to 136.5 produces contour lines that are 3 units apart.

1.5.6 Probing

Probing obtains dataset attributes by sampling one dataset (the input) with a set of one or more points (the probe), as shown in Fig. 1.34. Probing is also called *resampling*. Examples include probing an input dataset with a sequence of points along a line, on a plane, or in a volume. The result of the probing is a new dataset (the output) with the topological and geometric structure of the probe dataset and point attributes interpolated from the input dataset. Once the probing operation is complete, the output dataset can be visualized with any of the appropriate techniques described previously.

As Fig. 1.34 indicates, the details of the probing process are as follows. For every point in the probe dataset, the location in the input dataset (i.e., cell, subcell, and parametric coordinates) and interpolation weights are determined. Then the data values from the cell are interpolated to the probe point. Probe points

that are outside the input dataset are assigned a nil (or appropriate) value. This process repeats for all points in the probe dataset.

Probing can be used to reduce data or to view data in a particular fashion.

- Data is reduced when the probe operation is limited to a subregion of the input dataset or the number of probe points is less than the number of input points.

- Data can be visualized with specialized techniques by sampling on selected datasets. For example, using a line probe enables x-y plotting along a line, and using a plane probe allows surface color mapping or line contouring on the plane.

Probing must be used carefully or errors may be introduced. Undersampling data in a region can miss important high-frequency information or localized data variations. Oversampling data, while not creating error, can give false confidence in the accuracy of the data. Thus the sampling frequency should have a similar density as the input dataset, or if higher density, the visualization should be carefully annotated as to the original data frequency.

One important application of probing converts irregular or unstructured data to structured form using a probe volume of appropriate

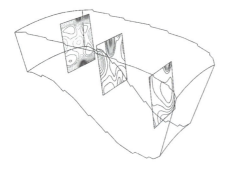

Figure 1.35 Probing data in a combustor. Probes are regular arrays of 50^2 points that are passed through a contouring filter.

resolution to sample the unstructured data. This is useful if volume rendering or another volume technique is to be used to visualize the data.

Fig. 1.35 shows an example of three probes. The probes sample flow density in a structured grid. The output of the probes is passed through a contour filter to generate contour lines. As this figure illustrates, we can be selective with the location and extent of the probe, allowing us to focus on important regions in the data.

1.5.7 Data Reduction

One of the major challenges facing the scientific visualization community is the increasing size of data. While just a short time ago data sizes of a gigabyte were considered large, terabyte and even petabyte data sizes are now available. Because the value of the visualization process is tied to its ability to effectively convey information about large and complex data, it is absolutely essential to find techniques to address this situation. A simple but effective approach is to use methods to reduce data size prior to the visualization process. The approaches taken depend on the type of data; for example, subsampling works well for structured data. Unstructured data (such as polygonal meshes) requires more sophisticated techniques. Since this topic is worth several books on

its own, we present some introductory approaches to data reduction. Note that the use of probing is also an excellent data-reduction tool.

1.5.7.1 Subsampling

Subsampling (Fig. 1.36) is a method that reduces data size by selecting a subset of the original data. The subset is specified by choosing a parameter n, specifying that every nth data point is to be extracted. For example, in structured datasets such as image data and structured grids, selecting every nth point produces the results shown in Fig. 1.36. Subsampling modifies the topology of a dataset. When points or cells are not selected, this leaves a topological "hole." Dataset topology must be modified to fill the hole. In structured data, this is simply a uniform selection across the structured *i-j-k* coordinates. In structured data, the hole must be filled in by using triangulation or other complex tessellation schemes. Subsampling is not typically performed on unstructured data because of its inherent complexity.

1.5.7.2 Decimation

Unstructured data can be reduced in size by applying a variety of decimation algorithms (also known as *polygon reduction* when applied to polygonal meshes). There are several approaches to decimation based on differing operations performed on the mesh (Fig. 1.37). Vertex removal deletes a vertex and all attached cells. The resulting hole is then triangulated.

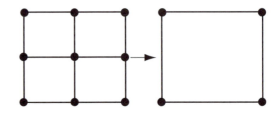

Figure 1.36 Subsampling structured data.

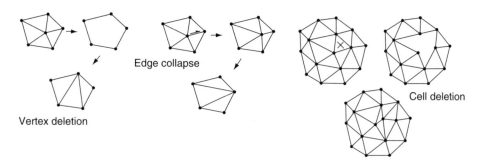

Figure 1.37   Decimating unstructured data.

Edge collapse results in merging two vertices into one. The position of the merged point is controlled by the particulars of the error metric and algorithm: choosing one of the two endpoints, or a point on the edge, is common. Some algorithms compute an optimal merge position based on minimizing error to the original data. Finally, some techniques may delete an entire cell (e.g., triangle) and attached cells and then retriangulate the resulting hole.

Decimation algorithms depend on the evaluation of an error metric to determine the operation to apply to the mesh. Simple approaches such as distance to an "average" plane work reasonably well. Probably the most widely used error metric is based on an accumulation of error represented by a quadric. The so-called quadric error metric measures the distance to a set of planes, each plane corresponding to an original triangle in the input mesh.

1.6   Bibliographic Notes

Color mapping is a widely studied topic in imaging, computer graphics, visualization, and human factors. [12,30,42]. You also may want to learn about the physiological and psychological effects of color on perception. The text by Wyszecki and Stiles [44] serves as an introductory reference.

Contouring is a widely studied technique in visualization because of its importance and popularity. Early techniques were developed for 2D data [43]. 3D techniques were developed initially as contour connecting methods [15]—that is, given a series of 2D

contours on evenly spaced planes, connecting the contours to create a closed surface. Since the introduction of marching cubes, many other techniques have been implemented [13,26,28]. A particularly interesting reference is given by Livnat et al. [22]. They show a contouring method with the addition of a preprocessing step that generates isocontours in near-optimal time.

Although we barely touched the topic, the study of chaos and chaotic vibrations is a delightfully interesting topic. Besides the original paper by Lorenz [24], the book by Moon [27] is a good place to start.

2D and 3D vector plots have been used by computer analysts for many years [16]. Streamlines and streamribbons also have been applied to the visualization of complex flows [41]. Good general information on vector visualization techniques is given by Helman and Hesselink [19] and Richter et al. [31].

Tensor visualization techniques are relatively few in number. Most techniques are glyph-oriented [10, 18]. We will see more techniques in later chapters.

Blinn [3], Bloomenthal [4,5], and Wyvill [45] have been important contributors to implicit modeling. Implicit modeling is currently popular in computer graphics for modeling "soft" or "blobby" objects. These techniques are simple, powerful, and becoming widely used for advanced computer graphics modeling.

Polygon reduction is a relatively new field of study. SIGGRAPH '92 marked a flurry of interest with the publication of two papers on this topic [32, 40]. Since then a number of valuable techniques have been published. One of the best techniques, in terms of quality of results, is given by Hoppe [21], although it is limited in time and space because it is based on formal optimization techniques. Other interesting methods include those by Hinker and Hansen [20] and Rossignac and Borel [32]. One promising area of research is multiresolution analysis, where wavelet decomposition is used to build multiple levels of detail (LODs) in a model [14]. The most recent work in this field stresses progressive transmission of 3D triangle meshes [21], improved error measures [17], and algorithms that modify mesh topology [29,36]. An extensive book on the technology is available that includes specialized methods for terrain simplification [25].

References

1. R. H. Abraham and C. D. Shaw. *Dynamics: The Geometry of Behavior*. Aerial Press, Santa Cruz, CA, 1985.

2. C. Upson, T. Faulhaber, Jr., D. Kamins, D. Laidlaw, and D. Schlegel. The application visualization system: a computational environment for scientific visualization. *IEEE Computer Graphics and Applications*, 9(4):30–42, 1989.

3. J. F. Blinn. A generalization of algebraic surface drawing. *ACM Transactions on Graphics*, 1(3):235–256, 1982.

4. J. Bloomenthal. Polygonization of implicit surfaces. *Computer Aided Geometric Design*. 5(4):341–355, 1982.

5. J. Bloomenthal, *Introduction to Implicit Surfaces*. San Francisco, Morgan Kaufmann, 1997.

6. H. Chernoff. Using faces to represent points in *K*-dimensional space graphically. *J. American Statistical Association*, 68:361–368, 1973.

7. H. Cline, W. Lorensen, and W. Schroeder. 3D phase contrast MRI of cerebral blood flow and surface anatomy. *J. Computer Assisted Tomography*, 17(2):173–177, 1993.

8. S. D. Conte and C. de Boor. *Elementary Numerical Analysis*. New York, McGraw-Hill, 1972.

9. *Data Explorer Reference Manual*. IBM Corp, Armonk, NY, 1991.

10. W. C. de Leeuw and J. J. van Wijk. A probe for local flow field visualization. In *Proceedings of Visualization '93*, pages 39–45, IEEE Computer Society Press, Los Alamitos, CA, 1993.

11. T. Delmarcelle and L. Hesselink. A unified framework for flow visualization. In *Computer Visualization Graphics Techniques for Scientific and Engineering Analysis* (R. S. Gallagher, ed.). Boca Raton, FL, CRC Press, 1995.

12. H. J. Durrett. *Color and the Computer*. Boston, Academic Press, 1987.

13. M. J. Durst. Additional reference to marching cubes. *Computer Graphics*, 22(2):72–73, 1988.

14. M. Eck, T. DeRose, T. Duchamp, H. Hoppe, M. Lounsbery, and W. Stuetzle. Multiresolution analysis of arbitrary meshes. In *Proceedings SIGGRAPH '95*, pages 173–182, 1995.

15. H. Fuchs, Z. M. Kedem, and S. P. Uselton. Optimal surface reconstruction from planar contours. *Communications of the ACM*, 20(10):693–702, 1977.

16. A. J. Fuller and M. L. X. dosSantos. Computer generated display of 3D vector fields. *Computer Aided Design*, 12(2):61–66, 1980.

17. M. Garland and P. Heckbert. Surface simplification using quadric error metrics. In *Proceedings SIGGRAPH '97*, pages 209–216, 1997.

18. R. B. Haber and D. A. McNabb. Visualization idioms: a conceptual model to scientific visualization systems. *Visualization in Scientific Computing* (G. M. Nielson, B. Shriver, L. J. Rosenblum, eds.). IEEE Computer Society Press, pages 61–73, 1990.

19. J. Helman and L. Hesselink. Representation and display of vector field topology in fluid flow data sets. *Visualization in Scientific Computing* (G. M. Nielson, B. Shriver, L. J. Rosenblum, eds.). IEEE Computer Society Press, pages 61–73, 1990.

20. P. Hinker and C. Hansen. Geometric optimization. In *Proceedings of Visualization '93*, pages 189–195, 1993.

21. H. Hoppe. Progressive meshes. In *Proceedings SIGGRAPH '96*, pp. 96–108, 1996.

22. Y. Livnat, H. W. Shen, and C. R. Johnson. A near optimal isosurface extraction algorithm for structured and unstructured grids. *IEEE Transactions on Visualization and Computer Graphics*, 2(1), 1996.

23. W. E. Lorensen and H. E. Cline. Marching cubes: a high-resolution 3D surface construction algorithm. *Computer Graphics*, 21(3):163–169, 1987.

24. E. N. Lorenz. Deterministic non-periodic flow. *J. Atmospheric Science*, 20:130–141, 1963.

25. D. Luebke, M. Reddy, J. Cohen, A. Varshney, B. Watson, and R. Huebner. *Level of Detail for 3D Graphics*. San Francisco, Morgan Kaufmann, 2002.

26. C. Montani, R. Scateni, and R. Scopigno. A modified lookup table for implicit disambiguation of marching cubes. *Visual Computer*, (10):353–355, 1994.

27. F. C. Moon. *Chaotic Vibrations*. New York, Wiley-Interscience, 1987.

28. G. M. Nielson and B. Hamann. The asymptotic decider: resolving the ambiguity in marching cubes. In *Proceedings of Visualization '91*,

pages 83–91, IEEE Computer Society Press, Los Alamitos, CA, 1991.

29. J. Popovic and H. Hoppe. Progressive simplicial complexes. In *Proceedings of SIGGRAPH '97*, pages 217–224, 1997.

30. P. Rheingans. Color, change, and control for quantitative data display. In *Proceedings of Visualization '92*, pages 252–259. IEEE Computer Society Press, Los Alamitos, CA, 1992.

31. R. Richter, J. B. Vos, A. Bottaro, and S. Gavrilakis. Visualization of flow simulations. *Scientific Visualization and Graphics Simulation* (D. Thalmann, ed.), pages 161–171. New York, John Wiley and Sons, 1990.

32. J. Rossignac and P. Borrel. Multi-resolution 3D approximations for rendering complex scenes. In *Modeling in Computer Graphics: Methods and Applications* (B. Falcidieno and T. Kunii, eds.), pages 455–465. Berlin, Springer-Verlag, 1993.

33. A. S. Saada. *Elasticity Theory and Applications*. New York, Pergamon Press, 1974.

34. W. Schroeder, J. Zarge, and W. Lorensen. Decimation of triangle meshes. *Computer Graphics (SIGGRAPH '92)*, 26(2):65–70, 1992.

35. W. Schroeder. A topology modifying progressive decimation algorithm. In *Proceedings of Visualization '97*. IEEE Computer Society Press, Los Alamitos, CA, 1997.

36. W. Schroeder, K. Martin, and W. Lorensen. *The Visualization Toolkit: An Object-Oriented Approach to 3D Graphics, 3rd Edition*. Clifton Park, NY, Kitware, Inc., 2003.

37. *SCIRun: A Scientific Computing Problem Solving Environment*. Scientific Computing and Imaging Institute (SCI), http://software.sci.utah.edu/scirun.html, 2002.

38. S. P. Timoshenko and J. N. Goodier. *Theory of Elasticity*, 3rd Ed. New York, McGraw-Hill, 1970.

39. E. R. Tufte. *The Visual Display of Quantitative Information*. Cheshire, CT, Graphics Press, 1990.

40. G. Turk. Re-tiling of polygonal surfaces. *Computer Graphics (SIGGRAPH '92)*, 26(2): 55–64, 1992.

41. G. Volpe. Streamlines and streamribbons in aerodynamics. Technical Report AIAA-89-0140, 27th Aerospace Sciences Meeting, 1989.

42. C. Ware. Color sequences for univariate maps: theory, experiments and principles. *IEEE Computer Graphics and Applications*, 8(5):41–49, 1988.

43. D. F. Watson. *Contouring: A Guide to the Analysis and Display of Spatial Data*. Pergamon Press, New York, 1992.

44. G. Wyszecki and W. Stiles. *Color Science: Concepts and Methods, Quantitative Data and Formulae*. New York, John Wiley and Sons, 1982.

45. G. Wyvill, C. McPheeters, and B. Wyvill. Data structure for soft objects. *Visual Computer*, 2(4):227–234, 1986.

PART II
Scalar Field Visualization: Isosurfaces

2 Accelerated Isosurface Extraction Approaches

YARDEN LIVNAT

Scientific Computing and Imaging Institute
University of Utah

2.1 Introduction

The *marching cubes* [7,15] method demonstrated that isosurface extraction can be reduced, using a divide-and-conquer approach to solving a local triangulation problem. In addition, the marching cubes method proposed a simple and efficient local triangulation using a lookup table. However, the marching cubes did not address the *divide* portion of the approach, i.e., how to efficiently search a large dataset for these small local triangulations. In fact, the marching cubes method checks each and every cell of the dataset.

In this chapter, we introduce the three main approaches to accelerate isosurface extraction (Section 2.2) and present two specific methods. Section 2.3 introduces the *span space* metaphor and uses it to devise a near-optimal search method in Section 2.4. Finally, Section 2.5 examines the view-dependent extraction approach and presents a particular implementation.

2.2 Isosurface Extraction Approaches

The various approaches to the acceleration of isosurface extraction fall into three main categories. Each approach is characterized based on the space in which it operates, namely, geometric, value, or image space decomposition. While some methods can be applied to structured and unstructured datasets, others lend themselves to only one, usually structured, grid.

2.2.1 Geometric Space Decomposition

Originally, only structured grids were available as an underlying geometry. Structured grids impose an order on the given cell set. By utilizing this order, methods based on the geometry of the dataset could take advantage of the coherence between adjacent cells.

2.2.1.1 Marching Cubes

Perhaps the best known isosurface extraction method to achieve high-resolution results is the *marching cubes* method introduced by Lorensen and Cline [7]. The marching cubes method concentrated on the approximation of the isosurface inside the cells rather than on efficient location of the involved cells. Toward this end, the marching cube method scans the *entire* cell set, one cell at a time. The novelty of the method is the way in which it decides for each cell *whether* the isosurface intersects that cell and, if so, *how* to approximate it.

2.2.1.2 Octrees

The marching cubes method did not attempt to optimize the time needed to search for the cells that actually intersect the isosurface. This issue was later addressed by Wilhelms and Van Gelder [14], who employed an octree, effectively creating a 3D hierarchical decomposition of the cell set. Each node in the tree was tagged with the minimum and maximum values of the cells it represents. These tags, and the hierarchical nature of the octree, enable one to trim off sections of the tree during the search and thus to restrict the search to only a portion of the original geometric space. Wilhelms and Van Gelder did not analyze the time complexity of the search phase of their algorithm. However, octree decompositions are known to be sensitive to the underlying data. If the underlying data contain some fluctuations or noise, most of the octree will have to be traversed. Livnat et al. [6] present an analysis of the octree algorithm and show that the algorithm has a worst-case complexity of $O(k + k \log n/k)$, where n is the number of cells in the dataset and k is the size of the extracted isosurface [6]. Finally, octrees have been applied primarily to structured grids, and they are not easily adapted to handle unstructured grids.

2.2.2   Value Space Decomposition

Decomposing the value space, rather than the geometric space, has two advantages. First, the underlying geometric structure is of no importance, so this decomposition works well with unstructured grids. Second, for a scalar field in three dimensions, the dimensionality of the search is reduced from three to two.

In general values, space decomposition methods exhibit worst-case complexity of $O(n)$. In Section 2.3, we introduce the span space metaphor, and in Section 2.4, we present a near-optimal isosurface extraction method based on the span space with a worst-case complexity of $O(k + \sqrt{n})$.

2.2.3   Image Space Decomposition

Today's large datasets pose new challenges. Datasets of several gigabytes can be found in many fields (e.g., medicine, flow simulation, and geosciences). The size of isosurfaces extracted from these datasets can reach several million polygons, many of which are less than one pixel in size. Two problems emerge due to the large number of polygons. First, due to the sheer number of cells containing an isosurface, the

Figure 2.1 (Left) The user view. (Right) The same isosurface from a 90-degree angle to the user view, illustrating the incomplete reconstruction. (See also color insert.)

computation of all the local triangulations can be very time-consuming, even if fast acceleration methods are used. Second, the huge number of polygons in the extracted isosurface can easily overwhelm even the most powerful graphics accelerators, leading to poor interaction. In other words, not only the size, n, of the datasets is very large, but also the size, k, of the extracted isosurface becomes a problem.

One current approach to the large-number-of-polygons problem is mesh reduction techniques [13,8]. The mesh reduction is applied to an isosurface either as a postprocess after the extraction phase or during the extraction phase itself [11]. However, mesh reduction is expensive and requires extracting the entire isosurface for examination. Furthermore, a change in the iso-value requires the *full* extraction of a new isosurface and the reapplication of the mesh reduction step.

Another approach is to employ ray-tracing techniques, which do not create an intermediate polygonal representation. Ray-tracing, nevertheless, does not take advantage of graphics hardware and requires a large number of CPUs to achieve interactivity [9].

View-dependent isosurface extraction [5], on the other hand, aims to reduce the search, construction, and rendering times, all at once. The key to this approach is accessing only cells that contain the *visible* portion of the isosurface from a given viewpoint, i.e., based on the image space. The approach is based on a hierarchical front-to-back traversal of the dataset while skipping the nonvisible

 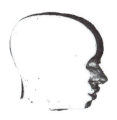

Figure 2.2 Extracted isosurface: a cut plane through the full and view-dependent isosurfaces extracted from the same viewpoint as in Fig. 2.1. Note the large internal structures that are part of the full isosurface but not part of the view-dependent isosurface. (See also color insert.)

sections of the dataset from the current viewpoint. Fig. 2.2 shows the potential savings of such an approach. Note the large section of the isosurface that represents the internal organs in the head yet is not part of the view-dependent isosurface.

2.3 The Span Space

Let $\varphi\colon G \to V$ be a given field and let $\mathcal{D}$ be a sample set over φ, such that

$$\mathcal{D} = \{d_i\} \qquad d_i \in D = G \times V \qquad (2.1)$$

where $G \subseteq R^p$ is a geometric space and $V \subseteq R$, for some $p \in Z$, is the associated value space. Also, let $d = |\mathcal{D}|$ be the size of the dataset.

2.3.1 Definition: Isosurface Extraction

Given a set of samples $\mathcal{D}$ over a field $\varphi\colon G \to V$, and given a single value $v \in V$, find

$$\begin{aligned} S &= \{g_i\} \quad g_i \in G \quad such \; that \\ \varphi(g_i) &= v \end{aligned} \qquad (2.2)$$

Note that S, the isosurface, *need not be topologically simple.*

Approximating an isosurface, S, as a global solution to Equation 2.2 can be a difficult task because of the sheer size, d, of a typical science or engineering dataset.

Data is often generated from 3D images or as solutions to numerical approximation techniques, such as from finite difference or finite

element methods. These methods naturally decompose the geometric space, G, into a set of polyhedral cells, C, where the data points define the vertices. While $n = |C|$, the number of cells, is typically an order of magnitude larger than d, the approximation of the isosurface over C becomes a manageable task. Rather than finding a global solution, one can seek a local approximation within each cell. Hence, isosurface extraction becomes a two-stage process: locating the cells that intersect the isosurface and then, locally, approximating the isosurface inside each such cell. We focus our attention on the problem of finding those cells that intersect an isosurface of a specified iso-value.

On structured grids, the position of a cell can be represented in the geometric space G. Because this representation does not require explicit adjacency information between cells, isosurface extraction methods on structured grids conduct searches over the geometric space, G. The problem as stated by these methods is defined in Section 2.3.2.

2.3.2 Approach: Geometric Search

Given a point $v \in V$ and given a set C of cells in G space where each cell is associated with a set of values $\{u_j\} \in V$ space, find the subset of C that an isosurface, of value v, intersects.

Efficient isosurface extraction for unstructured grids is more difficult, as no explicit order (i.e., position and shape) is imposed on the cells, only an implicit one that is difficult to utilize. Methods designed to work in this domain have to use additional explicit information or revert to a search over the value space, V. The advantage of the latter approach is that one needs only to examine the minimum and maximum values of a cell to determine whether an isosurface intersects that cell. Hence, the dimensionality of the problem reduces to two for scalar fields.

Many methods for isosurface extraction over unstructured grids, as well as some for structured grids, view the isosurface extraction problem as presented in Section 2.3.3.

2.3.3   Approach: The Interval Search

Given a point $v \in V$ and given a set of cells represented as intervals,

$$I = \{[a_i, b_i]\} \quad such\ that \quad a_i, b_i \in V \qquad (2.3)$$

find the subset I_s such that

$$I_s \subseteq I \quad and \quad a_i \leq v \leq b_i \quad \forall (a_i, b_i) \in I_s \qquad (2.4)$$

Posing the search problem over intervals introduces some difficulties. If the intervals are of the same length or are mutually exclusive, they can be organized in an efficient way suitable for quick queries. However, it is much less obvious how to organize an arbitrary set of intervals. Indeed, what distinguishes these methods from one another is the way they *organize* the intervals rather than the way they perform searches.

Our approach is not to view the problem as a search over intervals in V but rather as a search over points in V^2. We start with an augmented definition of the search space.

2.3.4   Definition: The Span Space

Let C be a given set of cells; define a set of points $P = \{p_i\}$ over V^2 such that

$$\forall c_i \in C \quad associate, \quad p_i = (a_i, b_i)$$

where $\qquad\qquad\qquad\qquad\qquad\qquad (2.5)$

$$a_i = \min_j \{v_j\}_i \quad and \quad b_i = \max_j \{v_j\}_i$$

and $\{v_j\}_i$ are the values of the vertices of cell i.

Though conceptually not much different from the interval space, the span space nevertheless leads to a simple and near-optimal search algorithm. In addition, the span space enables us to clarify the differences and commonalities between previous interval approaches as shown by Livnat et al. [6].

One key concept is that points in two dimensions exhibit no explicit relations between themselves, while intervals tend to be viewed as stacked on top of each other, so that overlapping intervals exhibit merely coincidental links. Points do not exhibit such arbitrary ties and in this respect lend themselves to many different organizations. However, as we shall show later, previous methods grouped these points in very similar ways because they looked at them from an interval perspective.

Using our augmented definition, the isosurface extraction problem can be stated as in Section 2.3.5.

2.3.5   Approach: The Span Search

Given a set of cells, C, and its associated set of points, P, in the span space, and given a value $v \in V$, find the subset $P_s \subseteq P$ such that

$$\forall (x_i, y_i) \in P_s \quad x_i < v \quad y_i > v \qquad (2.6)$$

We note that $\forall (x_i, y_i) \in P_s, x_i < y_i$ and thus the associated points will lie above the line $y_i = x_i$. A geometric perspective of the span search is given in Fig. 2.3.

2.4   Near-Optimal Isosurface Extraction (NOISE)

In this section we present an acceleration method that is based on the span space decomposition. Using the span space as our underlying domain, we employ a kd-tree as a means for simultaneously ordering the cells according to their maximum and minimum values.

2.4.1   Kd-Trees

Kd-trees were designed by Bentley [1] in 1975 as a data structure for efficient associative searching. In essence, kd-trees are a multidimensional version of binary search trees. Each node in the tree holds one of the data values and has two sub-trees as children. The sub-trees are constructed so that all of the nodes in one sub-tree, the *left* one for example, hold values that are less than the parent node's value, while the values in the *right* sub-tree are greater than the parent node's value.

Binary trees partition data according to only one dimension. Kd-trees, on the other hand, utilize multidimensional data and partition the data by alternating between each of the dimensions of the data at each level of the tree.

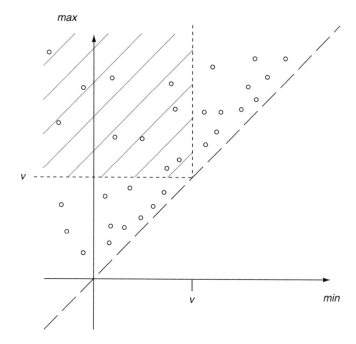

Figure 2.3 Search over the span space.

2.4.2 Search over the Span Space Using Kd-Trees

Given a dataset, a kd-tree that contains pointers to the data cells is constructed. Using this kd-tree as an index to the dataset, the algorithm can now rapidly answer isosurface queries. Fig. 2.4 depicts a typical decomposition of a span space by a kd-tree.

2.4.2.1 *Construction*

The construction of the kd-trees can be carried out recursively in optimal time $O(n \log n)$. The approach is to find the median of the data values along one dimension and store it at the root node. The data is then partitioned according to the median and recursively stored in the two sub-trees. The partition at each level alternates between the *min* and *max* coordinates.

An efficient way to achieve $O(n \log n)$ time is to recursively find the median in $O(n)$, using the

method described by Blum et al. [3], and to partition the data within the same time bound.

A simpler approach is to sort the data into two lists according to the maximum and minimum coordinates, respectively, in order $O(n \log n)$. The first partition accesses the median of the first list, the *min* coordinate, in constant time, and marks all of the data points with values less than the median. We then use these marks to construct the two subgroups, in $O(n)$, and continue recursively.

Although the above methods have complexity of $O(n \log n)$, they do have weaknesses. Finding the median in optimal time of $O(n)$ is theoretically possible yet difficult to program. The second algorithm requires sorting two lists and maintaining a total of four lists of pointers. Although it is still linear with respect to its memory requirement, it nevertheless poses a problem for very large datasets.

A simple and elegant solution is to use a Quicksort-based selection [12]. While this

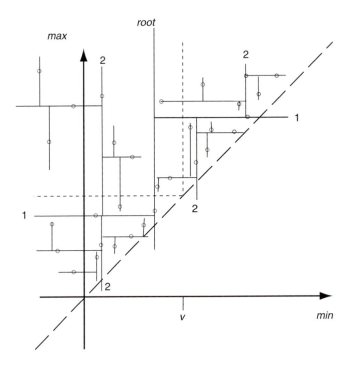

Figure 2.4 A typical decomposition of a span space by a kd-tree.

method has a *worst case* of $O(n^2)$, the *average* case is only $O(n)$. Furthermore, this selection algorithm requires no additional memory and operates directly on the tree.

It is clear that the kd-tree has one node per cell, or span point, and thus that the memory requirement of the kd-tree is $O(n)$.

2.4.2.2 Query

Given an iso-value v, we seek to locate all of the points in Fig. 2.3 that are to the *left* of the vertical line at v and are *above* the horizontal line at v. We note that we do not need to locate points that are *on* these horizontal or vertical lines.

The kd-tree is traversed recursively when the iso-value is compared to the value stored at the current node alternating between the minimum and maximum values at each level. If the node is to the left (above) of the iso-value line, then only the left (right) sub-tree should be traversed. Otherwise, *both* sub-trees should be

traversed recursively. For efficiency, we define two search routines, *search-min-max* and *search-max-min*. The dimension we currently check is the first named, and the dimension that we still need to search is named second. The importance of naming the second dimension will be evident in the next section, when we consider optimizing the algorithm.

Following is a short pseudo-code for the min-max routine.

```
search-min-max( iso-value, node )
{
    if ( node.min < iso-value ) {
        if ( node.max > iso-value )
            construct a polygon(s) from node
        search-max-min ( iso-value,
            node.right );
    }

    search-max-min ( iso-value,
        node.left );
}
```

Estimating the complexity of the query is not straightforward. Indeed, the analysis of the worst case was developed by Lee and Wong [4] only several years after Bentley introduced kd-trees. Clearly, the query time is proportional to the number of nodes visited. Lee and Wong analyzed the worst case by constructing a situation where all the visited nodes are not part of the final result. Their analysis showed that the worst-case time complexity is $O(\sqrt{n} + k)$. The average case analysis of a region query is still an open problem, though observations suggest that it is much faster than $O(\sqrt{n} + k)$ [2,12]. In almost all typical applications $\hat{k} \sim n^{2/3} > \sqrt{n}$, which suggests a complexity of only $O(k)$. On the other hand, the complexity of the isosurface extraction problem is $\Omega(k)$, because it is bound from below by the size of the output. Hence, the proposed algorithm, NOISE, is optimal, $\theta(k)$, for almost all cases and is near optimal in the general case.

2.4.3   Optimization

The algorithm presented in the previous section is not optimal with regard to both the memory requirement and the search time. We now present several strategies to optimize the algorithm.

2.4.3.1   Pointerless Kd-Trees

A kd-tree node, as presented previously, must maintain links to its two sub-trees. This introduces a high cost in terms of memory requirements. To overcome this, we note that, in our case, the kd-tree is completely balanced. At each level, one data point is stored at the node, and the rest are equally divided between the two sub-trees. We can, therefore, represent a pointerless kd-tree as a 1D array of the nodes. The root node is placed at the middle of the array, while the first $n/2$ nodes represent the left sub-tree and the last $(n - 1)/2$ nodes the right sub-tree, as shown in Fig. 2.5.

When we use a pointerless kd-tree, the memory requirements for our kd-tree, per node, reduce to two real numbers, for minimum and maximum values, and one pointer back to the original cell for later usage. Considering that each cell, for a 3D application with tetrahedral cells, has pointers to four vertices, the kd-tree

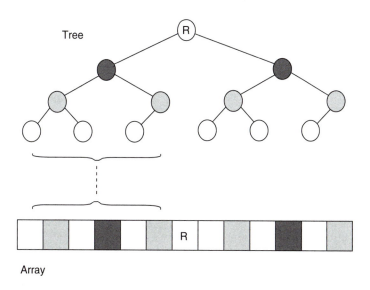

Figure 2.5   Pointerless kd-tree.

memory overhead is even smaller than the size of the set of cells.

The use of a pointerless kd-tree enables one to compute the tree as an offline preprocess and load the tree using a single read in time complexity of only $O(n)$. Data acquisition via CT/MRI scans or scientific simulations is generally very time consuming. The ability to build the kd-tree as a separate preprocess allows one to shift the cost of computing the tree to the data-acquisition stage. Hence, the impact of the initialization stage on the extraction of isosurfaces for large datasets is reduced.

2.4.3.2   Optimized Search

The search algorithm can be further enhanced. Let us consider again the min-max (max-min) routine. In the original algorithm, if the iso-value is less than the minimum value of the node, then we know we can trim the right sub-tree. Consider the case where the iso-value is greater than the node's minimum coordinate. In this case, we need to traverse *both* sub-trees. We have no new information with respect to the search in the right sub-tree, but for the search in the left sub-tree we *know* that the minimum condition is satisfied. We can take advantage of this fact by skipping over the odd levels from that point onward. To achieve this, we define two new routines: *search-min* and *search-max*. Adhering to our previous notation, the name *search-min* indicates that we are only looking for a minimum value.

Examining the search-min routine, we note that the maximum requirement is already satisfied. We do not gain new information if the iso-value is less than the current node's minimum and again only trim off the right sub-tree. If the iso-value is greater than the node's minimum, we recursively traverse the right sub-tree, but with regard to the left sub-tree we now know that all of its points are in the query's domain. We therefore need only to *collect* them. Using the notion of pointerless kd-tree as proposed in Section 2.4.3.1, any sub-tree is represented as a *contiguous* block of the tree's nodes. Collecting

all of the nodes of a sub-tree requires only sequentially traversing this contiguous block.

A pseudo-code of the optimized search for the odd levels of the tree, i.e., searching for minima, is presented in Fig. 2.6. The code for even levels, searching for maxima, is essentially the same and uses the same collect routine.

2.4.3.3   Count Mode

Extracting isosurfaces is an important goal, yet in a particular application one may wish only to know how many cells intersect a particular isosurface. Knowing the number of cells that intersect the isosurface can help one to make a rough estimate of the surface area of the isosurface on a structured grid and on a "well-behaved" unstructured grid. The volume encompassed by the isosurface can also be estimated if one knows the number of cells that lie inside the isosurface as well as the number of cells that intersect it.

The above algorithm can accommodate the need for such particular knowledge in a simple way. The number of cells intersecting the isosurface can be found by incrementing a counter rather than constructing polygons from a node and by replacing collection with a single increment of the counter with the size of the sub-tree, which is known without the need to traverse the tree. To count the number of cells that lie inside the isosurface, one need only look for the cells that have a maximum value below the iso-value.

The worst-case complexity of the count mode is only $O(\sqrt{n})$. A complete analysis is presented by Livnat et al. [6]. It is important to note that the count mode does not depend on the size of the isosurface. The count mode thus enables an application to quickly count the cells that intersect the isosurface and to allocate and prepare the appropriate resources *before* a full search begins.

2.4.4   Triangulation

Once a cell is identified as intersecting the isosurface, we need to approximate the isosurface

```
search_min_max( iso_value, node )
{
  if ( node.min < iso_value ) {
     if ( node.max > iso_value )
        construct polygon(s) from node;
     search_max_min( iso_value, node.right );
     search_max( iso_value, node.left );
  } else
     search_max_min( iso_value, node.left );
}
search_min( iso_value, node )
{
  if ( node.min < iso_value ) {
     construct polygon(s) from node;
     search_skip_min( iso_value, node.right );
     collect( node.left );
  } else
     search_skip_min( iso_value, node.left );
}
search_skip_min( iso_value, skip_node )
{
  if ( skip_node.min < iso_value )
     construct polygon(s) from skip_node;
  search_min( iso_value, skip_node.right );
  search_min( iso_value, skip_node.left );
}
collect( sub_tree )
{
  sequentially construct polygons for all nodes
  in this sub_tree
}
```

Figure 2.6   Optimized search.

inside that cell. Toward this goal, the marching cubes algorithm checks each of the cell's vertices and marks them as either *above* or *below* the isosurface. Using this information and a lookup table, the algorithm identifies the particular way the isosurface intersects the cell. The marching cubes method and its many variants are designed for structured grids, although they can be applied to unstructured grids as well.

Livnat et al. [6] have proposed an algorithm for unstructured grids of tetrahedral cells. We first note that if an isosurface intersects *inside* a cell, then the vertex with the maximum value *must* be above the isosurface, and the vertex with the minimum value *must* be below it.

To take advantage of this fact, we reorder the vertices of a cell according to their ascending values, say v1 to v4, *a priori*, in the initialization stage. When the cell is determined to intersect the isosurface, we need only to compare the iso-value against, *at most*, the two middle vertices. There are only three possible cases: only v1 is *below* the isosurface, only v4 is *above* the isosurface, or {v1, v2} are below and {v3, v4} are

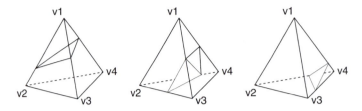

Figure 2.7 Triangulation. The vertices are numbered according to ascending values.

above (Fig. 2.7). Moreover, the order of the vertices of the approximating triangle(s), such that the triangle(s) will be oriented correctly with respect to the isosurface, is known in advance at no cost. We can take further advantage of the fact that there are only four possible triangles for each cell and compute their normals *a priori*. This option can improve the triangulation time dramatically, yet it comes with a high memory price tag.

2.5 View-Dependent Isosurface Extraction

The view-dependent extraction approach is based on the observation that isosurfaces extracted from very large datasets tend to exhibit high depth complexity for two reasons. First, since the datasets are very large, the projection

of individual cells tends to be subpixel. This leads to a large number of polygons, possibly nonoverlapping, projecting onto individual pixels. Second, for some datasets, large sections of an isosurface are internal and thus are occluded by other sections of the isosurface, as illustrated in Fig. 2.2. These internal sections, common in medical datasets, cannot be seen from any direction unless the external isosurface is peeled away or cut off. Therefore, if one can extract just the visible portions of the isosurface, the number of rendered polygons will be reduced, resulting in a faster algorithm. Fig. 2.8 depicts a 2D scenario. In a view-dependent method only the solid lines are extracted, whereas in non-view-dependent isocontouring, both solid and dotted lines are extracted.

Our view-dependent approach is based on a hierarchical traversal of the data and a marching cubes triangulation. We exploit coherency in

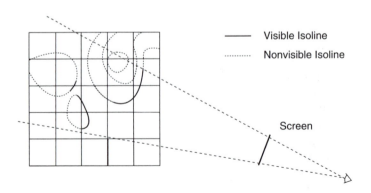

—— Visible Isoline

········ Nonvisible Isoline

Screen

Figure 2.8 A 2D scenario.

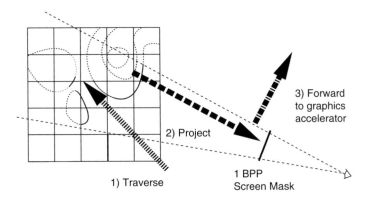

3) Forward
to graphics
accelerator

2) Project

1) Traverse

1 BPP
Screen Mask

Figure 2.9 The three-step algorithm.

the object, value, and image spaces, as well as balancing the work between the hardware and the software. The three-step approach is depicted in Fig. 2.9.

First, Wilhelms and Van Gelder's [14] algorithm is augmented by traversing down a hierarchical tree in a front-to-back order in addition to pruning empty sub-trees based on the min-max values stored at the tree nodes. The second step employs coarse software visibility tests for each [meta-] cell that intersects the isosurface. The aim of these tests is to determine whether the [meta-] cell is hidden from the view point by previously extracted sections of the isosurface (thus the requirement for a front-to-back traversal). Finally, the triangulations of the visible cells are forwarded to the graphics accelerator for rendering by the hardware. It is at this stage that the final and exact (partial) visibility of the triangles is resolved. A data-flow diagram is depicted in Fig. 2.10.

2.5.1 The Min/Max Tree

Wilhelms and Van Gelder [14] used an octree for their hierarchical representation of the underlying dataset. Each node of the octree contained the minimum and maximum values of its subtree. In order to reduce the memory footprint, the octree leaves were one level higher then the data cells. In other words, each leaf node represented the min/max values of 8 ($2 \times 2 \times 2$) data cells. Wilhelms and Van Gelder also introduced a new octree variant (BON tree) for handling datasets with sizes that are not a power of two.

The BON tree was adequate for relatively small datasets of total size less than 2^{28} or 256 MB. In addition, the use of 32-bit pointers for each node proves to be expensive when the data are 8 bits per node. In this case, the min/max values in each node require only 2 bytes but the pointer to the next node requires 4 bytes. The alignment of this pointer further increases the size of each node by 2 bytes, resulting in 8 bytes per node instead of only 2 bytes. As a result, the BON tree can consume as much memory as the dataset itself, and sometimes even more. For large datasets, this is too high a price to pay.

In the case of view-dependent extraction, where each node in the hierarchical tree has to be culled against the virtual framebuffer, each level in the hierarchy increases the cost of the traversal. The tradeoff is that deep hierarchy provides better culling, i.e., pruning of non-visible or empty sections, but with an increased cost in terms of both memory and processing time of each node.

To address these issues, we have implemented a shallow hierarchy [9]. Each level of the hierarchy can have a different number of nodes. The depth of the hierarchy can thus be adapted on a per-dataset case for optimum memory-vs.-time configuration.

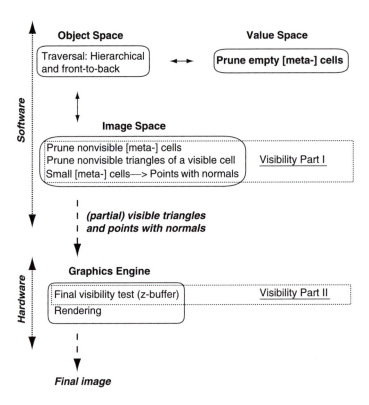

Figure 2.10 The algorithm data flow.

2.5.2 Visibility

Determining whether a meta-cell is hidden and thus can be skipped is fundamental to this algorithm. Toward this end, we create a virtual screen with only one bit per pixel. During the front-to-back traversal of the data we update this virtual screen with the projection of the triangles we extract. In effect, the virtual screen represents a dynamic visibility mask. At each stage of the traversal, the mask state represents the areas that are still visible from the user viewpoint.

Determining whether a meta-cell is visible is accomplished by projecting the meta-cell onto the virtual screen and checking if any part of it is visible, i.e., if any of the pixels it covers is not set. If the entire projection of the meta-cell is not visible, none of its children can be visible.

2.5.3 Hierarchical Framebuffer

It is important to quickly and efficiently classify a cell as visible. A hidden cell and all of its children will not be traversed further, and thus the potential savings can justify the time and effort invested in the classification. A visible cell, on the other hand, does not gain any benefit from this test, and the cost of the visibility test is added to the total cost of extracting the isosurface. As such, the cell-visibility test should not depend heavily on the projected screen area; otherwise, the cost would prohibit the use of the test for meta-cells at high levels of the octree—exactly those meta-cells that potentially can save the most.

To achieve fast classification we employ a hierarchical framebuffer. In particular, each node in the hierarchy represents 64 (8×8)

children. The branch factor of 64 was chosen such that it can be represented in one word and so that comparison against the projection of a meta-cell can be done efficiently.

2.5.3.1 Top-Down Visibility Queries

The main purpose of the hierarchical framebuffer is to accelerate the classification of a meta-cell as *visible*. As such, it is important to know for each node of the framebuffer hierarchy if any part of it might be visible. Therefore, a node in the hierarchy is marked as opaque if and only if *all* of its children are opaque.

Determining whether a meta-cell is visible can now be done in a top-down fashion. The meta-cell is first projected onto the framebuffer, and an axis-aligned bounding box is computed. This bounding box is then compared against the hierarchical framebuffer starting at the root node. The top-down approach accelerates the classification of the meta-cell as it can determine, at early stages, that some portion of the bounding box is visible.

2.5.3.2 Bottom-Up Updates

In order to keep the hierarchical framebuffer state current, we must update it as new triangles are extracted. However, a top-down update of the hierarchy is not efficient. If the extracted triangles are small, then each update of the hierarchy (i.e., rendering of a triangle) requires a deep traversal of the hierarchy. Such traversals are expensive and generally add only a small incremental change.

To alleviate the problem of projecting many small triangles down the hierarchy, we employ a bottom-up approach. Using this approach, the contribution of a small triangle is limited to only a small neighborhood at the lowest level, and thus only a few updates up the hierarchy will be necessary.

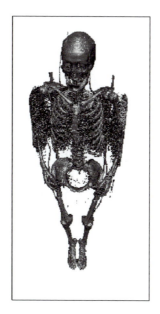
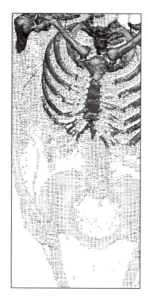

Figure 2.11 Rendering points. The left image was extracted based on the current viewpoint. The right image shows a close-up of the same extracted geometry. (See also color insert.)

2.5.4 Scan Conversion of Concave Polygons

Once a data cell is determined to both be visible and intersect the isosurface, we use the marching cubes method to triangulate the cell. In most cases, the marching cubes method creates more than one triangle, but it is not obvious which one of the triangles is in front of the other. As such, we need to render each one of them onto the framebuffer and forward all of them to the hardware for rendering.

Updating the hierarchical framebuffer one triangle at a time is not efficient, as the triangles from a single cell are likely to affect only a small section of the hierarchy and might even overlap. We thus employ a scan-conversion algorithm, which can simultaneously project a collection of triangles and concave polygons. The use of the scan-conversion algorithm is made particularly simple due to the bottom-up update approach. The projected triangles and polygons are scan-converted at screen resolution at the bottom level of the framebuffer hierarchy before the changes are propagated up the hierarchy. Applying the scan conversion in a top-down fashion would make the algorithm unnecessarily complex.

Additional acceleration can be achieved by eliminating redundant edges, projecting each vertex only once per cell, and using triangle strips or fans. To achieve these goals, the marching cubes lookup table is first converted into a triangle fans format. The usual marching cubes lookup table contains a list of the triangles (three vertices) per case.

2.5.5 Rendering Points

Another potential savings is achieved by using points with normals to represent triangles or (meta-) cells that are smaller than a single pixel. The use of points in isosurface visualization was first proposed as the *Dividing Cubes Method* by Lorensen and Cline [7]. Pfister et al. [10] also used points to represent surface elements (surfels) for efficient rendering of complex geometry.

In a view-dependent approach, during the traversal of the hierarchy, whenever a nonempty (meta-) cell is determined to have a size less than two pixels and its projection covers the center of a pixel, it is represented by a single point. Note that the size of the bounding box can be almost two pixels wide (high) and still cover only a single pixel. Referring to Fig. 2.12, we require

```
if (right - left < 2 ) {
    int L = trunc(left);
    int R = trunc(right)
if (L == R -1)
    // create a point at R + 0.5
else if (L == R)
    // too small: does not cover the center
    of a pixel
else
    // too large : covers more than two
    pixels
}
```

and similarly for the bounding box height.

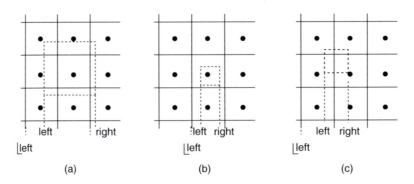

| | (a) | (b) | (c) |

Figure 2.12 Pixel center and the projected bounding box. A point is created for cases (a) and (b) but not for (c).

Figure 2.11 shows an example in which some of the projected cells are small enough that they can be rendered as points. On the left is the image as seen by the user, while on the right is a close-up view of the same extracted geometry (i.e., the user zoomed in but did not re-extract the geometry based on the new viewpoint). Notice that much of the image on the left is represented as points. Points are useful not only in accelerating the rendering of a large iso-surface but also in remote visualization because less geometry needs to be transferred over the network.

2.5.6 Fast Estimates of a Bounding Box of a Projected Cell

The use of the visibility tests adds an overhead to the extraction process that should be minimized. Approximating the screen area covered by a meta-cell, rather than computing it exactly, can accelerate the meta-cell visibility tests. In general, the projection of a meta-cell on the screen has a hexagon shape with non-axis-aligned edges. We reduce the complexity of the visibility test by using the axis-aligned bounding box of the cell projection on the screen, as seen

in Fig. 2.13. This bounding box is an overestimate of the actual coverage and thus will not misclassify a visible meta-cell, though the opposite is possible.

The problem is in how to find this bounding box quickly. The simplest approach is to project each of the eight vertices of each cell onto the screen and compare them. This process involves eight perspective projections and either two sorts (x and y) or 16 to 32 comparisons.

Our solution is to approximate the bounding box as follows. Let P be the center of the current meta-cell in object space. Assuming the size of the meta-cell is (dx, dy, dz), we define the eight vectors:

$$D = \left(\pm \frac{dx}{2}, \pm \frac{dy}{2}, \pm \frac{dz}{2}, 0 \right) \quad (2.7)$$

The eight corner vertices of the cell can be represented as

$$V = P + D = P + (\pm D_x, \pm D_y, \pm D_z)] \quad (2.8)$$

Applying the viewing matrix M to a vertex V amounts to

$$VM = (P + D)M = PM + DM \quad (2.9)$$

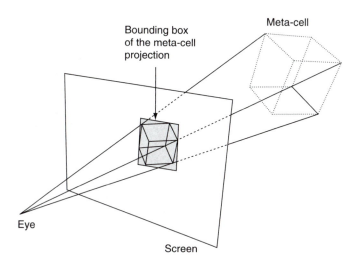

Figure 2.13 Perspective projection of a meta-cell, the covered area, and the bounding box.

After the perspective projection, the x screen coordinate of the vertex is

$$\frac{[VM]_x}{[VM]_w} = \frac{[PM]_x + [DM]_x}{[PM]_w + [DM]_w} \qquad (2.10)$$

To find the bounding box of the projected meta-cell, we need to find the minimum and maximum of these projections over the eight vertices in both x and y. Alternatively, we can overestimate these extreme values such that we may classify a nonvisible cell as visible but not the opposite. Overestimating can thus lead to more work but will not introduce errors.

The maximum x screen coordinate can be estimated as follows:

$$\begin{aligned}
\max\left(\frac{[VM]_x}{[VM]_w}\right) &\leq \frac{\max([PM]_x + [DM]_x)}{\min([PM]_w + [DM]_w)} \\
&\leq \frac{[PM]_x + \max([DM]_x)}{\min([PM]_w + [DM]_w)} \\
&\leq \frac{[PM]_x + [D^+M^+]_x}{\min([PM]_w + [DM]_w)}
\end{aligned}$$

where we define the $+$ operator to mean to use the absolute value of the vector or matrix elements.

Assuming that the meta-cells are always in front of the screen, we have

$$\begin{aligned}
V_z > 0 &\Rightarrow P_z - D_z^+ > 0 \\
&\Rightarrow [PM]_z - [D^+M^+]_z > 0
\end{aligned} \qquad (2.11)$$

Thus,

$$\max \frac{[VM]_x}{[VM]_w} = \begin{cases} \dfrac{[PM]_x + [D^+M^+]_x}{[PM]_w - [D^+M^+]_w} & \text{if numerator} \geq 0 \\[2ex] \dfrac{[PM]_x + [D^+M^+]_x}{[PM]_w + [D^+M^+]_w} & \text{otherwise} \end{cases}$$

Similarly, the minimum x screen coordinate can be overestimated as

$$\min \frac{[VM]_x}{[VM]_w} \leq \begin{cases} \dfrac{[PM]_x - [D^+M^+]_x}{[PM]_w + [D^+M^+]_w} & \text{if numerator} \geq 0 \\[2ex] \dfrac{[PM]_x - [D^+M^+]_x}{[PM]_w - [D^+M^+]_w} & \text{otherwise} \end{cases}$$

The top and bottom of the bounding box are computed similarly.

2.6 Summary

In Chapter 2, we classified the various approaches to the acceleration of isosurface extraction into three categories, namely geometric, value-based, and image-based. We also presented two particular acceleration methods. The NOISE method is based on the span space representation of the value space and exhibits a worst-case complexity of $O(k + \sqrt{n})$. The view-dependent method is based on a front-to-back traversal, dynamic pruning (based on a hierarchical visibility framebuffer), and point representation of distant meta-cells.

References

1. J. L. Bentley. Multidimentional binary search trees used for associative search. *Communications of the ACM*, 18(9):509–516, 1975.

2. J. L. Bentley and D. F. Stanat. Analysis of range searches in quad trees. *Info. Proc. Lett.*, 3(6):170–173, 1975.

3. M. Blum, R. W. Floyd, V. Pratt, R. L. Rivest, and R. E. Tarjan. Time bounds for selection. *J. of Computer and System Science*, 7:448–461, 1973.

4. D. T. Lee and C. K. Wong. Worst-case analysis for region and partial region searches in multidimensional binary search trees and balanced quad trees. *Acta Informatica*, 9(23):23–29, 1977.

5. Y. Livnat and C. Hansen. View-dependent isosurface extraction. In *Visualization '98*, pages 175–180. ACM Press, October 1998.

6. Y. Livnat, H. Shen, and C. R. Johnson. A near optimal isosurface extraction algorithm using the span space. *IEEE Trans. Vis. Comp. Graphics*, 2(1):73–84, 1996.

7. W. E. Lorensen and H. E. Cline. Marching cubes: A high resolution 3D surface construction algorithm. *Computer Graphics*, 21(4):163–169, 1987.

8. K. M. Oh and K. H. Park. A type-merging algorithm for extracting an isosurface from volumetric data. *The Visual Computer*, 12:406–419, 1996.

9. S. Parker, P. Shirley, Y. Livnat, C. Hansen, and P. P. Sloan. Interactive ray tracing for isosurface rendering. In *Visualization '98*, pages 233–238. IEEE Computer Society Press, 1998.

10. H. Pfister, M. Zwicker, J. van Baar, and M. Gross. Surfels: Surface elements as rendering primitives. In *Siggraph 2000, Computer Graphics Proceedings*, pages 335–342, 2000.

11. T. Poston, H. T. Nguyen, P. A. Heng, and T. T. Wong. 'Skeleton climbing': fast isosurfaces with fewer triangles. In *Pacific Graphics '97*, pages 117–126, Seoul, Korea, 1997.

12. R. Sedgewick. *Algorithms in C + +*. Boston, Addison–Wesley, 1992.

13. R. Shekhar, E. Fayyad, R. Yagel, and J. F. Cornhill. Octree-based decimation of marching cubes surfaces. In *Visualization '96*, pages 335–342. IEEE Computer Society Press, Los Alamitos, CA, 1996.

14. J. Wilhelms and A. Van Gelder. Octrees for faster isosurface generation. *ACM Transactions on Graphics*, 11(3):201–227, 1992.

15. G. Wyvill, C. McPheeters, and B. Wyvill. Data structure for soft objects. *The Visual Computer*, 2:227–234, 1986.

3   Time-Dependent Isosurface Extraction

HAN-WEI SHEN
Department of Computer Science and Engineering
The Ohio State University

New challenges for scientific visualization researchers have emerged over the past several years as the size of data generated from simulations experienced exponential growth. A major factor that is contributing to the exponential growth in the size of data is scientists' ability to perform time-varying simulations with finer temporal resolutions and a larger number of time-steps. To analyze complex dynamic phenomena from a time-varying dataset, it is necessary to navigate and browse the data in both the spatial and the temporal domain, select data at different resolutions, experiment with different visualization parameters, and compute and animate selected features over a period of time. To facilitate exploratory visual data analysis, it is very important that the visualization software be able to compute, animate, and track desired features at an interactive speed. This chapter discusses the topic of isosurface extraction for time-varying data. We will first discuss two isosurface cell search data structures that can minimize the storage overhead while keeping the search performance high. We will then discuss the work on extracting time-varying isosurfaces in 4D space that allows for smooth animation of isosurfaces in time. These high-dimensional isosurface extraction algorithms can also provide several additional benefits, such as computing the isosurface envelopes and interval volumes.

3.1   Space-Efficient Search Data Structure

One of the most commonly used approaches to computing isosurfaces is the marching cubes algorithm, proposed by Lorensen and Cline [1]. The marching cubes algorithm is simple and robust. However, the process of a linear search for isosurface cells can be expensive. To improve the performance, researchers have proposed various schemes that can accelerate the search process. Examples include Wilhelms and Van Gelder's [2] octrees, Livnat et al.'s [3] NOISE method, Shen et al.'s [4] ISSUE algorithm, Itoh and Koyamada's [5,6] Extrema Graph method, Bajaj et al.'s [7,8] Fast Isocontouring method, and Cignoni et al.'s [9] Interval Tree algorithm.

Inevitably, these acceleration algorithms incur overhead for storing extra search indices. For a steady scalar field, for example, only a single time-step of data is present; this extra space is often affordable, and the highly interactive speed of extracting isosurfaces can compensate for the overhead. However, for time-varying simulations, a typical solution can contain a large number of time-steps, and every simulation step can produce a great amount of data. The overall storage requirement for the search index structures can be overwhelming. Furthermore, when analyzing a time-varying scalar field, a user may want to explore the data back and forth in time, with the same or

different iso-values. This will require a significant amount of disk I/O for accessing the indices
for data at different time-steps when there is not
enough memory space for the entire time sequence. As a result, the performance gain from
the efficient isosurface extraction algorithm
could be offset by the I/O overhead.

Researchers have proposed various methods
for efficient searching of isosurfaces for time-
varying scalar fields. The main focus of the
research is to devise a new search-index structure for a time-varying field, so that the storage
overhead is kept small while the performance of
the isosurface extraction remains high. In the
following section, I describe two algorithms to
achieve this goal. One is to reduce the storage
overhead by trading the accuracy of time-varying isosurface searches using a data structure
called a *temporal hierarchical index tree*. The
other is to extend the octree data structure for
out-of-core processing.

3.1.1 Temporal Hierarchical Index Tree

Given a time interval $[i,j]$ and a time-varying
field, a cell's *temporal extreme values* in this
interval can be defined as

$$min_i^j = \text{MIN}(min_t), t = i..j$$
$$max_i^j = \text{MAX}(max_t), t = i..j$$

where MIN and MAX are the functions that
compute the minimum and the maximum
values, and min_t and max_t are the cell's extreme
values at the tth time-step; we call them the
cell's *time-specific extreme values*. To locate
the isosurface cells in the time-varying field,
one can approximate a cell's extreme values at
any time-step within the time span $[i,j]$ by the
cell's temporal extreme values, min_i^j and max_i^j,
and use them to create a single search index.
Using this approximated search index, an isosurface at a time-step t, $t \in [i,j]$ can be computed by first finding the cells that have min_i^j
smaller and max_i^j larger than the iso-value. The
actual scalar data of these cells at the specific
time t is then used to compute the geometry of
the isosurface. Using the approximated search

index can greatly reduce the storage space required, because only one index is used for all
the $j - i + 1$ time-steps. It also guarantees to
find all the isosurface cells, because

$$\text{if } t \in [i,j] \text{ and } min_t < V_{iso} \text{ and } max_t > V_{iso}$$
$$\Longrightarrow min_i^j < V_{iso} \text{ and } max_i^j > V_{iso}$$

where V_{iso} is the iso-value and t is the time-step
at which the query is issued.

Care should be taken when using the idea just
described because the temporal extreme values
provide only a necessary, not a sufficient, condition to qualify a cell as an isosurface cell. As a
result, many nonisosurface cells are visited as
well.

The goal of the *temporal hierarchical index
tree* data structure is to provide an adaptive
scheme that can reduce the storage overhead
incurred by the search index for isosurface extraction in time-varying fields without sacrificing
the performance too much. The underlying idea
of the data structure is to classify the cells
according to the amount of variation in the
cell's values over time. Cells that have a small
amount of variation are placed in a single node of
the tree that covers the entire time span. Cells
with a larger variation are placed in multiple
nodes of the tree multiple times, each for a
short time span. When generating an isosurface,
a simple traversal will retrieve the set of nodes
that contains all cell index entries needed for a
given time-step. The cells in each node can be
organized using existing algorithms developed
for generating isosurfaces from a steady dataset. In the following section, details about the
data structure are described.

3.1.1.1 Data Structure
The span space [3] is useful for analyzing the
temporal variation of a cell's extreme values. In
the span space, each cell is represented by a point
whose x coordinate represents its minimum value
and whose y coordinate represents its maximum
value. For a time-varying field, a cell has multiple
corresponding points in the span space, and each
point represents the cell's extreme values at one

time-step. To characterize a cell's scalar variation over time, the area over which the corresponding points spread in the span space provides a good measure—the wider these points spread, the higher is the cell's temporal variation. This variation can be quantified by using the *lattice subdivision* scheme of the span space [4], which subdivides the span space into $L \times L$ non–uniformly spaced rectangles, called *lattice elements*. To perform the subdivision, we first sort, in ascending order, all the distinct extreme values of the cells in the time-varying field within the given time interval and establish a list. We then find $L + 1$ scalar values, $\{d_0, d_1, \ldots, d_L\}$, in the list that can evenly separate the list into L sublists with an equal length. These $L + 1$ scalar values are used to draw $L + 1$ vertical lines and $L + 1$ horizontal lines to subdivide the span space. The list d_i is chosen in this way to ensure that cells can be more evenly distributed among the lattice elements. Fig. 3.1 is an example of the lattice subdivision.

Using the lattice subdivision, the temporal hierarchical index tree classifies the cells in a time-varying field based on the temporal variations of their extreme values. Given a time interval $[i,j]$ in the time-varying field, the root

node in the temporal hierarchical index tree, denoted as N_i^j, contains cells that have low scalar variations in the time interval $[i,j]$. We can determine whether a cell has a low temporal variation by inspecting the locations of the cell's $j - i + 1$ corresponding points in the span space. If all of the cell's corresponding points are located within an area of 2×2 lattice elements, a cell has a low temporal variation. This cell is then placed into the node N_i^j and is represented by its temporal extreme values, min_i^j and max_i^j. On the other hand, for cells that do not satisfy the criterion, the time interval $[i,j]$ is split in half, that is, into $[i, i + (j - i + 1)/2 - 1]$ and $[i + (j - i + 1)/2, j]$, and continues to classify the cells recursively into each of the two N_i^j subtrees that have roots $N_i^{i+(j-i+1)/2-1}$ and $N_{i+(j-i+1)/2}^j$. The temporal hierarchical tree has leaf nodes $N_t^t, t = i..j$. The leaf nodes contain cells that have the highest scalar variations in time so that the cells' time-specific extreme values are used. Cells that are classified into non-leaf nodes are represented by their temporal extreme values. The use of the temporal extreme values directly contributes to the reduction of the overall index size because the temporal extreme values are used to refer to a cell for more than one time-step. Fig. 3.2 shows an example of the temporal hierarchical index tree with a time interval [0,5].

To facilitate an efficient search for isosurface cells, a search index for each node of the

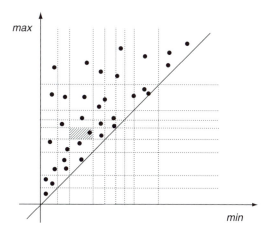

Figure 3.1 In this example, the span space is subdivided into 9×9 lattice elements. Each lattice element is assigned an integer coordinate based on its row and column number. The shaded lattice element in this figure has a coordinate (2,4).

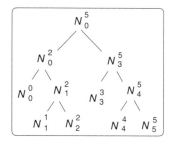

Figure 3.2 Cells in a time-varying field are classified into a temporal hierarchical index tree based on the temporal variations of their extreme values. In this figure, the tree is built from a time-varying field with a time interval [0,5].

temporal hierarchical tree is created. This can be done by using any existing isosurface extraction algorithm based on the value-partition paradigm. Here we propose to use a modified ISSUE algorithm [4] that can provide optimal performance. For every node N_i^j in the temporal hierarchical index tree, cells contained in the node are represented by their extreme values (min_i^j, max_i^j). To create the search index, we use the lattice subdivision described previously and sort cells that belong to the lattice elements of each row, excluding the lattice element at the diagonal line, into a list based on the cells' representative minimum values in ascending order. Another list in each row is created by sorting the cells' representative maximum values in descending order. For those lattice elements at the diagonal line, the interval tree method [9] is used to create one interval tree for each element.

3.1.1.2 Isosurface Extraction

Given the temporal hierarchical index tree, this section describes the algorithm that is used to locate the isosurface cells at run time. We first describe a simple traversal method to retrieve the sets of nodes that contain all cell index entries needed for a given time-step. We then describe the isosurface cell search algorithm used for the lattice search index built in each node.

Given an isosurface query at time-step t, the isosurface is computed by first locating the nodes in the tree that may contain the isosurface cells. This is done by recursively traversing from the root node N_i^j to one of its two child nodes, N_a^b, such that $a \leq t \leq b$ until the leaf node N_t^t is reached. Along the traversal path, the isosurface cell search is performed using a method that will be described later, at each encountered node. The tree is constructed so that every cell in the field exists in one of the nodes in the traversal path. These cells have their representative extreme values, temporal or time-specific, as the approximation of their actual extreme values at time-step t. Fig. 3.3 shows an example of the traversal path.

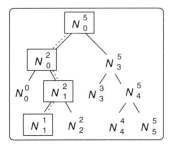

Figure 3.3   In this example, tree nodes that are inside the rectangular boxes are on the traversal path for an isosurface query at time-step 1.

At every node along the traversal path, the lattice search index built at the node is used to locate the candidate isosurface cells. Given an iso-value V_{iso}, the lattice element with integer coordinates $[I, I]$ that contains the point (V_{iso}, V_{iso}) in the span space is first identified. The isosurface cells are then located in the upper left corner that is defined by the vertical line $x = V_{iso}$ and the horizontal line $y = V_{iso}$ as shown in Fig. 3.4.

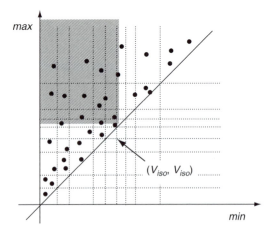

Figure 3.4   In this case, lattice element (4,4) contains the point (V_{iso}, V_{iso}). Isosurface cells are located in the shaded area.

The candidate isosurface cells can be collected from the following three categories:

1. For every list in the row R, $R = I + 1..L - 1$ that was sorted by the cells' minimum values, the cells from the beginning of the list are collected until the first cell is reached that has a representative minimum value that is greater than the iso-value.

2. For the list in row I that was sorted by the maximum values, the cells from the beginning of the list are collected until the cell is reached that has a representative maximum value that is smaller than the iso-value.

3. Collect the isosurface cells from the interval tree built at lattice element $[I, I]$. The method and its details are presented by Cignoni et al. [9].

After the candidate isosurface cells are located, we use the cells' actual data at time-step t to perform triangulation.

The above algorithm has optimal performance since the isosurface cells in categories 1 and 2 are collected without the need for any search. The interval tree method used in category 3 has an optimal efficiency of $O(logN)$, where N is the number of cells in the field. In addition, the number of cells in category 3 is also usually very small.

As mentioned previously, a candidate isosurface cell may not be an isosurface cell after all. These nonisosurface cells come from the non-leaf nodes in the temporal hierarchical index tree since a cell's time-specific extreme values, min_t and max_t, may not contain the given iso-value even though the approximated extreme values, i.e., the temporal extreme values min_i^j and max_i^j, do contain the iso-value. Although this problem will not cause a wrong isosurface to be generated, since the triangulation routine will detect the case and create no triangles from these cells, it does incur performance overhead. Actually, this performance overhead is an expected consequence of using temporal extreme values as the approximated extreme values for cells; we trade performance for storage space.

In fact, the performance overhead is bound by the resolution of the lattice subdivision in the span space. In the algorithm, a cell is placed into the node N_i^j in the temporal hierarchical index tree in such a way that its representing points at different time-steps within time interval $[i, j]$ always reside within an area of 2×2 lattice elements in the span space. Therefore, for any node N_i^j in the tree, the worst case for the number of the nonisosurface cells being visited is estimated as the number of cells in the two rows and two columns of the lattice elements at the boundary layers of the lattice elements that are searched for the candidate isosurface cells, as shown in the shaded area in Fig. 3.5. Therefore, the user-specified parameter L, in an $L \times L$ lattice subdivision, becomes a control parameter that is used to determine the tradeoff factor between the storage space and the isosurface extraction time.

3.1.1.3 Node Fetching and Replacement
Ideally, if the entire temporal hierarchical index tree resides in main memory, there is no I/O

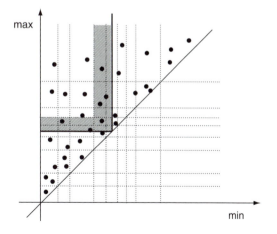

Figure 3.5 At every tree node, the nonisosurface cells being unnecessarily visited are confined within the two rows and two columns of the lattice elements, as shown in the shaded area. Increasing the resolution of the lattice subdivision can reduce the number of cells in this area, for the price of a larger temporal hierarchical index tree.

required when the user randomly queries for isosurfaces at different time-steps. However, the memory requirement is usually too high to make this practical. To address this issue, the temporal hierarchical index tree can be output to a file. When an isosurface at a time-step t is queried, the traversal path is followed along the tree as described previously, and only the nodes being visited are brought into main memory. Initially, all nodes on the traversal path must be read in. Subsequently, if the user queries for an isosurface at a different time-step, the algorithm traverses the search tree and brings in only those nodes that are not already in main memory. In fact, because the nonleaf nodes contain cell index entries that are shared by several time-steps, they are very likely to be in memory already. In this case, only the differential nodes, a small portion of the index tree, need to be read in from the disk. As a result, the amount of I/O required for a subsequent isosurface query can be considerably smaller. Fig. 3.6 gives an example.

Although it is always desirable to retain as many nodes in memory as possible in case the user needs to go back and forth in time when querying the isosurfaces, those nodes that are not in use have to be replaced when the memory limitation is exceeded. To determine the node

that must be replaced, the temporal hierarchical index tree algorithm includes a node replacement policy that assigns a priority to each node, based on its *depth* in the tree. The smaller the depth of a node is, the higher is its priority. For example, the root of a tree has a depth of zero, and therefore it has the highest priority. The reason is that the root node contains search index entries to those cells that have the lowest temporal variations, and, thus, these index entries are used by many time-steps. When a node has to be replaced, we select the node that has the lowest priority. If there are more nodes than one with the same priority, we remove the one that was least recently used (LRU).

3.1.2 Temporal Branch-on-Need (T-BON) Trees

Another method for time-varying isosurface extraction is the T-BON algorithm, proposed by Sutton et al. [10], which extends the branch-on-need octree (BONO) [2]. The octree is a widely used data structure for spatial subdivision in many graphics applications. For isosurface extraction, octrees allow for efficient pruning of data not containing isosurface patches. An octree is constructed by recursively bisecting each dimension of the volume, and thus subdividing the volume into eight octants each time, until the leaf nodes reach a predefined size. In each tree node, the minimum and maximum values of all the voxels in the corresponding volume block are stored. At run time, a user-supplied iso-value is used to compare against the extreme values stored in the tree nodes. If there is a node whose min/max values do not contain the iso-value, the entire branch under the node can be pruned. The efficiency of the octree relies on the fact that most of the regions in the volume do not contain the isosurface and thus can be quickly rejected. Compared with the value space partition algorithms [3,4,9], the space overhead for storing the octree is generally smaller, although it does not provide optimal search performance [3].

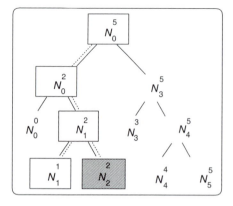

Figure 3.6 In this case, if the user changes the isosurface query from time-step 1 to time-step 2, only the node N_2^{-2} needs to be brought in from the disk.

To minimize the space overhead for the octree nodes even more, Wilhelms and Van Gelder [2] proposed the branch-on-need octree (BONO) data structure, which can reduce the number of tree nodes when the dimensions of the volume are not a power of two. In the BONO method, instead of subdividing the volume dimensions equally, "virtual zeros" are padded to the volume to force the dimensions of the volume to be powers of two, and then an equal subdivision is performed. This effectively makes the lower half of the subdivision in each dimension cover the largest possible power of two voxels. Fig. 3.7 shows a 2D illustration of the partition. Note that the zeros are only used to assist the partition, and thus no actual padding is performed. Wilhelms and Van Gelder [2] showed that the ratio of the BONO nodes to the data is less than .16.

In the T-BON algorithm, Sutton et al. [10] used the BONO data structure to index data at every time-step. Since it is assumed that the volume has the same dimensions in all the time-steps, only one BONO tree is needed to keep the spatial subdivision information. For the data values, the tree node stores a pair of min/max values for every time-step of the corresponding data block. The data associated with the leaf nodes of the tree is stored as separate blocks in disk, which are accessed at run time using a demand-paging technique similar to the one proposed by Cox and Ellsworth [11].

To reduce the amount of memory required at run time for keeping the T-BON structure, nodes are read into main memory only when necessary. Initially, the entire tree is stored in disk. When the user queries an isosurface at a particular time-step, a traversal of the T-BON begins by

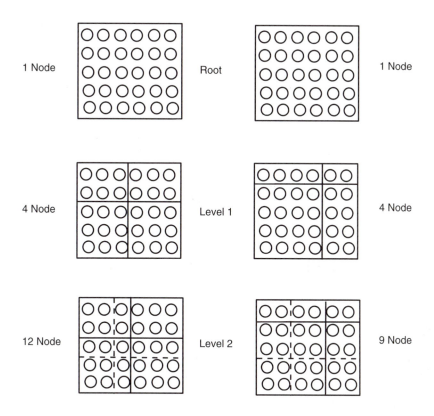

Figure 3.7   A 2D illustration of BONO partitioning.

bringing the root node to main memory. If the root's min/max values for the time-step intersect the iso-value, the children of the root are also read into main memory. This process is repeated recursively until all the tree nodes that intersect the isosurface are read. The data blocks at the time-step associated with the leaf nodes of the tree containing the isosurface are also read from the disk for surface extraction. To further minimize the input/output (I/O) overhead and the storage cost, the T-BON algorithm provides two additional features. One is to avoid redundant read of tree nodes and data blocks that have already been brought into memory. To ensure this, two lists are created to index the already in-core tree nodes and data blocks. The lists are deleted when the user requests isosurfaces at different time-steps. The T-BON algorithm also maintains a predefined memory footprint regardless of the total size of the underlying time-varying dataset, which is typically equal to one time-step of the tree and data.

Generally speaking, the T-BON algorithm provides a practical solution when the underlying time-varying data is large. This is because the storage overhead of the octree data structure is generally smaller compared to the algorithms that have theoretically better computation complexity but require more information for an efficient isosurface search. In addition, the octree is a more general data structure that can possibly be shared by other visualization algorithms such as volume rendering in a visualization system.

3.2 Extracting Isosurfaces in Four Dimensions

Displaying isosurfaces at discrete time-steps may not generate a smooth animation if the original data does not have enough temporal resolution. To address this problem, a straightforward method is to first interpolate the data linearly in time to create intermediate time-steps, and then apply a regular 3D isosurface extraction algorithm (such as the marching cubes method) to the interpolated data. Animating isosurfaces this way, however, can be expensive because the amount of data interpolations required will be in proportion to the number of intermediate surfaces needed for the animation, which can be quite large. Researchers have proposed algorithms to extract isosurfaces directly from the 4D (space plus time) data, which can be sliced down to a sequence of 3D isosurfaces at an arbitrary temporal resolution. In this case, no intermediate data need to be interpolated. In the following sections, we discuss two methods that perform 4D isocontouring. One is the recursive contour-meshing algorithms proposed by Weigle and Banks [12,13], and the other is a direct triangulation of hypercubes by Bhaniramka et al. [14].

3.2.1 Recursive Contour Meshing

Weigle and Banks [12,13] proposed an isocontouring method to extract isosurfaces from n-dimensional data. Given an n-cell, that is, a cell in n-dimensional space, their algorithm first splits the cell into a number of n-simplices. In 2D space, a 2-cell is a square, and a 2-simplex is a triangle. In 3D space, a 3-cell is a cube, and a 3-simplex is a tetrahedron. Fig. 3.8 shows examples of n-cells and n-simplices for n equal to 1, 2, 3, and 4. After splitting the n-cells to n-simplices, contours are computed from each of the n-simplices by looping through the faces of the simplices. Note that each face of a n-simplex is in fact an $(n-1)$-simplex. For example, the faces of a 3-simplex (tetrahedron) are 2-simplices, which are triangles. The faces of a 4-simplex are 3-simplices, which are tetrahedra. From the $(n-1)$-simplex faces, recursive contouring will be performed to form a polytope (generalization of a polyhedron) in the n-simplex. This polytope is the isocontour within the n-simplex.

Extracting isocontours from time-varying data can be seen as a special case of the above n-dimensional algorithm, where n is equal to four. In this case, the time-varying

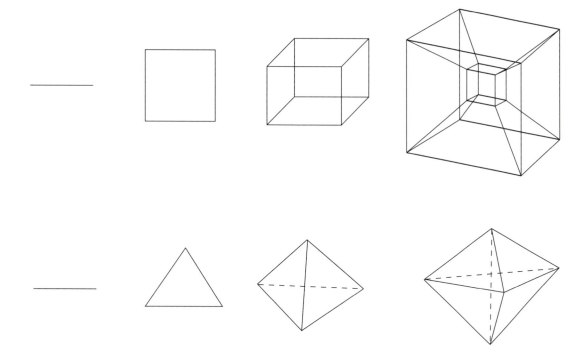

Figure 3.8 Examples of *n*-cells and *n*-simplices for *n* equal to 1, 2, 3, and 4.

field consists of a collection of 4-cells, or hyper-cubes. Each 4-cell can be subdivided into a number of 4-simplices. Each 4-simplex has five 3-simplex faces, or five tetrahedral faces. In each of the tetrahedra, we can compute the contour-polygons. These contour-polygons collectively form a polyhedron in four dimensions, which is the isocontour in the 4-simplex. The polyhedron can be further subdivided into a number of 3-simplices, which are tetrahedra in four dimensions.

When the 4D isosurfaces (3-simplices) are extracted using the above algorithm, to get a 3D isosurface at an arbitrary time, we can slice the 3-simplices according to the time values. That is, we can treat the time coordinate of each tetrahedron vertex as a scalar value, and then use the standard marching tetrahedra algorithm to perform triangulations, where the iso-value is the desired time.

3.2.2 High-Dimensional Triangulation Tables

The recursive contour-meshing algorithm described above can produce an excessive amount of simplices since the 4-cells need to first be subdivided into 4-simplices before contouring. This is analogous to breaking a cube into five or six tetrahedra in the 3D case and extracting the isosurface in each of the tetrahedra. Compared to the marching cubes algorithm, which uses a precomputed triangulation table, extracting iso-surfaces after simplicial decomposition will create more triangles.

To address the problem, Bhaniramka et al. [14] proposed a method to generate the tri-angulation table similar to the marching cubes table for any *n*-dimensional hypercubes. In their algorithm, midpoints of the edges of the hypercube that have one endpoint with a posi-tive label and one endpoint with a negative label are collected. A positive label is assigned to a vertex when its scalar value is greater than the iso-value, and the negative label means that the scalar value is smaller than the iso-value. Together with the other vertices that have a positive label (or negative label, if used consistently) in the hypercube, the convex hull of the points is created. This convex hull is an n-polytope lying in the hypercube. To extract the isosurface, any faces of the polytope lying on the boundary of the hypercube, i.e., any faces that share the vertices of the hypercube, are removed. The remaining faces comprise the isosurface in n dimensions, which can be written into the table. Note that the remaining faces of the polytope might not be simplices. In this case, triangulation needs to be performed. Care is taken by Bhaniramka et al. [14] to ensure a consistent triangulation between the adjacent faces of the hypercube. Fig. 3.9 illustrates a 2D example of the algorithm.

The above lookup table needs to be created only once for a given n dimension. Note that the size of the table increases exponentially as the dimension of the space increases. For an n-dimensional hypercube, there are 2^n vertices, and thus there are 2^{2^n} total possible cases in the table. For instance, for a 3D cube, there are $2^{2^3} = 256$ cases. For a 4D hypercube, there will be $2^{2^4} = 65,536$ cases. For a 5D cube, there will be 2^{2^5}, which will be over four billion cases in the table.

Once the triangulation table is created, isosurfaces in n dimensions can be generated using a method similar to the marching cubes algorithm. For a cell that intersects with the isosur-

face, if the value of a cell vertex is greater than the iso-value, we use one bit to encode the vertex and assign one to the bit; otherwise, it is zero. Since we have 2^n vertices, we can have 2^n in total possible values, where each value corresponds to a particular triangulation case. The triangulation table is used to decide which edges intersect with the isosurface.

3.2.3 Additional Applications

In addition to creating smooth animations, extracting 4D isosurfaces can have additional applications. For instance, Weigle and Banks [13] proposed a method to create the envelope of an isosurface in time. In their method, a 4D isosurface is first constructed within the desired time interval. Then, the silhouettes of the 4D isosurface can be found by locating points on the surface that have $df/dt = 0$, where f is the scalar function and t is the time. This is true because if we assume that the view vector in 4D is $(0,0,0,1)$ (along the time direction), then the silhouettes are the points that have that $(df/dx, df/dy, df/dz, df/dt)(0, 0, 0, 1) = 0$, which implies that $df/dt = 0$. If we project those points to 3D space, the silhouettes of the time-varying isosurface envelope can be visualized.

Another application of 4D isosurfaces is to extract interval volumes, a method proposed by Bhaniramka et al. [14]. An interval volume in 3D space in $[a,b]$ is defined by the points that have $a < f(x, y, z) < b$. Given a scalar field f, we can create one field $f_a = f - a$, give this field a time-step equal to 0, and then create another field $f_b = f - b$ and make this field a time-

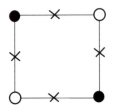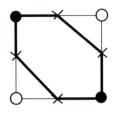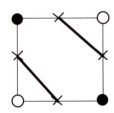

Figure 3.9   A 2D example of Bhaniramka et al.'s triangulation algorithm.

step equal to 1. Then, when creating a 4D iso-surface for f = 0 in the time range of [0,1], the isosurface will go through all the points $a < f(x,y,z) < b$ at time-steps between 0 and 1. If we perform a parallel projection of the 4D isosurface along the (0,0,0,1) direction, tri-angulation of the interval volume can be suc-cessfully obtained.

In general, extracting isosurfaces directly in 4D space allows for smooth animation and tracking of time-varying isosurfaces, which are typically difficult to achieve when utilizing a sequence of isosurfaces in 3D space. However, applying these algorithms for large-scale time-varying data might not be practical because the number of tetrahedra generated can be large. It is also difficult to store the 4D scalar fields in main memory, so additional care should be taken.

References

1. W. E. Lorensen and H. E. Cline. Marching cubes: A high resolution 3d surface construction algorithm. *Computer Graphics*, 21(4):163–169, 1987.
2. J. Wilhelms and A. Van Gelder. Octrees for faster isosurface generation. *ACM Transactions on Graphics*, 11(3):201–227, 1992.
3. Y. Livnat, H.-W. Shen, and C. R. Johnson. A near optimal isosurface extraction algorithm using the span space. *IEEE Transactions on Visualization and Computer Graphics*, 2(1), 1996.
4. H.-W. Shen, C. D. Hansen, Y. Livnat, and C. R. Johnson. Isosurfacing in span space with utmost efficiency (ISSUE). In *Proceedings of Visualization '96*, pages 287–294. IEEE Computer Society Press, Los Alamitos, CA, 1996.
5. T. Itoh and K. Koyamada. Automatic isosurface propagation using an extrema graph and sorted

boundary cell lists. *IEEE Transactions on Visualization and Computer Graphics*, 1(4), 1995.
6. T. Itoh, Y. Yamaguchi, and K. Koyamada. Volume thinning for automatic isosurface propagation. In *Proceedings of Visualization '96*, pages 303–310. IEEE Computer Society Press, Los Alamitos, CA, 1996.
7. C. L. Bajaj, V. Pascucci, and D. R. Schikore. Fast isocontouring for improved interactivity. In *1996 Symposium for Volume Visualization*, pages 39–46. IEEE Computer Society Press, Los Alamitos, CA, 1996.
8. M. van Kreveld, R. van Oostrum, C. L. Bajaj, D. R. Schikore, and V. Pascucci. Contour trees and small seed sets for isosurface traversal. In *Proceedings of 13th ACM Symposium on Comp. Geom.*, pages 212–219, 1997.
9. P. Cignoni, P. Marino, E. Montani, E. Puppo, and R. Scopigno. Speeding up isosurface extraction using interval trees. *IEEE Transactions on Visualization and Computer Graphics*, 3(2), 1997.
10. P. Sutton, C. Hansen, H.-W. Shen, and D. Schikore. A case study of isosurface extraction algorithm performance. In *Proceedings of Joint EUROGRAPHICS–IEEE TCCG Symposium on Visualization*, 2000.
11. M. Cox and D. Ellsworth. Application-controlled demand paging for out-of-core visualization. In *Proceedings of Visualization '97*, pages 235–244. IEEE Computer Society Press, Los Alamitos, CA, 1997.
12. C. Weigle and D. Banks. Complex-valued contour meshing. In *Proceedings of Visualization, '96*, pages 173–180. IEEE Computer Society Press, Los Alamitos, CA, 1996.
13. C. Weigle and D. Banks. Extracting iso-valued features in 4D scalar fields. In *1998 Symposium for Volume Visualization*, pages 103–108. IEEE Computer Society Press, Los Alamitos, CA, 1998.
14. P. Bhaniramka, R. Wenger, and R. Crawfis. Isosurfacing in higher dimensions. In *Proceedings of Visualization '00*, pages 267–273. IEEE Computer Society Press, Los Alamitos, CA, 2000.

4 Optimal Isosurface Extraction

PAOLO CIGNONI, CLAUDIO MONTANI, and ROBERTO SCOPIGNO
Istituto di Scienza e Tecnologie dell'Informazione
Consiglio Nazionale delle Ricerche

ENRICO PUPPO
Dipartimento di Informatica e Scienze dell'Informazione
Universitá degli Studi di Genova

4.1 Introduction

Like Chapter 3, this chapter is entirely dedicated to the topic of the fast extraction of one or more isosurfaces from a structured or unstructured volume dataset by means of a drastic reduction of the visited cells. In particular, we will show how the *interval tree* data structure, an optimally efficient search data structure proposed by Edelsbrunner [7] in 1980 to retrieve intervals of the real line that contain a given query value, can be effectively used for the fast location of cells intersected by an isosurface in a volume dataset.

The resulting search method can be applied to both structured and unstructured volume datasets, and it can be applied incrementally to exploit coherence between isosurfaces. In the case of unstructured grids, the overhead due to the search structure is compatible with the storage cost of the dataset, and local coherence in the computation of isosurface patches is exploited through a hash table. In the case of a structured dataset, a conceptual organization called the *chess-board approach* is adopted in order to reduce the memory usage and to exploit local coherence.

The use of these data structures for the fast extraction of isosurfaces from volume datasets was first presented by Cignoni et al. [4,5].

Let us introduce the topic by giving some basic definitions. A *scalar volume dataset* is a pair (V, W) where $V = \{v_i \in \mathbb{R}^3, i = 1, \ldots, n\}$ is a finite set of points spanning a domain $\Omega \subset \mathbb{R}^3$, and $W = \{w_i \in \mathbb{R}, i = 1, \ldots, n\}$ is a corresponding set of values of a scalar field $f(x,y,z)$, sampled at the points of V, i.e., $w_i = f(v_i)$. A mesh Σ subdividing Ω into polyhedral cells having their vertices at the points of V is also given (or computed from V, if the dataset is scattered): Σ can be made of hexahedra or tetrahedra, or it can be hybrid, i.e., made of tetrahedra, hexahedra, triangular prisms, and pyramids.

Given an *iso-value* $q \in \mathbb{R}$, the set $S(q) = \{p \in \Omega \mid f(p) = q\}$ is called an *isosurface* of field f at value q. For the purpose of data visualization, an isosurface $S(q)$ is approximated by a triangular mesh, defined piecewise on the cells of Σ: a cell $\sigma_j \in \Sigma$ with vertices $v_{j1}, \ldots v_{jh}$ is called *active* at q if $\min_i w_{ji} \leq q \leq \max_i w_{ji}$. An active cell contributes to the approximated isosurface for a patch made of triangles. Patches are obtained by joining points on the edges of active cells that intersect the isosurface (*active edges*), by assuming linear interpolation of the field along each edge of the mesh. Such intersection points are called *isosurface vertices*. In order to use smooth shading to render the isosurface, the surface normal at each surface vertex must also be estimated.

Therefore, the *isosurface extraction problem* consists of four main subproblems:

1. *Cell selection:* finding all active cells in the mesh Σ.

2. *Cell classification:* for each active cell, determining its active edges and how corres-

ponding isosurface vertices must be connected to form triangles.

3. *Vertex computation:* for each active edge, computing the 3D coordinates of its surface vertex by linear interpolation.

4. *Surface normal computation:* for each vertex of the isosurface, computing its corresponding surface normal.

In terms of computational costs, the impact of cell selection may be relevant to the whole isosurface extraction process, in spite of the simplicity of operations involved at each cell, because it involves searching the whole set of cells of Σ. Cell classification has a negligible cost because it is performed only on active cells and it involves only comparisons of values. Although vertex and normal computations are also performed only on active cells, they have a relevant impact, because they involve floating-point operations. Besides, such operations can also be redundant if the dataset is processed on a per-cell basis, because each active edge is shared by different cells.

In order to speed up such tasks, it can be worthwhile to use search structures and techniques that permit traversal of as few nonactive cells as possible during cell selection, and to avoid redundant vertex and normal computations. Speedup techniques can be classified according to the following criteria:

- *Search modality*, adopted in selecting active cells. There are three main approaches: in *space-based* methods, the domain spanned by the dataset is searched for portions intersected by the isosurface; in *range-based* methods, each cell is identified with the interval it spans in the range of the scalar field, and the range space is searched for intervals containing the iso-value; in *surface-based* methods, some facets of the isosurface are detected first, and the isosurface is traversed starting at such faces and moving through face/cell adjacencies.

- *Local coherence* (or coherence between cells). This refers to the ability of a method to

avoid redundancy in geometric computations by reusing the results obtained for an active face or edge at all its incident cells.

Since additional data structures may involve nonnegligible storage requirements, it is important to look for methods and structures that warrant a good tradeoff between time efficiency and memory requirements. The overhead due to auxiliary structures must be compared to the cost of storing a minimal amount of information necessary to support isosurface extraction, disregarding the computational complexity, which can be highly variable depending on whether the dataset considered is *structured* or *unstructured* (i.e., its connectivity is implicitly given, or it must be stored explicitly, respectively [20]). Therefore, evaluation criteria for a speedup method must take into account the following: its *range of applicability*, i.e., the type(s) of dataset (structured, unstructured, or both) for which the method is suitable; its *efficiency*, i.e., the speedup it achieves with respect to a nonoptimized reference method; and its *overhead*, i.e., the storage cost due to auxiliary structures.

The simplest speedup methods for cell selection are based on space partitions (*space-based* methods), and they are suitable only for structured data. Octrees (Velasco and Torres [23], Livnat and Hansen [13]), branch-on-need octrees (Wilhelms and Van Gelder [24], Sutton and Hansen [21]), and pyramids (Montani et al. [6]) are the most common search data structures used with this approach. Space-based techniques cannot be generalized easily to unstructured data, because spatial indices rely on the regular structure of the underlying dataset.

Range-based techniques apply to both structured and unstructured datasets, but they are generally more suitable for unstructured datasets because they cannot exploit the implicit spatial information contained in structured datasets and they have higher memory requirements. In the unstructured case, there is no implicit spatial information to exploit, and the higher storage cost for the input mesh highly reduces the overhead factor of auxiliary structures.

The methods proposed by Gallagher [8], Giles and Haimes [9], Shen and Johnson [19], Livnat et al. [14], Shen et al. [18], Cignoni et al. [4,5], and Chiang and Silva [2], for example, belong to this class. The work by Livnat et al. also introduces the *span space*, a 2D space where each point corresponds to an interval in the range domain; the span space is quite effective in the geometric understanding of the range-based methods. More details about range-based methods and the span-space scheme can be found in chapter 2 in this book, which is authored by Livnat.

Surface-based approaches rely essentially on two requirements: the ability to find an active cell (*seed*) for each connected component of the isosurface, and the ability to propagate the surface by traversing the mesh from cell to cell through adjacencies [20]. The works by Itoh et al. [10,11] and Bajaj et al. [1] present surface-based methods (see also Chapter 5).

In this chapter we address the application of speedup techniques in the various phases of isosurface extraction from both *structured* and *unstructured* datasets. A highly efficient technique for cell selection [2,4,5] that is based on the *interval tree* [7] is adopted. In the unstructured case, this technique is associated with the use of a hash table in order to exploit local coherence to avoid redundant vertex computations. In the structured case, a *chess-board approach* is adopted in order to reduce the memory requirements of the interval tree and to exploit local coherence intrinsic in the implicit structure of the dataset. In both cases, we adopt a precomputation of field gradients at data points in order to speed up the computation of surface normals. Moreover, we describe how the interval tree can be efficiently used to develop an incremental technique that exploits coherence between isosurfaces.

4.2 Selecting Cells Through Interval Trees

Let Σ be the input mesh. Each cell $\sigma_j \in \Sigma$ is associated with an interval I_j whose extremes a_j and b_j are the minimum and maximum field values at the vertices of σ_j, respectively. Since σ_j is active for an iso-value q if and only if its corresponding interval I_j contains q, the following general query problem is resolved:

"given a set $\mathcal{I} = \{I_1, \ldots, I_m\}$ of intervals on the real line of the form $[a_i, b_i]$ and a query value q, report all intervals of $\mathcal{I}$ that contain q."

The problem is effectively visualized using the span space [14]: each interval $I_i = [a_i, b_i]$ is represented as a point in a 2D Cartesian space using the extremes a_i and b_i as the x and y coordinates of the point, respectively. From a geometrical point of view, the problem of reporting all intervals that contain the query value q reduces to collecting the points in the span space lying in the intersection of the two half-spaces $min \le q$ and $max \ge q$.

An optimally efficient solution for the query problem above is obtained by organizing the intervals of $\mathcal{I}$ into an *interval tree* [17], a data structure originally proposed by Edelsbrunner [7], which is reviewed in the following section. For each $i = 1, \ldots, m$, let us consider the sorted sequence of values $X = (x_1, \ldots, x_h)$ corresponding to distinct extremes of intervals (i.e., each extreme a_i, b_i is equal to some x_j). The interval tree for $\mathcal{I}$ consists of a balanced binary search tree $\mathcal{T}$ whose nodes correspond to values of X, plus a structure of lists of intervals appended to nonleaf nodes of $\mathcal{T}$. The interval tree is defined recursively as follows. The root of $\mathcal{T}$ has a discriminant $\delta_r = x_r = x_{\lceil \frac{h}{2} \rceil}$, and $\mathcal{I}$ is partitioned into three subsets:

- $\mathcal{I}_l = \{I_i \in \mathcal{I} \mid b_i < \delta_r\}$
- $\mathcal{I}_r = \{I_i \in \mathcal{I} \mid a_i > \delta_r\}$
- $\mathcal{I}_{\delta_r} = \{I_i \in \mathcal{I} \mid a_i \le \delta_r \le b_i\}$

The intervals of $\mathcal{I}_{\delta_r}$ are arranged into two sorted lists $\mathcal{AL}$ and $\mathcal{DR}$, as follows:

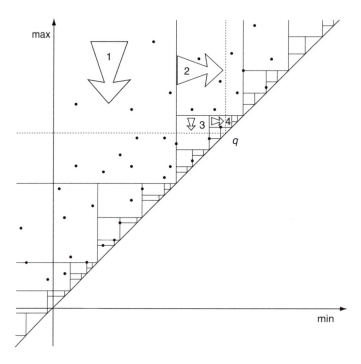

Figure 4.1 A graphical representation of an interval tree in the span space. By definition, the intervals lying on subdivision lines belong to the upper level of the tree. The tree search for a value q goes as follows: sectors with $\delta_r < q$ (intersected by the horizontal line $max = q$) are visited in a top-down order; sectors with $\delta_r > q$ (intersected by the vertical line $min = q$) are visited left to right.

- $\mathcal{AL}$ contains all elements of $\mathcal{I}_{\delta_\tau}$ sorted in Ascending order according to their Left extremes a_i.

- $\mathcal{DR}$ contains all elements of $\mathcal{I}_{\delta_\tau}$ sorted in Descending order according to their Right extremes b_i.

The left and the right subtrees are defined recursively by considering interval sets $\mathcal{I}_l$ and $\mathcal{I}_r$, and extreme sets $(x_1, \ldots, x_{\lceil \frac{h}{2} \rceil - 1})$ and $(x_{\lceil \frac{h}{2} \rceil + 1}, \ldots, x_h)$, respectively. The interval tree can be constructed in $O(m \log m)$ time by a direct implementation of its recursive definition. The resulting structure is a binary balanced tree with h nodes and a height of $\lceil \log h \rceil$, plus a collection of lists of type $\mathcal{AL}$ and $\mathcal{DR}$, each attached to a node of the tree, for a total of $2m$ list elements.

An example of the representation of an interval tree data structure in the span space is given by the subdivision in Fig. 4.1 (solid lines). It is noteworthy that, by construction, the last level of the tree is generally empty. The intervals of this level, if they exist, have to be null intervals (in our case, such intervals are, in fact, associated with cells having the same values at all vertices).

Given a query value q, the intervals containing q are retrieved by visiting tree T recursively, starting at its root:

- If $q < \delta_r$ then list $\mathcal{AL}$ is scanned until an interval I_i is found such that $a_i > q$; all scanned intervals are reported; the left subtree is visited recursively.

- If $q > \delta_r$ then list $\mathcal{DR}$ is scanned until an interval I_i is found such that $b_i < q$; all scanned intervals are reported; the right subtree is visited recursively.

- If $q = \delta_r$ then the whole list $\mathcal{AL}$ is reported.

The geometric interpretation of the search in the span space is also given in Fig. 4.1. The regions containing the active intervals are those to the left of and above dotted lines from q. Each sector of space (node of the tree) that contains the horizontal dotted line (i.e., such that $\delta_r \geq q$) is visited from the top down (scanning the $\mathcal{AL}$ list) until such a line is reached; each sector containing the vertical dotted line is visited left to right (scanning the $\mathcal{DR}$ list) until such a line is reached. Therefore, $\lceil \log h \rceil$ nodes of the tree are visited, and for each node, only the intervals reported in output, plus one, are visited. Hence, if k is the output size, then the computational complexity of the search is $O(k + \log h)$. Because $\log h$ is the minimum number of bits needed to discriminate between two different extreme values, no query technique could have a computational complexity smaller than $\Omega(\log h)$, hence the computational complexity of querying with the interval tree is optimally output-sensitive. It is interesting to note that the time complexity is independent of the total number m of intervals, i.e., on the input size: Indeed, it depends only on the output size and on the number of distinct extremes.

4.2.1 A General Data Structure

A general data structure for the interval tree can be devised by assuming that the set of input intervals is stored independently from the search structure, while each interval in the set can be accessed through a pointer. Therefore, for each element of $\mathcal{AL}$ and $\mathcal{DR}$ lists, we store only a pointer to its related interval. All such lists are packed into two arrays, one for lists of type $\mathcal{AL}$ and one for lists of type $\mathcal{DR}$, which will be called the *big $\mathcal{AL}$* and *big $\mathcal{DR}$* arrays, respectively. Lists are packed in a consistent order (e.g., by following a depth-first visit of the tree), in such a way that the $\mathcal{AL}$ and $\mathcal{DR}$ lists attached to a given node of the tree start at the same location in the two big arrays, respectively. For each node r of the tree, we store the following:

- The discriminant value δ_r.
- An index referring to the starting element of its lists in the two big arrays described above.
- The length of such lists (recall that both lists have the same length).

Because the tree is binary balanced, it can also be stored implicitly by using an array of nodes.

Therefore, if we assume a cost of a word for integers, pointers, and floating-point values, we will have that the bare tree requires $3h$ words, while the lists will require $2m$ words, for a total of $3h + 2m$. It should be taken into account that the cost of encoding the bare tree is expected to be small, at least in our application. Indeed, although in general we have $h \leq 2m$, in practice the intervals of $\mathcal{I}$ can have extremes only at a predefined, and relatively small, set of values; for instance, if data values are encoded by 16 bits, h is at most 65,536, while m can be several millions.

As for all other range-based methods, the storage cost of the interval tree is a crucial issue. In Section 4.3, we will address separately its application to unstructured and structured datasets, respectively, and we will discuss the way storage cost can be optimized by exploiting special characteristics of the two kinds of datasets, respectively.

4.2.2 Exploiting Global Coherence

The interval tree can also be used as an effective structure to address coherence between isosurfaces: active cells for a given iso-value q', sufficiently close to another iso-value q, can be extracted efficiently by exploiting partial information from the set of cells active at iso-value q. Following Livnat et al. [14], this problem can be visualized in the span space, as in Fig. 4.2a: assuming that active cells at iso-value q are known, the list of active cells at iso-values q' is obtained by eliminating all points lying in the right rectangular strip (dashed) and by adding all points lying in the bottom rectangular strip (gridded).

In order to perform this task, active cells at q must be stored in an *active list*, which is updated next to obtain the corresponding active list for iso-value q'. By using an interval tree, the active list can be maintained in a compressed form, as a path on the tree, namely, the path that is traversed when extracting active cells for iso-value q through the query algorithm described in Section 4.2. The path starts from the root node and has a length of log h. For each node in the path we just need to maintain one flag (1 bit) to discriminate whether the $\mathcal{AL}$ or the $\mathcal{DR}$ list was used, one index addressing the first interval that was *not* active in such a list, and one flag (1 bit) to denote whether the next node is the left or the right child of the current node. In the example of Fig. 4.1, the path is encoded as follows (by assuming that list locations are addressed starting at 0): $(\mathcal{DR},4,\text{right})$, $(\mathcal{AL},4,\text{left}),(\mathcal{DR},0,\text{right}),(\mathcal{AL},1,\text{null})$. It is evident that with a real dataset, the length of such a path is (on average) considerably smaller than the actual number of active cells.

The algorithm for computing active cells at iso-value q' scans the tree path and updates it by either adjusting the index associated to each node or recomputing the node completely. The traversal algorithm is described in detail by the pseudo-code in Fig. 4.3. The main principle of the algorithm is the following: as long as both q and q' lie on the same side of the discriminant of the current node, then the same list is used, and the same child will follow, while it is sufficient to adjust the interval index by moving it either backward or forward depending on whether $q > q'$ or $q < q'$ and $\mathcal{AL}$ or $\mathcal{DR}$ list is used. In the example of Fig. 4.2b, this happens for nodes 1 and 2 in the path; in this case, all intervals in the gridded part of the horizontal stripe are included simply by advancing the index in the first triple from 4 to 8, while all intervals in the dashed part of the vertical stripe are included simply by backtracking the index in the second triple from 4 to 1. As soon as a *branching* node is found, i.e., a node such that its discriminant lies between q and q', the search is continued independently of the rest of the active path at q.

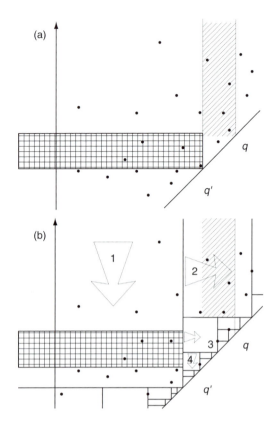

Figure 4.2   The active intervals at q' are obtained by taking active intervals at q, subtracting those in the dashed strip, and adding those in the gridded strip (a). Active list update: node 1 is updated by moving the index forward, in order to include points in the gridded strip; node 2 is updated by moving the index backward, in order to remove points in the dashed strip; tree traversal is repeated for nodes 3 and 4 (b).

Indeed, in this case, the other list for the current node must be used, while the rest of the path will certainly not be active at q'. This happens at node 3 in the example (compare it with Fig. 4.1), when the $\mathcal{DR}$ list was traversed for q, while the $\mathcal{AL}$ list must be traversed for q'. Note that after visiting such a node, the opposite branch of the tree (in the example, just the new node 4) must be visited.

In conclusion, we have that the update has a small overhead for encoding the list of active intervals, while it involves only traversing the intervals that make the difference between q and

```
UpdatePath(IntervalTree T, Path P, Real q,q')
begin
    r = root(T);
    (L,i,c) = first(P);
    δ_r = discriminant(r);
    while (q and q' are on the same side of δ_r)
        if (q > q') then
            if (L = AL) then
                while intervals not active at q' are found move i backward
            else
                while intervals active at q' are found move i forward
        else
            if (L = AL) then
                while intervals active at q' are found move i forward
            else
                while intervals not active at q' are found move i backward;
        r = child(r,c);
        δ_r = discriminant(r);
        (L,i,c) = next(P);
    end while;
    if r not empty then
        set flag L to the other list;
        set flag c to the other child;
        traverse list L to set value of i;
        discard the rest of the path;
        traverse T starting at child(r,c);
end;
```

Figure 4.3   Pseudo-code of the algorithm for active list update.

q', plus all of the intervals appended to the branching node (in the example, node 3). In the worst case (i.e., when q and q' lie on opposite sides of the discriminant of the root node), this algorithm is totally equivalent to performing the query from scratch on the interval tree.

4.3   Extraction of Isosurfaces from Structured and Unstructured Grids

As stated in Section 4.1, the isosurface extraction problem is not limited to the selection of active cells. Other important aspects (e.g., cell classification, vertex and normal computation) must be taken into account in order to ensure the efficiency of the whole extraction process. Moreover, the memory overhead of the auxiliary data structure used for cell selection

has to be considered in order to get a good tradeoff between time efficiency and memory requirements. While referring to the general method described in the previous section, we stress these aspects in the next subsections by distinguishing between *unstructured* datasets (composed of tetrahedra, hexahedra, prisms, or pyramids), in which connectivity must be encoded explicitly, and *structured* datasets (i.e., Cartesian, regular, rectilinear, curvilinear, and block-structured grids made of hexahedral cells), in which connectivity is implicit [20].

4.3.1   The Case of Unstructured Grids

In the case of unstructured datasets, the input mesh is encoded by an indexed data structure composed of the following elements:

- An array of vertices, where, for each vertex, we maintain its three coordinates and its field value.
- A list of cells, where, for each cell, we maintain its connectivity list. This list is made of four, five, six, or eight indices addressing its vertices in the vertex array, depending on whether the cell is a tetrahedron, a pyramid, a prism, or a hexahedron, respectively.

The indices in the connectivity list of each cell are sorted in ascending order, according to the field value of their corresponding vertices, in such a way that the minimum and maximum of the interval spanned by each cell will be given by the field values of the first and last vertex in the connectivity list, respectively. For a hybrid dataset, the list of cells can be encoded by using up to four different arrays, one for each type of cell. However, the list can be addressed as a single array, by assuming a conventional order (e.g., tetrahedra come first, then pyramids, prisms, and, last, hexahedra), and by using the length of each list as an offset.

Given a dataset composed of n points, t tetrahedra, p pyramids, s prisms, and k hexahedra, we have a storage cost of $4n + 4t + 5p + 6s + 8k$ for the whole dataset. Recall that $t + p + s + k = m$ is the total number of cells, that $3h + 2m$ is the cost of the interval tree, and that $h \leq n$. Therefore, we have a memory overhead for the interval tree variable between 25% and 50%, depending on the number of cells of each type, 25% being obtained for a dataset made only of hexahedra and 50% for a dataset made only of tetrahedra. Extreme values are not relevant, however, because in the first case the dataset would probably be a structured one, while in the second case further optimization can be adopted, as discussed next.

If the input mesh is a tetrahedralization, the cost of storing the input mesh is $4n + 4m$. Because all cells are of the same type, we can sort the whole array of tetrahedra according to the order of their corresponding intervals in the big $\mathcal{AL}$ array, described in Section 4.2.1. In this case,

we can avoid storing the big $\mathcal{AL}$ array explicitly, because this comes free from the list of tetrahedra. In this case, we need to maintain only the big $\mathcal{DR}$ array, with a total cost of $3h + m$, and hence less than 25% overhead.

4.3.1.1 Cell Classification

After active cells have been extracted, cell classification consists of testing the values of the cell's vertices with respect to the query value, in order to devise the topology of the isosurface patch inside the active cell. Cell classification is generally not a critical task in the isosurface extraction process. However, this step can be slightly improved by exploiting the fact that indices of vertices in the connectivity list of each cell can be stored in ascending value of field. In this case the first and last vertices of the list correspond to the minimum and maximum and therefore they are implicitly classified because the cell is active. This implies that cell classification can be performed with at most $\lfloor \log_2 (x - 2) \rfloor + 1$ tests for a cell with x vertices, by using bisection (i.e., either two or three tests in the worst case, depending on the type of cell).

4.3.1.2 Vertex and Normal Computation

A more critical task is the computation of vertices and normals. Due to the computational cost of this task, it is important to exploit the local coherence in order to avoid redundant computations. In the case of unstructured datasets, we adopt a dynamic hash-indexing technique. For each isosurface extraction, a hash table is built that is used to efficiently store and retrieve isosurface vertices and normals. The extracted isosurface is represented by adopting an indexed representation: an array of isosurface vertices, storing coordinates and normal vectors for each vertex, and an array of isosurface faces, each storing a connectivity list of three indices to the vertex array. Each isosurface vertex v is identified by the active edge $v_1 v_2$ of the input mesh Σ where v lies. The indices of v_1 and v_2 in the array of vertices of Σ are used to build the hash key for v:

$$key(v_1, v_2) = \mid XOR(v_1, v_2 * n_{\text{prim}}) \mid_{hashsize}$$

where n_{prim} is a sufficiently large prime number. The computational overhead due to the computation of hash indices is therefore small. When processing an edge during vertex computation, the hash table is inquired to know whether such computation has been done before and, if so, to retrieve the index of the interpolated vertex and normal in the corresponding array. Isosurface vertices and normals are computed explicitly, and inserted into the hash table, only if the hash search fails. In this way, each interpolation is done exactly once.

More details on how to carefully set the size of the hash table as a function of the number of active cells can be found in Cignoni et al. [5].

In order to speed up the computation of surface normals during isosurface extraction, we compute as a preprocessing step all field gradients at mesh vertices. Therefore, the surface normal at an isosurface vertex can be simply obtained by linear interpolation from the normalized gradients at the endpoints of the cell edge where the vertex lies. In the case of tetrahedral meshes, the gradient of the scalar field within each cell σ of the mesh is assumed to be constant; i.e., it is the gradient of the linear function interpolating the field at the four vertices of σ. Similar interpolating functions can be adopted in order to estimate the gradient within a single cell of the other types. Then, the gradient at each vertex v of the input mesh is computed as the weighted average of normalized gradients at all cells incident at v, where the weight for the contribution of a cell σ is given by the solid angle of σ at v. Note that this optimization on surface normal computation involves a further $3n$ storage cost, due to the need of maintaining gradients for all data points. The corresponding overhead is highly dependent on the ratio between the number of points n and the number of cells m. For a tetrahedral mesh, we have that on average $m \approx 6n$, and, therefore, the overhead will be less than 12.5%.

4.3.2 The Case of Structured Grids

In the case of structured datasets, i.e., grids based on a hexahedral decomposition and in which the connectivity information is implicit, we propose the use of a new conceptual organization of the dataset that both reduces the number of intervals to be stored in the interval tree and allows us to devise a dataset-traversal strategy that can efficiently exploit the local coherence. The resulting technique is, in practice, a compromise between a space-based approach and a range-based approach that tries to exploit the advantages of both. Though our proposal applies to every structured dataset, we will refer to regular ones in our discussion, for the sake of simplicity.

The number of intervals stored can be reduced on the basis of a simple but effective observation: in a marching cubes–like algorithm, the vertices of the triangular patches that form the extracted isosurface lie on the edges of the cells and, in a regular dataset, each internal edge is shared by four cells. Therefore, in order to be sure that every isosurface parcel will be detected, we only need to store the intervals of a minimal subset of cells that hold all the edges of the dataset.

Such a subset can be devised easily if we think of the volume dataset as a 3D chess board in which the *black* cells (Fig. 4.4) are those we are interested in. In other words, if $c[i,j,k]$ is a black cell, then its adjacent black cells are those that share a single vertex with $c[i,j,k]$. This conceptual arrangement presents some advantages:

- Given a regular $I \times J \times K$ dataset (i.e., a volume of $(I-1) \times (J-1) \times (K-1)$ cells), the black cells can be easily indexed as follows:

$$[2i + |k|_2, 2j + |k|_2, k],$$

with $i \in \{0, 1, \cdots, \lfloor \frac{I-2-|k|_2}{2} \rfloor\}, j \in \{0, 1, \cdots, \lfloor \frac{J-2-|k|_2}{2} \rfloor\}$, and $k \in \{0, 1, \cdots, K-2\}$, where terms $|k|_2$ (k modulo 2) make it possible to compute the indices for the even and odd layers of cells.

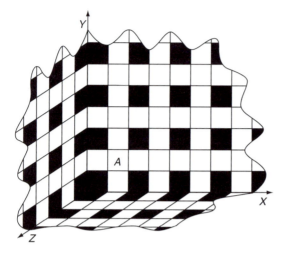

Figure 4.4 The chess-board arrangement: in the case of structured grids, the data structure used to speed up the isosurface extraction does not need to store the *min-max* intervals of all the cells of the volume. Because each internal edge belongs to four cells, only the intervals corresponding to the *black* cells (as in a 3D chess board) have to be maintained.

• The number of black cells in the dataset is 1/4 of the total number of cells, hence the number of intervals to be stored in the interval-tree data structure is 1/4 of the total. This implies not only lower memory occupancy but also shorter construction and traversal times for the auxiliary data structure.

Each black cell has (at most) 18 edge-connected white cells. For each active black cell, the adjacent white cells that are also active (because of isosurface intersections occurring at the edges of the black cell) are determined easily on the basis of the configuration of the current black cell (Fig. 4.5). Conversely, if a white cell is active, there must exist at least a black cell adjacent to it that is also active (special cases of white cells lying on the boundary of the dataset are discussed later). Therefore, once all active black cells have been located efficiently with an interval tree, all active white cells can be located

by searching, in constant time, the neighborhood of each active black cell.

The chess board reduces the number of intervals to be stored, but it does not help with the local coherence. This can be managed by maintaining, in a compact and easy-to-access data structure, the information already computed for vertices and normals of the isosurface. Such an auxiliary structure would require a relevant memory overhead, unless we maintain a sort of locality of the computations. This simple observation gives the key to a compromise between a space-based and a range-based approach: we need to visit the black cells on the basis of not only the intervals arrangement but also the topology of the grid.

In order to achieve this objective, we build an interval tree for each layer of cells (i.e., the cells formed by two adjacent slices of data), rather than building a single interval tree for the whole dataset. The interval tree for each layer stores the *min-max* intervals of the black cells in that layer. Each tree is then labeled with the *Tmin-Tmax* interval, where *Tmin* [*Tmax*] represents the minimum [maximum] of the *min* [*max*] values in the corresponding layer. Therefore,

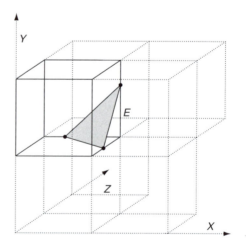

Figure 4.5 Isosurface extraction is propagated from each active *black* cell to the adjacent *white* cells that share one or more active edges.

we have a forest of interval trees, and, for each tree, we can know in constant time whether the tree contains active cells.

If the interval trees in the forest are visited according to the sequence of layers in the dataset, then, during the traversal of the kth tree, we need to maintain only a compact auxiliary data structure (called a *Vertex&Normal* data structure) for active cells of the three layers indexed by $k - 1$, k, and $k + 1$. The *Vertex&Normal* data structure stores information (e.g., vertices, normals, and visited cells) being computed at each active cell, and it avoids redundant geometrical computations. Advancing to the $(k + 1)$-th interval tree simply implies a circular shift of the indices of the layers in the *Vertex&-Normal* data structure. The extraction strategy and the exploitation of the local coherence (i.e., the runtime part of the method) can now be summarized as follows:

- *Interval tree selection:* given an iso-value q, the trees in the forest are tested in sequence in order to individuate the active trees, i.e., the trees for which $Tmin \leq q \leq Tmax$. Each active interval tree, say, the kth, is visited using the algorithm presented in Section 4.2.

- *Black cell processing:* for each active black cell, the marching cubes algorithm [15] is applied: on the basis of the configuration of the cell (determined with respect to q), we access the marching cubes lookup table, and we find the active edges of the current cell. By exploiting the *Vertex&Normal* data structure, we compute (and save) only the vertices and the normals not already computed during the processing of an adjacent white cell. On the basis of the configuration of the cell, we also select the adjacent active white cells where the isosurface extraction must be propagated. For example, if a vertex of the isosurface has been found on the edge E of the black cell $c[i, j, k]$ of the example in Fig. 4.5, then the edge-connected white cells $c[i + 1, j, k]$, $c[i + 1, j, k + 1]$, and $c[i, j, k + 1]$ will be examined.

- *Active white cells processing:* once a black cell has been processed, the algorithm examines the connected active white cells that have not been processed yet. For each of them, the marching cubes algorithm is applied as in the previous case. White cells already examined are individuated by means of simple flags in the *Vertex&Normal* data structures. Note that a *propagation* list for the white cells is not necessary because we individuate all the active white cells starting from one of the adjacent black cells.

- *Advancing:* the algorithm iterates on the next $(k + 1)$-th interval tree (if it is active) by a simple circular shift of the layers in the *Vertex&Normal* data structure: information for the $(k - 1)$-th layer is no longer necessary, and it is therefore rewritten by information on the $(k + 2)$-th layer.

A further remark is necessary for the white cells that lie on the boundary of the dataset. As shown in Fig. 4.4, some boundary edges of the dataset are not *captured* by black cells (e.g., the external edges of the cell labeled A in the figure). However, if all sizes of the dataset are even, no further information is needed for such edges. It is easy to see that if an isosurface cuts one or more edges of a white cell that do not belong to any black cell, the isosurface must also cut some edge of the same cell internal to the volume and hence shared by a black cell.

In case one or more of the sizes I, J, and K of the dataset are odd numbers, then part of the edges of at most $2(I - 1) + 2(J - 1) + 2(K - 1)$ cells (i.e., cells forming 6 of the 12 corners of the volume) are not captured (not even indirectly) by the black cells of the chess board (see Fig. 4.6). As shown in the figure, in these situations small isosurface subsections can be lost. To solve this problem we can add the following step to our algorithm:

- *Unreachable cells test:* Once an active tree has been visited and the corresponding active cells processed, the algorithm examines the (still not processed) white cells of the current

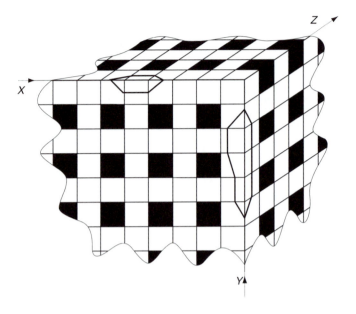

Figure 4.6 Some of the cells of a dataset with two odd sizes are not captured by the chess board. Small parts of the two isosurfaces could be lost.

layer whose edges are not captured by black cells.

An alternative solution to the previous step could be the insertion into the corresponding interval tree of the *black edges* of the unreachable white cells. However, the small number of cells to be tested separately does not justify the effort.

With the *chess-board* approach, the total asymptotic time for a query is, in the worst case, $O(\sqrt[3]{n}\log n + k)$, where k is the output size, by assuming a dataset with $\sqrt[3]{n}$ layers (i.e., $I = J = K$). Note that using a forest rather than a single interval tree adds an extra factor of $\sqrt[3]{n}$ to the optimal query time. Therefore, in this case we trade optimal asymptotic time for space. However, it should be noted that the $\sqrt[3]{n}\log n$ factor is usually negligible in practice, while the advantage that derives from exploiting local coherence is relevant.

As stated for the case of unstructured datasets, the complexity in space for the interval-tree data structures can be expressed in terms of

$3h + 2m$, with h the number of distinct interval endpoints and m the number of intervals to be stored. For a regular dataset with n data values, we have to store the intervals corresponding to the black cells, i.e., $m \cong n/4$ intervals. Because in real applications we usually have $h \ll n$, the requirement for $n/2 + 3h$ storage locations is very close to one-half of dataset size.

The ratio between the interval-tree memory requirements and the dataset occupancy becomes obviously more propitious in the case of nonregular structured datasets (e.g., the curvilinear ones).

Therefore, the *chess board approach* helps in solving the problem of the memory occupancy of the interval-tree data structure together with the problem of the local coherence.

4.4 Assessment

The presented data structures were proposed in Cignoni et al. [4,5] and tested on a number of different structured and unstructured datasets.

The speedup method for isosurface extraction based on the use of the interval-tree data structure considerably improves the performance of the traversal phase with respect to the standard marching tetrahedra and marching cubes algorithms. Optimal output-sensitive time complexity in extracting active cells is achieved. Memory overhead, according to the general interval-tree representation proposed, is $3h + 2m$ words, with h the number of distinct extremes of intervals and m the number of cells.

To reduce space occupancy, which becomes a critical factor in the case of high-resolution datasets, two different strategies, oriented to the two different data classes, can be used: (a) an optimized interval-tree representation for unstructured datasets, in order to reduce space occupancy to $3h + m$, and therefore enabling less than 25% overhead; and (b) a partial representation of the cell intervals, based on the *chess-board* approach, to reduce the number of intervals stored in the interval trees in the case of structured datasets. All of the active cells not represented directly are here detected by propagation. Although the reduced number of intervals are encoded, the speedups obtained are very similar to those obtained with the naïve interval-tree implementation, which encodes all of the intervals.

References

1. C. L. Bajaj, V. Pascucci, and D. R. Schikore. Fast isocontouring for improved interactivity. In *1996 Symposium on Volume Visualization Proc.*, pages 39–46, 1996.

2. Y. J. Chiang and C. T. Silva. I/O Optimal isosurface extraction. *Visualization '97 Conference Proceedings*, pages 293–300, 1997.

3. P. Cignoni, C. Montani, E. Puppo, and R. Scopigno. Multiresolution modeling and visualization of volume data. *Technical Report 95–22*, Istituto CNUCE–CNR, Pisa, Italy, 1995.

4. P. Cignoni, C. Montani, E. Puppo, and R. Scopigno. Optimal isosurface extraction from irregular volume data. In *1996 Symposium on Volume Visualization Proc.*, pages 31–38, 1996.

5. P. Cignoni, P. Marino, C. Montani, E. Puppo, and R. Scopigno. Speeding up isosurface extraction using interval trees. *IEEE Transactions on Visualization and Computer Graphics*, 3(2):158–170, 1997.

6. C. Montani, R. Scateni, and R. Scopigno. Decreasing isosurface complexity via discrete fitting. *Computer Aided Geometric Design*, 17:207–232, 2000.

7. H. Edelsbrunner. Dynamic data structures for orthogonal intersection queries. *Technical Report F59*, Inst. Informationsverarb., Tech. Univ. Graz, Graz, Austria, 1980.

8. R. S. Gallagher. Span filter: an optimization scheme for volume visualization of large finite element models. In *IEEE Visualization '91 Conf. Proc.*, pages 68–75, 1991.

9. M. Giles and R. Haimes. Advanced interactive visualization for CFD. *Computing Systems in Engineering*, 1:51–62, 1990.

10. T. Itoh, Y. Yamaguchi, and K. Koyamada. Volume Thinning for Automatic Isosurface Propagation. In *IEEE Visualization '96 Conf. Proc.*, pages 303–310, 1991.

11. T. Itoh and K. Koyamada. Automatic isosurface propagation using an extrema graph and sorted boundary cell lists. *IEEE Transactions on Visualization and Computer Graphics*, 1(4):319–327, 1995.

12. M. Laszlo. Fast generation and display of isosurfaces wireframe. *CVIGP: Graphical Models and Image Processing*, 54(6):473–483, 1992.

13. Y. Livnat and C. Hansen. View dependent isosurface extraction. *Visualization '98 Conference Proceedings*, Research Triangle Park, pages 175–180, 1998.

14. Y. Livnat, H. Shen, and C. Johnson. A near optimal isosurface extraction algorithm for structured and unstructured grids. *IEEE Transactions on Visualization and Computer Graphics*, 2(1):73–84, 1996.

15. W. Lorensen and H. Cline. Marching cubes: a high resolution 3D surface construction algorithm. *ACM Computer Graphics (SIGGRAPH '87 Conf. Proc.)*, 21(4):163–170, 1987.

16. C. Montani, R. Scateni, and R. Scopigno. Discretized marching cubes. In *Visualization '94 Conf. Proc.*, IEEE Computer Society Press, pages 281–287, 1994.

17. F. Preparata and M. Shamos. *Computational Geometry: an Introduction*. Springer-Verlag, 1985.

18. H. Shen, C. D. Hansen, Y. Livnat, and C. R. Johnson. Isosurfacing in span space with utmost efficiency (ISSUE). In *Visualization '96 Conf. Proc.*, pages 287–294, 1996.

19. H. Shen and C. Johnson. Sweeping simplices: a fast iso-surface extraction algorithm for un-

structured grids. In *IEEE Visualization '95 Conference Proceedings*, pages 143–150, 1995.

20. D. Speray and S. Kennon. Volume probes: interactive data exploration on arbitrary grids. *ACM Computer Graphics (1990 Symposium on Volume Visualization Proceedings)*, 24(5):5–12, 1990.

21. P. M. Sutton and C. Hansen. Accelerated isosurface extraction in time-varying fields. *IEEE Transactions on Visualization and Computer Graphics*, 6(2):98–107, 2000.

22. M. van Kreveld. Efficient methods for isoline extraction from a digital elevation model based on triangulated irregular networks. In *Sixth International Symposium on Spatial Data Handling Proc.*, pages 835–847, 1994.

23. F. Velasco and J. C. Torres. Cells octree: a new data structure for volume modeling and visualization. In *Vision, Modeling, and Visualization Conf. Proc.*, pages 151–158, 2001.

24. J. Wilhelms and A. Van Gelder. Octrees for faster isosurface generation. *ACM Transaction on Graphics*, 11(3):201–227, 1992.

25. G. Wyvill, C. McPheeters, and B. Wyvill. Data structures for soft objects. *The Visual Computer*, 2(4):227–234, 1986.

5 Isosurface Extraction Using Extrema Graphs

TAKAYUKI ITOH
IBM Japan Tokyo Research Laboratory

KOJI KOYAMADA
Kyoto University Center for the Promotion of Excellence in Higher Education

5.1 Introduction

Visualizing isosurfaces is one of the most effective techniques for understanding elements of scalar fields, such as the results of 3D numerical simulations and the results of 3D measurements in the medical field. An isosurface is usually approximated as a set of triangular facets [1] and displayed as a set of edges of triangles or as a set of filled triangles.

In the numerical simulation field, visualization tools that support a function for the continuous generation of isosurfaces with changing scalar values are used to understand the distribution of scalar fields. When a volume is huge and contains an enormous number of cells, the cost of generating an isosurface may be high. It may even prevent the user from understanding the distribution of the scalar field, because a long time is necessary to generate an isosurface from a huge volume of data. Fast isosurfacing methods are therefore needed to facilitate understanding of scalar fields.

In a basic isosurfacing procedure, all cells are visited, and those intersected by an isosurface, so-called *isosurface cells*, are extracted. Polygons inside the isosurface cells are then generated, and finally the positions and normal vectors of the polygon vertices are calculated. Some fast isosurfacing techniques focus on the acceleration of polygonization and the rendering processes, such as parallelization [2], graphics acceleration by generation of triangular strips [3], and geometric approximation [4]. However, it seems that the most effective technique for developing fast isosurfacing algorithms is the *reduction of visits to nonisosurface cells*. In our experience, the ratio of the number of nonisosurface cells to the number of isosurface cells is usually large.

Actually, many techniques have been proposed for reducing the number of nonisosurface cells that are visited. Algorithms that classify or sort cells according to their scalar values [5,6,7] have been proposed, but the number of cells visited in these algorithms is still estimated as $O(n)$, where n is the total number of cells. Algorithms that classify cells by using space subdivision [8,9] have also been proposed, but they are difficult to apply to unstructured volumes.

Many efficient isosurfacing algorithms have been developed that can be applied to unstructured volumes, and whose computation time for isosurfacing processes is much less than $O(n)$. They can be categorized into two approaches: range-based search methods and seed-set generation methods. The first approach uses an interval $[a,b]$ of a scalar range, where a is a cell's minimum value and b is its maximum value. The cell is intersected by an isosurface if the interval satisfies $a \leq C$ and $C \leq b$, where C is a constant value of the isosurface, the so-called *iso-value*. Such cells are efficiently extracted by traditional searching algorithms [10,11,12]. Livnat et al. [11] proposed an algorithm using a

Kd-tree method, whose isosurfacing cost is estimated as $O(n^{1/2} + k)$, where k is the number of isosurface cells (see Chapter 2 for more). Shen et al. [12] proposed an algorithm using a lattice classification, whose isosurfacing cost is estimated as $O(\log \frac{n}{L} + \frac{n^{1/2}}{L} + k)$, where L is a user-specified parameter. Cignoni et al. [10] proposed an algorithm using an interval tree, whose isosurfacing cost is estimated as $O(\log n + k)$ (see Chapter 4 for more). These can be applied to unstructured volumes, and they are more efficient than the previously mentioned methods, since the numbers of cells visited in the algorithms are much less than $O(n)$. However, these algorithms involve costs of over $O(n)$ for constructing substructures in preprocessing, since they use sorting processes. The lattice classification can be implemented without the sorting process; however, it is desirable to include the sorting process in order to make the ranges of classifications vary in order to have a similar number of cells in each lattice.

The second approach generates small groups of cells, the so-called *seed set*, which includes at least one isosurface cell for every isosurface [13,14]. This approach uses isosurface propagation algorithms, which recursively visit adjacent isosurface cells [15,16,17,18], starting from isosurface cells extracted from the seed set. Here, an adjacent cell means a cell that shares a face with the visited cell. When an isosurface consists of multiple disjoint parts, the approach extracts isosurface cells in all disjoint parts of the isosurface from the seed set. Bajaj et al. [13] reported a method for generating smaller groups of cells that sweeps cells in a grid space and removes many cells whose ranges of values are entirely shared by their adjacent cells. Kreveld et al. [14] reported a method for generating a contour tree that connects extreme and saddle points. The method solves the aforementioned problem, but it requires over $O(n)$ computation time for generating the contour tree.

Here let us summarize the key points for the development of fast isosurfacing algorithms that reduce the number of visits to nonisosurface cells:

- Application to unstructured volumes.
- Much less than $O(n)$ complexity for visiting isosurface cells.
- Reduction of complexity for preprocessing (hopefully $O(n)$).

This chapter introduces efficient isosurfacing techniques that can be applied to unstructured volumes as well as structured volumes. The techniques can be categorized as the second approach, which extracts seed sets and then propagates isosurfaces. Section 5.2 describes isosurface propagation techniques [15,17,18] that recursively visit adjacent isosurface cells, and it also describes acceleration of the propagation techniques [16], which eliminates the polygon–vertex identification process from the isosurface propagation algorithm. Section 5.3 describes the technique that uses an *extrema graph* [19]. An extrema graph connects extreme points in a volume by means of arcs, and cells through which the arcs pass are registered. The technique extracts seed sets from the cells registered in the extrema graph and cells on the boundary, and then generates isosurfaces by the propagation methods. Section 5.4 describes the technique that generates an *extrema skeleton* [20,21], which is an extension of an extrema graph [19], as a small seed set. An extrema skeleton consists of cells and connects all extreme points like an extrema graph. The technique generates an extrema skeleton by the *volume thinning* method, which is an extension of the thinning method used for image recognition, in the preprocessing. It then extracts isosurface cells from the extrema skeleton and generates an isosurface by the propagation method. Section 5.5 discusses the applications of the volume-thinning method for the extraction of numeric features of the volume date. Section 5.6 concludes this chapter.

5.2 Isosurface Propagation

5.2.1 Isosurface Propagation

An isosurface is efficiently generated by recursively visiting adjacent isosurface cells and skips the visits of nonisosurface cells. Such recursive polygonization algorithms were originally proposed for efficient polygonization of implicit functions [15,18], and have then been applied to volume datasets [17]. This chapter calls such techniques "isosurface propagation," because the process looks as if it propagates isosurfaces by generating polygons one by one.

Typical isosurface propagation algorithms extract isosurface cells by breadth-first traverses. The algorithms first insert several isosurface cells into a first-in-first-out (FIFO) queue. They then extract the isosurface cells from the FIFO and generate polygons inside the cells. At the same time, the algorithms insert isosurface cells adjacent to the extracted cells into the FIFO. They repeat the above process until the FIFO queue becomes empty, and finally the isosurface is constructed.

Fig. 5.1 shows an example of a typical isosurface propagation algorithm. When the algorithm first constructs the polygon P_1, it inserts four adjacent isosurface cells, C_2, C_3, C_4, and C_5, into the FIFO. When the algorithm extracts

these cells from the FIFO, it constructs four polygons, P_2, P_3, P_4, and P_5. When the algorithm constructs P_2, it similarly inserts adjacent isosurface cells C_6, C_7, and C_8 into the FIFO. The algorithm similarly constructs polygons P_6, P_7, and P_8 when it extracts C_6, C_7, and C_8 from the FIFO.

5.2.2 Polygon-Vertex Identification for Isosurface Polygons

When polygons in an isosurface are generated by a conventional isosurface generation technique such as the marching cubes method [1], all their polygon vertices lie on cell edges, and are mostly shared by several polygons. If a volume data structure contains all cell-edges data and then polygon vertices are linked to the cell edges, the technique can immediately extract the shared vertices by referring to cell edges of visited cells. However, cell-edge data is not usually preserved in a volume data structure, owing to the limited memory space. Therefore, a vertex-identification process should be implemented so

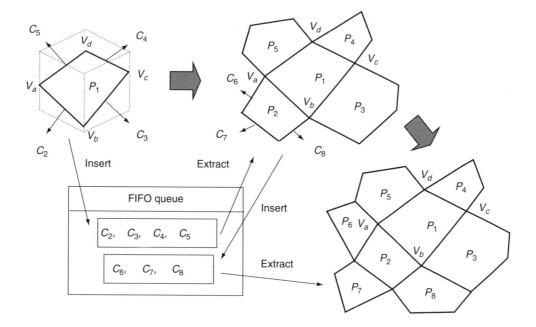

Figure 5.1 Isosurface propagation.

that polygons of isosurfaces can share polygon vertices.

An example of a vertex-identification process is as follows. Given an iso-value of an isosurface C, the implementation first determines the sign of $S(x,y,z) - C$ at nodes of a cell, where $S(x,y,z)$ denotes the scalar value at a node. If all the signs are equal, the cell is not an isosurface cell and the calculation of the action value is skipped. Otherwise, the process extracts isosurface cell edges from a cell. Here, a cell edge is represented as a pair of nodes. When a polygon vertex is generated on a cell edge, it is registered to the hash table with the pair of nodes that denotes the cell edge. When another isosurface cell that shares the same cell edge is visited, the polygon vertex is extracted from the hash table, by inputting the pair of nodes. In the implementation, all isosurface cell edges are registered to the hash table with the polygon vertices of an isosurface.

Fig. 5.2 shows an example of this process. When polygon P_1 is first constructed, vertices V_a, V_b, V_c, and V_d are registered in a hash table with pairs of nodes. For example, vertex V_b is registered with a pair of nodes, n_1 and n_2, that denote a cell edge that V_b lies on. When polygon P_2 is then constructed, vertex V_b is extracted from the hash table by inputting the pair of nodes, n_1 and n_2.

Note that the vertex identification process is still necessary in the isosurface propagation algorithm. For example, polygon vertices of P_1, V_a, V_b, V_c, and V_d in Fig. 5.1 are registered into a hash table when P_1 is generated. The polygon vertex V_b is then extracted from the hash table when P_2, P_3, and P_8 are generated.

5.2.3 Accelerated Isosurface Propagation Without Polygon-Vertex Identification

This vertex-identification process occupies a large part of computation time in isosurface generation methods. Recently an accelerated isosurface propagation technique that shares polygon vertices without using the vertex-identification process [16] has been proposed.

Fig. 5.3 shows the overview of the new isosurface propagation technique. The technique assumes that at least one isosurface cell is given. It first generates a polygon P_1 inside the given cell, and allocates its polygon vertices, V_a, V_b, V_c, and V_d. It then visits all cells that are adjacent to polygon vertices of P. In Fig. 5.3, cells that are adjacent to V_b are visited, and polygons P_2, P_3, and P_4 are generated. Note that polygon vertices of the three new polygons are not allocated at that time. It then assigns V_b to the three polygons. V_b is no longer required in this algorithm, because all polygons that share V_b have been generated at that time. It means that the search algorithm is not necessary for the vertex identification in the method. Similarly, in Fig. 5.3, cells that are adjacent to V_a are then visited. Polygons P_5 and P_6 are generated at that time, and V_a is

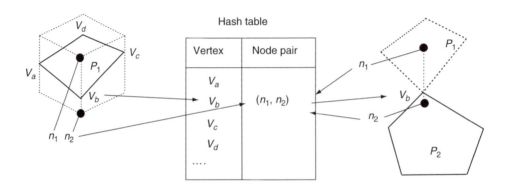

Figure 5.2 Polygon-vertex identification using a hash table.

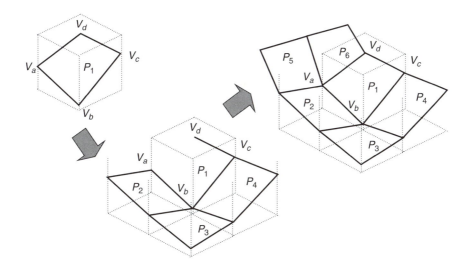

Figure 5.3   Isosurface propagation without the vertex identification process.

assigned to them. Fig. 5.4 shows the pseudo-code of the technique. The technique visits most of the isosurface several times (the `for` loop (3) in Fig. 5.4), and a polygon is constructed at the first visit. The cells are also visited when they are extracted from the FIFO (the `for` loop (1) in Fig. 5.4), and all polygon vertices of the polygons inside the extracted cells are set at the moment. The method processes a cell several times; however, experimental tests [16] showed that its computational time was less than that for the existing propagation technique.

5.3   Isosurface Generation Using an Extrema Graph

5.3.1   Overview

Isosurface propagation techniques have the great advantage of reducing the number of visited nonintersecting cells; however, they also have the problem that the starting isosurface cells must first be specified. Efficient automatic extraction of the starting cells was previously difficult, especially when the isosurface was separated into many disconnected parts. This section describes a technique that automatically extracts isosurface cells from all disconnected parts of the isosurface.

The technique introduced in this section [19] uses the following rule (see Fig. 5.5) as the key to searching for intersected cells:

There is at least one extreme point, or extremum, inside a closed isosurface (unless there is no interior point inside the isosurface, or the scalar field is flat inside the isosurface). There is at least one boundary cell intersected by an open isosurface.

According to this rule, at least one isosurface cell of a closed isosurface can be found by traversing cells across the graph connecting all extreme points in a volume. Also, at least one isosurface cell of an open isosurface can be found by traversing cells on the boundary of a volume. Fig. 5.5 shows the overview of the above idea.

The preprocessing of the technique consists of three technical components: (1) extraction of extreme points, (2) construction of an extrema graph, and (3) collection of boundary cells. The technique needs to execute the preprocessing only once, and then isosurfaces are quickly generated by referring to the extrema graph and the collection of boundary cells.

```
void Isosurfacing() {

  for (each given isosurface cell Cᵢ){insert Cᵢ into FIFO;}

  /* for-loop (1) */
  for (each cell Cᵢ extracted from FIFO){
    if(polygon Pᵢ in Cᵢ is not constructed){Construct Pᵢ in Cᵢ;}

    /* for-loop (2) */
    for (each intersected edge Eₙ){
      if(a polygon-vertex Vₙ on Eₙ is not added into Pᵢ) {
        Allocate Vₙ into Pᵢ in Cᵢ;
        Register Vₙ into Pᵢ in Cᵢ;

        /* for-loop (3) */
        for(each cell Cⱼ that shares Eₙ){
          if(Pⱼ in Cⱼ is not constructed){Construct Pⱼ in Cⱼ;}
          Register Vₙ into Pⱼ in Cⱼ;
          if (Cⱼ has never been inserted into FIFO){insert Cⱼ into FIFO;}
        } /* end for-loop(3) */
      } /* end if(there is not Vₙ) */
    } /* end for-loop(2) */
  } /* end for-loop(1) */

  for(each polygon-vertex Vₙ){Calculate position and normal vector;}

} /* end Isosurfacing() */
```

Figure 5.4   Pseudo-code of accelerated isosurface propagation technique.

5.3.2   Extraction of Extreme Points

In this chapter, extreme points are defined as grid-points whose scalar values are higher or lower than the values of all adjacent grid-points that are connected by cell edges. Here we interpreted the above definition as follows:

An extreme point is a grid-point whose scalar value is maximum or minimum in all cells that share the grid-point.

This definition is available for both structured and unstructured volumes.

The technique visits all cells once to extract extreme points and compares the scalar values of all grid-points for each cell. The technique

then marks all grid-points except the maximum-valued ones as "*not maximum.*" Similarly, it marks all grid-points except the minimum-valued ones as "*not minimum.*" Intermediate-valued grid-points are thus marked as both "*not minimum*" and "*not maximum.*" After comparing the values in all cells, the technique extracts only grid-points that have either a "*not maximum*" or a "*not minimum*" mark as extreme points.

Simply applying the above algorithm, all grid-points in a cloud of equivalued local minimum or maximum points are registered as extreme points; however, not all grid-points in the cloud are necessary for the extraction of seed cells. After all the cells in a volume have been

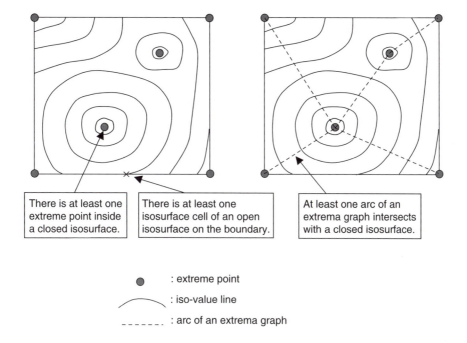

There is at least one extreme point inside a closed isosurface.

There is at least one isosurface cell of an open isosurface on the boundary.

At least one arc of an extrema graph intersects with a closed isosurface.

● : extreme point

⌒ : iso-value line

- - - - - : arc of an extrema graph

Figure 5.5 (Left) Rule between isosurfaces and extreme points. (Right) Extrema graph for extraction of isosurface cells.

visited, the technique recursively visits adjacent cells that share the extreme points again and erases unnecessary extreme points.

5.3.3   Construction of an Extrema Graph

This section describes an algorithm that constructs a group of arcs that connect a pair of extreme points to construct an extrema graph.

To generate an arc, the algorithm first selects an extreme point as a *"start"* point and then selects another extreme point as a *"goal"* point. Here, we assume the total cost of a graph is reduced when closer extreme points are connected by a graph, where we estimate the cost of a graph by the number of cells intersected by the arcs of the graph. In accordance with this assumption, several closer extreme points are enqueued as candidate goal points.

To generate an arc, the algorithm extracts one of the enqueued candidate goal points and calculates the vector of the arc connecting the start and goal points. Starting from a cell that includes the start point, the algorithm traverses

cells intersected by the arc through their adjacency, as shown in Fig. 5.6 (left). This process is repeated until the traverse arrives at the cell that touches the goal point, and the algorithm registers the traversed cells to the arc during the process. If the arc goes outside the volume, as shown in Fig. 5.6 (center), the algorithm terminates the traverse and starts a similar traverse after extraction of the next candidate goal point.

If no extreme points are connected with the start point, the algorithm extracts the closest candidate goal point again. Starting from a cell that includes the start point, the algorithm repeats the visit of adjacent cells that share the face whose sum of distances to the goal point is minimum. This process is repeated until it arrives at a cell that includes the goal point, and registers the visited cells to the arc, as shown in Fig. 5.6 (right). This distance-based process necessarily connects the chosen extreme points, but it is not always efficient because of the cost of calculating the distance for each grid-point. It should therefore be carried out after traversal of straight arcs.

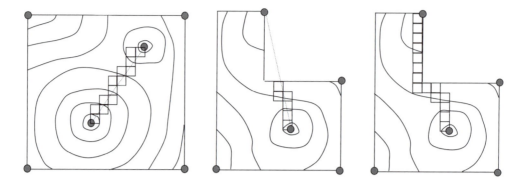

Figure 5.6   (Left) To generate a straight arc, two extreme points are selected as start and goal points, and cells intersected by the arc are traversed though their adjacency. (Center) It is possible that the arc goes outside the volume and cells cannot therefore be traversed. (Right) If no goal points can be connected by a straight arc, the algorithm connects the extreme points by polygonal cell traverse.

5.3.4   Collection of Boundary Cells

In this chapter, boundary cells are defined as a group of cells that have one or more faces not connected to any other cells. The proposed technique first collects the boundary cells, and then defines maximum and minimum values of the boundary cells. The technique finally generates two boundary cell lists; one preserves the boundary cells in the order of the maximum values, and the other preserves them in the order of the minimum values.

5.3.5   Isosurface Generation Using an Extrema Graph

Given an iso-value, the proposed technique searches isosurface cells by traversing one of the boundary cell lists and arcs of the extrema graph. If polygons inside the extracted isosurface cell have not yet been generated, the extracted cell is put into the FIFO queue and an isosurface is then propagated. The technique drastically reduces the number of visits of non-isosurface cells because the technique visits only cells from an extrema graph and a boundary-cell list, and then recursively visits isosurface cells but skips all other cells.

The computation time of preprocessing of the extrema graph method is estimated

as $O(n) + O(m^2 \log m + n^{1/3}m) + O(n^{2/3} \log n^{2/3})$, where m denotes the number of extreme points. In our experiments, the cost is shown to be linear in many cases. However, the process may be costly when a volume is very noisy and therefore has an enormous number of extreme points. The computation time for the isosurfacing process of the extrema graph method is estimated as $O(n^{2/3} + n^{1/3}m + k)$, where k is the number of isosurface cells. Here we estimate the complexity of algorithms with the following assumptions:

- The number of cells on the boundary of a volume is estimated to be $O(n^{2/3})$.

- The maximum degree of any node is bounded by a constant.

- The number of faces, edges, and nodes is $O(n)$ for all.

5.4   Isosurface Generation Using an Extrema Skeleton

5.4.1   Overview

The technique described in Section 5.3 has bottlenecks in computation time as follows:

- Its preprocessing may be costly when the number of m is large.

- Its isosurface generation is not always efficient when the number of boundary cells is large.

Here we consider the necessity of boundary cells to find isosurface cells. Unstructured volumes may have *through-holes* or *voids*. In this chapter, a through-hole is defined as a topological feature that causes a genus of a boundary. A void is defined as an empty space enclosed by a disjoint part of the boundary of a volume.

An isosurface separates a volume into several subdomains. There are extreme points in all of the subdomains, and an extrema graph therefore always connects all of the subdomains. An isosurface may also consist of several disjoint parts. If two adjacent subdomains share only one part of an isosurface, an arc of an extrema graph necessarily intersects the part. On the other hand, it is possible that the two subdomains share two or more disjoint parts of an isosurface if a volume has through-holes. In this case, arcs of the extrema graph do not always intersect all the parts of the isosurface. This means that an extrema graph may not find the intersection with several parts of an isosurface when a volume has through-holes. In other words, the extrema graph needs to preserve the topology of through-holes so that it intersects with all disjoint parts of every isosurface.

This chapter introduces an extension of the extrema graph, the *extrema skeleton*, which preserves the topology of through-holes so that it intersects with all disjoint parts of every isosurface [20,21]. It accelerates the isosurfacing process because boundary cell lists are no longer necessary to find seed cells from all the disjoint parts of isosurfaces. During the preprocessing to construct the extrema skeleton, the extended technique first extracts extreme points by using the algorithm described in Section 5.3.3, and then generates the extrema skeleton by the *volume thinning* method. An extrema skeleton is a set of cells that connects all extreme points while it intersects all disjoint parts of every isosurface.

The preprocessing of the technique introduced in this section consists of two technical components: (1) extreme point extraction and (2) volume thinning. The preprocessing is performed only once, and the extrema skeleton can be used while isosurfaces are generated over and over. Given an iso-value, the technique extracts isosurface cells from the extrema skeleton, and the isosurface is generated by the propagation technique.

The computation time of preprocessing is estimated as $O(n)$, because both the extrema point extraction and the volume-thinning processes visit most of the cells once. This is the main feature of the technique, since most of the fast isosurface generation techniques introduced in Section 5.1 require more than $O(n)$ computation time for their preprocessing. The computation time of the isosurfacing process is estimated as $O(n^{1/3}m + k)$, where k denotes the number of isosurface cells and m denotes the number of extreme points.

5.4.2 Volume Thinning for Extrema Skeleton Generation

The volume-thinning method [20,21] is an extension of the previously reported image-thinning method. While the image-thinning method that generates a skeleton of a painted area of an image consists of pixels, the volume-thinning method generates an extrema skeleton of a volume that consists of cells. The method initially assumes that a seed set of a volume contains all cells in the volume. It then marks each cell that touches an extreme point, as shown in Fig. 5.7a. The marked cells will never be eliminated from the seed set during the process. The method then visits unmarked cells on the boundary of the noneliminated cells and eliminates many of them from the seed set, as shown in Fig. 5.7b. Finally the seed set forms a one-cell-wide skeleton, while the cells in the skeleton preserve the connectivity of the marked cells. The visit procedure is similar to the image-thinning algorithm, which visits pixels at the boundary of a painted area and eliminates many of

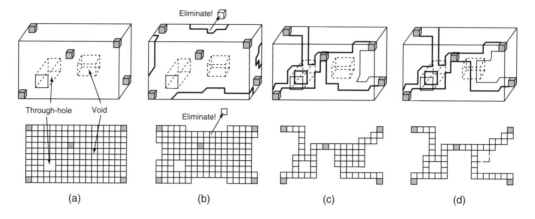

Figure 5.7 Overview of the volume-thinning method. (a) Cells that touch extreme points are marked. (b) Many cells are eliminated from a seed set. (c) An extrema skeleton with a bubble-like layer around a void. (d) An extrema skeleton after removal of the bubble-like layer.

them until a one-pixel-wide skeleton is generated. The skeleton preserves the topological features of the volume, such as through-holes and voids. Cycles of cells are generated around through-holes, since the skeleton contains the cycle of the through-holes. Bubble-like layers of cells are generated around the voids, since the skeleton retains any disjoint boundary faces around voids, as shown in Fig. 5.7c. The bubble-like layers occupy a large part of an extrema skeleton. However, we do not need to preserve the layers, because we need only the topology of the through-holes to find seed cells from all disjoint parts of isosurfaces. As shown in Fig. 5.7d, our implementation eliminates such bubble-like layers of cells to reduce the number of cells in the extrema skeleton.

The shape of the skeleton and the number of cells in it strongly depend on the order of visiting cells. However, we do not consider optimizing the order, because the optimization causes a more complicated implementation and only slightly reduces the number of seed cells.

Fig. 5.8 shows an example of how the volume-thinning process eliminates many of the cells from the extrema skeleton. The fourth image in Fig. 5.8 shows that there is a bubble-like layer, and the final image shows that most of the cells in the layer are then eliminated.

The initialization stage of the volume-thinning method first classifies cells according to the number of their adjacent cells n, where $n = 0, 1, 2, 3$ for tetrahedral cells and $n = 0, 1, 2, 3, 4, 5$ for hexahedral cells. It then allocates FIFO queues for C_n cells that are on the boundary of the noneliminated cells and inserts such cells into the FIFOs. At that time, C_4 tetrahedral cells or C_6 hexahedral cells

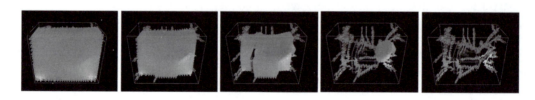

Figure 5.8 Example of the volume-thinning process. (See also color insert.)

are not inserted into FIFOs, but most of them will be inserted during the volume-thinning process, because their classification will be changed when their adjacent cells are eliminated from the seed set.

The main loop of the volume-thinning process extracts cells from FIFOs and checks the connectivity between the extracted cells and their adjacent cells, according to the conditions described by Itoh et al. [21]. The cells are eliminated from the seed set, if the connectivity of their adjacent cells can be preserved without them. At that time, the classifications of the eliminated cells are changed to C_0. When a cell is eliminated, the process changes the classifications of its adjacent noneliminated cells from C_n to C_{n-1} if $n > 0$, and inserts them into the C_{n-1} FIFO.

In our implementation, all C_1 cells are first extracted from a FIFO. An extracted C_1 cell is unconditionally eliminated from the seed set, since the connectivity of its adjacent noneliminated cell is not changed by the elimination. When the C_1 FIFO becomes empty, C_2 cells are extracted. When both C_1 and C_2 FIFOs become empty, C_3 cells are extracted. In the case of hexahedral cells, C_4 or C_5 cells are similarly extracted when all FIFOs up to C_3 or C_4, respectively, become empty.

When all FIFOs become empty, the noneliminated cells form a skeleton that connects all extreme points. Our implementation then eliminates bubble-like layers of cells around voids to reduce the number of cells in the extrema skeleton. It first finds disjoint parts of boundary faces of the extrema skeleton, and eliminates several cells that share the boundary faces, to prick the bubble-like layers. By inserting those several cells into FIFOs, the volume-thinning process eliminates many of the cells in the bubble-like layers. The detail of the implementation is described by Itoh et al. [21].

Fig. 5.9 shows a pseudo-code of the volume-thinning process for volumes consisting of tetrahedral cells. Pseudo-code for hexahedral cells is very similar, except it includes the processes for C_4 and C_5 cells after the processes for C_1 to C_3 cells.

5.5 Skeleton Generation for the Extraction of Numerical Features

The extrema skeleton, introduced in the previous section, has contributed to the high-performance processing searches for isosurfacing cells within volume datasets. Next, we would like to propose using the volume skeleton to search for volume datasets with some features among a very large number of volume datasets. In this case, we expect the skeleton to act as a search index, which is a simplified representation of a volume dataset.

To represent a volume dataset using the volume skeleton, the skeleton should retain the physical features of the original volume dataset. One possible skeleton is composed of all the volume cells intersecting the critical-point graph, which is an extension of our extrema graph. The critical-point graph is defined as the scalar topology of a scalar field of S, which is composed of the local maxima, the local minima, the saddle points of the scalar gradient field, and integral curves joining each of the critical points. Integral curves are defined as curves that are everywhere tangent to the gradient field. The initial positions of the calculations for the integral curves are set a small distance from the critical point along the appropriate eigenvector of the velocity gradient tensor [22]. This critical-point graph can also be used as a seed set for isosurface cells. Fig. 5.10 (left) shows an example of a critical-point graph, and Fig. 5.10 (right) shows an isosurface generated by using the critical-point graph as a seed set. A similar idea has been proposed by Bajaj et al. [13].

Since we might have a volume dataset containing a very large number of critical points, we will develop a technique for constructing a critical-point graph based on our volume-thinning algorithm. We extend our original volume-thinning algorithm so that the resulting skeleton can contain the saddle points and become parallel to the integral curves, and evaluate the degree to which the resulting skeleton coincides with the critical-point graph, so that the number of

```
void VolumeThinning(){

    /* Initialization */
    Classify cells that touch extreme points as C_-1,
    and other cells as C_1, C_2, C_3, and C_4;
    Insert C_1, C_2, and C_3 cells into FIFOs;
    Assign boundary ID to boundary faces and nodes;

    /* Main loop */
    while (1) {
        if( a C_1 cell ei is extracted ) { EliminateCell(e_i); }
          else if( a C_2 cell e_i is extracted ) {
          if(e_i cannot be eliminated ) continue; EliminateCell(e_i);
        } else if( a C_3 cell e_i is extracted ) {
            if(e_i cannot be eliminated ) continue; EliminateCell(e_i);
        } else if( there are bubble-like layers around voids ) {
            Select three cells to prick a layer;
              for(each above selected cells ei) { EliminateCell(ei); }
        } else break;
    } /* end of main loop */

    /* Postprocessing */
    Register non-C_0 cells to a seed set list;
}

void EliminateCell(e_i){
    Change the classification c_i into C_0;
    Assign the same boundary ID
         to all faces and nodes of e_i;
    for(each adjacent cell ei,j) {
        if( c_i,j = C_0 || c_i,j = C_-1) continue;
        Update the classification c_i,j;
        Insert e_i,j to a FIFO;
    }
}
```

Figure 5.9   Pseudo-code of the volume-thinning process for tetrahedral cells.

volume cells in the skeleton is increased compared to the original skeleton.

5.6   Conclusion

This chapter introduced techniques for fast isosurface generation. Section 5.2 introduced an isosurface propagation technique that recursively visits isosurface cells. Section 5.3 introduced an extrema graph that connects all extreme points in a volume. By traversing the cells intersected by the extrema graph and the

cells on the boundaries of those cells, the isosurface cells are extracted from all of the disjoint parts of an isosurface, and the isosurface is propagated efficiently. Section 5.4 introduced an extrema skeleton generated by a volume-thinning method, which connected all of the extreme points and preserved the topology of the volume. The extrema skeleton makes the traversal of the boundary cells unnecessary while propagating all disjoint parts of the isosurfaces. In contrast to many existing fast isosurface-generation techniques, the technique proposed in Section 5.4 satisfies the following three requirements:

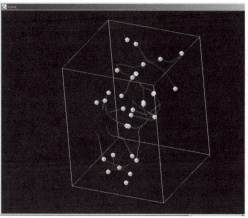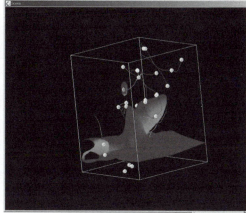

Figure 5.10 (Left) Example of a critical-point graph. (Right) Example of an isosurface generated by using the critical-point graph as a seed set. (See also color insert.)

- Can be applied to unstructured volumes.
- Is much lower than $O(n)$ complexity for visiting isosurface cells.
- Has $O(n)$ complexity for preprocessing.

We think that the volume-thinning method can be applied not only to fast isosurface generation but also to other purposes. Section 5.5 discussed the extension of the volume-thinning method for the extraction of numeric features as critical-point graphs.

References

1. W. E. Lorenson and H. E. Cline. Marching cubes: a high resolution 3D surface construction algorithm. *Computer Graphics*, 21(4):163–169, 1987.
2. C. D. Hansen and P. Hinker. Massively parallel isosurface extraction. *Proceedings of IEEE Visualization '92*, pages 77–83, 1992.
3. C. T. Howie and E. H. Blake. The mesh propagation algorithm for isosurface construction. *Computer Graphics Forum (Eurographics)*, 13(3):C65–C74, 1994.
4. J. W. Durkin and J. F. Hughes. Nonpolygonal isosurface rendering for large volume. *Proceedings of IEEE Visualization '94*, pages 293–300, 1994.
5. R. S. Gallagher. Span filtering: an optimization scheme for volume visualization of large finite element models. *Proceedings of IEEE Visualization '91*, pages 68–74, 1991.

6. M. Giles and R. Haimes. Advanced interactive visualization for CFD. *Computer Systems in Engineering*, 1(1):51–62, 1990.
7. H. Shen and C. R. Johnson. Sweeping simplices: a fast iso-surface extraction algorithm for unstructured grids. *Proceedings of IEEE Visualization '95*, pages 143–150, 1995.
8. D. Silver and N. J. Zabusky. Quantifying visualization for reduced modeling in nonlinear science: extracting structures from data sets. *Journal of Visual Communication and Image Representation*, 4(1):46–61, 1993.
9. J. Wilhelms and A. Van Gelder. Octrees for fast isosurface generation. *ACM Transactions on Graphics* 11(3):201–227, 1992.
10. P. Cignoni, P. Marino, C. Montani, E. Puppo, and R. Scopigno. Speeding up isosurface extraction using interval trees. *IEEE Transactions on Visualization and Computer Graphics*, 3(2):158–170, 1997.
11. Y. Livnat, H. Shen, and C. R. Johnson. A near optimal isosurface extraction algorithm using the span space. *IEEE Transactions on Visualization and Computer Graphics*, 2(1):73–84, 1996.
12. H. Shen, C. D. Hansen, Y. Livnat, and C. R. Johnson. Isosurfacing in span space with utmost efficiency (ISSUE). *Proceedings of IEEE Visualization '96*, pages 287–294, 1996.
13. C. L. Bajaj, V. Pascucci, and D. R. Schikore. Fast isocontouring for improved interactivity. *Proceedings of ACM Symposium on Volume Visualization '96*, pages 39–46, 1996.
14. M. Kreveld, R. Oostrum, C. L. Bajaj, V. Pascucci, and D. R. Schikore. Contour trees and

small seed sets for isosurface traversal. *Proceedings of 13th ACM Symposium of Computational Geometry*, pages 212–219, 1997.

15. J. Bloomenthal. Polygonization of implicit surfaces. *Computer Aided Geometric Design*, 5(4):341–355, 1988.

16. T. Itoh, Y. Yamaguchi, and K. Koyamada. Fast isosurface generation using the cell-edge centered propagation algorithm. *International Symposium on High Performance Computing (ISHPC)*, pages 547–556, 2000.

17. D. Speray and S. Kennon. Volume probe: interactive data exploration on arbitrary grids. *Computer Graphics*, 24(5):5–12, 1990.

18. G. Wyvill, C. McPheeters, and B. Wyvill. Data structure for soft objects. *The Visual Computer*, 2(4):227–234, 1986.

19. T. Itoh and K. Koyamada. Automatic isosurface propagation by using an extrema graph and sorted boundary cell lists. *IEEE Transactions on Visualization and Computer Graphics*, 1(4):319–327, 1995.

20. T. Itoh and K. Koyamada. Volume thinning for automatic isosurface propagation. *Proceedings of IEEE Visualization '96*, pages 303–310, 1996.

21. T. Itoh, Y. Yamaguchi, and K. Koyamada. Fast isosurface generation using an extrema skeleton and cell-edge-centered propagation. *IEEE Transactions on Visualization and Computer Graphics*, 7(1):32–46, 2001.

22. K. Koyamada and T. Itoh. Seed specification for displaying a streamline in an irregular volume. *Engineering with Computers*, 14:73–80, 1998.

6   Isosurfaces and Level-Sets

ROSS T. WHITAKER
School of Computing
University of Utah

6.1   Introduction

This chapter describes the basic differential geometry of isosurfaces and the method of manipulating the shapes of isosurfaces within volumes, called *level-sets*. Deformable isosurfaces, implemented with level-set methods, have demonstrated a great potential in visualization for applications such as segmentation, surface processing, and surface reconstruction. This chapter begins with a short introduction to isosurface geometry, including curvature. It continues with a short explanation of the level-set partial differential equations. It also presents some practical details for how to solve these equations using upwind-scheme and sparse-calculation methods. This chapter also presents a series of examples of how level-set surface models are used to solve problems in graphics and vision.

6.1.1   Motivation

This chapter describes mechanisms for analyzing and processing volumes in a way that deals specifically with *isosurfaces*. The underlying philosophy is to use isosurfaces as a modeling technology that can serve as an alternative to parameterized models for a variety of important applications in visualization and computer graphics. Level-set methods [1] rely on partial differential equations (PDEs) to model deforming isosurfaces. These methods have applications in a wide range of fields, such as visualization, scientific computing, computer graphics, and computer vision [2,3]. Applications in visualization include volume segmentation [4,5,6], surface processing [7,8], and surface reconstruction [9,10].

This chapter presents the mathematics and numerical techniques for describing the geometry of isosurfaces and manipulating their shapes in prescribed ways. It starts with a basic introduction to the notation and fundamental concepts and then presents the geometry of iso surfaces. It then describes the method of level-sets, i.e., moving isosurfaces, and presents the mathematical and numerical methods they entail.

6.1.2   Isosurfaces

6.1.2.1   Modeling Surfaces with Volumes

When considering surface models for graphics and visualization, one is faced with a staggering variety of options including meshes, spline-based patches, constructive solid geometry, implicit blobs, and particle systems. These options can be divided into two basic classes—explicit (parameterized) models and implicit models. With an implicit model, one specifies the model as a *level-set* of a scalar function,

$$\phi: \begin{array}{cc} U & \mapsto \mathbb{R} \\ x,y,z & k \end{array} \qquad (6.1)$$

where $U \subset \mathbb{R}^3$ is the domain of the volume (and the *range* of the surface model). Thus, a surface S is

$$S = \{x | \phi(x) = k\} \qquad (6.2)$$

The choice of k is arbitrary, and ϕ is sometimes called the *embedding*. Notice that surfaces defined in this way divide U into a clear inside and outside—such surfaces are always closed wherever they do not intersect the boundary of the domain.

Choosing this implicit strategy raises the question of how to represent ϕ. Historically, implicit models are represented using linear combinations of *basis* functions. These basis or potential functions usually have several degrees of freedom, including 3D position, size, and orientation. By combining these functions, one can create complex surface models. Typical models might contain several hundred to several thousand such primitives. This is the strategy behind the *blobby models* proposed by Blinn [11].

While such an implicit modeling strategy offers a variety of new modeling tools, it has some limitations. In particular, the global nature of the potential functions limits one's ability to model *local* surface deformations. Consider a point $x \in S$ where S is the level surface associated with a model $\phi = \sum_i \alpha_i$, and α_i is one of the individual potential functions that comprise that model. Suppose one wishes to move the surface at the point x in a way that maintains continuity with the surrounding neighborhood. With multiple, global basis functions one must decide which basis function or combination of basis functions to alter and at the same time control the effects on other parts of the surface. The problem is generally ill posed—there are many ways to adjust the basis functions so that x will move in the desired direction, and yet it may be impossible to eliminate the effects of those movements on other disjoint parts of the surface. These problems can be overcome, but the solutions usually entail heuristics that tie the behavior of the surface deformation to, for example, the choice of representation [12].

An alternative to using a small number of *global* basis functions is to use a relatively large number of *local* basis functions. This is the principle behind using a volume as an implicit model. A volume is a discrete sampling of the embedding ϕ. It is also an implicit model with a very large number of basis functions, as shown in Fig. 6.1. The total number of basis functions is fixed; their positions (grid-points) and extent are also fixed. One can change only the magnitude of each basis function, i.e., each basis function has only one degree of freedom. A typical volume of size $128 \times 128 \times 128$ contains over a million such basis functions. The shape of each basis function is open to interpretation—it depends on how one interpolates the values between the grid-points. A tri-linear interpolation, for instance, implies a basis function that is a piecewise cubic polynomial with a value of one at the grid-point and zero at neighboring grid-points. Another advantage of using volumes as implicit models is that for the purposes of analysis we can treat the volume as a continuous function whose values can be *set* at each point according to the application. Once the continuous analysis is complete, we can map the algorithm into the discrete domain using standard methods of numerical analysis. The sections that

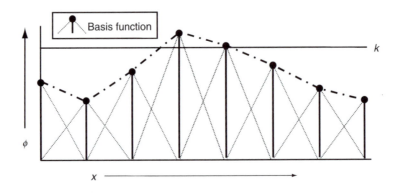

Figure 6.1 A volume can be considered as an implicit model with a large number of local basis functions.

follow discuss how to compute the geometry of surfaces that are represented as volumes and how to manipulate the shapes of those surfaces by changing the grey-scale values in the volume.

6.1.2.2 *Isosurface Extraction and Visualization*

This chapter addresses the question of how to use volumes as surface models. Depending on the application, however, a 3D grid of data (i.e., a volume) may not be a suitable model representation. For instance, if the goal is to make measurements of an object or visualize its shape, an explicit model might be necessary. In such cases it is beneficial to convert between volumes and other modeling technologies.

For instance, the literature proposes several methods for scan-converting polygonal meshes or solid models [13]. Likewise, a variety of methods exist for extracting parametric models of isosurfaces from volumes. The most prevalent method is to locate isosurface crossings along grid lines in a volume (between voxels along the 3 cardinal directions) and then to link these points together to form triangles and meshes. This is the strategy of *marching cubes* [14] and other related approaches. However, extracting a parametric surface is not essential for visualization, and a variety of direct methods [15,16] are now computationally feasible and arguably superior in quality. This chapter does not address the issue of extracting or rendering isosurfaces; it rather studies the geometry of isosurfaces and how to manipulate them directly by changing the grey-scale values in the underlying volume. Thus, we propose volumes as a mechanism for studying and deforming surfaces, regardless of the ultimate form of the output. There are many ways of rendering or visualizing them, and these techniques are beyond the scope of this discussion.

6.2 Surface Normals

The surface normal of an isosurface is given by the normalized gradient vector of ϕ. Typically,

we identify a surface normal with a point in the volume domain U. That is,

$$n(x) = \frac{\nabla\phi(x)}{|\nabla\phi(x)|} \quad \text{where } x \in D \tag{6.3}$$

The convention regarding the direction of this vector is arbitrary; the negative of the normalized gradient magnitude is also normal to the isosurface. The gradient vector points toward that side of the isosurface that has greater values (i.e., is brighter). When rendering, the convention is to use *outward-pointing* normals, and the sign of the gradient must be adjusted accordingly. However, for most applications any consistent choice of normal vector will suffice. On a discrete grid, one must also decide how to approximate the gradient vector (i.e., first partial derivatives). In many cases central differences suffice. However, in the presence of noise, especially when volume rendering, it is sometimes helpful to compute first derivatives using some smoothing filter (e.g., convolution with a Gaussian). Alternatively, when calculating high-order geometry, one should use a polynomial or spline, with the appropriate degree of continuity [17]. When using the normal vector to solve certain kinds of partial differential equations, it is sometimes necessary to approximate the gradient vector with discrete, one-sided differences, as discussed in Section 6.6.1.

Note that a single volume contains families of nested isosurfaces, arranged like the layers of an onion. We specify the normal to an isosurface as a function of the position within the volume. That is, $n(x)$ is the normal of the (single) isosurface that passes through the point x. The k value associated with that isosurface is $\phi(x)$.

6.3 Second-Order Structure

In differential geometric terms, the second-order structure of a surface is characterized by a quadratic patch that shares first- and second-order contact with the surface at a point (i.e., tangent plane and osculating circles). The *principal directions* of the surface are those associated with the quadratic approximation, and the *principal curvatures* k_1, k_2 are the curvatures in those directions.

As described by Kindlmann et al. [17], the second structure of the isosurface can be computed from the first- and second-order structure of the embedding, ϕ. All of the isosurface shape information is contained in a field of normals given by $\mathbf{n}(\mathbf{x})$. The 3×3 matrix of derivatives of this vector is

$$N = - [\mathbf{n}_x \ \mathbf{n}_y \ \mathbf{n}_z] \qquad (6.4)$$

The projection of this derivative onto the tangent plane of the isosurface gives the shape matrix, β. Let P denote the normal projection operator, which is defined as

$$P = \mathbf{n} \otimes \mathbf{n} = \frac{1}{\|\nabla \phi\|^2} \begin{pmatrix} \phi_x^2 & \phi_x \phi_y & \phi_x \phi_z \\ \phi_y \phi_x & \phi_y^2 & \phi_y \phi_z \\ \phi_z \phi_x & \phi_z \phi_y & \phi_z^2 \end{pmatrix} \qquad (6.5)$$

The tangential projection operator is $I - P$, and thus the shape matrix is

$$\beta = NT = TH_\phi T \qquad (6.6)$$

where H_ϕ is the Hessian of ϕ. The shape matrix β has 3 real eigenvalues, which are

$$e_1 = k_1, e_2 = k_2, e_3 = 0 \qquad (6.7)$$

The corresponding eigenvectors are the principal directions of the surface (i.e., in the tangent plane) and the normal, respectively.

The *mean curvature* is the mean of the two principal curvatures, which is one-half of the trace of β, which is equal to the trace of N:

$$H = \frac{k_1 + k_2}{2} - \frac{1}{2}\text{Tr}(N)$$

$$-\frac{\begin{matrix}\phi_x^2(\phi_{yy} + \phi_{zz}) + \phi_y^2(\phi_{xx} + \phi_{zz}) \\ + \phi_z^2(\phi_{xx} + \phi_{yy}) - 2\phi_x \phi_y \phi_{xy} \\ -2\phi_x \phi_z \phi_{xz} - 2\phi_y \phi_z \phi_{yz}\end{matrix}}{2(\phi_x^2 + \phi_y^2 + \phi_z^2)^{3/2}} \qquad (6.8)$$

The *Gaussian curvature* is the product of the principal curvatures:

$$K = k_1 k_2 = e_1 e_2 + e_1 e_3 + e_2 e_3 = 2\text{Tr}(N)^2 - \frac{1}{2}\|N\|$$

$$= \Big(\phi_z^2(\phi_{xx}\phi_{yy} - \phi_{xy}\phi_{xy}) + \phi_y^2(\phi_{xx}\phi_{zz} - \phi_{xz}\phi_{xz})$$

$$+ \phi_x^2(\phi_{yy}\phi_{zz} - \phi_{yz}\phi_{yz}) + 2(\phi_x\phi_y(\phi_{xz}\phi_{yz} - \phi_{xy}\phi_{zz})$$

$$+ \phi_x\phi_z(\phi_{xy}\phi_{yz} - \phi_{xz}\phi_{yy}) + \phi_y\phi_z(\phi_{xy}\phi_{xz} - \phi_{yz}\phi_{xx})) \Big)$$

$$\Big/ (\phi_x^2 + \phi_y^2 + \phi_z^2)^2 \qquad (6.9)$$

The total curvature, also called the deviation from flatness D [18], is the root sum of squares of the two principal curvatures, which is the Euclidean norm of the matrix β.

Notice these measures exist at every point in U, and at each point they describe the geometry of the particular isosurface that passes through that point. All of these quantities can be computed on a discrete volume using finite differences, as described in successive sections.

6.4 Deformable Surfaces

This section begins with mathematics for describing geometric surface deformations on parametric models. The result is an evolution equation for a surface. Any term in this geometric evolution equation can be reexpressed in a way that is independent of the parameterization. Finally, the evolution equation for a parametric surface gives rise to an evolution equation (differential equation) on a volume, which encodes the shape of that surface as a level-set.

6.4.1 Surface Deformation

A regular surface $\mathcal{S} \subset \mathbb{R}^3$ is a collection of points in 3D that can be represented *locally* as a continuous function. In geometric modeling, a surface is typically represented as a two-parameter object in a 3D space. That is to say, a surface is local to a mapping $\mathbf{S}$:

$$\begin{matrix} \mathbf{S}: V \times V \longmapsto & \mathbb{R}^3 \\ r & \mathcal{S} & x, y, z \end{matrix} \qquad (6.10)$$

where $V \times V \mathbb{R}^2$, and the bold notation refers specifically to a parameterized surface (vector-valued function). A deformable surface exhibits some motion over time. Thus $S = S(r, s, t)$, where $t \in \mathbb{R}^+$. We assume second-order-continuous, orientable surfaces; therefore, at every point on the surface (and in time) there is surface normal $n = n(r, s, t)$. We use S_t to refer to the entire set of points on the surface.

Local deformations of S can be described by an evolution equation, i.e., a differential equation on S that incorporates the position of the surface, local and global shape properties, and responses to other forcing functions. That is,

$$\frac{\partial S}{\partial t} = G(S, S_r, S_s, S_{rr}, S_{rs}, S_{ss}, \ldots) \qquad (6.11)$$

where the subscripts represent partial derivatives with respect to those parameters. The evolution of S can be described by a sum of terms that depends on both the geometry of S and the influence of other functions or data.

There is a variety of differential expressions that can be combined for different applications. For instance, the model could move in response to some directional *forcing* function [19,20], $F: U \mapsto \mathbb{R}^3$, that is,

$$\frac{\partial S}{\partial t} = F(S) \qquad (6.12)$$

Alternatively, the surface could expand and contract with a spatially varying speed. For instance,

$$\frac{\partial S}{\partial t} = G(S)n \qquad (6.13)$$

where $G: \mathbb{R}^3 \mapsto \mathbb{R}$ is a signed speed function. The evolution might also depend on the surface geometry itself. For instance,

$$\frac{\partial S}{\partial t} = S_{rr} + S_{ss} \qquad (6.14)$$

describes a surface that moves in a way that becomes more *smooth* with respect to its own parameterization. This motion can be combined with the motion of Equation 6.12 to produce a model that is pushed by a forcing function but maintains a certain smoothness in its shape and parameterization. There are myriad terms that depend on both the differential geometry of the surface and outside forces or functions to control the evolution of a surface.

6.5 Deformation: The Level-Set Approach

The method of level-sets, proposed by Osher and Sethian [21] and described extensively by Sethian [2] and Fedkiw and Osher [3], provides the mathematical and numerical mechanisms for computing surface deformations as time-varying iso-values of ϕ by solving a partial differential equation on the 3D grid. That is, the level-set formulation provides a set of numerical methods that describe how to manipulate the grey-scale values in a volume, so that the isosurfaces of ϕ move in a prescribed manner (Fig. 6.2).

We denote the velocity of a point on a surface as it deforms as dx/dt, and we assume that this motion can be expressed in terms of the position of $x \in U$ and the geometry of the surface at that point. In this case, there are generally two options for representing such surface movements implicitly.

Static: A single, static $\phi(x)$ contains a family of level-sets corresponding to surfaces as different times t. That is,

$$\phi(x(t)) = k(t) \Rightarrow \nabla\phi(x) \cdot \frac{\partial x}{\partial t} = \frac{dk(t)}{dt} \qquad (6.15)$$

To solve this static method requires constructing a ϕ that satisfies Equation 6.15. This is a boundary-value problem, and it can be solved somewhat efficiently, starting with a single surface using the fast marching method of Sethian [22]. This representation has some significant limitations, however, because (by definition) a surface cannot pass back over itself over time, i.e., motions must be strictly monotonic— inward or outward.

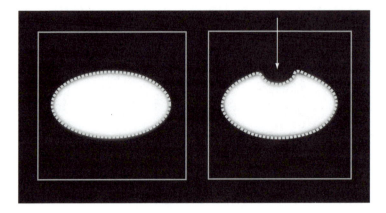

Figure 6.2   Level-set models represent curves and surfaces implicitly using grey-scale images. (Left) An ellipse is represented as the level-set of an image. (Right) To change the shape we modify the grey-scale values of the image.

Dynamic: The approach is to use a one-parameter *family* of embeddings, in which $\phi(x, t)$ changes over time, x remains on the k level-set of ϕ as it moves, and k remains constant. The behavior of ϕ is obtained by setting the total derivative of $\phi(x(t), t) = k$ to zero. Thus,

$$\phi(x(t), t) = k \Rightarrow \frac{\partial \phi}{\partial t} = -\nabla \phi \cdot \frac{dx}{dt} \qquad (6.16)$$

This approach can accommodate models that move forward and backward and cross back over their own paths (over time). However, to solve this requires solving the initial value problem (using finite forward differences) on $\phi(x, t)$—a potentially large computational burden. The remainder of this discussion focuses on the dynamic case, because of its superior flexibility.

All surface movements depend on position and geometry, and the level-set geometry is expressed in terms of the differential structure of ϕ. Therefore, the dynamic formulation from Equation 6.16 gives a general form of the partial differential equation on ϕ:

$$\frac{\partial \phi}{\partial t} = -\nabla \phi \cdot \frac{dx}{dt}$$
$$= -\nabla \phi \cdot F(x, D\phi, D^2\phi, \ldots) \qquad (6.17)$$

where $D^n\phi$ is the set of order-n derivatives of ϕ evaluated at x. Because this relationship applies to every level-set of ϕ, i.e., all values of k, this equation can be applied to all of U, and therefore the movements of *all* the level-set surfaces embedded in ϕ can be calculated from Equation 6.17.

The level-set representation has a number of practical and theoretical advantages over conventional surface models, especially in the context of deformation and segmentation. First, level-set models are topologically flexible; they can easily represent complicated surface shapes that can, in turn, form holes, split to form multiple objects, or merge with other objects to form a single structure. These models can incorporate many (millions of) degrees of freedom, and therefore they can accommodate complex shapes. Indeed, the shapes formed by the level-sets of ϕ are restricted only by the resolution of the sampling. Thus, there is no need to reparameterize the model as it undergoes significant deformations.

Such level-set methods are well documented in the literature [21,23] for applications such as computational physics [24], image processing [25,26], computer vision [5,27], medical image analysis [4,5], and 3D reconstruction [28,29].

For instance, in computational physics, level-set methods are a powerful tool for modeling moving boundaries between different materials (see Osher and Fedkiw [24] for a nice overview of recent results). Two examples are water–air and water–oil interfaces. In such cases, level-set methods can be used to compute deformations that minimize surface area while preserving volumes for surfaces that *split* and *merge* in arbitrary ways. The method can be extended to multiple, nonoverlapping objects [30].

Level-set methods have also been shown to be effective in extracting surface structures from biological and medical data. For instance, Malladi et al. [5] propose a method in which the level-sets form an expanding or contracting contour that tends to *cling* to interesting features in 2D angiograms. At the same time the contour is also influenced by its own curvature, and therefore remains smooth. Whitaker et al. [4,31] have shown that level-sets can be used to simulate conventional deformable surface models, and they demonstrated this by extracting skin and tumors from *thick-sliced* (e.g., clinical) MR data and by reconstructing a model of a fetus from 3D ultrasound. A variety of authors [26,32,33,34] have presented variations on the method and presented results for 2D and 3D data. Sethian [2] gives several examples of level-set curves and surfaces for segmenting CT and MR data.

6.5.1 Deformation Modes

In the case of parametric surfaces, one can choose from a variety of different expressions to construct an evolution equation that is appropriate for a particular application. For each of those parametric expressions, there is a corresponding expression that can be formulated on ϕ, the volume in which the level-set models are embedded. In computing deformations of level-sets, there can be no reference to the underlying surface parameterization (terms depending on r and s in Equations 6.10–6.14).

This has two important implications: (1) only those surface movements that are normal to the surface are represented—any other movement is equivalent to a reparameterization; (2) all of the derivatives with respect to surface parameters r and s must be expressed in terms of invariant surface properties that can be derived without a parameterization.

Consider the term $S_{rr} + S_{ss}$ from Equation 6.14. If r,s is an orthonormal parameterization, the effect of that term is based purely on surface shape, not on the parameterization, and the expression $S_{rr} + S_{ss}$ is twice the *mean curvature*, H, of the surface. The corresponding level-set formulation is given by Equation 6.8.

Table 6.1 shows a list of expressions used in the evolution of parameterized surfaces and their equivalents for level-set representations. Also given are the assumptions about the parameterization that give rise to the level-set expressions.

6.6 Numerical Methods

By taking the strategy of embedding surface models in volumes, we have converted equations that describe the movement of surface points to nonlinear, partial differential equations defined on a volume, which is generally a rectilinear grid. The expression $u_{i,j,k}^n$ refers to the nth time-step at position i,j,k, which has an associated value in the 3D domain of the continuous volume $\phi(x_i, y_j, z_k)$. The goal is to solve the differential equation consisting of terms from Table 5.1 on the discrete grid $u_{i,j,k}^n$.

The discretization of these equations raises two important issues. First is the availability of accurate, stable numerical schemes for solving these equations. Second is the problem of computational complexity and the fact that we have converted a *surface* problem to a *volume* problem, increasing the dimensionality of the domain over which the evolution equations must be solved.

The level-set terms in Table 6.1 are combined, based on the needs of the application, to create a partial differential equation on $\phi(x, t)$. The solutions to these equations are computed using finite differences. Along the time axis, solutions are obtained using finite *forward* differences, beginning with an initial model (i.e., volume) and stepping sequentially through a series of discrete time-steps (which are denoted as superscripts on u). Thus, the update equation is

$$u_{i,j,k}^{n+1} = u_{i,j,k}^n + \Delta t \Delta u_{i,j,k}^n \qquad (6.18)$$

The term $\Delta u_{i,j,k}^n$ is a discrete approximation to $\partial \phi / \partial t$, which consists of a weighted sum of terms such as those in Table 6.1. Those terms must, in turn, be approximated using finite differences on the volume grid.

6.6.1   Upwind Schemes

The terms in Table 6.1 fall into two basic categories: the first-order terms (items 1 and 2) and the second-order terms (items 3 through 5). The first-order terms describe a moving wave front with a space-varying velocity (expression 1) or speed (expression 2). Equations of this form cannot be solved using conventional central-difference schemes for spatial derivatives. Such schemes tend to overshoot, and they are unstable. To address this issue, Osher and Sethian [1] proposed an *upwind* scheme. The upwind method relies on a one-sided derivative that

looks in the upwind direction of the moving wave front, and thereby avoids the overshooting associated with finite forward differences.

We denote the type of discrete difference using superscripts on a difference operator, i.e., $\delta^{(+)}$ for forward differences, $\delta^{(-)}$ for backward differences, and δ for central differences. For instance, differences in the x direction on a discrete grid $u_{i,j,k}$ with domain X and uniform spacing h are defined as

$$\delta_x^{(+)} u_{i,j,k} \triangleq (u_{i+1,j,k} - u_{i,j,k})/h, \qquad (6.19)$$

$$\delta_x^{(-)} u_{i,j,k} \triangleq (u_{i,j,k} - u_{i-1,j,k})/h, \text{ and} \qquad (6.20)$$

$$\delta_x u_{i,j,k} \triangleq (u_{i+1,j,k} - u_{i-1,j,k})/(2h) \qquad (6.21)$$

where we have left off the time superscript for conciseness. Second-order terms are typically computed using the *tightest-fitting* central difference operators. For example,

$$\delta_{xx} u_{i,j,k} \triangleq (u_{i+1,j,k} + u_{i-1,j,k} \\ - 2u_{i,j,k})/h^2, \qquad (6.22)$$

$$\delta_{zz} u_{i,j,k} \triangleq (u_{i,j,k+1} + u_{i,j,k-1} \\ - 2u_{i,j,k})/h^2, \text{ and} \qquad (6.23)$$

$$\delta_{xy} u_{i,j,k} \triangleq \delta_x \delta_y u_{i,j,k} \qquad (6.24)$$

The discrete approximations to the first-order terms in Table 6.1 are computed using the upwind scheme proposed by Osher and Sethian [21]. This strategy avoids overshooting by approximating the gradient of ϕ using a one-sided difference in the direction that

Table 6.1   A list of evolution terms for parametric models has a corresponding expression on the embedding, ϕ, associated with the level-set models.

	Effect	Parametric evolution	Level-Set evolution	Parameter assumptions
1	External force	F	$F \cdot \nabla \phi$	None
2	Expansion/contraction	$G(x)N$	$G(x)\|\nabla \phi(x,\ t)\|$	None
3	Mean curvature	$S_{rr} + S_{ss}$	$H\|\nabla \phi\|$	Orthonormal
4	Gauss curvature	$S_{rr} \times S_{ss}$	$K\|\nabla \phi\|$	Orthonormal
5	Second order	S_{rr} or S_{ss}	$\left(H \pm \sqrt{H^2 - K}\right)\|\nabla \phi\|$	Principal curvatures

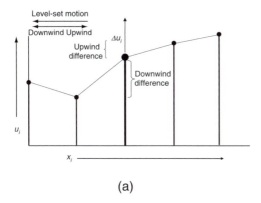
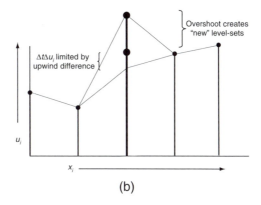

(a)   (b)

Figure 6.3 The upwind numerical scheme uses one-sided derivatives to prevent overshooting and the creation of new level-sets.

is upwind of the moving level-set, thereby ensuring that no *new* contours are created in the process of updating $u_{i,j,k}^n$ (Fig. 6.3). The scheme is separable along each axis (x, y, and z).

Consider Term 1 in Table 6.1. If we use superscripts to denote the vector components, i.e.,

$$\boldsymbol{F}(x, y, z) = (F^{(x)}(x, y, z), F^{(y)}(x, y, z),$$
$$F^{(z)}(x, y, z)) \tag{6.25}$$

the upwind calculation for a grid-point $u_{i,j,k}^n$ is

$$\boldsymbol{F}(x_i, y_i, z_i) \cdot \nabla \phi(x_i, y_j, z_k, t) \approx$$
$$\sum_{q \in \{x, y, z\}} F^{(q)}(x_i, y_i, z_i)$$
$$\begin{cases} \delta_q^+ u_{i,j,k}^n & F^{(q)}(x_i, y_i, z_i) < 0 \\ \delta_q^- u_{i,j,k}^n & F^{(q)}(x_i, y_i, z_i) > 0 \end{cases} \tag{6.26}$$

The time-steps are limited—the fastest-moving wave front can move only one grid unit per iteration. That is,

$$\Delta t_F \leq \frac{1}{\sum_{q \in \{x, y, z\}} \sup_{i, j, k \in X} \{|\nabla F^{(q)}(x_i, y_j, z_k)|\}} \tag{6.27}$$

For Term 2 in Table 6.1 the direction of the moving surface depends on the normal, and therefore the same upwind strategy is applied in a slightly different form.

$$G(x_i, y_j, z_k)|\nabla \phi(x_i, y_j, z_k, t)| \approx$$
$$\sum_{q \in \{x, y, z\}} G(x_i, y_i, z_i)$$
$$\begin{cases} \max^2(\delta_q^+ u_{i,j,k}^n, 0) + \min^2(\delta_q^- u_{i,j,k}^n, 0) \\ \quad G(x_i, y_i, z_i) < 0 \\ \min^2(\delta_q^+ u_{i,j,k}^n, 0) + \max^2(\delta_q^- u_{i,j,k}^n, 0) \\ \quad G(q)(x_i, y_i, z_i) > 0 \end{cases} \tag{6.28}$$

The time-steps are, again, limited by the fastest-moving wave front:

$$\Delta t_G \leq \frac{1}{3\sup_{i, j, k \in X} \{|\nabla G(x_i, y_j, z_k)|\}} \tag{6.29}$$

To compute the approximation of the update to the second-order terms in Table 6.1 requires only central differences. Thus, the mean curvature is approximated as

$$H_{i,j,k}^n =$$
$$\frac{1}{2} \left(\left(\delta_x u_{i,j,k}^n\right)^2 + \left(\delta_y u_{i,j,k}^n\right)^2 + \left(\delta_z u_{i,j,k}^n\right)^2 \right)^{-1} \times$$
$$\left[\left(\left(\delta_y u_{i,j,k}^n\right)^2 + \left(\delta_z u_{i,j,k}^n\right)^2 \right) \delta_{xx} u_{i,j,k}^n \right.$$
$$+ \left(\left(\delta_z u_{i,j,k}^n\right)^2 + \left(\delta_x u_{i,j,k}^n\right)^2 \right) \delta_{yy} u_{i,j,k}^n$$
$$+ \left(\left(\delta_x u_{i,j,k}^n\right)^2 + \left(\delta_y u_{i,j,k}^n\right)^2 \right) \delta_{zz} u_{i,j,k}^n$$
$$- 2\delta_x u_{i,j,k}^n \delta_y u_{i,j,k}^n \delta_{xy} u_{i,j,k}^n - 2\delta_y u_{i,j,k}^n \delta_z u_{i,j,k}^n \delta_{yz} u_{i,j,k}^n$$
$$\left. - 2\delta_z u_{i,j,k}^n \delta_x u_{i,j,k}^n \delta_{zx} u_{i,j,k}^n \right] \tag{6.30}$$

Such curvature terms can be computed by using a combination of forward and backward differences as described by Whitaker and Xue [35]. In some cases this is advantageous, but the details are beyond the scope of this chapter.

For the second-order terms, the time-steps are limited, for stability, by the diffusion number, to

$$\Delta t_H \leq \frac{1}{6} \qquad (6.31)$$

When combining terms, the maximum number of time-steps for each term is scaled by one over the weighting coefficient for that term.

6.6.2 Narrow-Band Methods

If one is interested in only a *single level-set*, the formulation described previously is not efficient. This is because solutions are usually computed over the entire domain of ϕ. The solutions $\phi(x, t)$ describe the evolution of an embedded family of contours. While this dense family of solutions might be advantageous for certain applications, there are other applications that require only a single surface model. In such applications the calculation of solutions over a dense field is an unnecessary computational burden, and the presence of contour families can be a nuisance because further processing might be required to extract the level-set that is of interest.

Fortunately, the evolution of a single level-set, $\phi(x, t) = k$, is not affected by the choice of embedding. The evolution of the level-sets is such that they evolve independently (to within the error introduced by the discrete grid). Furthermore, the evolution of ϕ is important only in the vicinity of that level-set. Thus, one should perform calculations for the evolution of ϕ only in a neighborhood of the surface $S_t = \{x | \phi(x,t) = k\}$. In the discrete setting, there is a particular subset of grid-points whose values control a particular level-set (Fig. 6.4). Of course, as the surface moves, that subset of grid-points must change to account for its new position.

Adalsteinson and Sethian [36] propose a *narrow-band* approach, which follows this line of reasoning. The narrow-band technique con-

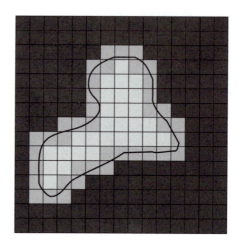

Figure 6.4   A level curve of a 2D scalar field passes through a finite set of cells. Only those grid-points nearest to the level curve are relevant to the evolution of that curve.

structs an embedding of the evolving curve or surface via a signed distance transform. The distance transform is truncated, i.e., computed over a finite width of only m points that lie within a specified distance to the level-set. The remaining points are set to constant values to indicate that they do not lie within the narrow band, or *tube*, as they call it. The evolution of the surface (they demonstrate it for curves in the plane) is computed by calculating the evolution of u only on the set of grid-points that are within a fixed distance to the initial level-set, i.e., within the narrow band. When the evolving level-set approaches the edge of the band (Fig. 6.5), they calculate a new distance transform (e.g., by solving the Eikonal equation with the fast marching method), which creates a new embedding, and they repeat the process. This algorithm relies on the fact that the embedding is not a critical aspect of the evolution of the level-set. That is, the embedding can be transformed or recomputed at any point in time, so long as such a transformation does not change the position of the kth level-set, and the evolution will be unaffected by this change in the embedding.

Despite the improvements in computation time, the narrow-band approach is not optimal

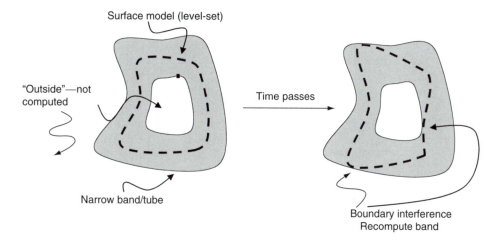

Surface model (level-set)

"Outside"—not
computed

Time passes

Narrow band/tube

Boundary interference
Recompute band

Figure 6.5   The narrow-band scheme limits computation to the vicinity of the specific level-set. As the level-set moves near the edge of the band, the process is stopped and the band recomputed.

for several reasons. First it requires a band of significant width ($m = 12$ in the examples of Adalsteinson and Sethian [36]) where one would like to have a band that is only as wide as necessary to calculate the derivatives of u near the level-set (e.g., $m = 2$). The wider band is necessary because the narrow-band algorithm trades off two competing computational costs. One is the cost of stopping the evolution and computing the position of the curve and distance transform (to sub-cell accuracy) and determining the domain of the band. The other is the cost of computing the evolution process over the entire band. The narrow-band method also requires additional techniques, such as smoothing, to maintain the stability at the boundaries of the band, where some grid-points are undergoing the evolution and nearby neighbors are static.

6.6.3   The Sparse-Field Method

The basic premise of the narrow-band algorithm is that computing the distance transform is so costly that it cannot be done at every iteration of the evolution process. Another strategy is to use an approximation to the distance transform that makes it feasible to recompute the neighborhood of the level-set model at

each time-step. Computation of the evolution equation is done on a band of grid-points that is only one point wide. The embedding is extended from the active points to a neighborhood around those points that is precisely the width needed at each time. This extension is done via a fast distance transform approximation.

This approach has several advantages. First, the algorithm does precisely the number of calculations needed to compute the next position of the level curve. It does not require explicit recalculation of the positions of level-sets and their distance transforms. Because the number of points being computed is so small, it is feasible to use a linked list to keep track of them. Thus, at each iteration the algorithm visits only those points adjacent to the k-level curve. This is important because for large 3D datasets, the very process of incrementing a counter and checking the status of all of the grid-points is prohibitive.

The *sparse-field* algorithm is analogous to a locomotive engine that lays tracks before it and picks them up from behind. In this way the number of computations increases with the surface area of the model rather than the resolution of the embedding. Also, the sparse-field approach identifies a single level-set with a specific set of points whose values control the position of that level-set. This allows one to compute external forces to an accuracy that is better than the grid spacing of the model, resulting in a modeling system that is more accurate for various kinds of *model-fitting* applications.

The sparse-field algorithm takes advantage of the fact that a k-level surface, S, of a discrete image u (of any dimension) has a set of cells through which it passes, as shown in Fig. 6.4. The set of grid-points adjacent to the level-set is called the *active set*, and the individual elements of this set are called *active points*. As a first-order approximation, the distance of the level-set from the center of any active point is proportional to the value of u divided by the gradient magnitude at that point. Because all of the derivatives (up to second order) in this approach are computed using nearest-neighbor differences, only the active points and their neighbors are relevant to the evolution of the level-set at any particular time in the evolution process. The strategy is to compute the evolution given by Equation 6.17 on the active set and then update the neighborhood around the active set using a fast distance transform. Because active points must be adjacent to the level-set model, their positions lie within a fixed distance to the model. Therefore, the values of u for locations in the active set must lie within a certain range. When active-point values move out of this *active range*, they are no longer adjacent to the model. They must be removed from the set, and other grid-points, those whose values are moving into the active range, must be added to take their place. The precise ordering and execution of these operations is important to the operation of the algorithm.

The values of the points in the active set can be updated using the upwind scheme for first-order terms and central differences for the mean-curvature flow, as described in the previous sections. In order to maintain stability, one must update the neighborhoods of active grid-points in a way that allows grid-points to enter and leave the active set without those changes

in status affecting their values. Grid-points should be removed from the active set when they are no longer the nearest grid-point to the zero crossing. If we assume that the embedding u is a discrete approximation to the distance transform of the model, then the distance of a particular grid-point $x_m = (i, j, k)$ to the level-set is given by the value of u at that grid-point. If the distance between grid-points is defined to be unity, then we should remove a point from the active set when the value of u at that point no longer lies in the interval $\left[-\frac{1}{2}, \frac{1}{2}\right]$ (Fig. 6.6). If the neighbors of that point maintain their distance of 1, then those neighbors will move into the active range just as x_m is ready to be removed.

There are two operations that are significant to the evolution of the active set. First, the values of u at active points change from one iteration to the next. Second, as the values of active points pass out of the active range, they are removed from the active set and other, neighboring grid-points are added to the active set to take their place. Whitaker [29] gives some formal definitions of active sets and the operations that affect them. These definitions show that active sets will always form a boundary between positive and negative regions in the image, even as control of the level-set passes from one set of active points to another.

Because grid-points that are near the active set are kept at a fixed value difference from the active points, active points serve to control the behavior of nonactive grid-points to which they are adjacent. The neighborhoods of the active set are defined in *layers*, $L_{+1}, \ldots L_{+N}$ and $L_{-1}, \ldots L_{-N}$, where the i indicates the distance (city-block distance) from the nearest active grid-point, and negative numbers are used for the outside layers. For notational convenience, the active set is denoted L_0.

The number of layers should coincide with the size of the footprint or neighborhood used to calculate derivatives. In this way, the inside and outside grid-points undergo no changes in their values that affect or distort the evolution

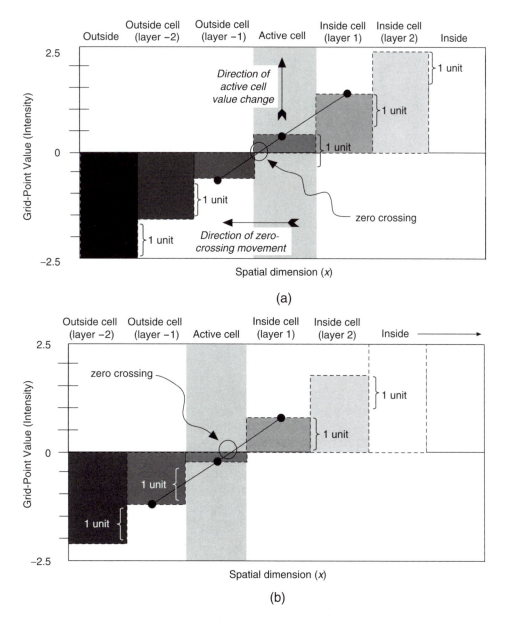

Figure 6.6 The status of grid-points and their values at two different points in time show that as the zero crossing moves, *activity* is passed from one grid-point to another.

of the zero set. Most of the level-set work relies on surface normals and curvature, which require only second-order derivatives of ϕ. Second-order derivatives are calculated using a $3 \times 3 \times 3$ kernel (city-block distance 2 to the corners). Therefore, only 5 layers are

necessary (2 inside layers, 2 outside layers, and the active set). These layers are denoted L_1, L_2, L_{-1}, L_{-2}, and L_0.

The active set has grid-point values in the range $\left[-\frac{1}{2}, \frac{1}{2}\right]$. The values of the grid-points in each neighborhood layer are kept 1 unit from the layer next closest to the active set (Fig. 6.6). Thus the values of layer L_i fall in the interval $\left[i - \frac{1}{2}, i + \frac{1}{2}\right]$. For $2N + 1$ layers, the values of the grid-points that are totally inside and outside are $N + \frac{1}{2}$ and $-N - \frac{1}{2}$, respectively. The procedure for updating the image and the active set based on surface movements is as follows:

1. For each active grid-point, $x = (i, j, k)$, do the following:

    (a) Calculate the local geometry of the level-set.

    (b) Compute the net change of u_x, based on the internal and external forces, using some stable (e.g., upwind) numerical scheme where necessary.

2. For each active grid-point x_j, add the change to the grid-point value and decide if the new value u_x^{n+1} falls outside the $\left[-\frac{1}{2}, \frac{1}{2}\right]$ interval. If so, put x on a list of grid-points that are changing status, called the *status list*; S_1 or S_{-1}, for $u_x^{n+1} > 1$ or $u_x^{n+1} < -1$, respectively.

3. Visit the grid-points in the layers L_i in the order $i = \pm 1, \ldots \pm N$, and update the grid-point values based on the values (by adding or subtracting one unit) of the next inner layer, $L_{i\mp 1}$. If more than one $L_{i\mp 1}$ neighbor exists, then use the neighbor that indicates a level curve closest to that grid-point, i.e., use the maximum for the outside layers and minimum for the inside layers. If a grid-point in layer L_i has no $L_{i\mp 1}$ neighbors, then it gets demoted to $L_{i\pm 1}$, the next level away from the active set.

4. For each status list $S_{\pm 1}, S_{\pm 2}, \ldots, S_{\pm N}$, do the following:

    (a) For each element x on the status list S_i, remove x from the list $L_{i\mp 1}$ and add it to the list L_i, or, in the case of $i = \pm(N + 1)$, remove it from all lists.

    (b) Add all $L_{i\mp 1}$ neighbors to the $S_{i\pm 1}$ list.

This algorithm can be implemented efficiently using linked-list data structures combined with arrays to store the values of the grid-points and their states, as shown in Fig. 6.7. This requires

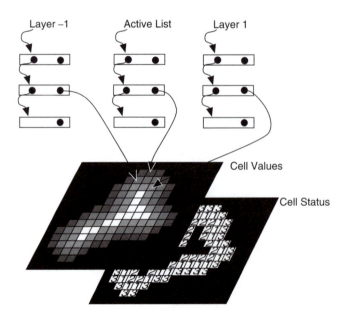

Figure 6.7 Linked-list data structures provide efficient access to those grid-points with values and status that must be updated.

only those grid-points whose values are changing, the active points and their neighbors, to be visited at each time-step. The computation time grows as m^{n-1}, where m is the number of grid-points along one dimension of x (sometimes called the resolution of the discrete sampling). Computation time for a dense-field approach increases as m^n. The m^{n-1} growth in computation time for the sparse-field models is consistent with conventional (parameterized) models, for which computation times increase with the resolution of the domain, rather than the range.

Another important aspect of the performance of the sparse-field algorithm is the larger time-steps that are possible. The time-steps are limited by the speed of the *fastest*-moving level curve, i.e., the maximum of the force function. Because the sparse-field method calculates the movement of level-sets over a subset of the image, time-steps are bounded from below by those of the dense-field case. That is,

$$\sup_{x \in \mathcal{A} \subset X} (g(x)) \leq \sup_{x \in X} (g(x)) \qquad (6.32)$$

where $g(x)$ is the space-varying speed function and $\mathcal{A}$ is the active set.

Results from previous work by Whitaker [29] have demonstrated several important aspects of the sparse-field algorithm. First, the manipulations of the active set and surrounding layers allow the active set to *track* the deformable surface as it moves. The active set always divides the inside and outside of the objects it describes (i.e., it stays closed). Empirical results show significant increases in performance relative to both the computation of the full domain and the narrow-band method, as proposed in the literature. Empirical results also show that the sparse-field method is about as accurate as both the full, discrete solution and the narrow-band method. Finally, because the method positions level-sets to sub-voxel accuracy, it avoids aliasing problems and is more accurate than these other methods when it comes to *fitting* level-set models to other surfaces. This sub-

voxel accuracy is an important aspect of the implementation and will significantly impact the quality of the results for the applications that follow.

6.7 Applications

This section describes several examples of how level-set surface models can be used to address problems in graphics, visualization, and computer vision. These examples are a small selection of those available in the literature.

6.7.1 Surface Morphing

This section summarizes the work of Breen and Whitaker [8], which describes the use of level-set surface models to perform 3D shape metamorphosis. The *morphing* of 3D surfaces is the process of constructing a series of 3D models that constitute a smooth transition from one shape to another (i.e., a homotopy). Such a capability is interesting for creating animations and as a tool for geometric modeling [37,38,39,40,41,42].

Level-set models provide an algorithm for 3D morphing, which is a natural extension of the mathematical principles discussed in previous sections. The strategy is to allow a free-form deformation of one surface (called the *initial* surface) using the signed distance transform of a second surface (the *target* surface). This free-form deformation is combined with an underlying coordinate transformation that gives either a rough global alignment of the two surfaces, or one-to-one relationships between a finite set of landmarks on both the initial and the target surfaces. The coordinate transformation can be computed automatically or using user input.

The distance transform gives the nearest Euclidean distance to a set of points, curve, or surface. For closed surfaces in three dimensions, the signed distance transform gives a positive distance for points inside and negative distance for points outside (one can also choose the opposite sign convention). If two connected shapes overlap, then the initial surface can expand or contract using the distance transform of the target. The steady state of such a deformation process is a shape consisting of the zero set of the distance transform of the target. That is, the initial object becomes the target. This is the basis of the proposed 3D morphing algorithm.

Let $D(x)$ be the signed distance transform of the target surface, B, and let A be the initial surface. The evolution process, which takes a model S from A to B, is defined by

$$\frac{\partial x}{\partial t} = ND(x) \tag{6.33}$$

where $x(t) \in S_t$ and $S_{t=0} = A$. The free-form deformations are combined with an underlying coordinate transformation. The strategy is to use a coordinate transformation as follows. A coordinate transformation is given by

$$x' = T(x, \alpha) \tag{6.34}$$

where $0 \le \alpha \le 1$ parameterizes a continuous family of these transformations that begins with identity, i.e., $x = T(x, 0)$. The evolution equation for a parametric surface is

$$\frac{\partial x}{\partial t} = ND(T(x, 1)) \tag{6.35}$$

and the corresponding level-set equation is

$$\frac{\partial \phi(x, t)}{\partial t} = |\nabla \phi(x, t)| D(T(x, 1)) \tag{6.36}$$

This process produces a series of transition shapes (parameterized by t). The coordinate transformation can be a global rotation, translation, or scaling, or it might be a *warping of the underlying 3D space* [40]. Incorporating user input is important for any surface-morphing technique, because in many cases finding the best set of transition surfaces depends on context. Only users can apply semantic considerations to the transformation of one object to another. However, this underlying coordinate transformation can, in general, achieve only some finite similarity between the *warped* initial model and the target, and even this may require a great deal of user input. In the event that a user is not able or willing to define every important

correspondence between two objects, some other method must *fill in* the gaps remaining between the initial and target surfaces. Lerios et al. [40] propose alpha blending to achieve that smooth transition—really just a fading from one surface to the other. Level-sets allow the use of the free-form deformations to achieve a continuous transition between the shapes that result from the underlying coordinate transformation.

Fig. 6.8 shows a 3D model of a jet that was built using a CSG modeling system. Lerios et al. [40] demonstrate the transition of a jet to a dart, which was accomplished using 37 user-defined correspondences, roughly 100 user-defined parameters. Fig. 6.9 shows the use of level-set models to construct a set of transition surfaces between a jet and a dart. The triangle mesh is extracted from the volume using the method of marching cubes [14].

The application in this section shows how level-set models moving according to the first-order term given in expression 2 in Table 6.1 can *fit* other objects by moving with a speed that depends on the signed distance transform of the target object. The application in the next section relies on expression 5 of Table 6.1, a second-order flow that depends on the principal curvatures of the surface itself.

Figure 6.8 A 3D model of a jet that was built using a CSG modeling system.

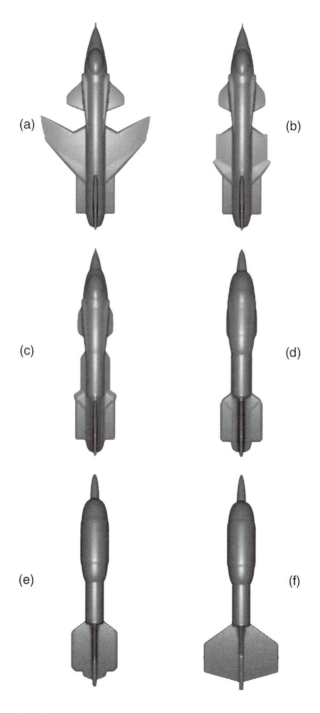

Figure 6.9 The deformation of the jet to a dart using a level-set model moving with a speed defined by the signed distance transform of the target object.

6.7.2 Surface Editing

This section gives a brief summary of the results in Museth et al. [44], who describe a system for surface editing based on level-sets. The creation of complex models for such applications as movie special effects, graphic arts, and computer-aided design can be a time-consuming, tedious, and error-prone process. One of the solutions to the model creation problem is 3D photography, i.e., scanning a 3D object directly into a digital representation (*http://www.tacc. utexas.edu/~reyes/tacc_personal/self/3D_fax.html*). However, the scanned model is rarely in a final desired form. The scanning process is imperfect and introduces errors and artifacts, or the object itself may be flawed. In order to overcome these difficulties, one can use a level-set approach to implementing operators for locally and globally editing closed surfaces.

Museth et al. [44] describe a number of surface-editing operators within the level-set framework by defining a collection of different level-set speed functions. The cut-and-paste operator gives the user the ability to copy, remove, and merge level-set models (using volumetric CSG operations) and automatically blends the intersection regions. The smoothing operator allows a user to define a region of interest and smooths the enclosed surface to a user-defined curvature value.

They also describe a point-attraction operator, in which a regionally constrained portion of a level-set surface is attracted to a single point. By defining line segments, curves, polygons, patches, and 3D objects as densely sampled point sets, the single-point attraction operator may be combined to produce a more general surface-embossing operator. Morphological operators, such as opening and closing, can also be implemented in a level-set framework [45]. Such operations are useful for performing global blending (closing) and smoothing (opening) on level-set models. Because all of the operators accept and produce the same volumetric representation of closed surfaces, the operators may be applied repeatedly to produce a series of surface-editing operations, as shown in Fig. 6.10.

6.7.3 Antialiasing Binary Volumes

This section presents a summary of results from a previous article [46] that addresses the question of using level-sets to reduce aliasing artifacts in binary volumes. Binary volumes are interesting for several practical reasons. First, in some cases, such as medical imaging, a volume dataset can be segmented to produce a set of voxels that correspond to some particular object (or anatomy). This segmentation can be manual, in which case

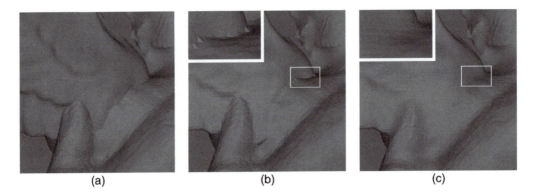

(a)	(b)	(c)

Figure 6.10 (a) Positioning the (red) wing model on the dragon model. (b) The models are pasted together (by CSG union operation), producing sharp, undesirable creases, a portion of which is expanded in the box. (c) The same region after automatic blending based on mean curvature. The blending is constrained to move only outwards. The models are rendered with flat-shading to highlight the details of the surface structure. (See also color insert.)

a user identifies (usually with the aid of a GUI) all of the voxels that belong to a certain object. The segmentation can also be automated, relying on methods such as pattern classification, flood fill, and morphological watersheds, which produce segmentations that are *hard*, i.e., binary. Binary volumes are also important when using a 3D imaging device that produces data with very high contrast. In such cases the measured data is *essentially* binary with regard to both the information it contains and the problems it presents in rendering. Binary volumes can be important when visualizing mathematical expressions, such as fractals, that cannot be evaluated as continuous functions. Binary volumes are also interesting because they require so little data, and, with the use of run length encoding, are very well suited to compression.

The strategy presented here is related to the work of Gibson [47], who uses a deformable-surface approach to reducing aliasing artifacts. Gibson dealt with the problem of extracting surfaces from binary volumes with an insightful strategy: *treat the binary data as a constraint on a surface that is subject to a regularization process.* She proposes a several-step algorithm called *constrained surface nets* that embodies this strategy. The algorithm begins by extracting a surface mesh from the volume. This initial mesh consists of a vertex in each cell (an eight-voxel neighborhood arranged in a cube) whose corners indicate a transition from inside to out. This mesh undergoes an iterative process of deformation, where each vertex moves to the mean of its neighbors but is prohibited from moving outside of its original cell. The resulting surface can be converted back into a volume by computing a discrete sampling of the distance transform to the set of triangles associated with the final mesh.

Constrained surface nets have some very useful properties. They are essentially the solution of a constrained minimization of surface area. Constrained surface nets are capable of creating flat surfaces through sequences of distant *terraces* or *jaggies*, which are not easily spanned by a filter or interpolating function.

The final solution is guaranteed to lie within a fixed distance of the original mesh, thus preserving small details, and even discontinuities, that can be lost through other anti-aliasing techniques such as those that rely on filtering [48,49] or surface approximation [50,51,52].

Using level-sets, one can implement the basic philosophy of constrained surface nets while eliminating the need for an intermediate surface mesh—i.e., operate directly on the volume. The result is a transformation of a binary volume to a grey-scale volume, and thus it is a kind of nonlinear filtering process. The zero set of the same volume that results from the algorithm has the desirable properties of the constrained surface net. However, the algorithm makes no explicit assumptions about the topology of the surface, but instead allows the topology of the surface to develop from the constrained minimization process.

Binary volumes are often visualized through treatment as implicit functions and rendering of the surfaces that correspond to the zero sets of interpolated version of $B : \mathcal{D} \mapsto \{-1, 1\}$, the binary volume. Alternatively, one could treat a binary volume, regardless of its origins, as a threshold (or binarization) of a discrete sampling of an embedding. That is,

$$\phi(x, y, z) \overset{\text{discretization}}{\longrightarrow} f_{i,j,k} \overset{\text{threshold}}{\longrightarrow} B_{i,j,k} \qquad (6.37)$$

From this point of view, the problem of *extracting* surfaces from a binary volume is really the problem of *estimating* either f or ϕ and extracting surfaces from one of those functions. However, the loss of information (i.e., projection) associated with the binary sampling leaves the inverse problem ill posed—that is, for a given binary volume, there are infinitely many embeddings from which it could have been derived.

The strategy in this section is to construct a discrete sampling of a ϕ that *could* have given rise to B. We say an estimate of the embedding, $\hat{\phi}$, is *feasible* if

$$\hat{\phi}(\boldsymbol{x}) B_{\boldsymbol{x}} \geq 0 \; \forall \; \boldsymbol{x} \in \mathcal{D} \qquad (6.38)$$

That is, B_x and $\hat{\phi}(x)$ must have the same sign at the grid-points of $\mathcal{D}$. This is the same as saying that the zero set of $\hat{\phi}$ must enclose all of those points indicated by the binary volume as *inside* and none of the points that are *outside*.

The ill-posed nature of the problem is addressed by imposing some criterion, a regularization, to which $\hat{\phi}$ must conform. In the case of surface estimation, a natural criterion is to choose the surface with minimal area. Often, but not always, surfaces with less area are qualitatively *smoother*. In the level-set formulation, the combined surface area of all of the level-sets of ϕ is the integral of the level-set density (which is the gradient magnitude of ϕ) over the domain $\mathcal{D}$. Thus, using level-sets, the constrained minimization problem for reconstructing surfaces from binary data is

$$\hat{\phi} = \arg\min_\phi \left[\int_D |\nabla\phi(x)| dx \right] \tag{6.39}$$

$$\text{such that } \phi(x)B_x \geq 0 \ \forall \ x \in \mathcal{D}$$

Using the method of undetermined multipliers, construct the Lagrangian:

$$F(\phi, \boldsymbol{\lambda}) = \int_D |\nabla\phi(x)| dx + \sum_{x_i \in \mathcal{D}} \lambda_i \phi(x_i) B_{x_i} \tag{6.40}$$

where $\lambda_i \leq 0$.

The Kuhn–Tucker [53] conditions describe the behavior of the solution. For points in the domain that are not on the grid, the level-sets of the solution are flat or hyperbolic (saddle points) with the principal curvatures offsetting one another. For points in the domain that fall on one of the grid-points, there are two cases: the level-sets of ϕ are convex at places with $B(x_i) > 0$ and concave where $B(x_i) < 0$, which is consistent with a solution that is stretched around the positive and negative constraints. Also, when the curvature is nonzero at a grid-point,

$$B(x_i)\phi(x_i) = \pm\phi(x_i) = 0 \tag{6.41}$$

which means that the zero set falls through grid-points of $\mathcal{D}$ except in those areas where the solution is flat.

This analysis leads to a gradient-descent strategy with an evolution parameter t. Starting with an initial estimate that is feasible, one can update ϕ in such a way that it minimizes the surface area but does not violate the constraints:

$$\frac{\partial\phi}{\partial t} = \begin{cases} 0 & \text{For } x = x_i \in D, \phi(x) = 0 \\ \text{and} & H(x)B_x > 0 \\ H(x) & \text{otherwise} \end{cases} \tag{6.42}$$

Because the solution must remain near the constraints, the full sparse-field solution is unnecessary, and one can instead use a static narrow band. The appropriate narrow-band algorithm is as follows:

1. Construct an initial solution $u^0_{i,j,k} = B_{i,j,k}$.

2. Find all of the grid-points in $u^0_{i,j,k}$ that lie adjacent to one or more grid-points of opposite sign. Call this set $\mathcal{A}_0$.

3. Find the set of all of the grid-points that are adjacent to $\mathcal{A}_0$ and denote it $\mathcal{A}_1$. Repeat this for $\mathcal{A}_2, \mathcal{A}_3, \ldots, \mathcal{A}_M$, to create a band that is $2M + 1$ wide. The union of these sets, $\mathcal{A} = \cup_{i=0}^M \mathcal{A}_i$, is the *active set*.

4. For each $(i, j, k) \in \mathcal{A}$ calculate $\Delta u^n_{i,j,k}$ using a central difference approximation to the mean curvature.

5. For each $(i, j, k) \in \mathcal{A}$ update the value of $u^{n+1}_{i,j,k}$ according to Equation 6.42.

6. Find the average change for points in the active set:

$$c^n = \left(\frac{1}{|\mathcal{A}|} \sum_\mathcal{A} \left| u^{n+1}_{i,j,k} - u^n_{i,j,k} \right|^2 \right)^{\frac{1}{2}} \tag{6.44}$$

7. If c^n is below some predefined threshold, then the algorithm is complete; otherwise, increment n and go to step 4.

Figs. 6.11a and 6.11b show the zero sets of grey-scale and binary volumes of a cube. The grey-scale volume is the distance transform. The cubes shown are rotated $22.5°$ around each axis (in order to create significant aliasing artifacts in the binary version), and each cube edge

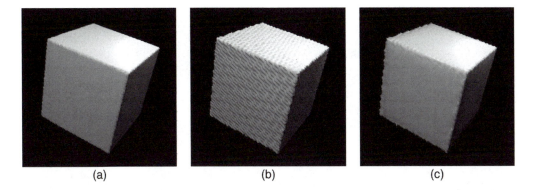

Figure 6.11   (a) An ideal grey-scale embedding of a cube (i.e., distance transform) results in a smooth, accurate isosurface. (b) A binary volume yields significant aliasing artifacts. (c) The surface estimation from the binary data with $M = 4$ and a stopping threshold of 0.002 shows a quality that is comparable to the ideal.

has a length of 50 grid units. Fig. 6.11c shows the zero set of the solution to the constrained minimization problem with a stopping threshold of 0.002 using a narrow band of $M = 4$. These results show significant improvements in the aliasing, especially along the flat faces where the minimum-surface-area approach is most appropriate. On the corners and edges, artifacts remain because the algorithm is trying to stretch the minimal surface across the constraints, which contain jaggies. The algorithm converges rapidly, in about 20 iterations. Experiments show that the choices of bandwidth and stopping threshold do not affect the results in any significant way, provided that the width is sufficiently large and the stopping threshold is sufficiently small.

Fig. 6.12 shows a series of before (left) and after (right) isosurface renderings. Generally the algorithm succeeds in reducing aliasing artifacts with a minimal distortion of the shapes. For some shapes, such as the low-resolution torus, the aliasing is reduced, but only marginally so, which demonstrates a fundamental limitation of the proposed algorithm: *the minimal surface criterion does not always get the solutions close to the ideal.* Instead, the solution is stretching across the rather coarse features formed by the binary volumes. This is especially bad in cases such as a torus, which

includes points for which one principal curvature is significantly greater than the other, causing the surface to *pucker* inward, leaving pronounced aliasing artifacts. On flat surfaces or those with higher resolution, the aliasing effects are virtually eliminated.

6.7.4   Surface Reconstruction and Processing

The ability to compute free-form surface deformations independent of topology or complexity opens up new possibilities in reconstructing and processing surfaces. For instance, in building 3D models from multiple laser range (ladar) images, one can express the likelihood of a closed surface as a function of an integral over the enclosed solid [29,54]. Using a gradient descent to minimize the likelihood gives rise to a data-driven deformation that fits a surface model to a collection of noisy ladar data. Combining the likelihood with an area or curvature-based prior and embedding the motion in the level-set framework generates a PDE for 3D surface reconstruction:

$$\frac{\partial \phi}{\partial t} = |\nabla \phi| G(\boldsymbol{x}, \boldsymbol{n}_\phi) + \beta |\phi| \mathcal{P} \qquad (6.45)$$

where G, the fitting term, depends on the set of input data and a sensor model, while $\mathcal{P}$, which

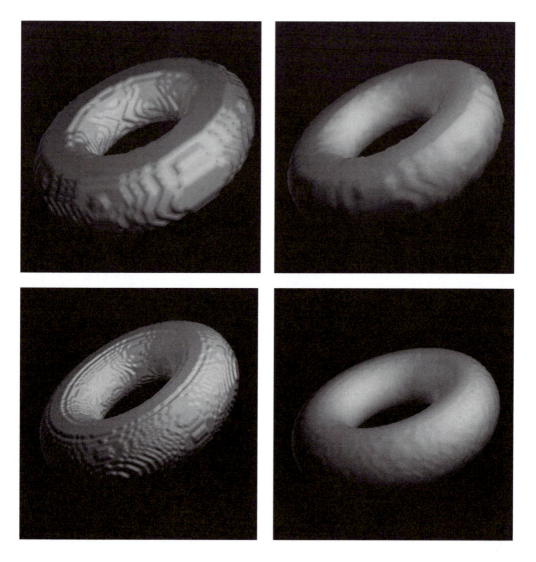

Figure 6.12 (Left) Binary input volumes. (Right) Results of surface estimation. (Top) Lower-resolution torus. (Bottom) Higher-resolution torus.

depends on derivatives of ϕ, is the first variation of the log prior. For example, $\mathcal{P}$ is the mean curvature in the case of a surface area prior. Fig. 6.13 shows a surface rendering of noisy range data and a 3D reconstruction using a level-set model and the curvature-based prior described by Tasdizen and Whitaker [55]. The ability to systematically combine data from different points of view and incorporate a

smoothing prior that preserves creases results in 3D reconstructions that exceed the accuracy of the laser range finder.

This same strategy applies to other imaging modalities. For instance, the problem of reconstructing 3D interfaces from tomographic projections leads to a formulation very similar to Equation 6.44, in which the data term depends on the set of input data and the shape of the

Figure 6.13   (a) A surface rendering of a noisy range image. (b) A 3D reconstruction obtained by fitting a level-set model to 12 noisy range images from different points of view.

surface estimate [9]. This is an important problem in situations where one is given limited or incomplete tomographic data, such as in transmission electron microscopy (TEM). Fig. 6.14 shows a TEM surface reconstruction of a spiny dendrite using this strategy. The application of level-sets for this problem is important, because the complex topology of this model changes during the fitting process.

A problem with relating the reconstruction is that of surface processing, which has gained importance as the number of 3D models grows. One would like to have the same set of tools for manipulating surfaces as exists for images. This would including cutting and pasting, blending, enhancement, and smoothing tools. This is especially important in visualization, where the 3D models are often derived from measured data

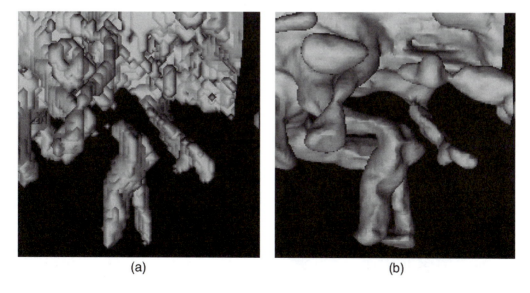

Figure 6.14   (a) An initial model constructed from a backprojection. (b) The model deforms to minimize the discrepancy with the projected data and forms new connections in the process.

and are therefore noisy or incomplete or somehow imperfect. Level-set models provide some mechanisms for filtering surfaces in a way that does not depend on a particular parameterization or topology. Tasdizen et al. [7] describe a strategy for filtering level-set surfaces that relies on processing a *fourth-order* geometric flow.

Using this strategy one can generalize a wide range of image-processing algorithms to surfaces. Fig. 6.15 shows a generalization of anisotropic diffusion [56] to surfaces in a way that enhances sharp creases. Fig. 6.16 shows a generalization of unsharp masking (a form of high-boost filtering), which brings out surface detail.

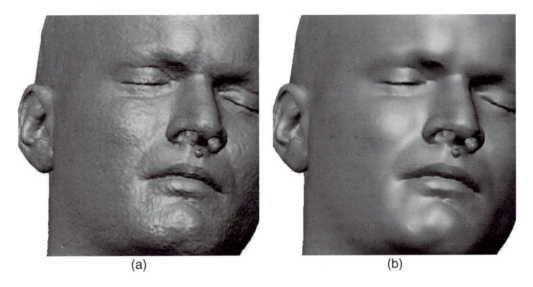

(a)   (b)

Figure 6.15   (a) A noisy isosurface obtained from an MRI volume. (b) Processing with feature-preserving smoothing alleviates noise while enhancing sharp features. (See also color insert.)

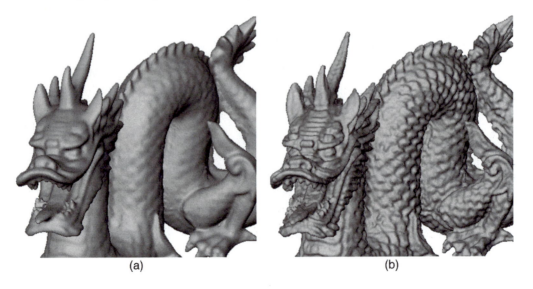

(a)   (b)

Figure 6.16   (a) A volumetric surface model is enhanced via (b) unsharp masking. (See also color insert.)

6.8   Summary

Volumes provide a powerful tool for modeling deformable surfaces, especially when one is dealing with measured data. With measured data, the shape, topology, and complexity of the surface are dictated by the application rather than the user. Implicit deformable surfaces, implemented as level-sets, provide a natural mechanism for processing such data in a manner that relieves the user of having to decide on an underlying parameterization. This technology easily handles the many degrees of freedom that are important to capturing the fine detail of measured data. Furthermore, the level-set approach provides a powerful mechanism for constructing *geometric flows* that results in output that depends only on the shape of input (and the resolution) and does not produce artifacts that are tied to an arbitrary, intermediate parameterization.

References

1. S. Osher and J. Sethian. Fronts propagating with curvature-dependent speed: algorithms based on Hamilton–Jacobi formulations. *Journal of Computational Physics*, 79:12–49, 1988.
2. J. A. Sethian. *Level Set Methods and Fast Marching Methods Evolving Interfaces in Computational Geometry, Fluid Mechanics, Computer Vision, and Materials Science.* Cambridge University Press, 1999.
3. R. Fedkiw and S. Osher. *Level Set Methods and Dynamic Implicit Surfaces.* Berlin, Springer-Verlag, 2002.
4. R. T. Whitaker. Volumetric deformable models: active blobs. In *Visualization In Biomedical Computing 1994*, pages 122–134, SPIE, 1994.
5. R. Malladi, J. A. Sethian, and B. C. Vemuri. Shape modeling with front propagation: a level-set approach. *IEEE Transactions on Pattern Analysis and Machine Intelligence*, 17(2):158–175, 1995.
6. R. Whitaker, D. Breen, K. Museth, and N. Soni. A framework for level-set segmentation of volume datasets. In *Proceedings of ACM Intnl. Wkshp. on Volume Graphics*, pages 159–168, 2001.
7. T. Tasdizen, R. Whitaker, P. Burchard, and S. Osher. Geometric surface processing via normal maps. *ACM Trans. on Graphics*, 22(4):1012–1033, 2003.
8. D. Breen and R. Whitaker. A level-set approach to 3D shape metamorphosis. *IEEE Transactions on Visualization and Computer Graphics*, 7(2):173–192, 2001.
9. R. Whitaker and V. Elangovan. A direct approach to estimating surfaces in tomographic data. *Medical Image Analysis*, 6:235–249, 2002.
10. K. Museth, D. Breen, L. Zhukov, and R. Whitaker. Level-set segmentation from multiple non-uniform volume datasets. In *IEEE Visualization 2002*, pages 179–186, 2002.
11. J. Blinn. A generalization of algebraic surface drawing. *ACM Trans. on Graphics*, 1:235–256, 1982.
12. S. Muraki. Volumetric shape description of range data using "blobby model." *Computer Graphics (SIGGRAPH '91 Proceedings)*, 25:227–235, 1991.
13. D. E. Breen, S. Mauch, and R. T. Whitaker. 3D scan conversion of CSG models into distance volumes. In *The 1998 Symp. on Volume Visualization*, pages 7–14, 1998.
14. W. Lorenson and H. Cline. Marching cubes: a high resolution 3D surface construction algorithm. *Computer Graphics*, 21(4):163–169, 1982.
15. M. Levoy. Display of surfaces from volume data. *IEEE Computer Graphics and Applications*, 9(3):245–261, 1990.
16. R. Drebin, L. Carpenter, and P. Hanrahan. Volume rendering. *Computer Graphics*, 22(4):65–74, 1988.
17. G. Kindlmann, R. Whitaker, T. Tasdizen, and T. Möller. Curvature-based transfer functions for direct volume rendering: methods and applications. In *Proc. IEEE Visualization 2003*, pages 513–520, 2003.
18. J. J. Koenderink. *Solid Shape.* Cambridge, MA, MIT press, 1990.
19. M. Kass, A. Witkin, and D. Terzopoulos. Snakes: active contour models. *International Journal of Computer Vision*, 1:321–323, 1987.
20. D. Terzopoulos and K. Fleischer. Deformable models. *The Visual Computer*, 4:306–331, 1988.
21. S. Osher and J. Sethian. Fronts propagating with curvature-dependent speed: algorithms based on Hamilton–Jacobi formulations. *Journal of Computational Physics*, 79:12–49, 1988.
22. J. A. Sethian. A fast marching level-set method for monotonically advancing fronts. *Proc. Nat. Acad. Sci.*, 93(4):1591–1595, 1996.
23. J. A. Sethian. *Level Set Methods: Evolving Interfaces in Gometry, Fluid Mechanics, Computer Vision, and Material Sciences.* Cambridge, England, Cambridge University Press, 1996.

24. S. Osher and R. Fedkiw. Level set methods: an overview and some recent results. Tech. Rep. 00–08, UCLA Center for Applied Mathematics, Department of Mathematics, University of California, Los Angeles, 2000.

25. L. Alvarez and J.-M. Morel. A morphological approach to multiscale analysis: from principles to equations. In *Geometry-Driven Diffusion in Computer Vision*, pages 4–21, 1994.

26. V. Caselles, R. Kimmel, and G. Sapiro. Geodesic active contours. In *Fifth International Conference on Computer Vision*, pages 694–699, IEEE Computer Society Press, 1995.

27. R. Kimmel and A. Bruckstein. Shape offsets via level sets. *Computer Aided Design*, 25(5):154–162, 1993.

28. R. Whitaker and D. Breen. Level-set models for the deformation of solid objects. In *The Third International Workshop on Implicit Surfaces*, pages 19–35, Eurographics, 1998.

29. R. T. Whitaker. A level-set approach to 3D reconstruction from range data. *International Journal of Computer Vision*, pages 203–231, 1998.

30. T. Chan and L. Vese. A multiphase level-set framework for image segmentation using the Mumford and Shah model. *International Journal of Computer Vision*, 50(3):271–293, 2000.

31. R. T. Whitaker. Algorithms for implicit deformable models. In *Fifth International Conference on Computer Vision*, IEEE Computer Society Press, 1995.

32. S. Kichenassamy, A. Kumar, P. Olver, A. Tannenbaum, and A. Yezzi. Gradient flows and geometric active contour models. In *Fifth Int. Conf. on Comp. Vision*, pages 810–815. Los Alamitos, CA, IEEE Computer Society Press, 1995.

33. A. Yezzi, S. Kichenassamy, A. Kumar, P. Olver, and A. Tannenbaum. A geometric snake model for segmentation of medical imagery. *IEEE Transactions on Medical Imaging*, 16:199–209, 1997.

34. L. Lorigo, O. Faugeraus, W. Grimson, R. Keriven, and R. Kikinis. Segmentation of bone in clinical knee MRI using texture-based geodesic active contours. In *Medical Image Computing and Computer-Assisted Intervention (MICCAI '98)*, pages 1195–1204, 1998.

35. R. Whitaker and X. Xue. Variable-conductance, level-set curvature for image denoising. In *IEEE International Conference on Image Processing*, pages 142–145, 2001.

36. D. Adalsteinson and J. A. Sethian. A fast level-set method for propagating interfaces, *Journal of Computational Physics*, pages 269–277, 1995.

37. J. Kent, W. Carlson, and R. Parent. Shape transformation for polyhedral objects. *Computer Graphics (SIGGRAPH '92 Proceedings)*, 26:47–54, 1992.

38. J. Rossignac and A. Kaul. AGRELS and BIPs: metamorphosis as a bezier curve in the space of polyhedra. *Computer Graphics Forum (Eurographics '94 Proceedings)*, 13(9):C-179–C-184, 1994.

39. J. Hughes. Scheduled fourier volume morphing. *Computer Graphics (SIGGRAPH '92 Proceedings)*, 26(7):43–46, 1992.

40. A. Lerios, C. Garfinkle, and M. Levoy. Feature-based volume metamorphosis. In *SIGGRAPH '95 Proceedings*, pages 449–456, 1995.

41. B. Payne and A. Toga. Distance field manipulation of surface models. *IEEE Computer Graphics and Applications*, 12(1):65–71, 1992.

42. D. Cohen-Or, D. Levin, and A. Solomivici. 3D distance field metamorphosis. *ACM Transactions on Graphics*, 17(2):117–140, 1998.

43. P. Getto and D. Breen. An object-oriented architecture for a computer animation system. *The Visual Computer*, 6(3):79–92, 1990.

44. K. Museth, D. Breen, R. Whitaker, and A. Barr. Level-set surface editing operators. In *ACM SIGGRAPH*, pages 330–338, 2002.

45. P. Maragos. Differential morphology and image processing. *IEEE Transactions on Image Processing*, 5(6):922–937, 1996.

46. R. Whitaker. Reducing aliasing artifacts in isosurfaces of binary volumes. In *IEEE Symp. on Volume Visualization and Graphics*, pages 23–32, 2000.

47. S. Gibson. Using distance maps for accurate surface representation in sampled volumes. In *1998 Symposium on Volume Graphics*, pages 23–30, ACM SIGGRAPH, 1991.

48. S. Wang and A. Kaufman. Volume-sampled 3D modeling. *IEEE Computer Graphics and Aplications*, 14(5):26–32, 1994.

49. S. Wang and A. Kaufman. Volume-sampled voxelization of goemetric primitives. In *Proceedings of the 1993 Symposium on Volume Visualization*, pages 78–84, ACM SIGGRAPH, 1993.

50. U. Tiede, T. Schiemann, and K. Höhne. High quality rendering of attibuted volume data. In *IEEE Visualization 1998*, pages 255–262, 1998.

51. M. I. Miller and B. Roysam. Bayesian image reconstruction for emission tomography incorporating Good's roughness prior on massively parallel processors. *Proc. Natl. Acad. Sci.*, 88:3223–3227, 1991.

52. T. McInerney and D. Terzopoulos. Deformable models in medical image analysis: a survey. *Medical Image Analysis*, 1(2):91–108, 1996.

53. L. Cooper and D. Steinberg. *Introduction to Methods of Optimization*. New York, W.B. Saunders Company, 1970.

54. R. Whitaker and J. Gregor. A maximum likelihood surface estimator for dense range data. *IEEE Transactions on Pattern Analysis and Machine Intelligence*, 24(10):1372–1387, 2002.

55. T. Tasdizen and R. Whitaker. Higher-order nonlinear priors for surface reconstruction. *IEEE Trans. on Pattern Recognition and Machine Intelligence*, 26(7):878–891, 2004.

56. P. Perona and J. Malik. Scale-space and edge detection using anisotropic diffusion. *IEEE Transactions on Pattern Analysis Machine Intelligence*, 12:629–639, 1990.

PART III

Scalar Field Visualization: Volume Rendering

7   Overview of Volume Rendering

ARIE KAUFMAN and KLAUS MUELLER
Center for Visual Computing
Stony Brook University

7.1   Introduction

Volume visualization is a method of extracting meaningful information from volumetric data using interactive graphics and imaging. It is concerned with volume data representation, modeling, manipulation, and rendering [31, 100,101,176]. Volume data are 3D (possibly time-varying) entities that may have information inside them, may not consist of tangible surfaces and edges, or may be too voluminous to be represented geometrically. They are obtained by sampling, simulation, or modeling techniques. For example, a sequence of 2D slices obtained from magnetic resonance imaging (MRI), computed tomography (CT), functional MRI (fMRI), or positron emission tomography (PET) is 3D-reconstructed into a volume model and visualized for diagnostic purposes or for planning of treatment or surgery. The same technology is often used with industrial CT for nondestructive inspection of composite materials or mechanical parts. Similarly, confocal microscopes produce data that is visualized to study the morphology of biological structures. In many computational fields, such as computational fluid dynamics (CFD), the results of simulations typically running on a supercomputer are often visualized as volume data for analysis and verification. Recently, the area of *volume graphics* [104] has been expanding, and many traditional geometric computer graphics applications, such as CAD and flight simulation, have been exploiting the advantages of volume techniques.

Over the years many techniques have been developed to render volumetric data. Since methods for displaying geometric primitives were already well established, most of the early methods involve approximating a surface contained within the data using geometric primitives. When volumetric data are visualized using a surface-rendering technique, a dimension of information is essentially lost. In response to this, volume rendering techniques were developed that attempt to capture the entire 3D data in a single 2D image. Volume rendering conveys more information than surface-rendered images, but at the cost of increased algorithm complexity and consequently increased rendering times. To improve interactivity in volume rendering, many optimization methods both for software and for graphics-accelerator implementations, as well as several special-purpose volume rendering machines, have been developed.

7.2   Volumetric Data

A volumetric dataset is typically a set V of samples (x,y,z,v), also called *voxels*, representing the value v of some property of the data, at a 3D location (x,y,z). If the value is simply a 0 or an integer i within a set I, with a value of 0 indicating background and the value of i indicating the presence of an object O_i, then the data is referred to as binary data. The data may instead be multivalued, with the value representing some measurable property of the data, including, for example, color, density, heat, or pressure. The value v may even be a vector, representing, for example, velocity at each location or results from multiple scanning modalities, such as anatomical (CT, MRI) and functional imaging (PERT, fMRI), or color (RGB) triples, such as the Visible Human cryosection dataset [91]. Finally, the volume data may be time-varying, in which case V becomes a 4D set of samples (x, y, z, t, v).

In general, the samples may be taken at purely random locations in space, but in most cases the set V is isotropic, containing samples taken at regularly spaced intervals along three orthogonal axes. When the spacing between samples along each axis is a constant, but there are three different spacing constants for the three axes, the set V is anisotropic. Since the set of samples is defined on a regular grid, a 3D array (also called the *volume buffer, 3D raster*, or simply *volume*) is typically used to store the values, with the element location indicating position of the sample on the grid. For this reason, the set V will be referred to as the array of values $V(x, y, z)$, which is defined only at grid locations. Alternatively, either rectilinear, curvilinear (structured), or unstructured grids are employed (e.g., Spearey and Kennon [240]). In a *rectilinear* grid the cells are axis-aligned, but grid spacings along the axes are arbitrary. When such a grid has been nonlinearly transformed while preserving the grid topology, the grid becomes *curvilinear*. Usually, the rectilinear grid defining the logical organization is called *computational space*, and the curvilinear grid is called *physical space*. Otherwise the grid is called *unstructured* or *irregular*. An unstructured or irregular volume data is a collection of cells whose connectivity has to be specified explicitly. These cells can be of arbitrary shapes, such as tetrahedra, hexahedra, or prisms.

7.3   Rendering via Geometric Primitives

To reduce the complexity of the volume rendering task, several techniques have been developed that approximate a surface contained within the volumetric data by way of geometric primitives, most commonly triangles, which can then be rendered using conventional graphics-accelerator hardware. A surface can be defined by applying a binary segmentation function $B(v)$ to the volumetric data, where $B(v)$ evaluates to 1 if the value v is considered part of the object, and evaluates to 0 if the value v is part of the background. The surface is then contained in the region where $B(v)$ changes from 0 to 1.

Most commonly, $B(v)$ is either a step function, $B(v) = 1, \forall v \geq v_{iso}$ (where v_{iso} is called the *iso-value*), or an interval $[v_1, v_2]$ in which $B(v) = 1, \forall v \in [v_1, v_2]$ (where $[v_1, v_2]$ is called the *iso-interval*). For the former, the resulting surface is called the *isosurface*, while for the latter, the resulting structure is called the *iso-contour*. Several methods for extracting and rendering isosurfaces have been developed; a few are briefly described here. The *marching cubes* algorithm [136] was developed to approximate an iso-valued surface with a triangle mesh. The algorithm breaks down the ways in which a surface can pass through a grid cell into 256 cases, based on the $B(v)$ membership of the 8 voxels that form the cell's vertices. By ways of symmetry, the 256 cases reduce to 15 base topologies, although some of these have duals, and a technique called *asymptotic decider* [185] can be applied to select the correct dual case and thus prevent the incidence of holes in the triangle mesh. For each of the 15 cases (and their duals), a generic set of triangles representing the surface is stored in a lookup table. Each cell through which a surface passes maps to one of the base cases, with the actual triangle vertex locations determined using linear interpolation of the cell vertices on the cell edges (Fig. 7.1). A normal value is estimated for each triangle vertex, and standard graphics hardware can be utilized to project the triangles, resulting in a relatively smooth shaded image of the iso-valued surface.

When rendering a sufficiently large dataset with the marching cubes algorithm, with an average of 3 triangles per cell, millions of triangles may be generated, and this can impede

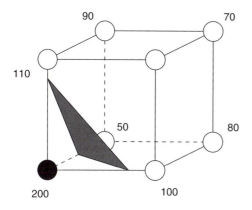

Figure 7.1 A grid cell with voxel values as indicated, intersected by an isosurface (iso-value = 125). This is base case #1 of the marching cubes algorithm: a single triangle separating surface interior (black vertex) from exterior (white vertices). The positions of the triangle vertices are estimated by linear interpolation along the cell edges.

the iso-value v_{iso} and isosurface to explore the different surfaces embedded in the data. By realizing that an isosurface can only pass through a cell if at least one voxel has a value above or equal to v_{iso} and at least one voxel has a value below or equal to v_{iso}, one can devise data structures that only inspect cells where this criterion is fulfilled (Fig. 7.2). Examples are the NOISE algorithm [135], which uses a kd-tree embedded into span space for quickly identifying the candidate cells (this method was later improved by Cignoni et al. [35], who used an interval tree), as well as the ISSUE algorithm [224]. Finally, since triangles are often generated that are later occluded during the rendering process, it is advisable to visit the cells in front-to-back order and only extract and render triangles that fall outside previously occluded areas [62].

interactive rendering of the generated polygon mesh. To reduce the number of triangles, one may either postprocess the mesh by applying one of the many mesh-decimation methods [63,88,220] or produce a reduced set of primitives in the mesh-generation process, via a feature-sensitive octree method [223] or discretized marching cubes [170]. The fact that during viewing many of the primitives may map to a single pixel on the image plane led to the development of screen-adaptive surface-rendering algorithms that use 3D points as the geometric primitive. One such algorithm is *dividing cubes* [36], which subdivides each cell through which a surface passes into subcells. The number of divisions is selected such that the subcells project onto a single pixel on the image plane. Another algorithm that uses 3D points as the geometric primitive is the *trimmed voxel lists* method [235]. Instead of subdividing, this method uses one 3D point (with normal) per visible surface cell, projecting that cell on up to three pixels of the image plane to ensure coverage in the image.

The traditional marching cubes algorithm simply marches across the grid and inspects every cell for a possible isosurface. This can be wasteful when users want to interactively change

7.4 Direct Volume Rendering: Prelude

Representing a surface contained within a volumetric dataset using geometric primitives can be useful in many applications; however, there are

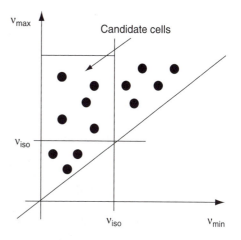

Figure 7.2 Each grid cell is characterized by its lowest (v_{min}) and its highest (v_{max}) voxel value, and represented by a point in span space. Given an iso-value v_{iso}, only cells that satisfy both $v_{min} \leq v_{iso}$ and $v_{max} \geq v_{iso}$ contain the isosurface and are quickly extracted from a kd-tree [135] or interval-tree [35] embedding of the span-space points.

several main drawbacks to this approach. First, geometric primitives can only approximate surfaces contained within the original data. Adequate approximations may require an excessive amount of geometric primitives. Therefore, a tradeoff must be made between accuracy and space requirements. Second, since only a surface representation is used, much of the information contained within the data is lost during the rendering process. For example, in CT-scanned data, useful information is contained not only on the surfaces, but within the data as well. Also, amorphous phenomena, such as clouds, fog, and fire, cannot be adequately represented using surfaces and therefore must have a volumetric representation and must be displayed using volume rendering techniques.

However, before moving to techniques that visualize the data directly without going through an intermediate surface-extraction step, we first discuss in the next section some of the general principles that govern the theory of discretized functions and signals, such as the discrete volume data. We also present some specialized theoretical concepts that are more relevant in the context of volume visualization.

7.5 Volumetric Function Interpolation

The volume grid V only defines the value of some measured property $f(x,y,z)$ at discrete locations in space. If one requires the value of $f(x,y,z)$ at an off-grid location (x,y,z), a process called *interpolation* must be employed to estimate the unknown value from the known grid samples $V(x,y,z)$. There are many possible interpolation functions (also called *filters* or *filter kernels*). The simplest interpolation function is known as zero-order interpolation, which is actually just a nearest-neighbor function. That is, the value at any location (x,y,z) is simply that of the grid sample closest to that location:

$$f(x,y,z) = V(round(x),\ round(y),\ round(z)) \tag{7.1}$$

which gives rise to a box filter (black curve in Fig. 7.4). With this interpolation method there is a region of constant value around each sample in V. The human eye is very sensitive to the jagged edges and unpleasant staircasing that result from a zero-order interpolation, and therefore this kind of interpolation gives generally the poorest visual results (see Fig. 7.3a).

Linear or first-order interpolation (the magenta curve in Fig. 7.4) is the next-best choice, and its 2D and 3D versions are called bi-linear and tri-linear interpolation, respectively. It can be written as three stages of seven linear interpolations, since the filter function is separable in higher dimensions. The first four linear interpolations are along x:

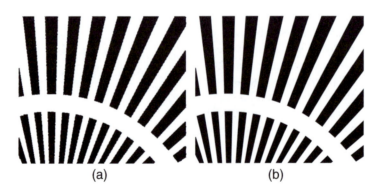

(a) (b)

Figure 7.3 Magnification via interpolation with (a) a box filter and (b) a bi-linear filter. The latter gives a much more pleasing result.

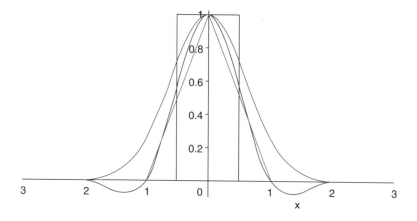

Figure 7.4   Popular filters in the spatial domain: box (black), linear (magenta), cubic (blue), Gaussian (red). (See also color insert.)

$$f(u, v_{0,1}, w_{0,1}) = (1 - u)$$
$$(V(0, v_{0,1}, w_{0,1}) + uV(1, v_{0,1}, w_{0,1})) \quad (7.2)$$

Using these results, 2 linear interpolations along y follow:

$$f(u, v, w_{0,1}) = (1 - v)f(u, 0, w_{0,1})$$
$$+ vf(u, 1, w_{0,1}) \quad (7.3)$$

One final interpolation along z yields the interpolation result:

$$f(x, y, z) = f(u, v, w)$$
$$= (1 - w)f(u, v, 0) + wf(u, v, 1) \quad (7.4)$$

Here u, v, w are the distances (assuming a cell of size 1^3, without loss of generality) of the sample at (x, y, z) from the lower, left, rear voxel in the cell containing the sample point (e.g., the voxel with value 50 in Fig. 7.1). A function interpolated with a linear filter no longer suffers from staircase artifacts (see Fig. 7.3b). However, it has discontinuous derivatives at cell boundaries, which can lead to noticeable banding when the visual quantities change rapidly from one cell to the next.

A second-order interpolation filter that yields a $f(x, y, z)$ with a continuous first derivative is the cardinal spline function, whose ID function is given by (see blue curve in Fig. 7.4):

$$h(u) = \begin{cases} (a+2)|u|^3 - (a+3)|u|^2 + 1 & 0 \le |u| < 1 \\ a|u|^3 - 5a|u|^2 + 8a|u| - 4a & 1 \le |u| \le 2 \\ 0 & |u| > 2 \end{cases}$$
$$(7.5)$$

Here, u measures the distance of the sample location to the grid-points that fall within the extent of the kernel, and $a = -0.5$ yields the Catmull–Rom spline, which interpolates a discrete function with the lowest third-order error [107]. The 3D version of this filter $h(u, v, w)$ is separable, i.e., $h(u, v, w) = h(u)h(v)h(w)$, and therefore interpolation in 3D can be written as a 3-stage nested loop.

A more general form of the cubic function has two parameters, and the interpolation results obtained with different settings of these parameters has been investigated by Mitchell and Netravali [165]. In fact, the choice of filters and their parameters always presents tradeoffs between the sensitivity to noise, sampling frequency ripple, aliasing (see below), ringing, and blurring, and there is no optimal setting that works for all applications. Marschner and Lobb [151] extended the filter discussion to volume rendering and created a challenging volumetric test function with a uniform frequency spectrum that can be employed to visually observe the characteristics of different

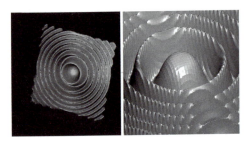

Figure 7.5 Marschner–Lobb test function, sampled into a 20^3 grid: (a) the whole function; (b) close-up, reconstructed, and rendered with a cubic filter. (See also color insert.)

filters (see Fig. 7.5). Finally, Möller et al. [167] applied a Taylor series expansion to devise a set of optimal nth order filters that minimize the $(n + 1)$-th order error.

Generally, higher filter quality comes at the price of wider spatial extent (compare Fig. 7.4) and therefore larger computational effort. The best filter possible in the numerical sense is the *sinc* filter, but it has infinite spatial extent and also tends to noticeable ringing [165]. Sinc filters make excellent, albeit expensive, interpolation filters when used in truncated form and multiplied by a window function [151,252], possibly adaptive to local detail [148]. In practice, first-order or linear filters give satisfactory results for most applications, providing good cost–quality tradeoffs, but cubic filters are also used. Zero-order filters give acceptable results when the discrete function has already been sampled at a very high rate, e.g., in high-definition function lookup tables [270].

All filters presented thus far are grid-interpolating filters, i.e., their interpolation yields $f(x, y, z) = V(x, y, z)$ at grid-points [254]. When presented with a uniform grid signal they also interpolate a uniform $f(x, y, z)$ everywhere. This is not the case with a Gaussian filter function (red curve in Fig. 7.4), which can be written as

$$h(u, v, w) = b \cdot e^{-a(u^2 + v^2 + w^2)} \qquad (7.6)$$

Here, a determines the width of the filter and b is a scale factor. The Gaussian has infinite con-

tinuity in the interpolated function's derivative, but it introduces a slight ripple (about 0.1%) into an interpolated uniform function. The Gaussian is most popular when a radially symmetric interpolation kernel is needed [268,183] and for grids that assume that the frequency spectrum of $f(x, y, z)$ is radially bandlimited [253,182].

It should be noted that interpolation cannot restore sharp edges that may have existed in the original function $f_{org}(x, y, z)$ prior to sampling into the grid. Filtering will always smooth or *lowpass* the original function somewhat. Nonlinear filter kernels [93] or transformations of the interpolated results [180] are needed to recreate sharp edges, as we shall see later.

A frequent artifact that can occur is *aliasing*. It results from inadequate sampling and gives rise to strange patterns that did not exist in the sampled signal. Proper prefiltering (bandlimiting) has to be performed whenever a signal is sampled below its *Nyquist limit*, i.e., twice the maximum frequency that occurs in the signal. Filtering after aliasing will not undo these adverse effects. Fig. 7.6 illustrates this by way of an example, and the interested reader may consult standard texts such as Wolberg [281] and Foley et al. [57] for more detail.

The gradient of $f(x, y, z)$ is also of great interest in volume visualization, mostly for the purpose of estimating the amount of light reflected from volumetric surfaces towards the eye (for example, strong gradients indicate stronger surfaces and therefore stronger reflections). There are three popular methods to estimate a gradient from the volume data [166]. The first computes the gradient vector at each grid-point via a process called *central differencing*:

$$\begin{bmatrix} g_x \\ g_y \\ g_z \end{bmatrix} = \begin{bmatrix} V(x - 1, y, z) \\ V(x, y - 1, z) \\ V(x, y, z - 1) \end{bmatrix} - \begin{bmatrix} V(x + 1, y, z) \\ V(x, y + 1, z) \\ V(x, y, z + 1) \end{bmatrix}$$
$$(7.7)$$

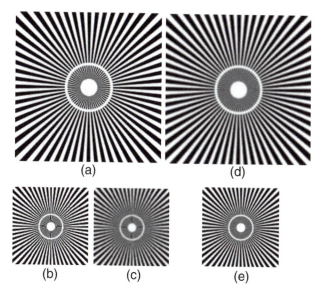

Figure 7.6 Anti-aliasing. (a) Original image. (b) Reduction by simple subsampling—disturbing patterns emerge, caused by aliasing the higher frequency content. (c) Blurring of (b) does not eliminate patterns. (d) Prefiltering (blurring) of the original image reduces its high-frequency content. (e) Subsampling of (d) does not cause aliasing, due to the prior bandlimiting operation.

and then interpolates the gradient vectors at a (x,y,z) using any of the filters described above. The second method also uses central differencing, but it does it at (x,y,z) by interpolating the required support samples on the fly. The third method is the most direct and employs a gradient filter [11] in each of the three axis directions to estimate the gradients. These three gradient filters could be simply the (u,v,w) partial derivatives of the filters described already or they could be a set of optimized filters [166]. The third method gives the best results since it performs only one interpolation step, while the other two methods have lower complexity and often have practical application-specific advantages. An important observation is that gradients are much more sensitive to the quality of the interpolation filter because they are used in illumination calculations, which consist of higher-order functions that involve the normal vectors, which in turn are calculated from the gradients via normalization [167].

7.6 Volume Rendering Techniques

In the next subsections various fundamental volume rendering techniques are explored. *Volume rendering* or *direct volume rendering* is the process of creating a 2D image directly from 3D volumetric data, hence it is often called *direct volume rendering*. Although several of the methods described in these subsections render surfaces contained within volumetric data, these methods operate on the actual data samples, without generating the intermediate geometric primitive representations used by the algorithms in the previous section.

Volume rendering can be achieved using an *object-order*, *image-order*, or *domain-based* technique. Hybrid techniques have also been proposed. Object-order volume rendering techniques use a *forward mapping* scheme where the volume data is mapped onto the image plane. In image-order algorithms, a *backward mapping* scheme is used where rays are cast from each pixel in the image plane through the volume data to determine the final pixel value. In a domain-based technique, the spatial volume data is first transformed into an alternative domain, such as compression, frequency, or wavelet, and then a projection is generated directly from that domain.

7.6.1   Image-Order Techniques

There are four basic volume rendering modes: *x-ray rendering, maximum intensity projection (MIP), isosurface rendering*, and *full volume rendering*, where the third mode is just a special case of the fourth. These four modes share two common operations: (1) They all cast rays from the image pixels, sampling the grid at discrete locations along their paths, and (2) they all obtain the samples via interpolation, using the methods described in the previous section. The modes differ, however, in how the samples taken along a ray are combined. In x-ray rendering, the interpolated samples are simply summed, giving rise to a typical image obtained in projective diagnostic imaging (Fig. 7.7a), while in MIP only the interpolated sample

with the largest value is written to the pixel (Fig. 7.7b). In full volume rendering (Figs. 7.7c and 7.7d), on the other hand, the interpolated samples are further processed to simulate the light transport within a volumetric medium according to one of many possible models. In the remainder of this section, we shall concentrate on the full volume rendering mode since it provides the greatest degree of freedom, although rendering algorithms have been proposed that merge the different modes into a hybrid image-generation model [80].

The fundamental element in full volume rendering is the volume rendering integral. In this section we shall assume the *low-albedo* scenario, in which a certain light ray only scatters once before leaving the volume. The low-albedo *optical model* was first described by Blinn [14] and Kajiya and Herzen [98], and then formally derived by Max [152]. It computes, for each cast ray, the quantity $I_\lambda(x, r)$, which is the amount of light of wavelength λ coming from ray direction r that is received at point x on the image plane:

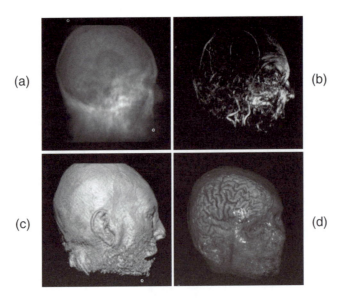

(a)

(b)

(c)

(d)

Figure 7.7   CT head rendered in the four main volume rendering modes: (a) x-ray; (b) MIP; (c) isosurface; (d) translucent. (See also color insert.)

$$I_\lambda(x,r) = \int_0^L C_\lambda(s)\mu(s)\exp$$
$$\left(-\int_0^s \mu(t)dt\right)ds \qquad (7.8)$$

Here L is the length of ray r. We can think of the volume as being composed of particles with certain mass density values μ (Max [152] calls them light extinction values). These values, as well as the other quantities in this integral, are derived from the interpolated volume densities $f(x,y,z)$ via some mapping function. The particles can contribute light to the ray in three different ways: via emission [215], transmission, or reflection [258], thus $C_\lambda(s) = E_\lambda(s) + T_\lambda(s) + R_\lambda(s)$. The latter two terms, T_λ and R_λ, transform light received from surrounding light sources, while the former, E_λ, is due to the light-generating capacity of the particle. The reflection term takes into account the specular and diffuse material properties of the particles. To account for the higher reflectivity of particles with larger mass densities, one must weight C_λ by μ. In low-albedo, we track only the light that is received on the image plane. Thus, in Equation 7.8, C_λ is

the portion of the light of wavelength λ available at location s that is transported in the direction of r. This light then gets attenuated by the mass densities of the particles along r, according to the exponential attenuation function.

$R_\lambda(s)$ is computed via the standard illumination equation [57]:

$$R(s) = k_a C_a + k_d C_l C_0(s)(N(s) \cdot L(s))$$
$$+ k_s C_l (N(s) \cdot H(s))^{ns} \qquad (7.9)$$

where we have dropped the subscript λ for reasons of brevity. Here, C_a is the ambient color, k_a is the ambient material coefficient, C_l is the color of the light source, C_0 is the color of the object (determined by the density–color mapping function), k_d is the diffuse material coefficient, N is the normal vector (determined by the gradient), L is the light direction vector, k_s is the specular material coefficient, H is the halfvector, and ns is the Phong exponent.

Equation 7.8 models only the attenuation of light from s to the eye (see Fig. 7.8). But the light received at s is also attenuated by the volume densities on its path from the light source to s. This allows us to develop the

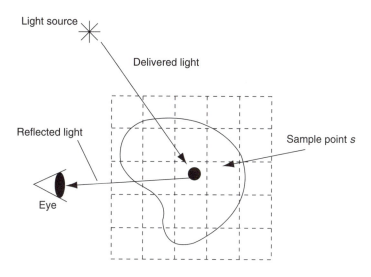

Figure 7.8 Transport of light to the eye. (See also color insert.)

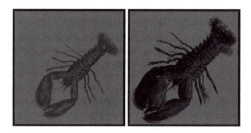

Figure 7.9 CT lobster rendered without shadows (left) and with shadows (right). The shadows on the wall behind the lobster and the self-shadowing of the legs create greater realism. (See also color insert.)

following term for C_1 in Equation 7.9, which is now dependent on the location s:

$$C_l(s) = C_L \exp\left(-\int_s^T \mu(t)dt\right) \qquad (7.10)$$

Here, C_L is the color of the light source and T is the distance from s to the light source (see Fig. 7.8). The inclusion of this term into Equation 7.9 produces volumetric shadows, which give greater realism to the image [191,290] (see Fig. 7.9). In practice, applications that compute volumetric shadows are less common, due to the added computational complexity, but an interactive hardware-based approach has been proposed [113,114].

The analytic volume rendering integral cannot, in the general case, be computed efficiently, if at all, and therefore a variety of approximations are in use. An approximation of Equation 7.8 can be formulated using a discrete Riemann sum, where the rays interpolate a set of samples, most commonly spaced apart by a distance Δs:

$$I_\lambda(\mathbf{x}, \mathbf{r}) = \sum_{i=0}^{L/\Delta s - 1} C_\lambda(i\Delta s)\mu(i\Delta s)\Delta s$$
$$\prod_{j=0}^{i-1} \exp(-\mu(j\Delta s)\Delta s) \qquad (7.11)$$

A few more approximations make the computation of this equation more efficient. First,

the *transparency* $t(i\Delta s)$ is defined as $\exp(-\mu(i\Delta s)\Delta s)$. Transparency assumes values in the range $[0.0, 1.0]$. The *opacity* $\alpha(i\Delta s) = (1 - t(i\Delta s))$ is the inverse of the transparency. Further, the exponential term in Equation 7.11 can be approximated by the first two terms of its Taylor series expansion: $t(i\Delta s) = \exp(-\mu(i\Delta s)\Delta s) \approx 1 - \mu(i\Delta s)\Delta s$. Then, one can write $\mu(i\Delta s)\Delta s \approx 1 - t(i\Delta s) = \alpha(i\Delta s)$. This transforms Equation 7.11 into the well known compositing equation

$$I_\lambda(\mathbf{x}, \mathbf{r}) = \sum_{i=0}^{L/\Delta s - 1} C_\lambda(i\Delta s)\alpha(i\Delta s) \cdot \prod_{i=0}^{i-1} (1 - \alpha(j\Delta s))$$
$$(7.12)$$

This is a recursive equation in $(1 - \alpha)$ and gives rise to the recursive *front-to-back compositing* formula [127,207]:

$$c = C(i\Delta s)\alpha(i\Delta s)(1 - \alpha) + c$$
$$\alpha = \alpha(i\Delta s)(1 - \alpha) + \alpha \qquad (7.13)$$

Thus, a practical implementation of the volumetric ray would traverse the volume from front to back, calculating colors and opacities at each sampling site, weighting these colors and opacities by the current accumulated transparency $(1 - \alpha)$, and adding these terms to the accumulated color and transparency to form the terms for the next sample along the ray. An attractive property of the front-to-back traversal is that a ray can be stopped once α approaches 1.0, which means that light originating from structures farther back is completely blocked by the cumulative opaque material in front. This provides for accelerated rendering and is called *early ray termination*. An alternative form of Equation 7.13 is the *back-to-front compositing* equation:

$$c = c(1 - \alpha(i\Delta s)) + C(i\Delta s)$$
$$\alpha = \alpha(1 - \alpha(i\Delta s)) + \alpha(i\Delta s) \qquad (7.14)$$

Back-to-front compositing is a generalization of the Painter's algorithm and does not enjoy speedup opportunities of early ray termination and is therefore less frequently used.

Equation 7.12 assumes that a ray interpolates a volume that stores at each grid-point a color vector (usually a (red,green,blue) = RGB triple) as well as an α value [127,128]. There, the colors are obtained by shading each grid-point using Equation 7.9. Before we describe the alternative representation, let us first discuss how the voxel densities are mapped to the colors C_o in Equation 7.9.

The mapping is implemented as a set of mapping functions, often implemented as 2D tables, called *transfer functions*. By way of the transfer functions, users can interactively change the properties of the volume dataset. Most applications give access to four mapping functions: (Rd), $G(d)$, $B(d)$, $A(d)$, where d is the value of a grid voxel, typically in the range of [0,255] for 8-bit volume data. Thus, users can specify semitransparent materials by mapping their densities to opacities < 1.0, which allows rays to acquire a mix of colors that is due to all traversed materials. More advanced applications also give users access to transfer functions that map $k_s(d), k_d(d), ns(d)$ and others. Wittenbrink et al. [280] pointed out that the colors and opacities at each voxel should be multiplied *prior* to interpolation to avoid artifacts on object boundaries.

The model in Equation 7.12 is called the *pre-classification model*, since voxel densities are mapped to colors and opacities *prior* to interpolation. This model cannot resolve high-frequency detail in the transfer functions (see Fig. 7.10 for an example), and it also typically gives blurry images under magnification [180]. An alternative model that is more often used is the *post-classification model*. Here, the raw volume values are interpolated by the rays, and the interpolation result is mapped to color and opacity:

$$I_\lambda(x,r) = \sum_{i=0}^{L/\Delta s-1} C_\lambda(f(i\Delta s), g(i\Delta s))$$
$$\alpha(f(i\Delta s)) \prod_{j=0}^{i-1}(1-\alpha(f(j\Delta s)))$$

(7.15)

The function value $f(i\Delta s)$ and the gradient vector $g(i\Delta s)$ are interpolated from $f_d(x,y,z)$ using a 3D interpolation kernel, and C_λ and α are now the transfer and shading functions that translate the interpolated volume function values into color and opacity. This

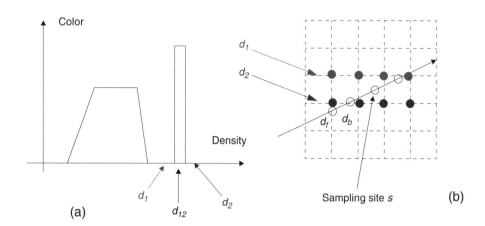

Figure 7.10 Transfer-function aliasing. When the volume is rendered preclassified, then both the red (density d_1, top row in (b)) and the blue (density d_2, bottom row) voxels receive a color of zero, according to the transfer function shown on the left. At ray sampling this voxel neighborhood at s would then interpolate a color of zero as well. On the other hand, in post-classification rendering, the ray at s would interpolate a density close to d_{12} (between d_1 and d_2) and retrieve the strong color associated with d_{12} in the transfer function. (See also color insert.)

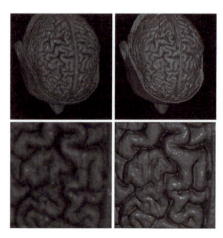

Figure 7.11 Pre-classified (left column) vs. post-classification (right column) rendering. The latter yields sharper images since the opacity and color classification is performed after interpolation. This eliminates the blurry edges introduced by the interpolation filter. (See also color insert.)

generates considerably sharper images (see Fig. 7.11).

A quick transition from 0 to 1 at some density value d_i in the opacity transfer function selects the isosurface $d_{iso} = d_i$. Thus, isosurface rendering is merely a subset of full volume rendering, where the ray hits a material with $d = d_{iso}$ and then immediately becomes opaque and terminates.

Post-classified rendering eliminates only some of the problems that come with busy transfer functions. Consider again Fig. 7.10a, and now assume a very narrow peak in the transfer function at d_{12}. With this kind of transfer function, a ray point sampling the volume at s may easily miss to interpolate d_{12}, but might have interpolated it had it just sampled the volume at $s + \delta_s$. Preintegrated transfer functions [55] solve this problem by precomputing a 2D table that stores the analytical volume rendering integration for all possible density pairs (d_f, d_b). This table is then indexed during rendering by each ray sample pair (d_b, d_f), interpolated at sample locations Δs apart (see Fig. 7.10b). The preintegration assumes a piecewise linear function within the density pairs and thus guaran-

tees that no transfer function detail falling between two interpolated (d_f, d_b) fails to be considered in the discrete ray integration.

7.6.2 Object-Order Techniques

Object-order techniques decompose the volume into a set of basis elements or *basis functions*, which are individually projected to the screen and assemble into an image. If the volume rendering mode is x-ray or MIP, then the basis functions can be projected in any order, since in x-ray and MIP the volume rendering integral degenerates to a commutative sum or MAX operation. In contrast, depth ordering is required when solving for the generalized volume rendering integral in Equation 7.8. Early work represented the voxels as disjoint cubes, which gave rise to the *cuberille* representation [68,83]. Since a cube is equivalent to a nearest neighbor kernel, the rendering results were inferior. Therefore, more recent approaches have turned to kernels of higher quality.

To better understand the issues associated with object-order projection, it helps to view the volume as a field of basis functions h, with one such basis kernel located at each grid-point where it is modulated by the grid-point's value (see Fig. 7.12, where two such kernels are shown). This ensemble of modulated basis functions then makes up the continuous object representation, i.e., one could interpolate a sample anywhere in the volume by simply adding up the contributions of the modulated kernels that overlap at the location of the sample value. Hence, one could still traverse this ensemble with rays and render it in image order. However, a more efficient method emerges when realizing that the contribution of a voxel j with value d_j is given by $d_j \cdot \int h(s)ds$, where s follows the line of kernel integration along the ray. Further, if the basis kernel is radially symmetric, then the integration $\int h(s)ds$ is independent of the viewing direction. Therefore, one can perform a preintegration of $\int h(s)ds$ and store the result into a lookup table. This table is called the kernel *footprint*, and the

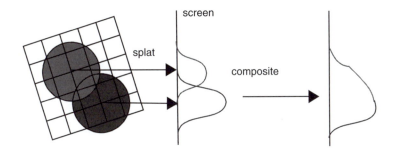

Figure 7.12   Object-order volume rendering with kernel splatting implemented as footprint mapping. (See also color insert.)

kernel projection process is referred to as *kernel splatting* or simply *splatting*. If the kernel is a Gaussian, then the footprint is a Gaussian as well. Since the kernel is identical for all voxels, we can use it for all voxels. We can generate an image by going through the list of object voxels in depth order and performing the following steps for each (see again Fig. 7.12): (1) Calculate the screen-space coordinate of the projected grid-point; (2) Center the footprint around that point and stretch it according to the image magnification factor; (3) Rasterize the footprint to the screen, using the preintegrated footprint table and multiplying the indexed values by the voxel's value [268,269,270]. This rasterization can either be performed via fast DDA procedures [147,174], or, in graphics hardware, by texture-mapping the footprint (basis image) onto a polygon [42].

There are three types of splatting: composite-only, axis-aligned sheet-buffered, and image-aligned sheet-buffered (IASB) splatting. The composite-only method was proposed first [269] and is the most basic one (see Fig. 7.12). Here, the object points are traversed in either front-to-back or back-to-front order. Each is first assigned a color and opacity using the shading equation (Equation 7.9) and the transfer functions. Then, each point is splatted into the screen's color and opacity buffers and the result is composited with the present image (Equation 7.13). In this approach, color bleeding and slight sparkling artifacts in animated viewing may be noticeable since the interpol-

ation and compositing operations cannot be separated due to the preintegration of the basis (interpolation) kernel [270].

An attempt to solve this problem gave way to the axis-aligned sheet-buffered splatting approach [268] (see Fig. 7.13a). Here, the grid-points are organized into sheets (basically the volume slices most parallel to the image plane), assigned a color and opacity, and splatted into the sheet's color and opacity buffers. The important difference is that now all splats within a sheet are added and not composited, while only subsequent sheets are composited. This prevents potential color bleeding of voxels located in consecutive sheets, due to the more accurate reconstruction of the opacity layer. The fact that the voxel sheets must be formed by the volume slices most parallel to the viewing axis leads to a sudden switch of the compositing order when the major viewing direction changes, and an orthogonal stack of volume slices must be used to organize the voxels. This causes noticeable popping artifacts where some surfaces suddenly reflect less light and others more [173]. The solution to this problem is to align the compositing sheet with the image plane at all times, which gives rise to the image-aligned sheet-buffered splatting approach [173] (see Fig. 7.13b). Here, a slab is advanced across the volume and all kernels that intersect the slab are sliced and projected. Kernel slices can be preintegrated into footprints as well, and thus this sheet-buffered approach differs from the original one in that

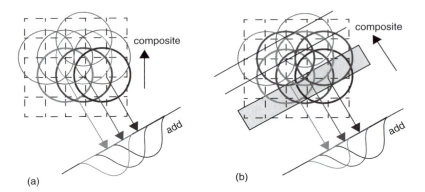

Figure 7.13   Sheet-buffered splatting. (a) Axis-aligned—the entire kernel within the current sheet is added. (b) Image-aligned—only slices of the kernels intersected by the current sheet-slab are added. (See also color insert.)

each voxel has to be considered more than once. The image-aligned splatting method provides the most accurate reconstruction of the voxel field prior to compositing and eliminates both color bleeding and popping artifacts. It is also best suited for post-classification rendering since the density (and gradient) field is reconstructed accurately in each sheet. However, it is more expensive due to the multiple splatting of a voxel.

The divergence of rays under perspective viewing causes undersampling of the volume portions farther away from the viewpoint (see Fig. 7.14). This leads to aliasing in these areas. As was demonstrated in Fig. 7.6, lowpassing can eliminate the artifacts caused by aliasing and replace them by blur (see Fig. 7.15). For perspective rendering, the amount of required lowpassing increases with distance from the viewpoint. The kernel-based approaches can achieve this progressive lowpassing by simply stretching the footprints of the voxels as a function of depth, since stretched kernels act as lowpass filters (see Fig. 7.14). EWA (Elliptical Weighted Average) Splatting [293] provides a general framework to define the screen-space shape of the footprints, and their mapping into a generic footprint, for generalized grids under perspective viewing. An equivalent approach for ray-casting is to split the rays in more distant volume slices to always maintain the proper

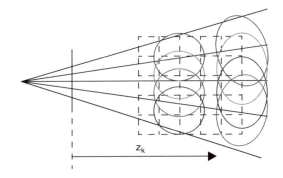

Figure 7.14   Stretching the basis functions in volume layers $z > z_k$, where the sampling rate of the ray grid is progressively less than the volume resolution. (See also color insert.)

sampling rate [190]. Kreeger et al. [118] proposed an improvement of this scheme that splits and merges rays in an optimal way.

A major advantage of object-order methods is that only the points (or other basis primitives, such as tetrahedral or hexagonal cells [273]) that make up the object must be stored. This can be advantageous when the object has an intricate shape, with many pockets of empty space [159]. While ray-casting would spend much effort traversing (and storing) the empty space, kernel-based or *point-based* objects will not consider the empty space, neither during rendering nor for storage. However, there are tradeoffs; since the rasterization of a footprint takes more time

Figure 7.15 Anti-aliased splatting. (Left) A checkerboard tunnel rendered in perspective with equal-sized splats. Aliasing occurs at distances beyond the black square. (Right) The same checkerboard tunnel rendered with scaled splats. The aliasing has been replaced by blur. (See also color insert.)

than the commonly used tri-linear interpolation of ray samples, since the radially symmetric kernels employed for splatting must be larger than the tri-linear kernels to ensure proper blending. Hence, objects with compact structure are more favorably rendered with image-order methods or hybrid methods (see next section). Another disadvantage of object-order methods is that early ray termination is not available to cull occluded material early from the rendering pipeline. The object-order equivalent is early point elimination, which is more difficult to achieve than early ray termination. Finally, image-order methods allow the extension of ray-casting to ray-tracing, where secondary and higher-order rays are spawned at reflection sites. This facilitates mirroring on shiny surfaces, inter-reflections between objects, and soft shadows.

There are a number of ways to store and manage point-based objects. These schemes are mainly distinguished by their ability to exploit spatial coherence during rendering. The lack of spatial coherence requires more depth sorting during rendering and also means more storage for spatial parameters. The least spatial coherence results from storing the points sorted by density [41]. This has the advantage that irrelevant points, being assigned transparent values in the transfer functions, can be quickly culled from the rendering pipeline. However, it requires that (x,y,z) coordinates, and possibly gradient vectors, are stored along with the points since neighborhood relations are completely lost. It also requires that all points be view-transformed before they can be culled due to occlusion or exclusion from the viewing pyramid. The method also requires that the points be depth-sorted during rendering, or at least tossed into depth bins [177]. A compromise is struck by Ihm and Lee [94], who sort points by density within volume slices only, which gives implicit depth-ordering when used in conjunction with an axis-aligned sheet-buffer method. A number of approaches exist that organize the points into run length encoded (RLE) lists, which allow the spatial coordinates to be incrementally computed when traversing the runs [108,182]. However, these approaches do not allow points to be easily culled based on their density value. Finally, one may also decompose the volume into a spatial octree and maintain a list of voxels in each node. This provides depth sorting on the node-level.

A number of surface-based splatting methods have also been described. These do not provide the flexibility of volume exploration via transfer functions, since the original volume is discarded after the surface has been extracted. They only allow a fixed geometric representation of the object that can be viewed at different orientations and with different shadings. A popular method is *shell-rendering* [259], which extracts from the volume (possibly with a sophisticated segmentation algorithm) a certain thin or thick surface or contour and represents it as a closed shell of points. Shell-rendering is fast since the number of points is minimized and its data structure used has high cache coherence. More advanced point-based surface rendering methods are QSplat [214], Surfels [201], and Surface Splats [292], which have been predominantly developed for point-clouds obtained with range scanners, but can also be used for surfaces extracted from volumes [293].

7.6.3   Hybrid Techniques

Hybrid techniques seek to combine the advantages of the image-order and object-order methods, i.e., they use object-centered storage for fast selection of relevant material (which is a hallmark of object-order methods) and they use early ray termination for fast occlusion culling (which is a hallmark of image-order methods).

The shear-warp algorithm [120] is such a hybrid method. In shear-warp, the volume is rendered by a simultaneous traversal of RLE-encoded voxel and pixel runs, where opaque pixels and transparent voxels are efficiently skipped during these traversals (see Fig. 7.16a).

Further speed comes from the fact that sampling only occurs in the volume slices via bi-linear interpolation, and that the ray grid resolution matches that of the volume slices, and therefore the same bi-linear weights can be used for all rays within a slice (see Fig. 7.16b). The caveat is that the image must first be rendered from a sheared volume onto a so-called base-plane, which is aligned with the volume slice most parallel to the true image plane (Fig. 7.16b). After completing the base-plane rendering, the base-plane image must be warped onto the true image plane and the resulting image is displayed. All of this com-

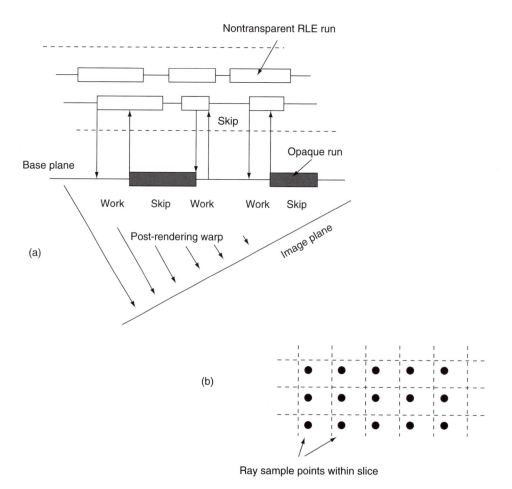

Figure 7.16   The shear-warp algorithm. (a) Mechanism; (b) interpolation scheme.

bined enables frame rates in excess of 10 frames/s on current PC processors, for a 128^3 volume. There are a number of compromises that had to be made in the process:

- Since the interpolation occurs within only one slice at a time, more accurate tri-linear interpolation reduces to less accurate bi-linear interpolation and the ray sampling distance varies between 1 and $\sqrt{3}$, depending on the view orientation. This leads to aliasing and staircasing effects at viewing angles near $45°$.

- Since the volume is run length encoded, one needs to use three sets of voxel encodings (but it could be reduced to two [249]), one for each major viewing direction. This triples the memory required for the runs, but in return, the RLE saves considerable space.

- Since there is only one interpolated value per voxel-slice 4-neighborhood, zooming can only occur during the warping phase and not during the projection phase. This leads to considerable blurring artifacts at zoom factors greater than 2. This post-rendering magnification, in fact, is a major source of the speedup for the shear-warp algorithm.

An implementation of the shear-warp algorithm is publicly available as the *volpack* package [90] from Stanford University.

7.6.4 Domain Volume Rendering

In domain rendering, the spatial 3D data is first transformed into another domain, such as the compression, the frequency, or the wavelet domain, and then a projection is generated directly from that domain or with the help of information from that domain. The frequency domain rendering applies the Fourier slice projection theorem, which states that a projection of the 3D data volume from a certain view direction can be obtained by extracting a 2D slice perpendicular to that view direction out of the 3D Fourier spectrum and then inverse-Fourier-transforming it. This approach obtains the 3D volume projection directly from the 3D

spectrum of the data, and therefore reduces the computational complexity for volume rendering from $O(N^3)$ to $O(N^2 \log(N))$ [50,149,256]. A major problem of frequency domain volume rendering is the fact that the resulting projection is a line integral along the view direction that does not exhibit any occlusion and attenuation effects. Totsuka and Levoy [256] proposed a linear approximation to the exponential attenuation [215] and an alternative shading model to fit the computation within the frequency-domain rendering framework.

The compression domain rendering performs volume rendering from compressed scalar data without decompressing the entire dataset and therefore reduces the storage, computation, and transmission overhead of otherwise large volume data. For example, Ning and Hesselink [187,188] first applied vector quantization in the spatial domain to compress the volume and then directly rendered the quantized blocks using regular spatial domain volume rendering algorithms. Fowler and Yagel [58] combined differential pulse-code modulation and Huffman coding and developed a lossless volume-compression algorithm, but their algorithm is not coupled with rendering. Yeo and Liu [288] applied the discrete cosine transform-based compression technique on overlapping blocks of the data. Chiueh et al. [33] applied the 3D Hartley transform to extend the JPEG still-image compression algorithm [261] for the compression of sub-cubes of the volume, and performed frequency domain rendering on the sub-cubes before compositing the resulting sub-images in the spatial domain. Each of the 3D Fourier coefficients in each sub-cube is then quantized, linearly sequenced through a 3D zig-zag order, and then entropy encoded. In this way, they alleviated the problem of lack of attenuation and occlusion in frequency domain rendering while achieving high compression ratios, fast rendering speed compared to spatial volume rendering, and improved image quality over conventional frequency domain rendering techniques. More recently, Guthe et al. [73] and Sohn and Bajaj [239] have used principles from

MPEG encoding to render time-varying datasets in the compression domain.

Rooted in time-frequency analysis, wavelet theory [34,46] has gained popularity in recent years. A wavelet is a fast-decaying function with zero averaging. The nice features of wavelets are that they have local property in both spatial and frequency domain and can be used to fully represent volumes with small numbers of wavelet coefficients. Muraki [181] first applied wavelet transform to volumetric data sets, Gross et al. [71] found an approximate solution for the volume rendering equation using orthonormal wavelet functions, and Westermann [266] combined volume rendering with wavelet-based compression. However, not all of these algorithms have focused on the acceleration of volume rendering using wavelets. The greater potential of wavelet domain, based on the elegant multiresolution hierarchy provided by the wavelet transform, is to exploit the local frequency variance provided by wavelet transform to accelerate the volume rendering in homogeneous areas. Guthe and Strasser [74] have recently used the wavelet transform to render very large volumes at interactive frame rates on texture-mapping hardware. They employ a wavelet pyramid encoding of the volume to reconstruct, on the fly, a decomposition of the volume into blocks of different resolutions. Here, the resolution of each block is chosen based on the local error committed and the resolution of the screen area the block is projected onto. Each block is rendered individually with 3D texture-mapping hardware, and the block decomposition can be used for a number of frames, which amortizes the work spent on the inverse wavelet transform to construct the blocks.

7.7 Acceleration Techniques

The high computational complexity of volume rendering has led to a great variety of approaches for its acceleration. In this section, we will discuss general acceleration techniques that can benefit software as well as hardware implementations. We have already mentioned a few acceleration techniques in the previous section, such as early ray termination [127], post-rendering warps for magnified viewing [120], and the splatting of preintegrated voxel basis functions [270]. The latter two gave rise to independent algorithms, that is, shear-warp [120] and splatting [270]. Acceleration techniques generally seek to take advantage of properties of the data, such as empty space, occluded space, and entropy, as well as properties of the human perceptual system, such as its insensitivity to noise over structural artifacts.

A number of techniques have been proposed to accelerate the grid traversal of rays in image-order rendering. Examples are the 3D DDA (Digital Differential Analyzer) method [1,59], in which new grid positions are calculated by fast integer-based incremental arithmetic, and the template-based method [284], in which templates of the ray paths are precomputed and used during rendering to quickly identify the voxels to visit. Early ray termination can be sophisticated into a Russian Roulette scheme [45] in which some rays terminate with lower and others with higher accumulated opacities. This capitalizes on the human eye's tolerance to error masked as noise [146]. In the object-order techniques, fast differential techniques to determine the screen-space projection of the points as well as to rasterize the footprints [147,174] are also available.

Most of the object-order approaches deal well with empty space—they simply don't store and process it. In contrast, ray-casting relies on the presence of the entire volume grid since it requires it for sample interpolation and addresses computation during grid traversal. Although opaque space is quickly culled, via early ray termination, the fast leaping across empty space is more difficult. A number of techniques are available to achieve this (see Fig. 7.17 for an illustration of the methods described in the following text). The simplest form of space leaping is facilitated by enclosing the object into a set of boxes, possibly hierarchical, and first quickly determining and testing the rays' inter-

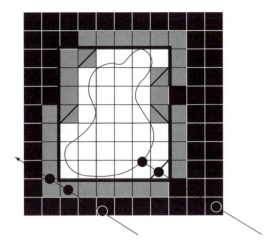

Figure 7.17 Various object approximation techniques. (Blue) Isosurface of the object. (Lightly shaded) Discretized object (proximity cloud = 0). (Red) Bounding box. (Green) Polygonal hull used in PARC. (Darker shaded areas) Proximity clouds with grey level indicating distance to the object. Note also that, while the right magenta ray is correctly sped up by the proximity clouds, the left magenta ray is missing, and the object is unnecessarily slowed down. (See also color insert.)

section with each of the boxes before engaging into more time-consuming volumetric traversal of the material in Kay and Kajiya [105]. A better geometrical approximation is obtained by a polyhedral representation, chosen crudely enough to still maintain ease of intersection. In fact, one can utilize conventional graphics hardware to perform the intersection calculation, where one projects the polygons twice to create two z-(depth) buffers. The first z-buffer is the standard closest-distance z-buffer, while the second is a farthest-distance z-buffer. Since the object is completely contained within the representation, the two z-buffer values for a given image plane pixel can be used as the starting and ending points of a ray segment on which samples are taken. This algorithm has been known as *PARC* (Polygon Assisted Ray Casting) [237] and it is part of the *VolVis* volume visualization system [4,5], which also provides a multialgorithm progressive refinement approach for interactivity. By using avail-

able graphics hardware, the user is given the ability to interactively manipulate a polyhedral representation of the data. When the user is satisfied with the placement of the data, light sources, and viewpoint, the z-buffer information is passed to the PARC algorithm, which produces a ray-cast image.

A different technique for empty-space leaping was devised by Zuiderfeld et al. [291] as well as Cohen and Shefer [37], who introduced the concept of *proximity clouds*. Proximity clouds employ a distance transform of the object to accelerate the rays in regions far from the object boundaries. In fact, since the volume densities are irrelevant in empty volume regions, one can simply store the distance transform values in their place, and therefore storage is not increased. Since the proximity clouds are the isodistance layers around the object's boundaries, they are insensitive to the viewing direction. Thus, rays that ultimately miss the object are often still slowed down. To address this shortcoming, Sramek and Kaufman [241] proposed a view-sensitive extension of the proximity-clouds approach. Wan [262] places a sphere at every empty voxel position, where the sphere radius indicates the closest nonempty voxel. They apply this technique for the navigation inside hollow volumetric objects, as occurring in virtual colonoscopy [87], and reduce a ray's space traversal to just a few hops until a boundary wall is reached. Finally, Meissner [160] suggested an algorithm that quickly recomputes the proximity cloud when the transfer function changes.

Proximity clouds only handle the quick leaping across empty space, but methods are also available that traverse occupied space faster when the entropy is low. These methods generally utilize a hierarchical decomposition of the volume where each nonleaf node is obtained by low-pass filtering its children. Commonly this hierarchical representation is formed by an octree [155] since these are easy to traverse and store. An octree is the 3D extension of a quadtree [218], which is the 2D extension of a binary tree. Most often a nonleaf node stores the average of its children, which is synonymous with a box filtering of the volume, but more sophisticated filters are possible. Octrees don't have to be balanced [274] nor fully expanded into a single root node or into single-voxel leaf nodes. The latter two give rise to a brick-of-bricks decomposition, where the volume is stored as a flat hierarchy of bricks of size n^3 to improve cache-coherence in the volume traversal. Parker et al. [194,195] utilize this decomposition for the ray-casting of very large volumes, and they also give an efficient indexing scheme to quickly find the memory address of the voxels located in the 8-neighborhood required for tri-linear interpolation.

When octrees are used for entropy-based rendering, the nonleaf nodes store either an entropy metric of its children, such as standard deviation [45], minimum-maximum range [274], or Lipschitz range [242], or a measure of the error committed when the children are not rendered, such as the root mean square or the absolute error [74]. The idea is either to have the user specify a tolerable error before the frame is rendered, or to make the error dependent on the time maximally allowed to render the frame, which is known as *time-critical rendering*. In either case, the rays traversing the volume will advance across the volume, but also transcend up and down the octree, based on the metric used, which will either accelerate or decelerate them on their path. A method called β-acceleration will also make this traversal sensitive to the ray's accumulated opacity so far. The philosophy here is that the observable error from using a coarser node will be relatively small when it is weighted by a small transparency.

Octrees are also easily used with object-order techniques, such as splatting. Laur and Hanrahan [124] have proposed an implementation that approximates nonleaf octree nodes by kernels of a radius that is twice the radius of the children's kernels, which gives rise to a magnified footprint. They store the children's average as well as an error metric based on their standard deviation in each parent node and use a preset error to select the nodes during

rendering. While this approach uses nonleaf nodes during rendering, other splatting approaches only exploit them for fast occlusion culling. Lee and Ihm [125] as well as Mora et al. [171] store the volume as a set of bricks, which they render in conjunction with a dynamically computed hierarchical occlusion map to quickly cull voxels within occluded bricks from the rendering pipeline. Hierarchical occlusion maps [289] are continuously updated during the rendering and thus store a hierarchical opacity man of the image rendered so far. Regions in which the opacity is high are tagged, and when octree nodes fall within such a region, all voxels contained in them can be immediately culled. If the octree node does not fall into a fully opaque region, then it has to be subdivided and its children are subjected to the same test. An alternative scheme that performs occlusion culling on a finer scale than the box-basis of an octree decomposition is to calculate an occlusion map in which each pixel represents the average of all pixels within the box-neighborhood covered by a footprint [177]. Occlusion of a particular voxel is then determined by indexing the occlusion map with the voxel's screen-space coordinate to determine if its footprint must be rasterized. One could attempt to merge these two methods to benefit both from the large-scale culling afforded by the octree-nodes and from the fine-scale culling of the average-occlusion map.

Hierarchical decomposition is not the only way to reduce the number of point primitives needed to represent an object for rendering. An attractive solution that does not reduce the volume's frequency content, by ways of averaging, is to exploit more space-efficient grids for storage. The most optimal regular lattices are the *face-centered cartesian (FCC)* lattices (see Fig. 7.19) [39]. The FCC lattices give the densest packings of a set of equal-sized spheres. If the frequency spectrum of the signal represented in the volume is spherical (and many of them are due to the sampling kernel used for volume generation), then they can be packed in the FCC lattice (see Fig. 7.18 for the 2D equiva-

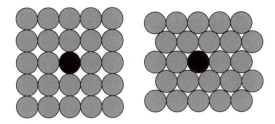

Figure 7.18 (Left) The cartesian grid vs. (Right) the hexagonal grid: two possible frequency-domain lattices. The latter provides the tightest packing of a discrete 2D signal's circularly bounded frequency spectrum. (Here, the dark circle contains the main spectrum, while the others contain the replicas or aliases.)

lent, the hexagonal lattice). The FCC lattice's dual in the spatial domain is the *body-centered cartesian (BCC)* lattice, and the spacing of samples there is the reciprocal of that in the frequency domain, according to the Fourier scaling theorem [17]. This BCC grid gives rise to two interleaved CC grids, each with a sampling interval of $\sqrt{2}$ and $1/(\sqrt{2})$ apart, which implies that a volume, when sampled into a BCC grid, only requires $\sqrt{2}/2 = 71\%$ of the samples of the usual cubic cartesian (CC) grid [182,253] (see Fig. 7.19 for an illustration of the

grid and Fig. 7.20 for images). The theorem extends to higher dimensions as well; for example, a time-varying (4D) volume can be stored in a 4D BCC at only 50% of the 4D CC samples. The BCC grids are best used in conjunction with point-based object-order methods, since these use the spherical (radially symmetric) filter required to preserve the spherical shape of the BCC grid-sampled volume's frequency spectrum. The reconstruction of a BCC grid with a tri-linear filter can lead to aliasing since the trilinear filter's frequency response is not radially symmetric and therefore will include higher spectra when used for interpolation.

A comprehensive system for accelerated software-based volume rendering is the UltraVis system devised by Knittel [115]. It can render a 256^3 volume at 10 frames/s. It achieves this by optimizing cache performance during both volume traversal and shading, which is rooted in the fact that good cache management is key to fast volume rendering, since the data are so massive. As we have mentioned before, this was also realized by Parker et al. [194,195], and it plays a key role in both custom and commodity hardware approaches as well, as we shall see later. The UltraVis system manages

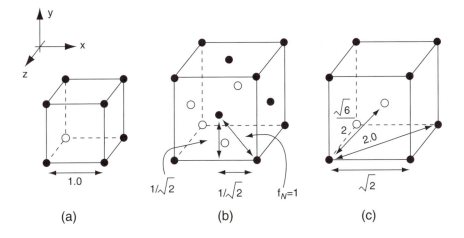

Figure 7.19 Various grid cells, drawn in relative proportions. We assume that the sampling interval in the CC grid is T = 1. (a) Cubic cartesian (CC) for cartesian grids (all other grid cells shown are for grids that can hold the same spherically bandlimited, signal content). (b) Face-centered cubic cartesian (FCC). (c) Body-centered cubic cartesian (BCC).

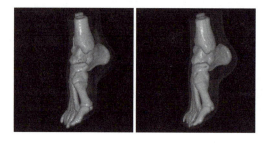

Figure 7.20   Foot dataset rendered on (left) a cubic cartesian (CC) grid and (right) a body-centered cubic cartesian (BCC) grid. The renderings are almost identical, but the BCC rendering took 70% of the time of the CC rendering. (See also color insert.)

the cache by dividing it into four blocks: one block each for volume bricks, transfer function tables, image blocks, and temporary buffers. Since the volume can only map into a private cache block, it can never be swapped out by a competing data structure, such as a transfer-function table or an image-tile array. This requires that the main memory footprint of the volume be four times as high since no volume data may be stored in an address space that would map outside the volume's private cache slots. By using a bricked volume decomposition in conjunction with a flock of rays that are

traced simultaneously across the brick, the brick's data will only have to be brought in once before it can be discarded when all rays have finished its traversal. A number of additional acceleration techniques give further performance.

Another type of acceleration is achieved by breaking the volume rendering integral of Equation 7.12 or 7.15 into segments and storing the composited color and opacity for each partial ray into a data structure. The idea is then to recombine these partial rays into complete rays for images rendered at viewpoints near the one for which the partial rays were originally obtained (see Fig. 7.21). This saves the cost for fully integrating all rays for each new viewpoint and reduces it to the expense of compositing a few partial segments per ray, which is much lower. This method falls into the domain of *image-based rendering (IBR)* [29,30,154,221] and is, in some sense, a volumetric extension of the *lumigraph* [69] or *lightfield* [129], albeit dynamically computed. However, one could just as well store a set of partial rays into a static data structure to be used for volumetric-style lumigraph rendering. This idea of using a cache of partial rays for accelerated rendering was exploited by Brady et al. [19,20] for the

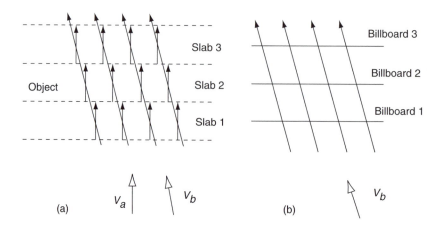

Figure 7.21   (a) The volume is decomposed into slabs, and each slab is rendered into an image from view direction V_a. The ray integrals for view direction V_b can now be approximated with higher accuracy by combining the appropriate partial ray integrals from view V_a (stored in the slab image). Interpolation is used to obtain partial integrals at nongrid positions. (b) The three billboard images can be composited for any view, such as V_b, shown here.

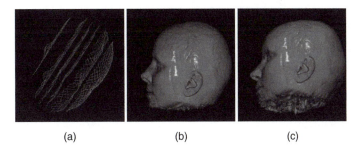

(a)   (b)   (c)

Figure 7.22   IBR-assisted volume rendering. (a) On-the-fly computed mesh derived from the slab's closest-voxel buffer; (b) head rendered from original viewpoint; (c) head rendered from a view 30° away. (See also color insert.)

volume rendering at great perspective distortions, such as found in virtual endoscopy applications [87]. Mueller et al. [178] stored the rays in form of a stack of depth-layered images and rendered these images warped and composited from novel viewpoints within a 30° view cone, using standard graphics hardware (see Fig. 7.22a). Since gaps may quickly emerge when the layers are kept planar, it helps to also compute, on the fly, a coarse polygonal mesh for each layer that approximates the underlying object, and then map the images onto this mesh when rendering them from a new viewpoint (see Figs. 7.22b and 7.22c). An alternative method that uses a precomputed triangle mesh to achieve similar goals for isosurface volume rendering was proposed by Chen et al. [28], while Yagel and Shi [286] warped complete images to nearby viewpoints, aided by a depth buffer.

7.8   Classification and Transfer Functions

In volume rendering we seek to explore the volumetric data using visuals. This exploration process aims to discover and emphasize interesting structures and phenomena embedded in the data, while de-emphasizing or completely culling away occluding structures that are currently not of interest. Clipping planes and more general clipping primitives [264] provide geometric tools to remove or displace occluding structures in their entirety. On the other hand, transfer functions that map the raw volume

density data to color and transparencies, can alter the overall look and feel of the dataset in a continuous fashion.

The exploration of a volume via transfer functions constitutes a navigation task, which is performed in a 4D transfer-function space, assuming three axes for RGB color and one for transparency (or opacity). It is often easier to specify colors in HSV (Hue, Saturation, Value) color space, since it provides separate mappings for color and brightness. Simple algorithms exist to convert the HSV values into the RGB triples used in the volume rendering [57]. Fig. 7.23 shows a transfer-function editor that also allows the mapping of the other rendering attributes in Equation 7.9.

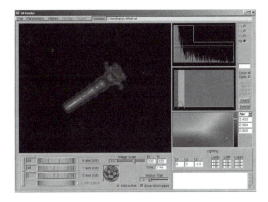

Figure 7.23   A transfer function editor with an HSV color palette and a mapping of densities to various material properties. (See also color insert.)

A generalization of the usual RGB color model has been pursued in *spectral volume rendering* [197], where the light transport occurs within any number of spectral bands. Noordmans [189] employed this concept to enable achromatic, elastic, and inelastic light scattering, which facilitates the rendering of inner structures through semitransparent, yet solid (i.e., nonfuzzy) exterior structures. Bergner et al. [12] described a spectral renderer that achieves interactive speeds by factoring the illumination term out of the spectral volume rendering integral and using post-illumination for the final lighting (a related technique, in RGB space and using a Fourier series approach, was presented by Kaneda et al. [99]). They describe a system that allows designers of a guided visualization to specify a set of lights and materials, whose spectral properties allow users to emphasize, de-emphasize, or merge specific structures by simply varying the intensity of the light sources.

Given the large space of possible settings, choosing an effective transfer function can be a daunting task. It is generally more convenient to gather more information about the data before the exploration via transfer functions begins. The easiest presentation of support data is in the form of 1D histograms, which are data statistics collected as a function of raw density, or some other quantity. A histogram of density values can be a useful indicator to point out dominant structures with narrow density ranges. A fuzzy classification function [48] can then be employed to assign different colors and opacities to these structures (Fig. 7.24). This works well if the data are relatively noise-free, the density ranges of the features are well isolated, and not many distinct materials, such as bone, fat, and skin, are present. In most cases, however, this is not the case. In these settings, it helps to also include the first and second derivative in the histogram-based analysis [109]. The magnitude of the first derivative (the gradient strength) is useful since it peaks at densities where interfaces between different features exist (Fig. 7.25). Plotting a histogram of first derivatives over density yields an arc that peaks at the interface density (Fig. 7.26). Knowing the densities at which feature boundaries exist narrows down the transfer-function exploration task considerably: One may now visu-

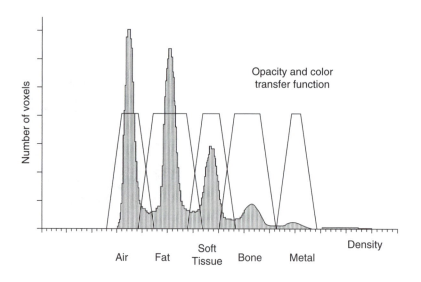

Figure 7.24   Histogram and a fuzzy classification into different materials.

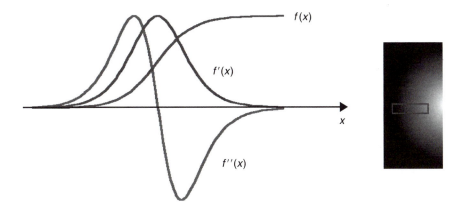

Figure 7.25   The relationship of densities and their first and second derivatives at a material interface.

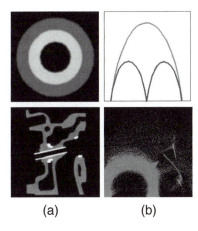

(a)   (b)

Figure 7.26   Histograms of (a) first- and (b) second-derivative strengths over density. In the concentric ring image (top row), the first arc is due to the background–outer ring interface, the second arc is due to the outer ring–inner ring interface, and the large arc is due to the background–inner ring interface that spans the widest density range. The second row shows the results of the same analysis for the CT engine volume.

alize these structures by assigning different colors and opacities within a narrow interval around these peaks. Levoy [127] showed that a constant width of (thick) surface can be obtained by making the width of the chosen density interval a linear function of the gradi-

ent strength (see Fig. 7.27). Kindlmann and Durkin [109] proposed a technique that uses the first and second derivative to generate feature-sensitive transfer functions automatically. This method provides a segmentation of the data, where the segmentation metric is a histogram of the first and second derivative. Tenginakai and Machiraju [251] extended the arsenal of metrics to higher-order moments and computed from them additional measures, such as kurtosis and skew, in small neighborhoods. These can provide better delineations of features in histogram space. Another proposed analysis method is based on maxima in cumulative Laplacian-weighted density histograms [198].

There are numerous articles (we can only reference a few here) on the topic of automatic segmentation of images and higher-dimensional datasets, using neural network-type approaches [142], statistical classifiers [222], region growing [117], the watershed algorithm [229], and many others. To that end, Tiede [255] describes an algorithm for rendering the tagged and segmented volumes at high quality. However, despite the great advances that have been made, automated segmentation of images and volumes remains a difficult task and is also, in many cases, observer and task dependent. In this regard, semi-supervised segmentation algo-

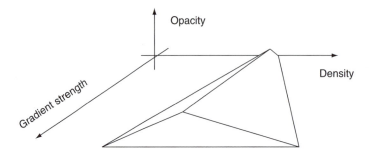

Figure 7.27   Gradient strength-dependent density range for isosurface opacities [127].

rithms where users guide the segmentation process in an interactive fashion have a competitive edge. There are two examples for such systems: the PAVLOV architecture, which implements an interactive region-grow to delineate volumetric features of interest [117], and the dual-domain approach of Kniss et al. [111,112], who embed Kindlmann's algorithm into an interactive segmentation application. Here, users work simultaneously within two domains, the histogram-coupled transfer-function domain and the volume rendering domain, to bring out certain features of interest. To be effective, an interactive (hardware-based) volume renderer is required, and the technique could embed more advanced metrics as well [251].

Another way to analyze the data is to look for topological changes in the iso-contours of the volume, such as a merge of split of two contours (see Fig. 7.28). These events are called *critical points*. By topologically sorting the critical points as a function of density, one can construct a *contour graph, contour tree*, or *Hyper Reeb Graph* that yields a roadmap for an exploration of the volume [7,26,60,119,227,250]. One can either use the contour graph to come up with an automatic transfer function (simply position an isosurface between two nodes), or one can use it to guide users in the volume-exploration process. A large number of critical points is potentially generated, especially when the data are noisy.

There has also been a significant body of work on more specific segmentation and volume-analysis processes, which aim to identify, track, and tag particular features of interest, such as vortices, streamlines, and turbulences [9,10, 233,234,279]. Once extracted, the features can then be visualized in form of icons, glyphs, geometry, or volumetric objects. These data-mining methods are particularly attractive for the exploration of very large datasets, where volume exploration with conventional means can become intractable.

All of the methods presented so far base the transfer function selection on a prior analysis of the volume data. Another suggested strategy has been to render a large number of images with arbitrary transfer-function settings and present these to the user, who then selects a subset of these for further refinement by application of genetic algorithms. This approach has been taken by the Design Galleries project [150], which is based, in part, on the method published by He et al. [81]. A good sample of all of the existing approaches (interactive trial-and-error, metric-based, contour graph, and design galleries) was squared off in a symposium panel [199].

7.9   Volumetric Global Illumination

In the local illumination equation (Equation 7.9), the global distribution of light energy is ignored and shading calculations are performed

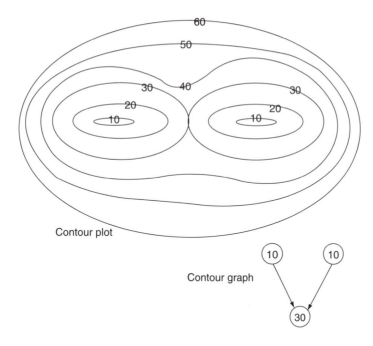

Figure 7.28 Simple contour graph. The first topological event occurs when the two inner contours are born at an iso-value of 10. The second topological event occurs at the iso-value at which the two inner contours just touch and give way to a single contour at iso-value = 30.

assuming full visibility of and a direct path to all light sources. While this is useful as a first approximation, the incorporation of global light visibility information (shadows, one instance of global illumination) adds a great deal of intuitive information to the image. This low-albedo [98,236] lighting simulation has the ability to cast soft shadows by volume density objects. Generous improvements in realism are achieved by incorporating a high-albedo lighting simulation [98,236], which is important in a number of applications (e.g., clouds [152], skin [75], and stone [47]). While some of these use hierarchical and deterministic methods, most of these simulations use stochastic techniques to transport lighting energy among the elements of the scene. We wish to solve the illumination transport equation for the general case of global illumination. The reflected illumination $I(\gamma, \omega)$ in direction ω at any voxel γ can be described as the integral of all incident radiation from directions ω', modulated by the phase function $q(\omega, \omega')$:

$$I(\gamma, \omega) = \int_V \int_\Gamma q(\omega, \omega')I(\gamma, \omega')d\omega'dv \qquad (7.16)$$

where Γ is the set of all directions and V is the set of all voxels v. This means that the illumination at any voxel is dependent upon the illumination at every other voxel. In practice, this integral equation is solved by finite-repeated projection of energy among voxels. This leads to a finite energy transport path, which is generally sufficient for visual fidelity.

In physics, equations of this sort are solved via Monte Carlo simulations. A large set of rays is cast from the energy sources into the volume, and at each voxel a "die is rolled" to determine how much energy is absorbed and how much energy is scattered and into what direction. After many iterations the simulation is stopped, and a final scattering of the radiosity volume is performed towards an arbitrarily positioned eye point. A practical implementation of this process is volumetric backprojection.

Backprojection is usually performed on a voxel-by-voxel basis, since this is the most obvious and direct method of computation. For example, in volumetric ray tracing [236], as illumination is computed for a volume sample, rays are cast toward the light sources, sampling the partial visibility of each. In computing high-albedo scattering illumination, Max [152] used the method of discrete ordinates to transport energy from voxel to voxel. For calculations of volumetric radiosity, voxels are usually regarded as discrete elements in the usual radiosity calculation on pairs of elements, thereby computing on a voxel-by-voxel basis [213, 236]. Particle tracing methods for global illumination track paths of scattered light energy through space starting at the light sources [97].

In many cases, the backprojection can be reorganized into a single sweep through the volume, processing slice by slice. Because sunlight travels in parallel rays in one direction only, Kajiya and Von Herzen [98] calculated the light intensity of a cloud-like volume one horizontal slice at a time. A similar technique was demonstrated as part of the Heidelberg ray-tracing model [157], in which shadow rays were propagated simultaneously slice-by-slice and in the same general direction as rendering. Dachille et al. [44] described a backprojection approach that scatters the energy in the volume by a multi-pass slice-by-slice sweep at random angles. He also devised a custom hardware architecture for a cache-efficient implementation of this algorithm.

Kniss et al. [113,114] proposed a single-pass algorithm that approximates the scattering of light within a volume by a recursive slice-blurring operation, starting at the light source. The profile of the blurring filter is determined by the user-specified phase function. The method exploits 3D texture-mapping hardware in conjunction with a dual image buffer and runs at interactive frame rates. One buffer, the repeatedly blurred (light) buffer, contains the transported and scattered light energy on its path away from the source, and the other (frame) buffer holds the energy headed for the eye and is attenuated by the densities along the path to the eye. At each path increment, energy is transferred from the light buffer to the frame buffer.

7.10 Rendering on Parallel Architectures

Much research towards parallel ray-casting has been reported in the literature, primarily due to the simplicity of the algorithm. To avoid volume data redistribution costs, researchers have proposed the distribution of data to processing nodes, where each node processes its own data for all frames or views. Each node generates a partial image with its data, which are then accumulated and composited into the final image [89,144,145,170,194,195].

Researchers have also investigated partitioning screen space into square tiles or contiguous scanlines, to be used as the basic task to be sent or assigned to processing nodes. For better load balancing, the task queue can be ordered in decreasing task size, such that the concurrency gets finer until the queue is exhausted [27]. Load balancing can also be achieved by having nodes steal smaller tasks from other nodes, once they have completed their own tasks [184,271]. Finally, time-out stamps for each node can be added, such that if the node cannot finish its task before the time-out, it takes the remnant of the task, repartitions it, and redistributes it [40].

A parallel shear-warp implementation on shared-memory architectures has been reported in Lacroute [121], with decent timing benchmarks. Amin et al. [2] ported the shear-warp algorithm onto a distributed memory architecture by partitioning in sheared volume space and using an adaptive load balancing. The parallel shear-warp implementation has been improved on distributed memory architectures by dividing the volume data after the shear operation into subvolumes parallel to an intermediate image plane of the shear-warp factorization [219].

Splatting and cell-projection methods have also been parallelized using a sort-last paradigm

[168]. The community has researched parallel splatting algorithms [133] that do not utilize occlusion-based acceleration. The volume data is distributed in either slices (axis-aligned planes) [54] or blocks [145] to processing nodes. Those are then rendered, in parallel, to partial images that are composited for the final image by the master node. Speedups can further be achieved by only passing the nonempty parts of the partial images [54] or by parallelizing the final compositing stage using a screen space partitioning [133]. Hierarchical data structures such as a kd-tree can be applied to facilitate prompt compositing and occlusion culling [145]. Machiraju and Yagel [147] report a parallel implementation of splatting, where the tasks are defined by subvolumes. Each processor is assigned a subvolume. The images are composited together in depth-sort order, also performed in parallel. This implementation splats all voxels, no matter if they are empty or occluded, while Huang [92] presents a parallel splatting algorithm that takes into account visibility and occlusion, which is considerably more challenging for load-balancing. PVR [230] is a parallel ray-casting kernel that exploits image-space, object-space, and time-space parallelism. See also [143] for a tutorial article on two highly scalable, parallel software volume rendering algorithms for unstructured grids.

7.11 Special-Purpose Rendering Hardware

The high computational cost of direct volume rendering makes it difficult for sequential implementations and general-purpose computers to deliver the targeted level of performance, although the recent advances in commodity graphics hardware have started to blur these boundaries (as we shall see in the next section). This situation is aggravated by the continuing trend towards higher and higher resolution datasets. For example, to render a dataset of 1024^3 16-bit voxels at 30 frames/s requires 2 GB of storage, a memory transfer rate of 60 GB/s and approximately 300 billion instructions per second, assuming 10 instructions per voxel per projection.

The same way as the special requirements of traditional computer graphics lead to high-performance graphics engines, volume visualization naturally lends itself to special-purpose volume-renderers that separate real-time image generation from general-purpose processing. This allows for stand-alone visualization environments that help scientists to interactively view their data on a single user workstation, augmented by a volume rendering accelerator. Furthermore, a volume rendering engine integrated in a graphics workstation is a natural extension of raster based systems into 3D volume visualization.

Several researchers have proposed special-purpose volume rendering architectures (Chapter 6) [67,96,102,116,152,162,163,192,245,246, 287].

Most recent research focuses on accelerators for ray-casting of regular datasets. Ray-casting offers room for algorithmic improvements while still allowing for high image quality. More recent architectures [84] include VOGUE, VIRIM, Cube, and VIZARD. The VolumePro board [200] is a commercial implementation of the Cube architecture.

VOGUE [116], a modular add-on accelerator, is estimated to achieve 2.5 frames/s for 256^3 datasets. For each pixel a ray is defined by the host computer and sent to the accelerator. The VOGUE module autonomously processes the complete ray, consisting of evenly spaced resampling locations, and returns the final pixel color of that ray to the host. Several VOGUE modules can be combined to yield higher-performance implementations. For example, to achieve 20 projections per second of 512^3 datasets requires 64 boards and a 5.2 GB/s ring-connected cubic network.

VIRIM [72] is a flexible and programmable ray-casting engine. The hardware consists of two separate units, the first being responsible for 3D resampling of the volume using lookup tables to implement different interpolation schemes. The second unit performs the ray-casting through the resampled dataset according to user-programmable lighting and viewing parameters. The underlying ray-casting model allows for arbitrary parallel and perspective projections and shadows. An existing hardware implementation for the visualization of $256 \times 256 \times 128$ datasets at 10 frames/s requires 16 processing boards.

The Cube project aims at the realization of high-performance volume rendering systems for large datasets and has pioneered several hardware architectures. Cube-1, a first-generation hardware prototype, was based on a specially interleaved memory organization [103], which has also been used in all subsequent generations of the Cube architecture. This interleaving of the n^3 voxels enables conflict-free access to any ray parallel to a main axis of n voxels. A fully operational printed circuit board (PCB) implementation of Cube-1 is capable of generating orthographic projections of 16^3 datasets from a finite number of predetermined directions in real time. Cube-2 was a single-chip VLSI implementation of this prototype [8].

To achieve higher performance and to further reduce the critical memory-access bottleneck, Cube-3 introduced several new concepts [203,205,206]. A high-speed global communication network aligns and distributes voxels from the memory to several parallel processing units, and a circular cross-linked binary tree of voxel combination units composites all samples into the final pixel color. Estimated performance for arbitrary parallel and perspective projections is 30 frames/s for 512^3 datasets. Cube-4 [202,204] has only simple and local interconnections, thereby allowing for easy scalability of performance. Instead of processing individual rays, Cube-4 manipulates a group of rays at a time. As a result, the rendering pipeline is directly connected to the memory. Accumulating compositors replace the binary compositing tree. A pixel-bus collects and aligns the pixel output from the compositors. Cube-4 is easily scalable to very high resolutions of 1024^3 16-bit voxels

and true real-time performance implementations of 30 frames/s.

EM-Cube [193] marked the first attempt to design a commercial version of the Cube-4 architecture. Its VLSI architecture features four rendering pipelines and four 64 Mbit SDRAMs to hold the volume data. VolumePro 500 was the final design, in the form of an ASIC, and was released to market by Mitsubishi Electric in 1999 [200]. VolumePro has hardware for gradient estimation, classification, and per-sample Phong illumination. It is a hardware implementation of the shear-warp algorithm, but with true tri-linear interpolation, which affords very high quality. The final warp is performed on the PC's graphics card. The VolumePro streams the data through four rendering pipelines, maximizing memory throughput by using a two-level memory block- and bank-skewing mechanism to take advantage of the burst mode of its SDRAMs. No occlusion-culling or voxel-skipping is performed. Advanced features such as gradient magnitude modulation of opacity and illumination, supersampling, cropping, and cut planes are also available. The system renders 500 million interpolated, Phong, illuminated, composited samples per second, which is sufficient to render volumes with up to 16 million voxels (e.g., 256^3 volumes) at 30 frames/s.

While the VolumePro uses a brute-force rendering mode in which all rays are cast across the volume, the VIZARD II architecture [162] implements an early ray-termination mechanism. It has been designed to run on a PCI board populated with four FPGAs, a DSP, and SDRAM and SRAM memory. In contrast to the VolumePro, it supports perspective rendering, but it uses a table-based gradient vector lookup scheme to compute the gradients at sample positions. The VIZARD II board is anticipated to render a 256^3 dataset at interactive frame rates. The VolumePro 1000 [282] is the successor of the VolumePro 500 and employs a different factorization of the viewing matrix, termed *shear-image* order ray-casting, which is a method of ray-casting that eliminates shear-warp's intermediate image and final warp step while preserving its memory-access efficiency. VolumePro 1000 uses empty space skipping and early ray termination, and it can render up to 10^9 samples/s.

The choice of whether one adopts a general-purpose or a special-purpose solution to volume rendering depends upon the circumstances. If maximum flexibility is required, general-purpose appears to be the best way to proceed. However, an important feature of graphics accelerators is that they are integrated into a much larger environment where software can shape the form of input and output data, thereby providing the additional flexibility that is needed. A good example is the relationship between the needs of conventional computer graphics and special-purpose graphics hardware. Nobody would dispute the necessity for polygon graphics acceleration despite its obvious limitations. The exact same argument can be made for special-purpose volume rendering architectures. The line between general-purpose and special-purpose, however, has become somewhat blurred in the past couple of years with the arrival of advanced, programmable commodity graphics processing units (GPUs). Although these boards do not, and perhaps never will, provide the full flexibility of a CPU, they gain more generality as a general computing machine with every new product cycle. In the following section, we will discuss the recent revolution in GPUs in light of their impact on interactive volume rendering and processing.

7.12   General-Purpose Rendering Hardware

Another opportunity to accelerate volume rendering is to utilize the texture-mapping capability of graphics hardware. The first such implementation was devised by Cabral et al. [24] and ran on SGI Reality Engine workstations. There are two ways to go about this. Either one represents the volume as a stack of 2D textures, one texture per volume slice, or as one single 3D texture, which requires more sophisticated

hardware. In the former case, three texture stacks are needed, one for each major viewing direction. An image is then rendered by choosing the stack that is most parallel to the image plane, and rendering the textured polygons to the screen in front-to-back or back-to-front order. If the machine has 3D texture capabilities, then one specifies a set of slicing planes parallel to the screen and composites the interpolated textures in depth order. The 3D texturing approach generally provides better images since the slice distance can be chosen arbitrarily, and no popping caused by texture-stack switching can occur. While the early approaches did not provide any shading, Van Gelder and Kim [65] introduced a fast technique to preshade the volume on the fly and then slice and composite a RGB volume to obtain an image with shading effects. Meißner et al. [161] provided a method to enable direct diffuse illumination for semi-transparent volume rendering. However, in this case, multiple passes through the rasterization hardware led to a significant loss in rendering performance. Instead, Dachille et al. [43] proposed a one-pass approach that employs 3D texture hardware interpolation together with software shading and classification. Westermann and Ertl [267] introduced a fast multi-pass approach to display shaded isosurfaces. Both Boada et al. [15] and LaMar et al. [122] subdivide the texture space into an octree, which allows them to skip nodes of empty regions and use lower-resolution textures for regions far from the viewpoint or of lower interest.

The emergence of advanced PC graphics hardware has made texture-mapped volume rendering accessible to a much broader community at less than 2% of the cost of the workstations that were previously required. However, the decisive factor stemming the revolution that currently dominates the field was the manufacturers' (e.g., NVidia, ATI, and 3DLabs) decision to make two of the main graphics pipeline components programmable. These two components are the vertex shaders (the units responsible for the vertex transformations in the

GLs Modelview matrix), and the fragment shaders, which are the units that take over after the rasterizer GLs Projection matrix. The first implementation that used these new commodity GPUs for volume rendering was published by Rezk-Salama et al. [209], who used the stack-or-textures approach since 3D texturing was not supported at that time. They overcame the undersampling problems associated with the large interslice distance at off-angles by interpolating, on the fly, intermediate slices using the register combiners in the fragment-shader compartment. Engel et al. [55] replaced this technique by the use of preintegrated transfer-function tables (see our previous section on transfer functions). The implementation can perform fully shaded semitransparent and iso-surface volume rendering at 1–4 frames/s for 256^3 volumes, using an NVidia GeForce3.

To compute the gradients required for shading, one must also load a gradient volume into the texture memory. The interpolation of a gradient volume without subsequent normalization is generally incorrect, but the artifacts are not always visible. Meißner and Guthe [158] use a shading cube texture instead, which eliminates this problem. Even the most recent texture-mapping hardware cannot reach the performance of the specialized volume rendering hardware, such as the VolumePro500 and the new VolumePro 1000, at least not when volumes are rendered by brute force. Therefore, current research efforts have concentrated on reducing the load for the fragment shaders. Level-of-detail (LOD) methods have been devised that rasterize lower-resolution texture blocks whenever the local volume detail or projected resolution allow them to do so [74,126]. Li and Kaufman [130,131] proposed an alternative approach that approximates the object by a set of texture boxes, which efficiently clips empty space from the rasterization.

Commodity graphics hardware also found much use for the rendering of irregular grids and for nonphotorealistic rendering, as will be discussed shortly. In addition, GPUs have also been extensively used for other nongraphics tasks, such as matrix computations [123], numerical simulations and computed [16,79,132], tomography [283]. These applications view the GPUs as general purpose SIMD machines, with high compute and memory bandwidth, and the latest feature, floating-point precision. It should be noted, however, that the limited capacity of the texture memory (currently 128 MB to 256 MB) and the slow CPU-GPU AGP bus bandwidth currently present the bottlenecks.

7.13   Irregular Grids

All the algorithms discussed above handle only regular gridded data. Irregular gridded data come in a large variety [240], including curvilinear data or unstructured (scattered) data, where no explicit connectivity is defined between cells (one can even be given a scattered collection of points that can be turned into an irregular grid by interpolation [186,153]). Fig. 7.29 illustrates the most prominent grid types.

For rendering purposes, manifold (locally homeomorphic to R^3) grids composed of convex cells are usually necessary. In general, the most convenient grids for rendering purposes are tetrahedral grids and hexahedral grids. One disadvantage of hexahedral grids is that the four points on the side of a cell may not necessarily lie on a plane, forcing the rendering algorithm to approximate the cells by convex ones during rendering. Tetrahedral grids have several advantages, including easier interpolation, simple representation (especially for connectivity information, because the degree of the connectivity graph is bounded, allowing for compact data-structure representation), and the fact that any other grid can be interpolated to a tetrahedral one (with the possible introduction of Steiner points). Among their disadvantages are that the sizes of the datasets tend to grow as cells are decomposed into tetrahedra and that sliver tetrahedra may be generated. In the case of curvilinear grids, an accurate (and naive) decomposition will make the cell complex contain five times as many cells.

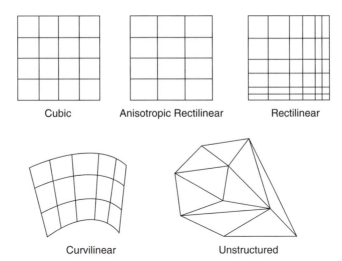

Figure 7.29   Various grid types in two dimensions.

As compared to regular grids, operations for irregular grids are more complicated and the effective visualization methods are more sophisticated in all fronts. Shading, interpolation, point location, etc., are all harder (and some even not well defined) for irregular grids. One notable exception is isosurface generation [136], which even in the case of irregular grids is fairly simple to compute given suitable interpolation functions. Slicing operations are also simple [240].

Volume rendering of irregular grids is a hard operation and there are several different approaches to this problem. The simplest and most inefficient is to resample the irregular grid to a regular grid. In order to achieve the necessary accuracy, a high-enough sampling rate has to be used that in most cases will make the resulting regular grid volume too large for storage and rendering purposes, not mentioning the time to perform the resampling. To overcome the need to fix the resolution of the regular grid to the smallest cell in the irregular grid, one can sample the data into a detail-adaptive octree whose local height is determined by the local granularity of the grid [126]. The octree decomposition also allows the grid to be rendered within a time-critical rendering framework.

One approach for rendering irregular grids is the use of feed-forward (or projection) methods, where the cells are projected onto the screen one by one, accumulating their contributions incrementally to the final image [153,228,273,275]. The projection algorithm that has gained popularity is the projected tetrahedra (PT) algorithm by Shirley and Tuchman [228]. It uses the projected profile of each tetrahedron with respect to the image plane to decompose it into a set of triangles. This gives rise to four classes of projections, which are shown in Fig. 7.30. The color and opacity values for each triangle vertex are approximated using ray integration through the thickest point of the tetrahedron. The resulting semitransparent triangles are sorted in depth order and then rendered and composited using polygonal graphics hardware. Stein et al. [243] sort the cells before they are split into tetrahedra, and they utilize 2D texture-mapping hardware to accelerate opacity interpolation and provide the correct per-pixel opacity values to avoid artifacts. While their method can handle only linear transfer functions without artifacts, Röttger et al. [211] introduced the concept of preintegrated volume rendering to allow for arbitrary transfer functions. They created a 3D texture map to provide hardware support in interpolating

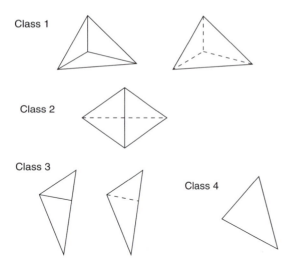

Figure 7.30 The four classes in tetrahedral projection.

along the ray between the front and back faces of a tetrahedral cell. In this texture map, two of the three coordinates correspond to values at the cell entry and exit points, with the third coordinate mapping to the distance through the cell. This texture map is then approximated using 2D texture mapping.

Cell-projection methods require a sorted list of cells to be passed to the hardware. Starting with Max et al.'s [153] and Williams's [276] works, there has been substantial progress in the development of accurate visibility-ordering algorithms [38,232]. A graphics hardware architecture was also proposed, but not yet realized, by King et al. [110], which can both rasterize and sort tetrahedral meshes in hardware.

An alternative technique to visualize irregular grids is by ray-casting [64,260]. Ray-casting methods tend to be more exact than projective techniques since they are able to "stab" or integrate the cells in depth order, even in the presence of cycles. This is generally not possible in cell-by-cell projection methods. Many ray-casting approaches employ the plane-sweep paradigm, which is based on processing geometric entities in an order determined by passing a line or a plane over the data. It was pioneered by Giertsen [66] for use in volume rendering. It is based on a

sweep plane that is orthogonal to the viewing plane (e.g., orthogonal to the y-axis). Events in the sweep are determined by vertices in the dataset and by values of y that correspond to the pixel rows. When the sweep plane passes over a vertex, an active cell list (ACL) is updated accordingly, so that it stores the set of cells currently intersected by the sweep plane. When the sweep plane reaches a y-value that defines the next row of pixels, the current ACL is used to process that row, casting rays corresponding to the values of x that determine the pixels in the row through a regular grid (hash table) that stores the elements of the ACL. This method has three major advantages: It is unnecessary to explicitly store the connectivity between the cells; it replaces the relatively expensive operation of 3D ray-casting with a simpler 2D regular grid ray-casting; and it exploits coherence of the data between scan lines. Since then, there have been a number of works that employ the sweep paradigm, most using a sweep plane that is parallel to the image plane. Some of these methods are assisted by hardware [267,285], while others are pure software implementations [22,56,231]. The ZSweep [56] algorithm is very fast and has excellent memory efficiency. It sweeps the plane from front to back and rasterizes the cell faces as they are encountered by the sweep plane. This keeps the memory footprint low since only the active cell set has be held in memory. Finally, Hong and Kaufman [85,86] proposed a very fast ray-casting technique that exploits the special topology of curvilinear grids.

7.14 Time-Varying and High-Dimensional Data

A significant factor contributing to the growth in the size of computational science datasets is the fact that the time-steps in the simulations have become increasingly finer in recent years. There have been significant developments in the rendering of time-varying volumetric datasets. These typically exploit time-coherency for compression and acceleration [3,74,141,225,247,

266], but other methods have also been designed that allow general viewing [6,13,76,77,78,106, 263] of high-dimensional (*n*-dimensional) datasets and require a more universal data decomposition.

In *n*-dimensional viewing, the direct projection from *n*-dimensional to 2D (for $n > 3$) is challenging. One major issue is that there are an infinite number of orderings to determine occlusion (for $n = 3$ there are just two, the view from the front and the view from the back). In order to simplify the user interface and to eliminate the amount of occlusion explorations a user has to do, Bajaj et al. [6] performed the *n*-dimensional volume renderings as an x-ray projection, where ordering is irrelevant. The authors demonstrated that, despite the lack of depth cues, much useful topological information of the *n*-dimensional space can be revealed in this way. They also presented a scalable interactive user interface that allows the user to change the viewpoint into *n*-dimensional space by stretching and rotating a system of *n* axis vectors.

On the other end of the spectrum are algorithms [13,263] that first calculate an *n*-dimensional hypersurface (a tetrahedral grid in 4D) for a specific iso-value, which can then be interactively sliced along any arbitrary hyperplane to generate an opaque 3D polygonal surface for hardware-accelerated view-dependent display. This approach is quite attractive as long as the iso-value is kept constant. However, if the iso-value is modified, a new iso-tetrahedralization must be generated, which can take on the order of tens of minutes [13].

Since 4D datasets can become quite large, a variety of methods to compress 4D volumes were proposed in recent years. Researchers used wavelets [73], DCT-encoding [141], RLE-encoding [3], and images [224,225]. All are lossy to a certain degree, depending on a set tolerance. An alternative compression strategy is the use of more efficient sampling grids, such as the BCC grids. Neophytou and Mueller [182] extended these grids for 4D volume rendering and used a 3D hyperslicer to extract 3D volumes for shaded and semitransparent volume visualization with occlusion ordering.

Early work on 4D rendering includes a paper by Ke and Panduranga [106], who used the hyperslice approach to provide views onto the on-the-fly computed 4D Mandelbrot set. Another early work is a paper by Rossignac [212], who gave a more theoretical treatment of the options available for the rendering of 4D hypersolids generated, for example, by time-animated or colliding 3D solids. Hanson et al. [76,77,78] wrote a series of papers that use 4D lighting in conjunction with a technique that augments 4D objects with renderable primitives to enable direct 4D renderings. The images they provided [77] are somewhat reminiscent of objects rendered with motion blur. The 4D isosurface algorithms proposed by Weigle and Banks [263] and Bhaniramka et al. [13] both use a marching cubes–type approach and generalize it into *n* dimensions.

Methods that focus more on the rendering of the time-variant aspects of 3D datasets have stressed the issue of compression and time-coherence to facilitate interactive rendering speeds. Shen and Johnson [226] used difference encoding of time-variant volumes to reduce storage and rendering time. Westermann [266] used a wavelet decomposition to generate a multiscale representation of the volume. Shen et al. [225] proposed the time-space partitioning (TSP) tree, which allows the renderer to cache and reuse partial (node) images of volume portions static over a time interval. It also enables the renderer to use data from subvolumes at different spatial and temporal resolutions. Anagnostou et al. [3] extended the RLE of the shear-warp algorithm [120] into four dimensions, inserting a new run block into the data structure whenever a change was detected over time. They then composited the rendered run block with partial rays of temporally unchanged volume portions. Sutton and Hansen [247] expanded the branch-on-need octree (BONO) approach of Wilhelms and Van Gelder [274] to time-variant data to enable fast out-of-core rendering of time-variant isosurfaces. Lum, Ma, and Clyne [141] advocated an algorithm that DCT-compresses time-runs of voxels into single scalars that are stored in a texture map. These texture maps, one per volume slice, are loaded into a GPU graphics board. Then, during time-animated rendering, the texture maps are indexed by a time-varying color palette that relates the scalars in the texture map to the current color of the voxel they represent. Although the DCT affords only a lossy compression, their rendering results are quite good and can be produced interactively. Another compression-based algorithm was proposed by Guthe and Straßer [74], who used a lossy MPEG-like approach to encode the time-variant data. The data were then decompressed on the fly for display with texture-mapping hardware.

7.15 Multichannel and Multimodal Data

So far, we have assumed that a voxel has a scalar density value from which other multivariate properties can be derived, e.g., via transfer-function lookup. We shall now extend this notion to datasets where the voxel data come originally in the form of multivariate vectors. In the context of this discussion, we shall distinguish between vectors of physical quantities, such as flow and strain, and vectors that store a list of voxel attributes. There is a large body of literature to visualize the former, including line integral convolution [23], spot noise [272], streamlines and stream-balls [21], glyphs, texture splats [42], and many more. In this section, we shall focus on the latter scenario, that is, volumes composed of attribute vectors. These can be (1) multichannel, such as the RGB color volumes obtained by cryosectioning the Visible Human [91] or multispectra remote sensing satellite data, or (2) multimodal, that is, volumes acquired by scanning an object with multiple modalities, such as MRI, PET, and CT.

The rendering of multimodal volumes requires the mixing of the data at some point in the rendering pipeline. There are at least three locations at which this can happen [25]. For the

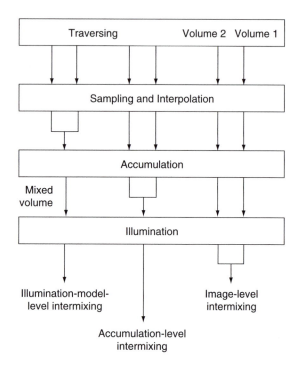

Figure 7.31 Levels in the volume rendering pipeline at which data mixing can occur.

following discussion, let us assume a set of two collocated volumes, but this is not a limitation. The simplest mixing technique is *image-level intermixing*, i.e., to render each volume separately as a scalar dataset and then blend the two images according to some weighting function that possibly includes the z-buffer or opacity channel (see Fig 7.32a). This method is attractive since it does not require a modification of the volume renderer, but as Fig. 7.32a (top) shows, it gives results of limited practical value since depth ordering is not preserved. This can be fixed by intermixing the rendering results at every step along the ray, which gives rise to *accumulation-level intermixing*. Here, we assign separate colors and opacities for each volume's ray sample, and then combine these according to some mixing function (see Fig. 7.32a (bottom)). A third method is *illumination-model level intermixing*, where one combines the ray samples before colors and opacities are

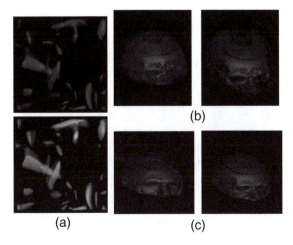

(a) (b) (c)

Figure 7.32 Multimodal rendering with data intermixing. (a) One time-step of a time-varying volume (magenta) and volume motion-blurred across 10 time-steps (blue). (Top) Image-level intermixing [182]. (Bottom) Accumulation-level intermixing [182]. (b) Accumulation-level intermixing of the Visible Man's CT and an MRI dataset. Here we assign blue if CT > MRI and green if MRI > CT. (Left) Gradients specified on CT while MRI is rendered as a point cloud. (Right) Surfaces rendered with gradient modulation [70]. (c) Accumulation-level intermixing of the Visible Man's CT and an MRI dataset, rendered in inclusive opacity mode, i.e., $\alpha = 1 - (1 - \alpha_{CT})(1 - \alpha_{MRI})$. (Left) Unweighted product of CT and MRI. (Right) More CT than MRI [70]. (See also color insert.)

computed. One could just use a weighted sum of the two densities to look up opacities and colors, or one could have one of the volumes act as an emission volume and the other as an attenuation volume. This would work quite naturally, for example, for the visualization of the emissive metabolic activities in a SPECT volume within the spatial context of a CT attenuation volume. Cai and Sakas [25] demonstrate this method in the scenario of dose planning in radiation therapy, where they visualize an emissive radiation beam embedded in an attenuating CT volume.

Multichannel data, such as RGB data obtained by ways of photographing slices of real volumetric objects, have the advantage that there is no longer a need to search for suitable color transfer functions to reproduce the original look of the data. On the other

hand, the photographic data do not provide an easy mapping to densities and opacities, which are required to compute normals and other parameters needed to bring out structural object detail in surface-sensitive rendering. One can overcome the perceptional nonlinearities of the RGB space by computing gradients and higher derivatives in the perceptually uniform color space $L^*u^*v^*$ [51]. In this method, the RGB data are first converted into the $L^*u^*v^*$ space, and the color distance between two voxels is calculated by their Euclidian distance in that color space. A gradient can then be calculated as usual via central differences but by replacing the voxel densities by the color distances. Although one cannot determine the direction of the normal vector with this method, this is not a limiting factor in practice. One can also derive more descriptive quantities, such as tensor gradients, since we are now dealing with vectors and not with densities in the gradient calculation. These can be used for segmentation, texture analysis, and others. Finally, opacities can be computed by using different functions of higher-level gradients to bring out different textural and structural aspects of the data [172].

7.16 Nonphotorealistic Volume Rendering

Nonphotorealistic volume rendering (NPVR) is a relatively recent branch of volume rendering. It employs local image processing during the rendering to produce artistic and illustrative effects, such as feature halos, tone shading, distance color blending, stylized motion blur, boundary enhancements, fading, silhouettes, sketch lines, stipple patterns, and pen-and-ink drawings [52,53,95,137,138,140,139,244,257]. The overall goal of NPVR is to go beyond the means of photorealistic volume rendering and produce images that emphasize critical features in the data, such as edges, boundaries, depth, and detail, to provide the user with a better appreciation of the structures in the data. This is similar to the goals of medical and other

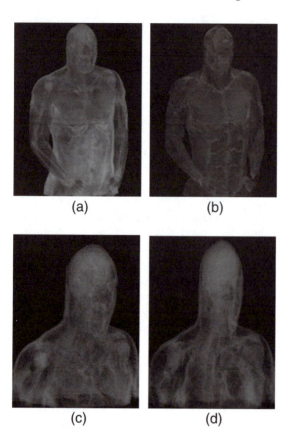

(a)   (b)

(c)   (d)

Figure 7.33 Rendering of multichannel (photographic) data. (a) The L* component (related to brightness). (b) The u* component (related to the chromatic change in red-green colors). (c) Color difference gradient computed in RGB color space. (d) Gradients computed in L*u*v* space, using the second derivative along the gradient direction to compute opacity. (Images from Gosh et al. [70]). (See also color insert.)

illustrators, as well as related efforts in general computer graphics [216,217,277,278]. Since the set of parameters that can be tuned in NPVR is even larger than the set for traditional volume rendering, interactive rendering of the NPVR effects is crucial, and indeed a number of researchers have proposed interactive implementations that exploit the latest generations of commodity-programmable graphics hardware [139, 244].

References

1. J. Amanatides and A. Woo. A fast voxel traversal algorithm for ray tracing. *Eurographics '87*, pages 1–10, 1987.
2. M. Amin, A. Grama, and V. Singh. Fast volume rendering using an efficient, scalable parallel formulation of the shear-warp algorithm. *Proc. Parallel Rendering Symposium '95*, pages 7–14, 1995.
3. K. Anagnostou, T. Atherton, and A. Waterfall. 4D volume rendering with the shear warp factorization. *Symp. Volume Visualization and Graphics '00*, pages 129–137, 2000.
4. R. Avila, L. Sobierajski, and A. Kaufman. Towards a comprehensive volume visualization system. *Proc. of IEEE Visualization '92*, pages 13–20, 1992.
5. R. Avila, T. He, L. Hong, A. Kaufman, H. Pfister, C. Silva, L. Sobierajski, and S. Wang. VolVis: A diversified system for volume visualization. *Proc. of IEEE Visualization '94*, 1994.
6. C. Bajaj, V. Pascucci, and D. Schikore. The contour spectrum. *Proc. IEEE Visualization '97*, pages 167–175, 1997.
7. C. Bajaj, V. Pascucci, G. Rabbiolo, and D. Schikore. Hypervolume visualization: A challenge in simplicity. *Proc. 1998 Symposium on Volume Visualization '98*, pages 95–102, 1998.
8. R. Bakalash, A. Kaufman, R. Pacheco, and H. Pfister. An extended volume visualization system for arbitrary parallel projection. *Proc. of Eurographics Workshop on Graphics Hardware '92*, 1992.
9. D. Banks and B. Singer. Vortex tubes in turbulent flows; identification, representation reconstruction. *Proc. of IEEE Visualization '94*, pages 132–139, 1994.
10. D. Bauer and R. Peikert. Vortex tracking in scale-space, *Joint EUROGRAPHICS–IEEE TCVG Symposium on Visualization (2002), VisSym '02*, 2002.
11. M. Bentum, B. B. A. Lichtenbelt, and T. Malzbender. Frequency analysis of gradient estimators in volume rendering. *IEEE Trans. on Visualization and Computer Graphics '96*, 2(3):242–254, 1996.
12. S. Bergner, T. Möller, M. Drew, and G. Finlayson. Interactive spectral volume rendering. *Proc. of IEEE Visualization '02*, pages 101–108, 2002.
13. P. Bhaniramka, R. Wenger, and R. Crawfis. Isosurfacing in higher dimensions. *Proc. of IEEE Visualization '00*, pages 267–273, 2000.
14. F. Blinn. Light reflection functions for simulation of clouds and dusty surfaces. *Proc. of SIGGRAPH '82*, pages 21–29, 1982.

15. I. Boada, I. Navazo, and R. Scopigno. Multiresolution volume visualization with a texture-based octree. *The Visual Computer*, 17(3):185–197, 2001.

16. J. Bolz, I. Farmer, E. Grinspun, and P. Schröder. The GPU as numerical simulation engine. *SIGGRAH'03*, pages 917–924, 2003.

17. R. Bracewell. *The Fourier Transform and its Applications*, 3rd Ed., McGraw-Hill, 1999.

18. M. Brady, K. Jung, H. Nguyen, and T. Nguyen. Interactive volume navigation. *IEEE Transactions on Visualization and Computer Graphics*, 4(3):243–256, 1998.

19. M. Brady, K. Jung, H. Nguyen, and T. Nguyen. Two-phase perspective ray-casting for interactive volume navigation. *Visualization '97*, pages 183–189, 1997.

20. M. Brady, W. Higgins, K. Ramaswamy, and R. Srinivasan. Interactive navigation inside 3D radiological images. *Proc. of Biomedical Visualization '99, Proc. of IEEE Visualization '99*, pages 33–40 and page 85 (color plate), 1995.

21. M. Brill, V. Djatschin, H. Hagen, S.V. Klimenko, and H.-C. Rodrian. Streamball techniques for flow visualization. *Proc. of IEEE Visualization '94*, pages 225–231, 1994.

22. P. Bunyk, A. Kaufman, and C. Silva. Simple, fast, and robust ray-casting of irregular grids. *Scientific Visualization*, pages 30–36, 1997.

23. B. Cabral and L. Leedon. Imaging vector fields using line integral convolution. *Proc. of SIGGRAPH '93*, pages 263–272, 1993.

24. B. Cabral, N. Cam, and J. Foran. Accelerated volume rendering and tomographic reconstruction using texture mapping hardware. *Symp. on Volume Visualization '94*, pages 91–98, 1994.

25. W. Cai and G. Sakas. Data intermixing and multi-volume rendering. *Computer Graphics Forum (Eurographics '99)*, 19(3):359–368, 1999.

26. H. Carr, J. Snoeyink, and U. Axen. Computing contour trees in all dimensions. *Computational Geometry Theory and Applications '02*, 24(2):75–94, 2003.

27. J. Challinger. Scalable parallel volume ray-casting for nonrectilinear computational grids. *Proc. of Parallel Rendering Symposium '93*, pages 81–88, 1993.

28. B. Chen, A. Kaufman, and Q. Tang. Image-based rendering of surfaces from volume data. *Workshop on Volume Graphics '01*, pages 279–295, 2001.

29. E. Chen. QuickTime VR: an image-based approach to virtual environment navigation. *Proc. of SIGGRAPH '95*, pages 29–38, 1995.

30. E. Chen and L. Williams. View interpolation for image synthesis. *Proc. of SIGGRAPH '93*, pages 279–288, 1993.

31. M. Chen, A. Kaufman, and R. Yagel (Eds.) *Volume Graphics*. Springer, London, 2000.

32. K. Chidlow and T. Möller. Rapid emission volume reconstruction. *Volume Graphics Workshop '03*, pages 15–26, 2003.

33. T. Chiueh, T. He, A. Kaufman, and H. Pfister. Compression domain volume rendering. *Technical Report 94.01.04, Computer science, SUNY Stony Brook, 1994*.

34. C. Chui. *An Introduction To Wavelets*. Boston, Academic Press, 1992.

35. P. Cignoni, C. Montani, E. Puppo, and R. Scopigno. Optimal isosurface extraction from irregular volume data. *Symposium on Volume Visualization '96*, pages 31–38, 1996.

36. H. Cline, W. Lorensen, S. Ludke, C. Crawford, and B. Teeter. Two algorithms for the 3D reconstruction of tomograms. *Med. Phys.*, 15:320–327, 1988.

37. D. Cohen and Z. Shefer. Proximity clouds: an acceleration technique for 3D grid traversal. *The Visual Computer*, 10(11):27–38, 1994.

38. J. Comba, J. Klosowski, N. Max, J. Mitchell, C. Silva, and P. Williams. Fast polyhedral cell sorting for interactive rendering of unstructured grids. *Computer Graphics Forum*, 18(3):369–376, 1999.

39. J. Conway and N. Sloane. *Sphere Packings, Lattices and Groups*, 2nd Ed. Berlin, Springer, 1993.

40. B. Corrie and P. Mackerras. Parallel volume rendering and data coherence. *Proc. of Parallel Rendering Symposium '93*, pages 23–26, 1993.

41. R. Crawfis. Real-time slicing of data space. *Proc. of IEEE Visualization '96*, pages 271–277, 1996.

42. R. Crawfis and N. Max. Texture splats for 3D scalar and vector field visualization. *Proc. of IEEE Visualization '93*, pages 261–266, 1993.

43. F. Dachille, K. Kreeger, B. Chen, I. Bitter, and A. Kaufman. High-quality volume rendering using texture mapping hardware. *SIGGRAPH/ Eurographics Workshop on Graphics Hardware '98*, pages 69–77, 1998.

44. F. Dachille, K. Mueller and A. Kaufman. Volumetric back-projection. *Volume Visualization Symposium '00*, pages 109–117, 2000.

45. J. Danskin and P. Hanrahan. Fast algorithms for volume ray tracing. *Workshop on Volume Visualization*, pages 91–98, 1992.

46. I. Daubechies. Ten lectures on wavelets. *CBMS-NSF Reg. Conf. Ser. Appl. Math. SIAM*, 1992.

47. J. Dorsey, A. Edelman, H. Jensen, J. Legakis, and H. Pederson. Modeling and rendering of weathered stone. *Proc. of SIGGRAPH '99*, pages 225–234, 1999.

48. R. Drebin, L. Carpenter, and P. Hanrahan. Volume rendering. *Proc. of SIGGRAPH '88*, 22(4):65–74, 1988.

49. D. Dudgeon and R. Mersereau. *Multi-dimensional Digital Signal Processing*. Englewood Cliffs, NJ, Prentice-Hall, 1984.

50. S. Dunne, S. Napel, and B. Rutt. Fast reprojection of volume data. *Proc. of IEEE Visualization in Biomed. Comput.*, pages 11–18, 1990.

51. D. Ebert, C. Morris, P. Rheingans, and T. Yoo. Designing effective transfer functions for volume rendering from photographics volumes. *IEEE Trans. on Visualization and Computer Graphics*, 8(2):183–197, 2002.

52. D. Ebert and P. Rheingans. Volume illustration: non-photorealistic rendering of volume models. *IEEE Transactions on Visualization and Computer Graphics*, pages 253–265, 2001.

53. D. Ebert and P. Rheingans. Volume illustration: non-photorealistic rendering of volume models. *Proc. of IEEE Visualization '00*, pages 195–202, 2000.

54. T. Elvins. Volume rendering on a distributed memory parallel computer. *Proc. of Parallel Rendering Symposium '93*, pages 93–98, 1993.

55. K. Engel, M. Kraus, and T. Ertl. High-quality pre-integrated volume rendering using hardware-accelerated pixel shading. *Proc. of SIGGRAPH Graphics Hardware Workshop '01*, pages 9–16, 2001.

56. R. Farias, J. Mitchell, and C. Silva. ZSWEEP: An efficient and exact projection algorithm for unstructured volume rendering. *ACM/IEEE Volume Visualization and Graphics Symposium*, pages 91–99, 2000.

57. J. Foley, A. Dam, S. Feiner, and J. Hughes. *Computer Graphics: Principles and Practice*, 2nd Ed. Addison-Wesley, 1996.

58. J. Fowler and R. Yagel. Lossless compression of volume data. *Symp. of Volume Visualization '94*, pages 43–50, 1994.

59. A. Fujimoto, T. Tanaka, and K. Iwata. ARTS: accelerated ray-tracing system. *IEEE Computer Graphics and Applications*, 6(4):16–26, 1986.

60. I. Fujishiro, Y. Takeshima, T. Azuma, and S. Takahashi. Volume data mining using 3D field topology analysis. *IEEE Computer Graphics & Applications*, 20(5):46–51, 2000.

61. A. Gaddipati, R. Machiraju, and R. Yagel. Steering image generation using wavelet based perceptual metric. *Computer Graphics Forum (Proc. Eurographics '97)*, 16(3):241–251, 1997.

62. J. Gao and H. Shen. Parallel view-dependent isosurface extraction using multi-pass occlusion culling. *ACM/IEEE Symposium on Parallel and Large Data Visualization and Graphics*, pages 67–74, 2001.

63. M. Garland and P. Heckbert. Surface simplification using quadric error metrics. *Proc. 24th Annual Conference on Computer Graphics and Interactive Techniques*, pages 209–216, 1997.

64. M. Garrity. Raytracing irregular volume data. *Computer Graphics*, pages 35–40, November 1990.

65. A. Van Gelder and K. Kim. Direct volume rendering via 3D texture mapping hardware. *Proc. of Vol. Rend. Symp. '96*, pages 23–30, 1996.

66. C. Giertsen. Volume visualization of sparse irregular meshes. *IEEE Computer Graphics and Applications*, 12(2):40–48, 1992.

67. S. Goldwasser, R. Reynolds, T. Bapty, D. Baraff, J. Summers, D. Talton, and E. Walsh. Physician's workstation with real-time performance. *IEEE Computer Graphics & Applications*, 5(12):44–57, 1985.

68. D. Gordon and R. Reynolds. Image-space shading of 3-D objects. *Computer Vision, Graphics, and Image Processing*, 29:361–376, 1985.

69. S. Gortler, R. Grzeszczuk, R. Szeliski, and M. Cohen. The lumigraph. *Proc. of SIGGRAPH '96*, pages 43–54, 1996.

70. A. Gosh, P. Prabhu, A. Kaufman, and K. Mueller. Hardware assisted multichannel volume rendering. *Computer Graphics International '03*, 2003.

71. M. Gross, R. Koch, L. Lippert, and A. Dreger. A new method to approximate the volume rendering equation using wavelet bases and piecewise polynomials. *Computers & Graphics*, 19(1):47–62, 1995.

72. T. Guenther, C. Poliwoda, C. Reinhard, J. Hesser, R. Maenner, H. Meinzer, and H. Baur. VIRIM: a massively parallel processor for real-time volume visualization in medicine. *Proc. of the 9th Eurographics Hardware Workshop*, pages 103–108, 1994.

73. S. Guthe, M. Wand, J. Gonser, and W. Strasser. Interactive rendering of large volume datasets. *Proc. of IEEE Visualization '02*, pages 53–60, 2002.

74. S. Guthe and W. Straßer. Real-time decompression and visualization of animated volume data. *Proc. of IEEE Visualization '01*, pages 349–372, 2001.

75. P. Hanrahan and W. Krueger. Reflection from layered surfaces due to subsurface scattering.

Computer Graphics (Proc. of SIGGRAPH '93), pages 165–174, 1993.

76. A. Hanson and P. Heng. 4D views of 3D scalar fields. *Proc. of IEEE Visualization '92*, pages 84–91, 1992.

77. A. Hanson and P. Heng. Illuminating the fourth dimension. *IEEE Computer Graphics & Applications*, 12(4):54–62, 1992.

78. A. Hanson and R. Cross. Interactive visualization methods for four dimensions. *Proc. of IEEE Visualization '93*, pages 196–203, 1993.

79. M. Harris, G. Coombe, T. Scheuermann, and A. Lastra. Physically based visual simulation on graphics hardware. *Proc. of 2002 SIGGRAPH/ Eurographics Workshop on Graphics Hardware*, pages 109–118, 2002.

80. H. Hauser, L. Mroz, G. Bischi, and M. Gröller. Two-level volume rendering-flushing MIP and DVR. *Proc. of IEEE Visualization '00*, pages 211–218, 2000.

81. T. He, L. Hong, A. Kaufman, and H. Pfister. Generation of transfer functions with stochastic search techniques. *Proc. of IEEE Visualization '96*, pages 227–234, 1996.

82. G. Herman and H. Liu. 3D display of human organs from computed tomograms. *Comput. Graphics Image Process*, 9(1):1–21, 1979.

83. G. Herman and J. Udupa. Display of 3D discrete surfaces. *Proceedings SPIE*, 283:90–97, 1981.

84. J. Hesser, R. Maenner, G. Knittel, W. Strasser, H. Pfister, and A. Kaufman. Three architectures for volume rendering. *Computer Graphics Forum*, 14(3):111–122, 1995.

85. L. Hong and A. Kaufman. Accelerated raycasting for curvilinear volumes. *Proc. of IEEE Visualization '98*, pages 247–253, 1998.

86. L. Hong and A. Kaufman. Fast projection-based ray-casting algorithm for rendering curvilinear volumes. *IEEE Transactions on Visualization and Computer Graphics*, 5(4):322–332, 1999.

87. L. Hong, S. Muraki, A. Kaufman, D. Bartz, and T. He. Virtual voyage: interactive navigation in the human colon. *Proc. of ACM SIGGRAPH '97*, pages 27–34, 1997.

88. H. Hoppe. Progressive meshes. *Proc. of SIGGRAPH '96*, pages 99–108, 1996.

89. W. Hsu. Segmented ray-casting for data parallel volume rendering. *Proc. of Parallel Rendering Symposium '93*, pages 93–98, 1993.

90. http://graphics.stanford.edu/software/volpack/

91. http://www.nlm.nih.gov/research/visible/visible_human.html

92. J. Huang, N. Shareef, R. Crawfis, P. Sadayappan, and K. Mueller. A parallel splatting algorithm with occlusion culling. *3rd Eurographics Workshop on Parallel Graphics and Visualization*, 2000.

93. J. Huang, R. Crawfis, and D. Stredney. Edge preservation in volume rendering using splatting. *IEEE Volume Vis. '98*, pages 63–69, 1998.

94. I. Ihm and R. Lee. On enhancing the speed of splatting with indexing. *Proc. of IEEE Visualization '95*, pages 69–76, 1995.

95. V. Interrante. Illustrating surface shape in volume data via principal direction-driven 3D line integral convolution. *Proc. of SIGGRAPH '97*, pages 109–116, 1997.

96. D. Jackel. The graphics PARCUM system: a 3D memory based computer architecture for processing and display of solid models. *Computer Graphics Forum*, 4(1):21–32, 1985.

97. H. Jensen and P. Christensen. Efficient simulation of light transport in sciences with participating media using photon maps. *Proc. of SIGGRAPH '98*, pages 311–320, 1998.

98. J. Kajiya and B. Herzen. Ray tracing volume densities. *Proc. of SIGGRAPH '84*, pages 165–174, 1994.

99. K. Kaneda, Y. Dobashi, K. Yamamoto, and H. Yamashita. Fast volume rendering with adjustable color maps. *1996 Symposium on Volume Visualization*, pages 7–14, 1996.

100. A. Kaufman. *Volume Visualization*, IEEE Computer Society Press Tutorial, Los Alamitos, CA.

101. A. Kaufman. Volume visualization. *ACM Computing Surveys*, 28(1):165–167, 1996.

102. A. Kaufman and R. Bakalash. CUBE – an architecture based on a 3-D voxel map. *Theoretical Foundations of Computer Graphics and CAD*, (R.A. Earnshaw, Ed.), Springer-Verlag, pages 689–701, 1985.

103. A. Kaufman and R. Bakalash. Memory and processing architecture for 3-D voxel-based imagery. *IEEE Computer Graphics & Applications*, 8(6):10–23, 1988. Also in Japanese, *Nikkei Computer Graphics*, 3(30):148–160, 1989.

104. A. Kaufman, D. Cohen, and R. Yagel. Volume graphics. *IEEE Computer*, 26(7):51–64, 1993.

105. T. Kay and J. Kajiya. Ray tracing complex scenes. *Proc. of SIGGRAPH '86*, pages 269–278, 1986.

106. Y. Ke and E. Panduranga. A journey into the fourth dimension. *Proc. of IEEE Visualization '89*, pages 219–229, 1989.

107. R. Keys. Cubic convolution interpolation for digital image processing. *IEEE Transactions. on Acoustics, Speech, and Signal Processing*, 29(6):1153–1160, 1981.

108. S. Kilthau and T. Möller. Splatting optimizations. *Technical Report, School of Computing Science, Simon Fraser University*, (SFU-CMPT-04/01-TR2001-02), 2001.

109. G. Kindlmann and J. Durkin. Semi-automatic generation of transfer functions for direct volume rendering. *Symp. Volume Visualization '98*, pages 79–86, 1998.

110. D. King, C. Wittenbrink, and H. Wolters. An architecture for interactive tetrahedral volume rendering. *International Workshop on Volume Graphics '01*, pages 101–112, 2001.

111. J. Kniss, G. Kindlmann, and C. Hansen. Interactive volume rendering using multidimensional transfer functions and direct manipulation widgets. *Proc. of IEEE Visualization '01*, pages 255–262, 2001.

112. J. Kniss, G. Kindlmann, and C. Hansen. Multidimensional transfer functions for interactive volume rendering. *IEEE Transactions on Visualization and Computer Graphics*, 8(3):270–285, 2002.

113. J. Kniss, S. Premoze, C. Hansen, P. Shirley, and A. McPherson. A model for volume lighting and modeling. *IEEE Transactions on Visualization and Computer Graphics*, 9(2):150–162, 2003.

114. J. Kniss, S. Premoze, C. Hansen, and D. Ebert. Interactive translucent volume rendering and procedural modeling. *Proc. of IEEE Visualization '02*, pages 109–116, 2002.

115. G. Knittel. The ULTRAVIS system. *Proc. of Volume Visualization and Graphics Symposium '00*, pages 71–80, 2000.

116. G. Knittel and W. Strasser. A compact volume rendering accelerator. *Volume Visualization Symposium Proceedings*, pages 67–74, 1994.

117. K. Kreeger and A. Kaufman. Interactive volume segmentation with the PAVLOV architecture. *Proc. of Parallel Visualization and Graphics Symposium '99*, pages 61–68, 1999.

118. K. Kreeger, I. Bitter, F. Dachille, B. Chen, and A. Kaufman. Adaptive perspective ray-casting. *Volume Visualization Symposium '98*, pages 55–62, 1998.

119. M. Kreveld, R. Oostrum, C. Bajaj, V. Pascucci, and D. Schikore. Contour trees and small seed sets for isosurface traversal. *Proc. of the 13th ACM Symposium on Computational Geometry*, pages 212–220, 1997.

120. P. Lacroute and M. Levoy. Fast volume rendering using a shear-warp factorization of the viewing transformation. *Proc. of SIGGRAPH '94*, pages 451–458, 1994.

121. P. Lacroute. Analysis of a parallel volume rendering system based on the shear-warp factorization. *IEEE Trans. of Visualization and Computer Graphics*, 2(3):218–231, 1996.

122. E. LaMar, B. Hamann, and K. Joy. Multiresolution techniques for interactive texture-based volume visualization. *Proc. of IEEE Visualization '99*, pages 355–361, 1999.

123. E. Larsen and D. McAllister. Fast matrix multiplies using graphics hardware. *Supercomputing '01*, pages 43–49, 2001.

124. D. Laur and P. Hanrahan. Hierarchical splatting: a progressive refinement algorithm for volume rendering. *Proc. of SIGGRAPH '91*, pages 285–288, 1991.

125. R. Lee and I. Ihm. On enhancing the speed of splatting using both object and image-space coherence. *Graphical Models and Image Processing*, 62(4):263–282, 2000.

126. J. Leven, J. Corso, S. Kumar, and J. Cohen. Interactive visualization of unstructured grids using hierarchical 3D textures. *Proc. of Symposium on Volume Visualization and Graphics '02*, pages 33–40, 2002.

127. M. Levoy. Display of surfaces from volume data. *IEEE Comp. Graph. & Appl.*, 8(5):29–37, 1988.

128. M. Levoy. Efficient ray tracing of volume data. *ACM Trans. Comp. Graph.*, 9(3):245–261, 1990.

129. M. Levoy and P. Hanrahan. Light field rendering. *Proc. of SIGGRAPH '96*, pages 31–42, 1996.

130. W. Li and A. Kaufman. Accelerating volume rendering with texture hulls. *IEEE/Siggraph Symposium on Volume Visualization and Graphics 2002 (VolVis'02)*, pages 115–122, 2002.

131. W. Li and A. Kaufman. Texture partitioning and packing for accelerating texture-based volume rendering. *Graphics Interface '03*, pages 81–88, 2003.

132. W. Li, X. Wei, and A. Kaufman. Implementing lattice Boltzmann computation on graphics hardware. *The Visual Computer*, 2003.

133. P. Li, S. Whitman, R. Mendoza, and J. Tsiao. ParVox—a parallel splatting volume rendering system for distributed visualization. *Proc. of Parallel Rendering Symposium '97*, pages 7–14, 1997.

134. B. Lichtenbelt, R. Crane, and S. Naqvi. *Volume Rendering*. Prentice-Hall, 1998.

135. Y. Livnat, H. Shen, and C. Johnson. A near optimal isosurface extraction algorithm for structured and unstructured grids. *IEEE Trans. on Vis. and Comp. Graph.*, 2(1):73–84, 1996.

136. E. Lorensen and H. Cline. Marching cubes: a high resolution 3D surface construction algorithm. *Proc. of SIGGRAPH '87*, pages 163–169, 1987.

137. A. Lu, C. Morris, D. Ebert, P. Rheingans, and C. Hansen. Non-photorealistic volume rendering using stippling techniques. *Proc. of IEEE Visualization '02*, pages 211–217, 2002.

138. A. Lu, C. Morris, J. Taylor, D. Ebert, P. Rheingans, C. Hansen, and M. Hartner. Illustrative interactive stipple rendering. *IEEE Transactions on Visualization and Computer Graphics*, pages 127–138, 2003.

139. E. Lum and K. Ma. Hardware-accelerated parallel non-photorealistic volume rendering. *International Symposium on Nonphotorealistic Rendering and Animation '02*, 2002.

140. E. Lum and K. Ma. Nonphotorealistic rendering using watercolor inspired textures and illumination. *Pacific Graphics '01*, 2001.

141. E. Lum, K. Ma, and J. Clyne. Texture hardware assisted rendering of time-varying volume data, *Proc. of IEEE Visualization '01*, pages 262–270, 2001.

142. F. Ma, W. Wang, W. Tsang, Z. Tang, and S. Xia. Probabilistic segmentation of volume data for visualization using SOM-PNN classifier. *Symposium on Volume Visualization '98*, pages 71–78, 1998.

143. K. Ma and S. Parker. Massively parallel software rendering for visualizing large-scale datasets. *IEEE Computer Graphics & Applications*, pages 72–83, 2001.

144. K. Ma. Parallel volume ray-casting for unstructured-grid data on distributed-memory architectures. *Proc. of Parallel Rendering Symposium '95*, pages 23–30, 1995.

145. K. Ma and T. Crockett. A scalable parallel cell-projection volume rendering algorithm for 3D unstructured data. *Proc. of Parallel Rendering Symposium '97*, 1997.

146. R. Machiraju, A. Gaddipati, and R. Yagel. Detection and enhancement of scale coherent structures using wavelet transform products. *Proc. of the Technical Conference on Wavelets in Signal and Image Processing V*, pages 458–469, 1997.

147. R. Machiraju and R. Yagel. Efficient feed-forward volume rendering techniques for vector and parallel processors, *Supercomputing '93*, pages 699–708, 1993.

148. R. Machiraju and R. Yagel. Reconstruction error and control: a sampling theory approach. *IEEE Transactions on Visualization and Graphics*, 2(3), 1996.

149. T. Malzbender and F. Kitson. A Fourier technique for volume rendering. *Focus on Scientific Visualization*, pages 305–316, 1991.

150. J. Marks, B. Andalman, P. A. Beardsley, W. Freeman, S. Gibson, J. Hodgins, T. Kang, B. Mirtich, H. Pfister, and W. Rum. Design galleries: a general approach to setting parameters for computer graphics and animation. *Proc. of SIGGRAPH '97*, pages 389–400, 1997.

151. S. Marschner and R. Lobb. An evaluation of reconstruction filters for volume rendering. *Proc. of IEEE Visualization '94*, pages 100–107, 1994.

152. N. Max. Optical models for direct volume rendering. *IEEE Trans. Vis. and Comp. Graph.*, 1(2):99–108, 1995.

153. N. Max, P. Hanrahan, and R. Crawfis. Area and volume coherence for efficient visualization of 3D scalar functions. *Computer Graphics*, 24(5):27–33, 1990.

154. L. McMillan and G. Bishop. Plenoptic modeling: an image-based rendering system. *Proc. of SIGGRAPH '95*, pages 39–46, 1995.

155. D. Meagher. Geometric modeling using octree encoding. *Computer Graphics and Image Processing*, 19(2):129–147, 1982.

156. D. Meagher. Applying solids processing methods to medical planning. *Proc. of NCGA '85*, pages 101–109, 1985.

157. H. Meinzer, K. Meetz, D. Scheppelmann, U. Engelmann, and H. Baur. The Heidelberg raytracing model. *IEEE Computer Graphics & Applications*, 11(6):34–43, 1991.

158. M. Meißner and S. Guthe. Interactive lighting models and pre-integration for volume rendering on PC graphics accelerators. *Graphics Interface '02*, 2002.

159. M. Meißner, J. Huang, D. Bartz, K. Mueller and R. Crawfis. A practical comparison of popular volume rendering algorithms. *Symposium on Volume Visualization and Graphics 2000*, pages 81–90, 2000.

160. M. Meißner, M. Doggett, U. Kanus, and J. Hirche. Efficient space leaping for ray-casting architectures. *Proc. of the 2nd Workshop on Volume Graphics*, 2001.

161. M. Meißner, U. Hoffmann, and W. Straßer. Enabling classification and shading for 3D texture mapping based volume rendering using OpenGL and extension. *Proc. of IEEE Visualization '99*, 1999.

162. M. Meißner, U. Kanus, and W. Straßer. VIZARD II: A PCI-card for real-time volume rendering. *Proc. of Siggraph/Eurographics Workshop on Graphics Hardware '98*, pages 61–67, 1998.

163. M. Meißner, U. Kanus, G. Wetekam, J. Hirche, A. Ehlert, W. Straßer, M. Doggett, and R. Proksa. A reconfigurable interactive volume rendering system. *Proc. of SIGGRAPH/ Eurographics Workshop on Graphics Hardware '02*, 2002.

164. J. Ming, R. Machiraju, and D. Thompson. A novel approach to vortex core detection. *Proc. of VisSym '02*, pages 217–225, 2002.

165. D. Mitchell and A. Netravali, Reconstruction filters in computer graphics. *Proc. of SIGGRAPH '88*, pages 221–228, 1988.

166. T. Möller, R. Machiraju, K. Mueller, and R. Yagel. A comparison of normal estimation schemes. *Proc. of IEEE Visualization '97*, pages 19–26, 1997.

167. T. Möller, R. Machiraju, K. Mueller, and R. Yagel. Evaluation and design of filters using a Taylor series expansion. *IEEE Transactions on Visualization and Computer Graphics*, 3(2):184–199, 1997.

168. S. Molnar, M. Cox, D. Ellsworth, and H. Fuchs. A sorting classification of parallel rendering. *IEEE Computer Graphics and Applications*, 14(4):23–32, 1994.

169. C. Montani, R. Scateni, and R. Scopigno. Discretized marching cubes. *Proc. of IEEE Visualization '94*, pages 281–287, 1994.

170. C. Montani, R. Perego, and R. Scopigno. Parallel volume visualization on a hypercube architecture. *Proc. of Volume Visualization Symposium '92*, pages 9–16, 1992.

171. B. Mora, J. Jessel, and R. Caubet. A new object-order ray-casting algorithm. *Proc. of IEEE Visualization '02*, pages 203–210, 2002.

172. C. Morris and D. Ebert. Direct volume rendering of photographic volumes using multi-dimensional color-based transfer functions, *EUROGRAPHICS IEEE TCVG Symp. on Visualization '02*, pages 115–124, 2002.

173. K. Mueller and R. Crawfis, Eliminating popping artifacts in sheet buffer-based splatting. *Proc. of IEEE Visualization '98*, pages 239–245, 1998.

174. K. Mueller and R. Yagel. Fast perspective volume rendering with splatting by using a ray-driven approach. *Proc. of IEEE Visualization '96*, pages 65–72, 1996.

175. K. Mueller and R. Yagel. Rapid 3D cone-beam reconstruction with the algebraic reconstruction technique (ART) by using texture mapping hardware, *IEEE Transactions on Medical Imaging*, 19(12):1227–1237, 2000.

176. K. Mueller, M. Chen, and A. Kaufman (Eds.) *Volume Graphics '01*. London, Springer, 2001.

177. K. Mueller, N. Shareef, J. Huang, and R. Crawfis. High-quality splatting on rectilinear grids with efficient culling of occluded voxels. *IEEE Transactions on Visualization and Computer Graphics*, 5(2):116–134, 1999.

178. K. Mueller, N. Shareef, J. Huang, and R. Crawfis. IBR assisted volume rendering. *Proc. of IEEE Visualization '99*, pages 5–8, 1999.

179. K. Mueller, T. Moeller, J. E. Swan, R. Crawfis, N. Shareef, and R. Yagel. Splatting errors and antialiasing. *IEEE Transactions on Visualization and Computer Graphics*, 4(2):178–191, 1998.

180. K. Mueller, T. Möller, and R. Crawfis. Splatting without the blur. *Proc. of IEEE Visualization '99*, pages 363–371, 1999.

181. S. Muraki. Volume data and wavelet transform, *IEEE Comput. Graphics Appl.*, 13(4):50–56, 1993.

182. N. Neophytou and K. Mueller. Space-time points: 4D splatting on efficient grids. *Symposium on Volume Visualization and Graphics '02*, pages 97–106, 2002.

183. N. Neophytou and K. Mueller. Post-convolved splatting. *Joint Eurographics–IEEE TCVG Symposium on Visualization '03*, pages 223–230, 2003.

184. J. Nieh and M. Levoy. Volume rendering on scalable shared-memory MIMD architectures. *Proc. of Volume Visualization Symposium*, pages 17–24, 1992.

185. G. Nielson and B. Hamann. The asymptotic decider: resolving the ambiguity in marching cubes. *Proc. of IEEE Visualization '91*, pages 29–38, 1991.

186. M. Nielson. Scattered data modeling. *IEEE Computer Graphics and Applications*, 13(1): 60–70, 1993.

187. P. Ning and L. Hesselink. Fast volume rendering of compressed data. *Proc. of IEEE Visualization '93*, pages 11–18, 1993.

188. P. Ning and L. Hesselink. Vector quantization for volume rendering. *Proc. of IEEE Visualization '92*, pages 69–74, 1992.

189. H. Noordmans, H. Voort, and A. Smeulders. Spectral volume rendering. *IEEE Transactions on Visualization and Computer Graphics*, 6(3):196–207, 2000.

190. L. Novins, F. X. Sillion, and D. P. Greenberg. An efficient method for volume rendering using perspective projection. *Computer Graphics*, 24(5):95–100, 1990.

191. M. Nulkar and K. Mueller. Splatting with shadows. *International Workshop on Volume Graphics '01*, 2001.

192. T. Ohashi, T. Uchiki, and M. Tokyo. A 3D shaded display method for voxel-based representation. *Proc. of EUROGRAPHICS '85*, pages 221–232, 1985.

193. R. Osborne, H. Pfister, H. Lauer, T. Ohkami, N. McKenzie, S. Gibson, and W. Hiatt. EM-cube: an architecture for low-cost real-time volume rendering. *Proc. of Eurographics Hardware Rendering Workshop '97*, pages 131–138, 1997.

194. S. Parker, M. Parker, Y. Livnat, P. Sloan, C. Hansen, and P. Shirley. Interactive ray tracing for volume visualization. *IEEE Transactions on Visualization and Computer Graphics*, 5(3):238–250, 1999.

195. S. Parker, P. Shirley, Y. Livnat, C. Hansen, and P. Sloan. Interactive ray tracing for isosurface rendering. *Proc. of IEEE Visualization '98*, pages 233–238, 1998.

196. V. Pascucci and K. Cole-McLaughlin. Efficient computation of the topology of level sets. *Proc. of IEEE Visualization '02*, pages 187–194, 2002.

197. S. Peercy. Linear color representations for full spectral rendering. *Computer Graphics*, 27(3):191–198, 1993.

198. V. Pekar, R. Wiemker, and D. Hempel. Fast detection of meaningful isosurfaces for volume data visualization. *Proc. of IEEE Visualization '01*, pages 223–230, 2001.

199. H. Pfister, B. Lorensen, C. Bajaj, G. Kindlmann, W. Schroeder, L. Avila, K. Martin, R. Machiraju, and J. Lee. The transfer function bake-off. *Proc. of IEEE Computer Graphics & Applications*, 21(3):16–22, 2001.

200. H. Pfister, J. Hardenbergh, J. Knittel, H. Lauer, and L. Seiler. The VolumePro real-time raycasting system. *Proc. of SIGGRAPH '99*, pages 251–260, 1999.

201. H. Pfister, M. Zwicker, J. Baar, and M. Gross. Surfels: surface elements as rendering primitives. *Proc. of SIGGRAPH '00*, pages 335–342, 2000.

202. H. Pfister, A. Kaufman, and F. Wessels. Towards a scalable architecture for real-time volume rendering. *Proc. of 10th Eurographics Workshop on Graphics Hardware '95*, pages 123–130, 1995.

203. H. Pfister, A. Kaufman, and T. Chiueh. Cube-3: a real-time architecture for high-resolution volume visualization. *Symposium of Volume Visualization '94*, pages 75–82, 1994.

204. H. Pfister and A. Kaufman. Cube-4: a scalable architecture for real-time volume rendering. *Proc. of Volume Visualization Symposium '96*, pages 47–54, 1996.

205. H. Pfister, F. Wessels, and A. Kaufman, Sheared interpolation and gradient estimation for real-time volume rendering. *Computer Graphics*, 19(5):667–677, 1995.

206. H. Pfister, F. Wessels, and A. Kaufman. Sheared interpolation and gradient estimation for real-time volume rendering. *Proc. of 9th Eurographics Workshop on Graphics Hardware '94*, 1994.

207. T. Porter and T. Duff. Compositing digital images. *Computer Graphics (Proc. Siggraph '84)*, pages 253–259, 1984.

208. R. Reynolds, D. Gordon, and L. Chen. A dynamic screen technique for shaded graphics display of slice-represented objects. *Computer Graphics and Image Processing*, 38:275–298, 1987.

209. C. Rezk-Salama, K. Engel, M. Bauer, G. Greiner, and T. Ertl. Interactive volume rendering on standard PC graphics hardware using multi-textures and multi-stage rasterization. *Proc. of SIGGRAPH/Eurographics Workshop on Graphics Hardware '00*, pages 109–118, 2000.

210. S. Roettger and T. Ertl. A two-step approach for interactive pre-integrated volume rendering of unstructured grids. *Proc. of VolVis '02*, pages 23–28, 2002.

211. S. Roettger, M. Kraus, and T. Ertl. Hardware-accelerated volume and isosurface rendering based on cell-projection. *Proc. of IEEE Visualization '00*, pages 109–116, 2000.

212. J. Rossignac. Considerations on the interactive rendering of 4D volumes. *Chapel Hill Workshop on Volume Visualization*, pages 67–76, 1989.

213. H. Rushmeier and E. Torrance. The zonal method for calculating light intensities in the presence of a participating medium. *Computer Graphics*, 21(4):293–302, 1987.

214. S. Rusinkiewicz and M. Levoy. QSplat: a multiresolution point rendering system for large meshes. *Proc. of SIGGRAPH '00*, 2000.

215. P. Sabella. A rendering algorithm for visualizing 3D scalar fields. *ACM SIGGRAPH Computer Graphics*, 22(4):51–58, 1988.

216. M. Salisbury, C. Anderson, D. Lischinski, and D. Salesin. Scale-dependent reproduction of pen-and-ink illustrations. *Proc. of SIGGRAPH '96*, pages 461–468, 1996.

217. M. Salisbury, M. Wong, J. Hughes, and D. Salesin. Orientable textures for image-based pen-and-ink illustration. *Proc. of SIGGRAPH '97*, pages 401–406, 1997.

218. H. Samet. *Application of Spatial Data Structures*. Reading, PA, Addison-Wesley, 1990.

219. K. Sano, H. Kitajima, H. Kobayashi, and T. Nakamura. Parallel processing of the shear-warp factorization with the binary-swap method on a distributed-memory multiprocessor system, *Proc. of Parallel Rendering Symposium '97*, pages 87–95, 1997.

220. W. Schroeder, J. Zarge, and W. Lorensen. Decimation of triangle meshes. *ACM SIGGRAPH Computer Graphics*, 26(2):65–70, 1992.

221. J. Shade, S. Gortler, Li-Wei He, and R. Szeliski. Layered depth images. *Proc. of SIGGRAPH '98*, pages 231–242, 1998.

222. N. Shareef, D. Wang, and R. Yagel. Segmentation of medical images using LEGION. *IEEE Transactions on Medical Imaging*, 18(1):74–91, 1999.

223. R. Shekhar, E. Fayyad, R. Yagel, and J. Cornhill. Octree-based decimation of marching cubes surfaces. *Proc of IEEE Visual Conf.*, pages 335–342, 1996.

224. H. Shen, C. Hansen, Y. Livnat, and C. Johnson. Isosurfacing in span space with utmost efficiency (ISSUE). *Proc. of IEEE Visualization '96*, pages 287–294, 1996.

225. H. Shen, L. Chiang, and K. Ma. A fast volume rendering algorithm for time-varying fields using a time-space partitioning tree. *Proc. of IEEE Visualization '99*, pages 371–377, 1999.

226. H. Shen and C. Johnson. Differential volume rendering: a fast volume visualization technique for flow animation. *Proc. Visualization '94*, pages 180–187, 1994.

227. Y. Shinagawa and T. Kunii. Constructing a reeb graph automatically from cross sections. *IEEE Computer Graphics and Applications*, 11(6):45–51, 1991.

228. P. Shirley and A. Tuchman. A polygonal approximation to direct scalar volume rendering. *Computer Graphics*, 24(5):63–70, 1990.

229. J. Sijbers, P. Scheunders, M. Verhoye, A. Linden, D. Dyck, and E. Raman. Watershed-based segmentation of 3D MR data for volume quantization. *Magnetic Resonance Imaging*, 15:679–688, 1997.

230. C. Silva, A. Kaufman, and C. Pavlakos. PVR: high-performance volume rendering. *IEEE Computational Science and Engineering*, pages 18–28, 1996.

231. C. Silva and J. Mitchell. The lazy sweep raycasting algorithm for rendering irregular grids. *IEEE Transactions on Visualization and Computer Graphics*, 3(2):142–157, 1997.

232. C. Silva, J. Mitchell, and P. Williams. An exact interactive time visibility ordering algorithm for polyhedral cell complexes. *Volume Visualization Symposium '98*, pages 87–94, 1998.

233. D. Silver and X. Wang. Tracking scalar features in unstructured datasets. *Proc. of IEEE Visualization '98*, pages 79–86, 1998.

234. D. Silver and X. Wang. Tracking and visualizing turbulent 3D features. *IEEE Transactions on Visualization and Computer Graphics*, 3(2), 1997.

235. L. Sobierajski, D. Cohen, A. Kaufman, R. Yagel, and D. Acker. A fast display method for volumetric data. *Visual Computer*, 10(2):116–124, 1993.

236. L. Sobierajski and A. Kaufman. Volumetric raytracing. *Symposium on Volume Visualization '94*, pages 11–18, 1994.

237. L. Sobierajski and R. Avila. A hardware acceleration method for volumetric ray tracing. *Proc. of IEEE Visualization '95*, pages 27–35, 1995.

238. L. Sobierajski, D. Cohen, A. Kaufman, R. Yagel, and D. Acker. A fast display method for volumetric data. *Visual Comput.*, 10(2):116–124, 1993.

239. B. Sohn, C. Bajaj, and V. Siddavanahalli. Feature based volumetric video compression for interactive playback. *VolVis '02*, pages 89–96, 2002.

240. D. Spearey and S. Kennon. Volume probes: interactive data exploration on arbitrary grids. *Computer Graphics*, 25(5):5–12, 1990.

241. M. Sramek and A. Kaufman. Fast ray-tracing of rectilinear volume data using distance transforms. *IEEE Transactions on Visualization and Computer Graphics*, 3(6):236–252, 2000.

242. B. Stander and J. Hart. A Lipschitz method for accelerated volume rendering. *Proc. of the Symposium on Volume Visualization '94*, pages 107–114, 1994.

243. C. Stein, B. Becker, and N. Max. Sorting and hardware assisted rendering for volume visualization. *Symposium on Volume Visualization '94*, pages 83–90, 1994.

244. A. Stompel, E. Lum, and K. Ma. Feature-enhanced visualization of multidimensional, multivariate volume data using non-photorealistic rendering techniques. *Proc. of Pacific Graphics '02*, pages 394–402, 2002.

245. M. Stytz and O. Frieder. Computer systems for 3D diagnostic imaging: an examination of the state of the art. *Critical Reviews in Biomedical Engineering*, pages 1–46, 1991.

246. M. Stytz, G. Frieder, and O. Frieder. 3D medical imaging: algorithms and computer systems. *ACM Computing Surveys*, pages 421–499, 1991.

247. P. Sutton and C. Hansen. Isosurface extraction in time-varying fields using a temporal branch-on-need tree (T-BON). *Proc. of IEEE Visualization '99*, pages 147–153, 1999.

248. E. Swan, K. Mueller, T. Moller, N. Shareef, R. Crawfis, and R. Yagel. An anti-aliasing technique for splatting. *Proc. of IEEE Visualization '97*, pages 197–204, 1997.

249. J. Sweeney and K. Mueller. Shear-warp deluxe: the shear-warp algorithm revisited. *Joint Eurographics–IEEE TCVG Symposium on Visualization '02*, pages 95–104, 2002.

250. S. Takahashi, T. Ikeda, Y. Shinagawa, T. L. Kunii, and M. Ueda. Algorithms for extracting correct critical points and constructing topological graphs from discrete geographical elevation data. *Computer Graphics Forum*, 14(3):181–192, 1995.

251. S. Tenginakai, J. Lee, and R. Machiraju. Salient isosurface detection with model-independent statistical signatures. *Proc. of IEEE Visualization '01*, pages 231–238, 2001.

252. T. Theußl, H. Hauser, and M. Gröller. Mastering windows: improving reconstruction. *Proc. of IEEE Symposium on Volume Visualization '00*, pages 101–108, 2000.

253. T. Theußl, T. Möller, and E. Gröller. Optimal regular volume sampling. *Proc. of IEEE Visualization '01*, 2001.

254. P. Thévenaz, T. Blu, and M. Unser. Interpolation revisited. *IEEE Transactions on Medical Imaging*, 19(7):739–758, 2000.

255. U. Tiede, T. Schiemann, and K. Hoehne. High quality rendering of attributed volume data. *Proc. of IEEE Visualization '98*, pages 255–262, 1998.

256. T. Totsuka and M. Levoy. Frequency domain volume rendering. *Proc. of SIGGRAPH '93*, pages 271–278, 1993.

257. S. Treavett and M. Chen. Pen-and-ink rendering in volume visualization. *Proc. of IEEE Visualization '00*, pages 203–210, 2000.

258. H. Tuy and L. Tuy. Direct 2D display of 3D objects. *IEEE Computer Graphics & Applications*, 4(10):29–33, 1984.

259. J. Udupa and D. Odhner. Shell rendering. *IEEE Computer Graphics and Applications*, 13(6):58–67, 1993.

260. S. Uselton. Volume rendering for computational fluid dynamics: initial results. *Tech Report RNR-91-026*, NASA Ames Research Center, 1991.

261. G. Wallace. The JPEG still picture compression standard. *Communications of the ACM*, 24(4):30–44, 1991.

262. M. Wan, Q. Tang, A. Kaufman, Z. Liang, and M. Wax. Volume rendering based interactive navigation within the human colon. *Proc. of IEEE Visualization '99*, pages 397–400, 1999.

263. C. Weigle and D. Banks. Extracting iso-valued features in 4-D datasets. *Proc. of IEEE Visualization '98*, pages 103–110, 1998.

264. D. Weiskopf, K. Engel, and T. Ertl. Volume clipping via per-fragment operations in texture-based volume visualization. *Proc. of IEEE Visualization '02*, pages 93–100, 2002.

265. R. Westermann. A multiresolution framework for volume rendering. *Proc. of Symp. on Volume Visualization '94*, pages 51–58, 1994.

266. R. Westermann. Compression domain rendering of time-resolved volume data. *Proc. of IEEE Visualization '95*, pages 168–174, 1995.

267. R. Westermann and T. Ertl. Efficiently using graphics hardware in volume rendering applications. *Proc. of SIGGRAPH '99*, pages 169–177, 1999.

268. L. Westover. Footprint evaluation for volume rendering. *Proc. of SIGGRAPH '90*, pages 367–376, 1990.

269. L. Westover. Interactive volume rendering. *Chapel Hill Volume Visualization Workshop*, pages 9–16, 1989.

270. L. Westover. SPLATTING: A parallel, feed-forward volume rendering algorithm. *Ph.D. Dissert.* UNC-Chapel Hill, 1991.

271. S. Whitman. A task adaptive parallel graphics renderer. *Proc. of Parallel Rendering Symposium '93*, pages 27–34, 1993.

272. J. Wijk. Spot noise-texture synthesis for data visualization. *Proc. of SIGGRAPH '91*, 25(4):309–318, 1991.

273. J. Wilhelms and A. Van Gelder. A coherent projection approach for direct volume rendering. *Proc. of SIGGRAPH '91*, 25(4):275–284, 1991.

274. J. Wilhelms and A. Van Gelder. Octrees for faster isosurface generation. *ACM Transactions on Graphics*, 11(3):201–227, 1992.

275. P. Williams. Interactive splatting of nonrectilinear volumes. *Proc. of IEEE Visualization '92*, pages 37–44, 1992.

276. P. Williams. Visibility ordering meshed polyhedra. *ACM Transaction on Graphics*, 11(2):103–125, 1992.

277. G. Winkenbach and D. Salesin. Computer-generated pen-and-ink illustration. *Proc. of SIGGRAPH '94*, pages 91–100, 1994.

278. G. Winkenbach and D. Salesin. Rendering parametric surfaces in pen and ink. *Proc. of SIGGRAPH '96*, pages 469–476, 1996.

279. T. Wischgoll and G. Scheuermann. Detection and visualization of closed streamlines in planar flows. *IEEE Transactions on Visualization and Computer Graphics*, 7(2):165–172, 2001.

280. C. Wittenbrink, T. Malzbender, and M. Goss. Opacity-weighted color interpolation for volume sampling. *Symposium on Volume Visualization '98*, pages 135–142, 1998.

281. G. Wolberg. *Digital Image Warping*. IEEE Computer Society Press, Los Alamitos, CA, 1990.

282. Y. Wu, V. Bhatia, H. Lauer, and L. Seiler. Shear-image ray-casting volume rendering. *ACM SIGGRAPH Symposium on Interactive 3D Graphics '03*, pages 152–162, 2003.

283. F. Xu and K. Mueller. A unified framework for rapid 3D computed tomography on commodity GPUs. *Proc. of IEEE Medical Imaging Conference '03*, 2003.

284. R. Yagel and A. Kaufman. Template-based volume viewing. *Computer Graphics Forum*, 11(3):153–167, 1992.

285. R. Yagel, D. Reed, A. Law, P.-W. Shih, and N. Shareef. Hardware assisted volume rendering of unstructured grids by incremental slicing. *Volume Visualization Symposium '96*, pages 55–62, 1996.

286. R. Yagel and Z. Shi. Accelerating volume animation by space-leaping. *Proc. of IEEE Visualization '93*, pages 62–69, 1993.

287. R. Yagel and A. Kaufman. The flipping cube architecture. *Tech. Rep.* 91.07.26, Computer Science, SUNY at Stony Brook, 1991.

288. B. Yeo and B. Liu. Volume rendering of DCT-based compressed 3D scalar data. *IEEE Trans. Visualization Comput. Graphics*, 1(1):29–43, 1995.

289. H. Zhang, D. Manocha, T. Hudson, and K. Hoff. Visibility culling using hierarchical occlusion maps. *Proc. of SIGGRAPH '97*, pages 77–88, 1997.

290. C. Zhang and R. Crawfis. Volumetric shadows using splatting. *Proc. of IEEE Visualization '02*, pages 85–92, 2002.

291. K. Zuiderveld, A. Koning, and M. Viergever. Acceleration of ray-casting using 3D distance transforms. *Visualization in Biomedical Computing '92*, pages 324–335, 1992.

292. M. Zwicker, H. Pfister, J. Baar, and M. Gross. Surface splatting. *Proc. of SIGGRAPH '01*, pages 371–378, 2001.

293. M. Zwicker, H. Pfister, J. Baar, and M. Gross. EWA volume splatting. *Proc. of IEEE Visualization '01*, 2001.

8   Volume Rendering Using Splatting

ROGER CRAWFIS, DAQING XUE, and CAIXIA ZHANG
The Ohio State University

8.1   The Theory

8.1.1   Reconstruction and Integration

In order to understand volume rendering and splatting, we must first have an appreciation of the continuous reconstruction of functions from discrete or sampled data. Typically, we are given the discrete data without any additional metadata that can be used to help reconstruct the original underlying function from which it was derived. Thus, many assumptions are made about the data, and different reconstruction kernels give rise to different continuous functions. Sampling theory will tell you that if you assume your function was periodic and rather smooth, in fact, so smooth that the fixed sampling of the function captured any rapid oscillations in the function, then using the well-known *sinc* function as your reconstruction operator will reproduce the original function. This, of course, is rarely true for most 3D objects, where there is an abrupt change from one material to another, such as from muscle to bone, or from air to tree bark. Even *fuzzy* phenomena rarely exhibit this *band-limited* behavior. Clouds have a fractal-like characteristic to them, and turbulent flow is chaotic. However, the data on the computer comes to us in two forms. The first possibility is that it was calculated using a simulation package, in which case a careful computational scientist has gone to great efforts to ensure a reasonable sampling. Most simulation packages attempt to control the maximum time step that can be taken and still produce reasonably accurate results. Extremely high gradients are still possible across a cell or between samples.

The second possibility is that the discrete data was acquired through some analog-to-digital device, where the continuous analog signal is usually low-pass filtered before being sampled, removing any high frequencies. A primary question to ask in visualizing such data is how we should represent the abstract 3D field. First of all, we need to ask what signal we are reconstructing. Should we reconstruct the output of the A-D converter, or make additional assumptions about the samples and try to reconstruct the original function?

The term *splat* means to spread flat or flatten out. Lee Westover [13] first used this term for the deposition of a single 3D reconstruction kernel being projected and integrated onto the image screen. In his paper, he refers to these as footprints, implying the extent to which a single voxel or reconstruction kernel covers the image plane. Fig. 8.1 shows the same situation in two dimensions. Here, a radially symmetric image-reconstruction kernel is being integrated along the direction perpendicular to the line AB. This converts the 2D function to a 1D function along the line. In this case, a *compact* reconstruction kernel is used, and the function quickly goes to zero (represented as black in the figure). This also allows the footprint function to go to zero rapidly. Finally, by using a radially symmetric kernel, the same 1D function is obtained regardless of the direction of integration.

Mathematically, we first center the reconstruction kernel over every discrete value in our 3D volume. This produces the continuous function $f(x,y,z)$:

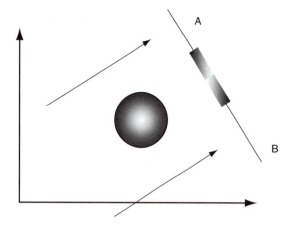

Figure 8.1 Flattening out of a reconstruction kernel to produce a 1D footprint or splat.

$$f(x, y, z) = \sum_i \sum_j \sum_k f_{i,j,k} \, h(x - i, y - j, z - k)$$

$$(8.1)$$

The particular choice of $h(x,y,z)$ is a well-studied subject and includes issues associated with practically implemented kernels as well as theoretical properties under ideal conditions. We will not address these issues here explicitly, other than to mention some design goals leading to the most effective kernels for splatting. The integral of our continuous function along any ray t is given by

$$\int_0^L f(x(t), y(t), z(t)) dt$$

$$= \int_0^L \sum_i \sum_j \sum_k f_{i,j,k} h(x(t) - i, \; y(t) - j, z(t) - k) dt$$

$$= \sum_i \sum_j \sum_k f_{i,j,k} \int_0^L h(x(t) - i, \; y(t) - j, z(t) - k) dt$$

$$(8.2)$$

This allows for a separation of the underlying discrete data values from the reconstruction kernel. A change of variables and a projection of our reconstruction kernel along t allow the computation of a single footprint for the reconstruction kernel centered at the origin.

Let

$$(u, v) = proj(x, y, z)$$

then $\qquad\qquad\qquad\qquad\qquad\qquad$ (8.3)

$$footprint_t(u, v) = \int_0^L h(x(t) - i, y(t) - j, z(t) - k) dt$$

This footprint can be either view-dependent, as denoted by the subscript t, or, for the case of radially symmetric kernels, view-independent. For now, we will consider orthographic projections. The continuous *footprint* function encodes all possible rays through a radially symmetric function, but it needs to be sampled differently for perspective projections. Thus, our final form for the integral of $f(x,y,z)$ over the image plane can be represented by the continuous function

$$\int_0^L f(x(t), y(t), z(t)) dt = \sum_i \sum_j \sum_k f_{i, j, k} footprint_t$$

$$(proj(x(t) - i, y(t) - j, z(t) - k))$$

$$(8.4)$$

This formula states that the integral is the summation of many translated copies of the footprint function, weighted by the discrete data. This integral formula was the basis for Westover's original volume renderings. We will now examine the illumination or volume rendering integral.

8.1.2   The Volume Rendering Integral

8.1.2.1   X-Rays and the Absorption-Only Model

For x-ray modeling, the function, $f(x,y,z)$, can be mapped to an extinction coefficient, τ, that controls the rate at which light is occluded. Max [3] derives the mathematical formula for this simple model. The differential change in the intensity along the ray can be written as

$$\frac{dI}{ds} = -\tau(s)I(s) \tag{8.5}$$

where s is a length parameter along a ray in the direction of the light flow.

This equation states that the change in the intensity (dI/ds) decreases (hence the negative multiplier) proportionally to the incoming intensity, as determined by the extinction coefficient. The analytical solution for this formula is simply

$$I(s) = I_0 e^{-\int_0^s \tau(t)dt} \tag{8.6}$$

This indicates the attenuation of the high energy source or backlight as it propagates from the background, $s = 0$, towards the eye. If τ is zero along the ray, then no attenuation occurs and the intensity at the pixel is I_0. If τ is a constant along the ray, then the attenuation is given by

$$I(s) = I_0 e^{-\int_0^s \tau dt} = I_0 e^{-\tau s} \tag{8.7}$$

Using a Taylor series expansion for the exponential and simplifying for the case where τs is small leads to the familiar compositing operator, over, from Porter and Duff [10]:

$$I_0 e^{-\tau s} = I_0\left(1 - \tau s + \frac{(\tau s)^2}{2!} - \frac{(\tau s)^3}{3!} + \cdots\right) \tag{8.8}$$

$$\approx I_0(1 - \tau s)$$

Here, τs represents the opacity, α, expressed as a function of the ray length. As Max [3] and Wilhelms and Van Gelder [14] point out, this relationship of increased opacity for longer ray-integration segments is crucial when considering

different sampling resolutions of the volume. This equation also points out that for volume rendering using relatively low-opacity values, the simple over operator is probably sufficient. Where volume rendering tries to replace surface-based rendering, the approximation is not very valid. However, in these cases, an x-ray model is also not desired.

So, how does this relate to splatting? Most other volume rendering techniques use the approximation above and a rather poor Riemann sum to approximate the x-ray integral. Splatting offers an analytical solution to this problem, provided that an analytical integration of the reconstruction kernel is obtainable. Combining Equations 8.4 and 8.6 for the ray from the background, $t=0$, to the eye at ray length L, produces the following formula:

$$I = I_0 e^{-\sum_i \sum_j \sum_k f_{i,j,k} footprint_t(proj(x(t)-i, y(t)-j, z(t)-k))} \tag{8.9}$$

To integrate the volume along an arbitrary viewing direction, w, using an x-ray model, we first transform the volume into eye space:

$$V(u, v, w) = \sum_{(s,t,r)\in Vol} \tilde{f}(s,t,r)h(s-u, t-v, r-w) \tag{8.10}$$

Here, $\tilde{f}(s,t,r) = \tilde{f}((i,j,k)M^T) = f(i,j,k)$, and M is the transformation matrix to the eye space. We calculate the integral, $D(u,v)$, of the volume, $V(u,v,w)$, along w as follows:

$$D(u, v) = \int \sum_{(s,t,r)\in Vol} \tilde{f}(s,t,r)h(s-u, t-v, r-w)dw$$

$$= \sum_{(s,t,r)\in Vol} \tilde{f}(s,t,r)\int h(s-u, t-v, r-w)dw \tag{8.11}$$

It can be rewritten as follows [15]:

$$D(u, v) = \sum_{(s,t)\in vol} \tilde{p}_w(s,t)\, footprint_w^h(s-u, t-v)$$

$$= \tilde{p}_w(u, v)^* \, footprint_w^h(u, v) \tag{8.12}$$

Here, $*$ is the convoluting operation, and $\tilde{p}_w(u, v)$ is the projection function of $\tilde{f}(s, t, r)$ along the w direction, as follows:

$$\tilde{p}_w(u, v) = \sum_{(s, t, r) \in vol} f(s, t, r)\delta(s - u, t - v) \quad (8.13)$$

Here, δ is a comb function. This implies that for x-ray models, we can pass a 2D convolution kernel (footprint) across the projected set of impulses.

8.1.2.2 Sabella Model: Absorption with Single Scattering

The absorption-plus-emission model is used by Sabella [11] and elaborated by Max [3]. The volume rendering integral for this model is

$$I(s) = I_0 e^{-\int_0^s \tau(r)dr} + \int_0^s g(t)e^{-\int_t^s \tau(r)dr} dt \quad (8.14)$$

Here, the background light is attenuated, as in the x-ray model, but new energy is scattered towards the eye along the ray according to the *glow function, g(t)*. This newly added energy is then attenuated based on the length of material that still exists between it and the eye. Integrating the function separately, as in the previous section, does not work for this model since the limits of integration are not fixed on the inner integral. See the paper by Mueller et al. [8] for more details on the inaccuracies involved in applying this scheme. The solution to this can be viewed by partitioning the integral.

$$I(s) = I_0 e^{-\int_0^s \tau(r)dr} + \int_0^{x_1} g(t)e^{-\int_t^s \tau(r)dr} dt$$
$$+ \ldots + \int_{x_{k-1}}^s g(t)e^{-\int_t^s \tau(r)dr} dt \quad (8.15)$$

Here, the limits of integration in the exponent do not change, since each slice's scattered energy still needs to be occluded by the entire portion of the volume remaining on the ray. By partitioning the inner integrals as well, we can factor out all but the most current partition.

$$I(x_n) = I_0 e^{-\int_0^{x_n} \tau(r)dr} + \int_0^{x_1} g(t)e^{-\left(\int_t^{x_1} \tau(r)dr + \int_{x_1}^{x_n} \tau(r)dr\right)} dt$$
$$+ \ldots + \int_{x_{n-1}}^{x_n} g(t)e^{-\int_t^{x_n} \tau(r)dr} dt$$
$$= I_0 e^{-\int_0^{x_n} \tau(r)dr} + e^{-\int_{x_1}^{x_n} \tau(r)dr} \cdot \int_0^{x_1} g(t)e^{-\int_t^{x_1} \tau(r)dr} dt$$
$$+ \ldots + \int_{x_{n-1}}^{x_n} g(t)e^{-\int_t^{x_n} \tau(r)dr} dt$$
$$= e^{-\int_{x_1}^{x_n} \tau(r)dr} \left(I_0 e^{-\int_0^{x_1} \tau(r)dr} + \int_0^{x_1} g(t)e^{-\int_t^{x_1} \tau(r)dr} dt \right)$$
$$+ \ldots + \int_{x_{n-1}}^{x_n} g(t)e^{-\int_t^{x_n} \tau(r)dr} dt \quad (8.16)$$

This leads to the recursion formula:

$$T_0 = 1$$
$$T_i = e^{-\int_{x_{i-1}}^{x_i} \tau(t)dt} T_{i-1} \quad (8.17)$$
$$I_i = T_i I_{i-1} + \int_{x_{i-1}}^{x_i} g(t)e^{-\int_t^{x_i} \tau(s)ds} dt$$

In Equation 8.4, the limits of integration are really assumed to be the full extent of the reconstruction kernel. Here, we may need to integrate over only a small portion of the reconstruction kernel. This requires many different footprint functions to be calculated, since the voxel location may be arbitrarily oriented in regards to the partitioning. Mueller et al. [8] provide details on this implementation, termed *image-aligned sheet-based splatting* or *IASB splatting*.

8.2 Image-Aligned Sheet-Based Splatting

In splatting, each voxel is represented by a 3D kernel weighted by the voxel value. The 3D kernels are integrated into a generic 2D footprint along the traversing ray from the eye. This footprint can be efficiently mapped onto the

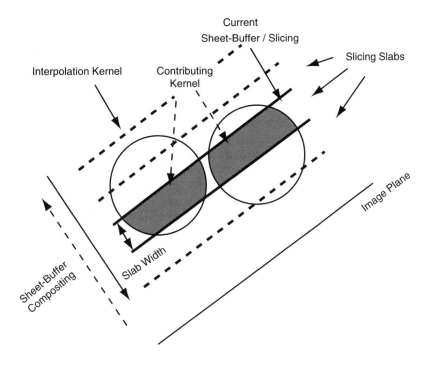

Figure 8.2   Image-aligned sheet-based splatting.

image plane; the final image is obtained by the collection of all projected footprints weighted by the voxel values. This splatting approach is fast, but it suffers from color bleeding and popping artifacts due to incorrect volume integration.

Mueller et al. [8] eliminate these problems by aligning the sheets to be parallel to the image plane (Fig. 8.2). All the voxel kernels that overlap a slab are clipped to the slab and summed into a sheet buffer. The sheet buffers are composited front-to-back to form the final image. While this significantly improves image quality, it requires much more compositing and several footprint sections per voxel to be scan-converted. Using a front-to-back traversal, this method can make use of the culling of occluded voxels by keeping an occlusion map and checking whether the pixels that a voxel projects have reached full opacity [2].

Traditionally, splatting classifies and shades the voxels prior to projection. Projecting the fuzzy color balls leads to a blurry appearance of object edges. Splatting using post-classification, which performs the color and opacity classification and shading process after the voxels have been projected onto the screen, was proposed by Mueller et al. [6] to generate images with crisp edges and well-preserved surface details. Now, after the projection of the voxels, each slice contains a small integral of the reconstructed function. After normalization, the integral represents the mean value of the function between two adjacent slices. For post-classification, we need a per-pixel transfer function to classify the function value at each pixel to the color and opacity values. The transfer function is designed by users to display the regions that they are interested in and display the regions in the way they like. Usually, we create a transfer function table that stores, for example, 256 entries of color and opacity values. Then, the normalized integral value at each pixel is used to look up the transfer function table to get the corresponding color and opacity for the pixel.

For the purpose of the shading calculation, we calculate per-pixel gradient using central differences in the projected image space. Three components of a pixel's gradient can be calculated using its six neighbor pixels in the three directions. Assume the viewing direction is along the z axis, and the image is in the x-y plane. Once we have three sheets of the function reconstructed, the z-component can be calculated using the following formula for orthographic projections:

$$f'_z = (f(p_{i,j}^{z+\Delta b}) - f(p_{i,j}^{z-\Delta b}))/(2 * \Delta b) \qquad (8.18)$$

where Δb is the distance between two adjacent slices, also called slab width.

The x and y components are calculated using the neighbor pixels in the current sheet. A simple way is to just use its nearest pixels to calculate the x and y components. But one problem encountered with this was the numerical errors for extremely close-up views. If the voxels project to a large screen space, then the difference between adjacent pixels becomes zero. To avoid this problem, we use a central difference operation with a step size equal to the number of pixels corresponding to the image-space voxel spacing. We can express the calculation of the x and y components of the gradient, using the following formula:

$$f'_x = (f(p_{i+\Delta p,j}^z) - f(p_{i-\Delta p,j}^z))/(2 * \Delta v)$$
$$f'_y = (f(p_{i,j+\Delta p}^z) - f(P_{i,j-\Delta p}^z))/(2 * \Delta v) \qquad (8.19)$$

where Δp is the number of pixels corresponding to the image-space voxel spacing, and Δv is the distance between two adjacent voxels in world space.

The normal at each pixel is obtained by normalizing the gradient (f'_x, f'_y, f'_z) and is used in the per-pixel shading calculation. Figs. 8.3 and 8.4 are two images for uncBrain and blood vessel datasets, respectively. The images are generated using IASB splatting with post-classification and per-pixel shading.

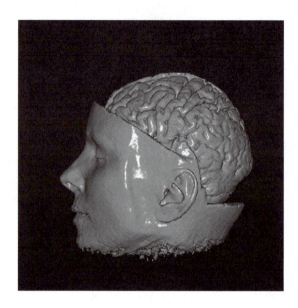

Figure 8.3 uncBrain. (See also color insert.)

8.3 Splatting with Shadows

Shadows are essential to realistic and informative images. We use sheet-based splatting with post-classification to keep track of the per-pixel contribution to the light attenuation and to generate per-pixel shadows.

8.3.1 Basic Shadow Algorithm Using Splatting

Visibility algorithms and shadow algorithms are essentially the same. The former determine the visibility from the eye, and the latter determine the visibility from the light source. However, it is hard to implement shadows, especially accurate shadows, in volume rendering, since the light intensity is continuously attenuated as the light traverses the volume. The fundamental problem, therefore, is not determining whether a point is visible from the light, but rather to determine the light intensity arriving at the point being illuminated.

In the shadow algorithm using splatting [16], we implement shadows by traversing the

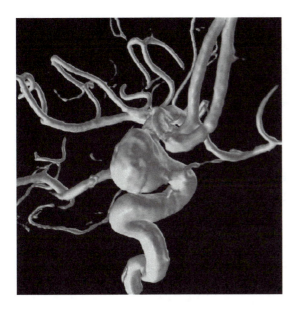

Figure 8.4 Blood vessel.

volume only once to generate per-pixel accurate shadows. The same splatting algorithm is used for both the viewer and the light source. For each footprint, while adding its contribution to the sheet buffer, as seen from the eye, we also add its contribution to a shadow buffer, as seen from the light source. In the sheet-based splatting, the light passing through the front sheets will be attenuated and cause shadows on the back sheets along the light rays. At the current sheet, the light intensity is attenuated by all front sheets. If the light source resides behind the object with respect to the viewer, then a back-to-front compositing order of the sheets is taken.

One limitation of IASB splatting to implement shadows is in dealing with light sources perpendicular to the eye vector. The image-aligned splatting makes it difficult to keep track of accurate opacities as seen from the perpendicular light source. To generate shadows using splatting, we propose a new non-image-aligned sheet-based splatting to keep track of accurate light attenuation [16]. We first calculate the halfway vector between the eye vector

and the light vector. Rather than slicing the reconstruction kernels via planes parallel to the image plane, we chop the volume by slices perpendicular to the direction of the halfway vector. We keep the image buffer aligned with the eye and the shadow buffer aligned with the light source (Fig. 8.5) to avoid sampling and resolution problems. This non-IASB splatting along the halfway vector will not have the popping artifacts mentioned for the volume-aligned sheet-based splatting in Mueller and Crawfis [8], since the splatting direction changes continuously with the eye vector and/or the light vector. Therefore, a consistent ray integration is generated with accurately reconstructed sheets.

For high-quality rendering, we use per-pixel post-classification and illumination. This implies the need to also support per-pixel shadowing. During the rendering, when we calculate the illumination for a pixel at the current sheet, we look up the accumulated opacity for the pixel from the shadow buffer by mapping the pixel to the shadow buffer (Fig. 8.5). The pixel (i,j) at the current image buffer is first transferred back to the point x in the eye space using the current sheet's z-value. It is then projected to the pixel (i',j') at the shadow buffer, aligned with the light source.

The light intensity arriving at the point x is calculated using the accumulated opacity stored at the corresponding pixel (i',j') on the shadow buffer:

$$I(x) = (1.0 - \alpha(x)) * I_{light} \qquad (8.20)$$

where $\alpha(x)$ is the accumulated opacity at x, which is the value at (i',j') on the shadow buffer, and I_{light} is the original intensity of the light source.

This shadow buffer has accumulated the energy loss from all the sheets in front of the current sheet. In this way, the light attenuation is accurately modeled. For a given point x, we get its $\alpha(x)$ by choosing its nearest pixel's opacity value in the shadow buffer, or using bi-linear interpolation of the opacity values of nearby pixels in the shadow buffer.

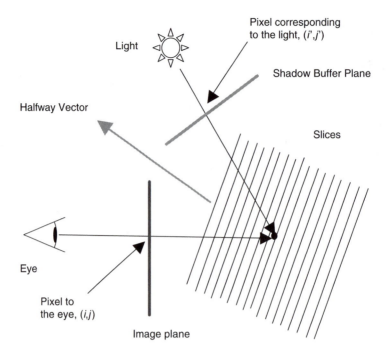

Figure 8.5   Non-image-aligned sheet-based splatting.

Since the shadow buffer is generated in lock-step with the image for each view, we can easily guarantee correct sampling of the shadow buffer.

Compared to splatting without shadows, two more buffers are needed in the shadow algorithm: a 2D shadow buffer to store the accumulated opacity from the light to the current sheet, and a working 2D sheet shadow buffer to which the current slab of voxel footprints is added. Then, a per-pixel classification is applied to the sheet shadow buffer, which is then composited into the accumulated shadow buffer.

The sheet-based splatting algorithm with shadows is demonstrated with the pseudo-code in Fig. 8.6.

8.3.2   Shadow Results

Using the above algorithm, we have implemented shadows for two different types of

1. Transform each voxel to the coordinate system having the halfway vector as the z-axis;
2. Bucket-sort voxels according to the transformed z-values;
3. Initialize opacity map to zero;
4. Initialize shadow buffer to zero;
5. For each sheet in front-to-back order
    6. Initialize image sheet buffer;
    7. Initialize shadow sheet buffer,
    8. For each footprint
        9. Rasterize and add the footprint to the current image sheet buffer;
        10. Rasterize and add the footprint to the current shadow sheet buffer;
    11. End for;
    12. Calculate the gradient for each pixel using central difference;
    13. Classify each pixel in the current image sheet buffer;
    14. Map pixel to the shadow buffer and get its opacity;
    15. Calculate the illumination to obtain the final color;
    16. Composite the current image sheet buffer to the frame buffer;
    17. Classify each pixel on current shadow sheet buffer and composite it to the accumulated shadow buffer;
18. End for;

Figure 8.6   Pseudo-code of the sheet-based splatting algorithm with shadows.

light sources: parallel light sources and point light sources.

The shadow of the rings composed of torus primitives is shown in Fig. 8.7. Notice how the per-pixel classification algorithm pro-duces sharp shadows. In Fig. 8.8, we have a scene of a smoky room with a floating cube inside.

Fig. 8.9 is the HIPIP (high-potential iron-sulfur protein) dataset, which describes

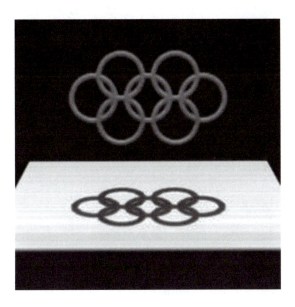

Figure 8.7   Rings with shadows. (See also color insert.)

Figure 8.8   A smoky room with a cube inside. (See also color insert.)

Figure 8.9   A scene of the HIPIP dataset. (Left) Without shadow; (Right) with shadow. (See also color insert.)

a one-electron orbital of a four-iron and eight-sulfur cluster found in many natural proteins. The data is the scalar value of the wave function 'psi' at each point. Shadows provide spatial-relationship information.

The splatting algorithm has been extended to support hypertextures. Fig. 8.10 shows the shadow of a hypertextured object, which is constructed using Perlin's turbulence function [9].

Figs. 8.11, 8.12, and 8.13 provide more results from the splatting algorithm with shadows. Fig. 8.11 is the Teddy bear, Fig. 8.12 is the Bonsai tree, and Fig. 8.13 is the uncBrain, with and without shadows. The insets (Figs. 8.13c and 8.13d) are close-up renderings in which precise curved shadows are generated. Again, notice that the shadows are calculated per pixel rather than per voxel.

These images are generated using a front-to-back rendering. If the light source is behind the objects, this algorithm proceeds as normal, but the compositing direction is changed from front-to-back to back-to-front. The room scene in Fig. 8.14 is an example of back-to-front

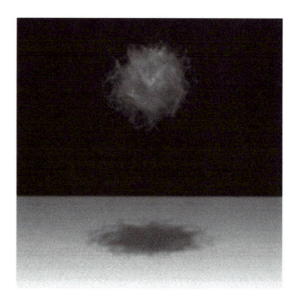

Figure 8.10 A hypertextured object with shadow. (See also color insert.)

rendering: light comes into the room through the window from the back. A desk and a chair reside in the room filled with a light haze, and they cast shadows.

Figure 8.11 Teddy bear. (Left) Without shadow; (Right) with shadow. (See also color insert.)

Figure 8.12   Bonsai tree. (Left) Without shadow; (Right) with shadow. (See also color insert.)

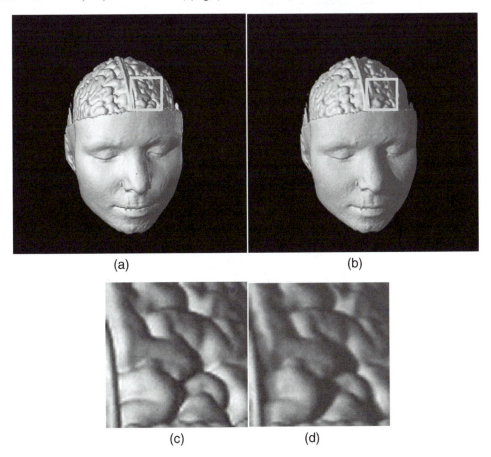

Figure 8.13   uncBrain. (a) Without shadow. (b) With shadow. (c) and (d) Close-up rendering of the specific patch. (See also color insert.)

Figure 8.14 Room scene (an example of back-to-front rendering). (See also color insert.)

When light is attenuated, the running time is longer than the time without shadows, because footprint evaluation and shadow-buffer compositing need to be done with respect to the light source. The algorithm with shadows takes less than twice the time of that without shadows.

8.3.3 Projective Textured Lights

Projective textures can be added for special effects. A light screen is used to get the effect of the "light window" or slide projector and map the light pattern to the scene. The range of the shadow buffer is determined by projecting the light screen to the shadow-buffer plane. The light screen is then given an initial image.

The projective textured lights are modeled as in Fig. 8.15. Now, the light intensity at point x depends not only on the light attenuation, but also on the light color.

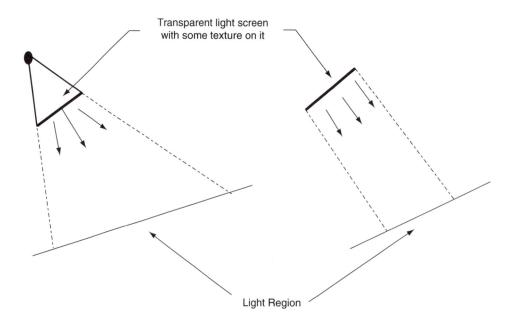

Figure 8.15 A schematic of projective textured light models. (Left) Point light; (Right) parallel light.

$$I(x) = I_{light} * light\_color(x) * (1.0 - \alpha(x)) \qquad (8.21)$$

A room scene in Fig. 8.16 is lit by a light with an image of the logo of The Ohio State University. Shadows are generated by the robot and the rings that reside in the room.

In Fig. 8.17, a parallel area light with a grid texture casts the grid pattern on a HIPIP scene. By controlling the grid pattern, we get some dimension information about the object.

Fig. 8.18 compares images with light beams passing through a semi-transparent cube. Three light beams with red, green, and blue colors enter the cube at the right top, traverse the cube, and come out from the left bottom. The left image is without consideration of light attenuation, while the right one is with light attenuation. The light intensity exiting the cube is the same as the original intensity entering the cube in the left image, while the resulting light intensity is lower than the original light intensity due to attenuation as the light traverses the cube in the right image.

Figure 8.16   A room scene for a light screen with an image of the OSU logo. (See also color insert.)

8.4   Future Work

Future work has progressed on extending the shadow algorithm to deal with extended light sources to generate soft shadows with penumbra and umbra [17], and extending the splatting algorithm to render mixed polygonal and volumetric objects. We are implementing the splatting using modern consumer-level graphics hardware to gain interactivity for volume rendering.

Acknowledgments

Our project was supported by the DOE ASCI program and the NSF Career Award received by Dr. Roger Crawfis. We also acknowledge the University of North Carolina (Chapel Hill) for providing the uncBrain dataset, the University of Erlangen-Nuremberg for providing the teddy

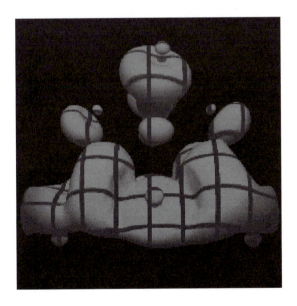

Figure 8.17   HIPIP with grid pattern. (See also color insert.)

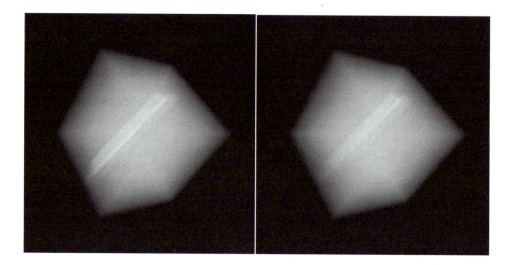

Figure 8.18 A scene with beams of light that pass through the semi-transparent cube. (Left) Without attenuation; (Right) with attenuation. (See also color insert.)

bear and bonsai tree datasets, and Philips Research in Germany for providing the blood vessel dataset.

References

1. R. Crawfis and J. Huang. High quality splatting and volume synthesis. *Data Visualization: The State of the Art* (F.H. Post, G.M. Nielson and G.-P. Bonneau, Eds.), pages 127–140, 2002.
2. J. Huang, K. Mueller, N. Shareef, and R. Crawfis. FastSplats: optimized splatting on rectilinear grids. *Visualization 2000*, pages 219–227, 2000.
3. N. Max. Optical models for direct volume rendering. *IEEE Transactions on Visualization and Computer Graphics*, 1(2):99–108, 1995.
4. M. Meissner, J. Huang, D. Bartz, K. Mueller, and R. Crawfis. A practical evaluation of popular volume rendering algorithms. *2000 Symposium on Volume Rendering*, pages 81–90, 2000.
5. K. Mueller, T. Moeller, J. E. Swan, R. Crawfis, N. Shareef, and R. Yagel. Splatting errors and antialiasing. *IEEE Transactions on Visualization and Computer Graphics*, 4(2):178–191, 1998.
6. K. Mueller, T. Moeller, and R. Crawfis. Splatting without the blur. *Proc. Visualization '99*, pages 363–371, 1999.
7. K. Mueller, N. Shareef, J. Huang, and R. Crawfis. High-quality splatting on rectilinear grids with efficient culling of occluded voxels. *IEEE Transactions on Visualization and Computer Graphics*, 5(2):116–134, 1999.
8. K. Mueller and R. Crawfis. Eliminating popping artifacts in sheet buffer-based splatting. *Proc. Visualization '98*, pages 239–245, 1998.
9. K. Perlin and E. M. Hoffert. Hypertexture. *Proc. SIGGRAPH '89*, pages 253–262, 1989.
10. T. Porter and T. Duff. Compositing digital images. *Computer Graphics*, 18(3):253–259, 1984.
11. P. Sabella. A rendering algorithm for visualizing 3D scalar fields. *Computer Graphics*, 22(4):51–58, 1988.
12. L. Westover. Interactive volume rendering. *Proceedings of Volume Visualization Workshop* (Chapel Hill, N.C., May 18–19), pages 9–16, 1989.
13. L. Westover. Footprint evaluation for volume rendering. *Proc. SIGGRAPH '90*, pages 367–376, 1990.
14. J. Wilhelms and A. Van Gelder. A coherent projection approach for direct volume rendering. *Computer Graphics*, 25(4):275–284, 1991.
15. D. Xue and R. Crawfis. Efficient splatting using modern graphics hardware. *Journal of Graphics Tools*, 8(3):1–21, 2003.
16. C. Zhang and R. Crawfis. Volumetric shadows using splatting. *Proc. Visualization 2002*, pages 85–92, 2002.
17. C. Zhang and R. Crawfis. Shadows and soft shadows with participating media using splatting. *IEEE Transactions on Visualization and Computer Graphics*, 9(2):139–149, 2003.

9 Multidimensional Transfer Functions for Volume Rendering

JOE KNISS, GORDON KINDLMANN, and CHARLES D. HANSEN
Scientific Computing and Imaging Institute
University of Utah

9.1 Introduction

Direct volume rendering has proven to be an effective and flexible visualization method for 3D scalar fields. Transfer functions are fundamental to direct volume rendering because their role is essentially to make the data visible: by assigning optical properties like color and opacity to the voxel data, the volume can be rendered with traditional computer graphics methods. Good transfer functions reveal the important structures in the data without obscuring them with unimportant regions. To date, transfer functions have generally been limited to 1D domains, meaning that the 1D space of scalar data value has been the only variable to which opacity and color are assigned. One aspect of direct volume rendering that has received little attention is the use of multidimensional transfer functions.

Often, there are features of interest in volume data that are difficult to extract and visualize with 1D transfer functions. Many medical datasets created from CT or MRI scans contain a complex combination of boundaries between multiple materials. This situation is problematic for 1D transfer functions because of the potential for overlap between the data-value intervals spanned by the different boundaries. When one data value is associated with multiple boundaries, a 1D transfer function is unable to render them in isolation. Another benefit of higher dimensional transfer functions is their ability to portray subtle variations in properties of a single boundary, such as thickness. When working with multivariate data, a similar difficulty arises with features that can be identified only by their unique combination of multiple data values. A 1D transfer function is simply not capable of capturing this relationship.

Unfortunately, using multidimensional transfer functions in volume rendering is complicated. Even when the transfer function is only 1D, finding an appropriate transfer function is generally accomplished by trial and error. This is one of the main challenges in making direct volume rendering an effective visualization tool. Adding dimensions to the transfer-function domain only compounds the problem. While this is an ongoing research area, many of the proposed methods for transfer-function generation and manipulation are not easily extended to higher-dimensional transfer functions. In addition, fast volume rendering algorithms that assume the transfer function can be implemented as a linear lookup table (LUT) can be difficult to adapt to multidimensional transfer functions due to the linear interpolation imposed on such LUTs.

This chapter provides a detailed exposition of the multidimensional transfer function concept, a generalization of multidimensional transfer functions for both scalar and multivariate data, as well as a novel technique for the interactive generation of volumetric shadows. To resolve the potential complexities in a user interface for multidimensional transfer functions, we introduce a set of direct manipulation widgets that make finding and experimenting with transfer functions an intuitive, efficient, and informative process. In order to make this process genuinely interactive, we exploit the fast rendering capabilities of modern graphics hardware, especially 3D texture memory and pixel-texturing operations. Together, the widgets and the hardware form the basis for new interaction modes that can guide users towards transfer-function settings appropriate for their visualization and data-exploration interests.

9.2   Previous Work

9.2.1   Transfer Functions

Even though volume rendering as a visualization tool is more than 10 years old, only recently has research focused on making the space of transfer functions easier to explore. He et al. [12] generated transfer functions with genetic algorithms driven either by user selection of thumbnail renderings or by some objective image-fitness function. The Design Gallery [23] creates an intuitive interface to the entire space of all possible transfer functions based on automated analysis and layout of rendered images. A more data-centric approach is the Contour Spectrum [1], which visually summarizes the space of isosurfaces in terms of metrics like surface area and mean gradient magnitude, thereby guiding the choice of iso-value for iso-surfacing, and also providing information useful for transfer-function generation. Another recent paper [18] presents a novel transfer-function interface in which small thumbnail renderings are arranged according to their relationship with the spaces of data values, color, and opacity.

The application of these methods is limited to the generation of 1D transfer functions, even though 2D transfer functions were introduced by Levoy in 1988 [22]. Levoy introduced two styles of transfer functions, both 2D and both using gradient magnitude for the second dimension. One transfer function was intended for the display of interfaces between materials, the other

for the display of iso-value contours in more smoothly varying data. The previous work most directly related to our approach for visualizing scalar data facilitates the semiautomatic generation of both 1D and 2D transfer functions [17,29]. Using principles of computer-vision edge detection, the semiautomatic method strives to isolate those portions of the transfer function domain that most reliably correlate with the middle of material-interface boundaries. Other work closely related to our approach for visualizing multivariate data uses a 2D transfer function to visualize data derived from multiple MRI pulse sequences [20].

Scalar volume rendering research that uses multidimensional transfer functions is relatively scarce. One paper discusses the use of transfer functions similar to Levoy's as part of visualization in the context of wavelet volume representation [27]. More recently, the VolumePro graphics board uses a 12-bit 1D lookup table for the transfer function, but also allows opacity modulation by gradient magnitude, effectively implementing a separable 2D transfer function [28]. Other work involving multidimensional transfer functions uses various types of second derivatives in order to distinguish features in the volume according to their shape and curvature characteristics [15,34].

Designing color maps for displaying non-volumetric data is a task similar to finding transfer functions. Previous work has developed strategies and guidelines for color map creation, based on visualization goals, types of data, perceptual considerations, and user studies [3,32,36].

9.2.2   Direct Manipulation Widgets

Direct manipulation widgets are geometric objects rendered with a visualization and are designed to provide the user with a 3D interface [5,14,31,35,38]. For example, a frame widget can be used to select a 2D plane within a volume. Widgets are typically rendered from basic geometric primitives such as spheres, cylinders, and cones. Widget construction is often guided by a constraint system that binds elements of a widget to one another. Each sub-part of a widget represents some functionality of the widget or a parameter to which the user has access.

9.2.3 Hardware Volume Rendering

Many volume rendering techniques based on graphics hardware utilize texture memory to store a 3D dataset. The dataset is then sampled, classified, rendered to proxy geometry, and composited. Classification typically occurs in hardware as a 1D table lookup.

2D texture-based techniques slice along the major axes of the data and take advantage of hardware bi-linear interpolation within the slice [4]. These methods require three copies of the volume to reside in texture memory, one per axis, and they often suffer from artifacts caused by under-sampling along the slice axis. Tri-linear interpolation can be attained using 2D textures with specialized hardware extensions available on some commodity graphics cards [6]. This technique allows intermediate slices along the slice axis to be computed in hardware. These hardware extensions also permit diffuse shaded volumes to be rendered at interactive frame rates.

3D texture-based techniques typically sample view-aligned slices through the volume, leveraging hardware tri-linear interpolation [11]. Other elements of proxy geometry, such as spherical shells, may be used with 3D texture methods to eliminate artifacts caused by perspective projection [21]. The pixel texture OpenGL extension has been used with 3D texture techniques to encode both data value and a diffuse illumination parameter that allows shading and classification to occur in the same lookup [25]. Engel et al. [10] showed how to significantly reduce the number of slices needed to adequately sample a scalar volume, while maintaining a high-quality rendering, using a mathematical technique of preintegration and hardware extensions such as dependent textures.

Another form of volume rendering graphics hardware is the Cube-4 architecture [30] and the subsequent Volume-Pro PCI graphics board [28]. The VolumePro graphics board implements ray-casting combined with the shear warp factorization for volume rendering [19]. It features tri-linear interpolation with supersampling, gradient estimation, and shaded volumes, and provides interactive frame rates for scalar volumes with sizes up to 512^3.

9.3 Multidimensional Transfer Functions

Transfer-function specification is arguably the most important task in volume visualization. While the transfer function's role is simply to assign optical properties such as opacity and color to the data being visualized, the value of the resulting visualization will be largely dependent on how well these optical properties capture features of interest. Specifying a good transfer function can be a difficult and tedious task for several reasons. First, it is difficult to uniquely identify features of interest in the transfer-function domain. Even though a feature of interest may be easily identifiable in the spatial domain, the range of data values that characterize the feature may be difficult to isolate in the transfer-function domain due to the fact that other, uninteresting regions may contain the same range of data values. Second, transfer functions can have an enormous number of degrees of freedom. Even simple 1D transfer functions using linear ramps require two degrees of freedom per control point. Third, typical user interfaces do not guide the user in setting these control points based on dataset-specific information. Without this type of information, the user must rely on trial and error. This kind of interaction can be especially frustrating since small changes to the transfer function can result in surprisingly large and unintuitive changes to the volume rendering.

Rather than classifying a sample based on a single scalar value, multidimensional transfer functions allow a sample to be classified based on a combination of values. Multiple data values tend to increase the probability that a feature can be uniquely isolated in the transfer-function domain, effectively providing a larger vocabulary for expressing the differences between structures in the dataset. These values are the axes of a multidimensional transfer function. Adding dimensions to the transfer function, however, greatly increases the degrees of freedom necessary for specifying a transfer function and the need for dataset-specific guidance.

In the following sections, we demonstrate the application of multidimensional transfer functions to two distinct classes of data: scalar data and multivariate data. The scalar-data application is focused on locating surface boundaries in a scalar volume. We motivate and describe the axes of the multidimensional transfer function for this type of data. We then describe the use of multidimensional transfer functions for multivariate data. We use two examples, color volumes and meteorological simulations, to demonstrate the effectiveness of such transfer functions.

9.3.1 Scalar Data

For scalar data, the gradient is a first-derivative measure. As a vector, it describes the direction of greatest change. The normalized gradient is often used as the normal for surface-based volume shading. The gradient magnitude is a scalar quantity that describes the local rate of change in the scalar field. For notational convenience, we will use f' to indicate the magnitude of the gradient of f, where f is the scalar function representing the data.

$$f' = \|f\| \tag{9.1}$$

This value is useful as an axis of the transfer function since it discriminates between homogeneous regions (low-gradient magnitudes) and regions of change (high-gradient magnitudes). This effect can be seen in Fig. 9.1. Fig. 9.1a shows a 1D histogram based on data value and identifies the three basic materials in the Chapel Hill CT Head: air (A), soft tissue (B),

and bone (C). Fig. 9.1b shows a log-scale joint histogram of data value versus gradient magnitude.

Since materials are relatively homogeneous, their gradient magnitudes are low. They can be seen as the circular regions at the bottom of the histogram. The boundaries between the materials are shown as the arches—air and soft tissue boundary (D), soft tissue and bone boundary (E), and air and bone boundary (F). Each of these materials and boundaries can be isolated using a 2D transfer function based on data value and gradient magnitude. Fig. 9.1c shows a volume rendering with the corresponding features labeled. The air–bone boundary (F), in Fig. 9.1, is a good example of a surface that cannot be isolated using a simple 1D transfer function. This type of boundary appears in CT datasets as the sinuses and mastoid cells. Often, the arches that define material boundaries in a 2D transfer function overlap. In some cases this overlap prevents a material from being properly isolated in the transfer function. This effect can be seen in the circled region of the 2D data value–gradient magnitude joint histogram of the human tooth CT in Fig. 9.2a. The background–dentin boundary (F) shares the same ranges of data value and gradient magnitude as portions of the pulp–dentin (E) and the background–enamel (H) boundaries. When the background–dentin boundary (F) is emphasized in a 2D transfer function, the boundaries (E) and (H) are erroneously colored in the volume rendering, as seen in Fig. 9.2c. A second derivative measure enables a more precise disambiguation of complex boundary configurations such as this. Some edge-detection algorithms (such as Marr–Hildreth [24]) locate the middle of an edge by detecting a zero crossing in a second derivative measure, such as the Laplacian. We compute a more accurate but computationally expensive measure, the second directional derivative along the gradient direction, which involves the Hessian (**H**), a matrix of second partial derivatives. We will use f'' to indicate this second derivative.

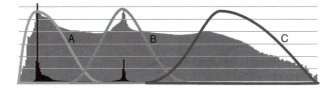

(a) A 1D histogram. The black region represents the number of data value occurrences on a linear scale; the grey is on a log scale. The colored regions (A, B, C) identify basic materials.

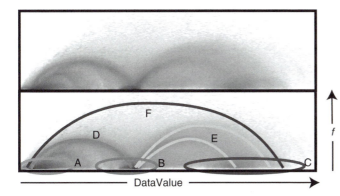

(b) A log-scale 2D joint histogram showing the location of materials (A, B, C) and material boundaries (D, E, F).

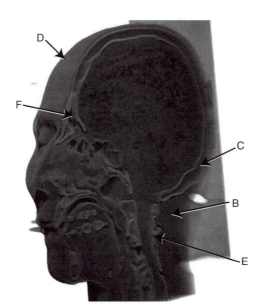

(c) A volume rendering showing all of the materials and boundaries identified above, except air (A), using a 2D transfer function.

Figure 9.1 Material and boundary identification of the Chapel Hill CT Head with data value alone (a) versus data value and gradient magnitude (f'), seen in (b). The basic materials captured by CT, air (A), soft tissue (B), and bone (C) can be identified using a 1D transfer function, as seen in (a). 1D transfer functions, however, cannot capture the complex combinations of material boundaries: air and soft tissue boundary (D), soft tissue and bone boundary (E), and air and bone boundary (F), as seen in (b) and (c). (See also color insert.)

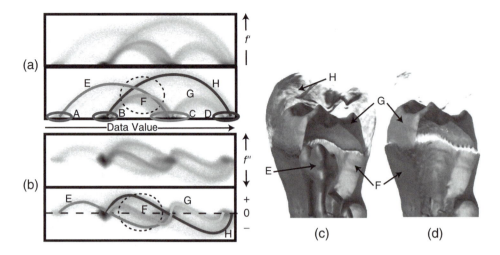

Figure 9.2 Material and boundary identification of the human tooth CT with (a) data value and gradient magnitude (f') and (b) data value and second derivative (f''). The background–dentin boundary (F) cannot be adequately captured with data value and gradient magnitude alone. (c) The results of a 2D transfer function designed to show only the background–dentin (F) and dentin–enamel (G) boundaries. The background–enamel (H) and dentin–pulp (E) boundaries are erroneously colored. Adding the second derivative as a third axis to the transfer function disambiguates the boundaries. (d) The results of a 3D transfer function that gives lower opacity to nonzero second-derivative values. (See also color insert.)

$$f'' = \frac{1}{\|f\|^2}(\nabla f)^{\mathrm{T}}\mathbf{H}f\nabla f \qquad (9.2)$$

More details on these measurements can be found in previous work on semiautomatic transfer function generation [16,17]. Fig. 9.2b shows a joint histogram of data value versus this second directional derivative. Notice that the boundaries (E), (F), and (G) no longer overlap. By reducing the opacity assigned to nonzero second-derivative values, we can render the background–dentin boundary in isolation, as seen in Fig. 9.2d. The relationship between data value, gradient magnitude, and the second directional derivative is made clear in Fig. 9.3. Fig. 9.3a shows the behavior of these values along a line through an idealized boundary between two homogeneous materials (inset). Notice that at the center of the boundary, the gradient magnitude is high and the second derivative is zero. Fig. 9.3b shows the behavior of the gradient magnitude and second derivative as a function of data value. This shows the curves as they appear in a joint histogram or a transfer function.

9.3.2 Multivariate Data

Multivariate data contains, at each sample point, multiple scalar values that represent different simulated or measured quantities. Multivariate data can come from numerical simulations that calculate a list of quantities at each time step, or from medical scanning modalities such as MRI, which can measure a variety of tissue characteristics, or from a combination of different scanning modalities, such as MRI, CT, and PET. Multidimensional transfer functions are an obvious choice for volume visualization of multivariate data, since we can assign different data values to the different axes of the transfer function. It is often the case that a feature of interest in these datasets cannot be properly classified using any single variable by itself. In addition, we can compute a kind of first derivative in the multivariate data in order to create more information about local structure. As with scalar data, the use of a first derivative measure as one axis of the multidimensional transfer function can increase the specificity with which we can isolate and visualize different features in the data.

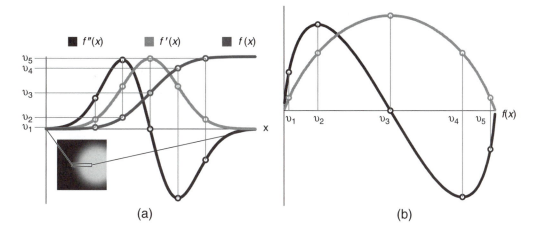

Figure 9.3   The behavior of primary data value (f), gradient magnitude (f'), and second directional derivative (f'') as a function of position (a) and as a function of data value (b). (See also color insert.)

One example of data that benefits from multidimensional transfer functions is volumetric color data. A number of volumetric color datasets are available, such as the Visible Human Project's RGB data. The process of acquiring color data by cryosection is becoming common for the investigation of anatomy and histology. In these datasets, the differences in materials are expressed by their unique spectral signatures. A multidimensional transfer function is a natural choice for visualizing this type of data. Opacity can be assigned to different positions in the 3D RGB color space.

Fig. 9.4a shows a joint histogram of the RGB color data for the Visible Male; regions of this space that correspond to different tissues are identified. Regions (A) and (B) correspond to the fatty tissues of the brain, white and gray matter, as seen in Fig. 9.4b. In this visualization, the transition between white and grey matter is intentionally left out to better emphasize these materials and to demonstrate the expressivity of the multidimensional transfer function Fig. 9.4c shows a visualization of the color values that represent the muscle structure and connective tissues (C) of the head and neck with the skin surface (D), given a small amount of opacity for

context. In both of these figures, a slice of the original data is mapped to the surface of the clipping plane for reference.

The kind of first derivative that we compute in multivariate data is based on previous work in color image segmentation [7,8,33]. While the gradient magnitude in scalar data represents the magnitude of local change at a point, an analogous first-derivative measure in multivariate data captures the total amount of local change, across all the data components. This derivative has proven useful in color image segmentation because it allows a generalization of gradient-based edge detection. In our system, we use this first-derivative measure as one axis in the multidimensional transfer function in order to isolate and visualize different regions of a multivariate volume according to the amount of local change, analogous to our use of gradient magnitude for scalar data.

If we represent the dataset as a multivariate function $\mathbf{f}(x,y,z): \mathbb{R}^3 \to \mathbb{R}^m$, so that

$$\mathbf{f}(x,y,z) = (f_1(x,y,z), f_2(x,y,z),$$
$$\ldots, f_m(x,y,z))$$

then the derivative $\mathbf{Df}$ is a matrix of first partial derivatives:

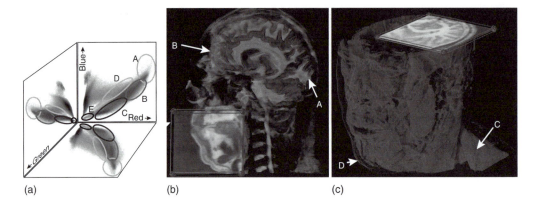

(a)   (b)   (c)

Figure 9.4 The Visible Male RGB (color) data. The opacity is set using a 3D transfer function, and color is taken directly from the data. The histogram (a) is visualized as projections on the primary planes of the RGB color space. (b) The white (A) and grey (B) matter of the brain. (c) The muscle and connective tissues (C) of the neck, showing skin (D) for reference. (See also color insert.)

$$
\mathbf{Df} = \begin{bmatrix}
\dfrac{\partial f_1}{\partial x} & \dfrac{\partial f_1}{\partial y} & \dfrac{\partial f_1}{\partial z} \\[2mm]
\dfrac{\partial f_2}{\partial x} & \dfrac{\partial f_2}{\partial y} & \dfrac{\partial f_2}{\partial z} \\
& \vdots & \\
\dfrac{\partial f_m}{\partial x} & \dfrac{\partial f_m}{\partial y} & \dfrac{\partial f_m}{\partial z}
\end{bmatrix}
$$

By multiplying $\mathbf{Df}$ by its transpose, we can form a 3×3 tensor $\mathbf{G}$ that captures the directional dependence of total change:

$$\mathbf{G} = (\mathbf{Df})^{\mathrm{T}}\mathbf{Df} \tag{9.3}$$

In the context of color edge detection [7,8,33], this matrix (specifically, its 2D analog) is used as the basis of a quadratic function of direction $\mathbf{n}$, which Cumani [7] terms the *squared local contrast* in direction $\mathbf{n}$:

$$S(\mathbf{n}) = \mathbf{n}^{\mathrm{T}}\mathbf{G}\mathbf{n}$$

$S(\mathbf{n})$ can be analyzed by finding the principal eigenvector (and associated eigenvalue) of $\mathbf{G}$ to determine the direction $\mathbf{n}$ of greatest local contrast, or fastest change, and the magnitude of that change. Our experience, however, has been that in the context of multidimensional transfer functions, it is sufficient (and perhaps preferable) to simply take the L2 norm of $\mathbf{G}$, $\|\mathbf{G}\|$,

which is the square root of the sum of the squares of the individual matrix components. As the L2 norm is invariant with respect to rotation, this is the same as the L2 norm of the three eigenvalues of $\mathbf{G}$, motivating our use of $\|\mathbf{G}\|$ as a directionally independent (and rotationally invariant) measure of local change. Other work on volume rendering of color data has used a non–rotationally invariant measure of $\mathbf{G}$ [9]. Since it is sometimes the case that the dynamic range of the individual channels (f_i) differ, we normalize the ranges of each channel's data value to be between zero and one. This allows each channel to have an equal contribution in the derivative calculation.

9.4   Interaction and Tools

While adding dimensions to the transfer function enhances our ability to isolate features of interest in a dataset, it tends to make the already unintuitive space of the transfer function even more difficult to navigate. This difficulty can be considered in terms of a conceptual gap between the spatial and transfer-function domains. The spatial domain is the familiar 3D space for geometry and the volume data being rendered.

The transfer-function domain, however, is more abstract. Its dimensions are not spatial (i.e., the ranges of data values), and the quantities at each location are not scalar (i.e., opacity and three colors). It can be very difficult to determine the regions of the transfer function that correspond to features of interest, especially when a region is very small. Thus, to close this conceptual gap, we developed new interaction techniques, which permit interaction in both domains simultaneously, and a suite of direct manipulation widgets that provide the tools for such interactions. Fig. 9.5 shows the various direct manipulation widgets as they appear in the system.

In a typical session with our system, the user creates a transfer function using a natural process of exploration, specification, and refinement. Initially, the user is presented with a volume rendering using a predetermined transfer function that is likely to bring out some features of interest. This can originate with an automated transfer function generation tool [16], or it could be the default transfer function described later in Section 9.6. The user would then begin exploring the dataset.

Exploration is the process by which a user familiarizes him or herself with the dataset. A clipping plane can be moved through the volume to reveal internal structures. A slice of the original data can be mapped to the clipping plane, permitting a close inspection of the entire range of data values. Sample positions are probed in the spatial domain, and their values, along with values in a neighborhood around that point, are visualized in the transfer-function domain. This feedback allows the user to identify the regions of the transfer function that correspond to potential features of interest, made visible by the default transfer function or the sliced data. Once these regions have been identified, the user can then begin specifying a custom transfer function.

During the specification stage, the user creates a rough draft of the desired transfer function. While this can be accomplished by manually adding regions to the transfer function, a simpler method adds opacity to the regions in the transfer function at and around locations queried in the spatial domain. That is, the system can track, with a small region of opacity in the transfer-function domain, the data values at the user-selected locations, while continually updating the volume rendering. This visualizes, in the spatial domain, all other voxels with similar transfer-function values. If the user decides that an important feature is captured by the current transfer function, he or she can add that region into the transfer function and continue querying and investigating the volume.

Once these regions have been identified, the user can refine them by manipulating control points in the transfer-function domain to better visualize features of interest. An important feature of our system is the ability to manipulate portions of the transfer function as discrete entities. This permits the modification of regions corresponding to a particular feature without affecting other classified regions.

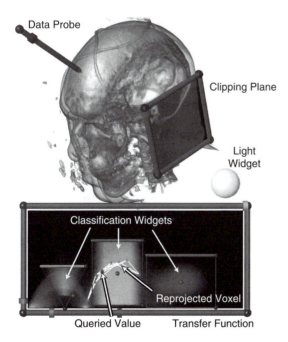

Figure 9.5   The direct-manipulation widgets. (See also color insert.)

Finally, this is an iterative process. Users continue the exploration, specification, and refinement steps until they are satisfied that all features of interest are made visible. In the remainder of this section, we introduce the interaction modalities used in the exploration and specification stages and briefly describe the individual direct manipulation widgets.

9.4.1   Probing and Dual-Domain Interaction

The concept of probing is simple: the user points at a location in the spatial domain and visualizes the values at that point in the transfer-function domain. We have found this feedback to be essential for making the connection between features seen in the spatial domain and the ranges of values that identify them in the transfer-function domain. Creating the best transfer function for visualizing a feature of interest is only possible with an understanding of the behavior of data values at and around that feature. This is especially true for multidimensional transfer functions where a feature is described by a complex combination of data values. The value of this dataset-specific guidance can be further enhanced by automatically setting the transfer function based on these queried values.

In a traditional volume rendering system, setting the transfer function involves moving the control points (in a sequence of linear ramps defining color and opacity), and then observing the resulting rendered image. That is, interaction in the transfer-function domain is guided by careful observation of changes in the spatial domain. We prefer a reversal of this process, in which the transfer function is set by direct interaction in the *spatial domain*, with observation of the transfer-function domain. Furthermore, by allowing interaction to happen in both domains simultaneously, we significantly lessen the conceptual gap between them, effectively simplifying the complicated task of specifying a multidimensional transfer function to pointing at a feature of interest. We use the term "dual-domain interaction" to describe this approach to transfer-function exploration and generation.

The top of Fig. 9.6 illustrates the specific steps of dual-domain interaction. When a position inside the volume is queried by the user with the data probe widget (a), the values associated with that position (multivariate values, or the data value, first and second derivative) are graphically represented in the transfer function widget (b). Then, a small region of high opacity (c) is temporarily added to the transfer function at the data values determined by the probe location. The user has now set a multidimensional transfer function simply by positioning a data probe within the volume. The resulting rendering (d) depicts (in the spatial domain) all the other locations in the volume that share values (in the transfer-function domain) with those at the data probe tip. If the features rendered are of interest, the user can copy the temporary transfer function to the permanent one (e), by, for instance, tapping the keyboard space bar with the free hand. As features of interest are discovered, they can be added to the transfer function quickly and easily with this type of two-handed interaction. Alternately, the probe feedback can be used to manually set other types of classification widgets (f), which are described later. The outcome of dual-domain interaction is an effective multidimensional transfer function built up over the course of data exploration. The widget components that participated in this process can be seen in the bottom of Fig. 9.6, which shows how dual-domain interaction can help volume render the CT tooth dataset. The remainder of this section describes the individual widgets and provides additional details about dual-domain interaction.

9.4.2   Data Probe Widget

The data probe widget is responsible for reporting its tip's position in volume space and its slider sub-widget's value. Its pencil-like shape is designed to give the user the ability to point at a feature in the volume being rendered. The other end of the widget orients the widget about its tip.

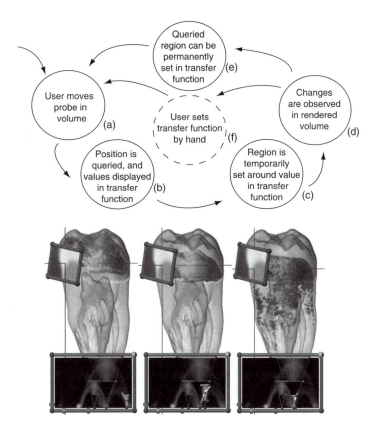

Figure 9.6   Dual-domain interaction. (See also color insert.)

When the volume rendering's position or orientation is modified, the data probe widget's tip tracks its point in volume space. A natural extension is to link the data probe widget to a haptic device, such as the SensAble PHANTOM, which can provide a direct 3D location and orientation [26].

9.4.3   Clipping Plane Widget

The clipping plane is responsible for reporting its orientation and position to the volume renderer, which handles the actual clipping when it draws the volume. In addition to clipping, the volume widget will also map a slice of the data to the arbitrary plane defined by the clip widget, and blend it with the volume by a constant opacity value determined by the clip widget's slider. It is also responsible for reporting the spatial position of a mouse click on its clipping surface. This provides an additional means of querying positions within the volume, distinct from the 3D data probe. The balls at the corners of the clipping plane widget are used to modify its orientation, and the bars on the edges are used to modify its position.

9.4.4   Transfer-Function Widget

The main role of the transfer-function widget is to present a graphical representation of the transfer-function domain, in which feedback from querying the volume (with the data probe or clipping plane) is displayed, and in which the transfer function itself can be set and altered. The balls at the corners of the transfer-function widget are used to resize it, as with a desktop window, and the bars on the edges are used to translate its position. The inner plane of the frame is a polygon texture-mapped with the lookup table containing the transfer function. A joint histogram of data, seen with the images in Section 9.3, can also be blended with the transfer function to provide valuable information about the behavior and relationship of data values in the transfer-function domain.

The data values at the position queried in the volume (via either the data probe or the clipping plane widget) are represented with a small ball in the transfer-function widget. In addition to the precise location queried, the eight data sample points at the corners of the voxel containing the query location are also represented by balls in the transfer-function domain, and are connected together with edges that reflect the connectivity of the voxel corners in the spatial domain. By "reprojecting" a voxel from the spatial domain to a simple graphical representation in the transfer-function domain, the user can learn how the transfer-function variables (data values at each sample point) are changing near the probe location. The values for the third, or unseen, axis are indicated by colorings on the balls. For instance, with scalar data, second-derivative values that are negative, zero, and positive are represented by blue, white, and yellow balls, respectively. When the projected points form an arc, with the color varying through these assigned colors, the probe is at a boundary in the volume as seen in Fig. 9.5. When the reprojected data points are clustered together, the probe is in a homogeneous region. As the user gains experience with this representation, he or she can learn to "read" the reprojected voxel as an indicator of the volume characteristics at the probe location.

9.4.5   Classification Widgets

In addition to the process of dual-domain interaction described above, transfer functions can also be created in a more manual fashion by adding one or more classification widgets to the main transfer-function window. Classification widgets are designed to identify regions of the transfer function as discrete entities. Each widget type has control points that modify its position or size. Optical properties, such as opacity and color, are modified by selecting the widget's inner surface. The opacity and color contributions from each classification widget are blended together to form the transfer function. We have developed two types of classification widget: triangular and rectangular.

The triangular classification widget, shown in Figs. 9.5, 9.6, and 9.8, is based on Levoy's "iso-value contour surface" opacity function [22]. The widget is an inverted triangle with a base point attached to the horizontal data value axis. The triangle's size and position are adjusted with control points. There are an upper and a lower threshold for the gradient magnitude, as well as a shear. Color is constant across the widget; opacity is maximal along the center of the widget, and it linearly ramps down to zero at the left and right edges.

The triangular classification widgets are particularly effective for visualizing *surfaces* in scalar data. More general transfer functions, for visualizing data that may not have clear boundaries, can be created with the rectangular classification widget. The rectangular region spanned by the widget defines the data values that receive opacity and color. Like the triangular widget, color is constant, but the opacity is more flexible. It can be constant or fall off in various ways: quadratically as an ellipsoid with axes corresponding to the rectangle's aspect ratio, or linearly as a ramp, tent, or pyramid.

As noted in the description of the transfer-function widget, even when a transfer function has more than two dimensions, only two dimensions are shown at any one time. For 3D transfer functions, classification widgets are shown as their projections onto the visible axes. In this case, a rectangular classification widget becomes a box in the 3D domain of the transfer function. Its appearance to the user, however, as 2D projections, is identical to the rectangular widget. When the third axis of the transfer function plays a more simplified role, interactions along this axis are tied to sliders seen along the top bar of the transfer function. For instance, since our research on scalar data has focused on visualizing boundaries between material regions, we have consistently used the second derivative to emphasize the regions where the second-derivative magnitude is small or zero. Specifically, maximal opacity is always given to zero second derivatives and decreases linearly towards the second-derivative extrema values. How much the opacity changes as a function of second-derivative magnitude is controlled with a single slider, which we call the "boundary emphasis slider." With the slider in its left-most position, zero opacity is given to extremal second derivatives; in the right-most position, opacity is constant with respect to the second derivative. We have employed similar techniques for manipulating other types of third-axis values using multiple sliders.

While the classification widgets are usually set by hand in the transfer-function domain, based on feedback from probing and reprojected voxels, their placement can also be somewhat automated. This further reduces the difficulty of creating an effective higher-dimensional transfer function. The classification widget's location and size in the transfer-function domain can be tied to the distribution of the reprojected voxels determined by the data probe's location. For instance, the rectangular classification widget can be centered at the transfer-function values interpolated at the data probe's tip, with the size of the rectangle controlled by the data probe's slider. The triangular classification widget can be located horizontally at the data value queried by the probe, with the width and height determined by the horizontal and vertical variance in the reprojected voxel locations. This technique produced the changes in the transfer function for the sequence of renderings in Fig. 9.6.

9.4.6 Shading Widget

The shading widget is a collection of spheres that can be rendered in the scene to indicate and control the light direction and color. Fixing a few lights in view space is generally effective for renderings; therefore, changing the lighting is an infrequent operation.

9.4.7 Color-Picker Widget

The color picker is an embedded widget that is based on the hue-lightness-saturation (HLS) color space. Interacting with this widget can be thought of as manipulating a sphere with hues mapped around the equator, gradually becoming black at the top and white at the bottom. To select a hue, the user moves the mouse horizontally, rotating the ball around its vertical axis. Vertical mouse motion tips the sphere toward or away from the user, shifting the color towards white or black. Saturation and opacity are selected independently using different mouse buttons with vertical motion. While this color picker can be thought of as manipulating an HLS sphere, no geometry for this is rendered. Rather, the triangular and rectangular classification widgets embed the color picker in the polygonal region, which contributes opacity and color to the transfer-function domain. The user specifies a color simply by clicking on that object and then moving the mouse horizontally and vertically until the desired hue and lightness are visible. In most cases, the desired color can be selected with a single mouse-click and gesture.

9.5 Rendering and Hardware

While this chapter is conceptually focused on the matter of setting and applying higher-dimensional transfer functions, the quality of interaction and exploration described would not be possible without the use of modern graphics hardware. Our implementation relies heavily on an OpenGL extension known as *dependent texture reads*. This extension can be used for both classification and shading. In this section, we describe our modifications to the classification portion of the traditional 3D texture-based volume rendering pipeline. We also describe methods for adding interactive volumetric shading and shadows to the pipeline.

Our system supports volumes that are stored as 3D textures with one, two, or four values per texel. This is due to memory-alignment restrictions of graphics hardware. Volumes with three values per sample utilize a four-value texture, where the fourth value is simply ignored. Volumes with more than four values per sample could be constructed using multiple textures.

9.5.1 Dependent Texture Reads

Dependent texture reads are hardware extensions that are similar but more efficient implementations of a previous extension known as *pixel texture* [10,13,25,37]. Dependent texture reads and pixel texture are names for operations that use color fragments to generate texture coordinates and replace those color fragments with the corresponding entries from a texture. This operation essentially amounts to an arbitrary function evaluation with up to three variables via a lookup table. If we were to perform this operation on an RGB fragment, each channel value would be scaled between zero and one, and these new values would then be used as texture coordinates of a 3D texture. The color values produced by the 3D texture lookup replace the original RGB values. The nearest-neighbor or linear-interpolation methods can be used to generate the replacement values. The ability to scale and interpolate color channel values is a convenient feature of the hardware. It allows the number of elements along a dimension of the texture containing the new color values to differ from the dynamic range of the component that generated the texture coordinate. Without this flexibility, the size of a 3D dependent texture would be prohibitively large.

9.5.2 Classification

Dependent-texture reads are used for the transfer-function evaluation. Data values stored in the color components of a 3D texture are interpolated across some proxy geometry (a plane, for instance). These values are then converted to texture coordinates and used to acquire the color and alpha values in the transfer-function texture per-pixel in screen space. For eight-bit data, an ideal transfer-function texture would have 256 color and alpha values along each axis. For 3D transfer functions, however, the transfer-function texture would then be $256^3 \times 4$ bytes. Besides the enormous memory requirements of such a texture, the size also affects how fast the classification widgets can be rasterized, thus affecting the interactivity of transfer-function updates. We therefore limit the number of elements along an axis of a 3D transfer function based on its importance. For instance, with scalar data, the primary data value is the most important, the gradient magnitude is secondary, and the second derivative serves an even more tertiary role. For this type of multidimensional transfer function, we commonly use a 3D transfer-function texture with dimensions $256 \times 128 \times 8$ for data value, gradient magnitude, and second derivative, respectively. 3D transfer functions can also be composed separably as a 2D and a 1D transfer function. This means that the total size of the transfer function is $256^2 + 256$. The trade-off, however, is in expressivity. We can no longer specify a transfer function based on the unique combination of all three data values. Separable transfer functions are still quite powerful. Applying the second derivative as a separable 1D portion of the transfer function is quite effective for visualizing boundaries between materials. With the separable 3D transfer function for scalar volumes, there is only one boundary-emphasis slider that affects all classification widgets, as opposed to the general case where each classification widget has its own boundary-emphasis slider. We have employed a similar approach for multivariate data visualization. The meteorological example used a separable 3D transfer function. Temperature and humidity were classified using a 2D transfer function and the multiderivative of these values was classified using a 1D transfer function. Since our specific goal was to show only regions with high values of $\|G\|$, we needed only two sliders to specify the beginning and ending points of a linear ramp along this axis of the transfer function.

9.5.3   Surface Shading

Shading is a fundamental component of volume rendering because it is a natural and efficient way to express information about the shape of structures in the volume. However, much previous work with texture-memory-based-volume rendering lacks shading. Many modern graphics hardware platforms support multitexture and a number of user-defined operations for blending these textures per pixel. These operations, which we will refer to as fragment shading, can be leveraged to compute a surface-shading model.

The technique originally proposed by Rezk-Salama et al. [6] is an efficient way to compute the Blinn–Phong shading model on a per-pixel basis for volumes. This approach, however, can suffer from artifacts caused by denormalization during interpolation. While future generations of graphics hardware should support the square root operation needed to renormalize on a per-pixel basis, we can utilize *cube map dependent texture reads* to evaluate the shading model. This type of dependent texture read allows an RGB color component to be treated as a vector and used as the texture coordinates for a cube map. Conceptually, a cube map can be thought of as a collection of six textures that make up the faces of a cube centered about the origin. Texels are accessed with a 3D texture coordinate (s,t,r) representing a direction vector. The accessed texel is the point corresponding to the intersection of a line through the origin in the direction of (s,t,r) and a cube face. The color values at this position represent incoming diffuse radiance if the vector (s,t,r) is a surface normal or specular radiance if (s,t,r) is a reflection vector. The advantages of using a cube map dependent texture read is that the vector (s,t,r) does not need to be normalized, and the cube map can encode an arbitrary number of lights

or a full environment map. This approach, however, comes at the cost of reduced performance. A per-pixel cube map evaluation can be as much as three times slower than evaluating the dot products for a limited number of light sources in the fragment shader stage.

Surface-based shading methods are well suited for visualizing the boundaries between materials. However, since the surface normal is approximated by the normalized gradient of a scalar field, these methods are not robust for shading homogeneous regions, where the gradient magnitude is very low or zero and its measurement is sensitive to noise. Gradient-based surface shading is also unsuitable for shading volume renderings of multivariate fields. While we can assign the direction of greatest change for a point in a multivariate field to the eigenvector (e_1) corresponding to the largest eigenvalue (λ_1) of the tensor $\mathbf{G}$ from Equation 9.3, e_1 is a valid representation of only orientation, not the absolute direction. This means that the sign of e_1 can flip in neighboring regions even though their orientations may not differ. Therefore, the vector e_1 does not interpolate, making it a poor choice of surface normal. Furthermore, this orientation may not even correspond to the surface normal of a classified region in a multivariate field.

9.5.4   Shadows

Shadows provide important visual queues relating to the depth and placement of objects in a scene. Since the computation of shadows does not depend on a surface normal, they provide a robust method for shading homogeneous regions and multivariate volumes. Adding shadows to the volume lighting model means that light gets attenuated through the volume before being reflected back to the eye.

Our approach differs from previous hardware shadow work [2] in two ways. First, rather than creating a volumetric shadow map, we utilize an off-screen render buffer to accumulate the amount of light attenuated from the light's point of view. Second, we modify the slice axis to be the direction halfway between the view and light directions. This allows the same slice to be rendered from both the eye and the light points of view. Consider the situation for computing shadows when the view and light directions are the same, as seen in Fig. 9.7a. Since the slices for both the eye and the light have a one-to-one correspondence, it is not necessary to precompute a volumetric shadow map. The amount of light arriving at a particular slice is equal to one minus the accumulated opacity of the slices rendered before it. Naturally, if the projection matrices for the eye and the light differ, we need to maintain a separate buffer for the attenuation from the light's point of view. When the eye and light directions differ, the volume would be sliced along each direction independently. The worst-case scenario happens when the view and light directions are perpendicular, as seen in Fig. 9.7b. In this case, it would seem necessary to save a full volumetric shadow map that can be resliced with the data

volume from the eye's point of view providing shadows. This approach, however, suffers from an artifact referred to as attenuation leakage. The visual consequences of this are blurry shadows and surfaces that appear much darker than they should due to the image-space high frequencies introduced by the transfer function. The attenuation at a given sample point is blurred when light intensity is stored at a coarse resolution and interpolated during the observer rendering phase. This use of a 2D shadow buffer is similar to the method described in Chapter 8 except we address slice-based volume rendering while they address splatting.

Rather than slice along the vector defined by the view direction or the light direction, we modify the slice axis to allow the same slice to be rendered from both points of view. When the dot product of the light and view directions is positive, we slice along the vector halfway between the light and view directions (Fig. 9.7c). In this case, the volume is rendered in front-to-

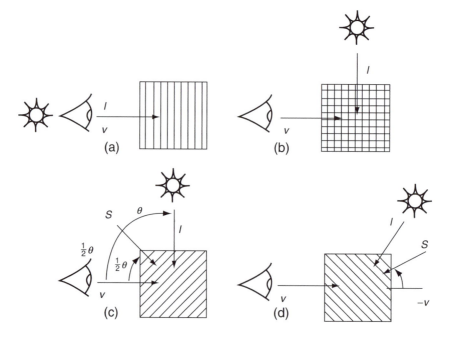

Figure 9.7   Modified slice axis for light transport.

back order with respect to the observer. When the dot product is negative, we slice along the vector halfway between the light and the inverted-view directions (Fig. 9.7d). In this case, the volume is rendered in back-to-front order with respect to the observer. In both cases the volume is rendered in front-to-back order with respect to the light. Care must be taken to ensure that the slice spacing along the view and light directions are maintained when the light or eye positions change. If the desired slice spacing along the view direction is d_v and the angle between v and l is θ, then the slice spacing along the slice direction is

$$d_s = \cos\left(\frac{\theta}{2}\right) d_v \qquad (9.4)$$

This is a multipass approach. Each slice is first rendered from the observer's point of view using the results of the previous pass from the light's point of view, which modulates the brightness of samples in the current slice. The same slice is then rendered from the light's point of view to calculate the intensity of the light arriving at the next layer.

Since we must keep track of the amount of light attenuated at each slice, we utilize an off-screen render buffer, known as a *pixel buffer*. This buffer is initialized to $1 - $ *light intensity*. It can also be initialized using an arbitrary image to create effects such as spotlights. The projection matrix for the light's point of view need not be orthographic; a perspective projection matrix can be used for point light sources. However, the entire volume must fit in the light's view frustum. Light is attenuated by simply accumulating the opacity for each sample using the over operator. The results are then copied to a texture that is multiplied with the next slice from the eye's point of view before it is blended into the frame buffer. While this copy-to-texture operation has been highly optimized on the current generation of graphics hardware, we have achieved a dramatic increase in performance using a hardware extension known as *render to texture*. This extension allows us to directly bind

a pixel buffer as a texture, avoiding the unnecessary copy operation.

This approach has a number of advantages over previous volume shadow methods. First, attenuation leakage is no longer a concern because the computation of the light transport (slicing density) is decoupled from the resolution of the data volume. Computing light attenuation in image space allows us to match the sampling frequency of the light transport with that of the final volume rendering. Second, this approach makes far more efficient use of memory resources than those that require a volumetric shadow map. Only a single additional 2D buffer is required, as opposed to a potentially large 3D volume. One disadvantage of this approach is that, due to the image-space sampling, artifacts may appear at shadow boundaries when the opacity makes a sharp jump from low to high. This can be overcome by using a higher resolution for the light buffer than for the frame buffer. We have found that 30–50% additional resolution is adequate.

As noted at the end of the previous section, surface-based shading models are inappropriate for homogeneous regions in a volume. However, it is often useful to have both surface-shaded and shadowed renderings regardless of whether homogeneous regions are being visualized. To ensure that homogeneous regions are not surface-shaded, we simply interpolate between surface-shaded and unshaded using the gradient magnitude. Naturally, regardless of whether a particular sample is surface-shaded, it is still modulated by the light attenuation providing shadows. In practice we have found that interpolating based on $1 - (1 - \|\nabla f\|)^2$ produces better results, since midrange gradient magnitudes can still be interpreted as surface features. Fig. 9.8 shows a rendering that combines surface shading and shadows in such a way. Fig. 9.1 shows a volume rendering using shadows with the light buffer initialized to simulate a spotlight. Fig. 9.2 shows volume rendering using only surface-based shading. Fig. 9.4 uses only shadows for illumination.

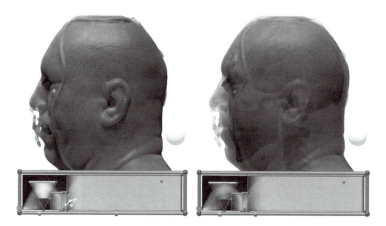

Figure 9.8   Volume renderings of the Visible Male CT (frozen) demonstrating combined surface shading and shadows. (See also color insert.)

9.6   Discussion

Using multidimensional transfer functions heightens the importance of densely sampling the voxel data in rendering. With each new axis in the transfer function, there is another dimension along which neighboring voxels can differ. It becomes increasingly likely that the data sample points at the corners of a voxel straddle an important region of the transfer function (such as a region of high opacity) instead of falling within it. Thus, in order for the boundaries to be rendered smoothly, the distance between view-aligned sampling planes through the volume must be very small. Most of the figures in this paper were generated with rates of about 3 to 6 samples per voxel. At this sample rate, frame updates can take nearly a second for a moderately sized ($256 \times 256 \times 128$) shaded and shadowed volume. For this reason, we lower the sample rate during interaction and rerender at the higher sample rate once an action is completed. During interaction, the volume rendered surface will appear coarser, but the surface size and location are usually readily apparent. Thus, even with lower volume sampling rates during interaction, the rendered images are effective feedback for guiding the user in transfer-function exploration.

While the triangular classification widget is based on Levoy's iso-contour classification function, we have found it necessary to have additional degrees of freedom, such as a shear. Shearing the triangle classification along the data value axis, so that higher values are emphasized at higher gradients, allows us to follow the center of some boundaries more accurately. This is a subtle but basic characteristic of boundaries between a material with a narrow distribution of data values and another material with a wide value distribution. This pattern can be observed in the boundary between soft tissue (narrow value distribution) and bone (wide value distribution) of the Visible Male CT, seen in Fig. 9.9. Thresholding the minimum gradient magnitude allows better feature discrimination.

While multidimensional transfer functions are quite effective for visualizing material boundaries, we have also found them to be useful for visualizing the materials themselves. For instance, if we attempt to visualize the dentin of the Human Tooth CT using a 1D transfer function, we erroneously color the background–enamel boundary, seen in Fig. 9.10a. The reason for this can be seen in Fig. 9.2a, where the range of data values that define the background–enamel boundary overlap with the dentin's data values. We can

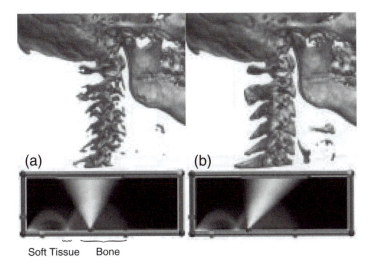

(a)

(b)

Soft Tissue   Bone

Figure 9.9   The soft tissue–bone boundary of the Visible Male CT. It is necessary to shear the triangular classification widget to follow the center of this boundary. (See also color insert.)

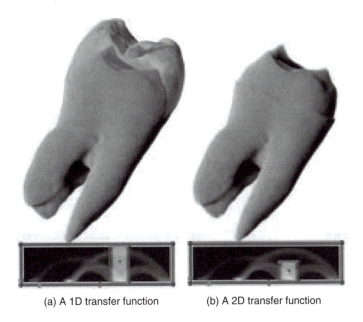

(a) A 1D transfer function   (b) A 2D transfer function

Figure 9.10   The dentin of the Human Tooth CT (a) shows that a 1D transfer function, simulated by assigning opacity to data values regardless of gradient magnitude, will erroneously color the background–enamel boundary. A 2D transfer function (b) can avoid assigning opacity to the range of gradient magnitudes that define this boundary. (See also color insert.)

easily correct this erroneous coloring with a 2D transfer function that gives opacity only to lower-gradient magnitudes. This can be seen in Fig. 9.10b.

A further benefit of dual-domain interaction is the ability to create feature-specific multidimensional transfer functions, which would be extremely difficult to produce by manual placement of classification widgets. If a feature can be visualized in isolation with only a very small and accurately placed classification widget, the best way to place the widget is via dual-domain interaction.

Dual-domain interaction has utility beyond setting multidimensional transfer functions. Dual-domain interaction also helps answer other questions about the limits of direct volume rendering for displaying specific features in the data. For example, the feedback in the transfer-function domain can show the user whether a certain feature of interest detected during spatial domain interaction is well-localized in the transfer function domain. If reprojected voxels from different positions in the same feature map to widely divergent locations in the transfer-function domain, then the feature is not well localized, and it may be hard to create a transfer function that clearly visualizes it. Similarly, if probing inside two distinct features indicates that the reprojected voxels from both features map to the same location in the transfer-function domain, then it may be difficult to selectively visualize one or the other feature.

Acknowledgments

The authors would like to thank Al McPherson from the ACL at LANL for fruitful and provocative conversations about volumetric shadows. This research was funded by grants from the Department of Energy (VIEWS 0F00584), the National Science Foundation (ASC 8920219, MRI 9977218, ACR 9978099), and the National Institutes of Health National Center for Research Resources (1P41RR12553-2). We would also like to thank NVIDIA, ATI, and SGI for making their latest generations of hardware available.

References

1. C. L. Bajaj, V. Pascucci, and D. R. Schikore. The contour spectrum. In *Proceedings IEEE Visualization 1997*, pages 167–173, 1997.

2. U. Behrens and R. Ratering. Adding shadows to a texture-based volume renderer. In *1998 Volume Visualization Symposium*, pages 39–46, 1998.

3. L. D. Bergman, B. E. Rogowitz, and L. A. Treinish. A rule-based tool for assisting color-map selection. In *IEEE Proceedings Visualization 1995*, pages 118–125, 1995.

4. B. Cabral, N. Cam, and J. Foran. Accelerated volume rendering and tomographic reconstruction using texture mapping hardware. In *ACM Symposium On Volume Visualization*, 1994.

5. D. B. Conner, S. S. Snibbe, K. P. Herndon, D. C. Robbins, R. C. Zeleznik, and A. van Dam. 3D widgets. In *Proceedings of the 1992 Symposium on Interactive 3D Graphics*, pages 183–188, 1992.

6. C. Rezk-Salama, K. Engel, M. Bauer, G. Greiner, and T. Ertl. Interactive volume rendering on standard PC graphics hardware using multi-textures and multi-stage rasterization. In *SIGGRAPH/Eurographics Workshop on Graphics Hardware 2000*, 2000.

7. A. Cumani, P. Grattoni, and A. Guiducci. An edge-based description of color images. *GMIP*, 53(4):313–323, 1991.

8. S. Di Zenzo. A note on the gradient of a multi-image. *Computer Vision, Graphics, and Image Processing*, 33(1):116–125, 1986.

9. D. Ebert, C. Morris, P. Rheingans, and T. Yoo. Designing effective transfer functions for volume rendering from photographic volumes. *IEEE TVCG*, 2002.

10. K. Engel, M. Kraus, and T. Ertl. High-quality pre-integrated volume rendering using hardware-accelerated pixel shading. In *SIGGRAPH/Eurographics Workshop on Graphics Hardware 2001*, 2001.

11. A. Van Gelder and K. Kim. Direct volume rendering with shading via 3D textures. In *ACM Symposium On Volume Visualization*, pages 23–30, 1996.

12. T. He, L. Hong, A. Kaufman, and H. Pfister. Generation of transfer functions with stochastic search techniques. In *Proceedings IEEE Visualization 1996*, pages 227–234, 1996.

13. W. Heidrich, R. Westermann, H.-P. Seidel, and T. Ertl. Applications of pixel textures in visualization and realistic image synthesis. In *Proceedings of the 1999 Symposium on Interactive 3D Graphics*, 1999.

14. K. P. Hernandon and T. Meyer. 3D Widgets for exploratory scientific visualization. In *Proceedings of UIST '94 (SIGGRAPH)*, pages 69–70, 1994.

15. J. Hladůvka, A. König, and E. Gröller. Curvature-based transfer functions for direct volume rendering. *Spring Conference on Computer Graphics 2000*, 16(5):58–65, 2000.

16. G. Kindlmann. Semi-automatic generation of transfer functions for direct volume rendering. Master's thesis, Cornell University, Ithaca, NY, 1999 (http://www.cs.utah.edu/~gk/MS).

17. G. Kindlmann and J. W. Durkin. Semi-automatic generation of transfer functions for direct volume rendering. In *IEEE Symposium On Volume Visualization*, pages 79–86, 1998.

18. A. König and E. Gröller. Mastering transfer function specification by using VolumePro technology. *Spring Conference on Computer Graphics 2001*, 17(4): 279–286, 2001.

19. P. Lacroute and M. Levoy. Fast volume rendering using a shear-warp factorization of the viewing transform. In *ACM Computer Graphics (SIGGRAPH '94 Proceedings)*, pages 451–458, 1994.

20. D. H. Laidlaw. Geometric model extraction from magnetic resonance volume data. PhD thesis, California Institute of Technology, 1995.

21. E. LaMar, B. Hamann, and K. I. Joy. Multiresolution techniques for interactive texture-based volume visualization. In *IEEE, Proceedings Visualization '99*, pages 355–361, 1999.

22. M. Levoy. Display of surfaces from volume data. *IEEE Computer Graphics & Applications*, 8(5):29–37, 1988.

23. J. Marks, B. Andalman, P. A. Beardsley, and H. Pfister. Design galleries: a general approach to setting parameters for computer graphics and animation. In *ACM Computer Graphics (SIGGRAPH '97 Proceedings)*, pages 389–400, 1997.

24. D. Marr and E. C. Hildreth. Theory of edge detection. *Proceedings of the Royal Society of London*, B 207:187–217, 1980.

25. M. Meissner, U. Hoffmann, and W. Strasser. Enabling classification and shading for 3D texture mapping based volume rendering using OpenGL and extensions. In *IEEE Visualization 1999*, pages 207–214, 1999.

26. T. Miller and R. C. Zeleznik. The design of 3D haptic widgets. In *Proceedings 1999 Symposium on Interactive 3D Graphics*, pages 97–102, 1999.

27. S. Muraki. Multiscale volume representation by a DoG wavelet. *IEEE Trans. Visualization and Computer Graphics*, 1(2):109–116, 1995.

28. H. Pfister, J. Hardenbergh, J. Knittel, H. Lauer, and L. Seiler. The VolumePro real-time raycasting system. In *ACM Computer Graphics (SIGGRAPH '99 Proceedings)*, pages 251–260, 1999.

29. H. Pfister, C. Bajaj, W. Schroeder, and G. Kindlmann. The transfer function bake-off. In *Proceedings IEEE Visualization 2000*, pages 523–526, 2000.

30. H. Pfister and A. E. Kaufman. Cube-4: a scalable architecture for real-time volume rendering. In *IEEE Symposium On Volume Visualization*, pages 47–54, 1996.

31. J. T. Purciful. 3D widgets for scientific visualization and animation. Master's thesis, University of Utah, 1997.

32. P. Rheingans. Task-based color scale design. In *Proceedings Applied Image and Pattern Recognition*. 1999.

33. G. Sapiro. Color Snakes. *CVIU* 68(2):247–253, 1997.

34. Y. Sato, C.-F. Westin, and A. Bhalerao. Tissue classification based on 3D local intensity structures for volume rendering. *IEEE Transactions on Visualization and Computer Graphics*, 6(2):160–179, 2000.

35. P. S. Strauss and R. Carey. An object-oriented 3D graphics toolkit. In *ACM Computer Graphics (SIGGRAPH '92 Proceedings)*, pages 341–349, 1992.

36. C. Ware. Color sequences for univariate maps: theory, experiments, and principles. *IEEE Computer Graphics and Applications*, 8(5):41–49, 1988.

37. R. Westermann and T. Ertl. Efficiently using graphics hardware in volume rendering applications. In *ACM Computer Graphics (SIGGRAPH '98 Proceedings)*, pages 169–176, 1998.

38. R. C. Zeleznik, K. P. Herndon, D. C. Robbins, N. Huang, T. Meyer, N. Parker, and J. F. Hughes. An interactive toolkit for constructing 3D widgets. *Computer Graphics*, 27(4):81–84, 1993.

10   Pre-Integrated Volume Rendering

MARTIN KRAUS and THOMAS ERTL
Visualization and Interactive Systems Group
University of Stuttgart

10.1   Introduction to Pre-Integrated Volume Rendering

The basic idea of pre-integrated volume rendering is the precomputation of parts of the volume rendering integral. In this sense, it is similar to volume rendering using splatting and even more similar to the accelerated evaluation of the volume rendering integral within a tetrahedron proposed by Max et al. [12]. However, pre-integrated volume rendering is a more general concept that can be applied to many volume rendering algorithms (Section 10.2) and supports several different rendering techniques (Section 10.4). There is also particular research interest in acceleration techniques for pre-integrated volume rendering, which are discussed in Section 10.3.

In this section, we describe the foundation of pre-integrated volume rendering, i.e., pre-integrated classification, and discuss its relation to the volume rendering integral and its numerical evaluation.

10.1.1   Volume Rendering Integral

In principle, any direct volume renderer performs an (approximate) evaluation of the *volume rendering integral* for each pixel, i.e., the integration of attenuated colors and extinction coefficients along each viewing ray. While this is obvious for ray-casting algorithms, many other volume rendering algorithms do not explicitly represent viewing rays. Instead, viewing rays are often defined implicitly by the positions of the eye point and the view plane. Also, the volume rendering integrals are not always evaluated one by one; rather, the integrals for all pixels may be evaluated simultaneously. For example, all object-order algorithms compute the contributions of parts of the volume to all ray integrals and then update the preliminary values of these integrals, usually by updating an intermediate image. In this larger sense, the evaluation of volume rendering integrals is common to all volume rendering algorithms.

We assume that a viewing ray $\mathbf{x}(\lambda)$ is parameterized by the distance λ to the eye point, and that color densities $color(\mathbf{x})$ together with extinction densities $extinction(\mathbf{x})$ may be calculated for any point in space $\mathbf{x}$. The units of color and extinction densities are color intensity per length and extinction strength per length, respectively. However, we will refer to them as colors and extinction coefficients when the precise meaning is clear from the context. The volume rendering integral for the intensity I of one viewing ray is then

$$I = \int_0^D color(\mathbf{x}(\lambda))$$

$$\times \exp\left(-\int_0^{\lambda} extinction(\mathbf{x}(\lambda'))\mathrm{d}\lambda'\right)\mathrm{d}\lambda$$

with the maximum distance D, i.e., there is no color density $color(\mathbf{x}(\lambda))$ for λ greater than D. In other words, color is emitted at each point $\mathbf{x}$ according to the function $color(\mathbf{x})$, and attenuated by the integrated function $extinction(\mathbf{x})$ between the eye point and the point of emission.

Unfortunately, this form of the volume rendering integral is not useful for the visualization of a continuous scalar field $s(\mathbf{x})$ because the calculation of colors and extinction coefficients is not yet specified. We distinguish two steps in the calculation of these colors and extinction coefficients: the *classification* is the assignment of a *primary color* and an extinction coefficient. (The term *primary color* is borrowed from OpenGL terminology [30] in order to denote the color *before* shading.) This classification is achieved by introducing transfer functions for color densities $\tilde{c}(s)$ and extinction densities $\tau(s)$, which map scalar values $s = s(\mathbf{x})$ to colors and extinction coefficients. In general, $\tilde{c}$ is a vector specifying colors in a color space, while τ is a scalar extinction coefficient.

The second step is called *shading* and calculates the color contribution of a point in space, i.e., the function *color*($\mathbf{x}$). The shading depends, of course, on the primary color, but it may also depend on other parameters, e.g., the gradient of the scalar field $\nabla s(\mathbf{x})$, ambient and diffuse lighting parameters, etc. (See also Section 10.4.2.) In the remainder of this section we will not be concerned with shading but only with classification; therefore, we choose a trivial shading, i.e., we identify *color*($\mathbf{x}$) with the primary color $\tilde{c}(s(\mathbf{x}))$ assigned in the classification. Analogously, *extinction*($\mathbf{x}$) is identified with $\tau(s(\mathbf{x}))$. The volume rendering integral is then written as

$$I = \int_0^D \tilde{c}(s(\mathbf{x}(\lambda)))$$
$$\times \exp\left(-\int_0^\lambda \tau(s(\mathbf{x}(\lambda')))d\lambda'\right)d\lambda \qquad (10.1)$$

10.1.2   Pre-Classification and Post-Classification

Direct volume rendering techniques differ considerably in the way they evaluate Equation 10.1. One important and very basic difference is the computation of $\tilde{c}(s(\mathbf{x}))$ and $\tau(s(\mathbf{x}))$. The scalar field $s(\mathbf{x})$ is usually defined by a mesh with scalar values s_i defined at each vertex $\mathbf{v}_i$ of the mesh in combination with an interpolation scheme, e.g., linear interpolation in tetrahedral cells or tri-linear interpolation in cubic cells.

The ordering of the two operations, interpolation and the application of transfer functions, defines the difference between *pre-classification*

and *post-classification*. Post-classification is characterized by the application of the transfer functions *after* the interpolation of $s(\mathbf{x})$ from the scalar values at several vertices (as suggested by Equation 10.1), while pre-classification is the application of the transfer functions *before* the interpolation step, i.e., colors $\tilde{c}(s_i)$ and extinction coefficients $\tau(s_i)$ are calculated in a preprocessing step for each vertex $\mathbf{v}_i$ and then used to interpolate $\tilde{c}(s(\mathbf{x}))$ and $\tau(s(\mathbf{x}))$ for the computation of the volume rendering integral. (The difference is also illustrated in Figs. 10.3a and 10.3b).

10.1.3   Numerical Integration

An analytic evaluation of the volume rendering integral is possible in some cases, in particular for linear interpolation and piecewise linear transfer functions [27,28]. However, this approach is not feasible in general; therefore, a numerical integration is usually required.

The most common numerical approximation of the volume rendering integral in Equation 10.1 is the calculation of a Riemann sum for n equal ray segments of length $d := D/n$. (See also Figs. 10.1 and 10.2 and Section IV.A in Max [10].) It is straightforward to generalize the following considerations to unequally spaced ray segments.

We will approximate the factor

$$\exp\left(-\int_0^\lambda \tau(s(\mathbf{x}(\lambda')))d\lambda'\right)$$

in Equation 10.1 by

$$\exp\left(-\sum_{i=0}^{\lfloor \lambda/d \rfloor} \tau(s(\mathbf{x}(id)))d\right) =$$
$$= \prod_{i=0}^{\lfloor \lambda/d \rfloor} \exp\left(-\tau(s(\mathbf{x}(id)))d\right) \approx \prod_{i=0}^{\lfloor \lambda/d \rfloor}(1 - \alpha_i)$$

where the *opacity* α_i of the ith ray segment is defined by

$$\alpha_i := 1 - \exp\left(-\int_{id}^{i(d+1)} \tau(s(\mathbf{x}(\lambda')))d\lambda'\right) \qquad (10.2)$$

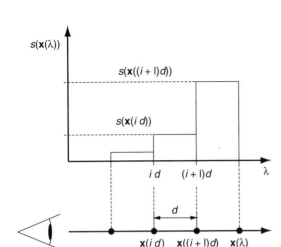

Figure 10.1 Piecewise constant approximation of the function $s(\mathbf{x}(\lambda))$ along a viewing ray.

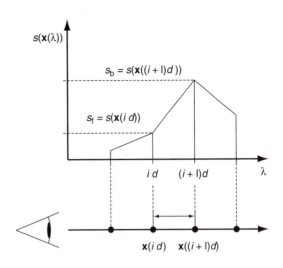

Figure 10.2 Piecewise linear approximation of the function $s(\mathbf{x}(\lambda))$ along a viewing ray.

and approximated by

$$\alpha_i \approx 1 - \exp(-\tau(s(\mathbf{x}(id)))d)$$

This approximation assumes a piecewise constant value of $s(\mathbf{x}(\lambda))$, as illustrated in Fig. 10.1. The result is often further approximated to

$$\alpha_i \approx \tau(s(\mathbf{x}(id)))d$$

We will call $1 - \alpha_i$ the *transparency* of the ith ray segment. Similarly, the color $\tilde{C}_i$ emitted in the ith ray segment is defined by

$$\tilde{C}_i := \int_{id}^{i(d+1)} \tilde{c}(s(\mathbf{x}(\lambda)))$$
$$\times \exp\left(-\int_{id}^{\lambda} \tau(s(\mathbf{x}(\lambda')))d\lambda'\right) d\lambda \qquad (10.3)$$

By neglecting the self-attenuation within the ith ray segment and again assuming a piecewise constant value of $s(\mathbf{x}(\lambda))$, $\tilde{C}_i$ may be approximated by

$$\tilde{C}_i \approx \tilde{c}(s(\mathbf{x}(id)))d \qquad (10.4)$$

Thus, the approximation of the volume rendering integral in Equation 10.1 is

$$I \approx \sum_{i=0}^{\lfloor D/d \rfloor} \tilde{C}_i \prod_{j=0}^{i-1} (1 - \alpha_j) \qquad (10.5)$$

Therefore, a back-to-front compositing algorithm will implement the equation

$$\tilde{C}_i' = \tilde{C}_i + (1 - \alpha_i)\tilde{C}_{i+1}' \qquad (10.6)$$

where $\tilde{C}_i'$ is the color accumulated from all ray segments j with $j \geq i$. The compositing equations for a front-to-back compositing algorithm are

$$\tilde{C}_i'' = \tilde{C}_{i-1}'' + (1 - \alpha_{i-1}'')\tilde{C}_i \text{ and}$$
$$\alpha_i'' = \alpha_{i-1}'' + (1 - \alpha_{i-1}'')\alpha_i \qquad (10.7)$$

where the color accumulated in the first i ray segments is denoted by $\tilde{C}_i''$ and the accumulated opacity is denoted by α_i''.

$\tilde{c}(s)$ is often specified by a product of two transfer functions $\tau(s)c(s)$. (See Max [10] for a comparison between a direct specification of the transfer function $\tilde{c}(s)$ and the specification of $c(s)$.) Substituting $\tilde{c}(s)$ by $\tau(s)c(s)$ in Equation 10.3 leads to the approximation

$$\tilde{C}_i \approx \tau(s(\mathbf{x}(id)))c(s(\mathbf{x}(id)))d$$

In this case, it is more common to use non-associated colors $C_i := \tilde{C}_i / \alpha_i$, i.e.,

$$\tilde{C}_i = \alpha_i C_i \text{ with } C_i \approx c(s(\mathbf{x}(id)))$$

This results in the approximation

$$I \approx \sum_{i=0}^{\lfloor D/d \rfloor} \alpha_i C_i \prod_{j=0}^{i-1} (1 - \alpha_j)$$

with the corresponding back-to-front compositing equation

$$\tilde{C}_i' = \alpha_i C_i + (1 - \alpha_i)\tilde{C}_{i+1}' \tag{10.8}$$

and the front-to-back compositing equations

$$\tilde{C}_i'' = \tilde{C}_{i-1}'' + (1 - \alpha_{i-1}'')\alpha_i C_i \text{ and}$$
$$\alpha_i'' = \alpha_{i-1}'' + (1 - \alpha_{i-1}'')\alpha_i \tag{10.9}$$

These compositing equations indicate that $\tilde{C}$ corresponds to a *premultiplied color* αC, which is also called *opacity-weighted color* (see [29]) or *associated color*. According to Blinn [1], associated colors have their opacities associated with them, i.e., they are regular colors composited on black. Blinn also notes that some intensity computations result in associated colors, although they are not explicitly multiplied by an opacity. In this sense, the transfer function $\tilde{c}(s)$ is, in fact, a transfer function for an associated color density.

The discrete approximation of the volume rendering integral will converge to the correct result for $d \to 0$, i.e., for high sampling rates $1/d$. According to the sampling theorem, a correct reconstruction is possible only with sampling rates greater than the Nyquist frequency. However, nonlinear features of transfer functions may considerably increase the sampling rate required for a correct evaluation of the volume rendering integral, as this sampling rate depends on the product of the Nyquist frequencies of the scalar field $s(\mathbf{x})$ and the maximum of the Nyquist frequencies of the two transfer functions $\tilde{c}(s)$ and $\tau(s)$, or of the product $c(s)\tau(s)$. (See Schulze et al. [19] and Section 2.4.3 in Kraus [7] for details.)

Thus, it is by no means sufficient to sample the volume rendering integral with the Nyquist frequency of the scalar field if nonlinear transfer functions are employed. Artifacts resulting from this kind of undersampling are frequently observed unless they are avoided by very smooth transfer functions, i.e., transfer functions with small Nyquist frequencies.

10.1.4   Pre-Integrated Classification

In order to overcome the limitations discussed above, the approximation of the volume rendering integral has to be improved. In fact, many improvements have been proposed, e.g., higher-order integration schemes, adaptive sampling, etc. However, these methods do not explicitly address the problem of high sampling frequencies required for nonlinear transfer functions. With *pre-integrated classification*, these high sampling frequencies are avoided by reconstructing a piecewise linear, continuous scalar function along the viewing ray and evaluating the volume rendering integral between each pair of successive samples of the scalar field by table lookups. This allows us to avoid the problematic product of Nyquist frequencies mentioned in the previous section, since the sampling rate for the reconstruction of the scalar function along the viewing ray is independent of the transfer functions.

For the purpose of pre-integrated classification, the sampled values of the 3D scalar field on a viewing ray define a 1D, piecewise linear scalar field, which approximates the original scalar field along the viewing ray. The volume rendering integral for this piecewise linear scalar field is efficiently computed by one table lookup for each ray segment. The three arguments of this table lookup for the ith ray segment from $\mathbf{x}(i)$ to $\mathbf{x}(id)$ are the scalar value at the start (front) of the segment $s_f := s(\mathbf{x}(id))$, the scalar value at the end (back) of the segment $s_b := s(\mathbf{x}((i + 1)d))$, and the length of the segment d (Fig. 10.2). For the purpose of illustration, we assume that the lengths of the segments are all equal to a constant d. In this case, the table lookup is independent of d.

More precisely spoken, the opacity α_i of the ith segment defined in Equation 10.2 is approximated by

$$\alpha_i \approx 1 - \exp\left(-\int_0^1 \tau((1-\omega)s_f + \omega s_b)d d\omega\right)$$
(10.10)

Thus, α_i is a function of s_f, s_b, and d (or of s_f and s_b if the lengths of the segments are equal).

The (associated) color $\tilde{C}_i$ defined in Equation 10.3 is approximated correspondingly:

$$\tilde{C}_i \approx \int_0^1 \tilde{c}((1-\omega)s_f + \omega s_b)$$
$$\times \exp\left(-\int_0^\omega \tau((1-\omega')s_f + \omega' s_b)d d\omega'\right)d d\omega$$
(10.11)

Analogously to α_i, $\tilde{C}_i$ is a function of s_f, s_b, and d.

Thus, pre-integrated classification will approximate the volume rendering integral by evaluating Equation 10.5.

$$I \approx \sum_{i=0}^{\lfloor D/d \rfloor} \tilde{C}_i \prod_{j=0}^{i-1}(1-\alpha_j)$$

with colors $\tilde{C}_i$ precomputed according to Equation 10.11 and opacities α_i precomputed according to Equation 10.10.

For nonassociated color-transfer functions, i.e., if $\tilde{c}(s)$ is substituted by $\tau(s)c(s)$, we will also employ Equation 10.10 for the approximation of α_i and the following approximation of the associated color $\tilde{C}_i^\tau$:

$$\tilde{C}_i^\tau \approx \int_0^1 \tau((1-\omega)s_f + \omega s_b)c((1-\omega)s_f + \omega s_b)$$
$$\times \exp\left(-\int_0^\omega \tau((1-\omega')s_f + \omega' s_b)d d\omega'\right)d d\omega$$
(10.12)

Note that pre-integrated classification always computes associated colors, whether a transfer function for associated colors $\tilde{c}(s)$ or for nonassociated colors $c(s)$ is employed.

In both cases, pre-integrated classification allows us to sample a continuous scalar field $s(\mathbf{x})$ without the need to increase the sampling rate for any nonlinear transfer function. Therefore, pre-integrated classification has the potential to

improve the accuracy (by less undersampling) and the performance (by fewer sampling operations) of a volume renderer at the same time.

Figure 10.3c summarizes the basic steps of pre-integrated classification and compares them with pre-classification (Fig. 10.3a) and post-classification (Fig. 10.3b).

10.2 Pre-Integrated Volume Rendering Algorithms

Pre-integrated classification is not restricted to a particular volume rendering algorithm; rather, it may replace the post-classification step of various algorithms as demonstrated by several publications in recent years. Instead of discussing these pre-integrated variants of well-known volume rendering algorithms in detail, this section attempts to give an overview of the literature on pre-integrated volume rendering algorithms.

10.2.1 Pre-Integrated Cell Projection

Cell projection is probably the most common method of exploiting graphics hardware for the rendering of tetrahedral meshes and unstructured meshes in general. Three different pre-integrated cell-projection algorithms are described in this section.

10.2.1.1 Shading Polyhedra

For the special case of a constant color per polyhedral cell, pre-integrated volume rendering is very similar to the shading of polyhedral cells suggested by Max et al. [12]. In our nomenclature, the algorithm scan converts the front and back faces of a polyhedron in order to compute the scalar values s_f on the front faces, s_b on the back faces, and the distance d between front and back faces (i.e., the thickness of the cell) for each pixel. Instead of tabulating the opacity $\alpha(s_f, s_b, d)$ of Equation 10.10, the algorithm employs a tabulated integral function of $\tau(s)$ to compute $\alpha(s_f, s_b, d)$. (This evaluation corresponds to

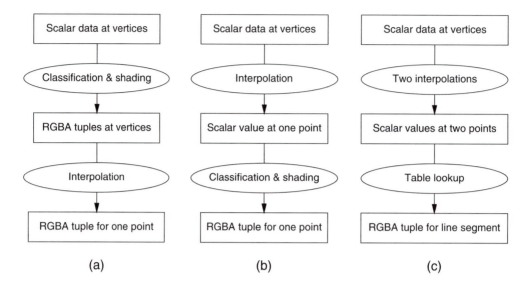

Figure 10.3   Data-flow schemes for (a) pre-classification, (b) post-classification, and (c) pre-integrated classification.

Equation 10.15.) For a constant transfer function for color $c(s) =: C$, Equation 10.12 is given by

$$\tilde{C}^{\tau}(s_f, s_b, d) = C \int_0^1 \tau((1 - \omega)s_f + \omega s_b)$$
$$\times \exp\left(- \int_0^{\omega} \tau((1 - \omega')s_f + \omega' s_b)d d\omega'\right) d d\omega$$
$$= C\left(1 - \exp\left(- \int_0^1 \tau((1 - \omega')s_f + \omega' s_b)d d\omega'\right)\right)$$
$$= C\alpha(s_f, s_b, d)$$

Thus, the computation of colors is basically free.

10.2.1.2   *Projected Tetrahedra Algorithm*

The first cell-projection algorithm that exploited graphics hardware efficiently was the Projected Tetrahedra (PT) algorithm by Shirley and Tuchman [20].

The original algorithm, which is restricted to tetrahedral cells, classifies each tetrahedron according to its projected profile and decomposes the projected tetrahedron into smaller triangles (Fig. 10.4). Colors and opacities are calculated only for the triangle vertices

using ray integration in software, while graphics hardware is employed to interpolate these colors and opacities linearly within the triangles. This, however, is an approximation that leads to rendering artifacts [11,12].

A pre-integrated variant of the PT algorithm was published by Röttger et al. [18]. In fact, pre-integrated volume rendering is particularly useful for tetrahedral cells, since scalar values are usually interpolated linearly within tetrahedra, i.e., the scalar field along a viewing ray varies linearly between two samples. Fig. 10.5 depicts the intersection of a viewing ray with a tetrahedral cell. More precisely spoken, this particular tetrahedron corresponds to a triangle generated by the PT decomposition. The goal is to render this tetrahedron by rasterizing its front face. For orthographic projections, s_f, s_b, and d vary linearly on the projected front face; thus, these parameters may be specified as texture coordinates at the vertices of the front face. Graphics hardware is then employed to linearly interpolate these texture coordinates and to perform a texture fetch in a 3D texture that contains pre-integrated colors $\tilde{C} = \tilde{C}(s_f, s_b, d)$ and

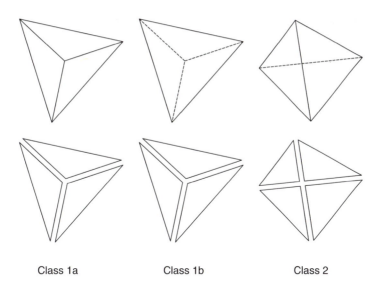

Class 1a   Class 1b   Class 2

Figure 10.4   Classification of nondegenerate projected tetrahedra (top row) and the corresponding decompositions (bottom row) [20].

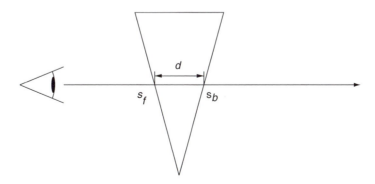

Figure 10.5   Intersection of a viewing ray with a tetrahedron corresponding to one of the triangles of the PT decomposition. s_f and s_b are the scalar values on the entry (front) face and exit (back) face, respectively; d denotes the thickness of the cell for this ray.

opacities $\alpha = \alpha(s_f, s_b, d)$ (see Equations 10.10–10.12). Thus, the texture fetch implements the lookup in a pre-integrated lookup table.

As discussed in Section 10.3.3, there are approximations that allow us to replace the 3D lookup table by a 2D lookup table. Implementations of these approximations for pre-inte-

grated cell projection with 2D textures are discussed by Röttger et al. [18] and Guthe et al. [4]. Moreover, the 3D texture can be replaced by a set of 2D textures with the help of a particular transformation of texture coordinates as demonstrated by Roettger and Ertl [16].

10.2.1.3   View-Independent Cell Projection

The main disadvantage of the *projected tetrahedra* algorithm is the need to perform the decomposition of tetrahedra in software. View-independent cell projection, on the other hand, allows the graphics hardware to perform the complete projection; thus, bandwidth requirements are dramatically reduced, provided that the tetrahedra data are stored on the graphics board.

A pre-integrated, view-independent projection of tetrahedra was discussed by Weiler et al. [22]. The basic idea is to rasterize only front faces of tetrahedra while the rasterization of back faces is avoided by back-face culling. For each pixel, the rasterization of a front face has to compute the entry and exit point of a viewing ray for the corresponding tetrahedron. Then, the scalar values s_f and s_b at these points, and the distance d between them, can be calculated. As demonstrated by Weiler et al. [22], all these computations can be performed by programmable graphics hardware. Based on s_f, s_b, and d, the color and opacity of the tetrahedron can be determined with the help of texture mapping, as discussed in Section 10.2.1.2.

10.2.2   Pre-Integrated Texture-Based Volume Rendering

Texture-based volume rendering rasterizes a stack of textured slices with either a stack of 2D textures (Fig. 10.6a, [15]) or one 3D texture (Fig. 10.6b, [2]).

Engel et al. [3] published a pre-integrated variant of texture-based volume rendering that may be characterized by the idea of rendering slabs instead of slices. For each rasterized pixel, pre-integrated classification requires the scalar data at the front and the back slice of each slab (Fig. 10.7a) between two adjacent slices (either object aligned or view aligned). Thus, the textures of the two slices have to be mapped onto one slice—either the front or the back slice; the latter case is illustrated in Fig. 10.7b. This mapping of the two textures onto one slice requires multitexturing and an appropriate calculation of texture coordinates. In this way, the scalar values of both slices are fetched for every pixel of the slice corresponding to one slab. These two scalar values are necessary for a third texture lookup operation, which fetches pre-integrated colors and opacities. As this texture lookup depends on previously fetched data, it is called a *dependent* texture lookup.

This approach is one of the most popular pre-integrated volume rendering algorithms; it has also been employed, for example, by Meißner et al. [13] and Roettger et al. [17].

10.2.3   Pre-Integrated Ray-Casting

Ray-casting is the most important image-order volume rendering algorithm. While the evaluation of the volume rendering integral discussed in Section 10.1.3 is less obvious for many other volume rendering algorithms, ray-casting usually computes just one volume rendering

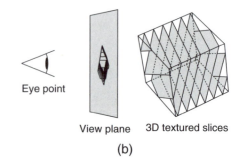

Eye point	Eye point
View plane 2D textured slices	View plane 3D textured slices
(a)	(b)

Figure 10.6   (a) Direct volume rendering with 2D textured slices. The slices are aligned to the volumetric object. (b) Direct volume rendering with 3D textured slices. The slices are usually aligned to the view plane.

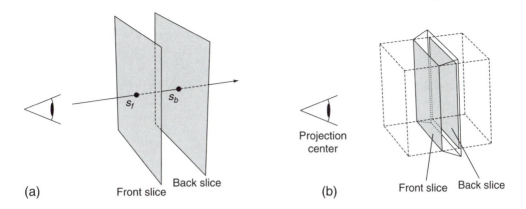

Figure 10.7 (a) A slab of the volume between two slices. The scalar value on the front (or back) slice for a particular viewing ray is called s_f (or s_b). (b) Projection of the front slice onto the back slice of a slab.

integral per pixel in order to determine the pixel's color. Therefore, the discussion of pre-integrated classification in Section 10.1.4 already covers its application to ray-casting algorithms.

A pre-integrated ray-casting system in software was described by Knittel [6], while a hardware solution with pre-integrated classification was presented by Meißner et al. [14]. Ray-casting algorithms have also been implemented in programmable graphics hardware. A pre-integrated variant for unstructured meshes was published by Weiler et al. [23]. As the computation of each cell's color and opacity is very similar to the view-independent cell projection described in Section 10.2.1.3, we will not discuss any details here. Roettger et al. [17] published an implementation of a pre-integrated ray-casting algorithm for uniform meshes in programmable graphics hardware. They also introduce an extension, called *adaptive pre-integration*, that chooses a step length depending on measures precomputed for each voxel.

10.2.4 Pre-Integrated Shear-Warp Algorithm

The shear-warp algorithm is a popular software volume rendering algorithm [8]. This algorithm is conceptually close to a software implementation of object-aligned texture-based volume rendering (see Section 10.2.2).

A pre-integrated variant of the shear-warp algorithm was published by Schulze et al. [19]. Similarly to texture-based volume rendering, the pre-integrated version renders slabs instead of slices; therefore, Schulze et al. introduce an additional slice buffer in order to have access to both scalar values, the value on the front slice and the value on the back slice of each slab.

10.3 Accelerated Pre-Integration

The primary drawback of pre-integrated classification in general is the required precomputation of the lookup tables that map the three integration parameters (scalar value at the front s_f, scalar value at the back s_b, and distance between samples d) to pre-integrated colors $\tilde{C} = \tilde{C}(s_f, s_b, d)$ defined by Equation 10.11 (or $\tilde{C}^\tau = \tilde{C}^\tau(s_f, s_b, d)$ defined by Equation 10.12) and opacities $\alpha = \alpha\,(s_f, s_b, d)$ defined by Equation 10.10. As these tables depend on the transfer functions, any modification of the transfer functions requires an update of the lookup tables. This might be no concern for games and entertainment applications, but it strongly limits the interactivity of applications in the domain of scientific volume visualization, which often depends on user-specified transfer functions.

Therefore, several techniques have been developed to accelerate the pre-integration step.

Note that many of these techniques not only accelerate the pre-integration but also reduce the size of pre-integrated lookup tables.

10.3.1 Local Updates of Lookup Tables

A local modification of transfer functions for a particular scalar value s does not require one to update the complete corresponding lookup table. In fact, only the entries $\tilde{C}(s_f, s_b, d)$ and $\alpha(s_f, s_b, d)$ with $s_f \leq s \leq s_b$ or $s_f \geq s \geq s_b$ have to be updated; i.e., about half of the pre-integrated lookup table has to be updated in the worst case.

10.3.2 Particular Optical Models

There are several interesting optical models that can be implemented with 2D lookup tables. As there is only one entry in these tables for each pair of s_f and s_b (instead of many entries for different values of d), their computation is usually far less costly.

10.3.2.1 Maximum Intensity Projection

Maximum intensity projection (MIP) computes the maximum intensity along each viewing ray. In our notation, $\tilde{C}(s_f, s_b, d)$ has to be the maximum intensity between scalar values s_f and s_b. As this maximum intensity $\tilde{C}(s_f, s_b, d)$ is independent of d, a 2D table is sufficient for $\tilde{C}$. Note that the back-to-front compositing for maximum intensity projection is given by the following (compare Equation 10.6):

$$\tilde{C}'_i = \max(\tilde{C}_i, \tilde{C}'_{i+1})$$

10.3.2.2 Emission Only

If there is no absorption (i.e., $\tau(s) = 0$ for all s), $\alpha(s_f, s_b, d)$ is always 0 and $\tilde{C}(s_f, s_b, d)$ becomes

$$\tilde{C}(s_f, s_b, d) = d \int_0^1 \tilde{c}((1 - \omega)s_f + \omega s_b) d\omega')$$

Thus, there is only a linear dependency on d, and a 2D lookup table for $\tilde{C}(s_f, s_b, d)/d$ is sufficient. (See also Section 10.3.3.)

10.3.2.3 Absorption Only

If there is no emission (i.e., $\tilde{c}(s) = 0$ for all s), $\tilde{C}(s_f, s_b, d)$ is 0. In this case, the opacity $\alpha(s_f, s_b, d)$ should be evaluated in two steps, as suggested by Max et al. [12]:

$$\tau'(s_f, s_b) = \int_0^1 \tau((1 - \omega)s_f + \omega s_b) d\omega \qquad (10.13)$$

$$\alpha(s_f, s_b, d) = 1 - \exp(-\tau'(s_f, s_b)d) \qquad (10.14)$$

Thus, the integration can be replaced by a 2D lookup table for $\tau'(s_f, s_b)$. In order to evaluate Equation 10.14, a 2D lookup table for

$$\alpha(\tau, d) = 1 - \exp(-\tau d)$$

can be employed as proposed by Stein et al. [21].

10.3.3 Pre-Integration for a Single Sampling Distance

In some cases, it is possible to reduce the dimensionality of pre-integrated lookup tables from three to two (only s_f and s_b) by assuming a constant sampling distance, i.e., a constant length of the ray segments. Obviously, this applies to ray-casting with equidistant samples. It also applies to 3D texture-based volume visualization with orthographic projection and is a good approximation for most perspective projections.

Even if different sampling distances occur, the dependency on the sampling distance is often approximated by a linear dependency, as suggested by Roettger et al. [17,18] or by a more accurate opacity correction, as suggested by Schulze et al. [19]: Assuming that opacities $\alpha(s_f, s_b, d')$ have been computed for a constant sampling distance d' according to Equation 10.14, the corrected opacity $\alpha(s_f, s_b, d)$ for a different sampling distance d may be computed by

$$\begin{aligned} \alpha(s_f, s_b, d) &= 1 - \exp(-\tau'(s_f, s_b)d) \\ &= 1 - \exp(-\tau'(s_f, s_b)d')^{d/d'} \\ &= 1 - (1 - \alpha(s_f, s_b, d'))^{d/d'} \end{aligned}$$

This correction can be efficiently computed by a lookup table for the mapping

$$\alpha \mapsto 1 - (1 - \alpha)^{d/d'}$$

If colors $\tilde{C}(s_f, s_b, d')$ have also been precomputed for a constant sampling distance d', they have to be corrected by the same factor $\alpha(s_f, s_b, d)/\alpha(s_f, s_b, d')$ [19].

Another way of exploiting Equations 10.13 and 10.14 is to precompute a 2D lookup table for $\tau'(s_f, s_b)$, as defined by Equation 10.13, and evaluate Equation 10.14 for $\alpha(s_f, s_b, d)$ only when it is required. Guthe et al. [4] show how to employ dependent textures for this task. As there is no corresponding formula for pre-integrated colors, Guthe et al. suggest a linear approximation, as mentioned above, or a more general polynomial approximation. However, the latter requires the computation of pre-integrated colors for several different segment lengths.

10.3.4   Pre-Integration Without Self-Attenuation of Segments

For small ray segments, the preintegration may be greatly accelerated by evaluating the integrals in Equations 10.10–10.12 with the help of integral functions for $\tau(s)$, $\tilde{c}(s)$ and $\tau(s)c(s)$, respectively. This technique neglects the self-attenuation of segments (as in Equation 10.4 for post-classification) and was published by Engel et al. [3]. A similar technique was first published by Max et al. [12]; see Section 10.2.1.1. More specifically, Equation 10.10 for $\alpha(s_f, s_b, d)$ can be rewritten as

$$
\begin{aligned}
\alpha(s_f, s_b, d) &\approx 1 - \exp\left(-\int_0^1 \tau((1-\omega)s_f + \omega s_b)d\,d\omega\right) \\
&= 1 - \exp\left(-\frac{d}{s_b - s_f}\int_{s_f}^{s_b} \tau(s)ds\right) \qquad (10.15) \\
&= 1 - \exp\left(-\frac{d}{s_b - s_f}(T(s_b) - T(s_f))\right)
\end{aligned}
$$

with the integral function $T(s) := \int_0^s \tau(s')ds'$, which is easily computed in practice, as the scalar values s are usually quantized.

Equation 10.11 for $\tilde{C}(s_f, s_b, d)$ may be approximated analogously; however, this requires one to neglect the attenuation within a ray segment. In fact, this is a common approximation

for post-classified volume rendering and is well justified for small products $\tau(s)d$, i.e., for small distances d.

$$
\begin{aligned}
\tilde{C}(s_f, s_b, d) &\approx \int_0^1 \tilde{c}((1-\omega)s_f + \omega s_b)d\,d\omega \\
&= \frac{d}{s_b - s_f}\int_{s_f}^{s_b}\tilde{c}(s)ds \qquad (10.16) \\
&= \frac{d}{s_b - s_f}(K(s_b) - K(s_f))
\end{aligned}
$$

with the integral function $K(s) := \int_0^s \tilde{c}(s')ds'$. For the nonassociated color transfer function $c(s)$, we approximate Equation 10.12 by

$$
\begin{aligned}
\tilde{C}^\tau(s_f, s_b, d) &\approx \int_0^1 \tau((1-\omega)s_f + \omega s_b) \\
&\quad \times c((1-\omega)s_f + \omega s_b)d\,d\omega \\
&= \frac{d}{s_b - s_f}\int_{s_f}^{s_b}\tau(s)c(s)ds \qquad (10.17) \\
&= \frac{d}{s_b - s_f}(K^\tau(s_b) - K^\tau(s_f))
\end{aligned}
$$

with $K^\tau(s) := \int_0^s \tau(s')c(s')ds'$.

Thus, instead of numerically computing the integrals in Equations 10.10–10.12 for each combination of s_f, s_b, and d, it is possible to compute the integral functions $T(s)$, $K(s)$, or $K^\tau(s)$ only once and employ these to evaluate colors and opacities according to Equations 10.15, 10.16, or 10.17 without any further integration. Moreover, 2D lookup tables are sufficient for $\tilde{C}(s_f, s_b, d)/d$ and $\tilde{C}^\tau(s_f, s_b, d)/d$. The opacity $\alpha(s_f, s_b, d)$ may also be evaluated with the help of 2D lookup tables, as mentioned in the dicussion of Equation 10.14.

10.3.5   Hardware-Accelerated Pre-Integration

Pre-integrated lookup tables are often implemented with hardware-accelerated texture mapping. In this case, the computation of pre-integrated lookup tables in graphics hardware as suggested by Roettger and Ertl [16] is particularly beneficial because the transfer of the lookup table from main memory into texture memory is avoided.

The basic idea is to evaluate the entries of a pre-integrated lookup table for all values of s_f and s_b (but a constant d) at the same time. These entries form a 2D image that is computed in the frame buffer and then transferred to texture memory before processing the next value of d. The RGBA color of each pixel is given by Equations 10.10 and 10.11 (or Equations 10.10 and 10.12). The integrals in these equations are discretized as discussed in Section 10.1.3. Note in particular that Equations 10.6 and 10.8 can be implemented by hardware-accelerated blending operations.

10.3.6   Incremental Pre-Integration

Incremental pre-integration is another acceleration technique for the computation of $\alpha(s_f, s_b, d)$, $\tilde{C}(s_f, s_b, d)$, and $\tilde{C}^\tau(s_f, s_b, d)$ (see Equations 10.10–10.12). Provided that a lookup table has been calculated for all entries with a sampling distance less than or equal to d', the entries for the next segment length $d = d' + \Delta d$ can be calculated by splitting the integrals into one part of length Δd and one part of length d'. The scalar value s_p at the split point is interpolated linearly between s_f and s_b, i.e., $s_p = (d' s_f + \Delta d s_b)/(\Delta d + d')$. As the integrals for these parts are already tabulated, the evaluation is reduced to table lookups and a blending operation. More specifically, $\alpha(s_f, s_b, d' + \Delta d)$ may be computed as

$$\alpha(s_f, s_b, d' + \Delta d) = \alpha(s_f, s_p, \Delta d) \\ + (1 - \alpha(s_f, s_p, \Delta d))\, \alpha(s_p, s_b, d')$$

$\tilde{C}(s_f, s_b, d)$ and $\tilde{C}^\tau(s_f, s_b, d)$ are given by

$$\tilde{C}(s_f, s_b, d' + \Delta d) = \tilde{C}(s_f, s_p, \Delta d) \\ + (1 - \alpha(s_f, s_p, \Delta d))\tilde{C}(s_p, s_b, d')$$

and

$$\tilde{C}^\tau(s_f, s_b, d' + \Delta d) = \tilde{C}^\tau(s_f, s_p, \Delta d) \\ + (1 - \alpha(s_f, s_p, \Delta d))\tilde{C}^\tau(s_p, s_b, d')$$

Incremental pre-integration was first published by Weiler et al. [23]. However, a very similar technique was independently developed by

Guthe for the ray caster described by Roettger et al. [17].

More recently, a further optimization for the generation of the pre-integrated lookup table, which is incremental in s_f and s_b instead of d, was suggested by Lum et al. [9].

10.4   Pre-Integrated Rendering Techniques

Pre-integrated volume rendering not only improves the accuracy of volume rendering algorithms but is also suitable for several particular rendering techniques. Some of these, namely isosurface rendering, volume shading, and volume clipping, are discussed in this section.

10.4.1   Isosurface Rendering

Rendering of isosurfaces is only a special case of direct volume visualization with appropriate transfer functions. For pre-integrated volume rendering, these transfer functions correspond to particularly simple, 2D pre-integrated lookup tables, which can be derived as follows.

10.4.1.1   Lookup Tables for Isosurface Rendering

In order to render the isosurface for an isovalue s_{iso}, the opacity transfer function $\tau(s)$ should be defined by $\tau(s) = 0$ for $s \neq s_{iso}$ and "$\tau(s_{iso}) = \infty$". Formally, we set $\tau(s) = \xi\delta(s - s_{iso})$ with a constant ξ and Dirac's delta function $\delta(x)$ [24]; multiple isosurfaces correspond to a sum of delta functions. As $\tilde{c}(s_{iso})$ and $c(s_{iso})$ are constant, we are only interested in the value of α as defined in Equation 10.10:

$$1 - \alpha = \exp\left(-\int_0^1 \tau((1-\omega)s_f + \omega s_b)d\,d\omega\right)$$
$$= \exp\left(-\int_0^1 \xi\delta((1-\omega)s_f + \omega s_b - s_{iso})d\,d\omega\right)$$
$$= \exp\left(-\int_0^1 \xi\left|\frac{1}{s_b - s_f}\right|\delta\left(\omega - \frac{s_{iso} - s_f}{s_b - s_f}\right)d\,d\omega\right)$$
$$= \exp\left(-\xi' H\left(\frac{s_{iso} - s_f}{s_b - s_f}\right)H\left(1 - \frac{s_{iso} - s_f}{s_b - s_f}\right)\right)$$

with $\xi' = \xi \left| \frac{1}{s_b - s_f} \right| d$ and the Heaviside step function $H(x)$ [24]. Thus, for $\xi \to \infty$ we obtain

$$\alpha = H\left(\frac{s_{\mathrm{iso}} - s_f}{s_b - s_f}\right) H\left(\frac{s_b - s_{\mathrm{iso}}}{s_b - s_f}\right)$$

which is independent of d. The dependency on s_f and s_b results in a checkerboard-like lookup table, which is visualized in Fig. 10.8 for three different values of s_{iso}. These 2D lookup tables are in fact special cases of the 3D lookup table defined by Equation 10.10. An alternative derivation of this result, which is closer to the previous work by Westermann and Ertl [25], is given by Röttger et al. [18].

The resulting lookup table may also be described as follows: its α-component, i.e., the opacity, has to be 1 for opaque isosurfaces if either s_f or s_b (but not both) is less than the iso-value, and 0 otherwise (see Fig. 10.8).

Usually, the RGB components of these lookup tables are constant and define the (uniform) color of an isosurface. However, if the colors of the two faces of an isosurface are different, the two rectangular regions in these lookup tables have to have different colors, as shown in Fig. 10.8c.

Unfortunately, the limited accuracy of the pre-integrated lookup table for isosurfaces will usually cause rendering artifacts. Holes in isosurfaces can be avoided by slightly modifying the lookup table, effectively "thickening" the isosurface. This eliminates artifacts for opaque isosurfaces; for partially transparent isosurfaces, however, this will visually enhance the sampling pattern by rasterizing some of the pixels of an isosurface more than once. Removing these artifacts for partially transparent isosurfaces in general is an open problem. For the special case of a single, semitransparent isosurface, these artifacts can be avoided by additional rendering passes [3,7]. More recent research suggests that at least some of these artifacts are due to the use of incorrect interpolation for perspective projection.

10.4.1.2 Multiple Isosurfaces

When rendering multiple isosurfaces, the 2D lookup tables of the individual isosurfaces have to be combined. An example of a combined lookup table is sketched in Fig. 10.9a, which shows the combination of the lookup tables from Fig. 10.8. The "visibility ordering" is easy to understand: for $s_f < s_b$ (upper left half) we view along the gradient of the scalar field; thus, isosurfaces for smaller iso-values occlude those for greater iso-values, and vice versa for $s_f > s_b$.

Assuming that the whole volume is rendered, the number of isosurfaces does not affect the rendering time. This feature is shared with, for example, Westermann's algorithm for multiple isosurfaces [26].

10.4.1.3 Shaded Isosurfaces

The shading calculation for isosurfaces is usually based on the normalized gradient of

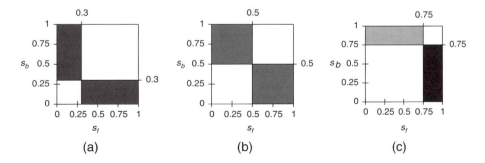

(a)   (b)   (c)

Figure 10.8 2D pre-integrated lookup tables for three iso-values: (a) iso-value 0.3, (b) iso-value 0.5, and (c) iso-value 0.75, with different shadings for the two faces of an isosurface.

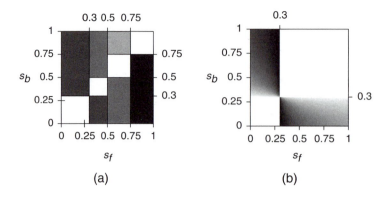

Figure 10.9   (a) The correct combination of the lookup tables from Fig. 10.8 in a single lookup table for multiple isosurfaces. (b) A modulation of the lookup table depicted in Fig. 10.8a with the weights $\omega = \frac{s_{\mathrm{iso}} - s_b}{s_f - s_b}$ for $s_{\mathrm{iso}} = 0.3$.

the scalar field, because gradients are perpendicular to isosurfaces. Note that shading of isosurfaces is closely related to volume shading, which is discussed in Section 10.4.2.

For a single pre-integrated isosurface, there are basically two different approaches to shading. The first is to perform the shading calculation at the sampling positions and to linearly interpolate the two resulting colors, say C_f at the start of the ray segment and C_b at the end. This interpolation results in the shaded color C_{iso} of the isosurface. The interpolation weight ω depends on the position of the intersection of the isosurface with the ray segment, i.e., it depends on the position of the iso-value s_{iso} within the interval between s_f and s_b:

$$C_{\mathrm{iso}} = \omega C_f + (1 - \omega)C_b \quad \text{with} \quad \omega = \frac{s_{\mathrm{iso}} - s_b}{s_f - s_b}$$

The second approach determines the normalized gradients at the sampling positions, say $\mathbf{g}_f$ and $\mathbf{g}_b$ (see also Fig. 10.10). The shading calculation for one ray segment is then based on a linearly interpolated and normalized gradient $\mathbf{g}_{\mathrm{iso}}$:

$$\mathbf{g}_{\mathrm{iso}} = \frac{\omega \mathbf{g}_f + (1 - \omega)\mathbf{g}_b}{|\omega \mathbf{g}_f + (1 - \omega)\mathbf{g}_b|} \quad \text{with} \quad \omega = \frac{s_{\mathrm{iso}} - s_b}{s_f - s_b}$$

In both cases a weight $\omega = \omega(s_f, s_b, s_{\mathrm{iso}})$ is required that may be tabulated in a lookup table; an example is depicted in Fig. 10.9b. The

first approach was employed for cell projection [18,25], while the second approach (without the normalization) was employed for texture-based volume rendering [3].

10.4.1.4 Mixing Isosurfaces with Semitransparent Volumes

Pre-integrated isosurfaces can be combined with semitransparent volumes either with separate or with combined lookup tables. In the former case, the semitransparent volume must be clipped in order to hide those parts of the volume that are occluded by isosurfaces (see Section 10.4.3). In the latter case, pre-integrated lookup tables are computed by allowing for iso-values in the evaluation of Equations 10.10, 10.11, or 10.12. That is, for opaque isosurfaces the ray integration has to be stopped as soon as one of the iso-values is reached [18]. Both approaches can be generalized to partially transparent isosurfaces.

10.4.2 Volume Shading

Volume shading applies illumination models (particularly Phong illumination) to volume rendering. For this purpose, the surface normal is usually replaced by the normalized gradient of the scalar field (see also Section 10.4.1.3). A detailed discussion of volume shading in the

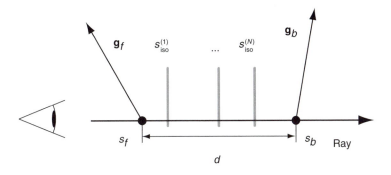

Figure 10.10 Some of the parameters used for shading a ray segment intersected by multiple isosurfaces.

context of pre-integrated volume rendering is given by Meißner et al. [13]. In fact, pre-integration is not restricted to colors and opacities but may also be applied, for example, to material properties. Here, however, we discuss only the pre-integrated interpolation of a gradient for multiple, semitransparent isosurfaces, i.e., for a ray segment with scalar values s_f and s_b, gradients $\mathbf{g}_f$ and $\mathbf{g}_b$, and n iso-values $s_{\mathrm{iso}}^{(1)}, \ldots, s_{\mathrm{iso}}^{(n)}$ within the interval $[s_f, s_b]$ according to Fig. 10.10.

As discussed in Section 10.4.1.3, the weight ω for the interpolation of a gradient for a single isosurface is given by $(s_{\mathrm{iso}} - s_b)/(s_f - s_b)$. In the case of multiple isosurfaces, there is one weight $\omega^{(i)}$ for each iso-value $s_{\mathrm{iso}}^{(i)}$:

$$\omega^{(i)} = \frac{s_{\mathrm{iso}}^{(i)} - s_b}{s_f - s_b}$$

These weights have to be weighted according to the transparencies $\alpha_{\mathrm{iso}}^{(i)}$ of the isosurfaces. Thus, the interpolation weight ω_{mult} for multiple isosurfaces is given by

$$\omega_{\mathrm{mult}} = \frac{\sum_{i=1}^{n}\left(\omega^{(i)}\alpha_{\mathrm{iso}}^{(i)}\Pi_{j=1}^{i-1}\left(1 - \alpha_{\mathrm{iso}}^{(j)}\right)\right)}{1 - \Pi_{j=1}^{n}\left(1 - \alpha_{\mathrm{iso}}^{(j)}\right)}$$

and the interpolated gradient is

$$\mathbf{g}_{\mathrm{iso}} = \frac{\omega_{\mathrm{mult}}\mathbf{g}_f + (1 - \omega_{\mathrm{mult}})\mathbf{g}_b}{|\omega_{\mathrm{mult}}\mathbf{g}_f + (1 - \omega_{\mathrm{mult}})\mathbf{g}_b|}$$

Analogously to the case of a single isosurface, the weights ω_{mult} depend on s_f and s_b, but not

on the sampling distance d. Therefore, they can be tabulated in a 2D lookup table.

10.4.2.1 Note from the Authors

After this chapter had been written, we learned about a completely different approach to volume shading for pre-integrated volume rendering suggested by Lum et al. [9]. We regret that we were not able to include a more detailed description of their method in this section.

10.4.3 Volume Clipping

A pre-integrated technique for volume clipping was published by Roettger et al. [17]. The basic idea is to define a second volumetric scalar field $\chi = \chi(\mathbf{x})$ that defines the volumetric clip region; for example, the volume is clipped where $\chi < 0.5$.

The pre-integration parameters (scalar values s_f' and s_b', and segment length d') of a clipped ray segment may be computed from the parameters of the unclipped segment (s_f, s_b, and d) and the values of χ at the start (χ_f) and at the end (χ_b) of the unclipped segment (Fig. 10.11). For example, in the case of $\chi_f < 0.5$ and $\chi_b > 0.5$ (Fig. 10.11a), the scalar value s_f' at the start of the clipped segment is closer to s_b than s_f and the length of the clipped segment d' is smaller than the original length d, while the scalar value at the end of the segment is unchanged, i.e., s_b equals s_b'. The actual calculation of s_f', s_b', and d' is straightforward:

$$s'_f = \omega_f s_f + (1 - \omega_f) s_b$$

$$\text{with } \omega_f = \begin{cases} \dfrac{\chi_b - 0.5}{\chi_b - \chi_f} & \text{if } \chi_f < 0.5 < \chi_b \\ 1 & \text{else.} \end{cases}$$

$$s'_b = \omega_b s_b + (1 - \omega_b) s_f$$

$$\text{with } \omega_b = \begin{cases} \dfrac{\chi_f - 0.5}{\chi_f - \chi_b} & \text{if } \chi_f > 0.5 > \chi_b \\ 1 & \text{else.} \end{cases}$$

$$d' = \omega_d d$$

$$\text{with } \omega_d = \begin{cases} \dfrac{\chi_b - 0.5}{\chi_b - \chi_f} & \text{if } \chi_f < 0.5 < \chi_b \\ \dfrac{\chi_f - 0.5}{\chi_f - \chi_b} & \text{if } \chi_f > 0.5 > \chi_b \\ 1 & \text{else.} \end{cases}$$

Since the weights ω_f, ω_b, and ω_d depend only on χ_f and χ_b, they may be tabulated in 2D lookup tables.

In summary, a ray segment is colored by first determining s_f, s_b, d, χ_f, and χ_b; then looking up $\omega_f = \omega_f(\chi_f, \chi_b), \omega_b = \omega_b(\chi_f, \chi_b)$, and $\omega_d = \omega_d(\chi_f, \chi_b)$; after that calculating s'_f, s'_b, and d' by linear interpolation; and eventually determining the color and opacity of the clipped ray segment depending on s'_f, s'_b, and d'. See Roettger et al. [17] for more details and a description of an implementation in the context of texture-based volume rendering.

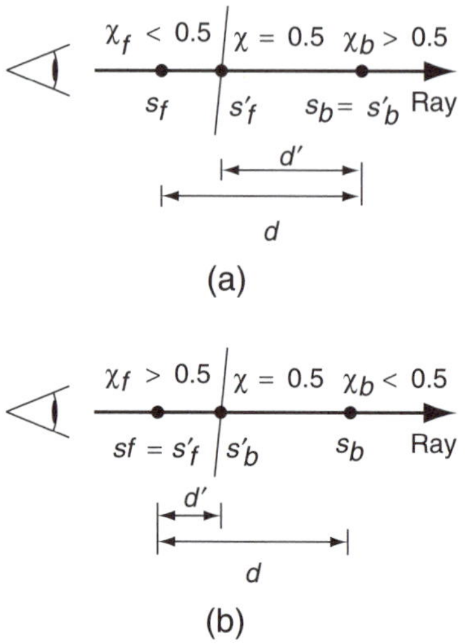

Figure 10.11 Two cases of clipping a pre-integrated ray segment: (a) clipping of the volume at the start (front) and (b) clipping of the volume at the end (back) of the segment.

10.5 Open Problems

Pre-integrated volume rendering is a relatively recent technique in volume graphics that is still evolving. While many researchers are convinced of its benefits (in particular for transfer functions with high frequencies), the discussion about the theoretical foundation of

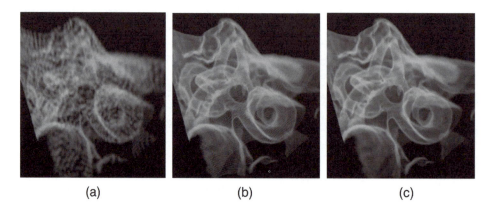

Figure 10.12 Volume visualization of a dataset of $128 \times 128 \times 30$ voxels showing tiny structures of the inner ear. These 128 textured slices were rendered with (a) preclassification, (b) post-classification, and (c) pre-integrated classification. The images appear courtesy of Klaus Engel.

these benefits is not yet closed. We have not joined this discussion here, as it is (and probably should be) focused on the appropriate sampling rate for volume rendering with *post-classification* [7,19].

Another open issue is the pre-integration of multidimensional transfer functions. Although the approach published by Kniss et al. [5] is feasible under many circumstances, a general solution to this problem still has to be found.

Further open problems related to pre-integrated volume rendering are, for example:

- Applications to other volume rendering algorithms, e.g., splatting.
- Additional acceleration techniques.
- Extensions to further illumination models, higher-order interpolation, etc.

Additional figures appear in the color insert section.

References

1. J. F. Blinn. Jim Blinn's corner—compositing, part I: theory. *IEEE Computer Graphics and Applications*, 14(5):83–87, 1994.
2. B. Cabral, N. Cam, and J. Foran. Accelerated volume rendering and tomographic reconstruction using texture mapping hardware. In *Proceedings 1994 Symposium on Volume Visualization*, pages 91–98, 1994.
3. K. Engel, M. Kraus, and T. Ertl. High-quality pre-integrated volume rendering using hardware-accelerated pixel shading. In *Proceedings Graphics Hardware 2001*, pages 9–16, 2001.
4. S. Guthe, S. Roettger, A. Schieber, W. Strasser, and T. Ertl. High-quality unstructured volume rendering on the PC platform. In *Proceedings Graphics Hardware 2002*, pages 119–125, 2002.
5. J. Kniss, S. Premoze, M. Ikits, A. Lefohn, C. Hansen, and E. Praun. Gaussian transfer functions for multi-field volume visualization. In *Proceedings Visualization 2003*, pages 497–504, 2003.
6. G. Knittel. Using pre-integrated transfer functions in an interactive software system for volume rendering. In *Proceedings Short Presentations EUROGRAPHICS 2002*, pages 119–123, 2002.
7. M. Kraus. Direct volume visualization of geometrically unpleasant meshes. Dissertation, University of Stuttgart, 2003.
8. P. Lacroute and M. Levoy. Fast volume rendering using a shear-warp factorization of the viewing transformation. In *Proceedings SIGGRAPH 94*, pages 451–458, 1994.
9. E. B. Lum, B. Wilson, and K.-L. Ma. High-quality lighting and efficient pre-integration for volume rendering. In *Proceedings Visualization Symposium (Vis Sym) '04*, pages 25–34, 2004.
10. N. Max. Optical models for direct volume rendering. *IEEE Transactions on Visualization and Computer Graphics*, 1(2):99–108, 1995.
11. N. Max, B. Becker, and R. Crawfis. Flow volumes for interactive vector field visualization. In *Proceedings Visualization '93*, pages 19–24, 1993.
12. N. Max, P. Hanrahan, and R. Crawfis. Area and volume coherence for efficient visualization of 3D scalar functions. *ACM Computer Graphics (Proceedings San Diego Workshop on Volume Visualization)*, 24(5):27–33, 1990.
13. M. Meißner, S. Guthe, and W. Straßer. Interactive lighting models and pre-integration for volume rendering on PC graphics accelerators. In *Proceedings Graphics Interface 2002*, pages 209–218, 2002.
14. M. Meißner, U. Kanus, G. Wetekam, J. Hirche, A. Ehlert, W. Straßer, M. Doggett, P. Forthmann, and R. Proksa. VIZARD II: A reconfigurable interactive volume rendering system. In *Proceedings Graphics Hardware 2002*, pages 137–146, 2002.
15. C. Rezk-Salama, K. Engel, M. Bauer, G. Greiner, and T. Ertl. Interactive volume rendering on standard PC graphics hardware using multi-textures and multi-stage rasterization. In *Proceedings Graphics Hardware 2000*, pages 109–118, 2000.
16. S. Roettger and T. Ertl. A two-step approach for interactive pre-integrated volume rendering of unstructured grids. In *Proceedings Volume Visualization and Graphics Symposium 2002*, pages 23–28, 2002.
17. S. Roettger, S. Guthe, D. Weiskopf, T. Ertl, and W. Strasser. Smart hardware-accelerated volume rendering. In *Proceedings Symposium on Visualization (VisSym) '03*, pages 231–238, 2003.
18. S. Röttger, M. Kraus, and T. Ertl. Hardware-accelerated volume and isosurface rendering based on cell-projection. In *Proceedings Visualization 2000*, pages 109–116, 2000.
19. J. P. Schulze, M. Kraus, U. Lang, and T. Ertl. Integrating pre-integration into the shear-warp

algorithm. In *Proceedings Volume Graphics 2003*, pages 109–118, 2003.

20. P. Shirley and A. Tuchman. A polygonal approximation to direct scalar volume rendering. *ACM Computer Graphics (Proceedings San Diego Workshop on Volume Visualization)*, 24(5):63–70, 1990.

21. C. M. Stein, B. G. Becker, and N. L. Max. Sorting and hardware assisted rendering for volume visualization. In *Proceedings 1994 Symposium on Volume Visualization*, pages 83–89, 1994.

22. M. Weiler, M. Kraus, M. Merz, and T. Ertl. Hardware-based view-independent cell projection. In *IEEE Transactions on Visualization and Computer Graphics*, 9(2):163–175, 2003.

23. M. Weiler, M. Kraus, M. Merz, and T. Ertl. Hardware-based ray-casting for tetrahedral meshes. In *Proceedings Visualization 2003*, pages 333–340, 2003.

24. E. W. Weisstein. *Eric Weisstein's world of mathematics*. http://mathworld.wolfram.com/

25. R. Westermann and T. Ertl. Efficiently using graphics hardware in volume rendering applications. In *Proceedings SIGGRAPH 98*, pages 169–177, 1998.

26. R. Westermann, C. Johnson, and T. Ertl. A level-set method for flow visualization. In *Proceedings Visualization 2000*, pages 147–154, 2000.

27. P. L. Williams and N. Max. A volume density optical model. In *Proceedings 1992 Workshop on Volume Visualization*, pages 61–68, 1992.

28. P. L. Williams, N. L. Max, and C. M. Stein. A high accuracy volume renderer for unstructured data. *IEEE Transactions on Visualization and Computer Graphics*, 4(1):37–54, 1998.

29. C. M. Wittenbrink, T. Malzbender, and M. E. Goss. Opacity-weighted color interpolation for volume visualization. In *Proceedings 1998 Symposium on Volume Visualization*, pages 135–142, 1998.

30. M. Woo, J. Neider, T. Davis, and D. Shreiner. *OpenGL Programming Guide: The Official Guide to Learning OpenGL, Version 1.2*, 3rd Ed. Addison-Wesley, 1999.

11 Hardware-Accelerated Volume Rendering

HANSPETER PFISTER
Mitsubishi Electric Research Laboratories

11.1 Introduction

Over the last decade, volume rendering has become an invaluable visualization technique for a wide variety of applications in medicine, biotechnology, engineering, astrophysics, and other sciences. Examples include visualization of 3D sampled medical data (CT, MRI), seismic data from oil and gas exploration, and computed finite element models. While volume rendering is very popular, the lack of interactive frame rates has long limited its widespread use. Fortunately, advances in graphics hardware have lead to interactive and even real-time volume rendering performance, even on personal computers.

High frame rates are essential for the investigation and understanding of volume datasets. Real-time rotation of 3D objects in space under user control makes the renderings appear more realistic due to kinetic depth effects [103]. Immediate visual feedback allows for interactive experimentation with different rendering parameters, such as transfer functions [88]. Dynamically changing volume data can now be visualized, for example, data from interactive surgical simulation or real-time 3D ultrasound. And the image quality of hardware-accelerated volume rendering rivals or equals that of the best software algorithms.

In this chapter, we review different hardware-accelerated methods and architectures for volume rendering. Our discussion will focus on direct volume rendering techniques. A survey of isosurface methods can be found elsewhere in this book. Direct volume rendering has the ability to give a qualitative feel for the density changes in the data. It precludes the need to segment the data [58]—indeed, it is particularly adept at showing structurally weak and "fuzzy" information.

Section 11.2 reviews the basics of direct volume rendering. It provides background and terminology used throughout this chapter. The following sections then describe several approaches to volume rendering that have been successfully accelerated by hardware. We will discuss ray-casting, 3D and 2D texture slicing, shear-warp rendering and its implementation on VolumePro, and splatting. We will focus on the underlying rendering algorithms and principles and refer to the literature for implementation details. The chapter ends with conclusions and an outlook on future work in Section 11.9.

11.2 Volume Rendering Basics

11.2.1 Volume Data

A volumetric dataset consists of information at sample locations in some space. The information may be a *scalar* (such as density in a computed tomography (CT) scan), a *vector* (such as velocity in a flow field), or a higher-order *tensor* (such as energy, density, and momentum in computational fluid dynamics (CFD)). The space is usually 3D, consisting of either three spatial dimensions or another combination of spatial and frequency dimensions.

In many applications the data is sampled on a *rectilinear grid*, represented as a 3D grid of volume elements, so-called *voxels*. Voxels are assumed to be zero-dimensional scalar values defined at integer coordinates. This is in contrast to an alternative definition, where a voxel is interpreted as a small unit cube of volume with a constant value. There are many good arguments as to why such a definition may lead to errors and confusion [101]. To describe voxels as volume points in 3D is consistent with signal-processing theory [76] and makes it easy to combine them with point-sampled surface models [137].

If all the voxels are spaced identically in each dimension, the dataset is said to be *regular*. Otherwise, the data is called *anisotropic*. Anisotropic volume data is commonly found in medical and geophysical applications. For example, the spacing of CT slices along the axis of the patient is determined by the (adjustable) speed of the table, while the spacing within a slice is determined by the geometry of the scanner. In addition, the gantry of a CT scanner may be tilted with respect to the axis of the patient. The resulting (rectilinear or anisotropic) data is called *sheared* because the axes are not at right angles.

Other types of datasets can be classified into *curvilinear grids*, which can be thought of as resulting from a warping of a rectilinear grid, and *unstructured grids*, which consist of arbitrarily shaped cells with no particular relation to rectilinear grids [104]. We restrict our discussion in this chapter to hardware-accelerated rendering of scalar voxels stored on a rectilinear volume grid, including anisotropic and sheared data.

11.2.2 Coordinate Systems and Transformations

Every volume rendering technique maps the data onto the image plane through a sequence of intermediate steps in which the data is transformed to different coordinate systems. We introduce the basic terminology in Fig. 11.1. Note that the terms *space* and *coordinate system* are synonymous. The volume data is stored in *source space*. The correction transformation C transforms source space to *object space*, correcting for anisotropy and shear in the volume data.

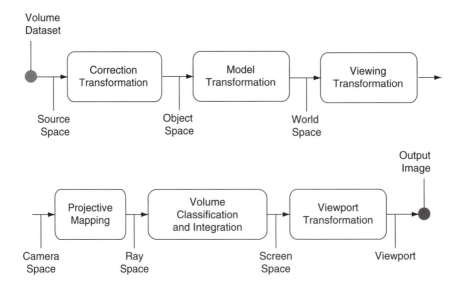

Figure 11.1   The volume rendering pipeline.

The model transformation M transforms object space to *world space*. This transformation allows one to place multiple volume and polygon objects in the scene. To render the scene from an arbitrary viewpoint, the world space is mapped to *camera space* using the viewing transformation V. The camera coordinate system is defined such that its origin is at the center of projection.

The volume rendering algorithm projects the data and evaluates the volume rendering integral. The details of this integration will be discussed in Section 11.2.3. For now, we use the projection transformation P to transform the data to *ray space* (Fig. 11.2). Ray space is a noncartesian coordinate system that enables an easy formulation of the volume rendering integral. In ray space, the viewing rays are parallel to a coordinate axis, facilitating analytical integration of the volume function. We denote a point in ray space by a vector $\mathbf{x} = (x_0, x_1, x_2)^T$, where the first two coordinates specify a point in screen space and can be abbreviated as $\hat{\mathbf{x}} = (x_0, x_1)^T$. The third coordinate x_2 specifies the Euclidean distance from the camera to a point on the viewing ray. Because the projection transformation P is similar to the projective transform used in rendering pipelines such as OpenGL, it is also called the *projective mapping*. For orthographic or parallel projection, P is the identity matrix.

Evaluating the volume rendering integral results in a 2D image in *screen space*. In the final step, this image is transformed to *viewport coordinates* using the viewport transformation VP. We will ignore the viewport transformation in the remainder of this chapter.

11.2.3 The Volume Rendering Integral

Volume rendering algorithms typically rely on the low-albedo approximation to how the volume data generates, scatters, or occludes light [10,44,68,125]. Effects of the light interaction are integrated along the viewing rays in ray space according to the *volume rendering integral*. The equation describes the light intensity $I_\lambda(\hat{\mathbf{x}})$ at wavelength λ that reaches the center of projection along the ray $\mathbf{x}$ with length L:

$$I_\lambda(\hat{\mathbf{x}}) = \int_0^L c_\lambda(\hat{\mathbf{x}}, \xi) g(\hat{\mathbf{x}}, \xi) e^{-\int_0^\xi g(\hat{\mathbf{x}}, \mu) d\mu} d\xi \qquad (11.1)$$

where $g(\mathbf{x})$ is the *extinction function* that models the attenuation of light per unit length along the ray due to scattering or extinction. $c_\lambda(\mathbf{x})$ is an *emission coefficient*, modeling the light added per unit length along the ray, including self-emission, scattered, and reflected light.

The exponential term can be interpreted as an *attenuation factor* that models the absorption of

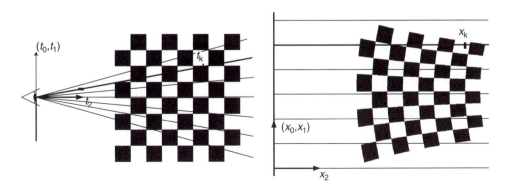

Figure 11.2   Transforming the volume from camera to ray space. (Left) Camera space; (Right) ray space.

light between a point along the ray and the eye. The product $c_\lambda(\mathbf{x})g(\mathbf{x})$ is also called the *source term* [68,125], describing the light intensity scattered in the direction of the ray $\mathbf{x}$ at the point x_2. In the remainder of this chapter we will omit the parameter λ, implying that Equation 11.1 has to be evaluated for different wavelengths separately.

We now assume that the extinction function is given as a weighted sum of coefficients g_k and reconstruction kernels $r_k(\mathbf{x})$:

$$g(\mathbf{x}) = \sum_k g_k r_k(\mathbf{x}) \tag{11.2}$$

This corresponds to the *source-attenuation* physical model [68] where the volume consists of individual particles that absorb and emit light. The reconstruction kernels r_k reflect position and shape of individual particles. The particles can be irregularly spaced and may differ in shape, so the model is not restricted to regular datasets.

Depending on how Equation 11.1 is evaluated, volume rendering algorithms can be divided into *backward-mapping* and *forward-mapping* methods. Backward-mapping algorithms shoot rays through pixels on the image plane into the volume data, and forward-mapping algorithms map the data onto the image plane.

11.2.4 Backward-Mapping Algorithms

Backward-mapping (or *image-order*) algorithms iterate over all pixels of the output image and determine the contributions of the integral to the current pixel [22,58,99]. Ray-casting is the most commonly used backward-mapping technique. It simulates optical projections of light rays through the dataset, yielding a simple and visually accurate mechanism for volume rendering.

The integral (Equation 11.1) can be evaluated using a Riemann sum approximation. By approximating the exponential function with the first two terms of its Taylor expansion ($e^{-x} \approx 1 - x$), we arrive at this equation:

$$I(\mathbf{\hat{x}}) = \sum_{l=0}^{L} \left(c(\mathbf{x}_l) \sum_k g_k r_k(\mathbf{x}_l) \prod_{j=0}^{l-1} (1 - \sum_m g_m r_m(\mathbf{x}_j)) \right) \tag{11.3}$$

The inner summation $\sum_k g_k r_k(\mathbf{x}_l)$ computes the sum of volume reconstruction kernels using Equation 11.2 at position $\mathbf{x}_l$ on the viewing ray. As described in Section 11.3.2, this is typically implemented by tri-linear interpolation. The product over j is the attenuation due to all sample points $\mathbf{x}_j$ that lie in front of the current position $\mathbf{x}_l$. The weighted sums of reconstruction kernels are typically replaced with the *opacity* α at the sample position. Thus we arrive at the familiar equation:

$$I(\hat{\mathbf{x}}) = \sum_{l=0}^{L} \left(c(\mathbf{x}_l)\alpha_l \prod_{j=0}^{l-1} (1 - \alpha_j) \right) \tag{11.4}$$

11.2.5 Forward-Mapping Algorithms

Forward-mapping (or *object-order*) algorithms iterate over the volume data and determine the contribution of each reconstruction kernel to the screen pixels. The traditional forward-mapping algorithm is splatting, introduced by Westover [124]. It convolves every voxel in object space with the reconstruction kernel and accumulates their contributions on the image plane.

Because of the linearity of integration, substituting Equation 11.2 into Equation 11.1 yields

$$I(\hat{\mathbf{x}}) = \sum_k g_k \left(\int_0^L c(\hat{\mathbf{x}}, \xi) r_k(\hat{\mathbf{x}}, \xi) \prod_j e^{-g_j \int_0^\xi r_j(\hat{\mathbf{x}}, \mu)d\mu} d\xi \right) \tag{11.5}$$

which can be interpreted as a weighted sum of projected reconstruction kernels.

To compute this integral numerically, splatting algorithms make a couple of simplifying assumptions. Usually, the reconstruction kernels $r_k(\mathbf{x})$ have local support. The splatting approach assumes that these local support areas do not overlap along a ray, and that the reconstruction kernels are ordered front to back. We also assume that the emission coefficient is constant in the support of each reconstruction kernel along a ray, hence we have $c_k(\hat{\mathbf{x}}) = c(\mathbf{x})$. Again, we approximate the exponential function with the first two terms of its Taylor expansion, thus $e^{-x} \approx 1 - x$. Finally, we ignore self-occlusion. Under these assumptions, we can rewrite Equation 11.5 to

$$\mathbf{I}(\mathbf{x}) = \sum_k \left(g_k c_k(\mathbf{x}) q_k(\hat{\mathbf{x}}) \prod_{j=0}^{k-1} (1 - g_j q_j(\hat{\mathbf{x}})) \right) \quad (11.6)$$

where $q_k(\hat{\mathbf{x}})$ denotes an integrated reconstruction kernel:

$$q_k(\hat{\mathbf{x}}) = \int_r r_k(\hat{\mathbf{x}}, x_2) dx_2 \quad (11.7)$$

The difference between backward and forward mapping is apparent when one compares Equations 11.3 and 11.6. In backward mapping, the evaluation of the volume rendering integral (Equation 11.1) is a Riemann sum along viewing rays. In forward mapping, we assume that the reconstruction kernels do not overlap and can be integrated separately using Equation 11.7. The volume rendering integral Equation 11.1 is then a sum of preintegrated reconstruction kernels, also called *footprints*.

11.3   The Volume Rendering Pipeline

Volume rendering can be viewed as a set of pipelined processing steps. *Pipelining* is an important concept in hardware design and for the design of efficient parallel algorithms with local communication. A pipeline consists of a sequence of so-called *stages* through which a computation and data flow. New data is input at the start of the pipeline while other data is being processed throughout the pipeline. In this section we look in more detail at the pipeline stages that are commonly found in volume rendering algorithms. The order in which these stages are arranged varies among implementations.

11.3.1   Data Traversal

A crucial step of any volume rendering algorithm is to generate addresses of *resampling locations* throughout the volume. The resampling locations in object space are most likely not positioned on voxel locations, which requires *interpolation* from surrounding voxels to estimate sample values at noninteger positions.

11.3.2   Interpolation

Interpolation at a resampling location involves a convolution of neighboring voxel values with a reconstruction filter (Equation 11.2). There is a wealth of literature that deals with the theory and application of appropriate reconstruction filters in computer graphics [32,130] and volume visualization [6,76,82]. In practice, due to the prohibitive computational cost of higher-order filters, the most commonly used filters for ray-casting are *nearest neighbor interpolation* and linear interpolation in three dimensions, also called *tri-linear interpolation*. Note that tri-linear interpolation is a nonlinear, cubic function in three dimensions [76]. This has consequences for the order of volume classification, as discussed below.

11.3.3   Gradient Estimation

To approximate the surface normals necessary for shading and classification requires the computation of a *gradient*. Given a continuous function $f(x, y, z)$, the gradient ∇f is defined as the partial derivative with respect to all three coordinate directions. Due to the sampled nature of volumetric data, the computation of this continuous gradient has to be approximated using discrete *gradient filters*.

Most gradient filters are straightforward 3D extensions of the corresponding 2D edge-detection filters, such as the Laplacian, Prewitt, or Zucker–Hummel [134] operators. The Sobel operator [102] is one of the most widely used gradient filters for volume rendering. In practice, and due to computational considerations, most volume rendering algorithms use the *central-difference gradient*, which is computed by local differences between voxel or sample values in all three dimensions [41]. Detailed analysis of several gradient filters for volume rendering can be found in several references [6,33,61,75].

11.3.4   Classification

Classification is the process of mapping physical properties of the volume, such as different material types, to the optical properties of the volume rendering integral, such as emission (color, RGB) and absorption (opacity, α). We distinguish between *pre- and post-classification*, depending on whether the voxel values of the volume are classified before or after interpolation.

11.3.4.1   Pre-Classification

In pre-classification, voxels may be mapped directly to RGBα values, which are then interpolated. Alternatively, voxels may be augmented by attributes that correspond to disjoint materials [24], which is common for medical image data that contains separate anatomical parts. Typically, these attributes are computed in a separate *segmentation* process using accurate statistical methods [23,120]. Such segmentation prevents *partial voluming* [43], one of the main sources of error in direct volume rendering. In partial voluming, single voxels ostensibly represent multiple materials, or tissue types in the case of medical data. Segmented volumes contain indices and associated probabilities for different material types, which can then be mapped to different colors and opacities during pre-classification [109].

However, the individual interpolation of color and opacity after pre-classification can lead to image artifacts [129]. The solution is to premultiply the color of each voxel with its opacity before interpolation. The resulting vector $(R\alpha, G\alpha, B\alpha, \alpha)$ is called *associated color* or *opacity-weighted color* [11,24]. If we denote

original colors with C, we will use the notation $\tilde{C} = C\alpha$ for associated colors. Wittenbrink et al. [129] present an efficient method to interpolate associated colors. Interpolation with associated colors is also necessary for preintegration techniques (see Section 11.4.3).

11.3.4.2   Post-Classification

In post-classification, the mapping to RGBα values is applied to a continuous, interpolated scalar field. Post-classification is easier to implement than pre-classification and mostly used in the hardware-accelerated algorithms described below. Note that post-classification does not require one to use associated colors, although it is still possible to do so. In that case, the transfer functions (see below) are stored for associated colors.

As discussed by Engel et al. [25], pre-classification and post-classification produce different results because classification is in general a nonlinear operation. The nonlinearity of tri-linear interpolation may lead to artifacts if it is applied to pre-classified data. For post-classification, the evaluation of the nonlinear classification function in a linearly interpolated scalar field produces the correct result [6,82]. However, as noted above, pre-classification remains a very important tool in medical imaging, and it is important that the hardware-accelerated volume rendering method be able to support both options.

11.3.4.3   Transfer Functions

The mapping that assigns a value for optical properties like $g(\mathbf{x})$ or $c(\mathbf{x})$ is called a *transfer function*. The transfer function for $g(\mathbf{x})$ is called the *opacity transfer function*, typically a continuously varying function of the scalar value s along the ray $g(\mathbf{x}) = T_o(s(\mathbf{x}))$. Often it is useful to include the gradient magnitude $|\nabla_s|$ as an additional parameter for classification. This approach has been widely used in the visualization of bone or other tissues in medical datasets or for the isosurface visualization of electron

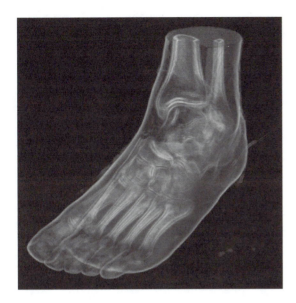

Figure 11.3 CT scan of a human foot, rendered on Volu-mePro 1000 with gradient magnitude modulation of opacity. Image courtesy of Yin Wu and Jan Hardenbergh, TeraRe-con, Inc. (See also color insert.)

density maps [58]. In the simplest case, the opacity is optionally multiplied with the gra-dient magnitude, which is also called *gradient* *magnitude modulation*, to emphasize surface boundaries or to minimize the visual impact of noisy data [30,87] (Fig. 11.3). Kindlmann et al. [47] and Kniss et al. [48] use higher-order derivatives for semiautomatic transfer-function design. Easy transfer-function design still re-mains one of the main obstacles to make volume rendering more accessible to nonexpert users [88].

The emission term $c(\mathbf{x})$ can also be specified as a transfer function of the scalar s: $c(\mathbf{x}) = T_c(s(\mathbf{x}))$. The simplest emission term is direction independent, representing the glow of a hot gas [68]. It may have red, green, and blue components, with their associated *color transfer* *functions* $f_{red}(s)$, $f_{green}(s)$, and $f_{blue}(s)$. More so-phisticated emission models include *multiple* *scattering* and *anisotropic scattering* terms [38,68]—mostly used for rendering of clouds—and *shading* effects.

11.3.5 Shading
Volume shading can substantially add to the realism and understanding of volume data (Fig. 11.4). Most volume shading is computed

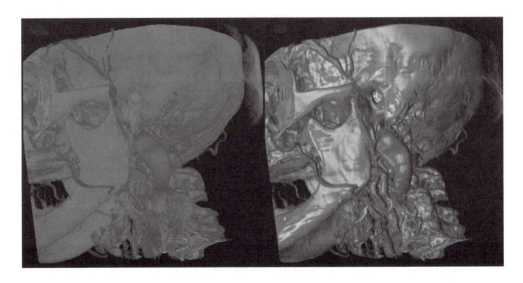

Figure 11.4 CT scan of a human head. (Left) Volume rendering with a simple emission model without shading. (Right) Including Phong shading. Images rendered on VolumePro 1000; courtesy of Yin Wu and Jan Hardenbergh, TeraRecon, Inc. (See also color insert.)

by the well-known Phong [89] or Blinn–Phong [9] illumination models. The resulting color is a function of the gradient, light, and view directions, as well as the ambient, diffuse, and specular shading parameters. It is typically added to the color that results from classification. Higher-order shading models, which include the physical effects of light-material interaction [17], are computationally too expensive to be considered for volume rendering. There are other illumination models for interactive systems [2].

Care has to be taken for sheared and anisotropic volume data. The different voxel spacing and alignment leads to incorrect gradients because the vector components are not orthonormal in object space. One remedy is to use full 3D convolution kernels for gradient computation, which may be prohibitive in real-time systems. The alternative is to use separable kernels and to apply *gradient correction* before classification and lighting calculations. For example, gradients can be transformed from voxel space to world space, including the correction transformation C. Shading is then computed using world-space versions of the eye and light vectors [87]. However, this transformation requires multiplication with a 3×3 matrix per gradient vector.

Alternatively, Wu et al. [131] describe an elegant and efficient solution for gradient correction using an intermediate lighting space. The product MC of model and correction transformation is decomposed into a shear-scale transformation L and a rotation R. Gradients are transformed to lighting space by $(L^{-1})^T$, similarly to how surface normals are transformed in polygon graphics [27], while light and eye vectors are transformed from world space to lighting space by R^{-1}. Note that this rotation preserves dot products, which enables one to precompute some shading calculations. The transformation $(L^{-1})^T$ is upper triangular and requires only six multiplications per gradient. For anisotropic but nonsheared volumes, this reduces to three multiplications per gradient.

11.3.6 Compositing

Compositing is the recursive evaluation of the numerical integration in Equations 11.4 and 11.6. It was first introduced in the context of digital image compositing, where it was formulated using the "*over*" operator [90]. The composition of n associated color samples $\tilde{C}_i = C_i \alpha_i$ is described by:

$$
\begin{aligned}
\tilde{C}(0, n-1) &= \sum_{x=0}^{n-1} \tilde{C}_x \prod_{t=0}^{x-1} (1 - \alpha_t) \\
&= \tilde{C}_0 + \tilde{C}_1 (1 - \alpha_0) + \tilde{C}_2 (1 - \alpha_0)(1 - \alpha_1) + \dots \\
&\quad + \tilde{C}_{n-1}(1 - \alpha_0)\dots(1 - \alpha_{n-2}) \\
&= \tilde{C}_0 \operatorname{over} \tilde{C}_1 \operatorname{over} \tilde{C}_2 \operatorname{over} \dots \tilde{C}_{n-1}
\end{aligned}
$$
(11.8)

Because of the associativity of the "over" operator, the composition of—for example—four samples $\tilde{C}_i$ can be computed in three different ways [127]:

Front-to-back: $\tilde{C} = (((\tilde{C}_1 \operatorname{over} \tilde{C}_2) \operatorname{over} \tilde{C}_3) \operatorname{over} \tilde{C}_4)$
Back-to-front: $\tilde{C} = (\tilde{C}_1 \operatorname{over}(\tilde{C}_2; \operatorname{over}(\tilde{C}_3; \operatorname{over} \tilde{C}_4)))$
Binary tree: $\tilde{C} = ((\tilde{C}_1; \operatorname{over} \tilde{C}_2); \operatorname{over}(\tilde{C}_3; \operatorname{over} \tilde{C}_4))$

The *front-to-back* or *back-to-front* formulations are used in most volume rendering algorithms. The last formulation as a *binary tree* is especially useful for parallel implementations algorithms, where partial results of segments along the ray can be computed on different processors [42,86,128]. The final composition of the partial results yields the same image as sequential compositing along the ray.

Compositing is expressed algorithmically using recursion. The front-to-back formulation is

$$
\begin{aligned}
\hat{t}_0 &= (1 - \alpha_0); \hat{C}_0 = \tilde{C}_0 \\
\hat{C}_i &= \hat{C}_{i-1} + \hat{t}_{i-1} \tilde{C}_i \\
\hat{t}_i &= \hat{t}_{i-1}(1 - \alpha_i)
\end{aligned}
$$
(11.9)

where $\hat{C}_i, \hat{t}_i$ indicate the results of the current iteration, $\hat{C}_{i-1}, \hat{t}_{i-1}$ the accumulated results of the previous iteration, and $\tilde{C}_i$ and α_i the associated sample color and opacity values at the current resampling location. Note that $\hat{t}$ is the accumulated *transparency*. Substituting $\hat{t}_i = (1 - \hat{\alpha}_i)$ leads to the less efficient—but more familiar—formulation with accumulated opacities $\hat{\alpha}_i$.

When compositing back to front, the accumulated transparencies of Equation 11.9 do not need to be maintained. However, they are useful if the final image is composited over a new background or for mixing volumes and polygons (see Section 11.4.5). The recursive back-to-front formulation for n sample points is

$$\hat{t}_n = (1 - \alpha_n); \hat{C}_n = \tilde{C}_n$$
$$\hat{C}_i = \hat{C}_{i-1}(1 - \alpha_i) + \tilde{C}_i \qquad (11.10)$$
$$\hat{t}_i = \hat{t}_{i-1}(1 - \alpha_i)$$

Because the extinction coefficient measures the volumetric light absorption per unit length, the opacity value must be adapted to the distance between interpolated samples. This scaling is called *opacity correction*. If opacities are defined for a distance d_{old}, and samples are spaced by a distance d_{new}, the scaling becomes [55]:

$$\alpha_{corrected} = 1 - (1 - \alpha_{stored})^{\frac{d_{old}}{d_{new}}} \qquad (11.11)$$

This can be efficiently implemented using a pre-computed lookup table that stores $\alpha_{corrected}$ as a function of α_{stored} and d_{new}.

As discussed by Schulze et al. [100], associated colors have to be corrected correspondingly:

$$\tilde{C}_{corrected} = \tilde{C}_{stored} \frac{\alpha_{corrected}}{\alpha_{stored}} \qquad (11.12)$$

Orthographic projections typically lead to constant sample distances throughout the volume. Maintaining constant sample distance for perspective projections leads to spherical shells of samples around the center of projection (Fig. 11.2). While some approaches have used spherical shell sampling [56] for uniform sample spacing, it is more common to correct opacities by evaluating Equations 11.11 and 11.12.

There are several alternatives to volumetric compositing that have proven useful. In *x-ray* or *weighted-sum* projections, the value of the pixel equals the sum of the intensities. *Maximum intensity projections* (MIPs) project the maximum intensity along the ray into a pixel. Other options include *first opaque projection and minimum intensity projection* [28].

11.4   Advanced Techniques

Given the high performance requirements of volume rendering, it becomes clear that a brute-force implementation requires an excessive amount of processing. It is therefore not surprising that many optimizations have been developed.

11.4.1   Early Ray Termination

Early ray termination is a widely used method to speed up ray-casting [3,22,59]. The accumulation of new samples along a ray is terminated as soon as their contribution to the currently computed pixel becomes minimal. Typically, the ray is terminated as soon as the accumulated ray opacity reaches a certain threshold, since any further samples along the ray would be occluded. More general methods terminate rays according to a probability that increases with increasing accumulated ray opacity [3,22] or decreases the sampling rate as the optical distance to the viewer increases [22]. The performance of early ray termination is dataset and classification dependent.

11.4.2   Space Leaping

Empty or transparent regions of the volume data may be skipped using precomputed data structures. In *content-based space leaping*, samples are skipped that are invisible by virtue of opacity assignment or filtering. Typically, the opacity transfer function is used to encode nontransparent areas of the data into hierarchical [22,59,69,106] or run-length encoded [54,93] data structures. Inherent problems of content-based space leaping are that its performance is classification dependent and that changes of the opacity transfer function lead to lengthy recomputation of the data structures.

In contrast, *geometry-based space leaping* skips empty space depending on the position of samples, not based on their values. Levoy [59] uses an octree data structure to skip empty subvolumes of the data during ray-casting. Avila et al. [4] use convex polyhedral shapes as bounding volumes and graphics hardware to efficiently skip space by rendering the bounding volume into the depth buffer. Other methods use precomputed distance functions to indicate the radial distance from each voxel in the data [105,135] or fast discrete line algorithms to progress quickly through empty regions [133].

11.4.3   Preintegration

The discrete approximation of Equation 11.1 will converge to the correct results only for high volume sampling rates, i.e., if the spacing between samples is sufficiently small. As discussed by Engel et al. [25], the Nyquist frequency for correct sampling is roughly the product of the Nyquist frequencies of the scalar field and the maximum of the Nyquist frequencies of the two transfer functions for $g(\mathbf{x})$ and $c(\mathbf{x})$. In other words, nonlinear transfer functions require very high sampling rates. Artifacts may still occur unless the transfer functions are smooth and the volume is band-limited.

To address the problem, Max et al. [67] and Roettger et al. [97] introduced a technique called *preintegration* (see Chapter 10). The volume rendering integral between two samples is a function of their interpolated values and the distance between the samples. For each combination of scalar values, the integration of the opacity or color transfer function is precomputed and stored in a 2D lookup table (for constant sampling distance). The computation of the integral can be accelerated by graphics hardware [35,96]. The value of the preintegrated opacity or color values is looked up during post-classification. Meissner et al. [73] also apply preintegration to the ambient, diffuse, and specular lighting parameters of the Phong shading model.

Despite preintegration, artifacts may still occur from high frequencies that are present in the scalar field [50]. Consequently, the distance between samples needs to be decreased to *oversample* the volume. Roettger et al. [98] suggest that four-times over-sampling yields the best quality-performance tradeoff in practice.

11.4.4   Volume Clipping

Volume clipping is an important operation that helps the understanding of 3D volumetric data. Clipping helps to uncover important details in the data by cutting away selected parts of the volume based on the position of *clip geometry*. The simplest clip geometries are one or more *clipping planes* that reveal slices and cross-sections of the volume data. *Cropping* is an easy way of specifying a rectilinear region of interest with multiple clipping planes parallel to the volume faces [87]. During cropping, cross-sections of the volume may be combined by taking intersections, unions, and inverses to define elaborate regions of visibility of the volume dataset. In its most general form, volume clipping uses arbitrary clip geometry that may be defined by a polygon mesh or by an additional volume (Fig. 11.9). Weiskopf et al. [119] provide a good overview of volume clipping methods.

11.4.5   Mixing Polygons and Volumes

The incorporation of polygonally defined objects into a volumetric scene is often important, especially in medical applications such as virtual endoscopy [29]. Volume data—such as CT or MR images—can be directly combined with synthetic objects, such as surgical instruments, probes, catheters, prostheses, and landmarks displayed as glyphs. In some instances, preoperatively derived surface models for certain anatomical structures such as skin can be more efficiently stored and better visualized as polygon meshes. A straightforward way of mixing volume and polygons is to convert the polygonal models into sampled volumes and then render them using a volume rendering method [46]. Another way is to simultaneously cast rays through both the polygonal and volume data, at the same sample intervals, and then composite the colors and opacities in depth sort order [60]. Bhalerao at al. [8] provide an overview of recent methods and propose a hardware-accelerated method for static views. All of the techniques described in this chapter are amenable to mixing volumes with polygons, although some with less efficiency than others.

We will now discuss several hardware-accelerated volume rendering algorithms for rectilinear grids: ray-casting, texture slicing, shear-warp and shear-image rendering, and splatting. Our discussion will focus on how these methods implement each stage of the volume rendering pipeline (Section 11.3) followed by an overview of the available extensions (Section 11.4). Throughout the chapter, we will rarely quote performance numbers, unless we are confident they will not change over time. For an additional comparison between most of these algorithms, including image-quality evaluations, see Meissner et al. [72].

11.5 Ray-Casting

Ray-casting is the most commonly used image-order technique. It simulates optical projections of light rays through the dataset, yielding a simple and visually accurate mechanism for volume rendering [58].

Data Traversal: Rays are cast from the viewpoint (also called center of projection) through screen pixels into the volume. The ray directions can be computed from the model and viewing transformations using standard computer-graphics techniques [27]. Empty space between the viewing plane and the volume can be skipped by rendering a polygonal bounding box of the volume into a depth buffer [4]. The ray starting points and their normalized directions are then used to generate evenly spaced resampling locations along each ray (Fig. 11.5).

Interpolation: At each resampling location, the data is interpolated, typically by tri-linear interpolation in *cells* of $2 \times 2 \times 2$ voxels. Note that tri-linear interpolation can be efficiently evaluated using caching of cell data among neighboring rays [87].

Gradient Estimation: Gradient vectors are usually precomputed at voxel locations and stored with the volume data. Alternatively, they are computed during rendering using the central difference gradient filter. The gradient vectors of a cell are interpolated to the nearest resampling locations. Optionally, the gradient magnitude is computed for gradient magnitude modulation during post-classification.

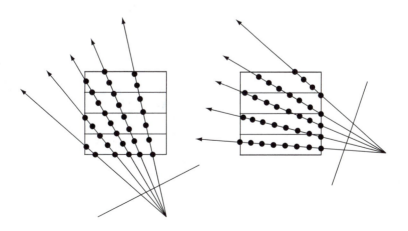

Figure 11.5 Ray-casting sample generation.

Classification: If scalar values were interpolated, color and opacity are assigned to each sample using a post-classification lookup. Optionally, the sample opacity is modulated by the gradient magnitude. If the data has been pre-classified, no further classification of the interpolated associated colors is necessary.

Shading: Using the gradient as the surface-normal approximation and the classified color as the emitted (or primary) color, a local shading model is applied to each sample. For example, the Phong illumination model for directional lights can be efficiently implemented using pre-computation and table lookup in so-called *reflectance maps* [114]. The reflectance map implementation supports an unlimited number of directional light sources, but no positional lights.

Compositing: All samples along the ray are composited into pixel values—typically in front-to-back order—to produce the final image. For higher image quality, multiple rays per pixel are cast and combined using high-quality image filters [27].

Ray-casting offers high image quality and is conceptually easy to implement. Unfortunately, these advantages come at the price of high computational requirements. The volume data is not accessed in storage order because of the arbitrary traversal direction of the viewing rays. This leads to poor spatial locality of data references, especially for sample and gradient interpolation.

There has been a lot of research on special-purpose hardware for volume ray-casting. There are general surveys on early work in this field [40,91]. The Cube project at SUNY Stony Brook resulted in VolumePro 500 (see Section 11.7.1), the first commercial real-time volume rendering engine, and various proposals for improvements [20,51,52] The VIZARD project [49,70] at the University of Tübingen led to the successful implementation of the VIZARD II hardware [74]. VIZARD II uses reconfigurable field-programmable gate arrays (FPGAs) for fast design changes and low-cost development. The system can be configured for high-quality

perspective ray-casting of volume data or for medical image reconstruction.

Roettger et al. [98] presented the first implementation of volume ray-casting on off-the-shelf graphics hardware (Fig. 11.6). All rays are processed in parallel in front-to-back order. The bounding box of the volume is rendered to provide starting locations for all rays. The parameters for ray traversal are stored in floating-point textures, which are subsequently updated. The optimal step size for each ray is precomputed and stored in a so-called importance volume. The step size depends on the pre-integrated emission and opacity value as well as second-order gradients. This technique, called *adaptive preintegration*, is a form of space leaping (see Section 11.4.2). Rays are terminated early or when they leave the volume by setting the z-buffer for the corresponding pixels

Figure 11.6 Hardware-accelerated ray-casting of a CT scan of a bonsai (128^3) with adaptive preintegration. Image courtesy of Stefan Roettger, University of Stuttgart, Germany. (See also color insert.)

such that further computation is avoided. Additional rendering passes are necessary to determine if all rays have terminated.

The number of rendering passes for this approach is $2n - 1$, where n is the maximum number of samples along a ray. Consequently, the performance is not interactive, but comparable to a software ray caster without preintegration. However, the image quality is higher than comparable texture slicing methods (see Section 11.6), and performance increases can be expected with future hardware improvements.

11.6   Texture Slicing

Texture slicing on programmable graphics processing units (GPUs) [62] is the predominant hardware-accelerated volume rendering method. Texture-based volume rendering approaches can be implemented using 3D or 2D texture-mapping functionality.

11.6.1   3D Texture Slicing

3D texture methods traverse the volume using *image-aligned texture slices* [1,13,19,34,126] (Fig. 11.7).

Data Traversal: The volume is stored in 3D texture memory and sampled during rendering by polygons parallel to the image plane.

The view-aligned polygons are generated on the CPU and clipped against the volume bounding box. The clipping operation, which has to be performed for each frame, requires an efficient algorithm [95]. The texture coordinates of each polygon vertex with respect to the 3D texture parameters in the range [0,1] are computed.

Interpolation: The polygons with associated 3D texture coordinates are projected by the graphics hardware using the standard transformations of the polygon graphics pipeline (see Section 11.2.2). The volume data is automatically resampled by the 3D texture hardware during polygon projection using tri-linear interpolation. Note that the sampling rate along viewing rays for orthographic projections is constant, whereas the sampling rate varies per ray for perspective projections [95]. This may lead to some artifacts, depending on the transfer functions and the data.

Gradient Estimation: Current GPUs do not support the computation of 3D gradients in hardware. Gradients are precomputed, scaled and biased into the range [0,1], and stored in the 3D texture. Typically, the RGB channel is used for the volume gradients, while the scalar values are stored in the alpha channel [121]. Since gradients are projected and interpolated the same way as scalar values, subsequent shading operations are computed per pixel in screen space.

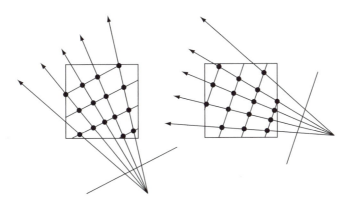

Figure 11.7   Image-aligned 3D texture slicing.

Gradients can also be stored using an integer encoding of the quantized gradient vector [26,30,31]. However, a nonlinear shading function is very sensitive to quantization errors of the gradients. Another alternative is to use a shading approximation without gradients by pairwise subtracting coplanar texture slices, one shifted in direction of the light source [84].

Classification: Most texture-mapping methods use post-classification (Fig. 11.8) by storing a 1D or higher transfer function as a texture [71]. The interpolated scalar value is stored as a texture, which is then used as a lookup coordinate into the transfer-function texture. This is also called *dependent texture lookup* because the texture coordinates for the second texture are obtained from the first texture. If pre-classification is used, a 3D texture with associated colors is stored in addition to a 3D gradient texture. Alternatively, paletted textures can be used, where the texture format is defined by an index to a color palette that maps scalar values to colors [7]. Opacity correction is applied based on the distance between texture slices. It can be implemented using a dependent lookup into a 1D floating point texture.

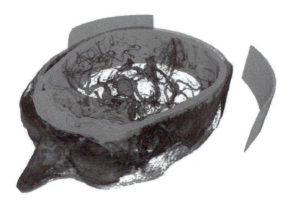

Figure 11.8 CT Angiography of a human brain ($512^2 \times 128$). Rendered on an ATI Radeon 9700 with 3D texture slicing and postclassification using dependent textures. Image courtesy of Christof Rezk-Salama, Siemens, Germany. (See also color insert.)

Shading: Before the introduction of programmable graphics hardware, shading of volumes stored as 3D textures was either ignored or performed in a preprocessing step [30]. However, preshaded volumes need to be recomputed whenever the light position and viewing direction change. A more fundamental problem is that classification and shading are in general nonlinear operations. Interpolation of the precomputed values degrades the image quality when compared to the evaluation of the nonlinear functions in a linearly interpolated scalar field [6,82].

Dachille et al. [21] use hardware for interpolation and compositing and compute shading on the CPU during volume rendering. Westermann and Ertl [121] introduced a hardware-accelerated ambient and diffuse shading technique for isosurface rendering. Meissner et al. [71] first expanded this technique for semitransparent volume data and then proposed an efficient technique to compute the full Phong illumination model using cube maps [73]. They also point out the need for normalized gradients in shading computations and propose an efficient solution for preintegrated classification. Engel et al. [25] compute diffuse and specular lighting for isosurface rendering. They observe that memory requirements can be reduced for static lighting by storing the dot products of light and gradient vectors per voxel in luminance-alpha textures. Behrens et al. [5] add shadows to the lighting model. Their multipass method works with 2D and 3D texture-mapping hardware.

Compositing: The resampled RGBα textures are accumulated into the frame buffer using back-to-front compositing [90]. If front-to-back compositing is used, accumulated opacities need to be stored for each pixel in the image.

Engel et al. [25] apply preintegration of opacity and color transfer functions to texture slicing. Their method produces high-quality images for semitransparent and isosurface volume rendering. The preintegration takes the scalar values at the entry and exit points and the distance between the slices into account.

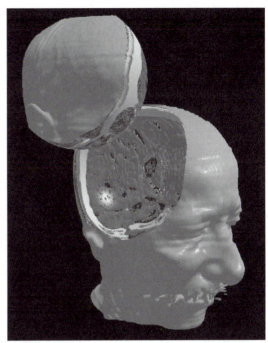

Figure 11.9 Volume clipping applied to (left) a CT scan of an engine ($256^2 \times 110$) and (right) an MRI scan of a human head (256^3). The cutting surface of the engine is enhanced by combining surface-based and volumetric shading. Image courtesy of Daniel Weiskopf, University of Stuttgart, Germany. (See also color insert.)

Weiskopf et al. [118,119] propose several techniques for volume clipping in 2D and 3D texture-slicing methods. Arbitrary clip geometry can be defined by polygons, voxelized geometry, or the isosurface of another volume. They also present a high-quality shading technique for clip boundaries (Fig. 11.9). Roettger et al. [98] extend some of their methods to work with preintegrated classification.

Kreeger and Kaufman [53] developed a texture-slicing method that renders opaque and translucent polygons embedded within volumes. Thin slabs of translucent polygons are rendered between volume slices and composited in the correct order.

A significant amount of texture memory is required to store the volume gradients. Typically, the storage increases by a factor of two to three. Visualization of very large volume data is also an issue for the limited memory of today's

GPUs. Meissner et al. [73] use lossy texture compression to compress the volume with a corresponding loss in image quality. LaMar et al. [56] propose a multiresolution framework based on an octree, where each node is rendered using 3D texture slicing. Weiler et al. [115] improve this algorithm to prevent discontinuity artifacts between different multiresolution levels. Guthe et al. [36] use a hierarchical wavelet decomposition, on-the-fly decompression, and 3D texture slicing. Their implementation is able to render very large datasets at interactive rates on PCs, although with a loss in image quality.

11.6.2 2D Texture Methods

Historically, 3D texture-mapping was not available on PC graphics cards, and 2D texture-mapping methods had to be used instead [13,94]. For example, Brady et al. [12] present

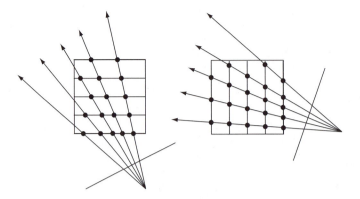

Figure 11.10   Object-aligned 2D texture slicing.

a technique for interactive volume navigation that uses ray-casting accelerated with 2D texture mapping. Despite the availability of 3D texture mapping on all modern GPUs, 2D texture slicing is still used for volume rendering today, and it outperforms 3D texture slicing for large volumes.

2D texture methods traverse the volume using *object-aligned* texture slices [13,19,34] (Fig. 11.10).

Data Traversal: Similar to 3D texture methods, each texture slice is defined by a polygon. In 2D texture slicing the polygons are always parallel to the face of the volume data that is most parallel to the viewing plane, which is also called the *base plane*. It can be easily determined by computing the minimum angle between viewing direction and face normals [95]. An arbitrary choice can be made in case of a tie at 45°. Each polygon vertex is assigned the texture coordinates of the corresponding 2D volume slice in texture memory. In contrast to 3D texture methods, three copies of the volume have to be stored in texture memory, one for each slicing direction.

Interpolation: The texture mapped slices are interpolated by the graphics hardware during projection of the polygon to screen space. Object-aligned 2D texture slicing requires only bi-linear instead of tri-linear interpolation, which leads to higher performance due to the coherent memory accesses.

The lack of interpolation between slices may lead to aliasing artifacts if the scalar field or transfer functions contain high frequencies [55]. Rezk-Salama et al. [94] improve the image quality by interpolating additional slices during rendering using multiple texture units in one pass (Fig. 11.11). The tri-linear interpolation of in-between slices is decomposed into two bi-linear interpolations (performed by 2D texture units in the graphics hardware) and one linear interpolation between slices (performed in the pixel shader of the GPU).

Gradient Estimation and Classification: Gradients and classification are computed similar as in 3D texture slicing. Precomputed gradients are stored in the RGB channel and bi-linearly interpolated to screen space during polygon rasterization. Classification can take place before or after interpolation.

Figure 11.11   CT scan of a carp (512^3) rendered on an ATI Radeon 9700 with 2D multitextures and post-classification. Image courtesy of Christof Rezk-Salama, Siemens, Germany. (See also color insert.)

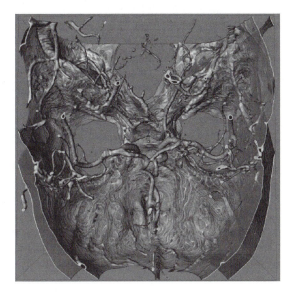

Figure 11.12 CT angiography of a human brain ($512^2 \times 128$). Transparent rendering of a nonpolygonal shaded isosurface with 2D multitextures on an NVIDIA GeForce-4Ti. Image courtesy of Christof Rezk-Salama, Siemens, Germany. (See also color insert.)

For opacity correction, Rezk-Salama et al. [94] show that scaling the opacities linearly according to the distance between samples is a visually adequate approximation. They also describe an algorithm for fast shaded isosurface display using multistage rasterization (Fig. 11.12). Engel et al. [25] improve the quality of 2D texture methods with preintegrated classification.

Shading and Compositing: The same methods are used for shading and compositing as in 3D texture slicing. For high image quality, the gradients need to be normalized by the fragment shader after projection [73]. The texture-mapped slices are composited onto the image plane using the texture blending modes of the graphics hardware.

When the viewing direction suddenly changes from one slicing direction to another, the sampling through the volume changes as well. This may lead to *popping artifacts*, which are sudden changes in pixel intensity with changes in slicing direction (Fig. 11.13). The problem is worse for anisotropic volume data. Note that 3D texture-slicing methods avoid popping artifacts by gradually adjusting the slice directions with the viewing angle. Rezk-Salama et al. [94] virtually eliminate popping artifacts in 2D texture slicing by interpolating and shifting in-between slices such that the sample spacing along viewing rays is practically constant independent of the viewing direction.

11.7 Shear-Warp Rendering

Shear-warp rendering algorithms resample the volume data from object space to the image co-

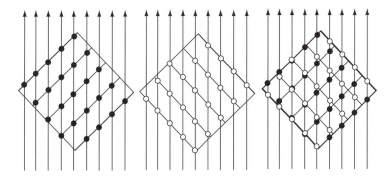

Figure 11.13 Popping artifacts in 2D texture slicing. The samples along rays may not be aligned after a small change in viewing angle leads to a change of slicing direction. The superimposition on the right shows that the location of resampling locations abruptly changes, which leads to sudden changes in pixel intensity. Figure suggested by Christof Rezk-Salama, Siemens, Germany.

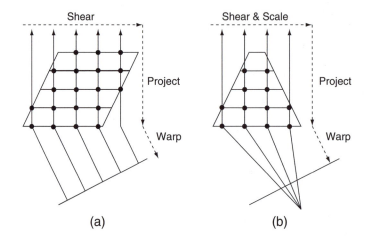

Figure 11.14   Shear-warp factorization. (a) Parallel projections; (b) perspective projections.

ordinate space so that the resampled voxels line up on the viewing axis in image space [24,110] (Fig. 11.14). The interpolated samples are then composited onto the viewing plane along axis-aligned viewing rays. The 3D affine transformation between object space and image space can be decomposed into three sequential 1D shear operations [37]. Alternatively, the viewing transformation can be decomposed into a shear and a 2D image warping operation [54,93]. Perspective projections require an additional transformation, typically in the form of a scale operation of the sheared data slices [54,112]. Shear-warp algorithms are very efficient due to the combination of object-order volume traversal and scan line–order resampling. More recently, they have been extended for improved image quality [108] and preintegration [100].

11.7.1   VolumePro 500

The VolumePro 500 system [87] is based on the Cube-4 architecture developed at SUNY Stony Brook [85]. Mitsubishi Electric licensed the technology, improved it [83], and started production shipments of the VolumePro 500 in 1999 [87]. The technology was subsequently acquired by TeraRecon Inc., which released the VolumePro 1000 system in 2002 [131].

Data Traversal: VolumePro 500 uses the standard shear-warp factorization [54,93] for orthographic projections. Instead of casting rays from image space, rays are sent into the dataset from pixels on the base plane. The ray-traversal mechanism ensures that ray samples are aligned on slices parallel to the base plane [132]. A key feature of the VolumePro architecture is the special memory address arithmetic called 3D skewing [45] and a highly optimized memory interface [83]. This enables one to efficiently read any blocks and axis-aligned voxel slices while storing only one copy of the volume data.

Interpolation: To prevent undersampling, VolumePro 500 uses tri-linear interpolation between volume slices. On-chip slice buffers and the axis-aligned processing order allow maximum memory-coherent accesses. The viewing rays can start at sub-pixel locations, which prevents popping artifacts during base-plane switches and allows over-sampling of the volume in the x, y, or z direction.

Gradient Estimation: VolumePro 500 has hardware for on-the-fly central-difference gradient estimation at each voxel. The gradients are then tri-linearly interpolated to resampling locations. The hardware includes gradient correction and gradient magnitude computation.

Contained in this first color section are Figures 1.3 through 21.5.

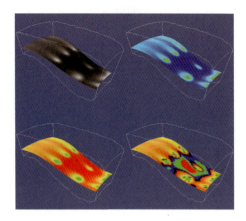

Figure 1.3 Flow density colored with different lookup tables. (Top left) Grey-scale; (top right) rainbow (blue to red); (lower left) rainbow (red to blue); (lower right) large contrast.

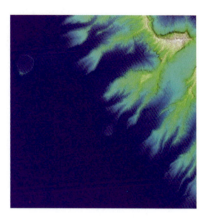

Figure 1.12 Computing scalars using normalized dot product. This part of the figure illustrates a technique applied to terrain data from Honolulu, HI.

Figure 1.13c Complex vector visualization.

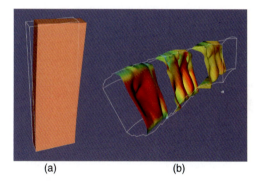

(a)   (b)

Figure 1.14 Warping geometry to show vector field. (a) Beam displacement; (b) flow momentum.

Figure 1.15b Surface plot of vibrating plate. Dark areas show nodal lines and bright areas show maximum motion.

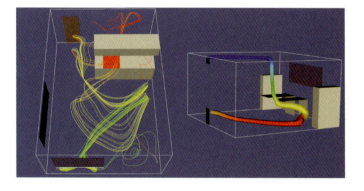

Figure 1.18 Flow velocity computed for a small kitchen (top and side view). Forty streamlines start along the rake positioned under the window. Some eventually travel over the hot stove and are convected upwards.

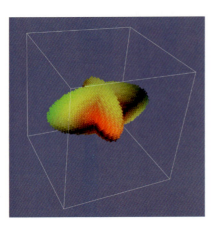

Figure 1.24b Two ellipsoids combined using the union operation used to select voxels from a volume. Voxels shrank 50%.

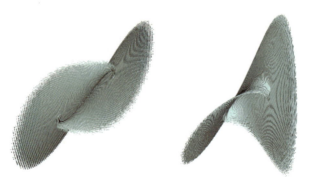

Figure 1.25 Visualizing a Lorenz strange attractor by integrating the Lorenz equations in a volume. The number of visits in each voxel is recorded as a scalar function. The surface is extracted via marching cubes using a visit value of 50. The number of integration steps is 10 million, in a volume of dimensions 200^3. The surface roughness is caused by the discrete nature of the evaluation function.

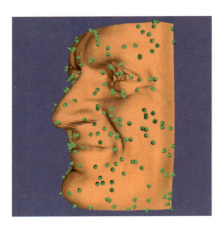

Figure 1.30 Glyphs indicate surface normals on a model of a human face. Glyph positions are randomly selected.

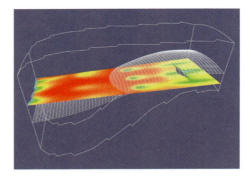

Figure 1.31 Cut through structured grid with plane. The cut plane is shown solid shaded. A computational plane of constant k value is shown in wireframe for comparison. The colors correspond to flow density. Cutting surfaces are not necessarily planes: implicit functions such as spheres, cylinders, and quadrics can also be used.

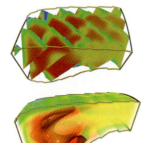

Figure 1.32 100 cut planes with opacity of 0.05, rendered back-to-front to simulate volume rendering.

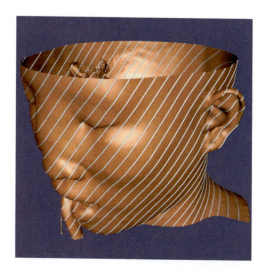

Figure 1.33 Cutting a surface model of the skin with a series of planes produces contour lines. Lines are wrapped with tubes for visual clarity.

Figure 2.1 (Left) The user view. (Right) The same isosurface from a 90-degree angle to the user view, illustrating the incomplete reconstruction.

Figure 2.2 Extracted isosurface: a cut plane through the full and view-dependent isosurfaces extracted from the same viewpoint as in Fig. 2.1. Note the large internal structures that are part of the full isosurface but not part of the view-dependent isosurface.

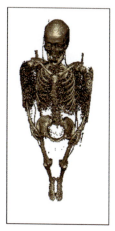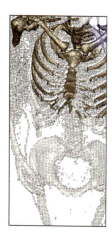

Figure 2.11 Rendering points. The left image was extracted based on the current viewpoint. The right image shows a close-up of the same extracted geometry.

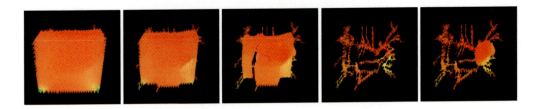

Figure 5.8 Example of the volume-thinning process.

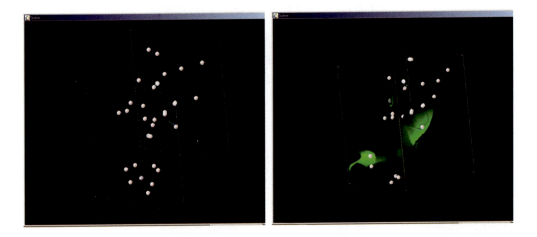

Figure 5.10 (Left) Example of a critical-point graph. (Right) Example of an isosurface generated by using the critical-point graph as a seed set.

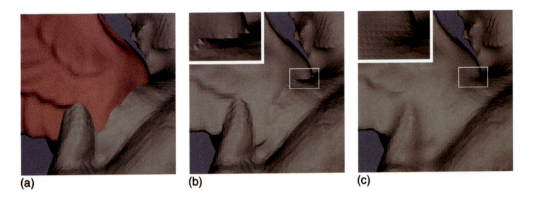

(a) **(b)** **(c)**

Figure 6.10 (a) Positioning the (red) wing model on the dragon model. (b) The models are pasted together (by CSG union operation), producing sharp, undesirable creases, a portion of which is expanded in the box. (c) The same region after automatic blending based on mean curvature. The blending is constrained to move only outwards. The models are rendered with flat-shading to highlight the details of the surface structure.

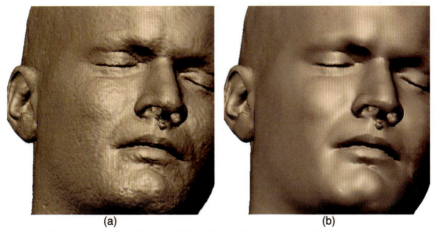
(a)   (b)

Figure 6.15 (a) A noisy isosurface obtained from an MRI volume. (b) Processing with feature-preserving smoothing alleviates noise while enhancing sharp features.

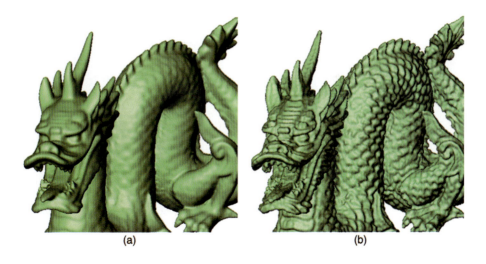
(a)   (b)

Figure 6.16 (a) A volumetric surface model is enhanced via (b) unsharp masking.

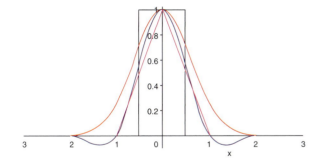

Figure 7.4 Popular filters in the spatial domain: box (black), linear (magenta), cubic (blue), Gaussian (red).

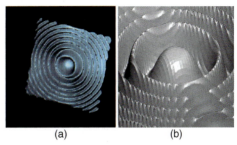
(a)   (b)

Figure 7.5 Marschner–Lobb test function, sampled into a 20^3 grid: (a) the whole function; (b) close-up, reconstructed, and rendered with a cubic filter.

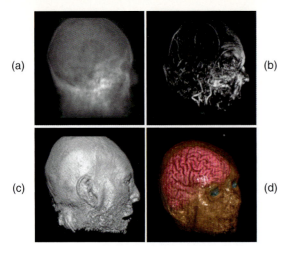

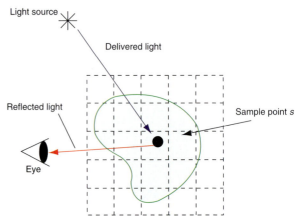

Figure 7.7 CT head rendered in the four main volume rendering modes: (a) x-ray; (b) MIP; (c) isosurface; (d) translucent.

Figure 7.8 Transport of light to the eye.

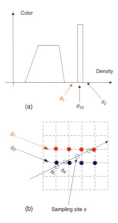

Figure 7.9 CT lobster rendered without shadows (left) and with shadows (right). The shadows on the wall behind the lobster and the self-shadowing of the legs create greater realism.

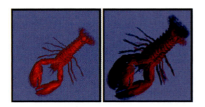

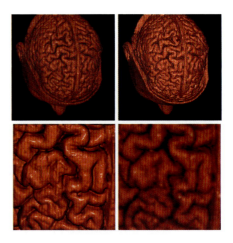

Figure 7.11 Pre-classified (left column) vs. post-classification (right column) rendering. The latter yields sharper images since the opacity and color classification is performed after interpolation. This eliminates the blurry edges introduced by the interpolation filter.

Figure 7.10 Transfer-function aliasing. When the volume is rendered preclassified, then both the red (density d_1, top row in (b)) and the blue (density d_2, bottom row) voxels receive a color of zero, according to the transfer function shown on the left. At ray sampling this voxel neighborhood at s would then interpolate a color of zero as well. On the other hand, in post-classification rendering, the ray at s would interpolate a density close to d_{12} (between d_1 and d_2) and retrieve the strong color associated with d_{12} in the transfer function.

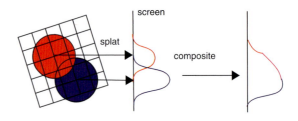

Figure 7.12 Object-order volume rendering with kernel splatting implemented as footprint mapping.

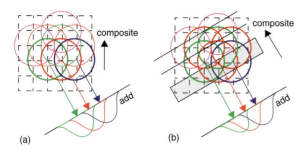

(a)

composite

add

(b)

composite

add

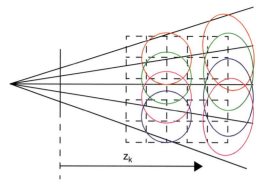

z_k

Figure 7.13 Sheet-buffered splatting. (a) Axis-aligned—the entire kernel within the current sheet is added. (b) Image-aligned—only slices of the kernels intersected by the current sheet-slab are added.

Figure 7.14 Stretching the basis functions in volume layers $z > z_k$, where the sampling rate of the ray grid is progressively less than the volume resolution.

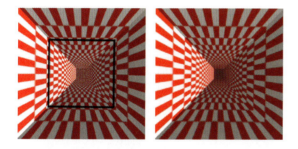

Figure 7.15 Anti-aliased splatting. (Left) A checkerboard tunnel rendered in perspective with equal-sized splats. Aliasing occurs at distances beyond the black square. (Right) The same checkerboard tunnel rendered with scaled splats. The aliasing has been replaced by blur.

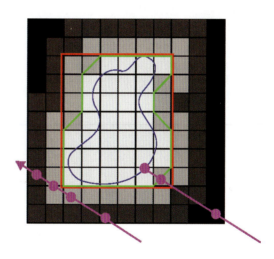

Figure 7.17 Various object approximation techniques. (Blue) Isosurface of the object. (Lightly shaded) Discretized object (proximity cloud = 0). (Red) Bounding box. (Green) Polygonal hull used in PARC. (Darker shaded areas) Proximity clouds with grey level indicating distance to the object. Note also that, while the right magenta ray is correctly sped up by the proximity clouds, the left magenta ray is missing, and the object is unnecessarily slowed down.

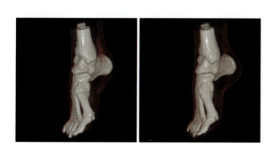

Figure 7.20 Foot dataset rendered on (left) a cubic cartesian (CC) grid and (right) a body-centered cubic cartesian (BCC) grid. The renderings are almost identical, but the BCC rendering took 70% of the time of the CC rendering.

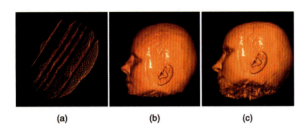

(a)   (b)   (c)

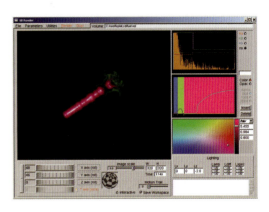

Figure 7.22 IBR-assisted volume rendering. (a) On-the-fly computed mesh derived from the slab's closest-voxel buffer; (b) head rendered from original viewpoint; (c) head rendered from a view 30° away.

Figure 7.23 A transfer function editor with an HSV color palette and a mapping of densities to various material properties.

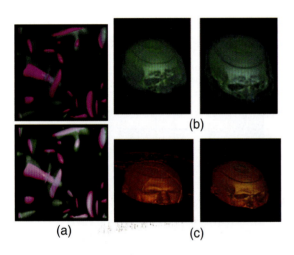

(a)   (c)

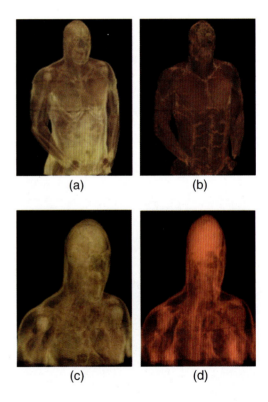

(a)   (b)

(c)   (d)

Figure 7.32 Multimodal rendering with data intermixing. (a) One time-step of a time-varying volume (magenta) and volume motion-blurred across 10 time-steps (blue). (Top) Image-level intermixing. (Bottom) Accumulation-level intermixing [182]. (b) Accumulation-level intermixing of the Visible Man's CT and an MRI dataset. Here we assign blue if CT > MRI and green if MRI > CT. (Left) Gradients specified on CT while MRI is rendered as a point cloud. (Right) Surfaces rendered with gradient modulation [70]. (c) Accumulation-level intermixing of the Visible Man's CT and an MRI dataset, rendered in inclusive opacity mode, i.e., $\alpha = 1 - (1 - \alpha_{CT})(1 - \alpha_{MRI})$. (Left) Unweighted product of CT and MRI. (Right) More CT than MRI [70].

Figure 7.33 Rendering of multichannel (photographic) data. (a) The L^* component (related to brightness). (b) The u^* component (related to the chromatic change in red-green colors). (c) Color difference gradient computed in RGB color space. (d) Gradients computed in $L^*u^*v^*$ space, using the second derivative along the gradient direction to compute opacity. (Images from Gosh et al. [70]).

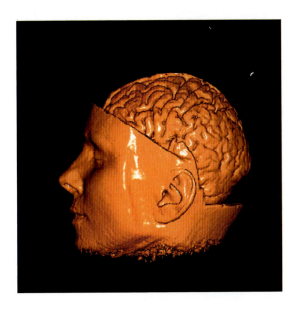

Figure 8.3 uncBrain.

Figure 8.7 Rings with shadows.

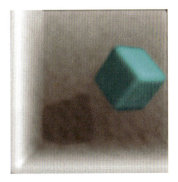

Figure 8.8 A smoky room with a cube inside.

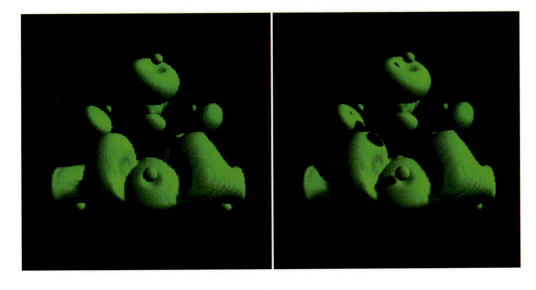

Figure 8.9 A scene of the HIPIP dataset. (Left) Without shadow; (Right) with shadow.

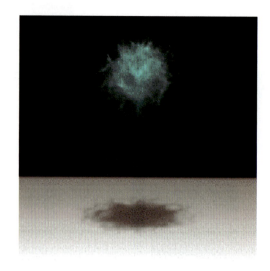

Figure 8.10   A hypertextured object with shadow.

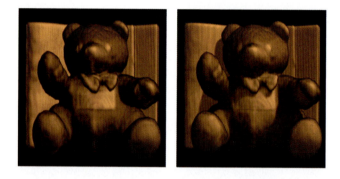

Figure 8.11   Teddy bear. (Left) Without shadow; (Right) with shadow.

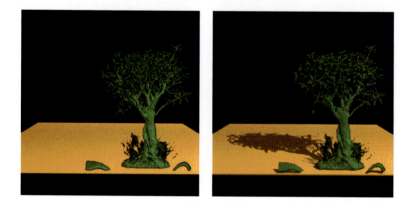

Figure 8.12   Bonsai tree. (Left) Without shadow; (Right) with shadow.

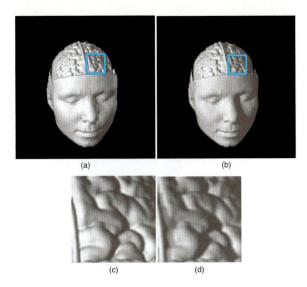

(a)   (b)

(c)   (d)

Figure 8.13 uncBrain. (a) Without shadow. (b) With shadow. (c) and (d) Close-up rendering of the specific patch.

Figure 8.14 Room scene (an example of back-to-front rendering).

Figure 8.16 A room scene for a light screen with an image of the OSU logo.

Figure 8.17 HIPIP with grid pattern.

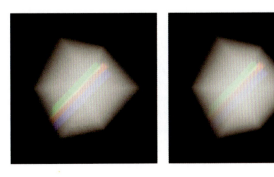

Figure 8.18 A scene with beams of light that pass through the semi-transparent cube. (Left) Without attenuation; (Right) with attenuation.

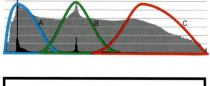

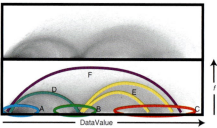

(a) A 1D histogram. The black region represents the number of data value occurrences on a linear scale; the grey is on a log scale. The colored regions (A, B, C) identify basic materials.

(b) A log-scale 2D joint histogram showing the location of materials (A, B, C) and material boundaries (D, E, F).

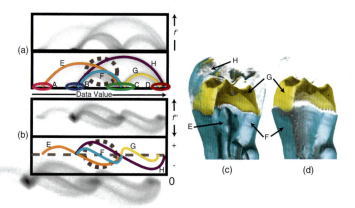

(c) A volume rendering showing all of the materials and boundaries identified above, except air (A), using a 2D transfer function.

Figure 9.1 Material and boundary identification of the Chapel Hill CT Head with data value alone (a) versus data value and gradient magnitude (f'), seen in (b). The basic materials captured by CT, air (A), soft tissue (B), and bone (C) can be identified using a 1D transfer function, as seen in (a). 1D transfer functions, however, cannot capture the complex combinations of material boundaries: air and soft tissue boundary (D), soft tissue and bone boundary (E), and air and bone boundary (F), as seen in (b) and (c).

Figure 9.2 Material and boundary identification of the human tooth CT with (a) data value and gradient magnitude (f') and (b) data value and second derivative (f''). The background–dentin boundary (F) cannot be adequately captured with data value and gradient magnitude alone. (c) The results of a 2D transfer function designed to show only the background–dentin (F) and dentin–enamel (G) boundaries. The background–enamel (H) and dentin–pulp (E) boundaries are erroneously colored. Adding the second derivative as a third axis to the transfer function disambiguates the boundaries. (d) The results of a 3D transfer function that gives lower opacity to nonzero second-derivative values.

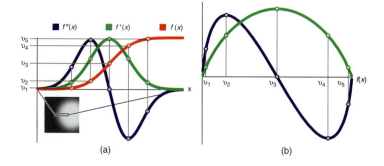

Figure 9.3 The behavior of primary data value (f), gradient magnitude (f'), and second directional derivative (f'') as a function of position (a) and as a function of data value (b).

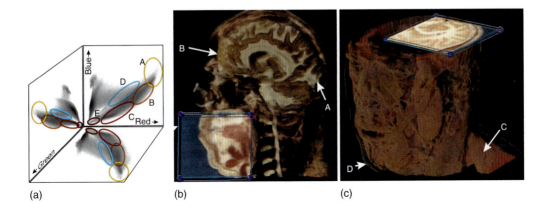

Figure 9.4 The Visible Male RGB (color) data. The opacity is set using a 3D transfer function, and color is taken directly from the data. The histogram (a) is visualized as projections on the primary planes of the RGB color space. (b) The white (A) and grey (B) matter of the brain. (c) The muscle and connective tissues (C) of the neck, showing skin (D) for reference.

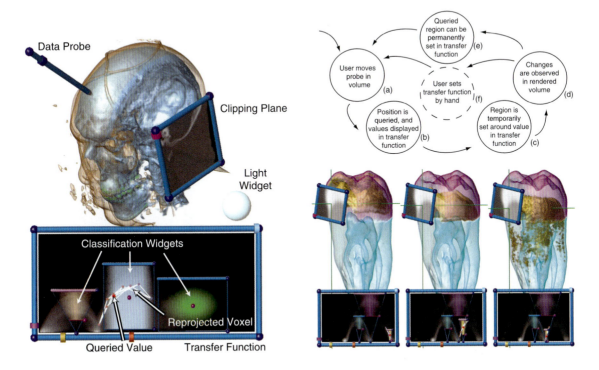

Figure 9.5 The direct-manipulation widgets.

Figure 9.6 Dual-domain interaction.

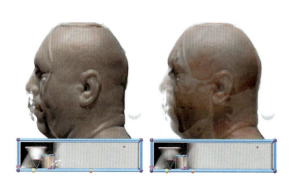

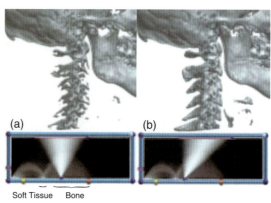

(a)   (b)

Soft Tissue   Bone

Figure 9.8 Volume renderings of the Visible Male CT (frozen) demonstrating combined surface shading and shadows.

Figure 9.9 The soft tissue–bone boundary of the Visible Male CT. It is necessary to shear the triangular classification widget to follow the center of this boundary.

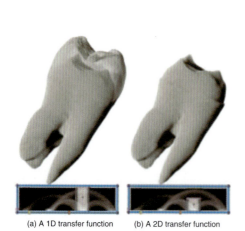

(a) A 1D transfer function   (b) A 2D transfer function

Figure 9.10 The dentin of the Human Tooth CT (a) shows that a 1D transfer function, simulated by assigning opacity to data values regardless of gradient magnitude, will erroneously color the background–enamel boundary. A 2D transfer function (b) can avoid assigning opacity to the range of gradient magnitudes that define this boundary.

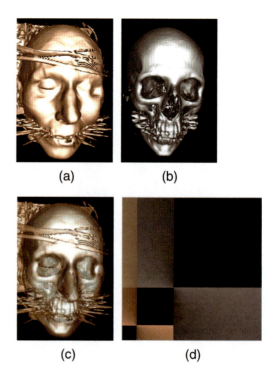

(a)   (b)

(c)   (d)

Figure 10.13 Pre-integrated volume rendering of isosurfaces. The isosurfaces in (a) and (b) were combined in (c) by combining the pre-integrated lookup tables, resulting in the lookup table visualized in (d). The images appear courtesy of Klaus Engel.

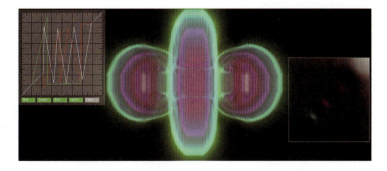

Figure 10.14 Screenshot of the pre-integrated volume renderer by Klaus Engel, available at *http://www.vis.uni-stuttgart.de/ ~engel/pre-integrated/*. The image appears courtesy of Klaus Engel.

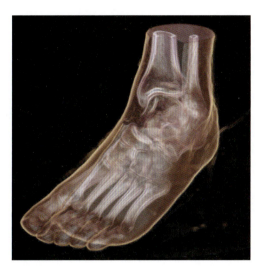

Figure 11.3 CT scan of a human foot, rendered on VolumePro 1000 with gradient magnitude modulation of opacity. Image courtesy of Yin Wu and Jan Hardenbergh, TeraRecon, Inc.

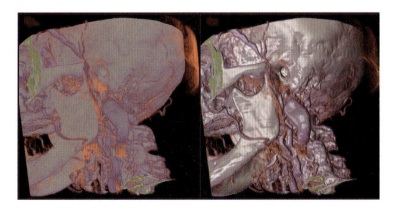

Figure 11.4 CT scan of a human head. (Left) Volume rendering with a simple emission model without shading. (Right) Including Phong shading. Images rendered on VolumePro 1000; courtesy of Yin Wu and Jan Hardenbergh, TeraRecon, Inc.

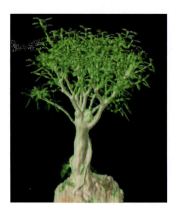

Figure 11.6 Hardware-accelerated ray-casting of a CT scan of a bonsai (128^3) with adaptive preintegration. Image courtesy of Stefan Roettger, University of Stuttgart, Germany.

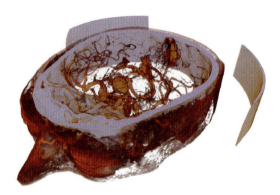

Figure 11.8 CT Angiography of a human brain ($512^2 \times 128$). Rendered on an ATI Radeon 9700 with 3D texture slicing and postclassification using dependent textures. Image courtesy of Christof Rezk-Salama, Siemens, Germany.

Figure 11.9 Volume clipping applied to (left) a CT scan of an engine ($256^2 \times 110$) and (right) an MRI scan of a human head (256^3). The cutting surface of the engine is enhanced by combining surface-based and volumetric shading. Image courtesy of Daniel Weiskopf, University of Stuttgart, Germany.

Figure 11.11 CT scan of a carp (512^3) rendered on an ATI Radeon 9700 with 2D multitextures and post-classification. Image courtesy of Christof Rezk-Salama, Siemens, Germany.

Figure 11.12 CT angiography of a human brain ($512^2 \times 128$). Transparent rendering of a nonpolygonal shaded isosurface with 2D multitextures on an NVIDIA GeForce-4Ti. Image courtesy of Christof Rezk-Salama, Siemens, Germany.

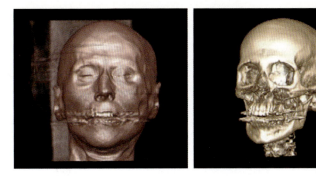

Figure 11.15 CT scan of a human head (256³) rendered on VolumePro 500 with Phong shading and different transfer functions.

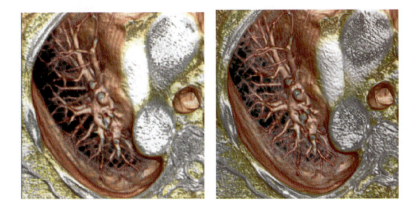

Figure 11.17 Comparison of (left) shear-warp (rendered by VolumePro 500) and (right) shear-image order (rendered by VolumePro 1000) ray-casting. Images courtesy of Yin Wu and Jan Hardenbergh, TeraRecon, Inc.

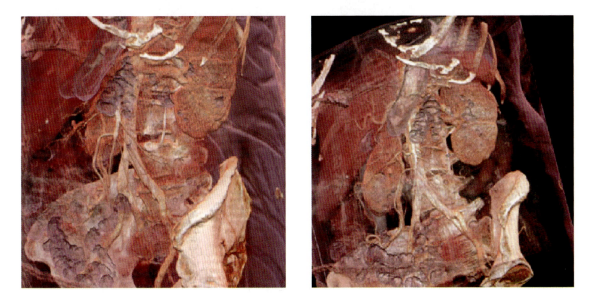

Figure 11.18 CT scans of a human torso, pre-classified with different transfer functions per material type. Images courtesy of Yin Wu and Jan Hardenbergh, TeraRecon, Inc.

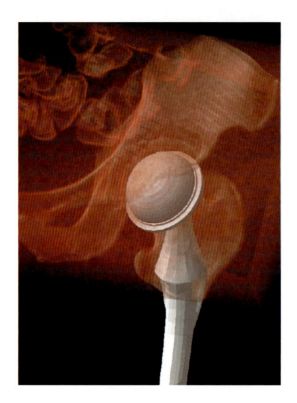
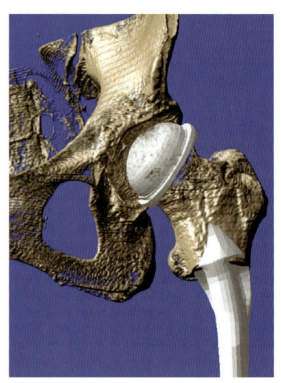

Figure 11.19 Embedding a polygon prosthesis into a CT scan of a human hip. Images courtesy of Yin Wu and Jan Hardenbergh, TeraRecon, Inc.

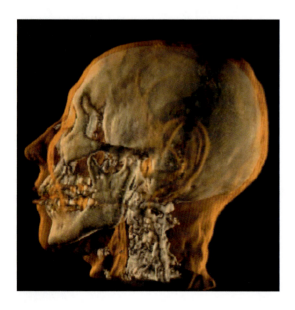

Figure 11.21 CT scan of a human head ($256^2 \times 129$) and of an engine ($256^2 \times 128$) rendered with hardware-accelerated EWA volume splatting on a GeForce FX Ultra 5900. Image courtesy of Wei Chen, Zhejian University, China, and Liu Ren, Carnegie Mellon University.

Figure 12.2 Glyph-based 3D flow visualization, combined with illuminated streamlines.

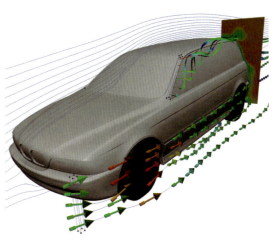

Figure 12.3 A combination of streamlines, streamribbons, arrows, and color coding for a 3D flow. Image courtesy of BMW Group and Martin Schulz.

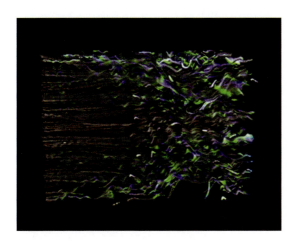

Figure 12.4 3D LIC with enhanced depth perception. Image courtesy of Victoria Interrante.

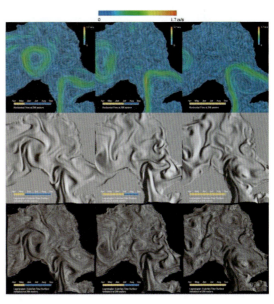

Figure 13.1 Application of flow textures to the advection of timesurfaces in the Gulf of Mexico [8]. (Top row) LEA on a slice; (middle row) timesurfaces viewed as shaded surfaces; (bottom row) flow texture superimposed on the timesurface. Each row represents three frames from an animation. Data courtesy of James O'Brien.

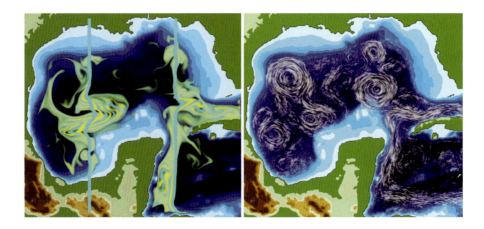

Figure 13.4 Flow in the Gulf of Mexico. (Left) Time lines visualized by dye advection; (Right) LEA with masking to emphasize regions of strong currents. Data courtesy of James O'Brien.

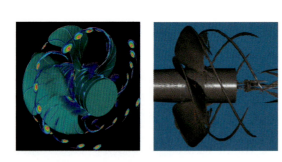

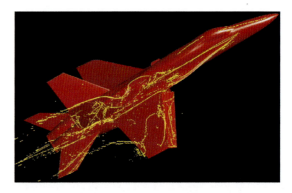

Figure 14.1 Swirl parameter. Images courtesy of Michael Remotigue, Mississippi State University.

Figure 14.2 Eigenvector approach (©1998 IEEE). Image courtesy of Robert Haimes, Massachusetts Institute of Technology.

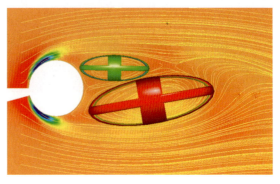

Figure 14.3 Parallel vector operator. Images courtesy of Martin Roth, Swiss Federal Institute of Technology, Zürich.

Figure 14.4 Winding-angle method. Image courtesy of I. Ari Sadarjoen. Delft University of Technology.

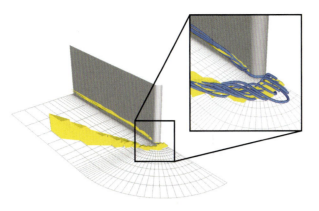

Figure 14.5 Combinatorial method (©2002 IEEE).

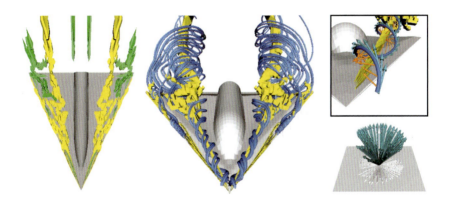

Figure 14.6 Geometric verification (© 2002 IEEE).

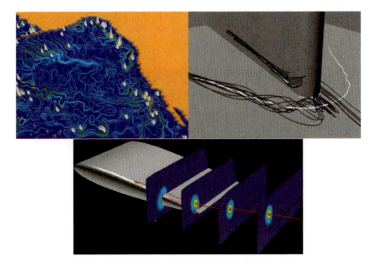

Figure 14.7 Visualization of vortices (© 1998 IEEE). Top left image courtesy of I. Ari Sadarjoen, Delft University of Technology. Top right image courtesy of Martin Roth, Swiss Federal Institute of Technology.

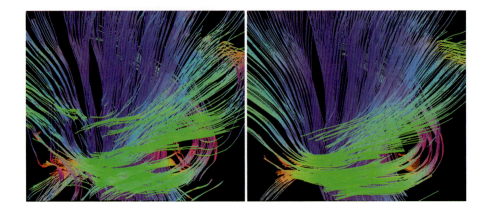

Figure 15.3 Comparison of nonfiltered (left) and MLS-filtered (right) fibers. Note the smoother and more regular behavior of the filtered fibers on the right image.

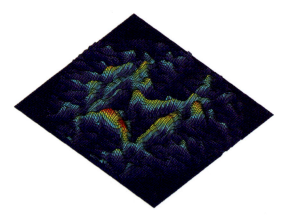

Figure 15.6 Height plot for anisotropy measure ("mountain" function) described in Section 15.2.6 for an axial slice of the data. The higher portions correspond to stronger anisotropy. See Equation 15.30.

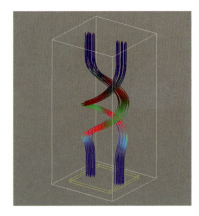

Figure 15.7 Double helix reconstructed using MLS method from artificial tensor data.

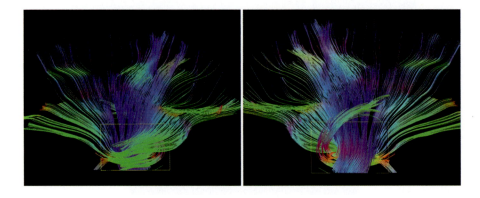

Figure 15.9 Right-hemisphere corona radiata shown from opposite directions. The yellow boxes show the seed region for the OTR fiber-tracing algorithm. Color coding indicates orthogonal directions in the amount of RGB (XYZ).

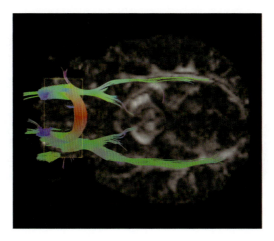

Figure 15.10 Brain structures: association fibers. Longitudinal and uncinate fasciculus of the optic tract. Color coding is the same as for Fig. 15.9.

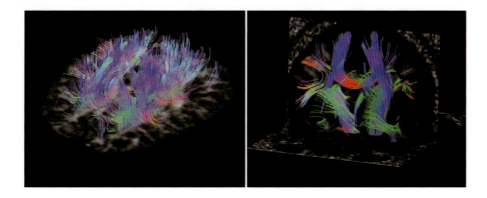

Figure 15.11 Brain structures: Fibers near the cortical surface and U-shaped fibers on the left; U-shaped fibers, parts of corona radiata, and corpus callosum are on the right.

Figure 15.12 Brain structures. (Left) A side view of the right hemisphere cingulum bundle on the background of corresponding c_l anisotropy; (Right) the same structure together with 3d models of the ventricle and CSF extracted by isosurfacing [13] on isotropic part c_s (see Equation 15.6) of the same DT-MRI dataset.

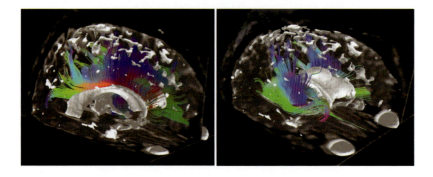

Figure 15.13 Brain structures: corpus callosum (left) and corona radiata (right) shown together with isotropic brain structures—ventricle, eye sockets, and pockets of CSF on the top of the brain. Cutting planes show isotropic c_s values.

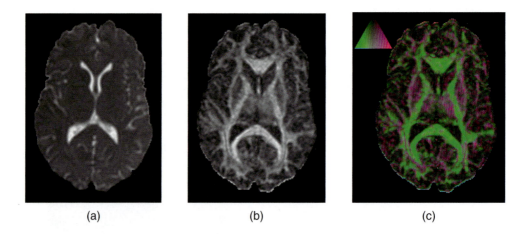

(a) (b) (c)

Figure 16.3 Different shape metrics applied to one slice of a brain DTI scan. (a) Tr: trace; (b) FA: fractional anisotropy; (c) C_L (green) and C_P (magenta).

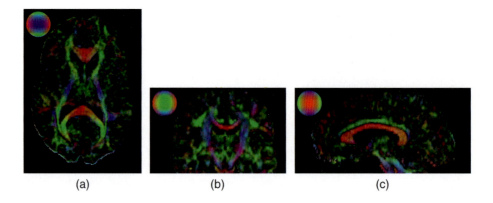

(a) (b) (c)

Figure 16.4 Eigenvector color maps shown on axis-aligned slices with three different slice orientations. The two axes within the slice are labeled with the anatomical name of the slice orientation. (a) Axial: x and y; (b) coronal: x and z; (c) sagittal: y and z.

Figure 16.5 Arrays of normalized ellipsoids visualize the diffusion tensors in a single slice.

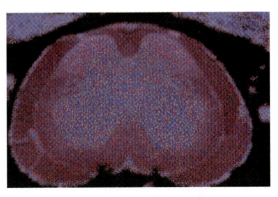

Figure 16.6 Brushstrokes illustrate the direction and magnitude of the diffusion. The background color and texture map show additional information.

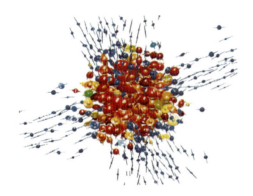

Figure 16.7 The composite shape of linear, planar, and spherical components emphasizes the shape of the diffusion tensor.

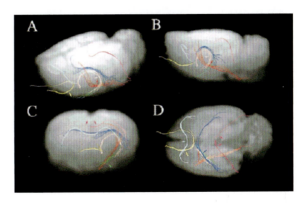

Figure 16.10 Tractography with streamlines. Image courtesy of Xue et al. [39].

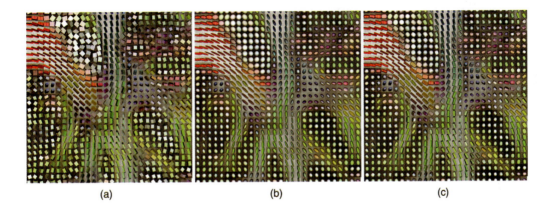

(a)   (b)   (c)

Figure 16.9 A portion of a brain DTI scan (also used in Figs. 16.3 and 16.4) as visualized by three different glyph methods. The overall glyph sizes have been normalized. (a) Boxes; (b) ellipsoids; (c) superquadrics.

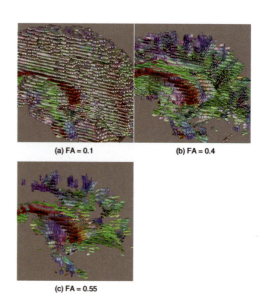

(a) FA = 0.1 (b) FA = 0.4

(c) FA = 0.55

Figure 16.11 Red streamtubes and green streamsurfaces show linear and planar anisotropy, respectively, together with anatomical landmarks for context.

Figure 16.12 Glyph-based visualization of a volumetric portion of a brain DTI scan (also used in Figs. 16.3, 16.4, and 16.9), with glyph culling based on three different fractional anisotropy thresholds.

Figure 16.13 Four different barycentric opacity maps and the corresponding renderings.

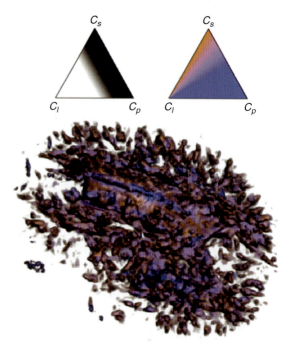

Figure 16.14 Assigning color and opacity with barycentric transfer function.

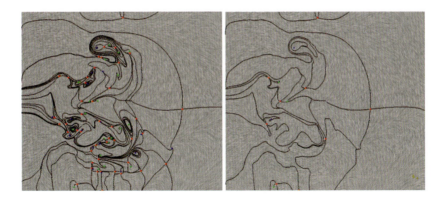

Figure 17.10 Turbulent and simplified topologies.

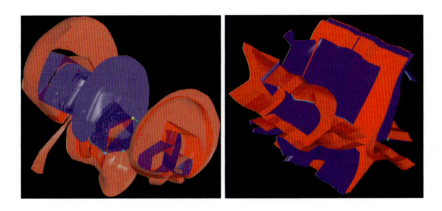

Figure 17.11 Unsteady vector and tensor topologies.

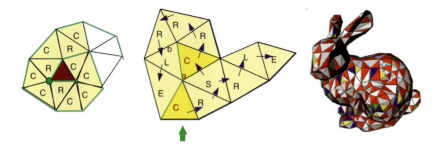

Figure 18.3 In this example of a typical compression situation, Edgebreaker starts with the darker triangle (left) and spirals out clockwise, filling the beginning of the clers string with CCCCRCCRCRC. It appends the tip of each C triangle to the vertex list. A typical situation where Edgebreaker finishes compression or closes a hole is shown in the center. It spirals counterclockwise, appending the label sequence CRSRLECRRRLE to the *clers* string and adding the vertices (a) and (b) to the vertex list. The triangles in the rabbit (right) have been shaded according to their Edgebreaker labels. Notice that half of the triangles are C (white) and about a third are R.

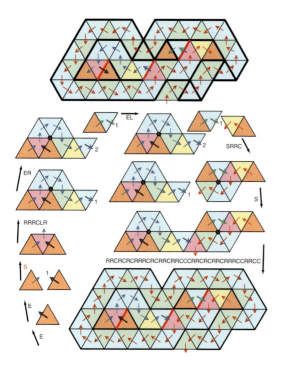

Figure 18.4 The connectivity of the remaining portion of the mesh shown on the top is encoded by Edgebreaker in the *clers* string: CCRRCCRRRCRRCRCRRCCCRRCRRCR CRRRCRCRCRRSCRRSLERERLCRRRSEE. The order in which the triangles are visited is shown by the arrows. The Spirale Reversi decompression receives the string reversed. Processing the first symbol (E) of the reversed string generates the first triangle (bottom left). The arrow leaving the previously reconstructed portion of the mesh indicates the gate where a new triangle will be attached. Then the next symbol (another E) puts the gate on a stack (1) and creates a new disconnected triangle with a new gate. Moving clockwise, the next symbol (S) makes a new triangle that joins the gate with the top of the stack. Reading the symbols RRC creates a right-turning fan that encloses a vertex (large dot), which will receive the reference number 1. Then the LR symbols are read. The next symbol (E) puts the gate on the stack (1) and creates an isolated triangle to which another one is attached as we read the next R symbol (top left). This creation, growth, and merging process continues as shown in the clockwise sequence.

Figure 18.7 (Left) The Topological Surgery approach merges concentric circles of triangles into a single TST. (Right) That TST and its dual VST have relatively few runs.

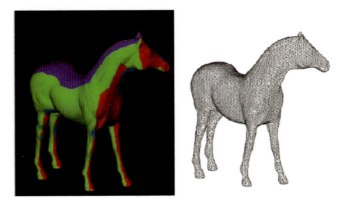

Figure 18.8 The reliefs produced by the Piecewise Regular Mesh (PRM) approach are shown (left) and resampled (right) into a nearly regular triangle mesh.

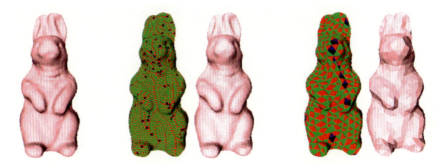

Figure 18.9 The original model (first from the left, courtesy of Cyberware) contains $t = 134,074$ triangles. A dense partitioning of its surface into triangloids (second) was produced by SwingWrapper. The corresponding retiled mesh (third) was generated by flattening of the triangloids. Its L^2 error is about 0.007% of the bounding box diagonal, and its 13642 triangles were encoded with a total of 3.5 bits per triangloid for both the connectivity and the geometry using Edgebreaker's connectivity compression combined with a novel geometry predictor, yielding a compressed file of $0.36t$ bits. A coarser partitioning (fourth) decomposes the original surface into 1505 triangloids. The distortion of the corresponding retiled mesh (fifth) is about 0.15%, and the total file size is $0.06t$ bits.

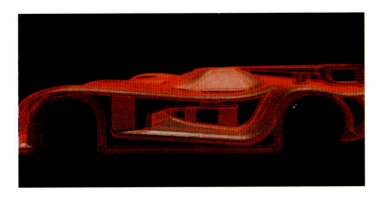

Figure 19.2 Race car reflection lines.

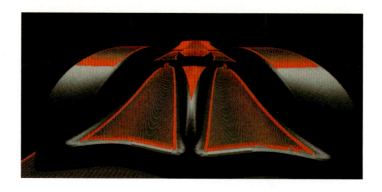

Figure 19.3   Race car focal analysis.

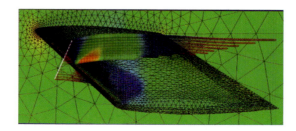

Figure 19.7   Airplane wing (streamballs).

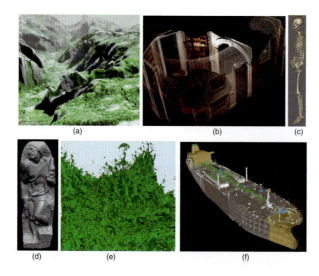

Figure 20.1   Large models from a variety of domains. (a) Yosemite Valley, California, 1.1 million triangles [74]. Copyright © 2001 IEEE. (b) Thomas Jefferson's Monticello, a fraction of the 19.5 million point samples from a single laser-rangefinder scan, from the Scanning Monticello project, courtesy of David Luebke (University of Virginia) and Lars Nyland (University of North Carolina at Chapel Hill). (c) Bones extracted from the Visible Female dataset, 9.9 million triangles before reduction, courtesy of Bill Lorensen of General Electric (d) St. Matthew, 372 M triangles [59]. Copyright © 2000 Digital Michelangelo Project, Stanford University. (e) Isosurface from DOE simulation of compressible turbulence, 500 million triangles (average depth complexity is 50), courtesy of Mark Duchaineau of Lawrence Livermore National Laboratory (f) Newport News double eagle tanker, 82 million triangles [29]. Copyright © 2001 Association for Computing Machinery, Inc.

Figure 20.2 Four discrete LODs of the armadillo model (2 million triangles) using normal maps [15]. LODs have 250,000, 63,000, 8,000, and 1,000 triangles, respectively. Copyright © 1998 Association for Computing Machinery, Inc.

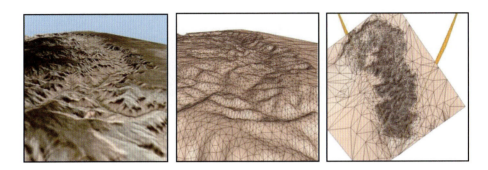

Figure 20.4 View-dependent rendering of the Grand Canyon, 10,013 triangles at 2 pixels of screen-space deviation [47]. Textured, wire-frame, and bird's-eye views. Copyright © 1998 IEEE.

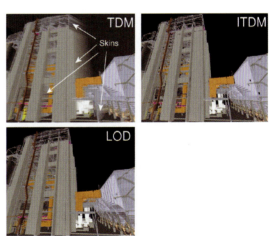

Figure 20.9 Comparison of Incremental Textured Depth-Meshes (ITDMs) [96] with regular TDMs and geometric levels of detail (LODs) on a 12.5M polygon power-plant model. ITDMs generate images almost as well as do static LODs, and at a frame rate 9 times faster. Moreover, ITDMs do not show the skin artifacts common in TDMs. The TDMs and ITDMs are used as a simplification of the far geometry. Copyright © 2003 Association for Computing Machinery, Inc.

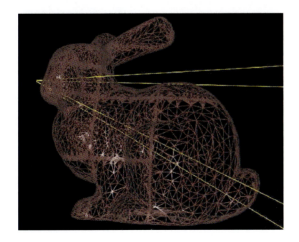

Figure 20.5 Using a 3D grid–based decomposition, HLOD performs a discrete approximation of the more fine-grained view-dependent hierarchy [29]. Copyright © 2001 Association for Computing Machinery, Inc.

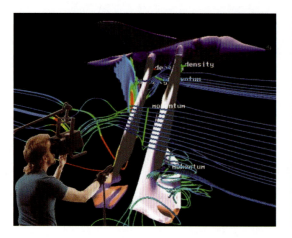

Figure 21.2 The Virtual Wind Tunnel, a virtual environment for the visualization of results of simulations arising in computational fluid dynamics. This example shows a variety of visualization widgets with visualizations including streamlines, local isosurfaces, and local cutting planes.

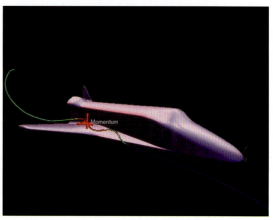

Figure 21.3 A point widget emitting a single streamline.

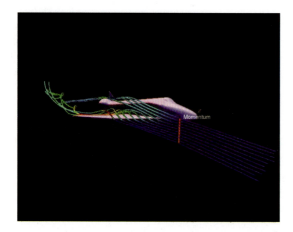

Figure 21.4 A line-visualization widget emitting a collection of streamlines.

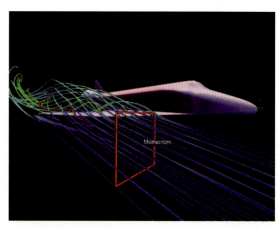

Figure 21.5 A plane-visualization widget emitting many streamlines.

The gradient magnitude is mapped by a lookup table to a user-specified piece-wise linear function. This function can be used to highlight particular gradient magnitude values or to attenuate the modulation effect. The lookup table is also used to automatically correct the gradient magnitudes in anisotropic volumes.

Classification: VolumePro 500 implements post-classification using a $4k \times 36$-bit classification lookup table that outputs 24-bit color and 12-bit α values. That precision is necessary for high accuracy during rendering of low-opacity volumes. Because of the uniform sample spacing, opacity correction can be applied in soft-ware for each frame. The opacity and color lookup tables can be dynamically loaded using double buffering in hardware.

Shading: The hardware implements Phong shading at each sample point at the rate of one illuminated sample per clock cycle. The diffuse and specular illumination are looked up in reflectance maps, respectively [111,114]. Each reflectance map is a precomputed table that stores the amount of illumination due to the sum of all of the light sources of the scene. Reflectance maps need to be reloaded when the object and light positions change

with respect to each other, or to correct the eye vector for anisotropic volumes (Fig. 11.15).

Compositing: The ray samples are accumulated into base-plane pixels using front-to-back alpha blending MIPs. The warping and display of the final image is performed by an off-the-shelf 3D graphics card using 2D texture mapping.

VolumePro 500 renders 256^3 or smaller volumes at 30 frames. Due to the brute-force processing, the performance is independent of the classification or data. In order to render a larger volume, the driver software first partitions the volume into smaller blocks. Each block is then rendered independently, and their resulting images are automatically combined to yield the final image. VolumePro 500 also provides various volume clipping and cropping features to visualize slices, cross-sections, or other regions of interest in the volume.

11.7.2 VolumePro 1000

The VolumePro 1000 system [131] uses a novel *shear-image order* ray-casting approach (Fig. 11.16).

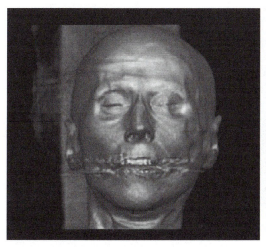
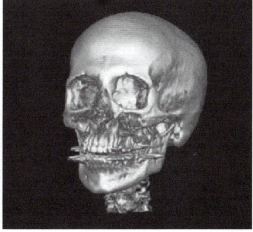

Figure 11.15 CT scan of a human head (256^3) rendered on VolumePro 500 with Phong shading and different transfer functions. (See also color insert.)

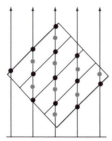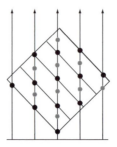

Figure 11.16 Shear-image order ray-casting. Grey samples are interpolated between original volume slices.

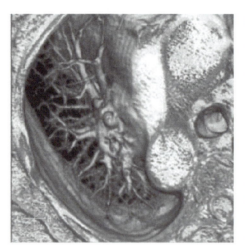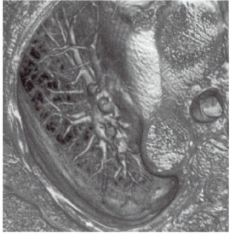

Figure 11.17 Comparison of (left) shear-warp (rendered by VolumePro 500) and (right) shear-image order (rendered by VolumePro 1000) ray-casting. Images courtesy of Yin Wu and Jan Hardenbergh, TeraRecon, Inc. (See also color insert.)

Data Traversal: Shear-image ray-casting casts rays directly through the centers of pixels but keeps the slices parallel to the base plane, similar to 2D texture-mapping methods. However, the 3D viewing transformation is explicitly decomposed into two matrices: a transformation from voxel coordinates to an intermediate coordinate system called *sample space*, and a transformation to adjust the depth values of sample points to reflect their distance from the image plane. A detailed derivation of these transformations is given by Wu et al. [131].

Sample space is coherent with image and voxel space, and the final image does not have to be warped because samples are aligned along viewing rays from image-plane pixels. This leads to higher image quality than in the traditional shear-warp factorization (Fig. 11.17).

Interpolation and Gradient Estimation: Similar to VolumePro 500, the resampling of the volume proceeds in object space for high memory coherence. VolumePro 1000 performs tri-linear interpolation of the volume data and computes gradient vectors in hardware. Similar to VolumePro 500 and the 2D texture slicing method of Rezk-Salama et al. [94], additional interpolated slices can be generated between original voxel slices. Since slices can be shifted

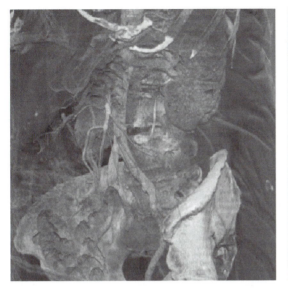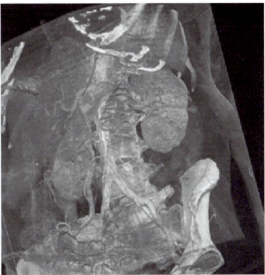

Figure 11.18 CT scans of a human torso, pre-classified with different transfer functions per material type. Images courtesy of Yin Wu and Jan Hardenbergh, TeraRecon, Inc. (See also color insert.)

with sub-pixel accuracy, this method avoids popping artifacts and keeps the ray spacing and sample spacing constant; it also does this for anisotropic and sheared volumes.

Classification: VolumePro 1000 uses a set of cascaded lookup tables that can be combined by a hierarchy of arithmetic–logic units [28]. Voxels can have up to four fields, and each field is associated with its own lookup table. The classification, and interpolation stage are cross-connected to allow the application to choose pre-or post-classification (Fig. 11.18). The hardware also supports opacity correction and gradient magnitude modulation of opacity.

Shading: VolumePro 1000 uses Phong shading hardware similar to that in VolumePro 500. Great care is taken to ensure correct gradient and Phong shading calculations for sheared and anisotropic data using lighting space [131] (see Section 11.3.5).

Compositing: In addition to the blending modes of VolumePro 500, the hardware also supports early ray termination for increased performance. VolumePro 1000 also implements geometry-based space leaping, volume clipping and cropping, and perspective projections using a variation of the shear-warp transformation. VolumePro 1000 is capable of rendering 10^9 samples per second.

For embedding of polygons into the volume data, the depth of volume samples can be compared with a polygon depth buffer (Fig. 11.19). The implementation uses multiple rendering passes: first, the polygons are rendered into the depth buffer. Next, rays are cast into the volume starting at the image plane and ending at the captured depth buffer. The color buffers of the polygon and volume rendering are then blended. In the second pass, rays are initialized with the result of the blending pass. They start at the depth buffer and end at the background to render the portion of the volume behind the polygon. The result is an image of the volume with embedded polygons. Volume-Pro 1000 also supports embedding of multiple translucent polygons using dual depth buffers [131].

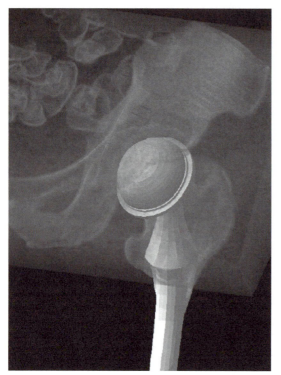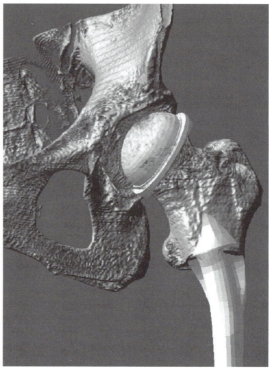

Figure 11.19 Embedding a polygon prosthesis into a CT scan of a human hip. Images courtesy of Yin Wu and Jan Hardenbergh, TeraRecon, Inc. (See also color insert.)

11.8 Splatting

Splatting, introduced by Westover [124], convolves every voxel in object space with a 3D reconstruction filter and accumulates the voxels' contribution on the image plane (see Chapter 8 and Fig. 11.20).

Data Traversal: Data traversal in splatting depends on the compositing method (see below). In its simplest form, voxels are traversed in object space and projected onto the screen (Fig. 11.20a). However, this leads to the wrong compositing order of the projected splats. Typically, traversal proceeds through the volume slice by slice, in approximate back-to-front order, similar to 2D texture slicing. For more advanced splatting methods, such as image-aligned sheet buffers, the traversal order is similar to 3D texture slicing (Fig. 11.20b).

Interpolation: Splatting is attractive because of its efficiency, which it derives from the use of preintegrated reconstruction kernels. For simple splatting, the 3D kernel can be preintegrated into a generic 2D footprint that is stored as a 2D texture.

Splatting also facilitates the use of higher-quality kernels with a larger extent than trilinear kernels. 3D Gaussian reconstruction kernels are preferable because they are closed under convolution and integration [138]. That is, the convolution of two Gaussians is another Gaussian, and the integration of a 3D Gaussian is a 2D Gaussian.

Additional care has to be taken if the 3D reconstruction kernels are not radially symmetric, as is the case for sheared, anisotropic, curvilinear, or irregular grids. In addition, for an arbitrary position in 3D, the contributions

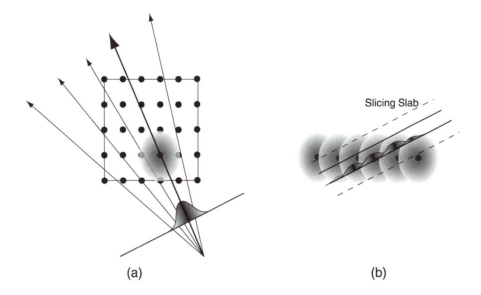

(a)   (b)

Figure 11.20   Splatting algorithm. (Left) In the splat-every-sample method, 3D reconstruction kernels are integrated into 2D footprints, projected, and composited onto the image plane. (Right) Image-aligned sheet buffers slice through the kernels. The contributions of 3D reconstruction kernels within a slab are added. The result of each slab is then composited onto the image plane.

from all kernels must sum up to one in the image. Zwicker et al. [137] discuss these issues in more detail and present a solution for Gaussian reconstruction kernels.

Gradient Estimation, Classification, and Shading: Typically, splatting uses pre-classification and preshading of the volume data. Each voxel stores the resulting RGBα values, which are then multiplied with the footprint before projection. Mueller et al. [80] propose a method for post-classification and shading in screen space. The gradients are either projected to screen space using so-called gradient splats, or they are computed in screen space using central differencing.

Compositing: Compositing is more complicated for splatting than for other volume rendering methods. While the principle is easy, it is more difficult to achieve high image quality.

The easiest compositing approach is called *splat-every-sample* (Fig. 11.20a). The 2D footprint of the kernel is multiplied by the scalar voxel value, projected to screen space, and

blended onto the image plane using graphics hardware [18]. However, this leads to visible artifacts, such as color bleeding from background objects, because of incorrect visibility determination [122].

To solve this problem, Westover [123] introduces *sheet buffer splatting*. 2D footprints are added (not composited) onto sheet buffers that are parallel to the base plane. Traversal proceeds in back-to-front order, and subsequent sheet buffers are composited onto the image plane. The approach solves color bleeding, but similar to 2D texture slicing, it introduces *popping artifacts* when the slice direction suddenly changes.

Mueller and Crawfis [77] proposed to use *image-aligned sheet buffers* (Fig. 11.20b). A *slicing slab* parallel to the image plane traverses the volume. The contributions of 3D reconstruction kernels between slab planes are added to the slab buffer, and the result is composited onto the image plane. This technique is similar to 3D texture slicing (see Section 11.6) and resolves the popping artifacts. But intersecting the slicing slab with the 3D reconstruction kernels has a high computational cost.

Mueller and Yagel [78] combine splatting with ray-casting techniques to accelerate rendering with perspective projection. Laur and Hanrahan [57] describe a hierarchical splatting algorithm enabling progressive refinement during rendering. Furthermore, Lippert [63] introduced a splatting algorithm that directly uses a wavelet representation of the volume data. For more extensions, see Chapter 8.

Westover's original framework does not deal with sampling-rate changes due to perspective projections. Aliasing artifacts may occur in areas of the volume where the sampling rate of diverging rays falls below the volume-grid sampling rate. The aliasing problem in volume splatting has first been addressed by Swan et al. [107] and Mueller et al. [79]. They use a distance-dependent stretch of the footprints to make them act as low-pass filters.

Zwicker et al. [137] develop EWA splatting along lines similar to the work of Heckbert [39], who introduced EWA filtering to avoid aliasing of surface textures. They extended his framework to represent and render texture functions on irregularly point-sampled surfaces [136] and to volume splatting [138]. EWA splatting results in a single 2D Gaussian footprint in screen space that integrates an elliptical 3D Gaussian reconstruction kernel and a 2D Gaussian low-pass filter. This screen-space footprint is analytically defined and can efficiently be evaluated. By flattening the 3D Gaussian kernel along the volume gradient, EWA volume splats reduce to surface splats that are suitable for high-quality isosurface rendering.

Ren et al. [92] derive an object-space formulation of the EWA surface splats and describe its efficient implementation on graphics hardware. For each point in object space, quadrilaterals that are texture-mapped with a Gaussian texture are deformed to result in the correct screen-space EWA splat after projection. A similar idea can be applied to EWA volume splatting, as shown in Fig. 11.21. The EWA

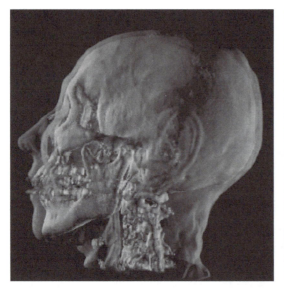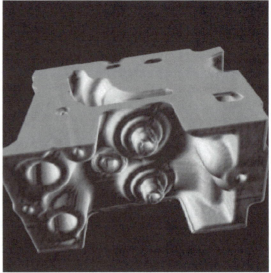

Figure 11.21   CT scan of a human head ($256^2 \times 129$) and of an engine ($256^2 \times 128$) rendered with hardware-accelerated EWA volume splatting on a GeForce FX Ultra 5900. Image courtesy of Wei Chen, Zhejian University, China, and Liu Ren, Carnegie Mellon University. (See also color insert.)

volume splat is evaluated in screen space by deforming a texture-mapped screen-space quadrilateral. The projection of samples and the deformation of the screen-space quads can be performed efficiently on modern GPUs [15].

11.9   Conclusions

Without a doubt, the availability of programmable graphics hardware on PCs has changed the field of hardware-accelerated volume rendering. It is has led to the great popularity of texture-slicing methods. More recently, it has become feasible to implement ray-casting on the GPU, including space leaping and early ray termination. The rapid progress of GPU hardware will address some remaining performance and image-quality issues soon. The recent introduction of procedural shading languages [66] will increase productivity and portability of code across hardware from different vendors.

A more serious issue is the continuing growth of volume data compared to the limited memory on the GPU and its low download–upload bandwidth. The availability of increasingly powerful computers and high-resolution scanners results in highly accurate and detailed data. For example, CT scanners now capture thousands of images with 512×512 resolution, supercomputers are producing terabytes of simulation data, and seismic scans for the oil and gas industry contain gigabytes or terabytes of data [113]. All of the GPU-accelerated algorithms presented in this chapter, such as texture slicing, multiply these memory requirements many times by storing gradients and other auxiliary volumes. Interesting directions to solve these problems are multiresolution techniques [36,56], compression-domain volume rendering [16], and image-based volume rendering (IBVR) [14,38,81].

On the high end, VolumePro remains the only commercially available solution. In its current incarnation it can store up to 1 GB of volume data on board; that memory size will undoubtedly increase with new releases. The business challenge is to make this hardware widely available in dedicated, high-end visualization systems, such as PACS or geophysical workstations, and 3D ultrasound systems. This challenge will increase with the continuing pressure from cheap, ever-stronger GPUs.

As hardware-accelerated techniques for rectilinear volumes mature, researchers focus their attention on the interactive or real-time rendering of unstructured volume data [35,116,117], time-varying data [65], and nontraditional, illustrative volume rendering [64]. If the rapid pace of innovation continues, the chapter on hardware-accelerated volume rendering will have to be expanded in the very near future.

Acknowledgments

I would like to thank the many people who provided images for this chapter, namely (in alphabetical order) Wei Chen, Tom Ertl, Jan Hardenbergh, Liu Ren, Christof Rezk-Salama, Stefan Roettger, Manfred Weiler, Daniel Weiskopf, and Yin Wu. A big thank you goes to Matthias Zwicker and Christof Rezk-Salama for stimulating and enlightening discussions. I also would like to thank the editors—Chuck Hansen and Chris Johnson—and Piper West for the opportunity to contribute to this book and for their tremendous patience. Finally, I would like to thank Lilly Charlotte Pfister for coming into the world—I could not have wished for a more beautiful excuse to procrastinate this project.

References

1.   K. Akeley. RealityEngine graphics. In *Computer Graphics, Proceedings of SIGGRAPH '93*, pages 109–116, 1993.

2.   T. Akenine-Möller and E. Haines. *Real-Time Rendering*, 2nd ed. A. K. Peters Ltd., 2002.

3.   J. Arvo and D. Kirk. Particle transport and image synthesis. In *Computer Graphics, Proceedings of SIGGRAPH '90*, pages 63–66, 1990.

4.   R. Avila, L. Sobierajski, and A. Kaufman. Towards a comprehensive volume visualization system. In *Proceedings of Visualization '92*, pages 13–20, 1992.

5. U. Behrens and R. Ratering. Adding shadows to a texture-based volume renderer. In *IEEE Symposium on Volume Visualization*, ACM Press, pages 39–46, 1998.

6. M. Bentum. *Interactive Visualization of Volume Data*. PhD thesis, University of Twente, Enschede, The Netherlands, 1995.

7. C. Berger, M. Hadwiger, and H. Hauser. A flexible framework for hardware-accelerated high-quality volume rendering. Tech. Rep. TR VRVIS 2003 001, Austria, 2003.

8. A. Bhalerao, H. Pfister, M. Halle, and R. Kikinis. Fast re-rendering of volume and surface graphics by depth, color, and opacity buffering. *Journal of Medical Image Analysis 4*, pages 235–251, 2000.

9. J. Blinn. Models of light reflection for computer synthesized pictures. *Computer Graphics 11*, Annual Conference Series, pages 192–198, 1977.

10. J. Blinn. Light reflection functions for simulation of clouds and dusty surfaces. In *Computer Graphics*, 16:21–29, 1982.

11. J. Blinn. Compositing: theory. *Computer Graphics & Applications* 14(5):83–87, 1994.

12. M. Brady, K. Jung, H. T. Nguyen, and T. P. Nguyen. Interactive volume navigation. *IEEE Transactions on Visualization and Computer Graphics*, 4(3):243–257, 1998.

13. B. Cabral, N. Cam, and J. Foran. Accelerated volume rendering and tomographic reconstruction using texture mapping hardware. In *1994 Workshop on Volume Visualization*, pages 91–98, 1994.

14. B. Chen, A. Kaufman, and Q. Tang. Image-based rendering of surfaces from volume data. In *Volume Graphics 2001*, pages 279–295, 2001.

15. W. Chen, L. Ren, M. Zwicker, and H. Pfister. Hardware-accelerated adaptive EWA volume splatting. In *Proceedings of IEEE Visualization 2004*, 2004.

16. T. Chiueh, T. He, A. Kaufman, and H. Pfister. Compression domain volume rendering. Tech. Rep. TR.94.01.04R, State University of New York at Stony Brook, 1994.

17. R. L. Cook and K. E. Torrance. A reflectance model for computer graphics. *ACM Transactions on Graphics*, 1(1):7–24, 1982.

18. R. Crawfis and N. Max. Direct volume visualization of 3D vector fields. *Workshop on Volume Visualization*, pages 55–50, 1992.

19. T. J. Cullip and U. Neumann. Accelerating volume reconstruction with 3D texture mapping hardware. Tech. Rep. TR93-027, University of North Carolina at Chapel Hill, 1993.

20. F. Dachille and A. Kaufman. Gi-cube: An architecture for volumetric global illumination and rendering. In *SIGGRAPH/EUROGRAPHICS Workshop On Graphics Hardware*, pages 119–128, 2000.

21. F. Dachille, K. Kreeger, B. Chen, I. Bitter, and A. Kaufman. High-quality volume rendering using texture mapping hardware. In *Eurographics/SIGGRAPH Workshop on Graphics Hardware*, pages 69–76, 1998.

22. J. Danskin and P. Hanrahan. Fast algorithms for volume ray tracing. In *Workshop on Volume Visualization*, A. Kaufman and W. L. Lorensen, Eds., pages 91–98, 1992.

23. J. Dengler, S. Warfield, J. Zaers, C. Guttmann, W. Wells, G. Ettinger, J. Hiller, and R. Kikinis. Automatic identification of grey matter structures from MRI to improve the segmentation of white matter lesions. In *Proceedings of Medical Robotics and Computer Assisted Surgery*, pages 140–147, 1995.

24. R. A. Drebin, L. Carpenter, and P. Hanrahan. Volume rendering. *Computer Graphics*, 22(4):65–74, 1988.

25. K. Engel, M. Kraus, and T. Ertl. High-quality pre-integrated volume rendering using hardware-accelerated pixel shading. In *Proceedings of the ACM SIGGRAPH/Eurographics Workshop on Graphics Hardware*, pages 9–16, 2001.

26. P. A. Fletcher and P. K. Robertson. Interactive shading for surface and volume visualization on graphics workstations. In *Proceedings of Visualization '93*, pages 291–298, 1993.

27. S. Francis and J. Hill. *Computer Graphics Using OpenGL*, 2nd Ed. Prentice Hall, 2000.

28. C. Gasparakis. Multi-resolution multi-field ray tracing: a mathematical overview. In *Proceedings of IEEE Visualization '99*, pages 199–206, 1999.

29. B. Geiger and R. Kikinis. Simulation of endoscopy. In *Comp. Vision Virtual Reality and Robotics in Medicine*, pages 227–281, 1995.

30. A. Van Gelder and K. Kim. Direct volume rendering with shading via 3D textures. In *ACM/IEEE Symposium on Volume Visualization*, pages 23–30, 1996.

31. A. S. Glassner, Ed. *Graphics Gems V*. New York, Academic Press, pages 257–264, 1990.

32. A. S. Glassner. *Principles of Digital Image Synthesis*. Morgan Kaufmann, 1995.

33. M. E. Goss. An adjustable gradient filter for volume visualization image enhancement. In *Graphics Interface '94*, pages 67–74, 1994.

34. S. Guan and R. Lipes. Innovative volume rendering using 3D texture mapping. In *Image*

Capture, Formatting, and Display, SPIE 2164, 1994.

35. S. Guthe, S. Roettger, A. Schieber, W. Strasser, and T. Ertl. High-quality unstructured volume rendering on the PC platform. In *Eurographics/ SIGGRAPH Graphics Hardware Workshop*, pages 119–125, 2002.

36. S. Guthe, M. Wand, J. Gonser, and W. Strasser. Interactive rendering of large volume data sets. In *Proceedings of IEEE Visualization '02*, pages 53–60, 2002.

37. P. Hanrahan. Three-pass affine transforms for volume rendering. *Computer Graphics*, 24(5):71–78, 1990.

38. M. J. Harris and A. Lastra. Real-time cloud rendering. In *Computer Graphics Forum (Eurographics 2001 Proceedings)*, 20, 2001.

39. P. Heckbert. *Fundamentals of Texture Mapping and Image Warping.* Master's thesis, University of California at Berkeley, Department of Electrical Engineering and Computer Science, 1989.

40. J. Hesser, R. Männer, G. Knittel, W. Strasser, H. Pfister, and A. Kaufman. Three architectures for volume rendering. In *Proceedings of Eurographics '95*, pages C-111–C-122, 1995.

41. K. H. Höhne and R. Bernstein. Shading 3D-images from CT using gray-level gradients. *IEEE Transactions on Medical Imaging*, 5(1):45–47, 1986.

42. W. M. Hsu. Segmented ray-casting for data parallel volume rendering. In *Proceedings of the 1993 Parallel Rendering Symposium*, pages 7–14, 1993.

43. J. Jacq and C. Roux. A direct multi-volume rendering method aiming at comparisons of 3-D images and models. *IEEE Transactions on Information Technology in Biomedicine*, 1(1):30–43, 1997.

44. J. T. Kajiya and B. P. V. Herzen. Ray tracing volume densities. In *Computer Graphics, SIGGRAPH '84 Proceedings*, 18:165–174, 1984.

45. A. Kaufman and R. Bakalash. Memory and processing architecture for 3D voxel-based imagery. *IEEE Computer Graphics & Applications*, 8(6):10–23, 1988.

46. A. Kaufman, R. Yagel, and D. Cohen. Intermixing surface and volume rendering. In *3D Imaging in Medicine: Algorithms, Systems, Applications*, K.H. Höhne, H. Fuchs, and S.M. Pizer, Eds., pages 217–227, 1990.

47. G. Kindlmann and J. Durkin. Semi-automatic generation of transfer functions for direct volume rendering. In *Proceedings 1998 IEEE Symposium on Volume Visualization*, pages 79–86, 1998.

48. J. Kniss, G. Kindlmann, and C. Hansen. Interactive volume rendering using multi-dimensional transfer functions and direct manipulation widgets. In *Proceedings of IEEE Visualization*, pages 255–262, 2001.

49. G. Knittel and W. Strasser. Vizard—visualization accelerator for real-time display. In *Proceedings of the SIGGRAPH/Eurographics Workshop on Graphics Hardware*, pages 139–146, 1997.

50. G. Knittel. Using pre-integrated transfer functions in an interactive software system for volume rendering. In *Eurographics 2002 Short Presentations*, pages 119–123, 2002.

51. K. Kreeger and A. Kaufman. Pavlov: A programmable architecture for volume processing. In *Proceedings of SIGGRAPH/Eurographics Workshop on Graphics Hardware*, pages 77–85, 1998.

52. K. Kreeger and A. Kaufman. Hybrid volume and polygon rendering with cube hardware. In *SIGGRAPH/Eurographics Workshop on Graphics Hardware*, pages 15–24, 1999.

53. K. Kreeger and A. Kaufman. Mixing translucent polygons with volumes. In *Proceedings of IEEE Visualization '99*, pages 191–198, 1999.

54. P. Lacroute and M. Levoy. Fast volume rendering using a shear-warp factorization of the viewing transform. In *Computer Graphics: Proceedings of SIGGRAPH '94*, pages 451–457, 1994.

55. P. Lacroute. *Fast Volume Rendering Using a Shear-Warp Factorization of the Viewing Transform.* PhD thesis, Stanford University, Departments of Electrical Engineering and Computer Science, 1995.

56. E. LaMar, B. Hamann, and K. Joy. Multiresolution techniques for interactive texture-based volume visualization. In *Proceedings of the 1999 IEEE Visualization Conference*, pages 355–362, 1999.

57. D. Laur and P. Hanrahan. Hierarchical splatting: a progressive refinement algorithm for volume rendering. In *Computer Graphics, SIGGRAPH '91 Proceedings*, pages 285–288, 1991.

58. M. Levoy. Display of surfaces from volume data. *IEEE Computer Graphics & Applications*, 8(5):29–37, 1988.

59. M. Levoy. Efficient ray tracing of volume data. *ACM Transactions on Graphics*, 9(3):245–261, 1990.

60. M. Levoy. A hybrid ray tracer for rendering polygon and volume data. *IEEE Computer Graphics & Applications*, 10(2):33–40, 1990.

61. B. Lichtenbelt, R. Crane, and S. Naqvi. *Introduction to Volume Rendering.* Los Angeles, Prentice Hall, 1998.

62. E. Lindholm, M. Kilgard, and H. Moreton. A user-programmable vertex engine. In *Computer*

Graphics: SIGGRAPH 2001 Proceedings, pages 149–158, 2001.

63. L. Lippert and M. Gross. Fast wavelet based volume rendering by accumulation of transparent texture maps. In *Computer Graphics Forum, Proceedings of Eurographics '95*, C-431–C-443, 1995.

64. A. Lu, C. Morris, J. Taylor, D. Ebert, C. Hansen, P. Rheingans, and M. Hartner. Illustrative interactive stipple rendering. *IEEE Transactions on Visualization and Computer Graphics*, 9(2):127–138, 2003.

65. K.-L. Ma. Visualizing time-varying volume data. *IEEE Transactions on Visualization and Computer Graphics*, 5(2):34–42, 2003.

66. B. Mark, S. Glanville, K. Akeley, and M. Kilgard. Cg: A system for programming graphics hardware in a C-like language. In *Proceedings of SIGGRAPH 2003*, 2003.

67. N. Max, P. Hanrahan, and R. Crawfis. Area and volume coherence for efficient visualization of 3D scalar functions. *Computer Graphics*, 24(5):27–34, 1995.

68. N. Max. Optical models for direct volume rendering. *IEEE Transactions on Visualization and Computer Graphics*, 1(2):99–108, 1995.

69. D. Meagher. Efficient synthetic image generation of arbitrary 3D objects. In *Proceedings of IEEE Computer Society Conference on Pattern Recognition and Image Processing*, 1982.

70. M. Meissner, U. Kanus, and W. Strasser. Vizard II, a PCI card for real-time volume rendering. In *Proceedings of SIGGRAPH/Eurographics Workshop on Graphics Hardware*, pages 61–68, 1998.

71. M. Meissner, U. Hoffmann, and W. Strasser. Enabling classification and shading for 3D texture mapping based volume rendering using OpenGL and extensions. In *Proceedings of the 1999 IEEE Visualization Conference*, pages 207–214, 1999.

72. M. Meissner, J. Huang, D. Bartz, K. Mueller, and R. Crawfis. A practical evaluation of popular volume rendering algorithms. In *IEEE Symposium on Volume Visualization*, pages 81–90, 2000.

73. M. Meissner, S. Guthe, and W. Strasser. Interactive lighting models and pre-integration for volume rendering on PC graphics accelerators. In *Proceedings of Graphics Interface 2002*, pages 209–218, 2002.

74. M. Meissner, U. Kanus, G. Wetekam, J. Hirche, A. Ehlert, W. Strasser, M. Doggett, and R. Proksa. A reconfigurable interactive volume rendering system. In *Proceedings of SIGGRAPH/Eurographics Workshop on Graphics Hardware*, 2002.

75. T. Möller, R. Machiraju, K. Mueller, and R. Yagel. A comparison of normal estimation schemes. In *Proceedings of IEEE Visualization '97*, pages 19–26, 1997.

76. T. Möller, R. Machiraju, K. Mueller, and R. Yagel. Evaluation and design of filters using a Taylor series expansion. *IEEE Transactions on Visualization and Computer Graphics*, 3(2):184–199, 1997.

77. K. Mueller and R. Crawfis. Eliminating popping artifacts in sheet buffer–based splatting. *IEEE Visualization '98*, pages 239–246, 1998.

78. K. Mueller and R. Yagel. Fast perspective volume rendering with splatting by utilizing a ray-driven approach. *IEEE Visualization '96*, pages 65–72, 1996.

79. K. Mueller, T. Moeller, J. Swan, R. Crawfis, N. Shareef, and R. Yagel. Splatting errors and antialiasing. *IEEE Transactions on Visualization and Computer Graphics*, 4(2):178–191, 1998.

80. K. Mueller, T. Moeller, and R. Crawfis. Splatting without the blur. In *Proceedings of the 1999 IEEE Visualization Conference*, pages 363–370, 1999.

81. K. Mueller, N. Shareef, J. Huang, and R. Crawfis. IBR assisted volume rendering. In *Proceedings of IEEE Visualization Late Breaking Hot Topics*, pages 5–8, 1999.

82. U. Neumann. *Volume Reconstruction and Parallel Rendering Algorithms: A Comparative Analysis*. PhD thesis, University of North Carolina at Chapel Hill, 1993.

83. R. Osborne, H. Pfister, H. Lauer, N. McKenzie, S. Gibson, W. Hiatt, and T. Ohkami. EM-Cube: An architecture for low-cost real-time volume rendering. In *Proceedings of the SIGGRAPH/Eurographics Workshop on Graphics Hardware*, pages 131–138, 1997.

84. M. Peercy, J. Airey, and B. Cabral. Efficient bump mapping hardware. In *Computer Graphics, Proceedings of SIGGRAPH '97*, pages 303–306, 1997.

85. H. Pfister and A. Kaufman. Cube-4—A scalable architecture for real-time volume rendering. In *1996 ACM/IEEE Symposium on Volume Visualization*, pages 47–54, 1996.

86. H. Pfister, A. Kaufman, and T. Chiueh. Cube-3: A real-time architecture for high-resolution volume visualization. In *1994 ACM/IEEE Symposium on Volume Visualization*, pages 75–83, 1994.

87. H. Pfister, J. Hardenbergh, J. Knittel, H. Lauer, and L. Seiler. The VolumePro real-time raycasting system. In *Proceedings of the 26th Annual Conference on Computer Graphics and*

Interactive Techniques (SIGGRAPH '99), pages 251–260, 1999.

88. H. Pfister, W. Lorensen, C. Bajaj, G. Kindlmann, W. Schroeder, L.S. Avila, K. Martin, R. Machiraju, and J. Lee. The transfer function bake-off. *IEEE Computer Graphics and Applications*, pages 16–22, 2001.

89. B. T. Phong. Illumination for computer generated pictures. *Communications of the ACM*, 1(18):311–317, 1975.

90. T. Porter and T. Duff. Compositing digital images. *Computer Graphics*, 18(3), 1984.

91. H. Ray, H. Pfister, D. Silver, and T. A. Cook. Ray-casting architectures for volume visualization. *IEEE Transactions on Visualization and Computer Graphics*, 5(3):210–223, 1999.

92. L. Ren, H. Pfister, and M. Zwicker. Object-space EWA surface splatting: A hardware accelerated approach to high quality point rendering. In *Computer Graphics Forum*, 21:461–470, 2002.

93. R. A. Reynolds, D. Gordon, and L.-S. Chen. A dynamic screen technique for shaded graphics display of slice-represented objects. *Computer Vision, Graphics, and Image Processing*, 38(3):275–298, 1987.

94. C. Rezk-Salama, K. Engel, M. Bauer, G. Greiner, and T. Ertl. Interactive volume rendering on standard PC graphics hardware using multi-textures and multi-stage rasterization. In *Eurographics/SIGGRAPH Workshop on Graphics Hardware*, pages 109–118, 2000.

95. C. Rezk-Salama. *Volume Rendering Techniques for General Purpose Graphics Hardware*. PhD thesis, University of Erlangen, Germany, 2001.

96. S. Roettger and T. Ertl. A two-step approach for interactive preintegrated volume rendering of unstructured grids. In *IEEE Symposium on Volume Visualization*. Boston, ACM Press, pages 23–28, 2002.

97. S. Roettger, M. Kraus, and T. Ertl. Hardware-accelerated volume and isosurface rendering based on cell-projection. In *Proceedings of IEEE Visualization*, pages 109–116, 2000.

98. S. Roettger, S. Guthe, D. Weiskopf, T. Ertl, and W. Strasser. Smart hardware-accelerated volume rendering. In *Eurographics/IEEE TCVG Symposium on Visualization 2003*, 2003.

99. P. Sabella. A rendering algorithm for visualizing 3D scalar fields. *Computer Graphics*, 22(4):59–64, 1988.

100. J. Schulze, M. Kraus, U. Lang, and T. Ertl. Integrating preintegration into the shear-warp algorithm. In *Proceedings of the Third International Workshop on Volume Graphics*, pages 109–118, 2003.

101. A. R. Smith. A pixel is not a little square, a pixel is not a little square, a pixel is not a little square! (and a voxel is not a little cube!) Tech. Rep., Microsoft, Inc., 1995.

102. I. Sobel. An isotropic $3 \times 3 \times 3$ volume gradient operator. Unpublished manuscript, 1995.

103. R. L. Sollenberg and P. Milgram. Effects of stereoscopic and rotational displays in a 3D path tracing task. In *Human Factors*, pages 483–499, 1993.

104. D. Speary and S. Kennon. Volume probes: Interactive data exploration on arbitrary grids. *Computer Graphics*, 24:5–12, 1990.

105. M. Sramek. Fast surface rendering from raster data by voxel traversal using chessboard distance. In *Proceedings of Visualization '94*, pages 188–195, 1994.

106. K. R. Subramanian and D. S. Fussell. Applying space subdivision techniques to volume rendering. In *Proceedings of Visualization '90*, pages 150–159, 1990.

107. J. E. Swan, K. Mueller, T. Möller, N. Shareef, R. Crawfis, and R. Yagel. An anti-aliasing technique for splatting. In *Proceedings of the 1997 IEEE Visualization Conference*, pages 197–204, 1997.

108. J. Sweeney and K. Mueller. Shear-warp deluxe: the shear-warp algorithm revisited. In *Eurographics/IEEE TCVG Symposium on Visualization*, pages 95–104, 2002.

109. U. Tiede, T. Schiemann, and K. Höhne. High quality rendering of attributed volume data. In *Proceedings of IEEE Visualization*, pages 255–262, 1998.

110. C. Upson and M. Keeler. V-BUFFER: Visible volume rendering. *Computer Graphics*, 22(4):59–64, 1988.

111. J. van Scheltinga, J. Smit, and M. Bosma. Design of an on-chip reflectance map. In *Proceedings of the 10th Eurographics Workshop on Graphics Hardware*, pages 51–55, 1995.

112. G. Vézina, P. Fletcher, and P. Robertson. Volume rendering on the MasPar MP-1. In *1992 Workshop on Volume Visualization*, pages 3–8, 1992.

113. W. Volz. Gigabyte volume viewing using split software/hardware interpolation. In *Volume Visualization and Graphics Symposium 2000*, pages 15–22, 2000.

114. D. Voorhies and J. Foran. Reflection vector shading hardware. In *Computer Graphics, Proceedings of SIGGRAPH '94*, pages 163–166, 1994.

115. M. Weiler, R. Westermann, C. Hansen, K. Zimmerman, and T. Ertl. Level-of-detail volume rendering via 3D textures. In *Volume*

Visualization and Graphics Symposium 2000, pages 7–13, 2000.

116. M. Weiler, M. Kraus, M. Merz, and T. Ertl. Hardware-based ray-casting for tetrahedral meshes. In *Proceedings of IEEE Visualization*, 2003.

117. M. Weiler, M. Kraus, M. Merz, and T. Ertl. Hardware-based view-independent cell projection. *IEEE Transactions on Visualization and Computer Graphics*, 9(2):163–175, 2003.

118. D. Weiskopf, K. Engel, and T. Ertl. Volume clipping via per-fragment operations in texture-based volume rendering. In *Visualization 2002*, pages 93–100, 2002.

119. D. Weiskopf, K. Engel, and T. Ertl. Interactive clipping techniques for texture-based volume visualization and volume shading. *IEEE Transactions on Visualization and Computer Graphics*, 9(3):298–313, 2003.

120. W. Wells, P. Viola, H. Atsumi, S. Nakajima, and R. Kikinis. Multi-modal volume registration by maximization of mutual information. *Medical Image Analysis*, 1(1):35–51, 1996.

121. R. Westermann and T. Ertl. Efficiently using graphics hardware in volume rendering applications. In *Computer Graphics, SIGGRAPH '98 Proceedings*, pages 169–178, 1998.

122. L. Westover. Interactive volume rendering. In *Proceedings of the Chapel Hill Workshop on Volume Visualization*, pages 9–16, 1989.

123. L. Westover. Footprint evaluation for volume rendering. In *Computer Graphics, Proceedings of SIGGRAPH '90*, pages 367–376, 1990.

124. L. A. Westover. *Splatting: A Parallel, Feed-Forward Volume Rendering Algorithm*. PhD thesis. The University of North Carolina at Chapel Hill, Technical Report TR91-029, 1991.

125. P. L. Williams and N. Max. A volume density optical model. *Workshop on Volume Visualization*, pages 61–68, 1992.

126. O. Wilson, A. V. Gelder, and J. Wilhelms. Direct volume rendering via 3D textures. UCSC-CRL-94-19, University of California at Santa Cruz.

127. C. Wittenbrink and M. Harrington. A scalable MIMD volume rendering algorithm. In *Eighth International Parallel Processing Symposium*, pages 916–920, 1994.

128. C. Wittenbrink and A. Somani. Permutation warping for data parallel volume rendering. In *Parallel Rendering Symposium, Visualization '93*, pages 57–60, 1993.

129. C. Wittenbrink, T. Malzbender, and M. Goss. Opacity-weighted color interpolation for volume sampling. In *Symposium on Volume Visualization*, pages 135–142, 1998.

130. G. Wolberg. *Digital Image Warping*. IEEE Computer Society Press, Los Alamitos, California, 1990.

131. Y. Wu, V. Bhatia, H. C. Lauer, and L. Seiler. Shear-image order ray-casting volume rendering. In *Symposium on Interactive 3D Graphics*, pages 152–162, 2003.

132. R. Yagel and A. Kaufman. Template-based volume viewing. *Computer Graphics Forum, Proceedings of Eurographics '92*, 11(3):153–167, 1992.

133. R. Yagel, D. Cohen, and A. Kaufman. Discrete ray tracing. *IEEE Computer Graphics & Applications*, pages 19–28, 1992.

134. S. W. Zucker and R. A. Hummel. A 3D edge operator. *IEEE Transactions on Pattern Analysis and Machine Intelligence*, 3(3):324–331, 1981.

135. K. Z. Zuiderveld, A. H. J. Koning, and M. A. Viergever. Acceleration of ray-casting using 3D distance transform. In *Proceedings of Visualization in Biomedical Computing*, 1808:324–335, 1992.

136. M. Zwicker, H. Pfister, J. Van Baar, and M. Gross. Surface splatting. In *Computer Graphics, SIGGRAPH 2001 Proceedings*, pages 371–378, 2001.

137. M. Zwicker, H. Pfister, J. Van Baar, and M. Gross. EWA splatting. *IEEE Transactions on Visualization and Computer Graphics*, 8(3):223–238, 2002.

138. M. Zwicker, H. Pfister, J. van Baar, and M. Gross. Ewa volume splatting. *IEEE Visualization 2001*, pages 29–36, 2001.

PART IV

Vector Field Visualization

12   Overview of Flow Visualization

DANIEL WEISKOPF
University of Stuttgart

GORDON ERLEBACHER
Florida State University

12.1   Introduction

Flow visualization is an important topic in scientific visualization and has been the subject of active research for many years. Typically, data originates from numerical simulations, such as those of computational fluid dynamics, and needs to be analyzed by means of visualization to gain an understanding of the flow. With the rapid increase of computational power for simulations, the demand for more advanced visualization methods has grown. This chapter presents an overview of important and widely used approaches to flow visualization, along with references to more detailed descriptions in the original scientific publications. Although the list of references covers a large body of research, it is by no means meant to be a comprehensive collection of articles in the field.

12.2   Mathematical Description of a Vector Field

We start with a rather abstract definition of a vector field by making use of concepts from differential geometry and the theory of differential equations. For more detailed background information on these topics, we refer to textbooks on differential topology and geometry [45,46,66,81]. Although this mathematical approach might seem quite abstract for many applications, it has the advantage of being a flexible and generic description that is applicable to a wide range of problems.

We first give the definition of a *vector field*. Let M be a smooth m-manifold with boundary, let N be an n-dimensional submanifold with boundary ($N \subset M$), and let $I \subset \mathbb{R}$ be an open interval of real numbers. A map

$$u : N \times I \rightarrow TM$$

is a *time-dependent vector field* provided that

$$u(x, t) \in T_x M$$

An element $t \in I$ serves as a description for time; $x \in N$ is a position in space. TM is a tangent bundle—the collection of all tangent vectors, along with the information of the point of tangency. Finally, $T_x M$ is the tangent space associated with x. The vector field maps a position in space and time, (x, t), to a tangent vector located at the same reference point x. For a *tangential time-dependent vector field*, the mapping remains in the tangent bundle TN and therefore does not contain a normal component. That is,

$$u : N \times I \rightarrow TN$$

For a nontangential vector field, a related tangential vector field can be computed by projection from $T_x M$ to $T_x N$, i.e., by removing the normal parts from the vectors.

Integral curves are directly related to vector fields. Let $u : N \times I \rightarrow TN$ be a continuous (tangential) vector field. Let x_0 be a point in N and $J \subset I$ be an open interval that contains t_0. The C^1 map

$$\xi_{x_0, t_0} : J \rightarrow N$$

with

$$\xi_{x_0, t_0}(t_0) = x_0 \text{ and } \frac{d\xi_{x_0, t_0}(t)}{dt} = u(\xi_{x_0, t_0}(t), t)$$

is an integral curve for the vector field with initial condition $x = x_0$ at $t = t_0$. The subscripts in the notation of ξ_{x_0, t_0} denote this initial condition. These integral curves are usually referred to as *pathlines*, especially in the context of flow visualization. If u satisfies the Lipschitz condition, the differential equation for ξ_{x_0, t_0} has a unique solution.

In all practical applications of flow visualization, the data is given on a manifold of two or three dimensions. To investigate a vector field on an arbitrary curved surface, the above formalism is necessary and, for example, the issue of describing the surface by charts has to be addressed. Very often, however, N is just a Euclidean space. This allows us to use a simpler form of the tangential vector field given by

$$u : \Omega \times I \longrightarrow \mathbb{R}^n, \qquad (\mathbf{x}, t) \mapsto \mathbf{u}(\mathbf{x}, t)$$

The vector field is defined on the n-dimensional Euclidean space $\Omega \subset \mathbb{R}^n$ and depends on time $t \in I$. We use boldface lowercase letters to denote vectors in n dimensions. The reference point x is no longer explicitly attached to the tangent vector. In this modified notation, the integral curve is determined by the ordinary differential equation

$$\frac{d\mathbf{x}_{\text{path}}(t;\mathbf{x}_0, t_0)}{dt} = \mathbf{u}(\mathbf{x}_{\text{path}}(t;\mathbf{x}_0, t_0), t) \qquad (12.1)$$

We assume that the initial condition $\mathbf{x}_{\text{path}}(t_0;\mathbf{x}_0, t_0) = \mathbf{x}_0$ is given at time t_0, i.e., all integral curves are labeled by their initial conditions $(\mathbf{x}_0, t_0)$ and parameterized by t. By construction, the tangent to the pathline at position x and time t is precisely the velocity $\mathbf{u}(\mathbf{x}, t)$. In more general terms, the notation $\mathbf{x}(t;\mathbf{x}_0, t_0)$ is used to describe any curve parameterized by t that contains the point $\mathbf{x}_0$ for $t = t_0$.

12.3 Particle Tracing in Time-Dependent Flow Fields

Integral curves play an important role in visualizing the associated vector field and in understanding the underlying physics of the flow.

There exist two important additional types of characteristic curves: *streamlines* and *streaklines*. In steady flows, pathlines, streamlines, and streaklines are identical. When the vector field depends explicitly on time, these curves are distinct from one another.

In a steady flow, a particle follows the streamline, which is a solution to

$$\frac{d\mathbf{x}_{\text{stream}}(t;\mathbf{x}_0, t_0)}{dt} = \mathbf{u}(\mathbf{x}_{\text{stream}}(t;\mathbf{x}_0, t_0))$$

In an unsteady context, we consider the instantaneous vector field at fixed time τ. The particle paths associated with this artificially steady, virtually frozen field are the streamlines governed by

$$\frac{d\mathbf{x}_{\text{stream}}(t;\mathbf{x}_0, t_0)}{dt} = \mathbf{u}(\mathbf{x}_{\text{stream}}(t;\mathbf{x}_0, t_0), \tau)$$

Here, t and t_0 are just parameters along the curve and do not have the meaning of physical time, in contrast to the physical time τ. A third type of curve is produced by dye released into the flow. If dye is released continuously into a flow from a fixed point $\mathbf{x}_0$, it traces out a streakline. For example, smoke emanating from a lit cigarette follows a streakline.

It is instructive to derive integrated equations for pathlines, streamlines, and streaklines. The solution to the ordinary differential equation (Equation 12.1) is obtained by formal integration, which gives the pathline

$$\mathbf{x}_{\text{path}}(t;\mathbf{x}_0, t_0) = \mathbf{x}_0 + \int_{t_0}^{t} \mathbf{u}(\mathbf{x}_{\text{path}}(s;\mathbf{x}_0, t_0), s)\, ds$$

Similarly, streamlines at time τ can be computed by

$$\mathbf{x}_{\text{stream}}(t;\mathbf{x}_0, t_0) = \mathbf{x}_0 + \int_{t_0}^{t} \mathbf{u}(\mathbf{x}_{\text{stream}}(s;\mathbf{x}_0, t_0), \tau)\, ds$$

To obtain the snapshot of a streakline at time t, a set of particles is released from $\mathbf{x}_0$ at times $s \in [t_{\min}, t]$ and the particles' positions are evaluated at time t:

$$\mathbf{x}_{\text{streak}}(s;\mathbf{x}_0, t) = \mathbf{x}_{\text{path}}(t;\mathbf{x}_0, s)$$

The streakline is parameterized by s, and $t_{\min}$ is the first time that particles are released.

12.4 Classification of Visualization Approaches

There exist many different vector-field visualization techniques, which can be distinguished according to their properties with respect to a number of categories. The following classification should be considered rather a collection of important issues than a complete taxonomy. These issues should be taken into account when choosing a visualization approach.

In one classification scheme, techniques are distinguished by the relationship between a vector field and its associated visual representation. Point-based direct visualization approaches take into account the vector field at a point, and possibly also its neighborhood, to obtain a visual representation. The vector field is directly mapped to graphical primitives in the sense that no sophisticated intermediate processing of data is performed. Another class is based on characteristic curves obtained by particle tracing. The third class thoroughly preprocesses data to identify important features, which then serve as a basis for the actual visualization.

Another type of property is the density of representation: the domain can be sparsely or densely covered by visualization objects. Density is particularly useful for subclassing particle-tracing approaches. Related to density is the distinction between local and global methods. A global technique essentially shows the complete flow, whereas important features of the flow could possibly be missed by a local technique.

The choice of a visualization method is also influenced by the structure of the data. The dimensionality of the manifold on which the vector field is defined plays an important role. For example, strategies that work well in 2D might be much less useful in 3D because of perception issues. The recognition of orientation and spatial position of graphical primitives is more difficult, and important primitives could be hidden by others. Dimensionality also affects performance; a 3D technique has to process substantially more data. If visualization is restricted to slices or more general hypersurfaces of a 3D flow, the projection of vectors onto the tangent spaces of the hypersurfaces has to be considered. Moreover, a distinction has to be made between time-dependent and time-independent data. A steady flow is usually much less demanding since frame-to-frame coherence is easy to achieve and streamlines, streaklines, and pathlines are identical. Finally, the type of grid has to be taken into account. Data can be provided, for example, on uniform, rectilinear, curvilinear, or unstructured grids. The grid type affects the visualization algorithms with respect to mainly data storage and access mechanisms or interpolation schemes. Ideally, the final visual representation does not depend on the underlying grid.

It should be noted that there is no "ideal" technique that is best for all visualization tasks. Therefore, it is often useful to combine different approaches for an effective overall visualization. Nevertheless, we focus on describing the methods individually. The techniques are roughly ordered according to their classification: point-based direct methods, sparse representations for particle-tracing techniques, dense representations based on particle tracing, and feature-based approaches. The other types of properties are discussed along with the descriptions of the individual methods.

This chapter is meant to provide an overview of flow visualization. For more detailed descriptions, we refer to [18,40,82,88,99] and to the other flow visualization chapters of this book.

12.5 Point-Based Direct Flow Visualization

The traditional technique of arrow plots is a well-known example for direct flow visualization based on glyphs. Small arrows are drawn at discrete grid-points, showing the direction of the flow and serving as local probes for the velocity field (Fig. 12.1a). In the closely related hedgehog approach, the flow is visualized by

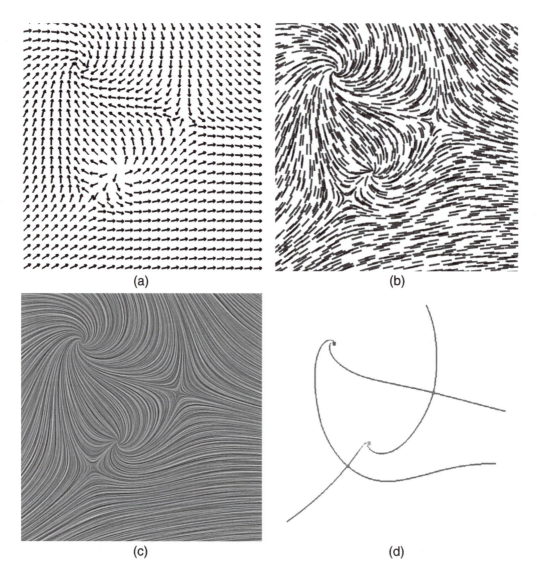

(a)

(b)

(c)

(d)

Figure 12.1   Comparison of visualization techniques applied to the same 2D flow: (a) arrow plot, (b) streamlets, (c) line integral convolution (LIC), and (d) topology-based. Image 12.1d courtesy of Gerik Scheuermann.

directed line segments whose lengths represent the magnitude of the velocity. To avoid possible distracting patterns for a uniform sampling by arrows or hedgehogs, randomness can be introduced in their positions [29]. Arrow plots can be directly applied to time-dependent vector fields by letting the arrows adapt to the velocity field for the current time. For 3D representations,

the following facts have to be considered: the position and orientation of an arrow is more difficult to understand due to the projection onto the 2D image plane, and an arrow might occlude other arrows in the background. The problem of clutter can be addressed by highlighting arrows with orientations in a range specified by the user [9], or by selectively seeding the arrows. Illumination and shadows serve to improve spatial perception; for example, shadowing can be applied to hedgehog visualizations on 2D slices of a 3D flow [65].

More complex glyphs [26] can be used to provide additional information on the flow at a point of the flow (Fig. 12.2). In addition to the actual velocity, information on the Jacobian of the velocity field is revealed. The Jacobian is presented in an intuitive way by decomposing the Jacobian matrix into meaningful components and mapping them to icons based on easily understandable metaphors. Typical data encoded into glyphs comprise velocity, acceleration, curvature, local rotation, shear, and convergence. Glyphs can also be used to represent information on the uncertainty of the vector-field data [134]. Glyph-based uncertainty visualization is also covered by Lodha et al. [69] and Pang et al. [86], who additionally discuss uncertainty representations for other visualization styles.

Another strategy is to map flow properties to a single value and apply techniques known from the visualization of scalar data. Typically,

the magnitude of the velocity or one of the velocity components is used. For 2D flow visualization, a mapping to color or to iso-lines (contour lines) is often applied. Volume-visualization techniques have to be employed in the case of 3D data. Direct volume rendering, which avoids occlusion problems by selective use of semitransparency, can be applied to single-component data derived from vector fields [30,120]; recent developments are specifically designed for time-dependent data [16,36,85].

12.6 Sparse Representations for Particle-Tracing Techniques

Another class of visualization methods is based on the characteristic lines obtained by particle tracing. Among these are the aforementioned pathlines, streamlines, and streaklines. In addition, *time lines*, constructed from particles released at the same time from different points along a curve, can be used. All these lines are quite intuitive because they represent some kind of transport along the flow. In this section, we

Figure 12.2 Glyph-based 3D flow visualization, combined with illuminated streamlines. (See also color insert.)

discuss sparse representations, in which the spatial domain is not densely covered.

A traditional particle-based approach computes characteristic curves and draws them as thin lines. Since many researchers handle time-independent vector fields, the notion of streamlines is used frequently. The visualization concepts can often be generalized to pathlines, streaklines, or time lines, even if not explicitly mentioned. Streamlines just serve as a role model for the other characteristic lines. Particles traced for a very short time generate short streamlines or *streamlets.*

Streamlines and streamlets can be used in 2D space, on 2D hypersurfaces of an underlying 3D flow, and for 3D flows. Hypersurfaces typically are sectional slices through the volume, or they are curved surfaces such as boundaries or other characteristic surfaces. It is important to note that the use of particle traces for vector fields projected onto slices may be misleading, even within a steady flow: a streamline on a slice may depict a closed loop even though no particle would ever traverse the loop. The problem is caused by the fact that flow components orthogonal to the slice are neglected during flow integration. For 3D flows, perceptual problems might arise due to distortions resulting from the projection onto the image plane. Moreover, issues of occlusion and clutter have to be considered. An appropriate solution is to find selective seed positions for particle traces that still show the important features of the flow, but do not overcrowd the volume; for example, a thread of streamlets along characteristic structures of 3D flow [71] can be used. The method of illuminated streamlines [136], based on illumination in diverse codimensions [1], improves the perception of those lines, and it also increases depth information and addresses the problem of occlusion by making the streamlines partially transparent. An example is shown in Fig. 12.2.

In 2D, particle traces are usually represented by thin lines, although the width of a line is sometimes modified to represent further infor-

mation. Fig. 12.1b shows an example with a collection of streamlets. In 3D applications, however, the additional spatial dimension allows more information to be encoded into the graphical representation through the use of geometric objects of finite extent perpendicular to the particle trace. Examples of such an extension of streamlines in 3D are *streamribbons* and *streamtubes.* A streamribbon is the area swept out by a deformable line segment along a streamline. The strip-like shape of a streamribbon displays the rotational behavior of a 3D flow. Fig. 12.3 shows a visualization of a 3D fluid simulation combining streamribbons, streamlines, arrows, and color coding [104]. An iconic streamtube [119] is a thick tube-shaped streamline whose radial extent shows the expansion of the flow. As a further extension of streamtubes, *dash tubes* [33] provide animated, opacity-mapped tubes. *Stream polygons* [103] trace arbitrary polygonal cross-sections along a streamline and thus are closely related to streamtubes and streamribbons. The properties of the polygons, such as the size, shape, and orientation, reflect properties of the vector field, including strain, displacement, and rotation. *Streamballs* [10] use their radii to visualize divergence and acceleration in a flow. Other geometric objects such as tetrahedra [110] may be used instead of spheres. Another extension of streamlines is provided by *stream surfaces*, which are everywhere tangent to the vector field. A stream surface can be modeled by an implicit surface [123] or approximated by explicitly connecting a set of streamlines along time lines. Stream surfaces present challenges related to occlusion, visual complexity, and interpretation, which can be addressed by choosing an appropriate placement and orientation based on *principal stream surfaces* [15] or through user interaction [47]. Ray casting can be used to render several stream surfaces at different depths [32]. *Stream arrows* [72] cut out arrow-shaped portions from a stream surface and thus provide additional information on the flow, such as flow direction and convergence or divergence. Stream surfaces can also be computed

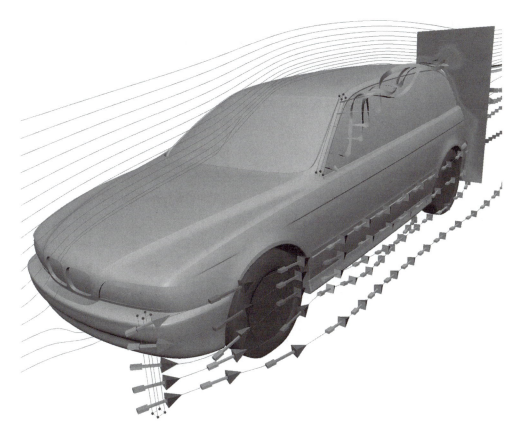

Figure 12.3   A combination of streamlines, streamribbons, arrows, and color coding for a 3D flow. Image courtesy of BMW Group and Martin Schulz. (See also color insert.)

and visualized based on surface particles [122], which are subject to less occlusion than is a full-bodied surface.

The generalization of the concept of particles released from single discrete points (for streamlines or streaklines) or from several points on a 1D line (for stream surfaces) leads to *flow volumes* [78]. A flow volume is a region of a 3D flow domain traced out by a 2D patch over time. The resulting volume can be visualized by volume rendering techniques. Since any flow volume can be (at least approximately) represented by a collection of tetrahedral cells, volume rendering techniques for unstructured grids can be applied, such as hardware-accelerated cell projection [93,107]. Flow volumes can

be extended to unsteady flows [6], yielding the analog of streaklines. Finally, *time surfaces* extend time lines to surfaces that are built from particles released from a 2D patch. The evolution of time surfaces can be handled by a level-set approach [131].

A fundamental issue of all particle-based techniques is an appropriate choice of initial conditions—seed-point positioning—in order to catch all relevant features of the flow. Two main strategies can be identified: interactive or automatic placement of seed points. The interactive approach leaves the problem to the user and, in this sense, simplifies the problem from an algorithmic point of view. Nevertheless, the visualization system should be designed to help the user identify appropriate seed points. For example, the virtual wind tunnel [11] is an early virtual-reality implementation of a flow-visualization system where particles can be interactively released by the user.

A useful approach for the automatic placement of seed points is to construct a uniform distribution of streamlines, which can be achieved for 2D vector fields [55,118] or for surfaces within curvilinear grids of a 3D flow [74]. The idea behind a uniform distribution of streamlines is that such a distribution very likely will not miss important features of the flow. Therefore, this approach can be regarded as a step towards a completely dense representation, which is discussed in the following section. Equally spaced streamlines can be extended to multiresolution hierarchies that support an interactive change of streamline density, while zooming in and out of the vector field [58]. Moreover, with this technique, the density of streamlines can be determined by properties of the flow, such as the magnitude of the velocity or the vorticity. Evenly spaced streamlines for an unsteady flow can be realized by correlating instantaneous streamline visualizations for subsequent time-steps [57]. Seeding strategies may also be based on vector-field topology; for example, flow structures in the vicinity of critical points can be visualized by appropriately setting the initial conditions for particle tracing [126].

Since all particle-tracing techniques are based on solving the differential equation for particle transport, issues of numerical accuracy and speed must be addressed. Different numerical techniques known from the literature can be applied for the initial value problem of ordinary differential equations. In many applications, explicit integration schemes such as nonadaptive or adaptive Runge-Kutta methods are used. The required accuracy for particle tracing depends on the visualization technique; for example, first-order Euler integration might be acceptable for streamlets but not for longer streamlines. A comparison of different integration schemes [111] helps to judge the tradeoff

between computation time and accuracy. Besides the actual integration scheme, the grid on which the vector field is given is very important for choosing a particle-tracing technique. Point location and interpolation depend heavily on the grid and therefore affect the speed and accuracy of particle tracing. Both aspects are detailed by Nielson et al. [82], along with a comparison between C-space (computational-space) and P-space (physical-space) approaches. The numerics of particle tracing is discussed, for example, for the tetrahedral decomposition of curvilinear grids [62], especially for the decomposition of distorted cells [94], for unstructured grids [119], for analytical solutions in piecewise linearly interpolated tetrahedral grids [83], for stiff differential equations originating from shear flows [111], and for sparse grids [113].

12.7   Dense Representations for Particle-Tracing Methods

Another class of visualization approaches is based on a representation of the flow by a dense coverage through structures determined by particle tracing. Typically, dense representations are built upon texture-based techniques, which provide images of high spatial resolution. A detailed description of texture-based flow visualization and, in particular, its support by graphics hardware is discussed in Chapter 13. A summary of research in the field of dense representations can be found in the surveys [40, 99].

The distinction between dense and sparse techniques should not be taken too rigidly because both classes of techniques are closely related by the fact that they form visual structures based on particle tracing. Therefore, dense representations also lead to the same intuitive understanding of the flow. Often, a transition between both classes is possible [125]; for example, texture-based techniques with only few distinct visual elements might resemble a collection of few streamlines and, on the other hand, evenly spaced streamline seeding can be used with a high line density.

An early texture-synthesis technique for vector-field visualization is *spot noise* [121], which produces a texture by generating a set of spots on the spatial domain. Each spot represents a particle moving over a short period of time and results in a streak in the direction of the flow at the position of the spot. Enhanced spot noise [27] improves the visualization of highly curved vector fields by adapting the shape of the spots to the local velocity field. Spot noise can also be applied on boundaries and surfaces [20,114]. A divide-and-conquer strategy makes possible an implementation of spot noise for interactive environments [19]. As an example application, spot noise was applied to the visualization of turbulent flow [21].

Line integral convolution (LIC) [14] is a widely used technique for the dense representation of streamlines in steady vector fields. An example is shown in Fig. 12.1c. LIC takes as input a vector field and a white-noise texture. The noise texture is locally smoothed along streamlines by convolution with a filter kernel. This filtering leads to a high correlation between the grey-scale values of neighboring pixels along streamlines and to little or no correlation perpendicular to streamlines. The contrast and quality of LIC images can be improved by postprocessing techniques, such as histogram equalization, high-pass filtering, or a second pass of LIC [84]. Both spot noise and LIC are based on dense texture representations and particle tracing and are, from a more abstract point of view, tightly related to each other [22]. The original LIC technique does not show the orientation and the magnitude of the velocity field, an issue that is addressed by variants of LIC. Periodic motion filters can be used to animate the flow visualization, and a kernel phase-shift can be applied to reveal the direction and magnitude of the vector field [31]. *Oriented Line Integral Convolution* (OLIC) [128] exploits the existence of distinguishable, separated blobs in a rather sparse texture and smears these blobs into the direction of the local velocity field by convolu-

tion with an asymmetric filter kernel to show the orientation of the flow. By sacrificing some accuracy, a fast version of OLIC (FROLIC) [127] is feasible. In another approach, orientation is visualized by combining animation and adding dye advection [105]. Multifrequency noise for LIC [64] visualizes the magnitude of the velocity by adapting the spatial frequency of noise.

Other visualization techniques achieve LIC-like images by applying methods not directly based on LIC. For example, fur-like textures [63] can be utilized by specifying the orientation, length, density, and color of fur filaments according to the vector field. The *integrate and draw* [90] approach deposits random grey-scale values along streamlines. *Pseudo LIC* (PLIC) [125] is a compromise between LIC and sparse particle-based representations and therefore allows for a gradual change between dense and sparse visualizations. PLIC uses LIC to generate a template texture in a preprocessing step. For the actual visualization, the template is mapped onto thin or thick streamlines, thus filling the domain with LIC-like structures. The idea of LIC textures applied to thick streamlines can be extended to an error-controlled hierarchical method for a hardware-accelerated level-of-detail approach [8].

LIC can be extended to nonuniform grids and curved surfaces, for example, to curvilinear grids [31] and to 2D unstructured or triangular meshes [4,76,112]. Multigranularity noise as the input for LIC [75] compensates for the nonisometric mapping from texture space to the cells of a curvilinear grid that differ in size. The projection of the normal component of the vector field needs to be taken into account for LIC-type visualizations on hypersurfaces [100].

Unsteady Flow LIC (UFLIC) [106] and its accelerated version [68] incorporate time into the convolution to visualize unsteady flow. The issue of temporal coherence is addressed by successively updating the convolution results over time. Forssell and Cohen [31] present a visualization of time-dependent flows on curvilinear surfaces. *Dynamic LIC* (DLIC) [109] is another extension of LIC, one that allows for time-dependent vector fields, such as electric fields. A LIC-like image of an unsteady flow can also be generated by an adaptive visualization method using streaklines, where the seeding of streaklines is controlled by the vorticity [98].

Since LIC has to perform a convolution for each element of a high-resolution texture, computational costs are an issue. One solution to this problem utilizes the coherence along streamlines to speed up the visualization process [42,108]. Parallel implementations are another way of dealing with high computational costs [13,137]. Finally, implementations based on graphics hardware can enhance the performance of LIC [43].

From a conceptional point of view, an extension of LIC to 3D is straightforward. The convolution along streamlines is performed within a volume; the resulting grey-scale volume can be represented by volume-visualization techniques, such as texture-based volume rendering. However, computational costs are even higher than in 2D, and therefore, interactive implementations of the filtering process are hard to achieve. Even more importantly, possibly severe occlusion issues have to be considered: in a dense representation, there is a good chance that important features will get hidden behind other particle lines. A combination of interactive clipping and user intervention is one possible solution [89]. Alternatively, 3D LIC volumes can be represented by selectively emphasizing important regions of interest in the flow, enhancing depth perception and improving orientation perception [48,49,50] (Fig. 12.4).

Another related class of dense representations is based on *texture advection*. The basic idea is to represent a dense collection of particles in a texture and transport that texture according to the motion of particles [77,80]. For example, the Lagrangian coordinates for texture transport can be computed by a numerical scheme for convection equations [7]. The *motion map* [56] is an application of the texture-advection concept for animating 2D steady flows. The motion map contains a dense representation of the flow and the information required for animation.

Lagrangian–Eulerian advection (LEA) [54] is a scheme for visualizing unsteady flows by integrating particle positions (i.e., the Lagrangian part) and advecting the color of the particles based on a texture representation (i.e., the Eulerian aspect). LEA can be extended to visualizing vertical motion in a 3D flow by means of time surfaces [38]. Texture advection is directly related to the texture-mapping capabilities of graphics hardware and therefore facilitates efficient implementations [53,129,130]. Another advantage of texture advection is the fact that both noise and dye advection can be handled in the same framework. Texture advection can also be applied to 3D flows [59,130].

Image-based flow visualization (IBFV) [124] is a recently developed variant of 2D texture advection. Not only is the (noise) texture transported along the flow, but a second texture is also blended into the advected texture at each time step. IBFV is a flexible tool that can imitate a wide variety of visualization styles. Another approach to the transport of a dense set of particles is based on *nonlinear diffusion* [28]. An initial noise image is smoothed along integral lines of a steady flow by diffusion, whereas the image is sharpened in the orthogonal direction. Nonlinear diffusion can be extended to the multiscale visualization of transport in time-dependent flows [12].

Finally, some 3D flow-visualization techniques adopt the idea of splatting, originally developed for volume rendering [132]. Even if some vector-splatting techniques do not rely on particle tracing, we have included them in this section because their visual appearance resembles dense curve-like structures. Anisotropic "scratches" can be modeled onto texture splats that are oriented along the flow to show the direction of the vector field [17]. *Line bundles* [79] use the splatting analogy to draw each data point with a precomputed set of rendered line segments. These semitransparent line bundles are composited together in a back-to-front order to achieve an anisotropic volume rendering result. For time-dependent flows, the animation of a large number of texture-

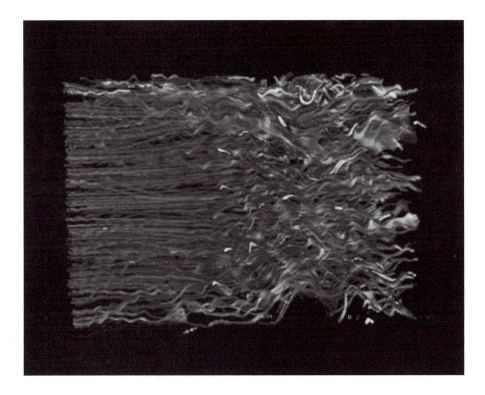

Figure 12.4   3D LIC with enhanced depth perception. Image courtesy of Victoria Interrante. (See also color insert.)

mapped particles along pathlines can be used [39]. For all splatting approaches, the density of representation depends on the number of splats.

12.8   Feature-Based Visualization Approaches

The visualization concepts discussed so far operate directly on the vector field. Therefore, it is the task of the user to identify the important features of the flow from such a visualization. Feature-based visualization approaches seek to compute a more abstract representation that already contains the important properties in a condensed form and suppresses superfluous information. In other words, an appropriate filtering process is chosen to reduce the amount of visual data presented to the user. Examples

for this more abstract data are flow topology based on critical points, other flow features such as vortices and shock waves, or aggregated flow data via clustering.

After features are computed, the actual visual representation has to be considered. Different features have different attributes; to emphasize special attributes for each type of feature, suitable representations must be used. Glyphs or icons can be employed for vortices or for critical points and other topological features. Examples are ellipses or ellipsoids to encode the rotation speed and other attributes of vortices. A comprehensive presentation of feature-extraction and visualization techniques can be found in the survey [88].

Topology-based 2D vector-field visualization [44] aims to show only the essential information of the field. The qualitative structure of a vector field can be globally represented by portraying its topology. The field's critical points and separatrices determine the nature of the flow. From a diagram of the topology, the complete flow can be inferred. Fig. 12.1d shows an example of a topology-based visualization. From a numerical point of view, the interpolation scheme is crucial for identifying critical points. Extended versions of topology-based representations make use of higher-order singularities [101] and C^1-continuous interpolation schemes [102]. Another extension is based on the detection of closed streamlines, a global property that is not detected by the aforementioned algorithms [133]. Topology-based visualization can also be extended to time-dependent vector fields by topology tracking [117]. The original topology-based techniques work well for datasets with small numbers of critical points. Turbulent flows computed on a high-resolution grid, however, may show a large number of critical points, leading to an overloaded visual representation. This issue is addressed by topology-simplification techniques [23,24,25,115,116,135], which remove some of the critical points and leave only the important features. For the visualization of 3D topology, suitable visual representations need to be used. For example, streamlines that are traced from appropriate positions close to critical points connect to other critical points or the boundary to display the topology, while glyphs can be used to visualize the various classes of critical points [37]. Topology can also serve as a means for determining the similarity between two different vector fields [3,67].

Vector-field clustering is another way to reduce the amount of visualization data. A large number of vectors of the original high-resolution field are combined into fewer vectors that approximately represent the vector field at a coarser resolution, leading to a visualization of aggregated data. An important issue is to establish appropriate error measures to control the way vectors are combined into clusters [70,114]. The extension of vector-field clustering to several levels of clustering leads to hierarchical representations [34,35,41]. In related

approaches, segmentation [28], multiscale visualization [12], or topology-preserving smoothing based on level-set techniques [131] reduce the complexity of the displayed vector fields.

An important class of feature-detection algorithms is based on the identification of *vortices* and their respective *vortex cores*. Vortices are useful for identifying significant features in a fluid flow. One way of classifying vortex-detection techniques is the distinction between point-based, local techniques, which operate directly on the vector data set, and geometry-based, global techniques, which examine the properties of characteristic lines around vortices [97]. Local techniques build a measure for vortices from physical quantities of a fluid flow. An overview of these techniques [2,91] and more recent contributions [5,52,61,92] are available. A mathematical framework [87] makes it possible to unify several vortex-detection methods. Point-based methods are rather simple to compute but are more likely to miss some vortices. For example, weak vortices, which have a slow rotational component compared to the velocity of the core, are hard to detect by these techniques. Geometry-based, global methods [51,95,96] are usually associated with higher computational costs but allow a more robust detection or verification of vortices. A more detailed description of vortex-detection techniques can be found in Chapter 14.

Shock waves are another important feature of a fluid flow because they can increase drag and even cause structural failure. Shock waves are characterized by discontinuities in physical quantities, such as pressure, density, and velocity. Therefore, shock-detection algorithms are related to edge-detection methods known from image processing. A comparison of different techniques for shock extraction and visualization is available [73]. Flow *separation* and *attachment*, which occur when a flow abruptly moves away from or returns to a solid body, are other interesting features of a fluid flow. Attachment and separation lines on surfaces in a 3D flow can be automatically extracted based on a local analysis of the vector field by means of phase plane analysis [60].

Vortex cores, shock waves, and separation and attachment lines are examples of features that are tightly connected to an underlying physical model and cannot be derived from a generic vector-field description. Therefore, a profound understanding of the physical problem is necessary to develop measures for these kinds of features. Accordingly, a large body of research on feature extraction can be found in the literature on related topics of engineering and physics. Since a comprehensive collection of techniques and references on feature extraction is beyond the scope of this chapter, we refer to the survey [88] for more detailed information.

Acknowledgments

We would like to thank the following people for providing us with images: Victoria Interrante (University of Minnesota) for the image in Fig. 12.4; Gerik Scheuermann (Universität Kaiserslautern) for Fig. 12.1d; BMW Group and Martin Schulz for Fig. 12.3.

The first author thanks the Landesstiftung Baden-Württemberg for support; the second author acknowledges support from NSF under grant NSF-0083792.

References

1. D. C. Banks. Illumination in diverse codimensions. In *Proceedings of ACM SIGGRAPH '94*, pages 327–334, 1994.
2. D. C. Banks and B. A. Singer. Vortex tubes in turbulent flows: identification, representation, reconstruction. In *IEEE Visualization '94*, pages 132–139, 1994.
3. R. Batra and L. Hesselink. Feature comparisons of 3-D vector fields using earth mover's distance. In *IEEE Visualization '99*, pages 105–114, 1999.
4. H. Battke, D. Stalling, and H.-C. Hege. Fast line integral convolution for arbitrary surfaces in 3D. In *Visualization and Mathematics* (H.-C. Hege and K. Polthier, Eds.), pages 181–195. Berlin, Springer, 1997.
5. D. Bauer and R. Peikert. Vortex tracking in scale-space. In *EG/IEEE TCVG Symposium on Visualization '02*, pages 233–240, 2002.
6. B. Becker, D. A. Lane, and N. Max. Unsteady flow volumes. In *IEEE Visualization '95*, pages 329–335, 1995.
7. J. Becker and M. Rumpf. Visualization of time-dependent velocity fields by texture transport. In *EG Workshop on Visualization in Scientific Computing*, pages 91–101, 1998.
8. U. Bordoloi and H.-W. Shen. Hardware accelerated interactive vector field visualization: a level of detail approach. In *Eurographics '02*, pages 605–614, 2002.
9. E. Boring and A. Pang. Directional flow visualization of vector fields. In *IEEE Visualization '96*, pages 389–392, 1996.
10. M. Brill, H. Hagen, H.-C. Rodrian, W. Djatschin, and S. V. Klimenko. Streamball techniques for flow visualization. In *IEEE Visualization '94*, pages 225–231, 1994.
11. S. Bryson and C. Levit. The virtual wind tunnel. *IEEE Computer Graphics and Applications*, 12(4):25–34, 1992.
12. D. Bürkle, T. Preußer, and M. Rumpf. Transport and anisotropic diffusion in time-dependent flow visualization. In *IEEE Visualization '01*, pages 61–67, 2001.
13. B. Cabral and C. Leedom. Highly parallel vector visualization using line integral convolution. In *SIAM Conference on Parallel Processing for Scientific Computing*, pages 802–807, 1995.
14. B. Cabral and L. C. Leedom. Imaging vector fields using line integral convolution. In *Proceedings of ACM SIGGRAPH '93*, pages 263–270, 1993.
15. W. Cai and P.-A. Heng. Principal stream surfaces. In *IEEE Visualization '97*, pages 75–80, 1997.
16. J. Clyne and J. M. Dennis. Interactive direct volume rendering of time-varying data. In *EG/IEEE TCVG Symposium on Visualization '99*, pages 109–120, 1999.
17. R. Crawfis and N. Max. Texture splats for 3D scalar and vector field visualization. In *IEEE Visualization '93*, pages 261–266, 1993.
18. R. Crawfis, H.-W. Shen, and N. Max. Flow visualization techniques for CFD using volume rendering. In *9th International Symposium on Flow Visualization*, pages 64/1–64/10, 2000.
19. W. C. de Leeuw. Divide and conquer spot noise. In *Supercomputing '97 Conference*, pages 12–24, 1997.
20. W. C. de Leeuw, H.-G. Pagendarm, F. H. Post, and B. Walter. Visual simulation of experimental oil-flow visualization by spot noise images from numerical flow simulation. In *EG Workshop on Visualization in Scientific Computing*, pages 135–148, 1995.

21. W. C. de Leeuw, F. H. Post, and R. W. Vaatstra. Visualization of turbulent flow by spot noise. In *EG Workshop on Virtual Environments and Scientific Visualization '96*, pages 286–295, 1996.

22. W. C. de Leeuw and R. van Liere. Comparing LIC and spot noise. In *IEEE Visualization '98*, pages 359–365, 1998.

23. W. C. de Leeuw and R. van Liere. Collapsing flow topology using area metrics. In *IEEE Visualization '99*, pages 349–354, 1999.

24. W. C. de Leeuw and R. van Liere. Visualization of global flow structures using multiple levels of topology. In *EG/IEEE TCVG Symposium on Visualization '99*, pages 45–52, 1999.

25. W. C. de Leeuw and R. van Liere. Multi-level topology for flow visualization. *Computers and Graphics*, 24(3):325–331, 2000.

26. W. C. de Leeuw and J. J. van Wijk. A probe for local flow field visualization. In *IEEE Visualization '93*, pages 39–45, 1993.

27. W. C. de Leeuw and J. J. van Wijk. Enhanced spot noise for vector field visualization. In *IEEE Visualization '95*, pages 233–239, 1995.

28. U. Diewald, T. Preußer, and M. Rumpf. Anisotropic diffusion in vector field visualization on Euclidean domains and surfaces. *IEEE Transactions on Visualization and Computer Graphics*, 6(2):139–149, 2000.

29. D. Dovey. Vector plots for irregular grids. In *IEEE Visualization '95*, pages 248–253, 1995.

30. D. S. Ebert, R. Yagel, J. Scott, and Y. Kurzion. Volume rendering methods for computational fluid dynamics visualization. In *IEEE Visualization '94*, pages 232–239, 1994.

31. L. K. Forssell and S. D. Cohen. Using line integral convolution for flow visualization: Curvilinear grids, variable-speed animation, and unsteady flows. *IEEE Transactions on Visualization and Computer Graphics*, 1(2):133–141, 1995.

32. T. Frühauf. Raycasting vector fields. In *IEEE Visualization '96*, pages 115–120, 1996.

33. A. Fuhrmann and E. Gröller. Real-time techniques for 3D flow visualization. In *IEEE Visualization '98*, pages 305–312, 1998.

34. H. Garcke, T. Preußer, M. Rumpf, A. Telea, U. Weikard, and J. J. van Wijk. A phase field model for continuous clustering on vector fields. *IEEE Transactions on Visualization and Computer Graphics*, 7(3):230–241, 2001.

35. H. Garcke, T. Preußer, M. Rumpf, A. Telea, U. Weikard, and J. van Wijk. A continuous clustering method for vector fields. In *IEEE Visualization '00*, pages 351–358, 2000.

36. T. Glau. Exploring instationary fluid flows by interactive volume movies. In *EG/IEEE TCVG Symposium on Visualization '99*, pages 277–283, 1999.

37. A. Globus, C. Levit, and T. Lasinski. A tool for visualizing the topology of 3D vector fields. In *IEEE Visualization '91*, pages 33–40, 1991.

38. J. Grant, G. Erlebacher, and J. O'Brien. Case study: visualizing ocean flow vertical motions using Lagrangian–Eulerian time surfaces. In *IEEE Visualization '02*, pages 529–532, 2002.

39. S. Guthe, S. Gumhold, and W. Straßer. Interactive visualization of volumetric vector fields using texture based particles. In *WSCG 2002 Conference Proceedings*, pages 33–41, 2002.

40. R. S. Laramee, H. Hauser, H. Doleisch, B. Vrolijk, F. H. Post, and D. Weiskopf. The state of the art in flow visualization: dense and texture-based techniques. *Computer Graphics Forum*, 23(2):143–161, 2004.

41. B. Heckel, G. Weber, B. Hamann, and K. I. Joy. Construction of vector field hierarchies. In *IEEE Visualization '99*, pages 19–25, 1999.

42. H.-C. Hege and D. Stalling. Fast LIC with piecewise polynomial filter kernels. *Mathematical Visualization* (H.-C. Hege and K. Polthier, Eds.), pages 295–314. Berlin, Springer, 1998.

43. W. Heidrich, R. Westermann, H.-P. Seidel, and T. Ertl. Applications of pixel textures in visualization and realistic image synthesis. In *ACM Symposium on Interactive 3D Graphics*, pages 127–134, 1999.

44. J. Helman and L. Hesselink. Representation and display of vector field topology in fluid flow data sets. *Computer*, 22(8):27–36, 1989.

45. M. W. Hirsch. *Differential Topology*, 6th Ed. Berlin, Springer, 1997.

46. M. W. Hirsch and S. Smale. *Differential Equations, Dynamical Systems, and Linear Algebra*. New York, Academic Press, 1974.

47. J. P. M. Hultquist. Interactive numerical flow visualization using stream surfaces. *Computing Systems in Engineering*, 1(2–4):349–353, 1990.

48. V. Interrante. Illustrating surface shape in volume data via principal direction-driven 3D line integral convolution. In *Proceedings of ACM SIGGRAPH 97*, pages 109–116, 1997.

49. V. Interrante and C. Grosch. Strategies for effectively visualizing 3D flow with volume LIC. In *IEEE Visualization '97*, pages 421–424, 1997.

50. V. Interrante and C. Grosch. Visualizing 3D flow. *IEEE Computer Graphics and Applications*, 18(4):49–53, 1998.

51. M. Jiang, R. Machiraju, and D. Thompson. Geometric verification of swirling features in

flow fields. In *IEEE Visualization '02*, pages 307–314, 2002.

52. M. Jiang, R. Machiraju, and D. Thompson. A novel approach to vortex core region detection. In *EG/IEEE TCVG Symposium on Visualization '02*, pages 217–225, 2002.

53. B. Jobard, G. Erlebacher, and M. Y. Hussaini. Hardware-accelerated texture advection for unsteady flow visualization. In *IEEE Visualization '00*, pages 155–162, 2000.

54. B. Jobard, G. Erlebacher, and M. Y. Hussaini. Lagrangian–Eulerian advection for unsteady flow visualization. In *IEEE Visualization '01*, pages 53–60, 2001.

55. B. Jobard and W. Lefer. Creating evenly spaced streamlines of arbitrary density. In *EG Workshop on Visualization in Scientific Computing*, pages 43–55, 1997.

56. B. Jobard and W. Lefer. The motion map: efficient computation of steady flow animations. In *IEEE Visualization '97*, pages 323–328, 1997.

57. B. Jobard and W. Lefer. Unsteady flow visualization by animating evenly spaced streamlines. In *Eurographics '00*, pages 31–39, 2000.

58. B. Jobard and W. Lefer. Multiresolution flow visualization. In *WSCG 2001 Conference Proceedings*, pages P34–P37, 2001.

59. D. Kao, B. Zhang, K. Kim, and A. Pang. 3D flow visualization using texture advection. In *IASTED Conference on Computer Graphics and Imaging 01 (CGIM)*, pages 252–257, 2001.

60. D. N. Kenwright. Automatic detection of open and closed separation and attachment lines. In *IEEE Visualization '98*, pages 151–158, 1998.

61. D. N. Kenwright and R. Haimes. Automatic vortex core detection. *IEEE Computer Graphics and Applications*, 18(4):70–74, 1998.

62. D. N. Kenwright and D. A. Lane. Interactive time-dependent particle tracing using tetrahedral decomposition. *IEEE Transactions on Visualization and Computer Graphics*, 2(2):120–129, 1996.

63. L. Khouas, C. Odet, and D. Friboulet. 2D vector field visualization using furlike texture. In *EG/IEEE TCVG Symposium on Visualization '99*, pages 35–44, 1999.

64. M.-H. Kiu and D. C. Banks. Multi-frequency noise for LIC. In *IEEE Visualization '96*, pages 121–126, 1996.

65. R. V. Klassen and S. J. Harrington. Shadowed hedgehogs: A technique for visualizing 2D slices of 3D vector fields. In *IEEE Visualization '91*, pages 148–153, 1991.

66. S. Lang. *Differential and Riemannian Manifolds*, 3rd Ed. New York, Springer, 1995.

67. Y. Lavin, R. Batra, and L. Hesselink. Feature comparisons of vector fields using Earth mover's distance. In *IEEE Visualization '98*, pages 103–109, 1998.

68. Z. P. Liu and R. J. Moorhead. AUFLIC: An accelerated algorithm for unsteady flow line integral convolution. In *EG/IEEE TCVG Symposium on Visualization '02*, pages 43–52, 2002.

69. S. K. Lodha, A. Pang, R. E. Sheehan, and C. M. Wittenbrink. UFLOW: visualizing uncertainty in fluid flow. In *IEEE Visualization '96*, pages 249–254, 1996.

70. S. K. Lodha, J. C. Renteria, and K. M. Roskin. Topology preserving compression of 2D vector fields. In *IEEE Visualization '00*, pages 343–350, 2000.

71. H. Löffelmann and E. Gröller. Enhancing the visualization of characteristic structures in dynamical systems. In *EG Workshop on Visualization in Scientific Computing*, pages 59–68, 1998.

72. H. Löffelmann, L. Mroz, E. Gröller, and W. Purgathofer. Stream arrows: enhancing the use of stream surfaces for the visualization of dynamical systems. *The Visual Computer*, 13(8):359–369, 1997.

73. K.-L. Ma, J. V. Rosendale, and W. Vermeer. 3D shock wave visualization on unstructured grids. In *1996 Volume Visualization Symposium*, pages 87–94, 1996.

74. X. Mao, Y. Hatanaka, H. Higashida, and A. Imamiya. Image-guided streamline placement on curvilinear grid surfaces. In *IEEE Visualization '98*, pages 135–142, 1998.

75. X. Mao, L. Hong, A. Kaufman, N. Fujita, and M. Kikukawa. Multi-granularity noise for curvilinear grid LIC. In *Graphics Interface*, pages 193–200, 1998.

76. X. Mao, M. Kikukawa, N. Fujita, and A. Imamiya. Line integral convolution for 3D surfaces. In *EG Workshop on Visualization in Scientific Computing*, pages 57–69, 1997.

77. N. Max and B. Becker. Flow visualization using moving textures. In *Proceedings of the ICASE/LaRC Symposium on Visualizing Time-Varying Data*, pages 77–87, 1995.

78. N. Max, B. Becker, and R. Crawfis. Flow volumes for interactive vector field visualization. In *IEEE Visualization '93*, pages 19–24, 1993.

79. N. Max, R. Crawfis, and C. Grant. Visualizing 3D velocity fields near contour surfaces. In *IEEE Visualization '94*, pages 248–255, 1994.

80. N. Max, R. Crawfis, and D. Williams. Visualizing wind velocities by advecting cloud textures. In *IEEE Visualization '92*, pages 179–184, 1992.

81. J. W. Milnor. *Topology from the Differentiable Viewpoint*. Charlottesville, VA, University Press of Virginia, 1965.

82. G. M. Nielson, H. Hagen, and H. Müller. *Scientific Visualization: Overviews, Methodologies, and Techniques*. Los Alamitos, CA, IEEE Computer Society Press, 1997.

83. G. M. Nielson and I.-H. Jung. Tools for computing tangent curves for linearly varying vector fields over tetrahedral domains. *IEEE Transactions on Visualization and Computer Graphics*, 5(4):360–372, 1999.

84. A. Okada and D. Kao. Enhanced line integral convolution with flow feature detection. In *Proceedings of IS&T/SPIE Electronic Imaging '97*, pages 206–217, 1997.

85. K. Ono, H. Matsumoto, and R. Himeno. Visualization of thermal flows in an automotive cabin with volume rendering method. In *EG/IEEE TCVG Symposium on Visualization '01*, pages 301–308, 2001.

86. A. T. Pang, C. M. Wittenbrink, and S. K. Lodha. Approaches to uncertainty visualization. *The Visual Computer*, 13(8):370–390, 1997.

87. R. Peikert and M. Roth. The "parallel vectors" operator—a vector field visualization primitive. In *IEEE Visualization '99*, pages 263–270, 1999.

88. F. H. Post, B. Vrolijk, H. Hauser, R. S. Laramee, and H. Doleisch. The state of the art in flow visualization: feature extraction and tracking. *Computer Graphics Forum*, 22(4): 775–792, 2003.

89. C. Rezk-Salama, P. Hastreiter, C. Teitzel, and T. Ertl. Interactive exploration of volume line integral convolution based on 3D-texture mapping. In *IEEE Visualization '99*, pages 233–240, 1999.

90. C. P. Risquet. Visualizing 2D flows: integrate and draw. In *EG Workshop on Visualization in Scientific Computing (Participant Edition)*, pages 57–67, 1998.

91. M. Roth and R. Peikert. Flow visualization for turbomachinery design. In *IEEE Visualization '96*, pages 381–384, 1996.

92. M. Roth and R. Peikert. A higher-order method for finding vortex core lines. In *IEEE Visualization '98*, pages 143–150, 1998.

93. S. Röttger, M. Kraus, and T. Ertl. Hardware-accelerated volume and isosurface rendering based on cell projection. In *IEEE Visualization '00*, pages 109–116, 2000.

94. I. A. Sadarjoen, A. J. de Boer, F. H. Post, and A. E. Mynett. Particle tracing in σ-transformed grids using tetrahedral 6-decomposition. In *EG Workshop on Visualization in Scientific Computing*, pages 71–80, 1998.

95. I. A. Sadarjoen and F. H. Post. Geometric methods for vortex extraction. In *EG/IEEE TCVG Symposium on Visualization '99*, pages 53–62, 1999.

96. I. A. Sadarjoen and F. H. Post. Detection, quantification, and tracking of vortices using streamline geometry. *Computers and Graphics*, 24(3):333–341, 2000.

97. I. A. Sadarjoen, F. H. Post, B. Ma, D. C. Banks, and H.-G. Pagendarm. Selective visualization of vortices in hydrodynamic flows. In *IEEE Visualization '98*, pages 419–422, 1998.

98. A. Sanna, B. Montrucchio, and R. Arina. Visualizing unsteady flows by adaptive streaklines. In *WSCG 2000 Conference Proceedings*, pages 84–91, 2000.

99. A. Sanna, B. Montrucchio, and P. Montuschi. A survey on visualization of vector fields by texture-based methods. *Recent Res. Devel. Pattern Rec.*, 1:13–27, 2000.

100. G. Scheuermann, H. Burbach, and H. Hagen. Visualizing planar vector fields with normal components using line integral convolution. In *IEEE Visualization '99*, pages 255–261, 1999.

101. G. Scheuermann, H. Hagen, H. Krüger, M. Menzel, and A. Rockwood. Visualization of higher order singularities in vector fields. In *IEEE Visualization '97*, pages 67–74, 1997.

102. G. Scheuermann, X. Tricoche, and H. Hagen. C^1-interpolation for vector field topology visualization. In *IEEE Visualization '99*, pages 271–278, 1999.

103. W. J. Schroeder, C. R. Volpe, and W. E. Lorensen. The stream polygon: A technique for 3D vector field visualization. In *IEEE Visualization '91*, pages 126–132, 1991.

104. M. Schulz, F. Reck, W. Bartelheimer, and T. Ertl. Interactive visualization of fluid dynamics simulations in locally refined Cartesian grids. In *IEEE Visualization '99*, pages 413–416, 1999.

105. H.-W. Shen, C. R. Johnson, and K.-L. Ma. Visualizing vector fields using line integral convolution and dye advection. In *1996 Volume Visualization Symposium*, pages 63–70, 1996.

106. H.-W. Shen and D. L. Kao. A new line integral convolution algorithm for visualizing time-varying flow fields. *IEEE Transactions on Visualization and Computer Graphics*, 4(2):98–108, 1998.

107. P. Shirley and A. Tuchman. A polygonal approximation to direct scalar volume rendering. In *Workshop on Volume Visualization '90*, pages 63–70, 1990.

108. D. Stalling and H.-C. Hege. Fast and resolution independent line integral convolution.

In *Proceedings of ACM SIGGRAPH '95*, pages 249–256, 1995.

109. A. Sundquist. Dynamic line integral convolution for visualizing streamline evolution. *IEEE Transactions on Visualization and Computer Graphics*, 9(3):273–282, 2003.

110. C. Teitzel and T. Ertl. New approaches for particle tracing on sparse grids. In *EG/IEEE TCVG Symposium on Visualization '99*, pages 73–84, 1999.

111. C. Teitzel, R. Grosso, and T. Ertl. Efficient and reliable integration methods for particle tracing in unsteady flows on discrete meshes. In *EG Workshop on Visualization in Scientific Computing*, pages 31–41, 1997.

112. C. Teitzel, R. Grosso, and T. Ertl. Line integral convolution on triangulated surfaces. In *WSCG 1997 Conference Proceedings*, pages 572–581, 1997.

113. C. Teitzel, R. Grosso, and T. Ertl. Particle tracing on sparse grids. In *EG Workshop on Visualization in Scientific Computing*, pages 81–90, 1998.

114. A. Telea and J. J. van Wijk. Simplified representation of vector fields. In *IEEE Visualization '99*, pages 35–42, 1999.

115. X. Tricoche, G. Scheuermann, and H. Hagen. Continuous topology simplification of planar vector fields. In *IEEE Visualization '01*, pages 159–166, 2001.

116. X. Tricoche, G. Scheuermann, and H. Hagen. A topology simplification method for 2D vector fields. In *IEEE Visualization '00*, pages 359–366, 2000.

117. X. Tricoche, T. Wischgoll, G. Scheuermann, and H. Hagen. Topology tracking for the visualization of time-dependent 2D flows. *Computers and Graphics*, 26(2):249–257, 2002.

118. G. Turk and D. Banks. Image-guided streamline placement. In *Proceedings of ACM SIGGRAPH '96*, pages 453–460, 1996.

119. S.-K. Ueng, C. Sikorski, and K.-L. Ma. Efficient streamline, streamribbon, and streamtube constructions on unstructured grids. *IEEE Transactions on Visualization and Computer Graphics*, 2(2):100–110, 1996.

120. S. P. Uselton. Volume rendering for computational fluid dynamics: initial results. Technical Report RNR-91-026, NASA Ames Research Center, 1991.

121. J. J. van Wijk. Spot noise—texture synthesis for data visualization. *Computer Graphics (Proceedings of AGM SIGGRAPH '91)*, 25:309–318, 1991.

122. J. J. van Wijk. Flow visualization with surface particles. *IEEE Computer Graphics and Applications*, 13(4):18–24, 1993.

123. J. J. van Wijk. Implicit stream surfaces. In *IEEE Visualization '93*, pages 245–252, 1993.

124. J. J. van Wijk. Image based flow visualization. *ACM Transactions on Graphics*, 21(3):745–754, 2002.

125. V. Verma, D. Kao, and A. Pang. PLIC: bridging the gap between streamlines and LIC. In *IEEE Visualization '99*, pages 341–348, 1999.

126. V. Verma, D. Kao, and A. Pang. A flow-guided streamline seeding strategy. In *IEEE Visualization '00*, pages 163–170, 2000.

127. R. Wegenkittl and E. Gröller. Fast oriented line integral convolution for vector field visualization via the Internet. In *IEEE Visualization '97*, pages 309–316, 1997.

128. R. Wegenkittl, E. Gröller, and W. Purgathofer. Animating flow fields: Rendering of oriented line integral convolution. In *Computer Animation '97*, pages 15–21, 1997.

129. D. Weiskopf, G. Erlebacher, M. Hopf, and T. Ertl. Hardware-accelerated Lagrangian–Eulerian texture advection for 2D flow visualization. In *Vision, Modeling, and Visualization VMV '02 Conference*, pages 77–84, 2002.

130. D. Weiskopf, M. Hopf, and T. Ertl. Hardware-accelerated visualization of time-varying 2D and 3D vector fields by texture advection via programmable per-pixel operations. In *Vision, Modeling, and Visualization VMV '01 Conference*, pages 439–446, 2001.

131. R. Westermann, C. Johnson, and T. Ertl. Topology-preserving smoothing of vector fields. *IEEE Transactions on Visualization and Computer Graphics*, 7(3):222–229, 2001.

132. L. Westover. Footprint evaluation for volume rendering. *Computer Graphics (Proceedings of ACM SIGGRAPH '90)*, 24:367–376, 1990.

133. T. Wischgoll and G. Scheuermann. Detection and visualization of closed streamlines in planar flows. *IEEE Transactions on Visualization and Computer Graphics*, 7(2):165–172, 2001.

134. C. M. Wittenbrink, A. T. Pang, and S. K. Lodha. Glyphs for visualizing uncertainty in vector fields. *IEEE Transactions on Visualization and Computer Graphics*, 2(3):266–279, 1996.

135. P. C. Wong, H. Foote, R. Leung, E. Jurrus, D. Adams, and J. Thomas. Vector field simplification—a case study of visualizing climate modeling and simulation data sets. In *IEEE Visualization '00*, pages 485–488, 2000.

136. M. Zöckler, D. Stalling, and H.-C. Hege. Interactive visualization of 3D-vector fields using illuminated stream lines. In *IEEE Visualization '96*, pages 107–113, 1996.

137. M. Zöckler, D. Stalling, and H.-C. Hege. Parallel line integral convolution. In *EG Workshop on Parallel Graphics and Visualisation*, pages 111–127, 1996.

13 Flow Textures: High-Resolution Flow Visualization

GORDON ERLEBACHER
Florida State University

BRUNO JOBARD
Université de Pau

DANIEL WEISKOPF
University of Stuttgart

13.1 Introduction

Steady and unsteady vector fields are integral to many areas of scientific endeavor. They are generated by increasingly complex numerical simulations and measured by highly resolved experimental techniques. Datasets have also grown in size and complexity, motivating the development of a growing number of visualization techniques to better understand their spatio–temporal structure. As explained in Chapter 12, they are often characterized by their integral curves, also known as pathlines. They can be best understood as the time evolution of massless particles released into the flow. In time-dependent flows, pathlines depend strongly on where the particles are released in space and time. Moreover, they can intersect when viewed in the physical domain, which often makes reliable interpretation of the flow field quite difficult and prone to error.

Rather than visualize a collection of pathlines within a single slice, it is advantageous to consider instead the instantaneous structure of the flow and its temporal evolution. For example, particles can be released along a planar curve and tracked. The time-dependent curve formed by the particles as they are convected by the flow is called a time line. Visualizations of streaklines, timesurfaces, etc., are other viable approaches based on integral curves. The extreme approach is to release a dense collection of particles, covering the physical domain, into the flow. At any point in time, due to flow divergence or strain, the particles form spatial patterns with varying degrees of complexity, and their temporal evolution becomes of interest. These patterns may become nonuniform under the effect of flow divergence, convergence, or strain. A dense, if not uniform, coverage is maintained by tracking a sufficiently large number of particles. If a property is assigned to each particle, and if the number of particles is sufficiently large, one can construct a continuous time-dependent distribution of the particle property over the physical domain.

Textures are a well known graphic representation with many useful properties, and they have been well supported on graphics hardware since the mid-1990s. They execute a wide variety of operations in hardware, including filtering, compression, and blending, and offer the potential to greatly accelerate many advanced visualization algorithms. Their main function is to encode detailed information without the need for large-scale polygonal models. In this chapter, we discuss flow-texture algorithms, which encode dense representations of time-dependent vector fields into textures and evolve those representations in time. These algorithms have the distinctive property that the update mechanism for each texel is identical. They are ideally suited to modern graphics hardware that relies on a *single instruction multiple data* (SIMD) architecture. Two algorithms can be viewed as precursors to flow textures: spot noise [18] and moving textures [14]. Spot noise tracks a dense collection of particles, represented as discs of finite radius. These two algorithms motivated the development of dense methods and their eventual mapping onto graphics hardware. A key component of flow-texture algorithms is a convolution operator

that acts along some path in a noise texture. This approach to dense vector-field representation was first applied to steady flows [5] and called line integral convolution (LIC). It has seen many extensions (see Chapter 12), some for unsteady flows, such as unsteady flow LIC (UFLIC) [16], dynamic LIC (DLIC) [17], Lagrangian–Eulerian advection (LEA) [10], image-based flow visualization (IBFV) [19], and others [3,14]. Fig. 13.1 illustrates the application of LEA to the evolution of unsteady vector fields on timesurfaces [8].

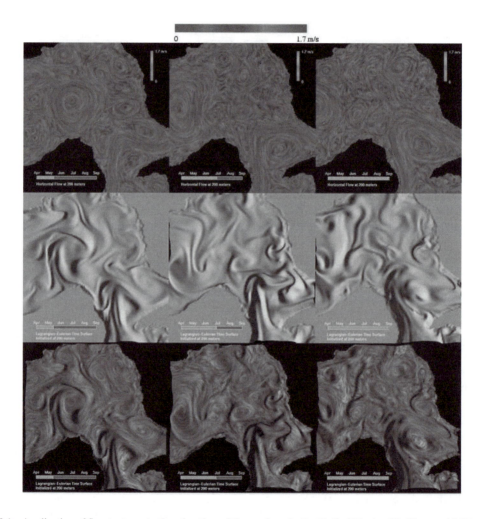

Figure 13.1   Application of flow textures to the advection of timesurfaces in the Gulf of Mexico [8]. (Top row) LEA on a slice; (middle row) timesurfaces viewed as shaded surfaces; (bottom row) flow texture superimposed on the timesurface. Each row represents three frames from an animation. Data courtesy of James O'Brien.

Rather than enumerate existing algorithms along with their advantages and disadvantages, we present a conceptual framework within which these algorithms can be described and understood. We limit ourselves to 2D flow fields, although the strategies presented below have straightforward (albeit much more expensive) extensions to 3D.

The chapter is structured as follows: Section 13.2 describes a framework that abstracts the salient features of flow-texture algorithms, in particular those related to temporal and spatial correlation. This is followed in Section 13.3 by a discussion of integration accuracy, texture generation, and the development of recurrence relations for easy mapping onto graphics hardware. Finally, we cover the details necessary to port the leading algorithms to current graphics hardware in Section 13.4.

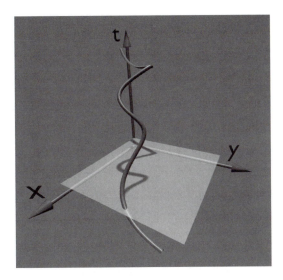

Figure 13.2 Illustration of a trajectory in the space–time domain V and a pathline viewed as its projection onto a spatial slice.

13.2   Underlying Model

Most of the existing flow-texture algorithms derive from different approximations to a physical model that we now describe. Consider a single particle in an unsteady flow $\mathbf{u}(\mathbf{r}, t)$ and its *trajectory*, henceforth defined as a curve in a 3D coordinate system with two spatial axes $\mathbf{r}$ and a time axis t. As illustrated in Fig. 13.2, the projection of the particle trajectory onto the (x, y) plane is a pathline of the underlying flow. The volume spanned by $\mathbf{r}$ and t is denoted by V. So that we can understand the temporal dependence of the flow, the 3D reference frame is densely filled with particle trajectories. The intersection of the volume with a 2D spatial plane then yields a dense collection of points (also referred to as particles). As this slice translates along the time axis, their spatial coordinates change continuously, thus ensuring temporal coherence. Local regions of flow strain or divergence create pockets of increased and reduced particle density, which form distinct macroscopic patterns. These patterns persist in time when collections of neighboring particles travel together.

All methods must address the following challenges:

1. Maintain a time-independent particle density without destroying the mechanisms responsible for changes in the particle density and without sacrificing temporal coherence; i.e., particle trajectories must be sufficiently long.

2. Develop a good strategy for adding and removing particles.

3. Introduce spatial correlation into each frame that encodes information from a short time interval to generate macroscopic structures representative of the flow.

4. These macroscopic structures should be correlated in time to simulate flow motion.

5. Steady flow should result as a special case of unsteady flow.

13.2.1   Particle Features

Particles released into the flow are subject to several constraints. They are identified by an invariant tag, along with one or more time-dependent properties. In most implementations that target visualization, one of these properties is color, often kept constant. To accentuate particle individuality, the colors can have a random distribution. Alternatively, to simulate the release of dye into the flow, all particles initially within a local volume are assigned the same color. Particles are constrained to lie on particle paths imposed by an underlying vector field. Finally, particles can enter and exit the physical domain, accumulate at sinks, or diverge from sources. Additional time-dependent properties, such as opacity, can be assigned to particles. These might be used to enhance the clarity of any macroscopic patterns that emerge.

At a given time, particles occupy a spatial slice through V and lie on separate trajectories. The challenge is to develop strategies that maintain a uniform distribution of particles for all time. We address this next.

13.2.2   Maintaining a Regular and Dense Distribution of Particles

Under the above constraints, one is left with the task of generating particle trajectories that sample space uniformly and densely, and properly take into account effects of inflow/outflow boundaries. In a direct approach, a large number of particles is seeded randomly across a spatial slice and tracked in time. Particle coordinates are maintained in some continuously updated data structure. Generally, one seeks to maximize the length of individual trajectories. However, this requirement must be balanced against the need for particle uniformity, which is easily destroyed as a result of boundary effects and finite velocity gradients. Particles near inflow boundaries are transported into the domain, leaving behind regions of reduced coverage. On the other hand, particles that exit the domain must be removed. Particle injection and destruction strategies are essential to counteract this effect. Control of particle density is also necessary within the interior of the physical domain.

We now present three strategies that achieve the above objectives, and we relate them to existing flow-texture algorithms:

1. All particles have a finite life span τ to ensure their removal from the system before nonhomogeneities become too severe. A new distribution of particles is generated at constant time intervals. The density of particles is chosen to achieve a dense coverage of the spatial domain. This approach is at the heart of UFLIC [16] and time-dependent spot noise [6]. In UFLIC, the time between consecutive particle injections is smaller than τ. On the other hand, they are injected at intervals of τ for the spot-noise approach. Aliasing errors are alleviated by having the initial particle injection distributed in time.

2. Keeping the previous approach in mind, we now seek to explicitly control the density of particles. The spatial slice is partitioned into bins within which the particle density will remain approximately constant. Particles are removed or injected into a bin when their number exceeds or falls below prescribed thresholds. We maximize the minimum length of trajectories by removing the oldest particles first. In this approach, a reduced number of particles is exchanged against more complex code logic. Accelerated UFLIC [13] and DLIC [17] are examples of this approach.

3. Particle tracking is abandoned in favor of an Eulerian point of view. The properties of a dense collection of particles on a spatial slice are sampled on a discrete grid. Their values are updated by following particle paths. This was the initial approach of Max and Becker [14]. More recent implementations are at the basis of LEA [10] and IBFV [19].

13.2.3   Temporal Correlation

To better understand the nature of temporal correlation, consider a single illuminated particle. This particle traces out a trajectory in space–time. Only a single dot is visible at any time. In the absence of temporal correlation, the resulting image would be a set of dots distributed randomly. Temporal correlation is thus a consequence of the continuity of particle paths.

The human visual system [4] can integrate the smooth motion of a single bright dot and infer its direction and approximate speed. There is a range of speed that maximizes the effectiveness of this correlation. If the particle moves too rapidly, the visual system is incapable of correlating successive positions, and the trajectory information is lost. If the motion if too slow, information about the particle motion cannot be established.

Whatever the visualization technique used, the objective is to simultaneously represent the structure of the flow in a spatial slice through spatial correlation and the temporal evolution of these structures along particle trajectories. In the next section, we discuss a general framework to introduce flow patterns within a spatial slice. This general framework subsumes all known flow-texture methodologies.

13.2.4 Spatial Correlation

To achieve spatial correlation, consider an intensity function $I(\mathbf{r}, t)$ defined over the 3D space–time volume V introduced earlier. We define a filtered spatial slice

$$D_t(\mathbf{r}) = \int_{-\infty}^{\infty} k(s) I(\mathcal{Z}(t - s; \mathbf{r}, t)) \, ds \qquad (13.1)$$

as a convolution along a trajectory $\mathcal{Z}(s; \mathbf{r}, t)$ through V. The subscript on D_t is a reminder that the filtered image depends on time. Trajectories are denoted by scripted variables and have three components: two spatial and one temporal. $\mathcal{Z}(s; \mathbf{r}, t)$ is parameterized by its first argument and passes through the point $(\mathbf{r}, t)$ when $s = t$. For generality, we perform the convolution along the entire time axis and rely on the definition of the filter kernel $k(s)$ to restrict the domain of integration. We must introduce some correlation into V. To this end, we define

the intensities in V to be either constant or piecewise continuous along another family of trajectories $\mathcal{Y}(s; \mathbf{r}, t)$. The structure of D_t is thus characterized by the triplet $[\mathcal{Y}, \mathcal{Z}, k]$.

13.2.4.1 Line Integral Convolution

LIC introduces coherence into a 2D random field through a convolution along the streamlines $\mathbf{x}(s; \mathbf{r}, t)$ of a steady vector field [5]. An equivalent formulation is possible within the 3D framework when $I(\mathbf{r}, t)$ is constant along lines parallel to the time axis with a random intensity across a spatial slice. Thus, $\mathcal{Y}(s; \mathbf{r}, t) = (\mathbf{r}, s)$ and $I(\mathbf{r}, t) = \phi(\mathbf{r})$, where $\phi(\mathbf{r}) \in [0, 1]$ is a random function. The convolution is computed along $\mathcal{Z}(s; \mathbf{r}, t) = (\mathbf{x}(s; \mathbf{r}, t), s)$, i.e., the streamlines of the steady vector field. With these definitions, spatially correlated patterns produced by LIC are defined by

$$D_t(\mathbf{r}) = \int_{-\infty}^{\infty} k(s) \phi(\mathbf{x}(t - s; \mathbf{r}, t)) \, ds$$

13.2.4.2 Image-Based Flow Visualization

IBFV [19] takes an approach similar to that of LIC, with two exceptions. First, the convolution trajectories are based on the pathlines $\mathbf{x}(s; \mathbf{r}, t)$ of $\mathbf{u}(\mathbf{r}, t)$. Second, the intensity function defined in LIC is modulated according to a time-dependent periodic function $w(t)$ along the trajectories $\mathcal{Y}(s; \mathbf{r}, t) = (\mathbf{r}, s)$ that define $I(\mathbf{r}, t)$. Thus, the volume intensity distribution becomes

$$I(\mathbf{r}, t) = w((t + \phi(\mathbf{r})) \bmod 1) \qquad (13.2)$$

where $\phi(\mathbf{x}) \in [0, 1]$ is now interpreted as a random phase. With the definitions in Equation 13.2, the convolution is given by Equation 13.1 with $\mathcal{Z}(s; \mathbf{r}, t) = (\mathbf{x}(s; \mathbf{r}, t), s)$.

Consider the case when a single trajectory from $\mathcal{Y}$ is illuminated with an intensity proportional to $w(t \bmod 1)$. The only points visible in $D_t(\mathbf{r})$ lie on trajectories that pass through $(\mathbf{r}, s)$ for some s within the support of the kernel. When the kernel is a monotonically decreasing function of $|s|$, the segment of the curve that corresponds to $s > 0$ is a streakline segment, with maximum intensity at **r** and decaying intensity away from **r**. It should be noted that in the limit of steady flow, the resulting streamline intensity remains unsteady, unless $w(t)$ is constant.

13.2.4.3   Lagrangian–Eulerian Advection

LEA [10] seeks to model the effect of a photograph taken with a long exposure setting. Under such conditions, an illuminated particle forms a streak whose points are spatially correlated. This result is modeled by performing the convolution along trajectories $\mathcal{Z}(s; \mathbf{r}, t) = (\mathbf{r}, s)$ parallel to the time axis with constant intensity along the trajectories $\mathcal{Y}(s; \mathbf{r}, t) = (\mathbf{x}(s; \mathbf{r}, t), s)$ associated with $\mathbf{u}(\mathbf{r}, t)$. The resulting display is then

$$D_t(\mathbf{r}) = \int_{-\infty}^{\infty} k(s) I(\mathbf{r}, t - s)\, ds \qquad (13.3)$$

To identify the patterns formed by the spatial correlation, it is expedient to rewrite $D_t(\mathbf{r})$ as a convolution along some curve through a spatial slice of V. Since $I(\mathbf{r}, t)$ is constant along trajectories of $\mathbf{u}(\mathbf{r}, t)$, it follows that $I(\mathbf{x}(s; \mathbf{r}, t), s) = I(\mathbf{x}(0; \mathbf{r}, t), 0)$ and

$$D_t(\mathbf{r}) = \int_{-\infty}^{\infty} k(s) I(\mathbf{x}(0; \mathbf{r}, t - s), 0)\, ds \qquad (13.4)$$

This is a convolution along the curve $\mathbf{x}(0; \mathbf{r}, s)$ in the $t = 0$ slice and parameterized by s. If

$k(s) = 0$ for $s > 0$, all points on this path lie on trajectories that pass through **r** at some time $s \leq 0$, which is a streakline of the flow $\mathbf{u}(\mathbf{r}, t)$ after time reversal. The curves $\mathbf{x}(0; \mathbf{r}, s)$, parameterized by s, are precisely the spatially correlated features we seek.

A comparison of Equations 13.1 and 13.4 shows that IBFV and LEA are, in a sense, dual to each other. IBFV constructs V from lines parallel to the time axis and convolves along trajectories, while LEA constructs V from trajectories and performs the convolution along lines parallel to the time axis. This duality is shown in Fig. 13.3.

From the above discussion, we find that it is important to distinguish carefully between the curves along which the convolution is computed and the resulting patterns formed in $D_t(\mathbf{r})$. In general, the projection of $\mathcal{Z}(s; \mathbf{r}, t)$ onto $D_t(\mathbf{r})$ differs from the resulting spatial patterns. If the kernel support is sufficiently small, the patterns are visually indistinguishable from short streamlines (streamlets). When $\mathbf{u}(\mathbf{r}, t)$ is steady, both IBFV and LEA produce streamlines.

13.2.4.4   Dynamic LIC

Not all vector fields derive from the velocity field of a fluid. For example, electric fields driven by time-dependent electric-charge dis-

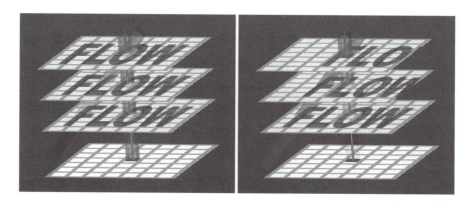

Figure 13.3   Duality between IBFV (left) and LEA (right). For IBFV, volume intensity is constant along vertical lines, and the convolution is along pathlines. For LEA, intensity is constant along pathlines, while convolution is along vertical lines.

tributions should be represented as successions of LIC images correlated in time. In DLIC [17], the motion of the electric field $\mathbf{u}(\mathbf{r}, t)$ is determined by a secondary flow field, $\mathbf{v}(\mathbf{r}, t)$, with pathlines $\mathbf{y}(s; \mathbf{r}, t)$. The generation of the sequence of LIC images is achieved by building $I(\mathbf{r}, t)$ from the pathlines of the motion field, i.e., $\mathcal{Y}(s; \mathbf{r}, t) = (\mathbf{y}(s; \mathbf{r}, t), s)$, and taking the convolution along the streamlines of $\mathbf{u}(\mathbf{r}, t)$, i.e., $\mathcal{Z}(s; \mathbf{r}, t) = (\mathbf{x}(s; \mathbf{r}, t), t)$. The filtered display becomes as follows:

$$D_t(\mathbf{r}) = \int_{-\infty}^{\infty} k(s) I(\mathcal{Z}(t - s; \mathbf{r}, t)) ds$$

The resulting structures are streamlines of $\mathbf{u}(\mathbf{r}, t)$ transported in time along the pathlines of $\mathbf{v}(\mathbf{r}, t)$.

13.3 Implementing the Dense Set of Particles Model

Several ingredients are necessary to implement a flow-texture algorithm:

1) The physical domain.

2) A time-dependent flow field defined over the physical domain (and perhaps a second vector field to determine trajectories in non-flows simulations).

3) A mechanism to generate noise texture (suitably preprocessed).

4) An integration scheme.

There are several issues that have to be addressed in all methods: 1) accuracy; 2) sampling; 3) inflow boundaries; 4) uniformity of the noise frequency; and 5) contrast (see Section 13.4).

13.3.1 Integration Scheme and Accuracy

The position $\mathbf{x}(t)$ of a single particle subject to a velocity field $\mathbf{u}(\mathbf{r}, t)$ satisfies

$$\frac{d\mathbf{x}(t)}{dt} = \mathbf{u}(\mathbf{x}(t), t) \tag{13.5}$$

(In what follows, we no longer refer to the starting location along a path unless necessary for clarity.) In practical implementations of flow-texture methods, the time axis is subdivided into uniform intervals Δt. Integration of Equation 13.5 over the time interval $[t_k, t_{k+1}]$ yields the relation

$$\mathbf{x}(t_{k+1}) = \mathbf{x}(t_k) + \int_{t_k}^{t_{k+1}} \mathbf{u}(\mathbf{x}(s), s)\, ds \tag{13.6}$$

between particle positions. At the discrete level, the particle position at t_k becomes $\mathbf{p}_k = \mathbf{p}(t_k)$. A first-order forward discretization of Equation 13.6 relates the positions of a particle at times t_k and t_{k+1}:

$$\mathbf{p}_{k+1} = \mathbf{p}_k + \Delta t\, \mathbf{u}(\mathbf{p}_k, t_k)$$

Similarly, a backward integration relates particle positions between times t_k and t_{k-1}:

$$\mathbf{p}_{k-1} = \mathbf{p}_k - \Delta t\, \mathbf{u}(\mathbf{p}_k, t_k)$$

A first-order scheme is sufficient when integrating noise textures; errors accumulate only over the length of a correlated streak. These errors are, in general, sufficiently small that they are not visible. When constructing long streaklines through dye advection (Fig. 13.4 and Section 13.3.3), this is no longer true. High-order integration methods are then necessary to achieve accurate results [15,10].

13.3.2 Sampling

The property field is sampled onto a texture that serves as a background grid. In a flow-texture algorithm, it is necessary to update the value of the particle property on each cell (i.e., each texel). The most general formulation is to compute a filtered spatial integration of the property at the previous time-step over the physical domain, according to

$$C_t(\mathbf{r}_{ij}) = \int_{\substack{\text{physical} \\ \text{domain}}} K(\mathbf{r}_{ij} - \mathbf{r}) C_{t-\Delta t}(\mathbf{x}(t - \Delta t; \mathbf{r}, t))\, d\mathbf{r}$$

$$\tag{13.7}$$

where $K(\mathbf{r})$ is some smoothing kernel. Different approximations to Equation 13.7 lead to trade-offs between speed and quality. Many subsampling schemes, along with supporting theory, are described by Glassner [7]. The simplest ap-

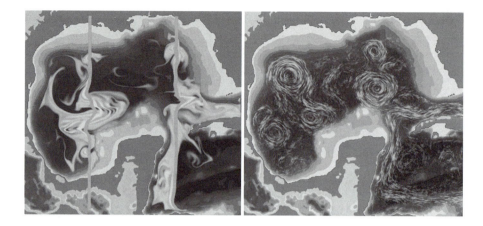

Figure 13.4 Flow in the Gulf of Mexico. (Left) Time lines visualized by dye advection; (Right) LEA with masking to emphasize regions of strong currents. Data courtesy of James O'Brien. (See also color insert.)

proximation, and the fastest to execute, is point sampling, which takes the form

$$C_t(\mathbf{r}_{ij}) = C_{t-\Delta t}(\mathbf{r}_{ij} - \Delta \mathbf{r}_{ij}) \tag{13.8}$$

where

$$\Delta \mathbf{r}_{ij} = \int_{t-\Delta t}^{t} \mathbf{u}(\mathbf{x}(s; \mathbf{r}_{ij}, t), s) \, ds$$

However, direct application of this formula may generate aliasing errors. LEA evaluates $C_{t-\Delta t}(\mathbf{r}_{ij} - \Delta \mathbf{r}_{ij})$ using nearest-neighbor interpolation; all other methods use bi-linear interpolation.

13.3.3 Texture Generation

Visual effects possible with flow-texture algorithms depend strongly on the textures that are advected. Noise textures lead to dense representations of time-evolving streaks; dye advection results from overlaying a smooth background texture with local regions of color and letting these regions advect with the flow. Some techniques, such as IBFV, explicitly construct a time-dependent noise texture. In all cases, it is essential to pay attention to the quality of the textures generated in order to minimize aliasing artifacts produced by improper relationships between the properties of the velocity field, the

spatial and temporal sampling rates, and the frequency content of the input textures.

13.3.3.1 Texture for Noise Advection

We begin by illustrating the relationship between filtering and sampling through a simple 1D example. Consider a uniform velocity field u and a noise texture $I(x)$ filtered by convolution:

$$D(x) = \int_{-\infty}^{\infty} k(s)I(x - us) \, ds \tag{13.9}$$

If $I(x)$ is sampled onto a uniform grid x_i with spacing Δx, a first-order approximation of Equation 13.9 yields

$$D(x) \approx \sum_{i=-\infty}^{\infty} k(i\Delta s)I(x - id)\Delta s$$

where $I(x - id)$ is estimated from $I_i = I(x_i)$ via some reconstruction algorithm and Δs is the time-sampling interval. By considering the properties of $D(x)$ when a single cell is illuminated, we derive a condition that relates Δs and $d = u\Delta s$, the distance traveled by a particle over the sampling interval. Using the Heaviside function $H(x) = 1$ for $x > 0$ and zero elsewhere, the texture has the representation

$$I(x) = H(x + \Delta x) - H(x)$$

which leads to a filtered texture

$$D(x) = \sum_{i=-\infty}^{\infty} k(i\Delta s)[H(x + \Delta x - id) - H(x - id)]\Delta s$$

We assume that the kernel is symmetric about $s = 0$. When $i = 0$, the term in brackets is a pulse of width Δx. When $i = 1$, the support of the pulse lies in $[d - \Delta x, d]$. To ensure overlap between these pulses, it is necessary that Δx exceed d. When this condition is violated, $D(x)$ is a series of disconnected pulses whose amplitudes follow $k(i\Delta s)$.

One way to avoid, or at least reduce, such aliasing effects is to ensure that high-frequency components are removed from $I(x)$ by a prefiltering step. Good filters have a short support in Fourier and physical space, making the Gaussian filter, of infinite support but exponential decay, a strong candidate. Multifrequency filtering has been proposed as a means to link the characteristics of spatial patterns in $D_t(\mathbf{r})$ to properties of the flow, such as velocity magnitude [12]. In the context of flow textures, it has been shown that if a 1D texture with a single illuminated texel is prefiltered with a triangular pulse of width greater than d, the resulting image is a smoothly decaying function that follows $k(s)$ [19]. Simple 2D filters are isotropic, so that spatial correlations introduced into the texture are independent of orientation. Unfortunately, control is lost over the width of the streaks formed in the final image. Anisotropic filters can be designed to prefilter the texture only in the direction of the instantaneous flow field, while leaving the direction normal to the streamline untouched. None of the techniques addressed in this paper implements such an approach, although LEA does use a LIC algorithm as an anisotropic filter to remove aliasing in a postprocessing step. Glassner [7] provides a thorough discussion of issues related to sampling, filtering, and reconstruction.

13.3.3.2   *Temporal Correlation*

As explained in Section 13.2.4, $I(\mathbf{r}, t)$ is an intensity distribution defined over a space–time domain. It is important to ensure that the temporal sampling interval Δs is consistent with the frequency content of the volume. IBFV and LEA address this problem differently. In IBFV [19], the intensity map is constructed independently of the underlying vector field. The temporal component is introduced through the periodic function $w(t \bmod 1)$ described by Equation 13.2. High contrast is achieved when $w(t)$ varies sharply over some small t interval. To avoid aliasing effects, $w(t)$ should be filtered so that the time discretization satisfies the Nyquist criterion. Since $w(t)$ is applied to all points with different initial phase, the filtering should be performed analytically. Van Wijk [19] advocates the use of $w(t) = (1 - t) \bmod 1$, which emulates the effects of Oriented LIC [20].

13.3.3.3   *Texture for Dye Advection*

Dye advection is treated similarly to noise advection, although the objective is different. When advecting noise, we seek to visualize streak patterns representative of the magnitude and direction of vectors in the velocity field. Instead, dye-advection techniques emulate the release of colored substance into a fluid and track its subsequent evolution over short or long time intervals. This process is simulated by pairing a connected region of the texture (referred to as dye) with a constant property value. When injecting dye, the texels located at the injection position are set to the color of the dye. Once released, the dye is advected with the flow.

13.3.4   Spatial Correlation

In this section, we discretize the convolution integrals (Equations 13.1 and 13.4) for IBFV and LEA respectively, and we transform them into simple recurrence relations that can be implemented efficiently. A kernel that satisfies the multiplicative relation $k(s)k(t - s) = k(t)$ for $s \leq t$ leads to a recurrence relation between the filtered display at two consecutive time-steps. The exponential filter $k(s) = \beta e^{-\beta s} H(s)$, normalized to unity, satisfies this relation. Although such a filter does not have compact support, the ease of implementing exponential filters through blending operations makes them very popular.

Using the normalized exponential filter, Equation 13.1 is discretized according to

$$D_n(\mathbf{r}_{ij}) = \beta \sum_{k=0}^{n} e^{-\beta s_k} \Delta s I(\mathcal{Z}(t_n - s_k; \mathbf{r}_{ij}, t_n)) \quad (13.10)$$

where the subscript n on D_n refers to t_n, $s_k = k\Delta s$, $t_n = n\Delta t$ and $t_0 = 0$. In general, $\Delta s = \Delta t$.

We next specialize the above relation to IBFV and LEA.

13.3.4.1 Image-Based Flow Visualization
IBFV uses trajectories

$$\mathcal{Z}(t_n - s_k; \mathbf{r}_{ij}, t_n) = (\mathbf{x}(t_n - s_k; \mathbf{r}_{ij}, t_n), t_n - s_k)$$

associated with $\mathbf{u}(\mathbf{r}, t)$ passing through the center $\mathbf{r}_{ij}$ of texel ij. Substitution into Equation 13.10 yields

$$D_n(\mathbf{r}_{ij}) = \beta \Delta s \sum_{k=0}^{n} e^{-\beta s_k} I(\mathbf{x}(t_n - s_k; \mathbf{r}_{ij}, t_n), t_n - s_k)$$

$$(13.11)$$

Although IBFV defines $I(\mathbf{r}, t)$ through Equation 13.2, we can derive a more general relationship valid for a time-dependent intensity function. Using the relation $\mathbf{x}(s; \mathbf{r}_{ij}, t_n) = \mathbf{x}(s; \mathbf{r}_{ij} - \Delta \mathbf{r}_{ij}, t_{n-1})$ and some straightforward algebra, a simple relation emerges between $D_n(\mathbf{r}_{ij})$ and $D_{n-1}(\mathbf{r}_{ij} - \Delta \mathbf{r}_{ij})$, namely

$$D_n(\mathbf{r}_{ij}) = e^{-\beta \Delta s} D_{n-1}(\mathbf{r}_{ij} - \Delta \mathbf{r}_{ij}) + \beta \Delta s I(\mathbf{r}_{ij}, t_n)$$

$$(13.12)$$

13.3.4.2 Lagrangian–Eulerian Advection
In the case of LEA, $\mathcal{Z}(t_n - s_k; \mathbf{r}_{ij}, t_n) = (\mathbf{r}_{ij}, t_n - s_k)$. The discretization

$$D_n(\mathbf{r}_{ij}) = \beta \Delta s \sum_{k=0}^{n} e^{-\beta s_k} I(\mathbf{r}_{ij}, t_n - s_k) \quad (13.13)$$

of Equation 13.3 can be recast into the recurrence relation

$$D_n(\mathbf{r}_{ij}) = e^{-\beta \Delta s} D_{n-1}(\mathbf{r}_{ij}) + \beta \Delta s I(\mathbf{r}_{ij}, t_n) \quad (13.14)$$

The second term is evaluated by following the trajectory $\mathcal{Y}(s; \mathbf{r}_{ij}, t_n) = (\mathbf{x}(s; \mathbf{r}_{ij}, t_n), s)$ from $t = t_n$ to the previous time t_{n-1}, i.e.,

$$I(\mathbf{r}_{ij}, t_n) = I(\mathbf{r}_{ij} - \Delta \mathbf{r}_{ij}, t_{n-1}) \quad (13.15)$$

13.3.4.3 Dynamic Line Integral Convolution
As explained in Section 13.2.4, DLIC constructs an intensity map based on the pathlines of the motion field $\mathbf{v}(\mathbf{r}, t)$, updates it according to Equation 13.15, and performs the LIC along the streamlines of $\mathbf{u}(\mathbf{x}, t)$. In the actual software implementation, a large number of individual particles, represented by discs of finite radius, is accurately tracked. The particles are allowed to overlap. Care is taken to ensure a uniform coverage of the physical domain, which in turn ensures good temporal correlation. The final LIC is of high quality. Note that LIC can also be implemented in graphics hardware [9].

13.3.5 Inflow Boundaries
All flow-texture algorithms have difficulties with inflow boundaries, i.e., points along the boundary where the velocity points into the physical domain (Fig. 13.5). The recurrence re-

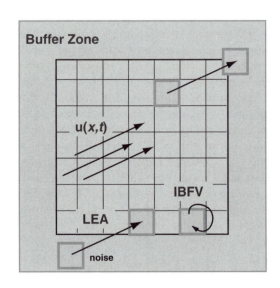

Figure 13.5 Inflow boundary treatment of LEA and IBFV.

lations derived above for LEA and IBFV clarify the issues. In LEA, as evident from Equation 13.14, particles that lie near an inflow boundary may lie outside the physical domain at the previous time. Should this occur, a random property value is assigned to that particle and blended into the display at the current time. IBFV, on the other hand, must access the value of the spatially correlated display at $\mathbf{r}_{ij} - \Delta\mathbf{r}_{ij}$. Simply replacing this value by a random value would destroy the spatial correlation near the inflow boundaries, which would then contaminate the interior of the domain. For such points, one possible update equation is

$$D_n(\mathbf{r}_{ij}) = e^{-\beta\Delta s}D_{n-1}(\mathbf{r}_{ij}) + \beta\Delta s I(\mathbf{r}_{ij}, t_n)$$

13.4   GPU-Based Implementation

In this section, we demonstrate how the different flow-visualization techniques described previously can be realized by exploitation of graphics hardware. The principal reason for using graphics processing units (GPUs) is to achieve a much higher processing speed, potentially two to three orders of magnitude higher than a comparable CPU-based implementation. Performance is an important issue because it might make the difference between a real-time GPU-based visualization, which allows for effective user interaction, and a noninteractive CPU version. We start with an abstract view on the capabilities of GPUs and how visualization techniques can benefit, followed by a discussion of specific details of IBFV and LEA implementations.

13.4.1   Generic GPU-Based Texture Advection

All implementations have to address the problem of how data structures and operations applied to them can be mapped to graphics hardware. From a generic point of view, the algorithmic structure of texture-advection techniques consists of the following steps (Fig. 13.6). First, (noise) textures and possibly other auxiliary data structures are

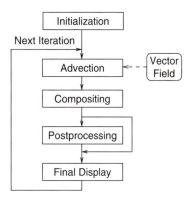

Figure 13.6 Flowchart for generic texture advection.

initialized. Then, each iteration step has to take into account the following:

1. The advection, based on the integration of pathlines.

2. A compositing operation, which combines information from the previous and current time-steps to introduce spatial coherence.

3. Optional postprocessing to improve image quality.

4. The final display on the screen.

The textures that represent the particles can be stored on the GPU by means of standard 2D or 3D textures, depending on the dimensionality of the domain. We assume that the particles are given on a uniform grid. In purely hardware-based implementations, the velocity field is also stored in a 2D or 3D texture. For hybrid CPU and GPU-based implementations, the vector field may be held in main memory and processed by the CPU. The advection step comprises both the integration of particle paths for one time-step (based on Equation 13.6) and the transport of the texture along these paths (based on Equation 13.8). Typically, an explicit first-order Euler integration is employed, which is executed either by the CPU for a hybrid approach or by the GPU. Texture transport is based on an appropriate specification of texture coordinates, which can be done either on a per-vertex basis (for hybrid CPU–GPU approaches) or on a per-fragment basis (i.e., purely GPU-based). The compositing operation is directly supported on the GPU by the blending of operations working on the framebuffer or within fragment programs.

Hardware-based implementations are very fast for the following reasons. GPUs realize a SIMD architecture, which allows efficient pipelining. In addition, bandwidth to texture memory is very high, which leads to a fast texture access. Finally, a transfer of visualization results to the graphics board for the final display is superfluous for GPU-based implementations. Since texture advection is compatible with the functionality of today's GPUs, a high overall visualization performance can be achieved.

However, the following issues have to be considered. First, the accuracy of GPUs is usually limited and might vary during the complete rendering pipeline. For example, color channels in the framebuffer or in textures have a typical resolution of only eight bits, whereas fragment processing may take place at higher precisions. Even floating-point accuracy within textures and fragment processing, provided by modern GPUs, is not comparable to double-precision numbers, available on CPUs. Second, the number of instructions might be limited. Therefore, the algorithms have to be designed to enable a rather concise implementation—sometimes at the cost of accuracy—or a less efficient multipass rendering technique has to be applied. Similarly, the number of indirection steps in fragment processing (i.e., so-called dependent texture lookups) may be restricted. A third issue concerns the choice of APIs for programming the GPU. OpenGL [2] and DirectX [1], the two widespread graphics APIs, have specific advantages and disadvantages. The main advantage of OpenGL is its platform (i.e., operating system) independence. However, at the time of writing, sophisticated features of the GPUs can only be accessed by GPU-specific OpenGL extensions. The situation might improve in the future with standardized interfaces for fragment

programs. The main advantages of DirectX are a GPU-independent API and the support for a fast render-to-texture functionality, which is extremely useful for multipass rendering and iterative processes like texture advection. On the downside, DirectX is available only on the Windows platform. Finally, for vertex or fragment programming, higher-level programming languages, such as NVidia's Cg (C for graphics) or the HLSL (high-level shading language) of DirectX, can be used to replace the assembler-like programming with OpenGL extensions or DirectX vertex- and pixel-shader programs. We try to keep the following discussion as API-independent as possible in order to allow a mapping to current and future programming environments.

13.4.2 Image-Based Flow Visualization

Visualization techniques based on the idea of IBFV implement 2D flow visualization according to the recurrence relation

$$D_n(\mathbf{r}) = (1 - \alpha)D_{n-1}(\mathbf{r} - \Delta t\, \mathbf{u}(\mathbf{r}, t_n)) + \alpha I_n(\mathbf{r})$$

(13.16)

which approximates Equation 13.12 by first-order integration of the particle path and by first-order approximation of the exponential function. Both D_n and I_n can be implemented by 2D textures. To alleviate the notation, D and I will henceforth refer to any of the D_n, I_n textures, and $\mathbf{r}_{ij}$ is replaced by $\mathbf{r}$. The first term of Equation 13.16 requires access to the texture D at the previous time-step at the old particle position $\mathbf{r} - \Delta t\, \mathbf{u}(\mathbf{r}, t_n)$. Three possible solutions can be employed for this advection.

The first solution implements advection on a per-fragment basis. Here, a fragment program computes the old position $\mathbf{r} - \Delta t\, \mathbf{u}(\mathbf{r}, t_n)$ by accessing the vector field stored in another 2D texture. Then a lookup in D_{n-1} is performed by interpreting the old position as a set of texture coordinates. This texture-fetch operation is an example of a dependent texture lookup. Fig. 13.7 shows the DirectX pixel-shader code for this advection. The fragments are generated by

```
ps.1.1           // We are fine with pixel shader V1.1
tex t0           // Accesses vector field u
texbem t1,t0     // Dependent tex lookup in D with shifted tex coords
mov r0, t1;      // Outputs results
```

Figure 13.7 Pixel-shader code for backward advection.

rendering a single quadrilateral that spans the spatial domain. Texture coordinates t_0 and t_1 are specified in such a way as to exactly cover the domain by the textures. The dependent texture lookup takes the result of the previously fetched vector field $u(r, t_n)$, scales this result by a constant value (a parameter that is specified outside the fragment program and that takes into account $-\Delta t$), and adds the scaled value to the texture coordinates t_0 to obtain the coordinates for the lookup in the texture D_{n-1}.

The second solution implements a similar backward advection on a per-vertex basis [14]. The domain is covered by a mesh, with velocities assigned at its vertices. On the CPU, the old positions $r - \Delta t\, u(r, t_n)$ are computed for the vertices by accessing the vector field in main memory. Once again, the old positions are interpreted as texture coordinates for a lookup in D. The domain is rasterized by drawing the entire mesh.

The third solution implements a forward advection on a per-vertex basis [19]. This technique differs from the previous one in that the vertex coordinates are changed instead of the texture coordinates, i.e., a forward Euler integration $r + \Delta t\, u(r, t_n)$ yields the vertex coordinates for a distorted triangle mesh. The results differ from backward integration because the Euler method is not symmetric with respect to the evolution of time. Nevertheless, first-order forward and backward integration are both first-order approximations of the true solutions, and forward mapping is an acceptable approximation to the advection in Equation 13.16.

In all three implementations, the texture lookup with shifted texture coordinates makes

use of bi-linear interpolation. An interpolation scheme is required because pathline integration usually results in texture coordinates that do not have a one-to-one correspondence to the texels. The artificial smearing-out by bi-linear interpolation does not invalidate the final images because their effects are continuously blended out by compositing, according to Equation 13.16.

This compositing operation can be achieved by a two-pass rendering [19]. The first pass writes the result of the above advection to the framebuffer. In the second pass, texture I is blended into the framebuffer by using α blending with weights α and $(1 - \alpha)$, respectively. Alternatively, the advected texture D and I can be combined by multitexturing within a fragment program. Finally, the framebuffer is saved in a texture to obtain the input for the subsequent iteration. In addition to noise-based visualization, IBFV emulates dye advection by interactively drawing additional dye sources into the framebuffer during each rendering step.

13.4.3 Lagrangian–Eulerian Advection

LEA is based on the recurrence relation in Equation 13.14, leading to

$$D_n(r) = (1 - \alpha)D_{n-1}(r) + \alpha I_{n-1}(r - \Delta t\, u(r, t_n))$$

$$(13.17)$$

where I is computed from the property of the particle at the previous time-step. When compared to IBFV, the roles of textures D and I are exchanged with respect to advection. A GPU-based algorithm [11,21] implements a per-fragment advection of I analogously to the fragment-based backward mapping for D in the case of IBFV. However, since no α blending is imposed onto the transported noise texture I, a bilinear interpolation would cause an unacceptable smearing and a fast loss of both contrast and high frequency. Therefore, a nearest-neighbor sampling is employed during the backward integration step. Unfortunately, a direct implementation of nearest-neighbor sampling would not permit subtexel motion of particles because a texel is virtually repositioned to the center of the respective cell after each iteration, i.e., small magnitudes of the velocity field would result in an erroneously still image [10]. The Lagrangian aspect of LEA makes possible subtexel motion: in addition to noise values, 2D coordinates of particles are stored in a texture; these coordinates are also updated during the particle integration and allow particles to eventually "jump" across texel boundaries, even at small velocities. Additional discussions concerning numerical accuracy on GPUs can be found in Weiskopf et al. [21].

Similarly to IBFV, textures D and I are combined by an α blending operation in the framebuffer or fragment program, and the framebuffer is saved in a texture to obtain the input for the subsequent iteration. Dye advection is included by a separate process that is based on per-fragment advection with bi-linear interpolation. The final image is constructed from advected noise and dye textures through blending.

13.4.4   Postprocessing

The results of IBFV and LEA can be improved by postprocessing. Since both techniques apply a summation of different noise textures by blending, image contrast is reduced. Postprocessing by histogram equalization [19] or high-pass filtering [16] could be applied to increase contrast. Aliasing artifacts occur in LEA when the maximum texture displacement is excessive (see Section 13.3.3). These artifacts can be avoided by imposing LIC as a postprocessing filter [10]. As specifically designed for dye advection, artificial blurring caused by bi-linear interpolation

can be reduced by applying a filter that steepens the fuzzy profile at the boundary of the dye, i.e., a nonlinear mapping of the unfiltered grey-scale values [10]. Finally, velocity masking can be used to change the intensity or opacity depending on the magnitude of the underlying velocity field [10,21]. In this way, important regions with high velocities are emphasized. Note that, with the exception of velocity masking [21], GPU-based implementations of the postprocessing methods discussed here have not yet been reported in the literature.

13.5   Conclusions

We have presented the leading flow-texture techniques within a single framework, first from a physical perspective based on a space–time domain filled with a dense collection of particle trajectories. This was followed by a formulation that explains how to derive the visualization techniques based on properly chosen convolutions within the volume. We feel that flow-texture algorithms are well understood in 2D planar flows, although some research remains to be done for flows on 2D curved manifolds. Flow-texture algorithms in 3D remain a formidable challenge. Straightforward extensions to 2D algorithms are prohibitively expensive. The display of dense 3D datasets introduces several perceptual issues, such as spatio–temporal coherence, depth perception, and orientation, that remain largely unsolved. Mechanisms for user interaction and navigation remain in their infancy. The holy grail of 3D flow-texture algorithms is the development of a real-time system to display high-quality dense representations with an interactively changing region of interest.

Acknowledgments

The first author thanks the National Science Foundation for support under grant NSF-0083792. The third author acknowledges support from the Landesstiftung Baden-Württemberg.

References

1. DirectX, http://www.microsoft.com/directx
2. OpenGL, http://www.opengl.org
3. J. Becker and M. Rumpf. Visualization of time-dependent velocity fields by texture transport. In *Visualization in Scientific Computing '98*, pages 91–102, 1998.
4. R. Blake and S.-H. Lee. Temporal structure in the input to vision can promote spatial grouping. In *Biologically Motivated Computer Vision 2000*, pages 635–653, 2000.
5. B. Cabral and L. Leedom. Imaging vector fields using line integral convolution. In *Proceedings of ACM SIGGRAPH '93*, pages 263–272, 1993.
6. W. C. de Leeuw and R. van Liere. Spotting structure in complex time dependent flows. In *Scientific Visualization, Proceedings: Dagstuhl '97*, pages 47–53, 1997.
7. A. S. Glassner. *Principles of Digital Image Synthesis*. San Francisco, Morgan Kaufmann, 1995.
8. J. Grant, G. Erlebacher, and J. J. O'Brien. Case study: visualization of thermoclines in the ocean using Lagrangian-Eulerian timesurfaces. In *IEEE Visualization '02*, pages 529–532, 2002.
9. W. Heidrich, R. Westermann, H.-P. Seidel, and T. Ertl. Applications of pixel textures in visualization and realistic image synthesis. In *ACM Symposium on Interactive 3D Graphics*, pages 127–134, 1999.
10. B. Jobard, G. Erlebacher, and M. Hussaini. Lagrangian–Eulerian advection for unsteady flow visualization. *IEEE Transactions on Visualization and Computer Graphics*, 8(3):211–222, 2002.
11. B. Jobard, G. Erlebacher, and M. Y. Hussaini. Tiled hardware-accelerated texture advection for unsteady flow visualization. In *Proceedings of Graphicon 2000*, pages 189–196, 2000.
12. M.-H. Kiu and D. C. Banks. Multi-frequency noise for LIC. In *Visualization '96*, pages 121–126, 1996.
13. Z. Liu and R. Moorhead. AUFLIC—An accelerated algorithm for unsteady flow line integral convolution. In *EG/IEEE TCVG Symposium on Visualization '02*, pages 43–52, 2002.
14. N. Max and B. Becker. Flow visualization using moving textures. In *Proceedings of the ICASE/LaRC Symposium on Visualizing Time-Varying Data*, pages 77–87, 1995.
15. H.-W. Shen, C. R. Johnson, and K.-L. Ma. Visualizing vector fields using line integral convolution and dye advection. In *1996 Volume Visualization Symposium*, pages 63–70, 1996.
16. H.-W. Shen and D. L. Kao. A new line integral convolution algorithm for visualizing time-varying flow fields. *IEEE Transactions on Visualization and Computer Graphics*, 4(2):98–108, 1998.
17. A. Sundquist. Dynamic line integral convolution for visualizing stream-line evolution. *IEEE Transactions on Visualization and Computer Graphics*, 9(8):273–282, 2003.
18. J. J. van Wijk. Spot noise—texture synthesis for data visualization. *Computer Graphics (Proceedings of ACM SIGGRAPH '91)*, 25:309–318, 1991.
19. J. J. van Wijk. Image based flow visualization. *ACM Transactions on Graphics*, 21(3):745–754, 2002.
20. R. Wegenkittl, E. Gröller, and W. Purgathofer. Animating flow fields: rendering of oriented line integral convolution. In *Computer Animation '97*, pages 15–21, 1997.
21. D. Weiskopf, G. Erlebacher, M. Hopf, and T. Ertl. Hardware-accelerated Lagrangian–Eulerian texture advection for 2D flow visualization. In *Vision, Modeling, and Visualization VMV '02 Conference*, pages 439–446, 2002.

14   Detection and Visualization of Vortices

MING JIANG and RAGHU MACHIRAJU
Department of Computer and Information Science
The Ohio State University

DAVID THOMPSON
Department of Aerospace Engineering
Mississippi State University

14.1   Introduction

In general, a feature can be defined as a pattern occurring in a dataset that is the manifestation of correlations among various components of the data. For many features that occur in scientific data, these correlations can be defined precisely. For other features, they are not well understood or do not lend themselves to precise definitions. Surprisingly, the swirling feature in flow fields, commonly referred to as a vortex, is an example of a feature for which a precise definition does not exist.

By most accounts [1–3], a vortex is characterized by the swirling motion of fluid around a central region. This characterization stems from our visual perception of swirling phenomena that are pervasive throughout the natural world. However, translating this intuitive description of a vortex into a formal definition has been quite a challenge.

Lugt [1] proposed the following definition for a vortex: *A vortex is the rotating motion of a multitude of material particles around a common center*. The problem with this definition is that it is too vague. Although it is consistent with visual observations, it does not lend itself readily to implementation in a detection algorithm. In light of this, Robinson [3] attempted to provide a more concrete definition of a vortex by specifying the conditions for detecting swirling flows in three dimensions:

> *A vortex exists when instantaneous streamlines mapped onto a plane normal to the vortex core exhibit a roughly circular or spiral pattern, when viewed from a reference frame moving with the center of the vortex core.*

The primary shortcoming of this second operational definition is that it is self-referential: the existence of a vortex requires *a priori* knowledge of the orientation and motion of its core.

Despite the lack of a formal definition, various detection algorithms have been implemented that can adequately identify vortices in most computational datasets. In this paper, we present an overview of existing detection methods; in particular, we focus on nine methods that are representative of the state of the art. Although this is not a complete listing of vortex-detection algorithms, the range of relevant issues covered by these nine methods is comprehensive in scope. The methods are these:

- *Helicity* method, by Levy et al. [4]
- *Swirl Parameter* method, by Berdahl and Thompson [5]
- *Lambda$_2$* method, by Jeong and Hussain [6]
- *Predictor–Corrector* method, by Banks and Singer [7]

- *Eigenvector* method, by Sujudi and Haimes [8]
- *Parallel Vectors* method, by Roth and Peikert [9]
- *Maximum Vorticity* method, by Strawn et al. [10]
- *Streamline* method, by Sadarjoen et al. [11]
- *Combinatorial* method, by Jiang et al. [12]

We first present three taxonomies for classifying these nine detection methods, in Section 14.2. We then describe each algorithm in Section 14.3, along with pseudo-code where appropriate. Next, we describe a recently developed verification algorithm for swirling flows in Section 14.4. In Section 14.5, we discuss the different visualization techniques for vortices. Finally, we conclude with highlights of future directions in this field.

14.2 Taxonomy

Almost every paper published on the subject of vortex detection has presented a classification of its predecessors in some fashion. One of the most comprehensive classifications of vortex-detection methods was proposed by Roth [13]. In this section, we present three taxonomies for classifying existing detection methods. These taxonomies are based on how the vortex is defined, whether the detection method is Galilean invariant, and the local or global nature of the identification process.

The first taxonomy classifies detection methods based on the definition of a vortex. A vortex can be defined either as a region or as a line. A region-based vortex definition specifies criteria for identifying contiguous grid nodes (or cells) that belong to either the vortex or its core. A line-based vortex definition, on the other hand, specifies criteria for locating vortex core lines. A set of contiguous line segments constitutes the vortex core line. In general, detection algorithms corresponding to region-based definitions are easier to implement and computationally cheaper than their line-based counterparts. Line-based algorithms must precisely locate points where the vortex core line intersects the grid cells. However, line-based algorithms provide more compact representations of vortices and can easily distinguish between individual vortices in close proximity. The latter is problematic for region-based approaches. Column 1 of Table 14.1 categorizes the nine detection methods based on this criterion.

The second taxonomy classifies detection methods based on whether they are Galilean (Lagrangian) invariant. Most detection methods

Table 14.1   Taxonomies of vortex-detection algorithms.

Method 1	Region/Line 2	Galilean 3	Local/Global 4
Helicity	Line	Not Invariant	Local
Swirl Parameter	Region	Not Invariant	Local
Lambda$_2$	Region	Invariant	Local
Predictor–Corrector	Line	Invariant	Global
Eigenvector	Line	Not Invariant	Local
Parallel Vectors	Line	Not Invariant	Local
Maximum Vorticity	Line	Invariant	Local
Streamline	Region	Not Invariant	Global
Combinatorial	Region	Not Invariant	Local

were designed under the assumption that steady flow fields and vortices move much more slowly than the average fluid particle within the flow. In a time-varying flow field, a vortex exhibits swirling motion only when viewed from a reference frame that moves with the vortex [1,3]. In order to detect vortices in unsteady (time-dependent) flows, it is necessary for the method to satisfy Galilean invariance. A detection method is Galilean invariant if it produces the same results when a uniform velocity is added to the existing velocity field. Thus, methods that do not depend directly on the velocity, such as pressure or vorticity, are Galilean invariant. This is an important property, especially in the context of tracking vortices in time-varying flow fields. Column 2 of Table 14.1 categorizes the nine detection methods based on this criterion.

The third taxonomy classifies detection methods based on the local or global nature of the identification process. A detection method is considered local if the identification process requires operations only within the local neighborhood of a grid cell. Methods that rely on the velocity gradient tensor are usually local methods. On the other hand, a global method requires examination of many grid cells in order to identify vortices. Methods that involve tracing streamlines in velocity or vorticity fields are considered global. From the definitions in the preceding section, it is apparent that a vortex is a global feature. It may be preferable to detect global features using global methods; however, on the basis of computation, global detection methods tend to be more expensive than local methods. However, in order to verify the accuracy of the detected results, a global approach is necessary. We describe this aspect in more detail in Section 14.4. Column 3 of Table 14.1 categorizes the nine detection methods based on this criterion.

14.3   Vortex-Detection Algorithms

14.3.1   Helicity Method

Levy et al. [4] introduced the use of normalized helicity, H_n, for extracting vortex core lines,

though they were not the first to identify the strong correlation between helicity and coherent structures in turbulent flow fields. H_n is a scalar quantity defined everywhere except at critical points:

$$H_n = \frac{v \cdot \omega}{|v||\omega|} \tag{14.1}$$

H_n is the cosine of the angle between velocity, v, and vorticity, ω. The underlying assumption is that near vortex core regions, the angle between v and ω is small. In the limiting case, where $v \| \omega$, $H_n = \pm 1$, and the streamline that passes through that point has zero curvature (it is a straight line). The authors suggested an approach to extract vortex core lines by first locating maximal points of H_n on cross-sectional planes, which are also points of minimal streamline curvature, and then growing the core line by tracing a streamline from the maximal points.

The sign of H_n indicates the direction of swirl (clockwise or counterclockwise) of the vortex with respect to the streamwise velocity component. It switches whenever a transition occurs between the primary and secondary vortices. The authors successfully used this feature with corresponding colors to distinguish between the primary and secondary vortices in the hemisphere-cylinder and ogive-cylinder datasets. However, the extracted core line may not always correspond to the actual vortex core line [13].

14.3.2   Swirl-Parameter Method

Berdahl and Thompson [5] presented a vortex-detection method based on the connection between swirling motion and the existence of complex eigenvalues in the velocity gradient tensor **J**. The authors introduced the intrinsic swirl parameter τ, defined by the ratio of the convection time t_{conv} (the time for a fluid particle to convect through the region of complex eigenvalues R_C) to the orbit time t_{orbit} (the time for a fluid particle to return to the same angular position). Thus,

$$t_{conv} = \frac{2\pi}{|Im(\lambda_C)|} \qquad t_{orbit} = \frac{L}{|v_{conv}|} \qquad (14.2)$$

where $Im(\lambda_C)$ is the imaginary part of the complex conjugate pair of eigenvalues, L is the characteristic length associated with the size of R_C, and v_{conv} is the convection velocity aligned along L. From Equation 14.2, τ can be written as

$$\tau = \frac{t_{conv}}{t_{orbit}} = \frac{|Im(\lambda_C)|L}{2\pi|v_{conv}|} \qquad (14.3)$$

When $\tau \to 0$, the fluid particle convects too rapidly through R_C to be "captured" by the vortex. Thus, τ is nonzero in regions containing vortices and attains a local maximum in the vortex core. For three dimensions, the length and orientation of L are unknown, because in general there is no single plane of swirling flow. The authors suggest using the plane normal to either the vorticity vector ω or the real eigenvector e_R, which are local approximations to the actual vortex core direction vector. The convective velocity v_{conv} is computed by projecting the local velocity vectors onto this plane:

$$v_{conv} = v - (v \cdot n)n \qquad (14.4)$$

where n is the plane normal computed from either ω or e_R.

Figure 14.1 illustrates the results when this method is applied to the propeller dataset. In the left image, the intensity of τ is described by a color map. In the right image, isosurfaces are generated showing the path of the tip vortex as well as a ring vortex that was shed from the propeller base. However, selecting the right threshold for τ in order to distinguish individual vortices is often difficult.

14.3.3 Lambda₂ Method

Jeong and Hussain [6] proposed a definition for a vortex that is commonly referred to as the λ_2 definition. They begin with the premise that a pressure minimum is not sufficient as a detection criterion. The problems are due to unsteady irrotational straining, which can create a pressure minimum in the absence of a vortex, and viscous effects, which can eliminate the pressure minimum within a vortex. To remove these effects, the authors decompose the velocity gradient tensor $\mathbf{J}$ into its symmetric part, the rate of deformation or strain-rate tensor $\mathbf{S}$, and its antisymmetric part, the spin tensor $\mathbf{\Omega}$, and consider only the contribution from $\mathbf{S}^2 + \mathbf{\Omega}^2$.

$$\mathbf{S} = \frac{\mathbf{J} + \mathbf{J}^T}{2} \qquad \mathbf{\Omega} = \frac{\mathbf{J} - \mathbf{J}^T}{2} \qquad (14.5)$$

They define a vortex as a connected region where $\mathbf{S}^2 + \mathbf{\Omega}^2$ has two negative eigenvalues. Because $\mathbf{S}^2 + \mathbf{\Omega}^2$ is real and symmetric, it has

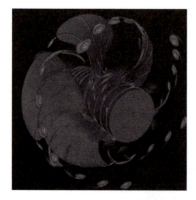
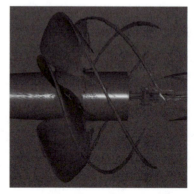

Figure 14.1 Swirl parameter. Images courtesy of Michael Remotigue, Mississippi State University. (See also color insert.)

only real eigenvalues. Let λ_1, λ_2, and λ_3 be the eigenvalues such that $\lambda_1 \geq \lambda_2 \geq \lambda_3$. If λ_2 is negative at a point, then that point belongs to a vortex core. Through several analytical examples and direct numerical-simulation datasets, the authors demonstrated the effectiveness of the λ_2 definition compared to others. However, in situations where several vortices exist, it can be difficult for this method to distinguish between individual vortices.

14.3.4 Predictor–Corrector Method

The vorticity-predictor, pressure-corrector method for detecting vortex core lines was proposed by Banks and Singer [7,14]. Their underlying assumption is that vortical motion is sustained by pressure gradients and indicated by vorticity ω. The algorithm extracts a skeleton approximation to the vortex core by tracing vorticity lines and then correcting the prediction based on local pressure minimum. In order to find the initial set of seed points for tracing vorticity lines, they consider grid-points with low pressure and high vorticity magnitude. However, as the authors pointed out, it is possible for a grid-point to satisfy both conditions without being part of a vortex core. An outline of the algorithm is provided in Algorithm 14.1.

For the predictor step, vorticity integration can be performed using fourth-order Runge-Kutta. The authors suggested, instead, a simplification whereby they relate the step size to the smallest dimension of the local grid cell. For the corrector step, the steepest-descent method is used to find the local pressure minimum, with the step size being, again, the smallest grid cell dimension.

Algorithm 14.1 terminates when the minimum pressure point is too far from the predicted point; however, the method is not guaranteed to terminate in every case, because the growing skeleton can form closed loops, which is not ideal for real vortices. Furthermore, special care must to be taken in order to minimize the number of skeletons approximating the same vortex core line, since the skeleton grown from each seed point may end up describing the same vortex core.

14.3.5 Eigenvector Method

The eigenvector method for detecting vortex core lines was first proposed by Sujudi and Haimes [8]. The method is based on critical-point theory, which asserts that the eigenvalues and eigenvectors of the velocity gradient tensor $\mathbf{J}$, evaluated at a critical point, define the local flow pattern about that point. As the authors

```
 1:  locate seed points with low pressure and high |ω|
 2:  for all seed points do
 3:    repeat
 4:      compute ωᵢ at current skeleton point
 5:      step in ωᵢ direction to predict next point
 6:      compute ωᵢ₊₁ at predicted point P_ω
 7:      locate minimum pressure P_p on plane ⊥ ω
 8:      if dist (P_ω, P_p) < threshold then
 9:        correct next point to P_p
10:      else
11:        terminate skeleton growth
12:      end if
13:      eliminate seed points within distance_r
14:    until skeleton exits domain or is too long
15:  end for
```

Algorithm 14.1 Predictor–corrector method.

```
1:  decompose grid cells into tetrahedral cells
2:  for all tetrahedral cells do
3:    linearly interpolate v to produce J
4:    compute all three eigenvalues of J
5:    if two eigenvalues are complex conjugates then
6:      compute eigenvector e_R for the real eigenvalue
7:      project v onto e_R → reduced velocity v_r
8:      compute the zero v_r straight line γ_z
9:      if γ_z intersects cell twice then
10:       add line segment to vortex core
11:     end if
12:   end if
13: end for
```

Algorithm 14.2 Eigenvector method.

point out, there are swirling flows that do not contain critical points within their centers. In order to handle these cases, velocity vectors are projected onto the plane normal to the eigenvector of the real eigenvalue (assuming the other two eigenvalues are complex conjugate pairs) to see if they are zero. If they are, then the point must be part of the vortex core. An outline of the algorithm is given in Algorithm 14.2.

Initially, all mesh elements are decomposed into tetrahedral cells. Linear interpolation of $\mathbf{v}$ within the cell follows, which induces a constant $\mathbf{J}$. The reduced velocity $\mathbf{v}_r$ is computed by subtracting the velocity component in the direction of $\mathbf{e}_R$; this computation is equivalent to projecting $\mathbf{v}$ onto the plane normal to $\mathbf{e}_R$. Finding the zero locations on the plane requires setting up a system of three equations using the linearly interpolated components of $\mathbf{v}_r$, which can be solved using any two of the three linearly independent equations. The solution is a straight line of zero $\mathbf{v}_r$.

This method was successfully applied to detecting vortex cores in numerous CFD applications [15,16]. Fig. 14.2 illustrates one such example taken from Kenwright and Haimes [16]. The yellow line segments represent the vortex cores extracted from a transient F/A-18 simulation dataset. However, as the authors pointed out, producing contiguous vortex core lines is not always possible because the under-

lying interpolant may not be linear or line segments may not meet up at shared faces. Modifications to the original algorithm are proposed in Haimes and Kenwright [17] to address this issue and improve its performance.

14.3.6 Parallel-Vectors Method

The parallel-vectors operator was first introduced by Roth and Peikert [9] as a higher-order method for locating vortex core lines. They recast the first-order eigenvector method into a parallel alignment problem between $\mathbf{v}$ and its first derivative $\mathbf{Jv}$ (i.e., reduced velocity is zero when $\mathbf{v}$ is parallel to the real eigenvector of $\mathbf{J}$). In order to better capture the slowly rotating curved vortices that are typical in turbomachinery flow fields, they use the second derivative of $\mathbf{v}$, which is defined as

$$\mathbf{w} = \frac{D^2\mathbf{v}}{Dt^2} = \frac{D(\mathbf{Jv})}{Dt} = \mathbf{JJv} + \mathbf{Tvv} \tag{14.6}$$

where $\mathbf{T}$ is a $3 \times 3 \times 3$ tensor. Essentially, a vortex core line is the locus where $\mathbf{v}$ is parallel to $\mathbf{w}$: $\{x: \mathbf{v}(x) \times \mathbf{w}(x) = 0\}$. An outline of the algorithm is given in Algorithm 14.3.

Due to discretization errors, excessive fluctuations may result from computing the higher-order derivatives. To avoid this, the authors recommend smoothing the vector-field data as a preprocessing step. Roth and Peikert [13,18] present other approaches for finding parallel

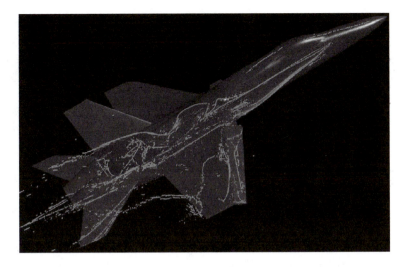

Figure 14.2 Eigenvector approach (©1998 IEEE). Image courtesy of Robert Haimes, Massachusetts Institute of Technology. (See also color insert.)

```
 1:  for all grid points do
 2:     calculate J and compute v' = Jv
 3:     calculate J' and compute W = Jv'
 4:  end for
 5:  for all grid faces do
 6:     find zero of function v × w
 7:     use Newton iterations starting from face center
 8:     if zero lies on face then
 9:        connect with straight line to previous zero
10:     end if
11:  end for
```

Algorithm 14.3 Parallel-vectors method.

vectors along with *post priori* criteria for removing line segments that might be of insufficient strength (speed of local rotation) or quality (angle between velocity at core and core line).

Fig. 14.3 illustrates the results for the Francis turbine runner dataset and the stator of a reversible pump-turbine dataset. The black line segments indicate the locations of detected vortex core lines. Note the existence of gaps in the detected core lines, which are mainly due to the large number of raw solution lines produced by the higher-order method [13].

14.3.7 Maximum-Vorticity Method

Strawn et al. [10] define a vortex core as a local maximum of vorticity magnitude $|\omega|$ in the plane normal to ω. This technique is applicable for free-shear flows, but not for shear layers, which have high $|\omega|$ but no local $|\omega|$ maxima. The motivation for this approach comes from situations where multiple vortices with the same orientation and overlapping cores are in close proximity. The resulting velocity field would only exhibit a single rotational center. To address this issue, the authors introduced

Figure 14.3 Parallel vector operator. Images courtesy of Martin Roth, Swiss Federal Institute of Technology, Zürich. (See also color insert.)

```
 1:   compute ω at all grid nodes
 2:   for all cell faces do
 3:      examine its 4 × 4 surrounding nodes
 4:      if∃ maximum |ω| in central nodes then
 5:        mark grid face as candidate face
 6:      end if
 7:   end for
 8:   for all candidate faces do
 9:      compute ∇|ω| using central difference at nodes
10:      compute solution points where ∇|ω| = 0
11:      if points are within face and are local maxima then
12:        mark them as vortex core points
13:      end if
14:   end for
```

Algorithm 14.4 Maximum-vorticity method.

the maximum-vorticity method, which is outlined in Algorithm 14.4.

For the preprocessing step, ω is transformed into computational space, where the search for $|\omega|$ maxima is done on a uniform grid. The gradient of $|\omega|$ is assumed to vary bi-linearly over the grid face. Finding the solution points where $\nabla|\omega| = 0$ requires solving a pair of quadratic equations derived from the bi-linear interpolation function. The authors also suggest using two thresholds in order to eliminate some of the weaker vortex centers. The first threshold eliminates cell faces with low $|\omega|$, and the second threshold eliminates cell faces whose normal may be misaligned with ω. This method

was successfully applied to distinguish individual vortices in the delta-wing dataset (primary, secondary, and tertiary vortices) and the V-22 tiltrotor blades dataset (tip and root vortices from each rotor blade).

14.3.8 Streamline Methods

Sadarjoen et al. [11] proposed an efficient algorithm for detecting vortices using the winding-angle method. This technique was first proposed by Portela [2] in a mathematically rigorous but computationally expensive fashion. Essentially, given a 2D streamline, the winding angle measures the amount of rotation of the streamline with respect to a point. Sadarjoen et al. [11,19,20] simplified the definition and proposed an efficient algorithm for extracting 2D vortices based on it. By their definition, the winding angle α_ω of a streamline is a measure of the cumulative change of direction of streamline segments.

$$\alpha_\omega = \sum_{i=1}^{N-2} \angle(\mathbf{p}_{i-1}, \mathbf{p}_i, \mathbf{p}_{i+1}) \qquad (14.7)$$

In Equation 14.7, $\mathbf{p}_i$ are the N streampoints of the streamline and $\angle(\mathbf{p}_{i-1}, \mathbf{p}_i \cdot \mathbf{p}_{i+1})$ measures the signed angle between the two line segments delimited by $\mathbf{p}_{i-1}, \mathbf{p}_i$, and $\mathbf{p}_{i+1}$; counterclockwise rotation is positive and clockwise rotation is negative. Therefore, a vortex exists in a region where $\alpha_\omega \geq 2\pi$ for at least one streamline. For slowly rotating vortices, the 2π winding criterion can be relaxed appropriately. An outline of the method is given in Algorithm 14.5.

Once the winding streamlines are marked, a clustering algorithm, based on the distance between center point and cluster, is used to group the streamlines that belong to the same vortex. The location of each cluster is taken to be the location of the vortex core. Various attributes of the vortex, such as shape and orientation, are used to quantitatively visualize the vortices. Fig. 14.4 depicts the results when the method is applied to a slice of the tapered-cylinder dataset. Elliptical icons are used to represent the shape of the extracted vortices, and the two colors (green and red) are used to represent the two different orientations.

Yet another streamline method is the curvature density center method for locating vortex cores in 2D flow fields [11,19,20]. Pagendarm et al. [21] extended this method for 3D flow fields. The underlying assumption behind this approach is that the center of curvature for each point on a winding streamline should form a tight cluster, within which the local maximum is identified as the vortex core. By our computing the curvature center at each sample point throughout the domain, a density field is formed whose peaks are the locations of vortex cores. As pointed out in [11,19,20], this approach lacks the robustness to work well for noncircular flows, such as the elliptically shaped vortices illustrated in Fig. 14.4.

14.3.9   Combinatorial Method

Jiang et al. [12] presented a method for extracting vortex core regions based on ideas from combinatorial topology. In this approach, a combinatorial labeling scheme based on *Sperner's Lemma* is applied to the velocity vector field in order to identify centers of swirling flows. The origin of Sperner's Lemma lies in the *fixed-point theory* of combinatorial topology. The connection between vortices and fixed points (i.e., critical points) is well known

```
 1:  select an initial set of seed points
 2:  for all seed points do
 3:     trace its streamline and compute αω
 4:     if |αω| ≥ 2π and initial point is near end point then
 5:        mark streamline as winding
 6:     end if
 7:  end for
 8:  for all winding streamlines do
 9:     compute its center point c (geometric mean)
10:     if c ∉ vortex clusters then
11:        add c to vortex clusters
12:     end if
13:  end for
```

Algorithm 14.5   Winding-angle method.

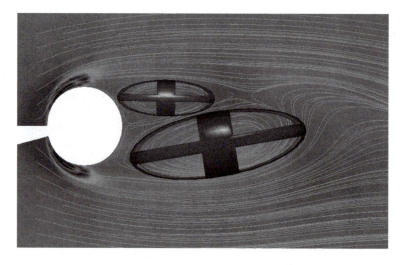

Figure 14.4   Winding-angle method. Image courtesy of I. Ari Sadarjoen. (See also color insert.)

[22,23]. Whereas Sperner's Lemma labels the vertices of a simplicial complex and identifies the fixed points of the labeled subdivision, the proposed method labels the velocity vectors at grid nodes and identifies the grid cells that are most likely to contain critical points.

Each velocity vector v is labeled according to the direction range in which it points. It is sufficient to examine the surrounding nodes of a grid cell for the existence of revolving velocity vectors. The number of direction ranges corresponds to the number of surrounding nodes. (For a quadrilateral mesh, there are four direction ranges, each spanning 90°.) For 2D flow fields, a grid cell belongs to a vortex core region if each of the four velocity vectors from the surrounding nodes points in a unique direction range, or satisfies the direction-spanning criterion. For 3D flow fields, it is necessary to approximate the local swirling plane at each grid cell and then project the surrounding velocity vectors onto this plane. An outline of the 3D algorithm is given in Algorithm 14.6.

The authors use a simple region growth algorithm along with Algorithm 14.6 in order to segment the individual vortex core regions.

What makes this method effective is its insensitivity to approximations to the local swirl plane normal **n**. Fig. 14.5 shows the results from this method on the blunt fin dataset. The yellow regions are detected vortex core regions, visualized using isosurfaces. The blue lines are the streamlines seeded near the detected vortex cores, and they serve to demonstrate the success of this approach by showing that the detected vortex cores actually lie in the center of the swirling flow. However, this approach can produce false positives [24].

14.4   Swirling Flow Verification

The main deficiency common to all these detection algorithms is not the false positives they may produce, but rather their inability to automatically distinguish between the false positives and the actual vortices. Imprecise vortex definitions and numerical artifacts are just two of the reasons these false positives occur. The fundamental problem is that most detection algorithms employ local operators (e.g., velocity-gradient tensor **J**) for detecting global features. As pointed out by Thompson et al. [25], these

```
 1:  for all grid cells do
 2:     compute swirl plane normal n at cell center
 3:     project v from surrounding nodes
 4:     for all v_p in swirl plane do
 5:        compute its angle α from local x-axis
 6:        label direction range for α
 7:     end for
 8:     if all direction ranges are labeled then
 9:        mark grid cell as vortex core
10:     end if
11:  end for
```

Algorithm 14.6   Combinatorial method.

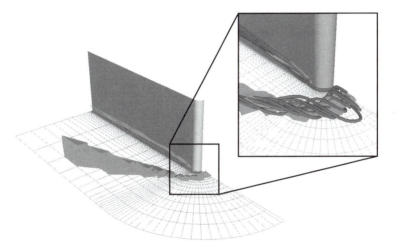

Figure 14.5   Combinatorial method (©2002 IEEE). (See also color insert.)

local operators are problematic because they do not incorporate the necessary global information into the detection process.

The most direct approach for verifying if a candidate feature is indeed a vortex is by visual inspection. The primary problem with this approach is that it requires human intervention, a process that is contrary to the automatic nature of the detection algorithms. The geometric verification algorithm proposed by Jiang et al. [24] addresses this issue by automating the verification process. By identifying the swirling streamlines surrounding a candidate vortex core, the

verification algorithm can arbitrate the presence or absence of the vortex most consistent with visual scrutiny.

As a postprocessing step, the verification algorithm can work with any detection algorithm. Given a candidate vortex core, the goal is to identify the swirling streamlines surrounding it by using various differential geometry properties of the streamlines. The algorithm was designed for 3D flow fields; in the 2D case, using the winding angle method discussed in Section 14.3.8 to verify planar swirling streamlines is sufficient. Identifying 3D swirling

```
 1:  uniformly distribute seed points at start position
 2:  for all seed points do
 3:    for i = 0 to N do
 4:      trace next streampoint
 5:      compute tangent vector t and probe vector
 6:      probe vortex core for swirl plane normal n
 7:      align n to z-axis and save transformation
 8:      apply transformation to t → tₐ
 9:      project tₐ on (x,y)-plane → tₚ
10:      if ∠(t_p^0, t_p^i) ≥ 2π then
11:        accept candidate vortex core
12:      end if
13:    end for
14:  end for
```

Algorithm 14.7   Geometric verification algorithm.

streamlines is nontrivial, since vortices can bend and twist in various ways. An outline of the verification algorithm for a candidate vortex core is given in Algorithm 14.7.

The verification algorithm begins by locating the upstream extent (tip) of the candidate vortex core. For candidate core lines, this is trivial; for candidate core regions, Jiang et al. [24] proposed a bounding box heuristic. The initial position is the tip of the candidate vortex core. Seed points are distributed uniformly on a circle in the swirl plane at the start position. Once the projected tangent vectors complete a full revolution around the z-axis in the (x,y)-plane (i.e., satisfy the 2π swirling criterion), the candidate vortex core is accepted as an actual vortex core.

Fig. 14.6 depicts the results for the delta-wing dataset. In the left image, the yellow regions are actual vortex cores and the green regions are false positives, artifacts from the combinatorial method. The middle image depicts the swirling streamlines surrounding the verified vortex cores. The right image shows the manner in which Algorithm 14.7 confirms that the identified candidate is indeed a vortex core. The cyan arrows represent the tangent vectors and the orange arrows represent the probe vectors. The bottom image on the right illustrates the projected tangent vectors revolving in the (x,y)-plane.

14.5   Visualization of Vortices

Methods used to visualize vortices are inextricably linked to the manner in which the vortices are detected. For example, line-based algorithms produce results that can best be visualized as line segments, as shown in Fig. 14.2. In contrast, results generated by region-type algorithms can best be visualized using color maps or isosurfaces, as shown in Fig. 14.1. Additionally, iconic representations, such as the elliptical icons shown in Fig. 14.4, can be used to quantitatively visualize various attributes of vortices.

By our seeding the streamlines near vortex cores, the swirling patterns that are generally associated with vortices can be visualized. This is one of the primary techniques used to ascertain the accuracy of detected results, either manually or automatically (see Section 11.4). Fig. 11.7 illustrates how some of the pioneers in this field leverage this technique to validate or invalidate results from detection algorithms. The top left image illustrates the Pacific Ocean dataset where streamlines (cyan lines) are seeded throughout the domain to show regions of winding streamlines. The intent [11] was to demonstrate the ineffectiveness of the curvature center density method. The density peaks (grey isosurfaces) do not correspond well to the

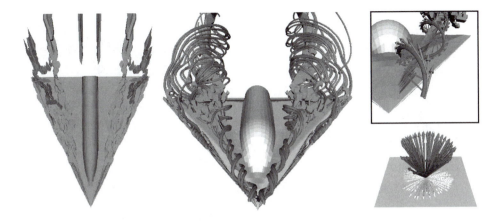

Figure 14.6   Geometric verification (© 2002 IEEE). (See also color insert.)

winding streamlines. The top right image depicts the vortical flow in the blunt-fin dataset. Vortex core lines (white lines) were extracted using the parallel-vectors method. In this case [9], the intent was to demonstrate the effectiveness of their method for extracting vortex core lines that correspond exactly to the center of swirling streamlines (black lines).

Besides seeding streamlines, the cutting plane technique is also an effective method to visualize vortices. Each cutting plane takes a sample slice of the dataset along a certain direction, and the visualization method can be isocontours of a scalar quantity or line integral convolution (LIC) [26] of velocity vectors. The bottom image of Fig. 14.7 depicts the wing-tip dataset where vortex core lines (red line segments) were extracted using the eigenvector method. Sample slices were taken along the detected vortex core [16] to demonstrate the correspondence between the isocontours and the extracted core line.

14.6   Conclusion

Throughout the past decade, there has been a steady stream of scholarly work on the subject of vortex detection. We presented an overview of nine detection algorithms that are representative of the state of the art. Each detection algorithm is classified based on how it defines

a vortex, whether it is Galilean invariant, and the local or global nature of its identification process. Although many of the algorithms share similarities, each has its own advantages and disadvantages. A recently developed verification algorithm that can be used in conjunction with any detection method, was also overviewed, as were various techniques for visualizing detected vortices.

Although much progress has been made towards detecting vortices in steady flow fields, there is still a paucity of methods that can do the same in unsteady (time-varying) flow fields. None of the detection methods described in this paper can adequately address all of the issues unique to unsteady vortical flows. A major challenge will be to develop efficient and robust vortex-detection and vortex-tracking algorithms for unsteady flow fields.

Acknowledgments

This work is partially funded by the National Science Foundation under the Large Data and Scientific Software Visualization Program (ACI-9982344), the Information Technology Research Program (ACS-0085969), an NSF Early Career Award (ACI-9734483), and a grant from the US Army Research Office (DAA-D19-00-1-0155).

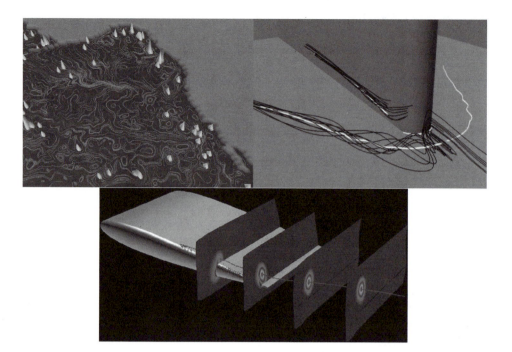

Figure 14.7 Visualization of vortices (© 1998 IEEE). Top left image courtesy of I. Ari Sadarjoen, Delft University of Technology. Top right image courtesy of Martin Roth, Swiss Federal Institute of Technology. (See also color insert.)

References

1. H. J. Lugt. *Vortex flow in nature and technology*. New York, Wiley, 1972.

2. L. M. Portela. Identification and characterization of vortices in the turbulent boundary layer. PhD thesis, Stanford University, 1997.

3. S. K. Robinson. Coherent motions in the turbulent boundary layer. *Ann. Rev. Fluid Mechanics*, 23:601–639, 1991.

4. Y. Levy, D. Degani, and A. Seginer. Graphical visualization of vortical flows by means of helicity. *AIAA J.*, 28(8):1347–1352, 1990.

5. C. H. Berdahl and D. S. Thompson. Education of swirling structure using the velocity gradient tensor. *AIAA J.*, 31(1):97–103, 1993.

6. J. Jeong and F. Hussain. On the identification of a vortex. *J. Fluid Mechanics*, 285:69–94, 1995.

7. D. C. Banks and B. A. Singer. A predictor–corrector technique for visualizing unsteady flow. *IEEE Trans. on Visualization and Computer Graphics*, 1(2):151–163, 1995.

8. D. Sujudi and R. Haimes. Identification of swirling flow in 3D vector fields. In *AIAA 12th Computational Fluid Dynamics Conference, Paper 95–1715*, 1995.

9. M. Roth and R. Peikert. A higher-order method for finding vortex core lines. In *IEEE Visualization '98*, pages 143–150, 1998.

10. R. C. Strawn, D. N. Kenwright, and J. Ahmad. Computer visualization of vortex wake systems. *AIAA J.*, 37(4):511–512, 1999.

11. I. A. Sadarjoen, F. H. Post, B. Ma, D. C. Banks, and H.-G. Pagendarm. Selective visualization of vortices in hydrodynamic flows. In *IEEE Visualization '98*, pages 419–422, 1998.

12. M. Jiang, R. Machiraju, and D. S. Thompson. A novel approach to vortex core region detection. In *Joint Eurographics–IEEE TCVG Symposium on Visualization*, pages 217–225, 2002.

13. M. Roth. Automatic extraction of vortex core lines and other line-type features for scientific visualization. PhD thesis, Swiss Federal Institute of Technology Zürich, 2000.

14. D. C. Banks and B. A. Singer. Vortex tubes in turbulent flows: identification, representation and reconstruction. In *IEEE Visualization '94*, pages 132–139, 1994.

15. D. N. Kenwright and R. Haimes. Vortex identification applications in aerodynamics. In *IEEE Visualization '97*, pages 413–416, 1997.

16. D. N. Kenwright and R. Haimes. Automatic vortex core detection. *IEEE Computer Graphics and Applications*, 18(4):70–74, 1998.

17. R. Haimes and D. N. Kenwright. On the velocity gradient tensor and fluid feature extraction. In *AIAA 14th Computational Fluid Dynamics Conference*, Paper 99–3288, 1999.

18. R. Peikert and M. Roth. The "parallel vectors" operator—a vector field visualization primitive. In *IEEE Visualization '99*, pages 263–270, 1999.

19. I. A. Sadarjoen and F. H. Post. Geometric methods for vortex extraction. In *Joint Eurographics–IEEE TCVG Symposium on Visualization*, pages 53–62, 1999.

20. I. A. Sadarjoen. *Extraction and visualization of geometries in fluid flow fields*. PhD thesis, Delft University of Technology, 1999.

21. H.-G. Pagendarm, B. Henne, and M. Rütten. Detecting vortical phenomena in vector data by medium-scale correlation. In *IEEE Visualization '99*, pages 409–412, 1999.

22. M. S. Chong, A. E. Perry, and B. J. Cantwell. A general classification of 3D flow fields. *Phys. Fluids, A*, 2(5):765–777, 1990.

23. A. E. Perry and M. S. Chong. A description of eddying motions and flow patterns using critical point concepts. *Ann. Rev. Fluid Mechanics*, 19:125–155, 1987.

24. M. Jiang, R. Machiraju, and D. S. Thompson. Geometric verification of swirling features in flow fields. In *IEEE Visualization '02*, pages 307–314, 2002.

25. D. S. Thompson, R. Machiraju, M. Jiang, J. Nair, G. Craciun, and S. Venkata. Physics-based feature mining for large data exploration. *IEEE Computing in Science & Engineering*, 4(4):22–30, 2002.

26. V. Verma, D. Kao, and A. Pang. PLIC: Bridging the gap between streamlines and LIC. In *IEEE Visualization '99*, pages 341–351, 1999.

PART V
Tensor Field Visualization

15 Oriented Tensor Reconstruction

LEONID ZHUKOV and ALAN H. BARR
Department of Computer Science
California Institute of Technology

15.1 Introduction

Directional tracking through vector fields has been a widely explored topic in visualization and computer graphics [5,19,20]. The standard streamline technique advects massless particles through the vector field and traces their location as a function of time. Analogously, a hyper-streamlines approach has been proposed to trace changes through tensor fields, following the dominant eigenvector direction [7]. These methods work best on "clean" datasets, which are usually produced as a result of simulations; these methods typically do not handle raw experimental data very well, due to noise and resolution issues.

Recently, attention has been given to the visualization of 2D [12] and 3D [10] diffusion tensor fields from DT-MRI data. Although these methods provide significant visual cues, they do not attempt to recover the underlying anatomical structures, which are the white matter fiber tracts (bundles of axons) found within the brain. (The white matter constitutes the "wiring" of the brain; the grey matter constitutes the computational components of the brain.)

Several previous endeavors have been made to recover the underlying structure by extracting fibers through the application of modified streamline algorithms. Examples include tensor lines [21] and streamtubes [6,24]. A direct fiber-tractography method has been developed [2].

Other work suggests separate regularization of eigenvalues and eigenvectors in the tensor fields before fiber tracing [14]. Another method uses level-sets with the front propagation approach [16]. These algorithms have had some success in recovering the underlying structures, but some problems still remain due to the complexity of the tensor field, voxelization effects, and the significant amount of noise that is omnipresent in experimental data. Recent work has concentrated on deriving a continuous tensor field approximation [15] and using signal-processing techniques (for example, Kalman filtering [8]) to clean up the data.

The goal of this chapter is to develop a stable tensor-tracing technique, which will allow the extraction of the underlying continuous anatomical structures from experimental diffusion tensor data. The proposed technique uses a moving local regularizing filter that allows the tracing algorithm to cross noisy regions and gaps in the data while preserving directional consistency.

This chapter is based on the results first presented in an IEEE Visualization 2002 paper by Zhukov and Barr [26].

15.2 Method

15.2.1 Diffusion Tensors

Diffusion-tensor magnetic resonance imaging (DT-MRI) [1] is a technique used to measure

Figure 15.1 Human brain pathways recovered from DT-MRI data using the oriented tensor reconstruction algorithm.

the anisotropic diffusion properties of the water molecules found within biological tissues as a function of the spatial position within the sample. Due to differing cell shape and cell-membrane properties, the diffusion rates of the water molecules are different in different directions and locations.

For instance, neural fibers are comprised mostly of bundles of long cylindrical cells that are filled with fluid and are bounded by less-water-permeable cell membranes. The average diffusion rate (at a spatial location) is fastest in the 3D axis direction along the length of the neuron cells, since more of the water molecules are free to move in this direction. The average diffusion rate is slowest in the two transverse directions, where the cell membrane interferes, reducing and slowing down the movement of the water molecules.

Other parts of the brain, such as the ventricles, are composed primarily of fluid without cell membranes. Here the average diffusion rate is larger and more uniform (almost the same in all directions).

The diffusion properties can be represented with a symmetric second-order tensor -3×3 matrix:

$$\mathbf{T} = \begin{pmatrix} T^{xx} & T^{xy} & T^{xz} \\ T^{yx} & T^{yy} & T^{yz} \\ T^{zx} & T^{zy} & T^{zz} \end{pmatrix} \tag{15.1}$$

The six independent values (the tensor is symmetric) of the tensor elements vary continuously with spatial location.

The 3D local axis direction of the neuron fibers will correspond to the dominant eigenvector of the tensor. There should be one large eigenvalue and two small eigenvalues. This can be seen from the physical interpretation of the diffusion tensor, which can be thought of as a vector-valued function whose input is the local 3D concentration gradient and whose output is the 3D directional vector flux of the water molecules. (Vector flux measures a quantity per unit area per time, in the direction perpendicular to the area.) The function is evaluated by multiplying the 3×3 matrix by the 3×1 concentration gradient, producing the 3×1 vector flux of the water molecules. Water will diffuse fastest in the direction along the axis of the neurons and slowest in the two transverse directions.

For the ventricles, a dominant eigenvector should not exist: the three eigenvalues of the tensor should have roughly the same value.

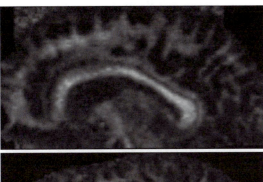

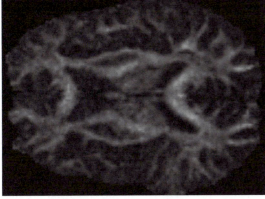

Figure 15.2 Sagittal and axial slices of anisotropy measure c_ℓ of the dataset. The lighter regions correspond to stronger anisotropy areas found in the white matter. See Equation 15.4.

Water will diffuse roughly at the same speed in all directions. Hence, we can use the diffusion tensor to distinguish tissues with a primary diffusion axis from parts that do not.

In this chapter, the experimental dataset contains sampled values of the diffusion tensor on a regularly spaced grid of $121 \times 88 \times 60$ (cubic) voxels. We will denote these given tensor values as $\mathbf{T}_{ijk}^{\alpha\beta}$, where α and β are the 3D tensor components $\{xx, xy, \ldots, zz\}$, and i, j, k are traditional integer indices into the regular grid volume. Also, when no upper indices are provided, the operations are assumed to be performed on the entire tensor component-wise $\mathbf{T} \equiv \mathbf{T}^{\alpha\beta}$, i.e., on each of the six independent values of the tensor.

15.2.2   Tensor Classification

Geometrically, a diffusion tensor can be thought of as an ellipsoid with its three axes oriented along the tensor's three perpendicular eigenvectors and semi-axis lengths proportional to the square root of eigenvalues of the tensor – mean diffusion distances [1].

In general, eigenvalues λ and eigenvectors $\mathbf{e}$ can be found as a solution to the eigen-equation

$$\mathbf{T}\mathbf{e}_i = \lambda_i \mathbf{e}_i \tag{15.2}$$

Since the tensor is symmetric, its eigenvalues are always real numbers, and the eigenvectors are orthogonal and form a Cartesian vector basis $\{\mathbf{e}_1, \mathbf{e}_2, \mathbf{e}_3\}$. This basis (frame of reference) can be used to represent the tensor in diagonal form and to specify directions with respect to the "world coordinate" system

$$\mathbf{T} = \{\mathbf{e}_1, \mathbf{e}_2, \mathbf{e}_3\} \begin{pmatrix} \lambda_1 & 0 & 0 \\ 0 & \lambda_2 & 0 \\ 0 & 0 & \lambda_3 \end{pmatrix} (\mathbf{e}_1, \mathbf{e}_2, \mathbf{e}_3)^T$$

$$\tag{15.3}$$

Using the ellipsoidal interpretation, one can classify the diffusion properties of tissue according to the shape of the ellipsoids, with extended ellipsoids corresponding to regions with strong linear diffusion (long, thin cells), flat ellipsoids to planar diffusion, and spherical ellipsoids to regions of isotropic media (such as fluid-filled regions like the ventricles). The quantitative classification can be done through the coefficients c_ℓ, c_p, c_s (linear, planar, spherical) first proposed in [22,23]:

$$c_\ell = \frac{\lambda_1 - \lambda_2}{\lambda_1 + \lambda_2 + \lambda_3} \tag{15.4}$$

$$c_p = \frac{2(\lambda_2 - \lambda_3)}{\lambda_1 + \lambda_2 + \lambda_3} \tag{15.5}$$

$$c_s = \frac{3\lambda_3}{\lambda_1 + \lambda_2 + \lambda_3} \tag{15.6}$$

These coefficients are normalized to the range of $[0..1]$ and could be interpreted as barycentric coordinates. For example, values of c_ℓ close to 1 choose the regions with strong linear ($\lambda_1 \gg \lambda_2 \approx \lambda_3$) diffusion.

15.2.3   Data Interpolation

We start by reconstructing a continuous tensor field in the volume through tri-linear interpol-

Figure 15.3   Comparison of nonfiltered (left) and MLS-filtered (right) fibers. Note the smoother and more regular behavior of the filtered fibers on the right image. (See also color insert.)

ation. In this scheme, the value of a tensor at any point inside the voxel is a linear combination of the eight values at its corners and is completely determined by them. Since the coefficients of this linear combination are independent of the tensor indices, the linear combination of the tensors can be done component-wise.

$$
\begin{aligned}
\mathbf{T}(x, y, z) = {} & \mathbf{T}_{ijk}(1 - x)(1 - y)(1 - z) \\
& + \mathbf{T}_{i+1,jk}\, x(1 - y)(1 - z) + \mathbf{T}_{i,j+1,k}(1 - x)y(1 - z) \\
& + \mathbf{T}_{ij,k+1}(1 - x)(1 - y)z + \mathbf{T}_{i+1,k+1}x(1 - y)z \\
& + \mathbf{T}_{i,j+1,k+1}(1 - x)yz + \mathbf{T}_{i+1,j+1,k}xy(1 - z) \\
& + \mathbf{T}_{i+1,j+1,k+1}xyz
\end{aligned}
$$

$$(15.7)$$

We can use tri-linear component-wise interpolation because symmetric tensors form a linear subspace in the tensor space: any linear combination of symmetric tensors remains a symmetric tensor, i.e., symmetric tensors are closed under linear combination (the manifold of symmetric tensors is not left). Component-wise interpolation is sufficient for our purposes; more sophisticated interpolation methods, however, would better preserve the eigenvalues along an interpolation path [14].

On the other hand, component-wise interpolation of eigenvectors and eigenvalues themselves would not lead to correct results, since a linear interpolation between two unit vectors is

not a unit vector anymore—the interpolated eigenvector value would leave the manifold of unit vectors. In addition, there can be a correspondence problem in the order of the eigenvalues.

Various types of tensor interpolation are discussed by, for example, Kindlmann et al. [11].

15.2.4   Regularization: Moving Least Squares

To perform a stable fiber tracing on experimental data, the data needs to be filtered. A simple global box or Gaussian filter will not work well, since it will blur (destroy) most of the directional information in the data. We want the filter to be adjustable to the data and we want to be able to put more weight on the data in the direction of the traced fiber, rather than between fibers. We also want the filter to have an adjustable local support that could be modified according to the measure of the confidence level in the data. Finally, we want the filter to preserve sharp features (ridges) where they exist but eliminate irrelevant noise if it can. Thus, the behavior of the filter at some voxel in space should depend on the "history of the tracing" (where it came from), so the filtering needs to be tightly coupled with the fiber-tracing process.

Due to these reasons, we chose to use a moving least-squares (MLS) approach. The idea behind this local regularization method is to find a low-degree polynomial that best fits the data, in the least-squares sense, in the small region around the point of interest. Then we replace the data value at that point by the value of the polynomial at that point $\mathbf{T}(x_p, y_p, z_p) \rightarrow \bar{\mathbf{T}}_p$. Thus, this is a *data approximation* rather than an interpolation method. The measure of "fitness" will depend on the filter location, orientation, and history of motion. The 1D MLS method was first introduced in signal-processing literature [9,17].

We start the derivation by writing a functional ε that measures the quality of the polynomial fit to the data in the world coordinate system. To produce ε, we integrate the squared difference between $\mathbf{F}$, which is an unknown linear combination of tensor basis functions, and $\mathbf{T}$, the known continuous tri-linear interpolated version of given tensor data. We integrate over all of the 3D space and use weighting function G to create a region of interest centered at chosen 3D point $\mathbf{r}_p$ with coordinates (x_p, y_p, z_p):

$$\varepsilon(\mathbf{r}_p) = \int_{-\infty}^{\infty} G(\mathbf{r} - \mathbf{r}_p; \mathbf{T}_p)[\mathbf{F}(\mathbf{r} - \mathbf{r}_p) - \mathbf{T}(\mathbf{r})]^2 d\mathbf{r}^3$$

(15.8)

The second argument $\mathbf{T}_p$ (value of the tensor at the center point) of the weighting function G determines the weighting function's size and orientation.

The square of the tensor difference in Equation 15.8 is a scalar, defined by the double-dot (component-wise) product [3,4]:

$$(\mathbf{F} - \mathbf{T})^2 = (\mathbf{F} - \mathbf{T}) : (\mathbf{F} - \mathbf{T})^T =$$
$$\sum_{\alpha\beta} (\mathbf{F}^{\alpha\beta} - \mathbf{T}^{\alpha\beta})(\mathbf{F}^{\beta\alpha} - \mathbf{T}^{\beta\alpha}) = \sum_{\alpha\beta} (\mathbf{F}^{\alpha\beta} - \mathbf{T}^{\alpha\beta})^2$$

(15.9)

Within the functional ε, the tensor function $\mathbf{F}$ is a linear combination of tensor basis functions that we will use to fit the data. The function G is the moving and rotating anisotropic filtering window, centered at the point $\mathbf{r}_p$ (Fig. 15.4).

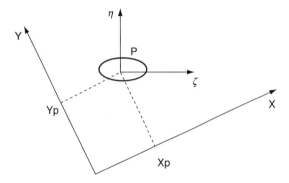

Figure 15.4   Coordinate systems used by linear transformation (Equation 15.10) to change from the $\{x, y, z\}$ coordinates in Equation 15.8 into the $\{\zeta, \eta, \theta\}$ coordinates in Equation 15.11.

We felt it was more convenient to perform the computations in the local frame of reference connected to a particular filtering window than to do so in world coordinates. It is straightforward to transform Equation 15.8 to a local frame of reference using a change of variables involving the following translation and rotation transformation:

$$\begin{pmatrix} \zeta \\ \eta \\ \theta \end{pmatrix} = \mathbf{R}_p^{-1} \begin{pmatrix} x - x_p \\ y - y_p \\ z - z_p \end{pmatrix} = \mathbf{R}_p^{-1}(\mathbf{r} - \mathbf{r}_p) \qquad (15.10)$$

where $\mathbf{R}_p = \{\mathbf{e}_1, \mathbf{e}_2, \mathbf{e}_3\}$ is a rotation matrix formed by the eigenvectors of the tensor $\mathbf{T}_p$ at the point $\mathbf{r}_p$. Then, in the local frame of reference $\{\zeta, \eta, \theta\}$, Equation 15.8 becomes

$$\varepsilon = \int_v [\mathbf{F}(\mathbf{R}_p\{\zeta, \eta, \theta\}) - \mathbf{T}(\mathbf{r}_p + \mathbf{R}_p\{\zeta, \eta, \theta\})]^2$$
$$G(\mathbf{R}_p\{\zeta, \eta, \theta\}; \mathbf{T}_p) d\zeta d\eta d\theta$$

(15.11)

Integration is performed over the parts of space where G has been chosen to be nonzero.

We now instantiate $\mathbf{F}$ and G in the local (rotated and translated) frame of reference. These functions can be thought of as functions of $\{\zeta, \eta, \theta\}$ only, since the rotation matrix $\mathbf{R}_p$ is independent of the integration variables.

For $\mathbf{F}$ we will use a polynomial function of degree N in the variables ζ, η, and θ:

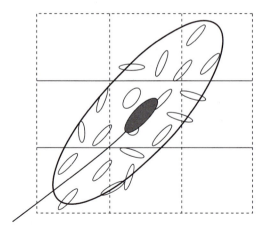

Figure 15.5 Oriented moving least-squares (MLS) tensor filter. The smallest ellipsoids represent the interpolated tensor data; the largest ellipsoid represents the domain of the moving filter $g(\ )$ described in Equation 15.8. The dark ellipsoid represents the computed filtered tensor. The filter travels continuously along the fiber line; the grid shows initial sampling of the tensor data.

$$F(R_p\{\zeta,\eta,\theta\}) \equiv f(A;\zeta,\eta,\theta) \tag{15.12}$$

where

$$f(A;\zeta,\eta,\theta) = \sum_{mnp}^{N} A_{mnp}\zeta^m\eta^n\theta^p \tag{15.13}$$

To instantiate G, we also use a function of variables ζ, η, and θ:

$$G(R_p\{\zeta,\eta,\theta\};T_p) \equiv g(\zeta,\eta,\theta;\lambda_p^i) \tag{15.14}$$

The function g is clipped to zero at some low threshold value to create a finite integration volume, and λ_p^i are eigenvalues of the tensor T_p.

Substituting the expressions in Equations 15.13 and 15.14 into Equation 15.11, we get

$$\varepsilon = \int_v \left[\sum_{mnp} A_{mnp}\zeta^m\eta^n\theta^p - T(r_p + R_p\{\zeta,\eta,\theta\})\right]^2$$
$$g(\zeta,\eta,\theta;\lambda_p^i)d\zeta d\eta d\theta \tag{15.15}$$

The least-squares fitting procedure reduces to minimization of functional ε with respect to

tensor elements $A_{rst}^{\alpha\beta}$. To minimize it, we differentiate Equation 15.15 with respect to each one of the A coefficients and tensor components, equate the result to zero, and linearly solve to find the unknown As:

$$\partial\varepsilon/\partial A_{rst}^{\alpha\beta} = 0 \tag{15.16}$$

This gives us the following linear system for the unknown As:

$$\sum_{mnp} M_{mnp,rst}A_{mnp}^{\alpha\beta} = B_{rst}^{\alpha\beta} \tag{15.17}$$

where elements of the matrix $M_{mnp,rst}$ and right-hand-side values of the system B_{rst} are computed through the following integrals:

$$M_{mnp,rst} = \int_V \zeta^{m+r}\eta^{n+s}\theta^{p+t}g(\zeta,\eta,\theta;\lambda_p^i)\,d\zeta d\eta d\theta \tag{15.18}$$

$$B_{rst}^{\alpha\beta} = \int_V T^{\alpha\beta}(r_p + R_p\{\zeta,\eta,\theta\})\zeta^r\eta^s\theta^t$$
$$g(\zeta,\eta,\theta;\lambda_p^i)d\zeta d\eta d\theta \tag{15.19}$$

These integrals can be evaluated numerically for any specific choice of g. (For Gaussian filter g, the integral in Equation 15.18 can be expanded over the entire domain and evaluated analytically using Gamma functions.)

Equation 15.17 is just a "regular" linear system to find the As. Written out component-wise for the tensors A, B, and M with contracted indexing

$$A_{mnp}^{\alpha\beta} \equiv A_{m+Nn+N^2p}^{\alpha\beta} = a_j^{\alpha\beta}$$
$$B_{rst}^{\alpha\beta} \equiv B_{r+Ns+N^2t}^{\alpha\beta} = b_i^{\alpha\beta} \tag{15.20}$$
$$M_{mnp,rst} \equiv M_{m+Nn+N^2p,\,e+Ns+N^2t} = M_{ij}$$

the system in Equation 15.17 becomes

$$\sum_j M_{ij}a_j^{\alpha\beta} = b_i^{\alpha\beta} \tag{15.21}$$

This type of system is also known as a system of "normal equations" for the least-squares optimization.

The optimization procedure allows us to compute the polynomial coefficients for best approximation of the tensor data within a region of a chosen point by a chosen degree of polynom. Then the tensor value at the point $\mathbf{r}_p$, which is the origin in the $\{\zeta, \eta, \theta\}$ frame of reference, can be easily calculated using Equations 15.13 and 15.20:

$$\bar{\mathbf{T}}_p^{\alpha\beta} = \sum_{mnp} \mathbf{A}_{mnp}^{\alpha\beta} \zeta^m \eta^n \theta^p \big|_{\zeta=\eta=\theta=0} = \mathbf{A}_{000}^{\alpha\beta} = a_0^{\alpha\beta}$$

$$(15.22)$$

It is important to notice that the value of $\mathbf{A}_{000}$ depends on the order of polynomial used for fitting.

We also notice that using a zero-order polynomial approximation (i.e., $N = 0$) is equivalent to finding the weighted average of a tensor function within the filter volume:

$$\bar{\mathbf{T}}_p = \int_V \mathbf{T}(\mathbf{r}_p + \mathbf{R}_p\{\zeta, \eta, \theta\}) g(\zeta, \eta, \theta; \lambda_p^i) d\zeta d\eta d\theta$$

$$(15.23)$$

The major advantage of the higher-order approximation is that it better preserves the absolute magnitude of the features in the areas that have maxima or minima, compared to simple averaging, which tends to lower the height of the maxima and the depth of the minima.

Finally, for the filter function g, we have chosen an anisotropic Gaussian weighting function G with axes aligned along the eigenvector directions and ellipsoidal semi-axes (the radii) proportional to the square root of corresponding eigenvalues.

$$g(\zeta, \eta, \theta; \lambda_p^i) =$$
$$\frac{1}{V} \exp\left(-\left(\zeta/(\sigma a)\right)^2 - \left(\eta/(\sigma b)\right)^2 - \left(\theta/(\sigma c)\right)^2\right)$$

$$(15.24)$$

with

$$a = \sqrt{\lambda_p^1}, \ b = \sqrt{\lambda_p^2}, \ c = \sqrt{\lambda_p^3}$$

$$(15.25)$$

The variable λ_p^1 is the largest eigenvalue of the diffusion tensor $\mathbf{T}_p$ at the location $\mathbf{r}_p$, λ_p^2 is the second largest, etc. The value σ is a parameter that can enlarge or contract all of the ellipsoid radii. It is important to notice that since we are

trying to trace fibers, i.e., to extract structures with very strong directional information, the filter is typically much more influenced by the data points "in front" and "behind" than those on the side. Thus, usually, $a \gg b, c$.

Also note that in Equation 15.14 for the filter function $g(\ )$, we have a choice of values for the diffusion tensor $\mathbf{T}_p$. In our algorithm, we use the filtered tensor value from the previous time step, $\bar{\mathbf{T}}_{p-1}$, to determine the weighting-function ellipsoid to use for the current time-step.

15.2.5   Streamline Integration

The fiber-tract trajectory $\mathbf{s}(\tau)$ can be computed as a parametric 3D curve through linear integration of the filtered principal eigenvector:

$$\mathbf{s}(\tau) = \int_0^\tau \bar{\mathbf{e}}_1(t) dt$$

$$(15.26)$$

where t is a parameter of the curve and has corresponding $t = t(x, y, z)$ values and $\bar{\mathbf{e}}_1$ is the MLS-filtered principal direction (unit) eigenvector as a function of position.

$$\bar{\mathbf{T}}\bar{\mathbf{e}}_1 = \bar{\lambda}_1 \bar{\mathbf{e}}_1$$

$$(15.27)$$

The discrete integration can be done numerically using explicit or implicit methods depending on the converging/diverging nature of the

Figure 15.6 Height plot for anisotropy measure ("mountain" function) described in Section 15.2.6 for an axial slice of the data. The higher portions correspond to stronger anisotropy. See Equation 15.30. (See also color insert.)

tensor field. The simplest approaches are forward (for diverging fiber fields):

$$\mathbf{r}_{new} = \mathbf{r}_{old} + \bar{\mathbf{e}}_1[\bar{\mathbf{T}}(\mathbf{r}_{old})]\Delta t \qquad (15.28)$$

or inverse Euler schemes (for converging fiber fields):

$$\mathbf{r}_{new} = \mathbf{r}_{old} + \bar{\mathbf{e}}_1[\bar{\mathbf{T}}(\mathbf{r}_{new})]\Delta t \qquad (15.29)$$

One can easily employ higher-order integration schemes, but they should still be chosen according to the local properties of the tensor field (converging or diverging) that are associated with the "stiffness" of the differential equation, bifurcations, and the desired geometry.

15.2.6 The "Mountain" Function

For the continuous tensor field, we use an anisotropy measure height function $c_\ell(x, y, z)$, defined using a continuous version of Equation 15.4

$$c_\ell(x, y, z) = \frac{\lambda_1 - \lambda_2}{\lambda_1 + \lambda_2 + \lambda_3} \qquad (15.30)$$

where λ_i are eigenvalues of $T(x, y, z)$. Metaphorically, we call this a "mountain function" because we initiate the fibers at the high points—the peaks of the mountain (the most highly directional portions of a region)—and grow them following the major eigenvector directions. The metaphor continues as the anisotropy measure decreases; we let the fibers grow until they go "under water" into the lakes (corresponding to a chosen lower value for the anisotropy measure); the low anisotropy values indicate an absence of fibers.

15.2.7 Fiber-Tracing Algorithm

The algorithm starts when the user selects a rectangular starting region. The fibers are traced starting only from the points where the anisotropy measure is bigger than the threshold, i.e., points that are high enough on the mountainside.

The initial direction will be determined by the "largest" eigenvector of locally filtered tensor

```
main
    for each P ∈ Seed − region
                ⎧ Tp = filter(T, P, sphere);
                ⎪ cl = anisotropy(Tp);
                ⎪ if (cl > eps)
         do ⎨                ⎧ e1 = direction(Tp);
                ⎪         then ⎨ trace1 = fibertrace(P, e1);
                ⎪                ⎪ trace2 = fibertrace(P, − e1);
                ⎩                ⎩ trace = trace1 + trace2;
    procedure FIBERTRACE (P, e)
        trace− > add(P);
        cl = anisotropy(Tp)
        while (cl > eps)
                ⎧ Pn = integrate_forward(P, e1, dt);
                ⎪ Tp = filter(T, Pn, ellipsoid, e1);
                ⎪ cl = anisotropy(Tp)
         do ⎨ if (cl > eps)                return(trace)
                ⎪                ⎧ trace− > add(Pn);
                ⎪         then ⎨ P = Pn;
                ⎩                ⎩ e1 = direction(Tp);
```

Algorithm 15.1 Fiber trace (seed region).

field. At this point the filter is not oriented. The tracing will proceed in two opposite directions along the "largest" eigenvector.

The tracing procedure integrates forward from the provided initial point and initial direction using the forward or the inverse Euler method. It then computes a filtered value of the tensor at the new point using the oriented filter (orientation and width of the filter are determined from the previous position: the filter is oriented along the "largest" eigenvector and is shaped according to the eigenvalues, with largest semi-axis along the "largest" eigenvector). If the anisotropy of the new point is greater than threshold value, the point is accepted and the tracing continues; otherwise, the tracing is finished. The tracing routine also chooses the direction of tracing consistent with previous steps— no turn of more than 90° is allowed.

We have also incorporated some simple mechanisms to ignore very short fibers and to stop tracing when the length of the fiber exceeds an allowed limit. The starting points are usually generated on a grid within user-defined regions. We use numerical integration to evaluate the integrals in Equations 15.18 and 15.19 inside the filter. We use SVD and LU factorization routines from the "Numerical Recipes" [17] to compute eigenvalues and eigenvectors in Equation 15.27 and solve the linear system (15.21). Evaluation of the tensor function **T** at the center of the filter (origin) requires only the first coefficient of the polynomial expansion in Equation 15.13, so we use only a single back-substitution procedure in LU factorization.

15.3   Discussion

15.3.1   Algorithm Validation

To validate our algorithm, we constructed an artificial tensor dataset that emulates a pair of "wound" fiber bundles (Fig. 15.7). We derived parametric equations that describe the bundle directions and control the size of the features (twist). We constructed a 3D tensor field by sampling the directional derivatives of the

Figure 15.7   Double helix reconstructed using MLS method from artificial tensor data. (See also color insert.)

bundle on a regular 3D grid. We tested various combinations of sampling and reconstruction parameters.

Accurate continuous reconstruction of the double helix was achieved when the sampling was at least one-third of the characteristic length (the scale of change); the integration step in reconstruction was one-fifth of the voxel size, and the MLS filter had a radius of two to three voxels. Increasing the radius of the reconstruction filter leads to oversmoothing and loss of features.

15.3.2   Human Brain Structure

The human brain structures consist of three physically different tissue types and materials: white matter, grey matter, and cerebrospinal fluid (CSF). A wealth of blood vessels permeate the brain tissue, to continually supply it with needed oxygen and nutrients.

CSF has the simplest microstructure. It consists primarily of cell-free fluid, contained within a few large hollow chambers in the brain called ventricles. The water molecules diffuse rapidly in all directions within the CSF.

Grey matter serves as the "computational" part of the brain. It consists of dense irregular groupings of structures, including cell bodies and many interdigitating and unaligned dendrites from many different nerve cells. In the grey matter, water molecules diffuse slowly in all directions because the water molecules are impeded in virtually all directions by the membranes of the nerve cells.

White matter serves as the "wiring" of the brain, and it allows neural signals to be communicated across long distances, from one part of the brain to another. In the cerebral cortex, there are about $3 \cdot 10^{10}$ neurons, which are highly interconnected with approximately 10^{14} synapses.

Physically, white matter consists of the axons of many microscopic nerve cells, each with its own cell membrane and coated with an insulating sheath of myelin. Each axon can be viewed as a long, electrically insulated microscopic tube of conductive cellular fluid, with a diameter on the order of approximately a few microns and with a length of a few centimeters to several feet (such as from the neck to the toe). In the white matter, the axons are aligned and are bundled together to create a long macroscopic fiber. Anatomically, the macroscopic white-matter fibers are immediately adjacent to one another and are quite delicate and fragile.

A diffusion-tensor MRI distinguishes between white matter, grey matter, and CSF due to the different microstructures within the tissues and fluids: it measures the movement of water molecules in different preferential directions for a given concentration gradient. Water molecules have difficulty in crossing cell membranes, and it is much easier for them to diffuse to different parts within the same cell than it is for them to cross cell membranes into a different cell.

In white matter, water molecules diffuse in a preferential direction, up and down the length of the fiber, because the axons are aligned. Along the length of the fiber, there are no impeding cell membranes, while there are many membranes separating the cells across the width of the fiber.

Due to these diffusion properties, white matter will be characterized by one large eigenvalue of the diffusion tensor, whose eigenvector is associated with the axial direction of the fiber and two small eigenvalues; the eigenvectors for these should be perpendicular to the direction of the fiber. Grey matter should have three small eigenvalues, while CSF should have three large eigenvalues.

15.3.3 Results

The DT-MRI dataset we used for this chapter has $121 \times 88 \times 60$ voxels, which provides a resolution of roughly $1\,mm^3$, which is sufficient to resolve fiber bundles. We used various orders of polynomial approximation from zero (average within the filter) up to third order (see Equation 15.13) to trace bundles of fibers that correspond to well known anatomical structures.

We present the results of fiber tracing on a human brain in the figures that follow. In these figures, the yellow boxes show the seed region for the fiber-tracking algorithm. Color coding indicates orthogonal directions in the amount of red for X, green for Y, and blue for Z.

White-matter fibers communicate information into and out of the major brain structures:

- The corona radiata (projection fibers) connect the cortical surface (grey matter) to the lower levels of the neural system and are visible in both hemispheres in Figs. 15.8 and 15.9.

- Long association fibers such as the longitudinal fasciculus and the cingulum connect cortical areas in the same hemisphere and are shown in Fig. 15.10.

- The cingulum bundle that connects the frontal lobe with the temporal-lobe regions and lies on top of the ventricle and above the corpus callosum fibers is clearly seen in Fig. 15.12.

- Short association fibers (U-shaped fibers), which are responsible for local communica-

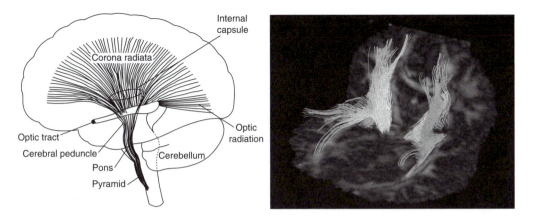

Figure 15.8 Brain structures: corona radiata. (Left) A diagram from Pritchard and Alloway [18]. (Right) The fibers are reconstructed from DT-MRI data using our oriented tensor reconstruction (OTR) algorithm. The corona radiata is visible in both hemispheres.

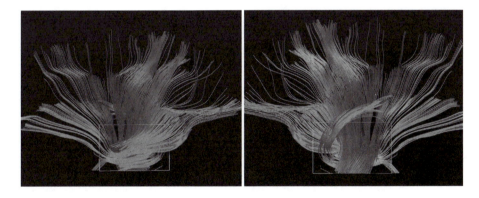

Figure 15.9 Right-hemisphere corona radiata shown from opposite directions. The yellow boxes show the seed region for the OTR fiber-tracing algorithm. Color coding indicates orthogonal directions in the amount of RGB (XYZ). (See also color insert.)

tion between the areas of cortical surface in the same hemisphere, are shown in Fig. 15.11.

- Anterior and posterior forceps of the frontal lobe are shown in Figs. 15.10 and 15.12.

- The corpus callosum, which is a commissural fiber and is a major communication band between the left and right hemispheres, is seen in Fig. 15.13 (also see Fig. 15.1).

Finally, the last figure, Fig. 15.13, combines white-matter fibers with other brain structures—the ventricle, eye sockets, and pockets of CSF on the top of the brain. The left figure shows corpus callosum fibers; the right figure shows corona radiata together with long association fibers. The surface models cover volumes of large uniform diffusion values and were obtained using isosurfacing [13] on isotropic part c_s (see Equation 15.6) of the same DT-MRI dataset. A more detailed discussion on isotropic and anisotropic tissue-model extrac-

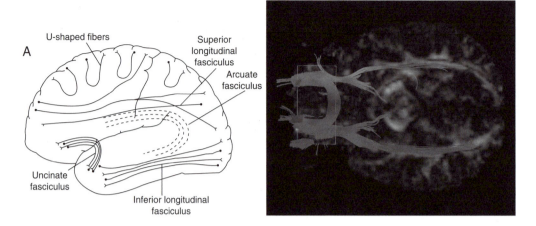

Figure 15.10 Brain structures: association fibers. (Left) A diagram from Pritchard and Alloway [18]. (Right) Longitudinal and uncinate fasciculus of the optic tract. Color coding is the same as for Fig. 15.9. (See also color insert.)

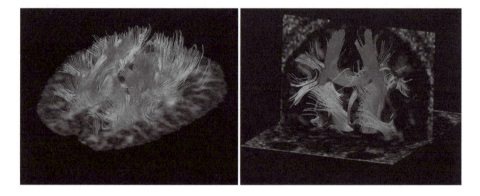

Figure 15.11 Brain structures: Fibers near the cortical surface and U-shaped fibers on the left; U-shaped fibers, parts of corona radiata, and corpus callosum are on the right. (See also color insert.)

tion from DT-MRI data using a combination of isosurfacing and level-set methods is given by Zhukov et al. [25].

15.4 Conclusions

In this paper we developed a new technique for tracing anatomical fibers from 3D DT-MRI tensor fields, recovering identifiable anatomical structures that correspond to many of the white matter brain-fiber pathways.

We found that simple component-wise interpolation of the tensors, forming a crude continuous approximate tensor field, worked well for extracting brain-fiber directions when combined with an MLS filtering approach.

The initial results seem promising; we feel that noninvasive DT-MRI brain-mapping techniques could have many important future diagnostic applications, ranging from research

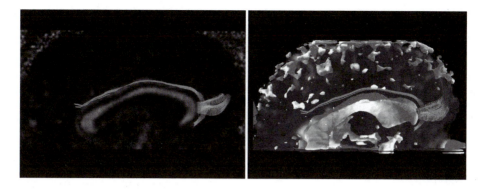

Figure 15.12   Brain structures. (Left) A side view of the right hemisphere cingulum bundle on the background of corresponding c_l anisotropy; (Right) the same structure together with 3d models of the ventricle and CSF extracted by isosurfacing [13] on isotropic part c_s (see Equation 15.6) of the same DT-MRI dataset. (See also color insert.)

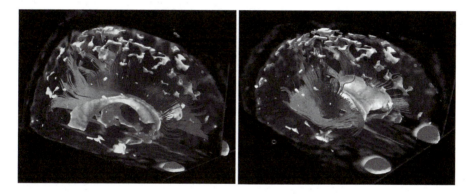

Figure 15.13   Brain structures: corpus callosum (left) and corona radiata (right) shown together with isotropic brain structures—ventricle, eye sockets, and pockets of CSF on the top of the brain. Cutting planes show isotropic c_s values. (See also color insert.)

techniques for determining the wiring in a live human brain to clinical applications for patients with disruption of fiber tracts due to brain injuries such as stroke or physical trauma.

Acknowledgments

We would like to thank Yarden Livnat for the initial discussion, Gordon Kindlmann and SCI Institute for the data, and David Breen for suggestions.

This work was supported by National Science Foundation grants #ASC-89-20219 and #ACI-9982273, and the National Institute on Drug Abuse and the National Institute of Mental Health, as part of the Human Brain Project.

References

1. P. J. Basser, J. Mattiello, and D. LeBihan. MR diffusion tensor spectroscopy and imaging. *Biophysical Journal*, 66:259–267, 1994.
2. P. J. Basser, S. Pajevic, C. Pierpaoli, J. Duda, and A. Aldroubi. *In vivo* fiber tractography using DT-MRI data. *Magnetic Resonance in Medicine*, 44:625–632, 2000.
3. P. J. Basser and C. Pierpaoli. Microstructural and physiological features of tissues elucidated by quantitative-diffusion-tensor MRI. *J. Magn. Res. Ser. B*, 111:209–219, 1996.

4. A. I. Borisenko and I. E. Tarapov. *Vector and Tensor Analysis*. New York, Dover, 1979.

5. B. Cabral and L. Leedom. Imaging vector fields using line integral convolution. In *Proceedings of SIGGRAPH 93*, pages 263–272, 1993.

6. M. J. da Silva, S. Zhang, C. Demiralp, and D. H. Laidlaw. Visualizing diffusion tensor volume differences. In *IEEE Visualization 01 Proceedings, Work in Progress*, 2001.

7. T. Delmarcelle and L. Hesselink. Visualization of second order tensor fields and matrix data. In *IEEE Visualization 92 Proceedings*, pages 316–323, 1992.

8. C. Gossl, L. Fahrmeir, B. Putz, L. M. Auer, and D. P. Auer. Fiber tracking from DTI using linear state space models: detectability of the pyramidal tract. *NeuroImage*, 16:378–388, 2002.

9. R. W. Hamming. *Digital Filters*. Prentice-Hall, 1983.

10. G. Kindlmann and D. Weinstein. Hue-balls and lit-tensors for direct volume rendering of diffusion tensor fields. In *IEEE Visualization '99 Proceedings*, pages 183–189, 1999.

11. G. Kindlmann, D. Weinstein, and D. Hart. Strategies for direct volume rendering of diffusion tensor fields. *IEEE Trans. on Visualization and Computer Graphics*, pages 124–138, 2000.

12. D. H. Laidlaw, E. T. Ahrens, D. Kremers, M. J. Avalos, R. E. Jacobs, and C. Readhead. Visualizing diffusion tensor images of the mouse spinal cord. In *IEEE Visualization '98 Proceedings*, pages 127–134, 1998.

13. W. Lorenson and H. Cline. Marching cubes: a high resolution 3D surface construction algorithm. In *Computer Graphics (Proceedings of SIGGRAPH '87)*, 21:163–169, 1987.

14. O. Coulon, D. C. Alexander, and S. R. Arridge. A regularization scheme for diffusion tensor magnetic resonance images. In *IPMI 2001, XVIth International Conference on Information Processing in Medical Imaging, in Lecture Notes in Computer Science*, 2082:92–105, 2001.

15. S. Pajevic, A. Aldroubi, J. Duda, and P. J. Basser. A continuous tensor field approximation of discrete DT-MRI data for extracting microstructural and architectural features of tissues. *Journ. Magn. Res.*, 154:85–100, 2002.

16. G. J. M. Parker, C. A. M. Wheeler-Kingshott, and G. J. Barker. Distributed anatomical brain connectivity derived from diffusion tensor imaging. In *IPMI 2001, XVIth International Conference on Information Processing in Medical Imaging*, 2082:106–120, 2001.

17. W. H. Press, S. A. Teukolsky, W. T. Vetterling, and B. P. Flannery. *Numerical Recipes in C.* Cambridge, England, Cambridge University Press, 1992.

18. T. C. Pritchard and K. D. Alloway. *Medical Neuroscience*. Fence Creek Publishing, 1999.

19. B. Steve and C. Levit. The virtual wind tunnel: an environment for the exploration of 3D unsteady flows. In *IEEE Visualization '91 Proceedings*, pages 17–24, 1991.

20. G. Turk and D. Banks. Image guided streamline placement. In *Proceedings of SIGGRAPH '96*, pages 453–460, 1996.

21. D. Weinstein, G. Kindlmann, and E. Lundberg. Tensorlines: advection-diffusion based propagation through diffusion tensor fields. In *IEEE Visualization '99 Proceedings*, pages 249–253, 1998.

22. C.-F. Westin, S. E. Maier, B. Khidhir, P. Everett, F. A. Jolesz, and R. Kikinis. Image processing for diffusion tensor magnetic resonance imaging. In *Proceedings of MICCAI '99*, pages 441–452, 1999.

23. C.-F. Westin, S. Peled, H. Gubjartsson, R. Kikinis, F. A. Jolesz, and R. Kikinis. Image processing for diffusion tensor magnetic resonance imaging. In *Proceedings of ISMRM '97*, 1997.

24. S. Zhang, C. Curry, D. Morris, and D. Laidlaw. Streamtubes and streamsurfaces for visualizing diffusion tensor MRI volume images. In *IEEE Visualization '00 Proceedings*, 2000.

25. L. Zhukov, K. Museth, D. Breen, A. H. Barr, and R. Whitaker. Level set modeling and segmentation of diffusion tensor magnetic resonance imaging brain data. *Journal of Electronic Imaging*, 12(1):125–133, 2003.

26. L. E. Zhukov and A. H. Barr. Oriented tensor reconstruction: tracing neural pathways from diffusion tensor MRI. In *IEEE Visualization 2002 Proceedings*, pages 387–394, 2002.

16   Diffusion Tensor MRI Visualization

SONG ZHANG and DAVID H. LAIDLAW
Department of Computer Science
Brown University

GORDON KINDLMANN
Scientific Computing and Imaging Institute
University of Utah

16.1   Introduction

Diffusion tensor magnetic resonance imaging (DT-MRI or DTI) is emerging as an important technology for elucidating the internal structure of the brain and for diagnosing conditions affecting the integrity of nervous tissue. DTI measurements of the brain exploit the characteristic microstructure of the brain's neural tissue, which constrains the diffusion of water molecules. The direction of fastest diffusion is aligned with fiber orientation in a pattern that can be numerically modeled by a *diffusion tensor*. Since DTI is the only modality for non-invasively measuring diffusion tensors in living tissue, it is especially useful for studying the directional qualities of brain tissue. Application areas include neurophysiology, neuroanatomy, and neurosurgery, as well as the diagnosis of edema (swelling), ischemia (brain damage from restricted blood flow), and certain types of brain tumors.

One of the fundamental problems in working with diffusion tensor data is its 3D and multivariate nature. Each sample point in a DTI scan can be represented by six interrelated values, and many features of interest are described in terms of derived scalar and vector fields that are overlaid logically on the original tensor field. Thus, the central tasks of DTI visualization include the following:

1. Determining which aspects of the tensor field will be graphically conveyed.

2. Determining where that information must be displayed and where it can be ignored.

3. Visually abstracting the DTI quantities into the graphics primitives by which the visualization is ultimately expressed.

This chapter introduces the acquisition and mathematics of diffusion tensor imaging and then surveys the current repertoire of visualization methods used with DTI. We finish by discussing some open questions.

16.2   Diffusion Tensor Imaging

Appreciating the origin and physical significance of any acquired scientific data is the first step in a principled approach to its visualization. This section briefly reviews the physical and mathematical underpinnings of diffusion tensor imaging.

Scientific understanding of the physical basis of diffusion converged at the beginning of the nineteenth century. In 1827, Robert Brown discovered Brownian motion, which underlies the thermodynamic model of diffusion: he observed that pollen grains suspended in water exhibit a zigzag "random walk." This motion was hypothesized by Desaulx in 1877 to arise when thermally energetic water molecules repeatedly collide with pollen grains. Einstein confirmed this hypothesis in 1905 as part of the development of his mathematical model of diffusion as a dynamically expanding Gaussian distribution

[14]. The path of the pollen grain suspended in water is a visible indicator of the similar Brownian motion of all liquid water molecules, whether in pure water, a porous medium, or biological tissue.

The intrinsic structural properties of many materials constrain diffusion so that diffusivity is *anisotropic*: greater in some directions than in others. If diffusion rates do not vary directionally, the diffusion is called *isotropic*. Biological tissues are often anisotropic because cell membranes and large protein molecules limit the motion of water molecules; Cooper et al. [9] call this *restricted diffusion*. Dissections and histological studies have shown that the grey matter of the brain is largely isotropic at the scale of MR scans, while the brain's white matter is more anisotropic because of the alignment of myelinated neuronal axons, which preferentially constrain water diffusion along the axon direction. Thus, via the mechanism of diffusion, the physical microstructure of white-matter tissue makes it possible to image the neural pathways connecting the brain. DTI imaging measurements have been validated within acceptable error on the fibrous muscle tissue of the heart [16,28].

Diffusion in biological tissue is measured by magnetic resonance imaging (MRI). In 1946, Purcell [27] and Block [7] independently discovered the nuclear magnetic resonance (NMR) effect. Water molecules contain hydrogen nuclei with uncoupled spins. In a strong magnetic field, the uncoupled spins cause the nuclei to align with and precess around the magnetic field direction, generating, in turn, a weak magnetic field aligned with the stronger ambient magnetic field. A second outside magnetic field can perturb this weak magnetic field so as to produce a magnetic resonance signal.

In 1950, Erwin Hahn discovered an important NMR signal called the spin echo [15] whose signal was perturbed by the diffusion of water molecules. Diffusion MR exploits this effect to measure hydrogen self-diffusivity *in vivo*. It is generally believed that the quantities measured with diffusion MR are a mixture of intracellular diffusion, intercellular diffusion, and the ex-change between the two sides of the cell membrane [29,30,31].

In 1973, Lauterbur described the principles of NMR imaging [20]. He encoded positioning information on NMR signals using gradient magnetic fields and an imaging-reconstruction algorithm, so that NMR imaging pinpoints the location where the signal is generated. This invention led to a new medical diagnostic instrument. In 1985, Bushel and Taylor combined the diffusion NMR and MR imaging techniques to create diffusion-weighted imaging [32]. A diffusion-weighted image (DWI) is a scalar-valued image that usually captures diffusion-rate information in one direction. In a DWI, the effect of diffusion on an MRI signal is an attenuation; the measured image intensity can be related to the diffusion coefficient by the following equation [21]:

$$\tilde{I}(x,y) = I_0(x,y)\exp(bD) \tag{16.1}$$

where $I_o(x,y)$ represents the voxel intensity in the absence of diffusion weighting, b characterizes the diffusion-encoding gradient pulses (timing, amplitude, shape) used in the MRI sequence, and D is the scalar diffusion coefficient.

Anisotropic diffusion information cannot be effectively represented in a scalar-valued DWI. In 1992, Basser et al. described the estimation of the diffusion tensor from the NMR spin echo [3]. The diffusion tensor, **D**, captures directional variation in the diffusion rate. **D** is a 3×3 positive symmetric matrix:

$$\mathbf{D} = \begin{pmatrix} D_{xx} & D_{xy} & D_{xz} \\ D_{yx} & D_{yy} & D_{yz} \\ D_{zx} & D_{zy} & D_{zz} \end{pmatrix} \tag{16.2}$$

b is also now a 3×3 matrix that represents the diffusion encoding. The equation becomes

$$\tilde{I}(x,y) = I_0(x,y)\exp\left(-\sum_{i=1}^{3}\sum_{j=1}^{3} b_{ij}D_{ij}\right) \tag{16.3}$$

A diffusion tensor has three real eigenvalues, λ_1, λ_2, and λ_3, each of which has a corresponding eigenvector, v_1, v_2, or v_3. A diffusion tensor is geometrically equivalent to an ellipsoid whose

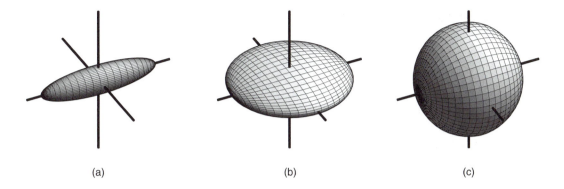

(a)   (b)   (c)

Figure 16.1   Ellipsoids represent diffusion tensors according to the eigensystem of the tensor: the eigenvalues are the radii of the ellipsoid, while the eigenvectors determine the axes' orientations. (a) Linear anisotropic diffusion; (b) planar anisotropic diffusion; (c) isotropic diffusion.

radii are the three eigenvectors of the diffusion-tensor matrix. This ellipsoid matches the shape to which water diffuses from a given data point in a fixed amount of time. The ellipsoids in Fig. 16.1 depict the three types of spatial diffusion.

In regions of complex-diffusion anisotropy, such as where two fiber bundles touch or cross each other, a tensor does not model the diffusion process accurately. These areas of ambiguity can be addressed with a diffusion model with more degrees of freedom than a second-order tensor. Tuch et al. acquire diffusion information in hundreds of directions to resolve ambiguity [33,34].

16.3   Approaches to Visualizing DTI Datasets

16.3.1   Overview

DTI visualization is challenging because the data has high information content and no single well established method exists to display the 3D patterns of matrix values that a tensor field contains. Another challenge for DTI visualization methods is keeping the results properly grounded in the specific application domain in which the data was originally acquired. Without such grounding, a visualization is unlikely to generate new hypotheses or provide answers about scientific problems.

Many DTI visualization approaches combine scalar, vector, and tensor methods by contracting the tensor to appropriate scalars or vectors for a particular application. These derived fields can be used in many ways in visualization applications and are, in some cases, sufficient alone. From a practical standpoint, the use of derived fields is also important in keeping the information in the visualization to its essential minimum. Much of 3D DTI visualization struggles with precisely this issue—that is, which regions in the tensor field are useful in the final visualization. The design of a visualization method is often a compromise between being informative and being legible. Part of what makes DTI visualization design an exciting research topic is that its rules and strategies are still being discovered.

16.3.2   Scalar Metrics

A common visualization approach involves contracting each tensor in a DTI to a scalar, thus reducing a DTI dataset to a scalar dataset. A carefully designed scalar metric can highlight useful information while reducing the effect of noise. Some scalar metrics are rotationally invariant, i.e., they do not depend on the coordinate system in which the tensor was measured. This property is often useful; without it, knowledge of the coordinate system must always be carried along with the metric for proper interpretation.

For instance, the trace of the diffusion tensor $T_r(D) = D_{11} + D_{22} + D_{33}$ measures the mean diffusivity and is rotationally invariant. It has been demonstrated that, after a stroke, images representing the trace of the diffusion tensor delineate the affected area much more accurately than images representing the diffusion in only one direction [35].

Many of the scalar metrics derived from DTI measure diffusion anisotropy in different ways. Douek et al. defined an anisotropic diffusion ratio (ADR) $ADR_{xz} = D_{xx}/D_{zz}$ for the anisotropy index [12]. Van Gelderen et al. [35] calculated a measure of diffusion anisotropy as the standard deviation of the three diffusion coefficients $A = \sqrt{\frac{1}{6}} \frac{\sqrt{(D_{xx}-D_{av})^2 + (D_{yy}-D_{av})^2 + (D_{zz}-D_{av})^2}}{D_{av}}$.

Both of these metrics are rotationally variant.

Basser et al. have calculated rotationally invariant anisotropy metrics from the diffusion tensor [4]: for relative anisotropy,

$$RA = \frac{\sqrt{(\lambda_1 - <\lambda>)^2} + \sqrt{(\lambda_2 - <\lambda>)^2} + \sqrt{(\lambda_3 - <\lambda>)^2}}{\sqrt{3}<\lambda>}$$

(16.4)

and for fractional anisotropy,

$$FA = \sqrt{\frac{3}{2}} \frac{\sqrt{(\lambda_1 - <\lambda>)^2} + \sqrt{(\lambda_2 - <\lambda>)^2} + \sqrt{(\lambda_3 - <\lambda>)^2}}{\sqrt{\lambda_1^2 + \lambda_2^2 + \lambda_3^2}}$$

(16.5)

where $<\lambda> = \frac{\lambda_1 + \lambda_2 + \lambda_3}{3}$.

Pierpaoli et al. showed that rotationally invariant metrics consistently show a higher degree of anisotropy than their variant analogs [25]. But because RA and FA are calculated over one diffusion tensor, they are still susceptible to noise contamination. Pierpaoli et al. then calculated an intervoxel anisotropy index, the *lattice index (LI)*, which locally averages inner products of diffusion tensors in neighboring voxels. LI has a low error variance and is less suscep-

tible to bias than are other rotationally invariant metrics.

Scalar anisotropy metrics such as FA and RA convey the anisotropy of a given diffusion distribution, but they do not convey whether the anisotropy is linear, planar, or some combination of the two. In terms of ellipsoid glyphs, cigar-shaped and pancake-shaped ellipsoids can have equal FA while their shapes differ greatly. Westin et al. [38] modeled diffusion anisotropy more completely with a set of three metrics that measure linear, planar, and spherical diffusion: $c_l = \frac{\lambda_1 - \lambda_2}{\lambda_1 + \lambda_2 + \lambda_3}$, $c_p = \frac{2(\lambda_2 - \lambda_3)}{\lambda_1 + \lambda_2 + \lambda_3}$, and $c_s = \frac{3\lambda_3}{\lambda_1 + \lambda_2 + \lambda_3}$, respectively. By construction, $c_l + c_p + c_s = 1$. Thus, these three metrics parameterize a barycentric space in which the three shape extremes (linear, planar, and spherical) are at the corners of a triangle, as shown in Fig. 16.2.

Fig. 16.3 shows one way to qualitatively compare some of the metrics described above by sampling their values on a slice of brain DTI data. Notice that the trace (Tr) is effective at distinguishing between the cerebrospinal fluid (where Tr is high) and the brain tissue (lower Tr) but fails to differentiate between different kinds of brain tissue. High fractional anisotropy FA, on the other hand, indicates white matter,

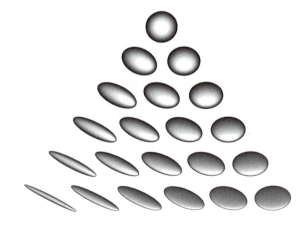

Figure 16.2 Barycentric space of diffusion tensor shapes.

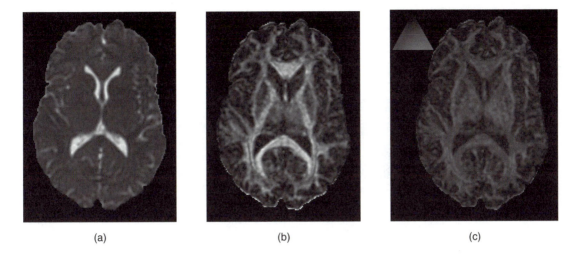

(a)   (b)   (c)

Figure 16.3 Different shape metrics applied to one slice of a brain DTI scan. (a) *Tr*: trace; (b) *FA*: fractional anisotropy; (c) C_L (green) and C_P (magenta). (See also color insert.)

because the directionality structure of the axon bundles permits faster diffusion along the neuron fiber direction than across it. *FA* is highest inside thick regions of uniformly anisotropic diffusion, such as inside the corpus callosum, the bridge between the two hemispheres of the brain. Finally, while both c_l and c_p indicate high anisotropy, their relative values indicate the shape of the anisotropy.

16.3.3 Eigenvector Color Maps

When diffusion tensors are measured with MRI, each tensor is represented by a 3×3 symmetric matrix; the matrix values are measured relative to the coordinate frame of the MRI scanner. Because they are real-valued, diffusion tensors have three real eigenvalues and three orthogonal eigenvectors. The eigenvalues are all nonnegative, because negative diffusivity is physically impossible. The eigenvectors define the orientation of the diffusion tensor. The *major* or *principal* eigenvector is associated with the largest eigenvalue and defines the direction of fastest diffusion. This direction can have significant physical meaning. In DTI scans of nervous tissue, for instance, the principal eigenvector is aligned with the coherent fibers.

A common visualization goal is to depict the spatial patterns of the principal eigenvector only in regions where it is meaningful, rather than depicting all the tensor information. Visualizing these patterns is often important in verifying that a given DTI scan has succeeded in resolving a feature of interest. A simple spherical color map of the principal eigenvector is the standard tool for this task, which first assigns an (R,G,B) color according to the (X,Y,Z) components of the principal eigenvector, v_1,

$$R = \text{abs}(v_{1x}), \quad G = \text{abs}(v_{1y}), \quad B = \text{abs}(v_{1z}),$$

and then modulates the saturation of the RGB color with an anisotropy metric. The direction of the principal eigenvector is numerically ill defined when the tensor is isotropic or has mostly planar anisotropy. In such cases, the visualization should not imply a particular direction with the hue of the RGB color. Thus, the saturation is modulated by c_l. Also, note that by design, the same color is assigned to v and −v. This also has a mathematical justification. The sign of eigenvectors is not defined. Numerical routines for their calculation may return either of two opposing vectors, and both should be visualized identically. Fig. 16.4 shows three

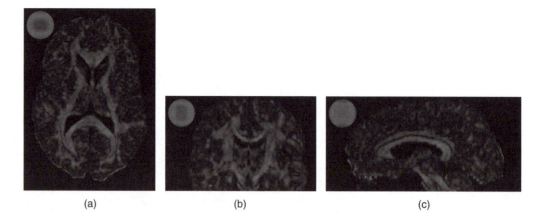

(a)   (b)   (c)

Figure 16.4 Eigenvector color maps shown on axis-aligned slices with three different slice orientations. The two axes within the slice are labeled with the anatomical name of the slice orientation. (a) Axial: x and y; (b) coronal: x and z; (c) sagittal: y and z. (See also color insert.)

examples of the eigenvector color map applied to the principal eigenvector.

16.3.4   Glyphs

A glyph is a parameterized icon that represents the data by its shape, color, texture, location, etc. Over the years, researchers have designed various glyphs for DTI visualization. We review and compare some of them here.

Diffusion ellipsoids are surfaces of constant mean-squared displacement of diffusing water molecules at some time τ after their release at the center of each voxel. Ellipsoids are a natural choice of glyph to summarize the information contained in a diffusion tensor [25]. The three principal radii are proportional to the eigenvalues and the axes are aligned with the three orthogonal eigenvectors of the diffusion tensor. The size of an ellipsoid can be associated with the mean diffusivity, and the preferred diffusion direction is indicated by the orientation of the ellipsoid. Arrays of ellipsoids can be arranged in the same order as the data points to show a 2D slice of DTI data. Laidlaw et al. [19] normalized the size of the ellipsoids to fit more of them in a single image (Fig. 16.5). While this eliminates the ability to show mean diffusivity,

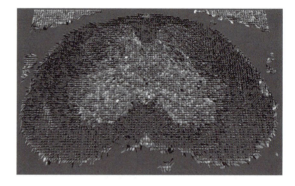

Figure 16.5 Arrays of normalized ellipsoids visualize the diffusion tensors in a single slice. (See also color insert.)

it creates more uniform glyphs that better show anatomy and pathology.

Laidlaw et al. also exploited the oil-painting concepts of brushstrokes and layering in visualization. They used 2D brushstrokes both individually, to encode specific values, and collectively, to show spatial connections and texture and to create a sense of speed corresponding to the speed of diffusion. Layering and contrast were used to create depth. The method clearly showed anatomy and pathology when applied to sections of spinal cords of mice with experimental allergic encephalomyelitis (EAE) (Fig. 16.6).

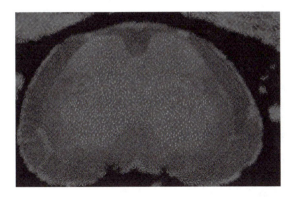

Figure 16.6 Brushstrokes illustrate the direction and magnitude of the diffusion. The background color and texture map show additional information. (See also color insert.)

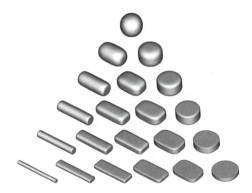

Figure 16.8 Superquadrics as tensor glyphs, sampling the barycentric space in Fig. 16.2.

In a still image, surface shading alone is often not enough to indicate the shape of an ellipsoid with only surface shading information. Westin et al. [37] used a composite of linear, planar, and spherical components to emphasize the shape of the diffusion ellipsoids. The components are scaled to the eigenvalues but can also be scaled according to the shape measures c_l, c_p, and c_s. Additionally, the color of the glyph is interpolated between the blue linear case, yellow planar case, and red spherical case (Fig. 16.7).

Kindlmann's approach adapted superquadrics, a traditional surface-modeling technique

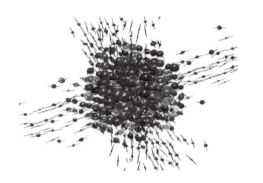

Figure 16.7 The composite shape of linear, planar, and spherical components emphasizes the shape of the diffusion tensor. (See also color insert.)

[1], as tensor glyphs. He created a class of shapes that includes spheres in the isotropic case and emphasizes the differences among the eigenvalues in the anisotropic cases. As shown in Fig. 16.8, cylinders are used for linear and planar anisotropy, and approximation to boxes represents intermediate forms of anisotropy. As with ellipsoid glyphs, a circular cross-section accompanies equal eigenvalues for which distinct eigenvectors are not defined.

The differences among the glyph methods can be appreciated by comparison of their results on a portion of a slice of a DTI brain scan, as shown in Fig. 16.9. The individual glyphs have been colored with the principal eigenvector color map. The directional cue given by the edges of box glyphs is effective in linearly anisotropic regions, but it can be misleading in regions of planar anisotropy and isotropy, since in these cases the corresponding eigenvectors are not numerically well defined. The rotational symmetry of ellipsoid glyphs avoids misleading depictions of orientation, with the drawback that different shapes can be hard to distinguish. The superquadric glyphs combine the best of the box and ellipsoid methods.

16.3.5 Tractography

In glyph-based methods, each glyph represents one diffusion tensor. Tractography, a term first

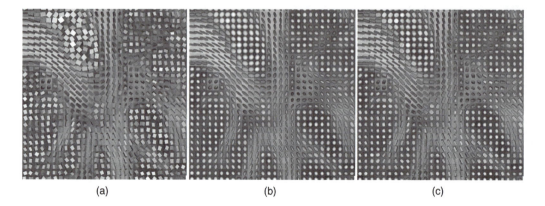

Figure 16.9   A portion of a brain DTI scan (also used in Figs. 16.3 and 16.4) as visualized by three different glyph methods. The overall glyph sizes have been normalized. (a) Boxes; (b) ellipsoids; (c) superquadrics. (See also color insert.)

applied to DTI analysis by Basser [2], yields curves of neural pathways, which are continuous and hard to represent with discrete glyphs. Streamlines and their derivatives are widely used for tractography results. Xue et al. show the result of fiber projection reconstruction by hand-selecting seeding points in a region of interest (ROI) and displaying the curves they generate [39]. The colors represent different groups of fiber structures. Zhang et al. used streamtubes and streamsurfaces to visualize the diffusion tensor field [41]. Streamtubes visualize fiber pathways tracked in regions of linear anisotropy. The trajectories of the streamtubes follow the major eigenvectors in the diffusion tensor field, the color along the streamtubes represents the magnitude of the linear anisotropy, and the cross-sectional shape represents the medium and minor eigenvectors. Streamsurfaces visualize regions of planar anisotropy. The streamsurfaces follow the expansion of major and medium eigenvectors in the diffusion tensor field, and the color represents the magnitude of planar anisotropy.

Zhang et al. used a culling algorithm to control the density of the scene's streamtubes [41] so that inside structures are visible and outside structures are still adequately represented. The metrics for the culling process include the trajectory length, the average linear anisotropy along a trajectory, and the similarity between a trajectory and the group of trajectories already selected. Combined with quantitative analysis, streamtubes and streamsurfaces can help elucidate structural heterogeneity in a DT-MRI dataset from a brain-tumor patient [40].

Trajectories calculated by integration have a serious drawback in regions where the white-matter structures change quickly: incorrect spurious connections can easily be generated [24]. Each diffusion tensor measurement is made over a small region. If the fiber direction is coherent throughout that region, then the measurement will be consistent. Otherwise, the tensor will be an amalgam of all the different values in the small region, and the major eigenvector may not point along a tract. These problems happen where tracts cross, diverge, or are adjacent to other tissues. Tractography is also sensitive to noise; a small amount of noise can cause significantly different results.

Some researchers have tried to address these problems by regularizing diffusion datasets [6,36,42] or direction maps [10,26]. Some researchers have explored new ways to find connectivity. Brun et al. [8] use sequential importance sampling to generate a set of curves, labeled with probabilities, from each seed point. Batchelor et al. generate a solution isosurface by solving for a diffusion-convection equation [5]. Parker et al. use front propagation in

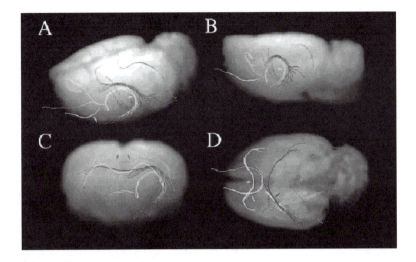

Figure 16.10   Tractography with streamlines. Image courtesy of Xue et al. [39]. (See also color insert.)

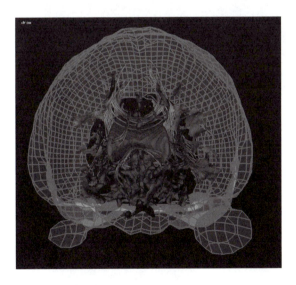

Figure 16.11   Red streamtubes and green streamsurfaces show linear and planar anisotropy, respectively, together with anatomical landmarks for context. (See also color insert.)

fast-marching tractography [23]. High-angular-resolution diffusion imaging is reported to ameliorate ambiguities in regions of complex anisotropy [33,34] and may ultimately be the best solution to this problem.

16.3.6   Volume Rendering

Glyphs and tractography communicate field structure with discrete geometry: the polyline or cylinder represents a fiber tract or the faceted surface of an polygonal ellipsoid, for example. Direct volume rendering, on the other hand, obviates the need for intermediate geometry by mapping and maps "directly" from measured field properties to optical properties like color and opacity, which are then composited and shaded [13,22]. The mapping is performed by a transfer function, which must be carefully designed to delineate and emphasize the features of interest, while not obscuring them with unimportant regions. In direct volume rendering of scalar data, the transfer function often maps from the scalar data values to opacity, although greater specificity and expressivity are possible with higher-dimensional and multivariate transfer functions [18]. Because transfer functions are applied without respect to field position, direct volume rendering has the potential to effectively convey large-scale patterns across the entire dataset.

Kindlmann et al. have explored various types of diffusion tensor transfer functions [17]; the present discussion focuses on transfer functions of tensor shape because of their intuitive definition and useful results. The barycentric space of tensor shapes in Figs. 16.2 and 16.8 captures two important degrees of freedom (from the total six) in a tensor field: degree and type of anisotropy. This space does not represent changes in overall size or orientation, but these are not crucial for visualizing the structure of white-matter fiber tracts. Fig. 16.3a indicates that the trace, Tr, does not vary significantly between gray and white matter, and, typically, the structural organization that distinguishes

white matter is interesting irrespective of its orientation.

Fig. 16.13 shows the results of using the barycentric shape space as the domain of transfer functions that assign opacity only. These renderings were produced with a brute-force renderer that samples each image ray at multiple points within the field, interpolates the tensor field componentwise, calculates the eigenvalues and the shape metrics (c_l, c_p, c_s), and then looks up the opacity for that sample. On the left-hand

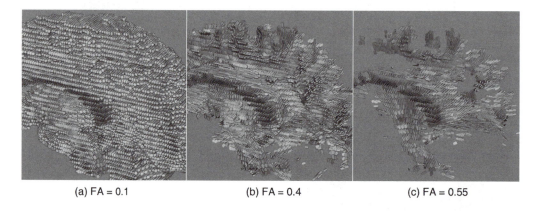

(a) FA = 0.1   (b) FA = 0.4   (c) FA = 0.55

Figure 16.12 Glyph-based visualization of a volumetric portion of a brain DTI scan (also used in Figs. 16.3, 16.4, and 16.9), with glyph culling based on three different fractional anisotropy thresholds. (See also color insert.)

Figure 16.13 Four different barycentric opacity maps and the corresponding renderings. (See also color insert.)

side of Fig. 16.13, the overall shape of the brain is seen when opacity is assigned to high c_s (isotropic) samples, while the shape of the white matter is visible when opacity is assigned to high c_l values. Arbitrary combinations of shape can be emphasized with this sort of transfer function (Fig. 16.13, right). Fig. 16.14 shows how specifying color as a function of barycentric shape can create a more informative rendering: the color variations indicate which portions of the field are more or less planarly anisotropic.

16.4   Open Issues

Interest in diffusion tensor visualization has been steadily increasing with the expanding applications of tensor imaging, increased computation and graphics capabilities for display, and advances in the visualization methods them-

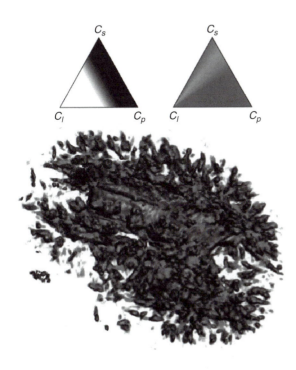

Figure 16.14 Assigning color and opacity with barycentric transfer function. (See also color insert.)

selves. Several important open issues and design challenges merit further research.

Visual design of glyphs: Glyphs must be parameterizable by at least as many variables as they will display. On the other hand, they should also be concise and compact, so that multiple glyphs viewed next to or on top of each other can still convey useful information. This is a daunting task for researchers looking to depict complex 3D patterns in the tensor field, but inspiration may be drawn from different artistic traditions of painting and technical illustration.

Seeding and culling schemes: While it is easy to survey information everywhere over a 2D domain, this is impossible in three dimensions because of occlusion. Glyph placement, trajectory placement, and selection of any visual abstraction are difficult problems of visual optimization, since the result must be legible in multiple contradictory ways. The interconnected nature of the white-matter fiber tracts in the brain further complicates this task. Solving this optimization may require level-of-detail (LOD) information and user-defined regions of interest.

Computational validation: This is perhaps the hardest aspect of scientifically *useful* visualization. An unresolved issue in DTI visualization is the extent to which the paths calculated by fiber tracking correspond to the paths of actual axons in the white matter. Locally, the fiber direction does correspond to tissue organization and the major "trunk lines" of connectivity are known from neuroanatomy. However, tractography methods can produce long and circuitous paths of purported connectivity whose actual validity is not, and cannot easily be, known. Scanning diffusion phantoms (with known connectivity) might address this, as would advances in histological preparations.

Display devices and interaction: DTI datasets are complicated and inherently 3D, and many visualizations involve large, complex graphical models. Recent advances in display may help boost the capabilities and applications of the visualization. Especially promising is the use of immersive virtual reality to display complex 3D fields of neural structures. But questions remain about the relative value of different display and interaction environments. For example, a user study comparing CAVE and fish-tank virtual reality systems shows that both types of display have pros and cons for certain tasks and goals [10].

Visual validation: No one visualization method stands out as the "gold standard" by which others are judged. Every method has advantages and limitations depending on the types of information it seeks to convey and the specific techniques used to convey it. Currently, most visualization methods are judged by their own inventors. User studies and other validation methods will both help appraise the methods in a more objective way and help evaluate different interaction environments.

Modeling: The interaction between water molecules and biological structures is vital for understanding the information contained in diffusion tensor images. It is generally believed that the sheath outside the neural axons causes most of the water-diffusion restrictions. Less is known about locations where fiber bundles diverge, cross, or kiss. What are the contributions of intracellular diffusion, intercellular diffusion, and exchange between the membranes? A physically realistic model may help us analyze the tensor data and then visualize the underlying structures in a meaningful way.

16.5 Summary

Since DTI technology emerged 10 years ago, the problems inherent in DTI acquisition, visualization, analysis, and application have spurred numerous multidisciplinary efforts. For scientific visualization students, these problem are especially intriguing because DTIs are large 3D multivariate datasets and thus present many visualization challenges. On the other hand, these problems have real-world origins, applications, and challenges: The datasets are noisy, resolution is never sufficient, and partial-volume effects limit the results. Perhaps the greatest challenge is the breadth of knowledge necessary to truly understand the entire process, from patient to imaging to computation to visual analysis and back to patient.

This chapter has given a brief survey of the last 10 years of visualization-related research, including some of the outstanding issues. We expect that over the next 10 years much more of the tremendous potential of this imaging modality will be realized.

References

1. A. Barr. Superquadrics and angle-preserving transformations. *IEEE Computer Graphics and Applications*, 18(1):11–23, 1981.
2. P. J. Basser, S. Pajevic, C. Pierpaoli, J. Duda, and A. Aldroubi. *In vivo* fiber tractography using DT-MRI data. *Magnetic Resonance in Medicine*, 44:625–632, 2000.
3. P. J. Basser, J. Mattiello, and D. LeBihan. Estimation of the effective self-diffusion tensor from the NMR spin echo. *J. Magn. Reson. B*, 103(3):247–54, March 1994.
4. P. J. Basser and C. Pierpaoli. Microstructural features measured using diffusion tensor imaging. *J. Magn. Reson. B*, pages 209–219, 1996.
5. P. G. Batchelor, D. L. G. Hill, D. Atkinson, F. Calamanten, and A. Connellyn. Fibre-tracking by solving the diffusion-convection equation. In *Proceedings of ISMRM 2002*, page 1135, 2002.
6. M. Bjornemo, A. Brun, R. Kikinis, and C.-F. Westin. Regularized stochastic white matter tractography using diffusion tensor MRI. In *MICCAI2002*, 2002.
7. F. Bloc. Nuclear induction. *Physical Review*, 70:460–474, 1946.
8. A. Brun, M. Bjornemo, R. Kikinis, and C.-F. Westin. White matter tractography using sequential importance sampling. In *ISMRM 2002*, 2002.
9. R. L. Cooper, D. B. Chang, A. C. Young, C. J. Martin, and B. Ancker-Johnson. Restricted diffusion in biophysical systems. *Biophysical Journal*, 14:161–177, 1974.
10. O. Coulon, D. C. Alexander, and S. R. Arridge. Tensor field regularisation for DT-MR images. In *Proceedings of British Conference on Medical Image Understanding and Analysis*, 2001.
11. C. DeMiralp, D. H. Laidlaw, C. Jackson, D. Keefe, and S. Zhang. Subjective usefulness of CAVE and fish-tank VR display systems for a scientific visualization application. In *IEEE*

Visualization '03 Poster Compendium, Seattle, WA, 2003.

12. P. Douek, R. Turner, J. Pekar, N. Patronas, and D. LeBihan. MR color mapping of myelin fiber orientation. *J. Comput. Assist. Tomogr.*, 15:923–929, 1991.

13. R. A. Drebin, L. Carpenter, and P. Hanrahan. Volume rendering. *Computer Graphics*, 22(4):65–74, 1988.

14. A. Einstein. Über die von der molekularkinetischen Theorie der Wärme geforderte bewegung von in ruhenden Flüssigkeiten suspendierten teilchen. *Annalen der Physik*, 17:549–560, 1905.

15. E. L. Hahn. Spin echoes. *Physical Review*, 80:580–594, 1950.

16. E. W. Hsu, A. L. Muzikant, S. A. Matulevicius, R. C. Penland, and C. S. Henriquez. Magnetic resonance myocardial fiber-orientation mapping with direct histological correlation. *Am. J. Physiol.*, 274:H1627–1634, 1998.

17. G. Kindlmann, D. Weinstein, and D. A. Hart. Strategies for direct volume rendering of diffusion tensor fields. *IEEE Transactions on Visualization and Computer Graphics*, 6(2):124–138, 2000.

18. J. Kniss, G. Kindlmann, and C. Hansen. Multidimensional transfer functions for interactive volume rendering. *IEEE Transactions on Visualization and Computer Graphics*, 8(3):270–285, 2002.

19. D. H. Laidlaw, E. T. Ahrens, D. Kremers, M. J. Avalos, C. Readhead, and R. E. Jacobs. Visualizing diffusion tensor images of the mouse spinal cord. In *Proceedings of IEEE Visualization 1998*, pages 127–134, 1998.

20. P. C. Lauterbur. Image formation by induced local interactions: examples employing nuclear magnetic resonance. *Nature*, 242:190–191, 1973.

21. D. LeBihan. Molecular diffusion nuclear magnetic resonance imaging. *Magn. Reson. Quant.*, 17:1–30, 1991.

22. M. Levoy. Display of surfaces from volume data. *IEEE Computer Graphics and Applications*, 8(5):29–37, 1988.

23. G. J. M. Parker, K. E. Stephen, G. J. Barker, J. B. Rowe, D. G. MacManus, C. A. M. Wheeler-Kingshott, O. Ciccarelli, R. E. Passingham, R. L. Spinks, R. N. Lemon, and R. Turner. Initial demonstration of *in vivo* tracing of axonal projections in the macaque brain and comparison with the human brain using diffusion tensor imaging and fast marching tractography. *NeuroImage*, 15:797–809, 2002.

24. C. Pierpaoli, A. S. Barnett, S. Pajevic, A. Virta, and P. J. Basser. Validation of DT-MRI tractography in the descending motor pathways of human subjects. In *Proceedings of ISMRM 2001*, page 501, 2001.

25. C. Pierpaoli and P. J. Basser. Toward a quantitative assessment of diffusion anisotropy. *Magn. Reson. Med.*, 36(6):893–906, 1996.

26. C. Poupon, C. A. Clark, V. Frouin, J. Regis, I. Block, D. LeBihan, and J.-F. Mangin. Regularization of diffusion-based direction maps for the tracking of brain white matter fascicles. *NeuroImage*, 12:184–195, 2000.

27. E. M. Purcell, H. C. Torrey, and R. V. Pound. Resonance absorption by nuclear magnetic moments in a solid. *Physical Review*, 69:37–43, 1946.

28. D. F. Scollan, A. Holmes, R. Winslow, and J. Forder. Histological validation of myocardial microstructure obtained from diffusion tensor magnetic resonance imaging. *Am. J. Physiol.*, 275:2308–2318, 1998.

29. G. J. Stanisz, A. Szafer, G. A. Wright, and R. M. Henkelman. An analytical model of restricted diffusion in bovine optic nerve. *Magn. Reson. Med.*, 37(1):103–11, 1997.

30. G. J. Stanisz and R. M. Henkelman. Tissue compartments, exchange and diffusion. In *Workshop on Diffusion MRI: Biophysical Issues*, pages 34–37, 2002.

31. A. Szafer, J. Zhong, and J. C. Gore. Theoretical model for water diffusion in tissues. *Magn. Reson. Med.*, 33:697–712, 1995.

32. D. G. Taylor and M. C. Bushell. The spatial mapping of translational diffusion coefficients by the NMR imaging technique. *Physics in Medicine and Biology*, 30:345–349, 1985.

33. D. S. Tuch, T.G. Reese, M. R. Wiegell, N. Makris, J. W. Bellireau, and V. J. Wedeen. High angular resolution diffusion imaging reveals intravoxel white matter fiber heterogeneity. *Magn. Reson. Med.*, 48(4):577–582, 2002.

34. D. S. Tuch, R. M. Weisskoff, J. W. Bellireau, and V. J. Wedeen. High angular resolution diffusion imaging of the human brain. In *Proceedings of the 7th Annual Meeting of ISMRM*, page 321, 1999.

35. P. van Gelderen, M. H. de Vleeschouwer, D. DesPres, J. Pekar, P. C. van Zijl, and C. T. Moonen. Water diffusion and acute stroke. *Magn. Reson. Med.*, 31:154–63, 1994.

36. D. M. Weinstein, G. L. Kindlmann, and E. C. Lundberg. Tensorlines: advection diffusion-based propagation through diffusion tensor fields. In *IEEE Visualization '99*, pages 249–254, 1999.

37. C.-F. Westin, S. E. Maier, H. Mamata, A. Nabavi, F. A. Jolesz, and R. Kikinis. Processing

and visualization for diffusion tensor MRI. *Medical Image Analysis*, 6:93–108, 2002.

38. C. F. Westin, S. Peled, H. Gubjartsson, R. Kikinis, and F. A. Jolesz. Geometrical diffusion measures for MRI from tensor basis analysis. In *Proceedings of ISMRM*, page 1742, 1997.

39. R. Xue, Peter C. M, van Zijl, B. J. Crain, M. Solaiyappan, and S. Mori. *In vivo* 3D reconstruction of rat brain axonal projections by diffusion tensor imaging. *Magn. Reson. Med.*, 42:1123–1127, 1999.

40. S. Zhang, M. E. Bastin, D. H. Laidlaw, S. Sinha, P. A. Armitage, and T. S. Deisboeck. Visualization and analysis of white matter structural asymmetry in diffusion tensor MR imaging data. *Magn. Reson. Med.*, 51(1):140–147, 2004.

41. S. Zhang, C. Demiralp, and D. H. Laidlaw. Visualizing diffusion tensor MR images using streamtubes and streamsurfaces. *IEEE Transactions on Visualization and Computer Graphics*, 9(4):454–462, 2003.

42. L. Zhukov and A. Barr. Oriented tensor reconstruction: tracing neural pathways from diffusion tensor MRI. In *IEEE Conference on Visualization '02*, pages 387–394, 2002.

17 Topological Methods for Flow Visualization

GERIK SCHEUERMANN and XAVIER TRICOCHE
University of Kaiserslautern

17.1 Introduction

Numerical simulations provide scientists and engineers with an increasing amount of vector and tensor data. The visualization of these large multivariate datasets is therefore a challenging task. Topological methods efficiently extract the structure of the corresponding fields to come up with an accurate and synthetic depiction of the underlying flow. Practically, the process consists of partitioning the domain of study into subregions of homogeneous qualitative behavior. Extracting and visualizing the corresponding graph permits conveyance of the most meaningful properties of multivariate datasets.

Vector and tensor fields are traditionally objects of major interest for visualization. They are the mathematical language of many research and engineering areas, including fundamental physics, optics, solid mechanics, and fluid dynamics, as well as civil engineering, aeronautics, turbomachinery, and weather forecasting. Vector variables in this context are velocity, vorticity, magnetic or electric field, and a force or the gradient of some scalar field like e.g., temperature. Tensor variables might correspond to stress, strain, or rate of deformation, for instance. From a theoretical viewpoint, vector and tensor fields have received much attention from mathematicians, leading to a precise and rigorous framework that constitutes the basis of specific visualization methods. In particular, Poincaré's work [20] at the end of the 19th century laid down the foundations of a geometric interpretation of vector fields associated with dynamical systems; the analysis of the phase portrait provides an efficient and aesthetic way to apprehend the information contained in abstract vector data. Nowadays, following this theoretical inheritance, scientists typically focus their study on the topology of vector and tensor datasets provided by numerical simulations or experimental measurements. A typical and very active application field is computational fluid dynamics (CFD), in which complex structural behaviors are investigated in the light of their topology [3,14,19,27]. It was shown, for instance, that topological features are directly involved in crucial aspects of flight stability like flow separation or vortex genesis [4]. Informally, the topology is the qualitative structure of a multivariate field. It leads to a partition of the domain of interest into subdomains of equivalent nature. Hence, extracting and studying this structure permits us to focus the analysis on essential properties. For visualization purposes, the depiction of the topology results in synthetic representations that transcribe the fundamental characteristics of the data. Moreover, it permits fast extraction of global flow structures that are directly related to features of interest in various practical applications. Further, topology-based visualization results in a dramatic decrease in the amount of data required for interpretation, which makes it very appealing for the analysis of large-scale datasets. These ideas are at the basis of the topological approach, which has gained an increasing interest in the visualization community during the last decade. First introduced for planar vector fields by Helman and Hesselink [9], the basic technique has been continuously extended since then. A significant milestone on the way was the work of Delmarcelle [5], which transposed the original vector method to symmetric, second-order tensor fields.

This chapter proposes an introduction to the mathematical foundations of the topological approach to flow visualization along with a survey of existing techniques in this domain. Note that the focus is on methods related directly to the depiction and analysis of the flow topology itself. In particular, visualization methods using topology for other purposes, like streamline seeding [43], data compression [40,41], smoothing [33], or modeling [42] are beyond the scope of this presentation.

The contents of this paper are organized as follows. Vector fields are considered first. Basic theoretical notions are introduced in Section 17.2. They result from the qualitative theory of dynamical systems, initiated by Poincaré. Nonlinear and parameter-dependent topologies are discussed in this section, along with the fundamental concept of bifurcation. Tensor fields are treated in Section 17.3. Following Delmarcelle's approach, we consider the topology of the eigenvector fields of symmetric, second-order tensor fields. It is shown that they induce line fields in which tangential curves can be computed, analogous to streamlines for vector fields. We explain how singularities are defined and characterized, and how bifurcations affect them in the case of unsteady tensor fields. This completes the framework required for the description of topology-based visualization of vector and tensor fields in Section 17.4. The presentation covers original methods for 2D and 3D fields, extraction and visualization of nonlinear topology, topology simplification for the processing of turbulent flows, and topology tracking for parameter-dependent datasets. Finally, Section 17.5 completes the presentation by addressing open questions and suggesting future research direc-

tions to further extend the scope of topology-based visualization.

17.2   Vector Field Topology

In this section, we propose a short overview of the theoretical framework of vector field topology, which we restrict to the requirements of visualization techniques.

17.2.1   Basic Definitions

We consider a vector field $\boldsymbol{v} : U \subseteq I\!R^n \times I\!R \to T I\!R^n \simeq I\!R^n$, which is a vector-valued function that depends on a space variable and on an additional scalar parameter, say time. The vector field $\boldsymbol{v}$ generates a *flow* $\phi_t : U \subseteq I\!R^n \to I\!R^n$, where $\phi_t := \phi(x, t)$ is a smooth function defined for $(\boldsymbol{x}, t) \in U \times (I \subseteq I\!R)$ satisfying

$$\frac{d}{dt}\phi(x, t)|_{t=\tau} = \upsilon(\tau, \phi(x, \tau)) \qquad (17.1)$$

for all $(\boldsymbol{x}, \tau) \in U \times I$. Practically we limit our presentation to the case $n = 2$ or 3 in the following. The function $\phi(\boldsymbol{x}_0, :) : t \to \phi(x_0, t)$ is an *integral curve* through $\boldsymbol{x}_0$. Observe that existence and uniqueness of integral curves are ensured under the assumption of fairly general continuity properties of the vector field. In the special but fundamental case of steady vector fields, i.e., fields that do not depend on the variable t, integral curves are called *streamlines*. Otherwise, they are called *pathlines*. The uniqueness property guarantees that streamlines cannot intersect in general. The set of all integral curves is called *phase portrait*. The qualitative structure of the phase portrait is called topology of the vector field. In the following we focus first on the steady case, and then we consider parameter-dependent topology.

17.2.2   Steady Vector Fields

The local geometry of the phase portrait is characterized by the nature and position of its *critical points*. In the steady case, these *singularities* are locations where the vector field is zero. Consequently, they behave as zero-dimensional integral curves. Furthermore, they are the only positions where streamlines can intersect (asymptotically). Basically, the qualitative study of critical points relies on the properties of the Jacobian matrix of the vector field at their position. If the Jacobian has full rank, the critical point is said to be linear or of *first order*. Otherwise, a critical point is nonlinear or of *higher order*. Next we discuss the planar and 3D cases successively. Observe that considerations made for 2D vector fields also apply to vector fields defined over a 2D manifold embedded in three dimensions, for example, the surface of an object surrounded by a 3D flow.

17.2.2.1 *Planar Case*

Planar critical points have benefited from great attention from mathematicians. A complete classification has been provided by Andronov et al. [1]. Additional excellent information is available in Abraham and Shaw [2] and Hirsch and Smale [12]. Depending on the real and imaginary parts of the eigenvalues, linear critical points may exhibit the configurations shown in Fig. 17.1. Repelling singularities act as *sources*, whereas attracting ones are *sinks*. *Hyperbolic* critical points are a subclass of linear singularities for which both eigenvalues have nonzero real parts. Thus, a center is nonhyperbolic. The analysis of nonlinear critical points, on the contrary, requires us to take into account higher-order polynomial terms in the Taylor expansion. Their vicinity is decomposed into an arbitrary combination of *hyperbolic, parabolic,* and *elliptic* curvilinear sectors (see Fig. 17.2). The bounding curve of a hyperbolic sector is called a *separatrix*. Back in the linear case, separatrices exist only for saddle points, where they are the four curves

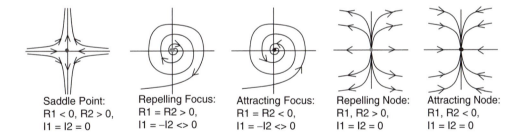

| Saddle Point:
R1 < 0, R2 > 0,
I1 = I2 = 0 | Repelling Focus:
R1 = R2 > 0,
I1 = –I2 <> 0 | Attracting Focus:
R1 = R2 < 0,
I1 = –I2 <> 0 | Repelling Node:
R1, R2 > 0,
I1 = I2 = 0 | Attracting Node:
R1, R2 < 0,
I1 = I2 = 0 |

Figure 17.1 Basic configurations of first-order planar critical points.

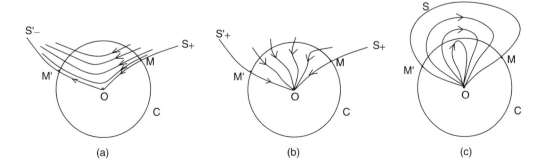

(a)   (b)   (c)

Figure 17.2 Sector types of arbitrary planar critical points. (a) Hyperbolic; (b) parabolic; (c) elliptic.

that reach the singularity, forward or backward in time. Thus, we obtain a simple definition of planar topology as the graph whose vertices are the critical points and whose edges are the separatrices integrated away from the corresponding singularities. This needs to be completed by *closed orbits* that are periodic integral curves. Closed orbits play the role of sources or sinks and can be seen as additional separatrices. It follows that topology decomposes a vector field into subregions where all integral curves have a similar asymptotic behavior: they converge toward the same critical points (respective closed orbits) both forward and backward. We complete our overview of steady planar topology by mentioning the *index* of a critical point introduced by Poincaré in the qualitative theory of dynamical systems. It measures the number of field rotations along a closed curve that is chosen to be arbitrarily small around the critical point. By continuity of the vector field, this is always a (signed) integer value. The index is an invariant quantity for the vector field and possesses several properties that explain its importance in practice. Among them we have the following:

1. The index of a curve that encloses no critical point is zero.

2. The index of a linear critical point is −1 for a saddle point and +1 for every other type (Fig. 17.3).

3. The index of a closed orbit is always +1.

4. The index of a curve enclosing several critical points is the sum of their individual indices.

17.2.2.2 3D Case

Similar theoretical results can be found for the analysis of the 3D case. However, only a few of them have concrete applications in scientific visualization so far. Therefore, we address only linear 3D critical points. Like the planar case, the analysis is based on the eigenvalues of the Jacobian. Two main possibilities exist: either the three eigenvalues are real or two of them are complex conjugates. Refer to Fig. 17.4 for a visual impression.

- *Three real eigenvalues.* One has to distinguish the case where all three eigenvalues have the same sign, where we have a 3D node (either attracting or repelling) from the case where only two eigenvalues have the same sign: the two eigenvectors associated with the eigenvalues of the same sign span a plane in which the vector field behaves as a 2D node and the critical point is a 3D saddle.

- *Two complex eigenvalues.* Once again there are two possibilities. If the common real part of both complex eigenvalues has the same sign as the real eigenvalue, one has a 3D spiral, i.e., a critical point (either attracting or repelling) that exhibits a 2D spiral structure in the plane spanned by the eigenspace related to the complex eigenvalues. If they have different signs, one has a second kind of 3D saddle.

Analogously to the planar case, a critical point is called *hyperbolic* in this context if the eigenvalues of the Jacobian have all nonzero real parts. Compared to 2D critical points,

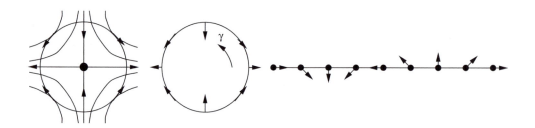

Figure 17.3 Simple closed curve of index −1 (saddle point).

separatrices in 3D are not restricted to curves; they can be surfaces, too. These surfaces are special types of so-called stream surfaces that are constituted by the set of all streamlines that are integrated from a curve. The linear 3D topology is thus composed of nodes, spirals, and saddles that are interconnected by curve and surface separatrices. Depending on the considered type, repelling and attracting eigenspaces can be 1D or 2D, leading to curves and surfaces (Figs. 17.4b and 17.4d). Surface separatrices emanate from 3D saddle points spanned by the eigenvectors associated with the two eigenvalues of the same sign.

17.2.3 Parameter-Dependent Vector Fields

The previous sections focused on steady vector fields. Now, if the considered vector field depends on an additional parameter, the structure of the phase portrait may transform as the value of this parameter evolves: position and nature of critical points can change along with the connectivity of the topological graph. These modifications—called *bifurcations* in the literature—are continuous evolutions that bring the topology from a stable state to another, structurally consistent, stable state. Bifurcations have been the subject of an intensive research effort in pure and applied mathematics [7]. The present section will provide a short introduction to these notions. Notice that the treatment of 3D bifurcations is beyond the scope of this paper, since they have not been applied to flow visualization up till now. We start with basic considerations about structural stability and then describe typical planar bifurcations.

17.2.3.1 *Structural Stability*

As said previously, bifurcations consist of topological transitions between stable structures. In fact, the definition of structural stability involves the notion of structural equivalence. Two vector fields are said to be *equivalent* if there exists a diffeomorphism (i.e., a smooth map with smooth inverse) that takes the integral curves of one vector field to those of the second while preserving orientation. *Structural stability* is now defined as follows: the topology of a vector field v is stable if any perturbation of v, chosen small enough, results in a vector field that is structurally equivalent to v. We can now state a simplified version of the fundamental Peixoto's theorem [7] on structural stability for 2D flows. *A smooth vector field on a 2D compact planar domain of* $I\!R^2$ *is structurally stable if and only if* (iff) *the number of critical points and closed orbits is finite and each is hyperbolic, and if there are no integral curves connecting saddle points.* Practically, Peixoto's theorem implies that a planar vector field typically exhibits saddle points, sinks, and sources, as well as attracting or repelling closed orbits. Furthermore, it asserts that nonhyperbolic critical points or closed orbits are unstable because arbitrarily small perturbations can

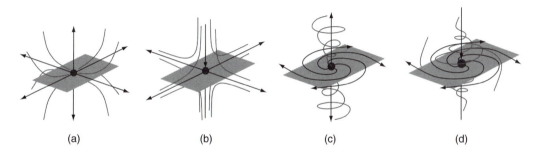

|(a)|(b)|(c)|(d)|

Figure 17.4   Linear 3D critical points. (a) 3D mode; (b) node saddle; (c) 3D spiral; (d) spiral saddle.

make them hyperbolic. Saddle connections, as far as they are concerned, can be broken by small perturbations as well.

17.2.3.2 Bifurcations

There are two major classes of structural transitions: local and global bifurcations.

Local Bifurcations: There are two main types of local bifurcations affecting the nature of a singular point in 2D vector fields. The first one is the so-called *Hopf bifurcation*. It consists of the transition from a sink to a source with simultaneous creation of a surrounding closed orbit that behaves as a sink, preserving local consistency with respect to the original configuration (Fig. 17.5). At the bifurcation point there is a center. The reverse evolution is possible too, as is an inverted role of sinks and sources. A second typical local bifurcation is called *fold bifurcation* and consists of the pairwise annihilation or creation of a saddle and a sink (respectively). This evolution is depicted in Fig. 17.6. Observe that the index of the concerned region remains 0 throughout the transformation.

Global Bifurcations: In contrast to the cases mentioned above, global bifurcations are not restricted to a small neighborhood of a singularity, but they entail significant changes in the overall flow structure and involve large domains by modifying the connectivity of the topological graph. Since global bifurcations remain a challenging mathematical topic, we mention here just a typical configuration exhibited by such transitions: the unstable saddle–saddle connection (see Peixoto's theorem). This is the central constituent of *basin bifurcations*, where the relative positions of two separatrices emanating from two neighboring saddle points are swapped through a saddle–saddle separatrix.

17.3 Tensor-Field Topology

Making use of the results obtained for vector fields, we now turn to tensor-field topology. We adopt for our presentation an approach similar to the original work of Delmarcelle [5,6] and focus on symmetric second-order real tensor fields that we analyze through their eigenvector fields. We seek here a framework that permits us to extend the results discussed previously to tensor fields. However, since most of the research done so far has been concerned with the 2D case, we put the emphasis on planar tensor fields and point out the generalization to 3D fields. A mathematical treatment of these notions can be found in Tricoche [28], where it is shown how covering spaces allow association of a line field with a vector field. In this section, we first introduce useful notations in the steady case and also show how symmetric second-order tensor fields can be interpreted as line fields. This makes possible the integration of tangential curves called tensor lines. Next, singularities are considered. We complete the presentation with tensor bifurcations.

17.3.1 Line Fields

17.3.1.1 Basic Definitions

In the following, we call *tensor* a symmetric second-order real tensor of dimension 2 or 3. This is a geometric invariant that corresponds

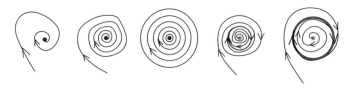

Figure 17.5 Hopf bifurcation.

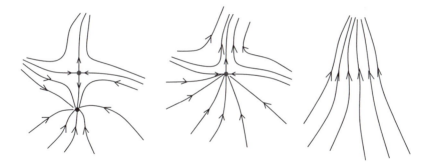

Figure 17.6   Pairwise annihilation.

to a linear transformation and can be represented by a matrix in a Cartesian basis. By extension, we define a tensor field as a map T that associates every position of a subset of the Euclidean space IR^n with a $n \times n$ symmetric matrix. Thus, it is characterized by $\frac{1}{2}n(n+1)$ independent, real scalar functions. Note that an arbitrary second-order tensor field can always be decomposed into its symmetric and antisymmetric parts. From the structural point of view, a tensor field is fully characterized by its *deviator field*, which is obtained by subtracting from the tensor its isotropic part, that is $D = T - \frac{1}{n}(\text{tr } T)I_n$, where tr T is the trace of T and I_n the identity matrix in IR^n. Observe that the deviator has trace zero by definition. The analysis of a tensor field is based on the properties of its eigensystem. Since we consider symmetric tensors, the eigenvectors always form an orthogonal basis of IR^n and the eigenvalues are real. It is a well known fact that eigenvectors are defined modulo a nonzero scalar, which means that they have neither inherent norm nor orientation. This characteristic plays a fundamental role in the following process. Through its corresponding eigensystem, any symmetric real tensor field can now be associated with a set of orthogonal eigenvector fields. We choose the following notations in three dimensions. *Let $\lambda_1 > \lambda_2 > \lambda_3$ be the real eigenvalues of the symmetric tensor field T (i.e., λ_1, λ_2, and λ_3 are scalar fields as functions of the coordinate vector x). The corresponding eigen-* *vector fields e_1, e_2, and e_3 are respectively called* major, medium, *and* minor eigenvector fields. In the 2D case, there are just major and minor eigenvectors. We now come to tensor lines, which are the object of our structural analysis.

17.3.1.2   Tensor Lines

A *tensor line* computed in a Lipschitz continuous eigenvector field is a curve that is everywhere tangent to the direction of the field. Because of the lack of both norm and orientation, the tangency is expressed at each position in the domain in terms of lines. For this reason, an eigenvector field corresponds to a *line field*. Nevertheless, except at positions where two (or three) eigenvalues are equal, integration can be carried out in a way similar to streamlines for vector fields by choosing an arbitrary local orientation. Practically, this consists of determining a continuous angular function θ^* defined modulo 2π that is everywhere equal to the angular coordinate θ of the line field, modulo π. Considering the set of all tensor lines as a whole, the topology of a tensor field is defined as the structure of the tensor lines. It is important to observe that the topology of a particular eigenvector field can be deduced from the topology of the other one through the orthogonality of the corresponding line fields.

17.3.2   Degenerate Points

Inconsistencies in the local determination of an orientation (as described previously) only occurs in the neighborhood of positions where several eigenvalues are equal. There, the eigenspace associated with the corresponding eigenvalues is no longer 1D. For this reason, such positions are singularities of the line field. To remain consistent with the notations originally used by Delmarcelle [5,6], we call them *degenerate points*, though they are typically called *umbilic points* in differential geometry. Because of the direction indeterminacy at degenerate points, tensor lines can meet there, which underlines the analogy with critical points.

17.3.2.1   Planar Case

The deviator part of a 2D tensor field is zero *iff* both eigenvalues are equal. For this reason, degenerate points correspond to zero values of the deviator field. Thus, D can be approximated as follows in the vicinity of a degenerate point P_0:

$$D(P_0 + dx) = \nabla D(P_0)dx + o(dx) \qquad (17.2)$$

Where $dx = (x, y)^T$, and where α and β are real scalar functions over $\mathbb{R}^2$,

$$D(x,y) = \begin{pmatrix} \alpha & \beta \\ \beta & -\alpha \end{pmatrix}, \text{ and } \nabla D(P_0)dx =$$

$$\begin{pmatrix} \dfrac{\partial \alpha}{\partial x}dx + \dfrac{\partial \alpha}{\partial y}dy & \dfrac{\partial \beta}{\partial x}dx + \dfrac{\partial \beta}{\partial y}dy \\ \dfrac{\partial \beta}{\partial x}dx + \dfrac{\partial \beta}{\partial y}dy & -\dfrac{\partial \alpha}{\partial x}dx - \dfrac{\partial \alpha}{\partial y}dy \end{pmatrix}$$

If the condition $\frac{\partial \alpha}{\partial x}\frac{\partial \beta}{\partial y} - \frac{\partial \alpha}{\partial y}\frac{\partial \beta}{\partial x} \neq 0$ holds, the degenerate point is said to be linear. The local structure of the tensor lines in this vicinity depends on the position and number of radial directions. If θ is the local angle coordinate of a point with respect to the degenerate point, $u = tan\theta$ is the solution of the following cubic polynomial:

$$\beta_2 u^3 + (\beta_1 + 2\alpha_2)u^2 + (2\alpha_1 - \beta_2)u - \beta_1 = 0$$

$$(17.3)$$

with $\alpha_1 = \frac{\partial \alpha}{\partial x}, \alpha_2 = \frac{\partial \alpha}{\partial y}$, and the same for β_i. This equation has either one or three real roots, which all correspond to angles along which the tensor lines reach the singularity. These angles are defined modulo π, so one obtains up to six possible angle solutions. Since we limit our discussion to a single (minor/major) eigenvector field, we are finally concerned with up to three radial eigenvectors. The possible types of linear degenerate points are trisectors and wedges (Fig. 17.7). In fact, the special importance of radial tensor lines is explained by their interpretation as separatrices. As a matter of fact, like critical points, the set of all tensor lines in the vicinity of a degenerate point is partitioned into hyperbolic, parabolic, and elliptic sectors. Separatrices are again defined as the bounding curves of the hyperbolic sectors (compare Fig. 17.7). In fact, a complete definition of the planar topology involves *closed tensor lines* too, although they are rare in practice. The analogy with vector fields may be extended by defining the tensor index of a degenerate point [5,28] that

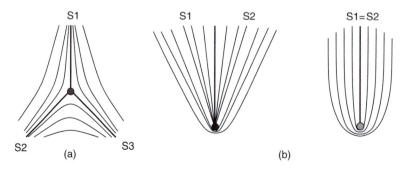

(a)   (b)

Figure 17.7   Linear degenerate points in the plane. (a) Trisector; (b) wedge point.

measures the number of rotations of a particular eigenvector field along a closed curve surrounding a singularity. Notice that tensor indices are half integers, due to orientation indeterminacy: trisectors have index $-\frac{1}{2}$ and wedges have index $\frac{1}{2}$. The nice properties of Poincaré's index extend here in a very intuitive fashion.

17.3.2.2 3D Degenerate Points
In the 3D case, singularities are of two types: eigenspaces may become 2D or 3D, leading to tangency indeterminacy in the corresponding eigenvector fields. To simplify the presentation, we restrict our considerations to the trace-free deviator $\mathbf{D} = (D_{ij})_{i,j}$. The characteristic equation is $-\lambda^3 + b\lambda + c = 0$, where $-c$ is the determinant of $\mathbf{D}$ and $b = \frac{1}{2}\sum_i D_{ii} + \sum_{i<j} D_{ij}^2$. The quantity $\Delta = (\frac{c}{2})^2 - (\frac{b}{3})^3$ determines the number of distinct real roots of the equation: $\Delta < 0$ yields three distinct real roots, while there are multiple roots *iff* $\Delta = 0$ (complex conjugate roots are impossible, since $\mathbf{D}$ is symmetric). Thus, degeneracies correspond to a maximum of Δ, which is everywhere negative, except at points, lines, or surfaces, where it is zero. Here a major difference with 3D vector field topology must be underlined: The loci of singularities are not restricted to points. Refer to Hesselink et al. [11] and Lavin et al. [16] for additional information.

17.3.3 Parameter-Dependent Topology
Again, the natural question that arises at this stage is the structural stability of topology under small perturbations of an underlying parameter. We restrict our considerations to the simplest cases in 2D of local and global bifurcations to remain in the scope of the methods to come. The observations proposed next are all inspired by geometric considerations.

Following the basic idea behind Peixoto's theorem, we see that the only stable degenerate points must be the linear ones. As a matter of fact, the asymptotic behavior of tensor lines in the vicinity of a degenerate point is determined by the third-order differential ∇D (thus leading to a linear degenerate point) except at locations where it becomes singular. This is, by essence, an unstable property, since arbitrary small perturbations in the coefficients lead to one of the linear configurations. Continuing our analogy with the vector case, we conclude that integral curves are unstable if they are separatrices for both of the degeneracies they link together. This is because a small-angle perturbation of the line field around any point along the separatrix suffices to break the connection. Using these elementary results, we review typical planar bifurcations.

17.3.3.1 Pairwise Creation and Annihilation
Since a wedge and a trisector have opposite indices, a closed curve enclosing them has index 0. This simple fact is the basic idea behind pairwise creations or annihilations. Indeed, the zero index computed along this closed curve shows that the combination of both degenerate points is structurally equivalent to a uniform flow. Therefore, a wedge and a trisector can merge and disappear: this is a *pairwise annihilation*. The reverse evolution is called a *pairwise creation*. Both are the equivalent of the fold bifurcations for critical points. An example is proposed in Fig. 17.8.

17.3.3.2 Homogeneous Mergings
When two linear degenerate points of the same nature merge, their half-integer indices are added and the resulting singularity exhibits a pattern corresponding to a linear critical point, e.g., two trisectors lead to a saddle point. However, according to what precedes, these new degenerate points are nonlinear and thus unstable. Details on that topic can be found in Delmarcelle [5].

17.3.3.3 Wedge Bifurcation
Both existing types of wedges have the same index, $\frac{1}{2}$. As a consequence, the transition from

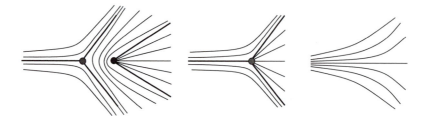

Figure 17.8 Pairwise annihilation of degenerate points.

one type to the other can occur without modifying structural consistency of the surrounding flow. From a topological viewpoint, this evolution corresponds to the creation (respectively disappearance) of a parabolic sector along with an additional separatrix.

17.3.3.4 *Global Bifurcations*

To finish this presentation of structural transitions in line fields, we briefly consider the simplest type of global bifurcation. It is intimately related to the unstable separatrices defined above. It occurs when the relative positions of two separatrices are changed through a common separatrix.

17.4 Topological Visualization of Vector and Tensor Fields

The theoretical framework described previously has motivated the design of techniques that build the visualization of vector and tensor fields upon the extraction and analysis of their topology. We start with a recall of the original method of Helman and Hesselink [9,10] for vector fields, later extended by Delmarcelle [5] to symmetric second-order tensor fields. After that, we focus on recent advances in topology-based visualization. New methods designed to complete the visualization of planar steady fields are discussed first. Nonlinear topology is addressed next. Techniques for reducing topological complexity in turbulent fields are considered, and the presentation ends with the topological visualization of parameter-dependent (e.g., unsteady) datasets. Throughout the presentation, we privilege a simultaneous treatment of vector and tensor techniques, making natural use of the profound theoretical relationships between their topologies.

17.4.1 Topology Basics
17.4.1.1 *Original Methods*

Helman and Hesselink pioneered topology-based visualization in 1989 [9]. They proposed a scheme for 2D vector fields, restricting the characterization of critical points to a linear precision. Remember that this leads to a graph where saddles, spirals, and nodes are the vertices and separatrices at the edges, integrated along the eigendirections of the saddle points. They extended their technique to tangential vector fields defined over surfaces embedded in 3D space [10]. Stream surfaces [13,22] are started along the separatrices. This was shown to permit the visualization of separation and attachment lines in some cases. At the same time, Globus et al. [8] suggested a similar technique to visualize the topology of 3D vector fields defined over curvilinear grids. They did not provide the separating surfaces associated with 3D saddles, but used glyphs to depict their structure locally and draw streamlines from them. Observe that all these methods require the integration of streamlines, which is typically carried out numerically, e.g., using Runge-Kutta with adaptive step size [21]. Analytical methods exist for piecewise linear vector fields over triangulations, respectively tetrahedrizations [18].

The topological approach for vector fields inspired Delmarcelle, who extended the original scheme to symmetric, second-order planar tensor fields [6] within his work on general techniques for tensor field visualization [5]. Here, too, analysis is restricted to linear precision. The missing quantitative information provided by eigenvalues is inserted into the representation by means of color coding. Existing attempts to generalize this method to 3D tensor fields suffer from the inherent difficulty of locating 3D degenerate points [11,16]. Typically they lead to complicated polynomial systems of high degree and lack surfaces to partition the domain conveniently.

17.4.1.2 Closed Orbits

The original topological method neglects the importance of closed orbits. As mentioned previously, these features play a major role in the flow structure, acting like sinks or sources and like additional separatrices. Moreover, they represent a challenging issue for numerical integration schemes used in practice, since every streamline that converges toward a closed orbit will result in an endless computation. A traditional but inaccurate way to solve this problem is to limit the number of integration steps. However, this does not permit us to distinguish a closed orbit from a slowly converging spiral, i.e., a spiral with high vorticity. Furthermore, this might be very inefficient if the number of iterations is set to a fairly large value to avoid a premature integration break. To overcome this deficiency, Wischgoll and Scheuermann [34] first proposed a method that properly identifies and locates closed orbits. Their basic idea is to detect on a cell-wise basis a periodic behavior during streamline integration. Practically, once a cell cycle has been inferred, a control is carried out over the edges of the concerned cells. This ensures that a streamline entering the cycle will remain trapped. If this condition is met and no critical point is present in the cycle, the Poincaré-Bendixon theorem [7] ensures that a

closed orbit is contained in it. Precise location is obtained by looking for a fixed point of the Poincaré map [2]. The method was generalized to 3D by Wischgoll and Scheuermann [35]. Note that the extension to tensor fields is straightforward even if closed tensor lines are rare in practice.

17.4.1.3 Local Topology

The definition of topology given previously does not specifically address vector fields defined over a bounded domain. As a matter of fact, the idea behind topology visualization is to partition the domain into subregions where all streamlines exhibit the same asymptotic behavior. If the considered domain is infinite, this is equivalent to looking for subregions where all streamlines reach the same critical points, both backward and forward, including a critical point at infinity [26]. Now, scientific visualization is typically concerned with domains spanned by bounded grids. In this case, the boundary must be incorporated into the topology analysis: outflow parts behave as sinks, inflow parts as sources, and the points separating them as half-saddles. Scheuermann et al. [23] proposed a method that identifies these regions along the boundaries of planar vector fields. It assumes that the restriction of the vector field to the boundary is piecewise linear. Half saddles are located and separatrices are started there, forward and backward, to complete the local topology visualization. Observe that the same principle can be applied in 3D: half saddles are no longer points but closed curves from which stream surfaces can be drawn.

17.4.1.4 Earth Mover's Distance

Lavin et al. [39] addressed the problem of vector field registration by automatically comparing two planar vector fields based on the characteristics of their critical points. The metric considered is called *Earth mover's distance* and is traditionally used for image retrieval. This technique was extended to 3D critical points by

Batra et al. [36]. Observe that in neither case is the connectivity of the topological graph taken into account by the metric.

17.4.2 Nonlinear Topology

The methods introduced so far are limited to linear precision in the characterization of singular points. We saw previously that nonlinear critical or degenerate points are unstable. However, when imposed constraints exist (e.g., symmetry or incompressibility of the flow), they can be encountered. To extend existing methods, Scheuermann et al. [25] proposed a scheme for the extraction and visualization of higher-order critical points in 2D vector fields. The basic idea is to identify regions where the index is larger than 1 (or less than -1). In such regions, the original piecewise linear interpolant is replaced by a polynomial approximating function. The polynomial is designed in Clifford algebra, based on theoretical results presented by Scheuermann et al. [24]. This permits us to infer the underlying presence of a critical point with arbitrary complexity, which is next modeled and visualized as shown in Fig. 17.9.

An alternative way to replace several close linear singularities by a higher-order one is suggested by Tricoche et al. [29]. It works with local grid deformations and can be applied to both tensor and vector fields in 2D. Moreover, it ensures continuity over the whole domain. Based on a mathematical background analogous to that of Scheuermann et al. [25], Mann and Rockwood presented a scheme for the detection of arbitrary critical points in three dimensions [17]. Geometric algebra is used to compute the 3D index of a vector field, which is obtained as an integral over the surface of a cube. Getting back to the original ideas of Poincaré in his study of dynamical systems, Trotts et al. [26] proposed a method to extract and visualize the nonlinear structure of a "critical point at infinity" when the considered vector field is defined over an unbounded domain.

17.4.3 Topology Simplification

Topology-based visualization usually results in clear and synthetic depictions that ease analysis and interpretation. Yet turbulent flows, like those encountered in CFD simulations, lead to topologies exhibiting many structures of very small scale. Their proximity and interconnection in the global picture cause visual clutter with classical methods. This drawback is

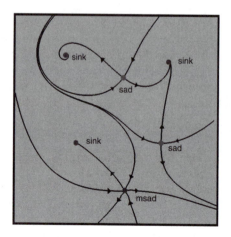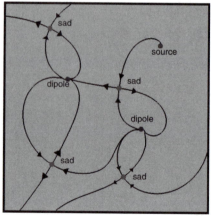

Figure 17.9 Nonlinear topologies.

worsened by low-order interpolation schemes, typical in practice, that confuse the results by introducing artifacts. Therefore, there is a need for postprocessing methods that permit clarification of the topologies by emphasizing the most meaningful properties of the flow and suppressing local details and numerical noise. The problem was first addressed by de Leeuw et al. [38] for planar vector fields. They proposed a scheme based on the pairwise pruning of interconnected critical points along with the corresponding edges. This pruning was first driven by a distance filter to obtain a multiscale topology visualization [37,38]. The same authors improved this original work later on by evaluating the importance of sinks and sources with respect to the surface of their inflow (respectively outflow) basins [15]. Since both methods are graph-based, the resulting simplified topology lacks a corresponding vector field description. Tricoche et al. [29] proposed an alternative approach for both vector and tensor fields defined in the plane. Close singularities are merged, resulting in a higher-order singularity that synthesizes the structural impact of small-scale features in the large picture. This reduces the number of singular points as well as the global complexity of the graph. The merging effect is achieved by local grid deformations that modify the vector field.

There is no assumption about grid structure or interpolation scheme. The same authors presented a second method that works directly on the discrete values defined at the vertices of a triangulation [31]. Angle constraints drive a local modification of the vector field that removes pairs of singularities of opposite indices. This simulates a fold bifurcation. Results are shown in Fig. 17.10 for a vortex breakdown simulation. A major advantage compared to the previous method is that the simplification process can be controlled not only by geometric considerations but by arbitrary user-prescribed criteria (qualitative or quantitative, local or region-based), specific to the considered application.

17.4.4 Topology Tracking

Theoretical results show that bifurcations are the key to understanding and thus properly visualizing parameter-dependent flow fields: they transform the topology and explain how the stable structures arise that are observed for discrete values of the parameter. Typical examples in practice are time-dependent datasets. This basic observation motivates the design of techniques that permit us to accurately visualize the continuous evolution of topology. A first at-

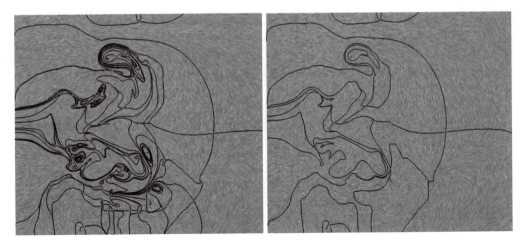

Figure 17.10 Turbulent and simplified topologies. (See also color insert.)

tempt was the method proposed by Helman and Hesselink [10]. The 1D parameter space is displayed in the third dimension (2D vector fields). However, the method is restricted to a graphical connection between the successive positions of critical points and associated separatrices, leading to a ribbon if consistency was preserved. Thus, no connection is made if a structural transition has occurred: bifurcations are missed. The same restriction holds for the transposition of this technique to tensor fields by Delmarcelle and Hesselink [6]. Tricoche et al. [30,32] attacked this deficiency. The central idea of their technique is to handle the mathematical space, made of the Euclidean space on one hand and the parameter space on the other hand, as a continuum. The vector or tensor data is supposed to lie on a triangulation that remains constant. A "space–time" grid is constructed by linking corresponding triangles through prisms over the parameter space. The choice of a suitable interpolation scheme permits an accurate and efficient tracking of singular points through the grid along with the detection of local bifurcations on the way. With the scheme of [34], closed orbits are tracked in a similar way. Again, the technique results in a 3D representation. The paths followed by critical points are depicted as curves. Separatrices integrated from saddles and closed orbits span smooth separating surfaces. These surfaces are further used to detect modifications in the global topological connectivity: consistency breaks correspond to global bifurcations. Examples are proposed in Fig. 17.11.

17.5   Future Research

So far, the major limitation of many existing topological methods is their restriction to 2D datasets. This is especially true in the case of tensor fields. In fact, the basic idea behind topology, i.e., the structural partition of a flow into regions of homogenous behavior, is definitely not restricted to two dimensions. However, the theoretical framework requires further research effort to serve as a basis for 3D visualization techniques. Now, in the simple case of linear precision in the characterization of critical points, topology-based visualization of 3D vector fields still lacks a fast, accurate, and robust technique to compute separating surfaces. This becomes challenging in regions of strong vorticity or in the vicinity of critical points, in particular for turbulent flows. In addition, topology-based visualization of parameter-dependent, 3D fields must overcome the limitations human beings experience in apprehending the information contained in 4D datasets.

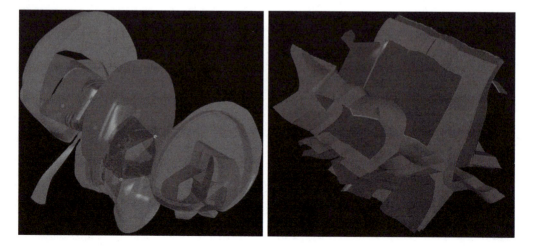

Figure 17.11   Unsteady vector and tensor topologies. (See also color insert.)

Dealing with time-dependent vector fields, there is a fundamental issue with topology. The technique described in Section 17.4.4 addresses the visualization of the unsteady streamlines' topology. Remember that streamlines are defined as integral curves in steady vector fields. In the context of time-dependent vector fields, they must be thought of as instantaneous integral curves, i.e., the paths of particles that circulate with infinite speed. This might sound like a weird idea. Actually, this is a typical way for fluid dynamicists to investigate the structure of time-dependent vector fields in practice. Observe that there is no restriction to this technique for the visualization of parameter-dependent vector fields, this parameter not being time. Nevertheless, if one is interested in the structure of pathlines, i.e., the paths of particles that flow under the influence of a vector field varying over time, one has to rethink the notion of topology. As a matter of fact, the asymptotic behavior of pathlines is not relevant for analysis since there is no longer infinite time for them to converge toward critical points. Thus, a new approach is required to define "interesting" behaviors. Furthermore, a structural equivalence relation must be determined between pathlines, upon which a corresponding topology can be built. This too seems to be a promising research direction in which to extend the scope of topological methods in the future.

References

1. A. A. Andronov, E. A. Leontovich, I. I. Gordon, and A. G. Maier. Qualitative theory of second-order dynamic systems. *Israel Program for Scientific Translations*, Jerusalem, 1973.

2. R. H. Abraham and C. D. Shaw. *Dynamics: the Geometry of Behavior*, I-IV. Aerial Press, 1982, 1983, 1985, 1988.

3. M. S. Chong, A. E. Perry, and B. J. Cantwell. A general classification of 3D flow fields. *Physics of Fluids*, A2(5):765–777, 1990.

4. U. Dallmann. Topological structures of 3D flow separations. DFVLR-AVA Bericht Nr. 221–82 A 07, *Deutsche Forschungs-und Versuchsanstalt für Luft-und Raumfahrt e. V.*, 1983.

5. T. Delmarcelle. *The Visualization of Second-Order Tensor Fields*. PhD Thesis, Stanford University, 1994.

6. T. Delmarcelle and L. Hesselink. The topology of symmetric, second-order tensor fields. *IEEE Visualization '94 Proceedings*, pages 140–147, 1994.

7. J. Guckenheimer and P. Holmes. Nonlinear oscillations, dynamical systems and linear algebra. New York, Springer, 1983.

8. A. Globus, C. Levit, and T. Lasinski. A tool for the topology of 3D vector fields. *IEEE Visualization '91 Proceedings*, pages 33–40, 1991.

9. J. L. Helman and L. Hesselink. Representation and display of vector field topology in fluid flow data sets. *Computer*, 22(8):27–36, 1989.

10. J. L. Helman and L. Hesselink. Visualizing vector field topology in fluid flows. *IEEE Computer Graphics and Applications*, 11(3):36–46, 1991.

11. L. Hesselink, Y. Levy and Y. Lavin. The topology of symmetric, second-order 3D tensor fields. *IEEE Transactions on Visualization and Computer Graphics*, 3(1):1–11, 1997.

12. M. W. Hirsch and S. Smale. *Differential Equations, Dynamical Systems, and Linear Algebra*. New York, Academic Press, 1974.

13. J. P. M. Hultquist. Constructing stream surfaces in steady 3D vector fields. *IEEE Visualization '92 Proceedings*, pages 171–178, 1992.

14. M. J. Lighthill. Attachment and separation in 3D flow. *Laminar Boundary Layers II*. Oxford, Oxford University Press, pages 72–82, 1963.

15. W. C. de Leeuw and R. van Liere. Collapsing flow topology using area metrics. *IEEE Visualization '99 Proceedings*, pages 349–354, 1999.

16. Y. Lavin, Y. Levy, and L. Hesselink. Singularities in nonuniform tensor fields. *IEEE Visualization '97 Proceedings*, pages 59–66, 1997.

17. S. Mann and A. Rockwood. Computing singularities of 3D vector fields with geometric algebra. *IEEE Visualization '02*, pages 283–289, 2002.

18. G. M. Nielson and I.-H. Jung. Tools for computing tangent curves for linearly varying vector fields over tetrahedral domains. *IEEE Transactions on Visualization and Computer Graphics*, 5(4):360–372, 1999.

19. A. E. Perry and M. S. Chong. A description of eddying motions and flow patterns using critical point concepts. *Ann. Rev. Fluid Mech.*, pages 127–155, 1987.

20. H. Poincaré. Sur les courbes définies par une équation différentielle. *J. Math.* 1:167–244, 1875, 2:151–217, 1876, 7:375–422, 1881, 8:251–296, 1882.

21. W. H. Press, S. A. Teukolsky, W. T. Vetterling, and B. P. Flannery. *Numerical Recipes in C*, 2nd ed. Cambridge, England, Cambridge University Press, 1992.

22. G. Scheuermann, T. Bobach, H. Hagen, K. Mahrous, B. Hahmann, K. I. Joy, and W. Kollmann. A tetrahedra-based stream surface algorithm. *IEEE Visualization '01 Proceedings*, pages 151–158, 2001.

23. G. Scheuermann, B. Hamann, K. I. Joy, and W. Kollmann. Visualizing local topology. *Journal of Electronic Imaging* 9(4), 2000.

24. G. Scheuermann, H. Hagen, and H. Krüger. An interesting class of polynomial vector fields. In *Mathematical Methods for Curves and Surfaces II*, pages 429–436. Nashville, TN, Vanderbilt University Press, 1998.

25. G. Scheuermann, H. Krüger, M. Menzel, and A. P. Rockwood. Visualizing nonlinear vector field topology. *IEEE Transactions on Visualization and Computer Graphics*, 4(2):109–116, 1998.

26. I. Trotts, D. Kenwright, and R. Haimes. Critical points at infinity: a missing link in vector field topology. *NSF/DoE Lake Tahoe Workshop on Hierarchical Approximation and Geometrical Methods for Scientific Visualization*, 2000.

27. M. Tobak and D. J. Peake. Topology of 3D separated flows. *Ann. Rev. Fluid Mech.*, 14:81–85, 1982.

28. X. Tricoche. *Vector and tensor topology simplification, tracking, and visualization*. Ph.D. thesis, Schriftenreihe FB Informatik 3, University of Kaiserslautern, Germany, 2002.

29. X. Tricoche, G. Scheuermann, and H. Hagen. Vector and tensor field topology simplification on irregular grids. *Data Visualization 2001—Proceedings of the Joint Eurographics – IEEE TCVG Symposium on Visualization*, pages 107–116, 2001.

30. X. Tricoche, G. Scheuermann, and H. Hagen. Tensor topology tracking: a visualization method for time-dependent 2D symmetric tensor fields. *Eurographics '01 Proceedings, Computer Graphics Forum*, 20(3):461–470, 2001.

31. X. Tricoche, G. Scheuermann, and H. Hagen. Continuous topology simplification of 2D vector fields. *IEEE Visualization '01 Proceedings*, 2001.

32. X. Tricoche, T. Wischgoll, G. Scheuermann, and H. Hagen. Topology tracking for the visualization of time-dependent 2D flows. *Computers & Graphics* 26:249–257, 2002.

33. R. Westermann, C. Johnson, and T. Ertl. Topology-preserving smoothing of vector fields. *IEEE Transactions on Visualization and Computer Graphics*, 7(3):222–229, 2001.

34. T. Wischgoll and G. Scheuermann. Detection and visualization of closed streamlines in planar flows. *IEEE Transactions on Visualization and Computer Graphics*, 7(2):165–172, 2001.

35. T. Wischgoll and G. Scheuermann. 3D loop detection and visualization in vector fields. In *"Mathematical Visualization" (VisMath 2002 Proceedings)*, 2003.

36. R. Batra and L. Hesselink. Feature comparisons of 3D vector fields using Earth mover's distance. In *Proceedings of IEEE Visualization '99*, pages 105–114, 1999.

37. W. de Leeuw and R. van Liere. Visualization of global flow structures using multiple levels of topology. In *Data Visualization '99, Eurographics*, pages 45–52, 1999.

38. W. de Leeuw and R. van Liere. Multi-level topology for flow visualization. In *Computers and Graphics*, 24(3):325–331, 2000.

39. Y. Lavin, R. Batra, and L. Hesselink. Feature comparisons of vector fields using Earth mover's distance. In *Proceedings of IEEE Visualization '98*, pages 103–110, 1998.

40. S. Lodha, J. Renteria, and K. Roskin. Topology preserving compression of 2D vector fields. In *Proceedings of IEEE Visualization 2000*, pages 343–350, 2000.

41. S. Lodha, N. Faarland, and J. Renteria. Topology preserving top-down compression of 2D vector fields using bintree and triangular quadtrees. *IEEE Transactions on Visualization and Computer Graphics*, pages 433–442, 2003.

42. H. Theisel. Designing 2D vector fields of arbitrary topology. In *Computer Graphics Forum*, 21(3):595–604, 2002.

43. V. Verma, D. Kao, and A. Pang. A flow guided streamline seeding strategy. In *Proceedings of IEEE Visualization 2000*, pages 163–170, 2000.

PART VI

Geometric Modeling for Visualization

18   3D Mesh Compression

JAREK ROSSIGNAC
College of Computing and Graphics, Visualization, Usability Center
Georgia Institute of Technology

18.1   Introduction

In this chapter, we discuss 3D compression techniques for reducing the delays in transmitting triangle meshes over the Internet. We first explain how vertex *coordinates*, which represent surface samples, may be compressed through quantization, prediction, and entropy coding. We then describe how the *connectivity*, which specifies how the surface interpolates these samples, may be compressed by compact encoding of the parameters of a connectivity-graph construction process and by transmission of the vertices in the order in which they are encountered during this process. The storage of triangle meshes compressed with these techniques is usually reduced to about a *byte per triangle*. When the exact geometry and connectivity of the mesh are not essential, the triangulated surface may be simplified or *retiled*. Although simplification techniques and the progressive transmission of refinements may be used as a compression tool, we focus on recently proposed retiling techniques designed specifically to improve 3D compression. They are often able to reduce the total storage, which combines coordinates and connectivity, to half a bit per triangle without exceeding a mean squared error of 1/10,000 of the diagonal of a box that contains the solid.

18.2   Background and Terminology

A triangle mesh is defined by a set of *vertices* and by its triangle-vertex *incidence* graph. The vertex description comprises *geometry* (3 coord-

inates per vertex) and optionally *photometry* (surface normals, vertex colors, or texture coordinates) [4,12,24], which will not be discussed here. *Incidence* (sometimes referred to as "topology") defines each triangle by the 3 integer indices that identify its vertices. For simplicity and elegance, we restrict our discussion in this chapter to *simple meshes*, which are homeomorphic to a triangulation of a sphere. However, most of the techniques presented here work for, or have been extended to, more general meshes with borders, handles, nonmanifold degeneracies, and nontriangular faces [4,15,24,29,33,46,50,64] and even to tetrahedral meshes [43,59,60].

In what follows, we assume that our triangle mesh is a connected manifold surface with no boundary and no handle and that it has v vertices, e edges, and t triangles. To simplify the formalism, we consider the *edges* to be the relatively open line segments that do not include their endpoints. Similarly, we use the term *face* to denote the relative interior of a triangle, excluding its edges and vertices. The *surface* of the mesh is the point-set union of its faces, edges, and vertices, which are all pairwise disjoint.

In order to prove the linear equation linking t and v and to facilitate the description of several compression approaches, we will use the following terminology. A *Vertex-Spanning Tree* (VST) of a triangle mesh is a subset of its edges, selected so that their union with all the vertices forms a tree (a connected cycle-free graph). Consider that a given VST has been selected. The edges it contains are called the *cut-edges*. The union of the cut-edges with all the vertices is called a *cut*. Because the VST is a tree, *there are $v - 1$ cut-edges*. The difference between the surface and its cut is called the *web*. Edges that are not cut-edges are called *hinge-edges*. The web is composed of all the faces and all the hinge-edges. Removing the cut, which has no loop, from the surface of a mesh will not disconnect it and will produce a web that is a (relatively open) triangulated 2D point set in three-space. Because by definition a simple mesh has no hole or handle, the web is simply connected and may be represented by an acyclic graph, whose nodes correspond to faces and whose links correspond to hinge edges. Thus there are $t - 1$ *hinge edges*. Note that by picking a leaf of this graph as the root and orienting the links, we can always turn it into a binary tree, which we call the *Triangle-Spanning Tree* (TST). It is a spanning tree of the dual of the graph made of the edges and vertices of the mesh. The TST defines a connected network of corridors through which one may visit all the triangles by walking across hinge-edges and never crossing a cut-edge. Because an edge is either hinge or cut, the total number of edges, e, is $v - 1 + t - 1$. Each triangle uses 3 edges and each edge is used by 2 triangles. Thus the number e of edges is also equal to $3t/2$. Combining these two equations yields $t = 2v - 4$, which shows that there are roughly twice as many triangles as vertices.

When 32-bit integers are used to represent triangle-vertex incidence references and 32-bit floats to represent vertex coordinates, an *uncompressed representation* of a simple mesh requires $12v$ bytes to store the geometry and $12t$ bytes (or equivalently $24v$–28 bytes) to store the incidence, which amounts to a total of $144t$ bits. Note that, surprisingly, the incidence information requires twice as much storage as the geometry.

18.3   Corner Table Representation

The *Corner Table* [50,51] is a simple data structure that simplifies the storage and pro-

cessing of manifold triangle meshes, whether they are simple or have holes and handles. We introduce it here and use it in this chapter to clarify the implementation details of compression and decompression techniques.

The *geometry* is stored in the *coordinate table*, G, where G[v] contains the triplet of the coordinates of vertex number v, and will be denoted v.g. Note that the order in which the vertices are listed in G is arbitrary, although once it is chosen, it defines the integer reference number associated with each vertex.

Triangle-vertex incidence defines each triangle by the three integer references to its vertices. These references are stored as consecutive integer entries in the V table. Note that each one of the $3t$ entries in V represents a *corner* (association of a triangle with one of its vertices). Let c be such a corner. Let c.t denote its triangle and c.v its vertex. Remember that c.v and c.t are integers in $[0, v - 1]$ and $[0, t - 1]$ respectively. Let c.p and c.n refer to the *previous* and *next* corners in the cyclic order of vertices around c.t.

Although G and V suffice to completely specify the triangles and thus the surface they represent, they do not offer direct access to a neighboring triangle or vertex. We chose to use the reference to the *opposite* corner, c.o, which we cache in the O *table* to accelerate mesh traversal from one triangle to its neighbors. For convenience, we also introduce the operators c.l and c.r, which return the *left* and *right neighbors* of c (Fig. 18.1).

Note that we do not need to cache c.t, c.n, c.p, c.l, or c.r because they may be quickly evaluated as follows: c.t is the integer division c.t DIV 3; c.n is c−2, when c MOD 3 is 2, and c+1 otherwise; c.p is c.n.n; c.l is c.n.o; and c.r is c.p.o. Thus, the storage of the connectivity is reduced to the O and V arrays.

We assume that all triangles have been consistently *oriented* so that c.n.v=c.o.p.v for all corners c. For example, one may adhere to the convention that when a triangle c.t is visible by a viewer outside of the solid (i.e., the finite set

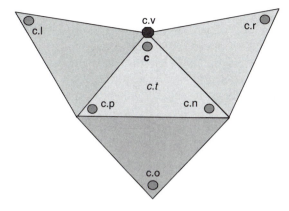

Figure 18.1 Corner operators for traversing a corner table representation of a triangle mesh.

that is bounded by the triangle mesh), the three vertices, c.p.v, c.v, and c.n.v, appear in clockwise order.

Assume that c.t.m is a Boolean set to TRUE when the triangle c.t has been visited. The procedure

```
visit (c) {
  if not c.t.m THEN {
    c.t.m:=true;
    visit(c.r);
    visit(c.l);
  }
}
```

will visit all the triangles in depth-first order in a TST.

Given the V table, the entries in O may be computed by the following:

```
for c:=0 to 3t−2 do
  for b: = c + 1 to 3t−1 do
    if (c.n.v==b.p.v) && (c.p.v==b.n.v)
    then{
      c.o:=b;
      b.o:=c
    }
```

A faster approach sorts the triplets {min(c.n.v,c.p.v), max(c.n.v,c.p.v), c} into bins. All entries in a bin have the same first record:

min(c.n.v,c.p.v), an integer in $[0, v − 1]$. There are rarely more than 20 entries in a bin. Then, we sort the entries in each bin by the second record: max(c.n.v,c.p.v). Now, pairs with identical first record and second record are consecutive and correspond to opposite corners, which are identified by the third record in each triplet. Thus, if a sorted bin contains consecutive entries (a,b,c) and (a,b,d), we set c.o:=d and d.o:=c.

Because it can be easily recreated, the O table need not be transmitted. Furthermore, the $31 − \log_2 v$ leading zeros of each entry in the V table need not be transmitted. Thus, assuming that floats are used for the coordinates, a compact but uncompressed representation of a triangle mesh requires $48t$ bits for the coordinates and $3t \log_2 v$ bits for the V table. Note that Edgebreaker (discussed below) encodes the full connectivity information contained in both V and O with a linear cost of less than $2t$ bits, and hence eliminates the need for the decompression modules on the client to recompute O from V.

18.4 Geometry Compression

The compression of vertex coordinates usually combines three steps: quantization, prediction, and statistical coding of the residues. We explain them briefly in this section.

Quantization truncates the vertex coordinates to a desired accuracy and maps them into integers that can be represented with a limited number of bits. To do this, we first compute a tight (min–max), axis-aligned bounding box around each object. The minima and maxima of the x, y, and z coordinates, which define the box, will be encoded and transmitted with the compressed representation of each object. Then, given a desired accuracy, e, we transform each x coordinate into an integer $i = INT((x − x_{min})/(e(x_{max} − x_{min})))$, which ranges between 0 and 2^B, where $B = \lceil \log_2 ((x_{max} − x_{min})/e) \rceil$ is the maximum number of bits needed to represent the quantized coordinate i. The y and z coordinates are quantized similarly. Choosing $e = \max((x_{max} - x_{min})/2^{12}, (y_{max} - y_{min})/2^{12}, (z_{max} - z_{min})/2^{12})$ yields $B \leq 12$ for all coordinates and ensures a sufficient geometric fidelity for most applications and most models. Thus, this *lossy* quantization step reduces the storage cost of geometry from $96v$ bits to less than *36v bits*.

The next, and most crucial, geometry compression step involves using a *vertex predictor*. Both the encoder and the decoder use the same predictor. Thus, only the *residues* between the predicted and the correct coordinates need to be transmitted. The *coherence* between neighboring vertices in meshes of finely tiled smooth surfaces reduces the magnitude of the residues.

Because most edges are short with respect to the size of the model, adjacent vertices are generally close to each other, and the differences between the coordinates are small. Thus, a new vertex may be predicted by a previously transmitted neighbor [8].

Instead of using a single neighbor, when vertices are transmitted in VST top-down order, a linear combination of the 4 *ancestors* in the VST may be used [62]. The 4 coefficients of this combination are computed to minimize the magnitude of the residues over the entire mesh and transmitted as part of the compressed stream.

The most popular predictor for single-rate compression is based on the *parallelogram* construction [64]. Assume that the vertices of c.t have been decoded. We predict c.o.v.g using c.n.v.g+c.p.v.g−c.v.g. The parallelogram prediction may sometimes be improved by predicting the angle between c.t and c.o.t from the angles of previously encountered triangles or from the statistics of the mesh.

Some of the residues may be large. Thus, good prediction by itself may not lead to compression. For example, if the coordinates have been quantized to B-bit integers, some of the coordinates of the corrective vector, c.o.v.g

−c.n.v.g −c.p.v.g +c.v.g, may require $B + 2$ bits of storage. Thus, parallelogram prediction could, in principle, expand storage rather than compress it. However, the distribution of the residues is usually biased towards zero, which makes them suitable for statistical compression [52].

In practice, the combination of these steps compresses vertex location data to about *7t bits*.

18.5   Connectivity Compression

As argued above, geometry may be encoded efficiently, provided that connectivity information is available during geometry decompression to locate previously decoded neighbors of each vertex. This section presents techniques for compressing the connectivity information from $3t \log_2 v$ bits to bt bits, where b is guaranteed never to exceed 1.80, and in practice is usually close to 1.0. As a result, meshes may be encoded with a total of about *8t bits* (*7t bits* for geometry, *1t bit* for connectivity).

Instead of retracing the chronological evolution of the research in the field of single-rate incidence compression, we first describe in detail Edgebreaker [48], which is arguably the simplest and one of the most effective single-rate compression approaches. The source code for Edgebreaker is publicly available [54]. Then, we briefly review several variants and other approaches using Edgebreaker's terminology to characterize their main differences and respective advantages or drawbacks.

18.5.1   Edgebreaker

The Edgebreaker compression visits the triangles in a spiraling (depth-first) TST order and generates the *clers* string of labels, one label per triangle, which indicate to the decompression how the mesh can be rebuilt by attaching new triangles to previously reconstructed ones (Fig. 18.2).

```
RECURSIVE PROCEDURE Compress (c) {          # compresses a simple t-meshes
  REPEAT {                                  # traverses TST, stopped by RETURN
    set(c.t.m);                             # mark triangle as visited
    IF !c.v.m                               # test whether tip vertex was visited
      THEN { encode(c.v.g);                 # store location of tip
        WRITE(clers, 'C');                  # append C to clers string
        set(c.v.m);                         # mark tip vertex as visited
        c:= c.r}                            # continue with the right neighbor
    ELSE IF c.r.t.u                         # test whether right triangle was visited
    THEN IF c.l.t.u                         # test whether left triangle was visited
      THEN { WRITE(clers, 'E'); RETURN }    # append code for E and pop or stop
      ELSE { WRITE(clers, 'R'); c:= c.l}    # append code for R, move to left triangle
    ELSE IF c.l.t.u                         # test whether left triangle was visited
      THEN { WRITE(clers, 'L'); c:= c.r}    # append code for L, move to right triangle
      ELSE { WRITE(clers, 'S');             # append code for S
        Compress(c.r); c:= c.l}} }          # recurse right, then continue left
```

The pseudo-code for the Edgebreaker compression is shown in the box above. The following explanations contain in parentheses the excerpts of the pseudo-code they reference. Edgebreaker works directly on the Corner Table and does not require any additional data structure, except for one bit per vertex and one bit per triangle to mark those that have already been processed. In particular, it does not require us to maintain linked lists of border edges. It traverses the mesh in depth-first order of a TST using iteration (REPEAT) and occasionally recursion (Compress) on corner indices. It marks all visited vertices (set(c.v.m)) and triangles (set(c.t.m)). The current triangle is identified by its tip corner (c). Note that the current triangle has been reached though the gate edge joining c.n.v with c.p.v. By testing the marks of the tip vertex of the current triangle and of neighboring triangles, it selects the label and appends it to the clers string.

If the tip vertex (c.v) has not yet been visited (lc.v.m), its location is encoded (encode(c.v.g)) using the parallelogram prediction and geometry compression, as explained earlier. The label C is appended to the clers string (WRITE(clers, C)) and the iteration moves to the right neigh-bor (c:=c.r). Note that the vertices are encoded in the order in which they are encountered by C-triangles during this traversal. This order does not usually reflect the order in which the vertices were listed in the original mesh. Similarly, the triangles are reordered during transmission. A dictionary mapping the original order on the server to the new order on the client may be kept on the server to reconcile vertex or triangle selections between one location and the other in subsequent processing.

When the tip of the current triangle has been previously visited, we distinguish four other types of triangles: L, R, S, and E (Fig. 18.2).

case L: When the left neighbor has been visited, but not the right one, we append the label L to the clers string and iterate on the right neighbor (c:=c.r).

case R: When the triangle on the right has been visited, but not the one on the left, we append R to the clers string and iterate on the left neighbor (c:=c.l).

case S: When both neighbors have not-visited status, we append S to clers, start a recursive process on the right

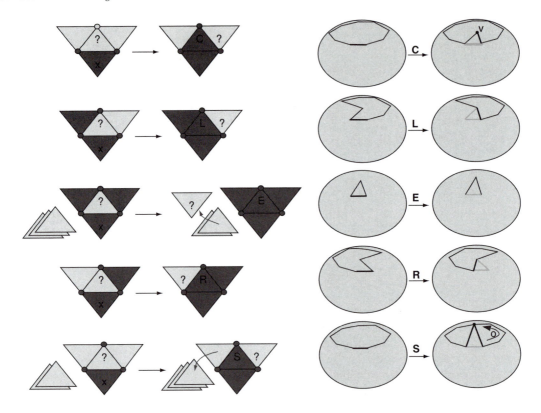

Figure 18.2 The five Edgebreaker situations (C, L, E, R, S) are illustrated top to bottom. On the left, we show the "before" and "after" states for each situation during compression. The current triangle is marked by "?". Previously visited triangles and vertices are darker. An X marks the triangle through which we came. We encode a C label when the tip vertex of the current triangle was not marked (top). Otherwise, the label depends on the status of the left and right neighbor triangles. When neither has been visited, we encode an S (bottom), after which compression goes right via a recursive call and then left. We show this symbolically by adding the left neighbor to a stack. When both neighbors have been visited previously, we encode and E and exit the procedure (possibly returning from a recursive call). The right column shows how decompression interprets the CLERS symbols to reconstruct the connectivity of the mesh. For each symbol in the *clers* string, Edgebreaker decompression attaches a new triangle to the gate edge (indicated by a thick line on the left figure, where the state before the insertion of the new triangle is shown). The gate for the next operation is placed as indicated by a thick line on the right column, which shows the state after the new triangle was inserted. Decoding a C symbol (top) creates a new vertex (v). When decoding an S symbol (bottom), the location of the tip of the new triangle is defined by the offset o from the gate around the bounding loop. The S operation puts the gate on the right edge of the new triangle and proceeds to fill the right hole using a recursive call. Then it sets the gate to the left edge of the new triangle and resumes the process. The offsets o for each S symbol may be computed from the *clers* string using the fact that C and S increment the edge count, L and R decrement it, and E reduces it by 3.

neighbor (*Compress*(c.r)), and then iterate on the left neighbor (c:=c.l).

case E: When both neighbors have been visited, we append E to the *clers* string and return from recursion or from the compression process (**RETURN**).

The connectivity of the first triangle is implicit. The initialization, detailed in the insert on the facing page, sets the visited tags (.m) to zero (not shown here). Then it encodes the first three vertices and marks them and their triangle. It calls the compression on one of the corners of that triangle (*Compress* (c.o)).

```
PROCEDURE initCompression (c){
        encode(c.p.v.g); encode(c.v.g);
        encode(c.n.v.g);                   # store first 3 vertices
        set(c.v.m, c.p.v.m, c.n.v.m,
        c.t.m);                            # mark first 3 vertices and triangle as
                                             visited
        Compress (c.o);}       # start compression at opposite corner
```

A typical execution of the compression process is illustrated in Fig. 18.3.

Because, except for the first two vertices, there is a one-to-one mapping between each C triangle and each vertex, the number of C triangles is $v - 2$. Consequently, the number of non-C triangles in a simple mesh is $t - (v - 2)$, which is also $v - 2$. Thus exactly *half of the triangles are of type C*. Hence, Edgebreaker guarantees that a compressed representation of the connectivity of a simple triangle mesh will never exceed $2t$ bits [48] if we use the following simple binary code for the labels $(C = 0, L = 110, E = 111, R = 101, S = 100)$.

Given that the sub-sequences CE and CL are impossible, a slightly more complex code [28] may be used to guarantee that the compressed file will never exceed $1.84t$ bits. This code uses $(C = 0, S = 10,$ and $R = 11)$ for symbols that follow a C, and one of the following 3 codes for symbols that do not follow a C:

- **Code 1:** C is 0, S is 100, R is 101, L is 110, E is 111
- **Code 2:** C is 00, S is 111, R is 10, L is 110, E is 01
- **Code 3:** C is 00, S is 010, R is 011, L is 10, E is 11

It was proven [28] that one of these 3 codes always takes less than $11t/6$ bits. A 2-bit switch header is used to identify which code is used for each model.

Further constraints exist on the *clers* string. For example, CCRE is impossible, because CCR increments the length of the loop, which must have been at least 3. By exploiting such constraints to better estimate the probability of the next symbol, a more elaborate code was

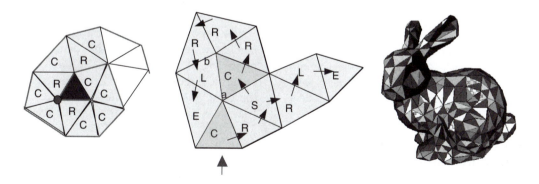

Figure 18.3 In this example of a typical compression situation, Edgebreaker starts with the darker triangle (left) and spirals out clockwise, filling the beginning of the clers string with CCCCRCCRCRC. It appends the tip of each C triangle to the vertex list. A typical situation where Edgebreaker finishes compression or closes a hole is shown in the center. It spirals counterclockwise, appending the label sequence CRSRLECRRRLE to the *clers* string and adding the vertices (a) and (b) to the vertex list. The triangles in the rabbit (right) have been shaded according to their Edgebreaker labels. Notice that half of the triangles are C (white) and about a third are R. (See also color insert.)

developed [16] that guarantees $1.778t$ bits when using a forward decoding [49] and $1.776t$ bits when using a reverse decoding scheme [25].

Hence, the Edgebreaker encoding of the connectivity of any mesh (homeomorphic to a sphere) may be compressed down to $1.78t$ bits. This brings it within 10% of the proven $1.62t$ theoretical lower bound for encoding planar triangular graphs, as established by Tutte [67], who by counting all planar triangulations of v vertices has proven that an optimal encoding uses at least $v \log_2 (256/7) \approx 3.245v$ bits for a sufficiently large v.

These recent developments constitute a significant advance in the study of short encodings of planar triangle graphs. They are often the best solution for compressing small or irregular meshes. For large and fairly regular meshes, better compression ratios may often be obtained. For example, one may encode CC, CS, and CR pairs as single symbols. Each odd C symbol will be paired with the next symbol. After an even number of C symbols, we use the following codes: $CR = 01, CC = 00$, $CS = 1101$, $R = 10$, $S = 1111$, $L = 1110$, $E = 1100$. This encoding guaranteed $2.0t$ bits, but usually yields between $1.3t$ and $1.6t$ bits [49].

Furthermore, by arranging symbols into words that each start with a sequence of consecutive Cs and by using a Huffman code [52], we often reduce storage to less than $1.0t$ bits. For example, $0.85t$ bits suffice for the Huffman codes of the Stanford Bunny. Including the cost of transmitting the associated 173-word dictionary brings the total cost to $0.91t$ bits. A gzip compression of the resulting bit stream reduces it by only 2%.

As shown earlier, the location of the next vertex may be predicted using previously decoded geometry and connectivity. Coors and Rossignac [6] have proposed to also predict the connectivity of the next triangle using the same information. In their Delphi system, compression and decompression perform the same geometric prediction of the location of the tip-vertex of the next triangle. Then they estimate the triangle connectivity, and thus its symbol in the

clers string produced by the Edgebreaker compression, by *snapping* the tip vertex to the nearest vertex in the active loop, if one lies sufficiently close. If no bounding vertex lies nearby, the next *clers* symbol is estimated to be a C. If the guess is correct, a single confirmation bit is sufficient. Otherwise, an entropy-based code is received and used to select the correct CLERS symbol from the other four possible ones (or the correct tip of an S triangle). Reported experiments indicate that, depending on the model, up to 97% of Delphi's guesses are correct, compressing the connectivity to $0.19t$ bits. When the probability of a wrong guess exceeds 40%, the Delphi encoding stops being advantageous.

Let us now discuss several approaches to the *decompression* of the *clers* string. All approaches attach a new triangle, one at a time, to a *gate* edge, which so far has only one incident triangle. The next symbol in the *clers* string defines where the tip of the new triangle is (Fig. 18.2). Symbol C indicates that the new triangle will have as a tip a new vertex. Note that the three vertices of the previously decoded triangle that is incident upon the gate have been previously decoded and may be used in a parallelogram prediction of the new vertex. Also note that the numbering of the vertices and hence their order in the G table of the reconstructed mesh reflects the order in which the vertices are instantiated as tips of C triangles.

Symbol L indicates that the tip vertex is immediately to the left of the gate along the boundary of the portion of the mesh decoded so far. R indicates that the tip is immediately to the right of the gate. E indicates that the new triangle will close a hole that must have exactly 3 vertices. S indicates that the tip of the new triangle is elsewhere on the boundary of the previously decoded portion of the mesh.

Consider the edge-connected components of the not-yet-decoded portion of the mesh. Let M be the component incident upon the gate. Because by definition of simple meshes M has no handle, an S triangle will always split it in two parts. Through a recursive call, Edgebreaker will first reconstruct the portion of M that is incident upon the right edge of S, as seen when entering the triangle through the gate. Then, upon return from the recursive call, the reconstruction of the rest of M will resume.

After the new triangle is attached, the gate is moved to the right edge of the new triangle for cases C and L. It is moved to the left edge for case R. When an S triangle is attached, the gate is first moved to the right edge of the S triangle and the right hole is then filled through a recursive call to decompression. Then the gate is moved to the left edge and the process resumes as if the S triangle had been an R triangle.

The only challenge in the Edgebreaker decompression lies in the location of the tips of the S triangles. Several approaches have been proposed, and they are briefly discussed below.

The integer reference number of the tip of each S triangle could be encoded using $\log_2 k$ bits, where k is the number of previously decoded vertices. A more economical approach encodes an *offset*, o, indicating the number of vertices that separate the gate from the tip in the current loop (Fig. 18.2). Because the current loop may include a large fraction of the vertices, one may still need up to $\log_2 k$ bits to encode the offset. Although the total cost of encoding the offsets is linear in the number of triangles [17], the encoding of the offsets constitutes a significant fraction of the total size of the compressed connectivity. Hence, several authors strove to minimize the number of offsets [2], mostly by using heuristics for selecting gates with a low probability of being the base of S triangles.

The breakthrough of Edgebreaker lies in the discovery that offsets need not be transmitted at all because they can be recomputed by the decompression algorithm from the *clers* string itself. The initial solution [48] is based on the observation that the attachment of a triangle of each type changes the number of edges in the current loop by specific amounts (Fig. 18.2). C increments the edge count. R and L decrement it. E removes a loop of three edges and thus decreases the edge count by 3. S splits the cur-

rent loop in two parts, but if we count the edges in both parts, it increments the total edge count. Each S label starts a recursive call that will fill in the hole bounded by the right loop and will terminate with the corresponding E label. Thus, S and E labels work as pairs of parentheses. Combining all these observations, we can compute the offset by identifying the substring of the *clers* string between an S and the corresponding E, and by summing the edge-count changes for each label in that substring. To avoid the multiple traversals of the *clers* string, all offsets may be precomputed by way of reading the *clers* string once and using a stack for locating the S of each E.

The elegant Spirale Reversi approach [25] to decompression of *clers* strings that have been created by the Edgebreaker compression avoids this preprocessing by reading the *clers* string backwards and building the triangle mesh in reverse order (Fig. 18.4). It assigns a reference number to a vertex not at its creation, but only when a C triangle incident upon it is created. The order in which vertices are assigned reference numbers by the Spirale Reversi decompression is reversed from the order in which they are first encountered by the Edgebreaker compression. Note that the vertices of new triangles are initially unlabeled, and they remain so until the corresponding C triangles are created.

A third approach, Wrap&Zip [49], also avoids the preprocessing necessary with Rossignac's method [48] and directly builds a corner table as it reads the *clers* string. It does not require the maintaining of a linked list of border vertices or edges. For each symbol, as a new triangle is attached to the gate, Wrap&Zip fills in the known entries to the V and O tables. Specifically, it fills in c.o for the tip corner, c, of the new triangle and for its opposite, c.o. It assigns vertex reference numbers to the tips of C triangles as they are created, by simply incrementing a vertex counter. It defers assignation of the reference numbers to other vertices until a Zip process matches them with vertices that already have a reference number. Thus, it

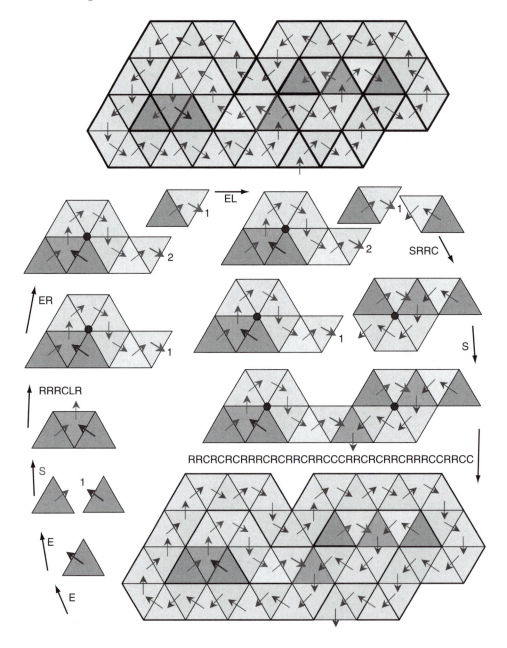

Figure 18.4   The connectivity of the remaining portion of the mesh shown on the top is encoded by Edgebreaker in the *clers* string: CCRRCCRRRCRRCRCRRCCCRRCRRCRCRRRCRCRCRRSCRRSLERERLCRRRSEE. The order in which the triangles are visited is shown by the arrows. The Spirale Reversi decompression receives the string reversed. Processing the first symbol (E) of the reversed string generates the first triangle (bottom left). The arrow leaving the previously reconstructed portion of the mesh indicates the gate where a new triangle will be attached. Then the next symbol (another E) puts the gate on a stack (1) and creates a new disconnected triangle with a new gate. Moving clockwise, the next symbol (S) makes a new triangle that joins the gate with the top of the stack. Reading the symbols RRC creates a right-turning fan that encloses a vertex (large dot), which will receive the reference number 1. Then the LR symbols are read. The next symbol (E) puts the gate on the stack (1) and creates an isolated triangle to which another one is attached as we read the next R symbol (top left). This creation, growth, and merging process continues as shown in the clockwise sequence. (See also color insert.)

produces a web, as defined earlier. The border edges of the web must be matched into pairs. The correct matching could be specified by encoding the structure of the cut [62,65]. However, as discovered by Rossignac and Szymczak [49], the information may be trivially extracted from the *clers* string by orienting the border edges of the web, as shown in Fig. 18.5. Note that these border orientations are consistent with an upward orientation of the cut-edges toward the root of the VST. The pseudo-code for the Wrap&Zip decompression algorithm is shown in the frame below. It was extended to meshes with handles by Lopes et al. [33].

```
PROCEDURE initDecompression {
  GLOBAL V[ ] = { 0,1,2,0,0,0,0,0, ...};       # table of vertex Ids for each corner
  GLOBAL O[ ] =
  {-1, − 3, − 1, − 3, − 3, − 3} . . .;          # table of opposite corner Ids for each corner
  GLOBAL T = 0;                                  # id of the last triangle decompressed
                                                 #   so far
  GLOBAL N = 2;                                  # id of the last vertex encountered
  DecompressConnectivity(1);                     # starts connectivity decompression

RECURSIVE PROCEDURE
DecompressConnectivity(c) {
  REPEAT {                                       # Loop builds triangle tree and zips it up
      T ++;                                      #   new triangle
      O[c] = 3T; O[3T] = c;                      #   attach new triangle, link opposite
                                                 #     corners
      V[3T + 1] = c.p.v; V[3T + 2] = c.n.v;      #   enter vertex Ids for shared vertices
      c = c.o.n;                                 # move corner to new triangle
      Switch decode(READ(clers) ) {              # select operation based on next symbol
      Case C: {O[c.n] = −1; V[3T] = ++N;}        # C: left edge is free, store ref to new
                                                 #   vertex
      Case L: {O[c.n] = −2; zip(c.n);}           # L: orient free edge, try to zip once
      Case R: {O[c] = −2; c = c.n}               # R: orient free edge, go left
      Case S: { DecompressConnectivity(c);
        c = c.n}                                 # S: recursion going right, then go left
      Case E: {O[c] = −2; O[c.n] = −2;
        zip(c.n); RETURN }}}}                    # E: zip, try more, pop

RECURSIVE PROCEDURE Zip(c) {                     # tries to zip free edges opposite c
  b = c.n; WHILE b.o>=0 DO b=b.o.n;              # search clockwise for free edge
  IF b.o!= −1 THEN RETURN;                       # pop if no zip possible
  O[c]=b; O[b]=c;                                #   link opposite corners
  a = c.p; V[a.p] = b.p.v;                        #   assign co-incident corners
  WHILE a.o>=0 && b!=a DO {a=a.o.p;
    V[ a.p]=b.p.v};
  c = c.p; WHILE c.o >= 0 && c!= b
    DO c = c.o.p;                                # find corner of next free edge on right
  IF c.o == −2 THEN Zip(c)}                      # try to zip again
```

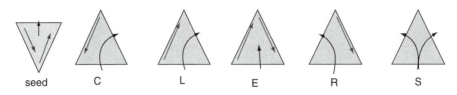

Figure 18.5   The borders of the web are oriented clockwise, except for the seed and the C triangles.

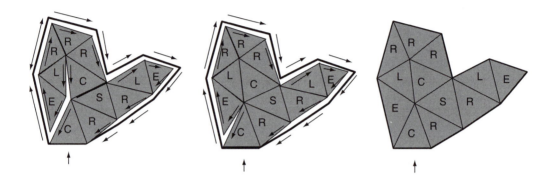

Figure 18.6   We assume that the part of the mesh not shown here has already been decoded into a web with properly oriented borders (exterior arrows). Building the TST (shown by the labeled triangles) for the substring CRSRLECRRRLE produces a web whose free borders are oriented clockwise for all non-C triangles and counterclockwise for C triangles (left). Each time Wrap&Zip finds a pair of edges oriented away from their common vertex, it matches them. The result of the first zip operation (center) enables another zip. Repeating the process zips all the borders and restores the desired connectivity (right).

The zipping part matches pairs of adjacent border edges that are oriented away from their shared vertices. Only the creation of L and E triangles opens new zipping opportunities. Zipping the borders of an E triangle may start a chain of zipping operations (Fig. 18.6). The cost of the zipping process is linear, since there are as many zipping operations as edges in the VST and the number of failed zipping tests equals the number of E or L triangles.

18.5.2   Other Approaches

The *cut-border machine* [18] has strong similarities to Edgebreaker. Because it requires the explicit encoding of the offset of S triangles and because it was designed to support manifold meshes with boundaries, the cut-border method is slightly less effective than Edge-

breaker. Reported connectivity compression results range from $1.7t$ to $2.5t$ bits. A context-based arithmetic coder further improves them to $0.95t$ bits [17]. Gumhold [16] proposes a custom variable-length scheme that guarantees less than $0.94t$ bits for encoding the offsets, thus proving that the cut-border machine has linear complexity.

Turan [65] noted that the connectivity of a planar triangle graph can be recovered from the structure of its VST and TST, which he proposed to encode using a total of roughly $12v$ bits. Rossignac [48] has reduced this total cost to $6v$ bits by combining two observations: (1) The binary TST may be encoded with $2t$ bits using one bit per triangle to indicate whether it has a left child and another one to indicate whether it has a right child. (2) The corresponding (dual) VST may be encoded with $1t$ bits, one bit per vertex indicating whether the node is a leaf and the other bit per vertex indicating whether it is the last child of its parent. (Remember that $2v = t + 4$.) This scheme does not impose any restriction on the TST. Note that for less than the $2t$-bits budget needed for encoding the TST alone, Edgebreaker [48] encodes the *clers* string, which describes not only how to reconstruct the TST, but also how to orient the borders of the resulting web so as to define the VST and hence the complete incidence. This surprising efficiency seems linked to the restriction of using a *spiraling* TST.

Taubin and Rossignac have noticed that a spiraling VST, formed by linking concentric loops into a tree, has relatively few branches. Furthermore, the corresponding dual TST, which happens to be identical to the TST produced by Edgebreaker, also generally has few branches (Fig. 18.7). They have exploited this regularity by *Run Length Encoding* (RLE) the TST and the VST. Each run is formed by consecutive nodes that have a single child. The resulting *Topological Surgery* 3D compression technique [61,62] encodes the length of each run, the structure of the trees of runs, and a marching pattern, which encodes each triangle run as a generalized *triangle strip* [10] using one bit per triangle to indicate whether the next triangle of the run is attached to the right or to the left edge of the previous one. An IBM implementation of the topological surgery compression has been developed for the VRML standard [63] for the transmission of 3D models across the Internet, thus providing a compressed binary alternative to the original VRML ASCII format [71], resulting in a 50-to-1 compression ratio. Subsequently, the topological surgery approach has been selected as the core of the Three-Dimensional Mesh Coding (3DMC) algorithm in MPEG-4 [38], which is the ISO/IEC multimedia standard developed by the Moving Picture Experts Group for digital television, interactive graphics, and interactive multimedia applications.

Instead of linking the concentric rings of triangles into a single TST, the *layered structure* color coded in Fig. 18.7 (left) may be preserved [4]. The incidence is represented by the total number of vertex layers, and by the triangulation of each layer. When the layer is simple, its triangulation may be encoded as a *triangle strip*, using one marching bit per triangle, as was originally done in the topological surgery approach. However, in practice, a significant number of overhead bits is needed to encode the connectivity of more complex layers. The topological surgery approach resulted from an attempt to reduce this additional cost by chaining the consecutive layers into a single TST (see Fig. 18.7).

Focusing on hardware decompression, Deering [8] encodes generalized triangle strips using a buffer of 16 vertices. One bit identifies whether the next triangle is attached to the left or the right border edge of the previous triangle. Another bit indicates whether the tip of the new triangle is encoded in the stream or is still in the buffer and can hence be identified with only 4 bits. Additional bits are used to manage the buffer and to indicate when a new triangle strip must be started. This compressed format is supported by Java 3D's Compressed Object node [26]. Chow [5] has provided an algorithm for compressing a mesh into Deering's format by extending the border of the previously visited part of the mesh by a fan of not-yet-visited triangles around a border vertex. When the tip of the new triangle is a previously decoded vertex no longer in the cache, its coordinates, or an absolute or relative reference to them, must be included in the vertex stream, significantly increasing the overall transmission cost. Therefore, the optimal encoding traverses a TST that is different from the spiraling TST of Edgebreaker in an attempt to reduce cache misses.

Given that there are 3 corners per triangle and that $t = 2v - 4$, there are roughly 6 times as many corners as vertices. Thus, the *average valence*, i.e., the number of triangles incident upon a vertex, is 6. In most models, the valence distribution is highly concentrated around 6. For example, in a subdivision mesh, all vertices

Figure 18.7   (Left) The Topological Surgery approach merges concentric circles of triangles into a single TST. (Right) That TST and its dual VST have relatively few runs. (See also color insert.)

that do not correspond to vertices of the original mesh have valence 6. To exploit this statistic, Touma and Gotsman [64] have developed a valance-based encoding of the connectivity, which visits the triangles in the same order as Edgebreaker does. As in Edgebreaker, they encode the distinction between the C and the S triangles. However, instead of encoding the symbols for L, R, and E, they encode the valence of each vertex and the offset for each S triangle. When the number of incident triangles around a vertex is one less than its valence, the missing L, R, or E triangle may be completed automatically. For this scheme to work, the offset must encode not only the number of vertices separating the gate from the tip of the new triangle along the border (Fig. 18.2), but also the number of triangles incident on the tip of the S triangle that are part of the right hole. To better appreciate the power of this approach,

consider the statistics of a typical case. Only one bit is needed to distinguish a C from an S. Given that 50% of the triangles are of type C and about 5% of the triangles are of type S, the amortized entropy cost of that bit is around $0.22t$ bits. Therefore, about 80% of the encoding cost lies in the valence, which has a low entropy for regular and finely tessellated meshes and in the encoding of the offsets. For example, when 80% of the vertices have valence 6, a bit used to distinguish them from the other vertices has entropy 0.72 and hence the sequence of these bits may be encoded using close to $0.36t$ bits. The amortized cost of encoding the valence of the other 20% vertices with 2 bits each is $0.40t$ bits. Thus, the valence of all vertices in a reasonably regular mesh may be encoded with $0.76t$ bits. If 5% of the triangles are of type S and each offset is encoded with an average of 5 bits, the amortized cost of the offsets reaches

0.25*t* bits. Note that the offsets add about 25% to the cost of encoding the C/S bits and the valence, yielding a total of 1.23*t* bits. This cost drops down significantly for meshes with a much higher proportion of valence-6 vertices.

Although attempts to combine the Edgebreaker solution that avoids sending the offsets with the valence-based encoding of the connectivity have failed, Alliez and Desbrun [2] managed to significantly reduce the total cost of encoding the offsets by reducing the number of S triangles. They use a heuristic that selects as gate a border edge incident upon a border vertex with the maximal number of incident triangles. To further compress the offsets, they sort the border vertices in the active loop according to their Euclidean distances from the gate and encode the offset values using an arithmetic range encoder [55]. They also show that if one could eliminate the S triangles, the valence-based approach would guarantee compression of the mesh with less than 1.62*t* bits, which happens to be *Tutte's lower bound* [67].

An improved Edgebreaker compression approach was proposed [56,57] for sufficiently large and regular meshes. It is based on a specifically designed context-based coding of the *clers* string and uses the Spirale Reversi decompression. For a sufficiently large ratio of degree-six vertices and a sufficiently large *t*, this approach is proven to guarantee a worst-case storage of 0.81*t* bits.

18.6 Retiling

Triangle mesh simplification techniques [20,36,44] reduce the sampling of the mesh, while decreasing its accuracy. Recently proposed approaches [11,32,37,41] combine ordered edge-collapse operations [21,22] with proximity-based or grid-based vertex clustering [35,45]. Both merge vertices and eliminate degenerate triangles. Simplification techniques have been developed to accelerate hardware-assisted 3D rasterization, and as such, they attempt to find the best compromise between

reducing the triangle count and reducing the error. The error resulting from such a simplification process may be estimated using the maximum [47] or the sum [12,13,14] of the squared distances between the new location of displaced vertices and the planes containing their incident triangles in the original model. It may also be evaluated [7] by sampling the original mesh and computing the distance between the samples and the simplified mesh. One may combine these simplification techniques with the 3D compression approaches described above to achieve a flexible lossy compression. To support the transmission of multiresolution scenes, several levels of detail (LODs) of each model in the scene should be generated through simplification and compressed. The level at which a model is downloaded may depend on the scale at which it is projected on the screen and on the total bit budget allocated to the transmission. After the initial transmission of such an approximation of the scene, compressed higher-resolution versions of selected models may be downloaded, either to improve the overall accuracy of the image or to adapt to the motions of the viewpoint.

When the complexity ratio between one LOD and the next is large, there is little or no savings in trying to reuse the lower LOD to reduce the transmission cost of the next level. When finer granularity refinements are desired, parameters to undo the sequence of simplifying edge-collapses in reverse order may be transmitted one by one [22], or grouped into 6 to 12 batched and compressed [23,39,40]. A transmission cost of 3.5*t* bits for connectivity and 7.5*t* bits for geometry was achieved by using one bit per vertex to mark which vertices must be split in each batch [39,40]. Marking the edges, instead of the vertices [68,69,70], allows the method to recover for free the connectivity of valence-6 portions of the mesh and in general reduces the cost of the progressive transmission of connectivity to between 1.0*t bits* and 2.5*t bits*. A series of simplification passes [1]—which do not take geometry into account, but each divide by 3 the number of vertices through a systematic removal of vertices of valence less than 7, a retriangulation of the resulting holes, and a subsequent removal of vertices of valence 3—permit us to encode connectivity upgrades with about $1.9t$ bits.

These simplification or progressive coding techniques encode a simplified version of the original connectivity graph and, optionally, the data necessary to fully restore it to its original form. When there is no need to preserve that connectivity, one may achieve better compression by producing a more regular sampling of the original mesh at the desired accuracy, so as to reduce the cost of connectivity compression, improve the accuracy of vertex prediction, and reduce the cost of encoding the residues. We review several such retiling techniques.

An early retiling technique [66] first places samples on the surface and distributes them evenly using repulsive forces derived from estimates of the geodesic distances [42] between samples. Then it inserts these samples as new vertices in the mesh. Finally, the old vertices are removed through edge-collapse operations that preserve topology.

The MAPS algorithm [31] was used [30] to compute a crude simplified model, which can be compressed using any of the single-rate compression schemes discussed above. Once received and restored by the decompression module, the crude model is used as the coarse mesh of a subdivision process. Each subdivision stage splits each edge into two and each triangle into four, by the insertion of one new vertex per edge, in accordance with the *Loop subdivision* [34] rules, which split each edge (c.n.v,c.p.v) by inserting a point (c.v.g+ c.o.v.g+3c.n.v.g+3c.p.v.g)/8 and then displace the old vertices toward the average of their old neighbors. After each subdivision stage, the client downloads a displacement field of *corrective vectors* and uses them to adjust the vertices, so as to bring the current-level subdivision surface closer to the desired surface. The distribution of the coefficients of the corrective vectors is concentrated around zero and their magnitude diminishes as the subdivision progresses. They are encoded using a wavelet transform and compressed using a modified version of the SPIHT algorithm [53] originally developed for image compression.

Instead of encoding corrective 3D vectors, the *Normal Mesh* approach [19] restricts each offset vector to be parallel to the surface normal estimated at the vertex. Only one corrective displacement value needs to be encoded per vertex, instead of three coordinates. A *Butterfly subdivision* [9] is used. It preserves the old vertices, and for each pair of opposite corners c and c.o splits the edge (c.n.v,c.p.v) by inserting a point (8c.n.v.g+8c.p.v.g+2c.v.g+2c.o.v.g+ c.l.v.g+c.r.v.g+c.o.l.v.g+c.o.r.v.g)/16. The corrective displacements of new vertices are compressed using the unlifted version of the Butterfly wavelet transform [9,72]. Further subdivision stages generate a smoother mesh that interpolates these displaced vertices. The challenge of this approach lies in the computation of a suitable crude simplified model and in handling situations where no suitable displacement for a new vertex exists along the estimated surface normal. The connectivity of the crude mesh and the constraint imposed by the regular subdivision process limit the way in which the retiling can adapt to the local shape characteristics, and thus may result in suboptimal compression ratios. For example, regular meshes may lead to suboptimal triangulations for surfaces with high curvature regions and saddle points, where vertices of valence different from 6 would be more appropriate.

In the *Piecewise Regular Mesh* (PRM) approach [58], the surface may be algorithmically decomposed into six reliefs, each one comprising triangles whose normals are closest to one of the six principal directions (Fig. 18.8, left). Each relief is resampled along a regular grid of parallel rays (Fig. 18.8, right). Triangles are formed between samples on adjacent rays and also, to ensure the proper connectivity, at the junction of adjacent reliefs. When the sampling rate (i.e., the density of the rays) is chosen so that the resulting PRM has roughly the same number of vertices as the original mesh, the PRM approximates the original mesh with the mean

Figure 18.8   The reliefs produced by the Piecewise Regular Mesh (PRM) approach are shown (left) and resampled (right) into a nearly regular triangle mesh. (See also color insert.)

squared error of less than 0.02% of the diameter of the bounding box. Because of the regularity of the sampling in each relief, the PRM may be compressed down to a total of about $2t$ bits, which accounts for both connectivity and geometry. PRM uses Edgebreaker compression [48] and the Spirale Reversi decompression [25] to encode the global relief connectivity and the triangles that do not belong to the regular regions. Edgebreaker produces the *clers* string, which is then turned into a binary string using the *context-based range coder*, which reduces the uncertainty about the next symbol for a highly regular mesh. The geometry of the reliefs is compressed using an iterated 2D variant of the differential coding. The regular retiling causes the entropy of the parallelogram rule residues to decrease by about 40% when compared to the entropy for the original models, because, on reliefs, two out of three coordinates of the residual vectors become zero. Since this approach does not require global parameterization, it may be used for

models with *complex topologies*. It is faster than the combination of the MAPS algorithm [31] and the wavelet mesh compression algorithm [19,30] while offering comparable compression rates.

By tracing geodesics, Swing Wrapper [3] partitions the surface of an original mesh M into simply connected regions called triangloids. From these, it generates a new mesh M'. Each triangle of M' is a linear approximation of a triangloid of M (Fig. 18.9). By construction, the connectivity of M' is fairly regular (96% of the triangles are of type C or R, and 82% of the vertices have valence 6) and can be compressed to less than a bit per triangloid using Edgebreaker. The locations of the vertices of M' are encoded with about 6 bits per vertex of M', thanks to a new prediction technique that uses a single correction parameter per vertex. Instead of using displacements along the surface normal or heights on a grid of parallel rays, Swing-Wrapper requires that the left and right edges of all C triangles have a prescribed length L, so

Figure 18.9   The original model (first from the left, courtesy of Cyberware) contains $t = 134,074$ triangles. A dense partitioning of its surface into triangloids (second) was produced by SwingWrapper. The corresponding retiled mesh (third) was generated by flattening of the triangloids. Its L^2 error is about 0.007% of the bounding box diagonal, and its 13642 triangles were encoded with a total of 3.5 bits per triangloid for both the connectivity and the geometry using Edgebreaker's connectivity compression combined with a novel geometry predictor, yielding a compressed file of $0.36t$ bits. A coarser partitioning (fourth) decomposes the original surface into 1505 triangloids. The distortion of the corresponding retiled mesh (fifth) is about 0.15%, and the total file size is $0.06t$ bits. (See also color insert.)

that the correction to the predicted location of their tip may be encoded using the angle of the hinge edge. For a variety of popular models, it is easy to create an M′ with 10 times fewer triangles than M. The appropriate choice of L yields a total file size of $0.4t$ bits and a mean squared error with respect to the original of about 0.01% of the bounding box diagonal.

18.7   Conclusion

The connectivity of triangle meshes homeomorphic to a sphere may always be encoded with less than $1.8t$ bits and usually requires less than $1.0t$ bits. Their quantized geometry may be compressed to about $7.0t$ bits. Compression and decompression algorithms are extremely simple and efficient. Progressive transmission doubles the total cost of connectivity. When the original connectivity need not be preserved, retiling the surface of the mesh to enhance regularity often reduces the storage cost to $0.5t$ bits with a mean squared error of less than 1/10,000th the size of the model. These statistics remain valid for meshes with relatively few holes and handles.

References

1. P. Alliez and M. Desbrun. Progressive encoding for lossless transmission of 3D meshes. *ACM SIGGRAPH Conference Proceedings*, 2001.
2. P. Alliez and M. Desbrun. Valence-driven connectivity encoding for 3D meshes. *EURO-GRAPHICS*, 20(3), 2001.
3. M. Attene, B. Falcidieno, M. Spagnuolo, and J. Rossignac. SwingWrapper: retiling triangle meshes for better Edgebreaker compression. Genova CNR-IMA Tech. Rep. No. 14/2001. In *ACM Transactions on Graphics*, 22(4), 2003.
4. C. L. Bajaj, V. Pascucci, and G. Zhuang. Single resolution compression of arbitrary triangular meshes with properties. *Computational Geometry: Theory and Applications*, 14:167–186, 1999.
5. M. Chow. Optimized geometry compression for real-time rendering. In *Proceedings of the Conference on Visualization '97*, pages 347–354, 1997.
6. V. Coors and J. Rossignac. Guess connectivity: delphi encoding in edgebreaker. GVU Technical Report, Georgia University of Technology, 2002.
7. P. Cignoni, C. Rocchini, and R. Scopigno. Metro: measuring error on simplified surfaces. *Proc. Eurographics '98*, 17(2):167–174, 1998.
8. M. Deering. Geometry compression. In *Proceedings of the 22nd Annual ACM Conference on Computer Graphics*, pages 13–20, 1995.

9. N. Dyn, D. Levin, and J. A. Gregory. A butterfly subdivision scheme for surface interpolation with tension control. *ACM Transactions on Graphics*, 9(2):160–169, 1990.

10. F. Evans, S. S. Skiena, and A. Varshney. Optimizing triangle strips for fast rendering. In *IEEE Visualization '96*, pages 319–326, 1996.

11. M. Garland and P. Heckbert. Surface simplification using quadric error metrics. *Proc. ACM SIGGRAPH '97*, pages 209–216, 1997.

12. M. Garland and P. Heckbert. Simplifying surfaces with color and texture using quadratic error metric. *Proceedings of IEEE Visualization*, pages 287–295, 1998.

13. M. Garland. *Quadric-Based Polygonal Surface Simplification*. PhD Thesis, Carnegie Mellon University, 1998.

14. M. Garland. QSlim 2.0 [Computer Software]. University of Illinois at Urbana-Champaign, UIUC Computer Graphics Lab, 1999. xhttp://graphics.cs.uiuc.edu/~garland/software/qslim.html

15. A. Gueziec, F. Bossen, G. Taubin, and C. Silva. Efficient compression of non-manifold polygonal meshes. In *IEEE Visualization*, pages 73–80, 1999.

16. S. Gumhold. Towards optimal coding and ongoing research. In *3D Geometry Compression*, Course Notes, SIGGRAPH 2000.

17. S. Gumhold. Improved cut-border machine for triangle mesh compression. In *Erlangen Workshop '99 on Vision, Modeling and Visualization*. IEEE Signal Processing Society, 1999.

18. S. Gumhold and W. Straßer. Real time compression of triangle mesh connectivity. In *Proceedings of the 25th Annual Conference on Computer Graphics*, pages 133–140, 1998.

19. I. Guskov, K. Vidimce, W. Sweldens, and P. Schroeder. Normal meshes. In *SIGGRAPH '2000 Conference Proceedings*, pages 95–102, 2000.

20. P. Heckbert and M. Garland. Survey of polygonal simplification algorithms. In *Multi-resolution Surface Modeling Course*, ACM SIGGRAPH Course Notes, 1997.

21. H. Hoppe, T. DeRose, T. Duchamp, J. McDonald, and W. Stuetzle. Mesh optimization. In *Computer Graphics: SIGGRAPH '93 Proceedings*, pages 19–25, 1993.

22. H. Hoppe. Progressive meshes. *Computer Graphics, Annual Conference Series*, 30:99–108, 1996.

23. H. Hoppe. Efficient implementation of progressive meshes. *Computers and Graphics*, 22(1):27–36, 1998.

24. M. Isenburg and J. Snoeylink. Face fixer: compressing polygon meshes with properties. In *SIGGRAPH 2000, Computer Graphics Proceedings*, pages 263–270, 2000.

25. M. Isenburg and J. Snoeyink. Spirale reversi: reverse decoding of the Edgebreaker encoding. *Computational Geometry*, 20(1):39–52, 2001.

26. Sun Microsystems. *Java3D API Specification*. http://java.sun.com/products/java-media/3D, 1999.

27. A. D. Kalvin and R. H. Taylor. Superfaces: polygonal mesh simplification with bounded error. *IEEE Computer Graphics and Applications*, 16(3):64–67, 1996.

28. D. King and J. Rossignac. Guaranteed 3.67V bit encoding of planar triangle graphs. *11th Canadian Conference on Computational Geometry (CCCG '99)*, pages 146–149, 1999.

29. D. King, J. Rossignac, and A. Szymczak. Connectivity compression for irregular quadrilateral meshes. *Technical Report TR-99-36*, GVU, Georgia Institute of Technology, 1999.

30. A. Khodakovsky, P. Schroeder, and W. Sweldens. Progressive geometry compression. In *SIGGRAPH 2000, Computer Graphics Proceedings*, pages 271–278, 2000.

31. A. W. F. Lee, W. Sweldens, P. Schroeder, L. Cowsar, and D. Dobkin. MAPS: multiresolution adaptive parametrization of surfaces. In *SIGGRAPH '98 Conference Proceedings*, pages 95–104, 1998.

32. P. Lindstrom. Out-of-core simplification of large polygonal models. *Proc. ACM SIGGRAPH*, pages 259–262, 2000.

33. H. Lopes, J. Rossignac, A. Safanova, A. Szymczak and G. Tavares. A simple compression algorithm for surfaces with handles. *ACM Symposium on Solid Modeling*, 2002.

34. C. Loop. Smooth spline surfaces over irregular meshes. *Computer Graphics, Annual Conference Series*, 28:303–310, 1994.

35. K-L. Low and T. S. Tan. Model simplification using vertex clustering. *Proc. Symp. Interactive 3D Graphics*, pages 75–82, 1997.

36. D. Luebke, M Reddy, J. Cohen, A. Varshney, B. Watson, and R. Hubner. *Levels of Detail for 3D Graphics*. Morgan Kaufmann, 2002.

37. D. P. Luebke. View-dependent simplification of arbitrary polygonal environments. Doctoral Dissertation, University of North Carolina at Chapel Hill, 1998. http://www.cs.virginia.edu/~luebke/simplification.html

38. ISO/IEC 14496-2. Coding of audio-visual objects: visual. 2001.

39. R. Pajarola and J. Rossignac. Compressed progressive meshes. *IEEE Transactions on*

Visualization and Computer Graphics, 6(1):79–93, 2000.

40. R. Pajarola and J. Rossignac. Squeeze: Fast and progressive decompression of triangle meshes. In *Proceedings of Computer Graphics International Conference*, pages 173–182, 2000.

41. J. Popovic and H. Hoppe. Progressive simplicial complexes. *Computer Graphics*, 31:217–224, 1997.

42. K. Polthier and M. Schmies. Geodesic flow on polyhedral surfaces. *Proc. Eurographics Workshop on Scientific Visualization*, 1999.

43. R. Pajarola, J. Rossignac, and A. Szymczak. Implant sprays: compression of progressive tetrahedral mesh connectivity. *IEEE Visualization 1999*, pages 24–29, 1999.

44. E. Puppo and R. Scopigno. Simplification, LOD and multiresolution: principles and applications. *Tutorial at the Eurographics '97 conference*, 1997.

45. J. Rossignac and P. Borrel. Multi-resolution 3D approximations for rendering complex scenes. *Geometric Modeling in Computer Graphics*. Berlin, Springer Verlag, pages 445–465, 1993.

46. J. Rossignac and D. Cardoze. Matchmaker. manifold breps for non-manifold r-sets. *Proceedings of the ACM Symposium on Solid Modeling*, pages 31–41, 1999.

47. R. Ronfard and J. Rossignac. Full range approximation of triangulated polyhedra. *Proc. Eurographics '96*, 15(3):67–76, 1996.

48. J. Rossignac. Edgebreaker: connectivity compression for triangle meshes. *IEEE Transactions on Visualization and Computer Graphics*, 5(1):47–61, 1999.

49. J. Rossignac and A. Szymczak. Wrap&Zip decompression of the connectivity of triangle meshes compressed with Edgebreaker. *Computational Geometry, Theory and Applications*, 14(1/3):119–135, 1999.

50. J. Rossignac, A. Safonova, and A. Syzmczak. 3D compression made simple: Edgebreaker on a corner-table. *Shape Modeling International Conference*, Genoa, Italy, 2001.

51. J. Rossignac, A. Safonova, and A. Szymczak. Edgebreaker on a corner table: a simple technique for representing and compressing triangulated surfaces. In *Hierarchical and Geometrical Methods in Scientific Visualization* (G. Farin, H. Hagen, and Hamann, Eds. Heidelberg, Germany, Springer-Verlag, 2002.

52. D. Salomon. *Data Compression: The Complete Reference*, 2nd Ed. Berlin, Springer Verlag, 2000.

53. A. Said and W. A. Pearlman. A new, fast, and effcient image codec based on set partitioning in hierarchical trees. *IEEE Trans. Circuits Syst. Video Technol.*, 6(3):243–250, 1996.

54. A. Safonova and J. Rossignac. Source code for an implementation of the Edgebreaker compression and decompression. http://www.gvu.gatech.edu/~jarek/edgebreaker/eb

55. M. Schindler. A fast renormalization for arithmetic coding. In *Proceedings of IEEE Data Compression Conference*, page 572, 1998.

56. A. Szymczak, D. King, and J. Rossignac. An Edgebreaker-based efficient compression scheme for regular meshes. In *Proceedings of 12th Canadian Conference on Computational Geometry*, pages 257–264, 2000.

57. A. Szymczak, D. King, and J. Rossignac. An Edgebreaker-based efficient compression scheme for connectivity of regular meshes. *Journal of Computational Geometry: Theory and Applications*, 2000.

58. A. Szymczak, J. Rossignac, and D. King. Piecewise regular meshes: construction and compression. *Graphical Models* 64:183–198, 2002.

59. A. Szymczak and J. Rossignac. Grow&Fold: compressing the connectivity of tetrahedral meshes. *Computer-Aided Design*, 32(8/9):527–538, 2000.

60. A. Szymczak and J. Rossignac. Grow&Fold: Compression of tetrahedral meshes. *Proc. ACM Symposium on Solid Modeling*, pages 54–64, 1999.

61. G. Taubin and J. Rossignac. Geometric compression through topological surgery. *IBM Research Report RC-20340*, 1996. http://www.watson.ibm.com: 8080/PS/7990.ps.gz

62. G. Taubin and J. Rossignac. Geometric compression through topological surgery. *ACM Transactions on Graphics*, 17(2):84–115, 1998.

63. G. Taubin, W. Horn, F. Lazarus, and J. Rossignac. Geometry coding and VRML. *Proceedings of the IEEE*, 96(6):1228–1243, 1998.

64. C. Touma and C. Gotsman. Triangle mesh compression. In *Graphics Interface*, pages 26–34, 1998.

65. G. Turan. On the succinct representations of graphs. *Discrete Applied Mathematics*, 8:289–294, 1984.

66. G. Turk. Retiling polygonal surfaces. *Proc. ACM SIGGRAPH '92*, pages 55–64, 1992.

67. W. Tutte. A census of planar triangulations. *Canadian Journal of Mathematics*, pages 21–38, 1962.

68. S. Valette, J. Rossignac, and R. Prost. An efficient subdivision inversion for wavemesh-based progressive compression of 3D triangle meshes.

IEEE International conference on Image Processing, 1:777–780, 2003.

69. S. Valette. Modeles de maillage deformables 2D et multiresolution surfacique 3D sur une base d'ondelettes. Doctoral dissertation, INSA Lyon, 2002.

70. S. Valette and R. Prost. A wavelet-based progressive compression scheme for triangle meshes: wavemesh. IEEE Transactions on Visualization and Computer Graphics, 10(2):123–129, 2004.

71. ISO/IEC 14772-1, The Virtual Reality Modeling Language (VRML), 1997.

72. D. Zorin, P. Schroeder, and W. Sweldens. Interpolating subdivision for meshes with arbitrary topology. *Computer Graphics*, 30:189–192, 1996.

19   Variational Modeling Methods for Visualization

HANS HAGEN and INGRID HOTZ
University of Kaiserslautern

19.1 Introduction

Variational modeling techniques are powerful tools for free-form modeling in CAD/CAM applications. Some of the basic principles are carrying over to scientific visualization. Others have to be modified and some totally new methods have been developed over the last couple of years. This chapter gives an extended survey of this area.

Surfaces and solids designed in a computer graphics environment have many applications in modeling, animation, and visualization. We concentrate in this chapter mainly on the visualization part. We start with a section on basics from differential geometry, which are essential for any variational method. Then we give a survey on variational surface modeling. Since the mid-1990s, a "modeling pipeline" has been state of the art, consisting of data generation, data enrichment, data reduction, modeling, and quality analysis. The last step is the visualization part of geometric modeling. We discuss this topic in Section 19.4. In this context, surface curves like geodesics and curvature lines play an important role. The corresponding differential equations are nonlinear, and in most cases numerical algorithms must be used. To be sure to visualize features at a high quality, we need algorithms with an inherent quality control. The next two sections present our geometric algorithms, which satisfy this demand. The last section discusses the streamball technique, a visualization tool for structural features in computational fluid dynamics.

19.2 Fundamentals from Differential Geometry

A parameterized C^T surface is a C^T-differentiable map $X: M \to S \subset I\!E^3$, where $X_1 := \frac{\partial X}{\partial u}$ and $X_2 := \frac{\partial X}{\partial w}$ are linearly independent.

The 2D linear subspace $T_p X$ of $I\!E^3$ generated by the span $\{X_1 \dot{X_2}\}$ is called the tangent space of X at a point $p \in S$. The unit normal field N is given by

$$N = \frac{[X_1, X_2]}{\|[X_1, X_2]\|} \qquad (19.1)$$

where $[\,,\,]: I\!E^3 \times I\!E^3 \to I\!E^3$ is the cross-product. The moving frame $\{X_1, X_2, N\}$ is called the Gaussian frame. The Gaussian frame is in general not an orthogonal frame. The bilinear form induced on $T_p X$ by the inner product of $I\!E^3$ by restriction is called the *first fundamental form* of the surface X. The matrix representation of the first fundamental form I_p with respect to the basis $\{X_1, X_2\}$ of $T_p X$ is given by

$$\begin{pmatrix} g_{11} & g_{12} \\ g_{21} & g_{22} \end{pmatrix} = \begin{pmatrix} \langle X_1, X_1 \rangle & \langle X_1, X_2 \rangle \\ \langle X_2, X_1 \rangle & \langle X_2, X_2 \rangle \end{pmatrix} \qquad (19.2)$$

where $\langle\,,\,\rangle: I\!E^3 \times I\!E^3 \to I\!R$ is the dot product. The first fundamental form I_p is symmetric, positive definite, and geometric invariant. Geometrically the first fundamental form allows measurements on the surface (length of curves, angles between tangent vectors, area of regions) without referring back to the space $I\!E^3$ in which the surface lies.

The linear mapping $L : T_pX \rightarrow T_pX$ defined by $L := -dN \circ dX^{-1}$ is called the Weingarten map. The bi-linear form II_p defined by $II_p(A, B) := \langle L(A), B \rangle$ for $A, B \in T_pX$ is called the *second fundamental form* of the surface. The matrix representation of II_p with respect to the basis $\{X_1, X_2\}$ of T_pX is given by

$$\begin{pmatrix} h_{11} & h_{12} \\ h_{21} & h_{22} \end{pmatrix} = \begin{pmatrix} \langle -N_1, X_1 \rangle & \langle -N_1, X_2 \rangle \\ \langle -N_2, X_1 \rangle & \langle -N_2, X_2 \rangle \end{pmatrix}$$
$$= \begin{pmatrix} \langle N, X_{11} \rangle & \langle N, X_{12} \rangle \\ \langle N, X_{21} \rangle & \langle N, X_{22} \rangle \end{pmatrix}$$

(19.3)

where $X_{ij}, i, j = 1, 2$ are the second partial derivatives. The Weingarten map L is self-adjoint. The eigenvalues κ_1 and κ_2 are therefore real, and the corresponding eigenvectors are orthogonal. The eigenvalues κ_1 and κ_2 are called the *principal curvatures* of the surface.

$$K = \kappa_1 \cdot \kappa_2 = \det(L) = \frac{\det(II)}{\det(I)}$$

(19.4)

is called the Gaussian curvature, and

$$H = \text{trace}(L) = \frac{1}{2}(\kappa_1 + \kappa_2)$$

(19.5)

is called the mean curvature.

Let $A := \Delta u \cdot X_1 + \Delta w X_2$ be a surface tangent vector with $\|A\| = 1$. If we intersect the surface with the plane given by N and A (Fig. 19.1), we get an intersection curve Y with the following properties:

$$\dot{y}(s) = A \quad \text{and} \quad e_2 = \pm N$$

(19.6)

where $\dot{y}(s)$ denotes the tangent vector of the space (and surface) curve $y(s)$ with respect to arc length, and e_2 is the principal normal vector of the space curve $y(s)$.

The implicit function theorem implies the existence of this normal section curve. To calculate the extreme values of the curvature of a normal section curve, we can use the method of Lagrange multipliers, because we are looking for the extreme values of the normal section curvature κ_n under the condition $\|\dot{y}(s)\| = 1$. As the result of these considerations, we obtain the following: Unless the normal section curva-

ture is the same for all directions, there are two perpendicular directions A_1 and A_2 in which κ_n obtains its absolute maximum values. These directions correspond to the principal curvatures κ_1 and κ_2.

19.3   Variational Surface Modeling

The process of creating a 3D CAD model from an existing physical model is called reverse engineering, and it is different from standard engineering, where a physical model is created from a CAD model. Both approaches have, from a mathematical point of view, certain principles in common. Since the mid-1990s, a five-step "modeling pipeline" has been the state of the art, and it consists of the following steps:

- Data generation and data reception: measurements and numerical simulations

- Data enrichment and improvement: filtering and clustering

- Data analysis and data reduction: structure recognition, testing of features, etc.

- Modeling: variational design, physically based modeling, etc.

- Quality analysis and surface interrogation: reflection lines, isophotes, variable offsetting, etc.

The last step is the scientific-visualization part of the modeling pipeline. We discuss this topic in detail in Section 19.4. After briefly discussing the topics of data reduction and segmentation, we concentrate in this chapter on the modeling step.

19.3.1   Data Reduction

Physical objects can be digitized using manual devices, CNC-controlled coordinate measuring machines, or laser range-scanning systems. In any case, we get large, unstructured datasets of arbitrary distributed points.

Let $P := \{p_i \in I\!\!R^3 | i = 1, \dots, n\}$ be a set of n distinct points. To reduce P to a smaller set

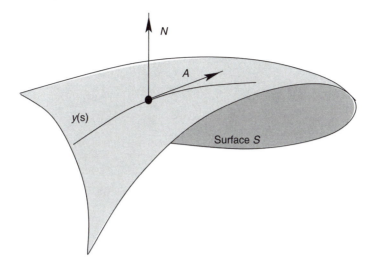

Figure 19.1   The intersection curve Y (See Equation 19.6).

$Q := \{q_i \in P | j = 1, \ldots, m\}$, a subdivision into m distinct clusters will be calculated. Clustering means grouping similar points by optimizing a certain criterion function. Subsequently, a single point out of each cluster is selected as a "representation point" for this cluster; these points build the so-called representation set Q. As a criterion to verify the quality of the subdivision, a function K_h is introduced, which assigns a numerical value to each cluster C_h.

$$K_n := \sum_{\substack{i,j \in I_h \\ i < j}} \|p_i - p_j\|^2 \qquad (19.7)$$

where $I_h := \{i | p_i \in C_h\}$ and $h = 1, \ldots, m$. This cost of a cluster is a measure of the distribution of the points in the cluster. An optimal subdivision of P is given by minimizing the cost. That is

$$\sum_{h=1}^{m} K_h = \sum_{h=1}^{m} \sum_{\substack{i,j \in I_n \\ i < j}} \|p_i - p_j\|^2 \to \min \qquad (19.8)$$

This expression is equivalent to

$$\sum_{h=1}^{m} \sum_{i \in I_h} \|p_i - S_h\|^2 \to \min \qquad (19.9)$$

S_h is the center of the cluster C_h. To find a global minimum of this expression is known to be NP-complete. Therefore, we have to use a heuristic method to find an "optimal" solution.

The Schreiber method is based on an interactive refinement strategy. The initial subdivision is the single cluster containing all points of P. In each step, the cluster with the highest cost is determined and divided into two new clusters, so that the cost is optimally reduced locally. For this purpose, a hyperplane is calculated orthogonal to the largest eigenvector of the covariance matrix of the points in a cluster. The optimal representation point for each cluster is the center point S_h of the cluster, but S_h is in general not a point of P. If this is a problem, the point of P nearest to S_k is used as the representation point. For more details on this algorithm, see Schreiber [4].

19.3.2   Segmentation

For a segmentation, all points have to be grouped in such a way that each group represents a surface (patch) of the final CAD model. The segmentations criterion is based on a curvature estimation scheme. The curvature at a point p can be estimated by calculating an approximating function for a local set of points around p. Hamann [3] uses the osculating paraboloids as approximating functions in his algorithm. Schreiber [5] extended this approach by using a general polynomial function to approximate a local set of points around p. First, a set of points neighboring p is determined by the Delaunay triangulation; this set of points is called the platelet. The platelet consists initially of all points that share a common edge of a triangle with p. This platelet is extended by adding all points that share a common edge with any platelet points. For a better curvature estimation, this extension is repeated several times.

19.3.3 Variational Design

The fourth step in the modeling pipeline is the final surface construction for a group of points. Analytical surfaces like planes, cylinders, and spheres can be created using standard CAD tools. Furthermore, the fillets with a constant radius in one parameter direction, as a connecting surface between two given surfaces, can be generated in a standard way, so the focus of this section is set to free-form modeling. This technique offers the possibility for a user to predefine boundary curves and to select neighboring surfaces for tangent or curvature continual transitions. Two approaches have become industry standards over the last couple of years: variational design and physically based modeling.

The variational design process of Brunnett et al. [6] combines a weighted least-squares approximation with an automatic smoothing of the surface. The chosen smoothing criterion minimizes the variation of the curvature along the parameter lines of the designed surface. This fundamental B-spline approach was extended for arbitrary degrees and arbitrary continuity conditions in both parameter directions, including given boundary information.

The following mathematical models can be used as variation principles:

$$(1 - w_s) \cdot \left\{ \sum_{k=1}^{n_p} w_p (F(u_k, v_k) - p_k)^2 \right\}$$

$$+ w_s \cdot \left\{ \sum_{i=1}^{n} \sum_{j=1}^{m} w_{u_g} \int_{v_j}^{v_{j+1}} \int_{u_i}^{u_{i+1}} w_{u_{ij}} \left\| \frac{\partial^3 F(u, v)}{\partial u^3} \right\| du\, dv \right.$$

$$+ w_{u_g} \int_{v_j}^{v_{j+1}} \int_{u_i}^{u_{i+1}} w_{v_{ij}} \left\| \frac{\partial^3 F(u, v)}{\partial u^3} \right\| du\, dv \right\} \rightarrow \min$$

$$(19.10)$$

where $F(u, v)$ is the representation of the surface, $\{p_k | k = 1, \ldots, n_p\}$ is the group of points, n and m are the numbers of segments in u and v directions, and $w_s, w_{u_g}, w_{v_g}, w_{u_{ij}}$, and $w_{v_{ij}} \in [0, 1]$ are the smoothing weights.

A successful alternative is to minimize the bending energy

$$\int_S \kappa_1^2 + \kappa_2^2 \, dS \rightarrow \min \qquad (19.11)$$

Variational design can to some extent be considered a part of *physically based modeling*. The starting point is always a specific physical demand of mechanic, electronic, aerodynamic, or similar origin.

Hamiltonian principle: Let a mechanical system be described by the functions $q_i, i = 1, \ldots, n$, where n is the number of degrees of freedom. The system between a fixed starting state $q_i(t_0)$ at starting time t_0 and a fixed final state $q_i(t_1)$ at final time t_1 moves in such a way that the functions $q_i(t)$ make the integral

$$I = \int_{t_0}^{t_1} L(q_i(t), \dot{q}_i(t)) \, dt =$$

$$\int_{t_0}^{t_1} \{ T(q_i(t), \dot{q}_i(t)) - U(q_i(t), \dot{q}_i(t)) \} dt \qquad (19.12)$$

stationary, compared with all functions $\bar{q}_i(t)$, which fulfill equal boundary conditions. T is the kinetic energy, U the potential energy, and $L = T - U$ the Lagrange function.

The functionals in variational design express an energy type. One very popular functional for surface generation describes the energy, which is stored in a thin, homogeneous, clamped plate with small deformations:

$$F(\eta) = \frac{h^3}{2n} \int \frac{E}{1-v^2} \left\{ \left(\frac{\partial^2 \eta}{\partial x^2} + \frac{\partial^2 \eta}{\partial y^2} \right)^2 \right.$$
$$\left. - 2(1-v) \left(\frac{\partial^2 \eta}{\partial x^2} \frac{\partial^2 \eta}{\partial y^2} + \frac{\partial^2 \eta}{\partial x \partial y} \right)^2 \right\} dx\ dy$$

$$(19.13)$$

The deduction of this term grounds on the equilibrium of volume and surface forces and uses some linearizations, where h denotes the thickness of the plate, η denotes the deformation, and E and v are material parameters. For more details and applications, see Hagen and Nawotki [2].

19.4   Curve and Surface Interrogation

One of the basic tasks in CAD/CAM is the design of free-form curves and surfaces. For quality-control purposes, it is necessary to inform the designer about certain properties and behaviors of the investigated curve or surface before entering the numerical control (NC) process. The detection of discontinuities or undesired regions becomes an indispensable tool for the interrogation.

We concentrate in this chapter on the two most important techniques:

- Reflection lines
- Variable offsets

The reflection-line method determines unwanted curvature regions by irregularities in the reflecting line pattern of parallel light lines.

Variable offset surfaces or focal surfaces are well known in the field of line congruences. They can be used to visualize many things: the pressure and heat distribution on an airplane, temperature, rainfall, ozone over the Earth's surface, etc. Given a set of unit vectors $E(u, v)$, a line congruence is defined in parameter form:

$$C(u, v, z) := X(u, v) + zE(u, v) \qquad (19.14)$$

where $X(u, v)$ represents the surface. In the special case of $E = N$ (surface normal) and $z = \kappa_1^{-1}$ or κ_2^{-1} we get the so-called focal surfaces

$$F_i(u, v) = X(u, v) + \kappa_i^{-1} \cdot N(u, v) \quad i = 1, 2 \quad (19.15)$$

Hagen and Hahmann [8] generalized this concept to achieve a surface-interrogation tool:

$$F(u, v) := X(u, v)$$
$$+ s \cdot f(\kappa_1, \kappa_2) \cdot N(u, v) \quad \text{with} \quad a \in \mathbb{R} \qquad (19.16)$$

The scalar function $f(\kappa_1, \kappa_2)$ is called the quality analysis function. For different applications, different functions are appropriate.

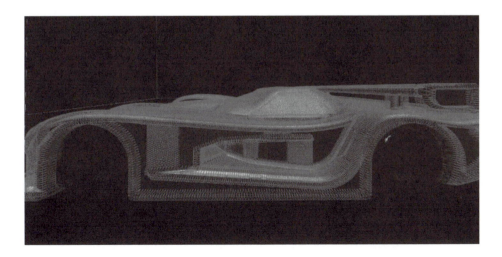

Figure 19.2   Race car reflection lines. (See also color insert.)

convexity test   — $f = \kappa_1 \cdot \kappa_2$
visualization of flat points   — $f = \kappa_1^2 + \kappa_2^2$

For more details see Hagen and Hahmann [8] and Hagen et al. [11]. We have just recently generalized these concepts to virtual environments [7].

19.5   Geodesics

Geodesics play an important role as shortest connections or straightest lines on curved surfaces in geometry and physics. In classical mechanics, geodesic curves can be interpreted as unaccelerated movements. In general relativity, geodesics are possible world lines of test particles. Geodesics are given by nonlinear differential equations for the surface coordinates. In general, these differential equations are solved with numerical standard algorithms, whereby the geometric origin of the problem is lost. We developed an alternative algorithm that is directly connected with the geometric definition of geodesics. Our approach allows a natural step-size adaptation without additional effort. This section consists of two parts. The first part presents the algorithm for the computation of the geodesics, and in the second part geodesics are

used to generate local nets on the surfaces as an application.

19.5.1   Mathematical Definition

There are two approaches to defining geodesics. The first is geodesics as locally shortest connections; the second is geodesics as straightest lines on a surface. For surfaces that are at least two times differentiable, the two definitions are equivalent. We consider only the second approach.

The covariant derivative of a vector field on a curved surface is the analog to the usual partial differentiation of a vector field defined on a plane. Let V be a vector field on the surface that assigns to each $p \in X$ a vector $V(p) \in T_p X$. Further let $\alpha(t) = (u(t), w(t))$ be a surface curve parameterized by arc length, with $\alpha(0) = p$ and $\dot{\alpha}(0) = A = \sum_i A_i X_i$. Restricting the vector field to the curve α

$$V(u(t), w(t)) \equiv V(t) = V_1(t)X_1 + V_2(t)X_2 \quad (19.17)$$

the projection of $(dV/dt)(0)$ on the tangent plane $T_p X$ is called the *covariant derivative* of the vector field V at p relative to the vector A. It is denoted by $\frac{DV}{dt} \equiv D_A V$. Related to the local basis $\{X_1, X_2\}$, we have the following:

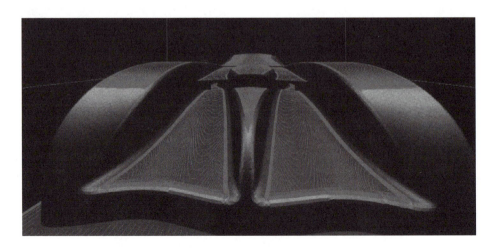

Figure 19.3   Race car focal analysis. (See also color insert.)

$$D_A V = \sum_{ij} A_i\, V_{i;j} X_j \quad \text{with} \quad V_{i;j}:$$

$$= \sum_{k=1}^{2} \frac{\partial V_i}{\partial X_j} + \Gamma^i_{jk} V_k \tag{19.18}$$

This definition depends not on the choice of α but only on the direction A. The coefficients Γ are the Christoffel symbols. They are a measure for the variation of the Frenet trihedron in the directions X_1 and X_2. The Christoffel symbols can be expressed in terms of the coefficients of the first fundamental form and its derivatives:

$$\Gamma^m_{kl} = \frac{1}{2} \sum_{i} g_{im}(g_{ik,l} + g_{li,k} - g_{kl,i}) \tag{19.19}$$

$$m, k, l, i \in \{1, 2\}$$

where $g_{ik,m} = \frac{\partial \langle X_i, X_k \rangle}{\partial x_m}$ with $x_1 = u$ and $x_2 = v$. A vector field V along a parameterized curve α is said to be parallel if $DV/dt = 0$ for all $t \in I$. A nonconstant, parameterized curve $\gamma : I \to S$ is said to be a *geodesic* if the field of its tangent vectors $\dot{\gamma}(t)$ is parallel along γ for all $t \in I$; that is,

$$D_t \dot{\gamma} \equiv 0 \tag{19.20}$$

In local coordinates a curve $\gamma(t)$ with $\gamma'(t) = \dot{u}(t)X_1 + \dot{w}(t)X_2$ is exactly a geodesic when it fulfills the following differential equations:

$$\ddot{u} + \Gamma^1_{11}\dot{u}^2 + 2\Gamma^1_{12}\dot{u}\dot{v} + \Gamma^1_{22}\dot{u}^2 = 0 \tag{19.21}$$

$$\ddot{u} + \Gamma^2_{11}\dot{u}^2 + 2\Gamma^2_{12}\dot{u}\dot{v} + \Gamma^2_{22}\dot{u}^2 = 0 \tag{19.22}$$

Given a point $p \in X$ and a vector $A \in T_p X, A \neq 0$, there exists always an $\varepsilon > 0$ and a unique parameterized geodesic $\gamma : (-\varepsilon, \varepsilon) \to X$ such that $\gamma(0)p$ and $\dot{\gamma}(0) = A$. One gets the same differential equation using a variational approach.

19.5.2   Geometric Construction of Geodesics

In general, the differential equations are the basis for the computation of geodesics. The most-used method for solving such systems of differential equations is the classical fourth-order Runge-Kutta formula. We choose another way, which is based on the geometric property of the geodesic being the straightest line on the surface. In contrast to the standard methods, this method allows a natural adaptation of the step length to the local geometry. This leads to a very accurate and efficient computation of geodesics on parameterized surfaces. Our approach has the advantage that it does not need the computation of Christoffel symbols at all.

For an arbitrary curve α on the surface, parameterized by its arc length, the algebraic value of the covariant derivative of its tangent vector field $D\dot{\alpha}(t)/dt = \kappa_g$ is called the *geodesic curvature of α at p*. Geodesics are thus characterized as curves whose geodesic curvature is zero. From a point of view external to the surface, the absolute value of the geodesic curvature κ_g at p is the absolute value of the tangential component of the vector $\ddot{\alpha}(t) = kn$, where κ is the curvature of α at p and n is the normal vector of α at p. It follows that $\kappa^2 = \kappa_g^2 + \kappa_n^2$. For geodesics, this leads to $\kappa^2 = \kappa_n^2$. This means that the normal of a geodesic corresponds to the surface normal. This allows us to construct the geodesic locally using the normal section. Explicitly, this means a projection of the vector $A \in T_p X$ on the surface.

In this way, the task of computing geodesics on a surface reduces to projection tangents on the surface. The main question is how to choose the length of the tangents to project to keep the fault small.

19.5.3   The Algorithm

We look at surfaces given by a parameterization $X(u, w) = (x(u, w), y(u, w), z(u, w))$, where x, y, z are differentiable functions of the parameters u and w.

1. A starting point and starting direction on the surface are chosen.

2. Depending on the parameter space (domain) and maximal step length, we

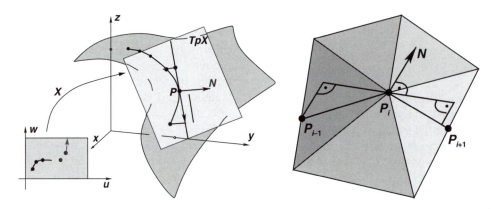

Figure 19.4   Computation of geodesics by tangent projection. (Left) On a parameterized surface; (Right) on a discrete surface.

compute a direction. The minimal step length is a fixed value.

3. Until the break conditions are fulfilled, the next point of the geodesic is determined by projecting (orthogonal to the tangent plane) the tangent on the surface. Because in general we do not have the inverse parameterization functions, the projection has to be computed numerically. If we use an angle corresponding to the direction of the geodesic in the parameter space as a variable, the projection becomes an iterative search for zero in one dimension.

Simultaneously, the step length is adapted to the geometry (the curvature and change of curvature) of the surface. These entities are represented by three angles.

Depending on the respective task, there are several conditions possible:

- The geodesic reaches the domain border (always active).
- The geodesic has a given length.
- The number of computed points exceeds a maximal point number.
- The geodesic intersects itself.
- The geodesic intersects another curve of a list of curves.

19.5.3.1   Computational Efficiency

Let us have a look at the computational effort of the computation of one new point. Besides some scalar products, the most time-consuming point is the projection step. The costs of one iteration are dominated by the computation of surface points, corresponding to an evaluation of the parameter functions. In our examples, the number of iterations varied from one to ten. This is not much in comparison with the standard fourth-order Runge-Kutta method with step adaptation, where one has to compute all six Christoffel symbols at least 11 times.

19.5.4   Geodesics on Discrete Surfaces

In computer graphics, we are often confronted with discrete surfaces. The definition of the geodesics needs not only continuity but at least two times differentiability. There does not exist a unique generalization. The two approaches as shortest and straightest curves are no longer equivalent.

Our projection algorithm needs, besides the computation of the surface normal, no derivatives of the surface. Therefore, it can easily be generalized to discrete surfaces. Because the projection of vectors to planes is very easy, the algorithm simplifies essentially. This generalization leads to an alternative definition of straightest geodesics, which is based on a discrete normal definition in the vertices.

19.5.5   Geodesic Nets

As an application, we consider nets built from geodesics. Because geodesics are very sensitive to local variations of the surface, they are not useful for global nets. But local nets allow an intuitive visualization and investigation of surfaces. A visualization of nets in the parameter space gives, without the embedded surface, an intuitive understanding of the geometric properties of the surface. We followed two different approaches.

19.5.5.1   *Pseudo-Orthogonal Nets*

These nets are constructed in the neighborhood of some point of interest. Two emphasized orthogonal geodesics, starting at this point, are computed and are used to determine the starting points and directions of two families of geodesics. The starting points are obtained by dividing the two start geodesics in parts of equal arc length. The start directions are orthogonal to these emphasized geodesics. (Orthogonal means orthogonality on the surface, not in the parameter space.) When the geodesics intersect themselves or another geodesic of the same family, the construction is

interrupted. In the ideal case, we get two bundles of orthogonal geodesics (e.g., on a plane or some isometric surface). But in general the intersections of the "orthogonal" families are not orthogonal or do not intersect at all. This means that we do not get nets in the classical sense.

The behavior of parallel geodesics differs depending on the Gaussian curvature. From the theorem of Gauss-Bonnet it follows that on an orientable surface of negative or zero curvature, two geodesics starting in one point cannot meet again in another point on the surface such that they are the boundary of a simple region. These nets give a good insight into the local geometry of the surface.

19.5.5.2   *Geodesic Polar Coordinates*

On a curved surface, one cannot find a coordinate system such that the Christoffel symbols $\Gamma^{\alpha}_{\beta\gamma} = 0$ on the entire surface. But for every point p there exists such a neighborhood. Special systems with this quality are the normal coordinate systems. Consider all geodesics passing one given point p, characterized by their tangential vectors $A \in T_p X, A \neq 0$. The points in the neighborhood of p are uniquely determined by the start direction of the geodesic passing them and the geodesic distance to p.

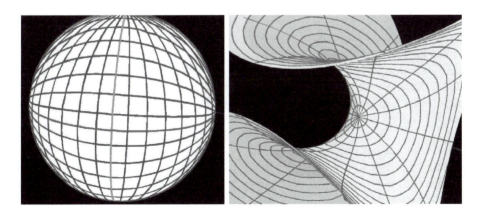

Figure 19.5   (Left) Pseudo-orthogonal net on a sphere. (Right) Geodesic polar coordinates on the Enneper minimal surface.

These so-called geodesic polar coordinates can be used to build a net on the surface around p. It consists of the geodesic circles, points with the same geodesic distance to p, and the radial geodesics. In some neighborhood of p, the family of the geodesic circles is orthogonal to the family of the radial geodesics.

The best-known example for usage of geodesic polar coordinates to visualize a surface is the Earth, where the geodesic polar coordinates correspond to the degrees of longitude and latitude. But in other cases they provide a very intuitive image of the surface. For more details, see Hotz and Hagen [9] and Hotz [10].

19.6 Curvature Lines

Curvature lines are curves that follow the maximum or minimum normal curvature of the surface. Because the curvature is an essential surface property, they are very important for the analysis of surfaces. Nets of curvature lines allow very good 3D impressions of the surface.

Using local coordinates, they are given by the following differential equation for the surface coordinates u and w:

$$(h_{12}g_{11} - h_{11}g_{12})\dot{u}^2 + (h_{22}g_{11} - h_{11}g_{22})\dot{u}\dot{v} + (h_{22}g_{12} - h_{12}g_{22})\dot{v}^2 = 0 \quad (19.23)$$

19.6.1 Geometric Construction of Curvature Lines

The standard method for the computation of the curvature lines is the solution of the differential equation with such numerical methods as Runge-Kutta. After the successful computation of geodesics with a geometric approach, we also investigated curvature lines under this aspect.

The definition of curvature lines as lines whose tangents are parallel to the principle directions leads directly to the property that the ruled surface built up by the curvature line and the surface normals is developable. This property is the basis for our algorithm. For a curva-

ture line $\alpha(t)$, this can be expressed in the following way:

$$\langle N(t), \dot{\alpha}(t) \rangle \times \dot{N}(t) = 0 \quad \text{for all } t \quad (19.24)$$

where $N(t)$ are the surface normals along the curvature line. Generalizing this condition to finite step length, we get

$$\text{Det}_{pi}(\Delta\alpha) := \frac{\langle N_i, \Delta\alpha \rangle \times \delta N}{\|N_i \times \Delta\alpha\|^2} = 0 \quad (19.25)$$

where N_i is the surface normal in the ith point $p_i, \Delta\alpha = p - p_i$ and $\Delta N = N(p) - N_i$. The geometric meaning of this determinate is the deviation of the surface normal in p from the plane spanned by the surface normal in p_i and $\Delta\alpha$. Using an angle variable to represent the direction of $\Delta\alpha$, the computation of a new point on the curvature line reduces to a search for zero in one dimension. Accuracy and efficiency of this algorithm are comparable to the Runge-Kutta method with adapted step length.

19.6.2 Nets of Curvature Lines

As an application, we consider nets built of curvature lines. Besides the geometric meaning of the curvature lines, as the direction of maximum or minimum normal curvature of the surface, nets are very useful for supporting the 3D effect of a surface. In contrast to geodesics, curvature lines are totally determined by the local geometry of the surface. There are exactly two orthogonal curvature lines passing every point, with the exception of the umbilical points. Umbilical points are surface points where the normal curvature is independent of the direction. These points are the singular points of the net, and they determine its topology. The neighborhood of umbilical points can be divided into areas where the curvature lines have characteristic behaviors. They can be separated by the radial curvature lines through the umbilical points.

These radial curvature lines correspond exactly to the zeros of the determinant. Thus, the determinant also allows an easy classification of the umbilical points and the detection of

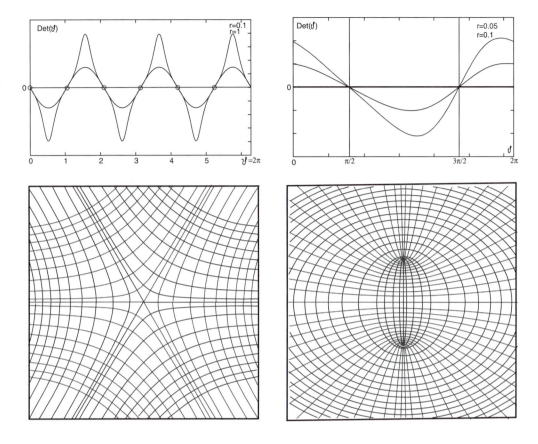

Figure 19.6 Two examples for the determinant function in the neighborhood of umbilical points. (Top left) The monkey saddle; (Top right) an elliptic paraboloid. Below them are the corresponding nets in the parameter space.

the separating directions. The separation lines are then used as a skeleton for the net computation [10].

19.7 Simulation-Based Solid Modeling

In computational fluid dynamics, visualization has become an important tool for gaining physical understanding of the investigated flow. The visualization is typically performed in a pipeline process similar to that in reverse engineering:

- Data generation: measurements and numerical simulations.
- Data preparation: filtering, feature recognition.

- Visualization mapping: generation of visualization primitives, mapping of physical quantities.

We concentrate on the visualization mapping part in this section. In CFD, a standard approach for visualizing flow situations consists of the construction of graphical objects whose shape delineates the structure of the flow field and which are used as a canvas for mapping physical parameters. The streamball technique [1] uses positions of particles in the flow for the construction of skeletons for implicit surfaces, which, by blending with each other, form 3D equivalents of streamlines, streamsurfaces, etc. The particle positions s_i are given by tracing flow lines through a flow field in

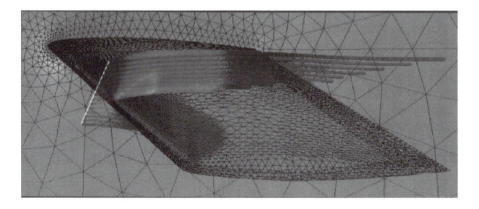

Figure 19.7   Airplane wing (streamballs). (See also color insert.)

discrete time-steps. Considering the set of all particle positions along a flow line as a skeleton S produces a discrete streamball. Connecting these particle positions by straight lines to form a polyline and using this as a skeleton creates a continuous streamball. Depending on skeletons S_j, a 3D potential field $G(x), x \in I\!\!R^3$ is generated by surrounding each skeleton with an individual potential field $F_j(x)$ and summing up the influences of all these individual fields:

$$G(x) = \sum_j F_j(x) \tag{19.26}$$

$F_j(x)$ can be seen as a composition of a monotonically decreasing influence function and a metric to measure the distance between a point x and the skeleton point.

As 3D objects, streamballs provide a variety of mapping techniques for the visualization of physical quantities of the flow. Two classes of mapping techniques associated with streamballs can be identified:

- Techniques changing shape (spherical, elliptical, etc.).
- Techniques changing appearance (color, transparency, etc.).

As an example, we visualize the heat distribution on an airplane wing. For more details, see Brill et al. [1].

References

1. M. Brill, H. Hagen, H. C. Rodrian, W. Djatschun, and S. Klimenko. Streamball techniques for flow visualization, *Visualization '94*, pages 225–231, 1994.
2. H. Hagen and A. Nawotki. Variational design and parameter optimized surface fitting. *Computing* 13:121–134, 1998.
3. B. Hamann. Visualization and modeling contours of trivariate functions. PhD thesis, Arizona State University, 1991.
4. T. Schreiber. A Voronoi diagram based adaptive k-means-type clustering algorithm for multidimensional weighted data in Nieri-Noltemeier. *Computational Geometry—Methods, Algorithms and Applications*, Springer Lecture Notes, pages 265–279, 1991.
5. T. Schreiber. Approximation of 3D objects. *Computing* 13:67–76, 1998.
6. G. Brunnett, H. Hagen, and P. Santarelli. Variational design of curves and surfaces. *Surv. Math. Ind.* 3:1–27, 1993.
7. H. Hagen, J. Guan, and R. Moorehead. Interactive surface interrogation using IPT technology, *IPT 2002 Symposium*, Orlando, FL, 2002.
8. H. Hagen and S. Hahmann. Generalized focal surfaces: a new method for surface interrogation. *Visualization '92*, pages 70–76, 1992.
9. I. Hotz and H. Hagen. Visualizing geodesics. *IEEE Visualization 2000*, pages 311–318, 2000.
10. I. Hotz. Geometrische algorithmen zur visualisierung diskreter und kontinuierlicher tensorfelder. PhD thesis, TU Kaiserslautern, 2003.
11. H. Hagen, T. Schreiber, and E. Geschwind. Methods for surface interrogation. *Visualization '90*, pages 187–193, 1990.

20 Model Simplification

JONATHAN D. COHEN
Johns Hopkins University

DINESH MANOCHA
University of North Carolina at Chapel Hill

20.1 Introduction

Interactive visualization is qualitatively different from generating images or sequences of images in batch mode; it provides a smooth, user-in-the-loop visualization process. This interactive process helps to deal with the challenging problem of finding the interesting aspects of some data through manipulations of a large space of visualization parameters. As the user changes visualization parameters, he or she sees immediate feedback, which guides further adjustments. This interactive discovery process is especially useful for modifying continuous parameters. For example, interactively controlling the virtual camera for 3D navigation of a complex dataset or modifying scale factors in color-mapping functions can assist the discovery process by allowing the user to maintain a mental context as the parameters change.

Unfortunately, rendering at interactive rates places a tremendous burden on the visualization system in terms of both computation and bandwidth requirements. We are fortunate today to have numerous hardware acceleration options, available at various levels of price and performance.

However, even with such hardware at our disposal, interactive visualization of today's large data remains a challenge. Recent advances in acquisition and modeling technologies have resulted in large databases of complex geometric models. These include large scanned datasets of real-world scenes, medical datasets, terrain models, and large synthetic environments created using modeling systems. These models are represented using polygons or higher-order primitives. Large models composed of millions of primitives are commonly used to represent architectural buildings, urban datasets, complex CAD structures, or real-world environments. The enormous size of these models poses a number of challenges in terms of interactive visualization and manipulation on current graphics systems.

Brute-force application of such hardware is insufficient for rendering data that fills the core memory of a modern computer at interactive rates, and data that is larger than core memory is now common. One useful approach to solving this problem is to build faster hardware (e.g., build a parallel machine such as a computer cluster). However, this is expensive and often has limited scalability. Achieving interactivity in the most general and cost-effective way requires a more sophisticated approach at the software level.

The fundamental concept of using hierarchical model representations to trade visual fidelity for interactivity was first proposed by Clark [12]. A variety of specific model representations and algorithms have been proposed for this purpose. The process of building these hierarchies automatically is known as *model simplification*, which builds a multiresolution hierarchy from a complex model and manages that new hierarchical representation during the interactive rendering process to enable this tradeoff. Using the model simplification approach, the user or the application can dynamically balance the need for interactivity against the need for high image quality during the image-generation process to manage both the computational and the bandwidth resources required at any given moment.

The last decade has seen a surge of research in the area of model simplification, producing a large number of research papers as well as several surveys [42,64], and recently a dedicated book on the subject [65]. In this article, we provide an overview of the concepts related to model simplification, a high-level look at the range of algorithms employed today, some applications, and open issues that still exist in this area.

20.2   Model Domains

The particular simplification algorithms and data structures that are appropriate for a given application are dependent on the type of input data to be visualized. Each class of data (shown in Fig. 20.1) has its own particular challenges, so it is reasonable for us to begin with a characterization of some typical applications and their associated data.

20.2.1   Terrain Visualization
Terrain visualization is a classic application of interactive graphics, with origins in flight simulation. Data may be acquired from satellite imaging and LIDAR airborne range imaging, with typical sample resolution ranging from 1 kilometer down to several meters, and current data sizes up to 933 million samples.

Terrains acquired in this way are often represented as regular height fields, stored as grey-scale images. However, this convenient and compact representation has some shortcomings, such as the inability to represent overhangs, caves, etc., as well as a general lack of adaptive sampling, so general triangle meshes (referred to as triangulated irregular networks, or TINs) may also be used.

Although the regularity of height field terrains can be used to specialize the algorithms, terrain data also poses a number of challenges for interactive visualization. The data sizes are often large

enough to require out-of-core processing [60,73]. In addition, they may be textured, requiring further management of out-of-core texture data [89]. Moreover, terrains are spatially large, spanning a huge range of depths in camera space and extending well beyond the rendered field of view. Such data requires crack-free, view-dependent adaptation of detail across its surface, taxing all of a computer's computational and bandwidth resources.

20.2.2   Visualization of 3D Scanned Models
Another large and growing source of rich models is laser range scanning. Though smaller in physical scale than terrains, these data provide wonderfully detailed representations of physical objects. The Stanford bunny model, comprising 70,000 triangles, is a classic test case for model simplification algorithms. More recent data are orders of magnitude larger, including statues from Stanford's Digital Michelangelo project [59], such as David, at 56 million triangles, and St. Matthew, at 372 million triangles.

Some simple models may be scanned in a single pass, and such models are often stored as cylindrical height fields. Models with more complex topological structure, or those spanning larger physical spaces, require multiple passes, as well as registration and merging of the data from each pass. These data are stored as more general triangle meshes or, in some cases, as points or point hierarchies [78].

The largest of these data require out-of-core processing [61]. Fortunately, the small spatial extent of most scanned models does not require view-dependent adaptation of detail. However, as more scanners are used in an outward-facing rather than an inward-facing fashion, scanning environments rather than objects, view-dependent adaptation becomes more essential.

20.2.3   Scientific and Medical Visualization
Scientific visualization spans a wide range of model types. Much of the data actually origin-

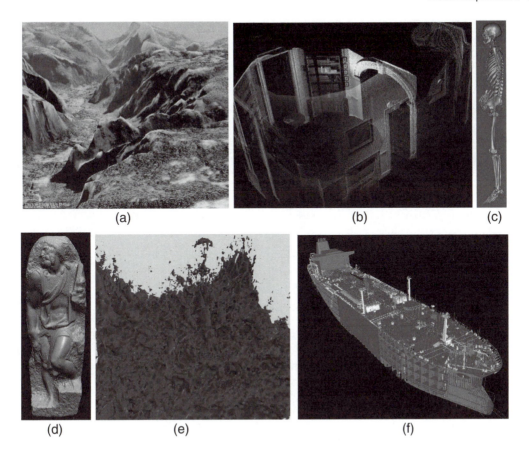

Figure 20.1 Large models from a variety of domains. (a) Yosemite Valley, California, 1.1 million triangles [74]. Copyright © 2001 IEEE. (b) Thomas Jefferson's Monticello, a fraction of the 19.5 million point samples from a single laser-rangefinder scan, from the Scanning Monticello project, courtesy of David Luebke (University of Virginia) and Lars Nyland (University of North Carolina at Chapel Hill). (c) Bones extracted from the Visible Female dataset, 9.9 million triangles before reduction, courtesy of Bill Lorensen of General Electric (d) St. Matthew, 372 M triangles [59]. Copyright © 2000 Digital Michelangelo Project, Stanford University. (e) Isosurface from DOE simulation of compressible turbulence, 500 million triangles (average depth complexity is 50), courtesy of Mark Duchaineau of Lawrence Livermore National Laboratory (f) Newport News double eagle tanker, 82 million triangles [29]. Copyright © 2001 Association for Computing Machinery, Inc. (See also color insert.)

ates as 3D volumetric data, represented as unstructured grids (tetrahedral meshes), structured grids (such as warped hexahedral grids), and regular grids (composed of volume elements, or voxels). Common origins include computational fluid dynamic and other simulations, as well as 3D medical imaging. The National Library of Medicine's Visible Female (see *http://www.nlm.nih.gov/research/visible/visible_human.html*) contains measured data at a resolution of 2048 × 1216 × 5600 (i.e., 14 billion voxels).

Similar data resolutions are now possible in simulation as well. The Department of Energy's ASCI project (see *http://www.llnl.gov/icc/sdd/img/viz.shtml*) performs material failure simulations in a simulation domain of 1 billion atoms, and turbulence simulations over a volumetric mesh of 8 billion samples (see *http://www.llnl.gov/CASC/asciturb*).

In some cases, these data are visualized using direct volume rendering algorithms. In other cases, an isosurface extraction algorithm is

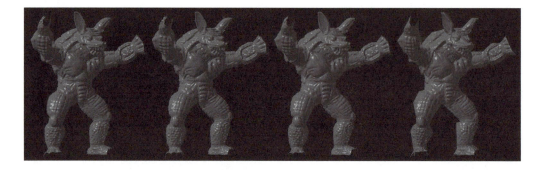

Figure 20.2   Four discrete LODs of the armadillo model (2 million triangles) using normal maps [15]. LODs have 250,000, 63,000, 8,000, and 1,000 triangles, respectively. Copyright © 1998 Association for Computing Machinery, Inc. (See also color insert.)

applied to generate a surface mesh for subsequent rendering and computation. These surface meshes may likewise have from 1 million to 1 billion elements. These large isosurfaces have many of the same size and spatial-extent challenges of terrain data but do not have the inherent regularity. In addition, these data may contain multiple data attributes and require strong quality guarantees on the rendered outputs so they may be effectively analyzed. Furthermore, the simulation data are not typically static, but are actually time-varying, often with a time resolution greater than their spatial resolution.

20.2.4   Computer-Aided Design and Synthetic Environments

CAD models and many large synthetic environments are substantially different from both terrain and scientific data. They are manmade, and the CAD models in particular are designed to very tight specifications. They exist to prototype equipment and perform simulations, as well as enable interactive visualization as part of the design process. CAD models may represent relatively simple machines, like automobiles, or complex machine systems, such as aircrafts, ships, factories, etc. They often exhibit high geometric complexity, comprising thousands to millions of individual parts that total millions to billions of primitives. Examples include the UNC power plant model, with 13 million triangles, the Newport News double-eagle tanker, with 82 million triangles, and the Boeing 777 aircraft model, with 190 million triangles.

When dealing with CAD models, we generally must maintain functional as well as spatial organization of the component parts. These parts exhibit a wide dynamic range of geometry, spanning large spatial extent but requiring close tolerances. Even without model simplification, such models can tax the accuracy of the standard Z-buffer rendering hardware. CAD models contain both sharp and smooth features, as well as irregular triangulations containing long, thin triangles. They are typically not "clean" data, exhibiting numerous degeneracies due either to human error or to numerous conversions between representations (such as the conversion from curved spline patches to triangles). These models tax all manner of geometric algorithms, including model simplification, and they are furthermore not the genre of model used to benchmark modern graphics card performance.

20.3   Types of Hierarchies

Before looking at how to build a simplification hierarchy, we should consider the various ways to store such hierarchies. In particular, we need to decide what granularity of multiresolution information we wish to keep around. We classify hierarchies as discrete, continuous, or view-

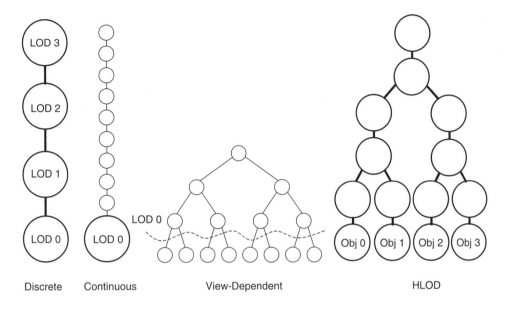

Figure 20.3 Simplification hierarchies at different granularities with different degrees of control. "Continuous" level of detail (LOD) is actually a large number of LODs, typically encoded by representing small changes from one LOD to the next. An LOD in the view-dependent hierarchy is represented as a *cut* across the tree. Hierarchical level of detail (HLOD) is similar to view-dependent, but it is much more coarse grained, allowing faster management and more efficient rendering. A cut across an HLOD hierarchy essentially selects a set of discrete LODs to represent a scene.

dependent according to this multiresolution granularity, as shown in Fig. 20.3.

20.3.1 Discrete

The discrete hierarchy is the simplest and most common form of simplification hierarchy. It encodes multiple levels of detail (LODs) of the original model at a very coarse granularity. Often these LODs are chosen such that each successive LOD has half the complexity of its predecessor.

This choice of granularity essentially doubles the storage requirements of the original high-resolution model. Each LOD may also contain some error value describing its quality.

A discrete hierarchy has a number of benefits. First, each level of detail may be easily compiled into an optimized form for efficient rendering. For example, this compilation might include such processes as triangle strip construction, reordering of vertices to optimize for hardware vertex cache size, and storage as indexed vertex arrays. The ability to perform this compilation is important, because we want the rendering of primitives from the hierarchy to be as efficient as the rendering of primitives from the original model (resulting in a potential speedup comparable to the reduction in primitives). Second, the management of one or more discrete hierarchies within the interactive application is not too computationally expensive. We just choose for each hierarchy the appropriate discrete level of detail given the current circumstances (see Section 20.5.1).

Discrete hierarchies are most useful when the data comprises one or more objects that are relatively small in spatial extent. In particular, if the range of depths of a model's primitives is small, then a single choice of resolution for the entire model may be appropriate. This is often the case for 3D scanned objects, virtual environments, and some CAD environments.

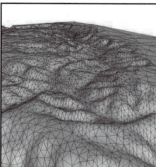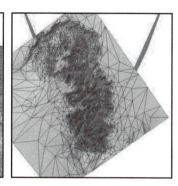

Figure 20.4 View-dependent rendering of the Grand Canyon, 10,013 triangles at 2 pixels of screen-space deviation [47]. Textured, wire-frame, and bird's-eye views. Copyright © 1998 IEEE. (See also color insert.)

20.3.2 Continuous

A continuous hierarchy (a widely accepted misnomer) describes a multiresolution hierarchy with so many levels of detail that it provides an effectively continuous progression [45]. Such a hierarchy is well suited for *delta encoding* by storing the individual small changes to the data rather than storing each level of detail as a complete, stand-alone model.

The main benefits of a continuous hierarchy are more exact choice of the number of primitives to use for an object, it enables more subtle transitions between levels of detail, and it provides a convenient representation for progressive transmission of data (e.g., over a network or from a hard disk). The basic continuous representation, which is a linear progression, does not allow selective refinement across the space of a particular model, so it is typically useful for the same class of models as the discrete hierarchy.

20.3.3 View-Dependent

The richest and most complex type of simplification hierarchy is the view-dependent hierarchy. It allows not only fine-grained selection of model complexity but also the ability to control the distribution of detail across the model at runtime. A view-dependent hierarchy is typically represented by a tree data structure [47,65,99] or a directed acyclic graph [21]. Like the continu-

ous hierarchy, the view-dependent hierarchy can represent fine-grained changes to the model. A complete approximation of the model is typically extracted from the view-dependent hierarchy on the fly by means of an appropriate *cut* through the hierarchy (as shown in Fig. 20.3). Raising some portion of the cut reduces the detail on the corresponding portion of the model, whereas lowering the cut results in local refinement.

In principle, view-dependent hierarchies allow the application to select just the right level of detail at appropriate portions of the model. Their ability to vary detail across the model is essential for a number of model domains. In particular, terrain visualization, large isosurface visualization, and visualization of some large-scale CAD structures (especially so-called *polygon soup* models, which have no connectivity or structural information) demand some form of view-dependent simplification. These applications include one or more models that cannot be easily decomposed into a set of smaller, independent objects and that have a wide span of depths.

View-dependent simplification involves additional runtime overhead, however. Adjustment of the detail across the model requires a large number of cost evaluations and comparisons across the model as opposed to the single decision required for a discrete hierarchy. The work performed in adapting the representation for a particular viewpoint may easily dominate the rendering time, overwhelming the benefit gained by view-dependent adaptation. In addition, maintaining optimized rendering structures as the model is dynamically changing remains a challenging problem, although some research has been done in this area [24,44].

20.3.4   Scenes

It is often possible to think of a model as a scene comprising multiple objects, each of which may be simplified independently. Each of these objects may be represented as a discrete, continuous, or view-dependent hierarchy, as appropriate. For example, spatially large objects might benefit from the use of a view-dependent hierarchy, whereas smaller objects might be usefully represented as discrete hierarchies.

The benefit to maintaining multiple hierarchies as opposed to merging all objects into a single hierarchy is that we can keep the efficiency of the discrete hierarchy and achieve some view-dependent adaptation by adjusting detail separately for each object. However, for very large collections of objects, the per-object overhead will eventually dominate. In this situation, it is useful to use hierarchical LOD, or HLOD [29]. HLOD looks like a discrete, view-dependent hierarchy. This LOD hierarchy is more coarse-grained than a typical view-dependent hierarchy and roughly corresponds to the notion of a hierarchical scene-graph structure. Each leaf node is the finest-resolution representation of each of the individual objects. Some number of discrete LODs is stored for each object, and these objects are eventually merged together. Each merged object may likewise have several discrete LODs before it is merged with another object, and so on. This coarse-grained representation scales well to complex scenes with large numbers of objects, reducing the per-object overhead for a given rendering frame both in the LOD selection process and in the actual number of primitives rendered. The LOD hierarchy also integrates

well with occlusion culling algorithms and has been used for interactive walkthrough of large and complex CAD environments [4,36].

In fact, given a set of discrete hierarchies for a scene, we can describe them trivially as an HLOD hierarchy by adding a virtual root node with each of the discrete hierarchies as a child. If we wish to create a more nontrivial hierarchical structure but do not wish to actually merge and simplify object geometries, we can cluster objects together to reduce the work of the selection process, while still rendering the objects individually at their selected levels of detail.

20.4   Building Hierarchies

Most simplification hierarchies are built in a bottom-up fashion, starting from the highest-resolution model and gradually reducing the complexity. For static models, this building process is a one-time preprocessing operation performed before interactive visualization takes place. The simplification operations performed to achieve the overall reduction may be ordered

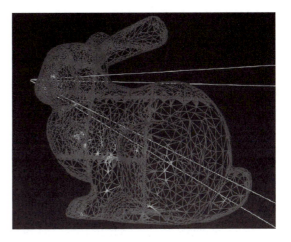

Figure 20.5   Using a 3D grid–based decomposition, HLOD performs a discrete approximation of the more fine-grained view-dependent hierarchy [29]. Copyright © 2001 Association for Computing Machinery, Inc. (See also color insert.)

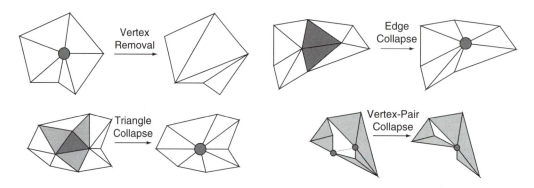

Figure 20.6 Several simplification operators for performing local decimation.

to produce either an *adaptive* or a *regular* hierarchy structure. An adaptive hierarchy retains more primitives in the areas requiring more detail, whereas a regular hierarchy retains a similar level of detail across all areas. A balanced tree structure is implied by a regular hierarchy, whereas the structure of an adaptive hierarchy is more unbalanced. In general, adaptive hierarchies take longer to build but produce fewer triangles for a given error bound. In some cases, the structure of a regular hierarchy allows optimizations for space requirements or specialized management algorithms.

20.4.1 Adaptive Hierarchies Using Decimation Operations

An adaptive hierarchy is typically built by performing a sequence of simplification operations in a greedy fashion, guided by a priority queue. The most common simplification operators in use today are various forms of vertex-merging operators. The *edge collapse* operator [48] merges together two vertices connected by an edge of the current model into a single vertex, discarding the primitives that become degenerate. Variants of the edge collapse include the *half-edge collapse* [53], which restricts the destination vertex to be one of the two source vertices; the *vertex-pair collapse* [33], which can merge any two connected or unconnected vertices; the *triangle collapse* [40], which merges three

connected vertices; and the *vertex cluster* [76], which merges together any number of vertices. Another well known operator is the *vertex removal* [82,91], which removes a vertex and its adjacent triangles from a mesh and then retriangulates the hole with new triangles. Most simplification algorithms employ some form of vertex merging, rather than vertex removal, because they are more easily computed, encoded, and smoothly interpolated. Half-edge collapses are often useful because they do not produce any new vertices, but they produce simplified surfaces with larger error than those produced using the edge collapse. Vertex-merging approaches have also been applied successfully to the simplification of tetrahedral meshes [10, 88,90,101]. Vertex removal is employed in some applications of terrain simplification and curved surface tessellation due to its close ties to incremental Delaunay triangulation.

Depending on the application, it may be desirable to preserve the topological structure of the original model, or it may be acceptable or even desirable to modify or simplify the topology. Operators such as edge collapse, triangle collapse, and vertex removal can be used to preserve local topological connectivity, and in some cases have been used to preserve global topology, preventing surface self-intersections [16]. Vertex-cluster operations often completely ignore topological structure, whereas vertex-pair operations can be used to make small topological modifications as they become reasonable. Other approaches may operate in either the volumetric domain [41,69] or the surface domain [26,27] to deliberately remove protrusions, indentations, and topological holes. Such topological reductions can improve model quality for a given primitive count or overcome topology-imposed lower limits on primitive counts. Although such reductions are useful for rendering acceleration, they may not be applicable to other applications of hierarchical models, such as scientific computations.

A number of error metrics have been applied to guide this priority-driven simplification process. These include *curvature estimates* [82,91], which remove vertices from areas of lower curvature first; *point–surface* metrics [48], which measure mean squared error between specific points on the original surface and their corresponding closest points on the simplified surface; *vertex–plane* metrics [33,75], which measure error between vertices of the simplified surface and their supporting planes from the original surface; *surface–surface* metrics, which bound maximum deviation between all points on the original and simplified surfaces [2,14,16,38,52]; *image-based* metrics [63], which compare images of the original objects with images of the simplified objects; and *perceptually based* metrics [66,83,95], which use simple models of human perception to decide which operations cause the least noticeable artifacts. Some metrics also measure error of nongeometric vertex attributes such as colors and normals [28,32,44,45,98], whereas other algorithms decouple these attributes from the simplification process [11,15]. In recent years, the vertex–plane, or *quadric error*, metrics have received the most attention due to their combination of easy implementation, fast execution, and good-quality results. However, these metrics do not provide guaranteed maximum error bounds on the models produced, so applications requiring such guarantees must use a different metric, combine it with a different metric [100], or measure such bounds as a post-process.

20.4.2 Regular Hierarchies

The image pyramid is a well known form of regular hierarchy. Each successive level of the image pyramid generally has half the resolution of the previous level in each spatial dimension. The representation is extremely compact because the parent–child relationships are implicit. In computer graphics, image pyramids are often stored as MIP-maps [94], which are used for efficient filtering of texture images.

Regular geometry hierarchies can leverage some of the same simplicity. Some terrain data takes the form of regularly sampled height fields. These regular terrains are often simplified using a quadtree [30,56,77] or binary triangle tree [22,35,71] scheme. Viewed from the top down, each starts with the entire model as a single square (or rectangle, etc.). The quadtree recursively subdivides the square into four quadrants until the final data resolution is reached. The binary triangle tree divides the initial square into two triangles, and then recursively splits each triangle in two by splitting the longest edge. Notice that neither scheme requires any error measurements to determine the structure of the hierarchy itself. However, some height error is typically stored with the hierarchy to allow view-dependent refinement during interactive visualization.

The rendering of 3D volumetric data can also employ regular hierarchies. Volume rendering of voxel data may be performed hierarchically using hierarchical splatting [55] or hierarchical 3D texture-based rendering [6,39,54,93]. Isosurface extraction may also be performed adaptively using a regular octree hierarchy [86].

One approach to dealing with irregular data is to resample it into a regular form. This can be particularly effective when combined with compression to reduce the new, redundant data. For example, a polygon mesh may be cut and unfolded into a square 2D domain and uniformly resampled for storage as a *geometry image* [37]. Although the result is not a height field, it has the same regular structure, allowing many techniques from terrain visualization to be applied. Unstructured 3D grids may also be

Before Simplification   After Simplification

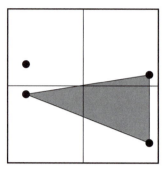

Figure 20.7   Cell-clustering simplification using a grid.

uniformly resampled using an octree to enable hierarchical 3D texture-based volume rendering [58].

Even when the original geometric data is not regular, we can still impose regularity on the hierarchy structure. The most common form of this is the *cell clustering* approach to model simplification [76]. To generate a discrete level of detail for some model using cell clustering, we partition the model's bounding box using a regular grid of some resolution. All the vertices of each cell are merged into a single vertex using a vertex cluster operator. The result is a model with a number of vertices less than or equal to the total number of cells. This algorithm can be very fast compared to adaptive, priority-driven algorithms, because we do not need to keep reevaluating the cost of operations as their neighboring operations are applied to the model.

We can replace the single-resolution regular grid of the cell-clustering algorithm with a multiresolution octree to generate a view-dependent hierarchy [65]. Although the hierarchy structure in such a scheme is still not adaptive, we can use such a hierarchy to provide view-dependent refinement within an interactive visualization application, also taking into account such effects as increased refinement around silhouettes. A hierarchical approach to cell clustering has also been used to create adaptive hierarchies in

reverse order, simplifying the root node first, and then splitting nodes according to a priority queue [7,9].

The cell-clustering approach has also proven to be useful for simplifying large models that do not fit in core memory [61,62]. If the output model is small enough to fit in core memory, we can make a single pass over the out-of-core list of input triangles, determining which become degenerate as they are quantized to the grid resolution. An optimized vertex position is determined for each cell by accumulation of a quadric error metric for each cell as the triangle list is traversed.

20.4.3   Hybrids

In many cases, we can benefit by combining adaptive and regular sampling in a hierarchy. For example, rendering a geometric simplification hierarchy with mip-mapped textures uses the geometric primitives to establish an adaptive sampling, whereas the texture maps provide image data stored as a regular hierarchy [15,80].

In general, some small amount of adaptive sampling may be used as a sort of normalization across the model, after which it becomes appropriate to perform sampling uniformly. This principle has been applied in a number of domains, including curved surface tessellation

Figure 20.8 Hybrid regular/adaptive simplification process for the David model (Stanford's Digital Michelangelo Project). (Left) Original model (8 million triangles). (Middle) Simplified using regular cell clustering (1,157 triangles). (Right) Simplified using multiphase approach to cell clustering followed by priority queue–driven edge collapses (1,000 triangles) [34]. Copyright © 2002 IEEE.

[8], hybrid point–triangle simplification [13], surface reparameterization [3,50,79], and conversion of meshes to semiregular subdivision meshes [23,57]. In each case, some base triangulation adaptively samples a surface, which is then relatively uniform within each element of the base structure.

A hybrid approach has recently been employed to improve the quality of out-of-core simplification [34]. Regular cell clustering is applied to bring the model to a small enough size to fit in core memory, and then an adaptive priority-queue algorithm is applied to continue the simplification down to the desired final polygon count. The results are shown to be significantly better than those obtained from performing a coarser cell clustering, but this algorithm assumes that the goal is to generate a model that is significantly smaller than the core memory size.

Another class of out-of-core simplification algorithms is the patch-based simplification algorithm. In a vein similar to the cell-clustering algorithms, a regular grid structure may be used to assist in the generation of a patch decomposition for a model, followed by an adaptive algorithm to simplify each patch in core memory. Various schemes are possible for dealing with simplification of patch boundaries. Multiple passes may be used, with a dual patch decomposition between passes [5], or patches may be clustered hierarchically at each pass, allowing for simplification of the boundaries that become interior at each successive pass [29].

20.4.4 Image-Based Simplifications

Image-based representations and impostors accelerate rendering by providing a drastic simplification of distant geometry. For example, many flight-simulation systems use images to represent terrains and other specialized models. Common image-based representations include point primitives, flat images, textured depth meshes, and depth images. Point primitives [72,78] work well for over-sampled datasets. Flat images [1,18,68,81,84,97] are mapped onto planar projections, but they only display the correct perspective when viewed from the location where the image was created. Textured depth meshes (TDMs) [17,51,87,97,102] replace the planar projections used by flat images with simplified meshes created from sampled depth values. Decoret et al. [19] extended TDMs to handle multiple layers. TDMs are height fields, and many algorithms have been proposed for simplification and view-dependent LOD control of such datasets [67]. Wilson and Manocha [96] have presented an algorithm for incrementally computing sample locations for generating TDMs and representing them incrementally.

20.5 Managing Hierarchies

Once we have precomputed one or more simplifications or scene hierarchies, one of the main goals is to use these hierarchical representations to improve the performance of an interactive visualization system. A visualization system

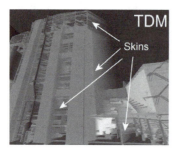

Figure 20.9 Comparison of Incremental Textured Depth-Meshes (ITDMs) [96] with regular TDMs and geometric levels of detail (LODs) on a 12.5M polygon power-plant model. ITDMs generate images almost as well as do static LODs, and at a frame rate 9 times faster. Moreover, ITDMs do not show the skin artifacts common in TDMs. The TDMs and ITDMs are used as a simplification of the far geometry. Copyright © 2003 Association for Computing Machinery, Inc. (See also color insert.)

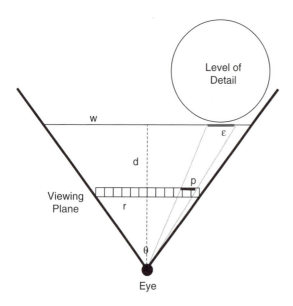

Figure 20.10 One way to project error from object space to screen space, enabling quality specification in *pixels of deviation*. The object space error, ε, is projected to the screen-space error, p, according to the viewing distance, d, as well as constant viewing parameters.

usually strikes a balance between manual detail control and fully automatic control. It will often permit the user to set several parameters controlling the target LOD and the prioritization used to reach that target. We next describe basic algorithms used to select the LOD during the interactive application as well as some issues related to managing the data for an entire scene.

20.5.1 LOD Selection

The problem of selecting an LOD is usually posed in one of two ways. In *quality-driven selection*, we use parameters corresponding to image quality, and the algorithm selects model representations that are as simple as possible while providing at least the specified quality. The quality specification could be in object space, screen space (Fig. 20.10), attribute space, perceptual space (including velocity and eccentricity [31,43,70]), etc. Such quality measures may be further scaled to account for user-applied semantic importance values. For view-dependent hierarchies, it is also possible to account for localized quality considerations such as preservation of silhouettes [65,98] and specular illumination effects [98]. Recent work has even applied perceptual metrics to account for the combination of screen-space geometric error, silhouette preservation, illumination, and texture-map content [95].

The other common way to pose the LOD selection problem is *performance-driven selection*. In this algorithm, the user or application specifies some parameter directly related to performance, such as number of primitives or frame time. It is then the job of the algorithm to select model representations that maximize the quality while meeting this performance budget. Notice that in this case, too, some measure of quality is required to perform prioritization.

Quality-driven selection is typically much simpler to implement than performance-driven selection, but it is insufficient to guarantee a particular level of interactivity. Given one or more hierarchies, each hierarchy may independently adjust its refinement level to just exceed the specified quality. For discrete hierarchies, this amounts to a form of switch statement to select the best LOD from the set of discrete LODs in the hierarchy. To find the best LOD, we can proceed from coarsest to finest or finest to coarsest, or we can maintain a current state from the previous frame as a starting position. Continuous hierarchies are no more complicated, although maintaining coherence becomes more important due to the increase in potential resolutions. View-dependent hierarchies are the most complex in practice. Each node may be treated as an LOD for the purpose of quality evaluation. A basic algorithm initializes a cut just beneath the root node. Each child node is tested against the quality threshold, and if it passes it is raised above the cut and its children are tested recursively, and this process is repeated. A small modification to this algorithm maintains the cut from the previous frame and allows it to move upward or downward as indicated by the quality of nodes above and below it.

One reason the performance-driven selection is more complex than the quality-driven selection is that the desired quality is not constant from frame to frame. In fact, one way to implement performance-driven selection is to employ a quality-driven selection algorithm, but increase or decrease the quality by an amount indicated by the performance of the preceding frames. This so-called *reactive* performance-management algorithm seems straightforward at first, but it has some fundamental limitations and can be difficult to tune in practice. It is a rather indirect way to control the performance, and because it relies only on old information, it is doomed to lag behind the needs of the current frame as new objects come into view.

Predictive performance-management algorithms, on the other hand, use a model of system performance to predict how a given representation of the input data should perform for the current frame. Such a predictive model requires some knowledge of the underlying rendering platform, and a number of parameters must be calibrated, but this approach has been demonstrated to work well. Given a total time budget and the predicted time required to render any particular model representation, we can employ a greedy algorithm to solve a variant of the well known Knapsack Problem (which is an NP-complete problem) [31]. We begin with the lowest quality representation of all hierarchies. If their sum does not exceed the time budget, we repeatedly refine the LOD with the lowest quality until we can no longer do so without exceeding the time budget. This algorithm has been demonstrated not only with discrete hierarchies [31], but with view-dependent hierarchies as well [22,65]. In the view-dependent setting, two priority queues are typically employed to prioritize refinements and reductions to the current cut.

For discrete and HLOD hierarchies, the time it takes to select the overall LOD, even in performance-driven mode, is typically small compared to the rendering time [29]. However, for the more fine-grained, view-dependent hierarchy, adapting the cut for the current frame can easily be the bottleneck. Typical implementations will decouple the updating of the cut for the current view from the actual rendering, either by using multiple threads [65] or by amortizing the update over some number of frames [46].

20.5.2   Scene Management

Given a large scene described as one or more hierarchies, an important challenge is to determine which portions of the scene need to be in core memory and at what LOD. We need at least the current cut of view-dependent hierarchy (or the current LODs of one or more discrete hierarchies) to make rendering feasible. We would also like to have in our memory cache the data needed for rendering the next one or more frames. This requires reasonable prediction of the viewing position and orientation for one or more frames in the future. This prediction is typically accomplished through a combination of velocity information and knowledge of the mode of user input for the specific application. For example, if a flight simulator limits the user to the controls of an actual aircraft, we can use that information to limit the range of positions and directions for the following frames. Based on this prediction information, we *prefetch* the data during frames before we expect to need the data.

In some cases, the user wants to avoid ever blocking the interactive application while it waits for data to arrive from disk, so the application must always have available *some* representation of any portion of the scene that could possibly be rendered during the next frame. One possible approach is to keep the coarsest LOD for the entire model loaded at all times as a fallback position. In some cases, it may be reasonable to keep all LODs that lie above the current cut, including the coarsest LODs, in memory at all times. Then it is always possible to coarsen any portion of the model on demand, and disk loads are only required for refinements.

Memory management is relatively straightforward with regular terrains [20,46,60], which are typically arranged in tiles, or hierarchical volume data [58]. The data for each LOD of each block can be loaded as the predictor says it may come into the field of view or otherwise change LOD. Scenes comprising multiple objects may be handled in much the same way, with the various LODs of the different objects loaded instead of the terrain tiles (predicting which objects will become visible due to disocclusion remains a challenging research problem today). One interesting way to think of this geometry cache is as a range of cuts surrounding the cut for the current frame. As we make predictions about the frames to come, we adapt these cache cuts, which tell us which geometry to load from disk.

In the case of a fine-grained, general-purpose view-dependent hierarchy, we will be loading the hierarchy structure as well as the actual geometry data from disk as we manage the cache [25]. For more coarse-grained hierarchies, such as the discrete hierarchy or an HLOD hierarchy, we may expect to fit the hierarchy structure entirely in core memory and manage just the geometry data with our caching algorithm [92,102].

20.5.3 Rendering Efficiency

Whatever the type of hierarchy or management algorithm, the goal is to achieve a higher level of rendering performance than we would achieve without simplification. If using LOD sacrifices rendering throughput to achieve primitive reductions, it can be a net slowdown rather than a speedup for all but the largest data. For discrete LODs and HLODs, this is not generally a problem because these coarse-grained, discrete structures may be optimized for whatever the current rendering platform requires. It is especially challenging, however, for continuous and view-dependent hierarchies, which are constantly making changes to the set of primitives to be rendered.

In terms of using current graphics hardware, the primary problem is still one of memory management. The fastest rendering is possible for data stored in *video memory*, local to the graphics processor. We can treat this memory as another level of cache for our geometry data and manage it much as we do the core memory cache.

A secondary factor in performance on today's hardware is the ability to use shared vertices and to reference these shared vertices with enough locality to maintain their state in an even smaller *vertex cache* on the graphics processor. Thus, maintaining this vertex locality is also an important consideration for achieving maximum performance. Although some work has been done in this area [44], it remains a challenging problem to maintain high rendering throughput in the presence of dynamically adapting, view-dependent LOD.

20.6   Conclusion

Model simplification has been an active area of research in the last decade. More than 100 papers have been published in this area, and the techniques are increasingly being used for interactive visualization of complex datasets arising from scientific visualization, virtual environments, real-world scenes, and computer gaming. Many tools are also available in both commercial and open-source form. A large list of such tools, and useful data sets, are currently available at *http://www.lodbook.com*.

A number of open problems remain in the area of model simplification for interactive visualization. Most noteworthy are those involving dynamic data. Almost all of the current research has focused on static data, whereas many application areas generate time-varying data, either by real-time acquisition or through simulations requiring simulation steering. Although some initial work has been done in this area [85], the field is still wide open.

References

1. G. D. Aliaga. Visualization of complex models using dynamic texture-based simplification. *Proceedings of IEEE Visualization '96*, pages 101–106, 1996.
2. C. Bajaj and D. Schikore. Error-bounded reduction of triangle meshes with multivariate data. *SPIE*, 2656:34–45, 1996.
3. L. Balmelli, F. Bernardini, and G. Taubin. Space-optimized texture maps. *Proceedings of Eurographics 2002*, pages 411–420, 2002.
4. B. Baxter, A. Sud, N. Govindaraju, and D. Manocha. Gigawalk: interactive walkthrough of complex 3D environments. *Proceedings of Eurographics Workshop on Rendering 2002*, pages 203–214 and 330, 2002.
5. F. Bernardini, H. Rushmeier, I. M. Martin, J. Mittleman, and G. Taubin. Building a digital model of Michelangelo's Florentine Pieta. *IEEE Computer Graphics and Applications*, 22(1):59–67, 2002.
6. I. Boada, I. Navazo, and R. Scopigno. Multi-resolution volume visualization with a texture-based octree. *The Visual Computer*, (17):185–197, 2001.
7. D. Brodsky and B. Watson. Model simplification through refinement. *Proceedings of Graphics Interface 2000*, pages 221–228, 2000.
8. J. Chhugani and S. Kumar. View-dependent adaptive tessellation of parametric surfaces. *Proceedings of 2001 Symposium on Interactive 3D Graphics*, pages 59–62 and 246, 2001.
9. P. Choudhury and B. Watson. Fully adaptive simplification of massive meshes. Technical Report, Northwestern University Department of Computer Science, 2002.
10. P. Cignoni, D. Constanza, C. Montani, C. Rocchini, and R. Scopigno. Simplification of tetrahedral meshes with accurate error evaluation. *Proceedings of IEEE Visualization 2000*, pages 85–92, 2002.
11. P. Cignoni, C. Montani, C. Rocchini, R. Scopigno, and M. Tarini. Preserving attribute values on simplified meshes by resampling detail textures. *The Visual Computer*, 15(10):519–539, 1999.
12. J. H. Clark. Hierarchical geometric models for visible surface algorithms. *Communications of the ACM*, 19(10):547–554, 1976.
13. J. Cohen, G. D. Aliaga, and W. Zhang. Hybrid simplification: combining multiresolution polygon and point rendering. *Proceedings of IEEE Visualization 2001*, pages 37–44 and 539, 2001.
14. J. Cohen, D. Manocha, and M. Olano. Simplifying polygonal models using successive mappings. *Proceedings of IEEE Visualization '97*, pages 395–402, 1997.
15. J. Cohen, M. Olano, and D. Manocha. Appearance-preserving simplification. *Proceedings of SIGGRAPH '98*, pages 115–122.
16. J. Cohen, A. Varshney, D. Manocha, G. Turk, H. Weber, P. Agarwal, F. Brooks, and W. Wright. Simplification envelopes. *Proceedings of SIGGRAPH '96*, pages 119–128, 1996.
17. L. Darsa, B. Costa, and A. Varshney. Walkthroughs of complex environments using image-based simplification. *Computers & Graphics*, 22(1):55–69, 1998.
18. P. Debevec, Y. Yu, and G. Borshukov. Efficient view-dependent image-based rendering with projective textures. *Proceedings of Eurographics Workshop on Rendering 1998*, pages 105–116, 1998.
19. X. Decoret, G. Schaufler, F. Sillion, and J. Dorsey. Multi-layered impostors for accelerated rendering. *Computer Graphics Forum*, 18(3):61–73, 1999.
20. L. DeFloriani, P. Magillo, and E. Puppo. VARIANT:asystemforterrainmodelingatvariableresolution.*GeoInformatica*,4(3):287–315,2000.

21. L. DeFloriani, P. Magillo, and E. Puppo. Building and traversing a surface at variable resolution. *Proceedings of IEEE Visualization '97*, pages 103–110.

22. M. Duchaineau, M. Wolinsky, D. E. Sigeti, M. C. Miller, C. Aldrich, and M. B. Mineev-Weinstein. ROAMing terrain: real-time optimally adapting meshes. *Proceedings of Visualization '97*, pages 81–88.

23. M. Eck, T. DeRose, T. Duchamp, H. Hope, M. Lounsbery, and W. Stuetzle. Multiresolution analysis of arbitrary meshes. *Proceedings of SIGGRAPH '95*, pages 173–182, 1995.

24. J. A. El-Sana, E. Azanli, and A. Varshney. Skip strips: maintaining triangle strips for view-dependent rendering. *Proceedings of IEEE Visualization '99*, pages 131–138, 1999.

25. J. A. El-Sana and Y. J. Chiang. External memory view-dependent simplification. *Computer Graphics Forum*, 19(3):139–150, 2000.

26. J. A. El-Sana and A. Varshney. Controlled simplification of genus for polygonal models. *Proceedings of IEEE Visualization '97*, pages 403–410, 1997.

27. J. A. El-Sana and A. Varshney. Topology simplification for polygonal virtual environments. *IEEE Transactions on Visualization and Computer Graphics*, 4(2):133–144, 1998.

28. C. Erikson and D. Manocha. GAPS: general and automatic polygonal simplification. *Proceedings of 1999 ACM Symposium on Interactive 3D Graphics*, pages 79–88, 1999.

29. C. Erikson, D. Manocha, and W. V. Baxter III. HLODs for faster display of large static and dynamic environments. *Proceedings of 2001 ACM Symposium on Interactive 3D Graphics*, pages 111–120, 2001.

30. J. S. Falby, M. J. Zyda, D. R. Pratt, and R. L. Mackey. NPSNET: hierarchical data structures for real-time 3D visual simulation. *Computers and Graphics*, 17(1):65–69, 1993.

31. T. A. Funkhouser and C. H. Sequin. Adaptive display algorithm for interactive frame rates during visualization of complex virtual environments. *Proceedings of SIGGRAPH '93*, pages 247–254, 1993.

32. M. Garland and P. Heckbert. Simplifying surfaces with color and texture using quadric error metrics. *Proceedings of IEEE Visualization '98*, pages 263–270, 1998.

33. M. Garland and P. Heckbert. Surface simplification using quadric error metrics. *Proceedings of SIGGRAPH '97*, pages 209–216.

34. M. Garland and E. Shaffer. A multiphase approach to efficient surface simplification. *Proceedings of IEEE Visualization 2002*, pages 117–124, 2002.

35. T. Gerstner. Multiresolution visualization and compression of global topographic data. *GeoInformatica*, 7(1):7–32, 2003.

36. N. Govindaraju, A. Sud, S.-E. Yoon, and D. Manocha. Interactive visibility culling in complex environments using occlusion-switches. *Proceedings of ACM Symposium on Interactive 3D Graphics 2003*, pages 103–112, 2003.

37. X. Gu, S. J. Gortler, and H. Hoppe. Geometry images. *Proceedings of SIGGRAPH 2002*, pages 355–361, 2002.

38. A. Guéziec. Locally toleranced surface simplification. *IEEE Transactions on Visualization and Computer Graphics*, 5(2):168–189, 1999.

39. S. Guthe and W. Strasser. Real-time decompression and visualization of animated volume data. *Proceedings of IEEE Visualization 2001*, pages 349–356, 2001.

40. B. Hamann. A data reduction scheme for triangulated surfaces. *Computer Aided Geometric Design*, 11:197–214, 1994.

41. T. He, L. Hong, A. Varshney, and S. Wang. Controlled topology simplification. *IEEE Transactions on Visualization and Computer Graphics*, 2(2):171–184, 1996.

42. P. Heckbert and M. Garland. Survey of polygonal simplification algorithms. *SIGGRAPH '97 Course Notes*, 1997.

43. L. E. Hitchner and M. W. McGreevy. Methods for user-based reduction of model complexity for virtual planetary exploration. *Proceedings of the SPIE—The International Society for Optical Engineering*, 1913:622–636, 1993.

44. H. Hoppe. Optimization of mesh locality for transparent vertex caching. *Proceedings of SIGGRAPH '99*, pages 269–276, 1999.

45. H. Hoppe. Progressive meshes. *Proceedings of SIGGRAPH '96*, pages 99–108, 1996.

46. H. Hoppe. Smooth view-dependent level-of-detail control and its application to terrain rendering. *Proceedings of Visualization '98*, pages 35–42, 1998.

47. H. Hoppe. View-dependent refinement of progressive meshes. *Proceedings of SIGGRAPH 97*, pages 189–198, 1997.

48. H. Hoppe, T. DeRose, T. Duchamp, J. McDonald, and W. Stuetzle. Mesh optimization. *Proceedings of SIGGRAPH '93*, pages 19–26, 1993.

49. H. Hoppe. New quadric metric for simplifying meshes with appearance attributes. *Proceedings of IEEE Visualization '99*, pages 59–66, 1999.

50. A. Hunter and J. D. Cohen. Uniform frequency images: adding geometry to images to produce

space-efficient textures. *Proceedings of IEEE Visualization 2000*, pages 243–250 and 563, 2000.

51. S. Jeschke and M. Wimmer. Textured depth mesh for real-time rendering of arbitrary scenes. *Proceedings of Eurographics Workshop on Rendering 2002*, pages 181–190 and 328, 2002.

52. R. Klein, G. Liebich, and W. Straßer. Mesh reduction with error control. *Proceedings of IEEE Visualization '96*, pages 311–318, 1996.

53. L. Kobbelt, S. Campagna, and H.-P. Seidel. A general framework for mesh decimation. *Proceedings of Graphics Interface '98*, pages 43–50, 1998.

54. E. C. LaMar, B. Hamann, and K. I. Joy. Multiresolution techniques for interactive texture-based volume visualization. *Proceedings of IEEE Visualization '99*, pages 355–362, 1999.

55. D. Laur and P. Hanrahan. Hierarchical splatting: a progressive refinement algorithm for volume rendering. *Proceedings of SIGGRAPH 1991*, pages 285–288, 1991.

56. Y. G. Leclerc and S. Q. Lau. TeraVision: a terrain visualization system. Technical Report 540. Menlo Park, CA, SRI International, 1994.

57. A. Lee, H. Moreton, and H. Hoppe. Displaced subdivision surfaces. *Proceedings of SIGGRAPH 2000*, pages 85–94, 2000.

58. J. Leven, J. Corso, J. Cohen, and S. Kumar. Interactive visualization of unstructured grids using hierarchical 3D textures. *Proceedings of IEEE/SIGGRAPH Symposium on Volume Visualization and Graphics 2002*, pages 37–44, 2002.

59. M. Levoy, K. Pulli, B. Curless, S. Rusinkiewics, D. Koller, L. Pereira, M. Ginzton, S. Anderson, J. Davis, J. Gensberg, J. Shade, and D. Fulk. The digital Michelangelo project: 3D scanning of large statues. *Proceedings of SIGGRAPH 2000*, pages 131–144, 2000.

60. P. Lindstrom and V. Pascucci. Visualization of large terrains made easy. *Proceedings of Visualization 2001*, pages 363–370 and 574, 2000.

61. P. Lindstrom. Out-of-core simplification of large polygonal models. *Proceedings of SIGGRAPH 2000*, pages 259–262, 2000.

62. P. Lindstrom and C. Silva. A memory insensitive technique for large model simplification. *Proceedings of IEEE Visualization 2001*, pages 121–126, 2001.

63. P. Lindstrom and G. Turk. Image-driven simplification. *ACM Transactions on Graphics*, 19(3):204–241, 2000.

64. D. P. Luebke. A developer's survey of polygonal simplification algorithms. *IEEE Computer Graphics & Applications*, 21(3):24–35, 2001.

65. D. P. Luebke and C. Erikson. View-dependent simplification of arbitrary polygonal environments. *Proceedings of SIGGRAPH '97*, pages 199–208, 1997.

66. D. Luebke and B. Hallen. Perceptually driven simplification for interactive rendering. *Proceedings of 2001 Eurographics Rendering Workshop*, pages 223–234, 2001.

67. D. Luebke, M. Reddy, J. D. Cohen, A. Varshney, B. Watson, and R. Huebner. *Level of Detail for 3D Graphics*. San Francisco, Morgan Kaufmann, 2002.

68. P. W. C. Maciel and P. Shirley. Visual navigation of large environments using textured clusters. *Proceedings of 1995 Symposium on Interactive 3D Graphics*, pages 95–102, 1995.

69. F. Nooruddin and G. Turk. Simplification and repair of polygonal models using volumetric techniques. Technical Report GIT-GVU-99-37, Georgia Institute of Technology, 1999.

70. T. Ohshima, H. Yamamoto, and H. Tamura. Gaze-directed adaptive rendering for interacting with virtual space. *Proceedings of 1996 IEEE Virtual Reality Annual International Symposium*, pages 103–110, 1996

71. R. Pajarola. Large scale terrain visualization using the restricted quadtree triangulation. *Proceedings of Visualization '98*, pages 19–26, 1998.

72. H. Pfister, M. Zwicker, J. van Baar, and M. Gross. Surfels: surface elements as rendering primitives. *Proceedings of SIGGRAPH 2000*, pages 335–342, 2000.

73. M. Reddy, Y. G. Leclerc, L. Iverson, and N. Bletter. TeraVision II: visualizing massive terrain databases in VRML. *IEEE Computer Graphics and Applications*, 19(2):30–38, 1999.

74. M. Reddy. Perceptually optimized 3D graphics. *IEEE Computer Graphics and Applications*, 21(5):68–75, 2001.

75. R. Ronfard and J. Rossignac. Full-range approximation of triangulated polyhedra. *Computer Graphics Forum*, 15(3):67–76 and 462, 1996.

76. J. Rossignac and P. Borrel. Multi-resolution 3D approximations for rendering complex scenes. Technical Report RC 17687–77951, IBM Research Division, 1992.

77. S. Röttger, W. Heidrich, P. Slussallek, and H.-P. Seidel. Real-time generation of continuous levels of detail for height fields. *Proceedings of 1998 International Conference in Central Europe on Computer Graphics and Visualization*, pages 315–322, 1998.

78. S. Rusinkiewicz and M. Levoy. QSplat: A multiresolution point rendering system for large

meshes. *Proceedings of SIGGRAPH 2000*, pages 343–352, 2000.

79. P. V. Sander, J. S. Gortler, J. Snyder, and H. Hoppe. Signal-specialized parametrization. *Proceedings of Eurographics Workshop on Rendering 2002*, pages 87–98 and 321, 2002.

80. P. V. Sander, J. Snyder, J. S. Gortler, and H. Hoppe. Texture mapping progressive meshes. *Proceedings of SIGGRAPH 2001*, pages 409–416, 2001.

81. G. Schaufler and W. Stuerzlinger. A 3D image cache for virtual reality. *Computer Graphics Forum*, 15(3):C227–C235, 1996.

82. J. W. Schroeder, J. A. Zarge, and W. E. Lorensen. Decimation of triangle meshes. *Proceedings of SIGGRAPH '92*, pages 65–70, 1992.

83. R. Scoggins, R. Machiraju, and R. J. Moorhead. Enabling level of detail matching for exterior scene synthesis. *Proceedings of IEEE Visualization 2000*, pages 171–178, 2000.

84. J. Shade, D. Lischinski, D. Salesin, T. DeRose, and J. Snyder. Hierarchical image caching for accelerated walkthroughs of complex environments. *Proceedings of SIGGRAPH '96*, pages 75–82, 1996.

85. A. Shamir, V. Pascucci, and C. Bajaj. Multiresolution dynamic meshes with arbitrary deformations. *Proceedings of IEEE Visualization 2000*, pages 423–430, 2000.

86. R. Shekhar, E. Fayyad, R. Yagel, and J. F. Cornhill. Octree-based decimation of marching cubes surfaces. *Proceedings of IEEE Visualization '96*, pages 335–344, 1996.

87. F. Sillion, G. Drettakis, and B. Bodelet. Efficient impostor manipulation for real-time visualization of urban scenery. *Computer Graphics Forum*, 16(3):207–218, 1997.

88. G. O. Staadt and M. H. Gross. Progressive tetrahedralizations. *Proceedings of IEEE Visualization '98*, pages 397–402, 1998.

89. C. C. Tanner, C. J. Migdal, and M. T. Jones. The clipmap: a virtual mipmap. *Proceedings of SIGGRAPH '98*, pages 151–158, 1998.

90. J. I. Trotts, B. Hamann, and K. I. Joy. Simplification of tetrahedral meshes with error bounds. *IEEE Transactions on Visualization and Computer Graphics*, 5(3):224–237, 1999.

91. G. Turk. Re-tiling polygonal surfaces. *Proceedings of SIGGRAPH '92*, pages 55–64, 1992.

92. G. Varadhan and D. Manocha. Out-of-core rendering of massive geometric environments. *Proceedings of IEEE Visualization 2002*, pages 69–76, 2002.

93. M. Weiler, R. Westermann, C. Hansen, K. Zimmerman, and T. Ertl. Level-of-detail volume rendering via 3D textures. *Proceedings of Volume Visualization and Graphics Symposium 2000*, pages 7–13, 2000.

94. L. Williams. Pyramidal parametrics. *Proceedings of SIGGRAPH '83*, pages 1–11, 1983.

95. N. Williams, D. Luebke, J. D. Cohen, M. Kelley, and B. Schubert. Perceptually guided simplification of lit, textured meshes. *Proceedings of ACM Symposium on Interactive 3D Graphics 2003*, pages 113–121, 2003.

96. A. Wilson and D. Manocha. Simplifying complex environments using incremental textured depth meshes. *Proceedings of SIGGRAPH 2003*, pages 678–688, 2003.

97. A. Wilson, K. Mayer-Patel, and D. Manocha. Spatially encoded far-field representations for interactive walkthroughs. *Proceedings of ACM Multimedia 2001*, pages 348–357, 2001.

98. J. C. Xia, J. A. El-Sana, and A. Varshney. Adaptive real-time level-of-detail-based rendering for polygonal models. *IEEE Transactions on Visualization and Computer Graphics*, 3(2):171–183, 1997.

99. J. C. Xia and A. Varshney. Dynamic view-dependent simplification for polygonal models. *Proceedings of IEEE Visualization '96*, pages 327–334, 1996.

100. S. Zelink and M. Garland. Permission grids: practical, error-bounded simplification. *ACM Transactions on Graphics*, 21(2):207–229, 2002.

101. Y. Zhou, B. Chen, and A. Kaufman. Multiresolution tetrahedral framework for visualizing regular volume data. *IEEE Visualization '97*, pages 135–142, 1997.

102. D. Aliaga, J. Cohen, A. Wilson, E. Baker, H. Zhang, C. Erikson, K. Hoff, T. Hudson, W. Stuerzlinger, R. Bastos, M. Whitton, F. Brooks, and D. Manocha. MMR: an interactive massive model rendering system using geometric and image-based acceleration. *Proceedings of 1999 Symposium on Interactive 3D Graphics*, 26–28:199–206 and 237, 1999.

PART VII

Virtual Environments for Visualization

21   Direct Manipulation in Virtual Reality

STEVE BRYSON
NASA Ames Research Center

21.1   Introduction

21.1.1   Direct Manipulation in Virtual Reality for Scientific Visualization

Virtual-reality interfaces offer several advantages for scientific visualization, such as the ability to perceive 3D data structures in a natural way [1,2,7,11,14]. The focus of this chapter is *direct manipulation*, the ability for a user in virtual reality to control objects in the virtual environment in a direct and natural way, much as objects are manipulated in the real world. Direct manipulation provides many advantages for the exploration of complex, multidimensional datasets, by allowing the investigator the ability to intuitively explore the data environment.

Because direct manipulation is essentially a control interface, it is better suited for the exploration and analysis of a dataset than for the publishing or communication of features found in that dataset. Thus, direct manipulation is most relevant to the analysis of complex data that fills a volume of 3D space, such as a fluid-flow dataset. Direct manipulation allows the intuitive exploration of that data, which facilitates the discovery of data features that would be difficult to find using more conventional visualization methods. Using a direct-manipulation interface in virtual reality, an investigator can, for example, move a "data probe" about in space, watching the results and getting a sense of how the data vary within their spatial volume.

Throughout this chapter, in order to focus the discussion, we will use the example of a data probe of a vector field in 3D space that emits streamlines of that vector field. The user is allowed to move the data probe anywhere in 3D space, and in response to that movement several operations must occur:

- Collision detection: the system must identify that the user has "picked up" or "moved" the data probe.

- Data access: for a given spatial position of the data probe, the system must locate the data (vector data, in our example) for that location and access that data, as well as all data involved in the visualization computation.

- Visualization computation: the geometry of the visualization (the streamline, in our example) must be computed.

- Graphical rendering: the entire graphical environment must be rerendered from the viewpoint of the user's current hand position.

In order for the user to have a sense that the object is moving with the user's hand position, this process must happen quickly, with low latency. Additional issues include the design of the data probes by the designer of the visualization environment and the possibility that the virtual environment may be implemented on a distributed system.

In this chapter we will explore the design and implementation of direct-manipulation interfaces useful for scientific visualization. The steps described above, combined with the low-latency human-factors requirements, place demands that result in several design challenges. After providing a simple model of the visualization process, we will develop implementation strategies tailored for scientific visualization in virtual environments, including issues of run-time software architectures, distribution, and control of time flow. These implementation strategies will be driven by consideration of the human factors of interaction.

21.1.2   The Data-Analysis Pipeline

In order to describe the special issues that arise in the implementation of a direct-manipulation based scientific visualization system in virtual reality, we require a conceptual model of a scientific visualization system. There are many ways to conceptualize scientific visualization, and we do not claim to present the most complete or optimal conceptualization. We have, however, found the following model very informative when considering implementation issues.

We consider the scientific visualization process as a pipeline, which in its most generic form starts with the data to be visualized. From this data visualization, primitives are extracted. These primitives may consist of vertices in a polygonal representation, text for a numerical display, or a bitmap resulting from, for example, a direct volume representation. Primitive extraction typically involves many queries for data values. The extracted primitives are then rendered to a display. This pipeline allows user control of all functions, from data selection through primitive extraction to rendering. We show this pipeline in Fig. 21.1.

Let us examine the operation of this pipeline in our example of streamlines of a vector field. Given a starting point of a streamline, data (the vectors at that point) are accessed by the streamline algorithm. The vector value is then added (sometimes with a complex high-accuracy algorithm) to the starting point, creating a line primitive. This process is iterated to build up a (typically curved) line with many vertices. These vertices are the streamline's extracted geometrical representation. They are then rendered in the visualization scene. The extraction of these primitives may involve significant computation even though the data may exist as a precomputed file. Computations like those in this example will turn out to be a significant issue in the implementation of scientific visualization in virtual environments.

21.1.3   Advantages of Direct Manipulation in a Virtual Environment

Direct manipulation in a virtual environment offers several advantages for many classes of scientific visualization. 3D interaction techniques common in virtual environments provide natural ways to control visualization selection and control in three dimensions. In addition, our experience has shown that one of the greatest advantages of scientific visualization in virtual environments is the inherent "near-real-

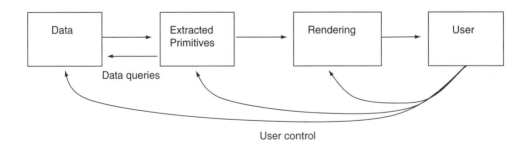

Figure 21.1   The data-analysis pipeline.

time" responsiveness required by a hand-tracked direct-manipulation based virtual environment. This responsiveness allows rapid queries of data in regions of interest. Maintaining this responsiveness in a scientific visualization environment is the most challenging aspect of such an application and will be one of the primary foci of this chapter. When designed well, the combination of 3D display, 3D interaction, and rapid response creates an intuitive environment for exploration and demonstration.

Fig. 21.2 shows the Virtual Wind Tunnel [1,5], an example of scientific visualization in virtual that makes extensive use of direct manipulation. The Virtual Wind Tunnel is used to investigate the results of simulations in computational fluid dynamics. This example exhibits the use of multiple visualization extracts in a single environment, all of which are interactive via visualization widgets.

21.1.4   How Scientific Visualization Differs from Other VR Applications

The design and development of a scientific visualization application within a virtual environment is different from most virtual reality application development. Scientific visualization environments are often abstract and involve large amounts of data access and computation in response to a query. For time-

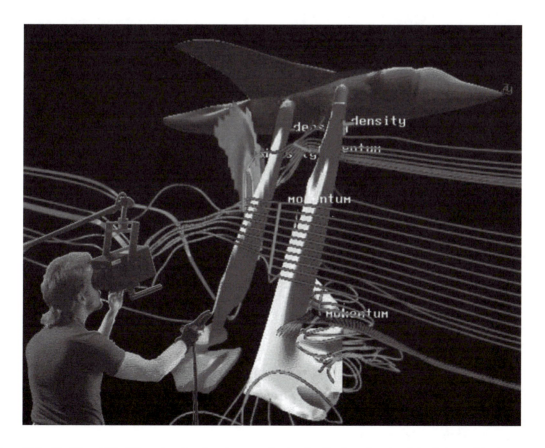

Figure 21.2   The Virtual Wind Tunnel, a virtual environment for the visualization of results of simulations arising in computational fluid dynamics. This example shows a variety of visualization widgets with visualizations including streamlines, local isosurfaces, and local cutting planes. (See also color insert.)

varying data, different senses of time arise, in which the data may evolve more slowly or even backwards relative to user time. Such differences between scientific visualization environments and more conventional VR applications can be generalized into the following areas:

- Greater flexibility in graphical representation: The inherently abstract nature of information implies opportunities for simpler, faster graphics, such as representing a streamline as a simple polyline. Conventional applications, such as training or entertainment, are typically more concerned with realistic graphical environments, and so may have less flexibility in the selection of graphical representation. Of course, this is not a universally applicable rule, as some representations of information such as direct volume rendering can be very graphically intensive.

- A large amount of numerical computation may be required: While scientific visualization typically addresses preexisting data, visualization extracts may themselves require considerable computation. Streamlines or isosurfaces in the visualization of continuous vector and scalar fields are well-known examples that require large amounts of computation. As more sophisticated data-analysis techniques are used in virtual environments, more computational demands can be expected.

- A large amount of data access may be required: Visualization extracts require access to data, and some extracts require more data access than others. Complete isosurfaces, for example, require traversing an entire dataset (at a particular time-step for time-varying data) for the computation of the dataset. Time-varying datasets can be extremely large, requiring access to hundreds of gigabytes of data in a single analysis session, albeit a single time-step at a time.

- There may be various senses of time: As will be discussed in Section 21.3.1, several senses

of time can arise in a scientific visualization system, particularly when addressing time-varying datasets. While some of these senses of time correspond to conventional time in other VE applications, completely new ways of thinking of time arise from the fact that a user may wish to manipulate time flow in a scientific visualization system.

These differences between a scientific visualization virtual environment and other applications have major impacts on the design of the virtual environment system. These impacts are the focus of this chapter.

21.2   Basics of Direct Manipulation

21.2.1   What is Manipulated: Visualization Widgets

The phrase "direct manipulation in virtual reality" refers to the ability of the user to pick up a virtual object much as the user would pick up an object in the real world. This type of manipulation contrasts with "indirect manipulation," where an object in the environment responds to controls generated by manipulation of some other object. A common example of indirect manipulation is a conventional graphical user interface (GUI), where buttons or sliders control some aspect of a visualization. While we focus on direct manipulation in this chapter, the strategies we describe to deliver fast responsiveness are beneficial to indirect manipulation as well.

A scientific visualization environment often does not mimic the real world, so there is some freedom to choose which objects should be manipulated and which objects should not. In some cases, it is not clear what direct manipulation of a visualization means. In our example, what should it mean to directly manipulate a streamline? While it conceptually makes sense to manipulate a streamline by grabbing it at any point, this can be very difficult if the data is time-varying and the streamline is rapidly and dramatically changing shape.

Experience has shown that while in some cases it makes sense to grab an object, in many cases it is easier for the user to grab tools that control a visualization rather than the visualization itself. These tools are spatially collocated with the visualization in some sense, so the user has the feeling of directly manipulating the visualization. We call these tools *visualization widgets*. This approach of directly manipulating visualization widgets rather than the visualizations themselves has several advantages:

- Unified interface: the user has to learn the interactive behavior of only a small set of widgets, rather than having to figure out how each type of visualization is manipulated.

- Affordance: a well designed widget will have a natural interpretation, facilitating the user's knowledge of what the widget does.

- Visualization grouping: it is often desirable to manipulate groups of visualizations. Such a group may consist of several streamlines in a row, or may be heterogeneous, combining streamlines, cutting planes, and isosurfaces under the control of one widget.

- Unchanging spatial position: in a time-varying visualization environment, the visualization widgets need not move with the data, which makes them easier for the user to pick up.

Of course, the use of visualization widgets does not preclude the ability to manipulate visualizations directly, but in our experience caution is advised. Visualization environments can become quite rich with many different objects in the environment. If many of these objects respond to user grab actions, then grabbing an object by mistake can become a significant issue. Restricting manipulation to a relatively small number of data widgets helps prevent this situation.

In order for the spatial location (and possibly orientation) of a visualization widget to control a visualization, that visualization must use spatial information in its specification. In our streamline example, this specification is natural: the widget controls the location of some point on the streamline. In our description of the streamline computation, that point is used to start both a forward and a backward integration of the vector field. The result is a streamline through that point. We will call the spatial location used to specify a visualization the *seed point* of that visualization.

Some visualizations, however, are not typically specified by a spatial location. An example of a non–spatially defined visualization is a traditional isosurface of a scalar field, which is specified by a scalar field value. The natural way to specify such a field value is through indirect manipulation, via a GUI. One way to convert such a visualization to a spatially specified *local isosurface* is given in Section 21.5.3. This is an example of the kind of creativity that may be required in the use of direct manipulation to control a visualization.

A visualization system need not use virtual reality to benefit from direct manipulation. Conventional mouse input can be used to directly manipulate objects in a visualization window, and nearly all of the considerations in this chapter apply.

21.2.2 Human-Factors Requirements

The act of reaching out and picking up an object is natural in the real world, but when the responsiveness of our hand is degraded by *inaccuracy* (it doesn't go where we tell it), *latency* (it doesn't go when we tell it), or a low *update rate* (we see our hand in snapshots), the act of picking up an object can become difficult. We can summarize this situation as follows:

- *Accurate tracking, fast response* to user commands, and *smooth feedback* are required for a direct manipulation interface to succeed in giving the user a sense of directly "picking up and moving" objects in the virtual environment.

In the presence of fast, smooth feedback, a certain amount of tracking inaccuracy is tolerable. Fast response and smooth feedback are, however, critical to the user's experience of directly manipulating objects in the virtual environment. In a scientific visualization environment, this places performance requirements on the data access, computation, and graphical rendering triggered by the manipulation of a visualization.

How fast the graphics and interaction responses must be turns out to be both application dependent and domain dependent. A virtual environment for training for real-world manual tasks such as a flight simulator requires response times of less than 1/30 of a second in order to mimic the response times of real-world objects. Information visualization, however, does not typically require fidelity to real-world time scales, so the performance requirements are driven by the human factors of manual control [12].

The human-factors issues that turn out to be important for scientific visualization in virtual environments are the following:

- Graphics update rate: how fast must the graphical display update rate (graphical animation rate) be to preserve a sense of object presence, the sense that the virtual object has a position in 3D space independent of the user? By graphical update rate, we mean the rate at which frames are drawn to the graphics framebuffer(s), not the display device's refresh rate.

- Interaction responsiveness: how quickly must objects in the environment respond to user actions in order to maintain a sense of presence and direct manipulation?

- Data display responsiveness: how fast must interactive data-display devices, such as a data probe, update to give the user a sense of exploring the data?

While these considerations are usually related, in a virtual environment they are distinct. The fast graphics-update rate is required by the vir-

tual-reality requirement of rendering the graphical scene from the user's current head position and orientation. Interaction responsiveness measures how well the interactive objects move with the user's hand, and it is a function of the graphical update rate, the latency of the input devices that measure the user's actions, and any computation that the interactive objects require to respond. This latter computation is trivial for visualization widgets. Data-display responsiveness measures how quickly the visualizations respond to the user's manipulation. The responsiveness of the interactive objects and the visualizations they control need not be the same; for example, the user can position a widget and subsequently watch the visualizations appear in response.

The relationships and differences between these time scales are subtle. The graphics update rate will limit the interaction and data-display responsiveness because the interactive displays cannot be presented faster than the graphics update rate. Update rate and responsiveness, however, are very different kinds of measures: update rate is measured in frames/s, while responsiveness is the *latency*, measured in seconds—the time interval between a user action and when the system's response is displayed. This latency is determined by all processes triggered by the user's action, from reading the user tracking devices, through processing the user's commands, through possible subsequent computation, to the display of the result.

Experience has shown that the limits on these time scales are the following:

- The graphics update rate must be greater than 10 frames/s. While faster update rates are desirable, 10 frames/s is sufficient to maintain a sense of object presence even though the discrete frames of the display are easily perceptible. Slower update rates result in a failure of the sense of object presence, compromising the enhanced 3D perception advantages of a virtual environment.

- Interaction responsiveness must be less than 0.1 s. While lower latencies and faster responsiveness are desirable, a latency of 0.1 s is fast enough to give the user a good sense of control of objects in the virtual environment. Longer latencies typically cause the user to experience unacceptable difficulty in selecting and manipulating objects in 3D space.

- Data-display responsiveness must be less than about 1/3 s. While faster responsiveness is desirable, a data-display latency of 1/3 s maintains a sense of "exploring" the environment, though the user may use slow movements to adapt to this relatively long latency. Longer latencies in data display require such slow movements on the part of the user that usability is lost.

The graphics update rate and the interaction responsiveness requirements are similar: only one graphics frame of latency is allowed in order to maintain good responsiveness for user interaction. The data-display responsiveness requirement, however, is less restrictive. The difference in latency requirements between interactivity and data displays is due to the fact that user interaction (selecting, acquiring, and moving objects) is a manual task driven by the human factors of manual control, while observing data display during movement is an intellectual task, in which the user observes what happens as a data probe is moved through space.

Because the interaction and data-display responsiveness requirements are different, the primary design strategy of a direct-manipulation system is to make the graphics and interaction processes independent of the visualization–computation processes. In such a system, the widgets can be made to update with 0.1 s latency even if the visualizations they control have a latency of 1/3 s.

A simple set of crosshairs in 3D space, with an associated streamline of a vector field, is an example of such a tool. In this example, the user can "pick up and move" the crosshairs, which will be very responsive (within the limits of the graphical update rate) due to their simple graphical nature. When these crosshairs are moved, they will trigger the computation of a streamline at their current location. If the process computing the streamline runs asynchronously from the process handling user interaction and graphical display, interaction responsiveness will be only slightly impacted by the computation of the streamline (assuming a preemptive multitasking operating system). This example shows that the difference in time scales between the interaction responsiveness and data-display responsiveness has strong implications for the runtime architecture of the scientific visualization system. These implications for the overall system architecture are discussed in Section 21.3.2.

Interactive time-varying environments potentially present a difficult challenge for the above requirements: all non-precomputed visualization extracts, not just the ones most recently manipulated, must be computed whenever the time-step changes. Furthermore, when the time-step changes, all computations must take place before any of the extracts can be displayed so that extracts from different data time-steps are not displayed at the same time. Experience has shown that it is acceptable for the data time-step to change quite slowly, so long as the 10 frames/s graphical update rate and the 1/3 s data-responsiveness requirements are met.

21.2.3   Input-Device Characteristics

Direct manipulation typically uses two pieces of information: the position and orientation of the user's hand, and some kind of user command indicating what action to perform at that location. In virtual reality systems, this information is provided by a 3D tracking device that delivers the hand position and (usually) orientation as well as a way to sense the user's command. The position and orientation data is often subject to noise, inaccuracies in position/orientation, and latency, which contribute (along with system latencies and frame rates) to degradation of the user's ability to directly manipulate objects in the virtual environment. Care must therefore be taken to use the highest-quality tracking device available, within practical constraints. Latency is minimized by use of the most recent tracking data, typically delivered by a high-frequency polling process.

Position-tracking data is typically defined as a 3D vector giving the user's hand position in some predefined coordinate system. The orientation data can take several forms. Commercial trackers typically return three orientation angles (either roll, pitch, and yaw or Euler angles), a 3×3 rotation matrix, or a quaternion [13]. A matrix or quaternion representation is preferable to orientation angles because orientation angles are unable to describe some orientations due to singularities (so called *gimbal lock*). These singularities are not present in a matrix representation or quaternions. In any case, the user's orientation and position data should be converted into a 4×4 graphics transformation matrix for use. This graphics transformation matrix should have the same form as transformation matrices used by the graphics system's matrix stack. The user data can then be used directly to transform any graphics objects being manipulated by the user.

Once the user's hand position and orientation are available as a graphics transformation matrix, it is simple to transform an object grabbed by the user so that the object appears to be rigidly attached to the user's hand. Define M_H as the graphics transformation giving tracking data for the user's hand in the current time frame, and M'_H as the graphics transformation giving that tracking data in the previous time frame. Similarly, let M'_O be the object's transformation matrix (from world coordinates) in the previous time frame. Our task is to compute the object's transformation matrix for the current frame M_O. Following the method of Warren Robinett, we have $M_O = M_H (M'_H)^{-1} M'_O$. This product is recomputed in each frame.

User commands are typically given via a device integrated with the hand-position tracker. Such a command device may be a simple set of buttons or the result of gesture recognition via an instrumented glove that measures the bend of the user's fingers. In any case, we assume that the device output is converted into a discrete command state such as "grab on" or "point on." When no command states are active, we have a neutral state.

Force-feedback devices are available that provide both user hand-tracking data as well as user feedback by restricting the movement of the tracking device via a mechanically exerted force. Such devices can be very powerful in direct manipulation, as they can give the experience of moving an object among other objects. Such a capability has shown great promise in some areas of scientific visualization, e.g., molecular modeling. Force feedback is an example of the many possibilities that are available in interface devices [15].

21.2.4 Collision Detection and User Commands

Collision detection in the context of direct manipulation refers to the ability to recognize when the user is able to "pick up" an object. In the real world we manipulate objects through a complicated interplay of muscle interaction, friction between complex surfaces, and gravity. Duplicating this interplay in a virtual environment is a highly nontrivial task. While some virtual-reality applications, such as task training, may benefit from detailed mimicking of real-world interaction, scientific visualization applications can usually take a much simpler approach.

The least complicated approach is to have a small number of active "grab points" contained in an interactive widget. These points become active when the user's hand position is within a predefined distance. Some kind of feedback is given to the user to indicate when the user's hand is close enough to a grab point to activate it. Then, if the user commands, for example, a grab state, then the grab point of the widget becomes grabbed and the widget reacts appropriately. When using this paradigm of inter-action, it is helpful to provide the user with unambiguous feedback as to the hand position. For example, representing the user's hand as a 3D set of crosshairs gives a better indication of the user's hand position than would showing an abstract hand shape.

More complex approaches can include using the user's hand-orientation information as part of the widget control, or using a more complex geometry-based collision-detection method.

21.2.5   Widget Design

The appropriate design of visualization widgets is critical to the success of a virtual-reality–based scientific visualization system. Unfortunately, space does not allow a complete review of this subject, so we will describe a simple approach that has proven effective in scientific visualization. For further examples, see Forsberg et al. [8] and Herndon and Meyer [10].

Inspired by the discussion at the end of Section 21.2.2, we discuss the design of simple widgets that can control a variety of visualiza-tions. These widgets are constructed based on some geometry defining the appearance of the widget, a set of defined grab points by which the widget may be manipulated, and a set of visual-ization seed points that are used to specify the visualizations controlled by the widget. Note that there can be several visualizations specified by the same seed point.

We give three examples of simple visualiza-tion widgets that differ in their spatial dimen-sions.

- Point widget (zero-dimensional; see Fig. 21.3): geometrical representation: a 3D set of crosshairs. Grab point: a single point located at the center of the crosshairs. Visu-alization seed point: the center of the cross-hairs.
- Line widget (1D; see Fig. 21.4): geometrical representation: line in 3D space. Grab

points: one at each end, each of which moves that end only (allowing control over the orientation and length of the line), and one at the line's center that, when grabbed, rigidly moves the entire line. Visualization seed points: a user-defined number are equally dis-tributed along the length of the line.

- Plane widget (2D; see Fig. 21.5): geomet-rical representation: a plane in 3D space. Grab points: one at each corner, each of which moves that corner only; one in the center of each edge, each of which moves that edge holding the opposite edge fixed; one in the center of the plane, which moves the plane rigidly. Visualization seed points: either a single seed point in the center of the plane (appropriate for local cutting plane visualizations, possibly re-stricted to the interior of the plane widget) or equidistributed on the plane, e.g., as an $n \times n$ array.

These widgets have a common design metaphor of simple geometry with grab points in the "ob-vious" places, and the visualization seed points have a natural association with the widget geometry. In all cases, the grab points provide the same graphical feedback to the user. These examples can be extended or generalized in obvious ways.

The visualization widgets and their associated visualizations will have properties that are typ-ically controlled via a conventional GUI inter-face. For immersive virtual environments, this GUI will usually be embedded in the 3D space of the visualization.

21.3   System Architecture Issues

There are several issues that arise in the design and implementation of virtual environments for information visualization. In this section we examine some of these issues in detail. First, however, we must classify the types of inter-action that may occur, which in turn depend on the time flow of the data.

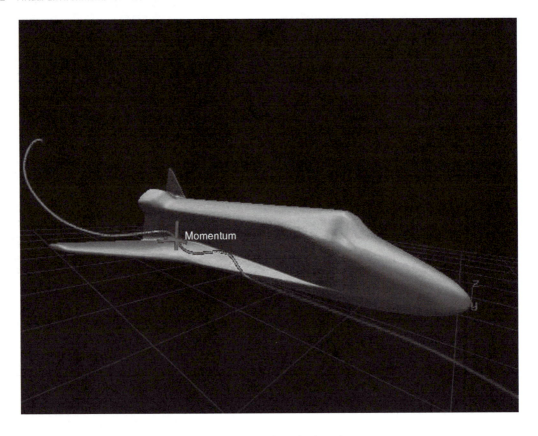

Figure 21.3 A point widget emitting a single streamline. (See also color insert.)

21.3.1 Classification of Interaction and Time Flow

There are two design questions that must be answered before a decision is made about the appropriate implementation of a scientific visualization application in virtual reality:

- Is the user allowed to interactively query the data at run time, generating new visualization extracts? If so, the system will likely have user-driven data accesses and computation to support extraction of new visualization geometry.

- Is the data time-varying? If so, there will be at least two senses of time in the virtual environment: user time and data time. The flow of data time can be put under user control so

that it can be slowed, stopped, reversed, or randomly accessed at specific time points.

These questions are independent, and both must be answered in order to determine which kind of implementation strategy will be appropriate. There are four combinations that arise from the answers to these two questions:

Noninteractive, non-time-varying data: This is the simplest scientific visualization environment, where the visualization geometry can be extracted ahead of time and displayed as static geometry in a head-tracked virtual environment. No data-access or computation issues occur in this case. The user may be able to rotate or move the static geometry. The design issues that arise in this case involve only collision detection and

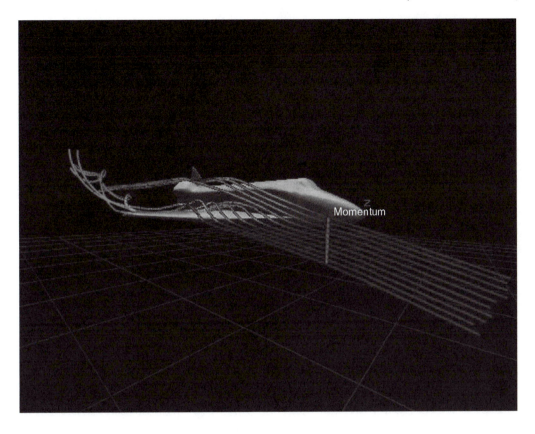

Figure 21.4   A line-visualization widget emitting a collection of streamlines. (See also color insert.)

possibly the design of widgets used to control the geometry.

Noninteractive, time-varying data: In this case, the visualization extract (VE) geometry can be precomputed as a time series, which can be displayed as a 3D animation in the virtual environment. The user may be given control over the flow of the animation to slow it down, stop it, or reverse its direction. Such user-controlled time flow implies an interface that may include rate controls to determine the speed and direction of the data time or a time-step control for the random access of the extracted geometry at a particular data time. The issues that arise in this case are common to most virtual-reality applications and so will not be considered further in this chapter.

Interactive, non-time-varying data: In this case, the data does not change in time, but the user specifies the visualization extracts at run time. In a virtual environment, such extracts may be specified via a direct-manipulation interface in which the user either specifies a point or manipulates an object in 3D space. The visualization extract may require significant computation, which will have an impact on the responsiveness of the system. This impact is discussed at length in Section 21.3.2. When visualization extracts do not change, they are typically not recomputed.

Interactive, time-varying data: For this type of environment, the data itself is changing with time, so any existing visualization extracts must be recomputed whenever the data time changes.

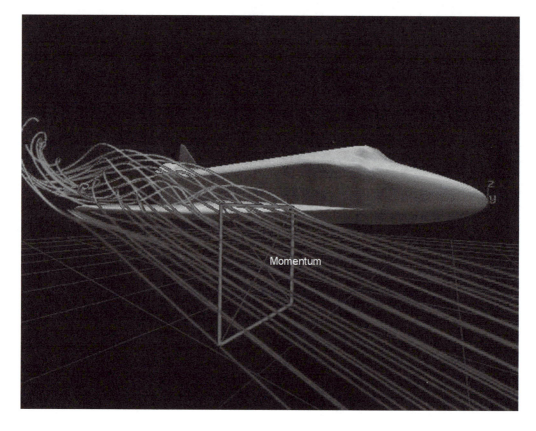

Figure 21.5 A plane-visualization widget emitting many streamlines. (See also color insert.)

This can result in a large amount of computation for each data time-step, the implications of which are discussed in Section 21.3.2. In addition, the user may be given control over the flow of data time, allowing it to run more quickly, more slowly, or in reverse. The user may wish to stop time and explore a particular time-step. When time is stopped, the system should act like an interactive, non-time-varying data environment, allowing visualization extracts to be computed in response to user commands.

21.3.2   System Architecture

The observations in the previous section imply that any implementation of an interactive scientific-visualization virtual environment in which

computation of visualization extracts takes place should contain at least two asynchronous processes: a *graphics and interaction process* and a *visualization extract-computation process*. More generally, the graphics and interaction tasks may be performed by a group of processes we shall call the *interaction (process) group*, one or more processes for graphics display, and possibly separate processes for reading and processing user tracking data. Similarly, the VE task may be performed by several processes called the *computation (process) group*, possibly operating on multiple processors in parallel. We choose these groupings because processes in the interaction group all have the same 10 frames/s and 0.1-s latency requirements, while the computation group has the 1/3-s latency requirement (see Section 21.2.2). This process structure decouples

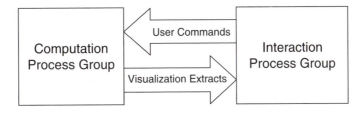

Figure 21.6   Runtime process architecture of a scientific visualization system for interactive and/or time-varying data.

display and computation, so a slow computation does not slow down the display process, and the speed requirements of the display process do not limit the computation.

The interaction process group passes user commands to the computation process group, which triggers the computation of visualization extracts. These resulting extracts are passed back to the interaction process group for display. This basic architecture is outlined in Fig. 21.6.

In an interactive time-varying environment, optimal computation and synchronization of the visualization extracts produced by the computation process group is a delicate issue, the resolution of which can be application dependent. An example of such a resolution is described by Bryson and Johan [4] and Bryson et al. [5].

21.4   Distributed Implementation

Distributed data analysis can be highly desirable for performance, computational steering, or collaborative purposes. The use of separate, asynchronous computation and interaction process groups communicating via buffers, as described in Section 21.3.2, facilitates a distributed implementation in which the process groups exist on separate, possibly remote machines communicating over a network.

21.4.1   Distribution Strategies

There are several strategies for the distribution of data analysis. These strategies are determined by the selection of where to place which oper-

ations in the data-analysis pipeline (Fig. 21.1). These strategies and their advantages and disadvantages are as follows:

- Remote data, local extraction and rendering: in this option, the data exists on a remote system, and individual data values are obtained over the network as required by the local VE algorithm. This strategy has the advantage of architectural simplicity; all visualization activities take place on the local system as if the data were local. This strategy has the disadvantage of requiring a network access each time data is required, which can be very time-consuming. There are techniques to overcome this disadvantage, such as clever prefetching, where data is delivered in groupings that anticipate expected future queries. For many applications, however, it will not be possible to use this strategy and meet the performance requirements described in Section 21.2.2.

- Remote data and extraction, local rendering: with this strategy, VE occurs on a remote system, typically the same system that contains the data. In an interactive system, the extraction computations occur in response to user commands passed from the user's local system. The results of the extraction, typically geometrical descriptions of 3D visualization objects, are transmitted to the user's local system for rendering. Architecturally, this strategy maps closely to the runtime architecture illustrated in Fig. 21.6, with the computation process group on a remote machine (or remote machines) and the display and interaction processes on the local system. This strategy has the advantage that the extraction algorithms are "close to the data," so data access does not suffer from a bottleneck. It also has the advantage of local rendering, which allows a head-tracking display for each participant, as required in virtual environments. The disadvantages of this strategy include the fact that response to user commands requires a round trip over the network and that the user's local system must be capable of rendering the extract's geometry.

• Remote data and extraction, distributed rendering: this strategy is a variation of the previous strategy (Remote Data and Extraction, Local Rendering). In this case, the rendering commands occur on the remote system and are passed as distributed rendering calls to the user's local system for actual rendering. A local client program is still required to read the user's trackers, as well as to process and send user commands. This strategy has advantages and disadvantages similar to those of the "remote data and extraction, local rendering" strategy, except that the network round-trip time is now part of the graphics-display loop. This may introduce unacceptable delays into head-tracking responsiveness.

• Remote data, extraction, and rendering: this strategy places all the operations of the data-analysis pipeline on the remote system(s), with the final rendered frames returned to the user's local system over the network. This strategy has the advantage of allowing very powerful remote systems to be used when the user has a low-power local system. The disadvantage is that the rendered frames can be large—for example, a 1024 × 1024 24-bit RGB-alpha display requires a 4 MB framebuffer, and two such framebuffers are required for stereo display. This implies an 80 MB transfer every second in our example to maintain the frame rate of 10 frames/s. This bandwidth requirement is beyond the capabilities of most available large-area networks. There are also serious issues of latency control in this strategy because the network time is part of the display-responsiveness loop. There are, however, situations where the local system is incapable of the kinds of rendering desired and this strategy may be the only viable option. Direct volume rendering is an example in which this strategy may provide the optimal choice.

These strategies are not exclusive: one may have remote VE taking place on a second, remote system while the data resides on a third system.

21.4.2   Remote Collaboration Strategies

Once a system is distributed, the opportunity arises for remote collaboration, where two or more noncollocated users examine the same data together. Strategies for remote collaboration are related to, but different from, distribution strategies. We briefly summarize the common remote-collaboration strategies:

• Distributed data: This collaboration strategy places copies of the data to be examined on all participants' client systems. Collaboration is implemented by passing either user commands or computed VE results among the participants' systems. The primary advantage of this strategy is that the software used is similar to stand-alone versions of the same systems. The main disadvantages include the difficulty of ensuring synchronization among the participants and the requirement that each participant's system be capable of storing the data and computing the visualization extracts (at least those that were locally generated). Many distributed collaborative VE systems, such as military training systems, utilize the distributed data-collaboration strategy.

• Data server: This collaboration strategy builds upon the remote data-distribution strategy. The data typically resides on a single remote system and is accessed by the participant's local system when needed. The visualization extracts are computed locally and are communicated in the same manner as in the distributed data-collaboration strategy explained previously.

- Visualization extract server: This collaboration strategy builds upon the remote extraction distribution strategy, in which the visualization extracts are computed on a remote system, typically the one where the data being examined is stored. The extracts are sent to each participant's system for local rendering. The advantages of this strategy include the following:

  - As there is only one set of extracts associated with each set of data, synchronization is greatly simplified.

  - Local rendering allows each participant to render the visualization extracts from a local point of view, which is required for head-tracked displays.

  - The extract server system can arbitrate conflicting user commands.

The visualization extract server-collaboration strategy has the disadvantage of poor scaling to large numbers of users, though this problem will be alleviated when reliable multicast technologies become available. The other disadvantages of this strategy are the same as those for the remote extraction distribution strategy.

- Scene replication: This collaboration strategy involves a privileged user whose view is presented to the other participants. This collaboration strategy is similar to the remote-rendering distribution strategy. This strategy has the same advantages and disadvantages as the remote data, extraction, and rendering distribution strategy, with the added disadvantage that all participants will see the same view, thereby precluding the use of head tracking for all participants.

21.5 Time-Critical Techniques

One of the prime requirements of virtual environments is responsiveness. In Section 21.2.2 we discussed the performance requirements for various aspects of a scientific visualization application within a virtual environment. These requirements can be difficult to meet, considering the possibly complex graphics and extensive computation required for the computation of VEs. One is often faced with a conflict between the requirements for a complete or accurate visualization and the requirements for responsiveness and fast graphical display rates. While accuracy is often critical in a scientific visualization environment, users often prefer fast response with a known degradation in accuracy for purposes of exploration. When a phenomenon of interest is found in the more responsive but less accurate mode, the user can request that this phenomenon be recomputed and displayed more slowly with higher accuracy. The automatic resolution of the conflict between accuracy and responsiveness, and the search for the appropriate balance, are collectively known as "time-critical design," the topic of this section.

21.5.1 The Time-Critical Philosophy

Time-critical design attempts to automate the process of finding a balance between required accuracy and responsiveness. This approach is very different from real-time programming, which guarantees a particular result in a specified time. Real-time programming typically operates in a fixed, highly constrained environment, whereas time-critical programs are typically highly variable. This variability is particularly evident in a scientific visualization environment, where the data and extracts computed and displayed may vary widely within a single user session. Time-critical design does not guarantee a particular result; it instead delivers the best result possible within a given time constraint. A successfully designed time-critical system will provide a graceful degradation of quality or accuracy as the time constraint becomes more difficult to meet.

Time-critical design for a particular aspect of a program begins with the creation of a cost–benefit metric for the task to be completed. The task is then parameterized in a way that controls both costs and benefits. When the costs and benefits of a task are known (as a function of the task's parameters) before that task is performed, the appropriate choice of parameters is selected to maximize the cost:benefit ratio. There are often many tasks to be performed in an environment, and the benefit of a particular task can be a function of the state of that environment. The solution of this problem is often approached as a high-dimensional constrained-optimization problem, maximizing the total cost:benefit ratio for the sum of the tasks to be performed, given the constraint of the total time allowed for all tasks.

As we shall see in Section 21.5.2, however, it is often very difficult to know the benefits and costs of a task before that task is performed. In such situations, hints provided by the user, or simple principles such as assuming equal benefit within a set of tasks, are often used.

21.5.2   Time-Critical Graphics

Time-critical techniques were pioneered in computer graphics [9], where objects were drawn with higher or lower quality depending on such benefit metrics as position in the field of view and distance from the user. Such implementations often used multiple representations of an object at varying levels of detail (LODs). In scientific visualization, however, many visualization extracts are already in a minimal form, such as streamlines defined as a set of points. In contrast, there are opportunities for graphical simplification in scientific visualization. For example, it may be the case that far more primitive elements are used to define a surface than are necessary for its display. An isosurface containing regions that are close to flat may have been derived with an algorithm that created many surface elements in that flat region.

Display of that surface would be faster if the flat region were represented by fewer surface elements. A surface might also be represented in simplified form until it became the focus of attention, in which case small variations from flatness would be important. Unfortunately, algorithms that identify such opportunities for surface simplification are computationally intensive and may therefore be unsuited for recomputation in every frame.

From similar considerations, we conclude that, unlike general computer graphics based on precomputed polygonal models, the use of time-critical graphics in scientific visualization will be highly dependent on the domain-dependent specifics of the visualization extracts. It may be very difficult, for example, to assign a benefit to a particular extract, especially when that extract may extend to many regions of the user's view. While simple benefit metrics such as the location of the extract on the screen may be helpful, one should keep in mind that the user's head may be pointed in one direction while the user's eyes are scanning the entire view. Such scanning is to be expected in a scientific visualization environment, where the scene may contain many related, extended objects.

From these considerations, few generalizations can be drawn about the use of time-critical graphics techniques in information visualization. Simple examples of time-critical graphics techniques that may be useful in scientific visualization include the following:

- Simplified representations, such as wireframe rather than polygonal rendering.
- Surfaces represented as 2D arrays, where simplified versions of the surface may be obtained by rendering every n points in the array.
- Techniques that have been developed for time-critical direct volume rendering [16].

A more general approach to time-critical graphics is to use multiresolution representations of the data or the resulting visualization extracts.

21.5.3   Time-Critical Computation

Computation of visualization extracts can provide several opportunities for time-critical design because such computation is often the most time-consuming aspect of a visualization system. As in the case of time-critical graphics, the specifics of time-critical computational design will be highly dependent on the nature of the extract computed. We can, however, make several general observations:

- Both the costs and the benefits of a visualization extract can be very difficult to estimate *a priori* based on its specification and control parameters, especially since the extent of an extract (e.g., where a streamline will be visible) is difficult to predict based on its specification alone.

- The cost of an extract can be roughly defined as the time required for its computation. Experience has shown that this cost does not vary widely between successive computations.

- The cost of a visualization extract may be most easily controlled through parameterization of the extent or resolution of the computation. Techniques that begin their computation at a particular location in space, such as a streamline emanating from a point in space, lend themselves well to control of their extent in space, which controls the time required for their computation. The costs of visualization techniques that rely on abstract or indirect specification, such as global isosurfaces, are more effectively controlled by variation of their resolutions.

- Other ways to control the cost of the visualization extract include the choice of computational algorithm and error metrics for adaptive algorithms. These control parameters have a more discrete nature and may be set by the user or set automatically via specific trigger criteria. For examples, see Bryson and Johan [4].

Given that the benefit of a visualization extract is hard to predict, one may treat all extracts as having equal benefit unless specified by the user.

In combination with the observation that costs do not change dramatically in successive computations, this allows the following simple solution to the problem of choosing the control parameters. For simplicity, we consider the situation in which all of the visualization extract's costs are controlled through limits on their extent. Here, each extract computation is assigned a time budget, and each extract's computation proceeds until its budget is used up. Then the time taken to compute all extracts is compared to the overall time constraint. Each extract's time budget is divided by a scale factor determined by the total actual time, divided by the total time constraint. This scale factor may have to take into account any parallel execution of the extract computations. If the time required to compute all the extracts is greater than the time constraint, this will result in smaller visualization extracts that will take less time to compute. If the extracts become too small, a faster but less accurate computational algorithm may be chosen. If the time required to compute all extracts is smaller than the time constraint, the time budget of each extract is increased, resulting in larger visualization extracts. A similar approach may be used to choose the resolution with which visualization extracts may be computed.

It may be evident from this discussion that the types of control parameters one may use in time-critical design will be highly dependent on the nature of the extract and how it is computed. Clever approaches can significantly enhance the time-critical aspects of a visualization technique. As an example, consider isosurfaces of a 3D scalar field. These isosurfaces are traditionally specified by selection of a value, resulting in a surface showing where that value is attained in the scalar field. Controlling the cost of a traditional isosurface by limiting the time of the computation will have unpredictable results: in the conventional marching cubes algorithm for computing isosurfaces, the entire dataset is traversed. If the marching cubes algorithm is stopped before completion and before regions of the field where the iso-value is attained are traversed, no isosurface will appear at all. It is possible to control the cost of the marching cubes algorithm by controlling the resolution with which the dataset is traversed, but this strategy does not provide fine control and may result in a significant degradation in quality. A different approach to isosurfaces, *local isosurfaces*, directly addresses this problem. Rather than choosing an iso-value when defining a local isosurface, the user chooses a point in the dataset; the iso-value is determined as the value of the scalar field at that point. The isosurface is then computed (via a variation on the marching cubes algorithm) so that it emanates from that point in space and is spatially local to the user's selection. The cost of a local isosurface is controlled by computation of the isosurface until that isosurface's time budget has been used up. Two examples of local isosurfaces can be seen in Fig. 20.2.

21.6   Conclusions

Direct manipulation in scientific visualization provides users with the ability to explore complex environments in a natural and intuitive way. In order to implement an effective scientific visualization application in a virtual environment, however, issues of responsiveness and fast updates must be addressed. These issues may be resolved via the use of appropriate system architectures, design based on human-factors issues, appropriate time control for time-varying data, implementation of time-critical techniques whenever possible, and appropriate choices for distributed implementations. The details of how these solutions are implemented will be highly dependent on the target domain and the specifics of the visualization techniques used.

References

1. S. Bryson. Virtual reality in scientific visualization. *CACM* 39(5):62–71, 1996.
2. S. Bryson and C. Levit. The virtual wind tunnel: an environment for the exploration of 3D unsteady flows. *Proceedings of Visualization '91*, San Diego, CA, 1991.
3. S. Bryson and M. Gerald-Yamasaki. The distributed virtual wind tunnel. *Proceedings of Supercomputing '92*, Minneapolis, MN, 1992.
4. S. Bryson and S. Johan. Time management simultaneity and time-critical computation in interactive unsteady visualization environments. *Visualization '96*, San Francisco, CA, 1996.
5. S. Bryson, S. Johan, and L. Schlecht. An extensible interactive framework for the virtual windtunnel. *VRAIS '97*, Albuquerque, NM, 1997.
6. D. B. Conner, S. S. Knibbe, K. P. Herndon, D. C. Robbins, R. C. Zeleznik, and A. van Dam. 3D widgets *Computer Graphics*, 25(2):183–188, 1992.
7. C. Cruz-Neira, J. Leigh, C. Barnes, S. Cohen, S. Das, R. Englemann, R. Hudson, M. Papka, L. Siegel, C. Vasilakis, D. J. Sandin, and T. A. DeFanti. Scientists in wonderland: a report on visualization applications in the CAVE virtual reality environment. *Proc. IEEE Symposium on Research Frontiers in Virtual Reality*, 1993.
8. A. S. Forsberg, K. P. Herndon, and R. C. Zeleznik. Aperture-based selection for immersive virtual environments. *Proc. 1996 ACM Symposium on User Interface and Software Technology (UIST)*, 1996.
9. T. A. Funkhouser and C. H. Sequin. Adaptive display algorithms for interactive frame rates during visualization of complex virtual environments. *Computer Graphics (SIGGRAPH '93)*, pages 247–254, 1993.
10. K. P. Herndon and T. Meyer. 3D widgets for exploratory scientific visualization. *Proc. 1994 Symposium on User Interface and Software Technology (UIST)*, pages 69–70, 1993.
11. S. G. Parker and C. R. Johnson. SCIRun: a scientific programming environment for computational steering. *Supercomputing '95*, 1995. See Chapter 31.
12. T. B. Sheridan and W. R. Ferrill. Man–machine systems. Cambridge, MA, MIT Press, 1974.
13. K. Shoemake. Animating rotations with quaternion curves. *Computer Graphics*, 19(3):1985.
14. D. Song and M. L. Norman. Cosmic Explorer: A virtual reality environment for exploring cosmic data, *Proc. IEEE Symposium on Research Frontiers in Virtual Reality*, pages 75–79, 1993.
15. R. M. Taylor, W. Robinett, V. L. Chi, F. P. Brooks, Jr., and W. Wright. The nanoManipulator: a virtual reality interface for a scanning tunnelling microscope. *Computer Graphics: Proceedings of SIGGRAPH '93*, 27:127–134, 1993.
16. M. Wan, A. E. Kaufman, and S. Bryson. High performance presence-accelerated ray casting. *IEEE Visualization*, pages 379–386, 1999.

22 The Visual Haptic Workbench

MILAN IKITS and J. DEAN BREDERSON
Scientific Computing and Imaging Institute
University of Utah

22.1 Introduction

Haptic feedback is a promising interaction modality for a variety of applications. Successful examples include robot teleoperation [57], virtual prototyping [63], painting [4], and surgical planning and training [30,55]. Such applications are augmented with force or tactile feedback for two reasons: (1) to increase the realism of the simulation, and (2) to improve operator performance, which can be measured by precision, fatigue level, and task completion times.

Even though a great variety of graphical visualization techniques have been developed in the past, effective display of complex multidimensional and multifield datasets remains a challenging task. The human visual system is excellent at interpreting 2D images. Understanding volumetric features, however, is difficult because of occlusion, clutter, and lack of spatial cues. Stereoscopic rendering, shadows, and proper illumination provide important depth cues that make feature discrimination easier. Using transparency reduces occlusion and clutter at the price of increasing ambiguity of the visualization.

In contrast to visual displays, haptic interfaces create a tightly coupled information flow via position sensing and force feedback. Such coupled information exchange results in more natural and intuitive interaction and utilizes some of the user's additional sensory-channel bandwidth. When users are presented with a proper combination of visual and haptic information, they experience a sensory synergy resulting from physiological reinforcement of the displayed multimodal cues [19].

Implementations of the traditional visualization pipeline typically provide a limited set of interactive data-exploration capabilities. Tasks such as finding and measuring features in the data or investigating the relationship between different quantities may be easier to perform with more natural data-exploration tools. To develop visualization and exploration techniques that further increase insight and intuitive understanding of scientific datasets, we designed and built an integrated immersive visual and haptic system, the Visual Haptic Workbench (VHW) [10]. In the following sections, we summarize our experiences with this system, discuss relevant issues in the context of developing effective visualization applications for immersive environments, and describe a haptic rendering technique that facilitates intuitive exploration modes for multifield volumetric datasets.

22.2 The Visual Haptic Workbench

The VHW is a testbed system developed primarily for haptic immersive scientific visualization. It is composed of a SensAble PHANToM 3.0L haptic device mounted above a Fakespace Immersive Workbench in an inverted configuration (Fig. 22.1). Head, hand, and stylus pose measurements are provided by a Polhemus Fastrak magnetic position tracker. Stereo images are generated by an Electrohome Marquee 9500LC projector and are reflected via folded optics onto the back of the nonlinear diffusion surface of the workbench. A pair of Stereographics Crystal-Eyes LCD shutter glasses, strobed at a 120 Hz refresh rate, is used for stereo viewing. In a typical

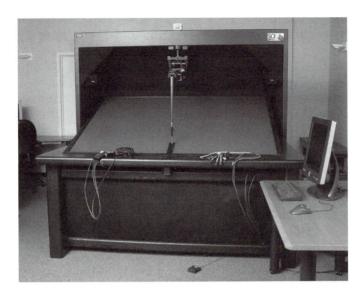

Figure 22.1 The Visual Haptic Workbench integrates a large workspace SensAble PHANToM with a Fakespace Immersive Workbench.

scenario, the user's dominant hand manipulates the PHANToM stylus to experience haptic feedback from the virtual scene, and the subdominant hand is used for system-control tasks such as navigating a menu interface. A pair of Fakespace Pinch Gloves and a pair of 5DT Data Gloves are provided for implementing more complex interaction techniques. Our custom additions to the workbench hardware include a "step to operate" footswitch instead of the original "push to interrupt" switch, which is used as a more convenient safety mechanism, a registration apparatus for placing the PHANToM in a fixed position on the surface of the workbench during encoder initialization, and an inexpensive 6DOF interaction device, the I^3Stick [9]. The system is constructed in such a way that it can be connected to a PC with the latest available graphics card without further modifications.

Compared to other similar systems, e.g., the UNC nano Workbench [21] or the CSIRO Haptic Workbench [68], our setup has the advantage of facilitating whole-arm interaction, using a wide-angle head-tracked visual display, and providing direct (1:1) correspondence between the visual and haptic workspaces. We found that tilting the workbench surface at a $20°$ angle both increases the visual range and aligns the hotspots of the workspaces [46]. Placing the PHANToM arm in front of the projection screen has the disadvantage of occluding the view, reducing the size of the available stereoscopic workspace. A related problem is that the low stiffness of the arm is apparent during hard surface contact, since the PHANToM end-effector may visually penetrate the virtual surface. To reduce these problems, we use a fixed offset between the actual endpoint and its virtual representation. Mounting the haptic device behind the screen would also solve these problems. Unfortunately, this is possible with front-projection screens only. Using front projection, however, further reduces the size of the visual workspace because the projection screen has to be located closer to the eyes than in other cases.

There are several important issues to consider when developing visualization applications for immersive environments. In the following subsections we discuss some of our observations and summarize what we have learned from our experiences with the Visual Haptic Workbench.

22.2.1   Calibration and Registration

Many immersive virtual-reality applications benefit from precisely calibrated system components. Greater accuracy is desired, though, to provide users with a more compelling experience and to increase precision and reduce frustration during 6DOF interaction tasks. Experimental test-bed systems require very accurate registration; otherwise, registration artifacts may be difficult to separate from other experimental factors [70]. Visual artifacts caused by registration errors include misalignment of real and virtual objects, as well as the notorious "swimming" effect, i.e., the motion and changing shape of stationary objects as the user moves around in the environment. It is also important to make sure that the generated visual and haptic cues match by precisely colocating the various workspaces of the system.

Registration error sources can be categorized according to whether they produce *geometric* or *optical* distortions. Geometric errors are the result of inaccurate tracking, system delay, misalignment of coordinate systems, and imprecise viewing and interaction parameters. Optical errors, caused by the limitations of the image-generation subsystem, are manifested as convergence and aliasing problems, display nonlinearities, and color aberrations. Haptic-rendering fidelity largely depends on the structural and dynamic characteristics of the haptic interface, the accuracy of its kinematic description, the update rate, and the control algorithm used to produce reaction forces and torques.

Possible geometric error sources for the VHW include tracker distortion and the unknown rigid-body transformations between the coordinate frames attached to the tracker transmitter, the PHANToM base, the display surface, and the eyepoints and interaction-device hotspots relative to the tracker sensors. Ideally, we want to reduce the discrepancies in these parameters so that the overall registration error does not exceed a few millimeters. In previous work, an external measurement device was used for coregistering the components of the Nano Workbench [21]. We have experimented with calibrating the magnetic tracker of our system using an optical device and found that it is possible to reduce measurement errors to a few millimeters within a large portion of the workspace [27]. We have also developed and evaluated a method for rapidly calibrating and registering the system components without using external metrology [26,29]. Our results indicate that to reach the desired level of accuracy, we need to replace the magnetic tracker with an accurate, low-latency optical solution.

22.2.2   Interaction Techniques

One of the "grand challenges" of using immersive environments for scientific exploration is "making interaction comfortable, fast, and effective" [72]. Designing and evaluating interaction techniques and user interfaces is an important area of virtual environment research [7]. Even though immersive virtual reality provides the possibility for more natural interaction, working with a computer-generated 3D world is difficult because the haptic cues that are part of the real world are missing from these environments. In the past, a variety of complex interaction techniques that are not particularly applicable for everyday use have been developed for immersive environments. In contrast, the desktop WIMP (Windows, Icons, Menus, Pointers) paradigm has been very successful due to its simplicity, robustness, and convenience. We found that the following guidelines should be observed when developing interaction techniques for immersive visualization applications [64]:

- Avoid complex and cumbersome devices, e.g., gloves.

- Use intuitive and simple interaction metaphors; reserve "magic" techniques for expert use, e.g., for shortcuts or text input [22,8].

- Utilize two-handed manipulation when possible, but provide ways to perform the same task with a single hand.

- Use physical and virtual constraints to increase precision and reduce fatigue.

Direct manipulation widgets provide a convenient means for exploring 3D datasets in desktop applications [15]. Typically, 3D widgets are implemented using the OpenGL picking and selection mechanism, or supported by the scenegraph API upon which the application is built [69,75]. Widgets are useful for immersive visualization, but the lack of physical constraints can make them cumbersome to use. We have developed a library of interface components for building applications that run in both desktop and immersive environments without modification (Fig. 22.2). Adding haptic feedback to the interface components is an area of future research.

22.2.3   Software Framework

Creating successful immersive applications is inherently an interactive process. An active area of virtual-environment research is making application development comfortable, fast, and effective. Application development and evaluation, however, usually happen in two different workspaces. Ideally, the developer should be able to run an application on every available platform without modifications, using an interface optimized to that particular platform [33] (Fig. 22.2). Previous efforts to create an immersive tool that could also run on the desktop required significant contributions from the programmer, because the interface remained platform-dependent [54]. Currently, most frameworks do not support this concept, and provide a *simulator mode* instead [5,16,37]. In this mode, a third-person view of the user is presented in such a way that immersive controls are mapped to a 2D desktop interface. Even though this mode is useful for examining how the user's actions affect the environment and vice versa, it prevents the developer from focusing on the content of the application.

22.2.4   Visualization Methods

Desktop visualization applications are event-based, and the majority of the events originate from the user. The viewer is in a fixed position, so the update rate and latency of interaction are less critical. In contrast, virtual-environment applications are built upon a continuous simulation with stringent requirements, including high update rate and low system latency, similar to those of computer games [12]. Thus, visualization techniques have to strike a balance between achievable quality and rendering

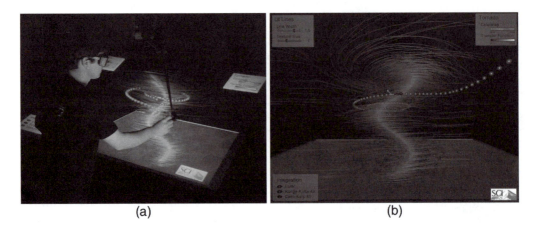

(a)   (b)

Figure 22.2   (a) A user explores a tornado dataset on the Visual Haptic Workbench. (b) Screenshot of the same application running in a desktop environment. Dataset courtesy of R. Crawfis, Ohio State University, and N. Max, Visualization Laboratory, Lawrence Livermore National Laboratory. (See also color insert.)

speed. Adaptations of traditional visualization methods have relied on multiresolution representations to maintain fluid interaction between the user and the application [13,20,71].

22.3 Haptic Rendering Techniques for Scientific Visualization

The majority of haptic rendering algorithms are geometric in nature, since they deal with the problem of interacting with various surface representations at real-time update rates. Surface rendering requires a suitable geometric model, typically combined with a bounding volume hierarchy, a rapid collision-detection technique, an incremental surface-tracing algorithm, and a model for generating contact forces from the probe-surface interaction. Surface-tracing algorithms exist for a variety of representations, including polygonal, parametric, and implicit surfaces. These algorithms rely on a combination of global and local distance queries to track the geometry closest to the interaction point. Haptic surface rendering has evolved from simple force-field methods [43] to constraint-based approaches that utilize a proxy point [58,77]. More recently, efficient techniques have emerged for haptic display of contact between complex polygonal objects [34,51].

Contact forces are usually modeled by interactions between the probe and a rigid viscoelastic surface. A virtual spring and damper are used to mechanically couple the probe with the proxy during contact. From the visualization point of view, surfaces are represented by a set of unilateral constraints that prevent the proxy from penetrating the object. Previous research has focused on improving the perceived crispness of surfaces and on augmenting them with various material properties to create realistic and convincing virtual objects [40,44,59,61,67].

Early work in haptic visualization used simple volumetric methods for exploring scalar and vector fields as well as molecular interactions [11,32]. The majority of previous methods for haptic display of volume data properties are based on a functional relationship between the reflected force and torque vectors, and the probe state and local data measures:

$$\vec{F} = \vec{F}(X, D, T) \tag{22.1}$$

where X denotes the state, typically position $\vec{x}$ and velocity $\dot{\vec{x}}$ of the haptic probe, D represents a set of local data measures at the probe position, and T stands for a set of haptic transfer functions and rendering parameters. We borrow the term *force-field rendering* for this class of techniques. The simplest examples in this category include density-modulated viscous drag for scalar data [3,52] and direct display of vector data [32,42]:

$$\vec{F}(\{\vec{x}, \dot{\vec{x}}\}, \{s(\vec{x})\}, \{k(s)\}) = -k(s(\vec{x}))\dot{\vec{x}} \tag{22.2}$$

$$\vec{F}(\{\vec{x}\}, \{\vec{v}(\vec{x})\}, \{k\}) = k\vec{v}(\vec{x}) \tag{22.3}$$

where the gain k is adjusted according to the scale and magnitude of the data measures and the capabilities of the haptic interface. Note that in Equation 22.2 we modulate viscous drag as a function of data value and in Equation 22.3 we apply a force directly proportional to the local field vector.

Even though this approach represents an important step in the evolution of haptic data-rendering techniques, it suffers from several limitations. First, it provides limited expressive power because it is difficult to display and emphasize features in a purely functional form. For example, we found that using complex transfer functions for rendering isosurfaces is less convincing than traditional surface-rendering approaches [3,31,39]. The reason for this is that the notion of *memory* is missing from these formulations [41,60]. Second, the device capabilities are captured implicitly in the rendering parameters. Applying a force as a function of the probe state can easily result in instability, especially when several rendering modes are combined. In general, it is very difficult and tedious to tune the behavior of the dynamic system formed by the force-field equation (Equation 22.1) and the motion equations of the haptic device by finding an appropriate set of rendering parameters.

Fortunately, haptic-rendering stability can be guaranteed with use of a virtual coupling network [1,14]. The coupler acts as a low-pass filter between the haptic display and the virtual environment, limiting the maximum impedance that needs to be exhibited by the device and preventing the accumulation of energy in the system [24,56]. Although the coupler is not part of the environment, the commonly used spring-damper form had been introduced implicitly in constraint-based surface-rendering algorithms. In the next section, we describe a similar approach to haptic rendering of directional information in volumetric datasets.

22.4 Data Exploration with Haptic Constraints

Constraints have been used successfully in both telerobotics and haptics applications. In early work, virtual fixtures or guides improved operator performance in robot teleoperation tasks [57]. More recently, a haptic-rendering framework was developed with algebraic constraints as the foundation [25]. Haptic constraints have helped guide users in a goal-directed task [23]. User interfaces can also benefit from guidance. Examples include a haptic version of the common desktop metaphor [47] and a more natural paradigm for media control [66].

We found that constraints provide a useful and general foundation for developing haptic-rendering algorithms for scientific datasets [28]. For example, constrained spatial probing for seeding visualization algorithms local to the proxy, e.g., particle advection, typically results in more cohesive insight than its unconstrained version. Volumetric constraints are obtained by augmentation of the proxy with a local reference frame, and control of its motion according to a set of rules and transfer functions along the axes of the frame. This approach has the advantage of providing a uniform basis for rendering a variety of data modalities. Thus, similar or closely related methods can be applied to seemingly unrelated datasets in such a way that the result is a consistent interaction experience. For example, to guide the user in vector-field data, the proxy can be constrained along a streamline such that any effort to move the probe in a direction perpendicular to the current orientation of the field results in a strong opposing force (Fig. 22.3b). However, if the user pushes the probe hard enough, the proxy could "pop over" to an adjacent streamline, allowing the user to move the probe in three dimensions and still receive strong haptic cues about the orientation of the flow. We can use an additional force component along the streamline to indicate the magnitude of the field. Alternatively, a secondary constraint can be added to convey information about the speed of the flow in the form of haptic "tickmarks." We found that such techniques result in intuitive feedback in exploration of vector-field data. A study on the effectiveness of various haptic rendering techniques for CFD datasets reached a similar conclusion [73].

Algorithms for constrained point-based 3DOF haptic rendering have been developed for scalar density data [6,41] as well as vector fields used in computational fluid dynamics (CFD) visualization and animation motion-control applications [18,73]. Haptic constraints have also been successfully used for displaying molecular flexibility [38]. Applications that require complex proxy geometry transform the proxy to a point shell to perform approximate 6DOF force and torque calculations using the individual point locations [42,45,53,56]. In recent work, a spherical approximation of tool–voxel interaction was used to speed up intersection calculations in a bone-dissection task [2].

22.4.1 Haptic Rendering with a Constrained Proxy Point

In general, haptic volume rendering algorithms based on a proxy point include four components

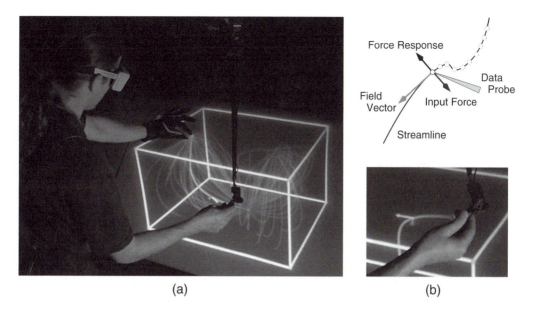

(a)

(b)

Figure 22.3 (a) A user explores a volumetric vector dataset. (b) The data probe is constrained along a streamline, resulting in intuitive haptic feedback. (See also color insert.)

that are executed at every iteration of the haptic servo loop (Fig. 22.4).

1. **Compute local data measures at current proxy location**: Data values and other measures, e.g., gradient or curvature information, are obtained from interpolation of data elements around the current proxy location. Typical methods include linear and tri-linear interpolation, although higher-order techniques may be more appropriate depending on the scale and resolution of the display [62]. Since haptic rendering is a local process, like particle advection, point-location algorithms for vector-field visualization on curvilinear and unstructured grids are readily applied [49]. A *local reference frame* ($\vec{e}_1$, $\vec{e}_2$, $\vec{e}_3$) is a key component of constraint-based techniques. Examples include the frame defined by the gradient and principal curvature directions in scalar data and the frame of eigenvectors in diffusion-tensor data. Note that the reference frame may be ill-defined or may not exist. Thus, an important

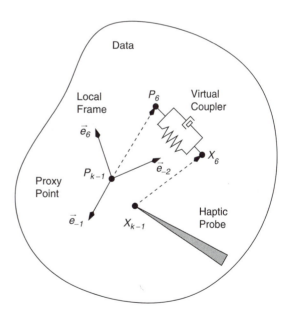

Figure 22.4 Components of constrained point-based 3DOF haptic data rendering. At time-step k, the state of the haptic probe has changed from X_{k-1} to X_k. The proxy state gets updated from P_{k-1} to P_k, from which the force response is computed using a virtual coupler. The p-Proxy update is based on data measures at the previous proxy location, as well as haptic transfer functions and rendering parameters.

requirement for the algorithm is to compute a stable force response even when transitioning into and out of homogeneous and ill-posed regions in the data. For example, in scalar volumes the reference frame is poorly defined in regions where the gradient vanishes. One way to achieve smooth transitioning is to modulate the force output as a function of gradient magnitude [41]. Another example is specifying transfer functions such that isotropic regions are handled properly in diffusion-tensor data. In this case, the transfer function has to be constructed in such a way that the force output either vanishes or degenerates to an isotropic point constraint.

2. **Evaluate haptic transfer functions to determine rendering parameters**: Similar to that of graphical visualizations, the goal of haptic transfer functions is to emphasize and combine features in the data. For example, a transfer function can be used to specify apparent stiffness and friction for isosurface regions based on data value and gradient magnitude [41]. In contrast to visual transfer functions, the design of haptic transfer functions is an unexplored area. Although it is possible to faithfully reproduce measured material properties [50], synthesizing them from different or abstract data remains difficult. In the examples presented in Section 22.5 we utilize stiffness and drag threshold transfer functions $\vec{k}$ and $\vec{\tau}$ to constrain the motion of the proxy along the axes of the local reference frame.

3. **Update proxy state**: In this step, the state of the proxy is updated according to simple motion rules. We have chosen a purely geometric approach that updates the proxy location based on probe motion and rendering parameters along the axes of the local frame:

$$\vec{p}_k = \vec{p}_{k-1} + \Delta\vec{p} = \vec{p}_{k-1} + \sum_{i=1}^{3} \Delta p_i\, \vec{e}_i \qquad (22.4)$$

where Δp_i is a function of probe position relative to the previous proxy location,

$\Delta x_i = (\vec{x}_k - \vec{p}_{k-1}) \cdot \vec{e}_i$. For example, surface-haptics algorithms locally constrain the proxy to the tangent plane by setting the normal component of change to zero. More sophisticated strategies incorporate the force response from previous steps as well as other stated variables. For example, physically based models assume the proxy has mass m and is moving in a medium with viscosity b [65]:

$$m\ddot{p}_i + b\,\dot{p}_i = F_i \qquad (22.5)$$

where F_i is the force component acting on the proxy point along $\vec{e}_i$. Friction effects can be incorporated by addition and moving of a static friction point within the constraint subspace [61].

Note that the linear approximation used in Equation 22.4 is not always appropriate for expressing a nonlinear constraint, such as staying on a surface or following a streamline. For example, when tracing volumetric isosurfaces, the first-order approximation obtained by projecting the probe point to the tangent plane defined by the gradient at the proxy location will result in the algorithm's quickly losing track of the surface. Thus, we find the new proxy location $\vec{p}_k$ by refining the initial estimate using Newton-Raphson iteration along the gradient direction [60]:

$$\Delta\vec{p} = -\frac{(s(\vec{p}) - s_0)\nabla s\,(\vec{p})}{|\nabla s\,(\vec{p})|^2} \qquad (22.6)$$

where s_0 is the target iso-value. The refinement is terminated when the step size $|\Delta\vec{p}|$ either is sufficiently small or reaches the maximum number of iterations permitted. Similarly, higher-order integration schemes, e.g., the fourth-order Runge-Kutta method, are necessary for computing the reference direction when following streamlines in vector data. For larger step sizes, supersampling and iteration of steps 1–3 may be required to ensure that constraints are satisfied accurately [6,60].

Linearized constraints can be applied in arbitrary order if the reference frame is orthogonal. For nonlinear constraints and nonorthogonal reference frames, the order of application defines which constraint is considered primary, which is considered secondary, etc. For example, to follow streamlines on a surface, we first move the proxy along the local field direction, then project it to the tangent plane of the surface. If the vector field has out-of-plane components, this order of steps corresponds to projecting the vector field onto the tangent surface. Reversing the order results in a different proxy location and creates a different haptic effect.

4. **Compute force response**: When using the spring-damper form of virtual coupling, the force response is computed from

$$\vec{F}_k = k_c\,(\vec{x}_k - \vec{p}_k) - b_c\,(\dot{\vec{x}}_k - \dot{\vec{p}}_k) \tag{22.7}$$

where k_c and b_c are chosen according to the device capabilities. The optimal choice maximizes the coupling stiffness without causing instability [1]. One problem is that these parameters may not be constant throughout the workspace. A choice that works well in the center may cause instability near the perimeter. Nevertheless, we can tune them by applying a point constraint at different locations in the workspace and determining which settings cause the device to become unstable on its own, i.e., without a stabilizing grasp. Analysis of the parameters could reveal the optimal operational region within the workspace of the device. In our implementation, we exclude the second term from Equation 22.7, since filtering velocity is difficult without high-resolution position measurements [14].

22.4.2 Motion Rules and Transfer Functions

Motion rules allow us to create various haptic effects that we can further modulate via haptic transfer functions. One effect simulates

plastic material behavior by generating increasing resistance between the probe and the proxy until a certain threshold is reached. At this point, the proxy is allowed to move towards the probe, keeping the reaction force at the same level. This effect is expressed succinctly by the following formula:

$$\Delta p_i = \text{sgn}(\Delta x_i)\max\left(|\Delta x_i| - \tau_i,\ 0\right) \tag{22.8}$$

This model yields free-space motion when $\tau_i = 0$:

$$\Delta p_i = \Delta x_i \tag{22.9}$$

and a *bilateral* constraint when $\tau_i > 0$. We use the term *drag threshold* for τ_i because it controls the difficulty of dragging the proxy along axis $\vec{e}_i$. Note that a stationary constraint is obtained when τ_i is sufficiently large, because it would take considerable effort to move the probe away from the proxy while resisting the increasing amount of force between them.

A *unilateral* constraint, which is the basis for surface-rendering algorithms, is obtained by considering the direction of travel along the axis:

$$\Delta p_i = \begin{cases} \Delta x_i & \text{if } \Delta x_i > 0 \\ \min\left(\Delta x_i + \tau_i,\ 0\right) & \text{if } \Delta x_i \le 0 \end{cases} \tag{22.10}$$

A bilateral *snap-drag* constraint changes the proxy location in discrete steps:

$$\Delta p_i = \begin{cases} \text{sgn}(\Delta \tau_i) & \text{if } |\Delta x_i| > \tau_i \\ 0 & \text{if } |\Delta x_i| \le \tau_i \end{cases} \tag{22.11}$$

The latter two rules are shown in Fig. 22.5, along with the resulting force responses.

We can influence proxy motion indirectly by scaling the force output according to stiffness transfer function $\vec{k}$:

$$F_{k,i} = k_i k_c\,(x_{k,i} - p_{k,i}) \tag{22.12}$$

where $0 \le k_i \le 1$. This reduces the force required for dragging the proxy. Note that setting either τ_i or k_i to zero produces no force output and creates frictionless motion along the axis. However, it yields two different proxy behaviors, since in the first case the proxy follows the

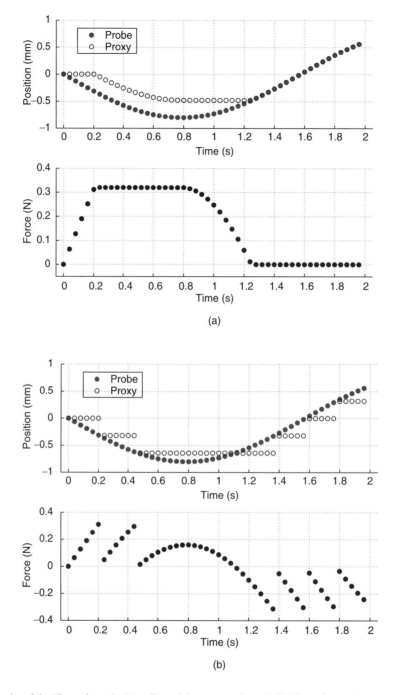

Figure 22.5 Examples of the 1D motion rule: (a) unilateral drag constraint and (b) bilateral snap-drag constraint. The motion of the probe and the proxy as a function of time are represented by the filled and empty circles, respectively. The resulting force responses are shown in the lower parts of the figures. Note that the sampling does not correspond to the haptic update rate.

probe exactly, while in the second case it lags behind by distance τ_i. Both parameters are necessary, because we want to express a range of effects, from subtle directional hints to stiff rigid constraints, in addition to both elastic and plastic material behavior.

22.5 Examples

In the following subsections, we describe how haptic constraints aid the user in two exploratory tasks: (1) investigating the relationship between cardiac muscle fibers and potential distributions, and (2) exploring the connectivity of brain white matter in diffusion-tensor MRI data.

22.5.1 Tracing Heart Muscle Fibers

Particle advection, i.e., integrating the motion of massless particles with velocities defined by the field, is a basic building block of vector-and tensor-field visualization techniques. The haptic equivalent is achieved by restriction of the motion of the proxy along the path of a single particle (Fig. 22.3a).

This method is easily modified to display orientation information on isosurfaces. Such a technique could be useful for investigating the relationship between heart muscle fiber orientations and potential distributions resulting from cardiac bioelectric finite-element simulations [48]. These simulations are typically carried out on a curvilinear grid that forms a number of epicardial and endocardial layers (Fig. 22.6). In our implementation, we reorganize the data to an unstructured tetrahedral grid by computing a Delaunay triangulation of the original data points. We assign a scalar value to the nodes in individual layers, in increasing order from inside to outside, such that isosurfaces of this scalar field correspond to muscle layers in the model. The gradient field computed from a central difference–approximation formula is used in the iterative refinement Equation 22.6 to make sure the proxy stays on the currently selected layer.

To avoid singularities when interpolating fiber orientation vectors within a tetrahedral element, we use component-wise linear interpolation of the tensor field, obtained by taking the outer product of the vectors with themselves. The major eigenvector of the interpolated tensor yields a smooth orientation field within a tetrahedral element, even when the vectors at the nodes point in completely different directions.

In this example, a local reference frame is formed by the interpolated fiber-orientation and gradient vectors. The snap-drag motion rule allows the user to explore a single layer and "pop through" to a neighboring layer by pushing against the surface. In this case, the drag threshold τ_i is not used for moving the proxy after breaking away from the current surface. Instead, we detect when the probe crosses a neighboring layer and set the proxy location to a numerical approximation of the intersection point. A secondary snap-drag rule constrains proxy motion along the fibers on the surface, allowing the user to switch to a nearby streamline in discrete steps. This method essentially creates a haptic texture on the surface composed of tiny valleys and ridges corresponding to the muscle fibers. See Fig. 22.6 for an illustration of this example.

22.5.2 Exploring Diffusion-Tensor Fields

Diffusion-tensor fields are difficult to comprehend because of the increased dimensionality of the data values and complexity of the features involved. Direct methods, such as glyphs and reaction-diffusion textures, work well on 2D slices, but they are less successful for creating 3D visualizations. Intermediate representations created by adaptations of vector-field techniques result in intuitive visual representations but fail to capture every aspect of the field [17,74]. Interactive exploration has helped users interpret the complex geometric models that represent features in the data [76]. Our goal is to aid the exploration process by adding

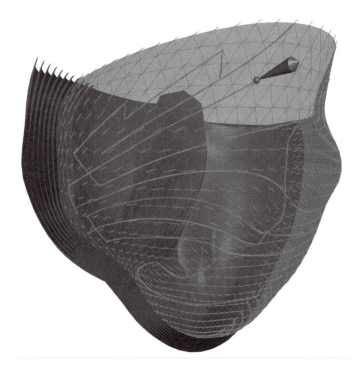

Figure 22.6   Exploring epicardial muscle fibers with haptic feedback. The probe is constrained to follow the local fiber orientation on the surface of a single layer. The user can "pop through" to a neighboring layer by pushing against the surface. Similarly, the user can choose a different fiber by pushing perpendicular to the currently selected fiber while staying on the surface. This effect feels as if the surface were textured with tiny valleys and ridges. The image shows the path of the proxy colored according to the magnitude of the applied force component perpendicular to the fiber orientation and tangent to the surface, from yellow to cyan, indicating increasing tension between the probe and the proxy. The dataset consists of about 30,000 nodes and 200,000 tetrahedral elements. Dataset courtesy of P. Hunter, Bioengineering Institute, University of Auckland. (See also color insert.)

haptic feedback that guides the user according to the local orientation and anisotropy of the field.

The rate and directionality of water diffusion in tissues is indicated by a second-order symmetric tensor. Anisotropy of the diffusion process can be characterized by the following barycentric measures [36]:

$$c_l = \frac{\lambda_1 - \lambda_2}{\lambda_1 + \lambda_2 + \lambda_3} \qquad (22.13)$$

$$c_p = \frac{2(\lambda_2 - \lambda_3)}{\lambda_1 + \lambda_2 + \lambda_3} \qquad (22.14)$$

$$c_s = \frac{3\lambda_3}{\lambda_1 + \lambda_2 + \lambda_3} = 1 - c_l - c_p \qquad (22.15)$$

where $\lambda_1 \geq \lambda_2 \geq \lambda_3$ are the sorted eigenvalues of the diffusion-tensor matrix. These measures indicate the degree of linear, planar, and spherical anisotropy, respectively. The associated eigenvectors $\vec{e}_1$, $\vec{e}_2$, $\vec{e}_3$ form an orthonormal frame corresponding to the directionality of diffusion. Regions with linear and planar anisotropy represent important features in the data, such as white-matter fiber bundles in brain tissue.

One way to use haptic feedback to indicate tensor orientation and degree of anisotropy is to control proxy motion such that it moves freely along the direction of the major eigenvector but is constrained in the other two directions. We found that setting the drag thresholds to a function of the anisotropy measures results in the desired feedback:

$$\tau_1 = 0 \qquad (22.16)$$

$$\tau_2 = \tau(c_l) \qquad (22.17)$$

$$\tau_3 = \tau(c_l + c_p) \qquad (22.18)$$

where $\tau(x)$ is a monotonically increasing function on $[0 \ldots 1]$. This choice ensures that the transfer functions yield a line constraint along the major eigenvector in regions with linear anisotropy ($c_l \gg c_p, c_s$), yield a plane constraint in regions with planar anisotropy ($c_p \gg c_l, c_s$), and allow free motion along all three directions in isotropic areas ($c_s \gg c_p, c_l$). Recall that the three indices sum to one, so when any one index dominates, the transfer functions emphasize the corresponding type of anisotropy. Alternatively, we can set the threshold to a constant value for all three directions and vary the stiffness similarly to Equation 22.18. In our implementation, we chose a linear ramp for $\tau(x)$, but other possibilities may be more appropriate.

The technique is illustrated in Fig. 22.7. We have observed that it takes little effort to trace out curves indicating fiber distribution and connectivity. Note that numerical methods for fiber tractography require careful specification of initial and stopping conditions and cannot be used straightforwardly for investigation of connectivity in regions of the data.

22.6 Summary and Future Work

We have designed and built a prototype system for synergistic display of scientific data. By developing and demonstrating initial applications, we have been able to refine our system and identify several important research issues in the context of building effective visualization applications for immersive environments. In the future, we plan to extend our collection of visualization techniques for the exploration of a variety of multidimensional and multifield datasets.

The presented approach for haptic data exploration has several desirable properties: it

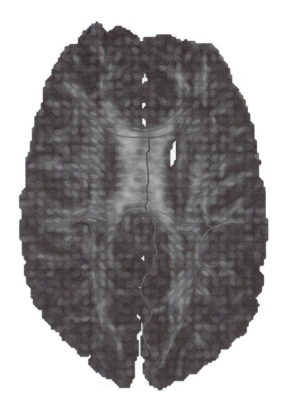

Figure 22.7 Exploring a 148×190 DT-MRI slice with haptic feedback. The ellipses represent local diffusion anisotropy and orientation. Lighter areas have higher associated anisotropy. The proxy path is colored according to the magnitude of the applied force, from yellow to red, indicating a larger tension between the probe and the proxy. The curves are tangent to the direction of the major eigenvector of the diffusion-tensor matrix in anisotropic areas. Dataset courtesy of G. Kindlmann and A. Alexander, W. M. Keck Laboratory for Functional Brain Imaging and Behavior, University of Wisconsin–Madison. (See also color insert.)

provides a unified rendering framework for different data modalities, allows secondary effects such as texture and friction to be easily realized, makes haptic transfer functions intrinsic to the algorithm, and allows control parameters to be tuned to the operational characteristics of the interface device.

A particular challenge we intend to address in the future is the issue of synthesizing useful haptic transfer functions from the underlying data. Investigating the synergistic relationship between visual and haptic transfer functions is another interesting research topic. A disadvantage of using the spring-damper form of virtual coupling is that it is too conservative, meaning that it may limit the efficient display of subtle haptic effects. We will experiment with a recent energy-based approach that uses a time-domain passivity observer and controller to adaptively adjust the coupling parameters [24]. In addition, we plan to extend the haptic rendering method to 6DOF devices. Transforming the constraint-based approach to a purely functional formulation would provide a very natural space for specifying rendering parameters. Finally, the real challenge for synergistic data display is validation. We intend to quantify the usability of our techniques and identify specific combinations that are useful to scientists who directly benefit from synergistic display of their datasets.

Acknowledgments

The authors are grateful to Gordon Kindlmann for enlightening and refreshing discussions as well as the Teem toolkit [35], which allowed efficient dataset probing and manipulation in this work. We thank Robert MacLeod for suggesting the heart visualization example and Hope M.F. Eksten for her help with various construction and fabrication projects. Support for this research was provided by NSF grant ACI-9978063, ARO DURIP grant DAAG-559710065, and the DOE Advanced Visualization Technology Center (AVTC).

References

1. R. J. Adams and B. Hannaford. A two-port framework for the design of unconditionally stable haptic interfaces. In *Proc. IEEE International Conference on Intelligent Robots and Systems*, pages 1254–1259, Victoria, BC, 1998.

2. M. Agus, A. Giachetti, E. Gobetti, G. Zanetti, and A. Zorcolo. Real-time haptic and visual simulation of bone dissection. *Presence: Teleoperators and Virtual Environments*, 12(1): 110–122, 2003.

3. R. S. Avila and L. M. Sobierajski. A haptic interaction method for volume visualization. In *Proc. IEEE Visualization*, pages 197–204, San Francisco, 1996.

4. W. Baxter, V. Scheib, M. C. Lin, and D., Manocha. DAB: Interactive haptic painting with 3D virtual brushes. In *Proc. ACM SIGGRAPH*, pages 461–468, Los Angeles, 2001.

5. A. Bierbaum, C. Just, P. Hartling, K. Meinert, A. Baker and C. Cruz-Neira. VR Juggler: A virtual platform for virtual reality application development. In *Proc. IEEE Virtual Reality*, pages 89–96, Yokohama, Japan, 2001.

6. D. J. Blezek and R. A. Robb. Haptic rendering of isosurfaces directly from medical images. In *Proc. Medicine Meets Virtual Reality*, pages 67–73, San Francisco, CA, 1999.

7. D. A. Bowman, E. Kruijff, J. J. LaViola, Jr., and I. Poupyrev. An introduction to 3D user interface design. *Presence: Teleoperators and Virtual Environments*, 10(1):96–108, 2001.

8. D. A. Bowman, C. J. Rhoton, and M. S. Pinho. Text input techniques for immersive virtual environments: an empirical comparison. In *Proc. Human Factors and Ergonomics Society Annual Meeting*, pages 2154–2158, 2002.

9. J. D. Brederson. The I^3 Stick: an inexpensive, immersive interaction device. *Technical Report UUCS-99–016*, School of Computing, University of Utah, 1999.

10. J. D. Brederson, M. Ikits, C. R. Johnson, and C. D. Hansen. The visual haptic workbench. In *Proc. PHANToM Users Group Workshop*, Aspen, CO, 2000.

11. F. P. Brooks, M. Ouh-Young, J. J. Batter, and P. J. Kilpatrick. Project GROPE—haptic displays for scientific visualization. In *Proc. ACM SIGGRAPH*, pages 177–185, Dallas, TX, 1990.

12. S. Bryson. *Virtual Reality Applications*, pages 3–15. Burlington, MA, Academic Press, 1995.

13. P. Cignoni, C. Montani, and R. Scopigno. MagicSphere: an insight tool for 3D data visualization. *Computer Graphics Forum*, 13(3): 317–328, 1994.

14. J. E. Colgate and J. M. Brown. Issues in the haptic display of tool use. In *Proc. IEEE International Conference on Intelligent Robots and Systems*, pages 140–145, Pittsburgh, PA, 1995.

15. D. B. Conner, S. S. Snibbe, K. P. Herndon, D. C. Robbins, R. C. Zeleznik, and A. van Dam. Three-Dimensional widgets. In *Proc. ACM Symposium on Interactive 3D Graphics*, pages 183–188, 1992.

16. C. Cruz-Neira, D. J. Sandin, and T. A. DeFanti. Surround-screen projection-based virtual reality: the design and implementation of the

CAVE. In *Proc. ACM SIGGRAPH*, pages 135–142, 1993.

17. T. Delmarcelle and L. Hesselink. Visualizing second order tensor fields with hyperstreamlines. *IEEE Computer Graphics and Applications*, 13(4):25–33, 1993.

18. B. R. Donald and F. Henle. Using haptic vector fields for animation motion control. In *Proc. IEEE International Conference on Robotics and Automation*, pages 3435–3442, 2000.

19. N. I. Durlach and A. S. Mavor, eds. *Virtual Reality: Scientific and Technological Challenges.* Washington, D.C., National Academy Press, 1994.

20. A. L. Fuhrmann and M. E. Gröller. Real-time techniques for 3D flow visualization. In *Proc. IEEE Visualization*, pages 305–312, 1998.

21. B. Grant, A. Helser, and R. M. Taylor II. Adding force display to a stereoscopic head-tracked projection display. In *Proc. IEEE Virtual Reality Annual International Symposium*, pages 81–88, 1998.

22. J. Grosjean and S. Coquillart. Command & control cube: a shortcut paradigm for virtual environments. In *Proc. Eurographics Workshop on Virtual Environments*, pages 39–45, 2003.

23. C. Gunn and P. Marando. Experiments on the haptic rendering of constraints: guiding the user. In *Proc. Advanced Simulation Technology and Training Conference*, Melbourne, Australia, 1999.

24. B. Hannaford and J.-H. Ryu. Time domain passivity control of haptic interfaces. *IEEE Trans. Robotics and Automation*, 18(1):1–10, 2002.

25. M. Hutchins. A constraint equation algebra as a basis for haptic rendering. In *Proc. PHANToM Users Group Workshop*, Aspen, CO, 2000.

26. M. Ikits. Coregistration of pose measurement devices using nonlinear least squares parameter estimation. Technical Report UUCS-00-018, University of Utah, 2000.

27. M. Ikits, J. D. Brederson, C. D. Hansen, and J. M. Hollerbach. An improved calibration framework for electromagnetic tracking devices. In *Proc. IEEE Virtual Reality*, pages 63–70, 2001.

28. M. Ikits, J. D. Brederson, C. D. Hansen, and C. R. Johnson. A constraint-based technique for haptic volume exploration. In *Proc. IEEE Visualization*, pages 263–269, 2003.

29. M. Ikits, C. D. Hansen, and C. R. Johnson. A comprehensive calibration and registration procedure for the visual haptic workbench. In *Proc. Eurographics Workshop on Virtual Environments*, pages 247–254, 2003.

30. Immersion Corporation. Medical simulators. http://www.immersion.com/

31. F. Infed, S. V. Brown, C. D. Lee, D. A. Lawrence, A. M. Dougherty, and L. Y. Pao. Combined visual/haptic rendering modes for scientific visualization. In *Proc. ASME Symposium on Haptic Interfaces for Virtual Environment and Teleoperator Systems*, pages 93–99, 1999.

32. H. Iwata and H. Noma. Volume haptization. In *Proc. IEEE Virtual Reality Annual International Symposium*, pages 16–23, 1993.

33. J. Kelso, L. E. Arsenault, and R. D. Kriz. DIVERSE: A framework for building extensible and reconfigurable device independent virtual environments. In *Proc. IEEE Virtual Reality*, pages 183–192, 2002.

34. Y. J. Kim, M. A. Otaduy, M. C. Lin, and D. Manocha. Six-degree-of freedom haptic display using localized contact computations. In *Proc. IEEE Symposium on Haptic Interfaces for Virtual Environment and Teleoperator Systems*, pages 209–216, 2002.

35. G. L. Kindlmann. The Teem toolkit, 2003. http://teem.sourceforge.net/

36. G. L. Kindlmann, D. M. Weinstein, and D. A. Hart. Strategies for direct volume rendering of diffusion tensor fields. *IEEE Trans. Visualization and Computer Graphics*, 6(2):124–138, 2000.

37. M. Koutek and F. H. Post. The responsive workbench simulator: a tool for application development and analysis. In *Proc. International Conference in Central Europe on Computer Graphics, Visualization and Computer Vision*, Plzen, Czech Republic, 2002.

38. A. Křenek. Haptic rendering of molecular flexibility. In *Proc. PHANToM Users Research Symposium*, Zurich, Switzerland, 2000.

39. D. A. Lawrence, C. D. Lee, L. Y. Pao, and R. Y. Novoselov. Shock and vortex visualization using a combined visual/haptic interface. In *Proc. IEEE Visualization*, pages 131–137, 2000.

40. D. A. Lawrence, L. Y. Pao, A. M. Dougherty, M. A. Salada, and Y. Pavlou. Rate-hardness: a new performance metric for haptic interfaces. *IEEE Trans. Robotics and Automation*, 16(4): 357–371, 2000.

41. K. Lundin. Natural haptic feedback from volumetric density data. Master's thesis, Linköping University, Sweden, 2001.

42. A. Mascarenhas, S. Ehmann, A. Gregory, M. Lin, and D. Manocha. *Touch In Virtual Environments: Haptics and the Design of Interactive Systems*, Chapter 5: Six degree-of-freedom haptic visualization, pages 95–118. Prentice-Hall, 2002.

43. T. H. Massie. Design of a three degree of freedom force-reflecting haptic interface. Bachelor's thesis, Massachusetts Institute of Technology, 1993.

44. T. H. Massie. Initial haptic explorations with the PHANToM: virtual touch through point interaction. Master's thesis, Massachusetts Institute of Technology, 1996.

45. W. A. McNeely, K. D. Puterbaugh, and J. J. Troy. Six degree-of-freedom haptic rendering using voxel sampling. In *Proc. ACM SIGGRAPH*, pages 401–408, 1999.

46. M. Meyer and A. H. Barr. ALCOVE: design and implementation of an object-centric virtual environment. In *Proc. IEEE Virtual Reality*, pages 46–52, 1999.

47. T. Miller and R. C. Zeleznik. An insidious haptic invasion: adding force feedback to the X desktop. In *Proc. ACM User Interface Software and Technology*, pages 59–64, 1998.

48. P. M. F. Nielsen, I. J. LeGrice, B. H. Smaill, and P. J. Hunter. Mathematical model of geometry and fibrous structure of the heart. In *American Journal of Physiology*, 260:H1365–H1378, 1991.

49. R. Y. Novoselov, D. A. Lawrence, and L. Y. Pao. Haptic rendering of data on unstructured tetrahedral grids. In *Proc. IEEE Symposium on Haptic Interfaces for Virtual Environment and Teleoperator Systems*, pages 193–200, 2002.

50. A. M. Okamura, J. T. Dennerlein, and R. D. Howe. Vibration feedback models for virtual environments. In *Proc. IEEE International Conference on Robotics and Automation*, pages 2485–2490, 1998.

51. M. A. Otaduy and M. C. Lin. Sensation preserving simplification for haptic rendering. *ACM Trans. Graphics*, 22(3):543–553, 2003.

52. L. Y. Pao and D. A. Lawrence. Synergistic visual/haptic computer interfaces. In *Proc. Japan/USA/Vietnam Workshop on Research and Education in Systems, Computation, and Control Engineering*, pages 155–162, 1998.

53. A. Petersik, B. Pflesser, U. Tiede, K. H. Höhne, and R. Leuwer. Haptic volume interaction with anatomic models at sub-voxel resolution. In *Proc. IEEE Symposium on Haptic Interfaces for Virtual Environment and Teleoperator Systems*, pages 66–72, 2002.

54. P. J. Rajlich. An object oriented approach to developing visualization tools portable across desktop and virtual environments. Master's thesis, Department of Computer Science, University of Illinois at Urbana-Champaign, 1998.

55. ReachIn Technologies AB. Laparoscopic Trainer. http://www.reachin.se/

56. M. Renz, C. Preusche, M. Pötke, H.-P. Kriegel, and G. Hirzinger. Stable haptic interaction with virtual environments using an adapted voxmap-pointShell algorithm. In *Proc. Eurohaptics*, 2001.

57. L. B. Rosenberg. Virtual fixtures: perceptual tools for telerobotic manipulation. In *Proc. IEEE Virtual Reality Annual International Symposium*, pages 76–82, 1993.

58. D. C. Ruspini, K. Kolarov, and O. Khatib. The haptic display of complex graphical environments. In *Proc. ACM SIGGRAPH*, pages 345–352, 1997.

59. S. E. Salcudean and T. D. Vlaar. On the emulation of stiff walls and static friction with a magnetically levitated input/output device. In *Proc. ASME Symposium on Haptic Interfaces for Virtual Environment and Teleoperator Systems*, pages 127–132, 1997.

60. J. K. Salisbury and C. Tarr. Haptic rendering of surfaces defined by implicit functions. In *Proc. ASME Symposium on Haptic Interfaces for Virtual Environment and Teleoperator Systems*, pages 61–67, 1997.

61. K. Salisbury, D. Brock, T. Massie, N. Swarup, and C. Zilles. Haptic rendering: programming touch interaction with virtual objects. In *Proc. ACM Symposium on Interactive 3D Graphics*, pages 123–130, 1995.

62. G. Sankaranarayanan, V. Devarajan, R. Eberhart, and D. B. Jones. Adaptive hybrid interpolation techniques for direct haptic rendering of isosurfaces. In *Proc. Medicine Meets Virtual Reality*, pages 448–454, 2002.

63. SensAble Technologies, Inc. FreeForm Modeling System. http://www.sensable.com/

64. L. Serra, T. Poston, N. Hern, C. B. Choon, and J. A. Waterworth. Interaction techniques for a virtual workspace. In *Proc. ACM Virtual Reality Software and Technology*, pages 79–90, 1995.

65. S. Snibbe, S. Anderson, and B. Verplank. Springs and constraints for 3D drawing. In *Proc. PHANToM Users Group Workshop*, Dedham, MA, 1998.

66. S. S. Snibbe, K. E. MacLean, R. Shaw, J. Roderick, W. L. Verplank, and M. Scheeff. Haptic techniques for media control. In *Proc. ACM User Interface Software and Technology*, pages 199–208, 2001.

67. M. A. Srinivasan and C. Basdogan. Haptics in virtual environments: taxonomy, research status, and challenges. *Computers and Graphics*, 21(4):393–404, 1997.

68. D. R. Stevenson, K. A. Smith, J. P. McLaughlin, C. J. Gunn, J. P. Veldkamp, and M. J. Dixon. Haptic workbench: a multisensory vir-

tual environment. In *Proc. SPIE Stereoscopic Displays and Virtual Reality Systems*, pages 356–366, 1999.

69. P. S. Strauss and R. Carey. An object-oriented 3D graphics toolkit. In *Proc. ACM SIG-GRAPH*, pages 341–349, 1992.

70. V. A. Summers, K. S. Booth, T. Calvert, E. Graham, and C. L. MacKenzie. Calibration for augmented reality experimental testbeds. In *Proc. ACM Symposium on Interactive 3D Graphics*, pages 155–162, 1999.

71. T. Udeshi, R. Hudson, and M. E. Papka. Seamless multiresolution isosurfaces using wavelets. Technical report ANL/MCS-P801-0300, Argonne National Laboratory, 2000.

72. A. van Dam, A. S. Forsberg, D. H. Laidlaw, J. J. LaViola, Jr., and R. M. Simpson. Immersive VR for scientific visualization: a progress report. *IEEE Computer Graphics and Applications*, 20(6):26–52, 2000.

73. T. van Reimersdahl, F. Bley, T. Kuhlen, and C. Bischof. Haptic rendering techniques for the interactive exploration of CFD datasets in vir-

tual environments. In *Proc. Eurographics Workshop on Virtual Environments*, pages 241–246, 2003.

74. D. M. Weinstein, G. L. Kindlmann, and E. Lundberg. Tensorlines: advection-diffusion based propagation through diffusion tensor fields. In *Proc. IEEE Visualization*, pages 249–253, 1999.

75. M. Woo, J. Neider, T. Davis, and D. Shreiner. *OpenGL Programming Guide: The Official Guide to Learning OpenGL, Version 1.2*. Addison Wesley, 1999.

76. S. Zhang, C. Demiralp, D. Keefe, M. DaSilva, D. H. Laidlaw, B. D. Greenberg, P. Basser, C. Pierpaoli, E. Chiocca, and T. Deisboeck. An immersive virtual environment for DT-MRI volume visualization applications: a case study. In *Proc. IEEE Visualization*, pages 437–440, 2001.

77. C. B. Zilles and J. K. Salisbury. A constraint-based god-object method for haptic display. In *Proc. IEEE International Conference on Intelligent Robots and Systems*, pages 146–151, 1995.

23   Virtual Geographic Information Systems

WILLIAM RIBARSKY
College of Computing
Georgia Institute of Technology

23.1   Introduction

Geospatial data are growing in diversity and size. Satellite imagery and elevation data at 30 M resolution are readily available for most of the Earth via Landsat and other sources. These sources also provide multispectral imagery at similar resolutions that distinguishes land use, vegetation cover, soil type, urban areas, and other elements. Higher-resolution aerial or satellite imagery for selected areas can be obtained. There are photographs at 1M resolution or better that cover most major cities, with insets at even higher resolution often available. There are also accurate digital maps. Tax assessment records and other geolocated records provide information about the uses of individual sectors of urban geography. GIS databases also provide geolocated access to names, addresses, and uses, and information about roads, bridges, buildings, and other urban features. Other GIS databases provide national, state, and local boundaries; paths of waterways and locations and extents of lakes; and boundaries of forests. In addition, techniques are now appearing that will lead to the automated and accurate collection of 3D buildings and streetscapes [20,62,66]. Most major U.S. and European cities have ongoing digital cities projects that collect these 3D models [32], although at the moment modeling is laborious. Among other things, these models are leading to a new, more detailed, and more comprehensive view of the city as it is now and as it is planned to be. Now detailed 3D, time-dependent atmos-pheric data are collected for extended areas. Sources include the 3D Doppler radar systems that cover the U.S. and Europe, and high-resolution weather, climate, or pollution simulations, all augmented by specialized satellite measurements. These weather data and simulations are at such a resolution and accuracy that detailed terrain elevation and coverage data can now be useful or necessary. For example, having detailed terrain-elevation models permits one to predict flood extents and the progress of flooding rather than just the flood heights (which is often all that is available widely). Elevation data are also a necessary input for high-resolution weather models. Other geolocated data, such as sources of industrial pollution, traffic congestion, and urban heat islands, can be important inputs for weather and pollution models.

Interactive visualization is of prime importance to the effective exploration and, analysis of the above integrated geospatial data. For systems dealing with geospatial data of any extent, the two capabilities of interactive visualization and integrated data organizations are inextricably intertwined. In this chapter, we will discuss both capabilities in the context of virtual geographic information systems (GISs). One main way in which a virtual GIS differs from a traditional GIS is that it supports highly interactive visualization of the integrated geospatial data. Visual navigation is a prime way of investigating these data, and queries are by direct manipulation of objects in the visual space. The visualization is thus a visual interface to the data that is supported by data retrieval and rendering mechanisms appropriate to multiscale, multiresolution data.

Virtual GIS systems are almost universally useful. Their use for the investigation of atmospheric phenomena and their effect on the land have already been mentioned. With appropriate urban data, virtual GIS can also be used for urban planning. Modern urban planning considers the issues of "smart growth" [14], where existing and already congested urban centers are redesigned for future development that concentrates work, school, shopping, and recreation to minimize car travel, congestion, and pollution while improving quality of life. Such projects are often infill projects with significant effects on the urban fabric. There are thus competing groups who often have significantly different objectives, groups including residents, businesses, developers, and local or state governments. This planning process is usually laborious and involves much negotiation and many plans vetted, modified, and discarded, missed opportunities, and results that often still don't satisfy the multiple groups. Interactive visualization is an essential new component for speeding the process, making alternatives clearer and more fully understandable, and reaching better results [19]. As mentioned above, comprehensive urban data combined with the visualization capability can also give a broader, more integrated, and more detailed view of the city and how multiple plans fit into it than was possible before.

There are many other uses for virtual GIS. For example, a highly detailed and interactive visualization system can be used for emergency planning and emergency response. Virtual GIS also has significant educational potential to show how cities fit with the wider environment, how the land fits with its natural resources, and how states and countries relate to each other. A virtual GIS with a sense of historical time can show, in context and in detail, the positions and movements of great battles, migrations of populations, development of urban areas, and other

events. Finally, there are many additional uses of virtual GIS, including tourism and entertainment, military operations, traffic management, construction (especially large-scale projects), various geolocated and mobile services, citizen–government relations (when complex civic projects are vetted), games based on real locations, and others. The dynamic nature of geospatial data collection provides all citizens with a unique capability to track the detailed change and development of urban areas, areas around waterways, farms, woodlands, and other areas.

In this chapter I will discuss key work in the development of current virtual GIS capabilities. I will then briefly discuss geospatial data-collecting organizations and multiresolution techniques. I will review interactive techniques for navigating and interacting with data at the wide range of scales in global geospatial systems. These will be for both tracked and untracked interaction and for a range of display environments, from PDAs to large projected screens. I will then discuss the application of virtual GIS to urban visualization and to 3D, time-dependent weather visualization. Finally, I will present some outstanding questions that should be addressed in the future.

23.2 Key Work in the Development of Virtual GIS

Initial work on interactive visualization of geospatial data focused on terrain visualization. Some systems supported virtual environments [28,42], including the capability for out-of-core visualization in a distributed networked environment [42]. These had limited terrain detail management. The first systems with enough capabilities to be virtual GIS systems were then developed [34,37]. These provided interactive visualization of terrain models including elevations and imagery, GIS raster layers, 3D buildings, moving vehicles, and other objects in both virtual (using a head-mounted display, or HMD) and Windows-based environments. About this time, an interactive system was developed [54] that integrated highly detailed phototextures, a terrain elevation map, road network features, and generic models for buildings with appropriate footprints, numbers of floors, and types. The result was an interactive visualization of an urban area that had good graphical quality and retained much information from the geographic sources. The initial work on virtual GIS was followed by the development of a system that integrated a standard GIS database to permit queries and display of GIS data through direct manipulation of the visualization [2]. The first reported global virtual GIS with the capability for handling scalable data, VGIS [18], will be discussed in detail below. Recently, commercial systems with the capability to handle global terrain data have appeared (*http://www.keyholecorp.com*). There also continue to be ongoing efforts to study the interface between computer graphics, visualization, and cartographic and geospatial data, such as the ACM Carto Project (*http://www.siggraph.org/~rhyne/carto/carto98.html*).

Recently there have been thrusts to extend, in both scale and type, the detailed models that can be included in virtual GIS and also to broaden its focus to include new kinds of geospatial data. In many cases, these thrusts are due to new acquisition tools (e.g., LIDAR, high-resolution photos, ground-based range images, 3D Doppler radar, etc.) that make available streams of data that can be turned into models of unprecedented scale and detail. Successfully integrating these models into a comprehensive, integrated virtual GIS remains a major challenge. Extended urban modeling efforts include the Virtual Los Angeles Project [5,32] and similar efforts to create extended, interactive urban landscapes (Fig. 23.1). Virtual LA has modeled hundreds to thousands of buildings in urban Los Angeles, as well as many street-level features. The buildings typically consist of a relatively small number of polygons with detail provided by textures that are usually culled from ground-level photoimagery. The visualization system uses scene graphs generated by SGI Performer or Multigen to navigate urban areas interactively at ground level.

Figure 23.1 Downtown Los Angeles, from the Virtual LA project [6]. (See also color insert.)

Extended environments such as Virtual LA (and the urban models contained in our VGIS system) clearly demonstrate an important point: that it is still a painstaking, time-consuming, hands-on process to produce the model collection. Fred Brooks [5] has said that modeling remains one of the major challenges in virtual environments research. One line of current research attacks the painstaking modeling problem while another line produces models of significantly greater detail. For example, a LIDAR system permits an airplane to quickly collect a height field, with lateral resolution better than 1 M and vertical resolution of inches, for a whole small city in a few hours. An image of extracted and refined models from LIDAR data acquired for the USC campus and environs is shown in Fig. 23.2. The refinement and modeling were accomplished with an almost completely automated set of tools [66]. Concurrently there are efforts to acquire ground-based urban range images [20,55] and place them into extended models. This research uses 3D recovery and reconstruction from multiple images or from imagery plus laser range-finding. In either case, 3D reconstruction of geometry plus appearance information produces urban scenes

Figure 23.2 High-resolution aerial imagery merged with georeferenced models automatically extracted from LIDAR data [66]. (See also color insert.)

(buildings and streetscapes including sidewalks, lampposts, trees, shrubs, etc.) that can resolve inch-size geometric features. On the other hand, there are efforts to automate and enrich the constructive modeling process with the use of 3D design tools rather than acquired data. Promising methods include procedural modeling tools that can produce an entire cityscape of generic buildings that conform either to an imagined or a real street layout [45]. Approaches to apply higher-level architectural-design principles and shape grammars [63] hold the promise to quickly construct buildings of much greater 3D architectural detail that conform to real or imagined structures. The organization into grammar rules means that details can be changed at will, perhaps even interactively (Fig. 21.3). This will be a boon to advanced urban planning. However, all of these approaches result in incomplete models. The acquisition approaches, in particular, are for only their domains (e.g., street-level façades for the ground-based methods versus footprints and top-level detail for the aerial methods) and must be combined for more complete models. Ultimately, all approaches should be combined to produce comprehensive and consistent urban models that will also have the advantageous capability to be changed and updated. As these models are created and extended, they must be prepared for interactive visualization, integrated with terrain data, and placed in a data organization that can handle their scale and complexity. This is the focus of the next sections.

23.3   Global, Comprehensive Organization of Geospatial Data

Everything on Earth has a time and a location. The latter is encoded in a global coordinate

Figure 23.3 3D urban models generated procedurally using an architectural grammar [63].

system such as latitude and longitude (lat/lon) or altitude. With the proper geodetic transformation, these coordinates give precise locations for terrain features or objects anywhere on Earth (including under the ocean). This is a fundamental aspect of GIS systems. Including the altitude coordinate can provide precise locations to objects or phenomena in the atmosphere or under the earth. For virtual GIS, we must organize data in this global coordinate system for multiresolution, interactive visualization. Once established, such a universal organization holds the promise of integrating and handling all geospatial data, whatever their type or source. Thus, an ultimate goal of virtual GIS is to go beyond the boundaries of traditional GIS and create a *virtual world* that encompasses all the knowledge of the real world. In this section we describe structures that can form the foundation of this virtual world and that, in particular, support interactive navigation and exploration in virtual GIS.

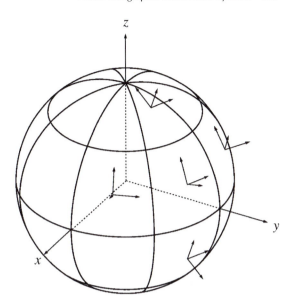

Figure 23.4 The Earth divided into 32 zones. The labeled axes correspond to Earth-centered, Earth-fixed Cartesian XYZ global coordinate systems for each zone.

23.3.1 Global Hierarchy

We have built a quadtree and shared a cache data structure that together constitute the basic components of a global data model. This is the data structure that is used in our VGIS system. The global structure is divided into 32 zones, each $45° \times 45°$ [10,16] (Fig. 23.4). Each zone has its own quadtree; all are linked so that objects or terrain-crossing quadrant boundaries can be rendered efficiently. We chose the number and extent of zones based on empirical observations of memory requirements, paging overhead, geometric accuracy, etc. A node in a quadtree corresponds to a raster tile of fixed dimensions and lat/lon resolution according to the level on which it appears in the quadtree. Quadnodes are identified by "quadcodes," which are built in a manner similar to the indices of representations of binary trees; that is, the children of a node with quadcode q are identified by $4q + 1$ through $4q + 4$. In addition, the quadcode contains a quadtree identifier that allows each quadcode to uniquely identify an area on the

globe. This structure is replicated in the underlying disk-management system so that files are aligned with the quadnodes in the set of linked quadtrees.

The quadtrees also define the boundaries of local coordinate systems. If a single, geocentric coordinate system were used, assuming a 32-bit single precision floating point were used to describe object geometries, the highest attainable accuracy on the surface of the Earth would be half a meter. Clearly, this is not sufficient to distinguish features with details as small as a few centimeters, e.g., features on a building façade. This lack of precision results in "wobbling" as the vertices of the geometry are snapped to discrete positions, which is present in other large-scale terrain systems, such as T_Vision [26]. We have developed an approach to overcome this problem [18]; we define a number of local coordinate systems over the globe, which have their origins displaced to the (oblate) spheroid surface that defines the Earth's sea level. (See Section 23.5 for a more detailed description of this spheroid structure.) The origins of the top-level coordinate systems are placed at the geographic centers (i.e., the means of the boundary longitudes and latitudes) of the quadtree roots. While the centroid of the terrain surface within a given zone would result in a better choice of origin in terms of average precision, we decided for the sake of simplicity to opt for the geographic center, noting that the two are very close in most cases. The z axis of each coordinate system is defined as the outward normal of the surface at the origin, while the y axis is parallel to the intersection of the tangent plane at the origin and the plane described by the North and South poles and the origin. That is, the y axis is orthogonal to the z axis and locally points due North. The x axis is simply the cross product of the y and z axes, and the three axes form an orthonormal area. This choice of orientation is very natural, as it allows us to approximate the "up" vector by the local z axis, which further lets us treat the terrain height field as a flat-projected surface with little error. Hence, the height-field level-of-detail (LOD) algorithm, which is based on vertical error in the triangulation, does not have to be modified significantly to take the curvature of the Earth into account. However, the delta values [39] must be computed in Cartesian rather than geodetic coordinates to avoid oversimplification of areas with constant elevation but that are curved, such as oceans. Fig. 23.4 shows the local coordinate systems for a few zones.

Using this scheme, the resulting worst-case precision for a $45° \times 45°$ zone is 25 cm—not significantly better than for the Earth-centered case. We could optionally use a finer subdivision with a larger number of zones to obtain the required precision, but this would result in a larger number of quadtrees, which is undesirable since the lowest-resolution datum that can be displayed is defined by the areal extent of the quadtree roots. Hence, too much data would be needed to display the lowest-resolution version of the globe. Instead, we define additional coordinate systems within each quadtree. In the

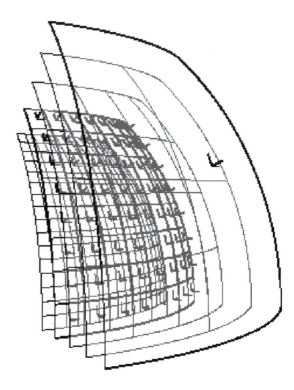

Figure 23.5   Nested coordinate systems in a quadtree. 8×8 smaller coordinate systems appear 3 levels below the root node.

current implementation, we have added 256×256 coordinate systems within each quadtree—one coordinate system per node, eight levels below each root node—resulting in a 1 mm worst-case precision. Fig. 23.5 illustrates a subset of these nested coordinate systems. The terrain and object managers keep track of which coordinate system to use among these thousands of systems and can even transition between coordinate systems for extended objects.

The general approach to using the hierarchical structure is illustrated in Fig. 23.6. In each zone, the quadtree is traversed to a certain level that depends on the type of geospatial data. Below this level, a nonquadtree detail-management scheme is used that depends on the detailed characteristics of the data. Thus, for example, buildings and terrain have different levels at which separate nonquadtree detail-

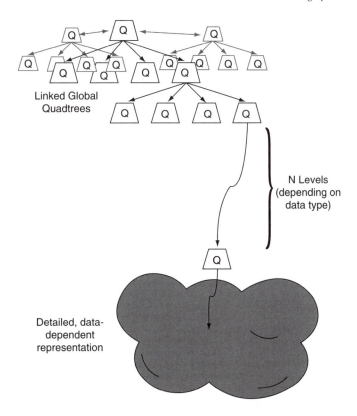

Linked Global
Quadtrees

N Levels
(depending on
data type)

Detailed, data-
dependent
representation

Figure 23.6 General global hierarchy.

management schemes take over. We describe these schemes further below.

This global hierarchical, nested structure will handle the Earth, everything on it, comprehensive atmospheric phenomena, and even subsurface data at levels of detail from global overviews to fine-resolution close-ups. Navigation between these extremes involves changes of sometimes 10 or more orders of magnitude. Note that this global structure does not require that the whole set of data reside in any one place. Rather, the structure is virtual and as long as one has appropriate lat/lon and altitude coordinates (and a procedure for organizing the particular type of data being handled), one knows where to store or retrieve the data. In this view, instead of server/client relations, there are peer relations. Anyone can collect data or share data, and servers are just peers that have

more data than others. This flexibility is quite useful for both distributed applications and mobile applications [35]. In addition, this structure is modular and can be efficiently incremented as new data become available. For example, terrain data (imagery and elevation) can be added in a time that is proportional to the amount of data added rather than to the total amount of data in the dataset. Furthermore, 3D weather data can be added in real time (in seconds, for a 3D Doppler radar system where the time between collection of successive radar volumes is typically 5–7 minutes [30,48]).

23.3.2 Caching and Paging

To conserve memory and promote efficiency, the view-dependent data associated with active nodes of the hierarchy are stored in a shared cache. This allows multiple managers for the various data types to access the data without having to replicate it. The shared cache consists of a set of hash tables, one for each data type (e.g., elevations, phototextures, weather data, buildings, moving objects, etc.), which have enough slots to hold all the quadnodes in the dataset. These slots are initially empty and are filled with geospatial data whenever a request is processed by a particular server. If a node is no longer needed by any of the managers, the space for it is deallocated. The quadcodes are used as hash keys for accessing nodes in the hash table. Since the hash table slots are initialized at startup, the managers know what nodes exist externally, so no invalid data requests are made to the server. To maintain scalability, we have developed a structure where only the high-level quadtree tables are loaded at startup, with an additional paging and caching mechanism to bring in more detailed portions of the quad-trees as they are needed. This reduces both startup time and the amount of memory needed to run VGIS. However, with the amount of memory available even on laptop systems and with preprocessed quadtree structures, we have found that fairly large datasets (multiple GB for the data and multiple MB for the hierarchical structure) can be handled without this paging mechanism, even on laptops.

To support parallelism and expandability, there are separate paging threads for the different types of geospatial data. Each thread has a server and a manager. The server loads pages from disk, while the manager decides which cells should be loaded (taking into account such things as user viewpoint and navigational speed) and passes it along for display or analysis. This communication path supports a demand-paging approach such as that of Cox and Ellsworth. When data are needed for a node in the quadtree, the manager allocates space in this shared cache and sends a message to the manager. Message priorities in this queue are changed dynamically according to the importance of the associated request as determined by the manager. Thus, requests that gradually become less important sift toward the end of the queue and get serviced only when no higher-priority requests remain in the queue.

We have found that this page-priority procedure sometimes falls short when handling global data. Users of such data frequently fly quickly from a global view, where the terrain elevation and imagery data are at 8 KM resolution, to views close to the ground, where the data are at 1 M resolution or higher and there may be hundreds or more buildings in view. If the user flies in too fast, the traversal of linked quadtrees by the terrain manager falls well behind the user's navigation. The process can stall in these circumstances, and the pages for the scene currently in view can take quite long to arrive.

Unfortunately, the system cannot just jump to the appropriate position in the quadtree. The quadtree has to be traversed to get important properties information, especially quadcell linking data, but also geospatial bounding boxes and other data, that are necessary to determine if the object or other data should be displayed or not. To address this problem we created separate sets of skeleton trees, one set for each geospatial data type [12,13]. This separate structure provides properties information, but is lightweight so it can be traversed quickly. Large segments of these skeleton trees can reside in main memory for fast access. With the flexibility of this scheme, we can skip one or more levels before paging in object data. A predictive mechanism is instituted based on user navigational speed and viewing direction to help predict where the terrain manager should skip. This method makes paging of data several times faster, especially on PCs. Full data for a particular node are retrieved only when they are needed for rendering or visual analysis.

It was stated at the beginning of this section that everything had a time and a location. We haven't yet discussed the time coordinate, and, indeed, in traditional GIS systems it is not handled with the generality of the spatial coordinates. However, this must change in the future, because time-dependent phenomena will become more prominent. This will be especially true for interactive visualization systems. One application where time is prominent is weather visual analysis. In this and other phenomena, one must consider the relevant time scales and how the data are to be used for each scale. With weather, for example, there is time within events and time between events. These scales should be handled with different queries of the data structure. Within an event, such as a storm front, one usually wants to follow the detailed dynamic behavior as an animation in time. Between events, one may want to do a different sort of query that gathers information about the number and types of storms during a period of time over an area. This could then be followed by detailed interaction with animations of individual storms. Time scales and how to handle them can emerge for all types of data in the virtual GIS. Buildings and urban streetscapes (including tree cover and the amount of concrete) will have time scales over which they change. Even terrain and natural features such as rivers will be dynamic over a long enough time scale.

23.4   Multiresolution Models

To support interactive visualization and efficient-networked data-passing, the previously mentioned global data structure must encompass multiresolution models. For large-scale data, whether terrain, buildings, weather, street and state boundaries, or something else, there must be a multiresolution model that fits the data type and integrates with the other data types for simultaneous display. These models must fit into the detailed, data-dependent representations of Fig. 23.6.

23.4.1   Terrain

Terrain models have received the most attention because they are the basis of virtual GIS systems and are also quite large. For example, the global terrain model in our VGIS system has, with high-resolution insets, nearly 100 GB of data (and it's growing). We will first review some

key work on the development of multiresolution models of terrain. Much of this work is in the context of broader efforts to develop multiresolution models of more general complex surfaces. Terrain is represented as a tessellated height field in one of two main forms: triangulated irregular networks (TINs) or regular grids. A number of different approaches have been developed to create TINs from height fields using Delaunay and other triangulations [21,50]. In addition, hierarchical triangulation representations have been proposed that lend themselves to multiresolution LOD algorithms [29]. Regular grids can also produce efficient hierarchical multiresolution representations [15,18,39,40]. A main advantage of TINs is that they can be set up to follow irregular terrain features (such as mountain ridges) and thus can represent details with fewer triangles than regular grids. On the other hand, regular grids, due to their implicit structure, are more compact, and their multiresolution hierarchical structure appears to be significantly less time-consuming to compute. In fact, global scale structures with very large terrain models have only been constructed for regular grids [18,40]. For these reasons, we have concentrated on regular grid terrain structures in VGIS. However, it is possible that a hybrid multiresolution terrain model could be constructed that was mainly regular but had TINs to efficiently represent fine features [61].

A multiresolution model can be made most efficient through application of a view-dependent criterion to determine the level of detail (LOD) for each of its features [29,39,64]. View dependence works by encoding geometry (and lately also appearance) errors, which occur in the transition from a higher to a lower LOD, into projected screen-space errors. The projection takes into account the position of the current viewpoint and thus the distance and orientation of the error. This error, expressed in pixels, is a natural measure of the perceptual fidelity of an approximated scene. Since different LODs are chosen dynamically (by frame) for different parts of a scene, view dependence can be quite efficient for complex terrain.

(A factor of a hundred or more reduction in displayed triangles is possible without any reduction in visual fidelity [39].) A final addition to interactive terrain visualization is on-the-fly occlusion culling [41]. This can reduce displayed triangles quite significantly when, for example, the user is flying between mountains or other terrain obstacles. View dependence and occlusion culling have now been extended to other types of multiresolution models, including those discussed next. Ultimately, the view-dependent multiresolution terrain is typically organized in quadtree-aligned triangle blocks for insertion into the data-dependent structure at the bottom of Fig. 23.6 [18,29,39]. Progressive edge collapse or vertex removal (for the case of simplification), depending on the method, produces the final list of triangles for rendering from a given viewpoint.

23.4.2   3D Structures

For reasons given in the first two sections, a comprehensive virtual GIS must go beyond interactive terrain visualization. We need an approach that handles general models that might be found in a scene, including highly detailed buildings, trees, statues, bridges, and other objects. General view-dependent approaches for triangle LODs have been developed by several researchers [17,29,40,66]. Recently, Qsplat, a point-based method based on multiresolution splats, has been developed [49]. This method permits fast construction of good-quality, view-dependent models and is especially useful for models with large amounts of small detail, such as those acquired with laser range finding. However, relatively flat or smoothly curving surfaces, such as those found in buildings, are not well represented by Qsplat. Ultimately, one would like a hybrid approach that combines both triangles and splats. This could be applied to trees or dense regions of detail on buildings. Initial research has been done in this direction [9], but more work must be done.

A promising general approach to good-quality mesh simplification is the quadric error

approach [22]. The basic method contracts arbitrary vertex pairs, not just along edges, to minimize surface error (that is, the error between the approximate surface and the original surface). Thus, unconnected regions of a model can be joined, which results in a better approximation both visually and in terms of geometric error than that obtained from topological simplification methods. The ability to handle nonmanifold surfaces makes the method attractive for the reconstructive (Fig. 23.2) or constructive (Fig. 23.3) models described in Section 23.2, which can often be topologically inconsistent. Recently we have extended this method to view-dependent rendering of models with an emphasis on collections of building models [31]. Here, the view-dependent metric has both geometry- and appearance-preserving components. The latter is derived from a measure of texture distortion from the original model as simplification is applied [8]. The view-dependent metric is a weighted sum of these two components, and the weights can be changed depending on the model and its appearance characteristics. This gives flexibility to the view-dependent simplification. This general approach can be applied to models from diverse sources, such as procedural models, models reconstructed from range images, architectural CAD models, and so on. Fig. 23.7 shows the application of the approach to a street façade model reconstructed from ground-based laser range images and associated photo textures [20]. Finally, the model must be constrained to simplify to a few textured planes. A rectangular office building, for example, should simplify to a textured rectangular box. In this way, the virtual GIS can make a smooth transition from, say, close-up street-level views to helicopter views over the urban area.

The models are now organized so that they can be inserted into the global hierarchy, which in the case of 3D structures is extended as in Fig. 23.8. Here the customized hierarchy for the building and streetscape geometry takes advantage of the natural organization of the urban setting. Buildings are grouped into blocks, which are typically separated by streets. Each

Figure 23.7 (Left) Original mesh with and without textures; (Right) same view with 1-pixel threshold. (See also color insert.)

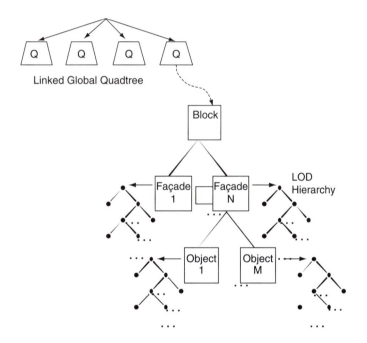

Figure 23.8 Hierarchical structure of 3D block geometry.

block is composed of a set of simple facades, to which textures and 3D details are attached as described in the previous paragraph. Accurately geolocated models from various sources, such as the ones in Figs. 23.2, 23.3, and 23.7, can be inserted into this structure. Attached to the façade are not only details such as window frames and doorways but also pieces of separ-

ated geometry that may be in front of the building, such as trees, lampposts, and sidewalks. It is efficient to construct an object tree for the block with the façades as first-level children and the separated objects as their children (Fig. 23.8). Both the façade and the separated objects have sub-trees to handle LODs in the manner given above. This whole structure is attached to the forest of quadtrees of our global terrain hierarchy (Fig. 23.6), giving buildings precise locations in the universal lat/lon coordinate system. This organizes the blocks and other 3D objects for fast view culling and scalable retrieval [13].

23.4.3   3D Weather

Atmospheric phenomena, such as weather, pollution, or climate models, can also be integrated into the virtual GIS. All these phenomena tend to be time-dependent, so time should also be a part of the structure. We will concentrate here on a structure for weather, but we expect the structures for other atmospheric phenomena to be similar. This structure is global in scale and will accept different 3D data formats. Thus we again start with the forest of quadtrees in Fig. 23.6. The quadtree extends to a certain level where a quadtree-aligned volume tree is inserted. Initial traversal of the quadtree is efficient because the atmosphere is a thin layer with respect to the extent of the terrain (Fig. 23.9). The quadnode is divided into Nx × Ny × Nz bins where x, y, z

are the longitude, latitude, and altitude directions, respectively. The bin sort is fast (O(n), where n is the number of volumetric data elements). This is a key step because the bins provide a structure that is quickly aggregated into a hierarchy for multiresolution detail management and for view frustum culling. However, the data element positions can be retained in the bins for full resolution rendering (and analysis), if desired. Such detail is needed for 3D Doppler radar analysis, as discussed further below. The hierarchy provides significant savings in memory space and retrieval cost, since only the data element coordinates for viewable bins at the appropriate LOD are retrieved. Note that the bins are not rectilinear in Cartesian space, a factor that may affect some analysis or volume rendering algorithms. In general, the bin widths in each direction are nonuniform (e.g., each of the bins in the, say, Nz direction may have a different width). This allows a useful flexibility in distributing bins, for example, when atmospheric measurements are concentrated near the ground with a fall-off in number at higher altitudes.

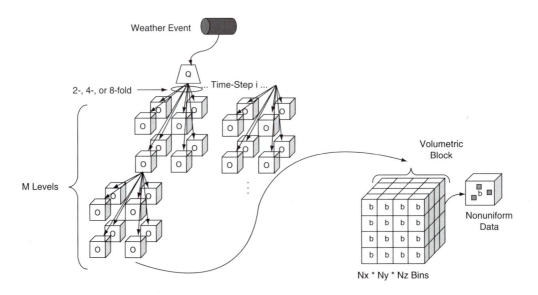

Figure 23.9   Hierarchy for 3D, time-dependent atmospheric data.

Each of the dimensions N in the x, y, z directions is a power of 2. This permits straightforward construction of a volume hierarchy that is binary in each direction. Our tests show that this restriction does not impose an undue limitation [48], at least for the types of atmospheric data we are likely to encounter. The number of children at a given node will be 2, 4, or 8. If all dimensions are equal, the hierarchy is an octree. Typically the average number of children is either 4 or 8. We restrict the hierarchy to the following construction (others are possible): Suppose that $Nx = 2m$, $Ny = 2n$, and $Nz = 2p$ where $m > n > p$. Then there will be p 8-fold levels (i.e., each parent at that level has 8 children), $n-p$ 4-fold levels, and $m-n$ 2-fold levels. If two out of three exponents are equal, there will be only 8-fold and 4-fold levels. The placement of 2-, 4-, and 8-fold levels within the hierarchy will depend on the distribution for the specific type of volumetric data. The octree structure and the number of levels depend on the type of data. The Doppler radar data, for example, will have more levels in the x, y directions than a global climate model, but the climate model will have more levels in the z direction.

Properties at parent nodes are derived from weighted averages of child properties. The parent also carries the following weighting factors: (1) the total raw volumetric data elements contained in the children; (2) the total filled bins contained in the children; and (3) the total bins contained in the children. The quadnode level is chosen such that there are between 1 and 10 K bins (i.e., leaf nodes in the volume hierarchy). This sets a reasonable balance between the costs of traversing global quadtree and volume hierarchies and the need to enable effective handling of volumetric data in the lon/lat and altitude dimensions. Note that the bin structure and volume hierarchy are static in space. We can efficiently apply this structure even to distributions of volumetric elements that move in space as long as the range of local spatial densities and the volume of the data do not change much over time.

For data incommensurate with the quadtree bins (e.g., Doppler radar data), the bin sizes are chosen such that there are at most a few volumetric elements in each bin. The reason for this choice is that we want a smooth transition between rendering of bin-based levels of detail and rendering of the raw data. The final step in the LOD process is the transition from the bins to the underlying raw data. Because the volume hierarchy permits fast traversal, this choice is efficient even for sparse data with holes and high-density clumps, as shown in Section 23.6.

To support interactive visualization, we further organize the volumetric structure as indicated at the bottom of Fig. 23.9. The volume tree structure goes down to a certain level, after which the bins are arranged in volumetric blocks. The block can be either a 3D array of bins or a list of filled bins, depending on whether the data distribution is dense or sparse. We have found in our applications so far that a block containing one bin gives good results [48]. (In other words, the volume tree goes all the way to single bin leaf nodes.) However, the multibin block structure is available if it should prove efficient for future data distributions. Ultimately, we expect that the data distribution will inform the visualization technique used. It may be sufficient to use traditional (continuous-field) volume visualization techniques for dense or uniform data, but different techniques may be better for sparse data. This structure is set up to handle simultaneous, overlapping datasets. These might include, for example, data from overlapping Doppler radar sets along with data from weather simulations, all of which might have different spatial distributions. In addition, a view-dependent technique for choosing the appropriate resolution will balance interactivity and visual quality, as it does for terrain and buildings. The application in Section 23.6 demonstrates the ability to handle overlapping datasets and view dependence.

As discussed in the last section, there are two levels of time that should be handled in this structure. One level is time within events and the other level is time between events. For example, time-dependence within a weather event such as a storm will be represented by time-steps that can be handled at the level of the volume tree, as shown in Fig. 23.9. At this point, an additional time structure could also be inserted [52] to provide further efficiency in rendering through temporal coherence. The structure that distinguishes between weather events (e.g., different storms) is at a higher level in the quadtree structure and contains lower quadnodes and volume trees for a particular event, as shown at the top of Fig. 23.9. These weather events are in turn embedded in a top-level quadtree structure. This multikey structure is designed to balance the benefits of spatial queries and queries based on events [46]. The goal is to efficiently bring forth displayable data for a query such as, "Show me the severe storms (of a certain magnitude of wind or rainfall) that occurred over this period of time in this region." Of course, the structure must reflect the nominal size of both a weather event and the region covered. Individual storm cells are too small to be considered independent events; storm fronts over an extent of a few hundred miles are probably more efficiently queryable. Even so, there must be annotations for important, but small, phenomena (such as tornado signatures) embedded in larger weather events. With the latter annotation, one could efficiently query for tornadoes over a certain region during a certain time period. Since time at the event level is a never-ending stream, there should also be a time structure imposed on the event structure so queries can be made efficiently over longer periods of time. All these structures should be the subject of further investigation and evaluation using real data. Identifying and extracting the weather events is also an area for further study. At present, we use the weather features (mesocyclones, storm cells, and tornado signatures) that come as part of the real-time Doppler radar analysis [16]. Similar features could be extracted from weather simulations or other observational data (e.g., ground flooding features). The weather features are discussed further in Section 23.6.

23.5   Interaction and Display

23.5.1   Interaction

Fast, intuitive, and effective interaction is at the core of an effective exploratory visualization system. For such a system, there are three main modes: *navigation*, *selection*, and *manipulation* [6,57]. The manifestation of these modes will depend on the type of system and the types of interaction devices (e.g., tracked, as in immersive virtual environments, or untracked). Nonetheless, whatever the type of interaction device, certain fundamental features of the interaction mode remain the same.

Navigation is of prime importance in a virtual GIS because the main way to get to a location in an extended geospatial database is to "fly" or "drive" there. As shown above, this act of navigation engenders a continuous act of data retrieval. The flying or driving, which involves a sense of moving past some details and getting closer to others (which then "unfold" to reveal inner details), is an example of *focus + context*. Focus + context is a main tenet of the exploratory visualization of extended information spaces [1], where one does not usually know exactly what one is looking for or where to find it. In virtual GIS, focus + context is achieved through continuous, scalable navigation and also through linked overview windows.

Our VGIS system has three types of navigation, which operate in both tracked and untracked environments: orbital mode, fly mode, and walk mode [59]. Through testing and experience we have found that navigational degrees of freedom (up to seven, including position, orientation, and scale) can be restricted depending on the mode while still retaining navigational flexibility. Orbital mode presents a third-person viewpoint and always looks down from above. Users can zoom in, pan, or rotate the scene (Fig. 23.10). This mode is good for positioning from a wide, even global, overview and then zooming in with continuous position selection updates. Since extreme changes in scale are often encountered (one can fly from global

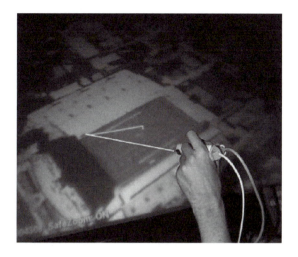

Figure 23.10 Rotation of detailed scene in orbital mode.

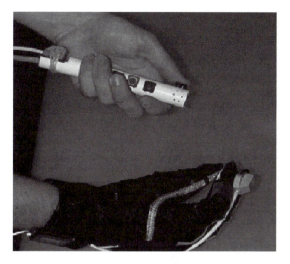

Figure 23.11 Button stick (top) and glove with finger contacts (bottom): two modes for 6 DoF interaction. (See also color insert.)

overviews to 1-foot resolution insets), one may not know where, say, Atlanta is in a global overview and must zoom in, adjusting on the fly as more detail is revealed. One may conversely be flying towards the Grand Canyon and notice an interesting feature along the way to stop and explore. These are examples of focus + context in action. Fly mode has the most degrees of freedom, and it simulates helicopter-like flight. The view direction is towards the horizon, and users can control position (latitude, longitude, height) plus pitch and yaw. We have removed roll as an unnecessary and also confusing degree of freedom. In walk mode, users are restricted to a ground-following altitude, which can be changed. Ground position (latitude, longitude), pitch, and yaw are also under user control. Through an interactive height adjustment, the user can also fly at a fixed height in this mode. When employing tracked interaction, a user can switch seamlessly between modes without menu or keyed selection [60]. The switch between orbital and fly modes, for example, is accomplished by turning the "button stick" controller (Fig. 23.11), which has a tracker attached to its shaft, from horizontal (where it is used like a pointer with a virtual ray emanating from its end) to vertical (where it operates like an airplane joy-

stick). The scene automatically switches from the top-down, orbital mode to fly mode at the same altitude.

Selection and manipulation modes are used to select objects (or regions) in 3D space and then manipulate them. In VGIS, a main mode is selecting at a distance via intersection with a virtual ray. One can also select via the cursor, which is especially effective in orbital mode but also can be used in fly mode. In this case, a virtual ray is cast from the viewpoint (either the head-tracked eyepoint or the default, screen-centered viewpoint in the untracked case) through the cursor to intersect the selected object. Because of the depth of scale, selection can be difficult since the scene may be cluttered with many objects that are small because they are far away. In addition, we can have atmospheric phenomena, which are volume rendered and thus do not have distinct surfaces to select. To handle the depth-of-scale problem, one can use interactivity and fly closer. Overview windows at different scales with selection capability in each can also be quite helpful (or alternatively act as movable "magic windows" for close-ups). The latter problem needs new techniques, such as selection of a distant point in space rather than an object. We are working on techniques to do this, which will also help with general object selection. Manipulation modes are restricted within VGIS [57,62]. One can move a selected object and change its orientation. One can do simple scaling in a body-oriented direction. Objects such as buildings and ground vehicles snap so they remain in an upright orientation. Such manipulations can be performed in either a tracked or an untracked environment. We use building manipulation, for example, as a step in our semi-automated urban modeling system [62], where newly modeled buildings are adjusted in VGIS with respect to other urban models and the street layout before final insertion into the database. In the future we expect to need additional manipulation capabilities for our urban planning applications.

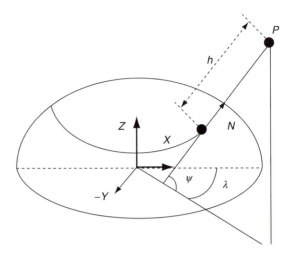

Figure 23.12   Two-parameter ellipsoidal coordinate system used for global space.

23.5.2   Navigation and Selection in a Global Space

It is worth discussing navigation and selection in greater detail to show how interactivity is maintained in a highly scalable global space. For full details refer to Wartell et al. [59,60]. The accurate global model uses a two-parameter ellipsoidal coordinate system commonly used in geodesy [56]. This coordinate system is based on a spheroid (Fig. 23.12). The two parameters are the spheroid's major semi-axis, a, along X and Y, and the minor semi-axis, b, along Z. In this system, longitude, λ, is equivalent to the longitude in spherical polar coordinates; however, latitude, ψ, is the angle between the surface normal and the equatorial plane. Altitude, h, is measured parallel to the normal between the point in question, P, and the underlying surface point. In this coordinate system, the quads in the global forest of quadtrees (Section 23.3) are bounded by meridians and parallels; the meridians provide East and West planar faces, while the parallels provide North and South conical faces (Fig. 23.13) Thus, the quads are triangles at the poles and quadrilaterals elsewhere. (Note that since meridians are

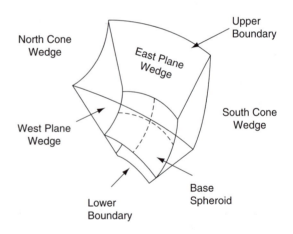

Figure 23.13   Quadrilateral height quad.

not geodesics, these are not true spheroidal triangles or quadrilaterals. We ignore this distinction here.) For surface features (terrain, buildings, vehicles, etc.), each quad is augmented with a height attribute equal to the maximum spheroidal height of the contained data; thus, it is a "spheroidal height quad" (Fig. 23.13).

Using this structure for surface features, we must provide an efficient method for finding arbitrary ray-to-surface intersections, since ray casting is a main mechanism for both free navigation and selection. Our basic algorithm traces the XY projection of a ray through the XY footprints of the spherical height quads. When the projection enters a height quad, the entering and exiting z coordinate of the ray is compared to the height of the quad. If the ray intersects the quad, the algorithm steps into the child quads at the next resolution level. Otherwise, the algorithm steps into the next quad at the same resolution level. Fig. 23.14 illustrates this process: it presents a side view with the ray in red and three levels of height quads. The red volume is the lowest level intersected by the ray. At a leaf node, the algorithm checks for intersection with a surface, e.g., terrain or building.

The basic algorithm is similar to ones used for fast, interactive traversal in Cartesian coordinates [7]. However, the spheroidal height quad complicates matters and requires a general 3D approach to ray–quad intersection. We construct this approach by considering the bounding surfaces of the quad. For a quadrilateral height quad, these consist of two plane wedges and two cone wedges (Fig. 23.13). The plane side boundaries are functions of the longitude λ, while the cone side boundaries are functions of the latitude ψ (Fig. 23.12). The upper boundary surface of the quad is not a spheroid. However, for a good approximation, we can make it a spheroid with major and minor axes that depend on h. Finally we can model the lower boundary surface of the quad as a sphere whose radius equals the distance from the spheroid center to the closest terrain vertex in the quad. This ensures that the sphere lies inside the true lower boundary and also is close to it. Details are in Wartell et al. [60].

When a user casts a ray, the algorithm first clips the ray to the volume bounded by a global upper boundary and a global lower boundary. The upper boundary is the spheroidal bounding surface with height equal to the maximum global surface feature height. The lower boundary is the minimum distance sphere for any terrain feature. Next the algorithm determines the zone in the forest of quadtrees that the ray first enters. Successive quad levels are stepped through, as described above, until either an intersection occurs or the ray exits the global boundaries. Since the upper boundary is curved, it is insufficient to check the height of the ray's entering and exiting intersections with the side boundaries. Instead, we must compute the ray's parameter values, t_in and t_out, at these side intersections; then we compute the ray's intersection parameters, t_0 and t_1, with the quad's upper boundary surface (Fig. 23.15). If and only if these two parameter intervals overlap will the ray have entered the height-quad volume. If the ray intersects the quad, the algorithm can determine which children to check by tracking the parent side boundaries that are intersected. Note that in the spheroidal case, a ray may intersect all four of a quad's children or may enter a quad twice (by intersecting a latitude cone side boundary twice, for example).

This ray-casting method can be used for any surface modeling method (e.g., voxel, bilinear patch, or triangles) or for any 3D objects upon the surface. However, the traversal of the individual elements is model-dependent. For

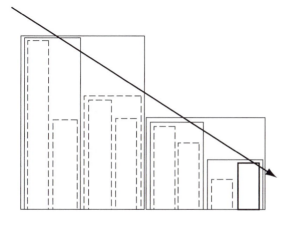

Figure 23.14   Side view of ray-traversing height quads.

Play

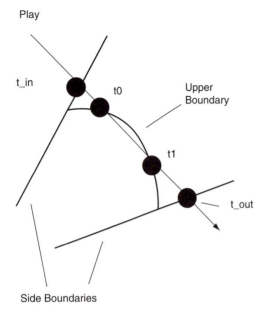

Figure 23.15   Side view of ray intersection.

example, for a regular triangle mesh, the height-quad tree does not typically recurse down to the smallest terrain element. Instead, the leaf quad will contain a fixed-size block of triangles [18,39], and the algorithm must separately trace the ray through this block. Triangle models pose the additional problem that they do not quite fit the concave spheroidal height quad boundaries (specifically the latitude cone boundary in Fig. 23.13). This containment problem can potentially cause the algorithm to miss a ray intersection. However, when using interactive pointing and grabbing in practice, it is our experience that this case never occurs, and we thus ignore it [60].

Now we turn to a discussion of how interactive display must be controlled in order to use these techniques most effectively in the global environment. We discuss interactive display in the context of zooming, since this is where changes of scale are greatest. One can also have panning, rotation, and other navigation modes.

23.5.3   Interactive Display

We will focus on head-tracked stereoscopic display for the highly scalable global environment [59]. This is the most complex display situation. Issues specific to head-tracked stereo display include the following: maintaining a good stereo effect as one moves closer to or farther away from objects (especially if the change of scale is great); the need to handle some degrees of freedom differently from in monoscopic interfaces; problems with stereo fusion at certain viewing positions; collapsing of the stereo effect due to clipping by the screen boundary or occlusion by user hands or bodies; and difficulties with accurate selection and detailed manipulation caused by the simultaneous stereo images. These issues apply to a range of stereoscopic display environments, including virtual workbench, CAVE, and wall-projected systems. We will now discuss these issues from the standpoint of a virtual workbench.

A natural default start position is a global view. To obtain this, we compute the radius, R, and offset, O, of the largest sphere that is contained in the default view volume. Fig. 23.16 shows a side view of the situation. We fix the sphere center to be in the projection plane in order to keep half of the planet above the physical display. This makes the planet as large as possible while permitting the user to reach as

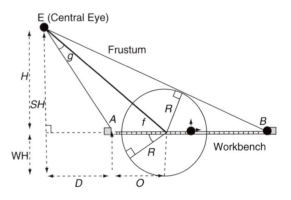

Figure 23.16   Side view of a virtual workbench.

much of the planet as possible directly with his or her hands.

WH is the height of the display surface. Points A and B represent the edges of the display area. Lowercase letters represent the illustrated angles. Computation is then done in a 2D coordinate system whose origin is at the display center. O and R are computed as follows:

$$O = \frac{H}{\tan e} - D, \quad R = O \sin a$$

Angles e and a can be easily calculated from the other labeled points in Fig. 23.16. The O and R values are then used to set the initial user scale and position.

To manage diplopia (double vision, in this case due to inability to fuse separate images in stereo display because of screen parallax), we initially dynamically adjusted the near clipping plane to clip out nonfusable terrain [53]. As a result, when a user moves his or her head too close to the planet, an upper portion is clipped out. We informally observed, however, that clipping the planet's top as the user leans in is as distracting as diplopia. We suggest that this occurs because real-world viewing can yield diplopia but cannot yield the effects of a near clipping plane. People experience diplopia when trying to view objects immediately in front of their faces. While real-world and virtual-world diploptic thresholds and causes differ, people are nonetheless more familiar with reaching and avoiding diploptic conditions than they are with seeing a near clipping plane suddenly slice through an object. As a result, we do not dynamically adjust the near clipping plane.

Zooming by scaling must augment zooming by camera translation within any interface that employs a head-tracked display, a stereoscopic display, or direct manipulation with a tracked device [59]. Thus, there are seven degrees of freedom for general interactions, as discussed above. With head-tracking, zooming out by moving the viewer away from an object will rapidly increase the sensitivity of the projected image to head position. Such sensitivity can be quite distracting. By using an independent scale

dimension, the system can scale down the object and preserve the object–viewer distance in order to avoid this problem. In the stereoscopic case, zooming by moving the viewer toward or away from an object can bring the object either far above the display surface or far below the display surface. In both cases the resulting screen parallax is likely to grow too large, causing user diplopia. Scaling the object while keeping the object near the projection plane solves the problem. Finally, direct manipulation using a tracked device will be difficult for large objects. Manipulating objects is easiest when the user can see the complete object and the object is within arm's reach of the viewer [44]. If, in order to see the complete object, the viewer must move away from the object, it is impossible to satisfy both of these requirements. Again, the solution is to scale so that the object is small enough to be brought close to the user and still be viewed in its entirety.

Thus, the zoom technique involves a user-controlled scale and translation plus an automatic translation. In addition, the zoom is toward a user-selected position using this ray-intersection technique. The position can be updated as one zooms in, which is quite useful when one is zooming over large changes in scale (e.g., from outer space to a section in a city). The scale and translation work as follows in an environment with 6 DoF input. (See Figs. 23.10 and 23.11 for a virtual workbench setup with tracked pointer input.) When the user presses the zoom button, the current 6 DoF pointer position is recorded. As the button is held and the pointer is moved toward or away from the projection plane, the magnitude of the displacement from the initial position is computed. The magnitude determines the zoom speed. The direction of zoom, either in or out, is determined by whether the pointer is displaced closer to or farther from the projection plane. To perform the zoom, we first scale the platform coordinate system up or down based on magnitude and direction of the pointer movement. This has the effect of changing the physical-world-to-virtual-world scale factor, making the perceived world shrink or grow. Next, if the pointer ray intersects the terrain, the platform coordinate system origin is simultaneously scaled about this intersection point. This causes the user to zoom about the selected point. This technique gives the user control of zoom speed and direction plus control of the zoom-in point.

In addition to this scale and translation activity, the user is automatically repositioned so that the planet appears to smoothly rotate about the selected terrain point. In detail, the planet rotates so that the planet normal at the selected terrain point becomes perpendicular to the projection plane. Without this automatic rotation, a zoom quickly brings too much of the planet out of the projection plane, leading to image-fusion problems and severe frame cancellation. Our viewing adjustment step (following) would only serve to push the planet deeper into the display plane, driving the target location farther away. The automatic rotation avoids these problems.

We need an automatic adjustment step to maintain good stereoscopic depth. Our goal is to take maximum advantage of stereoscopic depth cues while minimizing diplopia, frame cancellation, and image distortions. Maximizing stereoscopic depth cues entails keeping the terrain within 1.5 meters, the distance where stereo is strongest as a depth cue [11]. For an immersive virtual environment, this usually means keeping the terrain as close as possible, while still considering image fusability and frame cancellation. Also, keeping the terrain slightly above the display plane puts most objects within arm's reach and lets the user contact objects that are stereoscopically above the display. The adjustment step works in the following way. The rendering thread renders display lists containing terrain (and other object) geometry. This thread copies a sample of the right-eye depth buffer generated from the display list, and then the navigation thread examines this copy. The navigation thread scans the depth buffer and finds both the farthest point above the projection plane and the nearest point.

During the same loop, we also record the number of pixels, A, above the projection plane and the number of pixels, P, not equal to depth buffer clear-screen value. Two rules then apply. If A/P is less than a threshold (85%), we move the user along the projection-plane normal in order to bring the near point to a predetermined target height, TH, above the display plane. If A/P is greater than a threshold, then we move the user along the projection-plane normal so that the far point is flush with the display plane.

While the first rule simply draws the terrain peaks above the display, the second rule counters a problem. At certain scales and terrain formations, the first rule can bring an undesirable amount of the terrain above the display plane. For example, a particular dataset might contain a few peaks and then mostly flat land. At certain scales, the first rule would cause all of the flat terrain to be floating above the display. In effect, there is a large plane that extends far beyond the window limits hovering above the display. Even at target heights as small as 5 cm, the uniformity and extent of this plane create a strong frame-cancellation effect. In contrast for the same target height, if the terrain is more undulating, then the frame-cancellation effect is less disturbing. We surmise that this occurs because with undulating terrain only some of the terrain at the display edges is clipped by the view frustum, while with the flat planar terrain *all* terrain at the display edge is clipped. Given this situation, the more natural position for this problematic terrain is with the planar area flush with the display plane. The second rule catches such cases—where too much terrain is above the display plane—and pushes the terrain back down.

For the target height, TH, we use a constant value that empirically works well. While TH could be adjusted as a function of the nearest fusable image plane, this would cause the terrain to be pushed down into the display plane when the user leans down for a closer look. We informally observed that such behavior is more unnatural than diplopic conditions. While people experience real-world diplopia when peering too closely at an object, they do not experience inanimate objects autonomously moving away when closely examined. Therefore, TH is set to 10% of the standard user height above the workbench, a position within fusability constraints for the standard eye height [65]. Finally, we move the user at a logarithmic rate, rather than instantaneously, towards the target position. Since our rule pair leads to opposing motions and since the adjustment step is switched off above certain scales, a smooth transition is more appropriate to prevent abrupt transitions. Such instantaneous movements have been shown to adversely affect user spatial awareness [3].

Depending on user activities, depth of objects above the terrain might need to be considered. If the user wants to see aircraft, for instance, it is important to account for their depth values so that all of them are visible. On the other hand, if the user is focusing on the terrain, accounting for the aircraft could be problematic, as it could push the terrain far below the display plane when bringing the aircraft into view.

23.6 Some Applications

Several applications have been developed around the interactive capabilities of virtual GIS and its ability to integrate diverse data through visualization [2,6,18,27,32,34,42,47, 48,54]. Here I will highlight the global aspect of virtual GIS (as discussed in Section 23.3) by discussing weather visualization and urban visualization, two applications that use quite different data but use them in the context of the same general data organization. As shown below, these different data can even be displayed and analyzed together.

23.6.1 Weather Visualization

High-resolution weather data, either from computational models or from data sources such as 3D Doppler radar, are becoming more widely available and used. Methods for interactively

visualizing and analyzing these 3D data, such as volume rendering or isosurface extraction, are of importance. For example, Kniss et al. [33] have recently applied a multidimensional transfer function to the volume visualization and analysis of high-resolution weather models. Adjustment of the transfer function, which sets color and opacity values for the 3D scalar field being visualized, is important in analyzing the structure of the field. By having a transfer function that tracks both data values and gradients, Kniss et al. can better identify frontal zones and their characteristics. On the other hand, Gerstner et al. have recently used isosurfaces for multiresolution investigation of local rainfall over terrain [23]. The rainfall is derived from 3D radar measurements. Appropriately chosen isosurfaces show significant detail in the pattern of rainfall. (In Fig. 23.17, positioning of rain shafts and the shape and location of the rain cloud can be seen clearly.) Both of these techniques could be integrated into the general multisource, multiresolution structure described in the previous sections.

Figure 23.17 Isosurfaces of rainfall patterns over a terrain model. (See also color insert.)

Real-time analysis of time-dependent 3D Doppler radar datasets is of particular interest because the radar signatures will contain detailed information about the location, shape, direction, and intensity of precipitation patterns and wind shears. The latter are particularly useful in identifying severe storms, including tornadoes. The 3D structure of the storm in the Doppler radar profile can then provide important information, such as whether the storm extends to the ground (the most dangerous situation) or remains above the ground, as well as specifics of the 3D shape, that further indicate the severity and type of storm. Up to now, there has been little analysis of the full 3D structure of the Doppler data because there has been no effective means for visualization. Interactive, real-time 3D visualization of these events and structures, within the context of the global environment, has been a focus of both our work and others' work [23,30,48].

In our work, we have attacked both the nonuniform structure of the 3D Doppler radars and the scalability issue that arises because the U.S. is covered with more than 140 overlapping radars [23,48]. One would like to have overview visualizations and then zoom into a particular region and show any overlapping radars for that region. We have found the structure in Fig. 23.9 well suited for both scalability and nonuniform data. In addition, of course, it permits integrated visualization with high-resolution terrain models. As an example, Fig. 23.18 (left) shows a visualization of about 40 radars over the eastern U.S., while Fig. 23.18 (right) shows a zoomed-in view of a storm front captured by one of the radars. In the latter figure, storm cells are also visualized. These are features extracted from the Doppler data as it is collected, and they locate positions of possible severe storm activity [16]. A user can use these as guides to fly in for a closer look.

We require a visualization method that retains the nonuniform character of the data at highest resolution. Important storm features, such as tornadoes, may only be 1 KM or less across and thus cover only 1 or 2 gates in the Doppler radar sampling, so retaining a picture of the original data distribution is essential. However, typical volume visualization techniques apply to uniform grids, which would require resampling of the original data. Recently, extensions have been developed that apply to certain classes of irregular grids, but the arbitrary overlapping character of the radar areas makes it unlikely that these techniques would be generally applicable. We have instead opted for a multiresolution splatting technique. Splats can be treated independently and, if desired, without reference to the underlying data topology. Thus, splats can be applied to irregular volume datasets [43,67] and even to unstructured points with no explicit topology. It is then straightforward to set up a view-dependent

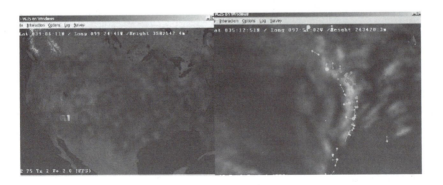

Figure 23.18   (Left) Overview of more than 40 radars in the eastern U.S.; (Right) close-up of storm front after zoom in. (See also color insert.)

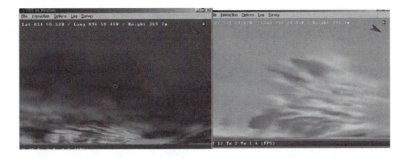

Figure 23.19   (Left) Ground-level view of the 3D structure of a severe storm cell. (Right) Close-up view of the same cell.

structure for the splats [30] where the highest resolution reveals the nonuniform distribution, and lower resolutions are based on the volume tree structure of Fig. 23.9. For the latter, a multiresolution splat hierarchy can be applied [36]. Some results from interactive navigation of the splat structure are shown in Fig. 23.19. Here the user has navigated in fly mode to a view along the ground. The 3D structure of a severe storm cell and its relation to the Earth's surface are clearly seen.

To permit the user to navigate more quickly to a weather event of interest and to investigate it closely, we have recently been working on 3D decision support tools. Here the user can cast rays to quickly select a volume in space, then grab it and pull it close, and finally travel around it (perhaps with background data turned off so the structure can be interacted with most quickly and seen most clearly). An initial version of these tools is shown in Fig. 23.20, where the user has placed a lens around a region of interest. The eventual goal is to get these tools into the hands of weather forecasters and other decision makers. These people must often make decisions in a few minutes. With interactive, comprehensive visualization, they will be able to get to weather events quickly and observe their time-dependent, 3D structures closely.

23.6.2   Urban Visualization and Planning

Urban visualization with accurate interactive visualization of buildings, shrubs and trees,

street layouts, and other city features has been an important area for virtual GIS. There are many applications, including urban planning, education, emergency response, tourism and entertainment, military operations, traffic management, construction (especially large-scale projects), various geolocated and mobile services, citizen–government relations (when complex civic projects are vetted), games based on real locations, and others. Recently urban planning has come to the fore as a prime application. This is due to the rise of "smart growth" principles applied to urban planning. Planning to counteract the effects of urban sprawl, including traffic congestion, increased pollution,

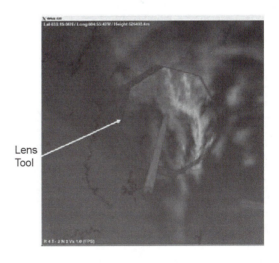

Lens Tool

Figure 23.20   3D decision support tools.

and lowered quality of life for citizens, is centering around higher-density environments with mixed uses and decreased commuting. Applying these smart-growth principles means carefully considering new developments in the context of the existing city. Furthermore, this planning is a negotiation between multiple groups that often have conflicting interests, such as local residents, developers, business owners, and government at all levels. Integrated visualization with efficient access to comprehensive data has been recognized as vital to optimizing this process [19]. The Virtual LA Project has provided initial evidence of the effectiveness of navigable visualization for planning [6,32] (Fig. 23.1). This work included developing and visualizing redevelopment alternatives for areas in Los Angeles, such as the Pico Union District and the Wiltshire-Vermont area. More recently the project team has helped develop plans for a mixed-use, master-planned community in Playa Vista, near Los Angeles.

Different levels of accuracy can apply to 3D urban visualization. At the lower end, one can have simple, textured polygonal models with accurate footprints, locations, and heights, and with textures extracted from photos of the actual buildings. This approach has been applied to extended urban environments with hundreds to thousands of buildings, as indicated in Figs. 23.1 and 23.2 and in Figs. 23.21 and 23.22. At the other end, as discussed at the beginning

Figure 23.22 The fly-in to downtown San Francisco. (See also color insert.)

of this chapter, one can have highly detailed models either reconstructed from ground-based or aerial acquisition (Fig. 23.7) or modeled using traditional CAD or procedural techniques (Fig. 23.3). In all cases, the modeling process is laborious, though recently tools have been developed for semiautomated reconstruction [62] and more automated procedural construction [63]. All of these cases can be successfully inserted in the global hierarchy customized as in Section 23.4 (Fig. 23.8) and efficiently paged in, culled, and visualized. Indeed, we have done this with extended collections of simpler models for cities such as Atlanta, Los Angeles, San Francisco, and Savannah. One can fly into any of these urban environments. As one gets closer, the buildings in the view frustum are automatically paged in, and one can then fly around the cityscape (Figs. 23.21 and 23.22). There is no obstacle to scaling up this collection of urban areas so that it contains 3D models for, say, all cities in the U.S. However, if the number of buildings concentrated in a single urban area grows into the tens of thousands or more, or if the amount of detail per building grows (as in Fig. 23.3), then new rendering techniques are needed to maintain interactivity. Urban visualizations such as those presented here are already being used for way-finding, emergency response, and other applications. In some cases,

Figure 23.21 The fly-in to a close-up view of downtown Atlanta. (See also color insert.)

traditional GIS databases have been attached and queried from within the visualization [2].

23.7   Conclusions and Future Work

I have described virtual geographic information systems, which have developed rapidly with the advent of 3D graphics cards, powerful desktop and laptop computers, and new algorithms for acquiring, handling, and visualizing geospatial data. As a result, the applications for interactive virtual GIS are rapidly growing. One of the current areas of intensive investigation is the handling of large amounts of data from diverse sources. These include access to much more extensive terrain models (e.g., 30 M elevation and image data for the entire U.S. plus higher-resolution insets, such as 1 M image data for the state of Georgia) plus LIDAR data for several urban areas. Lately these have been augmented by methods for ground-based collection of 3D urban data. All these different data could be integrated into a common framework by a global geospatial data organization. However, this data organization must be constructed for interactive visualization of scalably large amounts of data. This means that it must support multiresolution data structures appropriate for each type of data. These structures are best accessed for visualization through view-dependent methods that minimize the amount of detail in a scene while maintaining visual fidelity. This structure can even be efficiently extended to volumetric data associated with the Earth, such as atmospheric data or geophysical data having to do with subsurface structure. Collection and visualization of real-time weather data is one application that has been worked on extensively. Weather visualization brings with it the question of what to do with dynamic events. Here there is at least a two-level structure, both within events and between events. Methods have been developed to organize and visualize consecutive time-steps within events, but the between-event structure needs further development.

Fast, intuitive, and effective interaction is at the core of virtual GIS. The multiresolution models support interactivity, but we must also have appropriate interactive tools for navigating vast geospatial environments and interacting with objects within them. Effective navigation aids have been developed that take into account the large range of scales encountered. The navigation tools and selection and manipulation tools must interact with an accurate representation of the Earth. When an appropriate hierarchical traversal algorithm is provided with these tools, fast, precise, and dynamic interaction is possible.

A variety of display types and interaction modes are available for virtual GIS. In particular, since the extended geographic spaces are naturally immersive, it makes sense to develop immersive, stereoscopic displays. This has been done by careful consideration of the issues of stereo projection and the issue of dynamically handling scale in a head- and hand-tracked environment. Scale must be handled in such a way that objects in the environment can be reached easily through tracked interaction.

Several applications have been developed for virtual GIS, and its use in current and new applications is likely to grow significantly in the future. Among current applications are emergency planning and response, urban planning, environmental analysis, and weather analysis and forecasting. This chapter gave examples of ongoing applications in high resolution, real-time weather analysis, and urban visualization and planning.

23.7.1   Future Work

Virtual GIS is at the threshold of vast new data resources and new applications that will require a range of fundamental advances. The LIDAR data, 3D range images, and aerial and ground photoimages discussed here are just the beginning of what will be available. Ways of extracting models from these data, integrating them, building high-resolution terrains, and visualizing them must be significantly extended. In addition, time and dynamic character become factors in high-resolution models, because as the resolution increases, noticeable changes are more likely to occur over shorter periods of time. Thus, a time structure becomes necessary for all data in the virtual GIS. I have discussed the time structure for atmospheric data. This must be significantly extended in terms of an event structure so that it can be queried for weather or atmospheric events at particular time ranges or locations. There must also be a structure for changes in buildings, tree cover, terrain, etc. This dynamic character implies that there should be a mechanism for fast, dynamic updates, too. Significant new ways of manipulating the geospatial data will also be necessary. The results of these manipulations (e.g., marking the terrain or moving a building) must be dynamically added to the database. Such capabilities, especially for applications like urban planning, must be significantly extended. Urban-planning teams will want not only to navigate around 3D models in the context of the rest of the urban environment but also to move them, cut off or add wings and floors, change the external materials, change the landscaping, and so on. These manipulations must be easy to perform, and their results must be available interactively. Finally, the comprehensive and efficiently accessible accumulation of data, coupled with interactive visual analysis, opens the door to new methods of simulation and modeling. Already this is happening to some extent with simple terrain models and weather observations, which are being used as input for mesoscale weather models. However, in the future, much more will be possible. We have discussed using Doppler radar data (and higher-resolution terrain) as input to real-time, more accurate weather models. This is just one of many things that can be done. Flood models, including damage analysis, or high-resolution air-pollution analyses, could be run. In addition, many other types of simulations will be possible from traffic and pedestrian flow analyses to studies of risk and urban damage due to fire, earthquakes, or other disasters.

The availability of ever smaller, faster-networked computers means that mobile visualization will be a significant growth area. Work in this area is already starting [35], but much more remains to be done. It is now possible to carry around a large geospatial database in a wearable computer while interactively visualizing both the database and geolocated inputs from GIS, orientation trackers, and location-aware services. As the economy improves, the development of geolocated services and georeferenced applications will be significant.

References

1. P. P. Baudisch, N. Good, and P. Stewart. Focus plus context screens: combining display technology with visualization techniques. In *Proceedings of ACM UIST 01*, pages 31–40, 2001.
2. D. Bhaumik, N. L. Faust, D. Estrada, J. Linares. 3D urban GIS for Atlanta. *Proc. SPIE—The International Society for Optical Engineering*, 3085:115–124, 1997.
3. A. Bowman, D. D. Koller, and F. L. Hodges. Travel in immersive virtual environments: an evaluation of viewpoint motion control techniques. *Proc of IEEE VRAIS 97*, 1997.
4. D. Bowman. Interactive techniques for common tasks in immersive virtual environments: design, evaluation and application. PhD Thesis, Georgia Institute of Technology, 1999.
5. F. P. Brooks, Jr. What's real about virtual reality? *IEEE Computer Graphics and Applications*, 19(6):16–27, 1999.
6. R. Chan, W. Jepson, and S. Friedman. Urban simulation: an innovative tool for interactive planning and consensus building. *Proc. American Planning Association National Conference*, pages 43–50, 1998.
7. D. Cohen and A. Shaked. Photo-realistic imaging of digital terrains. *Eurographics '93*, 12(3):363–373, 1993.
8. J. Cohen, M. Olano, and D. Manocha. Appearance-preserving simplification of polygonal models. *Proc. SIGGRAPH '98*, pages 115–122, 1998.
9. J. Cohen, D. Aliaga, and W. Zhang. Hybrid simplification: combining multi-resolution polygon and point rendering. *Proc. IEEE Visualization '01*, pages 37–44, 2001.
10. M. Cox and D. Ellsworth. Application-controlled demand paging for out-of-core

visualization. *Proc. IEEE Visualization '97*, pages 235–244, 1997.

11. E. J. Cutting. How the eye measures reality and virtual reality. *High-Performance Computing and Human Vision I: Behavioral Research Methods, Instruments & Computers*, 29(1): 27–36, 1997.

12. D. Davis, T. F. Jiang, W. Ribarsky, and N. Faust. Intent, perception, and out-of-core visualization applied to terrain. *Proc. IEEE Visualization '98*, pages 455–458, 1998.

13. D. Davis, W. Ribarsky, T. Y. Jiang, N. Faust, and S. Ho. Real-time visualization of scalably large collections of heterogeneous objects, *IEEE Visualization '99*, pages 437–440, 1999.

14. A. Duany and E. Plater-Zyberk. *Suburban Nation: The Rise of Sprawl and the Decline of the American Dream*. North Point Press, 1999.

15. M. Duchaineau, M. Wolinsky, D. E. Sigeti, M. C. Miller, C. Aldrich, and M. B. Mineev-Weinstein. ROAMing terrain: real-time optimally adapting meshes. *Proc. IEEE Visualization '97*, pages 81–88, 1997.

16. M. D. Eilts, J. T. Johnson, E. D. Mitchell, S. Sanger, G. Stumpf, A. Witt, K. Hondl, and K. Thomas. Warning decision support system. *11th Inter. Conf. on Interactive Information and Processing Systems (IIPS) for Meteorology, Oceanography, and Hydrology*, pages 62–67, 1995.

17. J. El-Sana and E. Bachmat. Generalized view-dependent simplification. *Proc. IEEE Visualization '02*, pages 83–94, 2002.

18. N. Faust, W. Ribarsky, T. Y. Jiang, and T. Wasilewski. Real-time global data model for the digital earth. *Proc. International Conference on Discrete Global Grids*, 2000.

19. P. S. French. Overcoming the barriers to smart growth: regional benefits versus neighborhood concerns. *43rd ACSP Annual Conference*, 2001.

20. C. Früh and A. Zakhor. 3D model generation for cities using aerial photographs and ground level laser scans. *IEEE Computer Vision and Pattern Recognition Conference*, 2001.

21. M. Garland and P. S. Heckbert. Fast polygonal approximation of terrains and height fields. *Tech. Rep. CMU-CS-95-181*, Carnegie Mellon, 1995.

22. M. Garland and P. S. Heckbert. Surface simplification using quadric error metrics. *Proc. SIGGRAPH '97*, pages 209–216, 1997.

23. T. Gerstner, D. Meetschen, S. Crewell, M. Griebel, and C. Simmer. A case study on multiresolution visualization of local rainfall from weather measurements. *IEEE Visualization '02*, pages 533–536, 2002.

24. G. A. Grell, J. Dhudia, and A. Stauffer. A description of the fifth-generation Penn State/NCAR mesoscale model (MM5). NCAR Technical Note NCAR/TN-398+STR. NCAR, Boulder Colorado, 1994.

25. M. H. Gross, V. Kuhn, and N. M. Patrikalakis. A visualization and simulation system for environmental purposes. *Scientific Visualization of Physical Phenomena, Proc. CGI '91*, pages 639–654, 1991.

26. G. Grueneis, P. Mayer, J. Sauter, and A. T. Schmidt. Vision. *Visual Proc. of SIGGRAPH 95*, page 134, 1995.

27. L. E. Hitchner. Virtual planetary exploration: A very large virtual environment. *ACM SIGGRAPH '92 Tutorial on Implementing Immersive Virtual Environments*, 1992.

28. L. E. Hitchner and M. W. McGreevy. Methods for user-based reduction of model complexity for virtual planetary exploration. *Proc. SPIE 1993*, pages 1–16, 1993.

29. H. Hoppe. Smooth view-dependent level-of-detail control and its application to terrain rendering. *Proc. IEEE Visualization '98*, pages 35–42, 1998.

30. J. Jang, W. Ribarsky, C. Shaw, and N. Faust. View-dependent multiresolution splatting of non-uniform data. *Eurographics–IEEE Visualization Symposium '02*, pages 125–132, 2002.

31. J. Jang, W. Ribarsky, C. Shaw, and P. Wonka. Constrained view-dependent visualization of models. Submitted to *IEEE Visualization '03*. Report GIT-GVU-03-09.

32. W. Jepson, R. Liggett, and S. Friedman. Virtual modeling of urban environments. *Presence*, 5(1):72–86, 1996.

33. J. Kniss, C. Hansen, M. Grenier, and T. Robinson. Volume rendering multivariate data to visualize meteorological simulations: a case study. *Eurographics–IEEE Visualization Symposium '02*, pages 189–194.

34. D. Koller, P. Lindstrom, M. W. Ribarsky, L. Hodges, N. Faust, and G. Turner. Virtual GIS: a real-time 3D geographic information system. *Proc. IEEE Visualization '95*, pages 94–100, 1995.

35. D. M. Krum, W. Ribarsky, C. D. Shaw, L. Hodges, and N. Faust. Situational visualization. *Proc. ACM Symposium on Virtual Reality Software and Technology*, pages 143–150, 2001.

36. D. Laur and P. Hanrahan. Hierarchical splatting: a progressive refinement algorithm for volume rendering. *Proc. SIGGRAPH '91*, pages 285–288, 1991.

37. Y. G. Leclerc and S. Q. Lau Jr. TerraVision: a terrain visualization system. SRI International Technical Note No. 540, 1994.

38. B. Leibe, T. Starner, W. Ribarsky, D. Krum, L. Hodges, and Z. Wartell. The perceptive workbench towards spontaneous and natural interaction in semi-immersive virtual environments. *IEEE Virtual Reality 2000*, pages 13–20, 2000.

39. P. Lindstrom, D. Koller, W. Ribarsky, L. Hodges, N. Faust, and G. Turner. Real-time, continuous level of detail rendering of height fields. *Proc. SIGGRAPH '96*, pages 109–118, 1996.

40. P. Lindstrom and C. Silva. A memory insensitive technique for large model simplification. *Proc. IEEE Visualization '01*, pages 121–126, 2001.

41. B. Lloyd and P. Egbert. Horizon occlusion culling for real-time rendering of hierarchical terrains. *Proc. IEEE Visualization '02*, pages 403–409, 2002.

42. M. R. Macedonia and M. J. Zyda. NPSNET: a network software architecture for large scale virtual environments. *PRESENCE: Teleoperators and Virtual Environments*, 3:4, 1994.

43. X. Mao. Splatting of curvilinear volumes. *IEEE Trans. on Visualization and Computer Graphics*, 2(2):156–170, 1996.

44. M. R. Mine, F. P. Brooks, Jr., and C. H. Sequin. Moving objects in space: exploiting proprioception in virtual environment interaction. *Computer Graphics (SIGGRAPH '97)*, pages 19–26.

45. Y. Parish and P. Muller. Procedural modeling of cities. *Proc. SIGGRAPH '01*, pages 301–308, 2001.

46. B. Plale. Performance impact of streaming doppler radar data on a geospatial visualization system. *Technical Report* GIT-CC-01-07, 2001.

47. M. Reddy, Y. Leclerc, L. Iverson, and N. Bletter. TerraVision II: visualizing massive terrain databases in VRML. *IEEE Computer Graphics & Applications*, 19(2):30–38, 1999.

48. W. Ribarsky, N. Faust, Z. Wartell, C. Shaw, and J. Jang. Visual query of time-dependent 3D weather in a global geospatial environment. *Mining Spatio-Temporal Information Systems* (R. Ladner, K. Shaw, and Mahdi Abdelguerfi, Eds.) Amsterdam, Kluwer, 2002.

49. S. Rusinkiewicz and M. Levoy. Qsplat: a multiresolution point rendering system for large meshes. *Proc. SIGGRAPH 2000*, pages 343–352, 2000.

50. F. Schroder and P. Rossbach. Managing the complexity of digital terrain models. *Computers & Graphics* 18(6):775–783, 1994.

51. C. Shaw, W. Ribarsky, Z. Wartell, and N. Faust. Building the visual earth. Vol. 4744B, *SPIE 16th International Conference on Aerospace/Defense Sensing, Simulation, and Controls*, 2002.

52. H.-W. Shen, L.-J. Chiang, and K. L. Ma. A fast volume rendering algorithm for time-varying fields using a time-space partitioning (TSP) tree. *IEEE Visualization '99*, pages 371–378, 1999.

53. A. D. Southard. Viewing model for virtual environment displays. *Journal of Electronic Imaging,* 4(4):413–420, 1995.

54. M. Suter and D. Nuesch. Automated generation of visual simulation databases using remote sensing and GIS. *Proc. IEEE Visualization '95*, pages 86–93, 1995.

55. S. Teller. Toward urban model acquisition from geo-located images. *Proceedings Pacific Graphics '98*, pages 45–51, 1998.

56. P. Vanicek and E. Krakiwksy. Geodesy: the concepts. Amsterdam, North Holland, 1982.

57. R. Vanderpol, W. Ribarsky, L. F. Hodges, and F. Post. Evaluation of interaction techniques on the virtual workbench. *Proceedings of Eurographics Virtual Environments '99*, pages 157–168, 1999.

58. H. Veron, D.A. Southard, J. R. Leger, and J. L. Conway. Stereoscopic displays for terrain database visualization. *NCGA '90*, 1:16–25, 1990.

59. Z. Wartell, W. Ribarsky, and L. F. Hodges. Third-person navigation of whole-planet terrain in a head-tracked stereoscopic environment. *Proceedings of IEEE Virtual Reality '99*, pages 141–148, 1999.

60. Z. Wartell, W. Ribarsky, and L. F. Hodges. Efficient ray intersection for visualization and navigation of global terrain. *Eurographics— IEEE Visualization Symposium 99, Data Visualization '99*, pages 213–224, 1999.

61. Z. Wartell, E. Kang, T. Wasilewski, W. Ribarsky, and N. Faust. Rendering vector data over global, multiresolution 3D terrain. *Eurographics—IEEE Visualization Symposium 2003*, pages 213–222, 2003.

62. T. Wasilewski, N. Faust, and W. Ribarsky. Semi-automated and interactive construction of 3D urban terrains. *Proceedings of the SPIE Aerospace/Defense Sensing, Simulation & Controls Symposium*, 3694A:31–38, 1999.

63. P. Wonka, M. Wimmer, F. Sillion, and W. Ribarsky. Instant architecture. *ACM Transactions on Graphics*, 4(22):669–677, 2003.

64. J. C. Xia and A. Varshney. Dynamic view-dependent simplification for polygonal models. *Proc. IEEE Visualization '96*, pages 327–334, 1996.

65. Y.-Y. Yeh and L. D. Silverstein. Limits of fusion and depth judgement in stereoscopic color displays. *Human Factors*, 32(1):45–60, 1990.

66. S. You and U. Neumann. Automatic mosaic creation based on robust image motion estimation. *Proc. IASTED Signal and Image Processing*, 2000.

67. M. Zwicker, H. Pfister, J. van Baar, and M. Gross. EWA volume splatting. *Proc. IEEE Visualization '01*, pages 29–36, 2001.

24   Visualization Using Virtual Reality

R. BOWEN LOFTIN
Old Dominion University

JIM X. CHEN
George Mason University

LARRY ROSENBLUM
U.S. Naval Research Laboratory

24.1   Introduction

24.1.1   Purpose

This chapter provides a brief introduction to virtual reality, followed by a review of selected visualization applications implemented in a virtual-reality environment. The reader is provided with "pointers" to the major conferences and to more detailed compilations of research in virtual-reality-based visualization.

24.1.2   Virtual Reality

What we now often refer to as virtual reality was first proposed in the 1960s [16,38,39]. In addition to the term "virtual reality," many other terms such as "virtual environments," "synthetic environments," "virtual worlds," and "artificial reality" have been used. Virtual reality has the capability of providing sensory information of sufficient fidelity that the user can, in some cases, "suspend disbelief" and accept that he or she is actually somewhere else [10]. Further, the technology can also support perceptual interaction with the synthetic environment, enabling the user to transcend the role of passive observer and actively participate in shaping events [28]. A thorough, but somewhat dated, review can be found in the report of a National Research Council committee chartered to examine virtual reality in the mid-1990s [13].

For the purpose of this chapter, virtual reality will be defined as

The use of integrated technologies that provide multimodal display of and interaction with information in real time, enabling a user or users to occupy, navigate, and manipulate a computer-generated environment.

Key to an understanding of the potential of virtual reality in visualization is the recognition that virtual reality is not limited to *visual* displays and that it inherently provides those who use it with the means to navigate and manipulate the information that is displayed. An excellent entry into virtual-reality systems and applications is a recently published handbook on the field [37].

The visual element of virtual reality extends commonly available 2D computer graphics into the third dimension. To achieve true 3D graphics, a stereo image is produced by providing two slightly different views (images) of the same object for the user's two eyes. The provision of two separate images can be achieved by a number of methods. In the early days of virtual reality, a head-mounted display was the method of choice [16,38,39]. These displays used two image sources, one for each eye, for stereo viewing of the computer-generated environment. Subsequently, devices such as the ImmersaDesk [20,32] or the CAVE Automatic Virtual Environment [11] were used to produce stereo images. An ImmersaDesk or CAVE usually produces only one image on its display surface or surfaces but alternates between an image for the right eye and one for the left eye (commonly at 60 Hz). The viewer wears lightweight liquid crystal "shutter" or "active" glasses. These glasses are synchronized with the alternating displays so that the appropriate eye can view the image intended for that eye. These devices can also be used to produce stereo images by sending two images to the same display such that each image is polarized differently. The user then wears "passive" glasses with each lens polarized to match the image for its eye.

The principal hardware technologies required for producing a virtual reality are real-time graphics generators, stereo displays, 3D audio displays, tracking/interaction systems, and special display devices (e.g., haptic, vestibular, and olfactory displays). The development of effective virtual-reality applications, including those used for visualization, requires complex software environments. Commercial products are available, but many researchers rely on "home-grown" software that may not be widely supported. A small number of open-source systems have been created (e.g., VRJuggler and DIVERSE), as have application programmer interfaces (e.g., Java3D) and web-based graphics formats (e.g., VRML).

24.1.3   Characteristics of Virtual Environments

What does virtual reality offer visualization that conventional display technologies do not? While a number of items could be cited, three will be examined here: immersion, presence, and multimodal displays and interaction.

24.1.3.1   Immersion

"Immersion refers to what is, in principle, a quantifiable description of a technology. It includes the extent to which the computer displays are extensive, surrounding, inclusive, vivid and matching. The displays are more extensive the more sensory systems that they accommodate"

[35]. Immersion is depicted, in this definition, as a continuum. The argument is that the more sensory information provided and the more sensorially diverse that information, the more "immersion" a user will experience. Immersion has some properties that could directly enhance visualization. For example, immersion implies freedom from distractions. It also implies that a user's entire attention can be brought to bear on the problem at hand. Such characteristics should provide a user with an increased ability to identify patterns, anomalies, and trends in data that is visualized.

24.1.3.2   Presence

Our definition of presence is from Slater et al.:

> "Our general hypothesis is that presence is an increasing function of two orthogonal variables. The first variable is the extent of the match between the displayed sensory data and the internal representation systems and subjective world models typically employed by the participant. Although immersion is increased with the vividness of the displays . . . , we must also take into account the extent to which the information displayed allows individuals to construct their own internal mental models of reality. For example, a vivid visual display system might afford some individuals a sense of 'presence', but be unsuited for others in the absence of sound." [34]

> "The second variable is the extent of the match between proprioception and sensory data. The changes to the display must ideally be consistent with and match through time, without lag, changes caused by the individual's movement and locomotion—whether of individual limbs or the whole body relative to the ground." [35]

Presence can contribute to the "naturalness" of the environment in which a user works and the ease with which the user interacts with that environment. Clearly, the quality of the virtual reality—as measured by display fidelity, sensory richness, and real-time behavior—is critical to a sense of presence.

24.1.3.3   Multimodal Displays

Although the visual sense is arguably the most powerful sense in humans, it is important to note that most humans process inputs simultaneously from many senses (visual, auditory, haptic, vestibular, olfactory, and gustatory). For example, Fred Brooks and his colleagues at the University of North Carolina at Chapel Hill [30] have enabled chemists and biochemists to view, assemble, and manipulate large, complex molecules using a combination of visual and haptic displays. The well known conjecture that humans can simultaneously grapple with no more than seven, plus or minus two, separate items of information [27] does not specifically explore the human capacity for understanding multiple variables when expressed through multiple senses. In spite of this conjecture, the mapping of information onto more than one sensory modality may well increase the "human bandwidth" for understanding complex, multivariate data. Lacking a theory of multisensory perception and processing of information, the critical issue is determining what data best maps onto what sensory input channel. Virtual reality at least offers the opportunity to explore this interesting frontier to find a means of enabling users to effectively work with more and more complex information.

24.1.4   Virtual Reality and Visualization

Visualization is a tool that many use to explore complex (or even simple) data in a large number of domains. One can think of virtual reality as a specific means to achieve effective visualizations. As noted in the preceding section, virtual reality has a number of features that can contribute to the success of a visualization application, especially when that application must address high-dimensional data, high volumes of data, and/or highly complex data. One of the great potential strengths of virtual reality is its stated goal of giving users more accessible, more intuitive, and more powerful interaction capabilities. One way to grasp this concept is to imagine that a user is given a strange object that can be held in one hand. What is the natural thing for the user to do? First, the user will turn the object around to examine it—just as one can employ virtual reality

to provide users with different viewpoints on an object or scene. Next, the user may prod or poke the object to determine some of its properties or to elicit a behavior—just as one can employ virtual reality to provide users with gesture interfaces for direct interaction with virtual objects.

Virtual reality is a powerful display and interaction vehicle. It can structure abstract data and concepts, present the results of computations, and help researchers understand the unforeseen or find the unexpected.

24.2   A Visualization Sampler

24.2.1   Key Resources

Virtual reality has, by some measures, been available since the late 1980s through commercial entities as well as through academic, government, and industrial laboratories. From the time of the first available commercial systems, visualization has been linked to virtual reality. A few examples of visualization in the context of virtual reality can be found in SIGGRAPH proceedings, especially in the proceedings of more specialized conferences and workshops. Notable are the IEEE Visualization Conferences and their associated symposia and the IEEE Virtual Reality Conferences (known prior to 1999 as the Virtual Reality Annual International Symposium, or VRAIS). Many papers from these two conference series address specific applications of visualization using virtual reality, or discuss issues of data representation, human-computer interaction, or performance that are relevant to the success of such applications. In addition, the proceedings of the Eurographics Virtual Environments Workshop offer more examples of virtual reality applied to visualization. More recently, a series of meetings known as the Immersive Projection Technology Workshop (IPT) has been held (with the location alternating between Europe and the United States) to address virtual-reality technology development and applications. Again, many of the applications presented there illustrate the use of virtual reality in visualization.

A seminal workshop on visualization was held in July 1993 at the Fraunhofer Institute for Computer Graphics in Darmstadt, Germany. The proceedings of this conference [31] contains several useful papers, including a paper by Steve Bryson [7] entitled "Real-time exploratory scientific visualization and virtual reality" that sets forth many of the benefits and challenges of using virtual reality for visualization. In May 1994, a group of researchers met for the Dagstuhl Seminar on Scientific Visualization. The result of this seminar was the production of a book on scientific visualization [29] containing chapters written by some of the most respected research groups in the world. A number of chapters in this book treat virtual reality as an approach to visualization. The paper by Helmut Hasse, Fan Dai, Johannes Strassner, and Martin Göbel ("Immersive Investigation of Scientific Data") [17] is important in providing detailed examples of work up until that time. A more recent work [9] focuses on the use of virtual reality in information visualization.

Another very useful source comes from two workshops on virtual environments and scientific visualization that were sponsored by Eurographics, in Monte Carlo, Monaco, February 19–20, 1996 and in Prague, Czech Republic, April 23–23, 1996. The proceedings of these two workshops [15] provide a very good compilation of the work done through 1995.

24.2.2   Examples

Well over 100 extant publications address the application of virtual reality in visualization. Below are brief descriptions of specific projects that demonstrate the breadth of applicability of virtual reality to visualization. The examples below are not meant to be exhaustive or even to be a uniform sampling of the available literature. Inclusion or exclusion of a specific application does not imply a value judgment on the part of the authors of this chapter.

24.2.2.1   Archeology

A number of groups have used virtual reality to visualize archeological data. One group [1] used a CAVE to visualize the locations of lamps and coins discovered in the ruins of the Petra Great Temple site in Jordan.

24.2.2.2   Architectural Design

The Electronic Visualization Laboratory (EVL) at the University of Illinois in Chicago [23] has utilized virtual reality in architectural design and collaborative visualization to exploit virtual reality's capability for multiple perspectives on the part of users. These perspectives, including multiple mental models and multiple visual viewpoints, allow virtual reality to be applied in the early phases of the design process rather than during a walkthrough of the final design.

24.2.2.3   Battlespace Visualization

Work done at Virginia Tech and the Naval Research Laboratory [12,18] resulted in virtual-reality-based Battlespace visualization applications using both a CAVE and a projection workbench. The modern Battlespace extends from the bottom of the ocean into low earth orbit. Thus, 3D visualizations that support powerful direct interaction techniques offer significant value to military planners, trainers, and operators.

24.2.2.4   Cosmology

Song and Norman [36] demonstrated, as early as 1993, the utility of virtual reality as a tool for visualizing numerical and observational cosmology data. They have implemented an application that supports multiscale visualization of large, multilevel time-dependent datasets using both an immersive display and a gesture interface that facilitates direct interaction with the data.

24.2.2.5   Genome Visualization

Kano et al. [19] have used virtual reality to develop an application for pair-wise comparison between cluster sets generated from different gene expression datasets. Their approach displays the distribution of overlaps between two hierarchical cluster sets, based on hepatocellular carcinomas and hepatoblastomas.

24.2.2.6 Meteorology

Meteorologists typically use 2D plots or text to display their data. Such an approach makes it difficult to visualize the 3D atmosphere. Ziegler et al. [40] have tackled the problem of comparing and correlating multiple layers by using an immersive virtual environment for true 3D display of the data.

24.2.2.7 Oceanography

A multidisciplinary group of computer scientists and oceanographers [14] has developed a tool for visualizing ocean currents. The *c-thru* system uses virtual reality to give researchers the ability to interactively alter ocean parameters and communicate those changes to an ocean model calculating the solution.

24.2.2.8 Protein Structures

Protein structures are large and complex. Large-format virtual-reality systems support not only the visualization of such data, but the collaboration of small teams that analyze the data. One group [2] has visualized four geometric protein models: space-filling spheres, the solvent accessible surface, the molecular surface, and the alpha complex. Relationships between the different models are represented via continuous deformations.

24.2.2.9 Software Systems

Many computer programs now exceed one million lines of code. The ability to truly understand programs of such magnitude is rare. Visualizations of such systems offer a means of both comprehending the system and collaboratively extending or modifying it. An example of such a visualization is the work of

Amari et al. [3]. In this case, a visualization of static structural data and execution trace data of a large software application's functional units was developed. Further, the visualization approach supported the direct manipulation of graphical representations of code elements in a virtual-reality setting.

24.2.2.10 Statistical Data

A group at Iowa State University [4] has developed a virtual-reality-based application for the analysis of high-dimensional statistical data. Moreover, the virtual-reality approach proved superior to a desktop approach in terms of structural-detection tasks.

24.2.2.11 Vector Fields

Real-time visualization of particle traces using virtual environments can aid in the exploration and analysis of complex 3D vector fields. Kuester et al. [21] have demonstrated a scalable method for the interactive visualization of large time-varying vector fields.

24.2.2.12 Vehicle Design

The use of increasingly complex finite element (FE) simulations of vehicles during crashes has led to the use of virtual-reality techniques to visualize the results of the computations [22]. A program called VtCrash was designed to enable intuitive and interactive analyses of large amounts of crash-simulation data. The application receives geometry and physical-properties data as input and provides the means for the user to enter a virtual crash and to interact with any part of the vehicle to better understand the implications of the simulation.

24.2.2.13 Virtual Wind Tunnel

One of the earliest successful demonstrations of virtual reality as a visualization tool was the development of the Virtual Wind Tunnel at the NASA Ames Research Center [5,6,8]. Steve Bryson precomputed complex fluid flows around various aerodynamic surfaces. To view these flow fields, the user employed a tracked, head-mounted display (Fakespace's BOOM) that used relatively high-resolution color displays, one for each eye. These displays were attached to the head but were supported by a counterweighted boom to relieve the user of bearing the weight of the system. Optical encoders in the boom joints provided real-time, precise tracking data on the user's head position. Tools were developed to enable the user to explore the flow field using a tracked glove (a DataGlove) on one hand. For example, the user could use the gloved hand to identify the source point for streamlines that would allow visualization of the flow field in specific regions. A great deal of work went into developing both the precomputed data and the software that supported the visualization system. The software framework for the virtual wind tunnel was extensible and had interactive (i.e., real-time) performance. Fig. 24.1 shows Bryson examining the flow fields around an experimental vehicle.

Others, for example Severance [33], have extended the work of Bryson's group by fusing the data from several wind-tunnel experiments into a single, coherent visualization. Given the high cost of maintaining and operating wind tunnels and the limited regimes (of both wind speed and aerodynamic surface size), the virtual wind tunnel offers a significant potential to reduce the cost and expand the availability of wind-tunnel experiments.

Figure 24.1   Steve Bryson interacting with the Virtual Wind Tunnel developed at the NASA Ames Research Center. (See Fig. 21.2 in color insert.)

24.2.2.14 *Hydrocarbon Exploration and Production*

In recent years, virtual reality has had a growing impact on the exploration and production of hydrocarbons, specifically oil and gas. In 1997, there were only two large-scale visualization centers in the oil and gas industry, but by 2000, the number had grown to more than 20. In spite of this growth, the use of virtual reality technology was largely limited to 3D displays—interaction was still typically done via the keyboard and mouse. Lin et al. [24,25] created an application that supported more direct interaction between the user and the data in a CAVE. Not only could three to four users share an immersive 3D visualization, one of them could also interact directly with the data via natural gestures. Fig. 24.2 shows a user in a CAVE working with objects representing geophysical surfaces within a salt dome in the Gulf of Mexico. The user holds

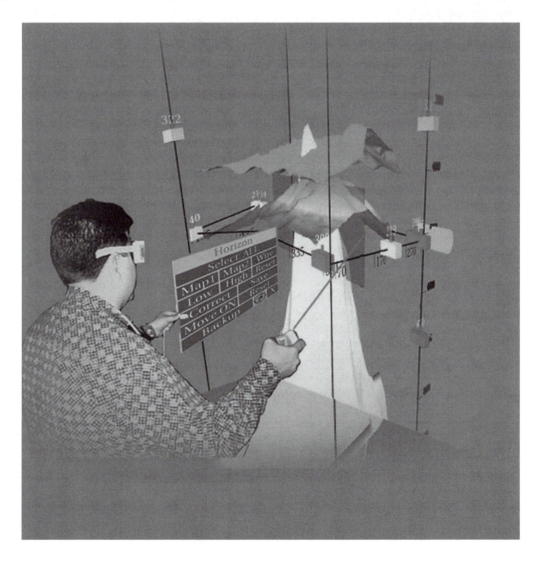

Figure 24.2   A user in a CAVE interacting with geophysical data describing a salt dome in the Gulf of Mexico. (See also color insert.)

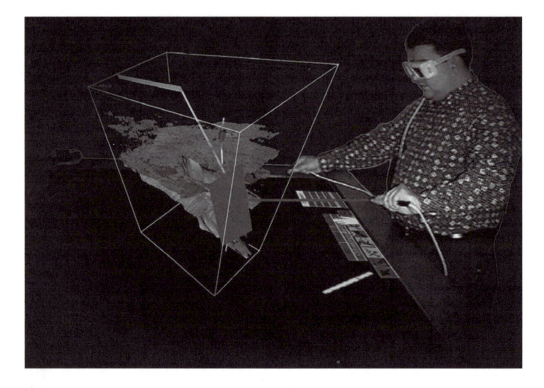

Figure 24.3   A user interacting with geophysical data on a projection workbench. (See also color insert.)

a tracked pointing/interaction device in the dominant hand (the right hand, in the figure) while a virtual menu is attached to the non-dominant hand. Studies were done to determine the effectiveness of this approach when compared with typical desktop applications.

Additional work was done by Loftin et al. [26] to implement powerful interaction techniques on a projection workbench. In Fig. 24.3, the user is again using both hands for navigation, menu interaction, and detailed manipulation of the data. The success of these efforts and similar work by other groups has led the oil and gas industries to make major investments in virtual-reality-based visualization systems. More importantly, these systems have had a demonstrable return on investment in terms of improved speed and success of decision making for both exploration and production activities.

24.3   Research Challenges

Although there is strong evidence that virtual-reality systems can offer significant advantages over conventional display systems for visualization applications, many challenges remain to be overcome. Below are some of the largest barriers to the use of virtual reality as the principal platform for visualization in many fields.

- Fidelity. Fidelity has two "faces"; in one sense, fidelity can refer to the resolution of data and/or displays of that data. One can think in terms of the resolution of a dataset or of a display. Another aspect of fidelity is in the data representation itself. A primary unanswered question is, "How much fidelity is enough?" That is, how much fidelity must an application and/or its display system have in order to achieve a specific outcome?

- Multimodal displays. While visual display technology is relatively mature and 3D audio displays are of very high quality, much work has to be done to advance the state of the art in displays for other senses (haptic, olfactory, vestibular, and gustatory) and in the integration of multimodal displays to provide a seamless sensory environment for the user. Perhaps the grandest challenge of all is the need for a robust theory of multi-sensory perception that guides the developer in mapping different types of data to different sensory modalities.

- Technical fragility. The hardware and software used in virtual reality is still fragile. The lack of a mass market has hindered manufacturers' desire to build more robust systems and to invest in the research needed to solve fundamental engineering problems (such as latency).

- Software inaccessibility. The available software systems, even those that are commercially available, are notoriously difficult to use. One needs great patience as well as a great deal of experience to become a proficient developer of virtual-reality applications of any real complexity.

- Cost. Large-scale virtual-reality systems cost a great deal. As long as it can cost over $1,000,000 to install the computational, display, and interaction technologies required for a sophisticated, multiuser system, many will choose not to use virtual reality as a means of implementing visualization applications.

References

1. D. Acevedo, E. Vote, D. H. Laidlaw, and M. S. Joukowsky. Archaeological data visualization in VR: analysis of lamp finds at the great temple of Petra: a case study. *Proceedings of the 2001 Conference on Virtual Reality*, pages 493–496, 2001.

2. N. Akkiraju, H. Edelsbrunner, P. Fu, and J. Qian. Viewing geometric protein structures from inside a CAVE. *IEEE Computer Graphics and Applications*, 16(4):58–61, 1996.

3. H. Amari, T. Nagumo, M. Okada, M. Hirose, and T. Ishii. A virtual reality application for software visualization. In *Proceedings of the 1993 Virtual Reality Annual International Symposium*, pages 1–6, 1993.

4. L. Arms, D. Cook, and C. Cruz-Neira. The benefits of statistical visualization in an immersive environment. In *Proceedings of the 1999 IEEE Virtual Reality Conference*, pages 88–95, 1999.

5. S. Bryson and C. Levit. The virtual wind tunnel. *IEEE Computer Graphics and Applications*, 12(4):25–34, 1992.

6. S. Bryson. The virtual wind tunnel: a high-performance virtual reality application. In *Proceedings of the 1993 Virtual Reality Annual International Symposium*, pages 20–26, 1993.

7. S. Bryson. Real-time exploratory scientific visualization and virtual reality. In *Scientific Visualization: Advances and Challenges*. (L. Rosenblum, R. A. Earnshaw, J. Encarnação, H. Hagen, A. Kaufman, S. Klimenko, G. Nielson, F. Post, and D. Thalman Eds.). London, Academic Press, 1994.

8. S. Bryson, S. Johan, and L. Schlecht. An extensible interactive visualization framework for the virtual wind tunnel. In *Proceedings of the 1997 Virtual Reality Annual International Symposium*, pages 106–113, 1997.

9. C. Chen. Information visualization and virtual environments. Berlin, Springer-Verlag, 1999.

10. J. C. Chung, M. R. Harris, F. P. Brooks, Jr., H. Fuchs, M. T. Kelley, J. Hughes, M. Ouh-Young, C. Cheung, R. L. Holloway, and M. Pique. Exploring virtual worlds with head-mounted displays. In *Proceedings of the SPIE Conference on Three-Dimensional Visualization and Display Technologies*, pages 42–52, 1990.

11. C. Cruz-Neira, D. J. Sandin, and T. A. DeFanti. Surround-screen projection-based virtual reality: the design and implementation of the CAVE. *Computer Graphics*, 27:135–142.

12. J. Durbin, J. E. Swan II, B. Colbert, J. Crowe, R. King, T. King, C. Scannell, Z. Wartell, and T. Welsh. Battlefield visualization on the responsive workbench. In *Proceedings of the 1998 IEEE Visualization Conference*, pages 463–466, 1998.

13. N. Durlach and A. Mavor (Eds). *Virtual Reality: Scientific and Technological Challenges*. Washington, DC, National Academy Press, 1995.

14. K. Gaither, R. Moorhead, S. Nations, and D. Fox. Visualizing ocean circulation models through virtual environments. *IEEE Computer Graphics and Applications*, 17(1):16–19, 1997.

15. M. Göbel (Ed.). Virtual environments and scientific visualization. *Proceedings of the Eurographics Workshops in Monte Carlo, Monaco, February 19–20, 1996 and in Prague, Czech Republic, April 23–23, 1996*. Berlin, Springer-Verlag, 1996.

16. M. R. Hall and J. W. Miller. Head-mounted electro-ocular display: a new display concept for specialized environments. *Aerospace Medicine*, pages 316–318, 1963.

17. H. Hasse, F. Dai, J. Strassner, and M. Göbel. Immersive investigation of scientific data. In *Scientific visualization: overviews, methods, and techniques*. (G. M. Nielson, H. Hagen, and H. Müller, Eds.). Los Alamitos, California, IEEE Computer Society, pages 35–58, 1997.

18. D. Hix, J. E. Swan II, J. L. Gabbard, M. McGree, J. Durbin, and T. King. User-centered design and evaluation of a real-time battlefield visualization virtual environment. In *Proceedings of the 1999 IEEE Virtual Reality Conference*, pages 96–103, 1999.

19. M. Kano, S. Tsutsumi, and K. Nishimura. Visualization for genome function analysis using immersive projection technology. In *Proceedings of the 2002 IEEE Virtual Reality Conference*, pages 224–231, 2002.

20. W. Krueger and B. Froehlich. The responsive workbench: a virtual work environment. *IEEE Computer Graphics and Applications*, 14(3):12–15, 1994.

21. F. Kuester, R. Bruckschen, B. Hamann and K. I. Joy. Visualization of particle traces in virtual environments. In *Proceedings of the 2001 ACM Symposium on Virtual Reality Software and Technology*, 2001.

22. S. Kuschfeldt, M. Schultz, T. Ertl, T. Reuding, and M. Holzner. The use of a virtual environment for FE analysis of vehicle crash worthiness. In *Proceedings of the 1997 Virtual Reality Annual International Symposium*, pages 209 and 1009, 1997.

23. J. Leigh, A. E. Johnson, C. A. Vasilakis, and T. A. DeFanti. Multi-perspective collaborative design in persistent networked virtual environments. In *Proceedings of the 1996 Virtual Reality Annual International Symposium*, pages 253–260, 1996.

24. C.-R. Lin, R. B. Loftin, and H. R. Nelson, Jr. Interaction with geoscience data in an immersive environment. In *Proceedings of the 2000 IEEE Virtual Reality Conference*, pages 55–62, 2000.

25. C.-R. Lin and R. B. Loftin. VR user interface: closed world interaction. In *Proceedings of the ACM Symposium on Virtual Reality Software & Technology 2000*, pages 153–159, 2000.

26. R. B. Loftin, C. Harding, D. Chen, C. Lin, C. Chuter, M. Acosta, A. Ugray, P. Gordon, and K. Nesbitt. Advanced visualization techniques in the geosciences. In *Proceedings of the Nineteenth Annual Research Conference of the Gulf Coast Section Society of Economic Paleontologists and Mineralogists Foundation*, Houston, Texas, December 5–8, 1999.

27. G. A. Miller. The magical number seven, plus or minus two: some limits on our capacity for processing information. *The Psychological Review*, 63:81–97, 1956.

28. M. Minsky, M. Ouh-Young, O. Steele, F. P. Brooks, Jr., and M. Behensky. Feeling and seeing: issues in force display. In *Proceedings of Symposium on 3-D Interactive Graphics*, pages 235–241, 1990.

29. G. M. Nielson, H. Hagen, and H. Müller. *Scientific Visualization: Overviews, Methods, and Techniques*. Los Alamitos, California, IEEE Computer Society, 1997.

30. M. Ouh-Young, M. Pique, J. Hughes, N. Srinivasan, and F. P. Brooks, Jr. Using a manipulator for force display in molecular docking. *IEEE publication CH2555-1/88*, pages 1824–1829, 1988.

31. L. Rosenblum, R. A. Earnshaw, J. Encarnação, H. Hagen, A. Kaufman, S. Klimenko, G. Nielson, F. Post, and D. Thalman (Eds). *Scientific Visualization: Advances and Challenges*. London, Academic Press, 1994.

32. L. Rosenblum, J. Durbin, R. Doyle, and D. Tate. The virtual reality responsive workbench: applications and experience. In *Proceedings of the British Computer Society Conference on Virtual Worlds on the WWW, Internets, and Networks*, Bradford, UK, 1997.

33. K. Severance, P. Brewster, B. Lazos, and D. Keefe. Wind tunnel data fusion and immersive visualization. In *Proceedings of the 2001 IEEE Visualization Conference*, pages 505–508, 2001.

34. M. Slater, M. Usoh, and A. Steed. Depth of presence in immersive virtual environments. *Presence: Teleoperators and Virtual Environments*, 3(2):130–144, 1994.

35. M. Slater, V. Linakis, M. Usoh, and R. Kooper. Immersion, presence and performance in virtual environments: an experiment with tri-dimensional chess. In *Proceedings of the ACM Virtual Reality Software and Technology (VRST) Conference*, pages 163–172, 1996.

36. D. Song and M. L. Norman. Cosmic explorer: a virtual reality environment for exploring cosmic data. In *Proceedings of the 1993 IEEE Symposium on Research Frontiers in Virtual Reality*, pages 75–79, 1993.

37. K. M. Stanney (Ed). *Handbook of Virtual Environments: Design, Implementation, and Applications. Mahwah, NJ, Lawrence Erlbaum Associates, 2002.*

38. I. E. Sutherland. Head-mounted 3D display. In *Proceedings of the 1998 Fall Joint Computing Conference*, 33:757–764, 1998.

39. D. Vickers. Head-mounted display terminal. In *Proceedings of the 1970 IEEE International Computer Group Conference*, pages 102–109, 1970.

40. S. Ziegler, R. J. Moorhead, P. J. Croft, and D. Lu. The MetVR case study: meteorological visualization in an immersive virtual environment. In *Proceedings of the 2001 IEEE Visualization Conference*, pages 489–492 and 596, 2001.

PART VIII
Large-Scale Data Visualization

25 Desktop Delivery: Access to Large Datasets

PHILIP D. HEERMANN and CONSTANTINE PAVLAKOS
Sandia National Laboratories

25.1 Introduction

Since the advent of modern digital computing, the rapid evolution of technology has been a major aspect of the high-performance computing field. Computer users, however, have changed less radically, and the desire to compute from their offices is constant. From punch cards and printouts through dial-in terminals and network-connected personal computers, users have enjoyed an increasing ability to interact with supercomputers from their offices.

The major issue in desktop interaction with supercomputers is the size of the data. For many years, the primary problem was the size of the results. Input decks and code were generally easily handled by desktops. Recently, setup elements including mesh definitions and the setting of initial and boundary conditions for highly detailed models have pushed problem setup needs beyond desktop capabilities. Nevertheless, the problem remains interacting with datasets beyond the capabilities available in an office.

For the context of discussion here, large data will be defined as datasets that are much greater than the memory capacity of the desktop machine. More formally, datasets can be described as follows:

$$D \gg 10 M_D$$

where D is the dataset of interest and M_D is the random access memory (RAM) of the desktop machine. For current high-end machines in the office, with 2 to 4 gigabytes (GB) of system memory, a large dataset would be hundreds of gigabytes to terabytes and perhaps even petabytes.

Computer network capabilities are also greatly strained by large data. Transferring very large datasets can require hours or days, and the reality of network errors and machine reboots can turn a planned 8-hour project into a 5-day marathon. Supercomputing resources are normally expensive, so these machines are usually shared. Often the user is accessing the supercomputer at a great distance over a widearea network (WAN). This adds to interaction latencies and may increase the chance of network failures.

Thus, the desktop delivery problem is one of engineering a solution to provide meaningful interaction with a large dataset using a relatively small computer connected by a relatively small network. The use of the word "relatively" is important because both the network and the desktop machine are truly superior to the supercomputer technology of the late 1980s. On all measures of performance (memory, disk, network speed, graphics processing), today's desktop is truly a last-generation supercomputer. The major problem is that the datasets of interest today still overwhelm their capabilities.

One final point of introduction is to highlight the true scale of modern large datasets. With commodity personal computers offering 100-GB disks and 1-GB memories, computer users are lulled into a false sense of large data size. A few GB of data doesn't seem so large when it is the camcorder video from a child's sporting event. But now answer the question, "How many flowering dandelions are on the soccer field?" If the videographer has recorded the entire game and zoomed in and out sufficiently, the number of dandelions is probably in the dataset. Finding them all, however, is the proverbial needle in the haystack.

Many users are considering or working on datasets a thousand times larger, in the terabyte range. These datasets are on the same scale as the U.S. gross domestic product (GDP) (about $10 trillion in 2001) [1]. At this scale, tables of unreduced data are uncompressible. Even simple statistical measures may have little meaning. Graphical representations are critical, requiring multiple views and representations to fully grasp the data. If one picture is worth a thousand words, then a terabyte is worth a billion pictures or one 10,000-hour movie.

For datasets of this size, interactivity is important or the data will never be fully explored. If the interactivity is too slow, only what is expected will be found, and the unexpected may never be noticed. The needles in the haystack will be found, but a correlation between their orientations may be missed. Therefore, a desktop must support a full suite of tools, including 3D graphics, 2D plotting, and statistical tools, and it must provide access to computing cycles that allow timely analysis.

This chapter will present an approach for delivering supercomputer results to offices. The approach was developed to visualize simulation results produced by the U.S. Department of Energy's Accelerated Strategic Computing Initiative (ASCI). Much of the technology was developed as a collaborative effort between Los Alamos, Lawrence Livermore, and Sandia national laboratories, teaming with leading universities and industry to develop technology as needed. The system design presented here is the instantiation at Sandia National Labs; similar systems have been implemented at Los Alamos and Lawrence Livermore national labs.

To begin, several background issues will be discussed, highlighting nonobvious issues to a successful solution. Next, a high-level framework will be discussed. The framework organizes the problem and helps to illuminate system design tradeoffs. The Sandia system design will be presented, followed by more detailed discus-

sion of each major component. Individual researchers may wish to skip Section 25.8, Large Enterprise Considerations, which is written to suggest ways to deliver high-performance supercomputer access to tens or hundreds of users in a large organization. The last section is a foray into the dangerous art of predicting the future, examining technology trends and the some of the desktop visualization research directions at Sandia.

25.2 Background

Supercomputer design centers around the all-important design point of maximizing floating point operations (FLOPs). Processing of large datasets, including desktop delivery, certainly makes use of available FLOPs, but bandwidth is a performance-limiting factor. Integer performance is important, but seemingly mundane issues often limit performance, like the number of copies required to move data through software stacks.

25.2.1 Computer Components

For desktop delivery systems, a critical factor is the bandwidth capability of the motherboard chip set. The chip set hosts the CPU; it also commonly provides the path to main memory and implements the interface buses and ports like PCI-X and AGP. For a high-performance desktop, it is highly desirable to be rapidly receiving data from a high-speed network while feeding data at full bandwidth to a high-performance graphics card. In practice, the chip sets are often overloaded. This is made apparent when running individual benchmarks of network and graphics performance, then running the benchmarks simultaneously. Desktop performance can be limited simply because the chip set cannot meet the combined demands.

Disks are the source of another common bottleneck. Physics dictate that moving heads and spinning disks cannot keep pace with CPUs running in the gigahertz (GHz). Disk drive manufacturers have made tremendous gains in disk drive capacity, but the bandwidth to the disk has not kept pace with the rate of capacity gains.

Networks, however, have made tremendous gains in performance. High-performance networks actually exceed the bandwidth performance of individual disks. Available networks severely challenge a CPU's ability to maintain pace, and for some local area network (LAN) situations, software compression techniques may actually slow the transfer. Consider a CPU running at 3 GHz feeding a 1-gigabit (Gbit) Ethernet network. With a properly tuned application and network, a 1-Gbit Ethernet line can deliver $70 + MB/s$. This means that the CPU has a budget of ~ 43 cycles per byte of data transferred. So in real-time applications like dynamic image delivery across a network, it may take longer to perform image compression than simply blast the image across the network. Compression can certainly improve network performance, but the time of compression must be balanced with the network speeds.

25.2.2   Networks

There are myriad networking technologies. From the desktop delivery perspective, there are two primary categories: 1) *internet protocol* (IP) networks and 2) *interconnects*. IP networks are the most widely deployed networks. They are the networks that comprise the Internet and are the foundations of most LANs and WANs. Interconnects are special networks designed to tightly couple computers together to build "computer clusters." Examples of interconnects are Myrinet [2], Quadrics [3], and Infiniband [4].

To understand why there are two types of networks and their impact on desktop delivery, one may benefit from a short primer on network architecture. Central to most networking is an architectural idea known as the OSI (Open System Interconnection) reference model. A central feature of the model is its layered structure: it has seven functional layers, from a physical transport layer at the bottom to the application at the top. Depending on the situation, some networks do not implement all the layers suggested by the OSI model.

Obviously, the physical layer is hardware, but often the middle layers are implemented in software. These software layers are sometimes referred to collectively as the "stack" or the "IP stack." The IP stack is the interface that provides an applications programming interface (API) to an application, coordinates network transfers with the operating system, and provides the driver software for the network interface card (NIC) hardware.

Knowledge of the IP stack is important because the number of CPU cycles required to execute the IP stack is substantial. As a general rule, it requires about a 1-GHz, 32-bit CPU to execute an IP stack rapidly enough to keep a 1-Gbit Ethernet full. This certainly varies with hardware, operating system, and network stack implementation, but in general it is true for optimized IP stacks. This leads to the common issue of computers that are unable to feed or receive data at the full network speed. Just because a fast network is connected to a computer does not mean that the computer can fully use it.

The second part of the short network primer is to understand *reliable* and *unreliable* communications. Networks, like IP networks, implement a connectionless, packet-based communication method. This means that at the lowest levels of the IP stack, packets of data are sent with no guarantee that they will arrive. Higher levels in the stack monitor the flow of packets and ask for packets to be resent if they are not properly transferred. Therefore, *reliable* communications are systems that have a positive means by which to ensure that data is properly transferred, and *unreliable* communications rely on the network being "good enough."

For common IP networks, there are two major protocols in wide use: transmission control protocol (TCP) and user datagram protocol (UDP). TCP ensures reliable communications and, together with its IP foundation, is referred to as TCP/IP. UDP does not provide reliable communications, but it provides fast transfer with less overhead than TCP (i.e., less CPU load to communicate data). *Unreliable* transmission is not necessarily worthless. For example, if playing a movie or animation across a network, a dropped or lost packet might leave a blemish in the image, but the next frame, in 33 milliseconds, will replace it. This is likely better than the alternative of a slow-playing animation.

Interconnect networks are designed to interconnect hundreds or thousands of computers normally occupying a single room. Taking advantage of the controlled conditions of a single room, the hardware can be specialized to the situation, and their network "stack" is designed to require much less processing. This results in two main advantages: lower latency and increased bandwidth. These allow the "cluster" of computers to communicate more efficiently and to operate together more readily on a single task. Leveraging the market base and resulting economies of scale for IP networking equipment, some clusters are built using IP networks. These systems, while certainly less expensive, pay the cost of slower communication and increased CPU utilization for communications.

To speed communications, increasing portions of the interconnect "stack" are being moved onto the NIC, which often includes a microprocessor. This reduces the CPU load for each cluster node, leaving more time for useful work instead of using main CPU cycles for communicating. These capabilities are beginning to appear in some IP NICs as well. This certainly reduces the CPU load, but latencies still remain higher than the specialized networks due to the heavier weight protocols.

Before leaving networks we want to mention Fibre Channel. Its primary application is to connect resources like disk drives or tape drives to a computer host. Its common protocol is small computer system interface (SCSI, pronounced "scuzzy"). Fibre Channel has the potential to reach several kilometers in distance, and Fibre Channel switches are able to support more complex network topologies than simple point-to-point. It is commonly used to attach large RAID disk systems at bandwidths reaching 100 MB/s per channel. Due to the available switches and distance possible, Fibre Channel can deliver high-performance disk access to desktops in a campus setting. This supports the possibility of delivering direct access to disk systems of tens or hundreds of terabytes without putting a rack of disks in an office.

25.2.3   Human Factors

Desktop delivery is obviously concerned with bringing information to a person in his or her office. Most of the discussion in the chapter presents methods for generating and delivering images to leverage the human visual system for analysis. The best means to use the human visual system is complex and beyond the scope of this chapter; however, large data analysis does present a major human-factors challenge.

The challenge is to provide interaction as rapidly as possible. Very large datasets can take a long time to analyze. If the desktop system requires minutes to create each image or view, the user will tend to look only where results are expected. He or she will not take the time to explore, thus leaving the large dataset not fully studied.

When thinking about user interactions with the desktop, it is helpful to consider several different rates. The highest desired rate is 20–30 Hz for animation playback. Slightly slower is the update rate for hand-eye coordination with the user interface, at a minimum of 10 Hz. This is the rate of cursor update for mouse movement or view rotation as the mouse is moved. The slowest is processing a user's request. For interactive systems, less than 10 seconds is the target, but with large data this may stretch to minutes. If processing takes longer than a few minutes, then batch preprocessing of the data for faster interaction is warranted. The key point is that the design goal is to analyze the data, and human needs should be considered.

25.2.4   Market Forces

Historically, supercomputing was the leading technology developer for computing. For decades, if a user needed the highest performance capabilities, they turned to supercomputing vendors to supply the need. In the last few years, however, an interesting twist has developed. Some components in the commodity computing markets are beginning to exceed the performance of the niche supercomputer components. In a sense this is nothing new, because Moore's Law has long been a major driving force for more capability in a given area of silicon real estate.

The market size, however, has become a major new factor. This is most visible in graphics cards. For years, the leading performance came from major workstation/supercomputer graphics vendors, like Evans Sutherland, Hewlett Packard, Silicon Graphics, and Sun Microsystems. Recently, commodity personal computer components have exceeded the supercomputer equipment's performance in several key areas. The drive behind this has been the tremendous market of computer gaming. The large markets attract and maintain larger research investments, resulting in faster hardware development.

For desktop delivery, it is critical to watch the markets. Failure to follow the trends can at best result in poor value for the investment dollar and at worst deploy a system that cannot be maintained or expanded due to obsolete components or a supplier that goes out of business. The bottom line is the successful long-term deployment of a desktop delivery system, and this calls for keeping an eye on the broad computer marketplace.

25.3   Framing the Problem

It is best to understand whatever problem you are trying to solve. From a high level, our problem can be stated as, "Provide a means for large dataset analysis in users' offices." This is a good goal, but more detail is needed to design a solution. At Sandia National Laboratories, a chart was developed to characterize the major activities in postprocessing results from large engineering physics simulations, but it is useful beyond the initial application.

The Sandia chart, Fig. 25.1, organizes the scientific visualization process from data sources on the top to displays on the bottom. *Data Sources* represents the input to a visualization process. This includes simulations, data archives, and experiments, but it can also include real-time data from sensors, video feeds, or any other source of static or dynamic data. The next step, *Data Services*, is the process of selecting what data is to be considered, converting it into a form more readily analyzed, filtering the data, or any other step that is necessary to prepare it for the next step. *Information Services* takes the data of interest from the *Data Services* section and converts it into a visual representation, something that can be rendered into an image. This process might be something along the lines of generating an isosurface or generating the polygons that represent the external surfaces of a finite element model. This information, usually polygons or voxels, is passed to the *Visualization Service* area for conversion to one or more images. The image data is at last presented to the user on a display device, which for the discussion here is an office desktop. Alongside the main data flow is *User Services*, the control aspect for the entire process. This documents the need to provide a means to tailor and control each step.

By its very nature, large data cannot reside at the user's desktop. The data, therefore, must be located remotely. Whether the data is down the hall or across the country, it is located separately. Similarly, the display must be located with the user in order to be useful. This separation between the display and the data must be accommodated somewhere, and Fig. 25.1 suggests some options.

Distance can be inserted between any two services or in the middle of a service. There are some arcane technical reasons to insert distance in the center of a service, but for the discussion here the focus will consider distance communication options between services.

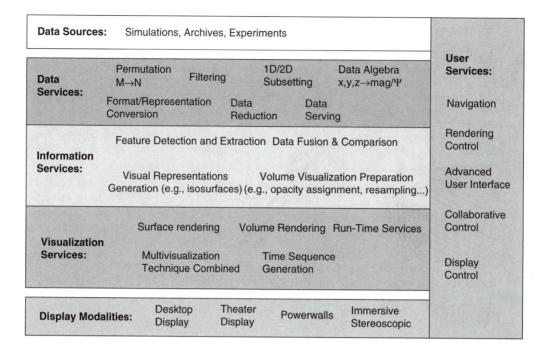

Figure 25.1 Scientific visualization logical architecture.

If the distance communication is between *Data Sources* and *Data Services*, this means simply moving the raw data. This has the drawback of generally moving the largest volume of data. Distance inserted between *Data Services* and *Information Services* is a raw data move with the benefit of moving only the data that is of interest. This down-select can greatly reduce the amount of data moved. Distance between *Information Services* and *Visualization Services* often provides a significant reduction in data because only the polygons or voxels that will contribute to the current image are communicated. The last possibility is to insert distance between *Visualization Services* and *Display Modalities*. This is simply image or animation delivery across a distance.

In fact, it is wise to design a system that supports as many of the options as possible, often placing distance between several services. This accommodates different user needs and provides options for pathological cases. An example of a pathological case is the common situation in which the data is reduced by a factor of 10 or more after conversion to a surface or voxel. However, Sandia researchers have experienced cases in which the polygon data was actually larger than the raw data.

For delivery to office desktops, the limited capabilities available in the office equipment strongly influence design decisions. Considering an office system with general computing and graphics capabilities, Fig. 25.2 gives design options.

This diagram illustrates the relationship between dataset size and a memory-limited forcing of the desktop mode. *Dataset Delivery*, *Geometry Delivery*, and *Image Delivery* modes correspond to inserting distance between *Data Sources* and *Data Services*, *Information Services* and *Visualization Services*, and *Visualization Services* and *Display Modalities*, respectively.

For the smallest datasets, or a dataset comprised of many relatively small pieces, the data

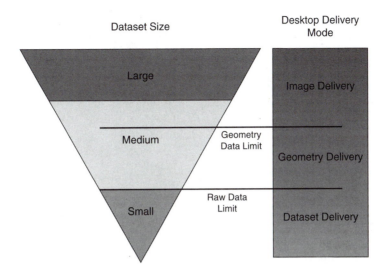

Figure 25.2   Effect of dataset size on desktop delivery mode.

can simply be brought to the desktop and analyzed completely on the desktop machine. A remote *Data Service* can be very useful for this mode by providing a data subset function to pull small pieces out of a large dataset. This operation actually places the distance between *Data Services* and *Information Services*.

When the data reaches a size that is too large to be readily handled, or breaking it into many small pieces is impractical, the Raw Data Limit (RDL) is reached. At this point, the *Information Service* is best performed on a server that has the memory capacity and I/O speeds to handle the data. The raw data is reduced to a renderable form, like polygons or voxels, and passed across the network. Normally, this provides a substantial reduction in data and maintains good interaction rates, since the rendering is handled locally. An example of a software tool that uses this mode of operation is Ensight from CEI [5]. Ensight uses a client–server model with a data server delivering polygon geometry across a TCP/IP network for rendering on the desktop machine.

For yet larger datasets, the data is too large for the geometry information to fit in the desk-

top machine. This limit is the Geometry Data Limit (GDL). Large data at this scale must be handled on large data and visualization server resources, with image data delivered to the desktop. This also brings the potential advantage of accelerated rendering supported from a high-performance rendering service. A major issue for remote rendering is that interaction tasks like zooming and rotation must be handled across the network, which adds latency between the user and the responding resource. Both hardware and software systems have demonstrated the feasibility of interacting effectively with remote rendering resources, but the latency must be carefully managed. An example of an image delivery system is a Sandia-developed hardware solution that provides high frame rate and rapid interaction across both LAN and WAN TCP/IP networks [6].

Other factors, like difficult-to-render images or computation-intensive *Information Service* tasks, can drive operations off the desktop to remote servers. Also, bringing dataset pieces to the desktop has a potential trap that does not deliver the best performance. The trap is that users bring the data across the network and place it on the local disk for processing. Desktop disk drives deliver sustained performance levels on the order of tens of megabytes per second. Remember that 1-Gb Ethernet can deliver 70+ Mb/s. Therefore, it could easily be faster to access data across the network from a fast RAID server than to process the data from the local disk.

25.4   System for Desktop Delivery

This section will present a reference design for production of large data desktop delivery. A *production system* is designed for a group of users (10 to 100) who need to access and analyze large data on a daily basis. Examples of organizations requiring this type of system are automobile or aerospace manufacturers, large research projects, and national laboratories. An individual or small group may not need the entire system, but the criteria for component selection are still applicable.

The first step in providing a practical production desktop delivery system is the realization that external services are needed to support the inadequacies of cost-effective desktop hardware. Large computer resources are generally necessary in order to handle the large data, and they tend to be expensive. The cost generally prohibits a design solution like deploying a large SMP or cluster resource to support each individual user. This means building a solution that shares large resources across many users.

If the large data is from a supercomputer or collected at one location, moving the users close to the data should be considered. Generally, this solution works only with a large corporation, but sabbaticals to another university or research location could be an option. Usually, the analyst or researcher desires to work from his or her current location, so a LAN or WAN is normally part of the design. Even colocating the users often necessitates a LAN to reach the users in situations such as college campuses.

If a large volume of data needs to be moved, bandwidths and loads of available networks need to be assessed. If the transfer times are excessive, physically moving tapes or disks may solve the problem. There is a common saying in supercomputing, "Never underesti-

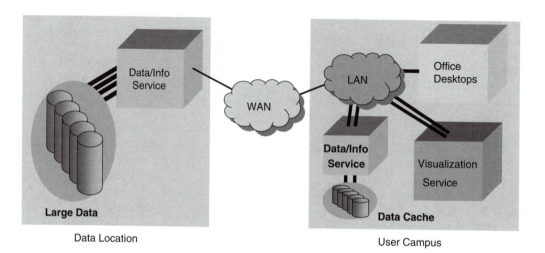

Figure 25.3   Visualization desktop delivery system.

mate the bandwidth of a station wagon full of tapes."

The scale of the data, however, suggests an alternative solution. Move the processing to the data rather than the data to the processing. This fundamental idea of moving the processing to the data means moving supercomputer capabilities to the location that has the best access to the data. That is the foundation of the design proposed below.

The reference-system design has two primary locations, the large data site and the user site. If both are on the same campus, then the WAN can be eliminated and the LAN supports the entire system. The services are located next to the large data or the user, depending on bandwidth and latency of communications channels.

The data/information service is located next to the large data store to maximize access speed and eliminate block transmission of the entire dataset. In Fig. 25.3, multiple parallel lines indicate that multiple parallel channels are necessary to provide sufficient bandwidth to the data. For example, to build a 1-GB/s channel for multiterabyte data store would require 10–15 Fibre Channel connections. Also, data and information services have the feature that their input is the large data but they perform a data reduction as a normal part of their operation. This provides a natural "compression" before transmission across a WAN or LAN.

The resources located at the user campus are driven primarily by network latency and bandwidth concerns. This locates the visualization service near the deskstop to facilitate interactive streaming of image data to the desktop. Internet web sites commonly stream image data across a WAN, but low-resolution images buffered into multisecond buffers are not a good solution for high-resolution interactive visualization. A smaller data/information service is placed at the user campus to provide high-bandwidth feeding of the visualization service and to provide high-speed network disk service to the desktop.

The next three sections will discuss in more detail the networks, servers, and desktops that support this design.

25.5 Networks

The primary focus of this section will be the LAN and WAN networks for the reference-system design. The connections within server systems to disks and between machines will be discussed in the next section on servers.

The most important item to consider when selecting the network is the marketplace. Certainly, the existing deployed networks at the user campus should be considered, but if the data is truly large, it is going to overwhelm the existing networks, so new capability will need to be deployed. Therefore, what equipment is currently available and the market directions for the next few years should be a primary concern. At present, this means TCP/IP networks for both the WAN and the LAN. In the WAN setting, TELCOs (long-distance telephone companies) are building toward their main growth market, Internet packet data. What is certain (and certainty is a rare thing in dynamic industries like communications) is that for many years to come there will be companies available to carry IP packets between sites.

The limiting factor in the WAN will most likely be the bandwidth cost. The TELCOs have the bandwidth available, but the yearly access fee must be considered. Access to an inexpensive or free high-performance research network is a possibility that should be considered for university or research-project systems.

Similarly, the LAN selection is largely based on the market directions. Here, the governing cost is often the equipment cost and the available cabling/fiber infrastructure. The office-to-LAN network cabling is not immediately obvious, but it can be a substantial expense. Nevertheless, it makes no sense to buy network switches and computer NICs if they cannot be connected.

The network equipment needed is the equipment to light up the fiber or send signals down the wire. This includes network switches, computer NICs, server NICs, and potentially routers to bring it all together. For desktop delivery of large data, this often is leading-edge equipment. Cutting-edge equipment can provide more capability, but often desktops will not be able to execute their IP stacks rapidly enough to fill the very fastest networks. The leading-edge equipment cost falls between the cutting-edge technology and the less-expensive commodity, network equipment. For large-data applications, it is generally worth the increased cost to purchase the leading-edge equipment, or the cost will be in wasted time for the large data analyst. The cutting-edge technology is worth considering, but only if the desktops or servers can utilize the extra bandwidth.

Parallel networks may be needed in locations where single channels provide insufficient bandwidth. This is generally not an issue in offices because the desktop machines have limited data-handling abilities. In machine rooms between servers, data archives, or other high-bandwidth needs, multiple parallel network channels are certainly a good solution.

It may be desirable to have parallel networks to an office if each network serves a different purpose. Extending a Fibre Channel network to an office can provide higher-speed and -capacity disk access than can be easily maintained in an office. If Fibre Channel is deployed to an office, an IP network connection is still desirable to provide access to the data and information services. Next-generation interconnects, like Infiniband, have the potential to provide access to both an IP network and a disk system, but it is too early to tell if and when combined features like this will reach the marketplace.

25.6 Servers

For the past 20 years, server hardware commonly implied a large shared-memory multiprocessor (SMP) computer. Recently, cluster systems built from numerous commodity computers have gained popularity. The major attraction of these systems is the high performance available for the price of the equipment. The cost of integration, however, is borne directly by the deploying organization rather than bundled in the cost of the machine.

Although clusters are gaining in popularity with researchers and with industry, their installation and operations still require considerable expertise. Also, the cluster environment requires some unique software tools and libraries to enable a broader range of applications beyond custom written applications for the environment.

When selecting components, the place to start is the motherboard. Commonly each node is a motherboard or system with 1 to 16 processors. One or two processors per motherboard are most commonly seen in commodity clusters. For data service nodes, CPU performance is important in the handling of tasks like IP stack execution or rapid processing of data.

For data serving nodes, many nodes have the responsibility of passing data from one network to another. For example, I/O nodes transfer data from disk to the interconnect fabric. This elevates the chip set to equal importance with the CPU. This requirement to process data through multiple ports simultaneously strains chip sets and CPUs alike. Only tests of the actual motherboard and the CPU(s) with the interfaces installed can determine if the computer system can handle the bandwidth needs.

A primary advantage of cluster systems in data-serving and rendering applications is their inherent scalability. Additional nodes can be added to increase the number of users supported, enhance the processing capabilities, or increase the disk bandwidth.

A high-performance interconnect is important as well. Large amounts of data are flowing through the cluster, and they all pass across the interconnect network. Bandwidth is certainly a key performance measure, since latency can often be masked by a pipelining software design. Nevertheless, a low-latency interconnect network does benefit fine-grain parallel processing and interactive rendering.

Last but not least, security should not be overlooked. Large datasets are generally valuable. Whether the data is a potential breakthrough for a researcher or a next-generation proprietary design, the large dataset was likely expensive to compute or collect and it should be protected. Often security can be greatly enhanced by use of the operating system's security options on each node. The cluster, however, does present additional challenges to ensure that nodes are cleaned up before being passed from one user to the next. Also, the cluster design can enhance security, and the designs presented next help support secure operation.

To aid cluster integrators, the remainder of this section will present reference designs for production deployment of a cluster as data server and a rendering engine. Los Alamos, Lawrence Livermore, and Sandia National Laboratories, teaming with university and industry partners, developed the software tools and cluster designs. The remainder of this section will present two cluster-reference designs.

The data server design has three major node types: login, processing, and I/O. The login nodes connect to both the LAN and the cluster interconnect. The login nodes are the only nodes that allow user login, and they spawn and coordinate jobs running on the cluster. The processing nodes are nodes that provide the processing for data subsetting, isosurfacing, or other data processing tasks. They are primarily distinguished as not having two jobs (like a login node) so that they have CPU cycles and chip set room to spare for processing. The I/O nodes move data onto and off of the cluster. They are the nodes connected to the disk system and are nodes that provide parallel streaming of data into the cluster.

The "Connect to Visualization Server" connection is only relevant at the user campus. At the data location, the LAN IO nodes are the means to pass data across the WAN to the user campus. The "Connect to Visualization Server" arrow is shown as a thick bus to illustrate the tight coupling desired between the data server and the visualization server. Here it represents extending the interconnect fabric between the two systems.

The software environment is parallel processing with MPI message passing between nodes.

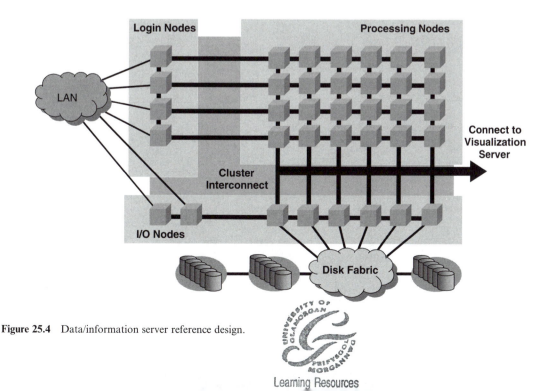

Figure 25.4   Data/information server reference design.

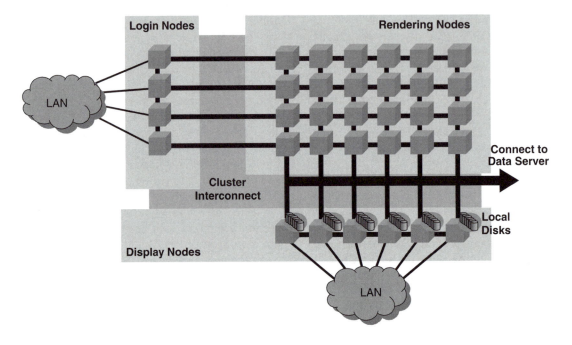

Figure 25.5   Visualization server design.

Microsoft Windows or Linux operating systems are viable on the nodes. A current disk system option is cluster wide NFS service with one file system per I/O node. Several high-performance parallel object-based file systems are in development that promise to provide a global parallel file system.

The visualization server is similar in design to the data server. The visualization server has three major node types: login, render, and display. The login nodes, like the data server, connect to the LAN and the cluster interconnect. The login nodes are the only nodes that allow user login, and they spawn and coordinate rendering activities on the cluster. The render nodes are nodes that contain commodity graphics cards to provide a scalable rendering capability. Presently, a render node's motherboard differs from the login and display nodes because they support an AGP (Advanced Graphics Port) for the graphics card.

The display nodes are designed to stream data off of the cluster to a user's desktop.

Each of the display nodes contains a local disk to cache results, and the disks can be used to ensure smooth playback of animations. The decision to play back animations from the display nodes or download and play them from a user's desktop depends on the balance of speed between the network and the desktop's disks.

The "Connect to Data Server" arrow indicates the tight coupling back to a local data server. As mentioned in the data-server discussion, a good option for the tight coupling is to extend the interconnect fabric between the two systems.

Like the data server, the software environment is parallel processing with MPI message passing between nodes. Microsoft Windows or Linux operating systems are viable on the nodes. Here the determining factor of the operating system is the availability of high-performance drivers for the graphics cards. If a Linux driver is not available, then Windows may be the only option.

The ASCI program has developed software to aid in constructing render clusters. Two primary tools are Chromium and ParaView. Chromium is a library that coordinates Open GL rendering on a cluster, allowing scalar and parallel Open GL applications access to cluster rendering. Chromium [7] was initially developed at Stanford under an ASCI contract, but it has transitioned to open-source development and availability. Similarly, ASCI contracted Kitware to extend the open-source Visualization Toolkit (VTK) to enable distributed-memory parallel processing. The resulting application, ParaView [8], is also open-source available. Other cluster tools, like Distributed Multihead X (DMX) [9] for cluster-driven tiled displays, are under development, and most will be released as open source.

25.7   Desktop Machine Selection

It is fitting that the section on desktop machine selection is buried numerous pages after discussions of markets, networks, and servers. When an office desktop is mentioned, thoughts usually turn only to the hardware sitting in the office. The truth, however, is that the desktop machine is only the tip of the iceberg, and most of the system necessary to bring large data to a desktop is hidden in network closets and machine rooms.

The first step in selecting a desktop is to carefully look at the markets. Visualizing and analyzing large data are, by definition, tasks well beyond the current desktop abilities. The consumer markets bring out doublings of performance regularly. If the desktop hardware and software are properly selected, then an advantage can be leveraged from the frequent release of improved technology. Doing this successfully allows the users to *ride the wave,* using the technology advances to propel the capability forward.

Today, the ride-the-wave technology is commodity personal computers. The commodity machines are matching the performance of traditional scientific workstations at a fraction of

their cost. Proper selection of a desktop machine centers on testing the machine with the applications that are intended to be run. Standard benchmarks are useful to narrow the search, but in the end, testing before buying is critical.

Because large data comprehension is usually highly dependent on visualization, the graphics card is a key component. The fastest vendor regularly changes in the marketplace, but nearly all vendors are targeting the PC platform. Here, benchmarks running the applications of importance to the users are critical. Different graphics cards from the same vendor and from different vendors vary greatly in their abilities. Application timings are the sure way to find the best one for users' applications.

The graphics-card vendors release significant improvements approximately every 6 months. Most vendors try to release a new card for the Christmas holiday, so a card purchased in October is likely to be surpassed by one purchased in December.

Because the vendors are targeting the PC, the graphics cards can be readily upgraded. The cards are relatively inexpensive ($300 to $1500, with a few above $2000). Due to the low cost, the graphics card can usually be upgraded at least annually. During the last few years, an annual upgrade has delivered nearly a doubling in performance each year. The more expensive cards provide extra performance in certain areas like line drawing. If the users' applications are not greatly accelerated by the extra performance, then buying the most expensive card may not make the most sense, since it will upgraded in a year.

Selecting the remainder of the desktop machine involves choosing a PC that can handle a high-performance graphics card and a fast NIC. Currently the PC will need an AGP port for the graphics card and PCI slot for the NIC. The chip set must be able to handle operations of both simultaneously. Tests should be run to confirm that the NIC and graphics card do not strongly conflict. It is also important to remember to include plenty of CPU cycles for executing an IP stack. Two CPUs can help, if the load is more than one can handle. Load the PC with as much RAM as possible. It is useful for processing locally and to help with the playback of animation.

From the large data perspective, the local disk is mainly for booting. A large disk is useful, however, because the PC will probably be used for more than just large data analysis. If local animation playback is desired, a small local RAID can be very useful. Small RAID controllers are even integrated on some motherboards.

The display selection can be summed up in one phrase, "Bigger is better." One large screen is good; multiple screens are often desirable as well. Construction blueprints are printed on huge paper for a reason; similarly, a large high-resolution screen allows viewing of more information.

In conclusion, buy the latest top-of-the-line PC. Load it with a fast graphics card, but be sure to test several to select the best one for your application. Put in as much memory as the slots can hold. Then connect it to the world with a leading-edge IP network.

25.8   Large Enterprise Considerations

Recognizing that the markets that supply desktops, servers, and networks are rapidly delivering improved technology, the challenge becomes deploying products rapidly and frequently to ride the wave. For a small team, the team generally just does it all themselves. In a large organization, the do-it-yourself option is possible, but it is also possible to engage the IT (Information Technology) organizations to help deploy to a large number of users.

Supporting a large number of users in addition to handling the large data has benefits and challenges. The larger user base generally brings larger budgets to help purchase servers, disk systems, and networks. The downside is that there are more users to support and the logistics of deploying 100 new desktops can be daunting. Also, large organizations are generally slow to adapt to change.

A slow rate of adoption for new computer technology can hinder the ability to ride the wave. With the availability of faster graphics cards every 6 months, a corporate practice like the 3-year refresh cycle for desktops can render desktops quickly obsolete. A good technique is to develop a deployment strategy that includes plans to ride the wave. At Sandia National Laboratories, by planning for graphics card refresh, the desktop delivery team has developed practices that permit upgrade of a user's graphics card in 20 minutes. This short update time allows annual updates of graphics capabilities.

Another strategy for a large organization to rapidly deploy and ride the wave is to team with vendors. Rather than purchase a desktop and have the internal IT team install the software, buy the desktop ready, with all the software installed, including user-specific tools. When you make the supplier a part of the team, more personnel are available to speed deployment or graphics card updates.

Large organizations generally have corporation-wide teams that supply networking to everyone. Similarly to the difficulties of deploying hundreds of desktops, deploying a new high-performance network can be difficult. A strategy to assist with deploying an advanced network is to deploy multilevel networks. The idea is to supply higher-speed networks to users that need the service. Telephone companies supply different levels of service, and it makes sense to have the same setup within large organizations. This allows the network budget to be focused on the users that need access to the large data. The corporate network team often likes this strategy too, because usually they enjoy the challenge of the leading-edge technology.

Finally, remember to scale the servers for the number of users in addition to the size of the data. Added login nodes may be needed on data and visualization clusters. Network backbones may need to be enlarged to handle multiple users accessing the large data simultaneously. The large data is still the first-priority problem, but a large user base complicates the problem.

25.9 Desktop Futures

Predicting the future is a dangerous practice, but there are some significant trends that are worth mentioning. The discussion here helps to illustrate the broad range of technologies that have the potential to greatly change the office desktop. The exercise also is useful in the assessment of current desktop ideas in order to avoid dead-end purchases.

The high-definition television (HDTV) deployment in the United States is not proceeding as rapidly as expected. Nevertheless, HDTV displays (i.e., next-generation TV sets) are dropping in price and the number of broadcast stations that carry this format is increasing. This growing marketplace is likely to foster technologies to deliver relatively high-resolution moving-picture content. As HDTV production and home display markets grow, the quality of both sending and receiving equipment will increase in quality and decrease in price. Because the television markets are so large, this technology segment could deliver components that exceed the capabilities of existing cluster-PC desktop systems. Thus, a desktop of the future may simply be an advanced "closed-circuit" HDTV system feeding flat-panel HDTV displays.

The next question should be, what about the human interface? How will a user interact with the system? The answer may be sitting in front of millions of 12-year-olds in game boxes like the XBox or the Playstation. There are two trends that point this way.

First, look at the effect of Moore's Law on computing. Mainframe computers, like the large CDC machines, gave way to minicomputers, like the DEC VAX. Microcomputers, like the PC and Apples, have superseded minicomputers. So the question becomes, what is next in the lineage? Game boxes are certainly a major option. The latest game boxes delivered graphics performance at their release that exceeded PC capabilities, and with a price of $300 or less, they could literally be disposable in an office setting.

Second, recent PC graphics cards have been bandwidth-limited because they run on PC plat-forms. Tests of rendering speed run from main memory are about half the speed of the same polygons loaded up into the card's memory. A solution is to build a special computer to properly feed the graphics chips, and game boxes are exactly these machines. Take a game box, add the increasing availability of broadband IP networks to homes, attach an HDTV display, and you have the perfect environment in which to deliver computing as a service rather than a system purchase. A game box has plenty of processing power to run common applications like a word processor, Internet browser, and annual tax software in addition to the usual suite of games. Provide this all as a service to a set-top box with no user maintenance, and the PC could be in trouble. If this proves true, the next-generation office hardware could easily be a game box.

Now let us turn to the future of displays. In the PC markets, the displays have been growing slowly in size and the resolution has crept up. However, neither the resolution nor the size of the display has seen the doubling like CPUs or graphics processing. The primary market movement has been toward flat-panel displays, but at similar sizes and resolutions as the CRT monitors.

The ASCI program recognized that some progress was needed in this area, and it funded IBM to modify a manufacturing line so that they could release a major new display. The result is the IBM T221, a 22.2″ flat-panel display delivering 9.2 million pixels. The T221 delivers extremely crisp images with its 200-pixels-per-inch resolution. This is certainly a welcome advance, and other manufacturers have been releasing large high-resolution flat panels.

For the future, there are three other technologies to watch in addition to the flat-panel market: projectors, OLEDs, and personal displays. Projectors are certainly improving in brightness and resolution and decreasing in size. With microelectronic technologies, small devices facilitate low costs, and perhaps projectors will ride some of this wave. A book-sized projector projecting from ceiling to wall might easily be more inexpensive than the large piece of glass or plastic that is part of a large flat-panel display.

Organic light-emitting diode displays (OLEDs) are another potential challenger to current flat panels. The display can be manufactured as a very thin sheet, leading to the potential of a "wallpaper display." The current displays are small and of relatively low resolution, but the technology is certainly worth the time to track.

Another area is personal displays. For years, a variety of researchers and companies have demonstrated direct imaging on the retina. To date, none have made a major market impact. Nevertheless, small devices can be made inexpensively, and a practical solution like a display integrated into a lightweight pair of glasses could be successful.

Also, a variety of technologies have been developed for 3D displays. Some have spinning screens or stacked flat-panel displays. Today the resolutions are relatively low, but a good 3D display has the potential to be very useful for large data visualization and analysis.

The existing and emerging technologies open opportunities to reengineer the office workspace. Henry Fuchs, working with other staff and students at the University of North Carolina at Chapel Hill, has been exploring future office environments for many years [10,11]. The goal is to apply display, teleconferencing, and high-performance graphics technology to deliver enhanced office work environments. Sandia National Laboratories

Figure 25.6 A picture of one of Sandia's "Office of the Future" test beds. (See also color insert.)

has been teaming with Fuchs' team and developing unique solutions for a future office environment.

Figure 25.6 shows the office test bed with a large high-resolution display for individual work and team meetings. A private system (the laptop) is available for private e-mail or it can act as an additional display when the user is working a large problem. On the back wall is a videoconference display and camera to expand interactions beyond the walls of the room. A goal of this "Office of the Future" test bed is to demonstrate that a next-generation office should use the technology to enhance office environments beyond a single display and a keyboard on a desk.

The future of large data should be considered. Large datasets today reach beyond a terabyte. Datasets of this size are certainly difficult to visualize and analyze. Through innovation, the computer technology is growing to meet the ever-increasing sizes. The human, however, has not changed much. Certain databases, like the Human Genome database, demonstrate the difficulty in making sense of all the data.

As datasets reach petabytes, it is likely that computer systems will play an increasing role in "data discovery." Scientific visualization supports information discovery by presenting information to the user to analyze, but with human attention and senses limited, new ways to better present large datasets will be needed. Scientific visualization took discovery past tables of numbers and curve fits. Perhaps scientific visualization will, in the future, evolve into the related field called Information Visualization (Info-Vis).

25.10 Conclusion

According to an old saying, "By the time an engineer completes an undergraduate engineering degree, he knows nothing. All he has learned is just enough that a senior engineer can now communicate with him." This chapter echoes that saying in some ways. Computer technology is advancing at such a rate that discussions of specific hardware and software will quickly become dated. A PC desktop of today may give way to a game box in each office. The core methods and concepts, however, will remain relevant. The goal of this chapter has been to provide a foundation that spans the major issues of office desktop delivery to aid construction of systems today and tomorrow.

Large data does not seem to be going away. As desktops grow in capability, our desire to tackle ever-larger problems advances too. For decades to come, people will likely be working in their offices or may be at home and looking to understand large databases. To provide these capabilities, full systems are likely to be needed. Networks, servers, and desktops must function together to bring large data access to the office.

Average consumers, with their combined spending power, are devouring new technology and driving rapid advances. Large data analysis and visualization must find ways to leverage these technologies or be left in the past with antiquated equipment. With careful observation of technology markets and some creative innovation, visualization and computer experts can find ways to solve the next generation of large-data visualization and analysis problems. Major technology changes are a way of life. Success is dependent on our harnessing this change to speed the understanding of complex phenomena hidden in the massive data.

References

1. United States Department of Commerce, Bureau of Economic Analysis, http://www.bea.gov/bea/dn/gdplev.xls
2. Myricom Home Page, http://www.myri.com
3. Quadrics Home Page, http://www.quadrics.com
4. Infiniband Trade Association, http://www.infinibandta.org/home
5. Computational Engineering International, http://www.ceintl.com/
6. Sandia News Center, "Be There Now," http://www.sandia.gov/news-center/news-releases/2002/comp-soft-math/remote-viz.html
7. The Chromium Project, http://sourceforge.net/projects/chromium

8. Paraview,   http://www.paraview.org/HTML/Index.html

9. Distributed Multihead X, https://sourceforge.net/projects/dmx/

10. R. Raskar, G. Welch, M. Cutts, A. Lake, L. Stesin, and H. Fuchs. The office of the future: a unified approach to image-based modeling and spatially immersive displays, *ACM SIGGRAPH 1998*, pages 179–188, 1998.

11. Henry Fuchs' home page, http://www.cs.unc.edu/~fuchs/

26 Techniques for Visualizing Time-Varying Volume Data

KWAN-LIU MA and ERIC B. LUM
University of California at Davis

26.1 Introduction

Our ability to study and understand complex, transient phenomena is critical to the solution of many scientific and engineering problems. Examples include data from the study of neuron excitement, crack propagation in a material, evolution of a thunderstorm, unsteady flow surrounding an aircraft, seismic reflection from geological strata, and the merging of galaxies. A typical time-varying dataset from a computational fluid dynamics (CFD) simulation can contain hundreds of time-steps, and each time-step can have more than millions of data points. Generally, multiple values are stored at each data point. As a result, a single dataset can easily require hundreds of gigabytes to even terabytes of storage space, which creates challenges for the subsequent data-analysis tasks.

The ability of scientists to visualize time-varying phenomena is absolutely essential in ensuring correct interpretation and analysis, provoking insights, and communicating those insights to others. For instance, by directly and appropriately rendering a time-varying dataset, we can produce an animation sequence that illustrates how the underlying structures evolve over time. In particular, interactive visualization is the key that allows scientists to freely explore in both spatial and temporal domains, as well as in the visualization parameter space, by changing view, classification, colors, etc. However, rendering of time-varying data requires the reading of large files continuously or periodically throughout the course of the visualization process, preventing it from achieving interactive rendering rates.

In this chapter, we describe how time-varying volume data can be efficiently rendered to achieve interactive visualization, with a focus on employing data encoding, hardware acceleration, and parallel pipelined rendering. Careful encoding of the data can not only reduce storage requirements but also facilitate subsequent rendering calculations. We provide a survey of encoding methods for time-varying volume data. Modern PC graphics cards can be used to render volume data at highly interactive rates. When rendering time-varying data, it is crucial to accelerate the loading of each time-step of the data into the video memory from either main memory or disk. A hardware decoding approach makes it possible to transport and render a much smaller, compressed version of the data instead, which enables interactive exploration in the spatial and temporal domains. Most rendering calculations can be straightforwardly parallelized to increase rendering rates while maintaining high image quality, but, again, the I/O issue must be addressed so that the volume data and image transfer rate can keep up with the rendering rate. A pipelined approach coupled with image compression is demonstrated to reach the highest possible efficiency.

26.2 Characteristics of Time-Varying Volume Data

When designing visualization techniques for time-varying data, one should take the characteristics of the data into consideration. Features of interest in a time series might exhibit a regular, periodic, or random pattern. A regular pattern is characterized by a feature that moves steadily through the volume. The feature's structure neither varies dramatically nor follows a periodic path. Features exhibiting a periodic pattern appear and disappear over time. Transient features of interest or features that fluctuate randomly are common, such as those found in turbulent flows. Generally, we can more easily detect and more efficiently render regular and periodic patterns.

It is also necessary to take into consideration the data value ranges of the datasets. For example, large value ranges must be treated with care; otherwise, many important features in later time-steps may become invisible. Fig. 26.1 plots the maximum and minimum values for each time-step of two datasets. The left one shows values of a turbulence flow dataset that consists of 81 time-steps. It has an extremely wide value range. The plot of the other dataset shows a much smaller value range that would help simplify some of the temporal-domain

visualization calculations. More detailed statistical information about a dataset, if available, could provide hints regarding how to derive more effective visualization in a more efficient manner.

Ideally, visualizing time-varying data should be done while data is being generated, so that users receive immediate feedback on the subject under study and so that the visualization results, rather than the much larger raw data, can be stored. Even though simulation-time visualization is a promising approach, considering the wide variety of possible data characteristics it is clear that exploratory data visualization must also be made available. That is, scientists need to have the capability to repeatedly explore the spatial, temporal, and parameter spaces of the data. In particular, most scientists run their large-scale simulations on parallel supercomputers operated at national supercomputer centers and also store the output data there. Data visualization and understanding are mostly done as post-processing tasks because either the simulation itself or the data transport is not real-time and thus does not

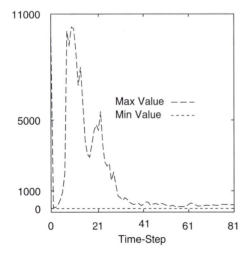
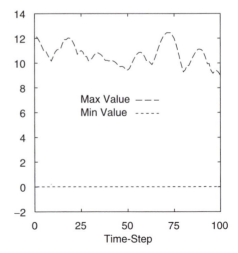

Figure 26.1 (Left) Maximum and minimum values at each time-step of a dataset from the study of the generation and evolution of turbulent structures in shear flows. Early time-steps contain values in a very large value range, which makes quantization more difficult. (Right) Maximum and minimum values at each time-step of a dataset from the study of coherent turbulent vortex structures. This dataset has a small value range, and the distribution of values is quite uniform.

allow interactive visualization. While this chapter focuses on postprocessing visualization techniques, some of the techniques we introduce are also applicable to simulation-time visualization.

26.3 Encoding

The size of a time-varying volume dataset can be reduced and therefore made more manageable through the use of either value-based encoding or physically based feature extraction methods. Value-based encoding methods transform (and potentially compress) data by exploiting coherence. Physically based methods extract data features, such as vortices or shocks, and represent them in a compact fashion. This section discusses value-based encoding techniques for time-varying volume data. Topics in feature extraction are covered by other chapters of this book.

First, the reduction in storage requirements can be used to fit more time-steps into a computer's main memory, enabling interactive browsing of the temporal domain of the data. Second, when the data must be transferred across a system bus, like from disk to main memory or from main memory to video memory, the transfer of reduced data can be done substantially faster than that of raw data. There are two basic approaches to value-based compression of time-varying data. The first approach is to treat time-varying volume data as 4D data. For example, the 4D data can be encoded with a 4D tree (an extension of an octree) and an associated error/importance model to control compression rate and image quality [28]. A more refined design is based on a *4th-root-of-2* subdivision scheme coupled with a linear B-spline wavelet scheme for representing time-varying volume data at multiple levels of detail [12]. Another way to treat 4D data is to slice or volume-render in the 4D space. The resulting hyperplane and hyperprojection present unique space–time features [29].

The second approach to time-varying data encoding is to separate the temporal dimension from the spatial dimensions. A simple difference encoding that exploits the data coherence between consecutive time-steps can result in a significant reduction, but it is limited to a sequential browsing of the temporal aspect of the data [25]. A more flexible encoding method couples nonuniform quantization with octree encoding for spatial-domain data compression, and difference encoding for temporal-domain compression [18]. Rendering can become optimal when neighboring voxels are fused into macrovoxels, if these voxels have similar values, and subtrees at consecutive time-steps are merged if they are identical. Fig. 26.2 displays the performance of such a renderer compared to a brute-force renderer for rendering a turbulent jet dataset. A subsequent design based on a hierarchical data structure called a *time-space partitioning* (TSP) tree [24] can achieve further improvement in utilization of both spatial and temporal coherence.

The TSP tree is a time-supplemented octree. The skeleton of a TSP tree is a standard complete octree, which means it recursively subdivides the volume spatially until all subvolumes reach a predefined minimum size. To store the temporal information, each TSP tree node itself is a binary tree. Every node in the binary time tree represents a different time span for the same subvolume in the spatial domain. The objective of the TSP design is to reduce the amount of data required to complete the rendering task and to reduce the volume-rendering time. Rendering with TSP trees essentially allows efficient traversals of regions with different spatial and temporal resolutions in order of increasing fidelity.

Several other techniques are also worth mentioning. One is based on wavelet compression to establish an underlying analysis model for characterizing time-varying data. Essentially, it involves separate wavelet encoding of each time-step to derive compressed multiscale tree structures [27]. By examining the

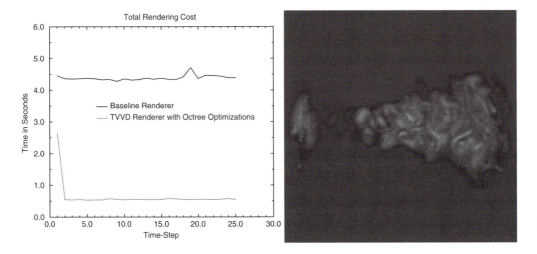

Figure 26.2   (Left) Rendering cost for a turbulent jet dataset using a ray-casting volume renderer. The time is the total time to process, including reading encoded data from disk, decoding when necessary, calculating the gradient, rendering, and compositing. (Right) One frame from the resulting animation. (See also color insert.)

resulting multiscale tree structures and wavelet coefficients, we can perform feature extraction, tracking, and further compression. It is also possible to compress time-varying iso-surfaces and associated volumetric features with wavelet transforms to allow fast reconstruction and rendering [26]. Finally, a technique based on shear-warp volume-rendering exploits temporal coherence to render only the changed parts of each slice and uses run-length encoding to compress the spatial domain of the data [1].

Whether one should treat time-varying volume data as 4D data depends on the characteristics of the data. For example, if the discrepancy between the temporal resolution and the spatial resolution is large, it could become very difficult to locate the temporal coherence in certain subdomains of the data; consequently, the time domain should be considered separately for encoding. Another problem with using 4D trees is that coupling spatial and temporal domains makes it difficult to locate regions with temporal coherence but not spatial coherence.

26.4   Interactive Hardware-Accelerated Rendering

Commodity PC graphics cards have been effectively used for volume-rendering static volumetric data [4,10,14]. Even though real-time rendering rates can be achieved, the volumetric data must be loaded into the video memory prior to the rendering. Since the access and transfer of data from main memory across the graphics bus to the graphics card is relatively slow compared to the direct access of video memory, the need to load each time-step of the data into video memory can limit rendering performance for time-varying volume data. Reducing the amount of data that must be transferred to the video memory seems to be the most effective way to remove this data-transport bottleneck. As discussed previously, time-varying data can be reduced in size and therefore made more manageable through the use of compression. However, one key requirement here is that the compressed volume data be uncompressed in the video memory. This section describes a technique allowing decoding and rendering of time-varying volume data in hardware, delivering the desirable interactive rendering rates [15].

26.4.1   Compression

The advantages of compressing volumetric data are twofold. First, the compression reduces the storage requirements needed for the data. This could allow a dataset to fit in main memory that might otherwise not fit, eliminating the need to transfer data from disk. The reduction in storage can also be used to fit relatively small compressed datasets entirely in texture memory, thus eliminating the need to transfer data across the graphics bus. The other to benefit of compression is a reduction in I/O. If a compressed volume fits entirely in main memory, the cost of transferring compressed data to the graphics card is lower than the cost of transferring uncompressed data. If a dataset does not fit into main memory, the transfer of compressed data from disk can be substantially faster than with uncompressed data, allowing for interactive visualization from disk.

We therefore treat video and main memory together as a two-level cache for volume-rendering. The compression of volumetric data not only increases the amount of data that can fit in each level but also decreases the I/O costs of transfers between these levels. Through the use of compression, and careful management of the time costs associated with the transfers between levels, it is possible to load texture maps representing volume data into video memory at rates suitable for interactive visualization.

If a compressed volume is to be rendered directly, from video memory, it must also be uncompressed using the graphics hardware. This is a significant constraint, since the operations supported in video hardware are extremely limited compared to those of a general-purpose CPU. Another constraint is imposed by the fact that it is very desirable to encode the scalar voxel values in terms of their scalar value rather than as a red, green, blue, alpha (opacity) set. Using scalar values and color indexed textures allows a user to manipulate the color palette to interactively change the opacity and color maps, permitting exploration of the data's transfer-function space. Storing voxels in terms of RGBA would require recompression of the entire dataset as parameters are changed, which can be impractical for very large datasets. In addition, storing a single scalar value, rather than four color scalars, reduces the amount of data by a factor of four.

26.4.2   Palette-Based Decoding

With the limitations of graphics hardware in mind, we need a method for the temporal encoding of indexed volumetric data that can quickly be decoded in hardware. A viable approach is to make extensive use of hardware support for the changing of color palettes without the reloading of textures. The cycling of color palettes can be used to create simple animations from static images. Similarly, we can use color-palette manipulation to allow a single scalar index to represent grid-points at several time-steps.

With paletted textures, a single scalar index is used to represent an RGB or RGBA color. The palette consists of a limited set of colors that sample the RGBA color space. Each of these colors is encoded in a single value, often a single byte. Consequently, we can encode a sequence of temporally changing scalar values into a single index. In this way, the value stored in each texel represents an approximation of a sequence of scalar values. Each index is therefore a sample in the space of possible time-varying scalar values. The scalar values that an indexed texel represents are decoded to its temporally changing values through the frame-to-frame manipulation of the palette. For each frame, the color for each palette entry is set to the color found in the transfer function for the scalar encoded by that index value during that frame, as shown in the following pseudo-code, which renders N time-steps using a single indexed texture. Note that 8-bit indexed textures are assumed.

```
{
    // stores color map from the transfer
    function
    Color color map [256];
    // stores the N time-varying scalars
    encoded by each
    // of the 256 possible texel values.
    This array is created
    // during the compression process
    int decoder [N] [256];
    // the color palette to be calculated
    for each time-step
    Color palette [256];

    for each time-step t (0 to N-1) {
        for each palette entry i (0 to 255) {
            palette[i] =color map[ decoder
            [t] [i] ] ;
        }
        // set the palette for the current
        frame
        setPalette (palette);
        renderTexture ();
    }
}
```

The textures are rasterized to the screen using linear interpolation. In contrast to dependent textures, for paletted textures, post-classified linear interpolation occurs in terms of RGBA values after the table lookup. If interpolation occurred in terms of palette indices, the resulting images would show severe artifacts since the mapping between palette indices and decoded scalar values is not linear.

26.4.3 Temporal Encoding

The encoding process consists of mapping sequences of scalars into single scalar indices. This operation can be approached as a vector quantization problem and can be solved using a variety of techniques. One method for quantizing the sequences we describe in this section uses transform encoding, specifically the *discrete cosine transform* (DCT) [8,9,23]. Transform encoding is a compression method that transforms

data into a set of coefficients that are then quantized to create a more compact representation. The transform by itself is reversible and does not compress the data. Rather, a transform is selected that puts more energy into fewer coefficients, thus allowing the less important, lower-energy coefficients to be quantized more coarsely, and thus requiring less storage.

The DCT is defined by

$$C(u) = \alpha(u) \sum_{x=0}^{N-1} f(x) \cos\left[\frac{(2x+1)u\pi}{2N}\right] \qquad (26.1)$$

and

$$\alpha(u) = \begin{cases} \sqrt{\frac{1}{N}} & \text{for } u = 0 \\ \sqrt{\frac{2}{N}} & \text{for } u = 1, 2 \ldots, N-1 \end{cases} \qquad (26.2)$$

where $C(u)$ are the transformed coefficients, N is the number of input samples, and $f(x)$ are the input samples. DCT is chosen because it is known to have good information-packing qualities and tends to have less error at the boundaries of a sequence [8].

In the encoding process, first a window size is selected, which will be the length of the time sequence that will be encoded into a single value. The longer the window size, the greater the compression that will be achieved at the expense of temporal accuracy. For each window of time-evolving scalars, the DCT is applied. The result is a set of coefficients equal in number to the size of the window used. The first coefficient stores the average value over the window and tends to be largest in value. The remaining coefficients store increasingly higher frequency components contained in the windowed sequence. These coefficients tend to represent decreasing amounts of signal as the frequency gets higher.

These coefficients are then quantized and combined into a single scalar value. Bits are adaptively allocated for each coefficient based on the variance of each coefficient [23]. Those coefficients of high variance are allocated more bits than those coefficients of low variance.

Using this technique, bits are allocated based on the temporal characteristics of the windowed sequence of the dataset. For example, a dataset with minimal amounts of movement would use fewer bits to store the temporal changes in the data, allowing more bits to be used to more precisely represent the stationary values in the sequence. On the other hand, a sequence with high speed motion (low temporal coherence) would use more bits to encode this motion at the expense of precision for the static values.

Once bit allocation for the transformed coefficients is determined, the coefficients are quantized to their respective precisions. Uniform quantization is not well suited for quantizing these coefficients, since they often have fairly nonuniform distributions. Instead, quantization occurs using Lloyd–Max quantization [13,21], which adaptively selects quantization levels that minimize mean squared error. The quantized coefficients are then combined into a single scalar, which is stored as an index in a paletted texture.

26.4.4 Texture Implementation

As described in the previous section, the quantization step can be adapted based on the characteristics of the dataset. Since the temporal properties can vary widely across a volume, to reduce the amount of error from quantization it is advantageous to independently compress small sections of a volume, each with its own set of palettes. The volume can be subdivided into 3D blocks stored as 3D textures, or as view-aligned slabs stored as 2D textures. Although the palette-based temporal compression technique can be applied to both 3D and 2D textures, if 3D textures are used and the volume is subdivided into blocks, the borders between those blocks can become visible. This is because the same scalar value can map to differing colors depending on its block. Decomposing the volume into view-aligned 2D slices has the advantage of having these quantization discontinuities occur along the view direction, which is made less noticeable by the volume-rendering integral. The 2D texture also has the advantage of being a feature more widely supported on more graphics cards, often with faster rasterization than with 3D textures. The use of 2D textures has the disadvantage of requiring three view-aligned copies of the dataset for viewing from an arbitrary direction.

Usually, when bit allocation occurs, most bits are used for storing the average value over a windowed sequence. As a result, when the transition occurs between two compressed sequences, the shift in average value can cause a perceived jump in the animation. With 2D textures, this can be fixed by interleaving the starting times of the time windows for each slice. Fig. 26.3 shows such an interleaving scheme. This decorrelates temporal transitions so that the jump occurs during every frame, but for interleaved slices in the volume, rather than the whole volume. This is analogous to interlaced video, except rather than being interlaced vertically, the textures are interleaved along the viewing direction. As with per-slice quantization, the volume-rendering integral helps to soften the interleaving effect.

For a transform window of length N, without interleaving, an entire new compressed volume must be loaded every N frames. Since the loading of data across the graphics bus is relatively slow, this can cause a noticeable drop in frame rate every N frames. This problem can be solved by loading $1/N$ of the next compressed volume every frame but requires storage of a copy of the next volume in texture memory. This, however, is not necessary if the textures are interleaved, since for every frame $1/N$ of the volume can be flushed from texture memory and replaced with a new texture. Thus, by amortizing data movement costs, interleaving allows for a more consistent frame rate without the expense of needing the texture memory to store a second compressed volume. If the user moves to an arbitrary frame nonsequentially, then all textures must be reloaded, and there is a drop in frame rate.

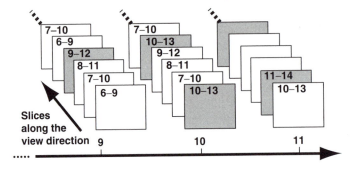

Figure 26.3 When 2D texture interleaving is utilized, for every time-step, every Nth 2D texture is replaced starting with the t modulo Nth texture slice, where t is the time-step and N is the compression ratio. In this example, N is four. The numbers on each slice indicate which time-steps the texture stores. The shaded slice is the slice that is updated at time t.

26.4.5 Performance

Here we present the performance of the renderer on a low-cost PC configured with an AMD 1.2 GHz Athlon processor, 768 megabytes of main memory, an Nvidia GeForce 3 based graphics card with 64 megabytes of texture memory, and an IDE level 0 Raid (4 drives). Using the hardware-accelerated decoding and rendering method, it is possible to render moderate-resolution, time-varying volumetric datasets at interactive rates.

The test results were from the rendering of a quasi-geostrophic (QG) turbulent flow dataset consisting of 1492 time-steps provided by researchers at the National Center for Atmospheric Research (NCAR). Its spatial resolution is $256 \times 256 \times 256$. The QG calculations simulate large-scale motions in the Earth's atmosphere and oceans and are representative in size and complexity of many earth sciences turbulent fluid-flow simulations. Fig. 26.4 shows selected time-steps of the data.

Table 26.1 shows frame rates for different compression cases using NCAR's QG dataset. Compressing each time-step of a 256^3 QG data set takes between 5 and 15 seconds, depending on the level of compression, using an approximation of Lloyd–Max Quantization [9]. The implementation uses 8-bit paletted textures,

although our technique could be applied to hardware that supports higher-precision textures for encoding strategies that allocate more bits to each transformed coefficient. The results were obtained when rendering the volume to a 512×512 window, with the volume occupying approximately one-third of the window area.

If a compressed dataset fits entirely in main memory, then the bottleneck in the rendering process is the transfer of textures from main memory to the graphics card. Compression helps with both of these limitations, not only increasing the number of time-steps that fit in main memory, but also decreasing the amount of time necessary for transferring data across the graphics bus. If only one set of axis-aligned textures is stored in main memory, then the number of time-steps that can be stored in memory increases by a factor of three at the expense of the user's ability to view the dataset from an arbitrary angle without swapping data from disk.

In the case of the 256^3 volumetric QG data, using a compression factor of four and 256 axis-aligned textured polygons, we can fit 140 time-steps into main memory and sustain a frame rate of approximately 25.8 frames/s. If 128 axis-aligned textured polygons are used instead, which requires only half the data to be trans-

Contained in this second color section are Figures 22.2 through 41.6.

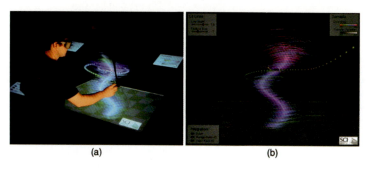

(a)   (b)

Figure 22.2 (a) A user explores a tornado dataset on the Visual Haptic Workbench. (b) Screenshot of the same application running in a desktop environment. Dataset courtesy of R. Crawfis, Ohio State University, and N. Max, Visualization Laboratory, Lawrence Livermore National Laboratory.

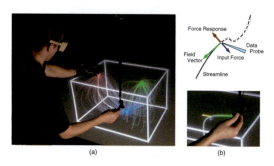

(a)   (b)

Figure 22.3 (a) A user explores a volumetric vector dataset. (b) The data probe is constrained along a streamline, resulting in intuitive haptic feedback.

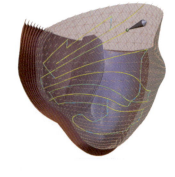

Figure 22.6 Exploring epicardial muscle fibers with haptic feedback. The probe is constrained to follow the local fiber orientation on the surface of a single layer. The user can "pop through" to a neighboring layer by pushing against the surface. Similarly, the user can choose a different fiber by pushing perpendicular to the currently selected fiber while staying on the surface. This effect feels as if the surface were textured with tiny valleys and ridges. The image shows the path of the proxy colored according to the magnitude of the applied force component perpendicular to the fiber orientation and tangent to the surface, from yellow to cyan, indicating increasing tension between the probe and the proxy. The dataset consists of about 30,000 nodes and 200,000 tetrahedral elements. Dataset courtesy of P. Hunter, Bioengineering Institute, University of Auckland.

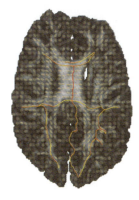

Figure 22.7 Exploring a 148×190 DT-MRI slice with haptic feedback. The ellipses represent local diffusion anisotropy and orientation. Lighter areas have higher associated anisotropy. The proxy path is colored according to the magnitude of the applied force, from yellow to red, indicating a larger tension between the probe and the proxy. The curves are tangent to the direction of the major eigenvector of the diffusion-tensor matrix in anisotropic areas. Dataset courtesy of G. Kindlmann and A. Alexander, W. M. Keck Laboratory for Functional Brain Imaging and Behavior, University of Wisconsin–Madison.

Figure 23.1 Downtown Los Angeles, from the Virtual LA project [6].

Figure 23.2 High-resolution aerial imagery merged with georeferenced models automatically extracted from LIDAR data [66].

Figure 23.7 (Left) Original mesh with and without textures; (Right) same view with 1-pixel threshold.

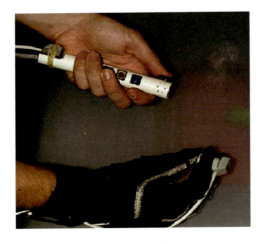

Figure 23.11 Button stick (top) and glove with finger contacts (bottom): two modes for 6 DoF interaction.

Figure 23.17 Isosurfaces of rainfall patterns over a terrain model.

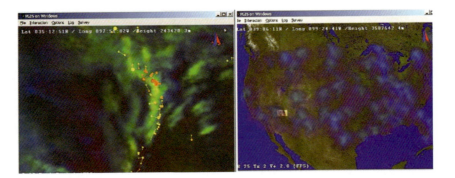

Figure 23.18 (Left) Overview of more than 40 radars in the eastern U.S.; (Right) close-up of storm front after zoom in.

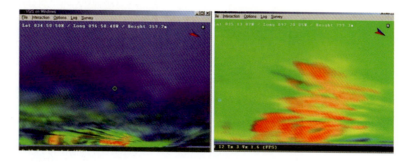

Figure 23.19 (Left) Ground-level view of the 3D structure of a severe storm cell. (Right) Close-up view of the same cell.

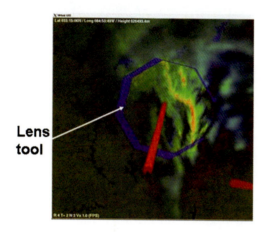

Figure 23.20 3D decision support tools.

Figure 23.21 The fly-in to a close-up view of downtown Atlanta.

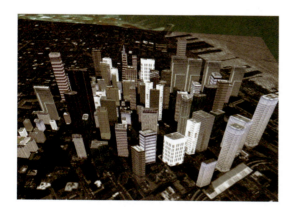

Figure 23.22 The fly-in to downtown San Francisco.

Figure 24.2 A user in a CAVE interacting with geophysical data describing a salt dome in the Gulf of Mexico.

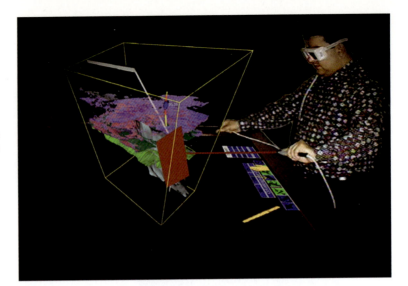

Figure 24.3 A user interacting with geophysical data on a projection workbench.

Figure 25.6 A picture of one of Sandia's "Office of the Future" test beds.

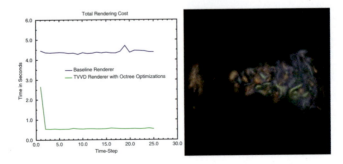

Figure 26.2 (Left) Rendering cost for a turbulent jet dataset using a ray-casting volume renderer. The time is the total time to process, including reading encoded data from disk, decoding when necessary, calculating the gradient, rendering, and compositing. (Right) One frame from the resulting animation.

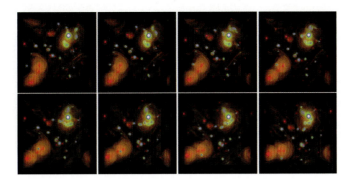

Figure 26.4 Visualizations of time-steps 1120 through 1155 of the QG dataset in 5-time-step intervals.

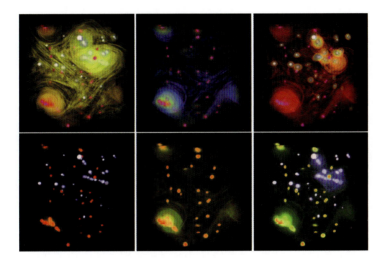

Figure 26.5 Selected visualizations of the QG dataset produced by varying transfer functions.

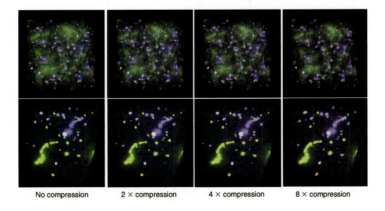

No compression   2 × compression   4 × compression   8 × compression

Figure 26.6 (Top) Visualizations of time-step 210 of the QG dataset at different compression levels. As the level of compression increases, some of the finer features become blurred. (Bottom) Visualizations of time-step 970.

Figure 27.3 The progression produced by the data streaming of eight subset stages of an entropy isosurface, colored by subset, from the Terascale Supernova Initiative.

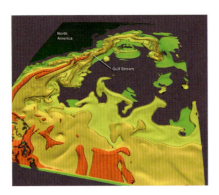

Figure 27.7 An example of the isosurfaces produced by the task-parallelism example.

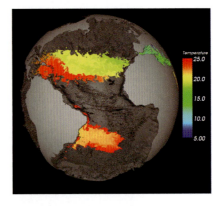

Figure 27.10 Salinity isosurface colored by temperature, produced using the Full POP dataset.

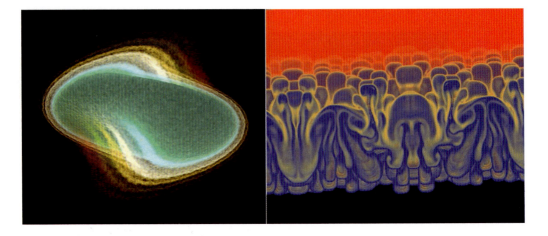

Figure 27.14 Two volume rendered images produced by TRex. The image on the left represents the particle density produced by a simulation of a linear accelerator model using 300 million particles. The image on the right shows a classic Raleigh-Taylor fluid dynamics simulation.

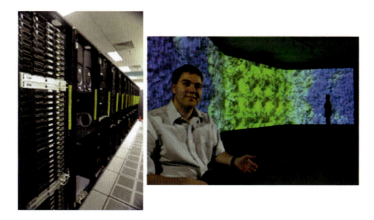

Figure 28.8 Sandia's 128-node graphics cluster (left) and 60-million-pixel display wall (right). The wall is driven by graphics-cluster technology. The image on the wall is a rendering of the 470-million-polygon isosurface extracted from the PPM dataset described in Section 28.2.

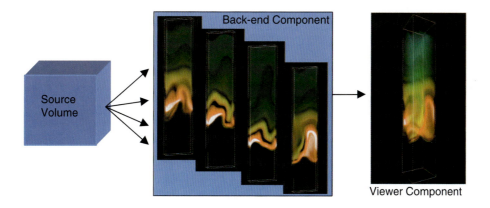

Figure 29.1 Visapult consists of a viewer component and a back-end component.

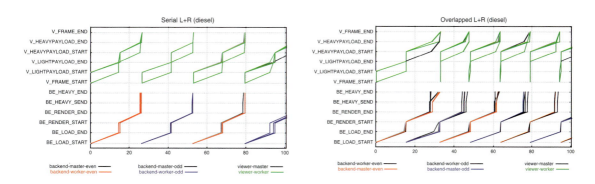

Figure 29.4 Visapult duty cycle with serial I/O and rendering.

Figure 29.5 Visapult duty cycle with overlapped I/O and rendering.

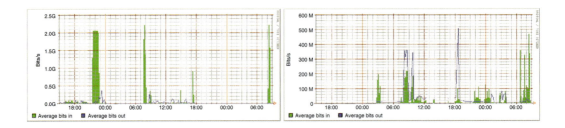

Figure 29.9 These are images of the MRTG performance graphs from the incoming data streams arriving from Esnet OC-48 and NCSA OC-12, respectively, for the SC 2001 Bandwidth Challenge. The challenge took place for approximately 7 minutes at 8:30 a.m. The performance MRTG graphs actually underreport the bandwidth utilization because they use a 5-minute average. The instantaneous bandwidth (which also included an additional ESNet OC-12) reached 3.3 Gbits/s. The 2+ Gbit peaks on the ESNet graph were practice runs using Visapult.

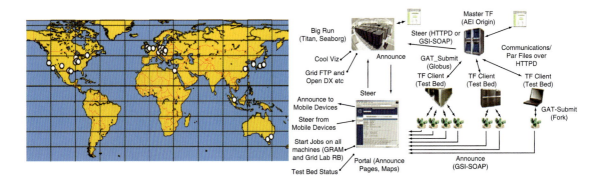

Figure 29.10 (Left) A map of sites participating in the SC 2002 Bandwidth Challenge Global Grid Testbed Collaboration. (Right) A the logical workflow of the Grid's task-farming scenario. The task farmer uses an application-level abstraction layer for Grid Services called the Grid Application Toolkit (GAT). For more information on the GAT, please refer to *http://www.gridlab.org*

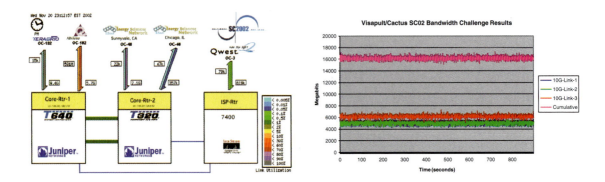

Figure 29.12 SCinet weather map.   **Figure 29.13** SC 2002 Bandwidth Challenge results.

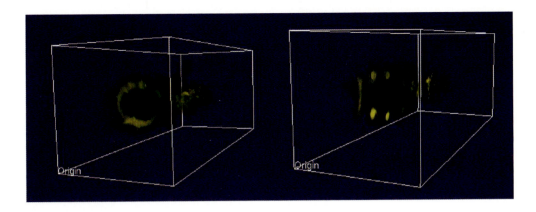

Figure 29.14 Visapult's new "Omniview" capabilities (left) produce much better visual fidelity than the original IBRAVR algorithm.

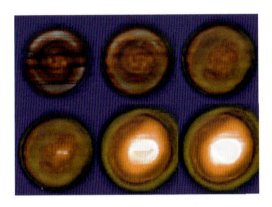

Figure 29.15 Visapult renderings of Cactus's 3D Wave Equation Evolution.

Figure 30.1 Rendering classes.

Figure 30.7 Streaming pieces of a sphere.

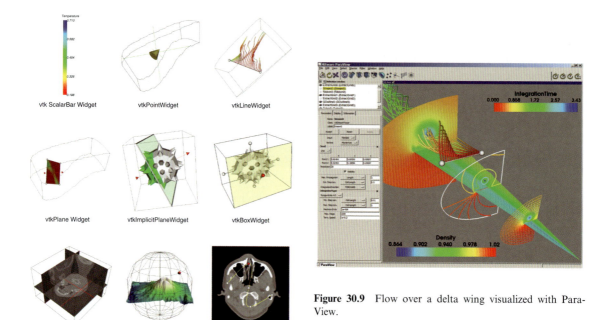

vtk ScalarBar Widget   vtkPointWidget   vtkLineWidget

vtkPlane Widget   vtkImplicitPlaneWidget   vtkBoxWidget

vtkImagePlaneWidget   vtkSphereWidget   vtkSplineWidget

Figure 30.9 Flow over a delta wing visualized with Para-View.

Figure 30.8 Palette showing some of VTK's 3D widgets.

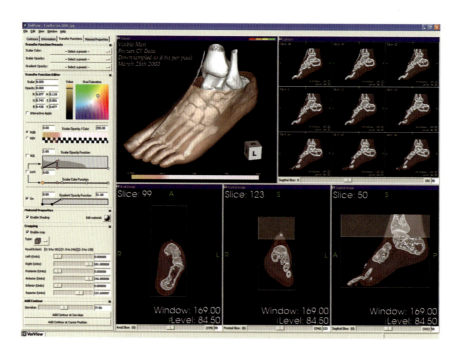

Figure 30.10 Visible Human dataset visualized with VolView.

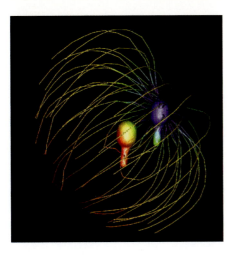

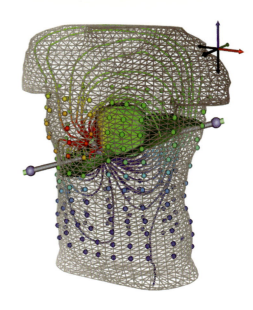

Figure 31.4 The simulation of two Aplysia motor neurons using the bridging capabilities between Genesis and SCIRun. First, Genesis solves the time-dependent Hodgkin-Huxley equations for each compartment in each cell. One result of the Genesis simulation is a calculation of neuron membrane current density, which is passed to SCIRun through a SQL database. SCIRun uses the current density to solve the forward field problem in the volume surrounding the cell. In this picture, streamlines show current flow within the volume; voltage is encoded by color (blue is negative; red is positive). Image courtesy of Chris Butson, University of Utah, Department of Bioengineering.

Figure 31.6 A visualization of a SolveMatrix convergence. In this case, an electric dipole is placed near the heart inside the Uintah torso model. The electric field set up by this dipole is visualized by color-mapped streamlines. Also visualized is the surface potential (indicated by color-mapped spheres on the surface of the torso) and a field potential isosurface (green surface inside the torso). This visualization was produced using the BioPSE forward-fem net with the movable dipole widget.

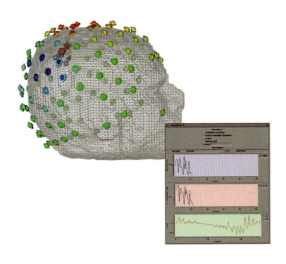

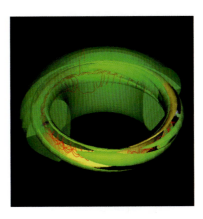

Figure 31.7 A visualization of an inverse EEG simplex search produced by the dipole-localization net. The accuracy of the solution at each electrode position is also shown (disks show measured voltage; spheres show computed voltage), the view window shows simplex dipoles as four arrows (connected with lines), and the test dipole is shown as a sphere. Also shown are the error metrics plotted for this particular solution.

Figure 31.8 A simulation of an experiment inside a Tokomak Fusion Reactor, visualized with the SCIRun Fusion package. This image is of NIMROD simulation data showing an isosurface of the n = 0 part of the pressure field (yellow), which shows the 1/1 structure, and an isosurface of the n = 2 part of the toroidal current field (green), which shows the developing 3/2 structure. Between the two isosurfaces is a streamline using the sum of the n = 0, n = 1, and n = 2 modes of the magnetic field (red). The underlying model consists of a toroidal grid with 737,280 nodes (in 10 arbitrary phi slices) with 22 time-slices. Image provided by Dr. Allen Sanderson and the National Fusion Collaboratory, with support from the US Department of Energy SciDAC initiative.

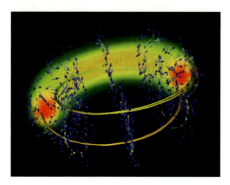

Figure 31.9 A simulation of an experiment inside a Tokomak Fusion Reactor, visualized with the SCIRun Fusion package. This is frame one of two from a time sequence showing the stochastic nature of the real-space magnetic field lines. A comparison of the two frames shows that the magnetic field lines start to diverge earlier as time progresses. The field lines are overlaid in a volume rendering of the pressure field, which provides visual cues to the location while rendering the plasma velocity-vector field. Image provided by Dr. Allen Sanderson and the National Fusion Collaboratory with support from the US Department of Energy SciDAC initiative.

Figure 31.10 A simulation of a heptane pool fire. The simulation was done using the Uintah derivative of SCIRun, and visualization was done using the ray-tracing package described in Section 31.4.4.2. This image is courtesy of the Center for the Simulation of Accidental Fire and Explosions (C-SAFE), working under funding from the Department of Energy as part of the Accelerated Strategic Computing Initiative (ASCI) Academic Strategic Alliance Program (ASAP).

Figure 31.11 A maximum intensity projection (MIP) of the 1-GB Visible Female dataset using ray-tracing. The MIP algorithm seeks the largest data value that intersects a particular ray. Ray-tracing allows interactive visualization of the MIP of this dataset.

Figure 31.12 Visualization of the Utah Electrode Array embedded in the cochlear nerve of a cat. The insert is a picture of the Utah Electrode Array. The 27-micron resolution of the CT scan allows definition of the cochlea, the modiolus (right), the implanted electrode array, and the lead wires (in purple) that link the array to a head-mounted connector. The resolution of the scan even allows definition of the shanks and tips of the implanted electrode array. Volume rendering also allows the bone to be rendered as translucent (as on the left half of this image), enabling the electrode to be clearly viewed. Thus, the combination of high-resolution scanning, image processing, and interactive visualization tools such as raytracing allows noninvasive verification of the implantation site in an anatomical structure that is completely encased in the thick temporal bone. Data provided by Dr. Richard Normann and Dr. Charles Keller, University of Utah.

Figure 32.2 Running a map in IRIS Explorer. The Module Librarian is at top left, and the Map Editor is at top middle. The map that is currently running is analyzing results from a simulation of oil transport through a permeable medium. The region is divided up into cells, each of which has a permeability value. The data is read into IRIS Explorer as a pyramid and is then culled using the **CullPyr** module to find only those cells whose data values are outside some range—this is an example of the filtering step in the visualization pipeline of Fig. 32.1. The **CullPyr** control panel, containing the widgets used to set its parameter values, is at bottom left, while the large display window comes from the **Render** module. The small window at top right is an editor for the headlight in the 3D scene; this is an Open Inventor control that is exposed in IRIS Explorer because it uses Open Inventor for 3D scene generation and display (Section 32.4.1).

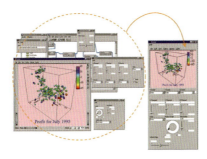

Figure 32.3 Building an application in IRIS Explorer by module grouping. On the left, a map is shown running in the map editor. The connections between the modules can be seen in the background at the top, and the control panels of several modules surround the large control panel from the **Render** module, which contains the visualization. The same application is shown on the right after the modules have been grouped together (and optionally compiled into a single executable). The group control panel contains widgets from the original modules' control panels–including the large window from **Render**. Here, IRIS Explorer is running in application mode which hides the development windows from view; the only thing appearing on-screen is the group control panel. This application is displaying stock market data as a 3D scatter plot.

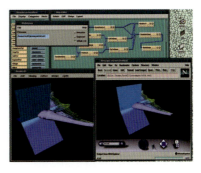

Figure 32.4 Using the **WriteGeom** module to output a scene from IRIS Explorer as VRML 2.0 (a.k.a. VRML97). The **Render** window is at bottom left, containing the scene, which consists of an isosurface, probe, and sheet geometry. In the map, the same inputs are passed to the **CombineGeom** module, which creates a single scene and passes it to **WriteGeom** for output to file. The file is rendered using a VRML browser in the bottom right-hand corner; here, the VRML is being displayed on the same machine as the original visualization, but publishing it on the Web would, of course, allow it to be displayed anywhere.

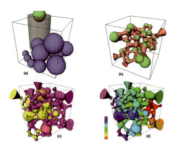

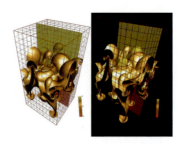

Figure 32.5 Using IRIS Explorer in pore-scale simulations of fluid flow [33]. (a) Creating a sphere pack by "dropping" particles at randomly chosen locations. The shaded volume is searched for the lowest position. (b) The network of nodes (green) and channels (pink) found in the sphere pack. The rendering uses simple primitives (cones) to approximate some of the geometrical information encoded in the pyramid. (c) Multiphase fluid flow through the network. Yellow fluid has displaced purple fluid in parts of the channels. Nodes, containing a mixture of the two phases, are colored according to pressure. (d) The same network as in (c), but now with all parts (nodes and channels) colored according to pressure.

Figure 32.6 Using IRIS Explorer to display the results of a simulation of the Rayleigh-Taylor instability in the interface between two fluids [35]. The upper fluid has a density of 3 units, the lower has a density of 1 unit, and the isosurface has been calculated for a threshold value of 2; thus, it follows the fluid–fluid interface irrespective of its shape. Here, the simulation cell has been added to orient the display, and a semi-transparent cutting plane has been used to show how density varies in a plane passing through the interface. Two views of the cell for a given time-step are displayed here that clearly show the characteristic loops, spikes, and bubbles in the surface.

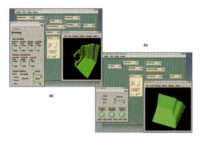

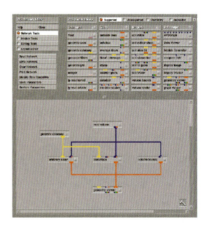

Figure 32.9 Performing a collaborative simulation of atmospheric dispersion using IRIS Explorer. The collaboration is envisioned as being between a numerical analyst—whose screen is displayed in (a)—and a chemist, who sees the screen shown in (b). The numerical analyst is running the simulation and viewing the error from the solution, while the chemist is viewing the computed solution as an isosurface.

Figure 33.1 The AVS network editor.

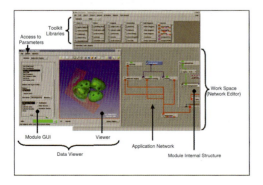

Figure 33.2 The AVS/Express visual editor.

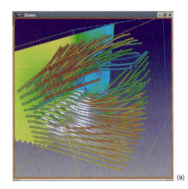

Figure 33.3 (a) Illuminated streamlines model.

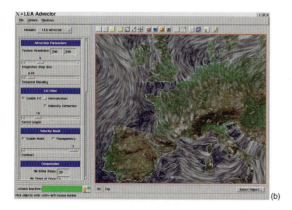
(b)

Figure 33.3 (b) Visualization of time-dependent wind fields with the IAC LEA module.

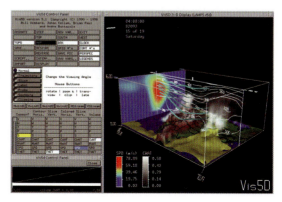

Figure 34.1 A Vis5D display combining terrain, isosurface, volume rendering, particle trajectories, pseudo-coloring, and contour curves.

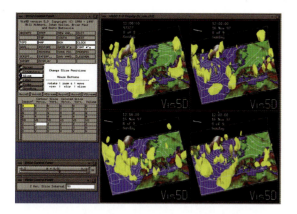

Figure 34.2 A Vis5D spreadsheet of four members of an ensemble forecast from the European Centre for Medium-Range Weather Forecasts.

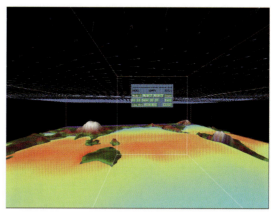

Figure 34.3 A planar simulation of a Cave5D virtual-reality display of coupled atmosphere and ocean simulations used for a demo at the Supercomputing '95 conference.

Figure 34.4 A VisAD display combining a GOES satellite image with map boundaries.

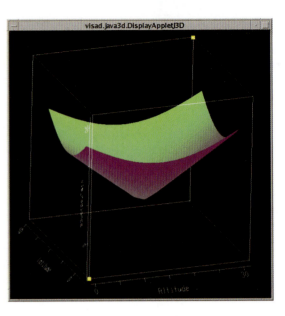

Figure 34.5 A simple demonstration using VisAD to build embedded 3D user interface components. Dragging the two large yellow points causes the display to rescale to keep the yellow points at opposite box corners.

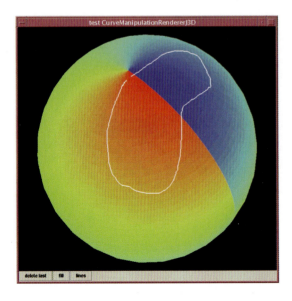

Figure 34.6 A more complex embedded user interface component built using VisAD. The user can draw freehand curves on the sphere's surface.

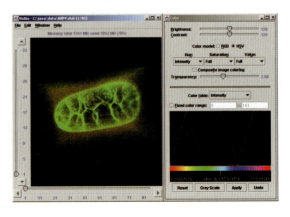

Figure 34.7 The VisBio system, using VisAD for volume rendering of a live *C. elegans* embryo stained with a fluorescent membrane probe. Imaging done by Dr. William Mohler of the University of Connecticut Health Center.

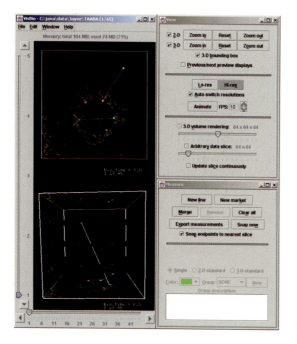

Figure 34.8 The VisBio system using VisAD for 3D and 4D measurements quantitating movement of mitochondria in different regions of a two-photon dataset of a two-cell hamster embryo labeled with a mitochondria-specific dye. Imaging done by Dr. Jayne Squirrell of the University of Wisconsin at Madison.

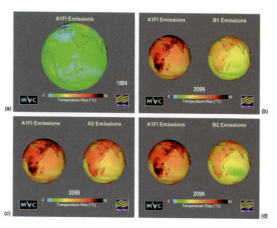

Figure 35.2.1 Comparison of SRES scenarios for global warming. Data courtesy of the Hadley Centre, UK Meteorological Office.

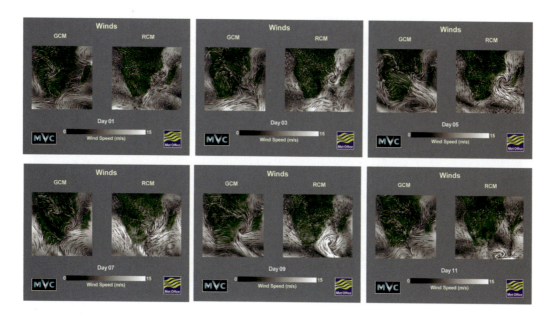

Figure 35.2.2 A series of frames from an animation comparing the Global Climate Model (GCM) and the Regional Climate Model (RCM) for simulation of a cyclone in the Mozambique Channel. Data courtesy of the Hadley Centre, UK Meterological Office.

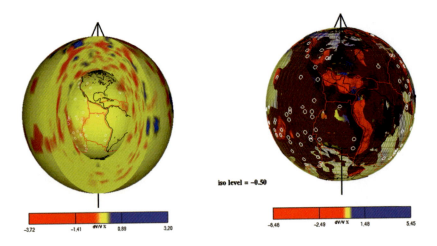

iso level = −0.50

Figure 35.3.1   3D images of seismic tomography data of the Earth's mantle.

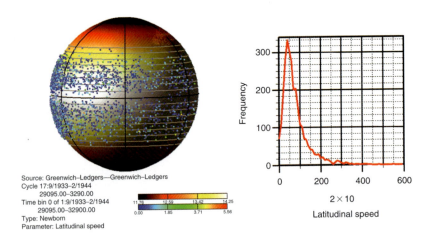

Source: Greenwich–Ledgers—Greenwich–Ledgers
Cycle 17:9/1933–2/1944
29095.00–3290.00
Time bin 0 of 1:9/1933–2/1944
29095.00–32900.00
Type: Newborn
Parameter: Latitudinal speed

Figure 35.3.2   The location bias for newborn sunspot observations.

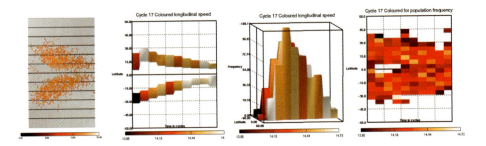

Figure 35.3.3   Plots based on the traditional butterfly plot. The left-most image is a 3D plot that is viewed as a traditional butterfly plot.

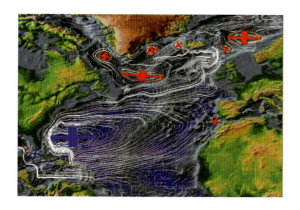

Figure 35.4.2 Simulation of the Atlantic Ocean, with stream-lines and vortex features visualized by ellipses. Red and blue ellipses rotate in opposite directions.

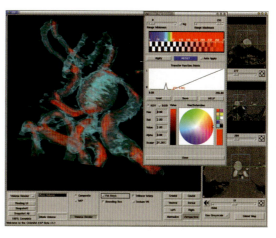

Figure 35.6.1 Blood flow visualized with software volume rendering.

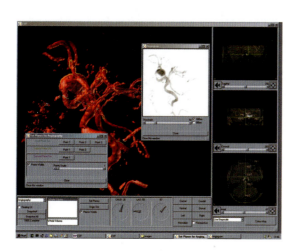

Figure 35.6.2 Angiography simulation.

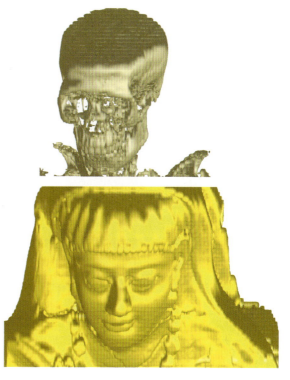

Figure 35.7.1 Mummy 1766.

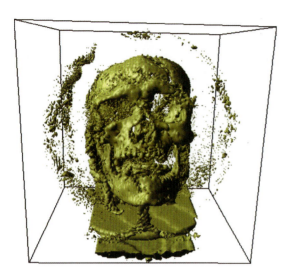

Figure 35.7.2 Visualization of Worsley Man showing the noise in the data.

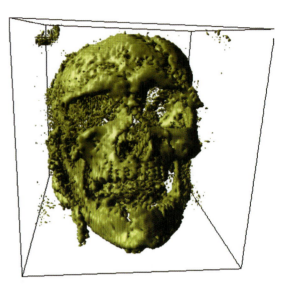

Figure 35.7.3 Result of crop on Worsley Man.

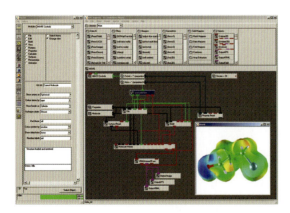

Figure 35.8.2 An example application constructed using standard and application-specific AVS/Express modules, with output structure and properties in the lower-right window.

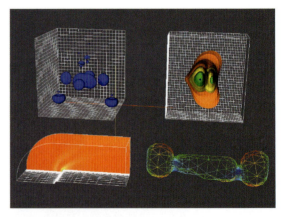

Figure 36.1 The figure shows the different types of datasets that VTK and ParaView can handle. The upper-left dataset is a uniform rectilinear volume of an iron potential function. The upper-right image shows an isosurface of a non-uniform rectilinear structured grid. The lower-left image shows a curvilinear structured grid dataset of airflow around a blunt fin. The lower-right image shows an unstructured grid dataset from a blow-molding simulation.

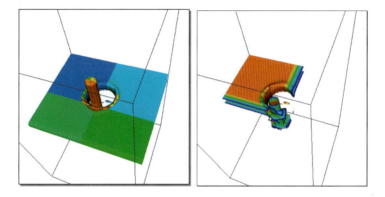

Figure 36.2 The image on the left was generated by a CTH dataset partitioned into eight pieces. Each piece was assigned a different color. The image on the right shows only one partition with six extra ghost levels. The cells are colored by ghost level. In practice, usually only one ghost level is necessary.

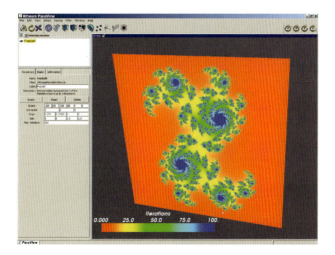

Figure 36.3 ParaView.

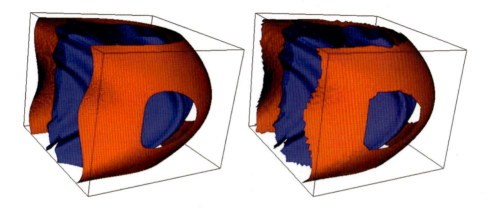

Figure 36.5 The full-resolution surface (left) has 821,495 triangles and renders in intermediate mode in 0.93 seconds on a GeForce2 GTS, and the decimated surface (right) has 35,400 triangles and renders in 0.04 seconds.

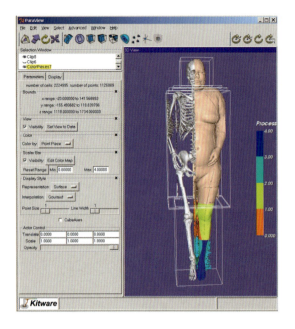

Figure 36.6 The Visible Woman dataset in ParaView.

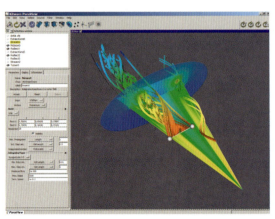

Figure 36.7 The delta wing dataset in ParaView.

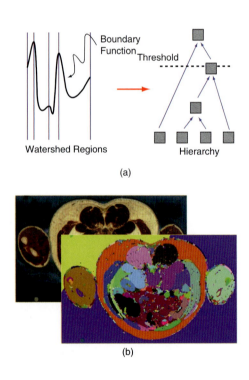

(a)

(b)

Figure 37.1 Watershed segmentation. (a) A 1D height field (image) and its watersheds with the derived hierarchy. (b) A 2D VHP data slice and a derived watershed segmentation. Example courtesy of Ross Whitaker and Josh Cates, University of Utah.

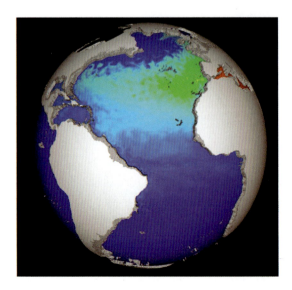

Figure 36.8 View of the Atlantic Ocean from a global one-tenth-of-a-degree simulation showing ocean temperature at a depth of 1140 meters generated by a ParaView batch script.

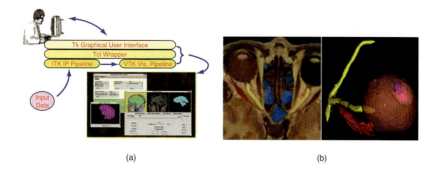

(a)   (b)

Figure 37.2 An ITK watershed segmentation application with visualization provided by VTK and with the user interface provided by Tcl/Tk. This example shows how ITK can be integrated with multiple packages in a single application. (a) Dataflow and system architecture for the segmentation user application. (b) Anatomic structures (the rectus muscle, the optic nerve, and the eye) highlighted with color overlays (shown in 2D but extending in the third dimension) and the resulting visualization. Example courtesy of Ross Whitaker and Josh Cates, University of Utah.

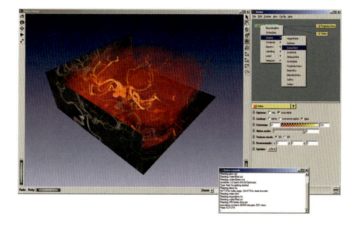

Figure 38.1 The amira user interface consists of the main window with the object pool and the control area, a large 3D graphics window, and a console window for messages and command input. The figure shows a medical image dataset from MR angiography, visualized by a projection view module and a volume rendering module.

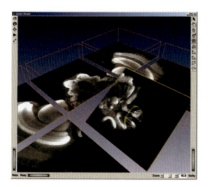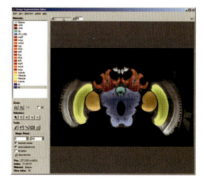

Figure 38.2 (Left) Multipart 3D confocal image stack of a bee brain visualized by slices. The orientation of an oblique slice can be easily adjusted with a trackball dragger. (Right) The amira segmentation editor. This component allows the user to identify and separate different objects in a 3D image stack. Here different parts of the bee brain have been segmented.

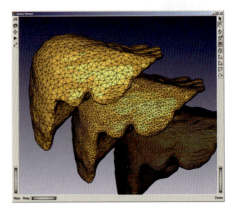

Figure 38.3 (Left) Reconstructed polygonal surface model of a bee brain; a volume rendering of the image data has been superimposed, showing how well the model matches the image data. (Right) Reconstructed model of a human liver, displayed at three different resolutions (1 cm maximum edge length, 0.5 cm maximum edge length, and surface at original resolution with 0.125 cm voxel size).

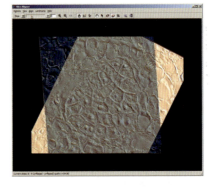

Figure 38.4 (Left) Tetrahedral finite-element grid of a human body embedded in a device for hyperthermia treatment. Only parts of the grid are shown, in order to reveal interior structures. (Right) The amira slice aligner, an interactive tool for aligning 2D physical cross-sections.

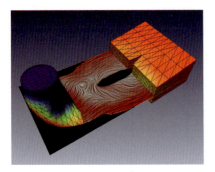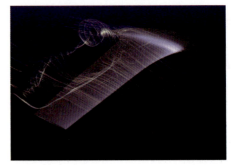

Figure 38.5 (Left) Visualization of a turbine flow using the fast line integral convolution method in a user-selected plane. (Right) Visualization of the 3D flow around an airfoil using the illuminated streamline technique.

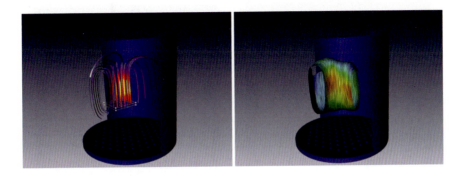

Figure 38.6 Visualization of fluid flow within a bioreactor. (Left) Streamribbons starting in the interactively positioned seed box. (Right) Streamsurface with the tangential flow depicted by line integral convolution (LIC).

 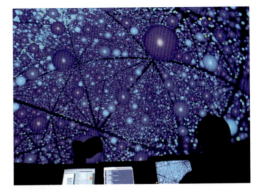

Figure 38.7 (Left) Coating on a car body investigated with amiraVR. For every module, a 3D version of its respective user interface can be used. This allows full control of amira from within an immersive environment. (Right) Atoms in a crystal lattice shown in a dome using amiraVR. The dome was illuminated by six laser projectors with partially overlapping images, with five arranged in a circle and one at the top. Image courtesy of Carl Zeiss Jena.

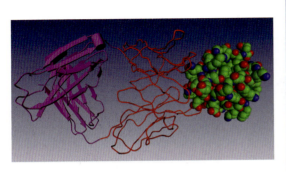 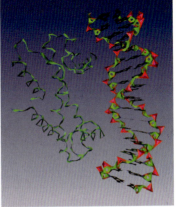

Figure 38.8 (Left) Complex consisting of a mouse antibody and an antigen of Escherichia Coli. The secondary structure representation is on the left side, and the pure backbone representation is in the middle. Van der Waals balls depict the antigen. (Right) Bond angle representation of a complex of the factor for inversion simulation (FIS) protein and a DNA fragment (colored according to atom types).

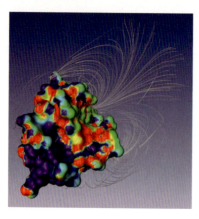 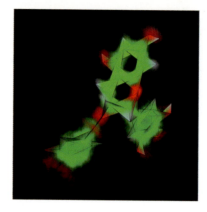

Figure 38.9 (Left) Solvent accessible surface of the ribonuclease T1, pseudo-colored according to the molecule's electrostatic potential; illuminated streamlines depict the electrostatic field. (Right) Configuration density and superimposed bond angle representation of Epigallocatechin-Gallat.

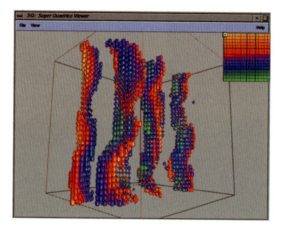

Figure 39.1 Visualization of a magnetohydrodynamics simulation of the solar wind in the distant heliosphere showing both velocity components and vorticity components of the vortex tubes.

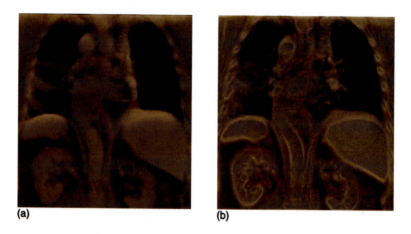

Figure 39.2 Comparison of (a) realistic rendering of an abdominal CT dataset with (b) a volumetric illustration rendering using silhouette and boundary enhancement.

Figure 39.3 Rendering of segmented kidney CT dataset showing selective volumetric illustration enhancement applied only to the right kidney to focus the viewer's attention.

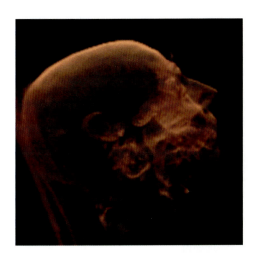

Figure 39.4 Interactive volume illustration rendering of a head dataset with silhouette and boundary enhancement.

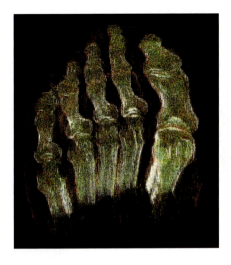

Figure 39.6 Volume stipple rendering of a foot dataset showing tone shading, distance color blending, silhouette and boundary enhancement, and silhoutte lines.

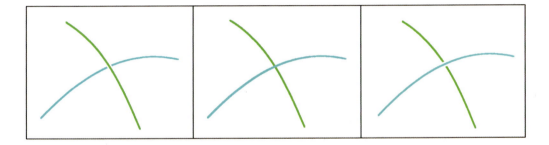

Figure 40.1 Two overlapping lines of roughly equivalent luminance but differing hue, shown in three subtly different depictions.

Figure 40.2 Since prehistoric times, artists have used gaps to indicate the passing of one surface behind another, as shown in this photograph taken in a Smithsonian museum exhibit reproducing the creation of the second "Chinese Horse" in the Painted Gallery of the cave of Lascaux. Photograph by Tomás Filsinger.

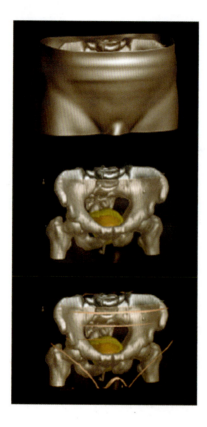

Figure 40.5 By selectively highlighting valley lines, we aim to enhance the perception of important shape features on a transparent surface in a visualization of 3D radiation therapy treatment planning data. (Top) The skin rendered as a fully opaque surface. (Middle) The skin rendered as a fully transparent surface, in order to permit viewing of the internal structures. (Bottom) The skin rendered as a transparent surface with selected opaque points, located along the valley lines. The bottom view is intended to facilitate perception/recognition of the presence of sensitive soft tissue structures that should be avoided by radiation beams in a good treatment plan. Data courtesy of Dr. Julian Rosenman. Middle and lower images © IEEE.

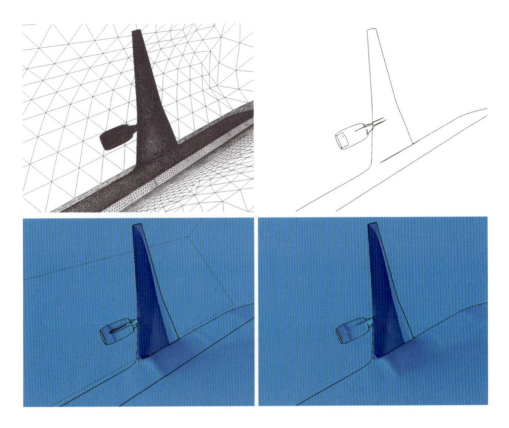

Figure 40.6 By highlighting feature lines on a surface mesh, we may enhance appreciation of the essential structure of the form, a goal that assumes particular importance under conditions where the use of surface shading is problematic. Clockwise from top left: the full original mesh; the silhouette and contour edges only; the silhouette and contour edges highlighted in a surface rendering in which polygon color is defined purely as a function of the value of a scalar parameter that is being visualized over the mesh; same as previous, except that the feature line set is augmented by crease edges determined by our algorithm to be locally perceptually significant. Data courtesy of Dimitri Mavriplis.

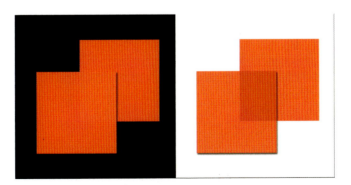

Figure 40.7 Examples of two different models of transparency. (Left) Additive transparency, exhibited by materials such as gauze that are intrinsically opaque but only intermittently present. (Right) Multiplicative or subtractive transparency, exhibited by materials, such as colored glass, that selectively filter transmitted light.

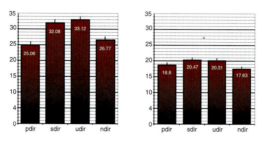

Figure 40.13 Pooled results (mean angle error) for all subjects, all surfaces, by texture type. (Left) flat presentation; the differences {pdir,rdir} < {sdir, udir} were significant at the 0.05 level. (Right) stereo presentation; only the differences {rdir < sdir, udir} were significant at the 0.05 level.

Figure 40.14 Representative stimuli used in our second experiment to investigate the relative extents to which differently oriented patterns have the potential to mask the perceptibility of subtle shape differences. The texture conditions are, from left to right, principal direction, swirly direction, and uniform direction.

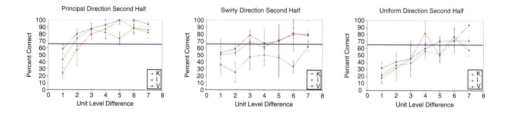

Figure 40.15 Summary results of our second experiment. Accuracy increased with increasing magnitude of shape difference in all cases, but it increased at a faster rate under the principal direction texture condition. Error bars represent 95% confidence intervals.

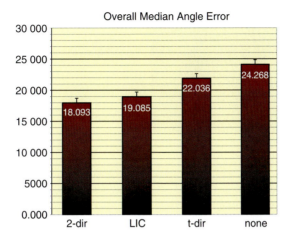

Figure 40.17   The results of the third cumulative experiment.

Figure 41.1   Part of the frieze from Cap Blanc.

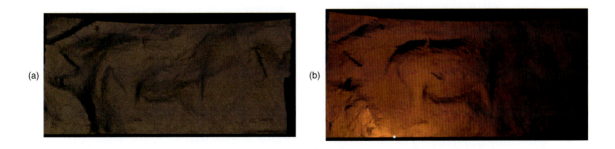

Figure 41.2   (a) The frieze lit by modern 55 w lighting. (b) The frieze lit by an animal tallow candle as used 15,000 years ago.

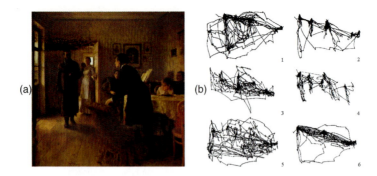

Figure 41.3 (a) Repin's picture *An Unexpected Visitor*. (b) Effects of task on eye movements. Repin's picture was examined by subjects with different instructions: 1. Free viewing, 2. Judge their ages, 3. Guess what they had been doing before the unexpected visitor's arrival, 4. Remember the clothes worn by the people, 5. Remember the position of the people and objects in the room, and 6. Estimate how long the unexpected visitor had been away from the family [30].

Figure 41.4 The scene visualized with a closeup of the mug showing the pencils and paintbrushes.

Figure 41.6 Visual angle covered by the fovea for mugs in the first two rooms at 2° (green circles) and blending circle at 4.1° (red circles).

(a)   (b)

Figure 41.5 (a) High-quality (HQ) image (Frame 26 in the animation). (b) Closeup of HQ and low-quality (LQ) rendered chairs.

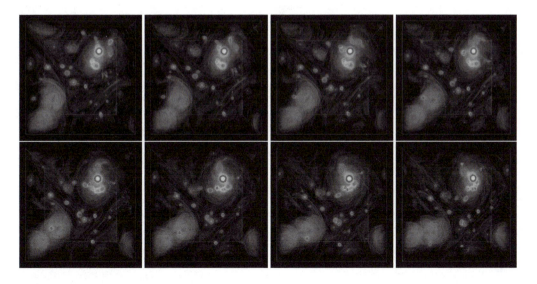

Figure 26.4   Visualizations of time-steps 1120 through 1155 of the QG dataset in 5-time-step intervals. (See also color insert.)

ferred and drawn, the frame rate doubles and we can render 280 time-steps from memory. Without compression, the same 140 time-steps no longer fit into main memory. A memory-resident subset of the uncompressed data can be rendered at only about 11.5 frames/s, compared to 25.8 frames/s with compression. We note that although the amount of data transferred with compression is one-fourth of that without, the frame rate does not scale linearly. This is due to the time required to rasterize the textured polygons to the screen. The performance would scale more linearly if a graphics card with a higher fill rate were used, or if the fill-rate requirements were reduced by projecting the volume to a lower-resolution display.

Often the temporal resolution of a dataset is too large to fit the desired number of time-steps into main memory, even with compression. In this case, it is necessary to load and render the volume from disk. Compression can substantially decrease the amount of data that must be loaded for each frame, resulting in a noticeably higher frame rate, as shown in Table 26.1. For example, all 1492 timesteps of the 256^3 QG

dataset can be rendered at 13.4 fps when compressed by a factor of eight, versus only 2.0 frames/s when rendered uncompressed from disk. Once the user finds a shorter temporal region of interest, that data can then be loaded into main memory and rendered at a faster frame rate or higher image fidelity. Fig. 26.5 shows select visualizations of the QG dataset for the same time-step using different transfer functions defined through interactive exploration.

Table 26.1   Frame rates for rendering the QG data with different compression levels

Compression ratio	FPS (Time-steps rendered)	
	In core	Out of core
8×	31.6 (280)	13.4 (1492)
4×	25.8 (140)	6.8 (1492)
2×	17.3 (70)	3.5 (1492)
1×	11.5 (35)	2.0 (1492)

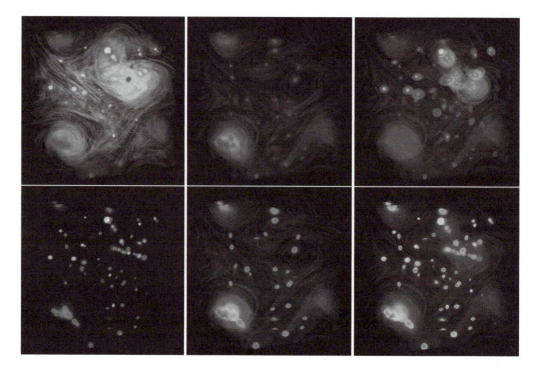

Figure 26.5   Selected visualizations of the QG dataset produced by varying transfer functions. (See also color insert.)

By changing the window size used in the encoding step, we can vary the compression ratio and quality. Table 26.2 shows the peak signal-to-noise ratio that results from compressing each dataset over 50 time-steps. Fig. 26.6 shows volumes that have been rendered using varying degrees of compression. As the amount of compression increases, some of the more subtle features as well as the faster-moving features can become blurred. Thus, there is a distinct trade-off between the compression ratio and rendering performance versus the quality of the compressed volume. This gives users a degree of flexibility in choosing compression ratios that best meet their needs. For example, if a scientist is interested in viewing a short time sequence at high quality, a lower compression ratio can be used. On the other hand, to view a very long sequence of data at high speeds, a higher compression rate can be selected. The scientist can combine compression ratios to pre-

Table 26.2   NCAR QG dataset error

Compression ratio	PSNR (dB)>
2×	41.1
4×	35.5
8×	32.1

view a dataset at a coarser temporal resolution and then view a specific time sequence of interest with less compression.

26.4.6   Discussion

The current approach is very scalable with respect to the temporal size of a dataset. With regard to the size of the dataset in the spatial domain, the amount of texture memory on a

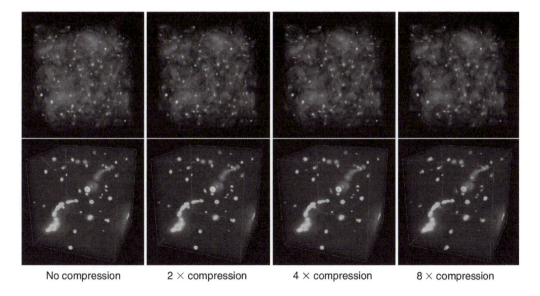

| No compression | 2 × compression | 4 × compression | 8 × compression |

Figure 26.6   (Top) Visualizations of time-step 210 of the QG dataset at different compression levels. As the level of compression increases, some of the finer features become blurred. (Bottom) Visualizations of time-step 970. (See also color insert.)

single card can be a limiting factor. Only compressing temporally does not reduce the amount of texture memory utilized to below that which would be required to render a single static volume. For out-of-core rendering, the cost of swapping textures from the graphics card to main memory is much lower than the cost of reading from disk; thus, texture memory capacity restraints become less of a concern. One way to support datasets whose spatial resolutions exceed what a single graphics card is capable of handling is to cluster multiple PCs together. Clustering effectively increases the amount of aggregate texture memory available, and, just as importantly, increases the aggregate bandwidth between all the levels in the storage hierarchy.

The use of compression here presents two potential shortcomings that are worth addressing. First, since the compression scheme is lossy, there is the potential for modest, but noticeable, image-quality degradation that increases with the degree of compression. However, a moderate loss of image fidelity due to compression or other optimization strategies is an acceptable tradeoff for enabling interactive exploration of temporal data, provided the gross features of evolving structures are preserved, as they are in the test cases. It is worth noting that many NCAR researchers commonly perform crude data reduction using simple zero-order subsampling in order to accommodate interactive exploration with the tools presently available to them. In essence, they have already demonstrated a willingness to sacrifice image quality to gain interactive exploration capabilities that are essential for maximizing scientific productivity. Once a feature of importance is detected in the reduced dataset, the full-resolution data may be further analyzed if necessary. Second, compression requires additional storage (for maintaining both the raw version and the compressed version of the data), and it takes time to perform the encoding. Similar to the loss of image fidelity, researchers are already bearing these costs by their use of subsampled data to achieve interactive rendering.

26.4.7   Future Work

Methods based on TSP trees reduce the amount of texture memory utilized by exploiting temporal and spatial coherence to reuse textures [7,24]. They represent several similar textures as a single static texture. The DCT-based encoding method stores several time slices in terms of lower precision averages and differences stored in a single texel. Through palette manipulation, these texels dynamically represent several time slices. This compressed encoding comes at the expense of the numerical precision used to store these averages and differences. The method exploits temporal coherence by using more bits to represent the average value over a set of slices, but it also reserves bits for storing the change over a set of slices. The method could be combined with TSP-based techniques to store textures at varying degrees of both spatial and temporal resolution. A preliminary study comparing the error that results from the use of the temporal compression method with the error produced by storing volumetric datasets at reduced spatial resolutions shows that for datasets with fine features, temporal compression produces significantly less error than spatial downsampling for a given bitrate. However, under some circumstances, spatial downsampling can produce less error, particularly when the volumetric features are relatively large and smooth. Often, the characteristics of a volume are mixed, clearly indicating a future research direction towards the combination of the temporal compression technique with varying degrees of spatial resolution. These results suggest that the DCT-based hardware-accelerated technique could be combined with the TSP tree–based methods to achieve better overall efficiency.

A 3D texture implementation of the DCT-based method has been done with lighting added [22]. Its performance then was not comparable to the 2D texture version, due to the higher cost of the rasterization of 3D textured polygons.

26.5   Parallel Pipelined Rendering

Many scientists are relying on centralized supercomputing facilities to conduct large-scale computational studies. The data produced by their simulations are stored in the mass storage devices at the supercomputer center. Due to the size of the datasets, it is convenient to also use a parallel computer at the supercomputer center for visualization calculations and to transfer the smaller images rather than the raw data to the scientist's desktop computer for viewing. This section discusses design criteria for realizing remote visualization of time-varying volume data using a parallel computer. The resulting design is a parallel pipelined renderer that achieves optimal processor utilization and fast image-display rates [16].

A typical parallel volume-rendering process consists of four steps. The data-input step reads data elements from disk and distributes them to the processors. Each processor receives a subset of the volume data. In the following rendering step, each processor renders the assigned subvolume into a 2D partial image that is independent of other processors. Next, a combining step, which generally requires interprocessor communication, composites the set of 2D partial images (according to the view position) to derive the final 2D projected image. The final step delivers the final image to a display or storage device.

When the degree of parallelism is small to modest, e.g., under 16 nodes, the major portion of the computational cost is attributed to *subvolume rendering*. However, when the degree of parallelism is high or when the dataset itself is large (say 512^3 or 1024^3 voxels per time-step), *3D data distribution* would become a significant performance factor.

Parallel rendering of time-varying data involves rendering multiple data volumes in a single task. There are three potential performance metrics: start-up latency, the time until the rendered image of the first volume appears;

overall execution time, the time until the rendered image of the last volume appears; and interframe delay, the average time between the appearance of consecutive rendered images. In conventional volume-rendering applications, since only one dataset is involved, start-up latency and overall execution time are the same, and interframe delay is irrelevant. When interactive viewing is desired, start-up latency and interframe delay play a crucial role in determining the effectiveness of the system. When visualization calculations are done in a batch mode, overall execution time should be the major concern. Different design tradeoffs should be made for different performance criteria.

Given a generic parallel volume-renderer and a P-processor machine, there are three possible approaches to managing the processors for rendering time-varying datasets. The first approach simply runs the parallel volume-renderer on the sequence of datasets one after another. At any point in time, the entire P-processor machine is dedicated to rendering a particular volume. That is, only the parallelism associated with rendering a single data volume, i.e., *intravolume* parallelism, has been exploited.

The second approach takes the exact opposite approach by rendering P data volumes simultaneously, each on one processor. This approach thus only exploits *intervolume* parallelism, and is limited by each processor's main memory space.

To attain the optimal rendering performance we should carefully balance two performance factors, resource utilization efficiency and parallelization overhead; this suggests exploiting both intravolume and intervolume parallelism. That is, instead of using all the processors to collectively render one volume at a time, a pipelined rendering process is formed by partitioning processors into groups to render multiple volumes concurrently. In this way, the overall rendering time may be greatly minimized because the pipelined rendering tasks are overlapped with the I/O required to load each volume into a group of processors (Fig. 26.7); moreover, parallelization overhead may

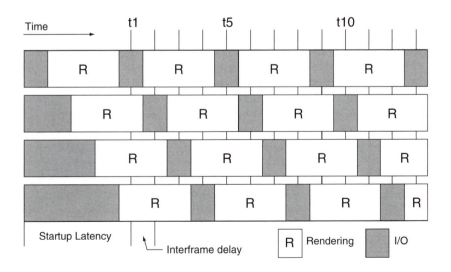

Figure 26.7 Processor grouping and pipelined rendering that overlaps rendering calculations with I/O can achieve optimal performance.

be reduced as a result of the partitioning of the processors.

The third approach is thus a hybrid one, in which P processor nodes are partitioned into L groups $(1 < L < P)$, each of which renders one volume (i.e., one time-step) at a time. The optimal choice of L generally depends on the type and scale of the parallel machine as well as the size of the dataset. The optimal partitioning strategy for minimizing the overall rendering time can be characterized with a performance model and revealed with an experimental study [5], which shows that the third approach indeed performs the best among the three for batch-mode rendering.

For remote, interactive visualization over a wide-area network, minimizing interframe delay becomes more important. The key is to not only find an optimal processor partitioning (as by studying the interplay between the rendering and I/O) but also develop an efficient image compression/transfer mechanism.

26.5.1   Image Transport

The mix of pipelined rendering with processor grouping makes possible overlapping data input, rendering, and image output, and therefore leads to optimal overall rendering performance in the absence of parallel I/O and high-speed network support. We assume that the volume dataset is local to the parallel supercomputing facility and is transmitted to the parallel renderer through fast local area networks (LANs). Note that the performance of a pipeline is determined by its slowest stage. The cost of the last stage of the pipeline—image output—cannot be ignored, since the resulting images must be assembled and transported to the desired display device(s) with minimal delay, possibly over a wide-area network. Fig. 26.8 shows such a setting for parallel pipelined rendering incorporating the processor grouping strategy.

It is relatively easy to implement the display program as an X-Window client such that X takes care of image transport, but the performance is not acceptable. Except with a very high-speed network or low-resolution images, a special display mechanism using compression and more clever buffering is required to deliver the desired frame rates.

For parallel rendering applications, we need image compression techniques that will compress with reasonable speed, exhibit good compression with short input strings, accept arbitrary orderings of input data, and decompress rapidly. The choice of a compression technique will also be influenced by factors such as rendering performance, network bandwidth, and image accuracy and quality. To date, ren-

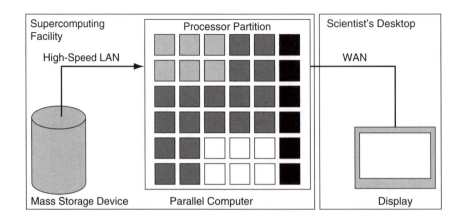

Figure 26.8   A parallel pipelined rendering system.

derer implementations exploiting image compression have mostly adopted relatively simple lossless schemes that rely on frame-differencing and run-length encoding, in part because they satisfy many of the desired criteria. While these techniques can usually deliver several frames per second over LANs, their compression ratios are highly dependent on image content and are insufficient for use on slower wide-area and long-haul networks.

It is therefore desirable to use lossy (visually lossless) compression methods capable of providing acceptable image quality for many applications, while retaining desirable properties such as efficient parallelizable compression, insensitivity to image organization, and, most importantly, rapid decompression.

26.5.1.1   A Framework for Image Transport

Compression can be done collectively by the rendering nodes or by a dedicated node for efficient image transport. Parallel compression would require decompression of multiple images at the receiving end, which usually uses a less powerful machine.

"Parallel compression" can be realized in two different ways. One is to have each processor compress a portion of the image independent of other processors. The other is to have all the processors collectively compress an image, which would require interprocessor communication. The latter would give the best compression results in terms of both quality and efficiency, but it is a less portable solution. Parallelizing compression calculations is itself an interesting research problem [25], and the goal is to minimize communication costs as much as possible. This section only discusses the former approach, which is easier to implement and can achieve the needed efficiency for fast image transport.

The compression-based image output stage is based on a framework consisting of three parts: renderer interface, display interface, and display daemon. The renderer interface provides each rendering node with image compression (if not done by the renderer) and communication to and from the display daemon. We can choose from a variety of image-compression methods and transport mechanisms according to visualization requirements.

The display interface provides three basic functions: image decompression, image assembly, and communication to and from the display daemon. The communication path can instruct the system to change the compression method, start the renderer, or pass a message directly to the renderer. The display daemon's main job is to pass images from the renderer to the display. It also allows the display to communicate with the renderer. In addition, the display daemon can accept any number of connections from the renderer interface and the display interface.

Interaction with the parallel renderer is provided by the display application. The user-interface tasks are split between the local controlling workstation and the parallel renderer. The display application drives the control panel display and passes events to the renderer though the display interface, and through the display daemon to all renderer interfaces in the form of "remote callbacks." The renderer responds with the appropriate action and may need to rerender the image.

26.5.1.2   Compression Methods

While image compression is a well developed field, progress is continuing due to the demand for higher compression performance from an increasing number of application areas. For example, JPEG-2000 [20] is an emerging standard for still image compression, and MPEG [11] is a standard for coding moving pictures. For remote time-varying visualization tasks, low cost is probably the most relevant selection criterion; low decompression cost is particularly important when considering our remote visualization setting because computing resources are generally low at the receiving end. This eliminates JPEG-2000 (based on wavelet transform) because of its relatively high computational complexity, even though it provides significantly lower distortion for the same bit rate. JPEG-2000 also requires more memory than JPEG.

Rendering time-varying data produces an animation sequence. MPEG, which is good for compressing existing videos, is not well suited for our interactive setting, in which each image is generated on the fly and is displayed in real time. Using MPEG is not completely impossible, but the overhead would be too high to make both the encoding and the decoding efficient in software.

Therefore, we consider three other more favorable compression methods: LZO, BZIP, and JPEG. LZO does lossless data compression, and it offers fast compression and very fast decompression. The decompression requires no extra memory. In addition, there are slower compression levels that achieve a quite competitive compression ratio while still decompressing at very high speed. In summary, LZO is well suitable for data compression or decompression in real time, which means it favors speed over compression ratio.

BZIP has very good lossless compression; it is better than *gzip* in compression and decompression time. BZIP compresses data using the Burrows-Wheeler block-sorting compression algorithm [3] and Huffman coding. Its compression is generally considerably better than that achieved by more conventional LZ77/LZ78-based compressors, and it approaches the performance of the PPM family of statistical compressors.

JPEG (*http://www.jpeg.org/*) is designed to compress full-color images of real-world scenes by exploiting known limitations of the human eye, notably the fact that small color changes are perceived less accurately than small changes in brightness. The most widely implemented JPEG subset is the "baseline" JPEG, which provides lossy compression, though the user can control the degree of loss by adjusting certain parameters. Another important aspect of JPEG is that the decoder can also trade off decoding speed against image quality by using fast but inaccurate approximations of the required calculations. Remarkable speedups for decompression can be achieved in this way.

Consequently, JPEG provides the flexibility to cope with the required frame rates. The newer "lossless JPEG," JPEG-LS, offers mathematically lossless compression. The decompressed output of the "baseline JPEG" can be visually indistinguishable from the original image. JPEG-LS gives better compression than original JPEG, but still nowhere near what one can get with a lossy method. Further discussion of compression methods can be found in many published reports and websites and is beyond the scope of this chapter.

26.4.2 Performance

A performance study of an implementation of the proposed parallel pipelined renderer was done by using parallel computers operated at the NASA Ames Research Center and a research laboratory in Japan with images displayed on a PC in a laboratory at the University of California at Davis (UCD). A parallel ray-casting volume-renderer [17] was used in the experimental study. This renderer is reasonably optimized and capable of generating high-quality images. Tests were performed to reveal the relationship between the overall execution time and the number of processor partitions (L). Fig. 26.9 displays the test results on a logarithmic scale along the x-axis; they show that an optimal partition does exist for each processor size. In this case, it is four for all three processor sizes (16, 32, and 64).

Table 26.3 compares the compressed image sizes for the three compression methods considered and also for a combination of them. When lossy compression is acceptable, JPEG is the choice because of the excellent compression it can achieve. Moreover, it is beneficial to use either LZO or BZIP to compress the output of JPEG; the result is additional compression, which may lead to the key reduction required for achieving the desired frame rates. Such a

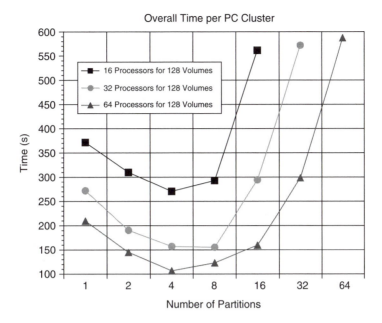

Figure 26.9   The overall execution time versus the number of partitions for three different processor sizes for rendering a 128-time-step turbulent jet dataset. The test results show that there is, indeed, an optimal partition for each processor size.

Table 26.3   Compressed image sizes (in bytes) with various compression methods

Method/image size	128^2	256^2	512^2	1024^2
Raw	49152	196608	786432	3145728
LZO	16666	63386	235045	848090
BZIP	12743	44867	152492	482787
JPEG	1509	3310	9184	28764
JPEG+LZO	1282	2667	6705	18484
JPEG+BZIP	1642	3123	7131	18252

two-phase compression approach was implemented in the display system of the parallel pipelined renderer. The compression rates achieved are 96% and up. Although using LZO adds only 1–2% to compression, reducing the transferred image size by another couple of kilobytes can effectively increase frame rates.

Fig. 26.10 compares the time via display mechanisms using X-Window and the compression-based setting. Four different image sizes were used, and it is clear that as the image size increases, the benefit of using compression becomes even more dramatic. In this set of tests, JPEG and LZO are used together to achieve the best compression rates. The cost of compression is between 6 milliseconds for 128^2 pixels and 500 milliseconds for 1024^2 pixels. The decompression cost is between 12 milliseconds and 600 milliseconds. Note that the decompression time is long because it was done on a single PC. Table 26.4 lists the actual frame rates observed during transmission of the resulting images from NASA Ames to UCD.

With parallel compression, as soon as a processor completes the sub-image that it is responsible for compositing, it compresses and sends the compressed sub-image to the display daemon. In this case, the step to combine the sub-images is waived. The daemon forwards

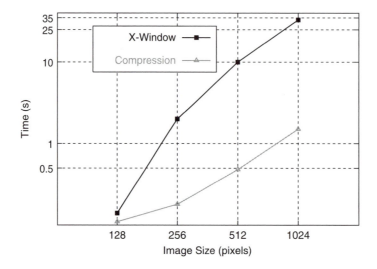

Figure 26.10   Average interframe delay from NASA Ames to UCD via X or the compression-based display daemon. Four different image sizes were considered.

Table 26.4   Actual frame rates (frames/s) from NASA Ames to UCD

Method/image size	128^2	256^2	512^2	1024^2
X-Window	7.7	0.5	0.1	0.03
Compression	9	5.6	2.4	0.7

the compressed sub-images it receives from all the processors to the display interface because compressing each image piece independent of other pieces would result in poor compression rates. Furthermore, with this approach, while compression time is reduced, decompression generally takes longer because of the overhead of processing multiple images. As shown in Fig. 26.11, the decompression time increases significantly with 16 or more processor (sub-image) cases. The plot also reveals that decompressing 2, 4, or 8 smaller sub-images is faster than decompressing a single larger image. Therefore, this set of test results suggests that a hybrid approach might give us the best performance.

That is to say, a small number of sub-images are combined to form a larger sub-image before compression. These combined sub-images are then compressed in parallel and delivered to the display interface for decompression and display.

Note that compression performance is also data- and transfer-function dependent. For large images, the image-transport and display time could be longer than the rendering time. Although the display daemon uses an image buffer to cope with faster rendering rates, a more effective compression mechanism is desired. On the other hand, a large dataset takes longer to render, making image-transport cost less of a concern. Even though the speed of the parallel computers (a PC cluster with 200 MHz processors and an SGI Origin 2000) and the data sizes (128 time-steps, $128^3 - 256^3$ voxels) used in the performance study cannot be compared to today's technologies, the test results obtained do adequately show the expected performance trends. Complete test results can be found in Ma and Camp [16].

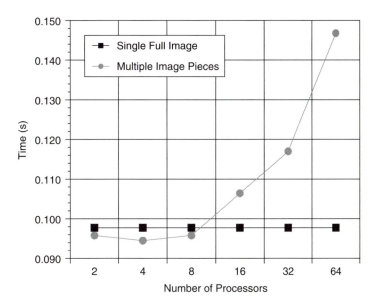

Figure 26.11 Time used to decompress all sub-images to be displayed for cases using up to 64 processors. The overall image size is 512×512 pixels.

26.5.3 Discussion

As scientific data resolution continues to increase, the demand for high-resolution imaging will also increase. We have shown how users of a supercomputing facility can perform remote, interactive visualization of time-varying data stored at the facility by using a combination of pipelining and image compression. In particular, the test results show that, in order to keep up with parallel rendering rates, image compression plays a key role.

Even though the pipelined setting hides most of the I/O cost, as rendering rates increase (as a result of using 3D texture graphics hardware), I/O will become a bottleneck again. Parallel I/O, if available, can be incorporated into the pipeline rendering process quite straightforwardly and can improve the overall system performance. A parallel renderer designed for the visualization of time-varying volume data from large-scale earthquake simulations employs not only parallel pipelined rendering to overlap the rendering calculations with I/O, but also I/O servers to further cut down the I/O cost [30].

The optimal number of I/O servers can be predetermined. The result is a highly efficient renderer with the I/O cost almost completely hidden, except for the pipeline startup delay.

Refined compression methods can also further improve the image transmission rates. In addition, one potential problem with lossy methods is that the loss could change between adjacent frames, as well as in the proposed setting, between adjacent image blocks, which could produce a flickering in the final animation. In the tests performed, we have not experienced such a problem so far, but a feasible solution would be to parallelize the more expensive but higher-performance lossless compression methods. The other would be to exploit frame (temporal) coherence as the frame-differencing technique demonstrated by Crockett [6].

If the user (client) side possesses some minimum graphics capability (e.g., a commodity PC graphics card), other forms of remote viewing can also be considered. Instead of just a single frame for each time-step, "compressed" subset data can be sent. This subset data can be either a reduced version of the data or a collection of prerendered images that can be processed very efficiently with the user-side graphics hardware. For example, Bethel [2] demonstrates remote visualization using an image-based rendering approach. The server side computes a set of images by using a parallel supercomputer, ships it to the user side, and allows the user to explore the data from viewpoints that can be reconstructed from this set of images.

26.6   Conclusions

We have shown that time-varying volume data can be efficiently visualized on a PC by the use of hardware-assisted decoding and rendering, and also on a parallel supercomputer using parallel pipelined processing. It is feasible to integrate some of the techniques presented in this chapter into an end-to-end high-performance solution that would allow scientists to explore both the temporal and the spatial domains of their data at the highest possible resolution. This new explorability, likely not presently available to most computational scientists, would help lead to many new discoveries. Computational scientists should reevaluate their current approaches to data analysis and exploration problems.

There is a growing trend to use adaptive, unstructured grids in large-scale scientific computing to model problems involving complex geometries. Even though this chapter is mainly concerned with data on rectilinear grids, some of the approaches introduced are applicable to irregular-grid volume data. For example, the parallel pipelined approach works equally well for the rendering of time-varying unstructured-grid data, as already demonstrated by Ma et al. [19] for the visualization of large-scale earthquake simulation data. Furthermore, the advanced features of the newer generation of graphics hardware would also allow us to efficiently render irregular-grid volume data directly. An interesting and challenging problem

will be the encoding of irregular-grid data to facilitate visualization calculations.

Acknowledgments

The authors are grateful for the funding support provided by the National Science Foundation and the Department of Energy. The authors would especially like to thank John Clyne, Kenji Ono, Gabriel Rosa, and Han-Wei Shen for the valuable discussions and assistance they have provided to us.

References

1. K. Anagnostou, T. Atherton, and A. Waterfall. 4D volume rendering with the shear warp factorization. In *Proceedings of Volume Visualization and Graphics Symposium 2000*, pages 129–137, 2000.
2. W. Bethel. Visapult: a prototype remote and distributed visualization application and framework. In *Proceedings of the Conference Abstracts and Applications, ACM SIGGRAPH 2000*, 2000.
3. M. Burrows and D. Wheeler. A block-sorting lossless data compression algorithm. Technical Report Center Research Report 124, Digital Equipment Corporation, Palo Alto, CA, 1994.
4. B. Cabral, N. Cam, and J. Foran. Accelerated volume rendering and tomographic reconstruction using texture mapping hardware. In *1994 Workshop on Volume Visualization*, pages 91–98, 1994.
5. T.-Z. Chiueh, C. K. Yang, T. He, H. Pfister, and A. Kaufman. Integrated volume compression and visualization. In *Proceedings of the Visualization '97 Conference*, pages 329–336, 1997.
6. T. W. Crockett. Design considerations for parallel graphics libraries. In *Proc. Intel Supercomputer Users Group 1994 Ann. North America Users Conf.*, pages 3–14, 1994.
7. D. Ellsworth, L. Chiang, and H.-W. Shen. Accelerating time-varying hardware volume rendering using TSP trees and color-based error metrics. In *Proceedings of 2000 Symposium on Volume Visualization*, pages 119–128, 2000.
8. R. Gonzalez and R. Woods. *Digital Image Processing*. Addison Wesley, 1992.
9. A. K. Jain. *Fundamentals of Digital Image Processing*. Prentice Hall, 1989.
10. J. Kniss, P. McCormick, J. McPherson, A. Ahrens, A. Keahey, and C. Hansen. Interactive

texture-based volume rendering for large data sets. *IEEE Computer Graphics and Applications*, 21(4):52–61, 2001.

11. R. Koenen. MPEG-4—multimedia for our time. *IEEE Spectrum*, 36(2):26–33, 1999.

12. L. Linsen, V. Pascucci, M. Duchaineau, B. Hamann, and K. Joy. Hierarchical representation of time-varying volume data with '4th-root-of-2' subdivision and quadrilinear B-spline wavelets. In *Proceedings of the 10th Pacific Conference on Computer Graphics and Applications*, page 346, 2002.

13. S. P. Lloyd. Least squares quantization in PCM. *IEEE Transactions on Information Theory*, IT-28:129–137, 1982.

14. E. Lum and K.-L. Ma. Hardware-accelerated parallel non-photorealistic volume rendering. In *Proceedings of 2nd International Symposium on Nonphotorealistic Rendering and Animation*, page 67, 2002.

15. E. Lum, K.-L. Ma, and J. Clyne. A hardware-assisted scalable solution of interactive volume rendering of time-varying data. *IEEE Transactions on Visualization and Computer Graphics*, 8(3):286–301, 2002.

16. K.-L. Ma and D. Camp. High performance visualization of time-varying volume data over a wide-area network. In *Proceedings of Supercomputing 2000 Conference*, article 29, 2000.

17. K.-L. Ma, J. S. Painter, C. Hansen, and M. Krogh. Parallel volume rendering using binary-swap compositing. *IEEE Computer Graphics Applications*, 14(4):59–67, 1994.

18. K.-L. Ma, D. Smith, M.-Y. Shih, and H.-W. Shen. Efficient encoding and rendering of time-varying volume data. Technical Report ICASE Reprot No. 98–22, Institute for Computer Applications in Science and Engineering, 1998.

19. K.-L. Ma, A. Stompel, J. Bielak, O. Ghattas, and E. Kim. Visualizing large-scale earthquake simulations. In *Proceedings of the Supercomputing 2003 Conference*, 2003.

20. M. W. Marcellin, M. J. Gormish, A. Bilgin, and M. P. Boliek. An overview of JPEG-2000. In *Proceedings of 2000 Data Compression Conference*, pages 523–541, 2000.

21. J. Max. Quantizing for minimum distortion. *IEEE Transactions on Information Theory*, IT-06:7–12, 1960.

22. G. Rosa, E. Lum, K.-L. Ma, and K. Ono. An interactive volume visualization system for transient flow analysis. In *Proceedings of Volume Graphics 2003 Workshop*, pages 137–144, 2003.

23. K. Sayood. *Introduction to Data Compression*. Morgan Kaufmann, 2000.

24. H.-W. Shen, L.-J. Chiang, and K.-L. Ma. A fast volume rendering algorithm for time-varying fields using a time-space partitioning (TSP) tree. In *Proceedings of Visualization '99*, pages 371–377, 1999.

25. H.-W. Shen and C. R. Johnson. Differential volume rendering: a fast volume visualization technique for flow animation. In *Proceedings of the IEEE Visualization '94 Conference*, pages 180–187, 1994.

26. B.-S. Sohn, C. Bajaj, and V. Siddavanahalli. Feature based volumetric video compression for interactive playback. In *Proceedings of Volume Visualization and Graphics Symposium 2002*, pages 89–96, 2002.

27. R. Westermann. Compression time rendering of time-resolved volume data. In *Proceedings of the Visualization '95 Conference*, pages 168–174, 1995.

28. J. Wilhelms and A. Van Gelder. Multidimensional trees for controlled volume rendering and compression. In *Proceedings of the 1994 Symposium on Volume Visualization*, pages 27–34, 1994.

29. J. Woodring, C. Wang, and H.-W. Shen. High dimensional direct rendering of time-varying volumetric data. In *Proceedings of the Visualization 2003 Conference*, pages 417–424, 2003.

30. H. Yu, K.-L. Ma, and J. Welling. A parallel visualization pipeline for terascale earthquake simulations. In *Proceedings of the Supercomputing 2004 Conference* (in press).

27 Large-Scale Data Visualization and Rendering: A Problem-Driven Approach

PATRICK MCCORMICK and JAMES AHRENS
Advanced Computing Laboratory
Los Alamos National Laboratory

27.1 Introduction

The performance of computing resources has grown dramatically over the last 10 years. This growth has allowed scientists to simulate physical processes in much greater detail than has ever been possible. Computed tomography systems have also been developed that produce extremely detailed scans. This increased resolution has in turn produced extremely large datasets that must be analyzed and visualized to complete the scientific process. In many cases, the same large computing resources required for scientific simulations are often needed to complete these visualization tasks. The goal of this chapter is to present a classification of the various algorithmic approaches that can be used to visualize and render these large datasets. This classification provides the fundamental building blocks for a solution that is based on techniques used in the design of parallel algorithms. These techniques focus on the systems level and are based on the parallel decomposition of the data and/or the required tasks. Before describing the classification in detail, we present the terminology and notation that will be used throughout the rest of the chapter.

In order to characterize our definition of large-scale data, we consider only situations in which the datasets are larger than can be processed by a single computer. In addition, we assume that some degree of interactivity, in the range of 5 to 30 frames/s, is desired to allow users to effectively explore and analyze the data. We have visualized datasets that range from hundreds of megabytes to petabytes in size. The ability to achieve interactive frame rates for such large datasets is an open research challenge. In many cases, new algorithms are required to appropriately leverage the available computing resources and effectively handle the unique properties of a particular dataset. The techniques presented in this chapter provide a foundation for the development of these algorithms. In many cases, these techniques will often work in combination with other methods (e.g., multiresolution representations) to achieve adequate interactive performance. In terms of an algorithmic specification, it is important to consider the characteristics of the various tasks that are required. For example, is the algorithm made up of several independent tasks, or is it composed of a series of dependent subtasks? By building upon the answer to this question and having a detailed knowledge of both the data and the computing environment, we can begin to describe a set of techniques for the visualization of large-scale data. There are four fundamental techniques that can be used to solve the large-data visualization problem:

- Data streaming.
- Task parallelism.

- Pipeline parallelism.
- Data parallelism.

It is also important to note that a combination of these techniques may be used to construct a solution. We will refer to these combinations as *hybrid* algorithms and will present the details of such solutions later in the text. The advantages and disadvantages of each of these four approaches can often be impacted by the available hardware (e.g., is the graphics pipeline implemented in software or hardware?). Although it is not possible to consider the use of every hardware configuration, we demonstrate the basic behavior of each technique and highlight the areas where hardware selection and optimization can play a crucial role in terms of performance.

We will use a data-flow depiction in the sections that follow to present the details of the various classifications. This representation is used primarily because of its convenience for illustration and does not necessarily imply that a strict data-flow implementation is required. At a high level of abstraction, a network represents an entire visualization application. Fig. 27.1 shows an example network that contains five nodes, A, B, C, D, and E, which we will refer to as modules. These modules represent the individual algorithmic steps that are used to construct the final application. For the purposes of

our discussion, we will assume that modules have a well-defined interface, for both inputs and outputs, and have no side effects. In addition, there are data connections between various modules in the network. In Fig. 27.1, the outputs of modules A, B, and C are all inputs into module D. Module D sends the result of its computation to module E. Fig. 27.2 shows a simple but more specific example of a module network. In this network, the *Read* module is responsible for reading a data file from disk, the *Isosurface* module is responsible for creating an isosurface of the data, and the *Render* module is responsible for producing an image of the isosurface. In the following sections, we consider each of the four algorithmic classifications, discuss their advantages and disadvantages, and present several case studies that show their use in practice.

27.1.1 Data Streaming

Data streaming is most commonly used to process independent subsets of a larger dataset, one subset at a time [1]. This is often the only feasible approach in situations where the size of a dataset exceeds the capacity of the available

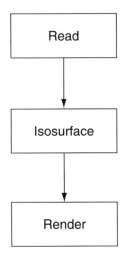

Figure 27.2 An example module network for isosurface computation and rendering.

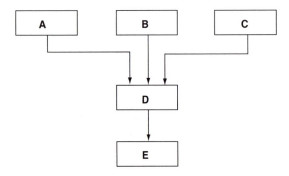

Figure 27.1 This figure shows a module network consisting of five modules (A, B, C, D, and E) and arrows depicting the data connections between modules.

computing resources (memory and swap space). For example, it is not uncommon for scientific datasets, especially time series, to easily exceed the amount of available memory. The key advantage of this approach is that any size dataset can be successfully processed. In some cases, streaming can also result in more effective cache utilization, and on busy computers, a reduction in virtual memory swapping. The drawback of this technique is that it often requires a substantial amount of execution time and does not allow for the interactive exploration of the data. The use of a high-throughput I/O subsystem is advantageous for improving the overall performance of streaming algorithms.

While the streaming of individual time-steps that fit into memory is a straightforward process, the streaming of subsets from a single time-step that exceeds system memory has different requirements. In order to produce the correct solution, the algorithms must be result-invariant—that is, the results must be consistent regardless of the number of subsets into which the data is split. This obviously requires that the algorithms be able to divide the original dataset into pieces. Many algorithms require a knowledge of the data values contained in neighboring cells in order to produce the correct result. For example, the classic marching cubes algorithm requires this knowledge [2]. It is therefore necessary to include these neighboring cells, known as ghost cells, in the individual subsets. While the determination of neighboring cells is a simple and straightforward process on regularly structured grids, it is more challenging with irregular grid structures. For an example of using streaming with irregular grid data, see Ahrens et al. [3]. In the next section, we consider the use of streaming to visualize two different sized datasets using a single desktop computer.

27.1.1.1   Case Study

It is not uncommon to discover that a dataset is too large for visualization and analysis using the single desktop PC found in many scientists'

offices. In this case study, we present the details of how this situation impacts the visualization and analysis of two different-sized datasets. The first dataset has been produced as part of the Terascale Supernova Initiative (TSI) project [4]. The goal of TSI is to develop models of core-collapse supernovae. Core-collapse supernovae produce the most powerful explosions known—they release 10^{53} erg of energy in the form of neutrinos at a rate of 10^{57} neutrinos per second. The second dataset is a larger version of the TSI data produced by resampling the data. These datasets are processed using a dual, 1.2 GHz Pentium 4 Xeon™ processor system with 2 GB of RAM and an NVIDIA Quadro FX graphics card (in an AGP 4x slot). The goal of this study is to understand the impact of dataset size on streaming performance as well as to highlight the advantages and disadvantages of the technique. By using the data streaming features available in the Visualization Toolkit [11], we can simulate computers with varying amounts of RAM. This is done by specifying the number of pieces into which the original dataset is subdivided for processing. One drawback to this approach is that it makes it difficult to mimic the exact behavior of the operating system's virtual memory system. Therefore, all of the results presented below assume that the system is mostly idle and that the amount of available RAM is at or near a maximum.

The TSI dataset contains single-precision floating-point values that represent the entropy from a 3D regular grid structure with dimensions of $320 \times 320 \times 320$. This represents roughly 32 million cells and requires approximately 131 MB of storage. The second dataset represents a $426 \times 426 \times 426$ resampled version of the first dataset. The storage space required for this dataset is approximately 310 MB. For our benchmarks, we will use reading the data file from disk, computing an isosurface of the entropy, and rendering the resulting polygonal data. This is the process depicted by the network shown in Fig. 27.2. The series of images in Fig. 27.3 show the progression of the streaming computations, with each subset

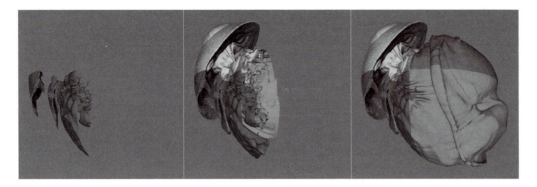

Figure 27.3   The progression produced by the data streaming of eight subset stages of an entropy isosurface, colored by subset, from the Terascale Supernova Initiative. (See also color insert.)

represented as a different color. The bar charts presented in Fig. 27.4 show the amount of time required to complete the streaming process as well as the maximum amount of memory required for the individual subset sizes. These performance measures provide what at first appear to be surprising results. The first result is that it takes longer to process a single piece of data than to process multiple subsets of data. This is primarily a result of the cache performance gains due to the smaller subsets' fitting into cache memory more efficiently. As a measure of this benefit, our performance studies indicated that there are approximately 80,000 second-level cache read misses during processing of the entire 320^3 dataset. In comparison, when processing the dataset subdivided into eight pieces, there were only 1,300 cache misses. Another interesting result shows that by combining the performance benefits of data streaming and the reduction in overall memory usage, it is possible to use a computer system with a much lower price point than might initially be expected to successfully process a large dataset. Based on the execution times presented in Fig. 27.4, it is clear that a straightforward approach to data streaming fails to meet the performance goal of 5 to 30 frames/s. The most important capability that data streaming has introduced is the ability to correctly process subsets of a large

dataset, which is critical for supporting further parallelism by allowing data decomposition. The advantages of this capability are presented in section 27.1.4.

27.1.2   Task Parallelism

With task parallelism, independent modules in an application execute in parallel. In Fig. 27.1, this would be achieved by having modules A, B, and C all execute at the same time. This requires that an algorithm be broken up into independent tasks and that multiple computing resources be available. The key advantage of this technique is that it enables multiple portions of a visualization task to be executed in parallel. The main disadvantage of this technique is that the number of independent tasks that can be identified, as well as the number of CPUs available, limits the maximum amount of parallelism. In addition, it can be difficult to load-balance the tasks, and therefore it can often be very challenging to take full advantage of the available resources. Task parallelism is used effectively in the movie industry, where several frames in an animated production are rendered in parallel. Specific hardware choices for improving the performance of task parallelism are dependent upon the details of the required tasks.

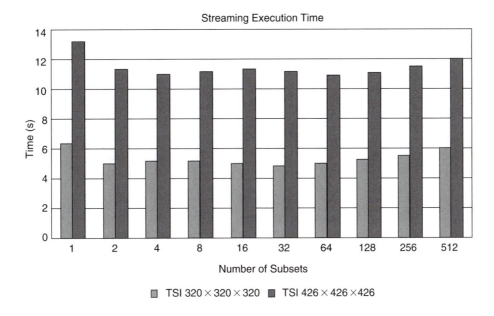

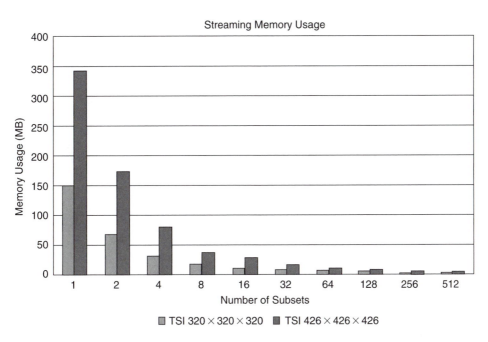

Figure 27.4   Performance and memory usage of the streaming algorithm as the number of pieces increases.

27.1.2.1 Case Study

In our experience, the use of task parallelism is rare when dealing with large datasets. However, one possible application well suited to task parallelism is that of comparative visualization. In this case study, we present the use of task parallelism in exploring the relationship between data values produced by the Parallel Ocean Program (POP) at Los Alamos National Laboratory. The largest dataset produced by POP represents the Earth's oceans at a resolution of one-tenth of a degree. This requires approximately 350 million cells, and a decade-long simulation generates about 6 terabytes of data. Ocean modeling plays an important role in the prediction of global warming and the exploration of the mechanisms that cause climate variability. The computational grid for this simulation has dimensions of $3600 \times 2400 \times 40$; therefore, a single-precision floating-point variable produced by the model requires approximately 1.4 GB of storage (given that one single-precision floating-point value requires four bytes of storage). In this example, we investigate the behavior of the Gulf Stream off the eastern coast of North America. The accurate simulation of the Gulf Stream is important for model validation and can be studied by extracting multiple isosurfaces of the water temperature.

For this example, we consider two possible approaches to breaking the processes of computation and visualization into multiple tasks. In the first, which we will call the *geometry approach*, the individual tasks are responsible for computing one or more isosurfaces of the data. In this situation, the resulting geometry is sent to a single process, which both renders and displays the results. In the second, which is called the *image approach*, the tasks are responsible for both the isosurface computation and the rendering of the resulting geometry. The resulting imagery is then similarly sent to a process for display. Fig. 27.5 presents both of these tasks as data-flow networks.

In order to focus on the Gulf Stream region of the global model, the *Read* module(s)

extract(s) a small portion of the full POP dataset. This region has dimensions of $150 \times 500 \times 40$ computational cells and contains two variables, which represent the salt content and the temperature of the ocean. To understand the performance characteristics of task parallelism, we consider the computation and rendering of 16 isosurfaces. These isosurfaces will be computed by both of the approaches depicted in Fig. 27.5, using a PC cluster comprised of dual-processor 800 MHz Pentium 3 Xeon systems with 1 GB of RAM, a Wildcat 4210 AGP graphics card, and a Myrinet 2000 connection [13]. The parallel infrastructure of the Visualization Toolkit is used to implement this example [11]. We consider the performance results for 2, 4, 8, and 16 processors. The isosurface computations are always evenly distributed among the processors. It is important to note that the *Render* module on the lefthand side of Fig. 27.5 always receives the geometry representing all 16 isosurfaces, regardless of the number of processors in use. In contrast, the *Display* module on the righthand side of Fig. 27.5 receives one image per task process; therefore, the more task processes, the more image data there is to receive. For this example, all images contain 1024×1024 RGBA pixels.

An important characteristic of both techniques is that they collect the results of several tasks into a single module. This situation normally limits the scalability of a parallel solution because the gathering point acts as a bottleneck for performance as the number of inputs, or the input data size, increases. In the case of the geometry-based tasks, the amount of data sent to the *Render* module is constant regardless of the number of processors. Overall, the geometry associated with these 16 isosurfaces requires approximately 36 MB of memory. The image-based tasks, however, send more data as the number of processors increases. When we reach 16 processors, we are sending approximately 16×4 MB of image data to the *Display* process, in comparison to the 4 MB required for a single process. Both techniques achieve performance improvements by distributing the

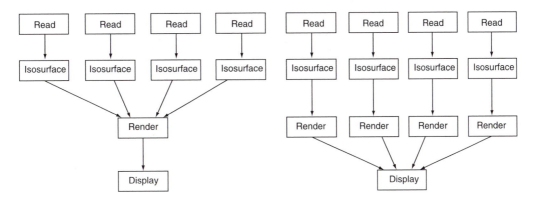

Figure 27.5   The two networks used in the task parallelism examples. The key difference between the two is the delivery of data to the display task. The display task on the left receives geometry, while the one on the right receives image data.

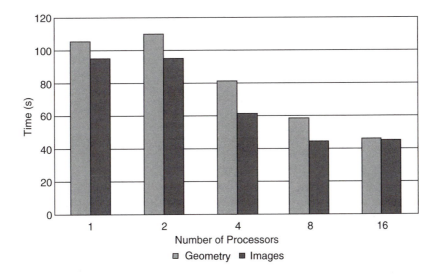

Figure 27.6   The total execution time required for each task-parallel approach on varying numbers of processors.

work of computation among different processors. The required execution times for both methods are shown in Fig. 27.6. Fig. 27.7 shows an example of isosurfaces depicting the Gulf Stream.

Besides the bottleneck in gathering the results on a single node, the other limitation of both task-parallel approaches is that they are limited to 16 total isosurface tasks. This puts a theoretical limit on our performance gains of a factor of 16 times faster than a single processor implementation. As the execution time results show, the task-parallel results are well below this limit. This performance limitation is the result of the cost of sending the data from each of the tasks to the display process as well as the load imbal-

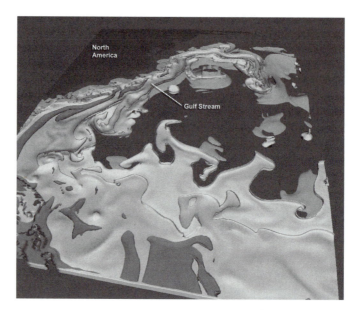

Figure 27.7 An example of the isosurfaces produced by the task-parallelism example. (See also color insert.)

ance between the various tasks. For example, the isosurface times for eight processors, with each processor computing two isosurfaces, range from 16.5 seconds to 29.3 seconds for both networks. Initially, the image network achieves better performance because the amount of data that must be sent to the display process is much lower than that required for the geometry process. In addition, this network also allows multiple rendering processes to run in parallel. The cost of sending the resulting image data, plus the overhead of reading the resulting images from the frame buffers on the graphics cards, introduces an additional overhead that impacts the initial advantages of this parallelism.

27.1.3 Pipeline Parallelism

Pipeline parallelism occurs when a number of modules in an application execute in parallel but on independent subsets of data (thus distinguishing this process from task parallelism). In Fig. 27.1, this would occur when modules A, D, and E are all operating on independent portions of the data. This approach is best suited for situations where there are multiple, heterogeneous tasks. The advantage of this approach is that it allows parallel use of the overall computing resources. For example, one process can be reading from disk, another process computing results using the CPU, and a third process rendering using a hardware-accelerated graphics card. The main disadvantage of this approach is that it can make it difficult to balance the execution time required by the individual stages; in an unbalanced pipeline, the slowest stage directly impacts the overall performance. In addition, the length of the pipeline directly limits the amount of parallelism that can be achieved (i.e., you must have as many processors as there are pipeline stages). In order to maximize performance, it is necessary to quickly move data from one stage of the pipeline to the next. Examples of this capability include the use of shared memory architectures and a high-speed interconnection network between processors. The following case study highlights the advantages and disadvantages of pipeline parallelism.

27.1.3.1   Case Study

In this case study, we consider a simple PC-based animation application that reads the image files from disk and displays them on a single monitor. In order to study both the advantages and the disadvantages of pipeline parallelism, we compare a serial version of this application to a multithreaded, pipelined implementation. In the serial implementation, a single process does all the work; in the pipelined version, one thread reads the data from disk and another displays the images (a classic producer–consumer model). In order to balance this two-process pipeline, we must be able to read an image from disk in approximately the same amount of time it takes us to display the image. The top graph in Fig. 27.8 shows the amount of time required for each process as the image data increases in size. It is important to note that the two tasks require similar time for small images but the read operation becomes more costly as the image size increases. This situation will create an unbalanced pipeline for large image sizes, as the slower read process will hold up the display process. Fig. 27.8 also shows the performance of the nonpipelined and the pipelined code. A more advanced use of pipelined parallelism is presented in the Hybrid Systems section.

27.1.4   Data Parallelism

With data parallelism, the code within each module of an application executes in parallel. Referring to Fig. 27.1, this occurs when the code within module A runs in parallel. This requires that a dataset be subdivided and multiple processes run the same algorithm on the resulting pieces concurrently. Data parallelism can be implemented as an extension of the data-decomposition technique described in the streaming section. In this case, the data is subdivided in the same fashion, but we have the extra step of assigning a processor to each of the resulting pieces. This approach is commonly referred to as a single program–multiple data (SPMD) model because each process executes the same program on different subsets of the data. The biggest advantage of this approach is that it can achieve a high degree of parallelism; solutions tend to scale well as the number of processors increases. When there is a large number of processors available, this approach is often one of the best ways to achieve increased performance. A possible drawback to this approach is that scalability can be limited by interprocess communication costs. In order to achieve the best possible performance with data parallelism, it is often important to consider the communication costs and data localities among the processors. In the best possible situations, there is no dependence between processors, and in the worst case every processor is required to share information with every other. Fortunately, many visualization algorithms have few communication dependencies between processors, and therefore data parallelism is often one of the most effective techniques for achieving increased performance.

27.1.4.1   Case Study

In this case study, we visualize a one-tenth-of-a-degree dataset generated by the Parallel Ocean Program (POP), previously introduced in Section 27.1.2. We investigate the salinity of the Atlantic by creating and viewing an isosurface of salinity colored by temperature. The computational grid for the dataset has dimensions of $3600 \times 2400 \times 40$ (referred to as Full); therefore, a single-precision floating point variable produced by the model requires approximately 1.4 GB of storage. Our visualization requires processing temperature and salinity variables. To study scalability and performance at different dataset sizes, we subsampled the full-resolution dataset to create two additional datasets: one with dimensions $360 \times 240 \times 20$ (referred to as Small), and one with dimensions $760 \times 480 \times 40$ (referred to as Medium). The isosurface algorithm processes approximately 12 MB for the Small dataset, 111 MB for the Medium dataset, and 2.8 GB for the Full dataset. With a data-parallel algorithm, the computational cost and memory requirements of the

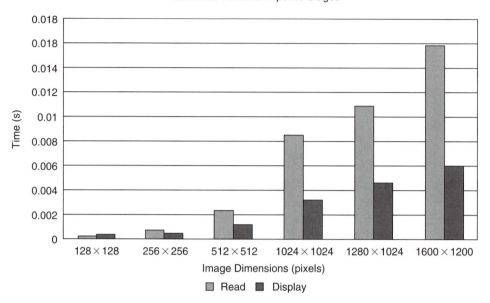

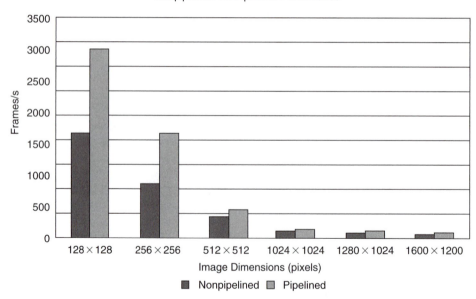

Figure 27.8   (Top) The amount of processing time required by each stage of the pipeline as image size increases. (Bottom) The frames/s performance comparison between nonpipelined and pipelined versions of the code.

isosurface computation are divided over the number of available processors. These isosurfaces will be computed using a PC cluster comprised of 64 nodes, 32 of which contain dual 800 MHz Pentium 3 Xeon processors and 32 of which contain 1 GHz Xeon processors (all systems contain 1 GB of RAM). The data-parallel infrastructure of the Visualization Toolkit is used to implement this example. Information about neighboring cells was automatically handled by the reader, and therefore the isosurfacing algorithm requires no interprocess communication. Fig. 27.9 shows the performance of a data-parallel isosurfacing algorithm on the three datasets. Notice that performance improves by a factor of two with each doubling of the number of processors for all dataset sizes. Notice also that due to the memory requirements of the full dataset and the resulting graphics primitives, visualization is only possible when we use 16 or more processors. Fig. 27.10 shows the resulting isosurface for the Full POP dataset.

27.1.5   Summary

The preceding sections have outlined a classification of techniques based on the decomposition of both tasks and data. The table shown in Fig. 27.11 presents a mapping from the characteristics of the problem to the supported solution. The dataset size column describes the quantity of data to visualize. A dataset is "Large" when it exceeds the resources of a single machine. The tasks column describes the type of work. *Homogenous* tasks are the same type of work applied to different data. *Independent* tasks can be run in parallel at the same time. *Sequential* tasks must run one after another in a fixed order. The resources column describes the number of resources available, and the solution column identifies the technique to use. For example, if you have a large dataset and only a single CPU, then data streaming is perhaps the only possible solution by which you can explore the entirety of the data.

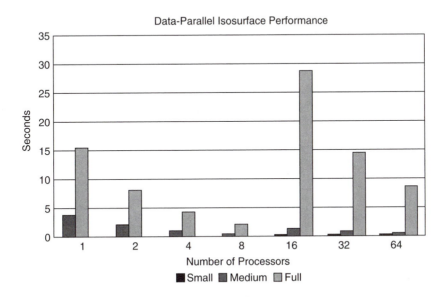

Data-Parallel Isosurface Performance

Figure 27.9   Execution times for data-parallel isosurfacing for the three different POP grid sizes. Due to its large size, the Full POP grid can be run only on 16 or more processors.

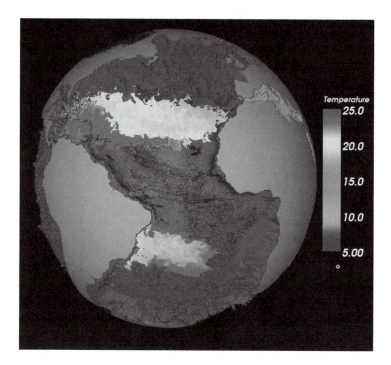

Figure 27.10   Salinity isosurface colored by temperature, produced using the Full POP dataset. (See also color insert.)

Dataset Size	Tasks	Resources	Solution
Large	Any	Single CPU	Data streaming
Large	Homogenous	Multiple CPUs	Data parallelism
Any	Independent	Multiple CPUs	Task parallelism
Any	Sequential	Multiple CPUs	Pipeline parallelism

Figure 27.11   Summary table of solution techniques.

27.2   Hybrid Systems

In several situations, it is possible to take advantage of many of the described techniques within the same application. For example, data streaming and data parallelism were used to visualize a dataset that was approximately a petabyte in size [3]. Although using multiple techniques can lead to a more complex implementation, it is often a very powerful way to fully utilize available resources. In this section, a case study that highlights the successful use of a combination of several approaches is presented.

27.2.1   Case Study

The specific performance characteristics and design details of a particular computer architecture can often play an important role in providing the foundation for an algorithmic solution. This case study presents an application in which this detailed knowledge was required to solve a large-data visualization problem. TRex is a hybrid system that uses data streaming, pipelining, and data parallelism [5]. The primary goal in designing and developing TRex was to support the rendering of large, full-resolution, time-varying, volume datasets at approximately 5 frames/s. Due to the computational expense associated with software-based volume rendering, it was necessary to leverage the performance of hardware-accelerated volume rendering using 3D textures [7,8]. The target platform for this application was an SGI Origin 2000 containing InfiniteReality2 (IR2) graphics engines and 128 processors. Although a single IR2 is capable of supporting hardware-accelerated volume rendering at rates beyond 5 frames/s, the amount of available texture memory is well below the amount required to support large datasets. It requires the texture memory of 16 IR2 pipes to render a $1024 \times 1024 \times 1024$ volume without texture paging. Therefore, the key issue in reaching our performance goal was not limitation from the graphics performance of the system, but finding a way to deliver the time-varying subsets of the data to the IR2 engines at a fast enough rate to sustain 5 frames/s.

As the section on pipeline parallelism showed, the process of reading data from disk can often be a bottleneck in the overall performance of an application. Initial I/O subsystem tests on the Origin 2000 revealed that by using a striped redundant array of inexpensive disks (RAID), and system routines that bypassed kernel interrupts, we could achieve a data rate of approximately 140 MB/s. Unfortunately, this rate was not fast enough to reach our goal of 5 frames/s for a 1024^3 dataset. To solve this problem, we were forced to look more closely at the architectural design of the Origin 2000 to see if it was possible to build a customized file system for streaming volume data from disk and then directly into texture memory.

Fig. 27.12 shows the topology of a 16-node, 32-processor Origin 2000 system where squares represent CPU nodes and circles represent router chips that handle communications between the nodes. Notice that each node contains the connections between the CPUs, memory, and I/O and graphics subsystems. In order to improve the performance of the I/O system for better graphics throughput, we collocated the IR2 pipes and the I/O controllers so that they share access of the same memory region, and therefore avoid the routing of data through the interconnection network. This step not only reduces the latency involved in accessing data in memory, it also provides a balance between the performance characteristics of all 16 graphics engines. The data-transfer rate between the memory and the graphics subsystem is approximately 300 MB per second. This gives a total data rate of 5 GB per second when taking all

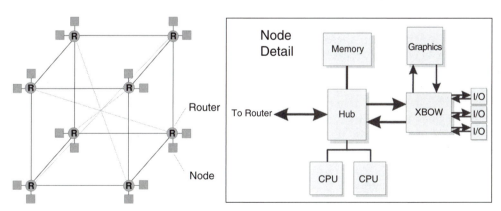

Figure 27.12 (Left) The interconnection topology of the SGI Origin 2000 architecture. (Right) The details of the nodes (represented as squares in the topology).

16 IR2 engines into consideration, and it is also the rate that the I/O subsystem must achieve to reach a balanced system performance. By placing four dual-fiber-channel controllers on each node's I/O subsystem and having them each drive 576 disks, we can achieve a data rate of approximately 70 MB per second per controller. This allows us to achieve 280 MB per second per node, which falls 20 MB per second below our goal. In comparison to our initial 140 MB-per-second data rate, we are now much closer to reaching a balanced system performance. In the remainder of this section, we present the details of the software application built upon this hardware configuration.

As was previously mentioned, it is necessary to break a large volume of data into individual subsets in order to efficiently render the data on the IR2 engines. In this case, the data decomposition has the advantage of increasing the total amount of texture memory available to our application. This decomposition provides the foundation for the data parallelism used in TRex. The network in Fig. 27.13 shows the overall series of parallel tasks required by TRex. The *Read* module takes advantage of the I/O configuration described above and reads individual subvolumes from disk. In order to guarantee peak performance, it is necessary to assign the read process to one of the processors within the same node as the graphics and I/O subsystems. Next, the *Render* module takes this data and downloads it as a 3D texture to the IR2, where it is then rendered. For adequate performance, this process must be assigned to the remaining processor within the node. In the next step, the *Composite* module takes the resulting images produced by each of the graphics engines and stitches them together to form the final image. In the final step, the *Display* module receives the composited image

and displays it to the user. By streaming multiple time-steps of data through the *Read* module, the second hybrid characteristic is incorporated into TRex. Even though the data has been divided into subvolumes, this use of streaming is a basic temporal approach that is easily achieved by reading one time-step after another from disk. To introduce the last hybrid characteristic into TRex, it is important to notice that each stage of this network produces output data that is independent from the data produced by the other modules. In addition, the streaming of temporal data from disk introduces the ability to have a constant flow of data. Therefore, it is possible to introduce pipeline parallelism into this network and overlap the execution of each stage. This requires that the compositing stage be separated from the other stages. It is often tempting to implement compositing within the graphics hardware, but in this situation, doing so would limit the length of the pipeline in TRex and thus reduce performance, as well as introduce overhead related to the reading and writing of multiple frame buffers—a task that is often much more expensive than might be expected. Therefore, TRex uses a software-based compositing technique that leverages any remaining processors that are not already involved in the reading or rendering portions of the pipeline.

By taking the specific details of a hardware architecture into consideration and leveraging the advantages of three of the techniques presented in this chapter, TRex allows for the exploration of 1024^3 datasets at five frames/s. The biggest drawback of this approach is that the size of data that can be processed is limited by the amount of texture memory that can be placed in the Origin 2000. Examples of the output produced by TRex are shown in Fig. 27.14.

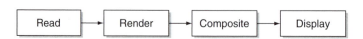

Figure 27.13   The parallel pipeline used in the TRex volume renderer.

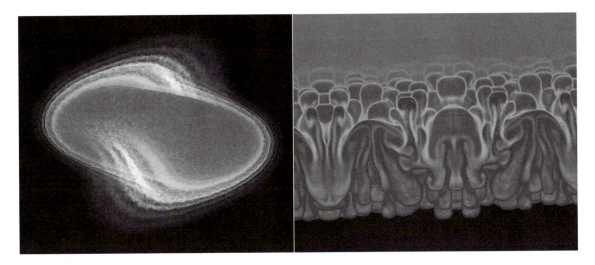

Figure 27.14   Two volume rendered images produced by TRex. The image on the left represents the particle density produced by a simulation of a linear accelerator model using 300 million particles. The image on the right shows a classic Raleigh-Taylor fluid dynamics simulation. (See also color insert.)

27.3   Rendering

In the previous sections, we have focused primarily on the large-data visualization process and not on the specific issues related to rendering the resulting graphics primitives. The rendering of these primitives is a computationally intensive process and is especially challenging when interactive rendering rates are desired. In this section, we present a review of the techniques that are commonly used to handle the rendering and display of large-scale datasets.

When we deal with gigabytes or petabytes of data, it is not uncommon for visualization algorithms to produce millions to billions of graphics primitives. It is beyond the ability of a single CPU and graphics accelerator to render these primitives at interactive rates. The use of data-parallel rendering methods is required to achieve adequate interactive performance. For an introduction to parallel rendering techniques, the reader is referred to Crockett [12]. The most commonly used classification of parallel rendering algorithms is based on the loca-

tions in the rendering pipeline where the graphics primitives are sorted from object space into screen space [6]. In summary, the following classifications are introduced: sort-first, sort-middle, and sort-last. We will briefly review each of these methods.

27.3.1   Sort-First

Sort-first algorithms begin by distributing graphics primitives at the start of the rendering pipeline. We assign primitives to processors by subdividing the output image and assigning a processor to handle each resulting region. Once the primitives have been assigned, each process completes the entire graphics pipeline to produce the final sub-image. The initial assignment of primitives to processors is the key step in sort-first algorithms. This step requires the transformation of the primitives into screen-space coordinates and introduces a computational overhead to the algorithm. The initial assignment to processors is generally done in an arbitrary fashion, but after each processor completes the transformation stage of the pipeline it reassigns primitives to the correct processors. Additional consideration must be given to primitives that overlap the subdivided regions of the final image.

Sort-first is advantageous because the processors implement the entire rendering pipeline. This has the advantage of allowing the use of a commodity-based, hardware-accelerated pipeline, and/or better system cache behavior. In addition, the communication requirements between the processors can be low, resulting in lower overhead and higher performance. The main disadvantage of sort-first is that the initial distribution of primitives among the processors can easily lead to a workload imbalance. Attempting to redistribute the primitives during the rendering process can also lead to poor scalability due to the large amounts of required messaging traffic.

27.3.2   Sort-Middle

In sort-middle algorithms, the redistribution of data occurs between the geometry-processing and scan-conversion stages of the rendering pipeline. In this case, it is common to describe the steps as two sets of operations. The first set of operations handles the geometry portion of the pipeline (transformations, lighting, etc.), and primitives are initially assigned in an arbitrary fashion. The second set of operations is assigned contiguous regions of the final output image, and these operations are responsible for rasterizing the screen-space primitives produced by the first set of processors. Note that it is possible to dedicate processors to each of these sets separately or to allow all the processors to perform both tasks. If separate processors are used, it is possible to create a set of parallel pipelines. This pipelined approach has the same advantages and disadvantages discussed in the pipeline section.

The split in the rendering pipeline is the biggest disadvantage of sort-middle, because it makes it difficult to leverage hardware-accelerated rendering. In addition, the cost of communicating between the stages of the pipeline scales with the number of primitives rendered. Finally, the algorithm can suffer from load imbalance when the primitives are not evenly distributed across the output image.

27.3.3   Sort-Last

The sort-last approach delays primitive sorting until the final stages of the rendering pipeline. Primitives can be initially assigned in an arbitrary fashion, and each processor renders its portion of the final image. All of these images are then composited to form the final complete image. As with sort-middle, it is possible to use two sets of processors. The first set of processors is responsible for completing the rendering pipeline, and the second set is responsible for creating the final image. This assignment also allows for the creation of a parallel pipeline that can be used to improve overall performance. Since the rendering stages of sort-last create full-resolution images, the interconnection network between processors must have very high bandwidth if interactive rendering is required. Other techniques can be used to reduce the amount of traffic required to complete the compositing stage [9,10].

The advantages of sort-last include the ability for processors to implement the entire rendering pipeline. It is easier to uniformly distribute primitives to the processors and thus avoid load-balancing issues. The main disadvantage of sort-last is the communication requirements introduced by sending large amounts of data between the processors. Techniques for reducing the cost of this compositing message traffic are still an area of active research.

27.3.4   Hardware

It is important to note that all of today's commodity graphics hardware takes advantage of parallelism in the form of pipeline and SIMD optimizations. This hardware-supported parallelism makes the use of the full rendering pipeline even more advantageous for both sort-first and sort-last techniques. It is extremely difficult to cost-effectively outperform today's commodity hardware with pure software-based rendering. For example, a single NVIDIA CineFX engine is capable of executing eight four-way operations per clock tick. With the graphics processor(s) running at 500 MHz, this is equivalent to 16 billion floating-point operations per second. The additional program-mability that is being built into the recent graphics hardware also has the potential to provide even more flexibility in the solution of the large-data visualization and rendering challenge.

27.4   Conclusions

This chapter presented a systems-based classification of algorithms for dealing with the visualization and rendering of large-scale datasets. By using these techniques as a foundation for the development of existing and future algorithms, we will be able to provide efficient and effective tools for application scientists to explore simulation results.

Acknowledgments

The authors would like to thank Tony Mezza-cappa and John Blondin for providing us with access to the terascale supernova data and Mat Maltrud for the one-tenth-of-a-degree POP dataset. Special thanks go to Berk Geveci and Charles Law for providing us with help in producing the testing code used in Section 27.1.1. We also acknowledge the Advanced Computing Laboratory at Los Alamos National Labora-tory; the examples presented in this chapter were computed using resources located at this facility.

References

1. C. C. Law, K. M. Martin, W. J. Schroeder, and J. E. Temkin. A multithreaded streaming pipeline architecture for large structured data sets, *Proc. IEEE Visualization 1999*, pages 225–232, 1999.
2. W. E. Lorensen and H. E. Cline. Marching cubes: a high resolution 3D surface recon-struction algorithm. *Computer Graphics*, 21(4): 163–169, 1987.
3. J. Ahrens, K. M. Martin, B. Geveci, and C. Law. Large-scale data visualization using paral-lel data streaming. *IEEE Computer Graphics and Applications*, 21(4):34–41, 2001.
4. Terascale Supernova Initiative, http://www. phy.ornl.gov/tsi/
5. J. Kniss, P. McCormick, A. McPherson, J. Ahrens, J. Painter, A. Keahey, and C. Hansen. TRex: interactive texture based volume rendering for extremely large datasets. *IEEE Computer Graph-ics and Applications*, 21(4):52–61, 2001.
6. S. Molnar, M. Cox, D. Ellsworth, and H. Fuchs. A sorting classification of parallel rendering. *IEEE Computer Graphics and Appli-cations*, 14(4):23–32, 1994.
7. B. Cabral, N. Cam, and J. Foran. Accelerated volume rendering and tomographic reconstruc-tion using texture mapping hardware. *ACM Symposium Volume Visualization* (A. Kaufman and W. Krueger, Eds.). New York, ACM Press, 1994.
8. O. Wilson, A. Van Gelder, and J. Wilhelms. Direct volume rendering via 3D textures. Tech. Report UCSC-CRL-94-19, Univ. of California at Santa Cruz, 1994.
9. J. Ahrens and J. Painter. Efficient sort-last rendering using compression based image com-positing. *Proc. of Second Eurographics Workshop on Parallel Graphics and Visualization*, 1998.
10. B. Wylie, C. Pavlakos, V. Lewis, and K. More-land. Scalable rendering on PC clusters. *IEEE Computer Graphics and Applications*, 21(4):62–69, 2001.
11. W. J. Schroeder, K. M. Martin, and W. E. Lorensen. *The Visualization Toolkit: An Object-Oriented Approach to 3D Graphics.* Upper Saddle River, NJ, Prentice Hall, 1996.
12. T. Crockett. An introduction to parallel render-ing. *Parallel Computing*, 23(7):819–843, 1997.
13. Myricom home page, http://www.myri.com

28 Issues and Architectures in Large-Scale Data Visualization

CONSTANTINE PAVLAKOS and PHILIP D. HEERMANN
Sandia National Laboratories

28.1 Introduction

"The purpose of computing is insight—not numbers." How many times has this quote from R. W. Hamming been used in the context of scientific visualization? Yet a steady stream of numbers, and other forms of data, continues to pour out from computations, experiments, real-time data capture, historical data capture, etc. Our ability to generate data continues to grow in leaps and bounds, and our ability to comprehend it all continues to encounter great challenges.

"I've never seen most of my data—I can't." This quote, from Professor George Karniadakis, appears with Hamming's quote in the Data and Visualization Corridors (DVC) technical report [1], published in 1998 as a result of a series of workshops on the manipulation and visualization of large-scale data that was sponsored by the Department of Energy (DOE) and the National Science Foundation (NSF). The two quotes together are quite telling, one stating a fundamental objective, the other capturing the pragmatic frustration of many a computational scientist.

Lest it appear that the situation is hopeless, we should point out that there has been progress. Indeed, the DVC report referred to above can and should be credited with bringing attention to the problem of large-scale data visualization, as well as with spurring additional research activity in this area. Among the efforts influenced by this report is the DOE's Advanced Simulation and Computing (ASCI) [2] Visual Interactive Environment for Weapons Simulations (VIEWS) program, which, in recent years,

has figured prominently in the experience base for the two authors of this chapter.

In this chapter, we take a high-level, end-to-end, systems view of the large-scale data visualization problem in high-performance computing environments. The chapter introduces many of the issues associated with the problem, as well as solution approaches (many of which receive more detailed treatment elsewhere in this handbook) and architectural features that are either proven or have promise. Successes in the ASCI VIEWS program are used to help provide credence to some of the architectural elements.

28.2 Characterizing the Problem

Today's supercomputers are capable of computing at performance levels in the tens of teraflops. The DOE's ASCI laboratories currently operate machines that provide 10–20 teraflops of peak computing power, and the Earth Simulator machine built by Japan's NEC Corporation provides a peak of 40 teraflops. Such machines are being used to perform simulations at unprecedented complexity and fidelity in diverse computational science and engineering applications, in areas including the biological sciences, environmental sciences, energy research, and defense. Terabyte datasets are no longer uncommon, and the existence of petabyte datasets is anticipated in the next few years.

The process of analyzing and visualizing large-scale scientific data is depicted in Fig. 28.1. The process includes a diverse group of data service functions that may be applied to

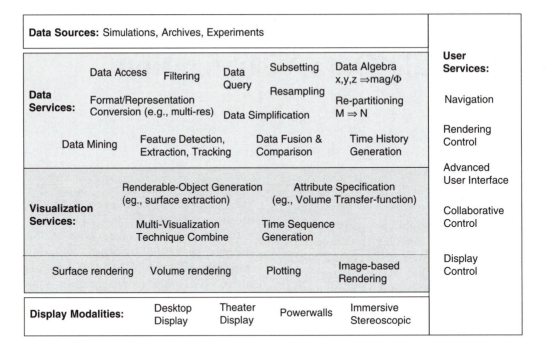

Data Sources: Simulations, Archives, Experiments						User Services:
Data Services:	Data Access Filtering Data Query		Subsetting Resampling	Data Algebra $x,y,z \Rightarrow mag/\Phi$		Navigation
	Format/Representation Conversion (e.g., multi-res)		Data Simplification	Re-partitioning $M \Rightarrow N$		Rendering Control
	Data Mining	Feature Detection, Extraction, Tracking	Data Fusion & Comparison	Time History Generation		Advanced User Interface
Visualization Services:		Renderable-Object Generation (eg., surface extraction)		Attribute Specification (eg., Volume Transfer-function)		Collaborative Control
		Multi-Visualization Technique Combine	Time Sequence Generation			Display Control
	Surface rendering Volume rendering		Plotting	Image-based Rendering		
Display Modalities:	Desktop Display	Theater Display	Powerwalls	Immersive Stereoscopic		

Figure 28.1 The data analysis and visualization process.

data prior to visualization, where renderable objects are generated and converted to images. It is important to understand that a substantial portion of the visualization process is often spent in the manipulation, transformation, and preparation of data for subsequent rendering. High-performance rendering by itself is inadequate. A complete environment for effective visualization of large-scale data must include a rich set of high-performance data services that can efficiently feed high-performance visualization services.

28.2.1 Enabling Data Exploration and Discovery

An ideal environment for computational scientific and/or engineering analysis would give very simple answers to complex questions, perhaps only a yes or no or a single number—in any case, a definitive answer to whatever the current question of interest is. In reality, however, we are faced with the problem of how to discover simple, definitive answers from a plethora of bits and bytes. Such discovery suggests the need for tools and environments that support the efficient, effective exploration of data. A robust interactive environment is needed that enables the scientist or analyst (the user) to receive timely and useful feedback in the search for answers.

Equally important, this environment must be accessible from the office, which is where day-to-day work is done. The user should be able to manipulate a complete set of available resources and tools from the desktop. At the same time, because of the size of datasets, the ability to process data demands that one have access to resources that far exceed the conventional power of an office computer or workstation. Ideally, such access should be provided in a manner that is as seamless as possible.

A rich set of data service functions is key to support of robust data exploration. These functions should be characterized by the ability to retrieve, manipulate, scrutinize, and interpret a portion of data that is of current interest to the end user. This implies the ability to efficiently extract specific data objects from large datasets. Analysis may also require the derivation of new data objects from pre-existing data objects; for large data, such a derivation may demand significant computational resources.

It is sometimes useful to change the form or representation of data in order to enhance the data-exploration process. One example is to convert data into a hierarchical or multiresolution form. Such a form allows data to be manipulated at various resolutions and/or levels of detail (LODs) very efficiently. Low-resolution data can be used for highly interactive analysis or data browsing. When immediate access to a complete multiresolution representation can be retained, visualization can be tailored to the current viewing parameters and the capabilities of particular visualization hardware.

Another useful data transformation is the conversion of data into a form that is optimally arranged for processing by key visualization algorithms. More specifically, certain representations organize data for fast searching and sorting to enable more interactive visualization processing (e.g., isosurface extraction). Such representations are also used to lay out data very carefully on disk for out-of-core methods.

Data transforms can be used to resample data onto a different type of grid. For example, it may be useful to sample data from a highly unstructured or irregular volumetric grid onto a structured, regular, possibly uniform grid in order to leverage certain high-performing hardware, tools, and/or algorithms that can operate only on structured grids.

The transformations described above generally convert data into forms whose sizes are on approximately the same order of magnitude as the original data. However, there are cases when a new representation incurs substantial overhead in size for data organization or partial replication, resulting in resource tradeoff decisions with regard to their use.

In contrast, certain other data-service functions can be used to simplify and reduce the amount of data for subsequent processing. Once in reduced form, it may be possible to process the data (e.g., visualize it) using relatively modest-capability resources, including those at the desktop. One example is enabled by multiresolution data representations—once a full multiresolution form has been computed, it is straightforward to extract a low-resolution representation of the data, which can be much smaller than the original data. An alternative is to use a sub-sampling function, which would compute a lower-resolution version of the data from a higher-resolution form upon demand. Other techniques for reducing data include the cropping of data to reduce the spatial domain, a reduction in dimensionality (such as extracting 2D slices from 3D data) and the extraction of surface data from volumetric data.

When surface geometries are extracted from very large data, the surface geometries themselves may also be very large. As with raw data, a variety of techniques can be applied to reduce the LOD needed to represent surface-geometry approximations—better techniques make use of error constraints to help ensure high-quality approximations.

One of the most challenging issues associated with large, complex data analysis is that of how to find what one wants, or, more importantly, needs, to see. Research is underway to develop tools that would help identify features and characteristics of data that are of special interest, ultimately helping the user discover information in the data. Such feature detection, extraction, and data mining can be performed as part of the associated computational simulation, or during postprocessing of results data. Unfortunately, features of interest tend to be application-specific, making it difficult to develop general solutions.

The strategic use of compression–decompression techniques can be considered to augment all other data service functions. Compression–decompression can be applied to raw data, extracted surface geometries, and images as appropriate in order to improve the data-transmission characteristics of the end-to-end system.

28.2.2 The Need for Scalable Solutions

The computing resources demanded by complex, high-fidelity simulation applications inherently imply the use of parallel computing. It should be no surprise that the prospect of analyzing data produced by such applications demands that data service and visualization service components also leverage extensive parallelism. Additionally, the data interfaces between parallel components throughout the system must support parallelism, because any sequential interfaces result in bottlenecks that degrade end-to-end system performance.

It is not uncommon for large simulation runs to produce datasets that have the following characteristics:

- At full resolution, the visualization data of current interest do not fit in the memory of a conventional office visualization system.

- The visualization data of current interest, at full resolution and level of complexity, are large enough that conventional office graphics packages are inadequate.

- The whole dataset does not fit on the local disk of a conventional office visualization system; indeed, it may not fit on the local disk of a large visualization system.

- The visualization data of current interest, at full resolution and level of complexity, are large enough that traditional high-performance graphics systems (e.g., a modest-sized SMP with modest graphics parallelism, 4–16 graphics pipelines) lack the rendering performance for interactive data exploration and visualization.

This suggests the need for shareable high-performance servers that can be applied, as needed, to analysis and visualization of data at scale. It also suggests the need for technology that can be scaled to significant performance levels, to hundreds (if not more) of processors and graphics pipelines. The scalability needed is a function of the maximal visualization or rendering capability needed for any one problem and the aggregate capability needed to service a required number of simultaneous users. Such systems are also useful for driving multiple displays in advanced visualization environments, such as powerwalls and CAVEs.

28.2.3 Some Data Facts and Observations

Table 28.1 presents certain statistics from a couple of actual simulation runs that were performed on ASCI supercomputing resources. Both runs shown were performed on the ASCI "White" platform, an IBM massively parallel supercomputer at Lawrence Livermore National

Table 28.1 Data statistics from example simulation runs

Computation	Number of cells	Number of variables	Number of time dumps	Size of one variable	Total size of vis dataset	Anecdotal isosurface data
PPM SC99 Gordon Bell (turbulence simulation)	8 billion	2	273	~8GB (~550MB compressed)	4.3TB (~300GB compressed)	One surface: ~470M triangles (~5GB compressed)
ASCI simulation (LANL on White)	468 million (maximum)	14–25	364	NA	21 TB	Example: ~32M triangles (~773MB) (~4% of total mesh size)

Laboratory. The ASCI run was performed by scientists from Los Alamos National Laboratory (LANL).

We offer the following tips:

• Note the size of the datasets.

• Note the size of the huge isosurface extracted from one of the time-steps in the PPM simulation. While somewhat pathological, it demonstrates that the size of extracted surface geometry can be very big, in this case substantially larger than its associated volumetric data. Also, compare the number of triangles to the number of pixels on a conventional display ($\sim$2M). This comparison suggests the utility of very high-resolution displays for certain visualizations, as well as the opportunity for techniques that would dynamically manage LOD to optimize rendering against display resolution.

• On the other hand, the ASCI simulation statistics show that extracting surface geometry can be an effective way to reduce data, especially for transmission to a remote visualization system. In this dataset, empirical statistics indicated that an extracted isosurface was nominally about 4% to 5% of the size of the entire 3D geometry.

• Even when datasets are very large, it is often possible to comfortably fit one or more full-size data objects on a desktop or laptop local disk.

Table 28.2 presents some simple statistics relating to the migration of visualization data. Statistics such as these are particularly relevant in the consideration of remote data analysis and visualization of large data, or visualization in high-performance computing environments that

are distributed (which most are, whether within a building, across a campus, or across a wide area). The key observation here is that migration of entire datasets can be very difficult, whereas the migration of select data subsets can be achieved in reasonable time.

28.2.4 Optimizing the End-to-End Process

The research community is attacking the problem of large-scale visualization by a variety of approaches. These approaches include the parallelization of visualization and graphics algorithms, data simplification, multiresolution data representation, out-of-core methods, image-based modeling and rendering, and data compression and decompression. While all of these approaches have their own merit for optimizing portions of the data-analysis and visualization process, they sometimes also have their costs. Costs can include time to compute a data transformation, use of a precious resource (e.g., supercomputing cycles), memory/storage overhead, and even loss of information, as some techniques use various forms of data interpolation to enable greater interactivity. It is important that all of these things be considered within the context of a complete end-to-end process. A successful end-to-end process for data analysis and visualization optimizes the end user's time and his or her ability to accurately complete an analysis.

28.3 An End-to-End Architecture for Large Data Exploration

In this section we present a functional system architecture (Fig. 28.2) for high-performance computing environments that can be used to

Table 28.2 Time to transfer various data parts from a single large dataset.

Data Transfer	Full 21 TB dataset	100 GB of selected data objects	1 GB selected data object
@ 100 MB/s	$\sim$60 hrs. ($\sim$2.5 days)	$\sim$17 min.	$\sim$10 sec.

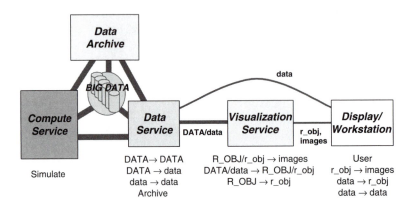

Figure 28.2   A functional system architecture for high-performance computing, data exploration, and visualization.

provide the services and end-to-end system characteristics described in Section 28.2. The architecture links the key components of computational resources, data archival, large high-performance online storage, data service resources, visualization service resources, and the end user's display and/or workstation interface. The diagram shown is functional in the sense that the functional components can be instantiated in a diversity of actual hardware configurations. In one case, all of the functions may exist within a single local system, such as a visualization supercomputer with locally attached storage and archival. In another case, all of the functional components may represent separate pieces of hardware that are attached to the network. In yet another possibility, some of the components may themselves actually be distributed. For example, the data service as a whole may consist of multiple data-manipulation engines, or the visualization service may be distributed across a network.

For convenience, the architecture diagram is annotated as follows:

- Objects shown in capital letters represent large objects, with objects in lower case representing smaller objects (e.g., "DATA" means big data, and "data" means smaller data).

- "R_OBJ" and "r_obj" refer to renderable objects (e.g., polygons, voxels, etc.).

- "$\rightarrow$" is an operator that implies a conversion, extraction, or reduction (e.g., "DATA $\rightarrow$ data" means large data is subsetted or reduced to produce smaller data; "R_OBJ $\rightarrow$ r_obj" means a large renderable object is somehow reduced to a smaller representation; "r_obj $\rightarrow$ images" means renderable objects are converted to images, that is, rendered).

Note that the data service component generally provides high-performance services that can access and process large data into other large or smaller forms, as well as deliver data to visualization resources, which can include the desktop workstation when data is adequately reduced. The data service component also provides access to archive services. The visualization service component provides services that are capable enough to produce renderable objects from large data and render them at high performance. The visualization service can also generate smaller renderable objects for downstream delivery to desktop visualization resources.

A key feature of the proposed architecture is in the shareable access to large data as it resides in one place, on a high-performance storage system. The notion is that data services can access data produced by large-scale simulations without moving the data—the data is accessible from the storage onto which it was written by the simulation. Data services also enable data object extraction, manipulation, and delivery, which, again, obviates the need to move and/or copy entire datasets. The long-standing brute-force habit of moving entire datasets around to get them to target systems for further analysis needs correcting. Users almost never need immediate access to all the data in a large dataset. When performing interactive analysis at any point in time, the user is typically interested in scrutinizing a relatively small percentage of the data. By providing more direct access to data subsets and data objects, we can deliver data more efficiently and we can make better use of available network bandwidth in the end-to-end system.

A very rich data service capability would also track and manage data across a highly distributed environment. Users would make requests to a virtual data service, and just the right data would be delivered to just the right place, transparently repackaging data as needed to enable efficient data migration and/or preparation for whatever subsequent processing is planned. The data service would also catalogue the locations of various pieces of data, using a relatively coarse caching scheme to leave data in temporary distributed locations where it was used, eliminating the need for redelivery whenever possible. Concepts like these are being explored for the Data Grid [3], although the Grid's Replica concept is based on managing replication at the file level rather than at a more abstract data level.

The desktop itself can receive data in one of three forms: (1) small- to modest-sized sets of data objects provided by data services; (2) small- to modest-sized sets of renderable objects that can be rendered at acceptable performance by the local graphics resources; and (3) images delivered from high-performance visualization services. This flexibility allows the remote end user to choose from respective usage modes to (1) make occasional data-service requests that deliver data subsets or special representations (e.g., multiresolution, out-of-core) to the desk-

top for local data analysis and visualization; (2) use back-end high-performance data and visualization services to occasionally extract and deliver graphical objects, such as polygonal surfaces, to the desktop for local interactive viewing; or (3) use back-end high-performance data and visualization services to render final images that are delivered to the end user's display (note that this mode, if used with the expectation of interactive frame rates, demands high quality of service and sustainable network bandwidth between the visualization service and the display and is the most latency-sensitive). All in all, the end user can leverage back-end high-performance resources as needed while also enabling the use of desktop resources, which are significant in today's personal computer systems with commodity graphics cards.

28.4 Commodity-Based Scalable Visualization Systems

The emergence of massively parallel computing has inevitably resulted in the ability to produce massive datasets. For many years, efforts to visualize the largest of datasets were confined to the use of software implementations of parallel visualization and rendering algorithms on general-purpose parallel machines. Traditional high-performance graphics systems did not offer the required graphics scalability, and, if they did so now, the cost of such systems would likely be prohibitive, given the cost of traditional high-performance graphics hardware.

In recent years, the video-game market has resulted in a graphics-hardware revolution. The performance of PC graphics cards for 3D graphics has been increasing at a rate that exceeds Moore's law for general-purpose processors. The result is that it is now possible to buy a commodity graphics card for a few hundred dollars that has greater rendering performance than that of a $100,000 graphics pipeline from 5 years ago. The availability of such low-cost graphics cards, together with cluster com-

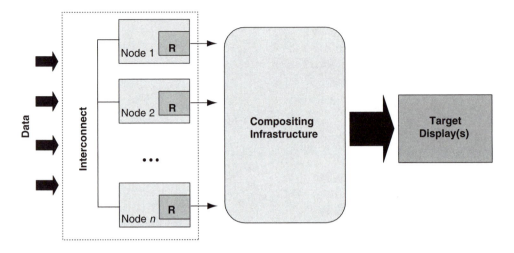

Figure 28.3 A generic clustered-rendering architecture. Boxes labeled "R" represent renderers, that is, graphics accelerator cards.

puting technology, has provided the opportunity to build scalable visualization systems [4] that are better matched with today's scalable supercomputers.

The architecture for a generic graphics cluster is shown in Fig. 28.3. The system integrates a number of computer nodes (generally PCs), each of which includes an attached graphics accelerator. The system uses a high-performance network-interconnect for communication among the nodes. The compositing infrastructure is the part of the system that accepts the image output contributions from all of the independent renderers in the system and composites them into a complete image. The compositing infrastructure may consist of special-purpose hardware and a dedicated network for compositing, or it may simply reuse the system's interconnect and general-purpose processors. Note that the cost of compositing contributes to the total time to deliver a final image, so the system's ability to deliver a new image is min-

imally bounded by the time-to-composite (TTC). For this reason, the parallel graphics community is still exploring ways to provide a dedicated compositing infrastructure [5] that reduces the TTC until it becomes negligible. This is particularly important for real-time interactive graphics applications, that is, applications that demand greater than 30 frames/s.

Each of the graphics cards in a clustered-rendering system embodies a full 3D rendering pipeline. Simplistically, a 3D rendering pipeline implements the following:

[Graphics Primitives] → **3D Geometry Processing** → **Rasterization** → [Display]

Use of many pipelines to render in parallel requires that data be sorted across the pipelines in one of three ways [6]: sort-first, sort-middle, or sort-last. The stages at which each of these sorts would occur in the overall pipeline are shown below:

[Graphics Primitives] → **3D Geometry Processing** → **Rasterization** → [Display]

↑   ↑   ↑

Sort-First   Sort-Middle   Sort-Last

Since sort-middle would impose a sort across the nodes right in the middle of the HW pipeline in the graphics cards, it is not a suitable candidate for use with commodity-based clustered systems, so we will ignore it. That leaves us with sort-first and sort-last, which are shown in more detail in Fig. 28.4.

In the case of sort-first, the image space is partitioned into N regions, where N is the number of renderers, and graphics primitives (e.g., polygons) are sorted according to their destination region on the final image. Each renderer renders all of the primitives that impact its associated image region. Compositing is a simple matter of collecting all the image parts. The sorting of graphics primitives is performed by the general-purpose processors on the cluster nodes, and primitives are redirected across the network-interconnect to their appropriate rendering nodes. The sorting work is load-balanced across the nodes somehow so that each does approximately the same amount of work. Some observations regarding sort-first:

- The amount of communication that has to happen across the interconnect grows as the size of the data grows, creating drawbacks for scalability. For applications where the data is static from frame to frame and successive images are relatively small perturbations of the images that preceded, such as with real-time viewing and/or animations, this can be relieved significantly using frame-to-frame coherence—the distribution of primitives from the previous frame can be used as the starting point for the next frame's sort.

- Any static partitioning of the image space can result in severe rendering-load imbalances when primitives have very high concentrations in certain regions—in the worst

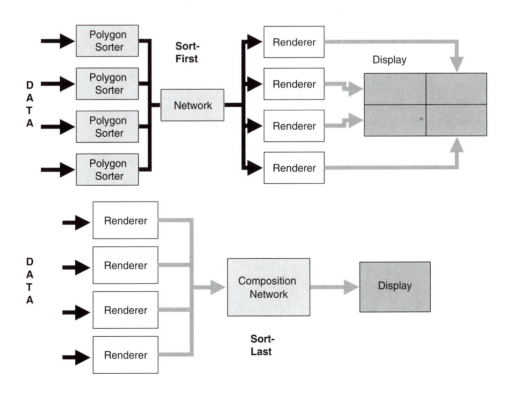

Figure 28.4 Sort-first and sort-last, the two basic approaches for implementing parallel rendering on a graphics cluster.

case, a single renderer (i.e., graphics card) renders all of the primitives.

- Sort-first can be applied straightforwardly to drive tiled displays (i.e., displays that are constructed from multiple displays, typically to achieve high resolution).

In the case of sort-last, each graphics node renders a portion of the graphics primitives onto the full image space (each renderer produces a full-resolution image), after which all of the separate image contributions are used, together with depth information from each renderer's depth buffer (i.e., Z-buffer), to complete a depth sort that produces the final image. Compositing is a more complex matter of performing the depth sort. If no special-purpose compositing infrastructure is present, compositing is performed by reading back the data needed from each of the graphics cards and completing the composition on the general-purpose processors, passing image and depth data around as needed. Some observations regarding sort-last:

- The amount of communication that has to happen across the interconnect is a function of image size, rather than data size, so it scales well as data sizes increase.

- Since each renderer normally renders a full-resolution image, in its simplest form, sort-last is limited to final images whose pixel resolution is no greater than the display resolution supported by the graphics-card hardware. Using sort-last to drive very high-resolution displays (e.g., tiled displays) takes extra effort.

- Proper handling of transparency is hard. Accurate transparency, regardless of sort, requires that primitives be rendered and accumulated in very precise order for each pixel in the final image. With sort-last, data would need to be partitioned spatially in a manner that allowed the resulting partial image contributions, once they themselves have been rendered properly, to be blended together in a predetermined order that is view-dependent.

- How data is distributed onto the parallel nodes for rendering is almost inconsequential, other than for purposes of trying to distribute the amount of rendering work equally. Even a round-robin approach can be used.

With either sorting approach, when the general-purpose processors and interconnect are relied upon to perform compositing, data must be read back from the graphics cards. While this capability generally exists, it may not be optimized by graphics-hardware drivers, as there is no such need in computer gaming. It is also true that any resources that are being used to retrieve data from the graphics cards are temporarily unavailable to contribute to the rendering of a subsequent frame. This data read-back problem is worse for sort-last because it requires read-back of full-size image data and depth-buffer data. The read-back problem as a whole is one reason that special-purpose compositing infrastructures are still being explored.

We have alluded to the use of cluster graphics systems for driving tiled displays. A common approach is to directly connect the display inputs for each of the display tiles to the display outputs from an equal number of cluster graphics cards. The compositing step must then ensure that the proper image parts end up on the proper graphics cards. If the number of graphics nodes equals the number of display tiles, and a naïve sort-first (i.e., a static image-space partitioning based on the physical display-tiles) is used, then the cost of compositing is reduced to zero. This works very well for small enough data, but not so well for large data, for reasons discussed above.

A promising emerging feature of commodity graphics cards is their programmability, which is increasing. Research [7] is showing that this programmability can be used to implement custom functionality that leverages the exceptional performance of graphics hardware. It is anticipated that further research and greater programmability over the next few years will result in innovative use of such hardware to accelerate various algorithms.

An important issue that has yet to be fully resolved for cluster technology in general is that of I/O. It is critical that applications running on a scalable visualization cluster be able to retrieve and write data at very high performance, using parallel I/O streams. The raw ability to connect to numbers of parallel I/O devices for high-performance I/O has been demonstrated—but delivering such raw capability in general-purpose ways that can be easily leveraged by parallel applications is another story. Parallel file systems may ultimately help provide the solution, but applications or application services will likely still have to organize and manage data judiciously to achieve desired performance scalability.

28.5 Interoperable Component-Based Solutions—Is There Any Hope?

Interoperability is a highly desirable attribute for software and systems. The ability to leverage a wide set of interoperable tools to get a job done is very powerful, to say the least. This ability is exemplified in the World Wide Web (WWW), where an endless, extensible set of plug-ins can be used to deliver functionality. Similarly, when applications can be constructed from reusable software components, new applications can be developed quickly and functionality can be easily extended. Of course, interoperability and component-based solutions ultimately rely on standards (not necessarily in a formal sense) and canonical interfaces, even if they are many, as in the WWW environment.

An end-to-end system such as we have described is based on the availability of a broad set of fully integrated services, many of which have high performance. Wouldn't it be nice if this set of fully integrated services could be realized using interoperable and component-based solutions? Clearly, the answer is yes, but when high performance is at stake, this is much easier said than done.

Interoperability is often achieved using the software equivalent of an adapter—if the data

is not already in the desired form, then convert it. This works fine when data is small—the time to convert is often negligible. When a dataset is large, however, such conversions are generally very costly. Occasional conversions of large data as part of the overall data-analysis and visualization process are useful only when the conversion results in a payoff for subsequent data analysis that substantially outweighs the cost of the conversion (e.g., perhaps a conversion to multiresolution form). Any large-data conversions embedded in recurrent processing, especially within parts of the system that are meant to deliver quick responses, are generally unacceptable. High performance is, by necessity, synonymous with minimal large-data movement, copying, and conversion.

Since conversions are to be avoided, why not define a single standard scientific data format that satisfies all application needs and performance requirements? All scientific codes could write the data format directly, and all data and visualization software could be implemented to process it directly. Some researchers have tried. Formats such as HDF [8] and NetCDF [9] have had success, but they are low level. They standardize a relatively small set of data constructs that applications then use to define higher-level data objects, with application-specific semantics (e.g., they standardize constructs such as an array, but they do not specify how to interpret the data in the array). A standard data format may be achievable in specific application domains or in-house environments, but experience suggests that this, too, is sometimes difficult.

Another conflict for interoperable high-performance visualization relates to the reliance of many visualization algorithms on data representations that are optimized for a specific algorithm. For example, suppose a certain isosurface extraction algorithm runs very fast when data is organized just the right way in memory. Use of this algorithm mandates the translation of data to produce the proper organization. In terms of performance, this may be acceptable if the new organization gets adequate reuse, but the overall performance of the algorithm must consider the data translation costs. In terms of interoperable component-based software, such data translations add to the complexity of the software system and can result in memory overhead (especially when multiple representations have to be kept in memory at the same time).

We have already observed the need for parallelism in large data analysis and visualization. This means that we need tools for the development of component-based parallel software. The Common Component Architecture (CCA) [10] is a cooperative R&D effort targeted at the development of such tools. CCA is a framework that enables the development of high-performance components—it does not provide the application components themselves. A use of CCA for constructing modular high-performance applications is depicted in Fig.

28.5 (for these purposes, assume a distributed-memory parallel machine). A set of interoperating components can be developed, from which various combinations can be loaded to form an application. When the application executes, within any one processor, the merged components execute as a single process and in a single memory space. Component interaction is achieved through direct calls to other components—each component "provides" (in CCA lingo) its own interface, which another component "uses" to make calls. We can implement a data object component that provides a canonical set of data objects, together with associated methods, that can be used in common by the set of interoperating components. Through management of the data objects, the data object component can point multiple components to single instantiations of data objects for data sharing.

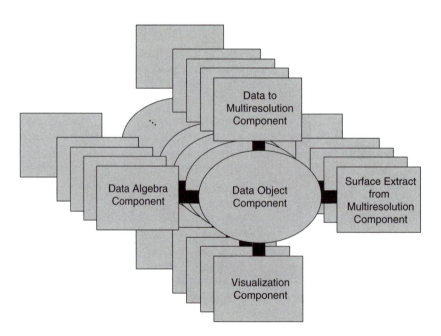

Figure 28.5 A parallel application constructed from CCA components. The components that make up the application are replicated across the set of processors. Each component runs as a parallel module, using message passing as needed to communicate between processors. A data object component can provide a sharable data space to all the components on a processor—the same data, in the same memory location, can be accessible to all of the components. This prevents unnecessary copying of the data from component to component.

Interoperability and component-based solutions collide with high performance in the following areas: the complexities associated with parallelism and parallel interfaces; unacceptability of overheads for embedded, recurring data copy and/or conversion (suggesting a strong reliance on a standard data representation); the challenges associated with defining general-purpose, consensus-based standard interfaces; and conflicts associated with algorithm-specific data representations. Is there hope? We think so, but much work is still needed.

28.6   The ASCI/VIEWS Program

The ASCI/VIEWS program was formed in 1998, although many of the activities that make up the program had their inception with the start of the ASCI program itself a few years earlier. The responsibility of the program is to provide "see and understand" solutions that support an advanced modeling and simulation-based capability for DOE's defense programs. The program is a tri-laboratory effort involving Sandia National Laboratories, Los Alamos National Laboratory, and Lawrence Livermore National Laboratory. The program has also engaged a broad community of collaborators in both industry and academia.

VIEWS has been working to develop scalable solutions for data analysis and visualization on a number of fronts. Some of these efforts are represented in Fig. 28.6 in the form of an end-to-end, layered software-stack, together with certain key partnerships.

It all starts with proper access to the data itself. The ASCI tri-labs are developing a set of scientific data-management tools and higher-level data services, which together with system-level scalable I/O, data transfer, and storage activities in other closely-cooperating parts of the ASCI program are targeted at providing the needed data access and manipulation capabilities. Sandia has a data services development project that is prototyping the use of CCA as a framework for delivering high-performance data-manipulation software.

At the level of visualization applications, VIEWS is working primarily with two tools or tool kits. CEI Inc.'s EnSight Gold [11] tool is a commercial tool that was adopted as a common

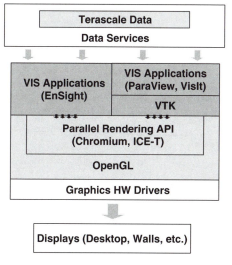

ASCI Tri-Labs:
• Sandia National Laboratories
• Lawrence Livermore National Laboratory
• Los Alamos National Laboratory

Partners (Past and Present):
• CEI (EnSightGold, parallel EnSight)
• Kitware (ParaView, parallel VTK)
• Stanford University (Chromium)
• NVIDIA (Linux graphics drivers)
• RedHat (Chromium and DMX)
• Hewlett Packard (Compositing Infrastructure & Scalable Graphics)
• Princeton University (Scalable Displays)
• IBM ("Bertha" displays)

Figure 28.6   ASCI/VIEWS has worked with partners to develop scalable end-to-end solutions for large-data visualization.

tool for the tri-labs in the early years of ASCI. Since then, VIEWS has guided and supported development of many features in EnSight Gold, including parallelization in the form of the "Server-of-Servers" capability. Efforts are still underway to develop a more complete parallelization of EnSight that would take fuller advantage of clustered visualization systems.

An important use of EnSight's distributed Server-of-Servers capability to support production data analysis for ASCI applications is shown in Fig. 28.7. As proposed during architectural discussions in Section 28.3, this capability allows us to visualize data across great distances without moving the dataset from where it was originally written (in this case, on the large parallel file system connected to the ASCI White platform at Lawrence Livermore National Laboratory). The Server-of-Servers part of EnSight runs on a part of the platform that is dedicated for VIEWS services and has direct access to the platform's parallel file system. It loads data in parallel, extracts surface geometries as directed by the end user from the client side, and sends those geometries together with associated data on the surface to the client.

Rendering occurs on the client side, so a static surface geometry can be viewed at will until different data is desired. Since the surface data is normally relatively small in comparison to the full dataset or the full 3D volume of data from which it is extracted, it can be transmitted quickly enough to support interactive data-visualization sessions. It should be noted that ASCI has deployed a dedicated wide-area network (WAN) between the tri-labs that supports aggregate data transmissions at OC-12 rates, or approximately 2.5 Gbits/s.

The Visualization Toolkit (VTK) [12] is an open-source toolkit, well known in the visualization community, that receives ongoing maintenance, support, and development from Kitware, Inc. VIEWS investments have supported the development of a parallel and distributed version of VTK, as well as the ParaView [14] application built on it. Lawrence Livermore National Laboratory has implemented its own custom end-user application on top of VTK, which they call VisIt. VTK is attractive to VIEWS as an open-source framework that can be extended directly by the tri-labs, as well as a framework for at least some of

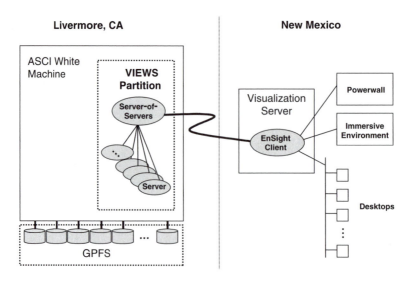

Figure 28.7   Long-haul distributed visualization using EnSight's Server-of-Servers.

the R&D performed at the laboratories in visualization algorithms and techniques, etc.

To provide parallel rendering for graphics clusters, the tri-labs have partnered with Stanford University and RedHat to develop and deliver the open-source Chromium [15,16] software, which is gaining popularity in the parallel-rendering community. Chromium provides an elegant framework for scalable rendering that enables customizable implementations. Chromium is based on a streaming-data approach that is inherently better suited to sort-first parallel rendering than to other types. Chromium has been integrated with EnSight and VTK. Researchers at Sandia National Laboratories have developed an alternative parallel-rendering library known as ICE-T, which uses sort-last principles [17]. ICE-T has been integrated with VTK. The two together have been used with a 128-node graphics cluster at Sandia to render a 470-million polygon dataset at rates exceeding 1 billion polygons/s. Both Chromium and ICE-T are based on the OpenGL standard. Indeed, Chromium can be used with legacy OpenGL applications to provide parallel rendering without any changes to the application.

When VIEWS first started working with commodity graphics cards, it was recognized that there were shortcomings for the graphics drivers in the Linux domain. This was addressed through funding of a couple of efforts to develop and/or optimize Linux OpenGL graphics drivers, including a partnership with NVIDIA. This has directly resulted in the creation of certain quality Linux drivers, as well as some heightened awareness of the desire for drivers in the Linux domain. At the same time, the sustainable availability of quality Linux-based graphics drivers is still a question mark, as the gaming industry has no market demand in the Linux arena. While VIEWS is strategically pursuing Linux-based clusters for scalable data and visualization services, the labs are also still exploring the use and integration of Windows-based systems.

VIEWS has also partnered to develop advanced display technologies. A partnership with Princeton University has helped explore and promote the development of cluster-driven powerwall technologies, or scalable displays [18]. A diversity of display walls, including a 60-million-pixel wall at Sandia, have been deployed at the ASCI tri-lab sites. Such a wall

Figure 28.8 Sandia's 128-node graphics cluster (left) and 60-million-pixel display wall (right). The wall is driven by graphics-cluster technology. The image on the wall is a rendering of the 470-million-polygon isosurface extracted from the PPM dataset described in Section 28.2. (See also color insert.)

is a resource for visualization at superior resolutions that are better matched to the fidelity in large-scale simulations, a resource for collaborative data visualization by a group of people, and a setting for high-quality presentation. A separate partnership with IBM was instrumental in the production of super-resolution office display technology, namely for the development of IBM's "Bertha" display, now more properly known by model numbers such as T221. The Bertha is a flat-panel display that delivers more than 9 million pixels on a 21-inch display.

28.7   Conclusion

In the simple analysis, the power of computers and their environments continues to grow, resulting in a propensity by their users to generate more and more data and larger and larger datasets. Can there be any doubt that the terabyte datasets of today will be dwarfed in the years to come? At the same time, computing is becoming a fundamental tool of modern science. Still, the ability to do effective science is ultimately limited by the ability to discover and understand results.

Together, these observations suggest a critical need for tools and environments that enable effective and efficient data analysis and visualization of large-scale data, in ways that need to become increasingly intuitive, even as the data itself continues to grow and become more complex. Features of such tools and environments include the following:

- The ability to process large data without moving it all around, and the ability to manipulate data at an object level rather than at a complete dataset level.

- A rich functional set of data and visualization services that enable data exploration and discovery.

- The ability to get the right data to the right place for further processing when needed.

- Wide availability of cost-effective, scalable, high-performance infrastructures that can be used to provide intensive data and visualiza-

tion services, proportionate to available computing resources (i.e., for simulation).

- Parallelism and the ability to support end-to-end parallelism throughout high-performance parts of the environment.

- The ability to apply a broad set of diverse tools, leveraging interoperability.

- The ability to leverage the interactivity and power of increasing desktop computing and visualization resources, which are dedicated to the scientist and analyst.

- The ability to drive the whole environment from the computational-science laboratory of choice, namely the office.

In the current state of technology, many piecemeal solutions are being developed and provided that target some niche aspect of large-scale data visualization. Efforts such as VIEWS are also working to construct end-to-end environments that integrate key solution approaches. As environments become more functional, one future challenge will be to enable the use of such environments in a relatively seamless, intuitive way.

Acknowledgments

Thanks to Randy Frank at Lawrence Livermore National Laboratory for access to PPM simulation data and statistics. Thanks to Bob Kares at Los Alamos National Laboratory for ASCI simulation statistics. Thanks to Brian Wylie and Rena Haynes for review. Thanks to many VIEWS coworkers and collaborators.

Further Reading

- Special issue on large-scale data visualization, *IEEE Computer Graphics and Applications*, 21(4), IEEE Computer Society, 2001.

- 2001 Symposium on parallel and large-data visualization and graphics, *Proceedings, IEEE*, Catalog Number 01EX520, ACM Order 429017, 2001.

- C. Silva, D. Bartz, J. Klosowski, W. Schroeder, and P. Lindstrom. *Tutorial M-9: High Performance Visualization of Large and Complex Scientific Datasets.* Course Notes, SC 2002, 2002.
- S. Whitman. *Multiprocessor Methods for Computer Graphics Rendering.* Jones and Bartlett Publishers, 1992.

References

1. P. H. Smith and J. van Rosendale. Data and visualization corridors. Report on the 1998 DVC workshop series, Technical Report CACR-164, California Institute of Technology, 1998.

2. National Nuclear Security Administration's Advanced Simulation and Computing Program, http://www.nnsa.doe.gov/asc/home.htm

3. B. Allcock, S. Tuecke, J. Bester, J. Bresnahan, A. L. Chervenak, I. Foster, C. Kesselman, S. Meder, V. Nefedova, and D. Quesnel. Data management and transfer in high-performance computational grid environments. *Parallel Computing Journal*, 28(5):749–771, 2000.

4. B. Wylie, C. Pavlakos, V. Lewis and W. Moreland. Scalable rendering on PC clusters. *IEEE Computer Graphics and Applications*, pages 62–70, 2001.

5. A. Heirich and L. Moll. Scalable distributed visualization using off-the-shelf components. *1999 IEEE Parallel Visualization and Graphics Symposium Proceedings*, pages 55–59, 1999.

6. S. Molnar, M. Cox, D. Ellsworth, and H. Fuchs. A sorting classification of parallel rendering. *IEEE Computer Graphics and Applications*, pages 23–32, 1994.

7. B. Wylie, K. Moreland, L. A. Fisk, and P. Crossno. Tetrahedral projection using vertex shaders. *Volume Visualization and Graphics Symposium 2002*, Proceedings, IEEE, 2002.

8. National Center for Supercomputing Applications' Hierarchical Data Format Group, http://hdf.ncsa.uiuc.edu/

9. Network Common Data Form, http://www.unidata.ucar.edu/packages/netcdf/

10. R. Armstrong, D. Gannon, A. Geist, K. Keahey, S. Kohn, L. McInnes, S. Parker, and B. Smolinski. Toward a common component architecture for high-performance scientific computing. *Proc. Eighth IEEE International Symposium on High Performance Distributed Computing*, IEEE, 1999.

11. Ensight Gold, http://www.ceintl.com/products/ensightgold.html.

12. Visualization Toolkit, http://www.vtk.org/

13. J. Ahrens, K. Brislawn, K. Martin, B. Geveci, C. C. Law, and M. Papka. Large-scale data visualization using parallel data streaming. *IEEE Computer Graphics and Applications*, 21(4):34–41, 2001.

14. ParaView, http://www.paraview.org/HTML/Index.html

15. The Chromium Project, http://sourceforge.net/projects/chromium

16. G. Humphreys, M. Houston, R. Ng, R. Frank, S. Ahern, P. Kirchner, and J. T. Klosowski. Chromium: a stream-processing framework for interactive rendering on clusters. *SIGGRAPH Proceedings*, ACM, 2002.

17. K. Moreland, B. Wylie, and C. Pavlakos. Sort-last parallel rendering for viewing extremely large data sets on tile displays. *Proc. 2001 Symposium on Parallel and Large-Data Visualization and Graphics*, IEEE, 2001.

18. K. Li, H. Chen, Y. Chen, D. W. Clark, P. Cook, S. Damianakis, G. Essl, A. Finkelstein, T. Funkhouser, T. Housel, A. Klein, Z. Liu, E. Praun, R. Samanta, B. Shedd, J. P. Singh, G. Tzanetakis, and J. Zheng. Building and using a scalable display wall system. *IEEE Computer Graphics and Applications*, 20(4):29–37, 2000.

29 Consuming Network Bandwidth with Visapult

WES BETHEL and JOHN SHALF
Lawrence Berkeley National Laboratory

29.1 Introduction

During the period of the Next Generation Internet Combustion Corridor project (1999–2000), there was a conundrum: the network research community declared there were no applications that could take advantage of high-speed networks, yet applications developers were confounded by poorly performing networks that impeded high-performance use. Our goal during that period was to create an extremely high-performance visualization application that not only was capable of pushing the performance envelope on wide-area networks (WANs) but was generally useful from a scientific research perspective. We wanted an application that would perform remote interactive visualization of large scientific datasets but that would not suffer from the delays inherent in network-based applications. We wanted our application to use multiple, distributed resources so that data, user, and required computing machinery need not be collocated. Such objectives are consistent with the needs of contemporary scientists: large datasets are often centrally located and have many remote users needing to view and analyze the data.

Our answer was to create the Visapult application. Visapult is a visualization application composed of visualization of multiple software components that execute in a pipelined-parallel fashion over WANs. By design, Visapult was tailored for use in a remote and distributed visualization context. The first use of Visapult was visualization of turbulent-flow simulation data computed on supercomputers at the Na-

tional Energy Research Scientific Computing (NERSC) Center by a researcher located at Sandia National Laboratories in Livermore, CA during the early part of 2000. Over time, we broadened our efforts to include a more careful study and more deliberate use of networking infrastructure, as well as refinement of the Visapult design to maximize performance over the network. Visapult is arguably the world's fastest performing distributed application, consuming approximately 16.8 gigabits per second (Gbits/s) in sustained network bandwidth during the SC 2002 Bandwidth Challenge over transcontinental network links. Visapult's performance is a direct result of architecture, careful use of custom network protocols, and application performance tuning.

In this chapter, we reveal the secrets used to create the world's highest-performing network application. In the first section, we present an overview of Visapult's fundamental architecture. Next, we present three short case studies that reflect our experiences using Visapult to win the SC Bandwidth Challenge in 2000, 2001, and 2002. We conclude with a discussion about future research and development directions in the field of remote and distributed visualization.

29.2 Visapult Architecture

Visapult is a highly specialized and extremely high-performance implementation of and extension to the Image-Based Rendering Assisted Volume Rendering (IBRAVR) proposed by Müller et al. [12]. Visapult consists of two software components, a *viewer* and a *back-end*. The viewer component implements the IBRAVR framework using OpenRM Scene Graph [3], an Open Source scene graph API. The viewer receives the source images it needs for the IBRAVR algorithm from the back-end component. Each of the viewer and back-end components is implemented as a parallel application, and they communicate using a custom TCP-based protocol to exchange information (control data) and images (payload data). Fig. 29.1 illustrates the architecture, which is discussed in substantially more detail in an earlier publication [2].

Looking more deeply into each of these components, we see that each of the back-end and viewer components is in turn a parallel application. The back-end is a distributed-memory parallel software component written using MPI. Each back-end processing element (PE) is responsible for reading and rendering a subset of a 3D volume of data and sending the resulting image to a peer listener in the viewer. The viewer is also a parallel application, but it uses threads and a shared memory model. The viewer creates one thread per back-end PE to receive image payload data. Each of these listener threads is responsible for updating a portion of a thread-safe scene graph. A detached rendering thread performs interactions with the user and invokes

the scene graph system's frame-based rendering function to generate new images.

Our earliest Visapult deployment tests in late 1999 used data sources that were resident on the same platform as the back-end PEs, and our attention was focused upon the wide-area networking behavior between the back-end and the viewer. We wanted to use computational resources located "close to" the data and provide interactive visualization capabilities to a remotely located user. With such requirements, our primary concern was the network link between the back-end and the viewer.

However, in preparing to use Visapult for the SC2000 Bandwidth Challenge, we wanted to stretch the limits of our design and use the emerging high-speed network backbones in order to test the hypothesis that pipelined parallelism over a WAN was useful for remote and distributed visualization. During early field tests, we had established that Visapult's pipelined architecture did indeed support interactivity on the desktop regardless of the performance of the underlying network. Early in Visapult's evolution, the principal developer routinely used a DSL connection between the back-end and the viewer. Even with a DSL line, the viewer was completely interactive on the desktop; the listener threads would accumulate the

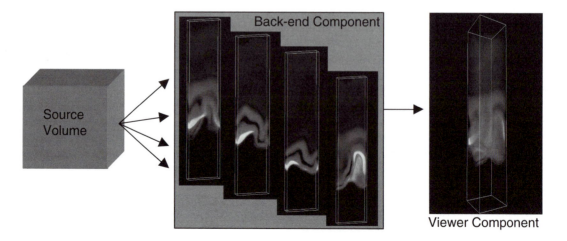

Figure 29.1 Visapult consists of a viewer component and a back-end component. (See also color insert.)

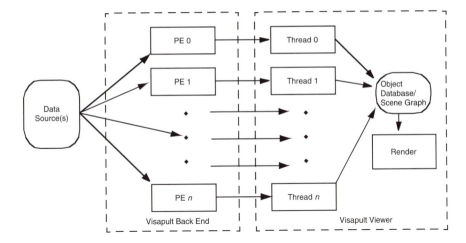

Figure 29.2 Visapult's end-to-end data parallelism: major blocks represent components that can be collocated or separated via long-haul network connections.

next frame's worth of partial prerendering before updating the scene graph. Meanwhile, the freely running rendering thread would continue to operate at desktop rates. This behavior is possible because of the substantial amount of data reduction that occurs as scientific data traverses the Visapult pipeline: incoming data is of size $O(n^3)$; the back-end renders data down to images, resulting in data of size $O(n^2)$.

We were motivated to find a way to successfully capitalize upon an emerging trend within the computational science community: a few high-powered resources are interconnected via high-speed fabric, and they provide service to users regardless of location. The idea is that data produced by simulations is stored on data caches located near the resource used to perform the simulation in network space. Similarly, subsequent visualizations are performed by remotely located users, leveraging resources that are interconnected by high-performance networks. Ideally, the user need not be aware of the vagaries of accessing remote resources; all such connections are automatically "brokered" by an intermediate agent. Visapult's design—using IBRAVR to reduce the data load between the back-end and the viewer by an order of magnitude—fits well within this model. With these objectives in mind, we begin to explore the use of high-speed, network-attached data caches.

29.3   Reading Data Over the Network Using TCP

To begin the process of using such data caches, we added support in Visapult to use the Distributed Parallel Storage System (DPSS) [16], created by the Data Intensive and Distributed Computing group at Berkeley Lab. The basic idea behind DPSS is "RAID-0 over the network." To the DPSS client, for example, each back-end PE, the DPSS looks like a file that is accessible through *open, seek*, and *read* calls. DPSS is a block-oriented system composed of commercial, off-the-shelf components and the DPSS library (Fig. 29.3). Using the DPSS library, a client application issues a "file open" call, which in turn results in many different files' being opened on the DPSS servers, all mediated by the DPSS master. Next, the client's "read" calls are routed to the DPSS master, but they are answered by the DPSS servers. The client application does not know that it is, in fact, talking to multiple machines on the network, since the DPSS libraries encapsulate those

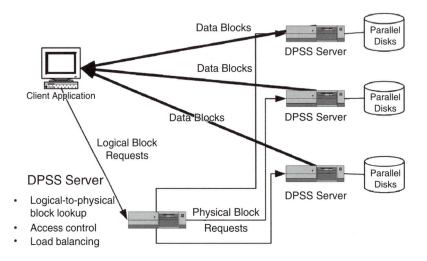

Figure 29.3 DPSS Architecture.

details. We'll discuss the performance characteristics of DPSS in the section on the SC00 Bandwidth Challenge.

During field testing, we used one DPSS system located in Berkeley and two separate computer systems located at Sandia National Laboratories. One Sandia system was an SGI Origin system, and the other was the "CPLANT" facility, which is a large, Alpha-based Linux cluster. Between the two sites was an experimental OC-48 NTON network connection. (NTON—the National Transparent Optical Network test bed— was an experimental WAN that consisted of dedicated OC-48 (and higher) segments connecting sites on the west coast of the United States. The NTON website, *http://www.ntonc.org*, is now defunct.) Despite the substantial amount of computing and NIC horsepower, we were not satisfied with performance results, which indicated we were consuming only a small fraction of a dedicated OC-48 link between two sites (about 10–15%).

Based upon initial performance data, we began to perform careful execution profiling of the Visapult application using a tool called Netlogger [15]. In order to use Netlogger's profiling capabilities, you must instrument your

application by inserting subroutine calls to the Netlogger library. When the instrumented application is executed, the Netlogger code invoked from the application sends data to a Netlogger host, which accumulates data (similar in fashion to the familiar syslog facility in Unix systems). After a run has been completed, Netlogger's analysis and display tool presents the profile data obtained during the run for visual inspection. Netlogger's strength lies in its ability to perform profiling and execution analysis of distributed software components.

In Fig. 29.4, we see the results obtained from profiling the Visapult back-end and viewer for a sample run. The graph on the top shows the execution profile of the Visapult viewer. The graph on the bottom shows the execution profile of the Visapult back-end. The horizontal axis of the graph is elapsed time, and the vertical axis represents instrumentation points in the code. For the purposes of this discussion, we are concerned only with the performance of the Visapult back-end, since the "heavy payload" network link was between DPSS and Visapult's back-end. The profile graph of the back-end is colored so that even-numbered frames are shown in red and odd-numbered frames in blue. The observation

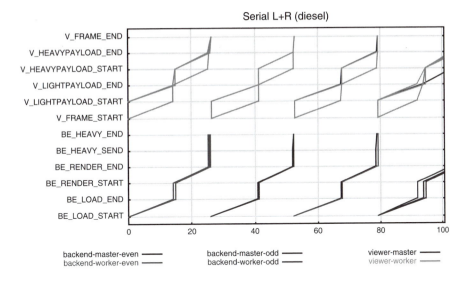

Figure 29.4   Visapult duty cycle with serial I/O and rendering. (See also color insert.)

we can make from the performance profile shown in Fig. 29.4 is that the network performance will never reach its theoretical maximum value because I/O is blocked while rendering occurs. Data is loaded into the back-end between the vertical axis labels "BE_LOAD_START" and "BE_LOAD_END." Then, back-end rendering occurs between "BE_RENDER_START" and "BE_RENDER_END." The result is "gaps" of time when there is no network I/O.

As a result of the performance analysis shown in Fig. 29.4, we modified the Visapult back-end so that rendering and network I/O were placed into separate threads of execution. (In MPI-parallel applications, it is possible for the MPI application to launch detached threads, but those threads typically cannot directly participate in MPI communication—only the parent/master threads can. The converse is not true.) By doing so, we were able to maintain a constant load on the network, thereby eliminating one source of network inefficiency in Visapult. Fig. 29.5 shows the profile analysis resulting from having I/O and rendering occur simultaneously in the back-end. So long as software rendering speed exceeds the time required to

load data over the network, the network link will be kept as full as possible. Furthermore, as long as the time required for rendering is less than the time required to send data over the network, the network will remain completely filled—or so we thought. The reality turns out to be quite different.

29.4   The Bandwidth Challenges and Results

In 2000, a new high-performance computing contest was announced: the High Performance Network Bandwidth Challenge. The primary objective of the competition was to use as much network bandwidth as possible within a window of time. Secondary objectives included most creative use of network bandwidth and so forth. The "real prize," however, was unabashed and gluttonous consumption of resources. We felt, based upon the results of our field tests and performance analysis, that we had a reasonably good chance of making a competitive showing at the Bandwidth Challenge (BWC) using Visapult. The sections that follow describe our experiences

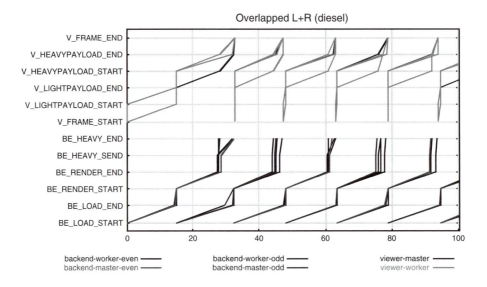

Figure 29.5   Visapult duty cycle with overlapped I/O and rendering. (See also color insert.)

in the SC BWC during the years 2000 to 2002. As we shall show, we learned many lessons during these competitions.

29.4.1   SC 2000 Bandwidth Challenge

For the SC 2000 BWC, we teamed with Helen Chen and Jim Brandt from the Networking Security Research Group at Sandia National Laboratories and with Brian Tierney, Jason Lee, and Dan Gunter of the Data Intensive and Distributed Computing group at Lawrence Berkeley National Laboratory (LBL) in Berkeley, CA. The resources we used included an eight-node DPSS system located at LBL, an OC-48 connection to SC00 in Dallas, TX, and a pair of hosts on the show floor. One host was an eight-node SGI Onyx2, located in the ASCI booth. The other was a small Linux cluster, located in the SC00 booth of Argonne National Laboratory (ANL). We ran two separate Visapult back-ends, one on each of these two platforms, and ran Visapult viewers in the LBL booth (Fig. 29.6).

Fig. 29.7 shows the performance of our application as measured in the SCinet NOC during our run. Despite having adequate computing power and a dedicated OC-48 link, we were able to consume only a fraction of the available network bandwidth. The "glitch" in the middle of our run occurred because of an application crash, which required us to restart the DPSS. During our 60-minute run, we achieved a peak bandwidth rate of about 1.48 Gbits/s, with a sustained average of about 582 Mbits/s.

The fundamental reason Visapult was not able to sustain greater than 25% of the theoretical network capacity on the OC-48 link between LBL and Dallas, TX was TCP, not application design. Visapult and DPSS were able to load the network with data, but the flow-control algorithm used by TCP resulted in poor throughput, even on an unloaded network. It is well known that single-stream TCP performance is poor on high-bandwidth networks. The response is to use a multistreamed approach, which is exactly what we did for our SC 2000 BWC entry. The TCP congestion-avoidance algorithm regulates the rate at

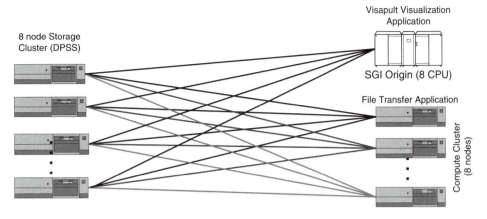

Network Throughput: 5-second peak 1.48 Gbits/sec (72 streams: 20.5
Mbits/stream); 60-minute sustained average 582 Mbits/sec

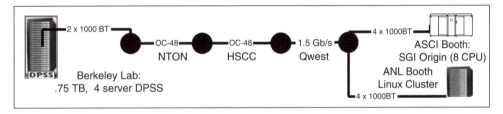

Figure 29.6   SC 2000 Bandwidth Challenge resource map.

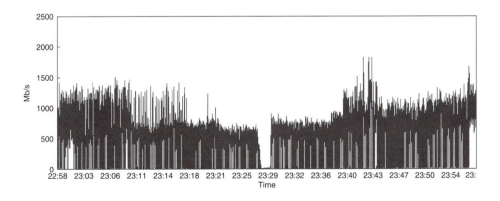

Figure 29.7   SC 2000 Bandwidth Challenge results.

which packets are dispatched onto the network in response to perceived congestion events. It performs such control on a per-stream basis, which means that any flow-regulation activities performed on one stream are independent of all other streams. In other words, a congestion event occurring on one stream, along with the resultant "TCP backoff" (reduction in bandwidth usage), will not affect peer TCP streams. This method creates a network stream that behaves as a "bully," because it doesn't respond as rapidly to congestion as its peers. As you can see from the performance graphs in Fig. 29.7, the result can best be characterized as "unstable." Such instability has a huge impact on the quality of interactive visualization applications as they stutter and halt, while the TCP stream's data rate "thrashes" in response to perceived congestion.

29.4.2 SC 2001 Bandwidth Challenge

Based upon the lesson we learned during SC 2000—namely, that TCP is inadequate as a high-performance network protocol—we tried a different approach for 2002. We modified Visapult to use a custom, application-level protocol based on the User Datagram Protocol (UDP) method of packet movement. The fundamental difference between TCP and UDP is that TCP guarantees delivery of packets in the order sent, whereas UDP makes no such guarantees. In UDP, packets could arrive in any order, or they might not arrive at all. In terms of performance, TCP uses mechanisms that control the rate at which packets are sent over the network. In contrast, UDP has no such flow regulation. The flow-congestion avoidance algorithm used by TCP is widely recognized as the reason TCP-based flows are unable to realize more than 25–30% of the theoretical line rate. We felt compelled to use UDP in order to achieve the maximum possible level of network performance.

UDP-based flows are "unregulated" in the sense that there is no flow control provided by the IP stack. Whereas TCP automatically adjusts its flow rate in response to environmental conditions, UDP flow-rate management must be performed by the application. One can mimic TCP's behavior using Madhavi and Floyd's TCP-friendly method [10], but in doing so, one will suffer the same corresponding performance loss that we've come to expect of TCP. Rather than attempting to infer congestion from packet-loss rates at run time, we de-

cided to carefully select a flow rate based on the measured end-to-end capacity of these dedicated links, allowing us to realize nearly 100% of the theoretical line rate. While abandoning TCP is considered "cheating" in some circles, we feel that we were merely addressing the problem at hand in the most direct manner possible. Under these conditions, we felt that we would be able to realize much higher rates of bandwidth utilization by employing UDP than would be possible with TCP, and our SC 2001 BWC entry was the field test for our hypothesis.

Unfortunately, replacing TCP with UDP was not as straightforward as replacing one set of subroutine calls with another. In our SC 2000 TCP-based implementation, Visapult would request a large block of data from the DPSS with a single *read* call. The large block of data represented one time-step's worth of data. Then, after the data for a given time-step had arrived, the back-end would begin rendering (and at the same time, the next time-step's worth of data was requested). One of the nice properties of TCP is that the application can ask that N bytes of data be moved from one machine to another. TCP breaks the N bytes of data into packet-sized chunks on the outbound side, and reassembles the N bytes of data from individual packets on the receiver side. In contrast, UDP-based applications are completely packet-oriented: there is no mechanism built into the IP stack that breaks large data blocks into packets on the sender side, nor is there one that reassembles packets into blocks on the receiver side. (In IPv4, large packets are automatically fragmented and reassembled. This activity is not supported in IPv6.) The maximum amount of data that can be sent by a UDP-based application with one transfer is one packet's worth of data. (The size of a network packet is defined by the Maximum Transmission Unit [MTU], which is the minimum of the configuration parameter of a Network Interface adapter and the minimum packet size supported by all intervening switches along a given path on the network. Typically, the Ethernet MTU is 1500 bytes. So-called "jumbo frames" use an MTU of 9000 bytes.) It is up to the application to partition large data blocks into packets for transmission by UDP, and then to reassemble the packet payload data into a large data block on the receiving end.

Another difference from our previous entry was the use of a live-running simulation as a data source rather than use of precomputed data stored on the DPSS disk cache. The simulation was built using the Albert Einstein Institute's Cactus framework (*http://www.cactuscode.org*), which models the collision of binary black holes by directly evolving Einstein's equations for general relativity. This particular simulation has substantial computational demands, including the need for exorbitant amounts of physical RAM and processing power in order to solve a very large set of equations. More information about general relativity is provided at *http://jean-luc.aei.mpg.de*.

In order to enable use of UDP in Visapult, we needed to do away with "frame boundaries," while at the same time using the existing back-end architecture, which provided for simultaneous network I/O and rendering. In addition, we needed to provide enough information in each UDP packet so that each packet could be treated independently. To that end, we encoded contextual information into the header of each packet. The information encoded into the header was sufficient to identify the location in the computational grid where the data payload should be placed. In this way, placement of the payload data, or the actual simulation data we wanted to visualize, was independent of the order in which packets arrived. The encoding scheme we used is described more fully elsewhere [13], and it resolved the problem of packets arriving out of order.

To a large extent, we can avoid creating conditions in which packets are dropped by regulating the rate at which UDP packets are placed onto the network. Generally speaking, packets are lost because of unrecoverable bit errors that are detected by the Cyclic Redundancy Check (CRC) or because of buffer overflows either at the endpoints or at congested switch interfaces in the network core. Our testing showed a very low rate of packet loss when we manually regulated the flow to match the known end-to-end

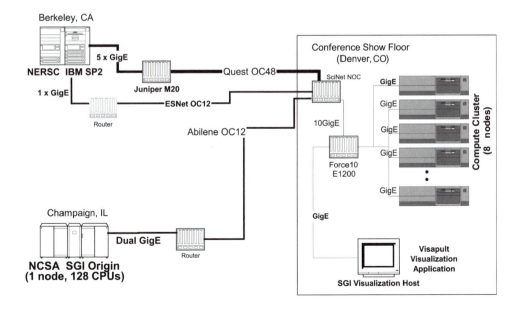

Figure 29.8   Resource map for the SC 2001 Bandwidth Challenge.

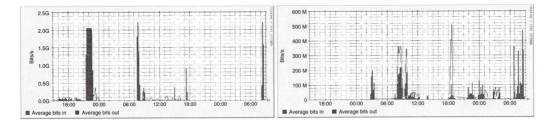

Figure 29.9   These are images of the MRTG performance graphs from the incoming data streams arriving from Esnet OC-48 and NCSA OC-12, respectively, for the SC 2001 Bandwidth Challenge. The challenge took place for approximately 7 minutes at 8:30 a.m. The performance MRTG graphs actually underreport the bandwidth utilization because they use a 5-minute average. The instantaneous bandwidth (which also included an additional ESNet OC-12) reached 3.3 Gbits/s. The 2+ Gbit peaks on the ESNet graph were practice runs using Visapult. (See also color insert.)

network capacity. This approach is exactly what we used in our custom UDP protocol between Cactus and Visapult. While we can minimize packet loss through careful flow regulation, packets are still sometimes dropped. In our experiments and testing, we find the impact of a few lost packets on the resulting visualization to be negligible. In fact, we observe that many well-accepted forms of media are predicated upon lossy compression methods. JPEG and MPEG compression, for example, typically produce 10–80% of information loss during the compression process, but the resulting images are visually quite acceptable. On the other hand, packet loss is entirely unacceptable in certain critical systems, such as medical applications. For the purposes of visualization for remote monitoring, however, consumers of this technology are well accustomed to lossy representations of information. An example of a Cactus/Visapult visualization complete with missing data is shown in the "Lessons Learned" section.

The demonstration involved a simple task-spawning scenario with the Cactus code. In this case, a large simulation typically runs for a few days on the 5-teraflop NERSC SP-2 simulating the merger of two black holes. During the course of the simulation, events such as the joining of the event horizons of the two black holes that require additional analysis occur. Rather than interrupt the current simulation, Cactus can spawn off the "horizon finder" to perform the specialized analysis on the SGI Origin supercomputers at NCSA, as a slave to the master simulation running at NERSC. Visapult's distributed-component architecture made it very simple to use our "visualization cluster" to process large quantities of live data arriving from multiple, widely distributed simulation resources in real time.

The SCinet network engineers who were monitoring the performance of our challenge entry were shocked to see that the ramp-up of the data stream was instantaneous, whereas a typical TCP-slow start would take several minutes to reach full line rate, even assuming absolutely no packet loss. Within 7 minutes, we were able to ramp up beyond 3 Gbits—double the performance of the nearest competing application (which was also UDP-based). For the first time in many years, we were able to fully utilize a dedicated high-bandwidth pipe—a feat that had been increasingly difficult to achieve, given TCP's inadequacies.

29.4.3   SC 2002 Bandwidth Challenge

Because of the success of our SC 2001 entry, along with results we obtained in the laboratory during the summer of 2002, our SC 2002 entry was intended to literally "flatten" any network we could get our hands on. (In July of 2002, we were able to completely fill a 10 G network link using the Cactus/Visapult combination. See *http://www.supercomputingonline.com/article.php?*

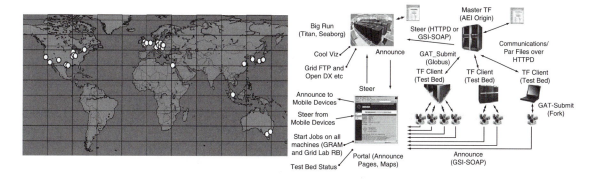

Figure 29.10 (Left) A map of sites participating in the SC 2002 Bandwidth Challenge Global Grid Testbed Collaboration. (Right) A the logical workflow of the Grid's task-farming scenario. The task farmer uses an application-level abstraction layer for Grid Services called the Grid Application Toolkit (GAT). For more information on the GAT, please refer to *http://www.gridlab.org*. (See also color insert.)

sid=2252 for more information.) As it turned out, garnering resources for the SC 2002 entry proved to be the real challenge. The result is reflected in the diversity of team members and resources we assembled for the SC 2002 run. The SC 2002 entry was a transglobal collaboration involving supercomputers, networking resources, and people from around the world drawn from a different project known as the Global Grid Testbed Collaboration (GGTC). (The GGTC was organized during the Global Grid Forum Applications Working Group Meeting in Chicago in the fall of 2002. The GGTC includes people and resources from five continents and over 14 countries that aggregated about 70 machines totaling approximately 7500 processors. Architectures ranged from a 5-teraflop SP2 system to a Sony Playstation 2 running Linux. See *http://scb.ics.muni.cz/static/SC2002/.)*

Our BWC application expanded on the task-spawning scenario of the SC 2001 entry. In the task-spawning scenario, the primary simulation spawns dozens of smaller subtasks during execution in order to perform a run-time parameter study with results that feed back to the master simulation. The master simulation responds by making adjustment to the spawned simulations—in effect, steering them. For instance, the master simulation can spawn a parameter

study to determine the effect of adjusting its courant factor. The result might show improvement in the evolution rate, but it might also show an unexpected impact on the accuracy and stability of the numerics. The primary simulation code could actually scan for available resources worldwide to launch slave simulations to explore all of those possibilities. Visapult fits into this framework as a means of visually inspecting the results of these slave simulations in real time using volume rendering.

For the purpose of the BWC, we focused our effort on five well connected sites: NCSA in Champaign, IL, NERSC in Oakland, CA, Argonne National Laboratory in Chicago, the University of Amsterdam, and Masaryk University in the Czech Republic (Fig. 29.11). During the actual challenge run, we opted to force manual launching at these strategic sites in order to reach full bandwidth within the 15 minutes we were allotted for the benchmarked run. The resulting run managed to push an unprecedented 16.8 Gbits/s of data to the LBL cluster on the SC show floor in Baltimore from these worldwide-distributed simulation sources. Fig. 29.12 shows the SCinet "weather map," which depicts the bandwidth consumed over each of the three inbound WANs used as part of our run. Note that the sum of bandwidths across the three WANs totals 17.2 Gbits/s. Our

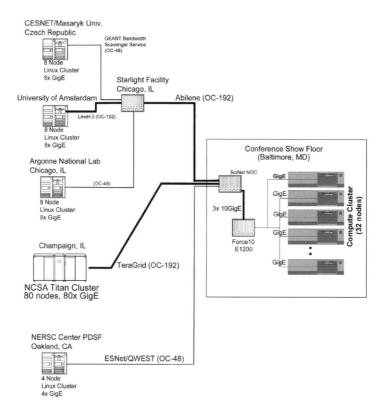

Figure 29.11   SC 2002 Bandwidth Challenge resource map.

metered average rate of 16.8 Gbits/s covers a 15-minute window, and we were ahead of the average rate at the time Fig. 29.12 was made.

In contrast, Fig. 29.13 shows the bandwidth measured between the SCinet NOC outbound to the three 10-Gbit connections to the LBL booth. Figs. 29.12 and 29.13 are two different views of the same network traffic. While Fig. 29.12 shows flow rates on each of the inbound WANs, Fig. 29.13 shows the flow rate on the LAN side of the SCinet NOC. During the SC 2002 run, we achieved a peak transfer rate of about 17.21 Gbits/s and had a minimum transfer rate of about 15.04 Gbits/s. During our 15-minute window, we transferred a total of approximately 138.26 Tbits of data from the remote resources into the Visapult back-end.

Visapult itself was extensively rewritten during the time between the SC 2001 and SC 2002 BWC events. Whereas the original Visapult IBR used slices exclusively in the Z direction to create the illusion of volume rendering, the SC 2002 version supported omnidirectional viewing by automatically changing the choice of compositing slices as a function of view angle (Fig. 29.14). The Visapult back-end was rewritten to accommodate block decomposition to support the new rendering capabilities and also to improve overall load balancing and flexibility in the visualization process.

The Visapult/Cactus protocol was enhanced significantly to support autonegotiation of load-balanced data-parallel UDP streams between the Cactus and the back-end, in contrast to the static mapping required by the SC 2001 version. Finally, the accuracy of the packet-rate regulation method used by Cactus was greatly im-

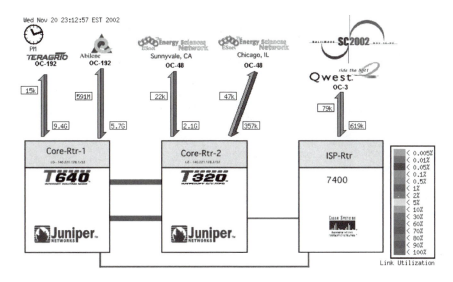

Figure 29.12   SCinet weather map. (See also color insert.)

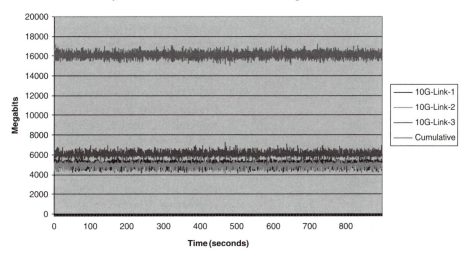

Figure 29.13   SC 2002 Bandwidth Challenge results. (See also color insert.)

proved by being moved from a static inter-packet delay mechanism to a method based on error diffusion. Errors from previous packet sends were carried forward to compute the interpacket delay of the next packet send, thereby achieving a high degree of accuracy in regulation of the flows despite the coarse granularity of available system timers.

Of critical importance is the fact that this bandwidth challenge was part of a larger fabric of interconnected Grid services and applications. For instance, all of the spawned simula-

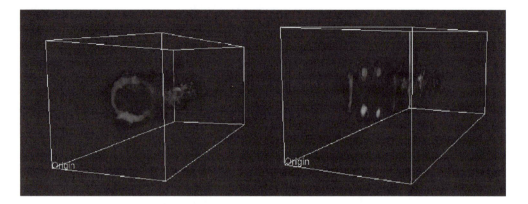

Figure 29.14 Visapult's new "Omniview" capabilities (left) produce much better visual fidelity than the original IBRAVR algorithm. (See also color insert.)

tion tasks were tracked automatically using a Grid Portal [17]. We can even follow links through the portal using a web browser to directly connect to the running simulation codes to monitor its progress, steer it, adjust network bandwidth, and restart or kill jobs running anywhere in the world. This infrastructure even allowed us to dynamically attach to running jobs to perform live/interactive visualization to the emerging results we obtained with Visapult. Such access was also available from handheld devices like iPAQ handheld computers and cell phones. Some testbed applications would even leave SMS messages on the scientists' cell phones to keep them apprised of their progress and location. The advances in the emerging Grid infrastructure and application-level services were nearly as exciting and dramatic as the increased network bandwidth and global connectivity that such applications engender. This event gave us a glimpse into a future of the Grid—a framework composed of independently developed components that can be tightly woven together into a pervasively accessible global infrastructure. Future development of Visapult and other RDV frameworks at LBL will focus on this larger view of component architectures for distributed high-performance computing on the Grid rather than on any individual stand-alone application.

29.5   Lessons Learned

There are two primary lessons we learned while evolving Visapult to make effective use of the network in a remote and distributed visualization context. First is that TCP, the most commonly used network transport protocol, is very ill suited for use in high-performance network applications. In order to effectively use high-bandwidth networks, you must avoid use of TCP. The ramifications of such a strategy are profound and far-reaching in terms of application design. Second, we learned that using UDP, which is an alternative to TCP for point-to-point network communication and is much better in terms of using available bandwidth, comes with a set of costs: applications must be resilient to dropped network packets, and applications designed for TCP will probably need to be retooled for use with UDP.

29.5.1   Don't Bet on TCP for High-Performance Network Applications

TCP's algorithm for regulating packet flow rates is having great difficulty meeting the needs of high-performance networking. If one takes a dramatically simplified view of the problems facing the aging TCP protocol, one could focus on two causal factors out of a field of many. First is the TCP congestion-avoidance algorithm's assumption that any packet loss indicates network congestion that requires rapid reduction in the flow rate (a fundamental assumption about the meaning of packet loss). The second is the slow rate at which TCP returns to full speed in response to packet loss. In order to understand why TCP is inadequate for today's high-performance networks, we must look at how the TCP congestion-avoidance algorithm works, and how that algorithm is inappropriate for use on modern, high-performance networks.

Two key parameters used to characterize network performance are latency and bandwidth. The network latency is affected by buffering in the network routers and switches as well as by the responsiveness of the hosts at the endpoint. However, it is primarily bounded by the fundamental limit of the speed of light. Bandwidth, on the other hand, has been increasing dramatically over the past decade—it has outpaced Moore's Law by four times. The TCP congestion-avoidance algorithm was created to deal with rate management on networks that peaked at 1–5 Mbits in the late 1980s. When a packet is lost in a transfer, TCP immediately assumes that the loss indicates congestion, and it "backs off" to half its current flow rate. The rate at which it can return to its original flow rate is bounded by the round-trip time (RTT) of the network. When the product of the bandwidth of the network and the RTT (the "bandwidth-delay product") is a small number, as it was in the late 1980s, the rate of recovery is not particularly noticeable.

With networks now pushing four orders of magnitude more bits per second with very similar latencies, the TCP congestion-avoidance algorithm is proving to be a hindrance. As bandwidth increases, so too does the exposure to the packet loss. Generally speaking, packets are lost because of unrecoverable bit errors that are detected by the Cyclic Redundancy Check (CRC) or because of buffer overflows at either the endpoints or congested switch interfaces in the network core. Given the comparatively large-bandwidth-delay products involved, the

stream cannot quickly recover from lost packets despite its increased sensitivity to the loss. In fact, using TCP on a 10 Gbit/s network connection, it would take $1\frac{2}{3}$ hours of continuous transmission without a single lost packet to achieve full line rate (Sally Floyd, *http://www. icir.org/ floyd/papers/draft-floyd-tcp-highspeed-02.txt*). Increasing the rate at which TCP can recover from packet-loss events makes the TCP algorithm most unstable, as it oscillates around its quiescent noncongestive data rate. Consequently, the Additive Increase Multiplicative Decrease (AIMD) algorithm employed for TCP congestion avoidance is necessary for stability from the control-theoretical standpoint. It also guarantees increasingly inefficient use of the network resources as the bandwidth-delay product increases. Based upon empirical evidence, the asymptotic behavior results in a tendency to send fewer and fewer packets, with the steady state reaching about 20–25% of the maximum line rate. As network bandwidth continues to increase, these efficiencies will continue to plummet.

Therefore, the time has come for a fundamental paradigm shift in the way that congestion is managed on the WAN. Some have proposed having the network infrastructure provide direct feedback to the TCP/IP stacks of the endpoint hosts regarding current congestive conditions on the network [6,9,14], but these methods require dramatic changes to the routers and switches that comprise our current network infrastructure. Other methods attempt to monitor the end-to-end performance of the network externally using probes or instrumented TCP kernels to provide this information [1,4,5] but can suffer from incomplete coverage or weak ability to understand anomalous behavior in the network core. Since we started using this fixed-data-rate method for our transport, we've seen a number of other efforts adopt a similar manual-rate-control methodology, including Tsunami and SABUL [7,11]. While we believe the latter approach is the most effective given current circumstances, it is not clear how it will scale as its use becomes more pervasive.

In the special case of HPC applications, it is not uncommon to see dedicated links, and such manual control is entirely reasonable. However, over time there will be an increasing need for brokers to mediate these fixed-bandwidth data streams.

29.5.2 How Well Does a Lossy Transport Mechanism Perform for Visualization?

When using TCP for data transport, one need not worry about data loss. TCP guarantees delivery of packets by ordering the packets and asking for retransmission if they are lost along the way. The result is that TCP is slow. When using UDP, which is very fast, for transport, we will occasionally lose packets. Just how much of an issue is the absence of a few packets?

The approach we used for encoding data into UDP packets results in portions of the data from the computational grid being copied into the packet. To maximize efficiency on the sender side, we copy data from adjacent memory locations into the data-payload portion of the packet. Typically, these regions consist of partial "scan lines" from the source volume. When a packet is dropped, we don't have data for that particular scan line when doing the visualization. Instead, what we have is data left over from the previous time-step that was loaded into memory on the receiver side. Keep in mind that we use a single-buffering approach for the receive buffer; there is no notion of "frame boundaries," which correspond to one time-step's worth of simulation data, so there is no advantage to using a double-buffered approach on the receive side. We simply receive data as fast as we can and asynchronously render it. Due to the asynchronous nature of the receive/render cycle, there is no design guarantee that, at any given point in time, the receive buffer will have data from just one simulation time-step.

Another advantage of this approach is the nearly immediate delivery of data to the application. TCP, in its attempt to enforce the ordered presentation of data, must actually hide received data from the application until all gaps in the packet stream are filled. If there is a gap in the stream because a packet is lost, then the receiver is blocked while the source is notified of the loss and the packet retransmitted. The result is that the receiver spends a good deal of time idle, waiting for data that has already arrived and is simply held by the OS in an opaque system buffer. For interactive graphics applications, the application performance is irregular, and it is very noticeable and annoying to the application. In contrast, UDP always delivers the information to the application almost immediately after it arrives. When using UDP, the application must have a graceful way to deal with loss and ordering issues. Visapult demonstrates that such management is possible, and it provides very reasonable results.

Fig. 29.15 shows six different snapshots of an evolving simulation. In the upper-left corner, the simulation has just started. You can see portions of the receive buffer that have not yet been filled in with simulation data—the background is visible through the volume rendering. Moving to the right, these regions begin to fill in with data while the simulation continues to evolve. This particular example was created using a rate-limited flow of approximately 10 Mbits/s over a 100 Mbits/s link. Since we were using only a fraction of the total available bandwidth, the likelihood of packet loss was quite small.

In the second row, the simulation has evolved even more. The middle and right-most images on the bottom row show the effects of how an asynchronous receive/render strategy can produce artifacts. In the middle image of the bottom row, the center of the field has a region of white that is partially covered with a light orange, which terminates abruptly. This is a good example of having data from (at least) two time-steps loaded into memory at once. After a few more updates, that artifact goes away, as seen in the bottom-right image.

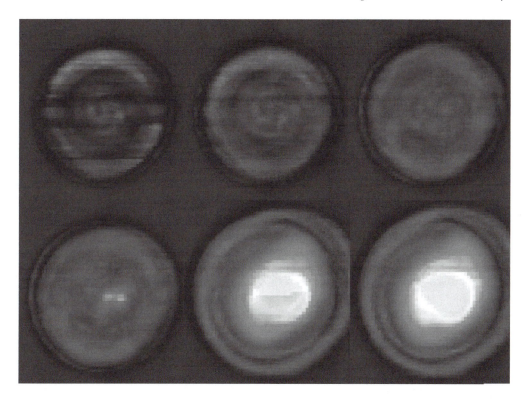

Figure 29.15   Visapult renderings of Cactus's 3D Wave Equation Evolution. (See also color insert.)

Keep in mind that the visualization we were performing is of a dynamic, evolving simulation. The artifacts seen in the images we present here are short lived. If a packet is lost, the resulting visualization has an artifact that can be characterized, in this case, as a "streak." Due to the nature of the fundamental algorithmic design, the resulting visualizations have artifacts caused by the presence of data from more than one time-step; it is difficult to ascertain whether any given visual artifact is caused by packet loss or by regular and normal software execution.

Mitigation of the type of artifacts we have described can occur on two fronts. The first is careful rate regulation of UDP flows. We perform such regulation as a general practice; the Cactus interface allows a user to specify the bit rate used to send packets from Cactus to Visa-

pult. The second is to impose some notion of "frame boundaries" upon the Cactus/Visapult rendering algorithm. Doing so would ensure that data from only one time-step is used to perform visualization. The disadvantage of imposing such a restriction is that overall end-to-end performance will be reduced. For the types of problems being studied, physicists are able to quickly ascertain whether a simulation run is proceeding in "the right direction" by examination of the large structures seen in the resulting visualization. Presence of a few visual artifacts does not cause them concern for general questions like "are my simulation parameters grossly incorrect?" For situations requiring more careful visualization and analysis work, lossless data transmission is entirely appropriate.

29.6   Future Directions for High-Performance Remote and Distributed Visualization

The work we have described in this chapter addresses only a small number of topics in high-performance, remote, and distributed visualization. Generally speaking, our work is similar in approach to related activities in distributed visualization: it involves a collection of custom software components using a custom protocol on dedicated resources. The future of research in remote and distributed visualization must include sweeping changes in basic infrastructure. For instance, toolkits based upon standard techniques should be employed to implement flow regulation underneath the application. These tools need access to standard mechanisms to obtain information about network-based resources, such as link bandwidth and computational and graphics capacity of a resource and so forth. Without such standards, each new research project must "reinvent the wheel" by creating these components that form the basis for remote and distributed visualization applications. With such an infrastructure, a new world of opportunity opens for transition of research prototypes into the hands of the visualization technology consumers.

The UDP method we have described so far relies on pacing of packets to meet, but not exceed, network capacity. Indeed, there is no reason that these fixed-data-rate methods cannot be applied to reliable transport protocols like TCP. Recent examples of fixed-rate reliable protocols include the University of Illinois protocol SABUL [7], which is used to support data mining, and more recent demonstrations of Indiana University's Tsunami file transport program [11]. In both cases, the protocols emphasize throughput for a stream-oriented protocol requiring considerable buffering. Such buffering is necessary in order to preserve the notion of the stream. Our group is proposing to separate the stream orientation from the notion of reliability in a Reliable Independent Packet Protocol (RIPP). Such decoupling is critically important for support of the needs of interactive visualization applications like Visapult.

As bandwidth-hungry fixed-data-rate methods like network video, SABUL, and Tsunami become more pervasive, we must consider ways to provide global mediation of their flow rates. It is clear that fixed-rate implementations of both reliable and unreliable protocols can be disruptive to commodity networks, and they are most appropriate for use on dedicated network links, Private Virtual Circuits (PVCs), Experimental Networks, or even scheduled access. (During our SC 2000 BWC run, the commercial service provider of the link between NTON and Dallas asked us to not exceed 1.5 Gbits/s in transfer rates lest we excessively interfere with commercial traffic. Our performance numbers during the SC 2000 run do not reflect any attempt on our part to limit bandwidth consumption.) While it is unreasonable to assume that all WAN connections will be dedicated links, it is quite reasonable to target an architecture where high-performance dedicated or schedulable links will exist between supercomputing centers and satellite data-analysis centers on high-performance production and experimental network backbones. More dynamic shared environments require continuous adjustment of the data rates. In this context, there are some concerns that selecting an appropriate packet rate for UDP-based methods to minimize loss is too tedious to be practical.

Ideally, the network switching fabric should provide detailed quality-of-service hints through informational packets to the endpoint hosts to indicate ideal send rates per MIT's XCP (eXplicit Congestion control Protocol) proposal [9], Core-Stateless Fair Queue (CSFQ) [14], or Explicit Congestion Notification (ECN) [6]. We could, for instance, have a TCP or UDP implementation that uses these hints to ignore packet loss if the switching fabric says that it is noncongestive but defaults to the standard congestion-avoidance algorithm when no such hints are available. In lieu of the wide availability of an intelligent network infrastructure, we are working on Grid-based bandwidth brokers that monitor current network conditions through SNMPv3 and provide feedback about network congestion through the Globus Metacomputing Directory Service (MDS) so as to avoid congestive loss. Even without intelligent switching fabric or explicit hints about congestion avoidance, we could create a system of peer-to-peer feedback/auto-negotiation by having endpoints multicast their path and current packet-rate information on a fixed set of designated paths. This allows hosts to negotiate amongst themselves for appropriate packet rates rather than involving a third party, like a bandwidth broker or the switching fabric itself. Ultimately, it is time to explore methods of coordinating fixed-data-rate flows as an alternative to current congestion-avoidance methods that attempt to infer congestion from packet-loss statistics. The latter methods have clearly reached their scalability limit!

The primary area of growth in considering custom UDP protocols is in the development of fault-tolerant/fault-resilient encoding techniques. The simplest approach provides fault tolerance by copying data from the previous time-step to fill in lost data. A more advanced methodology could use a wavelet or frequency domain encoding of the data so that any loss is hidden in missing spatial frequencies (similar to JPEG compression). For transport of geometric models, we can look at packet encodings that support progressively refined meshes using triangle bisection [8]. Such techniques make packet loss less visually distracting and eliminate the need for data retention on the sending side. Any reliable technique requires data to be retained at the source until its receipt is acknowledged. Given the large bandwidth-delay products involved for future Tbit networks, the window sizes necessary for reliable transport will be considerable. The buffering required to support TCP retransmission and large windows creates noticeable lag in the responsiveness of remote-visualization applications and produces low-bandwidth utilization rates. Fast response times are essential for creating the illusion of

locality, so low-latency connectionless techniques will be essential for Grid visualization and collaborative interfaces. Overall, there are many avenues to consider for information encoding that make performance-enhancing but unreliable delivery methods offer graceful degradation of visual quality in response to packet loss rather than simply settling for degradation in interactivity.

As mainstream visualization systems start to move outside the confines of the desktop workstation, we must begin to consider issues of the emerging global computing infrastructure—the emerging "Grid." Like most current remote and distributed visualization applications to date, ours is a one-off prototype that has little chance of interoperating with components or network protocols that we haven't explicitly programmed it to understand. Such interoperability requires standards for interfaces and protocols through community consensus. There must be considerable work on basic architectural frameworks that enable seamless movement from the desktop to the grid and that support sharing of components across RDV component systems and services developed by different groups who may not be working on the same code base. Without such a framework, the entire community will be so mired in issues of Grid management that there will be little time left to spend on visualization.

For instance, how does one actually go about the task of launching remote components in the emerging Grid infrastructure? In the case of our custom-built remote and distributed visualization applications, we typically use a manual process of staging the components on the machines we want to use for a distributed application. This approach is entirely impractical in Grid environments, where there may be hundreds or thousands of resources that constitute a single application. There must be a notion of a database or directory service that can distribute and keep track of components, both executables and running instances, on heterogeneous resources. Similarly, many contemporary distributed applications employ custom methods that signal each other to indicate state changes and to move data. In order to make use of distributed resources, which could be dynamic in location and availability, visualization frameworks must make use of Grid services that automate the process of resource location and component launching. Fledgling services of this form are provided in Grids based upon the Globus architecture (*http://www.globus.org*) in the form of Grid Information Services (GISs), which are essentially LDAP-based hierarchical directories of resources for a given Grid. Such object directories and indexing infrastructures are necessary to support minimal automation for launching distributed components for a Grid-based visualization architecture.

As we consider how components are managed for distributed environments, one key observation about moving to a Grid architecture is that it isn't just about components. Grid-based applications typically involve a fabric of components and persistent services woven together into complete applications. Services are loosely defined as persistent software components that are shared by more than one application or user. For example, an off-screen hardware renderer could be a component of a visualization that is time-shared by multiple users because there is only one of them available on the Grid.

Integrating the concept of services into remote and distributed visualization applications leads to hard problems in terms of mediating access and security. Should you have an external "broker" to ensure fair usage of available visualization services in a Grid, or should this be a function of the service itself? What do you do when the service fails to deliver its promised performance—a situation referred to as "contract violation"? Services must have an internal authorization and access control model in order to ensure that these access rules and behaviors are even minimally enforceable. Visualization researchers are not accustomed to dealing with these issues.

The issues described above are merely a starting point for attacking issues involved in hiding the incredible complexity of the Grid. Indeed, we are moving rapidly away from a computing model where you have complete and dedicated control from all of your computing resources (it is all on the motherboard, to your computer). The Grid is constantly changing in performance and capabilities, even as you are using it. Imagine what it would be like if, while you were trying to run a visualization application on some platform, the system administrator were constantly removing and adding components, even while you were working. That approximates your "Grid" experience. The pieces are distributed over the wide area, and your scope of control over them is extremely limited relative to your desktop. If we spend all of our time focusing on "stovepipe" visualization system designs, we will never be able to share any common infrastructure that is needed to tackle the problem of managing Grid resources! Now, more than ever, the visualization community must come together to work towards this common goal. One venue for discussing these topics is the Global Grid Forum (*http://www.gridforum.org*), where groups have the opportunity to discuss and ratify community standards that support compatibility between disparate remote and distributed visualization implementations and services. Participation by the visualization community in these forums is absolutely necessary in order to make wide area visualization and supercomputing possible.

29.7   Conclusion

Like many research projects, the Visapult effort has changed over time in response to new and unanticipated challenges. Visapult began as an effort to provide highly efficient and scalable software tools for data visualization to research scientists. In the relentless pursuit of ever-increasing levels of efficiency and performance, we learned a lot more about networking technology than we could have predicted at the outset. Our accomplishments with this project include winning the SC BWC for 3 years in a row and achieving unheard-of levels of bandwidth performance for a single application. Despite these successes, Visapult is still an island of capability in the larger ocean of remote and distributed visualization. Much work in remote and distributed visualization is still needed before the research efforts of different programs can begin to work together as a coherent whole.

References

1. D. Agarwal. Self-configuring network monitor project: an infrastructure for passive network monitoring. Presented at the CITRIS NorCal Network Research Meeting, Berkeley, CA, 2002.
2. W. Bethel, B. Tierney, J. Lee, D. Gunter, and S. Lau. Using high-speed WANs and network data caches to enable remote and distributed visualization. In *Proceedings of the IEEE/ACM 2000 Conference on Supercomputing (CDROM)*, Dallas, TX, 2000.
3. W. Bethel, R. Frank, and J. D. Brederson. Combining a multithreaded scene graph system with a tiled display environment. In *Proceedings of the 2002 IS&T/SPIE Conference on Electronic Imaging and Technology*, San Jose, CA, 2002.
4. The Data Intensive Distributed Computing Research Group, http://www.didc.lbl.gov/DMF/
5. T. Dunigan, M. Mathis, and B. Tierney. A TCP tuningdaemon. *Proceedings of IEEE supercomputing 2002 Conference*, LBNL-51022, 2002.
6. S. Floyd. TCP and explicit congestion notification. *ACM Computer Communication Review*, 24(5):10–23, 1994.
7. Y. Gu, X. Hong, M. Mazzucco, and R. Grossman. SABUL: a high performance data transfer protocol, 2002.
8. H. Hoppe. Progressive meshes. *Proc. 23rd Int'l. Conf. on Computer Graphics and Interactive Techniques SIGGRAPH '96*, pages 99–108, 1996.
9. D. Katabi and M. Handley. Congestion control for high bandwidth-delay product networks. *Proceedings of ACM Sigcomm 2002*, http://www.ana.lcs.mit.edu/dina/XCP/
10. J. Madhavi and S. Floyd. TCP-friendly UDP rate-based flow control. *Technical Note* 8, http://www.psc.edu/networking/papers/tcp_friendly.html, 1997.
11. The Advanced Network Management Lab, http://www.anml.iu.edu/anmlresearch.html
12. K. Müller, N. Shareef, J. Huang, and R. Crawfis. IBR-assisted volume rendering. In *Late Breaking Hot Topics, Proceedings of IEEE Visualization '99*, pages 5–8, 1999.
13. J. Shalf and E. W. Bethel. Cactus and Visapult: an ultra-high performance grid-distributed visualization architecture using connectionless protocols. *IEEE CG&A*, 2003.
14. I. Stoica, H. Zhang, and S. Shenker. Core-stateless fair queueing: achieving approximately fair bandwidth allocation in high speed networks. *Proceedings of ACM SIGCOMM '98*, 1998.
15. B. Tierney, W. Johnston, B. Crowley, G. Hoo, C. Brooks, and D. Gunter. The Netlogger methodology for high performance distributed systems performance analysis. *Proceedings of IEEE High Performance Distributed Computing conference (HPDC-7)*, LBNL-42611, 1998.
16. B. Tierney, J. Lee, B. Crowley, M. Holding, J. Hylton, and F. Drake. A network-aware distributed storage cache for data intensive environments. *Proceedings of IEEE High Performance Distributed Computing conference (HPDC-8)*, 1999.
17. G. Von Laszewski, M. Russell, G. Allen, G. Daues, I. Foster, E. Seidel, J. Novotny, and J. Shalf. The community software development with the astrophysics simulation collaboratory. *Concurrency in Computation: Practice and Experience*, 2002.

PART IX

Visualization Software and Frameworks

30   The Visualization Toolkit

WILLIAM J. SCHROEDER and KENNETH M. MARTIN
Kitware, Inc.

30.1   Introduction

The Visualization Toolkit (VTK) is an open-source, object-oriented software system providing a toolkit of functionality for 3D data visualization. Because visualization inherently involves capabilities from computer graphics, image processing, volume rendering, computational geometry, human/computer interaction, and other fields, VTK supports these capabilities in an integrated, robust manner. More than 850 separate C++ classes, including several hundred data processing filters, are included in the toolkit. While the core of VTK is implemented in portable C++ across all major computer hardware/software configurations, VTK is packaged as part of a sophisticated development environment that includes advanced interface, build, and test tools. (Section 30.2.9 covers language interfaces; see CMake [1] for information about the CMake build tool and DART [2] testing environment.) It is this combination of openness, extensive features, reliability, and comprehensive development environment that makes VTK one of today's premier visualization tools for academic, research, and commercial applications.

In this chapter, we provide insight into the key features of the toolkit. The focus is on architecture and system concepts; the details of algorithms and data representations are covered elsewhere in this handbook. In the remainder of this introductory section, we provide the motivation, history, and goals for VTK. The next section describes important architectural features. In the third section, we describe how VTK is used in large-data visualization, one of the key challenges facing visualization

systems of the future. We conclude with a brief survey of important applications developed using VTK.

30.1.1   Motivation and History

As visualization researchers, the three initial authors of VTK (Will Schroeder, Ken Martin, and Bill Lorensen) were interested in creating a software platform for developing and delivering visualization technology to their customers, as well as in sharing their knowledge with students, researchers, and programmers. Initial efforts focused on writing a textbook [3] with companion code and exercises. Work began in late 1993, and the system reached a usable state late in 1994. The first edition of the textbook appeared early in 1996 (it is now available as a third edition from Kitware). Since that time, a large community of users and developers has emerged, accelerating the growth of the system to its current state. While much of the initial work was done by the authors in their spare time, and many developers continue to volunteer their services, commercial firms such as Kitware and GE Research continue to invest and contribute code to this open-source system. Indeed, prestigious US National Laboratories such as Los Alamos, Livermore, and Sandia fund significant developments of the system, including contributing software modules created by their own researchers.

30.1.2   Goals

VTK benefited from early experience with the proprietary *ad hoc* graphics and visualization system LYMB [5]. This experience led to fundamental decisions about what a good visualization toolkit should be. Some key conclusions and thus goals for VTK include the following:

- The code should be open-source. Create a generous copyright for source distribution for commercial and noncommercial application. GPL was deemed too prohibitive; a BSD variant was chosen instead.

- Create a toolkit, not a system. Design the software to be embedded into applications, not to become the application. Modern software implementations require simultaneous usage of multiple software packages; the software must work well with other packages.

- The design should be object-oriented. C++ was deemed the best implementation language because of its wide acceptance and efficiency. Conservative usage of object-oriented features should be employed (e.g., avoiding multiple inheritance and excessive templating). This results in portable, simple code.

- The GUI is separate from the algorithms. A clean separation of the GUI allows a focus on algorithms and data representation. Furthermore, this approach facilitates offline processing without the need for a window manager, etc. The GUI often represents a sizable part of an application, and there are many choices for the GUI. We wanted to avoid this morass and provide hooks to the GUI instead.

- Support a hybrid compiled/interpreted architecture. Compiled environments (such as C++) are efficient. Interpreted environments (such as Tcl or Python) are great for prototyping and application development. We realized the best of both by implementing the core in C++ and automatically wrapping it with interpreted languages.

- Support a pipeline architecture. In visualization, a pipeline architecture (data-flow-based) works very well. This is because of the flexibility required by a visualization system to map data into different forms.

- Support parallel processing. Many visualization algorithms are computationally expensive and deal with large data. VTK supports portable multithreading for shared-memory parallel implementations and portable distributed parallel processing.

- The code must exist within a comprehensive build and test environment. Robust, reliable systems of the size and complexity of VTK require a formal environment for building and testing. We use a novel continuous and nightly process that compiles and tests the software on dozens of systems located around the world. The results are posted on a centralized web page (the so-called "quality dashboard" [2]) providing immediate feedback to developers.

- Create consistent code. We strive to maintain consistency in code style and implementation methodology. The goal is to make the code appear as if it was written by one person. Consistent code is easy to work with and understand because the developer is not encumbered by the need to decipher variable and method names. The code is also easier to read once the basic coding style is learned.

- Embed the documentation into the source code. Documentation is embedded into all VTK class header files. The documentation is automatically extracted (using the open-source Doxygen system [4]), and HTML is formatted to produce manual pages.

- Maintain toolkit focus. Successful software systems have a tendency to grow until they become unwieldy. Disciplined focus is required to keep the toolkit vital.

Implementation details addressing some of these goals are found in the following sections.

30.2 Overview

This section provides an overview of key toolkit capabilities. We begin with a general description of the architecture, followed by a description of the graphics, volume rendering, and visualization pipeline. An example then ties these concepts together. We end the chapter by describing important system features, such as the instantiation process through object factories, data representation, wrapping into other programming languages, callbacks, and memory management.

30.2.1 Architecture

The architecture of a software toolkit has significant impact on its success. A good overall design improves the capabilities, usability, extensibility, and maintainability of the toolkit. VTK's architecture is based on object-oriented principles [6] and implemented in C++. In VTK, an object is an abstraction that encapsulates the properties and behaviors of an entity in a system. Examples of objects in VTK include datasets, filters, readers, and writers. Using an object-oriented architecture can be confusing to people who are accustomed to older-style functional libraries. With a functional library, calling the function shows an immediate result. With an object-oriented design, instantiating an object typically has little impact on your data. Only by invocation of the methods on the object are results actually produced.

Even during adaptation of an object-oriented design, there are still a number of decisions to be made. For VTK we decided to not use multiple inheritance or public templated classes. While this limits some of the C++ features we can use, it greatly enhances our ability to work with other languages, such as Tcl, Python, and Java. For VTK we decided to have a common superclass for all objects called vtkObject. This class provides some common functionality for runtime interrogation, such as IsA( ), IsTypeOf( ), and GetClassName( ). It also includes func-tionality for printing, reference counting, call-back, and modification time management. It is too large a class to use for small data structures like a single three-component vector. In fact, VTK's design avoids the use of small classes. Instead of representing a polygonal mesh with a vector of polygon objects, VTK represents it with just two objects: an array of floats for the point coordinates and an array of connectivity. This allows VTK classes to have a common superclass without significantly impacting the optimal memory footprint for its data structures.

In an attempt to make VTK as easy to use as possible, a number of common conventions have been created to ensure code consistency. For example, in VTK all method names are fully spelled out—abbreviations are avoided unless they are common—and similar terminology is used for concepts. For example, in VTK, "ComputeRadius( )" is used. As a VTK developer you know that the method will never be called anything like compute_radius, CalcRadius, CalculateRadius, or compRad, because they do not follow coding conventions. This makes it much easier to remember method names and helps to ensure that the same method in different classes will have the same name. Likewise, there is a convention that all member variables (also known as instance variables) are protected or private and can only be accessed from other classes using Set( ) and Get( ) methods, such as SetVariable( ) or GetVariable( ). The result is a fairly simple programming interface where most every command is of the form instance->Method (arguments).

VTK's architecture supports cross-platform development and runs on almost any brand of UNIX, as well as Microsoft Windows and Mac OSX. This is achieved through the use of functional abstractions for both hardware and software components. For example, many filters in VTK support multithreaded operation on multiple CPU systems. Such filters interface with a generic class that handles multithreading on all platforms so that the filter's implementation does not need to consider such issues. When the differences between the platform hardware or software become significant, an abstract object-inheritance hierarchy is used to map the differences to a uniform API. For example, a class like vtkRenderWindow has the three subclasses vtkWin32OpenGLRenderWindow, vtkXOpenGLRenderWindow, and vtkCarbonOpenGLRenderWindow to support OpenGL-based render windows targeted at different window systems. The application developer will create and use what looks like a vtkRenderWindow while actually a more specific subclass of vtkRenderWindow is instantiated at run time. This same approach is used in other VTK subsystems, including parallel processing using MPI, sockets, or shared memory, and using different volume rendering techniques. Because VTK's architecture is consistent across the different classes, once you have learned to use one class you can apply the same approach to any other class.

30.2.2   Graphics

Now that the architecture has been discussed, it is worthwhile to see how it is applied to the graphics engine in VTK. A common source of confusion for new VTK users is understanding how it differs from OpenGL. OpenGL is a stack-based rendering API without any direct visualization algorithms. In contrast, VTK is a visualization toolkit with a wide range of visualization techniques, such as streamlines, glyphing, iso-contouring, clipping, and cutting. VTK does include support for rendering of its results, typically using OpenGL, but it does not compete with OpenGL. The two serve different purposes.

VTK provides a higher-level API than OpenGL that simplifies rendering-visualization results and is more consistent with the toolkit design. The API consists of objects such as cameras, lights, and actors. These are all contained within renderers that appear within a render window (Fig. 30.1). Taken together, these objects form a scene. When a render

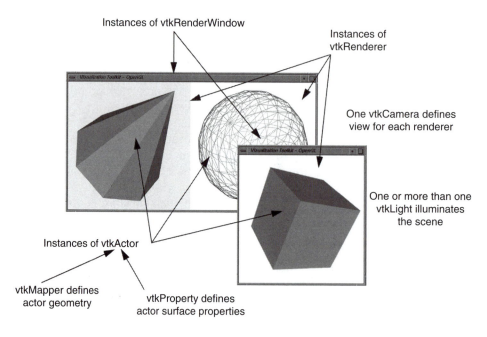

Figure 30.1   Rendering classes. (See also color insert.)

window is rendered, it sets up the viewports for each of its renderers and then passes control to them. They set up the camera and lights and then render each actor in turn. In VTK, each actor and its properties are independent from all other actors. Other approaches, such as scene graphs, are designed along the concept of a state machine, where changes in one part of the graph can impact other parts of the graph.

The VTK rendering process starts by performing view-frustum culling and render-time allocation based on screen coverage. This results in a subset of the actors' acting as targets for rendering with a time allocation provided to each. Each remaining actor then has three opportunities to render itself in a three-pass rendering process. The first opportunity is the "Opaque Geometry" pass, where all opaque geometry should be drawn. Next comes the "Translucent Geometry" pass, followed finally by the "Overlay" pass for annotations and markup. This three-pass structure has a number of advantages for techniques such as volume rendering. A texture-mapping volume renderer can simply draw its polygons in the translucent geometry pass. A ray casting–based volume renderer can also draw in the translucent geometry pass, but it first retrieves the color and z-buffers from the opaque geometry pass in order to properly blend the opaque geometry with the volume. The opaque geometry can also be used to perform early ray termination.

The objects in VTK's graphics API provide access to a wide range of functionality. Lights have parameters to simulate infinite lights or positional lights. Lights can also have color, spot angles, spot attenuations, and quadratic attenuation coefficients. Cameras support view angle, eye separation for stereo rendering, lens size for depth-of-field effects, and oblique angles for off-axis projections. There are various methods to adjust the camera's position and orientation, including roll, pitch, yaw, elevation, azimuth, dolly, and zoom. Both parallel and perspective cameras are supported. Actors have a matrix transformation that indicates their position and scale in world coordinates.

The actor's properties include ambient, diffuse, and specular colors, wire-frame versus surface rendering, and front- or back-face culling.

VTK uses mappers to map visualization data (see Section 30.2.4) to the graphics engine. In the simplest case, a mapper takes a polygonal mesh and calls the appropriate rendering functions to draw the mesh. The mapper handles issues such as mapping scalar values to colors with the help of lookup tables or transfer functions. It also manages display-list maintenance and normal generation for meshes that do not have normals.

To support interactive rendering rates, VTK can make use of multiple mappers per actor. In this case, a single actor (e.g., vtkLODActor) can have multiple mappers, each with their own input. When a render is requested of the actor, the mapper with the best quality that can be drawn within the allotted rendering time will be selected and rendered. One nice feature of this approach is that it provides the application programmer with significant flexibility in the design of the levels of detail (LODs). For example, a polygonal mesh might have LODs that include a bounding box, reduced point cloud, wire-frame outline, decimated mesh, or any other representation appropriate to the data. While VTK provides default LOD generation, the application programmer can override these and set specific LODs as appropriate to their application.

30.2.3 Volume Rendering

Many visualization algorithms produce and render polygonal data (points, lines, polygons, triangle strips, and other linear primitives). For example, iso-contouring produces dense triangle or triangle strip meshes. While these surface-rendering techniques work well in many applications, volume rendering is a more sophisticated rendering technique used to visualize the structure within complex 3D image datasets (i.e., volumes). VTK supports volume rendering using an object model similar to that of the surface-rendering architecture described previously (volume, volume mapper, and volume properties versus actor, mapper, and actor properties). In fact, VTK's rendering process seamlessly integrates image display, surface rendering, and volume rendering, which may be used in arbitrary combination. (Caveat: the surface geometry must be opaque when one is combining surface graphics with volume rendering.)

VTK supports several volume rendering methods through a variety of volume mappers. Software ray-casting (implemented using multi-threaded parallel processing) produces the highest-quality results. Different ray-casting strategies, such as isosurface, MIP, and composite, can be selected to obtain different effects. Volume rendering based on texture mapping takes advantage of modern computer graphics hardware to produce good-quality results at high frame rates. Other mappers, such as those supporting the VolumePro volume rendering hardware [7], are also available.

Like vtkActor, vtkVolume has a volume property associated with it. The volume property controls interpolation methods (nearest-neighbor or tri-linear), collects transfer functions for color and opacity (including gradient opacity), and enables volume shading. Lighting parameters such as ambient, specular, and diffuse may also be set through the volume property.

VTK tightly integrates images, volumes, and surface primitives into the rendering process and visualization pipeline. During the three-pass rendering process described previously, VTK renders its opaque surface geometry first, thereby populating the depth and color buffers. The volume renderer then composites its output image against these buffers, using the depth buffer to terminate unnecessary computation (e.g., terminate ray-casting). The final pass of the rendering process is used to draw annotation and/or images on top of the results of surface and volume rendering passes. A volume in VTK is a type of image, so volumes are fully integrated into the visualization pipeline; this provides the user with the ability to read, process, and then volume render data—along with

surface and image processing and display. This integration provides a rich environment for creating compelling visualizations.

30.2.4 Visualization Pipeline

Visualization is inherently a process of transformation: data is repeatedly transformed by a sequence of filtering operations to produce images. In this section we will describe the pipeline architecture used to accomplish this transformation.

The role of the graphics subsystem described previously is to transform graphical data into images. The role of the visualization pipeline is to transform information into graphical data. Another way of looking at this is that the visualization pipeline is responsible for constructing the geometric representation, which is then rendered by the graphics subsystem. VTK uses a data-flow approach—also referred to as the visualization pipeline—to transform information into graphical data.

There are two basic types of objects found in the visualization pipeline: data objects (vtkDataObject) and process objects (vtkProcessObject). Data objects represent and provide access to data. Process objects operate on the data and represent the algorithms in the visualization system. Data and process objects are connected together into directed networks (i.e., the visualization pipeline) that indicate the direction of data flow. The networks execute when data is requested; a sophisticated execution mechanism ensures that only those filters requiring execution are triggered.

At the most abstract level, the class vtkDataObject can be viewed as a general blob of data (i.e., an array of arrays). Data that has a formal geometric and topological structure is referred to as a dataset (class vtkDataSet). Fig. 30.2 shows the dataset objects supported in VTK. A vtkDataSet consists of a geometric and topological structure (points and cells) as illustrated by the figure, as well as additional attribute data, such as scalars or vectors. The attribute data can be associated

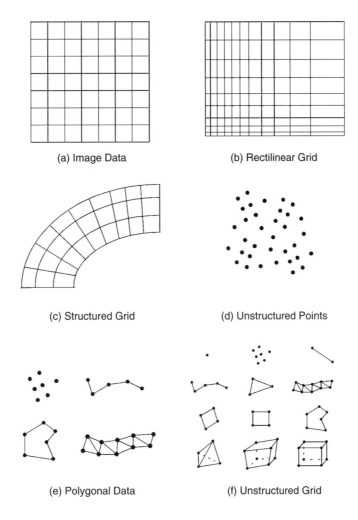

(a) Image Data

(b) Rectilinear Grid

(c) Structured Grid

(d) Unstructured Points

(e) Polygonal Data

(f) Unstructured Grid

Figure 30.2 Dataset and subclasses.

with the points and/or cells of the dataset. Cells are topological organizations of points; cells form the atoms of the dataset and are used to interpolate information between points. Figs. 30.3 and 30.4 show the 19 cell types supported by VTK. Fig. 30.5 shows the attribute data supported by VTK.

Process objects, also referred to generally as filters, operate on input data objects to produce new output data objects. The types of input and output are well defined, and filters will accept as input only those types that they can process. (Note: many filters accept as input vtkDataSet,

meaning they can process any type of data.) Source process objects are objects that produce data by reading (reader objects) or constructing one or more data objects (procedural source objects). Filters ingest one or more data objects, and they generate one or more data objects, on output. Mappers, which we have seen earlier in the graphics model, transform data objects into graphics data, which is then rendered by the graphics engine. A writer is a type of mapper that writes data to a file or stream.

Process and data objects are connected together to form visualization pipelines, as

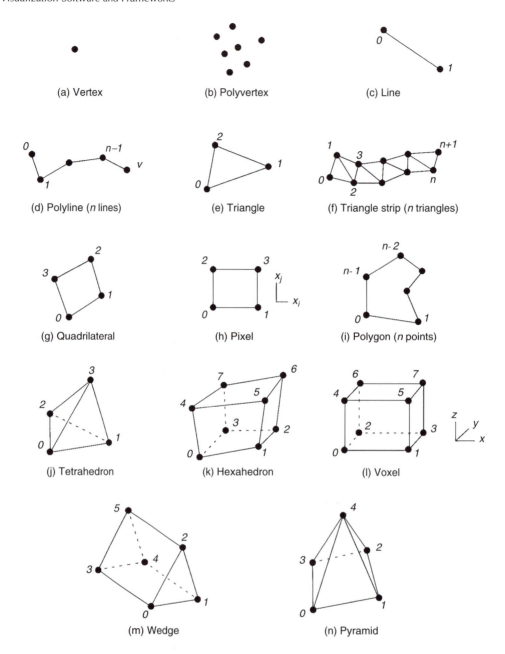

Figure 30.3 Linear cell types supported by VTK.

shown in Fig. 30.6. The pipeline topology is constructed using variations of the method.

```
aFilter->SetInput( anotherFilter->
                    GetOutput() );
```

It sets the input to the filter aFilter to the output of the filter anotherFilter. (Filters with multiple inputs and outputs have similar methods for setting input and output.) When data is requested, only those portions of the

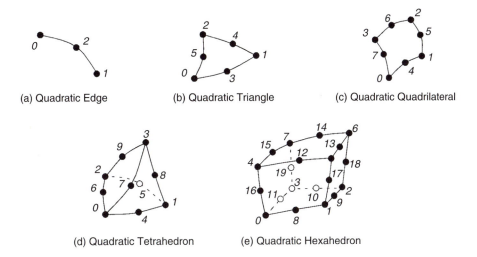

Figure 30.4   Higher-order cell types supported by VTK.

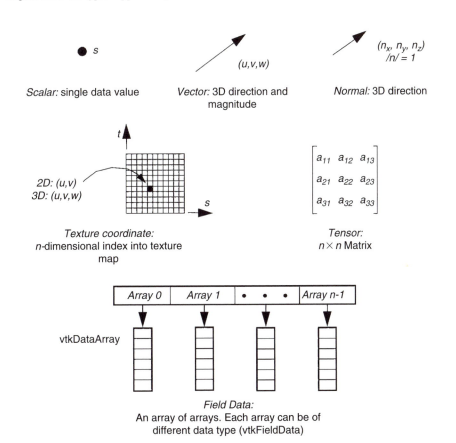

Figure 30.5   Attribute data supported by VTK.

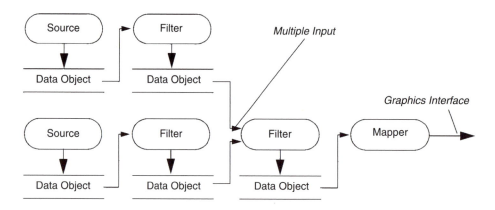

Figure 30.6   Pipeline topology.

pipeline necessary to bring the output up to data are executed. VTK uses a demand-driven, lazy evaluation scheme (i.e., execution is deferred) based on an internal modification time for each object. The modification time is compared to the time of last execution to determine whether the filter must execute again. Of course, any filters downstream of the executing filter must also execute.

In general, filters may accept as input any VTK data object type and may produce any of the VTK data objects. However, the image-processing pipeline in VTK operates only on vtkImageData. Because this regular data may be readily decomposed into subregions, the pipeline update mechanism supports special fea-

tures such as streaming and multithreaded parallel processing. This will be discussed in more detail in Section 30.3.

30.2.5   Example
VTK has a carefully designed interface that makes using the classes in the system relatively easy. The challenge to learning VTK (or any object-oriented system, for that matter) is learning the palette of objects and how they interact. The following C++ example shows typical usage of VTK. The example reads polygonal data from a data file, decimates it, smooths the surface, generates surface normals, and then displays the result.

```
vtkBYUReader *byu = vtkBYUReader::New();
byu->SetGeometryFileName(''fran_cut.g);

vtkDecimatePro *deci = vtkDecimatePro::New();
deci->SetInput( byu->GetOutput() );
deci->SetTargetReduction( 0.9 );
deci->PreserveTopologyOn();
deci->SetMaximumError( 0.0002 );

vtkSmoothPolyDataFilter *smooth = vtkSmoothPolyDataFilter::New();
smooth->SetInput (deci->GetOutput() );
smooth->SetNumberOfIterations( 20);
smooth->SetRelaxationFactor( 0.05 );
```

```
vtkPolyDataNormals *normals = vtkPolyDataNormals::New();
normals->SetInput( smooth->GetOutput() );

vtkPolyDataMapper *mapper = vtkPolyDataMapper::New();
mapper->SetInput( normals->GetOutput() );

vtkActor *actor = vtkActor::New();
actor->SetMapper (mapper);
actor-> GetProperty()->SetColor ( 1.0, 0.49, 0.25 );

vtkRenderer *ren1 = vtkRenderer::New();
vtkRenderWindow *renWin = vtkRenderWindow::New();
renWin->AddRenderer( ren1 );

vtkRenderWindowInteractor *iren = vtkRenderWindowInteractor ::New();
iren->SetRenderWindow( renWin );

ren1->AddActor( actor );
ren1->SetBackground( 1, 1, 1 );
renWin->SetSize( 500, 500 );
iren->Start();
```

The pattern outlined in this example is typical of many small VTK applications. The required filters are instantiated and connected together to form the visualization pipeline. (See the next section for details regarding instantiation with the object factory New( ) method.) The mapper terminates the pipeline, which is then connected to an actor (or subclass of vtkProp—the objects drawn in the scene). The actor is associated with a renderer, which is in turn associated with the render window. Interactors are often used to manage the event loop and provide simple mouse- and keyboard-based interaction. Note that in this example, lights, cameras, and properties are not created. VTK generally works intelligently, provides reasonable default values when necessary, and does things like instantiating objects such as lights when none have been explicitly created.

30.2.6 Object Factories and Instantiation

Earlier in this chapter we discussed how an application programmer would interact with a render window while at run time a more specific subclass of render window would be created and used. This is accomplished through the use of object factories. All objects in VTK are created using a special method called New( ) (note the different capitalization from the traditional C++ new method). Requiring all objects to use the New method enables a few different capabilities in VTK. The first is hardware abstraction. When the application creates a render window with the New( ) method, VTK determines at run time what rendering APIs were compiled in and examines environment variables to determine what subclass of the render window to create. On a Mac OSX system, invoking vtkRenderWindow::New( ) might create a vtkCarbonOpenGLRenderWindow, vtkCocoaOpenGLRenderWindow, or vtkXOpenGL-RenderWindow, depending on the system configuration. This allows the application programmer to write device-independent applications using VTK without the typical conditionals (and knowledge) surrounding all the device-dependent code.

Object factories can also be used in VTK to enable dynamic loading of new subclasses. VTK maintains a collection of object factories that are queried whenever a class is instantiated with New( ). Dynamically loaded factories can replace the existing implementation of a filter at run time with a new one that includes bug fixes, optimizations, enhancements, etc. For example, in a volume rendering application, an attempt might be made to load an object factory that contains a volume renderer supporting the VolumePro hardware. If the object factory successfully loads, then the application will suddenly start using the VolumePro hardware instead of the default volume renderer.

30.2.7 Data Representation

A key feature of any visualization system is the manner in which data is represented. In VTK, data is represented in native form. That is, data is not converted to a single canonical type (e.g., double). This approach avoids the memory penalty that would occur when, for example, a 1-byte char value was represented by an 8-byte double. In addition, conversion of data from one type to another can introduce precision errors.

The data types supported by VTK are limited to the native compiler types found in C++ (char, uchar, short, ushort, int, uint, long, ulong, float, and double). An additional ID type is used to index points and cells. This leads to a simple array-based data model (known in VTK as a data array). A data array is assumed to consist of a linear arrangement of tuples, each tuple having the same number of n components. A tuple may represent a scalar (1-tuple), point, vector, or normal (3-tuple), texture coordinate (1-, 2-, or 3-tuple), color (1-, 2-, 3-, or 4-tuple), 3×3 tensor (9-tuple), or arbitrary data array (n is arbitrary). The data is represented by contiguous arrays, which are easy to allocate, delete, and transport across the network.

Filters may access the data contained in the array using two different mechanisms. If performance is desired, and the number of types is limited, C++ templated functions are used to operate on the data. Alternatively, an abstract interface defined by the class vtkDataArray provides generic access to its concrete subclasses,

such as vtkUnsignedCharArray, vtkFloatArray, and so on. This more general approach avoids the complexities of templating but introduces overhead due to type conversion and virtual function access to the data. In VTK, both methods of data access are used—sometimes in combination—depending on the complexity and performance requirements of the filter.

30.2.8 Callbacks (Command/Observer Design Pattern)

VTK is designed to interoperate with other systems through a powerful callback mechanism. Callbacks are supported through the use of the command and observer design patterns [9]. Every object in VTK supports the AddObserver( ) method, which takes two key arguments: the event to observe and the command to invoke when the event is fired. These observers can be used in a variety of ways, depending on the events that a particular class invokes on itself. For example, all filters fire Start, Progress, and End events that can be used by a GUI to provide information on what filter is currently running and how much progress has been made on the computation. The progress events are particularly useful because they allow control to return to the application, at which point the application can check to see if the computation should be aborted and, if so, pass that information down to the filter. A similar event is used during rendering to abort a render operation when, for example, the user presses something on the GUI. This is particularly valuable for volume rendering with progressive refinement. The highest levels of quality can take a significant time to render, and the user may wish to abort the render and adjust some other properties instead of waiting for the render to complete.

While VTK is designed to be GUI independent, there is a connection between the render window and the GUI. Specifically, GUI events are typically used to drive camera operations, redraw the window, handle resizing, and manipulate 3D widgets. In VTK, there is a render window interactor class to handle these window-system events. Using VTK's Command/Observer design pattern, the render-window interactor simply receives window system events, translates them into VTK events, and then fires off these VTK-based events. There are a number of classes in VTK that register observers for these events. The most common are the interactor style (vkInteractor-Style) classes, which encapsulate different styles of mouse and keyboard interaction. An example of an interactor style is the camera trackball style. This style uses mouse events to manipulate the camera as if it were a trackball. There are a number of other styles that use different conventions for manipulating the camera, lights, and actors in VTK. (3D widgets make extensive use of events; they are discussed in Section 30.4.)

One advantage of using observers is that many objects can thereby listen to the same event. A mouse-down event could first be analyzed by a 3D widget to determine if it should respond to it. If so, it might perform the appropriate function and then abort further processing of the event, or leave the event for other observers to process. If not, the event would eventually be caught and handled by the currently selected interactor style, or any other registered observer of that event.

30.2.9 Wrapping

Part of the success of a toolkit depends on its accessibility. In many cases, the first test of accessibility is whether the toolkit can be used from the programmer's target development language. While C++ programmers can readily use VTK, programmers using the languages Tcl, Java, or Python can use it as well. VTK supports a thin-layer language binding that enables almost all of VTK's features to be used by these languages. The wrapping takes advantage of the consistency in VTK's coding style to determine what methods to wrap and how to handle reference counting and callbacks. The object-oriented nature of VTK makes writing

code in any of the languages straightforward. Consider the following example of creating a cone and setting its resolution in each of the four supported programming languages:

```
In C++
    vtkConeSource *cone =
      vtkConeSource::New();
    cone→ SetResolution ( 10 );

In Tcl
    vtkConeSource cone
    cone. SetResolution 10

In Python
    cone = vtk.vtkConeSource()
    cone.SetResolution( 10 )

In Java
    vtkConeSource cone =
      new vtkConeSource ();
    cone.SetResolution( 10 );
```

Wrapping the C++ code into additional programming languages has several important advantages beyond code accessibility. Interpreted languages like Tcl and Python do not require compilation or linking, support runtime extensibility, and provide many packages (such as GUI builders) to extend their capabilities. As a result, such languages provide a powerful prototyping environment while maintaining the efficiency of C++, since the algorithms are written in C++.

30.2.10 Memory Management

As described previously, the object factory New( ) method is used to instantiate classes in VTK. One of the benefits of this approach is that it supports reference counting. In order to minimize memory requirements and pass objects around the system, all objects in VTK are reference counted. When a new object is created with New( ), its reference count is initialized to 1. If another object uses that object, it will invoke Register( ) on it to increase its reference count. Likewise, when an object is done using another object, it will UnRegister( ) it, which decrements the reference count. When the reference count of an object falls to zero, the object is deleted and memory is freed.

A significant advantage of reference counting in VTK is that it allows data to be shared instead of duplicated. For example, consider a filter that converts a triangle mesh from independent triangles to triangle strips. If the input data to this filter has scalar values associated at each point, the output will have the same scalar values. In this case, the best solution is to pass the scalar values from input to output by reference instead of copying them. This is done by assigning them to the output, which will increase the scalar's reference count by one.

30.3   Methods in Large-Data Visualization

In VTK, special attention has been paid to handling large datasets. Large datasets present a number of problems, ranging from computation speed to communication to space limitations. VTK supports a number of techniques for handling large data, including data streaming and distributed parallel processing. The following sections cover these topics.

30.3.1   Data Streaming

In VTK, data streaming is implemented by breaking data into smaller pieces, processing the pieces (either serially or in parallel), and then combining the results [8]. The ability to stream data through a visualization pipeline offers two main benefits. The first is that visualization data that would not normally fit into memory or system swap space can be processed. The second is that visualizations can be run with a smaller memory footprint, resulting in higher cache hits and little or no swapping to disk. To accomplish this, the visualization software must support breaking the dataset into pieces and correctly processing those pieces. This requires that the dataset and the algorithms that operate on it be separable, mappable, and result invariant.

1. **Separable**. The data must be separable. That is, the data can be broken into pieces. Ideally, each piece should be coherent in geometry, topology, and/or data structure. The separation of the data should be simple and efficient. In addition, the algorithms in this architecture must be able to correctly process pieces of data.

2. **Mappable**. In order to control the streaming of the data through a pipeline, we must be able to determine what portion of the input data is required to generate a given portion of the output. This allows us to control the size of the data through the pipeline and to configure the algorithms.

3. **Result Invariant**. The results should be independent of the number of pieces and independent of the execution mode (i.e., single-threaded or multithreaded). This requires proper handling of boundaries and development of algorithms that are multithread safe across pieces that may overlap on their boundaries.

In VTK's demand-driven architecture, consumers of data—such as rendering engines or file writers—make requests for data that are fulfilled using a three-step pipeline update mechanism. The first step, UpdateInformation( ), is used to determine the characteristics of the dataset. This request is made by the consumer of the data and travels upstream to the source of the data. For structured data, the resulting information contains the native data type (such as float or short), the largest possible extent, expressed as $(i_{min}, i_{max}, j_{min}, j_{max}, k_{min}, k_{max})$, the number of scalar values at each point, and the pipeline-modification time. The native data type and number of scalar values at each point are used to compute how much memory a given piece of data requires. The largest possible extent is typically the size of the dataset on disk. This is useful in determining how to break the dataset into pieces and where the hard boundaries are (versus the boundaries of a piece). The pipeline-modification time is used to determine when cached results can be used.

Many algorithms in a visualization pipeline must modify the information during the UpdateInformation( ) pass. For example, a 2x image-magnification algorithm would produce a largest possible extent that was twice as large as its input. A volume-gradient algorithm would produce three components of output for every input component.

The second step, UpdateExtents( ), is used to propagate a request for data (the update extent) up the pipeline (to the data source). As the request propagates upstream, each algorithm must determine how to modify the request—specifically, what input extent is required for the algorithm to generate its requested update extent. For many algorithms, this is a simple one-to-one mapping, but for others, such as a 2x magnification or gradient computation using central differences, the required input extent is different from the requested extent. This is the origin of the requirement that the algorithms be mappable. A side effect of the UpdateExtents( ) pass is that it returns the total memory required to generate the requested extent. This enables streaming based on a memory limit. One simple streaming algorithm is to propagate a large update extent, and if that exceeds the user specified memory limit, then to break the update extent into smaller pieces until it fits. This requires that the dataset be separable. More flexible streaming algorithms can switch between dividing a dataset by blocks and dividing it by slabs and by what axis.

The final step, UpdateData( ), causes the visualization pipeline to actually process the data and produce the update extent that was requested in the previous step. These three steps require a significant amount of code to implement, but surprisingly their CPU overhead is negligible. Typically the performance speedup provided by better cache locality more than compensates for the additional overhead of configuring the pipeline. The exception is when boundary cells are recomputed multiple times because they are shared between multiple pieces. This is typical in neighborhood-based algorithms, and it creates a tradeoff between piece size (memory consumption) and recomputing shared cells (computation).

This entire three-step process is initiated by the consumer of the data, such as a writer that writes to disk or a mapper that converts the data into OpenGL calls. In both of these cases, the streaming is effective because the entire result is never stored in memory at one time. It is either written to disk in pieces or sent to the rendering hardware in pieces. We can also stream it in the middle of a visualization pipeline if there is an operation that requires a significant amount of input but produces a fairly small output.

The use of streaming within VTK is simple. Consider the following example. First an instance of an analytical volumetric source is created. It is then connected to a contour filter that is itself connected to a mapper. Normally the mapper would request all of its input data and then convert the data into graphics primitives, but this behavior can be changed by setting the memory limit on the mapper. The mapper will then initiate streaming if the memory consumption exceeds that limit. The only change required to support streaming in this example is the addition of the call to SetMemoryLimit( ) on the mapper.

30.3.2 Mixed Topologies

The preceding section described how to stream structured data. Streaming unstructured data or mixtures of structured and unstructured data poses several problems. First, an extent must be defined for unstructured datasets. With regularly sampled volumetric data, such as images, an extent defined as $(i_{min}, i_{max}, j_{min}, j_{max}, k_{min}, k_{max})$ can be used, but this does not work with unstructured data. With unstructured data, there are a few options. One is to use a geometric extent, such as $(x_{min}, x_{max}, y_{min}, y_{max}, z_{min}, z_{max})$, but it is an expensive operation to collect the cells that fit into that extent, and such an extent is difficult to translate into the extents used for structured data if they are not axis-aligned (consider a curvilinear grid).

A more practical approach is to define an unstructured extent as piece M out of N possible pieces. The division of pieces is made based on cells so that piece 2 of 10 out of a 1000-cell dataset would contain 100 cells. The approach for streaming based on a memory limit is the same as for structured data, except that instead of splitting the data into blocks or slabs, the number of pieces, N, is increased. This fairly basic definition of a piece dictates that there is not any control over what cells a piece will contain, only that it will represent about 1/N of the total cells of the dataset.

This raises the issue of how to support unstructured algorithms that require neighborhood information. The solution is to use ghost cells, which are not normally part of the current extent but are included because they are required by the algorithm. To support this, we extend the definition of an unstructured extent to be piece M of N with G ghost levels. This requires that any source of unstructured grid data be capable of supplying ghost cells. There is a related issue: some unstructured algorithms, such as contouring, operate on cells, while others, such as glyphing, operate on points. Points on the boundary between two different extents will be shared, resulting in duplicated glyphs when processed. To solve this we indicate which points in an extent are owned by that extent and which ones are ghost points. This way, point-based algorithms can operate on the appropriate points and yet still pass other points through to the cell-based algorithms that require them. In the end, both ghost cells and ghost points are required for proper processing of the extents.

Consider Fig. 30.7, which shows a piece of a sphere. The requested extent is shown in red, and two ghost levels of cells are shown in green and blue. The points are colored based on their ownership, so all red points are owned by the requested extent, and the green and blue points

Figure 30.7 Streaming pieces of a sphere. (See also color insert.)

indicate ownership of the points by other extents. Note that some cells use a mixture of points from different extents.

Now that extents have been defined for both structured and unstructured data, the conversion between them must be defined. For most operations that take structured data as input and produce unstructured data, a block-based division can be used to divide the structured data into pieces until there are N pieces as requested. If ghost cells are required, the resulting extent of the block can be expanded to include them. If ghost point information is required, it can be generated algorithmically based on the largest possible extent and some convention regarding which boundary points belong to which extent.

An extent can be converted from unstructured to structured in a similar manner, except that it is inappropriate for most algorithms that convert unstructured data to structured. Consider a Gaussian-splatting algorithm that takes an unstructured grid and resamples it to a regular volume. To produce one part of the resulting volume requires use of all the cells of the unstructured grid that would splat into that extent. With our definition of an unstructured extent, there is no guarantee that the cells in an extent are collocated or topologically related. So generation of one extent of structured output requires that all of the unstructured data be examined. While this could be done within a loop, our current implementation requires that during translation from a requested structured extent to an unstructured extent, the entire structured input be requested.

30.3.3 Distributed Parallel Processing

Most large-scale simulations make use of parallel processing, and the results are often distributed across many processing nodes. This distribution requires that the visualization system be capable of operating in such an environment. Parallelism support requires some of the same conditions as streaming, such as data separability and result invariance. It also requires methods for data transfer, asynchronous execution, and collection.

Data transfer is implemented in VTK by creation of input and output port objects that communicate between filters (i.e., algorithms) in different processes. Asynchronous execution is required so that one process is not unnecessarily blocked while waiting for input from another process. Consider a pipeline where one of its filters (FilterA) has two inputs and its first input is in another process. The first input requires an input and output port for managing the interprocess communication. Before FilterA executes, it must make sure that both of its inputs have generated their data. A naïve approach would be to simply ask each input to generate its data in order. The problem is that the second input of FilterA is idle while waiting for the first input to compute its data. To solve this problem, two modifications are made to the three-step pipeline execution process. The first modification is to add a nonblocking method to the update process. This method—TriggerAsynchronousUpdate( )—is used to start the execution of any inputs that are in other processes. Essentially, this method traverses upstream in the pipeline, and when it encounters a port, the port invokes UpdateData( ) on its input.

The second modification is to use the locality of inputs to determine in what order to invoke UpdateData( ) on each input. The locality of an input is a measure of how much of a pipeline resides within the current process. It ranges from 1.0 if the entire pipeline from that point back is within the current process to 0 if it is entirely in another process. This locality is computed as part of the UpdateInformation( ) call. So in the earlier example, TriggerAsynchronous Update( ) is sent to the first input of FilterA, which would cause it to start executing because it is in a different process. The second input would ignore the TriggerAsynchronousUpdate( ) call since there are no ports between it and FilterA. Then FilterA would call UpdateData( ) on its second input first, since it has the highest locality. Once it has completed executing, UpdateData( ) would be called on the first input, since it began executing after the Trigger-AsynchronousUpdate( ) call.

In addition to this parallel support in VTK, process initialization and communication calls have been encapsulated into a controller class so that the application programmer does not have to directly deal with them. Concrete sub-classes have been created for distributed-memory and shared-memory processes using MPI and pthreads. Likewise, a sort-last, parallel rendering class is provided that uses interprocess communication to collect and then composite parallel renderings into a final image. For smaller data that can be collected onto a single node, centralized rendering is supported by gathering the polygonal data together using ports connected to an append filter. Parallel rendering can also be implemented using polygon collection followed by parallel rendering such as WireGL.

Given our parallel data-streaming architecture, we can create a data-parallel program simply by writing a function that will be executed on each processor. Inside that function, each processor will request a different extent of the results based on its processor ID. Each processor can still take advantage of data streaming if its local memory is not sufficient, allowing this architecture to produce extremely large-scale visualizations.

30.4   VTK in Application

VTK is widely used in research and commercial applications. This section highlights two such applications and describes how the toolkit is integrated into the GUI.

30.4.1   Binding to the GUI

To facilitate the use of VTK in applications requiring a GUI, VTK includes support for a number of different widget sets and languages. For Tcl/Tk, a Tk widget is available so that a vtkRenderWindow can be directly embedded into a Tk application. A Python wrapper exists for the Tk rendering widget so that the same approach can be used with Tk-based Python

applications. For Java, VTK provides a Java AWT–based rendering canvas called vtkPanel. The vtkPanel can be used like any other AWT-based widget. Similarly, there is support to direct-embed a VTK render window into GUIs based on other toolkits including MFC, FLTK, Qt, and wxWindows. In all of these cases, the resulting window is a fully accelerated OpenGL window as long as the platform supports accelerated OpenGL. In most of these cases, the application programmer has the option of handling windowing system events in their application code or using the standard VTK event loop and command/observer mechanisms.

30.4.2   3D Widgets

The command/observer design pattern and GUI binding techniques described previously offer powerful capabilities by which we can build complex, interactive applications. However, this functionality is relatively low level and oriented towards skilled developers. Many tasks require complex interaction between the user and the visualization system. For example, positioning a data probe to obtain numerical values or clipping data with an oriented plane requires the user to manipulate and orient objects in the scene. Programming such functionality requires significant effort with the tools described previously, and common tasks have been encapsulated into the concept of a 3D widget. Similar to the pervasive 2D widgets found in modern user interfaces, 3D widgets have the added challenge of operating in one more depth dimension.

VTK offers a palette of more than a dozen widgets, as shown in Fig. 30.8. VTK's 3D widgets can be used to position points, lines, planes, boxes, and spheres. Specialized widgets are used to view orthogonal planes in volumes, manipulate splines, and position scalar bars (i.e., the color legend associated with scalar values). Most of the widgets provide positioning and transformation information. Many also provide implicit functions ($f(x, y, z) =$ constant), such as the planes and spheres used in VTK to

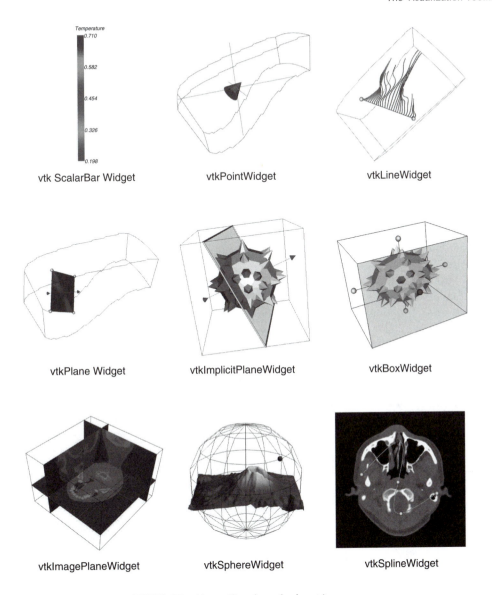

Figure 30.8 Palette showing some of VTK's 3D widgets. (See also color insert.)

cut, clip, and extract data. The widgets also provide primitives, such as points and lines that can be used to seed streamlines or probe data.

3D widgets are subclasses of the abstract class vtkInteractorObserver. This class observes mouse and keyboard events that are invoked by VTK's render-window class. The widgets

respond to registered events, modify themselves as appropriate, and invoke higher-level events that the application can observe. For example, selecting the endpoint of a line widget allows the user to interactively position the endpoint. As the point moves, InteractionEvents are invoked by the line widget. When the endpoint is released (on mouse up), an EndInteractionEvent is invoked. The application can respond to these events as appropriate—by, for example, seeding a streamline or producing an *x-y* plot along a line.

30.4.3   ParaView

ParaView is an open-source, turnkey application for general-purpose scientific visualization built on top of VTK [11]. Primary design goals include supporting large-data visualization using distributed parallel processing, as well as serving as a customizable, extensible platform (via XML-configured modules and Tcl/Tk scripts) that can be readily configured to address a variety of application areas. ParaView employs a data-parallel model with MPI and supports advanced rendering functionality including tiled display [10] support and the ability to automatically switch to using parallel composite rendering when data becomes large. ParaView makes ex-

tensive use of the VTK 3D widgets, including the point, line, plane, and scalar bar widgets. Fig. 30.9 is a screen shot of ParaView in action.

30.4.4   VolView

VolView is a commercial volume-visualization application built with VTK [12]. It provides an intuitive, interactive interface that enables researchers and clinicians to quickly explore complex 3D medical or scientific images. Novice users can easily generate informative images to include in patient reports and presentations. Data exploration and analysis are enhanced by tools such as filtering, contours, measurements, histograms, and annotation. VolView versions 2.0 and later support a plug-in architecture that enables complex image processing and segmentation tools to be added to the system at run time. These plug-ins can be written by the user and interfaced to VolView using a simple C-

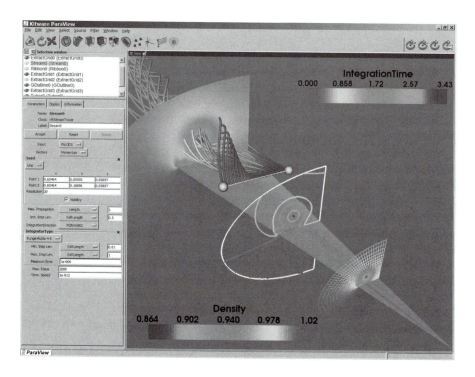

Figure 30.9   Flow over a delta wing visualized with ParaView. (See also color insert.)

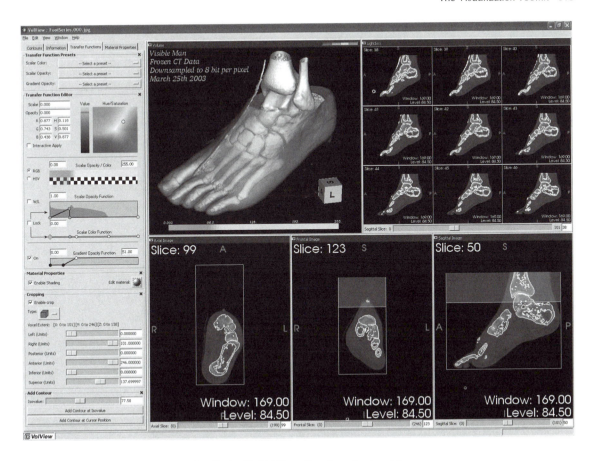

Figure 30.10   Visible Human dataset visualized with VolView. (See also color insert.)

language API. VolView takes advantage of VTK's integrated image processing, 3D surface graphics, and volume rendering capabilities to produce sophisticated visualizations for a variety of medical, scientific, geophysical, and imaging applications. Fig. 30.10 is a screenshot of VolView being used to visualize a portion of the Visible Human dataset [13].

References

1.   W. Hoffman and K. M. Martin. The CMake build manager. *Dr. Dobb's Journal*, 2003.
2.   DART software quality regression testing system, http://public.kitware.com/Dart/HTML/Index.shtml
3.   W. J. Schroeder, K. M. Martin, and W. E. Lorensen. *The Visualization Toolkit: An Object-Oriented Approach to Computer Graphics* (3rd Ed.) Kitware, Inc., 2003.
4.   The Doxygen documentation system, http://www.doxygen.org.
5.   W. E. Lorensen and B. Yamrom. Object-oriented computer animation. *Proceedings of IEEE NAECON*, 2:588–595, 1989.
6.   J. Rumbaugh, M. Blaha, W. Premerlani, F. Eddy, and W. Lorensen. *Object-Oriented Modelling and Design*. Englewood Cliffs, NJ, Prentice-Hall, 1991.
7.   VolumePro, http://www.rtviz.com/products/volumepro_prod.html
8.   C. C. Law, K. M. Martin, W. J. Schroeder, and J. E. Temkin. A multi-threaded streaming pipeline architecture for large structured data sets. *Proc. of Visualization '99*, 1999.

9. E. Gamma, R. Helm, R. Johnson, and J. Vlissides. *Design Patterns: Elements of Reusable Object-Oriented Software*. Addison Wesley, 1995.

10. K. Li, H. Chen, Y. Chen, D. W. Clark, P. Cook, S. Damianakis, G. Essl, A. Finkelstein, T. Funkhouser, A. Klein, Z. Liu, E. Praun, R. Samanta, B. Shedd, J. P. Singh, G. Tzanetakis, and J. Zheng. Early experiences and challenges in building and using a scalable display wall system. *IEEE Computer Graphics and Applications*, 20(4):671–680, 2000.

11. ParaView, http://www.paraview.org

12. VolView, http://www.kitware.com/products/volview.html

13. The Visible Human Project, http://www.nlm.nih.gov/research/visible/visible_human.html

31 Visualization in the SCIRun Problem-Solving Environment

DAVID M. WEINSTEIN, STEVEN PARKER, JENNY SIMPSON,
KURT ZIMMERMAN, and GREG M. JONES
Scientific Computing and Imaging Institute
University of Utah

31.1 Introduction to SCIRun

31.1.1 Motivation and History

Located at the crossroads of scientific applications, computer science, and numerical methods is the emerging field of computational science. With strongholds in applications ranging from chemistry to physics and genetics to astronomy, computational science is growing into prominence throughout the scientific world, taking a position next to "theoretical" and "experimental" as another branch of nearly every scientific discipline.

Each scientific discipline has its own terminology and its own specific problems of interest. But from a broader perspective, their similarities often outnumber their differences. Many problems of interest are based around a physical model of some system or domain; these problems often attempt to predict the result of well defined, equation-driven processes that take place within that domain; and the solutions to these problems are often most easily understood when recast into an interactive visual representation. Further, it is often not sufficient to run a single simulation of a system; the scientist typically wants to investigate and explore the problem space, setting up different initial conditions, system parameters, and so on, and then comparing the results.

Because of these common circumstances, it seems plausible that a general-purpose framework could be designed to assist scientists and engineers from a broad range of disciplines in investigating their respective computational science problems. Such a framework could be thought of as a "computational science workbench"; it would give a scientist a broad range of tools at hand for modeling, simulating, visualizing, and iteratively exploring a problem space. The framework would be an easy-to-use visual programming environment where the scientist could dynamically hook together computational components, just as a different scientist might hook together mechanical components in a laboratory. And, perhaps most importantly for scientists working on large-scale problems, the framework would have to be extremely efficient in the way it managed and processed data.

At the Scientific Computing and Imaging Institute (SCI) at the University of Utah, we set out to produce such a computational architecture beginning in the early 1990s. Our framework, called SCIRun (pronounced "ski-run"), was initially developed by a handful of graduate students, and it was targeted at the simulation of bioelectric field problems as its initial application [6,14]. Through the mid-1990s, SCIRun grew into a more robust platform as it was applied to more applications, including cognitive neuroscience and atmospheric simulation [4,15,23]. In 1997 and 1998, the SCI Institute was awarded Center grants from the National Institutes of Health (NIH) and the Department of Energy (DOE), respectively, to continue the research, development, and support of the SCIRun system.

31.1.2   Overview: Data-Flow Terminology

As an infrastructure, the SCIRun computational problem-solving environment is a powerful collection of high-performance software libraries. These libraries provide many operating system–type services, such as memory and thread management and interthread communication and synchronization, as well as development utilities, such as geometry, container, scene-graph, and persistent I/O classes.

While the SCIRun infrastructure is complex and is likely to be somewhat opaque to those who are not computer scientists, SCIRun's exterior layers are, in contrast, easy to use, extend, and customize. The SCIRun user-level programming environment, described earlier as a "computational workbench", is a visual data-flow environment that facilitates rapid development.

Fig. 31.1 shows an example of the SCIRun visual programming environment.

The boxes on the canvas are called *modules*, and the wires connecting them are called *datapipes*. Each module encapsulates a function or algorithm, and the datapipes carry input and output data between them. Taken as a whole, the group of modules and datapipes comprise a dataflow *network* or *net*. At run time, users can interactively instantiate, destroy, and reconnect new modules. In addition to datapipe I/O, each module also has the option of exposing additional input and output parameters through a graphical user interface (GUI). For example, as shown in Figs. 31.2 and 31.3, the SolveMatrix module is a linear solver that exposes input parameters, such as the solver method and the maximum error tolerance, and also reports

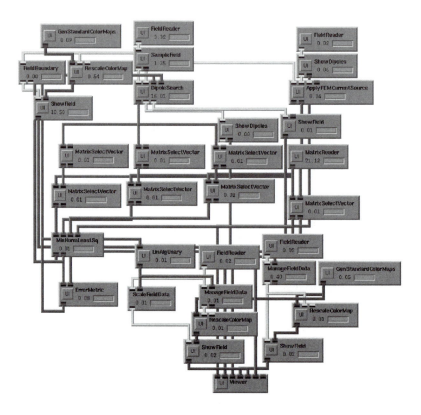

Figure 31.1   A SCIRun data-flow network. Each module encapsulates a function or algorithm, while the data pipes carry input and output data between the modules.

Figure 31.2 An example of a SCIRun module. The Solve-Matrix module is a linear solver that exposes input parameters such as the solver method and the maximum error tolerance and also reports output parameters such as convergence plots for iterative solvers. The "UI" button produces a module-specific interface, allowing the user to adjust parameters specific to that module.

output parameters, such as convergence plots for iterative solvers.

For the SolveMatrix module, we implemented several solvers natively within SCIRun. We have also, however, left placeholders for users to link in other solvers. This coupling of native support and optional hooks for extensibility has been a design pattern for SCIRun. In the SolveMatrix example (Fig. 31.3) we implemented Conjugate Gradient, Biconjugate Gradient, and Gauss-Seidel solvers; anyone downloading SCIRun will have immediate access to these methods. Then, in order to provide support for additional solvers, we created a *bridge* to the PETSc library. If a user chooses to download and install PETSc, he or she can configure SCIRun to use it, and the full set of PETSc solvers can then be leveraged within SCIRun. We have also applied this bridging mechanism to allow users to access the ImageMagick and MPEG libraries for saving images and movies, respectively. Additionally, this same bridging solution has been implemented to allow MATLAB users the ability to run their MATLAB scripts from within SCIRun. By leveraging other libraries and applications, we are able to stay focused on developing high-performance infrastructure and easy-to-use interfaces while still providing support for a wide range of application functionality. Shown in Fig. 31.4 is an example of one such bridge, where Genesis has been connected to SCIRun for the visualization of a combined genesis/SCIRun simulation of the bioelectric field between two Aplysia motor neurons.

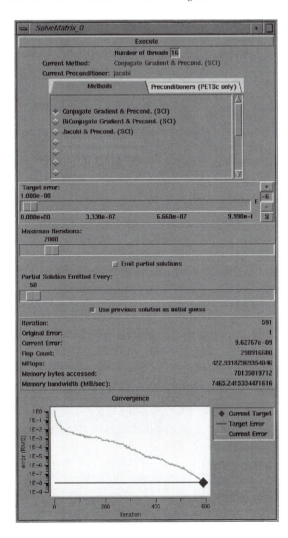

Figure 31.3 The SolveMatrix user interface (UI). This UI allows the user to interact with the model. In this case, the UI allows the user to chose various solvers, such as the Conjugate Gradient, Biconjugate Gradient, and Gauss-Seidel. The convergence of the solver is also displayed in the UI. In order to provide support for additional solvers, there is also a *bridge* to the PETSc library.

31.1.3 The Visualization Pipeline

A typical visualization algorithm, such as stream-line advection, works by computing sample positions, evaluating the value of the field at those positions, and creating a geometric representation for those values and positions. This

Figure 31.4 The simulation of two Aplysia motor neurons using the bridging capabilities between Genesis and SCIRun. First, Genesis solves the time-dependent Hodgkin-Huxley equations for each compartment in each cell. One result of the Genesis simulation is a calculation of neuron membrane current density, which is passed to SCIRun through a SQL database. SCIRun uses the current density to solve the forward field problem in the volume surrounding the cell. In this picture, streamlines show current flow within the volume; voltage is encoded by color (blue is negative; red is positive). Image courtesy of Chris Butson, University of Utah, Department of Bioengineering. (See also color insert.)

three-step process is common to many visualization methods: isosurfacing, streamlining, volume rendering, tensor-field rendering, surface-potential mapping, cutting-plane rendering, etc. Typically, a particular visualization algorithm will implement all three of these steps itself. Such an approach results in substantial coding inefficiencies. For example, the same geometric representations may be of interest in multiple visualization techniques (e.g., rendering pseudo-colored surfaces is common to surface-potential mapping, cutting-plane rendering, and often isosurfacing). In the spirit of modular pro-

gramming and reusable components, we have pipelined (or "networked") the majority of our visualization methods, and there are interchangeable modules available for each of the three stages.

31.2 SCIRun Visualization Tools

There exist a number of tools that are easily accessible within the SCIRun system. The following sections are a rundown of this toolset.

Tools for the Visualization of Scalar Fields
- Isosurface: Visualize isosurfaces of a volume or isocontours on a surface. Can use Marching Cubes or NOISE algorithm. Can specify a single iso-value, a list of iso-values, or a range and quantity for evenly spaced iso-values.
- Volume rendering/MIP (via 3D textures).
- Cutting plane (via 3D textures).
- Color-mapped geometry (ShowField): Visualizes the geometry that makes up a mesh inside a field. Where possible, the field takes its color from the data values that permeate the field.

Tools for the Visualization of Vector Fields
- Streamlines: The StreamLines module visualizes vector fields by generating curves that interpolate the flow of vectors in a field.
- Vector glyphs (ShowField): The ShowField module visualizes the geometry that makes up a mesh inside a field. Where possible, the field takes its color from the data values that permeate the field.
- ShowDipoles: The ShowDipoles model allows the user to edit vector positions and orientations via widgets.

Tools for the Visualization of Tensor Fields
- Glyphs: ellipsoids, colored boxes.
- Tensorlines.

Tools for Quantitative Visualization
- ShowLeads: The ShowLeads module graphs a matrix that has rows of potentials.
- ErrorMetric: The ErrorMetric module computes and visualizes error between two vectors.
- ShowField: The ShowField module visualizes the geometry that makes up a mesh inside a field. Where possible, the field takes

its color from the data values that permeate the field.
- ShowColorMap: The ShowColorMap module creates a geometry overlay containing the input color map and numerical values for its range.

31.3 Remote and Collaborative Visualization

In the last few years, scientists and researchers have given a great deal of attention to the area of remote visualization of scientific datasets within collaborative environments. *Remote visualization* refers to the process of running an application on one machine, often a supercomputer, and viewing the output on another machine in a different geographical location. *Collaborative visualization* is the use of tools (chat windows, annotations, synchronous viewing controls, etc.) that enable multiple geographically separated collaborators to directly exchange and simultaneously view information related to specific visualizations.

The recent interest in these tools has developed because researchers often use interactive viewing as their primary method of exploring large datasets. Researchers often need to extend this interactivity in order to collaborate remotely with colleagues. Additionally, the ability to use remote high-end computation resources is also driving the need for remote visualization tools. Certainly, visualization on grid based systems is also a driving demand for remote visualization tools.

Remote and collaborative visualization is by no means a new problem. In fact, many algorithms have been developed as solutions. General strategies available for achieving remote visualization fall into four rough categories:

1. Traditional XWindows remote display.
2. Image pixel streaming.

3. Geometry/texture rendering.

4. Some hybrid of these three methods.

While most remote-visualization tools based on the aforementioned methods successfully allow multiple parties to view images from different locations, most also face problems with efficiency and user interactivity at some level.

In general, popular recent approaches that have addressed these shortcomings focus on improving two different areas of remote visualization:

1. Increasing network bandwidth utilization.

2. Adjusting the amount of rendering performed on a local server versus on a remote client in order to optimally utilize resources, such as available bandwidth between the server and the client.

In particular, researchers and developers often use the client–server paradigm as a logical means to partition rendering responsibilities, efficiently utilizing valuable resources on both local and remote machines.

31.3.1 Current Work on Remote and Collaborative Visualization in SCIRun

Our work in implementing remote and collaborative visualization functionality in SCIRun has been largely experimental thus far. To this end, we built upon a prototype remote-visualization application that applies the client–server paradigm, along with several rendering methods, as an attempt to offer greater flexibility for remote viewing. In addition to using multiple rendering methods, we have experimented with different networking protocols for data transfer to compare efficiency and accuracy. While we have learned a great deal from our research with this prototype application, we have run into fundamental design problems that have prompted us to slight our remote visualization extension for redesign as a component in the design of the next generation of SCIRun, which will see the SCIRun system move toward a component architecture.

Presently, standard XWindows remote viewing is used for remote display of SCIRun.

31.3.2 Future Work on Remote and Collaborative Visualization in SCIRun

The future of remote and collaborative visualization in SCIRun is closely tied to the Common Component Architecture (CCA) that is planned for SCIRun2. Roughly, the long-term plan for SCIRun (SCIRun2) is to consider everything to be a component, including the computing modules and the user interface. Presumably, the remote client will be a component that uses a CCA protocol to communicate with other components.

As for our rendering method, our current plan is to utilize a hybrid of XWindows remote display and image streaming to transfer image data from the computing engine to the remote client.

In the process of designing the remote visualization component, we will adhere to a list of user-driven requirements. These include the following:

- Minimum x frames per second.
- UI that matches SCIRun local UI, both visually and functionally.
- Synchronized image manipulation for multiple remote viewers with locked controls.
- Exact representation of models, or level of detail (LOD) control-usability with "dumb" client machine.
- Usability with limited network bandwidth.
- Chat window.
- Annotation layer.
- Compliance with SCIRun2 architecture and communication protocols.
- Possible ability to record and replay sessions.

The end goal is to have a remote user interface that has the same appearance and functionality as the local SCIRun user interface, but with added collaborative tools and remote viewing capabilities.

31.4   SCIRun Applications

As mentioned above, one of the original applications of SCIRun was in bioelectric field problems. With the award of our NCRR grant from the NIH to create the Center for Bioelectric Field Modeling, Simulation, and Visualization, we have continued to focus on bioelectricity with the creation of modules, networks, and documentation to allow users to investigate both forward and inverse bioelectric field problems. In order to keep the core of SCIRun on a general-purpose level, we have created a separate package to house the components that are specific to bioelectricity. Taken together, the BioPSE Package and the SCIRun architecture comprise the BioPSE System [21], which is shown in Fig. 31.5. Similarly, the Uintah Package [1,9,13] is an extended set of functionality that targets combustion simulation. In fact, a number of grants have now leveraged the SCIRun core, adding specific components to address the needs of different various applications. These applications range from bioelectric fields to combustion simulation to magnetic

fusion. Each of these applications is briefly described below.

31.4.1   Modeling, Simulation, and Visualization of Bioelectric Fields

Our hearts and brains are electric organs. Electric activation at the cellular level causes the heart to beat, and it is the basis underlying our cognitive processes. However, unlike neurotransmitters and metabolic processes, electric patterns can be instantly detected at sites remote from the position of activation. By placing an ECG electrode on a patient's chest, we can "watch" the series of electrical events that make up a heart beat; by placing EEG electrodes on a patient's head, we can "watch" the electric activity as the patient thinks and reacts.

Cardiologists and neurologists are primarily interested in two types of bioelectric field problems: the forward problem and the inverse problem [19]. In the forward problem, the question at hand is, given a pattern of source activation, what level of electric activity would result

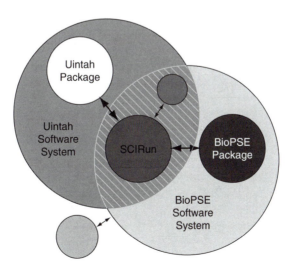

Figure 31.5   The relationship of the core infrastructure of SCIRun to the specialty packages BioPSE and Uintah. In order to keep the core of SCIRun general purpose, we have created a separate package to house the components that are specific to bioelectricity. Taken together, the BioPSE Package and the SCIRun architecture comprise the BioPSE System. Similarly, the Uintah Package is an extended set of functionality targeting combustion simulation.

through the rest of the domain [20] (see Fig. 31.6). Such studies are used in the investigation of internal implantable defibrillator designs. The inverse problem is typically more interesting, though it is unfortunately also less numerically stable: given a set of remote measurements, determine the position and pattern of source activation that gave rise to those remote measurements [17] (see Fig. 31.7).

The equations governing the flow of electricity through a volume conductor are well understood. The goal of the Bioelectric Problem Solving Environment (BioPSE) is to simulate those governing equations using discrete numeric approximations [22]. By building a computational model of a patient's body and then mapping conductivity values over the entire domain, we can accurately compute how activity generated in one region would be remotely measured in another region. The tools for modeling, simulating, and visualizing these bioelectric field phenomena comprise BioPSE.

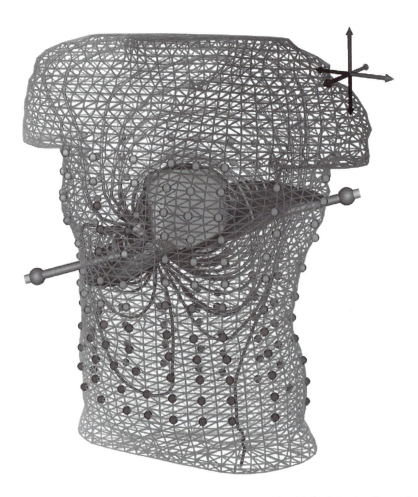

Figure 31.6   A visualization of a SolveMatrix convergence. In this case, an electric dipole is placed near the heart inside the Uintah torso model. The electric field set up by this dipole is visualized by color-mapped streamlines. Also visualized is the surface potential (indicated by color-mapped spheres on the surface of the torso) and a field potential isosurface (green surface inside the torso). This visualization was produced using the BioPSE forward-fem net with the movable dipole widget. (See also color insert.)

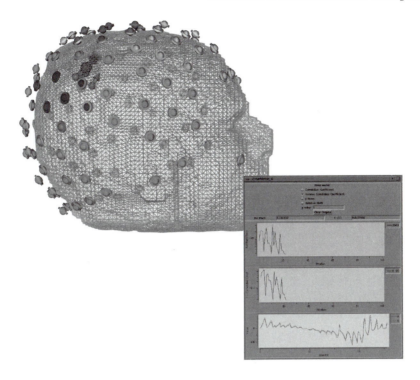

Figure 31.7   A visualization of an inverse EEG simplex search produced by the dipole-localization net. The accuracy of the solution at each electrode position is also shown (disks show measured voltage; spheres show computed voltage), the view window shows simplex dipoles as four arrows (connected with lines), and the test dipole is shown as a sphere. Also shown are the error metrics plotted for this particular solution. (See also color insert.)

31.4.2   Visualization for the Study of Magnetic Fusion

Alternate energy sources are becoming increasingly important as the world's finite resources, such as fossil fuels, are depleted. One promising source of unlimited energy is controlled nuclear fusion. Magnetic fusion in particular is a type of nuclear fusion in which scientists harness energy by using magnetic fields to confine fusion reactions taking place within hot plasma. Since magnetic fusion research is computationally intensive, many software tools are needed to support it. Visualization tools are particularly critical in helping fusion scientists analyze their data. As part of our work within the DOE and SciDAC–sponsored National Fusion Collaboratory (*http://www.fusiongrid.org/*), we have created the Fusion package in SCIRun in order to help meet this need [3,18].

Specifically, the Fusion package in SCIRun is designed to satisfy the goal of providing fusion scientists with visualization software tools that allow exploration of their data on a Linux workstation. The Fusion package consists of a set of SCIRun modules that, in concert with other standard SCIRun modules, allow the reading, visualization, and analysis of fusion data that is in MDSPlus format. The system provides fusion researchers with flexible visualization options and the feedback they need in order to properly adjust input parameters for the next iteration of data processing.

Currently, data generated using the NIMROD simulation code package [24] is being used as a test bed for development of the SCIRun Fusion package, with the extension of the SCIRun package to other data sources planned in the near future. The NIMROD package is publicly available (*http://www.nimrodteam.org/*) and was designed to study 3D, nonlinear electromagnetic activity in laboratory fusion experiments while allowing a large degree of flexibility in the geometry and physics models used in simulations. Using the visualization capabilities offered in the the SCIRun Fusion package makes it possible to visualize how the magnetic field moves within the pressure field represented in a NIMROD dataset. In this paradigm, the NIMROD simulations are done on supercomputers, and then SCIRun running on a Linux desktop is used to analyze the resulting data in order to appropriately adjust parameters for the next simulation.

An instantiation of the visualization pipeline created for NIMROD data includes the following steps. Once the NIMROD data is in SCIRun, a hexahedron mesh is built using the EditFusionField module, which takes into account toroidal geometry and then infuses it with the pressure values at the nodes using the ManageField module. Next, the data is passed downstream to several of the visualization modules available in SCIRun.

The simplest and most general visualization module is the ShowField module, which displays a visual representation of the pressure values. The user can also pass in a ColorMap to the second port, which will translate the pressure values into colors in the rendering.

The next visualization module in the pipeline is the Isosurface module. Given a scalar field, the isosurface module will extract triangular faces that approximate the ovule surface through the domain. Since the input is pressure, isobaric surfaces are generated using this module. As with the ShowField module, the pressure values are again mapped to color via an input ColorMap. With the help of a user interface, a user can choose from several different isosurface extraction algorithms and can set a number of extraction and display options.

Using this visualization pipeline, researchers are provided with the feedback needed to adjust their input parameters for the next iteration of data processing. In the case of the NIMROD

data, SCIRun makes it possible to visualize how the magnetic field moves within the pressure field of the dataset (see Figs. 31.8 and 31.9).

The long-term goals for the Fusion package are to develop a generalized visualization system that can support a wide variety of fusion data and to bring more of the computing portion of the fusion visualization pipeline into SCIRun itself.

31.4.3   The Simulation of Accidental Fires and Explosions

Our work in this area was funded by the DOE as part of the Accelerated Strategic Computing Initiative (ASCI) to form the Center for the Simulation of Accidental Fires and Explosions (C-SAFE). This work is primarily focused on the numerical simulation of accidental fires and explosions, especially within the context of handling and storage of highly flammable materials (Fig. 31.10). The objective of C-SAFE is to provide a system that comprises a problem-solving environment in which fundamental chemistry and engineering physics are fully coupled with nonlinear solvers, optimization, computational steering, visualization, and experimental data verification. For this work, a derivative of SCIRun coined *Uintah* has been developed [1,5,13,17]. The Uintah PSE has been built specifically to handle very large datasets, which are typical in the C-SAFE work. In this case, attempting to render an entire dataset can easily overwhelm the graphics hardware. To help us explore these datasets, we have incorporated into Uintah/SCIRun a multiresolution and a multipipe volume renderer.

31.4.3.1   *Multiresolution Volume Rendering*

Multiresolution techniques enable interactive exploration of large-scale datasets while providing user-adjustable resolution levels on a single graphics pipe. A user can get a feel for the entire dataset at a low resolution while viewing certain regions of the data at higher resolutions. A texture-map hierarchy is constructed in a way

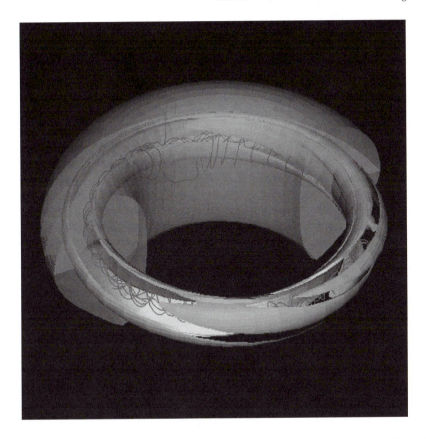

Figure 31.8 A simulation of an experiment inside a Tokomak Fusion Reactor, visualized with the SCIRun Fusion package. This image is of NIMROD simulation data showing an isosurface of the $n = 0$ part of the pressure field (yellow), which shows the 1/1 structure, and an isosurface of the $n = 2$ part of the toroidal current field (green), which shows the developing 3/2 structure. Between the two isosurfaces is a streamline using the sum of the $n = 0$, $n = 1$, and $n = 2$ modes of the magnetic field (red). The underlying model consists of a toroidal grid with 737,280 nodes (in 10 arbitrary phi slices) with 22 time-slices. Image provided by Dr. Allen Sanderson and the National Fusion Collaboratory, with support from the US Department of Energy SciDAC initiative. (See also color insert.)

that minimizes the amount of texture memory with respect to the power-of-two restriction imposed by OpenGL implementations. In addition, our hierarchical LOD representation guarantees consistent interpolation between different resolution levels. Special attention has been paid to the elimination of rendering artifacts that are introduced by noncorrected opacities at level transitions. Through adaptation of the sample slice distance with regard to the desired LOD, the number of texture lookups is reduced significantly, improving interaction.

31.4.3.2 Multipipe Volume Rendering

Multipipe techniques allow for interactive exploration of large-scale data at full resolution. Textures and color transfer functions are distributed among several rendering threads that control the rendering for each utilized graphics pipe or graphics display. On each draw cycle, view information and windowing information are stored in a shared data structure. While rendering is performed, compositing threads are supplied with the composite order for each partial image. Upon completion of the rendering, the rendering threads store the

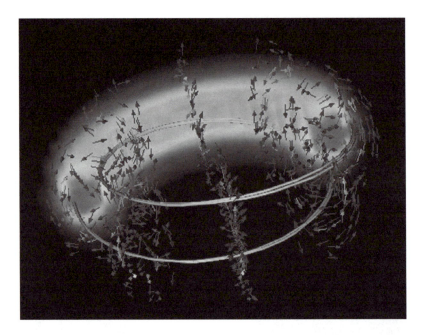

Figure 31.9 A simulation of an experiment inside a Tokomak Fusion Reactor, visualized with the SCIRun Fusion package. This is frame one of two from a time sequence showing the stochastic nature of the real-space magnetic field lines. A comparison of the two frames shows that the magnetic field lines start to diverge earlier as time progresses. The field lines are overlaid in a volume rendering of the pressure field, which provides visual cues to the location while rendering the plasma velocity-vector field. Image provided by Dr. Allen Sanderson and the National Fusion Collaboratory with support from the US Department of Energy SciDAC initiative. (See also color insert.)

resulting image in a local structure. When all renderers have completed, compositing threads copy the partial images to a final image buffer, using alpha blending techniques. Care is taken to prevent blending of artifacts in the final image by properly overlapping the texture data sent to each renderer and by premultiplying the colors in the transfer function by their corresponding alpha values.

31.4.4 Radiology and Surgical Planning

Most imaging systems currently used in medical imaging generate scalar values arranged in a highly structured rectilinear grid. These fields can be visualized by a variety of methods, including isosurface extraction, direct volume rendering, and maximum intensity projections (MIPs). The key difference between these techniques is that isosurfacing displays actual surfaces, while the direct volume rendering and MIP methods display some function of the values seen along a ray throughout the pixel. Ideally, the display parameters for each technique are interactively controlled by the user.

Interactivity is fast becoming a fundamental requirement of medical visualizations. While just a few years ago, radiologists and surgeons viewed inherently 3D images as 2D films, the use of interactive, 3D visualization tools has recently blossomed in medical imaging. This is certainly true in the field of surgical navigation. Interestingly, the radiological exams of today are generating very large datasets (tens to hundreds of megabytes (MB)), making interactivity a challenging requirement. While the commodity graphics card is making great strides, it is only recently that a texture memory of 256 MB has been introduced. The medical imaging community is continually increasing the resolution,

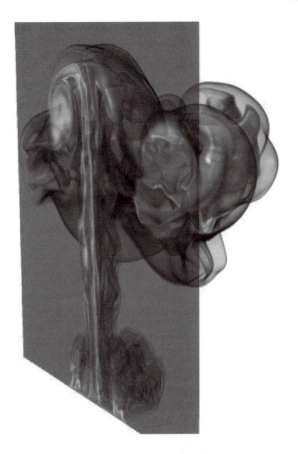

Figure 31.10 A simulation of a heptane pool fire. The simulation was done using the Uintah derivative of SCIRun, and visualization was done using the ray-tracing package described in Section 31.4.4.2. This image is courtesy of the Center for the Simulation of Accidental Fire and Explosions (C-SAFE), working under funding from the Department of Energy as part of the Accelerated Strategic Computing Initiative (ASCI) Academic Strategic Alliance Program (ASAP). (See also color insert.)

power, and size of their imaging tools, and consistently outpacing the graphics-card industry. Meanwhile, the medical imaging research community is now producing datasets in the multigigabyte range with new-generation small animal imagers.

Several visualization tools incorporated in SCIRun have been applied specifically to large medical datasets. Examples of these tools are volume bricking and ray-tracing.

31.4.4.1 Volume Bricking

While volume rendering can be performed in hardware on most modern graphics processing units (GPUs) via 3D texture mapping, the amount of memory available on a particular GPU (commonly referred to as "texture memory") limits the size of the models that can be volume-rendered. With large-scale data, it is not uncommon for a dataset to be several times larger than the available GPU memory. A common solution is to break the dataset up into smaller chunks, each of which is small enough to fit into the memory at hand. This process is known as bricking [12]. The bricks are then loaded into texture memory one at a time. After each brick is loaded, the corresponding texture is mapped to a series of polygons drawn perpendicular to the view. To avoid artifacts, bricks are sorted from farthest to nearest, based on the location of the viewpoint and the location of the brick. One must also take care to make sure that polygons drawn in neighboring bricks are aligned, in order to avoid artifacts at the brick boundaries. Using this volume-bricking approach, datasets that are many times larger than the available GPU memory can be processed and rendered nearly interactively.

31.4.4.2 Interactive Ray-Tracing

The basic ray-volume traversal method used in our ray-tracer allows us to implement volume-visualization methods that find exactly one value along a ray. Fundamentally, for each pixel of the image a ray is traced through a volume to compute the color for that pixel. The computational demand of ray-tracing is directly dependent upon the number of pixels (i.e., resolution of the viewing screen) being used, and it is less dependent on dataset size and allows both interactive isosurface extraction and MIP on very large datasets.

The ray-volume traversal method has been implemented as a parallel ray-tracing system that runs on both an SGI Reality Monster [12], which is a conventional shared-memory multiprocessor machine [12], and a Linux cluster with distributed memory [12]. To gain efficiency, several optimizations are used, including a volume-bricking scheme and a shallow data hierarchy. The graphics capabilities of the Reality Monster or cluster are used only for display of the final color image. This overall system is described in a previous paper [11]. Conventional wisdom holds that ray-tracing is too slow to be competitive with hardware z-buffers [25]. However, when one is rendering a sufficiently large dataset, ray-tracing should be competitive because its low time complexity ultimately overcomes its large time constant [7]. This crossover will happen sooner on a multiple CPU computer because of ray-tracing's high degree of intrinsic parallelism. The same arguments apply to the volume-traversal problem.

31.4.4.3 Examples of Ray-Tracing of Large Datasets

- The Visible Female: The Visible Female dataset, available through the National Library of Medicine as part of its Visible Human Project [10], was used to benchmark this ray-tracing method (Fig. 31.11). Specifically, we used the computed tomography (CT) data, which was acquired in 1 mm slices with varying in-slice resolution. This rectilinear data is composed of 1,734 slices of 512×512 images at 16 bits. The complete dataset is 910 MB. For the skin isosurface, we generated 18,068,534 polygons. For the bone isosurface, we generated 12,922,628 polygons. With these numbers of polygons, it would be challenging to achieve interactive rendering rates on conventional high-end graphics hardware. Our method can render a ray-traced isosurface of this data at multiple frames/s using a 512×512 image on multiple processors (for exact performance measures, see Parker et al. [12]).

- Small Animal Imaging: A recent example of ray-tracing is the work done by Dr. Richard Normann and his group at the University of Utah [8]. In this case, a relatively

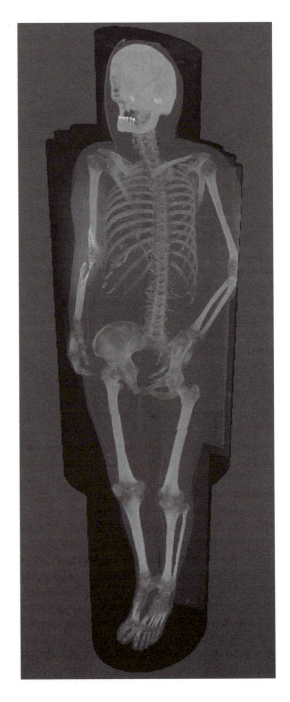

Figure 31.11 A maximum intensity projection (MIP) of the 1-GB Visible Female dataset using ray-tracing. The MIP algorithm seeks the largest data value that intersects a particular ray. Ray-tracing allows interactive visualization of the MIP of this dataset. (See also color insert.)

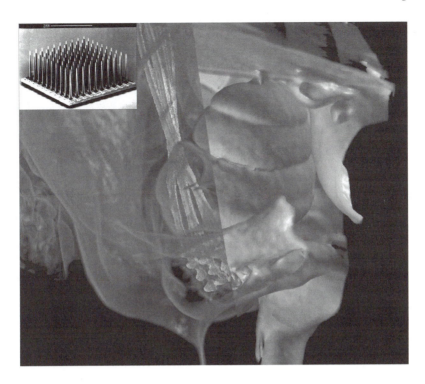

Figure 31.12   Visualization of the Utah Electrode Array embedded in the cochlear nerve of a cat. The insert is a picture of the Utah Electrode Array. The 27-micron resolution of the CT scan allows definition of the cochlea, the modiolus (right), the implanted electrode array, and the lead wires (in purple) that link the array to a head-mounted connector. The resolution of the scan even allows definition of the shanks and tips of the implanted electrode array. Volume rendering also allows the bone to be rendered as translucent (as on the left half of this image), enabling the electrode to be clearly viewed. Thus, the combination of high-resolution scanning, image processing, and interactive visualization tools such as ray-tracing allows noninvasive verification of the implantation site in an anatomical structure that is completely encased in the thick temporal bone. Data provided by Dr. Richard Normann and Dr. Charles Keller, University of Utah. (See also color insert.)

small dataset of 131 MB was rendered interactively across approximately 20 processors on an SGI Origin 3800. Again, no graphics hardware was required for the rendering, only to display the image. The imaging was used to examine an implantation of the Utah Electrode Array (Fig. 31.12, inset) into the cochlear nerve of a feline. In this case, the investigators used high-resolution CT imaging of the cat's head to verify the location of the electrode array in the cochlear nerve. The imaging was accomplished with a GE EVS-RS9 small-animal computed tomography (CT) scanner. There are distinct CT values for air, soft tissue, bone, and the elec-trode array, enabling the use of a combination of ray-tracing and volume rendering to visualize the array in the context of the surrounding structures, specifically the bone surface. Visualization results were improved when the voxels outside the electrode array were smoothed in order to better distinguish the bony structures.

The small-animal imaging systems, such as the CT scanner used in this work, are capable of producing extremely large datasets, in the 4- to 6-GB range. Additionally, the combination of multiple datasets from multiple imaging modalities, such as micro CT and micro positron emission tomography (PET) combinations, will compound the problem of dataset size. As scientists begin wanting to interact with these datasets, methods such as ray-tracing and distributed visualization will be at the forefront. It will certainly be quite some time before stand-alone, commodity graphics cards will be able to interactively handle this size dataset.

31.5 Getting Started in SCIRun

SCIRun and a variety of other software packages from the SCI Institute are available at the SCI software website, *http://software.sci.utah.edu/*. Documentation including download and installation instructions, sample datasets, and sample visualization networks are also available at the website. Following is a brief description of the SCIRun documentation.

31.5.1 Documentation

One of the greatest efforts in the transformation of our existing collection of research codes into a friendly environment for external users was the generation of the various forms of requisite documentation. We drafted documentation standards and investigated tools for integrating that documentation within and extracting it from our software system. The result is a hyperlinked "living document" that can be browsed from our website and is also included with our software distribution. The collection of documentation has been organized into a library of six manuals: an *Installation Guide*, a *Tutorial*, a *User Guide*, a *Developer Guide*, a *Reference Guide*, and *Frequently Asked Questions (FAQ)*.

31.5.1.1 The Installation Guide

The *Installation Guide* provides instructions for installing SCIRun from RPMs (Linux) and from Source Code (Linux and SGI), third-party libraries, the (optional) PETSc library, sample datasets, and SCIRun documentation. The PETSc library adds equation solvers to SCIR-

un's SolveMatrix module (see PETSc Installation).

31.5.1.2 The SCIRun Tutorial

The *SCIRun Tutorial* is an interactive introduction to SCIRun for new users. Since SCIRun is a large system, this tutorial provides a broad overview of SCIRun concepts and a core set of SCIRun user skills. There are seven chapters in this document. By the end of the tutorial, the user will have a grasp of data-flow programming, SCIRun architecture, and the specific modules and data types used along the way. The tutorial begins with Chapter 1, which demonstrates the construction of a simple yet functional SCIRun network. This demonstration is extended in Chapters 2 through 7 with additional functionality and complexity.

31.5.1.3 The User Guide

The *User Guide* describes how to get started using SCIRun. It includes a discussion of the data-flow programming paradigm and problem-solving environments. It explains basic concepts such as how to run and use SCIRun and how to write data-flow programs using SCIRun. The *User Guide* also includes descriptions of all the SCIRun modules, including data types used, functions performed, and explanations of the user interface elements.

31.5.1.4 The Developer Guide

The *Developer Guide* contains descriptions of the various SCIRun programming utilities, including our resource-management tools (memory, threads, persistent objects, exceptions, etc.). For each tool, we describe how the tool fits into SCIRun, the philosophy of why and when a developer would use that tool, and usage examples.

31.5.1.5 The Reference Guide

The *Reference Guide* contains the API specifications for all of the tools in SCIRun. This information is extracted directly from the source code using the Doxygen documentation system. For each class in SCIRun, the documentation contains a complete description of the class as well as cross-referenced hyperlinks to related classes.

31.5.1.6 Frequently Asked Questions (FAQ)

The last book in the documentation library is the *FAQ*. The *FAQ* has been subdivided into two sections: *Technical* and *User*. The *Technical* section contains answers to technical questions that arise during compilation and linking of SCIRun and its required third-party software. The *User* section contains questions and answers about the behavior of SCIRun, its modules, and the various messages the system issues.

31.5.2 Getting Help

Getting help in SCIRun is relatively easy. We have built a users' group email list; if you have questions at any time, you can email scirun-users@sci.utah.edu and either a SCIRun software engineer or another user will answer your questions.

References

1. J. D. de St. Germain, J. McCorquodale, S. G. Parker, and C. R. Johnson. Uintah: A massively parallel problem solving environment. In *Ninth IEEE International Symposium on High Performance and Distributed Computing*, pages 33–41, 2000.
2. D. E. DeMarle, S. G. Parker, M. Hartner, C. Gribble, and C. D. Hansen. Distributed interactive ray tracing for large volume visualization. *IEEE Symposium on Parallel Visualization and Graphics*, 2003.
3. D. P. Schissel for the National Fusion Collaboratory Project. An advanced collaborative environment to enhance magnetic fusion research. *Workshop on Advanced Collaborative Environments*, 2002.
4. C. R. Johnson, M. Berzins, L. Zhukov, and R. Coffey. Scirun: Applications to atmospheric diffusion using unstructured meshes. In *Numerical Methods for Fluid Dynamics VI*, (M. J. Baines, Ed.). Oxford University Press, 1998.
5. C. R. Johnson, S. Parker, D. Weinstein, and S. Heffernan. Component-based problem solving environments for large-scale scientific computing. *Journal on Concurrency and Computation: Practice and Experience*, (14):1337–1349, 2002.
6. C. R. Johnson and S. G. Parker. A computational steering model applied to problems in medicine. In *Supercomputing '94*, pages 540–549, 1994.
7. J. T. Kajiya. An overview and comparison of rendering methods, *A Consumer's and Developer's Guide to Image Synthesis*, 1988.
8. G. Kindlmann, R. A. Normann, A. Badi, J. Bigler, C. Keller, R. Coffey, G. M. Jones, and C. R. Johnson. Imaging of Utah electrode array, implanted in cochlear nerve. In *Digital Biology: The Emerging Paradigm*. NIH-Biomedical Information Science and Technology Initiative Consortium (BIS-TIC), 2003.
9. J. McCorquodale, J. D. de St. Germain, S. Parker, and C. R. Johnson. The Uintah parallelism infrastructure: a performance evaluation on the SGI Origin 2000. In *High Performance Computing 2001*, 2001.
10. National Library of Medicine Board of Regents. Electronic imaging: report of the board of regents, pages 90–2197, 1990.
11. S. Parker, W. Martin, P.-P. Sloan, P. Shirley, B. Smits, and C. Hansen. Interactive ray tracing. In *Symposium on Interactive 3D Graphics*, 1999.
12. S. Parker, M. Parker, Y. Livnat, P. Sloan, C. Hansen, and P. Shirley. Interactive ray tracing for volume visualization. *IEEE Transactions on Visualization and Computer Graphics*, 1999.
13. S. G. Parker. A component-based architecture for parallel multi-physics PDE simulation. *International Conference on Computational Science*, 3:719–734, 2002.
14. S. G. Parker, and C. R. Johnson. SCIRun: A scientific programming environment for computational steering. In *Supercomputing '95*, 1995.
15. S. G. Parker and C. R. Johnson. Scirun: applying interactive computer graphics to scientific problems. *SIGGRAPH (applications/demo)*, 1996.
16. O. Portniaguine, D. Weinstein, and C. Johnson. Focusing inversion of electroencephalography and magnetoencephalography data. In *3rd International Symposium On Noninvasive Functional Source Imaging*, 46:115–117, Innsbruck, Austria, 2001.
17. R. Rawat, S. G. Parker, P. J. Smith, and C. R. Johnson. Parallelization and integration of fire

simulations in the Uintah PSE. *Proceedings of the 10th SIAM Conference on Parallel Processing for Scientific Computing*, pages 12–14, 2001.

18. A. R. Sanderson and C. R. Johnson. Display of vector fields using a reaction-diffusion model. SCI Institute Technical Report UUSCI-2003-002, 2003.

19. R. Van Uitert, D. Weinstein, and C. R. Johnson. Volume currents in forward and inverse magnetoencephalographic simulations using realistic head models. *Annals of Biomedical Engineering*, 31:21–31, 2003.

20. R. Van Uitert, D. Weinstein, C. R. Johnson, and L. Zhukov. Finite element EEG and MEG simulations for realistic head models: quadratic vs. linear approximations. *Journal of Biomedizinische Technik*, 46:32–34, 2001.

21. D. Weinstein, P. Krysl, and C. Johnson. The BioPSE inverse EEG modeling pipeline. In *ISGG 7th International Conference on Numerical Grid Generation in Computation Field Simulations*, pages 1091–1100, 2000.

22. D. M. Weinstein, J. V. Tranquillo, C. S. Henriquez, and C. R. Johnson. BioPSE case study: modeling, simulation, and visualization of 3D mouse heart propagation. *International Journal of Bioelectromagnetism*, 5, 2003.

23. D. M. Weinstein, L. Zhukov, and C. R. Johnson. An inverse EEG problem solving environment and its applications to EEG source localization. In *NeuroImage (suppl.)*, page 921, 2000.

24. A. H. Glasser, C. R. Sovinec, R. A. Nebel, T. A. Gianakon, S. J. Plimpton, M. S. Chu, D. D. Schnack, and the NIMROD team. The NIMROD code: a new approach to numerical plasma physics. *Plasma Physics and Controlled Fusion* 41:A747, 1999.

25. E. E. Catmull. A subdivision algorithm for computer display of curved surfaces. Ph.D. Thesis, University of Utah, 1974.

32 NAG's IRIS Explorer

JEREMY WALTON
The Numerical Algorithms Group, Ltd.

32.1 Introduction

Data visualization can be defined as the gaining of insight by making a picture out of numbers, and the important role that it plays in the effective interpretation and analysis of numerical data has been recognized for a long time [1]. The type of data to be analyzed can vary from 1D time series (e.g., yearly changes in salary) to multidimensional vector-based datasets (e.g., air flow over an airplane wing). Although the simpler types of data can be effectively displayed using ubiquitous desktop applications such as Excel, more complex data require more sophisticated visualization techniques and applications.

Traditionally, a developer would write one of these applications in a high-level language such as Fortran or C. The application would calculate or *read* in the data and perhaps process or *filter* it in some way to isolate the component to be analyzed. The program would then call routines from a graphics library to *transform* the data into geometry (for example, a line chart, contour map, or scatter plot, etc.), which is then *rendered*. Examples of graphics libraries include OpenGL (which has been the *de facto* standard in 3D graphics for the past decade) [2], Open Inventor (an object-oriented toolkit that uses OpenGL for rendering), [3] and the NAG Graphics Library (a collection of Fortran routines for plotting and contouring) [4,5]. The use of libraries in a traditional programming environment has endured in the developer community—partly, no doubt, because of their incorporation within popular legacy codes that require maintenance. However, there is also some evidence [6] that the continued popularity of the NAG Graphics Library comes from a user requirement for underlying algorithms that are reliable, accurate, and well documented. These are the main features of the more widely known NAG Numerical Library (of which the NAG Graphics Library was originally a subset), and we highlight in Section 32.4.3 how the use of more reliable algorithms, such as those from the NAG library, can have a qualitative effect on the way data is transformed into geometry during the visualization process.

In spite of the fact that graphics libraries have great endurance, they can require a steep learning curve for the developer because of the granularity of the operations that they offer. The development cycle for applications built with a graphics library can be lengthy, which causes some disadvantages for visualization, particularly during the *exploratory* phase, when the user may have only a vague idea of what the picture of the data should look like. (The exploratory phase differs from the *presentation* phase, when the data is well understood by the user and visualization is being used to convey this understanding to others.) There may be a variety of techniques that could be applied to the data—for example, a 2D dataset can be transformed into a contour plot, a hidden line surface, or an illuminated surface. Even when the technique has already been selected, there will usually be a set of parameters such as color, texture, lighting, and orientation, and other attributes such as text labeling, that must be selected to produce the best picture of the data. Clearly, one technique will be more apposite for certain types of data than others, and the important things are to provide flexibility in the choice of technique for a given dataset and to see the results of its application with little delay. Moreover, this flexibility must be provided to the end users, since it is they who are the owners of the data and they who are making a value judgment about which is the technique producing the best-looking representation of their data [7].

Considerations such as these led to the design and development of so-called *modular visualization environments* (MVEs), which offer the user a collection of higher-level components that are connected together to build the application. The basis for MVE design is the *reference model* of the visualization process first set out by Upson et al. [8] and elaborated by Haber and McNabb [9]. This represents the formal organization of the sequence of steps in the visualization process noted above into a visualization *pipeline* (Fig. 32.1a).

Users have found the reference model to yield a rather intuitive representation of an application, especially when coupled with a *visual programming* interface, which gives a direct display of the flow of data and control. Here, components (usually referred to in this context as *modules*) are represented on the screen as graphical blocks, and connections between them as wires. Using this interface, the application is constructed as a network by selecting or replacing modules and by making or breaking connections via point-and-click actions. This provides a flexible route to application building. For example [10], it helps users to rapidly prototype new applications by interactively reconfiguring the modules or changing connections before rerunning the application. Because the MVE is dynamic, changes to the application are seen immediately; the network is always active during the editing process, leading to optimal prototyping efficiency and debugging. In addition, the way in which the modules can be used within many different applications provides an example of the reuse of code, another productivity benefit. If the system is *extensible*, the user can add his or her own modules; these can be used in the user's own applications and then shared with other users.

The popularity of the visualization reference model and the visual programming interface is reflected in the number of MVEs that have been developed around these bases, including apE [11], AVS [8], IRIS Explorer [12], IBM's Visualization Data Explorer [13] (now called Open DX), and SCIRun [14]. Although subsequent development of some MVEs has moved away from data flow (for example, AVS/Express and SCIRun are based on a data reference model), we believe that the model still has applicability, particularly in view of the way it has been recently used in the extension to distributed, collaborative applications running in a Grid environment (see Section 32.6).

Throughout the remainder of this chapter, we will repeatedly return to the visualization pipeline of Fig. 32.1 as new extensions to the model are introduced and demonstrated with their implementation in the IRIS Explorer visualization toolkit. In the following section, we introduce IRIS Explorer, describe its interface, its architecture, and some of its features, and indicate ways in which it can be extended through the addition of new modules. New modules are created with the aid of supplementary tools, which we describe in Section 32.2.2. In Section 32.3, we briefly outline the way in which an IRIS Explorer application can be distributed across a heterogeneous network of computers, together with options that are offered for the simplification of the application's interface. We also describe, in Section 32.3.3, its facility for building *collaborative* applications in which data and visualizations can be shared between workers on separate computers. Section 32.4 discusses some of the technology that underpins IRIS Explorer, and the way in which this serves to distinguish this system from other packages in this field. Section 32.5 highlights a number of user applications of the package in a variety of domains, while Section 32.6 describes some extensions to computational steering, along with current work that is concentrated on extending IRIS Explorer into a Grid-based environment, and shows how this is based on features of its architecture described in Section 32.3. The final

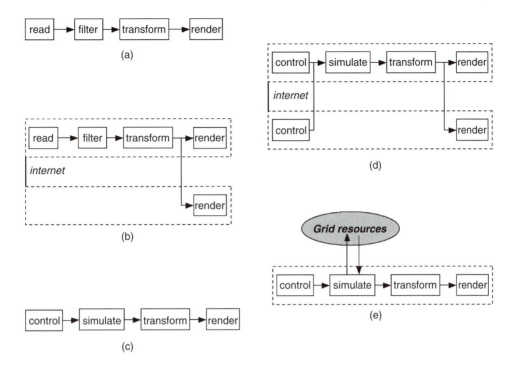

Figure 32.1   (a) The pipeline associated with the visualization reference model of Haber and McNabb [9]. (b) An extension to include collaboration (see Section 32.3.3). Data is passed from the original pipeline (top) over the network to a second pipeline (bottom). Collaboration could be introduced at any point in the pipeline; here, data has been shared following the transform step. (c) An extension to computational steering (see Section 32.6.1). (d) Adding collaboration to computational steering (see Section 32.6.1). Control of the simulation parameters is shared with the second pipeline, which also receives the visualization geometry for display. (e) Moving the simulation engine onto the Grid (see Section 32.6.2). The simulation is controlled by its interface, which is still part of the pipeline. Collaboration (although not shown in this sketch) is still an option here.

section, Section 32.7, collects some conclusions from this chapter.

32.2   Getting Started with IRIS Explorer

32.2.1   Creating Applications from Modules

As noted above, users of IRIS Explorer create applications by selecting and connecting software modules via a visual programming interface [15]. Each module is a routine that operates on its input data to produce some output, and typical functions of a module reflect the steps in the reference model of the visualization process in Fig. 32.1a. Thus, a module may, for example, do the following:

- *Read* data from a file, a database, or some other application that is running simultaneously.

- Modify or *filter* data by, say, clamping or normalizing values.

- *Transform* the data into geometric objects that can be displayed (such as slices, lines, surfaces, etc.).

- Output data to a file, or *render* geometry to a display device.

The behavior of each module is controlled by some set of *parameters* whose values may be set by the user while the application is running via a standard set of *widgets*—dials, sliders, text boxes, etc. Examples of parameters include the name of an input file, the threshold value for which an isosurface is to be calculated, or the number of contours to be displayed.

Data that is passed between modules is characterized by its type, and IRIS Explorer defines a small number of basic types, including the following:

- The *lattice*, a generalized array used for storing data values located at points on a regular mesh in multidimensional space.

- The *pyramid*, a hierarchical collection of lattices used for storing data values on irregular meshes, or for data associated with elements of higher dimensionality than points (i.e., edges, faces, or cells).

- The *parameter*, a single value used to control the behavior of a module.

- The *geometry*, used to store objects for display.

- The *pick*, used to store information about a selected point on a geometric object.

All these types are constructed using a data typing language and built-in compiler. The extensibility of the module set has been mentioned above (Section 32.1), and we will describe tools for adding new modules in Section 32.2.2. Here, we note that users can also extend the system by defining new data types [16,17] for the storage of data structures that cannot be fitted into one of the existing types. New data types are defined (in terms of existing types, if necessary) using the same methods as the definition of the basic types—via the typing language and compiler—and each user-defined type is given an automatically generated library of accessor functions to facilitate its use and to ensure the correctness of basic functions.

Users build their application in IRIS Explorer by connecting modules together into a network, or *map*. The constituent modules are first selected from the **Module Librarian** and then dragged onto the **Map Editor**, where the map is to be assembled. The user does this by connecting modules' inputs and outputs by a series of point-and-click actions. The connec-

tions define the way in which data flows through the map, and they provide a useful overview of its structure. Some of these elements of the architecture and design are illustrated in the screenshot of an IRIS Explorer session shown in Fig. 32.2.

Finally, we note that, although the visual programming interface helps users to visualize data without conventional programming, it is not a panacea. Visual displays of applications can become hard to navigate beyond a few dozen modules (although this can be simplified by *module grouping*—see Section 32.3.2), and MVEs that present only an interactive interface can have difficulty communicating with other applications. For these reasons, IRIS Explorer can also be driven by a textual scripting language that is based on a version of Scheme [15]. This is more useful for batch processing of long animation sequences, for the control of IRIS Explorer sessions lacking user interaction (for example, in the Visualization Web Server [18]—see Section 32.4.2), for interapplication communication (as used, for example, in collaborative sessions—see Section 32.3.3) and for automatic testing of IRIS Explorer applications.

32.2.2 Creating New Modules

Although the module suite provided with IRIS Explorer is extensive, users will eventually need new modules. This may be because they have existing code that they wish to incorporate into an IRIS Explorer map or because no module exists with the functionality they require. The modular nature of the environment makes it easy to add the missing pieces of their application, using the tools that are bundled with the IRIS Explorer system.

The IRIS Explorer **Module Builder** [17] is a graphical user interface (GUI) tool that the module developer uses to specify the data types to be input and output by the module, the interface to the module's computational function, and the selection and placement for the widgets in the module's control panel, all

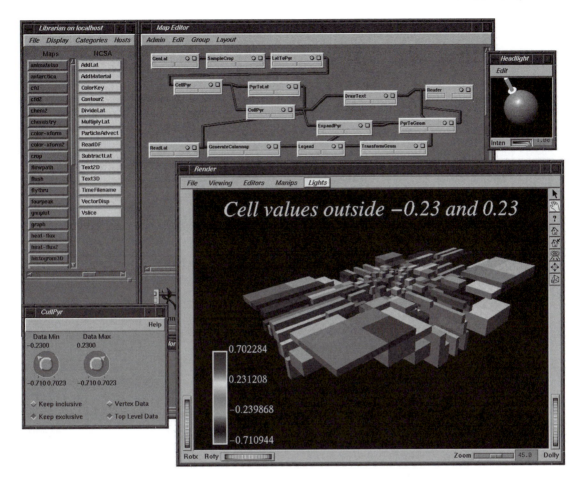

Figure 32.2   Running a map in IRIS Explorer. The Module Librarian is at top left, and the Map Editor is at top middle. The map that is currently running is analyzing results from a simulation of oil transport through a permeable medium. The region is divided up into cells, each of which has a permeability value. The data is read into IRIS Explorer as a pyramid and is then culled using the **CullPyr** module to find only those cells whose data values are outside some range—this is an example of the filtering step in the visualization pipeline of Fig. 32.1. The **CullPyr** control panel, containing the widgets used to set its parameter values, is at bottom left, while the large display window comes from the **Render** module. The small window at top right is an editor for the headlight in the 3D scene; this is an Open Inventor control that is exposed in IRIS Explorer because it uses Open Inventor for 3D scene generation and display (Section 32.4.1). (See also color insert.)

without programming. The user provides the module's computational function, writing in Fortran, C, or C++ and calling the IRIS Explorer API library to create and manipulate IRIS Explorer data types. In addition to this, the computational function can call other libraries, interface to files, or connect to external applications; the only specification of the function that the Module Builder requires is its

calling sequence. The source of practically all of the module suite is part of the IRIS Explorer distribution; this is freely available to users for modification or extension or as the starting point for the development of their own modules. In a similar way, many users have donated modules that they have developed to the module repository for use by others [19].

The output from the Module Builder is a *module resource* file, from which it automatically generates code wrappers around the computational function that handle module execution, intermodule communication (which may be heterogeneously networked—see Section 32.3.1) and data conversion between IRIS Explorer data types and user formats. The code wrappers and the computational function are compiled together to produce the module executable, which, together with the module resource file, constitutes the module that can be placed in the Module Librarian and invoked via the Map Editor (see Section 32.2.1).

One of the reasons a user may seek to add missing functionality is that IRIS Explorer is intended as a general-purpose visualization tool to be applicable across several domains (some of the areas where it has been used include computational chemistry [20], computational fluid dynamics (CFD) [21], and high-energy physics [22]). The consequences of this for IRIS Explorer and other MVEs are that the basic system modules perhaps cover the later steps of the visualization pipeline (render, transform, filter) better than the first step—data input—because this step, which usually forms the interface with other applications, is the most domain specific. Put another way, while it is to be expected that the basic modules would include one for isosurface calculation, for example, to find a module in that set for reading a specific computational chemistry data format (say) would perhaps be regarded as serendipitous. This preponderance of the requirement for data-reader modules is reflected in the additional tools provided for this purpose (although the general-purpose Module Builder can also be used for their creation).

The IRIS Explorer **DataScribe** extends the visual programming concept of the map editor (where applications are constructed from modules) to the construction of reader (or translator) modules from graphical elements representing scalars, arrays, and lattices [15]. DataScribe users manipulate these elements and make connections between them (using the same point-and-click actions as in the map editor) to construct a script that, along with the module resource file (also generated by DataScribe), comprises the module [23].

QuickLat is an alternative tool for building readers. Here, the user specifies the IRIS Explorer data that is to be output from the module, along with parameters such as file names. QuickLat then automatically generates a template (written in C or Fortran) for the module, which the user then completes by adding code to read data in from the data file. The template contains hooks into the IRIS Explorer data structures, which shield the user from calling the IRIS Explorer API for the creation and allocation of data types. Finally, the Module Builder is used to lay out the control panel of the module and build it.

32.3 Working with IRIS Explorer

In this section, we briefly describe a few features that arise naturally from IRIS Explorer's modular programming approach, which can be viewed as a process of breaking the application down into a small set of interconnecting units in the spirit of the visualization reference model.

32.3.1 Distributed Execution of Applications

If IRIS Explorer is being run in a networked environment, it might be useful to elect to run certain modules in the application on the local machine while others run on remote machines. Distribution of the application across a network of machines can be carried out in IRIS Explorer; moreover, since it is a multiplatform application, the network can be heterogeneous. (Originally produced under IRIX by SGI for their machines, development of IRIS Explorer was taken over by NAG in 1994, which ported it to other platforms. IRIS Explorer is currently available from NAG for Windows, Linux, Solaris, HP-UX, AIX, Digital Unix, and

IRIX.) Examples of situations where application distribution may be desirable include the following:

- Where some modules must be run on a high-performance machine.
- Where some degree of coarse-grained parallelism is being introduced into the application.
- Where a module must be run on a particular type of machine because it is only compiled for that architecture.
- Where a module has to be run on a special machine because it makes use of resources (for example, input files) that are only available on that machine.

A particular feature of the interface to remote modules is that (apart from a label identifying the machine on which they are running) it is exactly the same as that for local modules. Users make connections to modules and interact with their widgets in the same way irrespective of their location. This facility, which has been a part of IRIS Explorer since its original design, is currently being exploited in the extension of the system to a Grid environment (see Section 32.6.2).

32.3.2 Application Building via Module Grouping

Although the connections between the modules provide a convenient illustration of the form of the application, they can sometimes be distracting to an end user of the application, particularly if the application contains a large number of modules. Moreover, it can sometimes be hard to determine which widgets on the modules provide the most useful or relevant controls for the application. For these reasons, a simplified interface to any collection of modules in the application (including just one module or all the modules in the map) can be created on the fly through creation of a *module group*. More specifically, the user does the following:

- *Chooses* the modules that are to go into the group.
- *Selects* the module parameters that are to appear in the simplified interface.
- *Edits* the group's control panel by attaching suitable widgets to the parameters and positioning them in suitable locations on the window.

Grouping can be used to interactively change the interface of a single module in the map without requiring the user to return to the module-building stage. It can also be used to create an interface to an application consisting of a single control panel containing widgets attached to parameters from modules throughout the map (Fig. 32.3).

Although at first glance a group looks the same as a module, its underlying architecture is, in fact, the same as the original collection of separate modules. That is, grouping the modules only changes their user interface; underneath this, they still run as individual executables, each appearing to the operating system as a separate process, passing data via shared memory or sockets. This can carry some advantages over a monolithic process; for example, if one module in a map crashes, the remaining modules are generally unaffected. Nevertheless, the separate processes can sometimes represent a poor use of resources (such as memory and CPU), and the communications between the modules often appear to add significant overhead to the execution of the group, particularly if it is composed of a large number of small modules.

In the latest release of IRIS Explorer (5.0), these problems are addressed by the introduction of *group compiling*. In effect, this allows a new module to be built by combining existing ones. Once the group has been created and its control panel laid out, an additional option allows the user to compile the separate modules into a single executable. Once the compilation is complete, the user can replace the modules in the group with the new compiled group. The control panel of the compiled group is the same

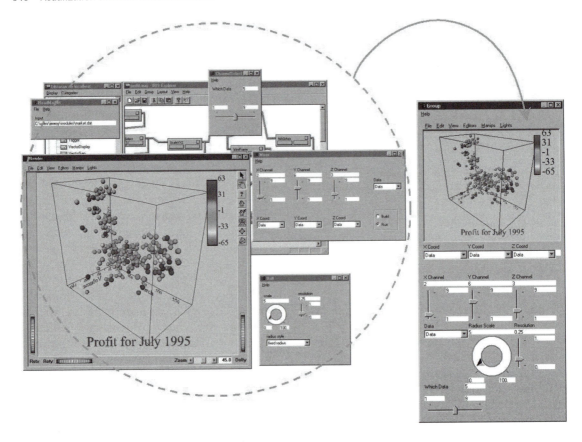

Figure 32.3   Building an application in IRIS Explorer by module grouping. On the left, a map is shown running in the map editor. The connections between the modules can be seen in the background at the top, and the control panels of several modules surround the large control panel from the **Render** module, which contains the visualization. The same application is shown on the right after the modules have been grouped together (and optionally compiled into a single executable). The group control panel contains widgets from the original modules' control panels–including the large window from **Render**. Here, IRIS Explorer is running in application mode which hides the development windows from view; the only thing appearing on-screen is the group control panel. This application is displaying stock market data as a 3D scatter plot. (See also color insert.)

as that of the old group, so the interface is unaffected. However, an examination of the machine's process space will quickly reveal that the compiled group is running as a single process, which consumes less CPU time and uses less memory than running all of the original separate modules. This yields improved performance, owing to the more efficient use of system resources.

As a final step towards application building, IRIS Explorer can be run in a mode where the development windows (module librarian, map editor) do not appear on the screen, leaving just the application windows for the user to interact with. Fig. 32.3 illustrates module grouping with an example taken from visualization of financial data.

32.3.3   Collaborative Visualization

Although our discussion so far has been based upon a single user model, there can arise situations (particularly in large or geographically disparate research teams) where data visualization and analysis must be performed *collaboratively*.

This problem was addressed by Wood et al. [24], who considered an extension of the visualization reference model that allowed communication between two or more pipelines (Fig. 32.1b) allowing users on multiple networked machines to collaborate. In the implementation of this model in IRIS Explorer, each user runs his or her own map but has the opportunity to make a connection to a *collaboratively aware* module that can pass data to and from the other collaborators' maps. Since any of the data types described in Section 32.2.1 can be shared, the users have a great deal of flexibility when setting up the collaboration. Thus, sharing parameters allows for joint control of modules; sharing lattices or pyramids enables different visualization techniques to be applied to the same datasets; and sharing geometry or pick data facilitates different views of, or interactions with, the same visualization. The choice of the point(s) at which to introduce collaboration is up to the users, based on their expertise and interests in the data. External constraints such as bandwidth and machine resources may also play a role in this choice—for example, it might be more efficient to work with local copies of the original data (which can be shared once at the start of the collaboration), or to share the filtered data (which will be smaller), or to collaborate on the geometry, as in the case of the scenario sketched in Fig. 32.1b (which might be smaller still), or just to share control of duplicated modules in collaborating maps.

In addition, modules, connections, and map fragments (groups of modules connected together) can be passed between collaborating maps. This is achieved using the scripting language mentioned in Section 32.2.1. To be more accurate, what is passed between the sessions are the commands to start modules, make or break connections, etc. Management of a collaborative session is performed by a central server, together with local servers on each machine that control the collaborative modules via IRIS Explorer scripts.

It is to be noted that collaborative modules use the same GUI as noncollaborative modules; no extra skill is required by the user. This is to be compared with the way in which the interaction model for modules running on remote machines is identical to that for local modules (see Section 32.3.1).

Wood et al. [24] illustrated the use of the collaborative visualization system with a scenario involving two doctors collaborating over the analysis of medical image data. Each works with a separate instance of IRIS Explorer, exchanging data and sharing control over the application. They can explore the data together, each introducing new visualization techniques as required and taking advantage of the other's complementary expertise. Finally, having created their collaborative map, they are able to group all the modules into an application for colleagues who require a simpler interface (see Section 32.3). We describe another collaborative scenario in somewhat more detail in Section 32.6.1 within the context of collaborative steering.

The collaborative functionality introduced by Wood et al. [24] proved to be so successful that it was incorporated into IRIS Explorer 4.0 [25] and subsequent releases, making this package one of the first commercial multiuser visualization systems in the world [48].

32.4 Underlying Software in IRIS Explorer

This section briefly describes some of the software technologies that underpin IRIS Explorer and the way in which each is utilized within the modules and the base system.

32.4.1 Open Inventor—An Object-Oriented 3D Toolkit

Open Inventor [3] is an object-oriented graphics library that provides a comprehensive environment for creating and manipulating 3D scenes in a database known as a *scene graph*. It provides a comprehensive software model for developing *interaction* with the scene and also defines a rather intuitive *file format* for scenes. Open Inventor is written in C++, and, as mentioned in Section 32.1, uses OpenGL [2] for rendering scenes.

IRIS Explorer uses Open Inventor to create and manipulate geometry—in fact, the IRIS Explorer geometry data type (see Section 32.2.1) *is* an Open Inventor scene graph. This means that the module developer in IRIS Explorer can take advantage of all the functionality in Open Inventor—and also OpenGL—when working with geometry. (See, for example, Fig. 32.2, in which the Open Inventor control for editing the direction and intensity of the headlight is used to enhance the scene displayed in the **Render** module.) In addition, scenes created within IRIS Explorer can be shared [26] with other Inventor-based applications, and vice versa. The close relationship between IRIS Explorer and Open Inventor has also been exploited in the incorporation of Inventor-based libraries into IRIS Explorer. An example is provided by 3D MasterSuite [27] from Template Graphics Software, which consisted of a collection of higher-order Inventor objects relating to efficient graphical plotting; these include axes, curves, text, and annotation. The inclusion of modules based on this library facilitated the production of higher-quality plots in IRIS Explorer.

The Open Inventor file format was the basis for the original version of the Virtual Reality Modeling Language (VRML), the description language for 3D scenes on the Web. Basing it on an extant format hastened the acceptance of VRML through the leverage of a large amount of existing content and applications, including IRIS Explorer. Because IRIS Explorer is based on Open Inventor, it is very easy to output any of its visualizations that as VRML and so publish it on the Web (see Fig. 32.4 for an example).

A number of examples of the use of VRML within IRIS Explorer have been reviewed elsewhere [28]. Here, we recall briefly the work of Brodlie and his coworkers on a Visualization Web Server [18] that provides a simplified, WWW-based interface to a visualization system. The user accesses a web page and fills in an HTML form describing the location of his or her dataset, the visualization techniques that are to be applied to the data, and any parameters that need to be selected for the techniques. This information is then passed over the Web to the visualization server, where it is used to assemble the application—in the form of a script—that is to be run by IRIS Explorer. The resultant geometry is passed back (via the server) as VRML to the user's client, where it is delivered to a browser. The use of the visualization web server has been illustrated [18] with an application to extract and display environmental data that is periodically posted to the Web. The system allows the user to select the data of interest (pollutant, location, time interval) and to display it in 3D—at any time and from anywhere on the Web.

32.4.2 Image Processing: The Image Vision Library

IRIS Explorer contains over 50 image-processing modules based on operators in SGI's ImageVision Library (IL) [29]. IL contains operators for input and output, color conversion, arithmetic functions, radiometric and geometric transforms, spatial and nonspatial domain transforms, and edge detection, amongst others. In addition, it has been designed to deal efficiently with large images by decomposing them into tiles and using a demand-driven *data pull* model of communication between operations. Thus, only the part of the image that is currently being processed is passed down the IL processing chain. In this way, chains of image processing commands can be performed in constant memory for any size image, assuming operations with local memory reference patterns (which most IL operators have).

In addition to these efficiencies, which are available for all platforms, IL takes advantage of specific hardware acceleration on SGI

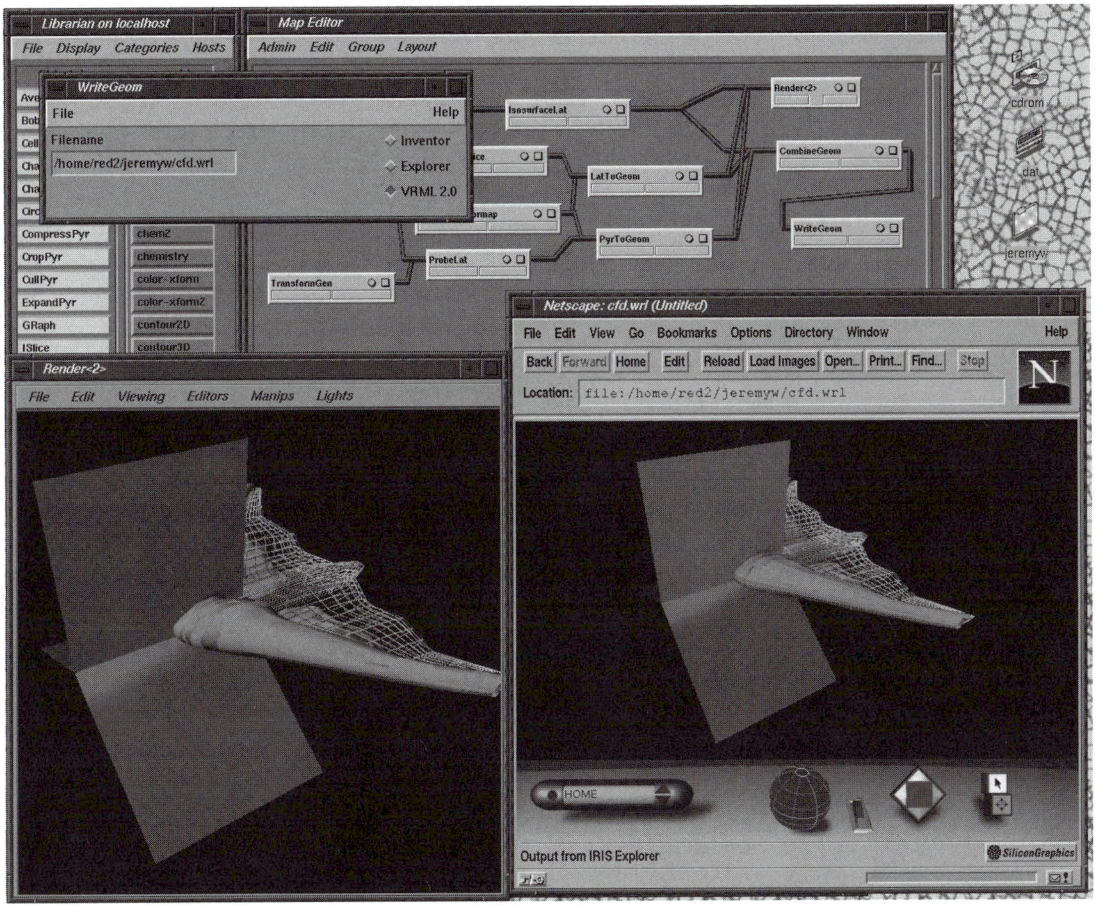

Figure 32.4   Using the **WriteGeom** module to output a scene from IRIS Explorer as VRML 2.0 (a.k.a. VRML97). The **Render** window is at bottom left, containing the scene, which consists of an isosurface, probe, and sheet geometry. In the map, the same inputs are passed to the **CombineGeom** module, which creates a single scene and passes it to **WriteGeom** for output to file. The file is rendered using a VRML browser in the bottom right-hand corner; here, the VRML is being displayed on the same machine as the original visualization, but publishing it on the Web would, of course, allow it to be displayed anywhere. (See also color insert.)

machines where available. The IL operators make use of high-speed graphics hardware for operations such as blurring, exponent computations, and rotation.

The way in which IRIS Explorer uses the IL in writing modules goes beyond the traditional module framework in order to preserve these performance efficiencies in the IL. Simply wrapping each IL operator into a module would interfere with both the tiling and the data pull model. Instead, IRIS Explorer provides a way to combine a map of image-processing modules into a single IL-based module that retains these efficiencies, while preserving IL hardware acceleration and reduced memory usage. Used in this way, the IRIS Explorer Map Editor becomes an efficient IL image-processing editor. (Note that this method of combining IL modules is distinct from the more generic group compilation described in Section 32.3.2, which works for any set of modules, not just IL ones.)

32.4.3   The NAG Libraries—for Reliable Numerics

The current proliferation of visualization software [8,11–14], with its promise of easy transformation of complex datasets into images, can leave unanswered questions of reliability and accuracy in output that is produced. Misleading results can arise when the algorithm used within the visualization is inappropriate, inaccurate, or unusable. Such questions about the reliability of visualization become even more important in the steering of computational processes via the real-time visualization of their results (see Section 32.6). Here, decisions about the future course of the calculation are taken based on their display, so it is important to ensure that it is as faithful to the data as possible.

Such questions highlight the importance of using good-quality algorithms in visualization. The NAG Numerical Library [30], which contains more than 1000 routines, has been used by a large community of users for almost 30 years, and it provides a straightforward way to access complex algorithms for a variety of problems. These include minimization, the solution of ordinary and partial differential equations, quadrature, statistical analysis, and the solution of linear and nonlinear equations. Using a reliable numerical library in applications can lead to savings in development time, and the testing of the accuracy of the library routines gives confidence in the solution.

The use of the NAG numerical routines within IRIS Explorer modules has included the interpolation algorithm in the API and the particle advection module. Particle advection generates the path of a particle released into a vector dataset by solving a differential equation. Although many algorithms can be used to obtain a solution, not all are equally accurate or applicable. Thus, for example, one of the simplest (the so-called Euler algorithm), although easy to implement, can produce incorrect results—sometimes dramatically so [31]. By contrast, the use of a more sophisticated algorithm from the NAG numerical library produces a more accurate solution [31,32]. It has

other advantages, also. Besides providing a high degree of confidence in the knowledge that the algorithm has been used in a very wide range of applications, it provides the module with a ready-made reference for documentation purposes [30], which is key to assessing the accuracy and applicability of the module.

32.5   Some Applications of IRIS Explorer

In this section, we recall some ways in which IRIS Explorer has been used in the visualization of data from a variety of sources and domains.

32.5.1   Oil Recovery Research

The extraction of oil from sand presents several technical challenges, largely owing to the high viscosity of the oil. This is a particular issue for the Athabasca tar sands in Alberta, Canada, and numerical simulations to help determine more efficient recovery methods are being carried out by London [33] and his colleagues at the Alberta Research Council. They are using IRIS Explorer as an application builder in the development of their simulations and for the visualization of their results [33].

Although commercial reservoir simulation products have some display functionality for production use, research cannot present a finite list of visualization requirements, which makes a flexible toolkit approach (such as that offered by IRIS Explorer) more useful in this context. By incorporating standard modules into specialized applications, and by adding suitable data reader modules (see Section 32.2.2), they have found that IRIS Explorer is a helpful complement to canned purpose-built packages.

To illustrate the use of IRIS Explorer here, we will describe one of the applications [33] in more detail. Pore-scale simulations are directed at exploring the complicated behavior of the flow of fluid through the irregular geometries representative of the channels and pores within the sand in most heavy oil deposits. The pore space is modeled as a flow network, paying particular attention to the pore-scale geometries and topologies. Because of the irregular structure, visualization is an essential part of this work, not only for viewing the final results but also for debugging the development and monitoring the various stages of modeling. IRIS Explorer has been used as the framework for all of these stages (Fig. 32.5).

The starting point for a model of the pore space is the creation of an artificial packing of spherical particles of various radii. A simple algorithm incorporating the idea of particle rearrangement in the presence of gravity is encoded in an IRIS Explorer module. The final packing is provided as an IRIS Explorer lattice that can easily be rendered as geometry using standard modules. In addition, the intermediate configurations can be usefully visualized in order to check on the packing algorithm, and these results can be assembled into an animation of the complete process.

Following the creation of the packing, the pore space is divided into *nodes*, where phases mix and flow without hindrance, and *channels*, where Stokes flow dominates and phases are separated by distinct interfaces. The resulting network is encoded into an IRIS Explorer pyramid, which stores details of the location of the particles, along with the nodes (with links to enclosing particles) and channels (with links

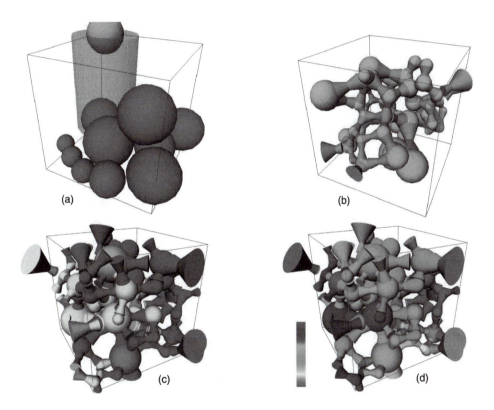

(a)   (b)

(c)   (d)

Figure 32.5   Using IRIS Explorer in pore-scale simulations of fluid flow [33]. (a) Creating a sphere pack by "dropping" particles at randomly chosen locations. The shaded volume is searched for the lowest position. (b) The network of nodes (green) and channels (pink) found in the sphere pack. The rendering uses simple primitives (cones) to approximate some of the geometrical information encoded in the pyramid. (c) Multiphase fluid flow through the network. Yellow fluid has displaced purple fluid in parts of the channels. Nodes, containing a mixture of the two phases, are colored according to pressure. (d) The same network as in (c), but now with all parts (nodes and channels) colored according to pressure. (See also color insert.)

to two nodes). Again, a standard module converts the pyramid to a geometry for viewing.

Once the visual check of the network is satisfactory, another module converts the pyramid to the ASCII data required by the simulator. Thus, except for some fluid property descriptions and boundary conditions, the entire setup is carried out graphically. The actual flow simulation is run offline, since it frequently takes days or weeks to complete.

The simulation output is encoded in a custom binary format, for which a special data reader is constructed. The simulator models single-phase *slugs* of fluid moving through the channels, with interfaces between them. This is accommodated in IRIS Explorer by addition of a slug layer to the pyramid, with links to the appropriate channels. Hence, the same module can be used to display both the input and the output data. A variety of views are available on output, including color-mapped pressures and flow rates.

The final stage of the analysis is the calculation of macroscopic properties, such as the permeability tensor. Again, this part of the simulation is carried out entirely within the IRIS Explorer map.

London [33] reports that the flexibility of IRIS Explorer cannot be matched by any application-specific visualization package, and that it facilitates the production of VRML and other common graphics files. He says, "The leverage gained by incorporating standard modules allows a single user to harness state-of-the-art visualization techniques without the services of a development team."

32.5.2 Large-Scale Computational Fluid Dynamics

Imagine half-filling a glass with water and then adding an equal amount of a light oil. The two liquids do not mix (they are immiscible), and, if the oil is the less dense of the two, it remains above the water, and the interface between the oil and the water is planar. Now imagine sealing the top of the glass (so that no liquid escapes) and turning it upside down. If the glass is wide enough, gravity will cause the water to fall into the lower part of the glass and the oil to rise into the upper part. The interface becomes distorted by spikes, bubbles, and other shapes, which is due to a phenomenon known as the *Rayleigh–Taylor instability*. This occurs whenever a light fluid is accelerated into a heavy one, and it has been studied widely because of its relationship to the onset of turbulence and mixing.

Cook [34] carried out a Rayleigh–Taylor simulation on the ASCI Blue-Pacific machine at Lawrence Livermore National Laboratories, using a mesh of dimensions $256 \times 512 \times 256$, and running for more than 3,000 time-steps. The simulation generated about a terabyte of data (each time-step generated about 260 MB), and Lombeyda et al. [35] used IRIS Explorer to visualize this, focusing attention on the way in which the interface between the fluids evolves in time. One way to display the interface is to calculate the density isosurface for a threshold value intermediate between the densities of the two fluids (Fig. 32.6).

Because the performance of the standard IRIS Explorer isosurface module with this dataset was not sufficient to extract and display isosurfaces dynamically in real time, Lombeyda et al. wrote a new module that spawned the isosurface extraction to run in parallel on a multiprocessor machine. Extraction was performed using a variation of the standard marching cubes algorithm [36], which kept track of visited cube edges (thus reducing calculation redundancy) and which was modified for parallel execution using simple data decomposition. Parallelization was explicated using p-threads, which could be invoked easily (using a simple one-line directive in the code) and had the additional advantage of working with sockets, which formed the basis of their communications framework.

Performance results were encouraging. Compared to execution using a single thread, they achieved a speedup factor of about 7.8 when using eight threads, indicating a high degree of parallelization. They could then load several

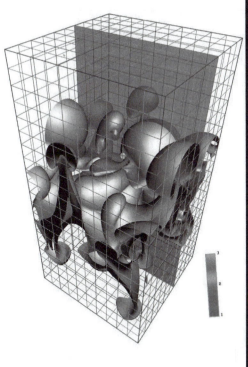
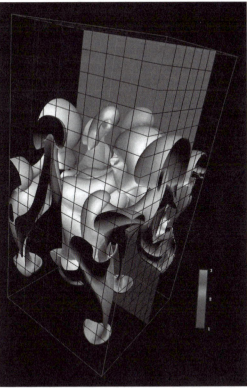

Figure 32.6 Using IRIS Explorer to display the results of a simulation of the Rayleigh-Taylor instability in the interface between two fluids [35]. The upper fluid has a density of 3 units, the lower has a density of 1 unit, and the isosurface has been calculated for a threshold value of 2; thus, it follows the fluid–fluid interface irrespective of its shape. Here, the simulation cell has been added to orient the display, and a semitransparent cutting plane has been used to show how density varies in a plane passing through the interface. Two views of the cell for a given time-step are displayed here that clearly show the characteristic loops, spikes, and bubbles in the surface. (See also color insert.)

frames of the dataset into the server, and, through the use of the module in the map, were able to dynamically request isosurfaces across different density thresholds and through different time-steps.

This work illustrates the way in which IRIS Explorer can be utilized to drive large computations, taking advantage of parallel processing where necessary, while using a standard graphics workstation to display the results. Finally, they noted that embedding their parallel isosurface module within IRIS Explorer promotes its reuse among the user community, many members of which have different needs and access to a variety of hardware configurations.

32.5.3 Biomechanical Modeling

More than 5000 heart-valve replacements are performed in the UK each year, and the design of improved artificial valves is an important area of research. IRIS Explorer has been used in the creation of an integrated visualization and design toolkit [37] that allows the valve designer to choose the shape and mechanical properties of a valve before carrying out numerical flow experiments to determine its behavior within the heart.

Early work [38] was devoted to the design of so-called *mechanical* valves, which have rigid leaflets that open and close to control the flow of blood. The rigidity makes them comparatively easy to model and manufacture and confers durability; however, it also causes disturbances in blood flow, leading to hemolysis. Accordingly, attention has been more recently switched to *flexible* valves, which give better hemodynamic performance, although the high bending stresses during the flexing of the leaflets carry the risk of earlier mechanical failure. This necessitates improved methods of design, modeling, and manufacture.

The use of the IRIS Explorer–based toolkit in the design of flexible heart valves has been described by Fenlon et al. [37]. They chose a simplified (2D) leaflet model that consists of a linked assembly of rigid dynamic elements, with each element connected to its neighbor by a

frictionless pivot, with stiffness and damping effects. The equation of motion of the leaflet is then that of an *n*-tuple pendulum, which can be integrated using a standard algorithm [30]. Coupling this to a simplified model of the fluid flow allows the calculation of the leaflet motion to be carried out in near real time. This, in turn, enables the rapid testing of leaflets with different physical properties (parameterized by the stiffness and damping terms), based on their dynamic and mechanical behaviors.

The simulation was embedded in IRIS Explorer as a module, and its output was visualized (Fig. 32.7) using a combination of custom-written modules and modules from the standard distribution. Fenlon et al. noted that building their

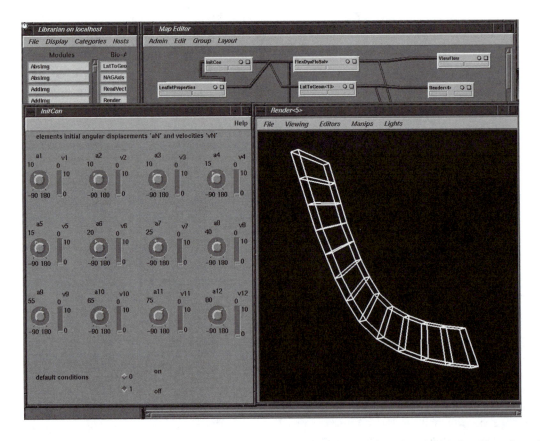

Figure 32.7 Setting the initial configuration of the valve leaflet [37]. The module control panel containing the widgets for setting the angles between the elements and their time derivatives is on the left, while the **Render** module on the right shows the leaflet in this configuration in a wireframe representation that clearly displays the elements in the leaflet.

toolkit in a framework like IRIS Explorer (rather than writing the package from scratch) meant that applications were easily customized through the replacement or addition of modules. For example, they found that the modular approach meant that, with some judicious design decisions, many modules could be shared between different flow models and experiments. In addition, one could explore alternative approaches such as different solvers by writing them as modules and then interchanging them in a particular application network.

Before the simulation was run, an IRIS Explorer map was used to set up the initial configuration of the valve leaflet and to assign values to its physical parameters. The configuration information was then fed into another map that solved for the leaflet motion and fluid flow, displaying results for the leaflet configuration as the simulation progressed in time. Finally, postprocessing of the results was performed by a third map, which displayed the flow around the leaflets in the form of vector arrows, streamlines, and vortex-wake surface plots. Other maps produced plots of the leaflet's curvature and an animation of the leaflet opening and closing (Fig. 32.8).

The key feature of this work was the way in which, by being embedded into IRIS Explorer, the simulation was tightly coupled with the visualization, allowing results to be displayed in near real time. (We continue with this

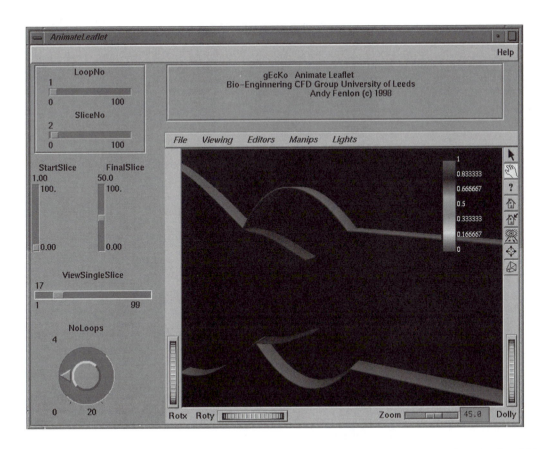

Figure 32.8 Control panel for the module (actually, a module group) used in the production of the leaflet animation. Widgets allowing the user to select the time-steps for the animation are on the left, and the display window (from the group's **Render** module) is on the right.

theme in Section 32.6, where we look in more detail at the use of IRIS Explorer in computational steering applications.) Changes in the physical properties of the leaflet (set using the toolkit framework) were immediately reflected in its mechanical response under flow, which, in turn, enables the toolkit user to examine a broad range of designs in the course of a single session.

32.6   Moving IRIS Explorer to the Grid

32.6.1   Steering and Collaboration

The visualization reference model in Fig. 32.1a is predicated on the display of a *static* data source—data is read from a file in the first step and then passed down the pipeline for display. Subsequent datasets could come from other files, or maybe from the same file if it is overwritten by an external application. Each time a new dataset is to be displayed, the user has to reexecute the visualization pipeline. This process could be automated by extending the reference model (Fig. 32.1c) to support a *dynamic* data source—for example, a numerical simulation of a time-dependent process. Here, the data source is incorporated into the visualization pipeline, its output is displayed as the calculation proceeds, and feedback from the visualization allows the user to control the course of the simulation by adjusting its control parameters; this is an example of so-called *computational steering*. In the same way that the reference model can be extended to incorporate steering, so can systems—such as IRIS Explorer—that are based upon it.

Before presenting an example of steering within IRIS Explorer, we will take advantage of the modular nature of the reference model and elaborate it still further by adding collaboration, as in Fig. 32.1d. Now the control parameters and the geometry are being shared, which enables one of the participants to control the

simulation (running on a remote machine), based on the visualization of its output—we return to this scenario later. Alternatively, if the output of the simulation can be shared as lattices or pyramids (see Section 32.3.3), then users can apply different visualization techniques in their separate locations rather than looking at a static display provided by one of the participants. This is a flexible approach that allows users with different backgrounds and motivations to explore the results of the simulation in their own way.

Collaborative computational steering within IRIS Explorer has been demonstrated by Walkley et al. [39], using a simulation of fluid transport aimed at modeling atmospheric dispersion [40]. This can be represented in simplified form as passive scalar convection of a single component, which may in turn be modeled using a standard streamline-upwind finite element method [41]. Realistic modeling requires expertise in both the physical processes involved and the numerical modeling of those processes, and Walkley et al. envision a collaboration between a chemist and a numerical analyst in the investigation of this model.

Typically, the collaborative researchers will have different motivations in the analysis of data from such a model. For example, the numerical analyst may wish to view the errors in the numerical solution at specific points, while the chemist will want to monitor the concentration of the component throughout the domain. The real-time display from the simulation can be helpful in debugging the simulation and in ensuring that only interesting areas of parameter space are explored.

Incorporating the simulation as a module into a map allows it to be interactively controlled via the widgets on its control panel. These may include parameters for the algorithm (e.g., error tolerance or time-step), or of the physical model (e.g., density or convection velocity). In a collaborative scenario, results generated in any part of the map can be shared with other users, while researchers with identical maps can jointly control any module by wiring in shared parameter modules at the appropriate point.

Fig. 32.9 shows a collaborative visualization session between the two users in the scenario described above. The numerical analyst (Fig. 32.9a) is hosting the application and viewing an estimate of the error in the solution. The mesh has been culled to show only those cells that have an error exceeding a specified tolerance (Fig. 32.2). The output from the simulation is shared with the chemist, whose view is shown in Fig. 32.9b, who is viewing an isosurface of

the computed solution. The parameters that control the simulation are also shared collaboratively, thereby allowing either user to steer the computation.

32.6.2   Bringing in the Grid

One collaborative scenario for steering is one in which the numerical application—which is generally computationally intensive—is run on the most powerful computer, while a remote user, who may have a relatively small computational resource, can control the simulation and receive output to display. In this case, it may be imprac-

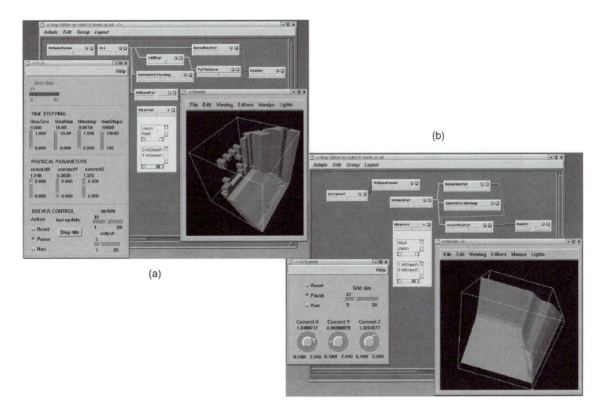

Figure 32.9 Performing a collaborative simulation of atmospheric dispersion using IRIS Explorer. The collaboration is envisioned as being between a numerical analyst—whose screen is displayed in (a)—and a chemist, who sees the screen shown in (b). The numerical analyst is running the simulation and viewing the error from the solution, while the chemist is viewing the computed solution as an isosurface. (See also color insert.)

tical for the remote user to monitor the simulation continuously, but it is also possible to use the collaborative modules to log in and out of a simulation. Indeed, the simulation can be run without human help, and the collaborative modules can be used to connect or reconnect a remote user when necessary. In the same way, computationally intensive visualization transformations (such as isosurface generation) can also be performed on the most suitable platform.

This transfer of the computationally intensive part of the application onto the most appropriate resource is reminiscent of the way in which IRIS Explorer modules can be run on remote machines (see Section 32.3.1). It is also suggestive of the philosophy behind the computational Grid, which represents a new paradigm for distributed computing. Recent research in computational steering and simulation has been focused on allowing all, or parts, of an application to be distributed in this way [42]. Current work with IRIS Explorer is focusing on extending the scenario above with the use of the Globus middleware [43] to allow the computing to be spread over a network of computational resources. In the scenario under development, depicted in Fig. 32.1e, we are using Globus to interface with Grid resources, while the user interacts with the simulation and performs steering and collaboration through the IRIS Explorer user interface. To do this, we separate the simulation's interface (which still appears in the map as a module control panel) from its computational engine (which runs on the Grid). (Note that this is reminiscent of the separation between the parallel isosurfacing code and its control described in Section 32.5.2.) Grid authentication and resource discovery are initiated by a specially written set of Globus modules, whose output is passed to the simulation interface module that spawns the engine onto the discovered resource; thus, much of the complexity associated with the Grid is hidden from the user.

Work to date includes the porting of the atmospheric dispersion model of Section 32.6.1

into the Grid environment; this formed the basis for a collaborative application demonstrating simulated pollutant flow over a terrain [44], and a scenario in which emergency services and scientists collaborated to track the progress of the pollution [45]. Other work [46] has taken an elastohydrodynamic lubrication simulation used in the modeling of solid deformation in gears and journal bearings and ported it to the IRIS Explorer steering environment running on the Grid. The simulation has been used within this environment by engineers for the investigation of solution profiles for particular lubricants under various operating conditions. Once again, the coupling of the simulation and visualization in a steering application yield benefits of efficiency in terms of the exploration of parameter space for the model.

While this work is ongoing [47,49], the IRIS Explorer Grid-based modules from the original pollutant demonstrator are available for download [44]. The porting of IRIS Explorer into a Grid-based environment has been of interest to other research projects on the Grid, many of which are currently investigating the toolkit as a candidate that may meet their visualization requirements in this environment.

32.7 Conclusions

This chapter has been an introduction to IRIS Explorer as a modular data-visualization system. We have described some of its features and indicated ways in which its modular architecture leads to applications in which the constituent modules may be distributed, compiled together, or made collaborative. We have also discussed some of the software that underlies the system and have selected a few user applications in diverse fields. Finally, we have indicated how the basic visualization reference model that forms the basis of this system can be extended to collaborative steering on the Grid, and we have given an account of work that has moved IRIS Explorer in this direction.

Acknowledgments

I am grateful to the numerous users and developers of IRIS Explorer responsible for much of the work described here, particularly Silhacene Aid, Ken Brodlie, Bob Brown, David Burridge, Hiro Chiba, Patrick Craig, Arnaud Desitter, Matt Dougherty, David Foulser, Alan Gay, Gary Griffin, Haruko Kanjo, Santiago Lombeyda, Mike London, Mick Pont, Jamie Shiers, Sarah Turner, Astrid van Maanen, Jason Wood, and Helen Wright.

References

1. B. McCormick, T. A. DeFanti, and M. D. Brown. Visualization in scientific computing. *Computer Graphics*, 21(6), 1987.
2. J. Neider, T. Davis, and M. Woo. OpenGL programming guide. Addison-Wesley, 1993.
3. J. Wernecke, The inventor mentor: programming object-oriented graphics with Open Inventor, release 2. Addison-Wesley, 1994.
4. K. W. Brodlie, D. L. Fisher, G. G. Tolton, and T. W. Lambert. The development of the NAG graphical supplement. *Computer Graphics Forum*, 1(3):133–142, 1982.
5. NAG Graphics Library, http://www.nag.co.uk/visual/GLGICH.asp
6. B. Ford, personal communication, 1998.
7. J. P. R. B. Walton. Get the picture—new directions in data visualization. In *Animation And Scientific Visualization: Tools & Applications* (R. A. Earnshaw and D. Watson, Eds.), pages 29–36, Academic Press, 1993.
8. C. Upson, T. Faulhaber, D. Kamins, D. Schlegel, D. Laidlaw, J. Vroom, R. Gurwitz, and A. van Dam. The application visualization system: a computational environment for scientific visualization. *IEEE Computer Graphics and Applications*, 9(4):30–42, 1989.
9. R. B. Haber and D. A. McNabb. Visualization idioms: a conceptual model for scientific visualization systems. In *Visualization In Scientific Computing* (B. Shriver, G. M. Nelson and L. J. Rosenblum, Eds.), pages 74–93, 1995.
10. H. D. Lord. Improving the application development process with modular visualization environments. *Computer Graphics*, 29(2):10–12, 1995.
11. D. S. Dyer. A dataflow toolkit for visualization. *IEEE Computer Graphics And Applications*, 10(4):60–69, 1990.
12. D. Foulser. IRIS Explorer: a framework for investigation. *Computer Graphics*, 29(2):13–16, 1995.
13. B. Lucas, G. D. Abram, N. S. Collins, D. A. Epstein, D. L. Gresh, and K. P. McAuliffe. An architecture for a scientific visualization system. In *Proceedings of Visualization '92* (A. E. Kaufmann and G. M. Neilson, Eds.), pages 107–114, 1992.
14. C. R. Johnson and S. G. Parker. Applications in computational medicine using SCIRun: a computational steering programming environment. In *Proceedings of Supercomputing '95* (H. W. Meuer, Ed.), pages 2–19, 1995.
15. The Numerical Algorithms Group, IRIS Explorer user's guide, http://www.nag.co.uk/visual/IE/iecbb/DOC/ html/unix-ieug5-0.htm
16. P. Craig. Implementing a statistical data type in IRIS Explorer. *IRIS Explorer Technical Report No. IETR/10* (NAG Technical Report No. TR3/97). http://www.nag.co.uk/doc/TechRep/PS/tr3_97.ps
17. The Numerical Algorithms Group, IRIS Explorer module writer's guide, http://www.nag.co.uk/visual/IE/iecbb/DOC/html/unix-iemwg5-0.htm, 2000.
18. J. D. Wood, K. W. Brodlie, and H. Wright. Visualisation over the world wide web and its application to environmental data. In *Proceedings of Visualization '96* (R. Yagel and G. M. Nielson, Eds.), pages 81–86, 1996.
19. User-Donated Module Repository, IRIS Explorer Center of Excellence, http://www.scs.leeds.ac.uk/iecoe/main_repository.html
20. O. Casher and H. S. Rzepa. A chemical collaboratory using Explorer EyeChem and the common client interface. *Computer Graphics*, 29(2):52–54, 1995.
21. D. Knight. CFD visualisation with IRIS Explorer 3.5. In *Render*, issue 6. NAG Ltd (1996). http://www.nag.co.uk/visual/IE/iecbb/Render/Issue6/CFDVis.html
22. J. Shiers. Analysis of high energy physics data using IRIS Explorer. In *Render*, issue 10. NAG Ltd., (1999). http://www.nag.co.uk/visual/IE/iecbb/Render/Issues10/issue10_4.html
23. J. P. R. B. Walton. Visualization of sphere packs using a dataflow toolkit. *J. Mol. Graphics*, 12(3):147–154, 1994.
24. J. D. Wood, K. W. Brodlie, and H. Wright. Collaborative visualization. In *Proceedings of Visualization '97* (R. Yagel and H. Hagen, Eds.), pages, 253–259, 1997. http://www.comp.leeds.ac.uk/vis/jason/vis97/vis97.html
25. The Numerical Algorithms Group, IRIS Explorer collaborative user's guide, http://www.

nag.co.uk/visual/IE/iecbb/DOC/html/unix-iecug 5-0.htm, 2000.

26. J. P. R. B. Walton and M. Dewar. See what I mean? Using graphics toolkits to visualise numerical data. In *Visualization and Mathematics: Experiments, Simulations and Environments* (H.-C. Hege and K. Polthier, Eds.), pages 279–299. Berlin, Springer, 1997. http://www. nag.co.uk/doc/TechRep/PS/tr8_96.ps

27. P. Barthelemy and R. Weideman. Open Inventor and MasterSuite. In *Proceedings of HEPVIS 96*, pages 87–98, 1996. http://www. cern.ch/Physics/Workshops/hepvis/hepvis96/ papers/p_barthelemy.ps.gz

28. J. P. R. B. Walton. World processing: data sharing with VRML. In *The Internet in 3D: Information, Images and Interaction* (R. A. Earnshaw and J. Vince, Eds.), pages 237–256, 1997.

29. SGI ImageVision Library, http://www.sgi.com/ software/imagevision/overview.html

30. NAG Numerical Libraries, http://www.nag.co.uk/ numeric/numerical_libraries. asp

31. J. P. R. B. Walton. Visualisation benchmarking: a practical application of 3D publishing. In *Proceedings of Eurographics UK 1996*, 2:339–351, 1996. http://www.nag.co.uk/doc/ TechRep/PS/tr9_96.ps

32. A. Lopes and K. W. Brodlie. Accuracy in 3D particle tracing, In *Mathematical Visualization: Algorithms, Applications and Numerics* (H.-C. Hege and K. Polthier, Eds.), pages 329–341. Berlin, Springer, 1998. http://www.scs. leeds.ac.uk/vis/adriano/acc_particle.ps.gz

33. M. London. Using IRIS Explorer in oil recovery research. In *Render*, issue 12. NAG Ltd, 2000. http://www.nag.co.uk/visual/IE/iecbb/Render/ Issue12/art7.html

34. A. W. Cook and P. E. Dimotakis. Transition stages of Rayleigh-Taylor instability between miscible fluids. *J. Fluid Mech.*, 443:69–99, 2001.

35. S. Lombeyda, J. Pool, and M. Rajan. Parallel isosurface calculation and rendering of large datasets in IRIS Explorer. In *Render*, issue 11. NAG Ltd, 1999. http://www.nag.co.uk/visual/ IE/iecbb/Render/Issue11/issue11_2.html

36. W. E. Lorensen and H. E. Cline. Marching cubes: a high resolution 3D surface reconstruction algorithm. *Computer Graphics*, 21(4):163–169, 1987.

37. A. J. Fenlon, T. David, and J. P. R. B. Walton. An integrated visualization and design toolkit for flexible prosthetic heart valves. In *Proceedings of Visualization 2000*, pages 453–456, 2000.

38. C. H. Hsu and T. David. The integrated design of mechanical bi-leaflet prosthetic heart valves. *Med. Eng. Phys.*, 18(6):452–462, 1996.

39. M. Walkley, J. Wood, and K. Brodlie. A distributed co-operative problem solving environment. In *Computational Science—ICCS 2002* (P. M. A. Sloot, C. J. K. Tan, J. J. Dongarra, and A. G. Hoekstra, Eds.), pages 853–861, 2002. http://www.comp.leeds.ac.uk/vis/kwb/e-science/paper098.pdf

40. G. Hart, A. Tomlin, J. Smith, and M. Berzins. Multi-scale atmospheric dispersion modelling by the use of adaptive gridding techniques. *Environmental Monitoring and Assessment* 52:225–228, 1998.

41. C. Johnson. *The Finite Element Method*. Wiley, 1990.

42. G. Allen, W. Benger, T. Goodale, H.-C. Hege, G. Lanfermann, A. Merzky, T. Radke, E. Seidel, and J. Shalf. The cactus code: a problem solving environment for the grid. In *Proceedings of Ninth IEEE International Symposium on High Performance Distributed Computing*, 2000.

43. Globus, http://www.globus.org

44. Covisa-G Collaborative Visualization, http:// www.visualization.leeds.ac.uk/CovisaG/

45. K. W. Brodlie, S. Mason, M. Thompson, M. Walkley, and J. D. Wood. Reacting to a crisis: benefits of collaborative visualization and computational steering in a grid environment. In *Proceedings of UK e-Science All Hands Conference*, 2002.

46. C. E. Goodyer and M. Berzins. Eclipse and ellipse: PSEs for EHL solutions using IRIS Explorer and SCIRun. In *Computational Science — ICCS 2002*, (P. M. A Sloot, C. J. K. Tan, J. J. Dongarra and A. G. Hoekstra, Eds.), pages 523–532. Berlin, Springer, 2002.

47. Visualization Middleware for e-Science, http:// www.visualization.leeds.ac.uk/gViz/

48. K. W. Brodlie, D. A. Duce, J. R. Gallop, J. P. R. B. Walton, and J. D. Wood. Distributed and collaborative visualization. *Computer Graphics Forum, 23(2):223–251, 2004.*

49. K. W. Brodlie, D. A. Duce, J. R. Gallop, M. Sagar, J. P. R. B. Walton, and J. D. Wood. Visualization in grid computing environments. In *Proceedings of Visualization 2004*, to appear.

33   AVS and AVS/Express

JEAN M. FAVRE and MARIO VALLE
Swiss National Supercomputing Center

33.1   Introduction

The introduction of the Application Visualization System (AVS) in 1989 marked a milestone in the visualization environment scene [1]. The computer industry had seen, at the end of the 1980s, the release of a class of hardware called "graphics superworkstations," while on the algorithmic side, the series of IEEE Visualization conferences had just started, bringing to light numerous visualization techniques and models. There remained, however, a large gap between the 3D graphics power and the software tools available. Graphics libraries available then (PHIGS+, GL, GKS, etc.) were too low-level and required a large programming investment that scientists were not ready to pay, while animations packages such as MOVIE.BYU were too specific and often restricted to the animation of geometric primitives.

AVS filled this gap by providing an interactive visualization framework that would allow scientists to quickly experiment with a wide range of visualization techniques, promoting the understanding of their data while relieving them of low-level programming tasks. The AVS environment was designed with the following goals:

- To enable quick application creation. To understand scientific data, it is critical to be able to explore and quickly experiment with visualization and analysis techniques. AVS provided a direct-manipulation interface to small software building blocks whose execution was entirely managed by the system.

- To shield the user from low-level programming, letting him or her concentrate on the visualization task and not on the programming system's idiosyncrasies. AVS generated a simple graphical user interface (GUI) for each modular component and took care of assembling all the graphics rendering components (e.g., views, cameras, and lights).

- To simplify integration and extension of the tool. Since visualization scenarios may change widely based on the application domain, and since visualization was—and continues to be—a rapidly evolving field, it was necessary for a visualization framework to provide a mechanism to easily include new user code in its environment.

- To provide a complete and integrated environment for all stages of the visualization, from data input and data transformation to graphics rendering.

- To develop a portable tool available on many heterogeneous platforms. AVS was based on standard graphics libraries, windowing systems, operating systems, and languages.

It was further recognized that a visualization tool should address the needs of three different categories of users:

- The *end users* want to visualize their data and experiment with different visualization techniques with as little programming as possible. They are typically scientists from numerical or experimental data production fields who use the tool without modification to support their scientific work.

- The *power users and visualization developers* are ready to extend the tool's functionalities by adding special visualization techniques and integrating their computational code. These users have graphics and visualization backgrounds and will enhance and extend the environment to better support their research work in visualization and the work of end users.

- The *application developers* prefer to work with reusable components to quickly build turnkey applications and deliver solutions specific to an application domain.

The architecture that resulted proved very successful as other similar environments (Iris Explorer, IBM Data Explorer, Khoros, VTK) followed this track in the 1990s. The AVS product line saw several versions of the tool that were eventually refined and improved in a fully object-oriented modular environment introduced in 1994 called "AVS/Express." This chapter will thus concentrate on this latest instance of the software product.

In the rest of the chapter, we begin by reviewing the design goals of AVS/Express, based on some of the limitations of AVS. We continue with a quick user tour to help illustrate AVS/Express' user interface environment and user-level functionalities. A detailed description of the most fundamental architecture aspects of AVS/Express, focusing on four areas (the object model, the execution model, the unified data model, and the rendering subsystem structure) follows. An example of a custom module is presented before the concluding paragraph.

33.2   Design Principles for AVS and AVS/Express

To address the requirements presented in the introduction, AVS introduced two paradigms essential to the creation of an open and modular environment. AVS reuses the concept of building blocks connected into a directed acyclic graph found in early animation and rendering systems. The power of this data flow–oriented

architecture is further strengthened by the addition of a visual programming interface that provides a one-to-one correspondence to the modules present in the visual workspace.

Because a crucial step in the visualization process is the extraction of data representations from raw data, intermediate data structures are necessary to hold the various filtered data prior to their conversion to geometric primitives. AVS introduced a data model—the *Field* data type—that would favor the interconnectivity of the modules and allow them to exchange typed data between each other.

At the time of the introduction of AVS, compliance with the standards was sought, and AVS supported Unix and VMS with PHIGS, GL, and Dore as rendering libraries.

Fig. 33.1 shows the Visual Editor area of the current version of AVS (AVS v5.6). The user creates the visualizations with a drag-and-drop interface, using modules from different libraries sorted out by application fields. In the working area, the modules can be connected together in a network-like topology, visually linking

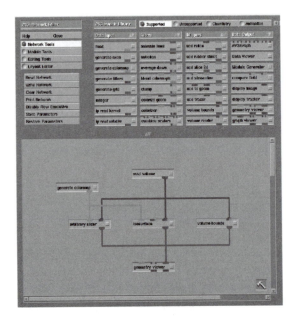

Figure 33.1   The AVS network editor. (See also color insert.)

modules via their input and output ports. The user is assisted during the network construction phase with color-coded ports that simplify the matching of typed inputs with typed outputs.

Each AVS module is a separate executable that runs in its own process. Its execution automatically starts as soon as all the required input data are present at its input ports. This style of program interface focuses on an interactive creation, connection, and deletion of modules, prototyping visualization scenarios in real time. The program structure is less important, as the user thinks in terms of flow and transformation of his data.

33.2.1   Lessons Learned from the Original AVS

The original AVS architecture had some drawbacks, addressed later by AVS/Express, as we will see in the remainder of the chapter. A pure data-flow architecture is inefficient from a memory usage point of view because each module makes a copy of the data. Further, the totally distributed nature of the running processes allowed simple and transparent connections to modules running on remote machines, but it proved very resource-consuming during sessions at a single desktop. The original Field data type was also not extensible. To support unstructured data grids, many modules had been duplicated. AVS's user interface was perhaps the strongest limitation. It was based on a set of proprietary widgets that do not resemble the accepted style for Unix applications' user interfaces. Finally, it was recognized that AVS was difficult to use as a framework for the creation of turnkey applications.

33.2.2   AVS/Express Design Goals

An enhanced data-flow architecture was central to the new architecture. The visual programming paradigm was a generally accepted concept, but the pure data-flow model underlying it had to be improved. To move beyond the creation of simple networks connecting no more than a few dozen modules, it was necessary to give the user an interpreted language interface (ASCII-based contents) and to place all objects in a hierarchy accessible visually or from the command language's interactive prompt.

It was also thought that the object-oriented nature of the system could be made directly available so that the user could use the software product as a true prototyping environment. The compile/link process necessary to create derived data representations and module implementations should be augmented to offer a consistent object-oriented view of the system, both during normal use and during new visualization prototyping activities.

In the modular approach to AVS, the module represented the atomic object of the lowest level. To favor a better extensibility of the application and to better use the system resources, a visualization developer needed access to data objects of widely different granularity, from the full 3D grid data level down to the basic numerical primitives (float, integer, etc.). A lot more flexibility and expressiveness would result from opening the object model.

Finally, a unifying data model that supported structured and unstructured data was necessary to make the system more intuitive. Data filtering and mapping operations would be simpler to use, and the user would leverage on his or her experience building visualization networks with different data characteristics.

33.3   AVS/Express Quick User Overview

When AVS/Express starts, it displays two windows: the Network Editor, which is the working area where the user creates visualization networks (Fig. 33.2, right), and the Data Viewer, which contains the rendering area, the module user interface, and the interface to the rendering subsystem (Fig. 33.2, left).

The Network Editor contains a list of libraries that group together modules with similar

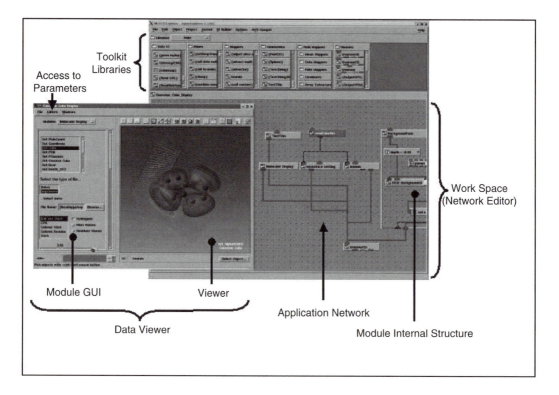

Figure 33.2 The AVS/Express visual editor. (See also color insert.)

functionality. A visualization application is called a network. Creation of a network consists of dragging modules from the module libraries to the Network Editor and then connecting them. Any module(s) can then be opened, exposing its (their) complete sub-object hierarchy for fine-grain editing tasks. Alternatively, objects can be grouped together into macros, reducing their visual footprint and hiding implementation details. The network editor supports strong-typed checking and will only allow connections between ports of compatible data types.

The Data Viewer window is composed of the viewer itself, the user interface area (in which the module control panels automatically appear when modules are instanced), and the various parameter editors that give access to the rendering subsystem. A toolbar allows quick access to the most common viewer functional-

ities (control over the mouse behavior for 2D and 3D viewing transformations, control over the light sources and the camera parameters, etc.).

33.3.1 Available Modules

AVS/Express offers an exhaustive list of modules. The modules are accessed trough the library section of the AVS/Express user interface. They can be divided into the following categories:

- Visualization-related modules. This category is further subdivided into Data Input/Output (I/O), Filters (which modify only their input data, not their geometry), Mappers (which create new data and geometry), and Field Mappers (which create gridded data from raw data arrays). Another grouping can be made between general 3D visualization techniques, imaging, and 2D charting. Modules generally come in two forms: one ready for immediate use with a user interface panel and ready-made connection ports, and a streamlined version providing only the raw computational method. This second set of modules is generally integrated into custom applications and gives the user a chance to provide a special user-interface panel.

- Modules for the creation of a user interface. These implement the classic interface widgets (buttons, sliders, frames, etc.) that support standard GUI styles. These modules allow the construction of a GUI in exactly the same way as a network of visualization modules. A GUI is immediately active as widgets are added to it. The AVS/Express environment offers a completely uniform prototyping and execution interface to all its components, shielding the user from platform differences.

- Modules for user interaction support. These include data value probes, interactive geometry editors, camera path editors, and so on.

- Modules for application development. This category includes modules for subprocess execution, application termination, and dynamic module loading.

Most standard visualization techniques are also provided as example applications. They are prebuilt networks to help the user familiarize himself or herself with the features of AVS/Express, and they can act as starting points for custom applications.

Modules are grouped into libraries or "kits" to simplify their retrieval. The kits list—like almost everything inside AVS/Express—can be modified to include custom-built libraries. Custom libraries are available from numerous sources. The largest of them is hosted at the International AVS Center (IAC), based in Manchester, UK (*http://www.iavsc.org*). It offers a complete repository of public-domain AVS/Express modules that complements the existing application libraries very well.

33.3.2 Module Examples

To illustrate the kind of visualization techniques available inside AVS/Express, we present two modules, one distributed with AVS/Express and the other one available through the IAC repository.

The first module is the *illuminated streamlines* module (Fig. 33.3a). The visualization of illuminated lines uses a dynamic texture map on 3D polylines to represent field streamlines to visualize continuous fluid dynamics fields. The method combines streamlines and particle animation into one hybrid technique, and it employs texture display on lines to represent full 3D lighting, motion, and 3D flow structure in one view. The technique exploits texture graphics systems in common use for games and achieves high graphics efficiency during animation [2,3].

The second module implements the Lagrangian-Eulerian Advection visualization technique (Fig. 33.3b). This is a recent technique to visualize dense representations of time-dependent vector fields. The algorithm produces animations with high spatio–temporal correlation at interactive rates. With this technique, every still frame depicts the instantaneous structure of the flow, whereas an animated sequence of frames reveals the motion that a dense collection of particles would take when released into the flow [4,5].

33.4 AVS/Express Architecture

The AVS/Express architecture can be introduced from two complementary points of view: that of the static structure and that of the dynamic behavior.

From a static view, AVS/Express is composed of objects. All objects live in a full hierarchical structure. There is no distinction between object types and object instances. Objects can be derived (sub-classed) from other objects or built by composition of other objects. The construct that fills the gap with the dynamic model is the *reference*. The AVS reference goes beyond

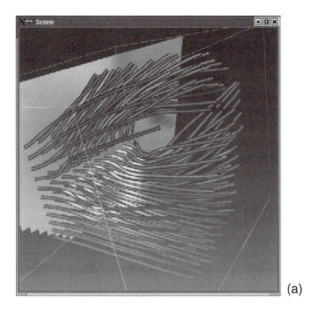

(a)

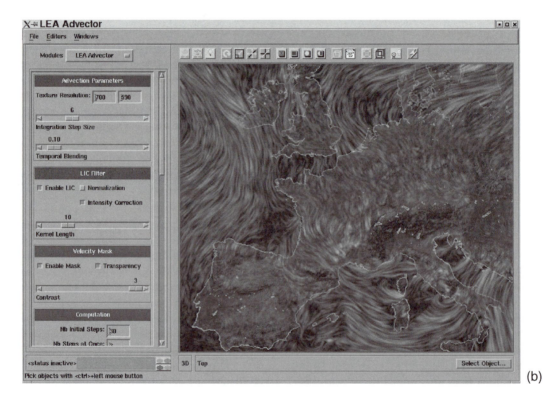

(b)

Figure 33.3 (a) Illuminated streamlines model. (b) Visualization of time-dependent wind fields with the IAC LEA module. (See also color insert.)

the standard C++ definition of a reference in that it also specifies the interaction between objects.

Dynamically, the AVS/Express architecture is a modified data-flow. In traditional data-flow architectures, the modules start execution when all the needed inputs are available, and they exchange data via copy operations through their connected ports. No global memory is used. AVS/Express modifies this architecture in two ways: it generalizes the set of events that makes a module execute, and it centrally manages the system memory, thus allowing the modules to exchange references to their data.

At the core of AVS/Express lies the Object Manager, which implements and controls this architecture. Everything is built on top of it, from the Field data type to the Rendering subsystem. The Object Manager exposes three interfaces: a drag-and-drop interface with visual programming via the Network Editor, a programmatic interface in AVS/Express "V" language, and an API that can be called from C, C++, and Fortran code.

33.4.1   The Object Architecture

The AVS/Express Object Manager is a programming environment that combines the strengths of object-oriented development methodology with the advantages of "visual programming" environments. It augments existing design and development methodologies by providing users with a high-level interface to glue together application objects developed with traditional object-oriented frameworks or structured programming techniques.

33.4.1.1   Base Types

The lowest-level unit of configurability in the system is a "base type." These "base types" implement the basic behaviors of program objects: storing and providing access to data values, storing and providing access to sub-objects, sub-classing operations, instantiating

and destroying objects, and propagating events through the system. Examples of base types include the "int," which stores an integer value, and the "group," which maintains a list of sub objects.

Base types are defined in an object-oriented way. All base types implement a common set of methods (some of these methods may return an error for specific base types). This common set of methods allows uniform access to the objects. For example, the Network Editor can add a new sub-object to an object without knowing which type of object it is dealing with.

An object stores a list of sub-objects. Sub-objects typically define an object hierarchy that can be traversed in the Network Editor or in the V language interface.

Many base types do not impose any specific behavior on the sub-objects, but other base types may treat the sub-objects as a list of attributes of the base type. This mechanism allows the user to treat everything as an object. For example, a "function" object can use its list of sub-objects as the list of parameters of the function itself.

33.4.1.2   Sub-Classing

The "sub-classing" and "instantiating" operations are both implemented using the *copy method* of an object. The copy method has the option of making a clean distinct copy, or it can make a *derived copy*, where the new object refers to the super-class to store information about its type. Thus, the implementer of the base type determines how the class hierarchy is maintained for objects of that particular type.

33.4.1.3   References and Notifications

Objects can have references to other objects. Normally, the reference is used to store the value of the object as an indirect pointer to another object. For example, an "int" object can store a reference to another object that will store its own value. In the V language example below, the operator => implements the reference. The value of *beta* is stored in *alpha*.

```
int alpha;
int beta => alpha;
```

The operator => has a second responsibility: allowing notifications of events. Any object can request a *notification* when a particular event occurs on an object in the system. Examples of events include "value changed", "object instantiated", and "connection made/broken".

Base types can propagate these "notification requests" through the references they manage. They might implement the "add notification request" method by performing this same method on the object they reference. Typically, however, they will only do this for "value changed" events.

Any request to be notified when the object *beta* is changed will be propagated so that the requestor is notified when the *alpha* value is changed as well. Thus, the => operator implements more than a one-time assignment of value. It manages a relationship of equality, which the Object Manager will guarantee at all times.

33.4.1.4   Object Manager Interfaces

The user may interact with the Object Manager via three complementary interfaces:

- The Network Editor: This is the visual programming interface. It allows class derivations, instantiations of objects, definitions of references (connections) and object editing.

- The V language: This is an interpreted declarative language. The experienced developer will often use it for faster prototype editing.

- The OM API: This is the lowest-level interface available for the C, C++, and Fortran codes.

All three interfaces allow the same access to the objects (Fig. 33.4). The developer will choose the prototyping interface that matches his or her degree of expertise in Object Manager Development.

33.4.1.5   Differences from C++

A comparison between the class interface and usage in C++ versus those in the AVS/Express

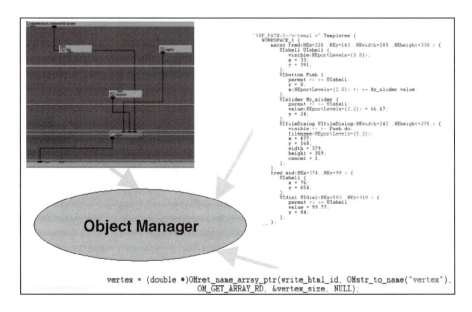

Figure 33.4   The Object Manager, a centralized control with three user interfaces.

Table 33.1   Different elements in C++ and AVS/Express

C++	AVS/Express Object Manager
class	module
member variable	sub object
method	method
public member variable	parameter
behavior defined by usage	behavior explicitly defined
constructor/destructor	module on_instance/ on_deinstance methods

Object Manager helps highlight their differences (Table 33.1).

The important difference is in the definition of the class (module) behavior. A module defines which events should trigger a method execution. There is no direct call of a method from a "main" program.

In C++, the "pixel" object created below must call its public method explicitly.

```
class screen_coords {
public:
  int u, v;
  map_coord (float x, float y);
};

class Point { float x, y;} ;
Point            wp;
screen_coords        pixel;
pixel.map_coord(wp.x, wp.y);
```

In AVS/Express, the module *screen_coords* specifies that the method will automatically execute whenever a "notify" event is received by either argument, *x* or *y*. The live object called "pixel" specifies connections (or references) to other sub-objects. (The notification mechanism will be explained further in a following paragraph). These sub-objects will be the source of the event notifications.

```
module screen_coords {
  int u, v;
  float x, y;
  method map_coord (x + notify,
                    y + notify);
};

world_position wp;
screen_coords pixel {
      x => wp.x;
      y => wp.y;
};
```

The AVS/Express developer can prototype a class derivation using the V language interpreter or the visual interface. In the Network Editor, the => operator is applied when making a connection between any two objects.

33.4.1.6 *Application Development Support*

Customizations and additions to the base AVS/ Express environment are saved in a project directory. A project can add to or supersede anything provided in the main distribution, and it can have descendents providing support for multideveloper projects.

At the code level, AVS/Express supports the application development with two specific tools: the "add module" wizard and the run-time generation support.

The wizard creates a code skeleton and all the support files needed to specify the integration of a user-defined module.

The run-time generation packages an application so that it can run on a machine without AVS/Express installed. The run-time also omits from the application all the code, libraries, and support files not needed by the application itself to achieve better performances.

33.4.2   The Execution Model

As stated earlier, no explicit procedural programming is available for or needed by the Object Manager. Control flow constructs like *for, while*, or *if* are embedded in the architecture using the same structure of standard processing modules without resorting to special language constructs. Two examples follow.

33.4.2.1   Control Flow Examples

Suppose the user wants to display a sequence of isosurfaces varying the threshold level from 0 to 255 with a step of 20. A *loop* module repeatedly executes, outputting a numerical value connected to the input threshold of the *isosurface* module. The Object Manager propagates the event "value changed" to the *isosurface* module, which will recompute a new isosurface. The renderer then reacts to the "value changed" event on its renderable object input and will refresh the display (Fig. 33.5).

In the second example, an if control structure can be emulated with the V language *switch ()* built-in function. *Switch(index, arg1, arg2,...)* returns one of its arguments, depending on the value of its index. If index is 1, it returns arg1; if index is 2, it returns arg2, and so on. The color of a GUI button can be changed based on the sign of an input value:

```
Uibutton  bicolor_button{
  color{
    backgroundColor => switch ((value >
      0) + 1, ``red'', ``green'');
    };
};
```

33.4.2.2   Event Scheduling

Every method definition declares the events it is interested in. The most important events are "module instantiation", "module de-instantiation", and "parameters value changes".

One advantage of using event scheduling for a particular method is that the object itself handles the expression of interest for a particular event. Thus, no outside user must explicitly manage the execution of code for that particular object; the event dispatcher handles the scheduling.

For example, the implementation of a user-interface widget typically handles the resize operation automatically. It registers an event handler to be called when its parent widget is resized. In response to this event, the callback function will appropriately resize the widget. The programmers using the widget in an application need not be aware of what is happening within the implementation of the widget. Without this event handler, the programmer would have to trap the conditions under which the widget needed to be resized, and it could handle this resize operation by explicitly calling a widget's method.

When modules are connected together, the Object Manager creates a dependency graph based on the modules' data access declarations. When an event occurs, e.g., a value changes in the user interface, the Object Manager will

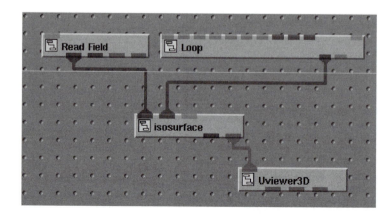

Figure 33.5   Animation of a parameter with the Loop module.

traverse this graph and check the *sequence number* maintained for every object. As a result, decisions will be made about which methods to fire and in what order.

A pair of push/pop context calls brackets this sequence of events. Those routines mark something similar to a transaction in a database system. This ensures that the objects are valid before event queuing, collapses a set of events directed to the same object, and ensures that no infinite loop occurs (e.g., if a method modifies the input that triggered the method execution).

This explicit execution model, with its central Object Manager, naturally integrates external events (e.g., sockets, X event loops) in the same execution model. The application developer can thus integrate an AVS/Express application inside another application or integrate other applications as part of an AVS/Express system.

33.4.2.3   *The Process Model*

AVS/Express maintains a collection of cooperating processes. A process can be running on any machine in a networked environment. Each process maintains a set of objects that includes any data and methods that make up the implementation of the object. Objects are allocated in a single namespace that spans all the processes in the system. A single master process distributes unique "process IDs" that are stored as part of the object to ensure its uniqueness in the system.

Given a particular object, the Object Manager can quickly obtain its process ID to ensure transparent interprocess communications. The connection is created dynamically the first time a process attempts to access information about a remote object. The system typically performs a procedure call on a particular object without needing to know whether the object is in the same process or in a different process. This multiprocess environment allows the user to separate a module in a different process (for example, to facilitate its debugging). At a later time, the module can integrate the major pro-

cess pool. Interprocess migration remains transparent, and performance will be better.

The most important benefit of this structure is that distributed execution can be easily implemented. The user defines a process and declares it as running on a particular machine. Then the user creates a module and assigns it to this process. When he or she instantiates this particular module and connects it into the application network, the system takes care of the distributed communication and remote process management. The resulting application network offers complete transparency for the modules' instantiation and execution.

33.5   The Field Data Type

The Field data schema encapsulates the data structures most commonly used in visualization applications. It covers a wide range of data, including images, volumes, finite difference and finite element solutions, and geometry. Its unifying property simplifies the integration between modules by making them operate on a single data type.

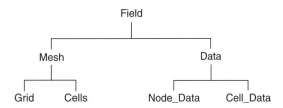

The Field data schema is defined using the V language and is interpreted by the Object Manager at run time. It is hierarchical, and the topmost object is referred to as the Field. Fields consist of a Mesh element and a Data element. A Mesh is a geometric description of the data domain that includes Grid and Cells. A Grid defines the spatial location of the nodes. Cells are used to specify node connectivity and the type of interpolation between nodes. Cells can have various properties, such as shape, order, etc. In the Field data schema, cells that have the same properties are combined into Cell Sets. Data represents scalar or vector values defined at specific domain locations, such as nodes and/ or cells. On 64-bit platforms, Fields can handle arrays of up to 2×10^9 elements.

The Field Access Library provides an API between the Field data schema and other components, such as the visualization modules and the Graphics Display Kit. It consists of an extensive set of functions that allow access to different subelements of Fields.

33.5.1   Mesh

The most general Mesh represents an unstructured domain; it can be viewed as a collection of cells or elementary volumes of different shapes covering the domain. Each cell is defined by a collection of bounding nodes that form a connectivity list. Node locations in space are defined by a coordinate array. The general Grid template groups those coordinates together with information about the number of nodes, dimensionality of the coordinate array, coordinate information, and spatial extents of a domain. Cell information is stored in the structure called Cells.

In the general case, a Mesh can have more than one cell type. For example, in structural analysis, the same Field can contain both 3D cells (such as hexahedrons) and 2D cells (such as triangles and quadrilaterals). The notion of `Cell_Set` efficiently represents different types of cells in a single Field. Cells in one `Cell_Set` have the same shape, order, topology, number of nodes, properties, and other attributes.

The following `Cell_Set` types are supported:

- Zero-dimensional: Point
- 1D: Line, Line2 (second-order cell), Polyline
- 2D: Triangle, Triangle2 (second-order cell), Quad, Quad2 (second-order cell), Polytri
- 3D: Tetrahedral, Tetrahedral2 (second-order cell), Hexahedral, Hexahedral2 (second-order

cell), Prism, Prism2 (second-order cell), Pyramid, Pyramid2 (second-order cell).

33.5.2   Structured Mesh Sub-Classes

Cell connectivity and node coordinates are explicitly defined in the general Mesh template. A specialized derived type of Mesh is a structured mesh (`Mesh_Struct`) that implies a certain cell connectivity rule and does not store any explicit connectivity information. This type of mesh is often used in computational fluid dynamics (CFD) analysis, where the domain is defined on a curvilinear grid (a topologically regular grid deformed in space).

In object-oriented terminology, `Mesh_Struct` is a sub-class of Mesh that has additional information about mesh topology. The data access methods defined in the parent class `Mesh` are available for `Mesh_Struct` without the need for overloading.

The class for a structured grid is derived from the general Grid class. The node connectivity array for the `Cell_Struct` is defined implicitly by using the V language function `get_connect_struct()`. An explicit evaluation would only take place when an application treated the `Mesh_Struct` object as a more general Mesh.

Sub-classes of the structured mesh are uniform meshes (`Mesh_Unif`) and rectilinear meshes (`Mesh_Rect`). They further specialize the definitions of node locations and do not store explicit node coordinates. A uniform mesh stores only two corner points representing grid extents in physical space. A rectilinear mesh stores coordinates for the boundary nodes for each dimensionality. Coordinate values in both cases are evaluated implicitly by the V language functions `get_coords_unif()` and `get_coords_rect()`. As for the connectivity array, these functions calculate coordinate arrays only in the case of a mesh being treated as a `Mesh_Struct` or of one of its parent classes.

The following class definitions complete the Field data schema for the following structured meshes:

- Cylindrical Uniform Mesh (2D and 3D)
- Cylindrical Rectilinear Mesh (2D and 3D)
- Spherical Uniform Mesh (3D)
- Spherical Rectilinear Mesh (3D)

33.5.3   Data

The Data class represents scalar or vector values defined at specific domain locations, such as nodes and/or cells. A Field can have different kinds of data: `Node_Data` and/or `Cell_Data`, each of them in turn may consist of multiple components, such as pressure, temperature, velocity, etc.

Data values can be of type *byte*, *char*, *short*, *int*, *float*, or *double*. Different components can have different data types. The Data class includes a notion of NULL data: a special value can be set and treated as absence of data at a particular node or cell.

`Cell_Data` is defined for each `Cell_Set`, meaning that different cell sets may have different data components.

33.5.4   Field Types and Visualization Methods

This variety of Field types fits well in an object-oriented design: methods can be associated and stored together with an object's template. This solves the problem of how to design data processing and visualization modules that work efficiently with all the different Field types.

Starting with the most general objects, such as `Mesh`, there are methods that are used in data processing and visualization modules (like interpolate_cell and find_cell) and methods associated with different cell types (like shape_functions, shape_derivatives, etc.).

The `Mesh` sub-classes (`Mesh_Struct`, `Mesh_Unif`, `Mesh_Rect`) overload the methods defined for their parent classes. Their methods are optimized for the particular sub-type and take advantage of information specific to the `Mesh` sub-class definitions.

33.5.5   Field Extensibility

The AVS/Express Field data structures allow user extensions for defining new mesh or cell types. By deriving new classes from the existing Field structures, cell types specific to certain application needs can be established. The Field data schema is defined using the V language and is interpreted by the Object Manager at run time, allowing prototyping without a compile/link. Thus, macros can be built using these definitions to map primitive coordinate, data, and connectivity arrays into groups that represent new cell types.

To allow display and processing of the new cells, a tessellation function can be registered. It applies a conversion algorithm that creates cell connectivity lists for the standard cell types.

A Field extension recently introduced in AVS/Express (v5.0) addresses the needs for a time-dependent field data type. It extends the Field structure to store data and/or coordinates at different time-steps. The whole hierarchy inherits the extension; thus, support for uniform, rectilinear, and structured meshes was easily derived. The "Read Field" module was updated to read this kind of Field, and some specific modules were added (i.e., the interpolate_time_step module allows generic substep interpolation). No rendering functions were added; time-dependent Fields are rendered one step at a time.

33.6   The Integrated Rendering Subsystem

The integrated rendering subsystem, also called the Data Display Kit, is the component of the AVS/Express system that provides the entire rendering and display functionality. Its structure is built in the Object Manager architecture framework and uses the same execution model as every other AVS/Express module. The choice of this structure makes the Data Display Kit highly extensible along three major directions: first, new data types may be added to the system and directly rendered; second, new renderers may be added; and third, the viewer's user interface may be replaced.

The most visible example of the extensibility of the Data Display Kit is its support of multiple renderers. On any supported platform, a software renderer is always provided. In addition, one or more of the following renderers are provided: OpenGL, XGL, PEXlib. In the AVS/Express Multipipe Edition, the SGI Multipipe Toolkit has been integrated as a new immersive rendering system. It is integrated transparently in the Data Display Kit and AVS/Express architecture. A single-button switch may transport the user from a desktop viewer to an immersive view (e.g., Reality Center, CAVE, etc.).

An example of another extension is the module available in the IAC AVS/Express public-domain repository that implements support for volume rendering for the VolumePro graphics subsystems of MERL. Direct rendering to OpenGL is used to take full advantage of graphics board API.

33.6.1 Rendering Features

The Data Display Kit has the ability to render not only 3D data but also 2D data, images, and volumes into a single view. Some toolkits focus on only one of these data types, or provide support for more than one data type but at less than optimal efficiency. The Data Display Kit supports all of these data types in a highly efficient manner.

The Data Display Kit renders data directly. It makes it unnecessary to convert data into a renderable form prior to passing it to the Display Kit. The toolkit will do this conversion on the fly. This renderable representation can be optionally cached for faster updates during direct interaction with the object. This provides the user with the typical tradeoff of memory versus speed. In immediate-mode rendering, the toolkit will convert data in real time. This allows large datasets or time-dependent data mappings to be rendered with minimal memory overhead. Alternatively, a display list can cache a renderable form of the data to accelerate interactive viewing of static objects.

Many of the typical attributes of a display list system, like inheritance of attributes, etc., are present in the toolkit.

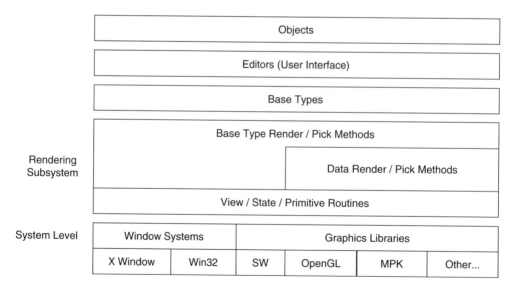

Figure 33.6 The Data Display Kit.

The Data Display Kit provides a number of powerful techniques to allow a high level of interaction in very complex scenes. These include alternate object rendering, accelerate mode, and usage of 2D primitives in 3D.

33.6.2 Architecture

The first level up from the operating-system level is the Data Display Kit rendering subsystem. This level provides all of the window system–dependent and graphics library–dependent routines that define a renderer in the toolkit. It also provides the data-independent routines that convert the raw data to geometric primitives. This level is independent of any other component of AVS/Express.

The second level above the operating-system level contains the base types that make up the Data Display Kit. These are objects like views, cameras, properties, etc. This level is largely independent of any other component of AVS/ Express, with only a minimal dependence on the framework. This layer and the rendering subsystem layer contain all of the functions that make up the high-level language API of the toolkit.

33.6.3 Multipipe Edition

The most notable extension to the Rendering Subsystem has been the integration of the SGI Multipipe Utility to implement a full parallel immersive rendering system within the AVS/Express Multipipe Edition [6].

The SGI Multipipe Toolkit (MPK) is designed as a new problem-solving API [7]. Its goal is to allow OpenGL applications to easily migrate to multipipe computing and graphics environments. The MPK system supports full portability, so the same application can run on a single-display low-end desktop system, or on an Onyx2 with multiple rendering engines. Unlike Inventor or Performer, MPK does not impose a scene-graph structure, so dynamic updating of scene contents, a capability often found in visualization applications, is handled in a more efficient way. The MPK system provides a transparent, callback-driven programming interface

and takes care of interprocess communication, parallel execution, and display-configuration issues.

33.7 Example of a New Module Derivation

AVS/Express applications and usage examples are covered in Chapter 35. We present here one low-level example, a specification of a custom GUI for an isosurface module.

This example demonstrates how an object can be customized for a particular purpose. Here, we modify the GUI of the isosurface module to present the user with three different interface options: a) a type-in widget to directly enter the threshold desired; b) a type-in widget to directly enter the percentage value between the minimum and the maximum of the scalar field selected (it ranges between 0 and 1); and c) a slider widget to animate that same percentage value (Fig. 33.7).

This object customization demonstrates many of the features of the AVS/Express design principles:

- The new object inherits its definition from a master copy. The derivation relationship is recorded, and the new object will inherit

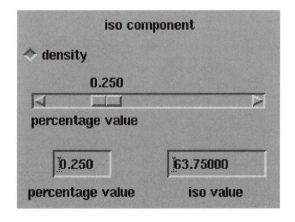

Figure 33.7 A custom isosurface GUI with percentage values.

from its parent once more when the software is upgraded.

- Customization is done by modifying inherited sub-objects and by adding new components. References to sub-objects in the object hierarchy are expressed with the $< -$. prefix, which tells the interpreter to search for the sub-objects one level higher than (and outside of) the current object.

- The numerical value of the isosurface threshold is specified via the V language $=>$ operator. Its value field will always be up to date. Special use of two other operators is done

here. The $< * >$ and $< + >$ allow the specification of a numerical formula that lets the user change any of the three widgets and have the value of the other two widgets be updated accordingly.

- Execution of the isosurface calculation is untouched; it remains under the responsibility of the update mechanism of the Object Manager. The corresponding V code (Fig. 33.8) declares *my_isosurface* as a sub-class of the standard isosurface from the MODS toolkit. The V code updates only the IsoUI component (its user interface) and leaves the rest *as is*.

```
MODS.isosurface my_isosurface {
  IsoParam {
    iso_level =>< -. IsoUI. UIfield.value;
  } ;
  IsoUI {
//add two float sub-objects
    float ndmin => in_fld.node_data[ param.iso_component] . min;
    float ndmax => in_fld.node_data [ param.iso_component] . max;
//add one typein widget whose value is in[ 0., 1.]
    UIfield PC {
      min = 0.;
      max = 1.;
    } ;
//modify the inherited sub-object UIiso_level
    UIiso_level {
      value => PC.value;
      title = ``percentage value'';
      min = 0.;
      max = 1.;
    } ;
//add a new typein whose value will verify at all times the
//relationship given
    UIfield UIfield {
      value => ( (<-. PC. value < * > (max - min) ) < + > min) ;
      min =>< -. ndmin;
      max =>< -. ndmax;
    } ;
  } ;
} ;
```

Figure 33.8 The V language to customize the isosurface module.

Note that the design of this new object via the object-oriented inheritance feature is done entirely during an interactive session, without the need for compilation and linking. Prototyping a system design is thus an extremely intuitive process.

33.8 Conclusions

AVS and AVS/Express were the first visualization environments to offer the visual programming paradigm in a visualization toolkit environment. AVS influenced many of the environments we see in use today. The framework offers full configurability, with the visual interface or with a fully interpretive language, and addresses the needs of many visualization and graphics tasks.

AVS/Express is a fully object-oriented environment where all objects are treated in a uniform fashion. From object instantiation to object derivation and execution, there is only one mode of development and usage. This applies to basic primitives, data and mesh object representations, GUI components, rendering components, and process and execution objects. This pervasiveness of the object-oriented architecture is the greatest strength of AVS/Express. Prototyping with it is very effective, thanks to many features. No compile/link cycle is necessary; all derived objects and modules are live in the environment.

The AVS/Express object orientation offers multiple levels of interaction to address the needs of different classes of users and programmers. Thus, one may choose different degrees of component integration levels and use or customize the environment as necessary. A basic user is able to reuse prebuilt visualization networks adapted to his or her own data without a single line of programming. A more thorough user will begin by customizing her or his environment via visual editions. Finally, the advanced programmer is able to rapidly prototype new components and deliver complete applications for use in a wide range of applications fields, from astrophysics to chemistry to computational fluid dynamics (CFD), and many other application fields.

AVS/Express goes well beyond a visualization toolkit which requires the additional knowledge of a GUI-building language such as Tcl/Tk or Java for advanced usage. It offers a fully functional interface ready to load data and produce renderings, but it can also be used via an API for specialized uses.

AVS and AVS/Express are based on a proven execution model: the data-flow paradigm. AVS/Express enhances this model with references and a transparent multiprocess execution model for different degrees of ease of use and efficiency. It is managed by a single event manager and task scheduler that maintains explicit control over all objects. Its default behavior is event driven: the network is executed after any single parameter change, but a demand-driven mode of execution is also possible. However, AVS/Express does not take full advantage of its centralized control to schedule parallel module execution. All modules, whether they are on a local process or on a remote process, are executed in a strictly synchronous manner.

However, with recent additions such as batch-mode rendering (off-screen rendering), multipipe rendering for immersive rendering, illuminated lines, volume rendering, two-pass transparency, and a CAVElib integration, AVS/Express is a modern programming environment with an advanced usage and prototyping environment based on object technology. Its customer base includes thousands of users worldwide in academic and industrial positions.

References

1. C. Upson, T. Faulhaber, Jr., D. Kamins, D. H. Laidlaw, D. Schlegel, J. Vroom, R. Gurwitz, and A. van Dam. The application visualization system: a computational environment for scientific visualization. *IEEE Computer Graphics and Applications*, 9(4):30–42, 1989.
2. D. Stalling, M. Zöckler, and H. C. Hege. Fast display of illuminated field lines. *IEEE Transac-*

tions on Visualization and Computer Graphics, 3(2), 1997.

3. I. Curington. Continuous field visualization with multi-resolution textures. *Proceedings of IEEE International Conference on Information Visualization*, London, July 14–16, 1999.

4. B. Jobard, G. Erlebacher, and Y. Hussaini. Lagrangian–Eulerian advection of noise and dye textures for unsteady flow visualization. *IEEE Transactions on Visualization and Computer Graphics*, 8(3):211–222.

5. B. Jobard, G. Erlebacher, and Y. Hussaini. Lagrangian–Eulerian advection for unsteady flow visualization. *Proceedings of IEEE Visualization 2001*, San Diego, CA, pages 53–60, 2001.

6. P. G. Lever, G. W. Leaver, I. Curington, J. S. Perrin, A. Dodd, N. W. John, and W. T. Hewitt. Design issues in the AVS/Express multi-pipe edition. *IEEE Computer Society Technical Committee on Computer Graphics*, Salt Lake City, 2000.

7. P. Bouchaud. Writing multipipe applications with the MPU SGI EMEA developer program document, http://www-devprg.sgi.de/devtools/tools/MPU/index.html see also the product description on: http://www.sgi.com/software/multipipe/sdk/

34   Vis5D, Cave5D, and VisAD

BILL HIBBARD
University of Wisconsin

34.1   Introduction

The Space Science and Engineering Center (SSEC) Visualization Project focuses on making advanced visualization techniques useful to scientists in their daily work. We accomplish this goal by making three scientific visualization systems (Vis5D, Cave5D, and VisAD) freely available over the Internet, and by using these systems as test beds for exploring and evaluating new techniques. It is important to distinguish between scientists who use advanced visualization techniques in their daily work and those who use them to create demos [11]. Use in ordinary work is the true measure of utility—demos often focus only on techniques that are merely new and dramatic but not useful. The systems described here are all freely available from the SSEC Visualization Project at *http://www.ssec.wisc.edu/~billh/vis.html*.

34.2   Vis5D

Vis5D grew out of work with the 4-D McIDAS [2,3] system during the 1980s, experimenting with animated 3D displays of various types of environmental data. We wrote Vis5D [5] in 1988 in response to three realities:

1. Our experiments indicated that data from simulation models are much easier to visualize than observational data.

2. Scientists do not trust depth information from binocular stereo. Instead, they need depth information from interactive 3D rotation.

3. The Stellar and Ardent commercial workstations that appeared in 1988 were the first with sufficient graphics performance for interactive rotation and real-time animation of Gouraud-shaded 3D scenes.

The input data to Vis5D are time sequences of regular 3D grids of values for multiple variables, for example, temperature, pressure, and fluid motion vector components. Such data are typically generated by atmosphere and ocean simulation models, and also by models and instruments in a variety of other scientific disciplines. The system takes its name from the fact that its input data can be stored in a 5D array, with three spatial dimensions, one time dimension, and one dimension for enumerating multiple variables.

Vis5D depicts scalar variables via isosurfaces, contour curves embedded on horizontal and vertical planes, pseudo-colors on horizontal and vertical planes, and volume rendering. The volume rendering technique uses three sequences of transparent planes, one each with planes perpendicular to the x, y, and z-axes. As the user rotates the view, the system switches to the sequence for the axis most closely aligned with the view direction. This widely used technique was first developed for the 4D McIDAS [4]. There is also an interactive point probe for scalar values, as well as an interactive vertical column probe used to generate traditional meteorological diagrams showing vertical thermodynamic structure. Vis5D depicts wind vectors via streamlines embedded on horizontal and vertical planes, particle trajectories, and vectors embedded on horizontal and vertical planes.

Fig. 34.1 shows some of these techniques combined in a single view, including the following:

1. A yellow isosurface of water vapor.

2. A white volume rendering of cloud water (i.e., transparent fog).

3. White contour curves of potential temperature embedded on a vertical plane.

4. A pseudo-color rendering of wind speed embedded on another vertical plane.

5. Cyan wind parcel trajectories.

6. A topographical map pseudo-colored by altitude and with embedded map boundaries.

All Vis5D rendering techniques are interactive. Users can interactively drag horizontal and vertical planes with the mouse for any graphics embedded in planes. Particle trajectory paths are integrated forward and backward through time-varying vector fields from points interactively selected by user mouse clicks. Users can interactively modify color tables for pseudo-colored graphics, including tables of transparency alpha values. Animation can be started, single-stepped, and stopped, running either forward or backward in time. The view volume can be interactively rotated, panned, and zoomed, and six clipping planes can be interactively dragged through the scene. Users also have interactive control over all sorts of numerical parameters of rendering techniques, including values for isosurfaces, spacing of contour curves and streamlines, and length-scaling of flow vectors and particle trajectories. Some types of graphical depictions of one variable can be pseudo-colored according to values of other variables, including isosurfaces, wind trajectories, and map topography.

A rectangular palette of buttons dominates the Vis5D user interface, with a row for each

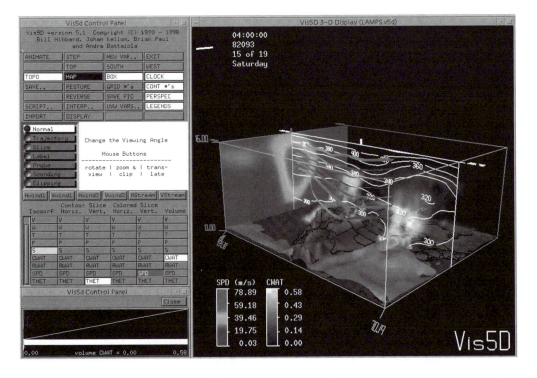

Figure 34.1   A Vis5D display combining terrain, isosurface, volume rendering, particle trajectories, pseudo-coloring, and contour curves. (See also color insert.)

scalar variable in the dataset and a column for each of six basic scalar rendering techniques. Clicking on a button adds the depiction of the selected variable by the selected rendering technique to the 3D display. This may cause the system to compute the geometries (i.e., vectors, triangles, and texture maps) for the rendering, if they are not already saved from a previous button click. Except for the volume rendering technique, which requires large amounts of memory, geometries are computed and saved for all time-steps in the dataset. This enables fast animation. It also enables users to click renderings of particular variables on and off quickly in the 3D scene, which can be very useful in cluttered scenes.

Memory management is a key problem for Vis5D and other interactive visualization systems. Interactive performance is best when the grids for all variables and all time-steps can be held in memory simultaneously, along with the computed geometries for each rendering technique. System responses are delayed whenever grids must be read from disk or geometries must be recomputed. When Vis5D loads a dataset, it computes the ratio of the dataset size to the amount of available memory, and, based on this ratio, chooses one of three memory management strategies:

1. If dataset size is significantly less than available memory, then the entire dataset is read into memory. When a variable and a rendering technique are selected, the geometries for all time-steps are computed and saved. If there is not enough memory to save the geometries, then the geometries for the least recently selected variable and rendering technique combination are discarded to make space for the new geometries.

2. If dataset size is many times larger than available memory, then grids are read from disk only as needed for computing geometries, and geometries are only computed and saved for the currently displayed time-step. This strategy of course makes

animation very slow, and it is appropriate when Vis5D is running from a script and computing an animation file (e.g., MPEG or animated GIF) for later viewing.

3. For datasets falling between the limits in cases 1 and 2, the system devotes part of the available memory to a cache of grids and devotes the rest of its memory to saving geometries for all time-steps for selected variables and rendering techniques. Grids and geometries are discarded from their caches on a least-recently-selected basis.

Vis5D uses a few other simple but important strategies for optimizing memory and performance:

1. Grids are generally stored in a compressed format of 1 byte per grid value, a format that suffices for most types of data. Options for 2 bytes and 4-byte floats are available for datasets that need more precision.

2. Geometries are stored compressed as scaled 1- and 2-byte integers.

3. When the user selects a variable and a rendering technique for display, geometries are computed first for the currently displayed time-step and then possibly computed for all other time-steps.

4. Vis5D maintains an internal work queue whose entries have the form (variable, rendering technique, time-step). On a symmetric (i.e., shared-memory) multiprocessor, Vis5D has one worker process per processor in order to exploit parallel processing. The worker processes remove entries from the work queue and compute appropriate rendering geometries from 3D grids.

Effective visualization requires information about the spatial and temporal locations of data, as well as the names of variables, units for numerical values, ways to indicate missing values, and so on. These pieces of information are called metadata. In 1988 there were no good standard file formats with adequate metadata for the 3D grids produced by weather and ocean simulation models. So we defined an internal file format for Vis5D that included the necessary metadata and used Vis5D's data-compression technique. Modelers write programs for converting their model output into this Vis5D internal format, and we supply C- and Fortran-callable libraries to make it easy.

The utility of visualization systems can be greatly enhanced by integration of data analysis operations. Vis5D includes capabilities to import data from various file formats into the Vis5D internal format, to resample grids spatially and temporally, to select a subset of variables from a file, and to merge data from multiple files. These capabilities are accessible within both a Vis5D visualization session and smaller stand-alone programs. Operations for listing and editing the contents of grid files are also available as small stand-alone programs. Users can also develop algorithms for deriving new grids from existing grids, either by simple formulas typed directly into the Vis5D user interface, or, for more complex operations, by writing Fortran functions that are dynamically linked to Vis5D.

Although weather modelers can use the increasing power of computers to increase the spatial and temporal resolution of their simulations, they have recently found more utility from increased computing power by running a statistical ensemble of simulations, each with slightly perturbed initial conditions. These ensembles give them a way to estimate the range of possible future outcomes due to the instability of nonlinear dynamics. Thus, we added to Vis5D the capability to read and display multiple sets. These can be displayed side by side in a rectangular spreadsheet of 3D displays, as shown in Fig. 34.2, or overlaid in the same 3D display. In the spreadsheet mode, the cells can be linked so that they animate, rotate, pan, and zoom in unison. Furthermore, selections of variables and rendering techniques made in one cell are mirrored in linked cells. While the capability to link displays of multiple 5D datasets would have justified changing the system's name to Vis6D, we stuck with the well known name Vis5D.

Vis5D also includes the rudimentary capability to overlay displays of simulation data with images from satellites and radars, as well as with randomly located observations from balloons, surface weather stations, etc. However, the visualization problems of observational data are very different from those of simulation data. Thus, most of our efforts in that direction have gone into the systematic approach in VisAD, described later in this chapter, rather than into retrofitting them in Vis5D.

Vis5D is an open-source system and has been freely available since 1989. In fact, it was probably the first open-source visualization system. From the start, users sent us modifications to fix bugs or add features that they wanted. Furthermore, large institutions expressed an interest in using Vis5D, but they wanted user interfaces that fit better with the look and feel of their own systems. In 1995, in order to make Vis5D easier to modify, we split the system into two parts separated by an application programming interface (API): a user interface part that calls the API functions, and a data-management and display part that implements those functions. This allows institutions to define custom user interfaces that call the Vis5D API for data-management and display functions. The system also includes links to an interpreter for the TCL scripting language and defines TCL commands for invoking the Vis5D API functions. This enables users to write TCL scripts to run Vis5D noninteractively. These scripts are often triggered by the completion of model runs, and they often generate images and animations that are loaded onto ftp and web servers. Alternatively, Vis5D TCL scripts are sometimes created and triggered from web forms, and in this case, output images and animations are returned to the web client. Vis5D also includes a mode in which it listens to a specified pipe file for the names of TCL script files, which it then executes.

Vis5D's API and TCL scripts provide a wide variety of ways to invoke its data-management and display functions from other systems. These are being used by institutions such as NASA,

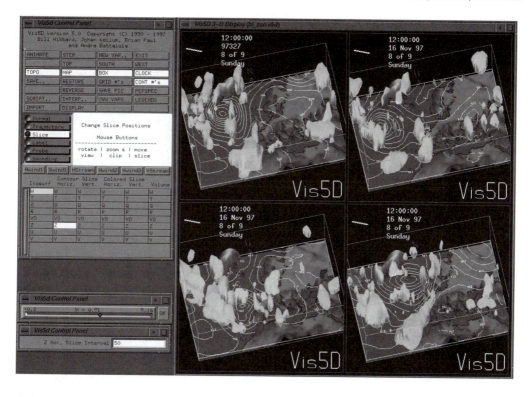

Figure 34.2   A Vis5D spreadsheet of four members of an ensemble forecast from the European Centre for Medium-Range Weather Forecasts. (See also color insert.)

the US EPA, the US Air Force, the US Navy, the US National Weather Service, the European Centre for Medium-Range Weather Forecasts, China's National Meteorological Center, and many others.

In the last few years, Vis5D development has ceased at the SSEC Visualization Project. However, it continues via the Vis5d+ project on SourceForge, the D3D project at the NOAA Forecast Systems Laboratory, the Cave5D system at Argonne National Laboratory, and many other derived systems. In the next section we describe Cave5D, an adaptation of Vis5D to immersive virtual reality.

34.3   Cave5D

Given its interactive 3D displays and its direct-manipulation user interface, Vis5D is a natural

for use in virtual reality. When Tom DeFanti and his collaborators at the University of Illinois–Chicago Electronic Visualization Lab developed their CAVE virtual reality system [1], we created Cave5D as a version of Vis5D for their CAVE. While demonstrating Cave5D at the Siggraph 94 VROOM, we met Glen Wheless and Cathy Lascara from Old Dominion University. They and we used Cave5D for demos at the Supercomputing 95 I-WAY [9,13]. Fig. 34.3 depicts a Cave5D display that combines ocean and atmosphere datasets from our I-WAY demo. During our demo, user wand clicks in the CAVE in San Diego triggered requests to an SP-2 supercomputer at Argonne National Laboratory, which sent new model data back to San Diego for display in the CAVE (during the week of the Supercomputing '95 conference, bandwidth into the San Diego Convention Center was greater than bandwidth into Manhattan). After

Figure 34.3   A planar simulation of a Cave5D virtual-reality display of coupled atmosphere and ocean simulations used for a demo at the Supercomputing '95 conference. (See also color insert.)

Supercomputing '95, we focused on VisAD, and Old Dominion took over development of Cave5D. I felt that immersive virtual reality would remain primarily a demo tool rather than a daily-use tool for many years and was happy to pass Cave5D on to Old Dominion.

User interfaces are a great challenge for virtual reality. Pop-up panels and large arrays of buttons and sliders do not feel natural in virtual reality. But Vis5D users need those large numbers of widget choices, in addition to the direct-manipulation choices (like scene rotation, dragging planes, and trajectory launch) that do feel natural in virtual reality. Thus, Cave5D has a modest array of buttons, such as those seen in Fig. 34.3, that hide many choices users would like to make. The Old Dominion folks added a script interpreter to Cave5D that enables users

to easily specify many of the Vis5D widget choices from a start-up script. They also integrated Cave5D with the Virtual Director system created by Donna Cox and Bob Paterson in order to support collaboration between multiple CAVEs with Cave5D.

I should note that Sheri Mickelson and John Taylor of Argonne National Laboratory have taken over support of Cave5D from Old Dominion. Cave5D has undoubtedly become the most widely used visualization software system for immersive virtual reality.

34.4   VisAD

Observational data is much more challenging than simulation data to visualize and analyze. To address this challenge, VisAD (the name stands for Visualization for Algorithm Development) is designed to deal with virtually any numerical or text data, to produce virtually any type of data depiction, to integrate interactive analyses with visualization, and to work with arbitrarily distributed computing resources [10]. It is written in pure Java (except for its use of the HDF libraries, which don't support Java) because Java is a language designed for a distributed environment. It is a component library that defines five basic kinds of components (names of Java classes, interfaces, and objects are in a monotype font):

1. `Data` components: These may be simple real numbers, text strings, vectors of real numbers and other values, sets in real vector spaces, functions from real vector spaces to other data spaces, or complex combinations of these. They are mostly immutable, in order to ensure thread safeness. The exception is that range values of functions (e.g., changing pixel intensities in an image) can be changed without necessitating a replacement of the entire function.

2. `Display` components: These contain visual depictions of one or more linked `Data` components. These may be 2D or 3D, may be a window on the screen or in a browser, or may be in an immersive virtual-reality system. `Display` components update data depictions in response to changes in linked `Data` components.

3. `Cell` components: These execute user-defined computations in response to changes in linked `Data` components (the name `Cell` is taken from spreadsheets).

4. `DataReference` components: These are mutable components used to connect `Display` and `Cell` components to `Data` components, which are often immutable. In "X = 3", the number 3 is immutable and plays the `Data` role, whereas X is mutable (i.e., it can be changed to a value other than 3) and plays the `DataReference` role.

5. User-interface components: These are traditional graphical user interface (GUI) widgets typically linked to `Data` or `Display` components.

`Display` and `Cell` components can be linked to `DataReference` (and hence to referenced `Data`) components on remote machines via Java Remote Method Invocation (RMI) distributed object technology. This facilitates collaborative visualization, in which a `DataReference` component on one machine is linked to `Display` components on the geographically distributed machines of multiple users. RMI is also used to enable applications to link groups of `Display` components on different machines, so interactive changes in any are reflected in all.

The core of VisAD's design is its data model, which is a mathematical description of the set of valid `Data` components. The data model grew out of our experience with a great variety of scientific data, and also out of the Siggraph 1990 Workshop on Data Structure and Access Software for Scientific Visualization, organized by Lloyd Treinish [12]. Participants included people who played leading roles in the development of AVS, IBM Data Explorer, HDF, netCDF, DOE CDM, and VisAD.

`Data` components contain numerical and text values, as well as metadata. The primary type of metadata is a data schema, in VisAD's `MathType` class, that defines names for primitive numerical and text values occurring in data, the way values are grouped into vectors, and functional dependencies among values. For example, a satellite image of Earth may be described as a functional dependence of `radiance` on pixel `line` and `element` coordinates, via the `MathType` (using the system's string representations for `MathType`s):

$$((\text{line, element}) \rightarrow \text{radiance})$$

This function is approximated by a finite sampling at discrete pixels. The sampling metadata of a function may be a regular or irregular set in a real vector space (typical image sampling is an integer lattice in the 2D space with coordinates line and element). This function may also include metadata that describes the Earth locations of pixels via an invertible coordinate transform:

$$\text{(line, element)} \leftrightarrow \text{(latitude, longitude)}$$

Any real values may include units. For example, latitude and longitude values may have units of degrees or radians. Function range values such as radiance may include metadata that indicates missing values (caused by instrument or computational failures), or metadata that defines estimates of errors.

A time sequence of images may have the following MathType:

$$\text{(time} \rightarrow \text{((line, element)} \rightarrow \text{radiance))}$$

This function will define some finite sampling of time values, and it may define units for time such as seconds since 1 January 1970.

A set of map boundaries may also be described using the MathType:

$$\text{set (latitude, longitude)}$$

VisAD defines various classes for sets in real vector spaces, for regular and irregular topologies, for different domain dimensions, and for sets restricted to sub-manifolds with dimensions smaller than their domains. For example, a set of map boundaries lies in a 1D sub-manifold of a 2D domain.

A more formal definition of the VisAD data model is provided by the MathType grammar below:

Any RealType may have associated Unit and ErrorEstimate objects and may be marked as missing. Any RealTupleType may have an associated CoordinateSystem object, defining an invertible transform to a reference RealTupleType. Any RealTupleType occurring in a FunctionType domain may have an associated finite sampling defined by a Set object. Unit conversions, coordinate transforms, and resampling are done implicitly as needed during computation and visualization operations on Data components, and missing data and error estimates are propagated in computations. This data model has proven robust for dealing with a wide variety of application requirements.

VisAD defines an architecture for interfacing its data model with various file formats and data-server APIs. This architecture presents applications with a view of a file (logical file, in the case of a server API) as a VisAD Data component. The architecture has been implemented for a number of commonly used scientific file formats and server APIs, including netCDF, HDF, FITS, BioRad, McIDAS, Vis5D, OpenDAP, and others. In some cases, entire files are read and used to create memory resident Data components. However, the architecture also includes support for format interfaces that transfer file data between disk and a memory cache as needed. Data transfers to the cache are implicit in application access to methods of the created Data component, and hence are transparent to applications.

The depictions of Data components linked to a Display component are defined by a set of

```
MathType        := ScalarType | TupleType | SetType | FunctionType
ScalarType      := RealType | TextType
RealType        := name
TextType        := name
TupleType       := (MathType, MathType, . . . , MathType)
TupleType       := RealTupleType
RealTupleType   := (RealType, RealType, . . . , RealType)
SetType         := set (RealTupleType)
FunctionType    := (RealTupleType→MathType)
```

ScalarMap objects linked to the Display. These are mappings from RealTypes and TextTypes to what are called DisplayReal-Types. For example, the depiction of a time sequence of images and a map boundary overlay in Fig. 34.4 is determined by the Scalar-Maps.

```
time      → Animation
latitude  → YAxis
longitude → XAxis
radiance  → RGB
```

Note the GUI widgets in Fig. 34.4 that allow the user to control time animation and the RGB color lookup table for radiance values. Each ScalarMap object has an associated Control object that provides a means to specify animation, color tables, contouring, flow rendering, 3 D to 2 D projection, and other parameters of its associated DisplayRealType. These Control objects can be linked to GUI widgets, as in Fig. 34.4, or can be manipulated by computations.

The system's intrinsic DisplayRealTypes include the following: XAxis, YAxis, ZAxis, Latitude, Longitude, Radius, CylRadius, CylAzimuth, CylZAxis, XAxisOff-set, YAxisOffset, ZAxisOffset (offset values are added to spatial coordinates), Red, Green, Blue, RGB (pseudo-color), RGBA (pseudo-color with transparency), Hue, Saturation, Value, HSV (pseudo-color to hue, saturation, and value), Cyan, Magenta, Yellow, CMY (pseudo-color to cyan, magenta, and yellow), Alpha, Flow1X, Flow1Y, Flow1Z, Flow2X, Flow2Y, Flow2Z (note two sets of flow coordinates), Flow1Elevation, Flow1Azimuth, Flow1Radial, Flow2Elevation, Flow2Azimuth, Flow2Radial (note also two sets of spherical flow coordinates), Animation, SelectValue (values not equal to a specified value are treated as missing in the depiction), SelectRange (values not in a specified range are treated as missing in the depiction), IsoContour, Text, Shape (values are sampled and used as indices in an array of icons), ShapeScale, LineWidth, PointSize, and LineStyle. System implementations for these DisplayRealTypes include just about every visualization technique.

ScalarMaps for some DisplayRealType (e.g., XAxis, RGB) allow applications to control the linear mapping from primitive numerical data values to DisplayRealType values

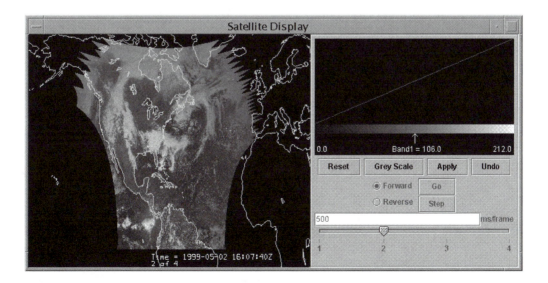

Figure 34.4 A VisAD display combining a GOES satellite image with map boundaries. (See also color insert.)

(e.g., a graphic coordinate in the case of XAxis, and a lookup table index in the case of RGB). If applications don't specify this mapping, then a system auto-scaling algorithm determines an optimal default mapping to keep data depictions visible (e.g., to ensure that longitude and latitude values are mapped to XAxis and YAxis values that are within the display screen).

ConstantMaps, which bind constant values to DisplayRealTypes, may be linked to Display components in just the way that ScalarMaps are. These allow applications to override default values for DisplayReal-Types, for example, to control locations and colors of data depictions when they are not determined by ScalarMaps of any Real-Types or TextTypes occurring in the data.

The generation of data depictions is automated based on an analysis of MathTypes, ScalarMaps, and other metadata. However, the system provides a way for applications to redefine that analysis and display-generation. When a Data component is linked to a Display component, an object of a sub-class of DataRenderer is used to analyze Math-Types, ScalarMaps, and other metadata, and then generate the depiction. There is a default sub-class for each supported graphics API (e.g., Java3D, Java2D), and these defaults can generate a visual depiction for just about any Data object and set of ScalarMaps. But applications have the option of defining and using nondefault subclasses of DataRenderer, and they may also add new DisplayRealTypes to describe parameters of those DataRenderer classes.

One important property of VisAD is that some of its DataRenderer sub-classes not only transform data into depictions but also invert the transform to translate user gestures on the depiction back into data changes. The default DataRenderer sub-classes do not translate user gestures into data changes, because in the general case of MathTypes and ScalarMaps there is no reasonable way to interpret user gestures as data changes. How-

ever, the VisAD system includes a number of nondefault DataRenderer sub-classes that do translate user gestures into data changes, and applications are free to define more. These first analyze a MathType and a set of ScalarMaps to make sure they are consistent with an interpretation of user gestures as data changes, and then they implement that interpretation. For example, a Data component with MathType

(latitude, longitude, altitude)

linked to a Display component with linked ScalarMaps

    latitude → YAxis
    longitude → XAxis
    altitude → ZAxis

will generate a data depiction as a simple point in 3D display space and allow the user to modify data values by dragging the point. An analysis by an object of the DirectManipu-lationRendererJ3D class verifies and implements this way of interpreting gestures.

Fig. 34.5 shows a display of a simple conical terrain surface in a 3D box with two large yellow points at opposite corners of the box. The yellow points are depictions of two 3-vector Data components linked to the Display component via objects of class DirectManipula-tionRendererJ3D, as described in the previous paragraph. These 3-vectors are linked to trigger the computation of a Cell compon-ent that modifies the linear mappings associated with the ScalarMaps to display spatial coord-inates in order to keep the depictions of the two 3-vectors at the corners of the display box. This little network of Data, Display, and Cell components and DirectManipulationRen-dererJ3D objects defines an embedded 3D GUI component for rescaling 3D display space.

As another example, in spatial data analysis applications it is often useful to apply analysis operations to restricted spatial regions. These regions may be defined in the data (e.g., within a map boundary), or they may be defined by users based on their judgement. For user defin-ition, we need a GUI component that enables

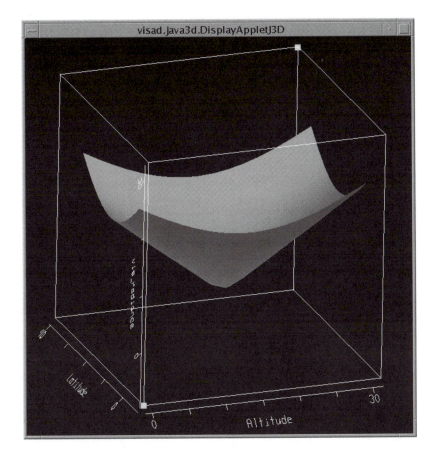

Figure 34.5 A simple demonstration using VisAD to build embedded 3D user interface components. Dragging the two large yellow points causes the display to rescale to keep the yellow points at opposite box corners. (See also color insert.)

users to draw the outlines of regions as freehand curves. In VisAD, the `CurveManipulation-RendererJ3D` class serves this purpose. It is a sub-class of `DataRenderer` that requires a `Data` component with `MathType` of the form

```
Set (x, y)
```

It also requires `ScalarMaps` of x and y to spatial `DisplayRealTypes`. These may be two of `XAxis`, `YAxis`, and `ZAxis`, or they may be two coordinates in a non-Cartesian spatial coordinate system. The `Data` component will lie on a 1D manifold embedded in the 2D domain with coordinates x and y. User

mouse movements are interpreted as samples along 1D curves in (x, y) space. According to the `ScalarMaps`, the curve is embedded on a 2D manifold in 3D display space. Fig. 34.6 is a snapshot of a curve being drawn on the 2D manifold on the surface of a sphere. In this case, the `ScalarMaps` are

```
x → Longitude
y → Latitude
```

Along with `Radius`, these `DisplayReal-Types` define a 3D spherical display coordinate system:

```
(Latitude, Longitude, Radius)
```

When a `DataReference` object is linked to a `Display` component, a number of `Constant-Maps` may be included that are applied only to the depiction of the referenced `Data` component. In the example in Fig. 34.6, a `Constant-Map` to `Radius` is used to specify which sphere defines the 2D manifold where curves are drawn.

Applications can use `CurveManipulationRendererJ3D` to draw on a nearly arbitrary 2D sub-manifold of 3D display space, by defining three `DisplayRealTypes` for a new coordinate system and defining a coordinate transform between these and Cartesian display coordinates (`XAxis`, `YAxis`, `ZAxis`). The 2D sub-manifold is defined by a `ConstantMap` that fixes the value of one of these `DisplayRealTypes`, and by `ScalarMaps` of x and y to the other two. This 3D GUI component can be used for freehand drawing in a wide variety of applications.

These examples illustrate the robust way that networks of VisAD components can be adapted to application requirements. Furthermore, many VisAD classes are designed to be extended so that they can meet almost any visualization requirements. These include the following:

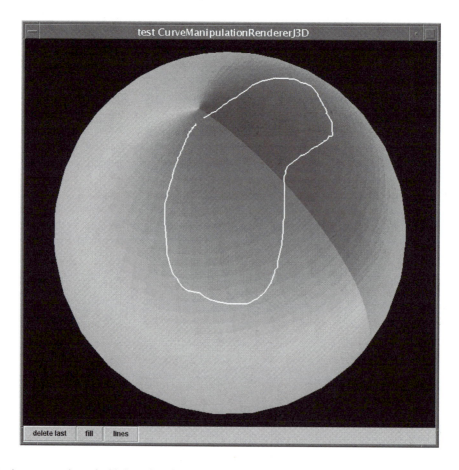

Figure 34.6 A more complex embedded user interface component built using VisAD. The user can draw freehand curves on the sphere's surface. (See also color insert.)

1. The `Set` class can be extended to define new sampling topologies of real vector spaces and to define new interpolation algorithms.

2. The `CoordinateSystem` class can be extended to define new coordinate-transformation algorithms.

3. New implementation classes for the `Function` interface (this is a sub-interface of `Data` corresponding to the `FunctionType` sub-class of `MathType`) can provide nonsampled approximations of functional dependencies, such as procedural definitions or harmonic sequences. Note that sampled approximations are classes that implement the `Field` interface (a sub-interface of `Function`).

4. New implementation classes for the `Form` interface define interfaces of the VisAD data model to new file formats and data-server APIs.

5. New implementation classes for the `Cell` component interface can be used to include application-defined computations in a network of VisAD components.

6. The `DataRenderer` class can be extended to customize the way that visual depictions are generated from `Data` components, including the interpretation of user gestures as data changes.

7. Various `Display` component classes can be extended to include implementations for new graphics APIs (the system includes implementations for the Java3D and Java2D APIs), or to adapt existing graphic APIs to new modes, such as immersive virtual reality (this has been done for the ImmersaDesk using the Java3D API).

In a recent development, a set of extensions of classes for `Data` and `Display` components have been defined to support visualization of large datasets distributed across the processors of a cluster. A large `Data` component is distributed by a partition of the samples of a `Field` across cluster processors. The partitioned `Field` does not have to be the top level of the `Data` component; it may occur as part of a containing `Data` organization. For example, large 3D grids from a weather model may be partitioned into sets of rectangles in latitude and longitude, with these 3D grids occurring as parts of larger time sequence `Data` components. This example corresponds to the way that weather models are usually partitioned across clusters, so defining `Data` components this way allows weather model output data to be visualized in the place where it is computed. A `Data` component on the user's visualization client connects via Java RMI to the `Data` components on the cluster processors, which hold the actual data values (which may be stored on disk via the caching architecture of a file format interface), and a `DataRenderer` object (its actual class is, of course, a sub-class of `DataRenderer`) on the user's visualization client connects via Java RMI to a set of `DataRenderer` objects on the cluster processors. Then, visualization computations, such as isosurface generation, are initiated from the visualization client, but the actual computations are distributed across the processor nodes. In order to prevent the generated geometries from swamping the memory of the visualization client, the client may request different rendering resolutions from different cluster processors. This enables users to visualize the overall dataset at low resolution and also to zoom into the data on any processor at full resolution. Because this cluster implementation is just another set of VisAD class extensions, it has the full generality of the VisAD data model and of VisAD's `ScalarMaps` for display definition. Furthermore, this approach makes it trivial to overlay depictions of large cluster data with depictions of data from other sources, such as geographical reference maps.

The VisAD library includes an interface to the Jython interpreter for Python, as well as Java method implementations that make VisAD data operations accessible via infix syntax from Python expressions, and a variety of Python functions that implement specialized graphics (e.g., histograms, scatter plots, image animation, etc.) via VisAD's components. The goal of this ongoing effort is to provide an easy way to use VisAD from Python scripts so that scientists and casual users do not have to learn Java.

The VisAD library is used by a number of popular visualization applications. One, the VisAD SpreadSheet, is distributed as part of the system. It is a general-purpose application that enables users to read files and perform simple computation and visualization operations. It includes a user interface for creating `ScalarMaps` and setting parameters in associated `Controls` (this user interface is also accessible from Python). The VisAD SpreadSheet is fully collaborative: users at different workstations can share the same displays and user interfaces, so interactions by one are shared by all.

VisAD supports more specialized applications for earth science, biology, astronomy,

and economics. The Unidata Program Center's Integrated Data Viewer (IDV) enables users to fuse environmental data from different servers and from different types of data sources (satellites, simulation models, radars, in situ observations, etc.) in displays with common geographical and temporal frames of reference. The IDV is being used for the National Science Foundation (NSF)-supported Digital Library for Earth Science Education (DLESE) as well as for a prototype environmental modeling project at the National Computational Science Alliance (NCSA). The Australian Bureau of Meterology is using VisAD as the basis for their Tropical Cyclone and Automated Marine Forecast systems and considering using it as the basis for all of its visualization applications. Ugo Taddei of the University of Jena is using VisAD to develop a number of hydrology applications. The National Center for Atmospheric Research

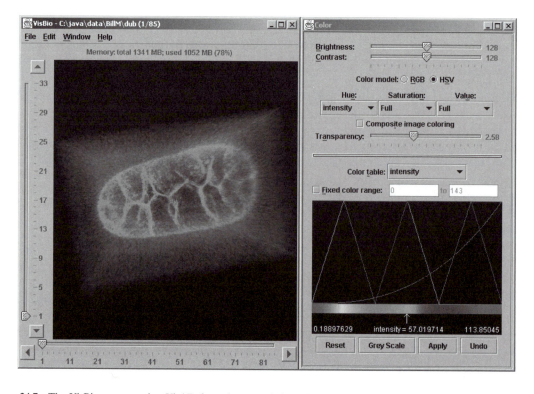

Figure 34.7 The VisBio system, using VisAD for volume rendering of a live *C. elegans* embryo stained with a fluorescent membrane probe. Imaging done by Dr. William Mohler of the University of Connecticut Health Center. (See also color insert.)

(NCAR) is using VisAD as the basis of its Visual Meterology Tool (VMET). The University of Wisconsin (UW) is developing a variety of applications for visualizing and analyzing data from hyperspectral (typically generating several thousand spectral bands) atmospheric observation instruments.

VisBio is a biological application of VisAD being developed by Curtis Rueden and Kevin Eliceiri under the direction of John White in his UW Laboratory for Optical and Computational Instrumentation. Fig. 34.7 shows a VisBio volume rendering of a live *C. elegans* embryo stained with a fluorescent membrane probe, and Fig. 34.8 shows 3D and 4D measurements with VisBio quantitating movement of mitochondria in different regions of a two-photon dataset of a two-cell hamster embryo labeled with a mitochondria-specific dye. The NuView system uses VisAD to enable

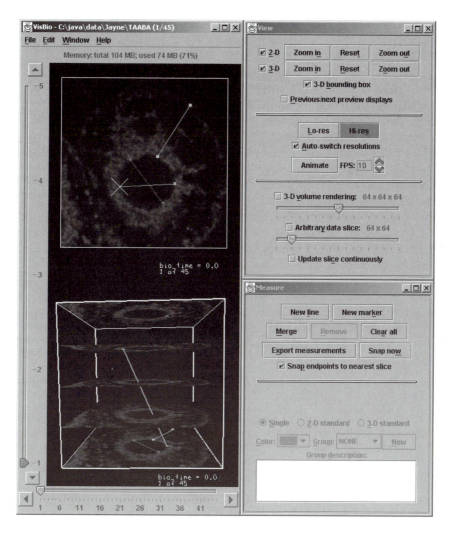

Figure 34.8   The VisBio system using VisAD for 3D and 4D measurements quantitating movement of mitochondria in different regions of a two-photon dataset of a two-cell hamster embryo labeled with a mitochondria-specific dye. Imaging done by Dr. Jayne Squirrell of the University of Wisconsin at Madison. (See also color insert.)

astronomers to visualize events from the AMANDA and IceCube neutrino detectors. There is a close collaboration among all the groups developing scientific applications of VisAD.

Acknowledgments

Marie-Francoise Voidrot, Andre Battaiola, and Dave Santek helped with the early development of Vis5D. Brian Paul did most of the Vis5D development starting in 1992, including development of Cave5D. Brian developed Mesa on his own while working in the SSEC Visualization Project, as an improvement over an earlier freeware implementation of OpenGL used by Vis5D. Johan Kellum took over most of the Vis5D development starting in 1997. Phil McDonald contributed changes to Vis5D that he made for the D3D system.

Charles Dyer supervised my University of Wisconsin Computer Science Ph.D. about the VisAD system [8]. Brian Paul helped with the first implementation of VisAD, in C [6,7]. John Anderson and Dave Fulker contributed major ideas to the Java implementation of VisAD. Curtis Rueden, Steve Emmerson, Tom Rink, Dave Glowacki, Tom Whittaker, Don Murray, James Kelly, Andrew Donaldson, Jeff McWhirter, Peter Cao, Tommy Jasmin, Nick Rasmussen, and Doug Lindholm helped write the Java implementation of VisAD. Ugo Taddei created the online VisAD tutorial.

References

1. C. Cruz-Neira, D. J. Sandin, and T. A. DeFanti. Surround-screen projection-based virtual reality: the design and implementation of the CAVE. *Proceedings of SIGGRAPH '93*, pages 135–142, 1993.

2. W. Hibbard. 4-D display of meteorological data. *Proceedings, 1986 Workshop on Interactive 3D Graphics*. Chapel Hill, pages 23–36, 1986.

3. W. Hibbard and D. Santek. Visualizing large data sets in the earth sciences. *IEEE Computer* 22(8):53–57, 1989.

4. W. Hibbard and D. Santek. Interactivity is the key. Chapel Hill Workshop on Volume Visualization, University of North Carolina, Chapel Hill, pages 39–43, 1989.

5. W. Hibbard and D. Santek. The Vis5D system for easy interactive visualization. *Proc. Visualization '90*, pages 28–35, 1990.

6. W. Hibbard, C. Dyer, and B. Paul. Display of scientific data structures for algorithm visualization. *Proc. Visualization '92*, pages 139–146, 1992.

7. W. Hibbard, C. Dyer, and B. Paul. A lattice model for data display. *Proc. Visualization '94*, pages 310–317, 1994.

8. W. Hibbard. Visualizing scientific computations: a system based on lattice-structured data and display models. PhD Thesis. Univ. of Wisc. Comp. Sci. Dept. Tech. Report #1226, 1995.

9. W. Hibbard, J. Anderson, I. Foster, B. Paul, R. Jacob, C. Schafer, and M. Tyree. Exploring coupled atmosphere-ocean models using Vis5D. *Int. J. of Supercomputer Applications*, 10(2):211–222, 1996.

10. W. Hibbard. VisAD: connecting people to computations and people to people. *Computer Graphics*, 32(3):10–12, 1998.

11. W. Hibbard. Confessions of a visualization skeptic. *Computer Graphics* 34(3):11–13, 2000.

12. L. A. Treinish. SIGGRAPH '90 workshop report: data structure and access software for scientific visualization. *Computer Graphics* 25(2):104–118, 1991.

13. G. H. Wheless, C. M. Lascara, A. Valle-Levinson, D. P. Brutzman, W. Sherman, W. L. Hibbard, and B. E. Paul. Virtual Chesapeake Bay: interacting with a coupled physical/biological model. *IEEE Computer Graphics and Applications*, 16(4):52–57, 1996.

35   Visualization with AVS

W. T. HEWITT, NIGEL W. JOHN, MATTHEW D. COOPER,
K. YIEN KWOK, GEORGE W. LEAVER, JOANNA M. LENG, PAUL G. LEVER,
MARY J. MCDERBY, JAMES S. PERRIN, MARK RIDING, I. ARI SADARJOEN,
TOBIAS M. SCHIEBECK, and COLIN C. VENTERS
Manchester Visualization Centre

35.1   Introduction

Manchester Visualization Center [1] (MVC) has been using the AVS family of software (AVS5 and AVS/Express) in research and development for more than a decade. This chapter aims to demonstrate the range of scientific visualization projects undertaken at MVC, from archaeological to flow-visualization to medical visualization. One of the case studies is a complete turnkey molecular visualization system.

It is also intended to demonstrate the flexibility of AVS products as a modular and extensible visualization system. None of these applications would be possible without the large repository of user-submitted modules available from the International AVS Centre [2], a supported open-source website.

35.2   Meterological Visualization: Global and Regional Climate Modeling

This work is part of an ongoing project to visualize numerous large-scale datasets for the Hadley Centre global climate group [3,4]. It is purely a visualization, animation, and presentation task to demonstrate the effects of global warming and the results of their climate models and simulations. The final animations were to be shown at the Conference of the Parties to the UN Framework Convention on Climate Change (CoP) [5,6,7,8] and distributed to various international news organizations.

35.2.1   Aims

MVC was approached to do this work based on its significant use of the AVS/Express environment and experience with visualization techniques. It was expected that MVC would be able to employ the most appropriate and latest visualization techniques to convey the message that global pollution needs to be significantly reduced.

The Hadley Centre required animated visualizations of their global climate simulations that would show dramatic and undeniable proof that global warming and its consequences were accelerating beyond control. This was evident in the data and the numerous 2D graphs and charts in the Hadley Centre's many publications, alongside of which were detailed descriptions of the processes and conclusions. Although useful, such publications lack impact and are accessible only to a small audience. To reach a wider audience and demonstrate the urgency of the situation, short and snappy animations that easily and readily showed what was happening were needed.

35.2.1.1   Background and Data

The data supplied was the result of both data collection from meterological stations around the world and subsequent simulations using the Hadley Centre Climate Models [9] (HadCM2 and HadCM3). MVC was not involved with the generation of the data. The data was delivered in the form of ASCII-based files with formatting information.

The Intergovernmental Panel on Climate Change (IPCC) [10] has published projections of future emissions in their Special Report on Emissions Scenarios (SRES). The basis for these scenarios is a number of "storylines" that describe the way in which the world will develop over the coming century. Assumptions are made about the future, including greater prosperity and increased technology. The levels of greenhouse gas emissions are generally lower than in previous IPCC scenarios, especially towards the end of the 21st century. The emissions of sulphur dioxide, which produce sulphate aerosols that have a cooling effect on climate, are substantially lower.

The SRES scenarios are based on recent projections of global population and span a range of potential economic futures. There are four families of scenarios: A1FI, A2, B1, and B2. Simplified, these represent a set of possible economic, technological, population, and globalization growth and decline predictions for the twenty-first century.

Between the present day and the end of the twenty-first century, the Hadley Centre predicts an average warming of over 4°C for the A1FI scenario, 3.5°C for the A2 scenario, and 2°C for the B2 scenario. From additional calculations, warming of just under 2°C is predicted for the B1 scenario. In some regions the warming peaked at a rise of 12°C.

35.2.2   Application Development

At first the modular system of AVS was used to import and preprocess data for use within AVS/Express. However, repeated reading of the ASCII files and regridding of the data proved to be computationally expensive and resulted in slow turnaround for development and production of the visualization animations. This was especially problematic for 3D oceanographic flow data with nonaligned vector components that could not be handled by the visualization modules within AVS/Express. However, given the modular nature of AVS/Express, it was easy to take the developed code within the modules and create independent

filter programs for offline execution. The ASCII files were processed offline to generate binary files that could be loaded quickly into AVS/Express, where data was correctly regridded, scaled, and generated. For example, to speed the production of animations, the scalar magnitude of the vectors was also generated offline.

Even though some of the processing pipeline was removed from the AVS/Express network application, there was still a significant workload remaining, which the modular system facilitated greatly. Filter and Mapper modules were still required to generate visualizations, control parameters, drive the animation loops, and output the animation frames.

35.2.2.1   Visualization
Each year the Hadley Centre has focused on specific themes, including the following:

- The global average temperature rise.
- Effects on sea-level rise due to gradual warming of the deep ocean.
- Severity of El Niño and subsequent climate changes.
- Breakdown of the Gulf Stream due to both deep ocean warming and desalination from the melting of the ice caps, and its effect on the UK.
- Reduction in the size of the polar ice caps.
- Impact on regional climate, e.g., cyclones in the Mozambique Channel and the Bay of Bengal and subsequent flooding.
- Decimation of South American rainforests and vegetation.

35.2.2.2   Output
In the early stages of the project, modules were used to generate labels, titles, counters, logos, and legends that were placed within the display output. These were later replaced, for two reasons:

1. Including multiple labels and counters within the display slowed the production rate.

2. As the quality of the output was increased, the quality of the aliased text labels was deemed too low for broadcast.

Additionally, we opted to generate visualization frames at twice the resolution necessary, and independently from the labels, etc. Labels and titles were then produced using software such as CorelDraw and Photoshop, and animated counters were generated using small AVS/Express networks, again at twice the required resolution. Further offline stages consisted of compositing the raw visualization images with the labels and titles. This process involved anti-aliasing of the raw frames and compositing using ImageMagick scripts. Finally, the raw frames were loaded into Adobe Premiere and assembled into both AVI (uncompressed broadcast quality) and MPEG-4 files. The AVI files were transferred to near-broadcast-quality DV tapes and also onto both Betacam-SP and VHS for further distribution.

35.2.3 Results

Fig. 35.2.1 shows changes in global temperature for four SRES emissions scenarios, known as A1Fi, A2, B1, and B2. The A1Fi scenario in Fig. 35.2.1a shows the simulation starting state with recorded data. Figs. 35.2.1b, 35.2.1c, and 35.2.1d show the unmitigated emissions of pollutants of A1Fi against the other scenarios, the result of which can be seen clearly in the year 2100, when some areas have risen by more

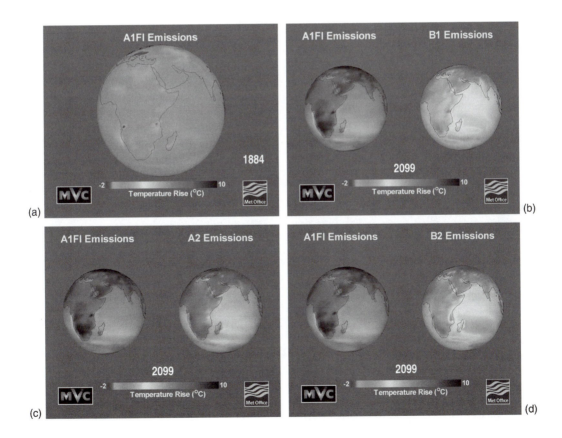

Figure 35.2.1 Comparison of SRES scenarios for global warming. Data courtesy of the Hadley Centre, UK Meteorological Office. (See also color insert.)

692 Visualization Software and Frameworks

than 12°C. The B1 scenario shows a lesser effect, where emissions have been reduced significantly.

The results for the scenarios show that surface warming is expected over most of the globe, with the largest increase at high northern latitudes. The melting sea ice causes less sunlight to be reflected and more to be absorbed at the surface, leading to large warming of the region. The patterns of temperature rise also show a sharp contrast between land and sea, where the land is warming approximately 80% faster.

Fig. 35.2.2 shows a comparison of two models, the Global Climate Model (GCM) and the Regional Climate Model (RCM). The GCM is of much smaller resolution than the RCM. In this visualization of a cyclone in the Mozambique Channel, it is clear that the GCM does not effectively simulate the cyclone. Fig. 35.2.2 shows the wind magnitude and direction for both models for a number of time-steps.

The higher-resolution model (RCM) clearly produces more effective results and will enable the Met Office to conduct more precise predictions in the future, when higher-resolution simulations can be used for global rather than regional climate prediction.

35.2.4 Conclusion

The animations produced have proven to be effective tools by which the Hadley Centre can dramatically demonstrate what the future may hold should the governments of this world not take heed of the warnings already evident. The animations have been well received at conferences and have led to the UK's adopting the proposal to reduce emissions levels to the lowest recommended.

Acknowledgments

The Hadley Centre, the U.K. Meteorological Office, Dr Geoff Jenkins, and Dr. Jason Lowe.

35.3 Earth Science Visualization: What Is Inside Stars and Planets

This section looks at the visualization of data that lies in or on a sphere. People who study the Earth, stars, or other planets use data gained from the surface to try to understand the processes that are occurring deep within. They can do this by a few methods:

- Using optical equipment to record and analyze surface features.

- Using spectrometry to analyze the elements contained in the body.

- Using seismic tomography/helioseismology, which inverts wave functions that pass through the body to gain internal images (the process is similar to medical scanning).

This type of research, whether in the field of earth science or in that of astrophysics, is carried out by computational scientists. These are scientists who model physical processes and encapsulate the basic physics of the system in a simplified mathematical model. The model is turned into a computer program. Typically, high-resolution models are computationally intensive and need large, high-end machines to run in a reasonable time scale.

Any computer model is a theoretical idealization and thus needs to be validated. Proof is made by comparison of real observational data and simulation data.

35.3.1 Problem

Data from the earth sciences and astrophysics has several common features:

- Observational data is sparse in places.

- Data lies in irregularly shaped cells.

- Data lies in a thick spherical shell that is located close to the center of the sphere, for example, the Earth's mantle.

- The features are self obscuring.

- The strength of a value is proportional to the radius.

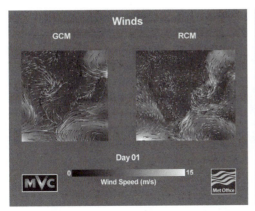

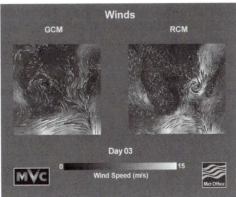

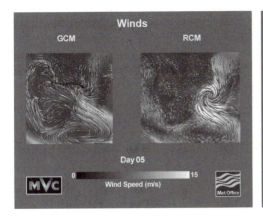

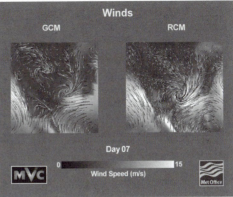

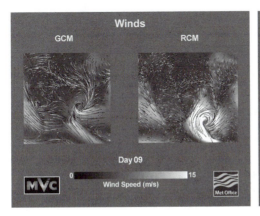

Figure 35.2.2   A series of frames from an animation comparing the Global Climate Model (GCM) and the Regional Climate Model (RCM) for simulation of a cyclone in the Mozambique Channel. Data courtesy of the Hadley Centre, UK Meterological Office. (See also color insert.)

- Points equally placed in latitude and longitude converge as the radial coordinate decreases from the surface to the center.

Here we look at two case studies that use visualization to validate and support computational models.

35.3.2   Seismic Tomography

Seismic tomography is the mathematical process that converts information in shock waves that pass through the Earth into a model of the materials within. Shock waves tend to travel through the Earth quickly where the internal rock is relatively hot and slowly where the rock is relatively cold. This gives information about tectonic plate movement and subduction; surface rock that is colder than its surroundings is the result of a tectonic plate being pushed down into relatively hot mantle below. Huw Davies, an earth scientist at the University of Cardiff, has proposed a model of mantle flow [12,13]. Initially he was using 2D visual analyses, where each spherical shell of data was projected into 2D and viewed as a series of images. He wished, however, to view all the data together as one 3D image (Fig. 35.3.1).

35.3.2.1   Implementation

The most important element of this work was the data reader. The data had to be read into AVS/Express in a way that allowed a number of analysis tools to be developed. The data reader was a C module that translated the data cells of the model into AVS/Express cells, while the analysis tools were developed as macros of AVS/Express components with individual GUIs. Example applications were produced and added to the libraries.

35.3.2.2   The Data Structure

A C structure defined the extents of the data cells in spherical coordinates. The data cells are placed in rings (latitudinal bands). The volumes of all cells are equal, so the number of cells in each band increases from pole to equator. The cells implicitly have a curved edge along each nonradial perimeter.

Systems like AVS/Express use Cartesian coordinates, not spherical coordinates, and all cells have straight edges. Each shell projected well into a 2D "image," but in 3D the cells did not tessellate. Gaps appeared between the latitudinal bands. This was distracting to the eye and caused holes to appear in underlying iso-

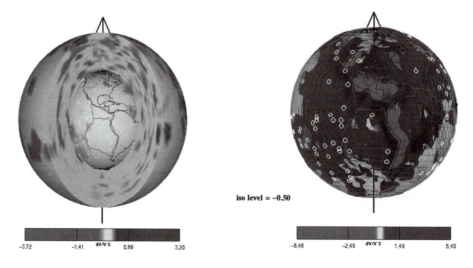

Figure 35.3.1   3D images of seismic tomography data of the Earth's mantle. (See also color insert.)

surfaces. The data needed to be resampled, and this process added a considerable number of cells.

35.3.2.3   *3D Display and Interaction*

A thick spherical shell that lies close to the core needs to be displayed differently from a thin shell that is far from the core (such as atmospheric data).

The bounds of the data show its limits, but a 3D shell has bounds very different from those of a 3D rectangular block. The outer bound is a sphere that surrounds the whole of the data, while the inner bound, another sphere, surrounds only the core. The data was organized so that the inner and outer shells of data could be separated from the main volume of data, allowing an isosurface to be displayed within the semitransparent outer shell and over the opaque inner shell. This data organization also allowed for the removal of internal data and when displayed as a volume made applications interactive on a Silicon Graphics 02.

Because of core heating, the data values are levels of magnitude higher near the core than at the crust. Features of interest are data that vary from the average. The 2D shells have an average different from that of the 3D whole, so a new normalization function was needed.

The topographic data of the Earth's mantle is strongly related to reference information, for instance, volcanic "hot spots," tectonic plate boundaries, and coastlines. It is vital that this data always be visible and unambiguously displayed. These features were displayed on either the outermost or the innermost shell. It was difficult to locate isosurfaces between these shells in monographic projection, but in stereographic projection, location of features became much easier.

35.3.2.4   *Further Work*

Simulation data needs to be of much higher resolution if plumes are to be detected around the core. Special techniques are needed to handle the large data. A preprocessing feature-extraction step could be used to help validate a model, and displaying selections of cells could aid debugging.

35.3.3   *Solar Physics*

The Sun has been studied throughout history. Sunspots are visible from the Earth and indicate the physical activity that occurs within the Sun. For 150 years, sunspot observations have been recorded by governmental observatories. Before this time, observations were not consistent, nor was a location recorded. Many revealing plots have been made of this data [15].

Sunspot activity is cyclical, with one obvious period of approximately 11 years. A cycle consists of a wave of sunspot activity moving from high latitudes to the equator. These cycles are thought to be an artifact of internal gyroscopic behavior. It is hypothesized that stochastic noise interacts with the Sun's large internal magnetic fields to cause tubes of magnetic flux to escape to the Sun's surface and cause sunspots. To validate the gyroscopic model, new visualization tools have been designed to do the following [11]:

- Analyze sunspot observations.
- Analyze computational-model data and add stochastic noise to produce hypothetical sunspots.
- Compare real data with theoretical model data.

Each sunspot observation is a recorded event. Separately, they are relatively unimportant, but how the behavior of the sunspots changes over the course of one cycle is quite important. This change defines a cycle of activity. Historically, the observations have been analyzed statistically to show significant trends both within a cycle and between cycles.

Scientists have defined several classes of sunspot useful for particular analyses:

- Single observations.

- One sunspot defined by a series of observations—has speed.
- Long lasting, high energy, multiple rotating sunspots—has speed.
- Collapsed sequences (location averages or newborns), where a series of observations is reduced to one observation.
- Groups of sunspots occur when a large sunspot splits into several; they may have the same location, but other parameters vary.

35.3.3.1 Implementation

Sunspot data is historically not stored in a date-ordered list [15]. The data is usually reordered and additional information (speed and a unique identifier) added in a preprocessing step. To do this, we defined an ideal observation as a C++ class.

Using V, AVS/Express's own internal language, a number of objects were created to handle sunspot data, for example, an ideal sunspot observation, an array of ideal observations, and a cycle of observations. Internally, AVS/Express generated C++ classes for each object and facilitated direct manipulation by C++ modules. Several analysis modules were written,

and these produced AVS/Express fields that are rendered in the viewer.

35.3.3.2 A Virtual Sun

Recently, Tuominen et al. [11,14] have used statistical analysis to show that data observed at either limb of the Sun (more than 60° from the viewer) is more likely than other data to be inaccurate. If the size, shape, and location of the spots are not accurately observed, then the statistical analysis of the observations could be biased. A virtual Sun with sunspots (Fig. 35.3.2) was produced to help the researchers understand the effects of observational bias or error.

35.3.3.3 Statistical Plots and Historic Analysis

Observations are traditionally converted into collapsed sequences and analyzed statistically to show significant trends; the classic example of this is the butterfly plot.

The butterfly plot is a scattergram of collapsed sequences in a 2D plot of time against latitude where cycles of activity look like the wings of a

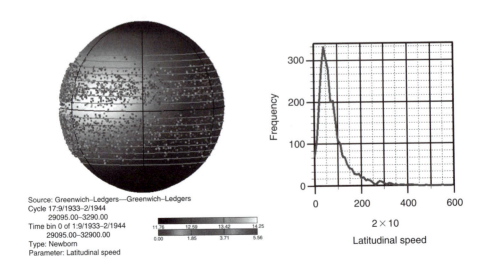

Figure 35.3.2 The location bias for newborn sunspot observations. (See also color insert.)

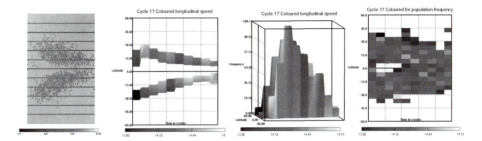

Figure 35.3.3   Plots based on the traditional butterfly plot. The left-most image is a 3D plot that is viewed as a traditional butterfly plot. (See also color insert.)

butterfly. 3D point plots of time against latitude against longitude were produced (Fig. 35.3.3). From the appropriate projection, these look identical to butterfly plots; however, color and glyphs are used to show other parameters, such as speed. Butterfly plots lose information because points are superimposed, but binning in location and time allows plots to be produced that show location, time, and population size more clearly.

35.3.3.4   Further Work

Adding computational steering will allow quicker validation of a model and give scientists a better understanding of their parameter space. Soon the first full cycle of activity will have been recorded by helioseismology. Visualization of this dataset and comparison to historic data would be valuable.

Acknowledgments

Huw Davies at the University of Cardiff, for his permission to use his tomographic data; John Brooke et al., for the data and collaborative help in producing a solar physics visualization toolkit.

35.4   Flow Visualization: Extracting and Visualizing Vortex Features

Vortices are important phenomena in computational fluid dynamics (CFD) and flow visualization, both in science and engineering practice,

and from a theoretical as well as from a practical viewpoint.

The problem with vortices is that they are easy to detect visually but difficult to extract algorithmically. Traditional methods based on physical quantities such as pressure, vorticity, or derived quantities have the problem that they are based on local physical properties describing an infinitesimal region, while vortices often span larger regions. In addition, it is often problematic to compare vortices in different datasets, e.g., in subsequent time-steps of the same simulation, because there is no explicit description.

There are two solutions to these problems:

1. To use a *geometric* method to detect the vortices, based on geometric properties of streamlines.

2. To treat vortices as *features.*

A feature is any object, pattern, or structure in data that is of interest and subject to investigation [16]. The advantage of this approach is that it enables quantification of (vortex) features using numerical parameters. This opens up many interesting possibilities, such as accurate comparison between different datasets e.g., simulations of the same model but with different boundary conditions, or different time-steps of the same simulation (an implementation known as time tracking) [18].

Here, an algorithm is presented for extracting and visualizing vortex features in CFD datasets. This algorithm was implemented in AVS5.

35.4.1   Algorithm

The method used here for detecting vortices is the so-called *winding angle* method [17]. This method works by calculating streamlines and selecting those that contain loops. These streamlines are then clustered in order to identify individual vortex features. These can be quantified by calculating numerical (statistical) attributes. Finally, the vortex features are visualized using icons. The complete process of extracting vortex features thus consists of five steps: streamline calculation, selection, clustering, quantification, and visualization. Each step is described below in more detail.

35.4.1.1   Streamline Calculation

First, streamlines are calculated with a density high enough for all vortical regions to be covered by the streamlines. Typically, streamlines are released from every grid-point, or every other grid-point, and integrated long enough that streamlines in vortical regions will make at least one full loop. In the AVS5 streamlines module, this can be achieved by setting the `sample_mode` parameter to "plane" and the `N. Segment`, `length`, and `step` parameters to values that are appropriate dependent upon the dataset. To optimize accuracy, the `advection_method` parameter is set to "Runge-Kutta".

35.4.1.2   Selection

The selection process chooses streamlines belonging to a vortex based on two criteria:

1. The winding angle of a streamline should be $k\pi$, with $k >= 1$.

2. The distance between the starting and ending points of the streamline should be relatively small.

The winding angle is illustrated in Fig. 35.4.1. Let S_i be a 2D streamline consisting of points $\mathbf{P}_{i,j}$ and line segments $(\mathbf{P}_{i,j}, \mathbf{P}_{i,j+1})$, and let $\angle (A,B,C)$ denote the angle between line segments AB and BC. Then, the winding angle

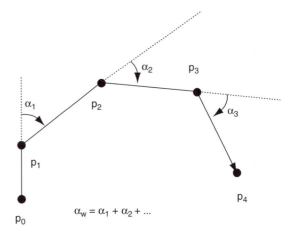

Figure 35.4.1   Winding angle.

$\alpha_{w,i}$ of streamline S_i is defined as the cumulative change of direction of the streamline segments: $\alpha_{w,i} = \sum_{j=1}^{n-1} \angle(\mathbf{P}_{i,j-1}, \mathbf{P}_{i,j}, \mathbf{P}_{i,j+1})$. A positive sign on an angle denotes a counterclockwise-rotating curve, and a negative sign denotes a clockwise-rotating curve. Hence $\alpha_{w,i} = \pm 2\pi$ for a fully closed curve; lower values might be used to find winding streamlines that do not make a full revolution.

Now the streamlines belonging to vortices have been selected, but they are still unrelated. This is solved in the clustering stage described in Section 35.4.1.3.

35.4.1.3   Clustering

The purpose of clustering is to group together those streamlines that belong to the same vortex. As it is easier to cluster points rather than streamlines, each streamline is mapped to a point by determination of the geometric mean of all points on the streamline. These center points are then clustered as follows. The first cluster is formed by the first point. For each subsequent point, it is determined which previous cluster lies closest. If the point is not within a predetermined radius of all the existing clusters, a new cluster is created. In this way, the selected streamlines are combined into a distinct number of groups.

Table 35.4.1   Vortex feature attributes

Streamline center:	$SC_i = 1/|S_i| \sum\limits_{j=1}^{|S_i|} P_{i,j}$
Cluster center:	$CC_k = 1/|C_k| \sum\limits_{l=1}^{|C_k|} SC_{k,1}$
Cluster covariance:	$M_k = \text{cov}(\Psi(C_k))$
Ellipse axis lengths:	$\lambda_k = \text{eig}(M_k)$
Ellipse axis directions:	$\mathbf{d_k} = \text{eigvec}(M_k)$
Vortex rotation direction:	$\mathbf{D_k} = \text{sign}(\alpha_{w,k})$
Vortex angular velocity:	$\omega_k = 1/|C_k|\Delta t \sum\limits_{l=1}^{|C_k|} \alpha_{w,1}$

Streamlines of the same group are considered to be part of the same vortex.

35.4.1.4   Quantification

Once the streamlines have been clustered, the vortices are quantified by a calculation of the numerical attributes of the corresponding streamline clusters. The shape of the vortices is approximated by *ellipses*. An ellipse is fitted to a set of points by calculation of statistical attributes, such as mean, variance, and covariance of the points [19]. In addition, we calculate specific vortex attributes, such as rotation direction and angular velocity. The number of points on a streamline S_i is denoted as $|S_i|$, and a cluster of streamlines as $C_k = \{S_{k,1}, S_{k,2}, \ldots\}$, where $S_{k,1}$ is streamline #1 in cluster k, the number of streamlines in that cluster as $|C_k|$, and all the points on all the streamlines in that cluster as $\Psi(C_k)$. The attributes in Table 35.4.1 can be calculated for each vortex.

35.4.1.3   Visualization

Visualization of the vortices is achieved with *icons*. Icons are abstract high-level objects for visualization [19]. Here, we use ellipse icons, as they reasonably approximate the shape of the vortical regions. The numerical statistical attributes are mapped to icon attributes. The first three attributes are used to calculate the axis lengths and directions of ellipses that approximate the size and orientation of the vortices. The rotation direction of a vortex is visualized by color, e.g., blue for clockwise vortices and red for counter-clockwise vortices.

35.4.2   Implementation

AVS5 was chosen to implement the algorithm described in the previous section for the following reasons:

- Many basic visualization tools are already built in. There is no need to to reinvent the wheel.

- High-quality rendering is built in, so there is no need to develop rendering engines or 3D viewers.

- If functionality is missing, it is easy to add custom modules. In this way, users (researchers, developers) can concentrate on their new visualization algorithms.

- A rich suite of modules has already been developed by thousands of users worldwide [2].

For this project, built-in modules were used for calculating streamlines, manipulating scalar and vector fields, doing color mapping, and rendering geometry. Custom-developed modules were used for creating and visualizing icons.

35.4.3   Results

The dataset used is a simulation performed at the Hadley Centre for Climate Research and Prediction of the UK Meteorological Office. One of the goals of the simulation was to predict the effects of air pollutant emissions on global warming and ocean currents. The simulation model is defined on a curvilinear grid of $288 \times 143 \times 20$ nodes spanning the globe with a resolution of 1.25 degrees longitude and latitude. From this grid, a part was selected covering the North Atlantic Ocean. The simulation spans a period from 1860 to 2099, with one time-step per year, but only one time-step was used (1999). At each node, the simulation calculated three velocity components plus the velocity magnitude and temperature. Fig. 35.4.2 shows a color visualization, where the geography is clearly illustrated by a topographic texture map.

The global flow patterns in the dataset are visualized by white streamlines released with a high density in a horizontal grid slice in the middle of the grid (slice 10 out of 20). Ellipse icons are used to visualize the vortex features. In addition to the size and orientation of the vortices, the colors indicate their rotation direction: red indicates counter-clockwise rotation, and blue indicates clockwise rotation. It can be seen that this method captures all vortices consisting of rotational streamlines, even weak ones.

In addition, numerical attributes were determined for the vortex features, as shown in Table 35.4.2. Notice the difference between the largest and the smallest vortex (approximate factor 25) and between the fastest and the slowest one (approximate factor 15). There does not seem to be any correlation between the size and the rotation speed of the vortices.

35.4.4   Conclusion

A technique was described for the extraction and visualization of vortex features. This technique allows us to not only find vortices more accurately than traditional techniques do, but also to explicitly describe them with numerical

Figure 35.4.2 Simulation of the Atlantic Ocean, with streamlines and vortex features visualized by ellipses. Red and blue ellipses rotate in opposite directions. (See also color insert.)

Table 35.4.2 Numerical attributes of the vortices found in the Atlantic Ocean

Number of clusters	14
Number of clockwise vortices	12
Number of counterclockwise vortices	2
Min. radius [km]	37.13
Max.radius [km]	920.91
Min. $\omega[s^{-1}]$	0.04
Max. $\omega[s^{-1}]$	0.71

parameters. This has the advantage of better comparison, better tracking of vortices in time, etc. AVS5 has proven to be a powerful tool for implementing these techniques.

Acknowledgments

U.K. Meteorological Office, Freek Reinders, and Frits Post.

35.5 Visualizing Engineering Data: Stresses and Deformations of Architectural Structures

The visualization of stresses and deformations of architectural structures under the load of gravity is an important way of ensuring the safety of constructions such as bridges or buildings. The statics of a building is a complex balance between the safety of the people using the construction and the costs of building it.

Visualizing engineering data in AVS/Express is demonstrated by the example of a dataset of unstructured cell data (UCD). The dataset represents half of the structure of a simple beam suspension bridge and its displacements under the load of gravity. The dataset was generated by DANFE, a finite-element analysis package. The other half of the bridge is a mirror image of the dataset.

The UCD data file contains the structural description of the bridge in 2336 nodes that form 1404 hexahedral cells. These cells are split into six separate cell sets with different material types defining to which cell set each belongs.

The data file also contains node data information (3 values per node) describing the x, y, and z components of the displacement of the bridge under load due to gravity. The values of the displacements are in the range from 1×10^{-05} to 1×10^{-09}, while the coordinates of the bridge elements themselves are in the region of 1×10^{00}. Consequently, the displacements are on the order of 100,000 times smaller than the coordinates. The x-axis of the bridge runs along the length of the bridge, the y-axis lies parallel to the plane of the bridge surface, and the z-axis lies parallel to the support towers.

35.5.1 Reading the Data

Using the ReadUCD module of AVS/Express, the data can be read directly into an AVS/Express field. The field contains all the cell data and node data that the file provides. The field is arranged in the 6 cell sets, which can be shown or hidden using the select_cells module. The different cell sets represent different material types for each part of the bridge structure. The bridge can be separated into six different material types describing the vertical towers, the tower basements, the cable, the two types of concrete, and the bedrock.

35.5.2 Glyphing
35.5.2.1 Converting the Data

To visualize the load of the gravity, the displacement values shall be shown as arrow glyphs. These arrows can be generated using the node data values of the UCD data files. The displacement data is stored in independent arrays for each vector component (i.e., X, Y, and Z). In order to use the data as vector data instead of scalar data, the three vector components have to be combined. To interleave three independent node data arrays into a single vector is not a trivial task in AVS/Express; it is done in three independent steps. The AVS/Express modules used are mentioned in parentheses. The first step is to extract the arrays from the AVS/Field and store them as plain float arrays (extract_data_array). The second step is to interleave the three arrays in order to get a single array in the structure $x_1 y_1 z_1 x_2 y_2 z_2 \ldots x_n y_n z_n$ (interleave_3_arrays). The final step is to generate the node data again using the node vector module (node_vector). The resulting node data can be combined with the field to be used with other modules (combine_mesh_data). The new field now contains vector node data that can be used with the glyph module to show arrow glyphs representing the displacements of the bridge cells.

35.5.2.2 Filtering the Data

Using the AVS/Express glyph module, arrows can be placed at each node of the cell sets. These arrows show the direction and the magnitude of the displacements in each point. One of the problems with showing an arrow at each node is that the number of arrows in the picture is hiding the information the arrows are supposed to show. A solution to reduce the number of glyphs is to show only those glyphs on a specified plane. This can be achieved using the slice_plane module. This module allows a plane to be specified that cuts through the dataset. Only data at points specified by this plane will be used for the glyphing. The problem in this approach is that there might be points of interest that are not on the plane.

The International AVS Centre [2] (IAC) provides a module repository that extends the functionality of AVS/Express. In this repository is a DVdownsize_scat module that allows reduction of the number of data points in a scattered field. This module filters a given percentage of node_data values. The remaining node data can be shown as before. The advantage of this module is that it reduces the number of glyphs in the picture without any interpolation. The selection of the data points to hide is not based on any statistical method. The user specifies a reduction

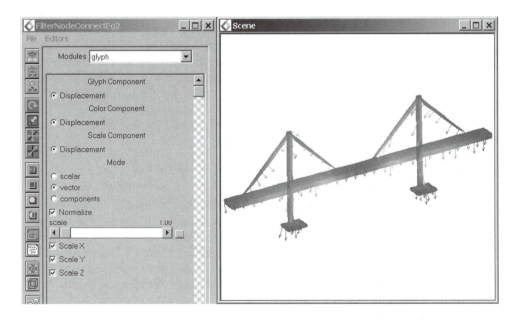

Figure 35.5.1 Glyphing arrows added to represent the displacement magnitude and direction on the bridge structure.

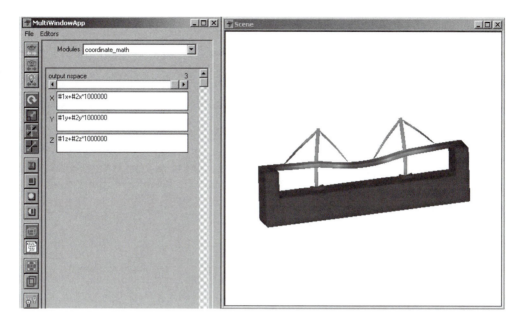

Figure 35.5.2 Deformation of the bridge structure to show displacements due to gravity.

factor, which defines how many points are taken out between two used points (i.e., if the down-size factor is specified as 2, every other node data value is omitted).

As we are only interested in the structures of the bridge, the bedrock cell sets should be hidden. The select_cells module allows individual cell sets to be hidden. One of the problems with hiding the cell sets is that node data values for hidden cell sets are still shown. The select_cells module does not filter node data, which causes a problem. Fortunately, there is an IAC module that has been written to take over this filtering task. The FilterNodeConnect module filters the node data values that belong to cells in unselected cell sets. Using this module, it is possible to hide not just the structures of the bridge but the glyphs connected to these structures as well.

Another helpful feature of AVS/Express is that it allows the user to set the opacity of solid objects. Therefore, glyphs inside the bridge structure can be shown in relationship to the object.

35.5.2.3 *Showing the Displacements*

The node data consists of displacement values that can be used to represent a visible deformation of the bridge structure. In order to achieve this we can make use of the coordinate_math module. This module provides two input boxes that allow user-defined mathematical formulae to combine the coordinates of several input meshes.

In order to use the coordinate_math module, the node data values have to be converted into coordinates. The simplest way of achieving this is to convert the interleaved arrays, which are produced as described in Section 35.5.2.1 into a point mesh. The point mesh does not need any connectivity. This mesh, together with the bridge mesh read by Read_UCD, is now provided to the coordinate_math module, which calculates new coordinates for the bridge.

The formulae for each of the coordinates in the coordinate_math module are $1x + 2x$ for the X-coordinate. The formulae for the Y and Z coordinates are similar.

The output from this network, with these parameters is, unfortunately, not very informative, since, as has been described earlier in this section, the displacements are several orders of magnitude smaller than the coordinates of the structure and so the displacements must be scaled in order to affect the structure coordinates sufficiently for the observer to discern the change. This can be done by simply altering the formulae for the coordinate_math module to $1x + 2x*1000000$, etc.

Acknowledgments

UK AVS and UNIRAS User Group (uAUug) and Priya Dey.

35.6   Medical Visualization: Endovascular Surgical Planning Tool

Medical visualization is one of the most widespread applications of scientific visualization. It has provided the driving force behind many developments in image processing and volume rendering. MVC has been involved in a number of projects to develop medical applications and tools, with AVS/Express being employed on several of these, both behind the scenes as a visualization engine for web-based systems and in the forefront to deliver applications to the clinical user. Described here is the development of an application that provides an endovascular surgical planning (ESP) tool for the coiling of cranial aneurysms [21].

35.6.1   Problem

An aneurysm is a ballooning of an area of arterial wall caused by the blood pressure exerted on it. If it bursts, it will cause hemorrhaging, and cranial aneurysms are often fatal. They can vary in size from millimeters to a few centimeters. Traditionally aneurysms are dealt with by major surgery to apply a clamp across the neck. Currently there is a move towards minor surgery, which involves threading a catheter into the femal artery, in the leg, and all the way up into the brain. Platinum coils are then

used to fill the aneurysm, causing clotting, effectively sealing the aneurysm, and preventing further possibility of bleeding.

To choose the correct size and shape of coils and to ascertain how best to position them, the clinician must gain a clear understanding of the anatomy of the aneurysm and how it relates to the artery to which it is attached. The size and orientation of the aneurysm neck are especially important. Our partners on this project at Manchester Royal Infirmary used a C-arm mounted x-ray machine to analyse the aneurysm. A contrast agent is injected into the arterial structure and a short x-ray movie is made showing its flow through the artery and into the aneurysm. These angiograms are 2D, and the spatial relationships have to be inferred using the clinician's knowledge and experience. The C-arm is also used during the procedure to allow the clinician to see the positioning of the coils.

Before the procedure, the best working views of the aneurysm must be found. These can take up to a dozen attempts to find, exposing the patient and clinicians to additional x-ray doses in the process.

35.6.2   Aim

In addition to the x-ray angiography, it was possible to take a magnetic resonance angiogram (MRA), which provided a data volume showing the intensity of blood flow. Though these were used on occasion to aid the clinician, the level of 3D visualization was limited by the scanner hardware and software to very small and slow maximum intensity projections (MIPs). The purpose of the project was to exploit these datasets to provide an easy-to-use rapid 3D visualization to generate detailed imagery of the artery and aneurysm anatomy. Secondly, it should provide a method to discern the best working views for use during the coiling procedure.

35.6.3   Application Development

The initial plan was that AVS/Express would only be used to create a prototype and that the final application would be coded using a lower-level package such as VTK. It was thought that the AVS/Express high-level environment would incur too much overhead for such a complex application. The Object Manager has control of module execution, which can make creating nonlinear applications difficult, as things that are trivial to achieve using a procedural programming language can be tricky to reproduce in a high-level environment. The time spent avoiding these situations can outweigh the benefits of the rapid feature and user interface development.

The main problem was importing the MRA data, which was in the form of a series of up to 100 DICOM [22] images, which, when correctly stacked and orientated, would create the volume dataset. A Read_DICOM project was found at the IAC [2]; however, this actually read the Papyrus format, a wrapper for a DICOM series. This was heavily reworked to read plain DICOM files, and a tool for scanning and selecting DICOM series was built. The work from this part of the project now forms the basis of the current Read_DICOM project at the IAC.

A prototype application was quickly put together to give the users a feel for what was achievable and to give the developers an idea of what the users wanted. This consisted of three orthogonal slice views and a large 3D main view. The functionality included data input, multiple crop volumes that could be created interactively, isosurfacing for each crop volume, and image capture. The user interface controls were kept compact by using pages for each area of functionality.

Use was made of relative coordinates for laying out the user interface, this being easy to do in AVS/Express. There are two advantages to such an approach: the application can be resized for use at different screen resolutions and window sizes, and if the user interface area changes size and shape, then minimum effort is required. User interface design is very important when developing applications for end users, especially clinicians who do not have a lot of time

to spend trying to learn complex pieces of software. Visualization terms need to be changed to clinical or descriptive labels. One solution is the use of icons, but they suffer from a lack of immediacy, since careful design is required and their overuse can look daunting. Familiar icons, however, such as those for opening file browsers, were used. To provide further help AVS/Express is able to display a message when the mouse pointer is over a widget.

It was decided to continue using AVS/Express, as the prototype had shown that by removal of extraneous functionality and by careful design, speed and usability could be maintained.

In the data used, called Time of Flight (ToF), the signal intensity is proportional to the blood flow. This method produces noisy data, and data processing had to be used to enhance the visualization. A 3D median filter and a 3D tangential smoothing method were developed in an image-processing package called TINA [23] and then integrated into AVS/Express. The tangential smoothing method calculates the data gradient at each voxel and then performs a Gaussian smoothing perpendicular to the gradient normal, thereby preserving surfaces.

Isosurfacing was enhanced by providing an interactive histogram to aid selection of level values. Even with preprocessing, isosurfaces may still contain a large number of artifacts. In order to deal with this, a segmentation tool was added that identifies the topologically separate pieces of the isosurface and ranks them by size. Pieces can then be displayed by making a simple selection from a list. Occluding objects, such as other arteries, can be removed as well. Surfaces can also be saved in simple ASCII format so that they can be reloaded into the application for comparison after the procedure.

The AVS/Express colormap editor required enhancement, and the volume rendering interface made no use of it, giving the impression that the volume-renderer produced poor results. By combining the two, proper transfer functions could be defined for volume rendering to produce insightful images. The renderer also supports different techniques (composite, MIP) and utilizes hardware texture support. Work from this project has been fed back into current versions of AVS/Express.

Volume rendering is the most appropriate method for visualizing ToF because it doesn't contain any explicit structural information, only the rate of blood flow. Judicious use of the transfer function allows information about the speed of blood in and around the aneurysm to be visualized (Fig. 35.6.1). This is important information in terms of placing coils so that they are not dislodged.

The simulation of angiograms was implemented using the volume renderer's ray-casting method with a simple step transfer function. The user can control the position of the step. This provided realistic angiograms with very little effort (Fig. 35.6.2).

The working views were obtained by first defining a center of rotation using three orthogonal planes. To coregister the patient with the C-arm, an isosurface showing the face of the patient is created and the intersection of the planes and the face is then displayed. A report (see below) can be given to the C-arm operator for this purpose. To obtain the actual working views, the user rotates the rendering of the appropriate arterial tree until the desired view is found. The rotation angles can then be read from the screen and applied to the C-arm. The screen turns red if there is a danger that the viewpoint will cause the C-arm to hit the table or the patient.

Exporting information from a medical application is very important for dissemination and reference materials. For this purpose, ESP can

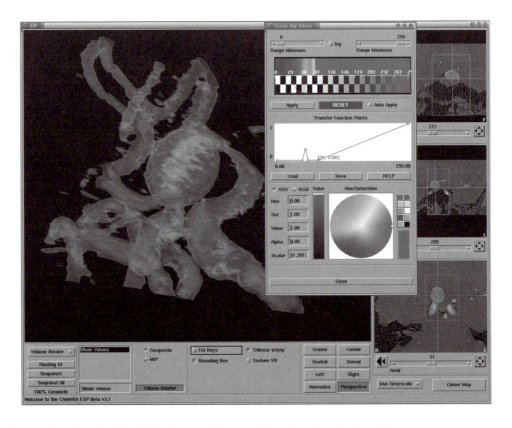

Figure 35.6.1   Blood flow visualized with software volume rendering. (See also color insert.)

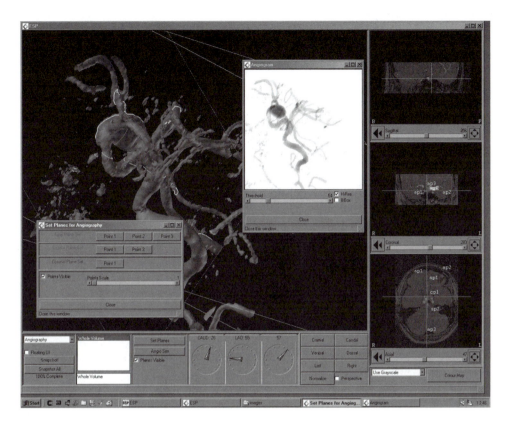

Figure 35.6.2 Angiography simulation. (See also color insert.)

produce AVIs through AVS/Express' key-frame animation macros. Images and VRML can also be exported. At a user's request, a simple reporting tool was added that allows images from the main view to be snapshot into a document. These can be laid out with comments and exported as postscript or HTML. Though Express is unsuitable for creating such a tool, it made a useful addition and demonstrated the flexibility of the package.

A clinician's time can be fractured, and so the ability to save the current state of the application is very useful. This was achieved in the ESP by creating a central parameters group, which contained the parameters for various sections of the application. A custom module was used to write this group to file in Express' V code and

could therefore be reloaded using standard objects for running V scripts. Using these session files, the majority of modifications to the data can be restored.

35.6.4 Conclusion

The project was very successful and produced an application that not only solved the specified task of aneurysm surgical planning but provided a useful medical visualization tool. The ESP application has been used in other areas, such as planning for cochlear implants. The project has pushed to the limits the size and complexity of a single application under AVS/Express and has exploited its strengths in rapid development and wealth of existing functionality. Where AVS/Express has proven weak, these problems have been overcome by the flexibility of the package to allow modification and extension of almost any part.

Acknowledgments

Prof. Alan Jackson, Dr. R. Laitt, and Anthony Lacy. The project was funded by the Sir Jules Thorn Charitable Trust.

35.7   Archaeological Visualization: Uncovering the Past

This section looks at the visualization of mummified humans. It is similar to medical visualization and uses many of the same types of techniques. However, there are a number of elements that make this a distinct topic:

- Mummies are rare and are the focus of specialist scientific research.
- Mummified tissues are physiologically different from live tissues.
- There may be physical or mechanical damage that occurred before, after, or at the time of death.

Research into mummies covers several disciplines: medicine (especially forensic medicine and art in medicine), anthropology, and archaeology. While the techniques used in visualization of mummies may be taken from the medical visualization field, the information obtained and the way it is used is very different.

35.7.1   The Problem

Due to the fact that mummies are rare, and because they are important for both current and future research, any postmortem investigation needs to be noninvasive. Dissection destroys tissues. Not only does it cause mechanical damage, but atmospheric water can damage the dehydrated tissues.

Medical scanning modalities such as Nuclear Magnetic Resonance (NMR) use the water con-

tent of tissues to produce images in which soft tissues show up best. Mummified remains, however, have very little water content, and consequently the success of this particular type of scan is reduced. Computed Tomography (CT) uses low-dose x-rays and is more suited to use with mummies, but it images dense tissues such as bone and teeth best.

The two case studies presented here show the special techniques needed to visualize mummified remains.

35.7.2   Mummy 1766

Mummy 1766 is an adult female Egyptian mummy from the first or second century AD. She was intentionally mummified after death, her body being enclosed in resin-coated bandages that were decoratively painted. She has an ornate gilded cover over the front of her head, breast, and feet.

Currently Mummy 1766 is part of a collection of mummies at the Manchester Museum, University of Manchester. Professor Rosalie David from the Museum has pioneered noninvasive testing of Egyptian mummies. Mummy 1766 has been key in this research for many years.

35.7.2.1   Implementation

The resultant work produced MPEG animations of Mummy 1766 for display as an exhibit at the museum in 1997. Volume rendering was used to reveal the mummy's hidden features.

A preprocessing segmentation step was implemented as a C program external to AVS/Express. The processed data was then volume-rendered, and transparency and rotation were used to show the following:

- The shape and structure of the gilded mask.
- The shape and structure of the skull and remaining facial soft tissues.
- The spatial relationship between the mask and the head.

Animation key frames were selected for both transparency and rotation through an AVS/Express GUI. MPEG animations were automatically generated.

35.7.2.2 Segmentation

During volume rendering, different objects and materials are distinguished from one another by the mapping of varying data intensity values to particular colors and transparencies. This relationship is normally referred to as a transfer function.

Ordinarily in AVS/Express, transfer functions are generated by the user and relate closely to the data histogram, with each peak presumed to be a separate material. A down side of this approach is that materials similar in density may have overlapping peaks, and so suitable transfer functions can be hard to find.

The mummy dataset suffered from this drawback; soft tissue had an intensity similar to that of the resin-coated bandages, and the intensity of the gilded decorative cover was similar to bone and teeth values.

An additional and more pressing problem was the complete occlusion of the head by the mask. Thus, in order to produce a meaningful visualization of the mummy's internal structure, the data had to first be segmented; the mask at least had to be distinguishable from the rest of the data.

To address this, a cutting contour was defined for each data slice in the axial plane, following the inside of the mask cover and extending to the edge of the dataset after the base of the mask was reached. Each contour was automatically generated by a custom C program, which employed a region-growing method [24]. In this way the dataset was split into two, with the mummy and bandages being separate from the mask.

A third dataset was created by recombination of the two segmented datasets, but the head and cartonnage data values were transposed, providing each data subset with its own distinct range of intensity values. This allowed a simple step function to be applied to the transparency map, in turn removing the gilding, decorations, mask, bandages, soft tissue, bone, and teeth.

The three datasets were then rendered to produce images and animations using AVS/Express.

35.7.2.3 Surface Rendering vs. Volume Rendering

The mummy's body was tightly wrapped in 47 layers of resin-coated bandages. Without the use of a view-dependent rendering, producing full-resolution isosurfaces of every material and then animating the transparency of each proved too computationally expensive for the machines of the time. Volume rendering, however, was a much more manageable approach. Animating

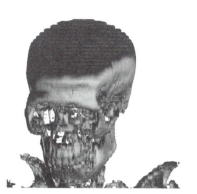

Figure 35.7.1 Mummy 1766. (See also color insert.)

the transparency of each material was important because it showed their spatial relationships, such as the location and orientation of the head behind the mask (Fig. 35.7.1).

35.7.3 Worsley Man

Worsley Man [25] is a bog body (a Celtic sacrifice) that was discovered in marshes in Salford, United Kingdom, in the 1950s. The aim of this work was to create a new computer graphics tool that would allow the rebuilding of a face from skeletal remains. This work was the start of a larger ongoing project.

In this section it is shown how the skull is extracted as a surface model so that muscle structure and facial features can be added later. The main components of this work are described below:

- Reading the data into AVS/Express. The scan was in Explicit DICOM format and therefore had to be converted to Implicit DICOM.

- Segmentation and the removal of unwanted features or noise. Due to dehydration, the physiology of the body changed, and materials from the bog became attached. Such extraneous data had to be segmented out from the dataset.

- An exhibit mount passes through the center of the skull; this, too, needed to be removed from the visualization.

35.7.3.1 Data Reading

To read the CT data into AVS/Express, the ReadDICOM module was used [2,22,27]. It should be noted that the ReadDICOM module can only read DICOM data in the Implicit format, and not the Explicit. Efilm [26] was used to convert the Explicit CT scan of Worlsey man into Implicit format.

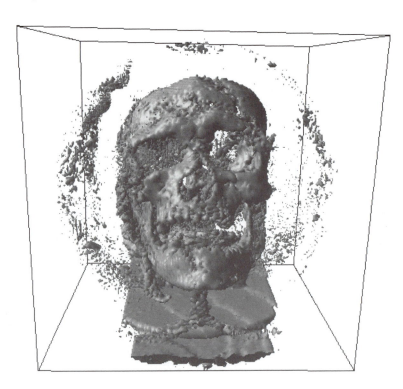

Figure 35.7.2 Visualization of Worsley Man showing the noise in the data. (See also color insert.)

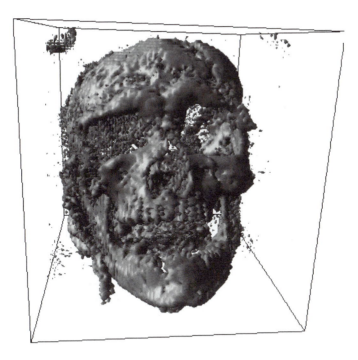

Figure 35.7.3   Result of crop on Worsley Man. (See also color insert.)

35.7.3.2   *Segmentation*

Segmentation was used to remove the noise around and inside of the skull (Fig. 35.7.2). A three-step segmentation method was employed:

1. The crop module removed the "noise" surrounding the skull (Fig. 35.7.3).

2. A region-growing module was used to segment the skull from the remaining data. A seed point was selected from the data by user selection. The region was then grown by comparison of the data values of the seed point, or the last point processed, to the current check point.

3. The IsoObjsReduce module was used to remove noise within the skull.

A low-resolution surface visualization of Worsley Man was produced, as high detail was not required for associated research. Several layers

of detail are added onto the skull to build up the face of Worsley Man.

Acknowledgments

Prof. Rosalie David and Dr. Caroline Wilkinson.

35.8   Molecular Visualization: Seeing 3D Chemical Structure

From the beginning of our understanding that 3D structure was important, scientists have explored a variety of approaches to rendering the 3D structure in 2D diagrams. They have manufactured physical models of molecular structures in the hope of understanding how the structure affects both their nature and their interaction with other molecules in their environment.

Computer-based 3D visualization has been in routine use in computational chemistry and biochemistry since the 1980s, when workstations sufficiently powerful to render structures interactively began to appear. The appearance of affordable stereo vision systems provided another important boost to such workstations, and their cost, though still very high, was easily justified by the benefits afforded. With modern, high-performance personal computers, 3D hardware graphics support has become the norm and so 3D molecular visualization has become an everyday tool for the computational chemist or biochemist.

35.8.1   Aim

The aim of this development has been to attempt to provide a widely usable and easily extensible framework for the visualization of molecular structures and properties. The work builds on the large number of valuable, powerful existing methods and techniques provided by an extensive visualization system such as AVS/Express. This gives the computational chemist tools that can be used for the visualization of molecular structures and properties. They can then be used to produce solutions tailored for specific needs by incorporating new methods of analysis, interaction, and, if necessary, visualization, while relying heavily upon core methods provided by the system itself.

35.8.2   Application Development

This development has been undertaken over several years and has been contributed to by a number of people who have worked on specific features. The initial work was carried out using UNIX on Hewlett Packard and Silicon Graphics workstations, but more recently development work has been carried out primarily on PC-type hardware using either the Linux or the Windows/NT platform. Portability has been an important issue throughout, and the system should be usable across most platforms supported by

AVS/Express, with minimal code rewriting required.

AVS/Express has proven a very useful development environment for this problem for a number of reasons, but three principal ones stand out. First, the user interface can be constructed, modified, and extended with great ease through a comprehensive widget set that can be readily combined in a hierarchical development scheme. Second, the internal data structure used by AVS/Express, the "field," can readily contain data in a form that is both comprehensible to the chemist and suitable for the AVS/Express environment for rendering and visualization. Third, the provided visualization methods can easily be integrated with the application-specific data to generate a single integrated display including structure and properties information.

35.8.2.1   Data Structures

The versatile internal data structure used by the AVS/Express system, usually referred to as a "field," can hold a wide variety of data types, including arbitrarily dimensioned tensor data in a space also of arbitrary dimension. This encompasses the ability to store scalar data values at points in a 3D space that is sufficient to hold the positions of atoms and specify their type in a 3D structure. More complex molecular components, such as protein residues, can also be specified, and orientation information can be held in a vector form at each point, allowing protein macromolecules to be built up from residue-based specifications.

The field also contains a facility to hold a mesh describing the way in which the points contained within the field define a multidimensional geometry. This feature can be used to specify how the atoms are interconnected, permitting full specification of the 3D molecular geometry. Thus, almost the most simple molecule viewer imaginable is that shown in Fig. 35.8.1, where, using a point mesh with atom data (radii and colors) to produce the balls and a line mesh with stick data (thicknesses and colors) to produce the sticks, a basic 3D

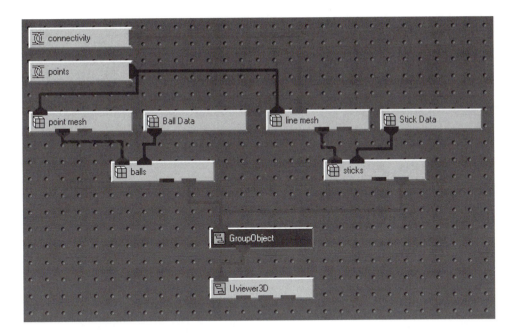

Figure 35.8.1   Almost the simplest AVS/Express molecule viewer imaginable.

molecular viewer is constructed using only 10 modules in only a few minutes of implementation time.

The rendering of this ball-and-stick model in AVS/Express is a straightforward process and is carried out automatically by the Uviewer3D when the correct additional entities are specified in the *balls* and *sticks* fields. Other standard entities can be set in the field and in the viewer to control the surface subdivision factor and the illumination and shading model of the surfaces of the spheres and tubes. Thus, high-quality renderings of large 3D structures can be trivially produced using AVS/Express.

35.8.2.2   Data Readers
Obviously, while this remarkably simple application is a fully functional molecule viewer, it would be somewhat tedious for the user to have to enter his or her geometry by hand into arrays of points, atom data, and connectivity just to produce basic 3D structure diagrams. As AVS/

Express allows the user to add new modules, a set of reader functions has been added that outputs data in a suitable form for inclusion in this viewer. Supported formats include standard Cartesian coordinates and z-matrices, as well as those for some of the most common chemistry software packages, such as Gaussian, CAD-PAC, and GAMESS. The addition of a new reader requires the developer to be able to implement an algorithm to extract the 3D structure as arrays of atomic coordinates, connectivity, and atom and bond properties, which are then connected to the correct points in the viewer. When this reader is combined with the simple application shown in Fig. 35.8.1, then the user has a full molecular viewer for the new data format.

35.8.2.3   Properties Data
Additional methods that use standard visualization techniques, such as direct volume rendering, surface extraction, and "glyphing" of data

from a variety of sources, can be incorporated in the structure viewer. Methods have been implemented for the visualization of properties data as produced by common computational chemistry software such as Gaussian. These methods allow the user to visualize molecular orbitals, electrostatic potentials, and electron density, overlaid on the 3D molecular structure, using volume- and isosurface-rendering techniques. In each case, little additional work was required to combine suitable readers with the standard visualization modules provided by AVS/Express in order to create this molecular viewing network.

35.8.2.4   Interactivity and User Interfacing

AVS/Express provides a comprehensive user interface development kit with an extensive widget set. This permits the user to develop a flexible and changeable user interface with relative ease. Sections of the interface can be switched on and off through simple toggle controls. This allows the development of modal interfaces that maximize the available space for the display and optimize the availability of controls for the current task. The modular nature of the user interface also permits the developer and user to make changes to suit new data types and new applications without reimplementing the processing and rendering controls.

Along with the obvious manipulations (zooming, translating, and rotating the display), AVS/Express provides facilities for the selection of drawn objects in the scene, returning object identifiers and position information. In this way, modules have been developed that permit the user to select particular atoms, bonds, and regions of the structure, which can then be manipulated. Using this feature, macros have been constructed for building and editing structures and for the selection of regions of macromolecules within which particular display features can be applied. Thus, regions of interest, such as active sites of macromolecules, can be displayed in more detail, while the less interesting parts of

the structure can be deselected and either displayed at lower resolution or not rendered at all. This allows the user to concentrate her or his attention where it is most appropriate.

35.8.2.5   Development Approach

The hierarchical development approach in AVS/Express helps in the development of applications that are readily modifiable and extensible. Interdependencies can be avoided, and the module and macro set developed is left open to the user and future developers, who may combine them in new ways. The example application, shown in Fig. 35.8.2, contains approximately 100 separate modules in a hierarchy of macros. The application includes a substantial user interface (in the left-most window) with controls for handling the loading, building, editing, and display of structures and properties with a wide range of visualization approaches and fine control over the rendering process.

35.8.3   Conclusion

We have produced a broad-based module set with capabilities for visualization of molecular structure and properties. This has created a framework that can be easily extended by users and developers to incorporate new methods and approaches as they are developed. Existing applications developed using this module set provide capabilities not only for visualization but also for interactive construction and editing of large and complex molecules from single atoms, residues, and libraries of structure fragments.

Today's chemist is presented with a wide variety of visualization software from numerous suppliers, and, since the essential need has been long identified, the competition has been fierce and development extensive. That said, new methods for data analysis are being developed daily within research groups, and there is still a great scope within which the user can develop new and interesting approaches to both properties and structure visualization. These goals could be assisted by the availability of a

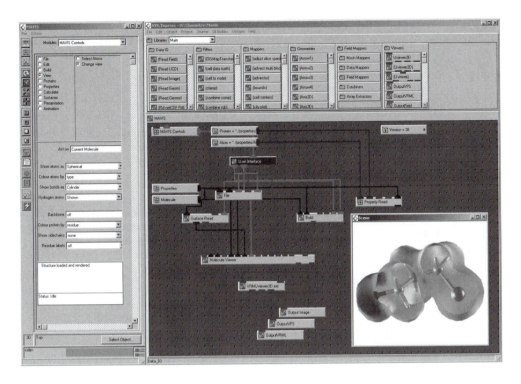

Figure 35.8.2   An example application constructed using standard and application-specific AVS/Express modules, with output structure and properties in the lower-right window. (See also color insert.)

standard, portable visualization environment within which new methods and techniques can be easily incorporated, and we hope to continue to develop these ideas further in the future.

Acknowledgments

Dr. Christopher Parkinson, Prof. Ian Hillier, and Dr. Neil Burton.

35.9   Visualization with AVS: Conclusion

This chapter has presented examples from several different scientific disciplines, all with different requirements. In all cases, the end users have been provided with added understanding of their data through the use of visualization techniques. There is no doubt that the flexibility of AVS/Express has contributed to this by providing many useful elements:

- The ability to develop upon existing modules or use modules from the IAC repository [2].

- Rapid prototyping of applications.

- An extensible user interface that can be constructed and controlled by the application.

MVC [1] is also authoring new high-performance features for the AVS software environment. The Multipipe Edition (MPE) enables transparent deployment of an AVS visualization network on multiple projector-based virtual environments. Both the SGI MultiPipe and CAVElib APIs are supported. Decoupled head tracking and control of peripheral devices are also provided. The rendering decomposition used by MPE has produced a good performance increase. This is being further improved by the Parallel Support Toolkit (PST), which provides a framework for heterogeneous distributed computation using a suite of parallelized performance-visualization modules. The framework will integrate the control of data decomposition, distribution, large dataset handling and LoD, data streaming, and asynchronous communication for facilitating computational steering. Finally, a version of AVS/Express that supports a haptic joystick has been developed. Users can feel and interact with their visualization, whether by feeling the surface of a porous rock or the pull of gravity around the Earth's surface or by some different method.

References

1. Manchester Visualization Centre. Manchester Computing, The University of Manchester, http://www.man.ac.uk/MVC

2. International AVS Centre, Manchester Visualization Centre. http://www.iavsc.org

3. The Hadley Centre for Climate Prediction and Research, U.K. Meteorological Office, http://www.metoffice.com/research/hadleycentre/index.html

4. Climate Research. http://www.metoffice.com/corporate/scitech0102/9_climate_research/index.html

5. United Nations Framework Convention on Climate Change. http://unfccc.int/index.html

6. Fifth Session of the UNFCCC Conference of the Parties (CoP5). http://cop5.unfccc.int/

7. Sixth Session of the UNFCCC Conference of the Parties (CoP6). http://cop6.unfccc.int/

8. Seventh Session of the UNFCCC Conference of the Parties (CoP7). http://unfccc.int/cop7/

9. Climate Models. http://www.metoffice.com/research/hadleycentre/models/modeltypes.html

10. Intergovernmental Panel on Climate Change (IPCC). http://www.ipcc.ch/

11. J. Leng and J. M. Brooke. Visualization of historic databases of sunspot observations and solar dynamo simulation. *UKHEC visualization case study*, 2001. http://www.ukhec.ac.uk/ publications

12. J. Leng, J. Brooke, W. T. Hewitt, and H. Davies. Visualization of seismic data in geophysics and astrophysics. *UKHEC visualization case study*, 2001. http://www.ukhec.ac.uk/ publications

13. J. Leng, J. Brooke, W. T. Hewitt, and H. Davies. Visualization of seismic data in geophysics and astrophysics. *Proceedings of SGI Users Conference*, Krakow, Poland.

14. P. Pulkkinen. Solar differential rotation and its generators: computational and statistical studies. Ph.D. dissertation by Pentti Pulkinen, published by the Finnish Meteorological Institute, Helsinki, 1998.

15. E. Tufte. *Envisioning Information*, 2nd Ed. Graphics Press, 2001.

16. K. F. J. Reinders. Feature-based visualization of time-dependent data. Ph.D. Thesis, Delft University of Technology, 2001.

17. I. A. Sadarjoen, F. H. Post, B. Ma, D. C. Banks, and H. G. Pagendarm: Selective visualization of vortices in hydrodynamic flows. *IEEE Proc. Visualization '98*, pages 419–423, 1998.

18. I. A. Sadarjoen and F. H. Post. Detection, quantification, and tracking of vortices using streamline geometry, *Computers and Graphics* 24:333–341, Elsevier Science, 2000.

19. T. van Walsum, F. H. Post, D. Silver and F. J. Post. Feature extraction and iconic visualization. *IEEE TVCG* 2(2):111–119, 1996.

20. Visualization cookbook using AVS/express, http://www.iavsc.org/training/cookbooks/index.html

21. J. Perrin, T. Lacey, A. Jackson, R. Laitt, and N. W. John. A visualization system for the clinical evaluation of cerebral aneurysms from MRA data. *EUROGRAPHICS 2001 Proceedings*, 2001.

22. DICOM, medical data format, http://medical.nema.org/

23. TINA, image analysis library, http://www.tina-vision.net/

24. K. Moltenbrey. Unraveling the mysteries of the mummy. *Computer Graphics World*, pages 51–53, 2002.

25. The Boothstown Website (Worsley Man), http://freespace.virgin.net/tony.smith/malkins.htm

26. Efilm, http://www.efilm.com

27. Papyrus 3.6 toolkit, courtesy of Digital Imaging Unit, University Hospital of Geneva.

28. C. I. Parkinson, M. D. Cooper, W. T. Hewitt, and I. H. Hillier. MAVIS: An interactive visualization tool for computational chemistry calculations in a distributed networked environment, In *Proceedings of the Pacific Symposium on Biocomputing*, 1998.

36 ParaView: An End-User Tool for Large-Data Visualization

JAMES AHRENS
Los Alamos National Laboratory

BERK GEVECI and CHARLES LAW
Kitware, Inc.

36.1 Introduction

This chapter describes the design and features of a visualization tool called ParaView (*http://www.paraview.org*), a tool that allows scientists to visualize and analyze extremely large datasets. The tool provides a graphical user interface (GUI) for the creation and dynamic execution of visualization tasks. ParaView transparently supports the visualization and rendering of large datasets by executing these programs in parallel on shared or distributed memory machines. ParaView supports hardware-accelerated parallel rendering and achieves interactive rendering performance via level-of-detail (LOD) techniques. The design balances and integrates a number of diverse requirements, including the ability to handle large data, ease of use, and extensibility by developers. This chapter describes the requirements that guided the design, identifies the importance of those requirements to scientific users, and discusses key design decisions and tradeoffs.

Sensors and scientific simulations are generating unprecedented volumes of data, making visualization by traditional solutions difficult or even impossible. To address the simulation scientists' visualization needs, we spoke with simulation scientists and gathered a set of requirements. The high-level requirements that guided the design of ParaView are (a) support for an efficient workflow and (b) support for the visualization and analysis of extremely large datasets. The challenge was to create a design and implementation that met both of these complex requirements and balanced the conflicts between them.

36.1.1 Workflow Requirements

Visualization is one task of the many that simulation scientists encounter. Other simulation tasks include theoretical work, programming, problem setup, analysis, and data management. Therefore, the first workflow requirement is tool ease of use. That is, how long it takes to create results and what visualization domain knowledge is required to run the tool will determine whether and how often the tool is used. A coarse approximation of the simulation scientists' visualization workflow includes two modes: an exploratory mode, in which an interactive GUI based tool is used to explore a dataset, and a batch mode, in which a scripting or programming language is used to write and execute a program that creates an animation. The second workflow requirement is support for both modes. This coarse approximation can be refined further by identification of how data is input (during the simulation run or after processing of the simulation) and what type of interface is used (GUI, scripting, VR) [1]. Additional workflow requirements include tool portability, accessibility, and extensibility. Portability is required because of the diverse collection of resources available to scientists to run their simulations and visualizations. Tool accessibility is the ability to quickly gain access to, set up, possibly modify, and run the tool. Open-source projects are more accessible because the package is typically available on the Internet, and any necessary tool modifications can be made quickly because the source is available. We define extensibility as the ability to easily add new functions and graphical interfaces to the tool.

36.1.2   Large-Data Visualization Requirements

The ability to handle large data is also a critical requirement. We define large data as data that exceeds the resource limits (i.e., the elements of the storage hierarchy—memory, disk, tape) of a single machine. The first aspect of the large-data handing requirement is a functional one: can the data be visualized at all? Techniques such as data streaming (i.e., incrementally processing the data) and parallelism can be used to process large datasets. Workflow requirements, such as portability, mandate that the tool execute on both shared and distributed-memory parallel machines. The second aspect of the large-data handling requirements is performance: can the data be processed quickly? Techniques such as multiresolution representations and parallelism can be used to improve both visualization and rendering performance.

36.2   Related Work

There are a number of visualization packages available for use by scientists. Each of these packages meets a subset of the identified requirements. In this section, we will discuss a few of these packages, specifically AVS [2], OpenDX [3], SCIRun [4], and Ensight [5], identifying their strengths and describing which requirements they meet. ParaView was designed to meet all of the identified workflow and large-data visualization requirements.

36.2.1   Workflow Requirements

Ensight and ParaView use GUIs to execute visualization tasks. AVS, OpenDX, and SCIRun use data-flow program graph editors to compose programs. Data-flow program graph editors were thought to provide a good tradeoff between the needs of visualization developers and those of end users: for developers, they provide the ability to create complex program graphs, and for end users they provide a graphical interface to create these graphs. In practice, learning visual programming with data-flow graphs is considered by many scientists a significant barrier to creating visualization tasks, and thus they consider GUI-based interfaces easier to use. OpenDX, SCIRun, and ParaView are all open-source packages, which makes them easily accessible and extensible. These packages offer interactive and batch interaction modes. SCIRun provides support for computational steering—the ability to interact with and visualize data from a running simulation. In contrast to these other packages, ParaView uses a general-purpose scripting language, Tcl, for batch commands. The advantages of using a general purpose scripting language include the availability of general-purpose computing functionality, robust documentation, and support for the scripting language that is independent of the visualization tool.

36.2.1   Large Data Visualization Requirements

All of these packages are portable to most architectures when run on a single machine. Differences arise in their portability to parallel architectures. AVS, OpenDX, and SCIRun all support parallel execution on shared-memory machines. They also all rely on a centralized executive to allocate memory and execute programs. This reliance makes it difficult to port these packages to distributed-memory machines. Ensight uses a client/server architecture where the client renders geometry and the server executes visualization and analysis tasks. Ensight currently provides shared-memory implementations of both the client and the server. Ensight also has a distributed-memory implementation of the server. ParaView is portable to both shared and distributed-memory machines, and it is the only listed package that can incrementally process data.

The ability to process datasets that are larger than the available computing resources is a key consideration when one is processing extremely large datasets, since resource availability changes over time. (ParaView's data-streaming feature is available in batch mode).

36.3   Design

Para View is designed as a layered architecture. The foundation is the visualization toolkit (VTK) [6,7], which provides data representations, algorithms, and a mechanism to connect these representations and algorithms together to form a working program. The second layer is the parallel extensions to the VTK. This layer extended VTK to support streaming of all data types and parallel execution on shared- and distributed-memory machines. (These extensions are currently part of the toolkit, but they were added after the original design of the toolkit was complete.) The third layer is Para-View itself. ParaView provides a GUI and transparently supports the visualization and rendering of large datasets via hardware acceleration, parallelism, and LOD techniques. Each layer meets a subset of the requirements and adds additional functionality to the layer below it.

36.3.1   The Visualization Toolkit

The VTK is the foundation of the ParaView architecture. VTK provides data representations for a variety of grid types, including structured (uniform and nonuniform rectilinear grids as well as curvilinear grids), unstructured, polygonal, and image data. Examples of these grid types are shown in Fig. 36.1. VTK provides hundreds of visualization and rendering algo-

rithms that process these data types, including isosurfacing, cutting/clipping, glyphing, and streamlines. VTK also provides algorithms for polygon rendering and volume rendering as well as a keyboard and mouse-based interaction model. Algorithms are organized into data-flow program graphs, and a demand-driven data-flow execution model is used to run these programs. Core functionality in VTK is written in C++. To use the toolkit, VTK offers both a C++ library interface and a set of scripting interfaces, including Java, Python, and Tcl interfaces. The library interface provides the best performance. The scripting interfaces offer the advantage of rapid prototyping of programs. Once a day and continuously (i.e., whenever a developer commits a change), tests are run using an open-source testing framework called Dart, which improves the toolkit's reliability. The toolkit provides the basis for Para-View's portability, accessibility, full range of features, and support for interactive and scripting usage modes. More details on VTK can be found in Chapter 30.

36.3.2   Parallel and Distributed Visualization Toolkit

Additional functionality was added to VTK to support data streaming and parallel computation [8]. Both depend upon the ability to break a dataset into smaller pieces. Data streaming incrementally processes these smaller pieces one at a time. Thus, a user can process an extremely large dataset with computing resources that cannot store the entire dataset (either in memory or on disk). Data streaming requires that all VTK data types be separable into pieces and that the toolkit algorithms correctly process these pieces. To process pieces in a dataflow pipeline, a mapping must be defined that specifies for each algorithm what portion of the input data is required to generate a portion of the output data. With this information, algorithms can generate only a portion of their output for a given input. Each algorithm must ensure that program results are invariant regardless of how the dataset is broken into pieces. These requirements are met when the user creates a partitioning of both structured and unstructured grid types and provides ghost levels, which are points/cells that are shared between processes and are used by algorithms that require neighborhood information. A piece of a structured grid is defined by its extent, which describes a contiguous range of elements in each dimension (i.e., in three dimensions, a sub-block of a complete block). VTK's unstructured grid types use an "element of a collection" scheme (i.e., piece M of N). A procedure for converting between grid types has also been defined, in which each structured extent piece maps to one unstructured piece. Additional policies take care of handling boundary conditions and creating ghost levels for all grid types. This data-streaming ability supports data parallelism. Instead of processing pieces one at a time, each processor processes a different piece in parallel. Examples of dataset partitioning and the creation of ghost levels are shown in Fig. 36.2, which shows a CTH nonuniform rectilinear grid dataset that was processed in eight pieces. The original dataset contained cell-centered attributes. Volume fraction attributes for both the projectile and the plate were first interpolated to vertices before an isosurface filter was used to extract the material surfaces. Both the vertex interpolation and the normal generation method require ghost cells to ensure partition-invariant results.

Parallel communication and control classes encapsulate details of process initialization and communication libraries, such as a shared-memory implementation or MPI. The streaming and parallel computing features can be accessed both through a C++ library interface and through a set of scripting interfaces. These feature extensions provide the basis for ParaView's large data functionality and performance requirements.

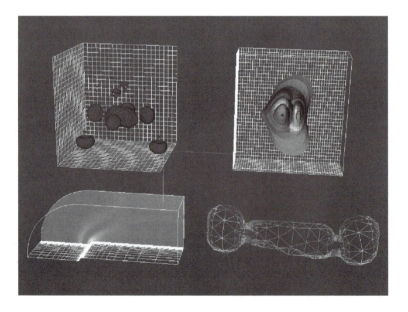

Figure 36.1 The figure shows the different types of datasets that VTK and ParaView can handle. The upper-left dataset is a uniform rectilinear volume of an iron potential function. The upper-right image shows an isosurface of a non-uniform rectilinear structured grid. The lower-left image shows a curvilinear structured grid dataset of airflow around a blunt fin. The lower-right image shows an unstructured grid dataset from a blow-molding simulation. (See also color insert.)

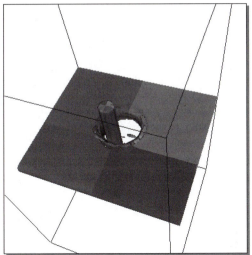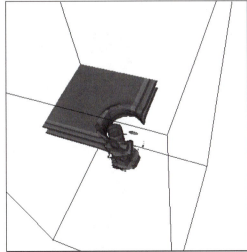

Figure 36.2 The image on the left was generated by a CTH dataset partitioned into eight pieces. Each piece was assigned a different color. The image on the right shows only one partition with six extra ghost levels. The cells are colored by ghost level. In practice, usually only one ghost level is necessary. (See also color insert.)

36.3.3 ParaView

ParaView provides a GUI for the interactive exploration of large datasets. It builds this functionality on parallel and distributed VTK. An overview of the tool from a user perspective is presented first, followed by a technical description of how the tool's functionality is achieved.

36.3.3.1 Overview

A sample ParaView session is shown in Fig. 36.3. There are several regions in the user interface, including the Menu Bar along the top of the application, the Toolbar just below the Menu Bar, the Left Panel on the left side, and the Display Area on the right side. Each of these areas is described in more detail below.

- **Menu Bar:** The top menu bar provides menu buttons for loading and saving data, creating sources and filters, viewing other windows, displaying help, and other standard functions.

- **Toolbar:** The toolbar contains buttons for resetting the camera, switching between 2D

and 3D interaction modes, and changing the center of rotation. In addition, the Toolbar contains shortcut icons to instantiate some commonly used filters.

- **Left Panel:** The top portion of this panel contains the selection or navigation window. The selection window provides a list of instantiated sources and filters. The navigation window provides a data-flow program graph representation of the user's task. The area below the selection/navigation window is where the properties of sources and filters are set, and we refer to it as a property sheet. Property sheets contain such module settings as the current isosurface values computed by the isosurface module.

- **Display Area:** The display area is where the 3D representation of the scene is rendered. Mouse and keyboard interaction are provided in this area.

To add new filters, the user selects a source or filter from the Source or Filter menu on the Menu Bar. Sources include various types of readers or

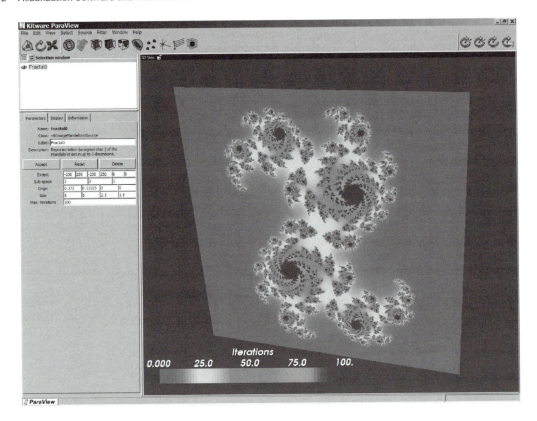

Figure 36.3   ParaView. (See also color insert.)

computer-generated sources. A sample of the possible filters includes the following:

- *Contours* and *isosurfaces* can be extracted from all data types using scalars or vector components. The results can be colored by any other variable or processed further. When possible, structured data contours/isosurfaces are extracted with fast and efficient algorithms that make use of the structured data layout.

- Vector fields can be inspected by applying *glyphs* (currently arrows, cones, and spheres) to the points in a dataset. The glyphs can be scaled by scalar, vector component, or vector magnitude and can be oriented using a vector field.

- A subregion of a dataset can be extracted by *cutting* or *clipping* with an arbitrary

plane, specifying a *threshold* criteria to exclude cells, and/or specifying a volume of interest (for structured data types only).

- *Streamlines* can be generated using constant step or adaptive integrators. (A parallel implementation of streamlines is not currently available; this feature is under development.) The results can be displayed as points, lines, tubes, and ribbons, and they can be processed by a multitude of filters.

- The points in a dataset can be *warped* or displaced with scalars or with vectors.

- With the *array calculator*, new variables can be computed using existing point or cell field arrays. Many scalar and vector operations are supported.

- Data can be *probed* on a point or along a line. The results are displayed either graphically or as text and can be exported for further analysis.

ParaView provides many other data sources and filters by default, including edge extraction, surface extraction, reflection, decimation, extrusion, and smoothing. Any VTK source or filter can be added to ParaView if the user provides a simple XML description for its user interface for its property sheet.

The Source and Filter menus are dynamically updated to contain a list of sources/filters that can input the output of the currently selected module. The selected module is either the last module created or the one most recently selected from the Selection/Navigation window. Once a module is chosen, a new instantiation of the module is created and connected to the selected module, and the module's property sheet is displayed. In this manner, a data-flow program graph is created. In order to manipulate or view the properties of a module, the module is selected and its property sheet shown, and the user can view or edit the listed values.

36.3.3.2 *Meeting the Workflow Requirements*

ParaView is simple because it minimizes the knowledge of data-flow programming required by users to use the tool. Specifically, a user can specify simple tasks (e.g., creating a source and applying simple filters) without needing to be aware of data-flow programming. This is because ParaView's default behavior is to add new modules to the last module created. When the user wants to change this behavior, for example, by applying another filter to the source data, he or she can use the Selection Window to reset the location where the new module will be added. ParaView also simplifies the choice of modules by listing only modules that accept the correct data type for insertion.

For advanced users who wish to create complex program graphs, the program graph is available for manipulation in the Navigation window. ParaView is designed so that visualized results dominate the GUI real estate and the manipulation of program graphs is relegated to a much smaller area. This allows scientists to focus on their visual analysis and not on visual programming, which is typically of secondary importance to them.

When modules are instantiated in ParaView, they create visual output in the display area that provides immediate visual feedback to the user about the data and the effect of the applied module. For example, as shown in Fig. 36.3, when the user creates a 2D source, in this case, a fractal source, ParaView automatically creates a color mapping of the data. This feature improves ease of use because it provides default settings, freeing the user from having to create them. This feature does have a down side: it hampers the ability to stream data, since every module instantiation would cause ParaView to stream visual results. For now, we have chosen to permit data streaming only in batch mode. Solutions to this problem include offering the option of turning on and off interactive data streaming and offering the option of turning on and off the immediate feedback feature.

Users can change the parameters of some modules directly by interacting with the 3D view shown in the Display Area using 3D manipulators. For example, the user can manipulate the seed line of a streamtrace filter by clicking on a control point and dragging the line to the new location. There are also 3D manipulators for probing a dataset with a point or line and cutting or clipping a dataset with a sphere or plane. These 3D manipulators improve ease of use by allowing users to quickly apply visualization modules to datasets by using the mouse to select location parameters, instead of setting the parameters, numerically in the user interface. When the manipulator is adjusted interactively in the 3D view, the numerical values of the location parameters are updated in the user interface and the user can then fine-tune these values.

ParaView is portable to most architectures. To achieve portability, only packages that work across many platforms were used in developing ParaView. For example, to achieve a portable user interface, Tk was chosen. Tk is the GUI companion to the Tcl scripting language. The application framework at the core of ParaView is a unique blend of Tcl/Tk and C++. Tk is used as the widget set, but C++ objects, which provide higher-level user interface components, are created to encapsulate the widgets. Like VTK objects, these C++ user interface objects are automatically wrapped in Tcl.

ParaView's user interface can be modified and extended both statically, with XML configuration files, and dynamically, at run time, using the Tcl scripting interface. All ParaView modules and their corresponding user interfaces are initialized by parsing XML-based configuration files. These files contain the input/output types of the modules, an icon name to be displayed on the toolbar, a list of widgets to be displayed on the module's parameter page, the corresponding module parameters, and, in the case of reader modules, information about the file type. For example, Fig. 36.4 presents the XML description listed in the ParaView default configuration file and corresponding user interface for isoline/isosurface modules.

```
<Module name=''Contour''
    class=''VTKPVContour''
    module_type=''Filter''
    root_name=''Contour''
    button_image=''PVContourButton''
    output=''VTKPolyData''
    input=''VTKDataSet'' >
<Filter class=''VTKPVContour
    Filter''/>
<InputMenu id=''im'' label=''Input''
    trace_name=''Input''
    input_name=''PVInput''
    input_type=''VTKDataSet'' / >
```

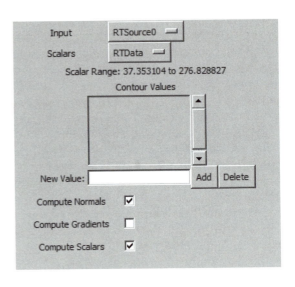

Figure 36.4 XML description for the isoline/isosurface module in ParaView and corresponding user interface generated by XML.

```
<ArrayMenu id=''am''
    input_name=''Input''
    attribute_type=''Scalars''
    label=''Scalars''
    input_menu=''im''
    number_of_components=''1'' / >
<ScalarRangeLabel array_menu=
    ''am'' />
<ContourEntry label=''Contour
    Values'' trace_name=''Contour
    Values'' / >
<LabeledToggle label=''Compute
    Normals'' variable=
    ''ComputeNormals'' / >
<LabeledToggle label=''Compute
    Gradients'' variable=
    ''ComputeGradients'' / >
<LabeledToggle label=''Compute
    Scalars'' variable=
    ''ComputeScalars'' / >
</Module>
```

Since all ParaView widgets have corresponding Tcl representations, the user can modify the GUI at run time by loading scripts or typing commands at a command console. This allows the user to, for example, add new widgets, create dialog windows, or load additional libraries at run time. These features can be used to customize which modules are loaded and how they are presented to the user. A visualization developer can edit the ParaView GUI configuration file and write custom scripts to customize ParaView for use by specific application users. For example, for the climate modeling community, a configuration file could be written to add a suite of climate analysis modules and customize the existing modules, such as contouring, to meet community conventions. ParaView meets the accessibility requirement because it is available in open-source form.

Recall that a coarse approximation of the simulation scientists' workflow includes two modes: an interactive mode, in which an interactive GUI-based tool is used to explore a dataset, and a batch mode, in which a program is executed to create an animation. ParaView supports both of these modes. ParaView's interactive mode was detailed in the overview section. Every interaction with the ParaView GUI can be saved in a session file, since every interaction has a corresponding Tcl command. The session file can be reloaded to reproduce the session. Furthermore, since the session file is a Tcl script, it can be edited or modified and then reloaded to obtain different results. In addition to dataset exploration, the interactive mode is also used to create programs to run in batch mode. ParaView also supports the ability to save the current program graph as a VTK script. A series of queries allows the user to customize the script. A session script differs from a VTK script in that a session script saves every ParaView interaction (i.e., every interaction used to create a program graph), and a VTK script saves only a program graph.

36.3.4 Meeting the Large-Data Visualization Requirements

ParaView supports large data visualization via techniques that include parallel processing, level-of-detail (LOD) rendering, and data streaming.

36.3.4.1 Parallelism and Data Streaming

ParaView supports parallelism, using either shared-memory processes or distributed-memory processes via MPI. When ParaView is run in parallel, data is automatically read, processed, and, if necessary, rendered in a data-parallel manner. ParaView's parallel software architecture includes three types of processes: a client process that runs the GUI, and two types of server processes: root and slave processes. The client process communicates with the root server process by broadcasting commands. The root server process receives these commands and rebroadcasts them to all slave servers. A command is a script; currently, a Tcl script is used. In the future, the user may be able to select the scripting language to use, for example, Python or Java. All servers run an interpreter and use it to execute received commands. This communication mechanism is used to create a copy of the same program graph on each process. The program graphs on the servers manipulate pieces of the full dataset for data-parallel processing, and the program graph on the client is used to store program states, such as parameter settings for modules. ParaView's user interface elements update the client's program graph, and the changes are sent as scripts to the root and slave servers. For example, when a user creates a filter module (such as an isosurface), a script is created that instantiates, sets parameters, and appends the isosurface module to a program graph. This script is then communicated to and interpreted by both the client and the server processes. All processes use the same naming convention, and thus one script works for all processes' program graphs. All program graphs are initialized with a rendering module.

To implement parallel algorithms, communication between processes is handled internally by the modules in the program graphs. For example, all rendering modules communicate to implement a parallel rendering algorithm. Although the client can be considered a centralized executive, ParaView's design supports independent process decisions and actions as much as possible. For example, the decision to allocate memory occurs locally. Furthermore, interprocess communication is limited to program instantiation and execution commands and parallel algorithms.

ParaView supports data streaming in batch mode. When the user writes a batch script, an option prompts the user to indicate whether he or she would like to stream the data and what memory limit he or she is bounded by. Streaming and data parallelism are effective techniques for processing large datasets, and they effectively fulfill ParaView's large-data visualization requirement. We have used these techniques to efficiently and effectively isosurface and color a collection of datasets that ranged in size from tens of gigabytes to approximately a petabyte in size [8].

36.3.4.2 Level of Detail and Parallel Rendering

ParaView's rendering module supports both LOD and parallel rendering techniques [9] to facilitate the interactive rendering of very large datasets. Interactive rendering of large datasets is an extremely challenging problem, and therefore, we applied a number of techniques to achieve good performance.

LOD techniques increase rendering performance at the expense of image quality. Two dif-

ferent LOD techniques are used in ParaView: geometric LOD and rendered-image LOD. The geometric LOD technique creates a model with reduced geometric resolution. In general, models with fewer geometric elements render faster than those with more elements. When the user is interacting with the model (i.e., rotating, translating, zooming), a reduced-resolution model is used in order to render quickly. When the interaction is complete, the full-resolution model is rendered. Fig. 36.5 shows an example of the full- and reduced-resolution models. VTK's quadric clustering algorithm is used to simplify surfaces. This algorithm preserves data attributes, so the LOD models have the same visual appearance as the original data. Timing results for the decimation algorithm are given in Table 36.1. Decimation can introduce significant visual artifacts into the rendered image. However, we have concluded that these artifacts are acceptable during interactive rendering. Decimation can also work well with the geometry redistribution technique discussed later in this section. The smaller decimated models can easily be collected and rendered locally.

The rendered-image LOD technique involves rendering to a small image and using pixel replication to magnify the image for final display. It is essentially lossy compression of the image

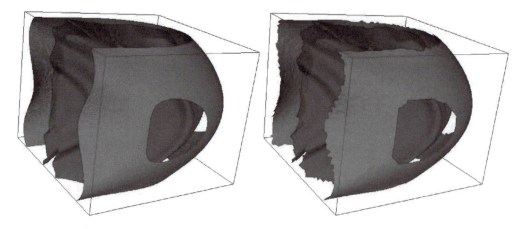

Figure 36.5 The full-resolution surface (left) has 821,495 triangles and renders in intermediate mode in 0.93 seconds on a GeForce2 GTS, and the decimated surface (right) has 35,400 triangles and renders in 0.04 seconds. (See also color insert.)

Contained in this third color section are Figures 41.7 through 47.14.

(a)

(b)

Figure 41.7 (a) A 2-second eye scan for an observer counting the teapots. The crosses are fixation points and the lines are the saccades. (b) Perceptual difference between selective-quality (SQ) and high-quality (HQ) images using Myszkowski's Visual Difference Predictor [20]. The red denotes areas of high perceptual difference. The superimposed blue line is the eye-scan path for another observer counting the teapots.

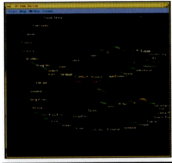

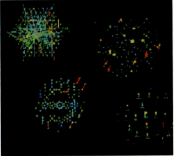

Figure 42.1 Node and link network displays.

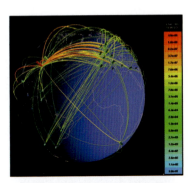

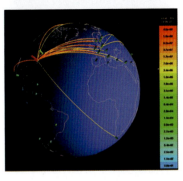

Figure 42.4 (Top) One frame from an animation showing worldwide internet traffic over a 2-hour period. (Bottom) Thresholding the link statistics with a slider (not shown) highlights the most important links and decreases display clutter.

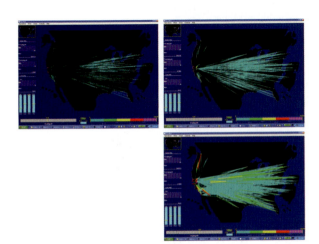

Figure 42.3 Three frames from a SeeNet network visualization animation.

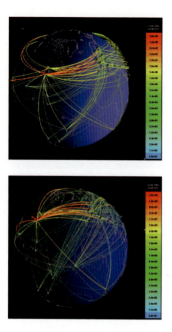

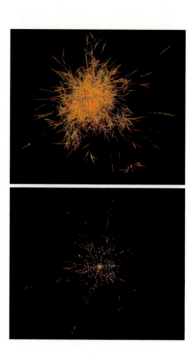

Figure 42.5 (Top) Rectangular latitude–longitude paths. (Bottom) Making the globe translucent reveals obscured arcs.

Figure 42.6 Images from a StarGraph visualization showing network structure for an IP network. (See also color insert.)

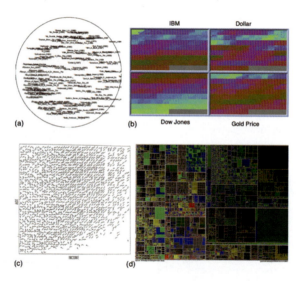

Figure 43.2 Some popular information visualization techniques. (a) Geometrically transformed displays: Interactive visualization of high-dimensional data using the hyperbolic plane [36]. Genre separation in movie space (red "x" marks science fiction, black "Δ" marks animation, and green "+" movies belonging to both genres) (© ACM). (b) Dense pixel displays: Recursive Pattern [4]—based on a generic back-and-forth recursive arrangement schema to represent each data value as a colored pixel and each attribute in separate sub-windows (example visualization shows the stock prices for Dow Jones, Gold, IBM, and US Dollar are depicted for almost 7 consecutive years, 7 vertical bars correspond to the 7 years (level (3)-patterns) and the subdivision of the bars to the 12 month within each year (level (2)-patterns), the coloring maps high attribute values (stock prices) to light colors and low attributes values (stock prices) to dark colors) (c) Iconic displays: Stick Figures [24,23]—visualization of multidimensional data using properties of angle and/or length of the limbs (US Census Data Median Household Income and Age of Householder). (d) Stacked displays: TreeMaps [9,31]—splitting the screen into rectangles in alternating horizontal and vertical directions in each level (example visualization shows a hierarchical file system of a large hard disk).

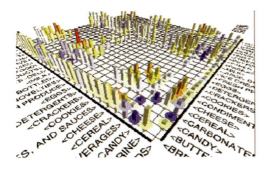

Figure 43.4. MineSet's Association Rule Visualizer [17] maps the left- and right-hand sides of the rules to the x- and y-axes of the plot and shows the confidence as the height of the bars and the support as the height of the discs; color of the bars shows the interestingness of the rule (example visualization shows market basket data for customer buying patterns). © SGI.

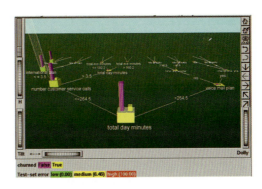

Figure 43.6 MineSet's Decision Tree Visualizer [17] displays decision trees as 3D landscapes; each node contains bars whose height, color, and disk correspond to important parameters. © SGI.

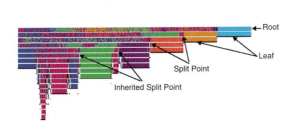

Figure 43.7 Visual Classification [3] shows attribute values by colored pixels arranged in bars (here we see a decision tree for DNA segment data with 19 attributes). © ACM.

Figure 43.8 Visualization based on a projection into 3D space [39]: 3D cluster-guided projection, where the 3D subspace is determined by centroids of 4 clusters 0, 1, 3, 5. © ACM.

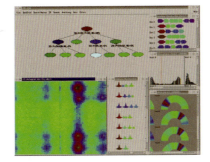

Figure 43.10 *HD-Eye* screenshot [14] showing different visualizations of projections and the separator tree. Clockwise from the top: separator tree, iconic representation of 1D projections, 1D projection histogram, 1D color-based density plots, iconic representation of multidimensional projections, and color-based 2D density plot (example visualization shows a large molecular-biology dataset). © IEEE.

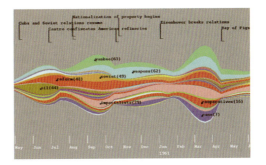

Figure 43.11 ThemeRiver [13]: visualization of thematic changes in documents (example visualization shows Castro data from November 1959 through June 1961). © IEEE.

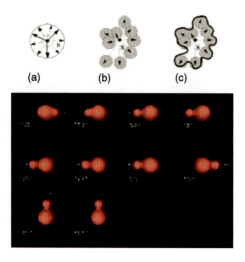

Figure 43.12 Shape-based Visual Interface for Text Retrieval [27]: shape-based visualization of query results for the key words lion, sheep, mouse, and wolf. © ACM.

Figure 44.1 Early GCM.

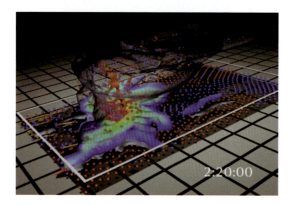

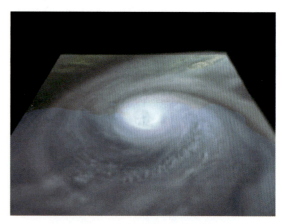

Figure 44.2 A transparent isosurface is combined with components of other earlier animation segments and includes moving balls representing air motion, a colored horizontal slice through the storm showing the region of precipitation, and a twisting ribbon rising up through the storm. The animators referred to this as the party scene. The rate of twist of the ribbon is related to the magnitude of the streamwise vorticity present in air rising through the storm. Such vorticity is analogous to the rate of rotation of a spiraling football as it moves from the quarterback to the receiver. To fully understand the behavior illustrated, it is important to look at the animation. For example, the colored balls released at regular intervals in a horizontal plane are colored blue when they are sinking and orange when they are rising. In the early portion of the animation, the alternating regions of blue and orange move away from the storm, revealing the presence of wave motion. The various visualization idioms used here are still used today.

Figure 44.5 A volume visualization of a Hurricane Opal simulation was created using Renderman by NCSA's David Bock with land terrain generated by NCSA's Rob Stein. The volume rendered view is based on the model water vapor field.

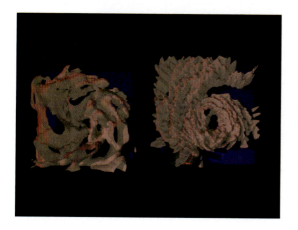

Figure 44.7 Isosurface visualization of Typhoon Herb and radar return signal, with simulated data on the left and observed data on the right. The vertical axis is the height of the domain, and the domain is roughly over Taiwan.

Figure 44.9 Visualization of a developing tornado revealed through the use of thousands of circulating particles.

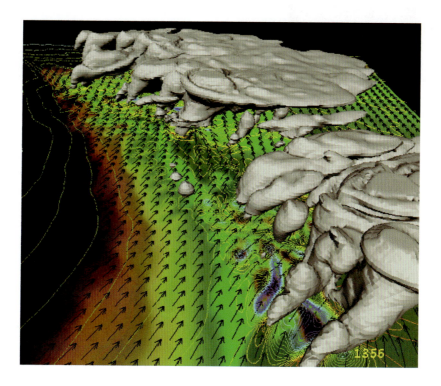

Figure 44.10 Model results from the simulation of the April 19, 1996 tornado outbreak in Illinois.

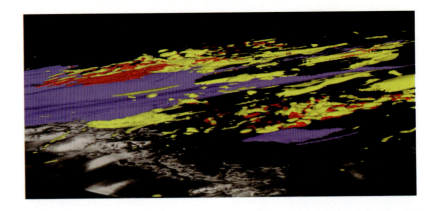

Figure 44.11 Simulation of clear air turbulence with jet stream in blue, enstrophy in red, and fast moving air parcels in yellow. The visualization was created using a combination of IDL, Vis5D, and the Persistence of Vision (POV) raytracer run on a 128-processor SGI Origin system.

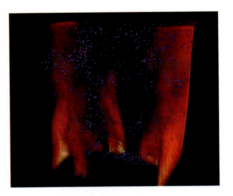

Figure 44.12 Volume visualization of a forest fire and "burning brand" lofting (particles). Image courtesy NCAR.

Figure 44.13 Visualization of turbulence and heat fields evolving in a fire model. Image courtesy NCAR.

Figure 44.15 Visualization of water vapor from an experimental version of the Community Climate System Model (CCM3) at T170 resolution. Water vapor is rendered in white. Image courtesy NCAR.

Figure 44.18 Tiled wall display of ensemble WRF model simulation. Image courtesy NCSA.

Figure 44.19 GeoWall in a geology labroom. Image courtesy Paul Morin, University of Minnesota.

Figure 44.20 A classroom visualizing weather data on NOAA's Science on a Sphere (SoS) display. Image courtesy NOAA.

Figure 44.21 The AccessGrid, hosting the CHEF portal, hosting the NOAA Live Access Server (LAS), hosting data and visualization for the Vegetation and Ecosystem Mapping Project (VEMAP).

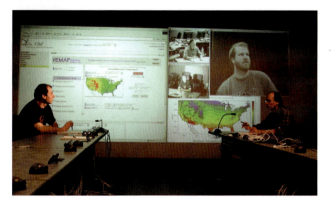

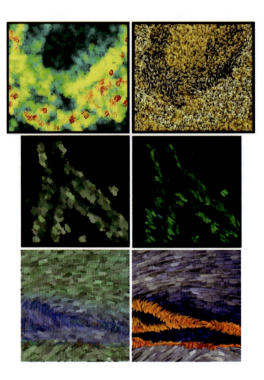

Figure 45.2 Meier layered strokes to build up computer paintings, much as painters layer their strokes to build up an oil painting. The stroke layers are shown here as they accrete. Here, the layers are organized around form and lighting, but other organizing principles can work in other contexts.

Figure 45.3 Several early "painterly" visualizations. We experimented with varying the visual representation of underlying data by changing stroke shapes, texture, color, size, and placement. The left and right image in each pair use the same underlying data.

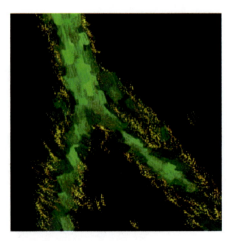

Figure 45.4 Scumbling, or lightly painting over an already-painted region, is an example of a painting technique we mimicked. We used small strokes to emulate the small bits of paint left behind.

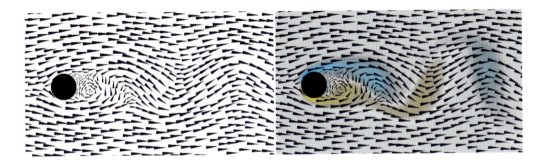

Figure 45.5 Typical visualization methods for 2D flow past a cylinder at Reynolds number 100. On the left, we show only the velocity field. On the right, we simultaneously show velocity and vorticity. Vorticity represents the rotational component of the flow. Clockwise vorticity is blue; counterclockwise is yellow.

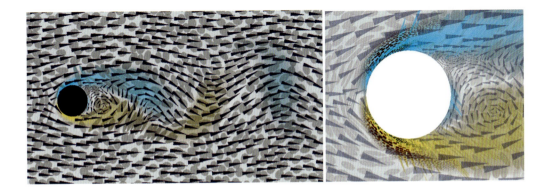

Figure 45.6 (Left) Visualization of 2D flow. Velocity, vorticity, and rate of strain (including divergence and shear) are all encoded in image layers. (Right) Additional values for turbulent charge and turbulent current for Reynolds number 100 flow are added to the visualization. A total of nine values are simultaneously displayed.

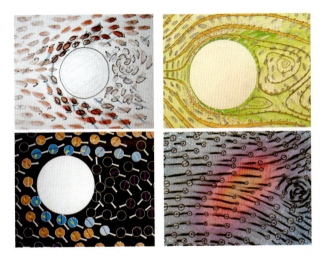

Figure 45.7 Students in a joint computer science/art scientific visualization class generated creative multivalued 2D flow visualizations.

Figure 45.8   Examples of 2D flow visualizations developed by students in a SIGGRAPH 2001 course.

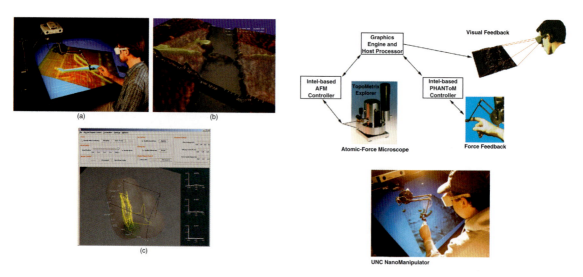

Figure 46.1   (a) UNC nanoManipulator being used to manipulate carbon nanotubes. (b) UNC NIMS overlays SEM image on AFM surface display. (c) UNC's 3D Force Microscope views and manipulates biological samples.

Figure 46.3   nanoManipulator system diagram. Separate threads control data acquisition, visual rendering, and force display.

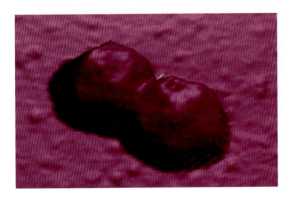

Figure 46.6 Two icosahedral adenovirus capsids. The nano-Manipulator was used to press a dimple into the one on the right.

Figure 46.7 The nanoManipulator was used to push a carbon nanotube on top of two others; haptic feedback enabled the scientist to peel apart the lower tubes.

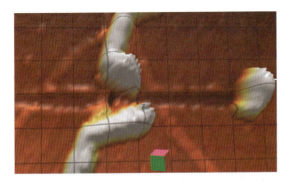

Figure 46.8 Here, the nanoManipulator system is being used to measure the rupture strength of a fibrin fiber.

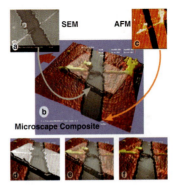

Figure 46.9 The NIMS system combines the display of SEM and AFM data from the same specimen.

Figure 46.15 Photograph of four magnetic poles in a 3DFM surrounding a specimen mounted between two cover slips.

Figure 46.16 3DFM user interface prototype. The current position of the bead is indicated by the green wire-frame sphere. Its trajectory is shown by the yellow trace. The brownish translucent surface shows the volume carved out during travel. The grey-scale plane shows live video from an optical microscope centered on the bead.

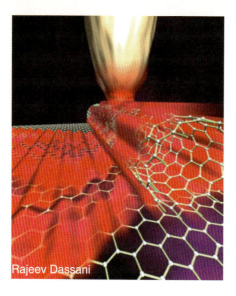

Rajeev Dassani

Figure 46.17 Artistic conception of constrained electron flows between carbon nanotube and surface. Charge is injected from the AFM probe above the nanotube.

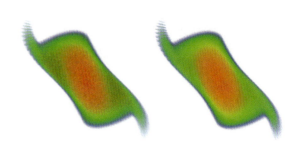

Figure 47.1 Comparison of (left) a volume rendering and (right) a mixed volume/point rendering of the phase plot (x, P_x, y) of frame 170. The volume rendering has a resolution of 256^3. The mixed rendering, with a volumetric resolution of 64^3 and 2 million points, provides more detail than the volume rendering while displaying at a much higher frame rate.

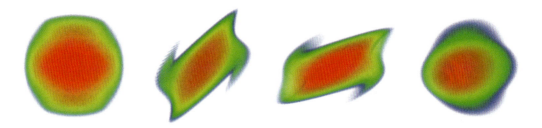

Figure 47.2 Selected distributions for time-step 180. From left to right: (x, y, z), (x, Px, y), (x, Px, z), and (Px, Py, Pz).

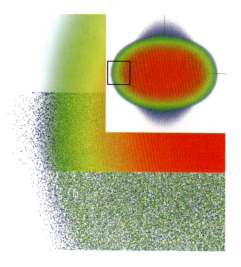

Figure 47.4 Portions of a hybrid rendering on a sphere-like (x, y, z) distribution showing (top) the volume-rendered portion, (middle) the combined hybrid rendering, and (bottom) the point-rendered portion alone. The front half of the sphere has been clipped; the points obscured by the volume rendering are on the far side. The points shown here are completely opaque so that they are more visible.

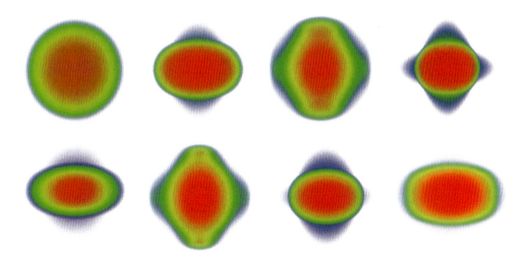

Figure 47.5 Selected time-steps from a simulation over 350 time-steps for the x-y-z distribution of the data. (Top) frames 1, 50, 100, 150. (Bottom) frames 200, 250, 300, 350.

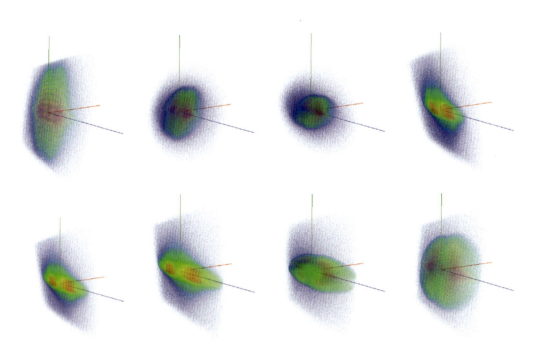

Figure 47.6 Selected frames from the visualization of a misoriented particle beam using the hybrid rendering technique.

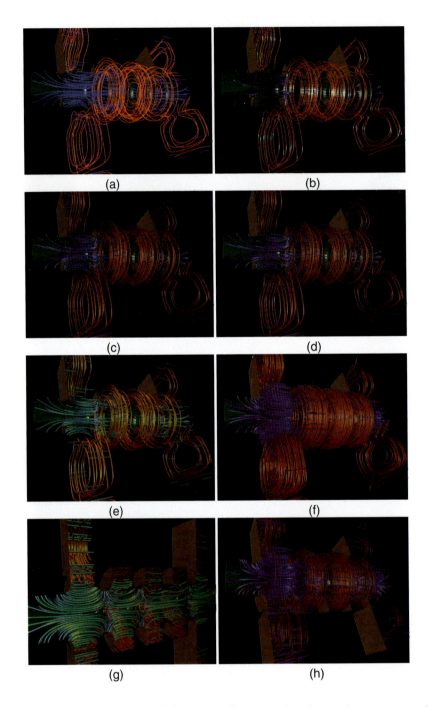

Figure 47.10 Visualization of an electromagnetic field corresponding to a section of an accelerator structure. (a) Conventional line drawing; (b) illuminated streamline technique; (c) conventional streamtube technique; (d) self-orienting surface technique; (e) with enhanced lighting; (f) dense lines; (g) cutting away; (h) use of transparency.

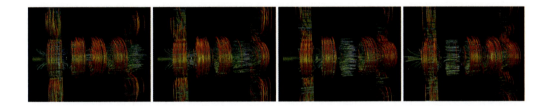

Figure 47.11 Selected time-steps that show RF waves propagating in through the input ports (left) and out through the output ports (right).

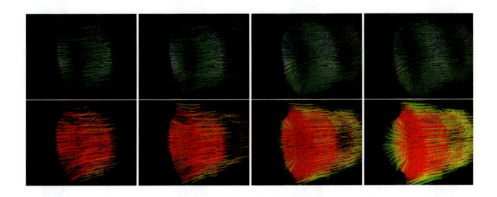

Figure 47.12 From left to right, incremental loading and rendering of electric field lines. (Top) with the line opacity proportional to local field strength. (Bottom) with enhancement using both color and transparency.

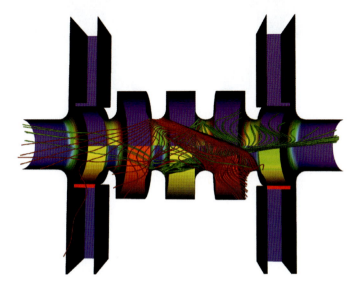

Figure 47.13 Particle tracking visualization.

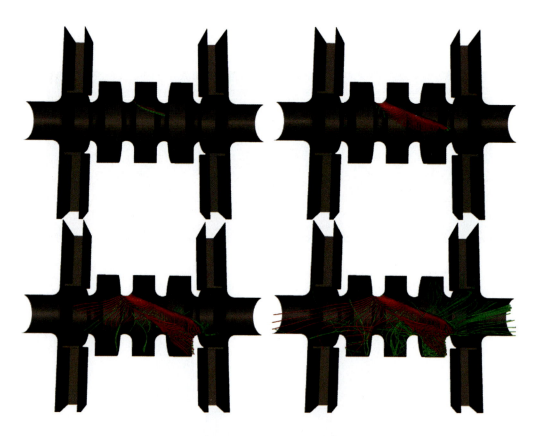

Figure 47.14 Primary particles released at evenly spaced time intervals follow the paths shown in red. When these particles hit the walls of the accelerator structure, they can liberate secondary particles, which follow the paths shown in green. (See also color insert.)

Table 36.1   Time to decimate based on number of processors

Number of Processors	Time To Decimate
1	6.25 Seconds
2	3.515 Seconds
4	1.813 Seconds

that has no compression cost and minimal decompression cost. Although the resulting visual artifacts are noticeable, they do not significantly impair interaction with the dataset. This technique reduces rendering times through all the steps listed below. Initial rendering times can be faster because fill areas are smaller. The time to read and composite image buffers is reduced as well because it is directly proportional to the area of the rendered image.

Using data parallelism in the renderer is also critical for high performance. The renderer is always in one of two states: an interactive state, when the user is interacting with the GUI, and a still state, when the user is not. The steps of the rendering algorithm are described below. Note that ParaView's geometry and image data can be either serial or parallel, and this may change during execution. The algorithm below is applied to each geometric object to be rendered.

1. **If (rendering state is interactive) then apply geometric and image LOD algorithm**:

   a. The geometric LOD algorithm is applied when an estimate of the time to render the object (based on the number of points in the object) exceeds a user-modifiable threshold. When the threshold is exceeded, rendering occurs with a reduced-resolution version of the object. If a reduced-resolution version does not exist, one is created.

   b. The image LOD algorithm is applied when the time to render the last frame exceeds a user-modifiable threshold. Using the previous time frame as an estimate, a new image size is calculated in order to reduce rendering time to below the threshold.

2. **If (geometry data is parallel) then apply parallel geometry load redistribution algorithm**: The result of the LOD algorithm is geometry data. If there is parallel geometry data, it can be redistributed from its current location on the processes to a subset of the processes. For example, if the geometry is small enough (i.e., after it is reduced by Step 1a), it can be more efficient to collect and render the geometry on a single process. This avoids the cost of parallel image compositing (Step 4). In the future, this step will also be used to balance geometric load across processes for more efficient performance.

3. **Rendering**: The result of the redistribution algorithm is geometry data. A rendering operation then renders this geometry to create an image and depth buffer result of the image-size set in Step 1b. Rendering can be serial or parallel, hardware- or software-based, and occur either onscreen or offscreen.

4. **If (image data is parallel) then apply parallel image compositing**: If there is parallel imagery, then this image data is composited together using the depth buffer to resolve conflicts and to create a final image. ParaView currently supports a binary tree compositing, with the option of using run-length encoding to losslessly compress images for speed [10]. With large window sizes and many processes, this communication time can be the major factor limiting rendering performance. Compositing transmission time grows linearly with render-window area and scales logarithmically with the number of processes. This is why both Step 1a and Step 1b offer methods (collecting geometry to a single process or compositing using smaller images) to either skip or speed up this compositing step.

Notice that different paths through these steps are possible. For example, a reduced LOD model can be rendered locally when the renderer is in the interactive state, and the full-resolution version of the same model can be rendered in parallel when the renderer is in the still state. Having the ability to render at different resolutions and speeds allows the user to interactively focus on an area of interest and then study the details of a full-resolution image, and it also meets the large-data rendering requirement.

36.4   Results

This section presents visualization results generated by ParaView for several application areas.

Fig. 36.6 shows isosurfaces of the Visible Woman dataset. The $512 \times 512 \times 1734$ dataset, which is 900 MB, is composed of seven sections.

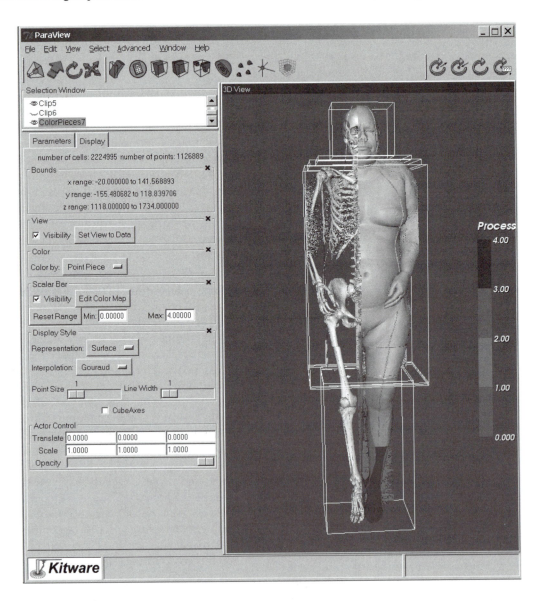

Figure 36.6   The Visible Woman dataset in ParaView. (See also color insert.)

Each section is a uniform rectilinear grid generated by a CT scan. Two isosurfaces were extracted, one for bone and one for skin. The skin isosurface was clipped in order to reveal the bone isosurface. One block of the skin was colored by process ID to show the data partitioning. ParaView was run with four server processes in this example. This example also demonstrates ParaView's ability to process multiblock datasets. Many structured datasets divide the domain into blocks. Each block is configured to get the best resolution sampling for its domain. Since some datasets can have hundreds of blocks, it is important to group these blocks into a single entity that can be filtered in one step. ParaView has group and ungroup filters, which simplify processing of multiblock datasets.

Fig. 36.7 shows streamlines generated by ParaView using a dataset of air flow around a delta wing. This example also shows the 3D line widget used to seed the streamline algorithm. The streamline filter propagates integration across partition boundaries and can execute in parallel. The delta wing and the contour surface were obtained by extracting sub-grids from the original curvilinear dataset. Since the actual dataset contains only half the wing, due to symmetry, a reflection filter was applied to all surfaces. Both surfaces were colored by mapping the stagnation energy through a default lookup table.

Fig. 36.8 shows the results of a batch script on the results from the Parallel Ocean Program (POP). The $3600 \times 2400 \times 40$ structured grid models the Earth's oceans at a resolution of one-tenth of a degree. Isosurfaces and extracted geometry are also used to represent land masses. A clip plane colored by temperature at a depth of 1140 meters is also shown. It is worth noting that climate-specific visualization tools are unable to process datasets of this magnitude.

Figure 36.7   The delta wing dataset in ParaView. (See also color insert.)

Figure 36.8   View of the Atlantic Ocean from a global one-tenth-of-a-degree simulation showing ocean temperature at a depth of 1140 meters generated by a ParaView batch script. (See also color insert.)

36.5   Conclusions

This chapter has presented the design of Para-View, an end-user tool for large-data visualization. ParaView provides a GUI for visualizing large datasets using techniques that include data parallelism, LOD, and streaming to meet its workflow and large-data visualization requirements. In the future, there are many directions in which to extend ParaView, including the incorporation of data streaming into the user interface and rendering support of extremely large datasets for tiled display walls. ParaView is an open-source tool, and members of the visualization community are invited to add new features.

Acknowledgments

This work was supported by grants from the US Department of Energy ASCI VIEWS program. We acknowledge the Advanced Computing Laboratory of the Los Alamos National Laboratory, where we performed portions of this work on its computing resources.

References

1. R. Knight, J. Ahrens, and P. McCormick. Improving the scientific visualization process with multiple usage modalities. LAUR-001619.

2. C. Upson, T. Faulhaber, Jr., D. Kamins, D. H. Laidlaw, D. Schlegel, J. Vroom, R. Gurwitz, and A. van Dam. The application visualization system: a computational environment for scientific visualization. *IEEE Computer Graphics and Applications*, 9(4):30–42, 1989.

3. G. Abrams and L. Trenish. An extended dataflow architecture for data analysis and visualization. *Proc. IEEE Visualization 1995*, pages 263–270, 1995.

4. S. G. Parker, D. M. Weinstein, and C. R. Johnson. The SCIRun computational steering software system. *Modern Software Tools in Scientific Computing* (E. Arge, A. M. Brauset, and H. P. Langtangen, eds.), Birkhauser Boston, Cambridge, Mass., 1997.

5. K. Misegades. EnSight's parallel processing changes the performance equation, http://www.ceintl.com/products/papers.html

6. W. Schroeder, K. Martin, and W. Lorensen. The design and implementation of an object-oriented toolkit for 3D graphics and visualization. *Proc. IEEE Visualization 1996*, pages 263–270, 1996.

7. W. J. Schroeder, K. M. Martin, and W. E. Lorensen. *The Visualization Toolkit: An Object-Oriented Approach to 3D Graphics.* Prentice Hall, Upper Saddle River, NJ, 1996.

8. J. Ahrens, K. Brislawn, K. Martin, B. Geveci, C. C. Law, and M. Papka. Large-scale data visualization using parallel data streaming. *IEEE Computer Graphics and Applications*, 21(4):34–41, 2001.

9. S. Molnar, M. Cox, D. Ellsworth, and H. Fuchs. A sorting classification of parallel rendering. *IEEE Computer Graphics and Applications*, 4(4):23–31, 1994.

10. J. Ahrens and J. Painter. Efficient sort-last rendering using compression-based image compositing. *Proc. of the Second Eurographics Workshop on Parallel Graphics and Visualization*, pages 145–151, 1998.

37   The Insight Toolkit: An Open-Source Initiative in Data Segmentation and Registration

TERRY S. YOO
Office of High Performance Computing and Communications
The National Library of Medicine, National Institutes of Health

37.1   Introduction

Unlike the other visualization systems and frameworks described in this text, the Insight Toolkit (ITK) does not, by itself, perform visualization. ITK is an open-source library of software components for data segmentation and registration, supported by federal funding through the National Library of Medicine (NLM) and National Institutes of Health (NIH) and available in the public domain. ITK is designed to complement visualization and user interface systems to provide advanced algorithms for filtering, segmentation, and registration of volumetric data. Created originally to provide preprocessing and analysis tools for medical data, ITK is being used for image processing in a wide range of applications from handwriting recognition to robotic computer vision. This chapter begins with an overview and introduction to ITK, its history, its design principles, and the resulting architecture of the collection image processing software components. The chapter also provides sources for obtaining additional information and locations for downloading ITK source code.

37.2   Background

In August of 1991, the NLM began the acquisition of the Visible Human Project (VHP) Male and Female datasets. The VHP male dataset contains 1,871 digital axial anatomical images (15 GB), and the VHP female dataset contains 5,189 digital images (39 GB) [1]. Researchers have complained that they are drowning in data, due in part to the sheer size of the image information and the available anatomical detail. Also, the imaging community continues to request data with finer resolution, a pressure that will only compound the existing problems posed by large data for analysis and visualization. The NLM and its partner institutes and agencies seek a broadly accepted, lasting response to the issues raised by the segmentation and registration of large 3D medical data.

In 1999, the NLM Office of High Performance Computing and Communications, supported by an alliance of NIH Institutes and Centers (ICs) and federal funding agencies, awarded six contracts for the formation of a software development consortium to create and develop an application programmer interface (API) and the first implementation of a segmentation and registration toolkit, subsequently named the Insight Toolkit (ITK). The final deliverable product of this group has been a functional collection of software components, compatible for direct insertion into the public domain via Internet access through the NLM or its licensed distributors. Ultimately, NLM hopes that this public software toolkit will serve as a foundation for future medical image understanding research. The intent is to amplify the investment being made through the Visible Human Project, and future programs for medical image analysis, by reducing the reinvention of basic algorithms. We are also hoping to empower young researchers and small research groups with the kernel of an image analysis system in the public domain.

A complete discussion of the ITK approach to segmentation and registration of medical data is beyond the scope of this chapter. Instead, we will concentrate on covering the user requirements and design principles that guided the construction of ITK. An understanding of this motivating perspective will help explain why certain choices were made in the construction of these tools. Later, we will explain some of the aggressive software engineering practices that the Insight Software Consortium has embraced. The supporting software systems for cross-platform builds, software wrapping for multiple language bindings, and automated build and regression testing will also be covered, followed by a brief tour of ITK. The chapter closes with a discussion of the integration of visualization systems with ITK, where to find additional information, and some observations on the future of the ITK.

37.3 An ITK Example

A critical problem in medical imaging is actually generating segmented data from which visualizations can be constructed. ITK targets the preliminary stages of visualization, namely segmentation (the partitioning of datasets into coherent, cohesive objects or structures) and data registration (the alignment or mapping of multiple, related datasets to a common coordinate frame or orientation). From its inception, ITK has been designed to address problems of large, multidimensional data, and it routinely considers data that is tens of gigabytes in size, sometimes three channel RGB 24-bit elements, and of three and sometimes four

spatial and/or temporal dimensions. Finally, while ITK does not itself generate visualizations, its components were designed to flexibly interface with a variety of graphical user interface (GUI) frameworks as well as visualization systems.

The example below (courtesy of Ross Whitaker and Josh Cates, University of Utah) applies a watershed segmentation technique to the VHP Female dataset. This 24-bit color dataset often presents complex challenges in visualization because of its density and size and the multiple channels of the voxel elements. The watershed technique creates a series of "catchment basins," representing local minima of image features or metrics (an example is shown in Fig. 37.1a). The resulting watersheds can be hierarchically organized into succesively more global minima, and the resulting graph (a schematic of a simple graph is shown in Fig. 37.1b), which links all of the individual watershed regions, can be generated as a preprocessing step. The graph hierarchy can later be used to navigate the image, interactively exploring the dataset through dynamic segmentation of the volume by selectively building 3D watershed regions from the 4D height field. ITK can thus be linked to a GUI as well as to a visualization back-end in order to create a functioning interactive segmentation and visualization system for complex data (Fig. 37.2a).

Fig. 37.2b shows the results of the semiautomated segmentation system. A user has defined the rectus muscle (highlighted on the left in red), the left eye (highlighted in pink), and the optic nerve (highlighted in yellow). These tissue types can be compared with the corresponding anatomy reflected on the right side of the subject, revealing the difference in the highlighting color and the relative fidelity of the segmentation. It should also be noted that, after preprocessing to find the primary watershed regions, navigating the hierarchy to build up regions is a natural and quick process with a GUI. The watersheds and their related hierarchies extend in three dimensions, making this segmentation example a native 3D application.

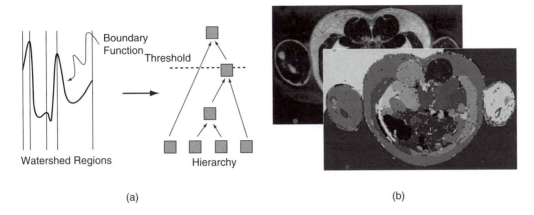

(a)   (b)

Figure 37.1   Watershed segmentation. (a) A 1D height field (image) and its watersheds with the derived hierarchy. (b) A 2D VHP data slice and a derived watershed segmentation. Example courtesy of Ross Whitaker and Josh Cates, University of Utah. (See also color insert.)

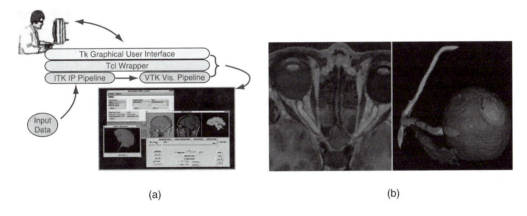

(a)   (b)

Figure 37.2   An ITK watershed segmentation application with visualization provided by VTK and with the user interface provided by Tcl/Tk. This example shows how ITK can be integrated with multiple packages in a single application. (a) Data-flow and system architecture for the segmentation user application. (b) Anatomic structures (the rectus muscle, the optic nerve, and the eye) highlighted with color overlays (shown in 2D but extending in the third dimension) and the resulting visualization. Example courtesy of Ross Whitaker and Josh Cates, University of Utah. (See also color insert.)

37.4   Background and History of the Toolkit

The NLM, in partnership with the National Institute for Dental and Craniofacial Research (NIDCR), the National Eye Institute (NEI), the National Science Foundation (NSF), the National Institute for Neurological Disorders and Stroke (NINDS), the National Institute of Mental Health (NIMH), the National Institute on Deafness and Other Communication Disorders (NIDCD), the National Cancer Institute (NCI), and the Department of Defense's Telemedicine and Advanced Technology Research Center (TATRC), has founded the Insight Software Consortium to support the creation of a public resource in high-dimension data-processing tools. The initial emphasis of this effort was to provide public software tools in 3D segmentation and deformable and rigid registration, capable of analyzing the head-and-neck anatomy of the VHP data. The eventual goal is for the consortium to provide the cornerstone of a self-sustaining software community in 3D, 4D, and higher-dimensional data analysis. The consortium is committed to open-source code and public software, including open interfaces supporting connections to a broad range of visualization and GUI platforms.

The Insight Software Consortium, including partners in academia and in industry, was formed to carry this work forward. The original developers include General Electric Global Research, Kitware, Inc., Insightful, Inc., the University of North Carolina at Chapel Hill, the University of Pennsylvania, Rutgers University, the University of Tennessee, Harvard Brigham and Women's Hospital, the University of Pittsburgh, Columbia University, and the University of Utah. Subsequently, an additional group of developers has been funded by the NLM as part of an expansion effort to grow the collection of algorithms and applications platforms. The expansion of the group has added the Imperial College of London, the Mayo Clinic, the Carnegie Mellon Robotics Institute, the Scientific Computing and Imaging (SCI) Institute of the University of Utah, Cognitica Corp., the University of Iowa, and Georgetown University to the list of supported developers.

The prime contractors and their subcontractors comprise the software research consortium, with the principal investigators of the prime contractors serving as the governing board. Together with the NLM Office of High Performance Computing and Communications as the executive member, the Insight Software Consortium is working to deliver a software toolkit to improve and enable research in volume imaging for all areas of health care.

The joint initiative is a continuing project that is now beginning to generate results and have an impact in the public and commercial sectors. After a 3-year development period, including a brief beta test of the toolkit, the developers' consortium released the first public version of the toolkit, ITK 1.0, in October 2002.

Subsequent releases have included at least three major revisions that updated the code, improved the toolkit organization, and added new examples and algorithms. The users' mailing list includes hundreds of users located in dozens of countries. ITK is beginning to appear in commerical efforts from small startup organizations to large corporations. The expansion effort is beginning to integrate ITK into established software systems, such as SCIRun from the University of Utah and Analyze from the Mayo Clinic. The effort is attracting new participants, and the number of active users is growing rapidly.

37.5 Design Principles

Even before the first meeting of the software developer consortium, the NLM declared that its intention was to create a public resource, an archive of working, operational algorithms that would be maintainable and sustainable through the coming years as new computing technologies emerged. We intentionally solicited 3D expertise to help develop multidimensional image-processing capabilities. We also purposely did not include visualization or user interface development in the program. From its inception, this project was intended as a companion and adjunct to visualization systems; it was never intended to supplant or compete with them. The purpose was to create open-source components to aid research and provide a collection point for the transfer of ideas and information.

As experience with ITK grows, we often receive questions from new developers and users asking why ITK has been structured the way that it is built. The data-flow architecture, generic programming practices, and object-oriented class organization are derived from our design principles and their subsequent software engineering decisions. This section and the next section explain the roots of the design of ITK, providing background that may answer many of the "why" questions about the design and programming of ITK. New and prospective users may wish to skip to the sections describing ITK and its contents and come back to these sections just as reference.

37.5.1   Open-Source Community Software

The Insight developers take the strong position that an essential vehicle for communicating software ideas is open access to source code. We recognize that many research communities become invested in a software system or product, often reducing the flexibility of software and sometimes the mobility of their personnel. Moreover, if new software is built on proprietary software foundations, the resulting new ideas are often difficult to port to other environments. We have created a public toolkit to encourage portable software ideas; open source is required to promote nonproprietary portable implementations as a means of communicating advanced ideas in medical image analysis.

37.5.2   A Working Archive

It has been said that computer scientists communicate the majority of their ideas in source code. Indeed, computer languages are now being accepted in some univerisities in lieu of foreign language requirements. Open-source software is considered an asset in computer science research and education. One of the most effective means of learning complex programming techniques is to read the program itself. Access to source code also encourages researchers to build upon the existing software base, making incremental refinements to the existing methods while reducing redundant work in basic programming of established methods. Thus, open-source software can assist small research groups that lack the means to support a large software infrastructure.

Finally, since much computer science research is evaluated by comparison of the accuracy, precision, robustness, and efficiency of multiple programs, public software resources provide shared implementations that can be used as benchmarks; if that software is source-code accessible, differences in programming techniques can be normalized and more effective comparisons can be made. By committing this project to a policy of open-source software, the Insight Software Consortium will provide all future software accompanied with the written source files as part of the toolkit.

37.5.3   Multiple Platforms, Multiple Language Bindings

The deliverables for this suite of contracts did not include user interface development, nor did they include visualization products. These programmatic decisions were not undertaken lightly, and they have their roots in the need for project focus and funding realities. Fortunately, the absence of user interface and visualization support reinforces a principle of versatility for the toolkit. By not providing a focus for user interface development, each team is encouraged to connect the toolkit to its own research environment, providing a *de facto* testbed for the API. Systems such as the Vesalius project viewers (Columbia), 3Dviewnix (University of Pennsylvania Radiology), and the Visualization Toolkit (VTK) (Kitware) have all already been promoted as visualization environments. Recent additional awards have been made to adapt ITK for the Analyze and the SCIRun software environments. Beyond these individual connections to the API, the consortium is actively pursuing the wrapping of toolkit functions to enable library calls in languages such as TCL, Perl, Python, and Java, facilitating access to the new tools from a variety of other systems.

NLM adheres to a policy of portable code. Unix (Linux, Irix, and Solaris), Mac OSX, and Windows-based environments are supported. Nightly builds and testing ensure the compilation and function of the same source code across multiple operating systems and CPUs.

37.5.4   Supportable Code

The Project officers recognized early that other similar programs have failed for lack of computer and user support. Although the toolkit may eventually be sustained by an active user community, NLM emphasized careful design of a software suite that can be supported. We accepted proposals that had strong components in systems architecture and systems integration. The technical evaluation team considered experience in supporting large collaborative software development programs a strong asset. Formal engineering practices, such as bug lists, enforced accountability for bug fixes, regression testing, etc., were adopted early in the framework of this program.

User training, courseware, tutorials, textbooks, and continuing software maintenance are being developed as part of the software development initiative. A Software Guide is already available online, with a growing collection of examples available on many different algorithms in source-code form. Continuing maintenance and support await future funding, pending the success of the initiative, but NLM considers the support of the software a necessary element in the long-term success of this project.

37.5.5 Pipelining, Streaming, and Out-of-Core Computation

As a group, the Insight developers agreed that the trend in datasets is toward large data and that this trend will often outstrip the capacity of main memory to accomodate it. Moreover, multiple processors in one frame are becoming increasingly available in consumer-level computing platforms. The result is a specification for the Insight toolkit to capitalize on these technological trends by requiring the design to support pipelining, streaming, and out-of-core computation. This requirement positions ITK to take advantage of complex multiprocessing hardware when available and to accomodate datasets that will not fit in main memory. The stated intention of the group is to make the code that is necessary to support these features as unobtrusive as possible for the casual ITK programmer.

37.5.6 3D

A pervasive and insidious exaggeration in visualization and image analysis research is that methods are easily generalized from two to three dimensions. NLM is emphasizing the development and distribution of 3D segmentation and registration methods through this initiative. We recognize that 2D and $2\frac{1}{2}$ D methods have been developed and are available as products as well as freeware. The growth of the size of medical datasets is making these methods untenable. When medical volume data is represented as a stack of 1000 slices, visiting each slice or even every tenth slice in a routine segmentation pass is unrealistic. In addition, registration of multiple medical datasets is inherently a 3D problem. Any incorporation and distribution of 2D methods in the segmentation and registration toolkit will be incidental in pursuit of 3D (and higher-dimensional) methods for image processing.

37.6 Software Engineering Infrastructure

Many users of ITK often wonder why the software is structured in its current form. The consequences of our adoption of the founding principles for the Insight Toolkit have led to a fertile experiment in software engineering and program architecture. All of our design decisions have been based on the needs and requirements of the users and the developers as assessed by our group in our early architectural planning meetings. This section describes the reasons behind our design choices, grounded in the principles described above.

37.6.1 Toolkits as a Software Engineering Philosophy

The promotion of this development effort as a toolkit distinguishes it from most previous sponsored programs: this project is not simply a software repository or library, nor will it result in a single programming environment or application system. All programs included in this effort will have a common, unified architecture, will adhere to particular coding standards and conventions within the consortium, and will comply with common style sheets that have been approved by the consortium members. If multiple software objects require similar functions or capabilities, those common elements will be built into base classes shared by the entire toolkit. The resulting toolkit is modular, linkable to other code, and easily incorporated into other programming structures.

37.6.2   Data-Flow Architecture

The overall pipeline, data-flow architecture for ITK is a direct consequence of our desire to accomodate multiprocessor hardware and the requirement to permit out-of-core streaming implementations that can process very large datasets. A conventional procedure-based approach in which whole datasets are loaded, processed, and then written to output was not feasible for this design. Moreover, pipelining and streaming require a very different software organization, one that is accomodated best by a data-flow design. Thus, programming in ITK is mostly the creation of pipelines of software components; the "execute pipeline" call is often the last command of the program.

Many other systems use a data-flow architecture as a result of the same types of requirements or design motivations. AVS, VTK, IBM Data Explorer, and others all use a data-flow organization, some with easy-to-use graphical languages that allow users to have a visual representation of the software pipelines. The original development did not originally have a user interface component; however, the ITK classes, objects, and methods were designed with sufficient support to make graphical programming languages possible.

37.6.3   C++, Templates, and Generic Programming

Requirements for the code to be both supportable and compact created the need for the ITK

developers to adopt generic programming practices. Often times in image processing, algorithms must be implemented multiple times to accomodate the different atomic data types that comprise the input image and volume datasets. Thus, core code can bloat with examples of addImageInt, addImageFloat, addImageDouble, etc., not only creating a library or toolkit that is uncomfortably large but also increasing the maintenance burden whenever changes must be made to an algorithm, or, worse, when changes must be made to foundation classes that will propagate throughout all of the different instantiations and variations of an algorithm. Algorithmic variation is natural and necessary in this type of endeavor, but the multiple implementations based on the native atomic data type are an artificial variation that does not reflect fundamental differences in the mathematics of the algorithm itself.

Moreover, multiple implementations of an algorithm, method, or operator should not be required to accomodate images of varying dimensions. For instance, a toolkit should not need an "addImage2D" method and an "addImage3D" method, as well as a separate "addImage4D" method, in order to cover the simple image-processing task of adding two images, where those images might be 2D, 3D, or 4D. However, this idea of implementations being essentially dimensionless affects algorithm expression at the flow-control level, disrupting the basic notion of nested iterative loops (nested "for" loops).

ITK is built around the concepts of object factories, generic iterators, and templated generic objects. These concepts allow the details of flow control and atomic typing to be deferred to compile time, where the actual code for executing the process or algorithm will be instantiated and only those types and data structures needed for a particular application will be invoked, expressed, and compiled. Programming ITK is not the subject of this chapter. Interested readers should seek the ITK Software Guide [2]. The essential notion that we hope to convey here is that ITK is written using advanced features of C++, including a heavy dependence on templates. In order to invoke a feature or method, a programmer first declares the dimension and type of a generic object, such as a template. Subsequently, the pipelines built using those objects as input and output types, including file readers, writers, and imaging filters, will conform to the data being manipulated.

The result is a programming process that does require some acclimation; novice users will often be surprised and have to study the concepts before they can become comfortable with the style and structure. However, the ultimate result is that programmers and algorithm designers can operate on images without the burden of detail regarding the dimension or elemental types involved, leading to compact, readable, and supportable, yet versatile, high-level source code.

37.6.4 CMake and Dart—Tools for Building on Multiple Platforms

The need to run ITK on multiple platforms and have the source code remain supportable and maintainable creates particular challenges. Although there is a large install base of workstations running Microsoft operating systems on Intel-type CPUs, much of the image processing, segmentation, and registration algorithm development is still conducted on Unix-type systems. It is necessary, then, to accomodate the range of system types used by developers (Cygwin or Visual C++ on Windows, GNU C++ on Unix/Linux/Irix/Solaris/MacOSX, etc.) on a diverse spectrum of hardware that includes serial and multiprocessing architectures.

ITK developers chose to create a single base of source code that could be compiled for different system configurations. Older software development projects have experienced significant divergence in internal versions, as some features are available on PC systems but not available for Macintoshes, for example. ITK is constructed from a single software effort that is continually checked so that it can be compiled and run on the range of system types supported by the project. At no point in the development

and extension of ITK do we intend to have core features available only on some systems.

In order to support this mission, ITK has crafted new software engineering superstructure systems, including CMake and DART. CMake is a system for configuring ITK (and other toolkit software) for compilation under different programming environments. DART is a distributed build-and-test system capable of continuously checking the source code for compilation and run-time errors and reporting them on a web-accessible dashboard that illuminates bugs, design problems, and critical programming faults as they happen. The driving concepts behind these tools are not new; the tools are new implementations inspired by similar programs previously available. The refinements of these ideas, however, have led them to be reincorporated into older programming projects including VTK, demonstrating that open-source programming can leverage itself to benefit more than just the narrow constituencies of individual projects.

Both of these tools are essential to the operation of a distributed programming group and the maintenance of common, multiplatform core software. A complete discussion of these tools cannot be presented here, and references for finding more information are listed at the end of this section. Please note that these tools have been created as versatile components, capable of integration in projects other than ITK. Arising from a sponsored program in open-source software, both CMake and DART are available in the public domain in source-code form, inviting adoption and refinement by a broad community of programming professionals.

37.6.5 Multiple Language Bindings and Cable

Although ITK is written in C++, we recognize that the use of the toolkit may be predominantly handled in other programming environments. Interpretive languages such as Tcl/Tk, Python, Perl, and Java empower programmers with a more flexible setting for faster prototyping of complex tools. Since the basic premise of toolkits as software engineering tools is to create software components that are easily assembled into complex methods, supporting multiple languages is an essential mission for ITK.

Our heavy dependence on templates to create a generic programming structure later raises problems when we try to support toolkit calls to other languages and other programming environments. The basic underlying problem in linking programs across languages is to be able to use a parsing of the high-level language to provide sufficient information to be able to invoke a subroutine call at the machine level from a different language or programming stratum (such as an interpreter). However, our generic programs manipulate templates rather than atomic or user-defined data types, and the machine-level expression of these generic types does not occur until compilation. It is therefore difficult to observe the templated, generic code alone to create toolkit, library, or system calls for other languages.

The solution is Cable. Cable is an integrated series of programs and modifications that permits the wrapping of templated (and non-templated) C++ code so that subroutines, methods, and procedures can be called from other languages in compiled and interpreted systems. Cable begins with a complete template instantiation process followed by a routine parsing of the resulting C++ source code. This step has been enabled by a permanent modification to the GNU C++ compiler, an addition that has been accepted and embraced by the Free Software Foundation (the trustees for the GNU project). The parsed language structure is then written in an intermediate form as XML files, describing the parameters, types, and calling order of the subroutines to be wrapped. Using the XML descriptions, code can be generated that wraps the ITK calls with stub programs in Tcl, Python, and soon Java and Perl, as well as other languages, permitting access to ITK algorithms from a rich cross-section of programming and teaching settings.

Cable is described in depth in a technical article [3], and it serves as an important addition to the arsenal of tools for object-oriented programming. Like CMake and DART, Cable is open-source and is available in the public domain in source-code form. It is offered without restrictions to give programmers the capability to support multiple language bindings for all C++ library software.

37.6.6 Implications of Good Software Engineering Infrastructure

Along with the significant design and software engineering approaches pioneered by the ITK developers, we have adopted other programming practices, such as the inclusion of in-line documentation using Doxygen, a system for generating manual pages and online information directly from source files. These practices, as well as the practical derivation of concepts from user requirements to software design, have led to a strong, cohesive software infrastructure. Added to the array of developer support tools such as CMake, DART, and Cable, the resulting code is reasonably compact, is therefore more maintainable, and permits the nimble response of our developers to meet challenges in software integration and redesign.

37.7 The Elements of ITK

This section describes what classes and methods are available in ITK. We describe the contents of ITK briefly as primary classes, basic filters, the segmentation infrastructure, and the registration framework. Within the ITK release (or its associated supplements), users will find examples, applications, and algorithm-validation studies complete with test data. This treatment is necessarily neither exhaustive nor particularly deep. The great joy about open-source code, however, is that if you want a truly deep presentation of the software, you're permitted to read the source directly.

One common failing of open-source initiatives is that documentation is often insufficient to support a large body of application developers in creating derivative work. More complete documentation is available in the ITK Software Guide and online at *http://www.itk.org*.

37.7.1 Primary Classes

Basic data representation in ITK begins with three primary classes upon which many ITK methods are constructed. They are *image, pointset*, and *mesh*. An additional primary class, *spatial objects*, is emerging as an important adjunct to the existing classes for describing the meta-information regarding classes, objects, and types and their relations to one another. The three primary classes are detailed below.

- **Image**: an *image* is a multidimensional array type for holding pixel elements. Using generic programming principles, the image can be instantiated to contain pixels of arbitrary types. ITK supports scalar, vector, and RGB pixel types, and the toolkit has provisions for user-specified pixel types. In addition to arbitrary dimensions and pixel types, the image class also contains provisions for managing pixel spacing, permitting images of noncubic voxels to be easily stored and manipulated.

- **Pointset**: The *pointset* class is a basic class intended to represent spatial locations. As the name implies, a pointset is a collection of points in N-dimensional space. As with the image class, each pointset is defined and instantiated from a templated class, and points may be associated with values, also determined by instantiation with types. Pointsets are the base class for the ITK mesh class.

- **Mesh**: A *mesh* is an ITK class intended to represent shapes. Built on the pointset class, a mesh also contains connectivity information linking points, effectively converting them into vertices of a mesh. Since this class is derived from pointsets, it inherits properties from that class; thus, vertices in a mesh may have values attached to them.

The mesh and pointset classes can be declared to be either *static* or *dynamic*. The former char-

acteristic is used when the number of points or mesh vertices can be known in advance, yielding a fixed-memory implementation. Dynamic meshes or pointsets are used for modeling and manipulations applications in which points can be inserted and deleted and which thus require a dynamic memory management method such as a list structure.

37.7.2 Basic Filters

Although ITK was developed as a segmentation and registration toolkit for medical images, no method can exist without the capabilities of basic filters for noise suppression, contrast enhancement, local derivative measurement, and generation of multilocal features such as gradient magnitude. ITK contains an array of "basic" filters for the manipulation, extraction, scaling, filtering, smoothing, noise removal, and feature measurement of multidimensional data. Almost all of these filters can be applied to datasets of arbitrary dimension with a multitude of pixel/voxel types (from byte scalars to multichannel RGB values).

ITK naming conventions attempt to capture all of the information regarding a particular method in its name. This makes for awkward initial implementation sometimes, but it saves significant time in finding and debugging toolkit elements. Novice users appreciate the descriptive filenames, despite the burden of significant typing.

ITK contains a powerful collection of filter-based tools, from simple mean and median filtering to complex distance-mapping tools, grey-scale mathematical morphology operators, and boundary-preserving noise-removing anisotropic diffusion filters. To cover the spectrum of available filters would be prohibitive here. Given the descriptive nature of ITK labels, we will list just a sample of the filters available as examples in ITK, portraying a view of the breadth and sophistication of available ITK algorithms:

- MeanImageFilter
- MedianImageFilter
- DiscreteGaussianImageFilter

- SmoothingRecursiveGaussianImageFilter
- BilateralImageFilter
- BinaryMinMaxCurvatureFlowImageFilter
- BinaryThresholdImageFilter
- BinomialBlurImageFilter
- CurvatureFlowImageFilter
- DanielssonDistanceMapImageFilter
- GradientMagnitudeImageFilter
- MathematicalMorphologyBinaryFilters
- MathematicalMorphologyGrayscaleFilters
- MinMaxCurvatureFlowImageFilter
- ResampleImageFilter
- SigmoidImageFilter
- CurvatureAnisotropicDiffusionImageFilter
- GradientAnisotropicDiffusionImageFilter
- RGBCurvatureAnisotropicDiffusionImage-Filter
- VectorGradientAnisotropicDiffusionImage-Filter

All of these examples and more are illuminated in the ITK Software Guide [2]. The examples' underlying filters and their source code are freely available with the ITK toolkit.

37.7.3 Segmentation Infrastructure

Segmentation is the process of partitioning images into coherent (often contiguous) regions based on some property of similarity among the individual pieces. In medical images or volumes, these separations are often fuzzy and made more complex by the multichannel, multimodal properties of the data. Traditional approaches to segmentation have followed structural pattern-recognition techniques based on region growing and edge assembly or have followed statistical pattern-recognition techniques based on Bayesian classifiers and parametric and non-parametric partitioning of scatter plots and feature spaces. In recent years, the line between the structural and statistical approaches has become increasingly blurred as good ideas are

synthesized and hybridized from the mathematical and statistical foundations of both areas.

The result has been an explosion of segmentation approaches that incorporate the concepts of filtering and registration. An open-source approach such as the one offered by ITK encourages this type of crossover between subdisciplines, sharing the needed access to the basic implementation of each algorithm for incorporation into a complex stream of ideas comprising new techniques. ITK contains a sampling of some of this myriad of different methods, and it houses the components for assembling as-yet undiscovered combinations of methods that may solve new problems in image understanding.

Some of the related methods found in ITK include the following:

- Threshold methods
- Region growing
- Watershed methods
- Level-set segmentation
- Fuzzy connectedness and Voronoi hybrid method
- Gibbs prior and deformable model hybrid method
- Statistical pattern recognition

While it would be impossible to cover segmentation here (such a topic requires far more than one chapter), we can briefly present an example that can illustrate the power of an integrated toolkit. Since multiple methods are represented in ITK, they can be combined into hybrid segmentation systems that amplify the strengths of more than one approach while reducing their weaknesses. Thus, statistical methods can be combined with region-growing ones, or in the case below, region-growing techniques (such as Fuzzy Connectedness) with boundary-based deformable models.

When combined, FuzzyConnectedness and VoronoiDiagramClassification result in a segmentation system that requires minimal manual intialization. A FuzzyConnectedness filter can generate a prior for a Voronoi partitioning classifier, which can later be refined with a deformable model to generate smooth 3D boundary surfaces of segmented objects. Fig. 37.3 shows a schematic of how multiple methods can be serially combined into an integrated approach for segmentation. Fig. 37.4 shows the input MRI slice and an output binary mask that is the result of the combined method. This method can be extended to 3D, or the stacked 2D results can be used with a deformable 3D model to generate coherent 3D structures.

Other segmentation methods are available in ITK. The opening example shows a 3D watershed technique to segment RGB data. The nearly plug-and-play framework of interchangable filtering and segmentation elements increases the design space of segmentation implementations almost combinatorially. A more complete description of the segmentation methods, native implementations in ITK, and software components for alternative algorithms are all given in the ITK Software Guide.

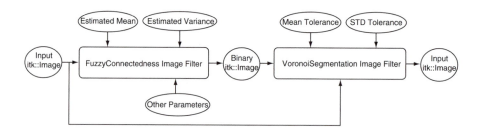

Figure 37.3 A pipeline showing the integration of FuzzyConnectedness with Voronoi Diagram Classification. This process can be used as an input to a deformable model for additional processing and analysis. Figure courtesy of Celina Imielinska, Columbia University; Jayaram Udupa, University of Pennsylvania; and Dimitris Metaxas, Rutgers University.

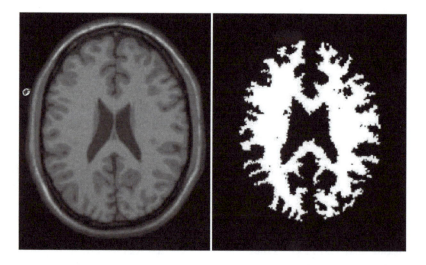

Figure 37.4 Segmentation results for the hybrid segmentation approach, showing the input image and the binary mask resulting from the segmentation pipeline. Figure courtesy of Celina Imielinska, Columbia University; Jayaram Udupa, University of Pennsylvania; and Dimitris Metaxas, Rutgers University.

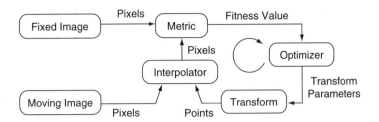

Figure 37.5 The basic components of the registration framework are two input images, a transform, a metric, an interpolator, and an optimizer. Registration is usually an interative mechanism with incremental refinement using the metric to measure improvement in the match between the two input images. Courtesy of Luis Ibáñez and Will Schroeder, Kitware, Inc.

37.7.4 Registration Framework

The approach to image registration in ITK is shown in Fig. 37.5. This basic model can be elaborated to create user-supervised interactive, deformable registration tools or simplified to describe automated affine registration. ITK supports pointset-to-pointset registration approaches (e.g., iterative closest point), image-to-image methods (e.g., maximum entropy or mutual information methods), model-to-image fitting, and hierarchical, multiresolution, and deformable techniques. Some examples of represented algorithms include the following:

- Iterative closest point (ICP).
- Deformable registration.
- Mutual Information, an implementation of the Viola–Wells algorithm.
- Demons Deformable Registration, an implementation of Thirion's demons.

To cover all of the registration methods in ITK would be impossible here. However, one important example is the multimodality Viola–Wells Mutual Information registration approach, which uses an image-to-image metric for matching the moving image to the fixed one. Fig. 37.6 shows two input images for the mutual information algorithm, a T1-weighted MRI and an offset Proton Density (PD)–weighted MRI of the same patient. Although it may seem specious to use two MRIs of the same subject, a TI-weighted scan is taken using a

separate sequence from a PD-T2 MRI scan, and significant patient motion can occur between the two sequences. The shift shown in the example is more dramatic than is normally seen in combined acquisitions.

Mutual Information is an iterative method for incrementally modifying a transform to match two images. In this example, taken from the ITK software distribution, the method is implemented with a gradient descent optimizer, but it could use one of the many other solvers available in the toolkit. After 200 iterations, the process was stopped and the results were analyzed. Fig. 37.7 shows the resulting match between the moving PD-weighted image and the T1-weighted fixed image. A checkerboard representation is used to superimpose both images, with the image in the middle showing the initial mismatch and the image on the right showing the final result. Notice that after alignment, outlines and structures continue smoothly between the squares extracted from alternating images.

The individual registration approaches are supplemented by a broad cross-section of available metrics and a collection of interpolation schemes, as well as an array of optimizers. As with the hybrid segmentation approaches, the design space of registration implementations is enhanced by the interchangeable components almost combinatorially. A more complete description of the registration framework, native implementations in ITK, and software compon-

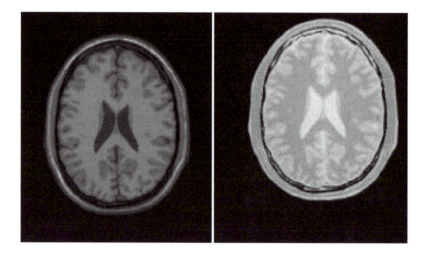

Figure 37.6 T1 MRI (fixed image) and Proton Density MRI (moving image) provided as input to the Viola–Wells Mutual Information registration method. Courtesy of Bill Lorensen and Jim Miller, GE Global Research.

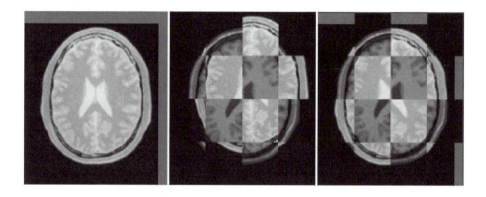

Figure 37.7 Mapped moving image (left) and composition of fixed and moving images before (center) and after (right), using the Viola–Wells Mutual Information registration method. Courtesy of Bill Lorensen and Jim Miller, GE Global Research.

ents for alternative algorithms are all given in the ITK Software Guide.

37.8 More on ITK

ITK has been supported by funding from a coalition of federal agencies and institutes at the NIH, the NSF, and the Department of Defense. Continued support is planned in order to foster growth and establish a secure user base for ITK. Maintenance, documentation, and courseware are all planned or are already underway as part of ITK development.

37.8.1 How to Use ITK

ITK is available via anonymous CVS download or directly from the online website (*http://www.itk.org*). CVS is the online source-code control system used by the consortium to manage the contributions from multiple developers. If you wish to download the latest developmental versions of ITK, you will need to acquire a CVS executable from a reputable location. CVS is maintained by an open-source organization that is not affiliated with the Insight software developers.

In addition to downloading ITK, a new user will also need an executable version of CMake to configure the ITK build. Executable versions of CMake can be found through the "related software" link from the ITK home page. Alternately, CMake source files can be downloaded and built directly from the source for your target system. CMake should then be used to build ITK for your local hardware configuration.

To aid in this process, a "getting started" document is available online (see Section 37.8.2), as is the *ITK Software Guide*, which will help walk you through all of the examples as well as some basic orientation to the ITK internal structure. Executable examples, including test data, are all available as part of the ITK release.

Beyond the stand-alone process of reading, filtering, and writing data files, ITK is being integrated with a variety of user-interface systems and visualization platforms to enhance and empower users with a powerful collection of new tools. Some of those systems and interfaces being supported include the following:

- SCIRun (Utah Scientific Computing and Imaging Institute)
- Analyze (Mayo Clinic)
- VTK (Kitware)
- FLTK
- Tcl/Tk

The upcoming test for ITK will be to prove that as an Application Programmers Interface (API), it can not only support new application environments for visualization but also can be retrofitted and integrated into systems already in use.

37.8.2 Information and Documentation

This text is only an overview of the ITK software. It provides a brief description of the motivations for the building of this API and the design decisions that led to its particular software architecture. Further information on downloading, getting started with, and programming ITK can be found online. Interested users should check the following:

http://www.itk.org: a website has been established where users may find software releases of ITK, access to the testing dashboard showing the error status of the latest software build, and links to related software.

Users' mailing list: Anyone can subscribe to the itk-users mailing list. This is the primary forum for asking questions about the toolkit, seeking answers to installation and design questions, and sharing ideas and information about ITK. At the current time, there are hundreds of users subscribed to the mailing list in dozens of countries. Check the website (see above) for subscription instructions.

ITK Software Guide: The *ITK Software Guide* is one of the most important resources for ITK users. It comprises more than 300 pages of illustrated examples, programming tips, software design notes, and parameter selection advice, and is located online and available in the public domain as an open-source product. Users can find this and the other "Getting Started" instructions at *http://www.itk.org/ HTML/Documentation.htm*

Many open-source programming efforts can fail from lack of support and documentation. Developers have been mistaken in the past to believe that all users would be comfortable directly reading source code as their primary means of studying the algorithms, mathematics, and science behind the software. In addition to the web pages and the ITK Software Guide, the ITK developers are preparing course material for teaching image processing using ITK, as well as a reference book describing the principles and practice of segmentation and registration methods embodied in ITK. We intend to continue to support this project through additional documentation, courseware, tutorials, and software maintenance, thus helping ensure its longevity and usefulness for the community it is intended to serve.

37.9 Conclusion

The ITK is an experiment in building a segmentation and registration API to support the processing and, by default, the visualization of complex high-dimensional data. As with other forms of computing, the quality of the product or output of the process depends quite strongly on the quality of the input given. It follows that the quality and precision of the processing that feeds a visualization system have a profound impact on the effectiveness and power of the images, animations, and other products of that system.

Beyond the creation of a working archive of segmentation and registration tools, the Insight Toolkit is intended to be a means of communicating ideas among research professionals. It is a functional example of an aggressive software engineering approach in generic programming among a distributed development group. ITK

has been constructed to integrate into a variety of visualization systems, and so can be crafted to support the field in many different ways.

Effective visualization depends on the ability of the designer to abstract and elevate the important information hidden in the data and present it for maximum comprehension and retention of the ideas so exposed. The Insight software developers have crafted ITK to aid in the process, providing precision software components for a broad audience of researchers to help partition, sculpt, mask, and illuminate information buried within multidimensional datasets. Renowed computer scientist Richard Hamming has been quoted as saying, "The purpose of computing is *insight*, not numbers." This aphorism is aptly applied to the field of visualization, and so we adopt it to our cause.

References

1. M. J. Ackerman. The Visible Human Project. *Proceedings of the IEEE* 86(3):504–511, 1998.
2. L. Ibáñez and W. Schroeder (Eds.). *The ITK Software Guide*. Clifton Park, NY, Kitware, Inc., 2003.
3. B. King and W. Schroeder. Automated wrapping of complex C++ code. *C/C++ User's Journal* 21(1), 2003.

38 amira: A Highly Interactive System for Visual Data Analysis

DETLEV STALLING, MALTE WESTERHOFF, and HANS-CHRISTIAN HEGE
Zuse Institute Berlin

38.1 Introduction

What characteristics should a good visualization system hold? What kinds of data should it support? What capabilities should it provide? Of course, the answers depend on the particular task and application. For some users a visualization system may be nothing more than a simple image viewer or plotting program. For others, it is integrated software dedicated to their personal field of work, such as a computer algebra program or a finite-element simulation system. While in such integrated systems visualization is usually just an add-on, there are also many specialized systems whose primary focus is upon visualization itself.

On the one hand, there are many self-contained special-purpose programs written for particular applications. Examples include flow-visualization systems, finite-element post-processors, and volume rendering software for medical images. On the other hand, several general-purpose visualization systems have been developed since scientific visualization became an independent field of research in the late 1980s. These systems are not targeted to a particular application area; they provide many different modules that can be combined in numerous ways, often adhering to the data-flow principle and providing means for "visual programming" [1,4,5,7,15,17,23].

In these ways, custom pipelines can be built to solve specific visualization problems. Although these visualization environments are very flexible and powerful, they are usually more difficult to use than special-purpose software. In addition, a major drawback induced by the data-flow principle or pipelining approach is the lack of sophisticated user interaction. Any operation that requires manual interaction, such as segmenting a medical image into different regions, editing a polygonal surface model or molecular structure, or cropping and selecting different parts of a complex finite-element model, is difficult to incorporate into a pipeline of modules that is executed automatically. One may argue that these interactive tasks are not visualization problems per se. But it is a matter of fact that such operations require visual support and are essential for solving problems in many application areas.

In order to close the gap between the ease of use, power, and interactivity of monolithic special-purpose software and the flexibility and extensibility of data-flow-oriented visualization environments, the software system amira has been designed. Initially developed by the scientific visualization group at the Zuse Institute Berlin (ZIB), today amira is available as a commercial product together with several extensions and add-ons [2]. One major focus of the software is the visualization and analysis of volumetric data, which is common in medicine, biology, and microscopy. With amira, volumes can be displayed and segmented, 3D polygonal models can be reconstructed, and these models can be further processed and converted into tetrahedral volume grids. Due to its flexible design, amira can also perform many other tasks, including finite-element postprocessing, flow visualization, and visualization of molecules. In this chapter we discuss the fundamental concepts, techniques, and features of amira.

38.1.1 Design Goals

The development of amira was driven by the following design goals:

- *Ease of use*. Simple visualization tasks such as extracting an oblique slice from a 3D image or computing an isosurface should not require more than a few mouse clicks. Untrained users should be able to get results as quickly as possible.

- *Flexibility*. The system should be able to work with a large number of different data types and with multiple datasets at once. Complex operations requiring a combination of multiple modules should be possible, too.

- *Interactivity*. Techniques or components requiring heavy user interaction, both in 2D and in 3D, should be easy to integrate. Examples are image segmentation, surface editing, and alignment operations.

- *Extensibility*. Users should be able to add new features to the system, e.g., new input/output (I/O) routines, new modules, or even new data types or interactive editors. Existing components should be customizable to some extent.

- *Scripting interface*. The system should be programmable via a scripting language, enabling batch processing, in order to simplify user-specific routine tasks and presentations and to facilitate customization of the software.

- *Multiplatform support*. Different hardware platforms and operating systems should be supported, in particular, Windows, Linux, and other Unix variants. Also, 64-bit code should be supported in order to process large datasets efficiently.

- *State-of-the-art algorithms*. Modern visualization techniques such as direct volume rendering and texture-based flow visualization should be implemented. All techniques should be optimized for both performance and image quality.

In the following section we first describe the general concepts we have chosen to use in order to meet these goals. Next, we will illustrate how the system can be applied in different fields of work. We do this by identifying common tasks and showing which methods are provided to solve these tasks. Finally, we discuss some amira extensions, most prominently amiraVR, an extension that allows the software to operate on large tiled displays and within immersive virtual-reality environments.

38.2 General Concepts

In Fig. 38.1, a snapshot of the amira user interface is shown. When the software is started, three windows are invoked: the main window, containing the "object pool" and the control area; a graphics window, where visualization results will appear; and the console window, where messages are printed and additional commands can be typed in. Datasets can be easily imported via the file browser or via drag-and-drop. After a dataset has been imported, it is represented as a small icon in the object pool. The dataset can be visualized by an appropriate display module chosen from a context-sensitive pop-up menu over the data icon. This pop-up menu lists only modules that can be connected to the particular data object. The output of a display module is immediately shown in the graphics window. Thus, a dataset can often be visualized with a single mouse click once it has been imported. Also, different visualization techniques can be combined with each other without any limitations.

Having discussed the basic modes of operation, we now consider the concepts behind amira in more detail.

38.2.1 Object Orientation

In amira, datasets and modules are considered to be objects. These objects are represented visu-

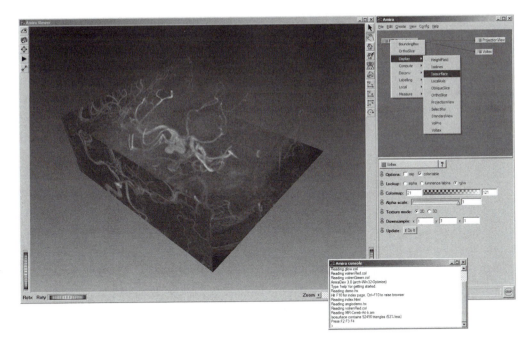

Figure 38.1   The amira user interface consists of the main window with the object pool and the control area, a large 3D graphics window, and a console window for messages and command input. The figure shows a medical image dataset from MR angiography, visualized by a projection view module and a volume rendering module. (See also color insert.)

ally as icons in the object pool. Looking at the object pool, one can easily observe which objects exist and how they are related to each other. If an object is selected with the mouse, additional information and corresponding user-interface elements are displayed in the work area. These interface elements, known as "ports" in amira, allow users to interact with a module. For example, the threshold of an iso-surface module or the position of a slicing module can be adjusted with corresponding sliders. In addition, every object provides a script interface. One can query an object for its properties, or one can interact with an object by calling it with certain commands.

A major advantage of the object concept is that objects can be easily modified or edited by other components. For this purpose powerful editors are provided, allowing modification of datasets in a highly interactive way. Examples are the amira segmentation editor, the landmark editor, and the surface editor. Objects can be modified not only by editors but also by other components, mainly compute modules. This is not easily possible in a data-flow-oriented system, because in such systems data is usually not represented to users as an independent object.

38.2.2   Inheritance and Interfaces

The object-oriented approach in amira also makes use of class inheritance. For example, all data objects are derived from a common data class. This class provides methods to dupli-cate and save the object or to associate arbitrary parameters with it. A more specialized data class represents spatial data objects, i.e., data objects that are embedded in 3D space. This class provides methods to query the bounding box of the data object as well as to set or get an optional affine transformation matrix. Another base class represents fields, i.e., data objects that can be evaluated at any point in a 3D domain. This class provides methods by which to query the component range of the field, or to evaluate it at an arbitrary point in a transparent way, i.e., without needing to know how the field is actually represented. The field may be defined on a regular grid, on an unstructured grid, or even procedurally, by specifying an arithmetic expression. New fields defined in other ways may also be easily added. All display modules defined for the base class can be used automatically for these new fields. In amira the generic evaluation methods of the field class are used (for example) by data-probing modules, which plot field values at a point or along a line; by flow visualization modules, which need to compute trajectories in a vector field; or by slicing modules, which need to resample a field on a 2D grid.

However, in some cases it is not possible to derive data classes with common properties from a common base class. For example, in amira there are separate base classes for scalar fields and for vector fields. At the same time, scalar fields and vector fields defined on the same type of grid, e.g., on a regular Cartesian grid, have many things in common. For this reason, amira provides a mechanism called interfaces. These are classes that describe common properties of objects that do not necessarily have a common base class. For example, both regular scalar fields and regular vector fields provide a lattice interface. In this class, the number of data values per node is a variable. Thus, a module or export routine using the lattice class interface can be automatically applied to both regular scalar fields and regular vector fields, or to any other object that provides this interface.

38.2.3 User-Interface Issues and 3D Interaction

A primary design goal of amira is ease of use. Of course, this is a somewhat subjective and loosely defined requirement. We try to accomplish this goal using several different strategies. First, context-sensitive pop-up menus are provided, offering only those modules that actually can be connected to a given data object. Next, fewer but more powerful modules are preferred, compared to a larger number of simpler entities. For example, in order to display one slice of a 3D image in a data-flow-oriented visualization system, users must often first extract a 2D sub-image, then convert this into geometry data, and finally display the geometry using a render module. In amira all of this is done by a single OrthoSlice module.

By default, visualization modules show their results directly in the main graphics window. In many cases one simply chooses a module from a data object's pop-up menu and immediately receives a visual result. Modules that may need more time for preprocessing usually provide an additional *DoIt* button, which must be pressed in order to generate a result. In this way it is possible, for example, to first adjust the threshold of an isosurface module or to first select the color map of a volume rendering module before starting any computation. Optionally, *DoIt* buttons can be "snapped" to an "on" position, thus facilitating automatic updates. Another point improving the clarity of the user interface is to reduce the number of open windows and avoid overlapping windows so far as possible. The controls of amira objects are shown in a single scrolled list once an object has been selected. Although multiple objects can be selected at once, usually this is not the case. Typically, the user interface is organized as a list of so-called ports, where each port comprises a single line with a label, followed by some buttons, text fields, or sliders. If required, important ports can be "pinned," which makes them visible even if the corresponding object has been deselected.

In addition to the standard controls, many modules also provide a means for 3D interaction in the graphics window. For example, a slice may be picked and translated in 3D. In order to choose a different orientation, a trackball icon can be activated, which in turn can be picked and rotated. The positions of landmarks and other points can be defined by simply clicking on other objects. Similarly, information about points or faces of a grid or surface, as well as associated data values, can be obtained by clicking on them. Parts of a 3D model or individual triangles can be selected using a lasso tool, i.e., by drawing a contour in the graphics window.

38.2.4   Software Technology

Flexibility and extensibility of amira are ensured by a specific modular software architecture in which multiple related modules are organized into different packages. Each package exists in the form of a shared library (or DLL under Windows) that is linked to the main program at run time. For each package there is a resource file specifying which data classes, modules, editors, or I/O routines are defined in that package. This way, only shared libraries providing code that is actually being used need to be loaded. This keeps the executable size small but still makes it possible to have an almost unlimited number of different components in the system. In order to extend the functionality of the system, developers can add custom packages or even replace existing packages as needed.

amira is written in C++, and it requires only a few external standard libraries. For 3D graphics support amira uses the TGS Open Inventor toolkit, which is a well established and proven scene-graph layer [13]. Open Inventor provides multiplatform support, multithreaded rendering, and many advanced display nodes. In addition, several Open Inventor custom nodes have been added to amira for improved performance and image quality. All of these nodes have been directly implemented using OpenGL. The graphical user interface (GUI) of amira is built using Qt, which is a multiplatform widget library [16]. With Qt it is possible to use the same code base for all supported platforms. Currently supported platforms are Windows, Linux, IRIX, HP-UX, and Solaris, all with both 32-bit and 64-bit code. A Mac OS X version is planned for the near future.

The ability to run in 64-bit mode is important because it allows one to process very large datasets as they become more and more frequent in many areas of science and engineering.

38.2.5   Scripts and Script Objects

The script interface of amira is based on the scripting language *Tcl*, which is also an established industry standard [14]. The standard set of Tcl commands has been extended by many amira-specific commands. In particular, the name of each object in the amira object pool can be used as a command name, allowing the user to interact with that particular object. Furthermore, the name of each port of a data object or module can be used as an argument or subcommand for that object. All ports provide Tcl methods to set or get their respective values. For example, to set the threshold of an isosurface module in a script (i.e., the value of the port representing the threshold), one would use the command

```
Isosurface threshold setValue 100
```

Scripts allow one to simplify routine tasks or to run complex presentations. When an amira network is saved, a Tcl script is generated that, when executed, restores the current state. Tcl code can also be bound to certain function keys or to entries in the context menu of a data object. This menu lists all modules that can be connected to an object. It is even possible to modify the default settings of any amira module using Tcl code. Finally, Tcl expressions can be used to decide at run time whether a particular module can be connected to some data object or whether a particular export routine can be called for that object. In this way, the default rules for matching object types and interface names can be overwritten.

Besides standard scripts, amira also supports *script objects*. Script objects are similar to ordinary modules but are implemented completely in Tcl. Usually they provide at least three Tcl procedures: a constructor, a compute routine, and a destructor. The constructor is called when the object is initialized. Any number of standard GUI components, i.e., ports, can be created and initialized here. The compute routine is invoked whenever one of the ports is changed. Finally, the destructor is called when the script object is deleted. Script objects are well suited to solve specific problems using high-level commands. Often multiple standard modules are combined in a script object in order to generate a result.

38.2.6   Affine Transformations

In many applications, alignment or registration of multiple datasets is important. Therefore, visualization environments should allow the user to easily transform individual datasets spatially with respect to others. Such transformations should be applicable to any spatial data object, i.e., any data object embedded in 3D space. For this reason, in amira an optional transformation matrix can be set for all data objects derived from the spatial data's base class. This transformation matrix is automatically applied to any display module visualizing the data object.

The transformations can be defined interactively using the "transformation editor." This editor allows easy transformation in 3D using so-called Open Inventor draggers, such as a 3D tab box or a 3D virtual trackball. In addition, absolute or relative translations, rotations, and scaling operations can be applied using text input. It is important to note that the data themselves are not modified by such a transformation. In order to actually apply the transformation, a separate module is provided. This module transforms the point coordinates of a vertex-based data object (such as a surface or a tetrahedral grid) and resets the transformation matrix to the identity matrix. Voxel-based data objects, such as 3D images, must be resampled onto a new grid. This can be done using several different interpolation filters, either taking the original bounding box or using an enlarged one that encloses the complete transformed dataset.

38.2.7   Parameters

Another concept that amira users have found very helpful across many different applications is the ability to add arbitrary parameters to a data object. Parameters are identified by a unique name and are associated with some value. The value may be a simple number or a tuple of numbers, a string, some binary data, or a subfolder containing an additional list of parameters. In this way, a hierarchy of parameters can be defined. Parameters are used to store additional information for a dataset, which may be contained in specific file formats. For example, in the case of confocal images, the wavelength of the emitted light and a description of the particular optics are often encoded. For medical data, the patient name or a patient ID is usually stored together with many additional parameters. Some parameters are interpreted by certain amira modules. For example, a parameter called *DataWindow* is used to indicate the default greylevel window of an image dataset. Similarly, a parameter called *Colormap* specifies the name of the default colormap used to visualize the dataset. This way it is easy to associate additional information with existing data types, which then may be interpreted by custom modules. Parameters can be edited interactively using the amira parameter editor. In addition, they can also be set and evaluated via the Tcl command interface.

38.3   Features and Applications

In this section we wish to illustrate how amira can be applied in different areas of science and engineering. Though the spectrum of applications is rather wide, there are often similar tasks to be solved. Therefore, we will identify common requirements and show how these are addressed by amira.

38.3.1   Visualization of 3D Image Data

3D image data are important in medicine, biology, and many other areas. Sources of 3D images include CT or MRI scanners, ultrasound devices, 3D confocal microscopes, and even conventional microscopes (which usually require the specimen to be physically cut into sections). The main characteristic of a 3D image is its regular structure, i.e., voxels arranged in a 3D array. Many modules in amira require a 3D image to have uniform or stacked coordinates, although rectilinear and curvilinear coordinates can be represented as well. In the case of uniform coordinates, all voxels have the same size. In the case of stacked coordinates, the distance between subsequent slices in the z-direction may vary.

The most basic approach for investigating 3D images is to extract individual 2D slices. In amira, two modules are provided for this, *OrthoSlice* and *ObliqueSlice* (Fig. 38.2, left). The first module extracts axis-aligned slices, while the second displays arbitrarily oriented slices. In the latter case, the data must be resampled onto a 2D plane. This can be done using different interpolation kernels. Another useful module is *ProjectionView*, which computes a maximum-intensity projection on the xy-, xz-, and yz-plane. In order to grasp the 3D structure of an image dataset, isosurfaces can be computed or direct volume rendering can be applied. For the latter, two different modules are provided. One utilizes the standard texture capabilities of modern graphics cards, while the other one makes use of special-purpose hardware (VolumePro 1000 from TeraRecon, Inc.). In any case a suitable colormap must be chosen to define how the image data are mapped to color and opacity. With the exception of isosurfaces, all methods can be applied not only to grey-scale images but also to RGBA color images and multichannel images.

38.3.2   Image Segmentation

Image segmentation denotes the process of identifying and separating different objects in a 3D image. What constitutes an object depends on the application. Image segmentation is a prerequisite for geometry reconstruction from image data and for more advanced analysis of image data. Consequently, it is an important feature in an image-oriented 3D visualization system such as amira.

In amira, segmentation results are represented by labels. For each voxel, a label is stored specifying to which object or material this voxel belongs. In general, image segmentation cannot be performed fully automatically, and human intervention is necessary. For this reason, amira provides a special-purpose component, the *segmentation editor* (Fig. 38.2, right). The editor offers a variety of different tools for manual and semiautomatic segmentation, in both 2D and 3D. In the simplest case, regions can be selected using a lasso, a brush, or thresholding. More advanced tools such as 2D or 3D region growing or a live-wire method are also provided. In region growing, the user selects a seed point and adjusts the lower and upper bound of a greylevel interval. All connected voxels within this interval are then selected. In the live-wire tool, the user selects a starting point on a boundary and then drags the cursor roughly around the outline [3]. The minimum cost contour from the seed point to the current cursor position is displayed in real time. The cost is based on the image gradient and Laplacian, such that computed paths cleanly follow region boundaries.

Although segmentation is primarily performed in 2D, a 3D view of the currently selected regions is available at any time. For this purpose a fast point-based rendering technique is applied. Noisy regions or regions that have been falsely selected by a 3D threshold or region-growing operation can be easily cleared by marking them in the 3D view using the lasso tool. Another approach to reducing the amount of work needed for image segmentation is to interpolate segmentation results between subsequent slices. Optionally, the interpolated results can be automatically adapted to the image data using a "snakes" technique [9]. Furthermore, shape interpolation from a few segmented orthogonal slices is provided by a 3D wrapping

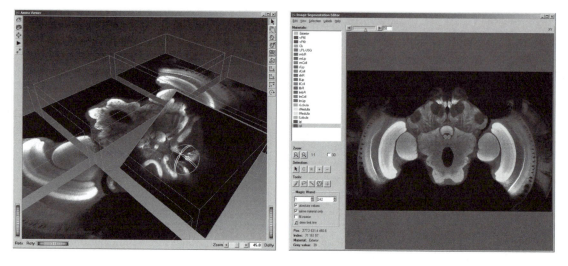

Figure 38.2 (Left) Multipart 3D confocal image stack of a bee brain visualized by slices. The orientation of an oblique slice can be easily adjusted with a trackball dragger. (Right) The amira segmentation editor. This component allows the user to identify and separate different objects in a 3D image stack. Here different parts of the bee brain have been segmented. (See also color insert.)

tool. The segmentation editor also provides a number of different filters, e.g., denoising and smoothing filters, and/or morphological filters for erosion, dilation, opening, and closing operations. Various other experimental (research-stage) amira modules exist, providing additional image-segmentation methods, e.g., based on statistical shape models [10].

38.3.3   Geometry Reconstruction

After a 3D image has been segmented, i.e., after every voxel has been assigned to some material, a polygonal surface model can be created. Several algorithms have been described that attempt to construct a surface model by connecting contours in neighboring slices in the appropriate way. However, these algorithms are not fail-safe, especially if multiple different materials are involved. In this case, nonmanifold surfaces must be created, i.e., surfaces with edges where more than two triangles join. In amira a robust and fast surface reconstruction algorithm is applied that triangulates all grid cells individually, similar to the marching cubes algorithm for computing isosurfaces [12]. This algorithm guarantees that the resulting

surfaces are free from cracks and holes, that no triangles intersect each other, and that all regions assigned to different materials are well separated from each other. If additional weights are defined (by prior calculations) for each voxel, a smooth surface can be reconstructed. The weights are computed by applying a Gauss filter to the binary labels, so that a nonbinary smooth result is obtained. A disadvantage of this technique is that small details of the segmented dataset may be lost. Therefore, a constrained smoothing method is also provided that ensures that the final surface is still consistent with the original labeling. A similar but more computationally expensive method has been described by Whitaker [25]. An example of a smooth 3D model reconstructed by amira is shown in Fig. 38.3 (left).

38.3.4   Surface Simplification and Editing

Surfaces reconstructed from a segmented 3D image usually have a large number of triangles. In fact, the polygon count of the triangular surface is on the order of the voxel size. For many purposes, the number of triangles needs to be reduced, i.e., the surface needs to be sim-

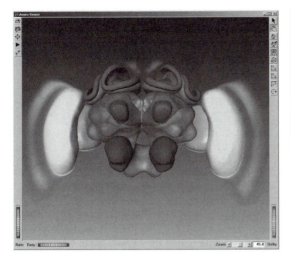
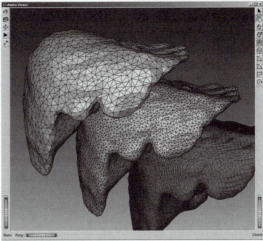

Figure 38.3 (Left) Reconstructed polygonal surface model of a bee brain; a volume rendering of the image data has been superimposed, showing how well the model matches the image data. (Right) Reconstructed model of a human liver, displayed at three different resolutions (1 cm maximum edge length, 0.5 cm maximum edge length, and surface at original resolution with 0.125 cm voxel size). (See also color insert.)

plified. In amira this can be done using an advanced simplification algorithm based on edge contraction. The method tries to reduce the error induced by the simplification process as far as possible while simultaneously optimizing triangle quality. In order to control the maximal deviation, a quadric error metric is used (as proposed by Garland and Heckbert [6]). In all cases, intersecting triangles are strictly avoided. The result of a simplified surface is shown in Fig. 38.3 (right).

Simplification is not the only surface-editing operation that can be performed in amira. For other operations an interactive *surface editor* is provided. Among other tasks, this editor allows the user to iteratively smooth or refine the surface (in whole or part); to cut parts out of a surface and to copy them into other surfaces; or to define boundary conditions on the surface. The latter is important when performing numerical simulations on the surface or on a tetrahedral finite-element grid derived from it. The surface editor also provides several tools for modifying the surface at a fine-grained level. In particular, individual edges can be flipped, subdivided, or contracted, and points can be

moved. Also, tests can be performed to check whether the surface has intersections, holes, or inconsistently oriented triangles. Finally, triangles with a bad aspect ratio or with small dihedral angles can be found.

All these operations allow the user to interactively modify an arbitrary surface in such a way that a good tetrahedral grid can be generated afterwards. Grid generation itself is implemented as a separate module in amira using an advancing front algorithm [8,11]. The grid quality can be improved by a subsequent smoothing or relaxation step. An example of a tetrahedral grid generated by amira is shown in Fig. 38.4 (left).

38.3.5 Alignment of Physical Cross-Sections

With the previously described techniques, polygonal surface models and tetrahedral volume grids can be reconstructed from 3D image stacks recorded by CT scanners, MR scanners, or confocal microscopes. Another common approach in microscopy is to physically cut an object into slices and to image each slice separately. In order

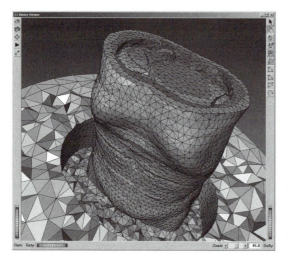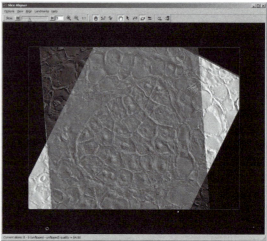

Figure 38.4   (Left) Tetrahedral finite-element grid of a human body embedded in a device for hyperthermia treatment. Only parts of the grid are shown, in order to reveal interior structures. (Right) The amira slice aligner, an interactive tool for aligning 2D physical cross-sections. (See also color insert.)

to reconstruct geometries from such data, the individual slices usually need to be aligned with respect to each other. For this purpose another tool is provided in amira, the *slice aligner* (Fig. 38.4, right). The slice aligner supports interactive, semiautomatic, and automatic alignment. The tool displays two slices of a 3D image stack at once. Different view modes can be selected that help to visually distinguish the slices. For example, one image can be displayed in green and the other one in red, or the colors of one image can be inverted. The image slice then can be manually translated or rotated. Semiautomatic alignment via landmarks is possible, too. Fully automatic prealignment can be achieved by matching the centers of gravity as well as the principal axes of the two images. Once this has been done, a multilevel optimization algorithm can be called that attempts to minimize the pixelwise difference of the two images.

38.3.6   Multiple Datasets and 3D Registration

In biomedical applications, users often work with multiple datasets. For example, one wants to compare images of multiple individuals, images of the same individual recorded at different times, or images of the same object taken with different imaging modalities, such as CT and MR. In all of these cases it is crucial that it be possible to visualize multiple datasets simultaneously. This requirement is met by amira in a natural way. In order to compare multiple datasets, one can use (for example) semitransparent displays. amira supports high-quality transparency with depth sorting and opacity enhancement at silhouettes. Other techniques include colorwash displays, where images of multiple datasets are superimposed on 2D slices, or computation and visualization of difference images. For surfaces it is also possible to compute the distance between the vertices of one surface and the nearest point on some other surface. The result can be visualized using conventional pseudo-coloring.

When multiple corresponding datasets are to be used, the problem of registering or aligning these datasets with each other becomes relevant. Here, the two major techniques are rigid and elastic registration. In the case of rigid transformation, the dataset will be only translated, rotated, and possibly scaled. Such transformations can be easily encoded in an affine transformation matrix, which is supported for all amira data objects. Thus, a manual rigid registration can be performed using amira's *transformation editor*. Another possibility is to make use of landmarks. Corresponding landmarks can be defined in both datasets with amira's *landmark editor*. Afterwards, a rigid transformation can be computed that minimizes the squared distance between each pair of landmarks. Finally, a voxel-based automatic registration can also be computed. This method attempts to optimize a quality measure indicating the difference between both images. Several different quality measures are implemented, including the sum over squared pixel differences and a mutual information measure [24]. The latter is suitable for registration of multimodal images, e.g., CT or MR, when there is no one-to-one correspondence between the grey values in the two images. Sometimes it is more appropriate to align reconstructed surfaces than image data. In amira this is supported by an iterative method that automatically tries to find corresponding vertices and then minimizes the squared distance between these points.

In contrast to rigid registration, elastic registration is usually much more difficult to define. In addition, it requires image data to be resampled on a new axis-aligned grid. Currently, amira supports only an elastic registration method based upon landmarks. This method computes a Bookstein spline that exactly matches corresponding landmarks and smoothly interpolates in between. This approach can be applied to both 3D images and triangular surfaces. An automatic voxel-based elastic registration method is currently under development.

38.3.7 General Data Processing and Data Analysis

In addition to the specific tools we have described, amira also provides other more general utilities for data processing. Probably one of the most important is a resampling module for reducing or enlarging the resolution of a 3D image or other dataset defined on a regular grid. Some care must be taken when choosing a filter kernel for resampling. In amira, several different kernels are supported, ranging from fast box and hat filters to a high-quality Lanzcos filter (which approximates a sinc function, the optimal filter from sampling theory), for finite images. Other tools are provided for cropping a dataset, for enlarging it by replicating boundary slices, and for changing the primitive data type of a dataset. For images, the most common primitive data types are bytes and 16-bit shorts, either signed or unsigned. Simulation data is usually encoded using 32-bit floating-point numbers. In addition, amira supports 32-bit signed integers and 64-bit floating-point numbers. While a scalar field has only one such component, any number of other components is also possible. For example, a vector field usually has three components. In amira a module is provided to extract one component from such a field, and to combine multiple components from different sources into a new field. Another valuable tool is the *arithmetic module*, used for combining multiple datasets by evaluating a user-defined arithmetic expression per voxel or per data value. In this way it is possible to, for example, subtract two datasets, compute the average, scale the data values, or mask out certain regions using boolean operations.

Another class of utility modules is related to statistical data analysis. This includes simple probing modules, which evaluate a dataset at some discrete points or plot it along a user-defined line segment. Moreover, a histogram of the data values can be computed, possibly restricted to some region of interest. Other modules are provided to compute statistical quantities such as volume, mean grey value, standard deviation, and so on for different regions encoded in a segmented label field, and also to compute volume-dose diagrams, or to count and statistically analyze the connected components in a binary-labeled 3D image.

38.3.8 Finite-Element Postprocessing

Most of the tools described in the previous sections were related to the processing of 3D image data, or, more generally, to data defined on regular 3D grids. However, other data types are also important, in particular, finite-element data defined on unstructured grids. amira supports the generation of triangular surfaces and tetrahedral volume grids suitable for numerical simulations. Such simulations are typically performed using some external code, but the results can again be visualized in amira. This task is known as finite-element postprocessing. Besides tetrahedral grids, amira also supports hexahedral grids. Most of the general-purpose visualization techniques and analysis tools can also be applied to data on unstructured grids—for example, slice extraction, computation of isolines or isosurfaces, direct volume rendering (implemented via a cell-projection algorithm), data probing, or computation of histograms. In addition, scalar quantities can be visualized using color coding of the grid itself. In the case of mechanical simulations, deformations are often computed. Such data can be visualized with displacement vectors or by applying the displacement vectors to the initial grid sequentially such that an animation sequence is obtained. All of these methods can also be applied to visualize results from numerical simulations in biomedicine—e.g., simulations of mechanical loads in bones or of heat transport in tissue—or to visualize results from numerical simulations in engineering and related disciplines.

38.3.9 Flow Visualization

Flow visualization has evolved into an independent field of research in scientific visualization. Since flow fields are often generated by numerical computations, it can also be considered a special form of finite-element postprocessing. Beyond engineering domains such as computational fluid dynamics (CFD), where (for example) virtual wind tunnel experiments are performed, flow visualization techniques are important also in biomedicine—e.g., for analyzing a simulated flow in blood vessels or air flow in a nasal pathway.

Flow-visualization techniques have been reviewed in depth in other chapters of this book. Therefore, here we list the different methods supported in amira without presenting algorithmic details.

Likely the simplest method for visualizing a vector field is to draw small arrows attached to discrete points. Arrows can be drawn on a slice, within the volume, or upon a surface in amira. More highly resolved and comprehensible visual representations can be obtained using texture-based methods. amira supports fast line integral convolution, both on slices and on surfaces with arbitrary topology [2,20] (Fig. 38.5, left). Probably the most popular approach to reveal the structure of a flow field in 3D is to draw streamlines. amira includes support for illuminated streamlines (Fig. 38.5, right)—i.e., streamlines that are rendered as line primitives with a lighting applied to them [22]. This allows rapid rendering of many streamlines while at the same time highlighting the 3D structure of the field. Another method based on streamline computation is the display of streamribbons (Fig. 38.6 [left]). Like streamlines, streamribbons also show the swirl and torsion of flow fields. A further extension supported by amira is the streamsurface (Fig. 38.6 [right]). A streamsurface is spanned by multiple streamlines starting from some user-defined seed shape or rake. Streamsurfaces are commonly started from a straight line or from a line traced along the normal or binormal direction of the vector field. All of these stream visualization techniques are highly interactive. While seed-point distributions can be automatically calculated, users can also select and interactively manipulate seed points and structures, thus supporting the investigation of the flow field and highlighting of different features. Each of these techniques again supports 3D interaction, allowing the user to pick and move

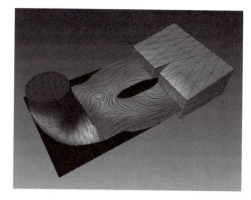
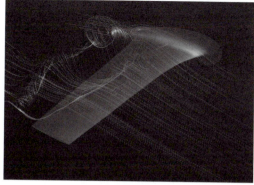

Figure 38.5 (Left) Visualization of a turbine flow using the fast line integral convolution method in a user-selected plane. (Right) Visualization of the 3D flow around an airfoil using the illuminated streamline technique. (See also color insert.)

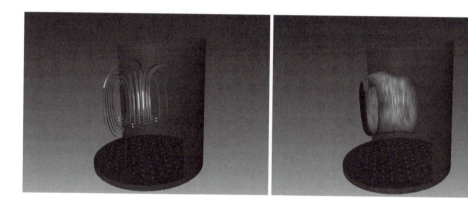

Figure 38.6 Visualization of fluid flow within a bioreactor. (Left) Streamribbons starting in the interactively positioned seed box. (Right) Streamsurface with the tangential flow depicted by line integral convolution (LIC). (See also color insert.)

the seed volume or seed shape directly within the 3D viewer.

38.4 amiraVR and Other Extensions

The modular structure of amira makes it possible to extend the system in various ways and to provide extensions that address more specific application areas. Some major extensions are directly available as optional products. Among these, the most prominent is amiraVR, which allows amira to operate on a large tiled display or in a multiwall virtual environment.

38.4.1 amiraVR

High-resolution multiwall displays have received considerable attention in scientific visualization since the turn of the century. Two major approaches have held special interest. The first is flat multitile displays, often called "power walls." Here, the goal is to create a very high-resolution display, usually by combining several projectors in one rear projection system.

With such a display, fine details in high-resolution datasets can be visualized and shown to a small- or medium-sized group of observers. The other approach is to construct an immersive environment for virtual-reality applications. Usually, such environments incorporate multiple screens in a nonplanar configuration. In order to compute correct views, the actual position of the observer needs to be tracked. Nontracked observers usually see somewhat distorted images and artifacts at the boundaries between neighboring screens.

For performance reasons, the images for the different parts of a tiled display or for the different screens of a VR environment should be rendered in parallel, if possible. The simplest approach from a software perspective is to use a multiprocessor shared-memory machine with multiple graphics pipes. This architecture is implemented by SGI Onyx systems and also by other workstations from vendors such as Sun or HP. To support such an architecture, it must be possible to perform the actual rendering in parallel using multiple threads. This is supported by amira. amira's rendering process involves the traversal of an Open Inventor scene graph and the calling of render methods for each node in this graph. Although early versions of Open Inventor were not thread-safe originally, this is currently the case (since the v3.1 release by TGS).

The use of amiraVR requires specification of the display configuration—i.e., the actual setup of the display system and (optionally) the tracking system. When the configuration file is read, additional graphics windows are opened on the graphics pipes as specified. In the case of a tiled display, the modules can be controlled via their usual interface with the 2D mouse. For user interaction within an immersive environment, 3D versions of all the standard GUI elements of amira data objects and modules are provided. In addition, a user-defined 3D menu can be displayed. Interactive elements such as slices or draggers, which can be picked in the viewer window using the 2D mouse, also react to events generated by a tracked 3D "mouse." Thus, slices can be easily translated and rotated in 3D, or seed volumes for flow visualization can be easily adjusted. All objects that can be picked with the 3D mouse, including menus, provide some visual feedback. This is important to make interaction in VR feasible. With these mechanisms, data can be visualized

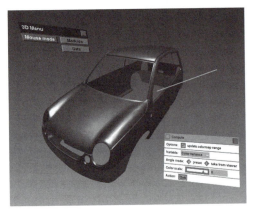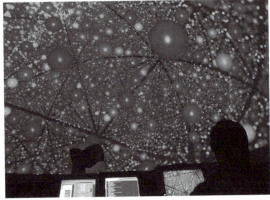

Figure 38.7 (Left) Coating on a car body investigated with amiraVR. For every module, a 3D version of its respective user interface can be used. This allows full control of amira from within an immersive environment. (Right) Atoms in a crystal lattice shown in a dome using amiraVR. The dome was illuminated by six laser projectors with partially overlapping images, with five arranged in a circle and one at the top. Image courtesy of Carl Zeiss Jena. (See also color insert.)

in a VR environemnt in a way similar to that of the desktop GUI. All modules and networks can be loaded without modification. This allows users (for instance) to prepare visual demonstrations for large display systems or VR environments on a PC or notebook computer.

38.4.2   Developer Version

For a modern visualization system, it is crucial that the user be able to add new functionality. We have already stated that this is possible within amira and simplified by amira's modular and object-oriented design. The amira developer version provides all of the header files and documentation required to derive new modules from existing classes. It also provides a unified make system, which creates either makefiles or project files for integrated development environments such as Microsoft Visual Studio. New modules based on Tcl code, *script objects*, do not require the developer version and can be implemented from within amira's base version.

38.4.3   Molecular Visualization

For the application domains of chemistry, biochemistry, and molecular biology, the amira extension amiraMol has been developed. This provides tools for the analysis of complex molecules, molecular trajectories, and molecular conformations. The extension is useful for inorganic and organic chemistry, but its emphasis is on the analysis of biomolecules.

The central goal of molecular biology is to elucidate the relationship between the sequence, structure, properties, and function of biomolecules. Such knowledge allows one to understand biological processes and pharmaceutical effects, as well as to identify and optimize drug candidates. Since bioactivity is guided by molecular shape and molecular fields, amiraMol provides special means for analyzing the dynamic shapes of molecules as well as the corresponding molecular fields. Standard and novel representations are available for visualizing biomolecules. Arbitrary grouping hierarchies on the mol-

ecule's topology can be defined for coloring, masking, and selection.

amiraMol supports common molecular file formats, such as PDB, Tripos, Unichem, and MDL. For trajectories from molecular dynamics simulation, amira provides its own native data format, but it also supports CHARMM's dcd format. Standard techniques for molecular visualization available in amiraMol are wire frame, ball-and-stick, van der Waals spheres, and secondary structure representations (Fig. 38.8, top). Beyond these, a novel technique called bond-angle representation has been developed (Fig. 38.8, bottom) that displays a triangle for every group of three atoms connected by two bonds. This representation requires only a few geometric primitives and provides a comprehensible view of the 3D structure.

Several color schemes can be used to enhance the molecular representations. They permit the user to color the atoms according to a number of attributes, such as atomic number, charge, hydrophobicity, radius, or the atom's index.

To take into account structural information beyond atoms and bonds, groups can be defined. A group is a combination of atoms and other groups that can (for example) represent a residue, a secondary structure, or an α-chain. The groups are organized into levels, such as the level of residues or the level of chains. The user can define arbitrary new groups and levels by using expressions. Groups may contain not only atoms but also groups of arbitrary and possibly different levels. Atoms can be colored according to their membership in a group of a chosen level.

In order to support easy investigation of molecular structures, amiraMol offers a *selection browser*. This displays all groups in a chosen level of the hierarchy. Furthermore, additional information, such as type and membership of the groups, can be displayed. Groups can be selected by being clicked on in the browser, by use of expressions, or through interaction within the viewer. Groups that were selected in the browser will also be highlighted in the viewer, and vice versa. Apart from selecting, the selection browser offers the possibility to hide arbitrary parts in all of the above-mentioned representations, so that the user can concentrate on particular regions of interest.

Shape complementarity is an important aspect in molecular interactions. Shape properties are relevant for manual docking of ligands to proteins and for automated docking procedures. The characterization of molecular shapes is therefore very useful for molecular modeling.

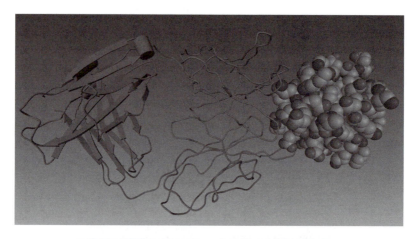

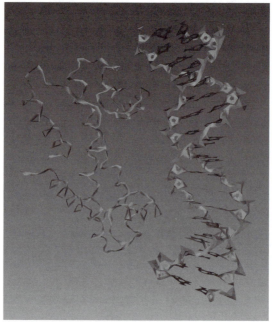

Figure 38.8 (Top) Complex consisting of a mouse antibody and an antigen of Escherichia Coli. The secondary structure representation is on the left side, and the pure backbone representation is in the middle. Van der Waals balls depict the antigen. (Bottom) Bond angle representation of a complex of the factor for inversion simulation (FIS) protein and a DNA fragment (colored according to atom types). (See also color insert.)

In addition to the previous techniques, amira-Mol offers algorithms for generating triangular approximations of solvent-excluded and solvent-accessible surfaces [18]. Additionally, van der Waals surfaces and interfaces between arbitrary parts of a molecule or between different molecules can be computed. All triangular surfaces can be color-coded by arbitrary scalar fields.

Of course, these molecular representations can be combined with all other visualization techniques available in amira. For instance, the electrostatic field can be depicted using vector-field visualization techniques (Fig. 38.9, left) or electron density isocontours can be computed.

However, molecules are not static; they are in constant motion. Typically they fluctuate for long periods around a certain "meta-stable" shape. Less frequently, they also undergo larger shape changes. Taken together, individual molecular configurations with similar shapes are called conformations. Conformation analysis is used to determine the essential shapes of

molecules and the probabilities of transitions between these shapes. amiraMol offers an extensive set of tools for visual analysis of trajectories from molecular dynamics simulations. This allows users to determine representatives of conformations and to depict conformations [19]. In Fig. 38.9 (right), a representative molecular shape is displayed, together with the shape density of the corresponding conformation.

38.5   Summary

We have presented the general design concepts behind amira, a 3D visualization and geometry reconstruction system. We have also given an overview of the different techniques and algorithms implemented in this system. It was shown that the combination of different concepts such as object orientation, a simple and well-structured user interface, the integration of highly interactive components such as the segmentation editor, intuitive 3D interaction

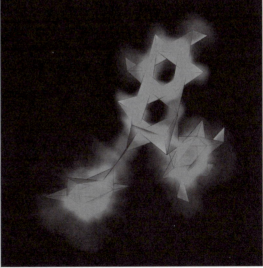

Figure 38.9   (Left) Solvent accessible surface of the ribonuclease T1, pseudo-colored according to the molecule's electrostatic potential; illuminated streamlines depict the electrostatic field. (Right) Configuration density and superimposed bond angle representation of Epigallocatechin-Gallat. (See also color insert.)

techniques, a powerful scripting interface, and a broad range of advanced and innovative algorithms for visualization and data processing have yielded a powerful software system that can be usefully applied to many problems in medicine, biology, and other scientific disciplines.

Acknowledgments

We would like to sincerely thank the many students, researchers, and software developers who contributed to the amira suite: Maro Bader, Werner Benger, Timm Baumeister, Daniel Baum, Philipp Beckmann, Robert Brandt, Martina Bröhan, Liviu Coconu, Frank Cordes, Olaf Etzmuß, Andrei Hutanu, Ralf Kähler, Ralf Kubis, Hans Lamecker, Thomas Lange, Alexander Maye, André Merzky, Olaf Paetsch, Steffen Prohaska, Hartmut Schirmacher, Johannes Schmidt-Ehrenberg, Martin Seebaß, Georg Skorobohatyj, Brygg Ullmer, Tino Weinkauf, Natascha Westerhoff, Gregor Wrobel, and Stefan Zachow. Special thanks go to Brygg Ullmer for his editing suggestions, which improved grammar, vocabulary, and style. We also would like to express our gratitude to Peter Deuflhard, who supported this project over many years. Furthermore, we thank all research collaboration partners and early users who helped us with their requirements to shape the system and make it practically useful.

References

1. G. Abram and L. A. Treinish. An extended data-flow architecture for data analysis and visualization. In *Visualization '95 proceedings*, pages 263–270. IEEE Computer Society Press, 1995.
2. amira—Advanced 3d visualization and volume modeling. Software and user's guide available from www.amiravis.com.
3. W. A. Barrett and E. N. Mortensen. Interactive live-wire boundary extraction. *Medical Image Analysis*, 1(4):331–341, 1997.
4. D. S. Dyer. A dataflow toolkit for visualization. *IEEE Computer Graphics & Applications*, 10(4):60–69, 1990.
5. D. Foulser. Iris explorer: a framework for investigation. *ACM Computer Graphics*, 29(2):13–16, 1995.
6. M. Garland and P. S. Heckbert. Surface simplification using quadric error metrics. In *SIGGRAPH 97 Conference Proceedings*, pages 209–216. ACM SIGGRAPH, Addison Wesley, 1997.
7. C. Gunn, A. Ortmann, U. Pinkall, K. Polthier, and U. Schwarz. An extended data-flow architecture for data analysis and visualization. In *Visualization and Mathematics*, pages 249–265. Springer Verlag, 1997.
8. H. Jin and R. I. Tanner. Generation of unstructured tetrahedral meshes by advancing front technique. *Int. J. Numer. Methods. Eng.*, 36:217–246, 1993.
9. M. Kass, A. Witkin, and D. Terzopoulos. *Snakes: Active Contour Models*. Academic Publishers, 1987.
10. H. Lamecker, T. Lange, and M. Seebass. Segmentation of the liver using a 3d statistical shape model. ZIB preprint 2002, submitted.
11. R. Löhner and P. Parikh. Generation of 3D unstructured grids by the advancing-front method. *Int. J. Numer. Methods. Fluids*, 8:1135–1149, 1988.
12. W. Lorensen and H. Cline. Marching cubes: a high resolution 3D surface construction algorithm. *Computer Graphics*, 21(4):163–169, 1987.
13. Open Inventor from TGS, http://www.tgs.com.
14. J. K. Ousterhout. *Tcl and the Tk Toolkit*. Addison-Wesley, 1994.
15. S. G. Parker, D. M. Weinstein, and C. R. Johnson. The SCIRun computational steering software system. In E. Arge, A. M. Bruaset, and H. P. Langtangen, Eds., *Modern Software Tools for Scientific Computing*, pages 5–44. Boston, Birkhauser (Springer-Verlag), 1997.
16. QT white paper, http://www.trolltech.com
17. J. Rasure and C. Williams. An integrated data flow visual language and software development environment. *Journal of Visual Languages and Computing*, 2:217–246, 1991.
18. M. F. Sanner, A. J. Olson, and J.-C. Spehner. Reduced surface: An efficient way to compute molecular surfaces. *Biopolymers*, 38:305–320, 1995.
19. J. Schmidt-Ehrenberg, D. Baum, and H.-C. Hege. Visualizing dynamic molecular conformations. In R. J. Moorhead, M. Gross, and K. I. Joy, Eds., *Proceedings of IEEE Visualization 2002*, pages 235–242, 2002.
20. D. Stalling. Fast texture-based algorithms for vector field visualization. PhD thesis, Zuse Institute Berlin (ZIB), 1998.

21. D. Stalling and H.-C. Hege. Fast and resolution independent line integral convolution. In *Proceedings of SIGGRAPH 95, Computer Graphics Proceedings*, Annual Conference Series, pages 249–256, 1995.

22. D. Stalling, M. Zöckler, and H.-C. Hege. Fast display of illuminated field lines. *IEEE Transactions on Visualization and Computer Graphics*, 3(2):118–128, 1997.

23. C. Upson, T. A. Faulhaber, Jr., D. Kamins, D. Laidlaw, D. Schlegel, J. Vroom, R. Gurwitz, and A. van Dam. The application visualization system: a computational environment for scientific visualization. *IEEE Computer Graphics and Applications*, 9(4):30–42, 1989.

24. W. M. Wells, P. Viola, H. Atsumi, S. Nakajima, and R. Kikinis. Multi-modal volume registration by maximisation of mutual information. *Medical Image Analysis*, 1(1):35–51, 1996.

25. R. Whitaker. Reducing aliasing artifacts in isosurfaces of binary volumes. In *IEEE Volume Visualization and Graphics Symposium*, pages 23–32, 2000.

PART X
Perceptual Issues in Visualization

39   Extending Visualization to Perceptualization: The Importance of Perception in Effective Communication of Information

DAVID S. EBERT
School of Electrical and Computer Engineering
Purdue University

39.1   Introduction

The essence of any human–computer interface is to convey information. Visualization has been a good information communication tool for more than two decades. While many visualization systems provide users with a better understanding of their data, they are often difficult to use and are not a reliable, accurate tool for conveying information, which limits their acceptance and use. As scientists, medical researchers, and information analysts face drastic growth in the size of their datasets, the efficient, accurate, and reproducible communication of information becomes essential. These problems become even worse when the datasets under investigation are multivariate and/or vector datasets.

Therefore, we need a fundamental change in the development of visualization techniques. Traditional visualization must evolve into *perceptualization of information*: conveying information through multiple perceptual channels and perceptually tuned rendering techniques. The choice of visual rendering techniques should be driven by characteristics of human perception, since perceptual channels are the communication medium. This chapter summarizes some basics of human perception and show several examples of how this perceptual basis can be used to drive new perceptualization techniques.

Throughout history, humans have tried to effectively convey important information through the use of images, using techniques such as illustrations, photographs, and detailed technical drawings. These techniques harness the enormous bandwidth and preattentive processing of the human visual system. Many disciplines of study have evolved to further these techniques, including photography, technical and medical illustration, and now visualization. While illustration and photography are both successful at capturing and conveying information, they utilize different techniques and characteristics of the human visual system to convey information. Photography concentrates on conveying information with light and color variation, while illustration additionally tries to effectively convey information by omitting unimportant details, enhancing the most significant components of the image, simplifying complex features, and exposing hidden features [28]. Scientific illustrations have been used for centuries because of their effective communicative ability [12].

The effectiveness of illustration techniques, the amazing power of the human perceptual system, and the enormous data deluge facing information analysts, medical researchers, and scientific researchers have motivated us to explore the possible extension of visualization techniques to *perceptualization*. The goal of perceptualization is to concisely convey information to the user through the creation of effective perceptual human inputs (visual, proprioceptive, and haptic).

39.2 Overview of Human Perception

Humans are aware of their environment through the human perceptual system, which consists of several perceptual channels/modalities: visual perception, auditory perception, haptic and tactile perception, olfactory and gustatory perception (smell and taste), and kinesthesis (perception of body movement/position). Proprioception, the body's innate sense of where joints and body segments are in space, is one form of kinesthesis that is sometimes used for effective interaction and 3D space perception.

These perceptual channels take physical stimuli and generate information that is transmitted to the brain. Some of the information is processed at a very low level in parallel without conscious thought, referred to as preattentive processing. Other information requires attention, or conscious thought, to perceive the information. In visualization, the visual perceptual channel is the most widely used communication channel, while the auditory and haptic channels are being incorporated to convey additional information or as redundant forms of communication to increase accuracy or speed of communication. Understanding the way humans perceive information is, therefore, vital to the effective conveyance of information through perceptualization. For a basic understanding of the human visual system and human visual perception, the texts by Glassner [9] and Ware [27] are very good references.

39.2.1 Preattentive Visual Processing

Preattentive processing is essential in grouping large sets of information, including visual stimuli. An understanding of which visual features are preattentive can be used to more effectively design visualization systems. According to Treisman, preattentive processing is visual processing that is apparently accomplished automatically and simultaneously for the entire visual field of view [26]. Many studies have investigated which visual stimuli are preattentive. One common procedure is to measure the response time to find a target in a set of "distracters." If a stimulus is preattentive, the response time should be independent of the number and types of "distracters" presented with the stimulus. Another method is to display a group of elements, with one element different from the rest in some way, for a short period of time (commonly 250 milliseconds) and then determine whether the viewer was able to pick out the unique element.

Cleveland [3] cites experimental evidence that shows that the most accurate method to visually decode a quantitative variable in 2D is to display position along a scale. This is followed in decreasing order of accuracy by interval length, slope angle, area, volume, and color. Bertin offers a similar hierarchy in his treatise on thematic cartography [2]. More recent experiments have shown that humans can preattentively perceive 3D shape [18]. Ware [27] provides the following classification of preattentively processed features:

- Form
  - line orientation, length, width, and colinearity
  - size
  - curvature
  - spatial grouping
  - added marks
  - numerosity (number of items)
- Color: hue and intensity
- Motion: flicker and direction of motion
- Spatial position:
  - 2D position
  - Stereoscopic depth
  - convexity/concavity (shape from shading)

This information on preattentive processing capabilities is very useful for visualizing multifield datasets and glyph visualization. However, only a small number of distinct values can be perceived preattentively via each feature. Also, it is not necessarily the case that perceiving seven different values among five preattentive features means that you can preattentively perceive over 16,000 distinct combinations. The interference between features is not currently a well understood area. For an example of using preattentive features for multivariate visualization, see Healey's work [10,11].

39.3 Examples of Perceptualization Research

A basic understanding of perception can be used to develop systems that are perceptually significant and more powerful than traditional visualization approaches. Below are two examples of perceptualization systems we have developed.

39.3.1 Minimally Immersive Perceptualization

We have developed techniques to effectively convey information from large multidimensional, multifield datasets. Our interactive visualization system, the Stereoscopic Field Analyzer (SFA), successfully harnesses human perception to increase the quantity and clarity of the information conveyed [7,8,24]. As described by Ebert et al. [8], the system provides a minimally immersive interactive visualization tool that increases the understanding of both structured and unstructured volumetric data while being affordable on desktop PCs. Our system takes advantage of the priority structure of human visual perception [2,3], stereopsis, motion, and proprioception to create meaningful visualizations from scientific and information analysis data. The basic volume rendering mechanism is glyph-based volume rendering, with projected volume rendering, thin slab volume rendering, and contour visualization available on an interactive cutting plane [24]. SFA uses a glyph's location

(3 attributes), 3D size (1–3 attributes), color (1 attribute), orientation (1 vector attribute), and opacity (1 attribute) to encode up to nine data variables per glyph. We have also developed new techniques for automatic glyph shape generation that allow perceptualization of data through shape variation. Location, color, size, orientation, and opacity are more significant perceptual cues than shape [3]; however, shape variation can also be effectively used to convey related scalar variables, especially in an interactive system.

Our use of glyphs is related to the idea of marks as the most primitive component that can encode useful information [2]. Senay and Ignatius point out that shape, size, texture, orientation, transparency, hue, saturation, brightness, and transparency are retinal properties of marks that can encode information [22,23]. Since size and spatial location are more significant cues than shape, the importance mapping of data values should be done in a corresponding order. In decreasing order of data importance, data values are mapped to location, size, color, opacity, and shape. In our experience, shape is very useful for local area comparisons among glyphs: seeing local patterns, rates of change, outliers, and anomalies.

Glyph shape is a valuable visualization component because of the human visual system's preattentive ability to discern shape. Shapes can be distinguished at the preattentive stage [18], using curvature information of the silhouette contour and, for 3D objects, curvature information from surface shading [14]. Unlike an arbitrary collection of icons, curvature has a visual order since a surface of higher curvature looks more jagged than a surface of low curvature. Therefore, generating glyph shapes by maintaining control of their curvature will maintain a visual order. This allows us to generate a range of glyphs that interpolate between extremes of curvature, thereby allowing the user to read scalar values from the glyph's shape. Preattentive shape recognition allows quick analysis of shapes and provides useful dimensions for comprehensible visualization.

We have chosen to use a procedural approach for the generation of glyph shape to meaningfully encode two data variables/fields. Our goal for glyph shape design is to allow the automatic mapping of data to shape in a comprehensible, easily controllable manner. Superquadrics are a natural choice to satisfy this goal. Superquadrics [1] are extensions of quadric surfaces where the trigonometric terms are each raised to exponents. Of the four types of superquadrics, we have chosen superellipsoids due to their familiarity. With superellipsoids, the exponents of the geometric terms allow continuous control over the shape characteristics (e.g., the "roundness" or "pointiness") of the shape in the two major planes that intersect to form the shape, allowing a very simple, intuitive, comprehensible, abstract schema of shape specification. By using superquadrics, we can provide the appropriate visual shape cue, discerning data fields mapped to glyph shape, while not distracting from the cognition of global data patterns. We rely on the ability of superquadrics to create graphically distinct, yet related, shapes. The system allows the mapping of one independent variable to both glyph exponent or two related variables to each glyph exponent to ensure the understandability of the shapes.

SFA uses a pair of 3D magnetic trackers and stereo glasses to provide a minimally immersive desktop visualization system, where the user sits in front of a graphics console that has a screen, keyboard, mouse, and two 3D sensors. Each 3D sensor has buttons for more complete 3D interaction and interrogation of the data. The combination of two-handed interaction and stereo viewing allows us to harness the user's proprioceptive sense to convey information of data spatialization.

We have successfully applied SFA for CFD visualization, as well as information visualization of document corpora and intrusion detection data. The power of shape visualization can be seen in Fig. 39.1, which is a magnetohydrodynamics simulation of the solar wind in the distant heliosphere. In this simulation, the data is a $64 \times 64 \times 64$ grid containing the vector vor-

ticity and velocity for the simulation. Opacity is used to represent vorticity in the j direction, so that the six vortex tubes (only four are visible) represent zones in space where this vorticity is somewhat larger than zero. Glyph shape is based inversely on the velocity in the j direction. Positive velocities are displayed as larger, rounder to cuboid shapes, and negative velocities are displayed as spiky, star-like shapes. Zero velocity is represented by the diamond shape. The overall columnar pattern of the data is not disturbed by the introduction of the shape mapping, but the velocity variation can still be seen as we traverse the lengths of the tubes. In this case, values close to zero in terms of j vorticity (still fluid) have been masked out.

39.3.2 Volume Illustration: Creating Effective Perceptualization by Incorporating Illustration Techniques

As mentioned earlier, technical and medical illustrators have developed techniques over the past several centuries to very compactly and effectively convey the important information in an illustration. Traditional visualization techniques create complex images that may be difficult to interpret and do not have the expressiveness of illustrations. We are developing techniques to capture the enhancement and expressive capability of illustrations. Illustration techniques use characteristics of visual perception, such as edge detection, strokes, contrast sensitivity, clutter reduction, and focus to more effectively convey information.

We are extending nonphotorealistic rendering (NPR) [21,28,29] to volume visualization and creating images that are very effective for training, education, and presentation of medical and scientific data. We have introduced the volume illustration approach, combining the familiarity of a physics-based illumination model with the ability to enhance important features using NPR rendering techniques. Since features to be enhanced are defined on the basis of local and global volume characteristics rather than volume sample value, the application of volume

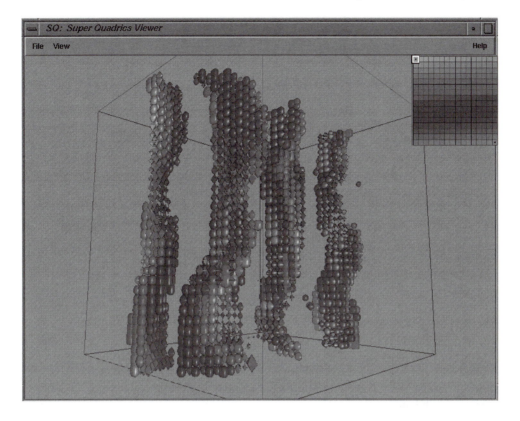

Figure 39.1   Visualization of a magnetohydrodynamics simulation of the solar wind in the distant heliosphere showing both velocity components and vorticity components of the vortex tubes. (See also color insert.)

illustration techniques requires less manual tuning than does the design of a good transfer function. Volume illustration provides a flexible unified framework for enhancing structural perception of volume models through the amplification of features and the addition of illumination effects [6,19].

Several researchers have applied NPR techniques to the display of 2D and surface data. Laidlaw et al. used concepts from painting to create visualizations of 2D data, using brush-stroke-like elements to convey information [16] and a painterly process to compose complex visualizations [15]. Treavett et al. have developed techniques for pen-and-ink illustrations of surfaces within volumes [25]. Interrante applied principles from technical illustration to

convey depth relationships with halos around foreground features in flow data [13]. Saito converted 3D scalar fields into a sampled point representation and visualized selected points with a simple primitive, creating an NPR look [20]. However, with few exceptions, the use of NPR techniques has been confined to surface rendering.

Our earlier work proposing volume illustration [6] was the first system to apply NPR techniques to volume rendering. More recently, several researchers have developed additional systems to advance volume illustration to interactive rendering [4,5].

There are several important questions that should be answered by a volume illustration system.

- What information should be displayed?
- What techniques should be used to display the information?
- How should it be implemented in the rendering process?

Advanced feature analysis and technical illustration principles can be used to determine the important information that should be displayed by the system. We have developed the toolbox of illustrative techniques described below to display and represent the information and have taken several approaches to incorporating these techniques into the rendering process.

39.3.2.1 Toolbox of Illustration Techniques

There are three categories of illustration techniques that we have explored: feature enhancement, depth and orientation enhancement, and regional enhancement [19]. Our feature-enhancement techniques include techniques to enhance volumetric boundaries, volumetric silhouette regions, and volume sketching. Depth and orientation enhancement includes distance color blending

(e.g., aerial perspective), oriented fading, feature halo rendering, and tone shading. Regional enhancement refers to the illustration technique of selectively enhancing and showing details in only certain regions of the image to quickly draw the user's focus. Details of the implementation of these techniques can be found in Rheingans and Ebert [19]. The results achievable by these enhancements can be seen by comparison of the abdominal CT renderings in Figs. 39.2a and 39.2b. Fig. 39.2a has no enhancement, whereas Fig. 39.2b has boundary and silhouette enhancement performed. Also, comparing the left kidney to the right kidney in Fig. 39.3 shows the effectiveness of regional enhancement.

39.3.2.2 Illustrative Rendering Approaches

Our volume illustration techniques are fully incorporated into the volume rendering process, utilizing viewing information, lighting information, and additional volumetric properties to provide a powerful, easily extensible framework for volumetric enhancement. By incorporating the enhancement of the volume

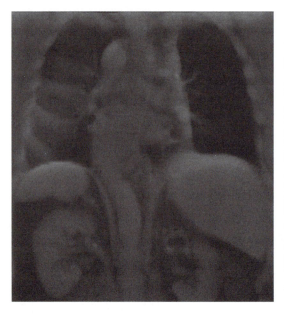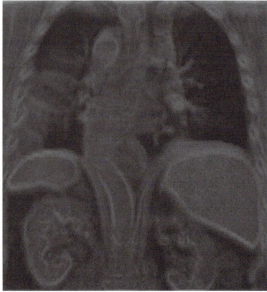

Figure 39.2 Comparison of (a) realistic rendering of an abdominal CT dataset with (b) a volumetric illustration rendering using silhouette and boundary enhancement. (See also color insert.)

Figure 39.3   Rendering of segmented kidney CT dataset showing selective volumetric illustration enhancement applied only to the right kidney to focus the viewer's attention. (See also color insert.)

sample's color, illumination, and opacity into the rendering system, we can implement a wide range of enhancement techniques. The properties that can be incorporated into the volume illustration procedures include the following:

- Volume sample location and value
- Local volumetric properties, such as gradient and minimal change direction
- View direction
- Light information

The viewing and light information allow global orientation information to be used in enhancing local volumetric features. Combining this rendering information with user-selected parameters provides a powerful framework for volumetric enhancement and modification for illustrative effects. It is important to remember that these techniques are applied to continuous volume properties and not to hard surfaces extracted from volume data.

We have extended our original atmospheric volume ray-casting illustration system to two other rendering techniques. First, we have applied these illustration approaches to interactive hardware texture-based volume rendering to achieve illustration effects at interactive rates. An example of these results can be seen in Fig. 39.4. Tone shading, silhouette enhancement, boundary enhancement, and distance color blending can easily be achieved at interactive rates using advanced programmable graphics hardware.

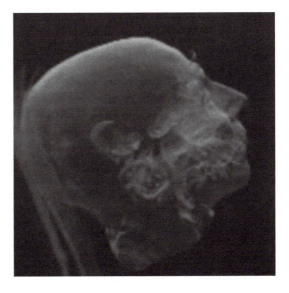

Figure 39.4   Interactive volume illustration rendering of a head dataset with silhouette and boundary enhancement. (See also color insert.)

Second, we have applied these basic principles to create interactive stipple rendering of volumetric datasets for quick previewing and exploration. Stipple drawing is a pen-and-ink illustration technique where dots are deliberately placed on a surface of contrasting color to obtain subtle shifts in value. Traditional stipple drawing is a time-consuming technique. However, points have many attractive features in computer-generated images. Points are the minimum element of all objects and are the simplest and quickest element to render.

By mimicking traditional stipple drawing, we can interactively visualize modestly sized simulations. The use of volume illustration techniques provides the stipple volume-renderer with its interactivity and illustrative expressiveness. In this system, a Poisson disk approximation is initially used for placement of random points within each voxel to simulate the traditional stipple point placement. Volume illustration and feature enhancement techniques are then applied to adjust the number of points that are drawn per voxel. Finally, silhouette curves are added to create more effective renderings. A complete description of this system can be found in the paper by Lu et al. [17]. Figs. 39.5 and 39.6 show the effective results achievable by this system.

39.4   Conclusions

This chapter has provided an introduction to the use of perception for visualization and the need to create perceptually significant visualizations to attack the data deluge facing scientists,

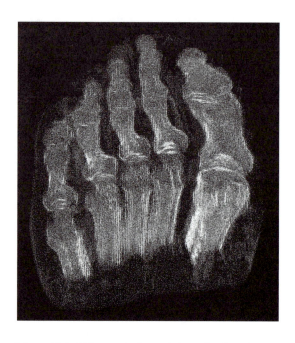

Figure 39.6   Volume stipple rendering of a foot dataset showing tone shading, distance color blending, silhouette and boundary enhancement, and silhoutte lines. (See also color insert).

doctors, and information analysts. An overview of two perceptualization systems and the effective results achievable by these techniques was presented.

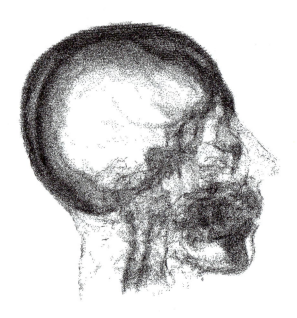

Figure 39.5   Head volume stipple rendering with silhouette, boundary, and distance enhancement and silhouette curves.

Acknowledgments

Many individuals were involved in the research described in this chapter. The SFA system was developed in collaboration with Cindy Starr, Chris Shaw, D. Aaron Roberts, Amen Zwa, Jim Kukla, Ted Bedwell, Chris Morris, and Joe Taylor. The volume illustration and stippling work has been performed collaboratively with Chris Morris, Aidong Lu, Chuck Hansen, Penny Rheingans, and Joe Taylor. This work was supported in part by grants from the NASA AIRSP program and from NSF grants NSF ACI-0081581, ACI-0121288, IIS-0098443, ACI-9978032, MRI-9977218, and ACR-9978099.

References

1. A. Barr. Superquadrics and angle-preserving transformations. *IEEE Computer Graphics & Applications*, 1(1):11–23, 1981.

2. J. Bertin. *Semiology of Graphics*. The University of Wisconsin Press, 1983.

3. W. S. Cleveland. *The Elements of Graphing Data*. Wadsworth Advanced Books and Software, Monterey, Ca., 1985.

4. B. Csébfalvi and M. Gröller. Interactive volume rendering based on a "bubble model". In *GI 2001*. pages 209–216, 2001.

5. B. Csébfalvi, L. Mroz, H. Hauser, A. König, and M. Gröller. Fast visualization of object contours by non-photorealistic volume rendering. *Computer Graphics Forum*, 20(3):452–460, 2001.

6. D. S. Ebert and P. L. Rheingans. Volume illustration: Non-photorealistic rendering of volume models. *IEEE Visualization 2000*, 2000.

7. D. S. Ebert and C. D. Shaw. Minimally immersive flow visualization. *IEEE Transactions on Visualization and Computer Graphics*, 7(4):343–350, 2001.

8. D. S. Ebert, C. D. Shaw, A. Zwa, and E. L. Miller. Minimally-immersive interactive volumetric information visualization. In *Proceedings Information Visualization '96*, 1996.

9. A. Glassner. *Principles of Digital Image Synthesis*. Morgan Kaufmann Publishers, 1995.

10. C. G. Healey and J. T. Enns. Building perceptual textures to visualize multidimensional datasets. In *IEEE Visualization '98*, pages 111–118, 1998.

11. C. G. Healey and J. T. Enns. Large datasets at a glance: Combining textures and colors in scientific visualization. *IEEE Transactions on Visualization and Computer Graphics*, 5(2):145–167, 1999.

12. E. Hodges. *The Guild Handbook of Scientific Illustration*. Guild of Natural Science Illustrators, John Wiley and Sons, 1989.

13. V. Interrante and C. Grosch. Visualizing 3d flow. *IEEE Computer Graphics & Applications*, 18(4):49–53, 1998.

14. V. Interrante, P. Rheingans, J. Ferwerda, R. Gossweiler, and T. Filsinger. Principles of visual perception and its applications in computer graphics. In *SIGGRAPH 97 Course Notes, No. 33*. ACM SIGGRAPH, 1997.

15. R. M. Kirby, H. Marmanis, and D. H. Laidlaw. Visualizing multivalued data from 2d incompressible flows using concepts from painting. In *IEEE Visualization '99*, pages 333–340, 1999.

16. D. H. Laidlaw, E. T. Ahrens, D. Kremers, M. J. Avalos, R. E. Jacobs, and C. Readhead. Visualizing diffusion tensor images of the mouse spinal cord. In *IEEE Visualization '98*, pages 127–134, 1998.

17. A. Lu, C. Morris, D. S. Ebert, C. Hansen, P. Rheingans, and M. Hartner. Illustrative interactive stipple rendering. *IEEE Transactions on Visualization and Computer Graphics*, 9(2):127–138, 2003.

18. A. J. Parker, C. Christou, B. G. Cumming, E. B. Johnston, M. J. Hawken, and A. Zisserman. The analysis of 3D shape: Psychophysical principles and neural mechanisms. In Glyn W Humphreys (Ed.), *Understanding Vision*, Chapter 8. Blackwell, 1992.

19. P. L. Rheingans and D. S. Ebert. Volume illustration: Non-photorealistic rendering of volume models. *IEEE Transactions on Visualization and Computer Graphics*, to appear.

20. T. Saito. Real-time previewing for volume visualization. *Proceedings of 1994 IEEE Symposium on Volume Visualization*, pages 99–106, 1994.

21. M. P. Salisbury, S. E. Anderson, R. Barzel, and D. H. Salesin. Interactive pen-and-ink illustration. In *Proceedings of SIGGRAPH 94*, Computer Graphics Proceedings, Annual Conference Series, pages 101–108, 1994.

22. H. Senay and E. Ignatius. A knowledge-based system for visualization design. *IEEE Computer Graphics and Applications*, 14(6):36–47, 1994.

23. H. Senay and E. Ignatius. Rules and principles of scientific data visualization. *ACM SIGGRAPH HyperVis Project*, www.siggraph.org/education/materials/HyperVis/percept/visrules.htm, 1996.

24. C. D. Shaw, J.A. Hall, D.S. Ebert, and A. Roberts. Interactive lens visualization techniques. *IEEE Visualization '99*, pages 155–160, 1999.

25. S. Treavett, M. Chen, R. Satherley, and M. Jones. Volumes of expression: Artistic modelling and rendering of volume datasets. In *Computer Graphics International 2001*, pages 99–106, 2001.

26. A. Treisman. Features and objects in visual processing. *Scientific American*, 255(2):114–125, 1986.

27. C. Ware. *Information Visualization: Perception for Design*. Morgan Kaufmann Publishers, 2000.

28. G. Winkenbach and D. Salesin. Computer-generated pen-and-ink illustration. In *Proceedings of SIGGRAPH 1994*, Computer Graphics Proceedings, Annual Conference Series, pages 91–100, 1994.

29. G. Winkenbach and D. Salesin. Rendering parametric surfaces in pen and ink. In *Proceedings of SIGGRAPH 1996*, Computer Graphics Proceedings, Annual Conference Series, pages 469–476, 1996.

40   Art and Science in Visualization

VICTORIA INTERRANTE
Department of Computer Science and Engineering
University of Minnesota

40.1   Introduction

Visualization research and development involves the design, implementation, and evaluation of techniques for creating images that facilitate the understanding of a set of data. The first step in this process, visualization design, involves defining an appropriate representational approach, determining the vision of what one wants to achieve. Implementation involves deriving the methods necessary to realize the intended results—developing the algorithms required to create the desired visual representation. Evaluation, or the objective assessment of the impact of specific characteristics of the visualization on application-relevant task performance, is useful not only to quantify the usefulness of a particular technique but also, more powerfully, to provide insight into the means by which a technique achieves its success, thus contributing to the foundation of knowledge upon which we can draw to create yet more effective visualizations in the future. In this chapter, I will discuss the art and science of visualization design and evaluation, illustrated with case study examples from my research. For each application, I will describe how inspiration from art and insight from visual perception can provide guidance for the development of promising approaches to the targeted visualization problems. As appropriate, I will include relevant details of the algorithms developed to achieve the referenced implementations, and where studies have been done, I will discuss their findings and the implications for future directions of work.

40.1.1   Seeking Inspiration for Visualization from Art and Design

Visualization design, from the creation of specific effective visual representations for particular sets of data to the conceptualization of new, more effective paradigms for information representation in general, is a process that has the characteristics of both an art and a science. General approaches to achieving visualizations that "work" are not yet straightforward or well defined, yet there are objective metrics that we can use to determine the success of any particular visualization solution. In this section I will discuss ways in which the practices and products of artists and designers can help provide inspiration and guidance to our efforts to develop new, more effective methods for communicating information through images.

Design, as traditionally practiced, is a highly integrative activity that involves distilling ideas and concepts from a great variety of disparate sources and assembling them into a concrete form that fulfills a complex set of objectives. It is an inherently creative process that defies explicit prediction or definition, yet whose results are readily amenable to comprehensive evaluation. Across disciplines, from graphic arts to architecture, the art of design is primarily learnt through practice, review, and the careful critical study of work by others, and expertise is built up from the lifelong experience of "training one's eyes." Providing a good environment for design is critical to enabling and facilitating the process of design conceptualization. Creative insights are difficult to come by in a vacuum—designers typically surround themselves in their work area with sketches, images, models, references, and other materials that have the potential to both directly and indirectly provide inspiration and guidance for the task at hand. In addition, designers rely heavily on the ability to quickly try out ideas, abstracting, sketching out, and contemplating multiple approaches before settling upon a particular design solution.

In visualization research, we can take a similar approach to the problem of design conceptualization—drawing inspiration from the work of others and from the physical world around us, and experimenting with new combinations and variants of existing techniques for mapping data to images. We can also benefit from establishing fertile design environments that provide rich support for design conceptualization and varied opportunities for rapid experimentation. Finally, it can sometimes be useful to work with traditional materials to create approximate mockups of potential design approaches that allow one to preview ideas to avoid investing substantial effort in software development in what ultimately prove to be unproductive directions.

Turning now from the process of design to the product, there is, in a wide variety of fields—from art to journalism, from graphic design to landscape architecture—a long history of research in visual literacy and visual communication through drawings, paintings, photographs, sculpture, and other traditional physical media that we have the potential to be able to learn from and use in developing new visualization approaches and methods to meet the needs of our own specific situations.

In computer graphics and visualization, as in art, we have complete control not only over what to show but also over how to show it. Even when we are determined to aim for a perfectly physically photorealistic representation of a specified model, as in photography, we have control over multiple variables that combine to define the "setting of the scene"

that creates the most effective result. In many cases this not only includes selecting the viewpoint and defining the field of view, setting up the lighting, and determining the composition of the environment, but also extends to choosing the material properties of the surfaces of the objects that we wish to portray. For practical reasons of computational efficiency or because of the limitations of available rendering systems, we often choose to employ simplified models of lighting and shading, which can also be considered to be a design decision. In addition, we may choose to use non–physically based "artificial" or "artistic" enhancement to emphasize particular features in our data, and we may selectively edit the data to remove or distort portions of the model to achieve specific effects.

Through illustration we have the potential to *interpret* physical reality, to distill the essential components of a scene, accentuate the important information, minimize the secondary details, and hierarchically guide the attentional focus. In different media, different methods are used to draw the eye to or away from specific elements in an image, and in each medium, different styles of representation can be used to evoke different connotations.

When seeking to develop algorithms to generate simplified representations of data or models, it is useful to consider where artists tend to take liberties with reality. They have similar motivations to avoid the tedium and difficulty of accurately representing every detail in a photorealistic manner, but at the same time they need to represent enough detail, with enough accuracy, to meet the expectations of the viewer and communicate the subject effectively. Numerous texts on methods of illustration present various artists' insights on this subject [e.g., 1,2,3,4]. Vision scientists have also considered this question, from the point of view of seeking to understand how the brain processes various aspects of visual input, and it is interesting to note the connection between the findings in perception and common practices in artistic representation. For example, people are found to be highly insensitive to the colors of shadows [9], being willing to interpret as shadows patches whose luminance and general geometry are consistent with that interpretation, regardless of hue, and to be broadly tolerant of inconsistencies among shadows cast from different objects in a scene [5], despite the significant role that cast shadows play in indicating objects' positions in space [6]. Although there is much about the processes of vision and perception that remains unknown, research in visual perception has the potential to make explicit some of the intuition that artists rely upon to create images that "work."

40.1.2 Drawing Insight for Visualization Design from Research in Visual Perception

In addition to seeking inspiration from art for the design of effective methods for conveying information through images, it is possible to use fundamental findings in human visual perception to gain insight into the "science behind the art" of creating successful visual representations. This can be useful because, although it is often (but not always) possible from informal inspection to determine how well a single, particular visualization meets the needs of a specific, individual application, or to comparatively assess the relative merits of alternative visualization solutions for a particular problem, it is much less straightforward to achieve a comprehensive understanding of the reasons that one particular visualization approach is more successful than another, and even more difficult to uncover the theoretical basis for why certain *classes* of approaches are likely to yield better results than others. From a fundamental understanding of the strengths and weaknesses, abilities and limitations, and basic functional mechanisms of the human visual system, we have the potential to become better equipped to more accurately predict which sorts of approaches are likely to work and which aren't, which can be of immense benefit in helping us determine how to guide our research efforts in

the most promising directions and to avoid dead ends.

Mining the vision research literature for insights into a particular visualization problem can be a daunting task. The field of visual perception is broad and deep and has a very long and rich history, with research from decades past remaining highly relevant today. The application domains targeted in visualization are typically far more complex than the carefully controlled domains used in perception studies, and extreme caution must be exercised in hazarding to generalize or extrapolate from particular findings obtained under specific, restricted conditions. Also, the goal in vision research—to understand how the brain derives understanding from visual input—is not quite the same as the goal in visualization (to determine how best to portray a set of data so that the information it contains can be accurately and efficiently understood). Thus, it is seldom possible to obtain comprehensive answers to visualization questions from just a few key articles in the vision/perception literature, but it is more often necessary to distill insights and understanding from multiple previous findings and then supplement this knowledge through additional studies.

Vision scientists and visualization researchers have much to gain from successful collaboration in areas of mutual interest. Through joint research efforts, interdisciplinary teams of computer scientists and psychologists have already begun to find valuable answers to important questions in computer graphics and visualization, such as determining the extent to which accurately modeling various illumination and shading phenomena, such as cast shadows and diffuse interreflections, can facilitate a viewer's interpretation of spatial layout [7,8].

In the remainder of this chapter, I will describe in more concrete detail the application of inspiration from art and insights from visual perception to visualization design and evaluation in the context of selected case-study examples from my research.

40.2 Case Study 1: Effectively Portraying Dense Collections of Overlapping Lines

Numerous applications in fields such as aerospace engineering involve the computation and analysis of 3D vector fields. Developing techniques to effectively visualize such data presents some challenges. One promising approach has been to provide insight into portions of the data using bundles of streamlines. However, special steps must be taken to maintain the legibility of the display as the number of streamlines grows. Stalling et al. [11] proposed a very nice technique, based on principles introduced by Banks [12], for differentially shading 1D streamlines along their length according to the local orientation of the line with respect to a specified light source. Still, the problem of preventing the local visual coalescing of similarly oriented adjacent or proximate overlapping lines remained. Attempts to visually segregate the individual lines by rendering them in a spectrum of different colors rather than in the same shade of grey meet with only partial success, because it is still difficult to accurately and intuitively appreciate the existence and the magnitude of the depth separation between overlapping elements in the projected view. The problem is that, in the absence of indications to the contrary, objects are generally perceived to lie on the background over which they are superimposed [10]. In some situations, cast shadows can be used to help disambiguate depth distance from height above the ground plane [13], but this technique is most effective when applied to simple configurations in which individual objects can be readily associated with their shadows. A more robust solution to the problem of effectively portraying clusters of overlapping and intertwining lines (Fig. 40.1) is inspired by examples from art and is explained by research in visual perception.

For thousands of years, artists have used small gaps adjacent to occluding edges as a visual device to indicate the fact that one surface should be understood to be passing behind another. This convention can be observed in visual representations dating as far back as the Paleolithic paintings within the caves of Lascaux (Fig. 40.2).

Recent research in visual perception [14] helps explain why the technique of introducing gaps to indicate occlusion is so effective. In ordinary binocular vision, when we view one surface in front of another using both of our eyes, we see different portions of the farther surface occluded by the nearer surface in the views from each eye (Fig. 40.3). Called "da Vinci stereopsis," in deference to the recognition of this phenomenon by Leonardo da Vinci, as reported by Wheatstone [15], the perception of these *inter-ocularly unpaired regions*, which he likened to "shadows cast by lights centered at each eye," has been shown to be interpreted by the visual system as being indicative of the presence of a disparity in depth between two surfaces.

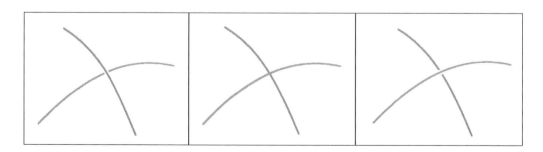

Figure 40.1 Two overlapping lines of roughly equivalent luminance but differing hue, shown in three subtly different depictions. (See also color insert.)

Figure 40.2 Since prehistoric times, artists have used gaps to indicate the passing of one surface behind another, as shown in this photograph taken in a Smithsonian museum exhibit reproducing the creation of the second "Chinese Horse" in the Painted Gallery of the cave of Lascaux. Photograph by Tomás Filsinger. (See also color insert.)

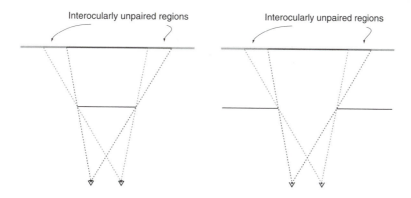

Figure 40.3 When one surface is viewed in front of another using both eyes, a different portion of the more distant surface is occluded in the view from each eye. Psychologists have found evidence that the presence of these interocularly unpaired regions evokes a perception of depth disparity by our visual system [14].

The use of gaps to clarify depth ordering in computer-generated line drawings dates back to the earliest days of computer graphics—one of the first "haloed" line-drawing algorithms was presented by Appel *et al.* at SIGGRAPH in 1979. In 1997, we [16] developed an algorithm for using "visibility-impeding halos" to clarify the depiction of what were effectively clusters of streamlines in volume-rendered images of 3D line integral convolution textures (Fig. 40.4). Our decision to take this approach, and the direction of the particular path that we followed in developing it, were guided both by the inspiration from examples in art and by the insights from research in visual perception.

Figure 40.4 A side-by-side comparison, with and without visibility-impeding halos, illustrating the effectiveness of this technique for indicating the presence of depth discontinuities and facilitating appreciation of the extent of depth disparities between overlapping lines in the 3D flow. Data courtesy of Dr. Chester Grosch.

Our implementation was based on a simple modification to the basic LIC algorithm [17,18] allowing the efficient computation of a matched pair of streamline and surrounding halo textures. We automatically defined a subtle and smoothly continuous 3D visibility-impeding "halo" region that fully enclosed each streamline in the original 3D texture by performing the LIC simultaneously over two input textures containing identically located spots of concentric sizes. Because the streamline tracing only had to be done once for the pair of volumes, the overhead associated with the creation of the halos was kept to a minimum. Halos were implemented, during ray-casting volume rendering, by decreasing the contribution to the final image of any voxel encountered, after a halo was previously entered and subsequently exited, by an amount proportional to the largest previously encountered halo opacity. (Because each line was necessarily surrounded everywhere by its own halo, it was important to allow the voxels lying between the entrance and exit points of the first-encountered halo to be rendered in the normal fashion.) It should be noted that this particular implementation assumes a black background, and will fail to indicate the existence of depth discontinuities between lines that pass closely enough that their halos overlap in 3-space, even if the lines themselves do not actually intersect.

40.3 Case Study 2: Using Feature Lines to Emphasize the Essential 3D Structure of a Form

The goal of an effective graphic representation is to facilitate understanding. When creating computer-generated visual representations of models or surfaces, we desire to portray the data in a way that allows its most important features to be easily, accurately, and intuitively understood. In applications where the information that we need to effectively communicate is the 3D shape of a rendered surface, there is particular value in seeking both inspiration from the practices of artists and insight from research in visual perception.

40.3.1 Nonphotorealistic Rendering in Scientific Visualization

A photographic depiction will capture the exact appearance of an object as it is actually seen, with subtle, complex details of coloration and texture fully represented at the greatest possible level of detail and accuracy. Despite the option to use photography, however, a number of scientific disciplines have historically retained artists and illustrators to prepare hand-drawn images of objects of study [20,23,24]. One of the reasons for this is that in a drawing it is possible to create an idealized representation of a subject, in which structural or conceptual information is clarified in a way that may be difficult or impossible to achieve in even the best photograph. In some fields, such as cartography [22] and archaeology [21,25], specific stylizations are used to encode particular information about important features of the subject within the graphical representation.

When drawings are prepared for visualization purposes, in many cases they are created using photographs of the subject as reference. Through comparing these drawings with the photographs from which they were derived, we have the ability to gain insight into the selective process used by the artist to identify and emphasize the important features in the data while minimizing the visual salience of extraneous detail. We can also observe at what points and in what respects the artist takes special care to remain faithful to details in the original depiction and where he chooses to exercise artistic license. Through these observations, and by consulting the relevant literature in the instruction of illustration techniques, we can derive inspiration for the development of visualization algorithms that aspire to achieve similar effects.

Searching for deeper insight into the process of determining how to design effective pictorial representations, it can be useful to carefully consider all aspects of the questions of not only *how* but also *why* and *when* a hand-drawn interpretation of a subject is preferable to a photograph. In some fields, such as zoology [26], it is not uncommon to find a wide range of representational modalities—from simple outline drawings to detailed, shaded, colored drawings to photographs—used at different times, for different purposes. Apart from the influence of various practical concerns, the definition of an optimal representation will depend on the particular purpose for which it is intended to be used.

The critical issue, I believe, is that drawings are intended to be *idealizations* of actual physical reality. Recognizing a subject from a drawing can require some translation. The success of the representation hinges on the extent to which it captures precisely the necessary information for successful comprehension, which depends not only on the vision of the artist in defining what that information is and on his skill in portraying it faithfully, but also on the needs and experience level of the observer. With photographs, there is no interpretation required—every subtle detail of color and texture, shading and shadow, is exactly represented for some specific configuration of subject, viewpoint, and lighting. Psychologists and others have conducted numerous studies to examine questions of whether, when, and under what circumstances it might be possible to increase perceptual efficiency by reducing the visual complexity of a pictorial representation [27,28,29,30]. Results are mixed, with the superiority of performance using drawings vs. photographs depending both on the task and on the quality and level of detail in the drawing. However, a common theme in all cases is the potential danger inherent in using highly simplified, schematic, or stylized representations, which can, rather than facilitating perception, instead increase cognitive load by imposing an additional burden in interpretation. Studies have found that face recognition from outline drawings is exceptionally poor [31], but that adding "mass," even through simple bi-level shading, improves recognition rates considerably [32]. Face recognition rates from outline drawings can also be improved by distorting outline drawings in the style of a caricature [33].

Carefully constructed nonphotorealistic representations have significant potential advantages for use in visualization. They allow us the possibility to increase the dynamic range of the attentional demands on the observer by allowing greater differentiation in the *salience* of the visual representation, and enable guiding of the attentional focus of the observer through highlighting critical features and deemphasizing the visual prominence of secondary details. However, the success of such efforts depends critically on the ability to correctly define what to emphasize and how to emphasize it, in order to best support the task objectives.

Nonphotorealistic representations also have the potential to facilitate visualization objectives by allowing greater differentiation in the *specificity* of the visual representation. By introducing the possibility of ambiguity, through representational methods that are optimized for *indication* rather than *specification*, one has the potential to portray things in a way that facilitates flexible thinking. This is critical for applications involving design development, such as architecture, where there is a need for visualization tools that facilitate the ability to work with ideas at an early stage of conceptual definition, when a precise physical instantiation of the model has not yet been determined and one needs to foster the envisioning of multiple possibilities. Nonphotorealistic rendering also allows the expression of multiple styles, potentially establishing various "moods" that can influence the subjective context within which information is perceived and interpreted. Finally, in some applications, ambiguity has the potential to be successfully used as a visual primitive to explicitly encode the level of accuracy or confidence in the data [35].

40.3.2 Critical Features of Surface Shape That Can Be Captured by Lines

Gifted artists, such as Pablo Picasso, have demonstrated the ability to capture the essence of a form in just a few critical strokes [36]. In visualization, if we would like to achieve a similar effect, we need to determine an algorithm for automatically defining the set of feature lines that we wish to represent. Both inspiration from art and insight from research in visual perception can be useful in helping to guide these efforts.

According to the Gestalt theory of visual perception, the process of visual understanding begins with the separation of figure from ground. The lines that define this separation are the *silhouettes*, and their use is ubiquitous in artists' line drawings. The silhouette can be imagined as the boundary of the shadow that would be cast by an object onto a planar back-ground from a parallel light source oriented in the direction of the line of sight, with the locus of silhouette points indicating the outline of the object. Closely related to the silhouettes is the set of lines called *contours* [38]. These are formed by the locus of points where the surface normal is orthogonal to the line of sight, and they typically correspond to boundaries across which there is a C^0 discontinuity in depth. On a polygonally defined object, the contour curves can be identified as the locus of all edges shared by a front-facing polygon and a back-facing polygon, plus all boundary edges. Complicating efforts to create effective visualizations using highlighted contour edges is the problem that their definition is viewpoint-dependent. Under conditions of stereo viewing, the set of surface points belonging to the contour curve defined by the view from the left eye will not, in general, be the same as the set of surface points belonging to the contour curve defined by the view from the right eye. If models are rendered in such a way as to imply a correspondence between highlighted contour curves shown in each view, the result will be to impede the stereo matching process and interfere with the observer's ability to accurately perceive the object's 3D shape [39]. In single images, however, highlighting silhouette and contour edges can be an effective way to emphasize the basic structure of the form [40]. There is abundant evidence from psychophysical research that our visual system is adept at inferring information about the 3D shape of an object from the dynamic deformation of its silhouette as it rotates in depth [42] and from the curvature characteristics of the outline of its 2D projection in static flat images [41].

Although silhouettes and contours are important shape descriptors, alone they are not ordinarily sufficient to unambiguously describe the essential shape of a 3D object. In addition to using silhouette and contour curves, artists often also include in their line drawings lines that indicate other sorts of visual discontinuities. These include discontinuities in shape, shading, color, texture, and function. Recognizing this, methods have been developed for automatically generating a line drawing style representation from a 2D image based on local functions of the pixel intensities [43]. Similarly, view-dependent methods have been developed to identify shape-emphasizing feature lines in projected 3D data based on the detection of C^1 or C^2 discontinuities in depth distance (from the viewpoint to the first encountered surface) between neighboring image points [40].

In our research, we have sought to identify supplementary characteristic lines that can be used to highlight additional perceptually or intrinsically significant shape features of arbitrary 3D objects, independent of any particular viewpoint. For several reasons, the most promising candidates for this purpose are the valley and sharp ridge lines on a smoothly curving object and the sharp creases, either convex or concave, on a polygonally defined surface. These lines derive importance first of all because they give rise to a variety of visual discontinuities. Prominent shading discontinuities are likely to be found at sharp creases on any object; on smoothly curving forms, specular highlights are relatively more likely to be found along or near ridges, where the surface normals locally span a relatively larger range of orientations, and sharp valleys have a higher probability of remaining in shadow. Miller [44] demonstrated impressive results using "accessibility shading" to selectively darken narrow concave depressions in polygonally defined models. Equally important was their association with shading discontinuities, crease lines, and valley lines in particular, which derived perceptual significance from their ability to specify the structural skeleton of a 3D form in an intuitively meaningful way. Since 1954 [45], psychologists have found evidence that the points of curvature extrema play a privileged role in observers' encodings of the shape of an object's contour. Recent research in object perception and recognition has suggested that people may mentally represent objects as being composed of parts [46,47], with the objects perceived as subdividing into parts along their valley lines [48,49].

Drawing upon this inspiration from art and insight from visual perception, we developed two distinct algorithms for highlighting feature lines on 3D surfaces for the purposes of facilitating appreciation of their 3D shapes. The first is an algorithm for highlighting valley lines on smoothly curving isointensity surfaces defined in volumetric data [51] (Fig. 40.5). The basic

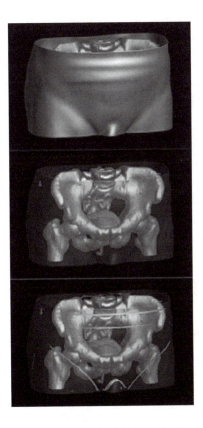

Figure 40.5 By selectively highlighting valley lines, we aim to enhance the perception of important shape features on a transparent surface in a visualization of 3D radiation therapy treatment planning data. (Top) The skin rendered as a fully opaque surface. (Middle) The skin rendered as a fully transparent surface, in order to permit viewing of the internal structures. (Bottom) The skin rendered as a transparent surface with selected opaque points, located along the valley lines. The bottom view is intended to facilitate perception/recognition of the presence of sensitive soft tissue structures that should be avoided by radiation beams in a good treatment plan. Data courtesy of Dr. Julian Rosenman. Middle and lower images © IEEE. (See also color insert.)

algorithm works as follows. On a smoothly curving surface, valley lines are mathematically defined as the locus of points where the normal curvature is locally minimum in the direction of least negative curvature [38]. One can determine, for each point on a surface, whether it should be considered to lie on a ridge or valley line by computing the principal directions and principal curvatures of the surface at that point and then checking to see if the magnitude of the strongest principal curvature assumes an extreme value compared to the magnitudes of the corresponding principal curvatures at the surrounding points on the surface in the corresponding principal direction. To compute the principal directions and principal curvatures at any point in a volumetric dataset, one can begin by taking the *greylevel gradient* [50], in our case computed using a Gaussian-weighted filter over a $3 \times 3 \times 3$ voxel neighborhood, to indicate the surface normal direction, $\vec{e_3}$. Choosing any two arbitrary orthogonal directions $\vec{e_1}$ and $\vec{e_2}$ that span the tangent plane defined by $\vec{e_3}$ gives an orthogonal frame at the point. It is then straightforward to obtain a *principal* frame by computing the Second Fundamental Form,

$$A = \begin{bmatrix} \tilde{\omega}_1^{13} & \tilde{\omega}_1^{23} \\ \tilde{\omega}_2^{13} & \tilde{\omega}_2^{23} \end{bmatrix}, \text{ where } \tilde{\omega}_j^{i3} \text{ is determined by}$$

the dot product of $\overline{e_i}$ and the first derivative of the gradient in the $\overline{e_j}$ direction, and then diagonalizing it to obtain $D = \begin{bmatrix} \kappa_1 & 0 \\ 0 & \kappa_2 \end{bmatrix}$ and

$$P = \begin{bmatrix} v_{1u} & v_{2u} \\ v_{1v} & v_{2v} \end{bmatrix}, \text{ where } A = PDP^{-1} \text{ and}$$

$|\kappa_1| > |\kappa_2|$. The principal directions $\vec{e_1}$ and $\vec{e_2}$ are then given by $v_{1u}\vec{e_1} + v_{1v}\vec{e_2}$ and $v_{2u}\vec{e_1} + v_{2v}\vec{e_2}$ respectively. In our implementation, we highlighted what we determined to be perceptually significant valley lines by increasing the opacities of candidate valley points by an amount proportional to the magnitude of the principal curvature at the point, if that curvature exceeded a fixed minimum value.

Our second algorithm [52] was developed for applications involving the visualization of surface mesh data (Fig. 40.6). In this case the goal was to determine which mesh edges were most important to show in order to convey a full and accurate impression of the 3D shape of the surface in the absence of shading cues. In addition to silhouette and contour edges, our algorithm marked for display selected *internal crease edges* where the angle between neighboring triangles was locally relatively sharp, in comparison with the angles subtended across other edges in the immediate vicinity.

40.4 Case Study 3: Clarifying the 3D Shapes, and Relative Positions in Depth, of Arbitrary Smoothly Curving Surfaces via Texture

The final case study I describe in this chapter concerns the development of visualization techniques intended to facilitate the accurate and intuitive understanding of the 3D shapes and relative positions in depth of arbitrary smoothly curving, transparent surfaces that are not easily characterized by a small number of feature lines. Examples of such surfaces arise in many applications in visualization, from medical imaging to molecular modeling to aerospace engineering, where scientists seek insight into their data through the visualization of multiple level surfaces in one or more 3D scalar distributions. In striving to accomplish this goal I again found great value in drawing upon both inspiration from the practices used by accomplished artists and insight from fundamental research in visual perception.

40.4.1 Cues to 3D Shape and Depth

As a first step in determining how to most effectively convey the 3D shapes and depths of smoothly curving transparent surfaces in computer-generated images, it is useful to briefly consider the questions of 1) how we ordinarily infer shape and depth information from visual input and 2) why the shapes and depths of transparent surfaces can be difficult to adequately perceive, even in everyday experience. Since a

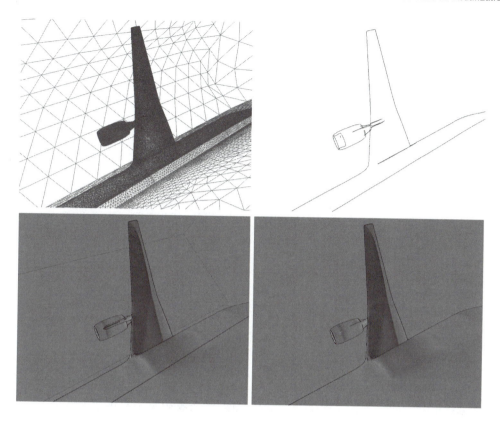

Figure 40.6 By highlighting feature lines on a surface mesh, we may enhance appreciation of the essential structure of the form, a goal that assumes particular importance under conditions where the use of surface shading is problematic. Clockwise from top left: the full original mesh; the silhouette and contour edges only; the silhouette and contour edges highlighted in a surface rendering in which polygon color is defined purely as a function of the value of a scalar parameter that is being visualized over the mesh; same as previous, except that the feature line set is augmented by crease edges determined by our algorithm to be locally perceptually significant. Data courtesy of Dimitri Mavriplis. (See also color insert.)

full discussion of shape and depth perception is beyond the scope of this chapter, I focus here on the most important, relevant aspects of scene construction that are typically under the control of the visualization developer: viewpoint and projection, lighting, and the definition of surface material properties.

After occlusion, which unambiguously specifies the depth-order relationships of overlapping surfaces, linear perspective is one of the strongest available pictorial cues to depth. Under linear perspective, lines that are parallel in 3D appear in a projected image to converge towards one or more vanishing points as they recede into the distance, and objects accordingly appear smaller/increasingly skewed, with increasing distance from the viewpoint/increasing eccentricity from the center of projection [53]. In an orthographic projection, or a perspective projection subtending a very narrow field of view, these convergence cues and size gradients are forfeited. The selection of an appropriate vantage point with respect to an object is also a consideration of some importance. In choosing preferred views for objects, observers appear to use task-dependent strategies [55]. Nongeneric viewpoints, in which accidental alignments or occlusions of particular object features fortuitously occur, have the greatest potential to be misleading [54].

Lighting is a complex and important factor that can influence the perception of a scene in multifaceted ways, as is well understood in fields related to the cinema and stage. For visualization purposes, in addition to effects of shadows on depth perception, which were mentioned earlier, we need to consider how best to control lighting parameters to facilitate an accurate perception of shape from shading. It has long been recognized that people are accustomed to things being lit from "above" [56], and that among the deleterious effects of directing light toward a surface from below is an increased chance of fostering depth reversal, where convexities are perceived as concavities and vice versa [57]. Recent research suggests more specifically a bias toward the assumption that lighting is coming from the above-left, possibly attributable to cerebral lateralization [58]. Somewhat less dramatic but also significant are the shape-enhancing effects of defining lighting to be at an oblique angle, as opposed to head-on [59], whereby shading gradients on surfaces receding in depth are enhanced.

It perhaps is in the definition of surface material properties that the greatest hitherto untapped potential lies for facilitating shape and depth perception in visualizations of surface data. Before addressing this topic, however, I would like to discuss the remaining issue of why transparent surfaces are so difficult to adequately perceive, which will help suggest how material properties might be selected to best advantage.

Plain transparent surfaces clearly provide impoverished cues to shape and depth. Shape-from-shading information is available only through the presence of specular highlights, which, in binocular vision, are perceived not to lie on the curved surface of an object but to float behind the surface if the object is locally convex and in front of the surface if it is locally concave [60]. Cues to the presence of contours, where the surface normal is orthogonal to the line of sight, are provided by the distorting effects of refraction; however, this comes at a significant cost to the clarity of visibility of the underlying mater-

ial. When artists portray transparent surfaces, they similarly tend to rely heavily on the use of specular highlights (Lucas Cranach) [61], specular reflection (Don Eddy) [63], and/or refractive distortion (Janet Fish) [62].

While it would be misleading to downplay the potential merits of employing a fully physically correct model of transparency for visualization applications—refraction provides a natural way to emphasize silhouettes and contours, and might also provide good cues as to the thickness of a transparent layer—it is at the same time clear that the visualization goal of enabling the effective simultaneous understanding of multiple layers will not be met simply by achieving photorealism in the surface rendering. Something must be done to supplement the shape and depth cues available in the scene. The solution that we came up with in our research was to "stick something onto" the surface, in the form of carefully designed, subtle, uniformly distributed texture markings, that can provide valuable cues to the surface shape, along with explicit cues to the surface depth, in a way that is not possible through reliance on specular highlights.

Before moving on to the discussion of surface textures, a final point in regard to the rendering of transparent surfaces bears mentioning. There are several alternative models for surface transparency, corresponding to different types of transparent material. The first model (Fig. 40.7a) corresponds to the case where you have an opaque material, such as a gauze curtain, that is very finely distributed, so that over a finite area it is both partly present and partly absent. This is the type of transparency represented by the "additive model": $I = I_f \cdot \alpha_f + I_b \cdot (1 - \alpha_f)$, where α_f is the (wavelength-independent) opacity of the transparent foreground material and I_f is its intensity, while I_b is the intensity of the background material. If a surface rendered using this model is folded upon itself multiple times, the color in the overlap region will, in the limit, converge to the color of the material; shadows cast by this material will be black. A second model (Fig. 40.7b)

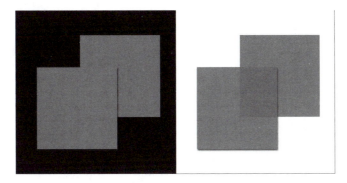

Figure 40.7 Examples of two different models of transparency. (Left) Additive transparency, exhibited by materials such as gauze that are intrinsically opaque but only intermittently present. (Right) Multiplicative or subtractive transparency, exhibited by materials, such as colored glass, that selectively filter transmitted light. (See also color insert.)

corresponds to the case where you have a semi-transparent material that impedes the transmission of certain wavelengths of light. This type of transparency can be approximated by a "multiplicative model": $I = \alpha_f I_b$, where α_f is the (wavelength-dependent) transmissivity of the foreground material. Multiple layers of this material will, in the limit, converge to looking black; shadows cast by this material will be colored. Other approaches, incorporating both additive and multiplicative combinations of foreground and background material, are also possible.

40.4.2   Using Texture on Surfaces to Clarify Shape

Having determined to attempt to clarify the 3D shapes of external transparent surfaces through the addition of sparsely distributed texture markings, we now ask what sort of markings we should add. If we could define the *ideal* texture pattern to apply to an arbitrary smoothly curving surface in order to enable its 3D shape to be most accurately and effectively perceived, what would the characteristics of that texture pattern be? To answer this question we again turn for inspiration to the observation of the practices of artists and illustrators, and for insight to the results of research in psychology on the perception of shape from texture.

In stipple drawings, artists carefully control the density of pen markings in order to achieve a desired distribution of tone. With line drawings, in addition to achieving variations in tone, there is the additional concern of carefully controlling the *directions* of the lines in order to emphasize the surface shape. Although there is no precise or universally recognized convention for defining the directions of pen strokes in line drawings, artists and illustrators have frequently noted the importance of using stroke direction appropriately, and have variously cited advantages for effectively conveying shape, in using lines that "follow the contours of the form" [64] or that run "at right angles to the length of the form" [65].

The significance of texture's effect on shape (slant) perception was first emphasized in the research literature and formally studied by James Gibson [66]. Using two different wallpaper patterns on large flat boards, viewed through a circular aperture, he found that slant perception was not only significantly more accurate under either texture condition than under the control condition of no texture, but also that accuracy was greatest in the case of the more "regular" texture pattern. In subsequent studies comparing the effects of different aspects of "texture regularity," researchers found evidence that regularity in element size, element shape, and element placement all had a positive effect in improving slant perception accuracy [67]. Ultimately, it was determined that linear convergence cues (from perspective projection) play the dominant role in slant perception from texture [68,69]. Looking at the effects of texture on shape perception in the case of curved surfaces, Cutting and Millard [70] found evidence primarily for the importance of "texture compression" cues, manifested as the changes in the projected shapes of circular texture elements. In later studies, Cumming et al. [71] found support for these findings. Evaluating the relative impacts of veridical size, density, and compression gradients on curvature perception accuracy under stereo viewing conditions, they found that perceived curvature was least when element compression was held constant, while the presence or absence of appropriate gradients in element size and/or spacing had little effect on curvature perception accuracy. These results are important because they provide clear evidence that the particular characteristics of a surface texture pattern can significantly affect shape perception, even in the presence of robust, veridical cues to shape from stereo.

Still, it remains unclear what sort of texture we should choose to apply to a surface in order to facilitate perception of its shape. Stevens [72], informally observing images from the Brodatz [73] texture album pasted onto a cylindrical form and viewed monocularly under conditions that minimized shading cues, reported obtaining a compelling impression of surface curvature only in the cases of the wire mesh and rattan textures, the most regular, synthetic patterns. However, observing an extensive variety of line-based texture patterns projected onto a complicated, doubly curved surface, obliquely oriented so as to exhibit contour occlusions, Todd and Reichel [74] noted that a qualitative perception of shape from texture seems to be afforded under a wide range of texturing conditions. Computer vision algorithms for the estimation of surface orientation from texture generally work from assumptions of texture isotropy or texture homogeneity.

Rosenholtz and Malik [75] found evidence that human observers use cues provided by deviations from both isotropy and homogeneity in making surface orientation judgments. Stone [76] notes that particular problems are caused for perception by textures that are not "homotropic" (in which the dominant direction of the texture anisotropy varies across the texture pattern). Optical artists such as Brigit Riley have exploited this assumption to create striking illusions of relief from patterns of waving lines. Mamassian and Landy [77] found that observers' interpretations of surface shape from simple line drawings are consistent with the following biases under conditions of ambiguity: to perceive convex, as opposed to concave, surfaces; to assume that the viewpoint is from above; and to interpret lines as if they were oriented in the principal directions on a surface. Knill [78] suggests that texture patterns with oriented components, which under the assumption of texture pattern homogeneity are constrained to follow parallel geodesics on developable surfaces, may provide more perceptually salient cues to surface shape than isotropic patterns. Finally, Li and Zaidi [79,80] have shown that observers can reliably discriminate convex from concave regions in front-facing views of a vertically oriented sinusoidally corrugated surface only when a perspective projection is used and the texture pattern contains patterns of oriented energy that follow the first principal direction. However, in more recent studies, considering a wider range of viewpoints, they have found indications that the texture conditions necessary to ensure the veridical perception of convexity vs. concavity are more complicated than previously believed [81].

Because of historical limitations in the capabilities of classical texture-mapping software and algorithms, with few exceptions nearly all studies investigating the effect of surface texture on shape perception that have been conducted to date have been restricted either to the use of developable surfaces—which can be rolled out to lie flat on a plane—or to the use of procedurally defined solid texture patterns, whose characteristics are in general independent of the geometry of the surfaces to which they are applied. For several years we have believed that important new insights into texture's effect on shape perception might be gained through studies conducted under less restrictive surface and texture pattern conditions. In the final part of this section I will describe the algorithms that we derived for the controlled synthesis of arbitrary texture patterns over arbitrary surfaces and the results of the studies we have recently undertaken in pursuit of a deeper understanding of how we might best create and apply custom texture patterns to surfaces in scientific datasets in order to most effectively facilitate accurate perception of their 3D shapes.

In our first studies [82], involving indirect judgments of shape and distance perception under different texture conditions on layered transparent surfaces extracted from radiation therapy treatment planning datasets (Fig. 40.8), we created a variety of solid texture patterns by scan-converting individual texture elements (spheres, planes, or rectangular prisms) into a 3D volume at points corresponding to evenly distributed locations over a pre-identified isosurface. We found clear evidence

that performance was better in the presence of texture, but we did not find a significant main effect of texture type. As expected, "sticking something onto the surface" helped, but the question of how best to define helpful texture markings remained open. Unfortunately, none of the textures we were able to achieve using this discrete element approach yet resembled anything one might find in an artist's line drawing.

Shortly afterward, we developed an improved method for synthesizing a more continuous surface texture pattern that everywhere followed the first principal direction over an arbitrary doubly curved surface [83]. To achieve this texture pattern we began by scattering a number of discrete high-intensity point elements evenly throughout an otherwise empty 3D volume, according to an approximate Poisson distribution. The location of each point element was determined by dividing the volume into uniformly sized cells and randomly selecting a location within each cell, under the restriction that no location could be selected that was within a predetermined distance from any previously chosen point. In a separate process, we precomputed the vector field of the first principal directions of the iso-level surfaces defined at every

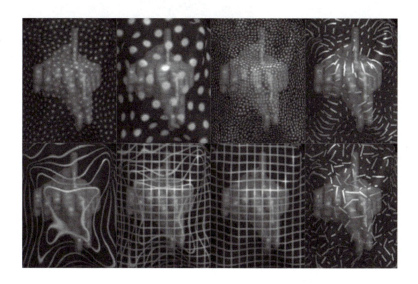

Figure 40.8 A variety of sparse textures applied to the same external transparent surface.

Figure 40.9 (Left) A texture of evenly distributed points over three slices in a volume. (Right) The results after advecting the texture according to the first principal direction vector field using 3D line integral convolution.

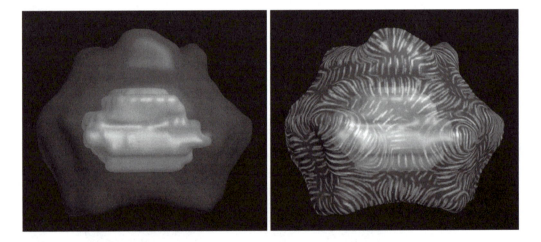

Figure 40.10 A 3D line integral convolution texture applied to a transparent iso-intensity surface in a 3D radiation-therapy treatment planning dataset. Data courtesy of Dr. Julian Rosenman.

point in the 3D volume dataset, using the principal direction computation approach described in Section 40.3.2 of this chapter. Finally, we used 3D line integral convolution to "comb out" the distributed point texture along the lines of curvature defined by the principal direction vector field (Fig. 40.9), to obtain a single solid texture that could be applied to any iso-level surface in the volume data (Fig. 40.10). In order to avoid artifacts in the pattern at the

points where the first and second principal directions switched places, we forced the filter kernel length to reduce to zero in the vicinity of these umbilic points.

Although these results were encouraging, important tasks remained. The first was to objectively evaluate the relative effectiveness of the new LIC-based principal direction texturing approach, and in particular to rigorously examine the impact of texture orientation on shape

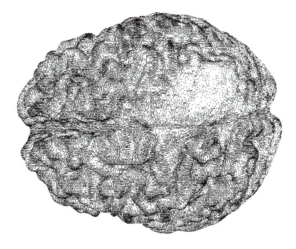

Figure 40.11 A line drawing of a brain dataset, generated by Steve Haker [84], in which tiny straight strokes are oriented in the first principal direction at vertices in the surface mesh.

perception in the case of complicated doubly curving surfaces. The second was to pursue development of a flexible, robust principal direction texturing method that could be applied to polygonal datasets (Fig. 40.11). The two principal challenges in that regard were to define a method for obtaining accurate estimates of the principal directions at the vertices of an arbitrary polygonal mesh and to determine how to synthesize an arbitrary texture pattern over an arbitrary doubly curved surface, in a way that avoided both seams and stretching and such that the dominant orientation in the texture pattern everywhere locally followed the first principal direction vector field.

While several methods have been previously proposed for estimating principal directions and principal curvatures at points on meshes [85,86], we have found in practice that all exhibit unexplained large errors in some cases. Recently, we set out to investigate the sources of these errors, and in the process we developed a new method for principal direction estimation that appears to produce better results [87]. Unfortunately, space does not permit a full description of that approach here, but the essential insights are these: 1) large errors can, and do, occur at points that are far from being umbilic; 2) errors

are most problematic when the underlying mesh parameterization is not regular; and 3) we can use the vertex location and (approximate) surface normal information available at the neighboring points to a vertex to achieve a least squares *cubic* surface fit to the mesh at that point, which appears to offer better potential for a more accurate fit than when the surface is locally restricted to be quadratic. The open questions that remain are how to robustly resolve problems that arise due to the first and second principal directions switching places on either side of critical points; how to gracefully determine an appropriate texture or stroke direction across patches of umbilic points, where the principal directions are undefined; and how to balance the concern of emphasizing shape with the concern of minimizing the salience of extraneous detail. Not all surface perturbations are worth drawing attention to, and, depending on the application, it may be desirable to enforce certain smoothness criteria before using a computed principal direction vector field for texture definition purposes.

In the first of our most recent experiments [93] intended to gain insights into methods for using texture effectively for shape representation, we investigated the effect of the presence and direction of luminance texture pattern anisotropy on the accuracy of observers' judgments of 3D surface shape. Specifically, we sought to determine 1) whether shape perception is improved, over the default condition of an isotropic pattern, when the texture pattern is elongated in the first principal direction; and 2) whether shape perception is hindered, over the default condition of an isotropic pattern, when the texture is elongated in a constant or varying direction other than the first principal direction. We had five participants, using a surface attitude probe [88], make judgments about local surface orientation at 49 evenly spaced points on each of six different smoothly curving surfaces, under each of four different luminance texture conditions. All of the texture patterns for this study were created via 3D line integral convolution, using either a random/isotropic vector field (*rdir*), a first principal direction vector field (*pdir*), a vector field following a constant uniform direction (*udir*), or a vector field following a sinusoidally varying path (*sdir*). Sample stimuli are shown in Fig. 40.12. The experiment was repeated under two different viewing conditions, flat and stereo. Charts summarizing the results are shown in Fig. 40.13. In the flat viewing condition, we found that

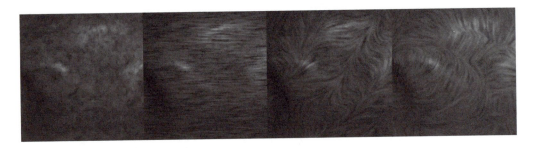

Figure 40.12 Representative examples of the sample stimuli used in our first recent experiment investigating the effect of texture orientation on the accuracy of observers' surface shape judgments. From left to right: Isotropic (rdir), Uniform (udir), Swirly (sdir), and Principal Direction (pdir).

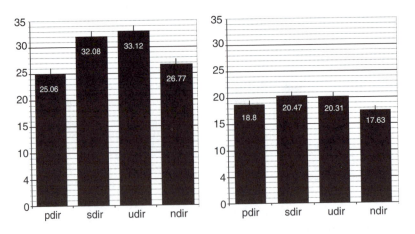

Figure 40.13 Pooled results (mean angle error) for all subjects, all surfaces, by texture type. (Left) flat presentation; the differences {pdir,rdir} < {sdir, udir} were significant at the 0.05 level. (Right) stereo presentation; only the differences {rdir < sdir, udir} were significant at the 0.05 level. (See also color insert.)

performance was significantly better in the cases of the pdir and rdir patterns than in the cases of the sdir and udir patterns. Accuracy was significantly improved in the stereo vs. the flat viewing condition for all texture types. Performance remained marginally better in the cases of the isotropic and principal direction patterns than under the other two texture conditions, but significance was achieved only in the rdir case. These results are consistent with the hypothesis that texture pattern anisotropy can impede surface shape perception when the elongated markings are oriented in a way that is different from the principal direction, but they do not support the hypothesis that principal direction textures will facilitate shape perception to a greater extent than will isotropic patterns.

In a follow-up study [94], we repeated the experiment using displacement textures instead of luminance textures, and we found the same pattern of results. However, two important questions were raised by this work. First, why does shape perception seem to be most accurate in the principal direction orientation condition, when there is little ecological justification for a texture pattern being oriented in the principal directions across a doubly curved surface? Is it because, from a generic viewpoint, the contours traced by a principal direction texture have the greatest potential to reveal the surface curvature to a maximum extent, while the contour traced out by the texture flow along any other direction at that point and for the same view will be intrinsically more flat, which may represent a loss of shape information that is not recoverable? Second, on arbitrary doubly curved surfaces, there are two orthogonal directions in which the normal curvature generically assumes a nonzero extremum. Although these directions can be reliably classified into two types, the first principal direction and the second principal direction, there is not a clear algorithm for determining which of these two directions a singly oriented directional texture should follow at any point in order to minimize artifacts due to the apparent turning of the texture pattern in the surface. Is it possible that the effectiveness of

the pdir texture used in this first experiment was compromised by these "corner" artifacts, and that we might be able to more effectively facilitate shape perception using an orthogonally bidirectional principal direction oriented pattern—one that has 90° rotational symmetry?

In order to investigate these questions, we needed to conduct further studies and to develop a more general texture synthesis method capable of achieving a wider variety of oriented patterns over surfaces.

Inspired by Efros and Leung's [90] algorithm for synthesizing unlimited quantities of a texture pattern that is nearly perceptually identical to a provided 2D sample, we developed a fast and efficient method for synthesizing a fitted texture pattern, without visible seams or projective distortion, over a polygonal model, such that the texture pattern orientation is constrained to be aligned with a specified vector field at a per-pixel level [91]. Our method works by partitioning the surface into a set of equally sized patches, then using a two-pass variant of the original Efros and Leung method to synthesize a texture for each patch, being careful to maintain pattern continuity across patch boundaries, and performing the texture lookup, at each point, in an appropriately rotated copy of the original texture sample, in order to achieve the desired local pattern orientation.

Using this system to render a new set of textured surface stimuli, we undertook a second experiment [92] intended to evaluate the information carrying capacities of two different base texture patterns (one singly oriented and one doubly oriented), under three different orientation conditions (pdir, udir, and sdir) and two different viewing conditions (upright and backward slanting). In a four-alternative forced choice task, over the course of 672 trials, three participants were asked to identify the quadrant in which two simultaneously displayed B-spline surfaces, illuminated from different random directions, appeared to differ in their shapes. We found that participants were consistently able to

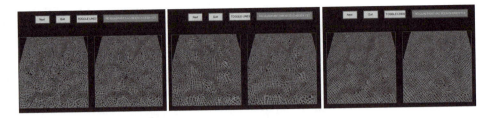

Figure 40.14 Representative stimuli used in our second experiment to investigate the relative extents to which differently oriented patterns have the potential to mask the perceptibility of subtle shape differences. The texture conditions are, from left to right, principal direction, swirly direction, and uniform direction. (See also color insert.)

more reliably perceive smaller shape differences when the surfaces were textured with a pattern whose orientation followed one of the principal directions than when the surfaces were textured either with a pattern that gradually swirled in the surface or with a pattern that followed a constant uniform direction in the tangent plane. We did not find a significant effect of texture type (performance was only marginally better overall in the two-directional case) or of surface orientation (performance was only marginally better overall in the tilted vs. front-facing case), nor evidence of an interaction between texture type and surface orientation. Sample stimuli and summary results are shown in Figs. 40.14 and 40.15. These findings support the hypothesis that anisotropic textures *not* aligned with the first principal direction may support shape perception more poorly, for a generic view, than principal direction–oriented anisotropic patterns, which can provide cues as to the maximal amount of surface normal curvature in a local region. However, this study did not yield much insight into the potential effects on shape perception of principal-direction texture type.

In our third recent experiment [89], we focused on the question of whether some principal-direction texture patterns might be more effective for conveying shape than others, and, if so, what the particular characteristics of these principal direction patterns might be. Five participants each adjusted surface attitude probes to provide surface orientation estimates at two different locations over five different surfaces under each of four different texture conditions. With five repeated measures, we had a total of 200 trials per person. We compared performance under the control condition of no texture to performance under three different texture type conditions: a high-contrast, one-directional line pattern that everywhere followed the first principal direction (1dir); a lower-contrast, one-directional line integral convolution pattern that similarly followed the first principal direction (lic); and a medium-high-contrast, two-directional grid pattern that was everywhere in alignment with both principal directions (2dir). All patterns had equivalent mean luminance. Sample stimuli are shown in Fig. 40.16. We used the statistical software package MacAnova, developed by Prof. Gary Oehlert from the Department of Statistics at the University of Minnesota, to perform a three-way (within subjects) mixed analysis of variance (ANOVA) to evaluate the statistical significance of the results. We found significant main effects of probe location ($p = 0.0000264$) and texture type ($p = 0.0002843$), and a significant two-way interaction between texture type and probe location ($p < 0.00000001$). We did not find a significant main effect of subject id ($p = 0.18$) nor a significant interaction between subject and texture type ($p = 0.62$). We used Tukey's Honestly Significant Difference (HSD) method to perform post-hoc pairwise comparisons of the means of the angle errors under the different texture conditions. We found that the following differences were statistically significant at the 0.01 level: 2-dir < 1-dir, 2-dir < None, 1-dir <

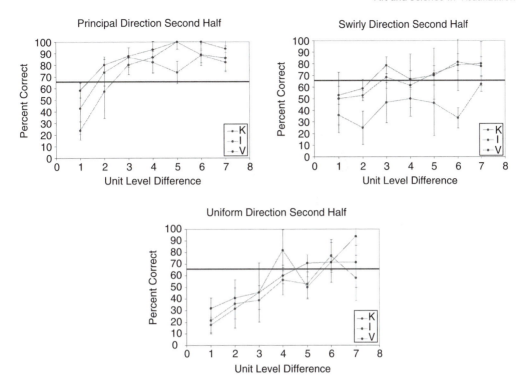

Figure 40.15   Summary results of our second experiment. Accuracy increased with increasing magnitude of shape difference in all cases, but it increased at a faster rate under the principal direction texture condition. Error bars represent 95% confidence intervals. (See also color insert.)

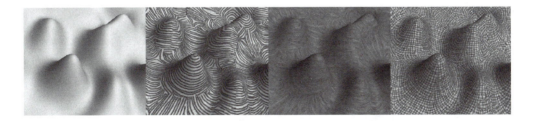

Figure 40.16   A test surface from our third experiment, in the control condition of no texture (left), and (from left to right) under the three studied principal direction texture conditions: 1dir, lic, and 2dir.

None, and LIC < None. The difference between performance in the 2-dir and LIC conditions was not statistically significant at the 0.01 level, nor was the performance difference between the LIC and the 1-dir conditions. Charts summarizing these results are shown in Fig. 40.17.

Through all of the efforts summarized in this section, we have realized that determining the characteristics of a texture pattern that is best able to facilitate surface shape perception is not as straightforward an undertaking as it first might seem. Conducting controlled experiments

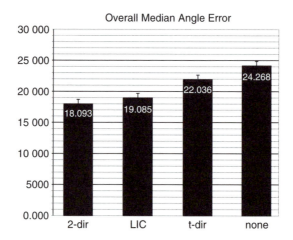

Figure 40.17 The results of the third cumulative experiment. (See also color insert.)

is a delicate and time-consuming business, and success is never guaranteed. However, through our efforts, we have been able to answer many important questions about the suitability of principal-direction-oriented patterns for shape representation, and the open questions that remain provide a welcome challenge for future work.

40.5 Conclusions

The process of creating an effective visual representation of a set of data is both an art and a science, requiring extensive efforts in visualization design, implementation, and evaluation. For visualization design, there are significant potential benefits in seeking inspiration from previous graphical work in art, illustration, visual communication, and design, and in seeking insights from research in vision and visual perception. The process of implementation—figuring out how to develop the algorithms necessary to translate our vision of the results we want to achieve into a reality—(though dealt with only lightly in this chapter) are of extreme importance, and this has historically been where the field of visualization has seen its greatest successes. Evaluation, through informal obser-

vation or, more rigorously, through controlled observer experiments, can be critical in clarifying our understanding of the strengths and weaknesses of alternative visualization approaches and for assessing the practical merits of a particular visualization approach for a specific task. Most importantly, evaluation helps us to better understand not only what works and what doesn't, and by how much, but also to gain insight into *why*. Based on this insight, we are better equipped to go back to the design stage and work on developing yet more effective approaches to meet our visualization objectives.

References

1. A. Loomis. *Creative Illustration.* The Viking Press, 1947.
2. E. W. Watson. *The Art of Pencil Drawing.* Watson-Guptill Publications, 1968.
3. A. L. Guptill. *Rendering in Pen and Ink.* Watson-Guptill Publications, 1976.
4. S. McCloud. *Understanding Comics: The Invisible Art.* Harper Perennial, 1994.
5. Y. Ostrovsky, P. Cavanagh, and P. Sinha. Perceiving illumination inconsistencies in scenes. *AI Memo #2001-029, MIT*, November 2001.
6. D. Kersten, P. Mamassian, and D. C. Knill. Moving cast shadows induce apparent motion in depth. *Perception*, 26(2):171–192, 1997.

7. H. Hu, A. A. Gooch, W. B. Thompson, B. E. Smits, J. J. Rieser, and P. Shirley. Visual cues for imminent object contact in realistic virtual environments. *Proceedings of IEEE Visualization 2000*, pages 179–185.

8. C. Madison, W. Thompson, D. Kersten, P. Shirley, and B. Smits. Use of interreflection and shadow for surface contact. *Perception & Psychophysics*, 63(2):187–194, 2001.

9. P. Cavanagh and Y. G. Leclerc. Shape from shadows. *Journal of Experimental Psychology: Human Perception and Performance*, 15(1):3–27, 1989.

10. J. J. Gibson. *The Perception of the Visual World*. Houghton-Mifflin, 1950.

11. D. Stalling, M. Zöckler, and H.-C. Hege. Fast display of illuminated field lines. *IEEE Transactions on Visualization and Computer Graphics*, 3(2):118–128, 1997.

12. D. C. Banks. Illumination in diverse codimensions. *Computer Graphics Proceedings, Annual Conference Series*, pages 327–334, 1994.

13. A. Yonas, L. T. Goldsmith, and J. L. Hallstrom. Development of sensitivity to information provided by cast shadows in pictures. *Perception*, 7(3):333–341, 1978.

14. K. Nakayama and S. Shimojo. Da Vinci stereopsis: depth and subjective contours from unpaired image points. *Vision Research*, 30(11):1811–1825, 1990.

15. C. Wheatstone. On some remarkable, and hitherto unobserved, phenomena of binocular vision. *Philosophical Transactions of the Royal Society of London*, 128:371–394, 1838.

16. V. Interrante and C. Grosch. Visualizing 3D flow. *IEEE Computer Graphics and Applications*, 18(4):49–53, 1998.

17. B. Cabral and L. C. Leedom. Imaging vector fields using line integral convolution. *Computer Graphics Proceedings, Annual Conference Series*, pages 263–270, 1993.

18. D. Stalling and H.-C. Hege. Fast and resolution independent line integral convolution. *Computer Graphics Proceedings, Annual Conference Series*, pages 249–256, 1995.

19. A. Appel, F. J. Rohlf, and A. J. Stein. The haloed line effect for hidden line elimination. *Proceedings of SIGGRAPH '79*, pages 151–157, 1979.

20. E. R. S. Hodges. *The Guild Handbook of Scientific Illustration*. Van Nostrand Reinhold, 1989.

21. B. D. Dillon, Ed. *The Student's Guide to Archaeological Illustrating*. Institute of Archaeology, University of California, Los Angeles, 1981.

22. E. Imhof. *Cartographic Relief Presentation*. De Gruyter, 1982.

23. J. L. Ridgway. *Scientific Illustration*. Stanford University Press, 1938.

24. W. E. Loechel. *Medical Illustration: A Guide for the Doctor-Author and Exhibitor*. Charles C. Thomas, 1964.

25. L. R. Addington. *Lithic Illustration: Drawing Flaked Stone Artifacts for Publication*. The University of Chicago Press, 1986.

26. G. R. Allen and D. R. Robertson. *Fishes of the Tropical Eastern Pacific*. University of Hawaii Press, 1994.

27. T. A. Ryan and C. B. Schwartz. Speed of perception as a function of mode of representation. *American Journal of Psychology*, 69:60–69, 1956.

28. D. Fussel and A. Haaland. Communicating with pictures in Nepal: results of practical study used in visual education. *Educational Broadcasting International*, 11(1):25–31, 1978.

29. K. Hirsh and D. A. McConathy. Picture preferences of thoracic surgeons. *Journal of BioCommunications*, pages 26–30, 1986.

30. I. Biederman and G. Ju. Surface versus edge-based determinants of visual recognition. *Cognitive Psychology*, 20(1):38–64, 1988.

31. G. Davies, H. Ellis, and J. Shepherd. Face recognition accuracy as a function of mode of representation. *Journal of Applied Psychology*, 63(2):180–187, 1978.

32. V. Bruce, E. Hanna, N. Dench, P. Healey, and M. Burton. The importance of 'mass' in line drawings of faces. *Applied Cognitive Psychology*, 6(7):619–628, 1992.

33. G. Rhodes, S. Brennan, and S. Carey. Identification and ratings of caricatures: implications for mental representations of faces. *Cognitive Psychology*, 19(4):473–497, 1987.

34. L. Anderson, J. Esser, and V. Interrante. A virtual environment for conceptual design in architecture. *Ninth Eurographics Workshop on Virtual Environments/Seventh International Workshop on Immersive Projection Technology*, May 2003.

35. V. Interrante. Harnessing rich natural textures for multivariate visualization. *IEEE Computer Graphics and Applications*, 20(6):6–11, 2000.

36. P. Picasso. *Study of a Bull's Head*, 5 November 1952.

37. O. Monga and S. Benayoun. Using partial derivatives of 3D images to extract typical surface features. *Computer Vision & Image Understanding*, 61(2):171–189, 1995.

38. J. J. Koenderink. *Solid Shape*. MIT Press, 1990.

39. V. Interrante. *Illustrating Transparency: Communicating the 3D Shape of Layered Transparent Surfaces via Texture*. PhD Dissertation, UNC-Chapel Hill, 1996.

40. T. Saito and T. Takahashi. Comprehensible rendering of 3-D shapes. *Computer Graphics*, 24(4):197–206, 1990.

41. W. Richards, J. Koenderink, and D. Hoffman. Inferring three-dimensional shapes from two-dimensional silhouettes. *Journal of the Optical Society of America, A, Optics and Imaging Science*, 4:1168–1175, 1987.

42. H. Wallach and D. N. O'Connell. The kinetic depth effect. *Journal of Experimental Psychology*, 45(4):205–217, 1953.

43. D. E. Pearson and J. A. Robinson. Visual communication at very low data rates. *Proceedings of the IEEE*, 73(4):795–812, 1985.

44. G. Miller. Efficient algorithms for local and global accessibility shading. *Computer Graphics Proceedings, Annual Conference Series*, pages 319–326, 1994.

45. F. Attneave. Some informational aspects of visual perception. *Psychological Review*, 61(3):183–193, 1954.

46. I. Biederman. Human image understanding: recent research and a theory. In *Human and Machine Vision*, Azriel Rosenfeld, ed., Academic Press, pages 13–57, 1985.

47. J. M. H. Beusmans, D. D. Hoffman, and B. M. Bennett. A description of solid shape and its inference from occluding contours. *Journal of the Optical Society of America A*, 4(7):1155–1167, 1987.

48. D. D. Hoffman and W. A. Richards. Parts of recognition. *Cognition*, 18(1-3):65–96, 1984.

49. M. L. Braunstein, D. D. Hoffman, and A. Saidpour. Parts of visual objects: an experimental test of the minima rule. *Perception*, 18(6):817–826, 1989.

50. K. H. Höhne and R. Bernstein. Shading 3D-Images from CT using gray-level gradients. *IEEE Transactions on Medical Imaging*, 5(1):45–47 (with a correction in 5(3):165), 1986.

51. V. Interrante, H. Fuchs, and S. Pizer. Enhancing transparent skin surfaces with ridge and valley lines. *Proceedings of IEEE Visualization '95*, pages 52–59, 1995.

52. K.-L. Ma and V. Interrante. Extracting feature lines from 3D unstructured grids. *Proceedings of IEEE Visualization '97*, pages 285–292, 1997.

53. J. M. Kennedy and I. Juricevic. Foreshortening gives way to forelengthening. *Perception*, 31(7):893–894, 2002.

54. W. T. Freeman. The generic viewpoint assumption in a framework for visual perception. *Nature*, 368(6471):542–545, 1994.

55. D. I. Perrett and M. H. Harries. Characteristic views and the visual inspection of simple faceted objects and smooth objects: 'tetrahedra and potatoes'. *Perception*, 17(6):703–720, 1988.

56. M. Luckiesh. *Light and Shade and Their Applications*. Van Nostrand, 1916.

57. V. S. Ramachandran. Perceiving shape from shading. *Scientific American*, 259(2):76–83, 1988.

58. P. Mamassian and R. Goutcher. Prior knowledge on the illumination position. *Cognition*, 81(1):B1–B9, 2001.

59. A. Johnson and P. J. Passmore. Shape from shading I: surface curvature and orientation. *Perception*, 23(2):169–189, 1994.

60. A. Blake and H. Bülthoff. Shape from Specularities: computation and psychophysics. *Philosophical Transactions of the Royal Society of London, B*, 331:237–252, 1991.

61. M. J. Friedländer and J. Rosenberg. *The Paintings of Lucas Cranach*. Cornell University, 1978.

62. V. Katz. *Janet Fish: Paintings*. New York, Harry N. Abrams, 2002.

63. D. Kuspit. *Don Eddy: the art of paradox*. Hudson Hills Press, 2002.

64. F. W. Zweifel. *A Handbook of Biological Illustration*. University of Chicago Press, 1961.

65. E. J. Sullivan. *Line: An Art Study*. Chapman & Hall, 1922.

66. J. J. Gibson. The perception of visual surfaces. *American Journal of Psychology*, 63:367–384, 1950.

67. H. R. Flock and A. Moscatelli. Variables of surface texture and accuracy of space perceptions. *Perceptual and Motor Skills*, 19:327–334, 1964.

68. F. Attneave and R. K. Olson. Inferences about visual mechanisms from monocular depth effects. *Psychonomic Science*, 4:133–134, 1966.

69. M. L. Braunstein and J. W. Payne. Perspective and form ratio as determinants of relative slant judgments. *Journal of Experimental Psychology*, 81(3):584–590, 1969.

70. J. E. Cutting and R. T. Millard. Three gradients and the perception of flat and curved surfaces. *Journal of Experimental Psychology: General*, 113(2):198–216, 1984.

71. B. G. Cumming, E. B. Johnston, and A. J. Parker. Effects of different texture cues on curved surfaces viewed stereoscopically. *Vision Research*, 33(5/6):827–838, 1993.

72. K. A. Stevens. The information content of texture gradients. *Biological Cybernetics*, 42:95–105, 1981.

73. P. Brodatz. *Textures: A Photographic Album for Artists and Designers*. Dover, 1966.

74. J. T. Todd and F. D. Reichel. Visual perception of smoothly curved surfaces from double-projected contour patterns. *Journal of Experimental Psychology: Human Perception and Performance*, 16(3):665–674, 1990.

75. R. Rosenholtz and J. Malik. Surface orientation from texture: isotropy or homogeneity (or both)? *Vision Research*, 37(16):2283–2293, 1997.

76. J. V. Stone. Shape from local and global analysis of texture. *Philosophical Transactions of the Royal Society of London, B*, 339:53–65, 1993.

77. P. Mamassian and M. S. Landy. Observer biases in the 3D interpretation of line drawings. *Vision Research*, 38(18):2817–2832, 1998.

78. D. C. Knill. Contour into texture: information content of surface contours and texture flow. *Journal of the Optical Society of America, A*, 18(1):12–35, 2001.

79. A. Li and Q. Zaidi. Perception of three-dimensional shape from texture is based on patterns of oriented energy. *Vision Research*, 40(2):217–242, 2000.

80. A. Li and Q. Zaidi. Information limitations in perception of shape from texture. *Vision Research*, 41(12):1519–1533, 2001.

81. A. Li and Q. Zaidi. Limitations on shape information provided by texture cues. *Vision Research*, 42(7):815–835, 2002.

82. V. Interrante, H. Fuchs, and S. Pizer. Conveying the 3D shape of smoothly curving transparent surfaces via texture. *IEEE Transactions on Visualization and Computer Graphics*, 3(2): 98–117,1997.

83. V. Interrante. Illustrating surface shape in volume data via principal direction-driven 3D line integral convolution. *Computer Graphics Proceedings, Annual Conference Series*, pages 109–116, 1997.

84. A. Girshick, V. Interrante, S. Haker, and T. LeMoine. Line direction matters: an argument for the use of principal directions in 3D line drawings. *First International Symposium on Nonphotorealistic Animation and Rendering*, pages 43–52, 2000.

85. G. Taubin. Estimating the tensor of curvature of a surface from a polyhedral approximation. *Proceedings of the 5th International Conference on Computer Vision* (ICCV '95), pages 902–907, 1995.

86. M. Desbrun, M. Meyer, P. Schroder, and A. H. Barr. Discrete differential geometry operators in nD. Preprint, July 22, 2000.

87. J. Goldfeather and V. Interrante. A novel cubic-order algorithm for approximating principal direction vectors. *ACM Transactions on Graphics*, 23(1):45–63, 2004.

88. Koenderink, Jan J., A. van Doorn, and A. M. L. Kappers. Surface perception in pictures. *Perception*, 52:487–496, 1992.

89. S. Kim, H. Hagh-Shenas, and V. Interrante. Showing shape with texture: two directions seem better than one. *Human Vision and Electronic Imaging VIII*, SPIE 5007, 2003.

90. A. A. Efros and T. K. Leung. Texture synthesis by non-parametric sampling. *Proceedings of the International Conference on Computer Vision*, 2:1033–1038, 1999.

91. G. Gorla, V. Interrante, and G. Sapiro. Texture synthesis for 3D shape representation. *IEEE Transactions on Visualization and Computer Graphics*, to appear.

92. V. Interrante, S. Kim, and H. Hagh-Shenas. Conveying 3D shape with texture: recent advances and experimental findings. *Human Vision and Electronic Imaging VII*, SPIE 4662:197–206, 2002.

93. V. Interrante and S. Kim. Investigating the effect of texture orientation on the perception of 3D shape. *SPIE Conference on Human Vision and Electronic Imaging VI*, SPIE 4299:330–339, 2001.

41 Exploiting Human Visual Perception in Visualization

ALAN CHALMERS and KIRSTEN CATER
University of Bristol

41.1 Introduction

Most visualizations serve some specific visual task, for example, investigating the air flow over a wing tip or looking at the perception of medieval pottery under medieval lighting conditions. In the majority of cases, objects or data relevant to the task can be identified in advance—for example, the wing tip or the medieval pottery. During the actual visualization, the viewer's visual system must focus its attention on these objects and data in order to complete the task. The human visual system is good, but it is not perfect. While focusing on these objects, the viewer will simply fail to notice other large parts of the scene. It is this feature of human visual perception that can be exploited in visualizations to save significant computational time by not computing to such a high resolution those parts of the visualization that the human viewer will fail to notice.

In this chapter we first discuss an application in which the visualization task is to recreate high-fidelity archaeological sites on a computer. The visualization task is to recreate the sites not as they are at present but the way they may have been visually perceived at the time they were created, including the lighting conditions of the time. We then go on to discuss how, knowing the nature of the visualization task being performed by the user *a priori*, we can exploit visual attention, and by selectively rendering the data, reduce overall computation time substantially without compromising the observer's perceived visual quality.

41.2 Visualizing the Past

Human visual perception can be exploited in many applications, including the visualization of archaeological sites. 3D computer reconstruction provides us with a means of visualizing past environments, allowing us a glimpse of the past that might otherwise be difficult to appreciate. This is especially true for sites that have been severely damaged or even destroyed in the passage of time. Many of the visualizations generated for this purpose look realistic, but if they are to provide any meaningful representations of the past, then these visualizations must be quantifiably real. That is, the archaeologist must be confident that what he or she sees in the generated images is comparable to what a human would have perceived in the real scene in the past [5]. Failure to guarantee this level of realism leads to the very real danger of our visualizations in fact misrepresenting the past.

The prehistoric site rock shelter site of Cap Blanc, France, is a good example of how high-quality visualization can provide meaningful insights for archaeologists. Cap Blanc, overlooking the Beaune Valley in the Dordogne, contains perhaps the most dramatic and impressive example of Upper Palaeolithic haut-relief carving. A frieze of horses, bison, and deer, some overlaid on other images, was carved some 15,000 years ago into the limestone as deeply as 45 cm, covering 13 m of the wall of the shelter. Since its discovery in 1909 by Raymond Peyrille, several descriptions, sketches, and surveys of the frieze have been published, but they appear to be variable in their detail and accuracy.

In 1999, a laser scan was taken of part of the frieze at 20 mm precision [26], using an eye-safe laser to ensure that there was no possibility of damage to the site. Fig. 41.1 shows part of the frieze from Cap Blanc. Some 55,000 points were obtained and converted into a triangular mesh. Using detailed photographs as textures (each with a rock-art chart to enable color calibration) and appropriate lighting values, the model was then rendered in *Radiance*. Fig. 41.2a shows the horse illuminated by a simulated 55W incandescent bulb (as in a low-power floodlight), which is how visitors view the actual site today. In Fig. 41.2b, the horse is illuminated by an animal-fat tallow candle as it may have been viewed 15,000 years ago. The difference between the two images is significant; the candle illumination gives a warmer glow to the scene, as well as increasing the shadows.

Visualization was used to investigate whether the dynamic nature of flame, coupled with the careful use of 3D structure, may have been used by our prehistoric ancestors to create animations in the cave-art sites of France some 15,000 years ago. The visualization showed that the shadows created by the moving flame do indeed appear to give the horse motion. We will never know for certain whether the artists of the Upper Palaeolithic were in fact creating animations 15,000 years ago, but the reconstructions do show that the effect is certainly possible. There is other intriguing evidence to support this hypothesis. As can be seen in the figures, the legs of the horse are not present in any detail. This has long been believed to be due to erosion, but this does not explain why the rest of the horse is not equally eroded. The possibility exists that the legs were deliberately not carved in any detail, thereby accentuating any motion by creating some form of motion blur. Furthermore, traces of red ochre have been found on the carvings. It is interesting to speculate whether the application of red ochre at key points on the horse's anatomy may also have been used to enhance any motion effects. High-fidelity visualization provides us with an opportunity to explore such scenarios.

41.3 Visual Perception

A major challenge in visualization is to achieve high-quality images at interactive rates. Even with the ready availability of modern high-performance graphics cards, the complexity of the models being considered and the high-fidelity requirements of the images mean that rendering such images is still simply not possible in a reasonable time frame, let alone real time. For

Figure 41.1 Part of the frieze from Cap Blanc. (See also color insert.)

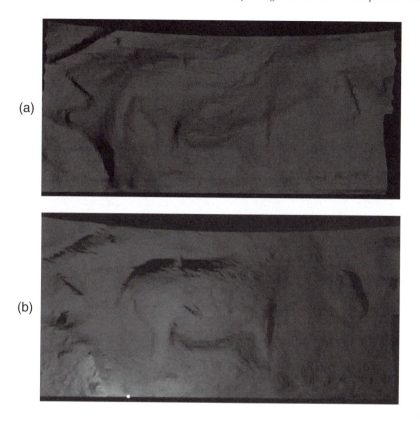

Figure 41.2 (a) The frieze lit by modern 55 w lighting. (b) The frieze lit by an animal tallow candle as used 15,000 years ago. (See also color insert.)

example, each of the frames of the Cap Blanc visualization took many minutes to render on a high-performance PC. Visual perception approaches offer one possibility of maintaining perceptual quality, but at reduced computational cost. This is achieved by taking into account that it is the human who will ultimately be looking at the resultant images, and while the human eye is good, it is not perfect. Exploiting knowledge of the human visual system can save significant rendering time by computing at a substantially lower quality those parts of a scene that the human will fail to notice. Researchers into flight simulation have long studied what parts of a scene or image are most likely to be noticed in an interactive setting.

Most of this research has attempted to exploit gaps in low-level visual processing, similar to JPEG and other image-compression schemes [1]. In this section we consider both low-level and high-level visual perception.

41.3.1 Visual Attention

Visual attention is the process by which we humans select a portion of the available visual information for localization, identification, and understanding of objects in an environment. It allows our visual system to process visual input preferentially by shifting attention about an image, giving more attention to salient locations and less attention to unimportant regions.

There are two general processes, called *bottom-up* and *top-down*, that determine where humans locate their visual attention [11]. A *bottom-up* visual process is purely stimulus driven, for example, a candle burning in a dark room, a red ball amongst a large number of blue balls, or a person's lips and eyes, the most mobile and expressive elements of the face. Here, the visual attention is captured automatically without volitional control. There is, in addition, another aspect to the human visual system: the *top-down* process. Here the brain directs the eyes to focus on one or more objects that are relevant to the observer's goal when studying the scene, for example, looking for a street sign or searching for a target in a computer game. Research has shown that conspicuous objects in a scene that would normally attract the viewer's attention may be deliberately ignored if they are irrelevant to the visual task at hand [2]. This is referred to as *inattentional blindness*.

Faced with a complex scene, the human visual system depends, at its primary level, on the retina to cope with this wealth of information [7]. The retina converts the information about the scene from light waves into neural signals that the brain can process. At the center of the retina is the fovea, which consists exclusively of densely packed color-sensitive cones. The fovea provides the highest spatial and chromatic resolutions in the retina. However, the visual angle covered by the fovea is only approximately $2°$, about the size of eight letters on a typical page of text or the size of your thumbnail held at arm's length. If detailed information is needed from an area of the visual environment, it can only be obtained by redirecting the eye so that the relevant area falls sequentially on the fovea.

Yarbus was one of the first to study how the eye moves when looking at complex images [30]. He successfully demonstrated that humans do not scan a scene in a raster-like fashion; rather, the eyes jump to foveate a new point of interest in the scene, called a saccade. What Yarbus also noted was that these saccades were linked to the task or question the viewer had been asked

about the scene. Yarbus illustrated this by asking several observers to answer a number of different questions concerning the depicted situation in Repin's picture *An Unexpected Visitor*. This resulted in substantially different saccade patterns, each one being easily construable as a sampling of those picture objects that were most informative for the answering of the question, as shown in Fig. 41.3.

What Yarbus, and subsequently many others, showed was that while performing a task, once an initial eye saccade has found the appropriate object to locate it on the fovea, the eye subsequently performs a *smooth pursuit movement* to keep the object in foveal vision [22]. This means that the image of a successfully tracked object is nearly stationary on the retina, while untracked objects are experienced as smeared and unclear because of their motion on the retina. To experience this, Palmer [22] suggests a simple example: *Place your finger on this page and move it fairly quickly from one side of the page to another. As soon as you track your moving finger, the letters and words appear so blurred you are unable to read them, but your finger is clear.* Even when you stop moving your finger, only the words located within the visual angle of your fovea become sharp and thus readable.

41.3.1.1 The Bottom-Up Process

Previous applications of visual perception to visualization have concentrated on the bottom-up visual processes. This work has included using knowledge of the human visual system to improve the quality of the displayed image, for example [8,9,19,20,23,24], and reducing the level of detail (LOD) in a scene without affecting the viewer's perception of the objects within the scene [14,15,25]. In addition, Maciel and Shirley proposed a visual navigation system that uses texture-mapped primitives to represent clusters of objects to maintain high and approximately constant frame rates [16].

Saliency models have been also been developed to simulate where people may focus

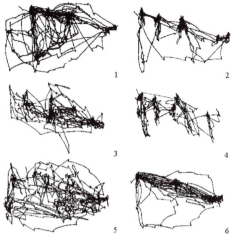

Figure 41.3 (a) Repin's picture *An Unexpected Visitor*. (b) Effects of task on eye movements. Repin's picture was examined by subjects with different instructions: 1. Free viewing, 2. Judge their ages, 3. Guess what they had been doing before the unexpected visitor's arrival, 4. Remember the clothes worn by the people, 5. Remember the position of the people and objects in the room, and 6. Estimate how long the unexpected visitor had been away from the family [30]. (See also color insert.)

their attention in images. Visual psychological researchers such as Yarbus [30], Itti and Koch [10], and Yantis [29] showed that the visual system is highly sensitive to features such as edges, abrupt changes in color, and sudden movements. This low-level human visual processing has been exploited in computer graphics by Yee [31] and Yee et al. [32] to accelerate global illumination computation in prerendered animations by using a model of visual attention to locate regions of interest in a scene and to modulate spatio–temporal sensitivity. They created a spatio–temporal error tolerance map, constructed from data based on velocity-dependent contrast sensitivity, and a saliency map for each frame in the animation. The saliency map is obtained by combining the conspicuity maps of intensity, color, orientation, and motion. An Aleph map is then created by combining the spatio–temporal error-tolerance map with the saliency map. The resulting Aleph map is then used as a guide to indicate where less rendering effort should be spent in computing the lighting solution, and thus significantly reduces the overall computational time to produce animations.

Knowledge of peripheral vision has also been used to reduce overall computation time, by allowing researchers to render those areas outside the observer's fovea visual angle at a lower resolution [27]. Watson et al. used such visual fidelity for simplifying polygonal models [28]. McConkie and Loschky measured viewers' image quality judgements and their eye movement parameters and found that photographic images filtered with a window radius of 4.1° produced results statistically indistinguishable from those of a full high-resolution display [12,13,18]. However, their research showed that the image needs to be updated after an eye saccade, within 5 milliseconds of a fixation, otherwise the observer will detect the low resolution. These high update rates were achievable only through use of an extremely high-temporal-resolution eye tracker and by prestoring all possible multiresolutional images that were to be used.

41.3.1.2   The Top-Down Process

As Yarbus showed, the choice of task is important in helping us predict the eye-gaze pattern of the viewer [30]. It is precisely this knowledge of the expected eyegaze pattern that allows us to reduce the rendered quality of objects outside the area of interest without affecting the viewer's overall perception of the quality of the rendering.

The significance of exploiting the top-down visual process for visualization tasks was illustrated by Cater et al. [2]. They showed that conspicuous objects in a scene that would normally attract the viewer's attention are ignored if they are irrelevant to the visualization task at hand. The task considered was for each user to count the number of pencils that appeared in a mug on a table in a room as they moved on a fixed path through four such rooms; it is shown in Fig. 41.4.

In their experiments there were three renderings of the animations, the only difference being the quality to which the animation was rendered, low quality (LQ), high quality (HQ), or selective quality (SQ). Each frame for the HQ animation took, on average, six times longer to render than the frames for the LQ animation. Fig. 41.5a shows the HQ rendered scene, while Fig. 41.5b shows the difference between the HQ and LQ

images. The SQ animation was created by using the LQ frames with HQ rendering substituted in the visual angle of the fovea ($2°$) centered on the pencils, shown by the green circle in Fig. 41.6. The higher quality is blended to the lower quality at $4.1°$ visual angle (the red circle in Fig. 41.6) [18].

In the experiment, a total of 160 subjects were studied, and each subject saw two animations. Fifty percent of the subjects were asked to count the pencils in the mug, while the remaining 50% were simply asked to watch the animations. On completion of the experiment, participants were asked to fill in a detailed questionnaire that asked them to compare the two animations they had just seen.

Their results showed that when observers were simply watching the animation, they could easily detect if there was a change in rendering quality between the two animations. This was also the case when the participants were performing the counting pencils task but were shown the HQ and the LQ animations. Of interest is that when the participants were performing the task and saw the SQ animation,

Figure 41.4 The scene visualized with a closeup of the mug showing the pencils and paintbrushes. (See also color insert.)

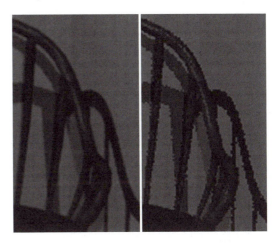

Figure 41.5   (a) High-quality (HQ) image (Frame 26 in the animation). (b) Closeup of HQ and low-quality (LQ) rendered chairs. (See also color insert.)

Figure 41.6   Visual angle covered by the fovea for mugs in the first two rooms at 2° (green circles) and blending circle at 4.1° (red circles). (See also color insert.)

they perceived no difference in quality from the HQ animation, i.e., the observers thought that they were seeing the same HQ animation twice. This shows that when observers perform a task within an animation, their visual attention becomes fixed exclusively on the area of the task at hand, and they consistently fail to notice significant differences in rendering quality.

Cater et al. have since gone on to show not only that the task has to be a visual one [3] but also that the same principle can be applied even if the task is carried out all over the image and not restricted to a single location, for example, counting teapots located all around a scene. They demonstrate that by using what they term a *task map*, you can selectively render the scene, rendering only the areas related to the task (in their example, the small area located around each teapot) to a higher quality. The rest of the scene can then be rendered at a substantially lower quality.

By placing the teapots all over the scene, we forced the viewer to scan the whole image and thus fixate on LQ as well as HQ areas (Fig. 41.7). Cater et al. showed that, even though this is the case, the observer visualizes only the quality of the objects related to the task, and not the rest of the scene. Thus, by selectively rendering animations according to the task, users can save significant computational time without having a significant effect on the viewer's perception of the scene.

41.4   Conclusions

For visualization applications in which the task is known *a priori*, the computational savings made by exploiting visual attention can be dra-

Figure 41.7   (a) A 2-second eye scan for an observer counting the teapots. The crosses are fixation points and the lines are the saccades. (b) Perceptual difference between selective-quality (SQ) and high-quality (HQ) images using Myszkowski's Visual Difference Predictor [20]. The red denotes areas of high perceptual difference. The superimposed blue line is the eye-scan path for another observer counting the teapots. (See also color insert.)

matic. Lower-quality peripheral vision and inattentional blindness are fundamental features of the human visual system. We can use these to our advantage to meet the demand for HQ images of increasingly complex content in less time.

High-level *task maps* and low-level *saliency maps* do indeed tell us where an observer will be looking in an image. This knowledge enables us to selectively render those parts attenuated to in high quality while the rest of the images can be rendered in low quality. The time taken to selectively render the image is significantly lower than the time taken to render the whole image in high quality. The key is that when the user is performing a visualization task within the scene, he or she will simply fail to notice this difference in quality.

Visual perception can thus be exploited to significantly lower overall rendering time while maintaining the same perceptual quality of the resultant images. Such perceptually HQ images have the potential to provide powerful real-time visualization tools to a wide range of disciplines, including archaeology.

References

1. M. R. Bolin and G. W. Meyer. A Perceptually based adaptive sampling algorithm. In *Proceedings of SIGGRAPH 1998*, ACM, pages 299–309, 1998.

2. K. Cater, A. G. Chalmers, and P. Ledda. Selective quality rendering by exploiting human inattentional blindness: looking but not seeing. In *Proceedings of Symposium on Virtual Reality Software and Technology 2002*, ACM, pages 17–24, 2002.

3. K. Cater and A. G. Chalmers. Maintaining perceived quality for interactive tasks. In *IS&T/ SPIE Conference on Human Vision and Electronic Imaging VIII*, SPIE 5007, 2003.

4. A. Chalmers, C. Green, and M. Hall. *Firelight: Graphics and Archaeology*. Electronic Theatre, SIGGRAPH '00, New Orleans, 2000.

5. A. Chalmers and K. Devlin. Recreating the past (SIGGRAPH 2002 Course #27), SIGGRAPH 2002, San Antonio, 2002.

6. S. Daly. Engineering observations from spatio-velocity and spatiotemporal visual models. In *IS&T/SPIE Conference on Human Vision and Electronic Imaging III*, SPIE 3299, 152–166, 1998.

7. J. E. Dowling. *The Retina: An Approachable Part of the Brain.* Cambridge, Belknap, 1987.

8. J. Ferwerda, S. N. Pattanaik, P. Shirley, and D. P. Greenberg. A model of visual adaptation for realistic image synthesis, In *Proceedings of SIGGRAPH 1996*, ACM, pages 249–258, 1996.

9. D. P. Greenberg, K. E. Torrance, P. Shirley, J. Arvo, J. Ferwerda, S. N. Pattanaik, E. Lafortune, B. Walter, S-C. Foo, and B. Trumbore. A framework for realistic image synthesis. In *Proceedings of SIGGRAPH 1997* (Special Session). ACM, pages 477–494, 1997.

10. L. Itti and C. Koch. A saliency-based search mechanism for overt and covert shifts of visual attention, In *Vision Research*, 40(10–12), 1489–1506, 2000.

11. W. James. *Principles of Psychology*, New York: Holt, 1890.

12. L. C. Loschky and G. W. Mcconkie. Gaze contingent displays: maximizing display bandwidth efficiency. ARL Federated Laboratory Advanced Displays and Interactive Displays Consortium, Advanced Displays and Interactive Displays Third Annual Symposium, pages 79–83, 1999.

13. L. C. Loschky, G. W. McConkie, J. Yang, and M. E. Miller. Perceptual effects of a gaze-contingent multi-resolution display based on a model of visual sensitivity. In the ARL *Federated Laboratory 5th Annual Symposium—ADID Consortium Proceedings*, pages 53–58, 2001.

14. D. Luebke and B. Hallen. Perceptually driven simplification for interactive rendering. In *Proceedings of 12th Eurographics Workshop on Rendering*, pages 221–223, 2001.

15. D. Luebke, M. Reddy, B. Watson, J. Cohen, and A. Varshney. Advanced issues in level of detail (SIGGRAPH 2001 Course #41). *SIGGRAPH 2001 Proceedings*. Los Angeles, CA, pages 12–17, 2001.

16. P. W. C. Maciel and P. Shirley. Visual navigation of large environments using textured clusters. In *Proceedings of Symposium on Interactive 3D Graphics*, pages 95–102, 1995.

17. A. Mack and I. Rock. *Inattentional Blindness.* Massachusetts Institute of Technology Press, 1998.

18. G. W. McConkie and L. C. Loschky. Human performance with a gaze-linked multi-resolutional display. *ARL Federated Laboratory Advanced Displays and Interactive Displays Consortium, First Annual Symposium*, pages 25–34, 1997.

19. A. McNamara, A. G. Chalmers, T. Troscianko, and I. Gilchrist. Comparing real and synthetic scenes using human judgements of lightness. In B. Peroche and H. Rushmeier (eds), *12th Eurographics Workshop on Rendering*, pages 207–219, 2000.

20. K. Myszkowski, T. Tawara, H. Akamine, and H.-P. Seidel. Perception-guided global illumination solution for animation rendering. In *Proceedings of SIGGRAPH 2001*, ACM, pages 221–230, 2001.

21. J. K. O'Regan, H. Deubel, J. J. Clark, and R. A. Rensink. Picture changes during blinks: looking without seeing and seeing without looking. *Visual Cognition*, 7(1):191–212, 2000.

22. S. E. Palmer. Vision Science—Photons to Phenomenology. Massachusetts Institute of Technology Press, 1999.

23. S. N. Pattanaik, J. Ferwerda, M. D. Fairchild, and D. P. Greenberg. A multiscale model of adaptation and spatial vision for realistic image display. In *Proceedings of SIGGRAPH 1998*, ACM, pages 287–298, 1998.

24. M. Ramasubramanian, S. N. Pattanaik, and D. P. Greenberg. A perceptually based physical error metric for realistic image synthesis, In *Proceedings of SIGGRAPH 1999*, ACM Press / ACM SIGGRAPH, New York. Computer A. Rockwood, Ed., Graphics Proceedings, Annual Conference Series, ACM, pages 73–82, 1999.

25. M. Reddy. Perceptually modulated level of detail for virtual environments. Ph.D. Thesis (CST-134-97), University of Edinburgh, 1997.

26. K. A. Robson Brown, A. G. Chalmers, T. Saigol, C. Green, and F. D'Errico. An automated laser scan survey of the upper palaeolithic rock shelter of Cap Blanc. *Journal of Archaeological Science 28*, pages 283–289, 2001.

27. B. Watson, A. Friedman, and A. McGaffey. An evaluation of level of detail degradation in head-mounted display peripheries. *Presence*, 6(6), pages 630–637, 1997.

28. B. Watson, A. Friedman, and A. McGaffey. Measuring and predicting visual fidelity, In *Proceedings of SIGGRAPH 2001*, ACM, pages 213–220, 2001.

29. S. Yantis. Attentional capture in vision. In A. Kramer, M. Coles, M. and G. Logan, (Eds.), *Converging Operations in the Study of Selective*

Visual Attention. American Psychological Association, pages 45–76, 1996.

30. A. L. Yarbus. Eye movements during perception of complex objects. In L. A. Riggs, Ed., *Eye Movements and Vision*, Plenum Press, New York, Chapter VII, pages 171–196, 1967.

31. H. Yee. Spatiotemporal sensitivity and visual attention for efficient rendering of dynamic environments. MSc. Thesis, Program of Computer Graphics, Cornell University.

32. H. Yee, S. Pattanaik, and D. P. Greenberg. Spatiotemporal sensitivity and visual attention for efficient rendering of dynamic Environments. In *ACM Transactions on Computer Graphics*, 20(1):39–65, 2001.

PART XI
Selected Topics and Applications

42   Scalable Network Visualization

STEPHEN G. EICK
SSS Research;
National Center for Data Mining
University of Illinois

42.1 Introduction

Many problems can be represented as networks
and analyzed using network visualization. Unfortunately, however, the sizes of datasets that
are easily collected overwhelm existing visualizations. The problem is that network visualizations become visually confusing and cluttered.
This chapter defines the concept of visual scalability for networks, illustrates it with three
examples, and proposes techniques to increase
network visualization scalability.

Many analysis problems involve understanding network data. Familiar examples include
monitoring electronic communications, tracking
money flows, understanding travel patterns, correlating personal contacts, and analyzing pur-
chasing correlations. Network data can be
generated by a single process or by overlapping
processes. For networks that represent a social
organization, datasets may be associated with
interconnected individuals involving multiple
events linked in both time and space.

At its most basic level, a network consists of
nodes and links. The nodes and links can represent physical objects, such as people or machines,
or nonphysical objects, such as meetings, events,
or hypertext pages. Statistics (possibly time-
varying) and events at discrete points in time
may be tied to the nodes and links. These statistics may be raw measurements, such as the locations of individuals, group membership, counts
of e-mail communications, computed aggregates, or imputed attributes.

The analysis challenges are many and broad.
The obvious tasks include understanding overall structure, traffic flows, changes, important
nodes, and key links. More subtle tasks involve
isolating a small signal from a massive amount
of background activities. This task, frequently
called outlier detection, is extremely difficult.
For example, in a financial analysis involving
monetary flows, it may be well known from
aggregates that transactions within the flow correspond to illegal activity. However, identifying
exactly which transactions are illegal may be
essentially impossible.

Our focus in this chapter is on the scalability of
network visualizations. As with many classes
of data, our ability to collect massive volumes of
network data overwhelms existing analysis techniques. This is particularly true for real-world
data analysis problems, where the data is heterogeneous, noisy, incomplete, highly fragmented,
time-varying, and at multiple levels of abstraction.

The new ideas involve the concept of visual
scalability [8] and its application to network data.
Here, we introduce the problem of visual scalability, illustrate it by analyzing the scalability
of three visualizations, and discuss techniques
to increase visual scalability.

42.2 Network Background

As is standard in the information visualization
community, we view a network as a graph in

which nodes represent entities of interest and edges represent a relationship. In the social network literature [11,15], the nodes correspond to individuals or actors and the edges correspond to a type of relationship, e.g., common group membership, telephone communications, or common friends. A network may represent a single type of relation among the nodes (simplex) or more than one kind of relation (multiplex). Each edge or relation may be directed (i.e., originates with a source node and reaches a target node), or it may be undirected (i.e., represents co-occurrence, copresence, or a bonded tie between the pair of nodes). Each edge may have an associated strength or weight that may be nominal, binary, signed, ordinal, or valued (measured on an interval or ratio level).

The basic properties of a network can be defined in terms of their relationship to graph theory. These include the following:

1. *Conductivity*—whether there is a path between two nodes.
2. *Path*—a sequence of connections.
3. *Size*—the number of nodes in the network.
4. *Density*—the proportion of all links that could logically be present.
5. *Reachability*—whether there is a path from the source to the target node.
6. *Distance*—the number of connections between two nodes along a path.
7. *Geodesic distance*—the distance of the shortest path between two nodes.
8. *Network diameter*—the maximum geodesic distance between any pair of nodes in a network.
9. *Flow*—the number of unique paths between two nodes.
10. *Cohesion*—a measure of the coupling between two nodes.

For any individual node, standard measures include the following:

1. *Node degree*—the number of connections to other nodes.
2. *Closeness*—whether the distance between two nodes is low.
3. *Betweenness*—a measure of alternative paths to other nodes.
4. *Centrality*—the measure of alternative paths.

In the social network literature, these measures are related to the distribution of power in a network. Central nodes with many relations are more powerful than nodes on the edge of the network.

42.2.1 Substructure

Every network consists of subgroups that determine its macrostructure. At the most basic level, a *clique* is a subset of a network in which the nodes are more closely and intensely tied to one another than they are to other members of the network. For example, in human networks, cliques form on the basis of age, gender, race, ethnicity, and religion. The smallest clique consists of two nodes and is called a *dyad*.

For some analysis tasks, a clique, where every node has a direct tie to every other, is too strong. However, the clique idea can be generalized to a 2-clique, where the 2-clique includes every member of the group at a distance two. For example, this definition corresponds to "friends of friends" in the group. In general, the following rules apply:

- *N-cliques* are substructures where all nodes are at a distance *n*.
- *N-clans* are *n*-cliques where the total diameter of the clique is constrained by a maximum.
- *K-plexes* are generalizations of *n*-cliques where nodes are in the *k*-plex if they are in the structure, if they have direct ties to *n-k* members of the group.

- *K-cores* are maximal groups of nodes, all of which are connected to some number (k) of other members of a group. The difference between *n*-cliques and *n*-clans is that cliques are often stringier, whereas clans are tighter.

Components of a graph are structures that are connected within themselves but disconnected from the remainder of the network. A *cutpoint* in a network is a node that, if removed, would cause the network to divide into disconnected components. Cutpoints are very useful for bridging disconnected groups. Similarly, *lambda sets* are links or bridges that, if removed, would disconnect the network.

42.3 Visual Metaphors for Networks

Perhaps the most conventional way to visualize a network is to use node and link diagrams (Fig. 42.1). Nodes encode the items, and links encode the network edges. Visual characteristics (and sometimes even nonvisual characteristics such as sound, position, shape, color, size, texture, transparency, drawing style, and thickness) encode node and link atributes.

Although node and link displays are the most common way to represent network data, there are other possibilities. Bertin [2,3] suggests a dynamic matrix (as shown in Fig. 42.2). The idea in a matrix representation is that each row and each column corresponds to a node, and the glyph at the (i,j)th cell encodes to the (directed) edge from node i to node j. Visual characteristics of the glyph are tied to attributes. If the glyphs are 3D, this visual representation is called a CityScape [12]. Although we will not address it in this manuscript, the predominant visual problem with matrix displays is occlusion. Taller bars towards the front of the matrix obscure smaller bars towards the rear.

42.3.1 Problems with Node and Link Displays

There are three fundamental reasons why node and link displays fail on larger networks:

- **Display Clutter.** Node and link diagrams become cluttered and visually confusing as the size of the network increases. The predominant reason for the perceptual clutter is

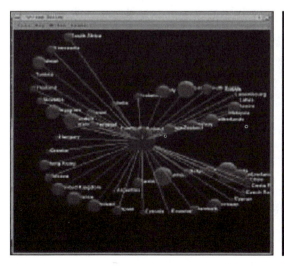
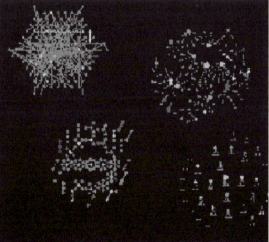

Figure 42.1 Node and link network displays. (See also color insert.)

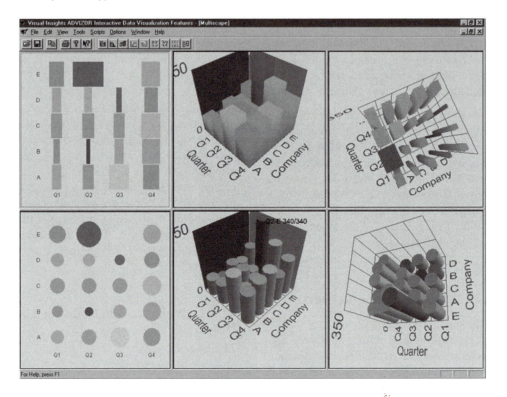

Figure 42.2 Matrix display with the glyph at cell (i,j) encoding the link from i to j.

all the line crossings, particularly crossings at nonright angles.

- **Node Positioning.** The interpretation and readability of the diagram are highly dependent on the node positions. The same network drawn with the nodes in different positions may lead to a totally different interpretation of the visualization.

- **Perceptual Tension.** Closely positioned nodes are interpreted perceptually as being related, but they are connected by a short line. Conversely, distant nodes are interpreted as perceptually unrelated, but they are connected by a long, visually dominant line. The most effective network visualization tools exploit the tradeoff between visual real estate of the connecting line and human perceptual grouping.

There are a variety of strategies for addressing these three key issues, which are related to

graph drawing. Graph drawing, which focuses on the visualization structure rather than the node and edge attributes, involves positioning the nodes on a 2D plane and drawing connections that satisfy one or more aesthetic constraints. The criteria typically include, for example, minimizing the number of edge crossings, maximizing the depiction of symmetry, minimizing the number of edge bends, maximizing the edge orthogonality, and maximizing the minimum angle between edges leaving a node [14].

Within the graph-drawing community, there has been an extensive multiyear research effort to develop efficient techniques for drawing graphs [4]. There is also an annual contest in which software packages compete to produce the best drawings. In the commercial sector, there are several companies offering graph-drawing packages, and there are also many research efforts.

42.4   Visual Scalability

Visual scalability [8], for our purposes, is the capability of the visualization to effectively display large network datasets, in terms of either the number of nodes, the number of edges, the number of node or edge attributes, or other similar elements. Mathematically,

$$\text{visual scalability} = F(\text{visual factors, network data}) \tag{42.1}$$

The research goals are to understand F, to characterize factors affecting F, and to create software with a high F. A significant research challenge is to invent measures of visual scalability. For example, possible visual scalability metrics for networks include the following:

1. Number of distinct nodes visible.

2. Number of distinct links visible.

3. Number of "features" visible in the displayed graph (e.g., "star," "spoke," "hub," or "chain")

4. Number of time periods displayed.

5. Number of connected components.

6. Number of comparisons made.

The ultimate measure, however, is the number of insights gained from the visualization. Unfortunately, this measure is exceedingly difficult to quantify.

42.4.1   Factors Affecting Visual Scalability

There are seven factors that affect network visualization scalability.

1. **Human perception.** The consumer of a network visualization is a person. Thus, the ultimate limiting factor is human perception. There is no value in producing visualizations with higher resolution than humans can perceive.

2. **Display resolution.** Visualizations are viewed on display devices. Thus, the resolution of the output display, whether it is a monitor, flat panel, or wall, can be measured by pixels, display size, and display fidelity. According to Wegman [16], about 6.5 million distinct pixels are visible on a 17-inch monitor at normal viewing distance.

Table 42.1 shows the standard pixel resolutions for currently available display devices. Resolution has been increasing more slowly than processing power and disk sizes. From the late 1990s to the early 2000s, typical workstation monitor resolution increased by a factor of only four, from 800 × 600 to 1600 × 1200. Over the same period, CPU speeds and hard-disk sizes have increased by two orders of magnitude. In the mid-2000s, even the most powerful monitor contains two orders of magnitude fewer pixels than the human eye's resolution, suggesting that increased visual scalability can be attained by improved monitors. However, wall-sized displays, such as AT&T's 4000 × 4000 pixel display, are within one magnitude of the limits of human perception [1,13].

3. **Visual metaphor.** As illustrated in Figs. 42.1 and 42.2, the choice of visual metaphor has a dramatic influence on scalability. Some metaphors are scalable and others are not. A typical network visualization might display a sparse network with

Table 42.1   Pixel resolutions of different devices.

Computer	Typical resolution	Pixels displayed
Ultralight	640 × 480	307,200
Laptop	800 × 600	480,000
Portable PC	1024 × 768	786,432
Desktop PC	1280 × 1024	1,310,720
Graphics workstation	1600 × 1200	1,920,000
Wall display	4000 × 4000	16,000,000

hundreds of nodes, with the best showing perhaps 100,000 nodes. Visualizing dense networks is extremely difficult.

4. **Node positioning algorithm.** Some node positioning algorithms lead to informative displays and others do not. There is no one optimal best positioning function. The most effective network visualization tools provide a suite of positioning algorithms that highlight various aspects of network structure.

5. **Interactivity.** Perhaps the most powerful technique to increase scalability uses interactivity. The key idea is dynamically updating the display to overcome the human perceptual constraints associated with fixed images. For users to perceive smooth continuous motion, the display updates must occur within one-tenth of a second. Even with fast graphics cards, achieving smooth motion requires programming tricks such as dropping out details while panning, drawing the most important nodes on top, and choosing graphical symbols that overplot gracefully.

6. **Data structures and algorithms.** Efficient data structures and algorithms are needed to achieve interactive display rates. This involves accessing the data, indirect computations (e.g., aggregation), and also direct computations. Precomputing a quad tree for node positions, for example, makes it possible to identify nodes at interactive rates.

7. **Computational infrastructure.** The computational infrastructure includes the speed of the CPU, the graphics card, the disk capacity, and network access. Network visualization issues for high-performance workstations are quite different from those involved with thin web-based displays.

42.5   Examples

In this section, we analyze the scalability of three network visualizations and suggest techniques for improving scalability. Two of the visualizations involve network traffic, and the third focuses on network structure.

42.5.1   Time-Varying Network Data

This example, described by Becker et al. [5], visualizes blocking statistics from a network of 110 switches collected every 5 minutes. The visualization technique positions the switches geographically on a map and draws color- and thickness-coded lines between the switches, with the color and thickness redundantly encoding link attributes. The node glyphs may also encode one or two attributes using circles and rectangles.

Fig. 42.3 shows three frames from a SeeNet animation. Although not shown in the figure, SeeNet addresses the display clutter problem by providing a set of interactive controls, e.g., line shortening, thresholding, zooming, and panning.

Figure 42.3   Three frames from a SeeNet network visualization animation. (See also color insert.)

42.5.1.1 Scalability Analysis

For each of the 150 time periods, there are $110 * 109 = 11,990$ directed link measurements. For each of the 30-odd attributes, there are as many as $11,990 * 150 = 1,785,000$ integer counts (7.2 MB). Zero counts are excluded from the display. The "bigness" in this example involves many time periods, many links, and many attributues, but few nodes. The software animates to display multiple time periods.

42.5.2 Worldwide Internet Traffic

The second example, described by Cox et al. [6], is also a visualization of time-varying network traffic. In this case, the dataset consists of IP traffic flows among 50 countries by 2-hour period for a 7-day period.

Fig. 42.5 shows three frames from an animation. Each country is represented by a box-shaped glyph that is both scaled and colored to encode the total packet count for all links emanating from the country. The glyphs are positioned at the locations of the countries' capitals and extend perpendicular to the surface of the globe. The color-coded arcs between the countries show the intercountry traffic, with

the higher and redder arcs indicating the larger traffic flows. The globe is illuminated by a light that is positioned to indicate the angle of the sun for the frame of the time series data that is displayed.

Drawing a world map on the surface of the sphere in Fig. 42.5 converts it into a globe, thereby providing spatial context for the location of each of the nodes. Our map contains only the continental outlines, avoiding possibly excessive detail that would obscure network information. The surface of the globe in Fig. 42.5 is an opaque blue and thus obscures those portions of arcs and nodes that lie behind and within the sphere. By interactively varying the translucency of the surface, the user may control how much of the display is obscured.

The boxes on the surface of the globe in Fig. 42.5 encode the node statistic by scaling in the radial direction. The visual effect is pleasing; the boxes appear to be small towers standing on the sphere, with the tower height and color of the tower tied to the statistic. The largest boxes correspond to the nodes that have the largest statistics, thereby focusing attention on the important nodes. Other glyphs and data encodings are possible. We have experimented

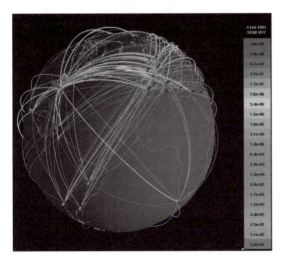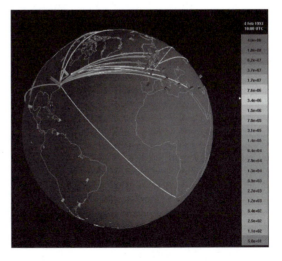

Figure 42.4 (Left) One frame from an animation showing worldwide internet traffic over a 2-hour period. (Right) Thresholding the link statistics with a slider (not shown) highlights the most important links and decreases display clutter. (See also color insert.)

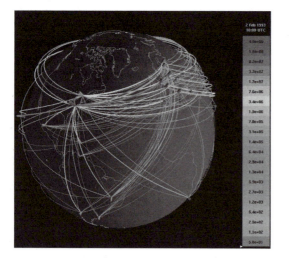 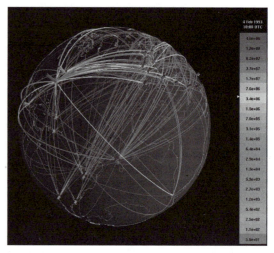

Figure 42.5   (Left) Rectangular latitude–longitude paths. (Right) Making the globe translucent reveals obscured arcs. (See also color insert.)

with cylinders and pyramids, negative scaling (where the glyph descends toward the center of the sphere), and encoding of two values using the glyph height and position above or below the surface. Our preliminary results show that the latter technique of information encoding can be readily understood by viewers.

The arcs are the analogs of the lines in the traditional 2D node and link displays. They touch the spherical surface at each end and reach a maximal radial height in the center. Tying the height of the arc to the link statistic ensures that the most important arcs corresponding to the largest values of the statistic are always visible and are never obscured by arcs corresponding to lesser values.

The first two examples use animation. However, we have found three problems with network animations. First, network animations work well for visualizing time-varying data that evolves smoothly. Changes are jarring, and our attention is automatically drawn towards them. The reason for this is that motion is a preattentive visual cue. Too many changes, however, are confusing. Thus, for network animations to be perceptually effective, the software needs to eliminate high-frequency spatial clutter. Specifically, node positions must change

gradually, and the link shapes must evolve smoothly.

Second, network animations involve inherent limits with human spatial memory. Human short-term visual memory is limited, and thus we find it difficult to see changes between frames. Visualization software can help overcome this problem by showing changes directly and thereby eliminating the need to remember previous frames and perform mental comparisons.

Third, the human attention span is limited, and it is easy for the user to become fatigued and miss important features in the animation. The visualization software needs an engaging user interface to help users concentrate.

42.5.2.1   Scalability Analysis

This dataset consists of $50 * 49 * 84 = 205{,}800$ counts (800 KB). Although smaller than the telephone network traffic, this dataset is still sizeable, and the interesting aspect involves visualizing a fully connected network where every link is nonzero. The problem with fully connected networks is that line crossings usually make the visualization unusable. The globe visual metaphor addresses this problem by drawing the arcs in 3D so that they no longer cross.

42.5.3 Visualizing Network Structure

This example shows a visualization of a network dataset that illustrates network conductivity (courtesy of Bill Cheswick at Bell Labs). The raw data consist of IP numbers (nodes) and pairs of connected IP numbers (edges). There are attributes such as date, degree, network class, etc., associated with the nodes and edges.

Fig. 42.6 shows a stargraph visualization where the nodes are positioned using a node and spring layout [9,10]. This layout positions related nodes closely together. The interesting aspect of this visualization is not the positioning algorithms but the interactive techniques for navigating through the network. The software provides filters, smooth pan and zoom, node details in linked windows, flexible ways to rebind visual characteristics, and show nodes at a graph distance of n from a designated node.

42.5.3.1 Scalability Analysis

The network visualized in Fig. 42.6 consists of 123,000 nodes and 86,000 edges. Altogether, there are 209,000 distinct entities that are displayed simultaneously on the screen. To achieve scalability, this visualization uses interactivity, filtering, and node placement.

42.6 Discussion

Our engineering goal is to develop software that usefully displays networks with thousands to hundreds of thousands of nodes and links. Beyond this limit, computational techniques must be used to reduce network sizes. Our strategy for increasing scalability is multifaceted. Our approach is to create a suite of techniques that involves both algorithms and visualization that can be independently combined to increase scalability. The techniques include the following:

- **Fast multipass adaptive algorithms for node placements.** For example, in previous work we developed a node placement algorithm consisting of four parts [10,17]. At startup, basic graph data structures such as a minimal spanning are precomputed and the raw processor speed of the graphics hardware calculated. When network nodes are repositioned, initially an O(n) algorithm does rough positioning followed by an O(n log n) algorithm, and finally an O(n2) for fine positioning. The cutpoints between the

Figure 42.6 Images from a StarGraph visualization showing network structure for an IP network. (See also color insert.)

Figure 42.7 Zoombar that supports 1000:1 zooming using a two-step linear zoom.

algorithms are determined using a heuristic that depends on the computational speed of the processor and graphics rendering rates. The object is to ensure that the layout code runs reasonably fast on any class of processor.

- **Engineering the software to hit human response rates.** As we indicated above, it is important that software be responsive to human performance constraints.

- **Using multiresolution visual metaphors** that overplot gracefully and can be made more scalable by rendering progressively less detail as the scale increases. A simple example of this is to render the glyphs corresponding to nodes as 3D spheres when zoomed in and 2D circles or even tiny points when zoomed out.

- **Labeling algorithms** that use variable font sizes and using positioning tricks to avoid overplotting.

- **Zooming and panning.** Although well known, our zoombars provide a zoom of approximately 1000:1 using double thumbs and a two-step linear zoom [7] (Fig. 42.7).

- **Fast algorithms for interactive operations.** A common interactive operation is to identify and label entities pointed to by the mouse. For example, using a standard data structure like quadtrees for identification and line-crossing algorithms for selections, it is possible to interactively manipulate networks with 100,000 nodes on standard desktop machines.

Each individual technique contributes to scalability. Taken together, the results can be striking.

Acknowledgments

Although the focus of this chapter is on scalability, the examples build on a sequence of research projects that I have undertaken with a variety of authors over the last decade. Contributers include Rick Becker and Alan Wilks (Section 42.5), Ken Cox and Taosong He (Section 42.5), Graham Wills (Section 42.5), and Alan Karr (Section 42.4).

References

1. J. Abello, E. Gansner, E. Koutsofios, and S. North. Large-scale network visualization. *ACM Computer Graphics*, 1999.
2. J. Bertin. *Graphics and Graphic Information Processing*. Berlin, Walter de Gruyter & Co., 1981.
3. J. Bertin. *Semiology of Graphics*. University of Wisconsin Press, Ltd., London, England, 1983.
4. G. Di Battista, P. Eades, R. Tamassia, and I. G. Tollis. *Graph Drawing – Algorithms for the Visualization of Graphs*. Prentice Hall, 1999.
5. R. A. Becker, S. G. Eick, and A. R. Wilks. Visualizing network data. *IEEE Transactions on Visualization and Graphics*, 1(1):16–28, 1995.
6. K. C. Cox, S. G. Eick, and T. He. 3D geographic network displays. *ACM Sigmod Record*, 25(4):50–54, 1996.
7. S. G. Eick. Visual discovery and analysis. *IEEE Transactions on Computer Graphics and Visualization*, 6(1):44–59, 2000.
8. S. G. Eick and A. F. Karr. Visual scalability. *Journal of Computational Graphics and Statistics*, 11(1):22–43, 2002.
9. S. G. Eick and G. J. Wills. Navigating large networks with hierarchies. In *Visualization '93 Conference Proceedings*, pages 204–210, 1993.
10. S. G. Eick and G. J. Wills. High interaction graphics. *European Journal of Operational Research*, 81:445–459, 1995.
11. R. A. Hanneman. Introduction to social network methods. http://faculty.ucr.edu/~hanneman.
12. W. C. Hill and J. D. Hollan. Deixis and the future of visualization excellence. In *IEEE Visualization '91 Conference Proceedings*, pages 314–320, 1991.
13. E. Koutsofios, D. A. Keim, and S. North. Visualization of large-scale telecommunications

data. *IEEE Computer Graphics and Applications*, pages 33–35, 1999.

14. H. C. Purchase. Graph drawing aesthetics. *Journal of Visual Languages and Computing*, 13(5):501–516, 2002.

15. S. S. Wasserman and K. Faust. *Social Network Analysis*. Cambridge University Press, 1994.

16. E. J. Wegman. Huge data sets and the frontiers of computational feasibility. *Journal of Computational and Graphical Statistics*, 4(4):281–295, 1995.

17. G. J. Wills. Nicheworks—interactive visualization of very large graphs. In *Graph Drawing '97 Conference Proceedings*. New York, Springer–Verlag, 1997.

43 Visual Data-Mining Techniques

DANIEL A. KEIM and MIKE SIPS
University of Konstanz

MIHAEL ANKERST
The Boeing Company

43.1 Introduction

Never before in history has data been generated at such high volumes as it is today. Exploring and analyzing the vast volumes of data has become increasingly difficult. Information visualization and visual data mining can help to deal with the flood of information. The advantage of visual data exploration is that the user is directly involved in the data-mining process. There are a large number of information visualization techniques that were developed in the early 2000s to support the exploration of large datasets. In this chapter, we provide an overview of information visualization and visual data-mining techniques and illustrate them using a few examples.

The progress made in hardware technology allows today's computer systems to store very large amounts of data. Researchers from the University of Berkeley estimate that every year about 1 exabyte (1 million terabytes) of data is generated, of which a large portion is available in digital form. This means that in about 2007 more data will be generated than in all of human history to date. The data is often automatically recorded via sensors and monitoring systems. Even simple transactions of everyday life, such as paying by credit card or using the telephone, are typically recorded by computers. Usually many parameters are recorded, resulting in data with high dimensionality. The

data is collected because people believe that it is a potential source of valuable information, providing a competitive advantage (at some point). Finding the valuable information hidden in the data, however, is a difficult task. With today's data-management systems, it is possible to view only small portions of the data. If the data is presented textually, the amount of data that can be displayed is in the range of some 100 data items, but this is like a drop in the ocean when you are dealing with datasets containing millions of data items. Having no possibility to adequately explore the large amounts of data that have been collected because of their potential usefulness, the data becomes useless and the databases become data "dumps." Information visualization focuses on datasets lacking inherent 2D or 3D semantics and therefore also lacking a standard mapping of the abstract data onto the physical screen space. There are a number of well known techniques for visualizing such datasets, such as x-y plots, line plots, and histograms. These techniques are useful for data exploration but are limited to relatively small and low-dimensional datasets. In the early 2000s, a large number of novel information visualization techniques were developed that allowed visualizations of multidimensional datasets without inherent 2D or 3D semantics. Good overviews of the approaches can be found in a number of books [8,28,38]. The techniques can be classified based on three criteria [20]

An earlier version of this paper with focus on visualization techniques and their classification has been published in *Visual Data Analysis: An Introduction* (D. Hand and M. Berthold, Eds.).

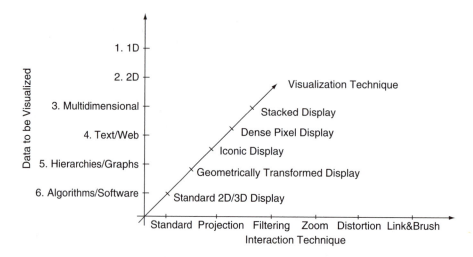

Figure 43.1   Classification of information visualization techniques.

(Fig. 43.1): the data to be visualized, the visualization technique, and the interaction technique used.

The *data type to be visualized* [32] may be 1D data, such as temporal (time-series) data; 2D data, such as geographical maps; multidimensional data, such as relational tables text, hypertext news articles, and web documents; or hierarchies and graphs, such as telephone calls and web documents, algorithms, and software.

The *visualization technique* used may be classified as standard 2D/3D displays, such as bar charts and x-y plots; geometrically transformed displays, such as hyperbolic plane [36] (Fig. 43.2a) and parallel coordinates [18]; icon-based displays, such as Chernoff faces [9] and stick figures [23,24] (Fig. 43.2c); dense pixel displays, such as the recursive pattern [4] (Fig. 43.2b) and circle segments [5]; and stacked displays, such as treemaps [19,31] (Fig. 43.2d) and dimensional stacking [37]. The third dimension of the classification is the *interaction technique* used. Interaction techniques allow users to directly navigate and modify the visualizations, as well as select subsets of the data for further operations. Examples include dynamic projection, interactive filtering, interactive zooming, interactive distortion, interactive linking, and brushing. Note that the three dimensions of our classification—data type to be visualized, visualization technique, and interaction technique— can be assumed to be orthogonal. Orthogonality means that any of the visualization techniques may be used in conjunction with any of the interaction techniques for any data type. Note also that a specific system may be designed to support different data types and that it may use a combination of visualization and interaction techniques. More details can be found in Keim and Ward [21].

43.2   Methodology of Visual Data Mining

First, the data analyst typically specifies some parameters to restrict the search space; data mining is then performed automatically by an algorithm, and finally the patterns found by the automatic data-mining algorithm are presented to the data analyst on the screen. For data mining to be effective, it is important to include the human in the data exploration process and combine the flexibility, creativity, and general knowledge of the human with the enormous storage capacity and the computational power of today's computers. Since there is a huge

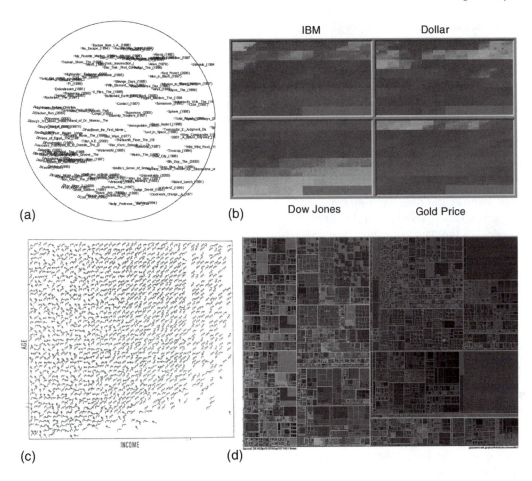

Figure 43.2   Some popular information visualization techniques. (a) Geometrically transformed displays: Interactive visualization of high-dimensional data using the hyperbolic plane [36]. Genre separation in movie space (red "x" marks science fiction, black "Δ" marks animation, and green "+" movies belonging to both genres) (© ACM). (b) Dense pixel displays: Recursive Pattern [4]—based on a generic back-and-forth recursive arrangement schema to represent each data value as a colored pixel and each attribute in separate sub-windows (example visualization shows the stock prices for Dow Jones, Gold, IBM, and US Dollar are depicted for almost 7 consecutive years, 7 vertical bars correspond to the 7 years (level (3)-patterns) and the subdivision of the bars to the 12 month within each year (level (2)-patterns), the coloring maps high attribute values (stock prices) to light colors and low attributes values (stock prices) to dark colors) (c) Iconic displays: Stick Figures [24,23]—visualization of multidimensional data using properties of angle and/or length of the limbs (US Census Data Median Household Income and Age of Householder). (d) Stacked displays: TreeMaps [9,31]—splitting the screen into rectangles in alternating horizontal and vertical directions in each level (example visualization shows a hierarchical file system of a large hard disk). (See also color insert.)

amount of patterns generated by an automatic data-mining algorithm in textual form it is almost impossible for the human to interpret and evaluate the pattern in detail and extract interesting knowledge and general characteristics. Visual data mining aims at integrating the human in the data-mining process as well as applying human perceptual abilities to the analysis of large datasets available in today's computer systems. Presenting data in an interactive, graphical form often fosters new insights, encouraging the formation and validation of new hypotheses to the end of better problem-solving and gaining deeper domain knowledge.

Visual data exploration usually follows a three-step process: *overview* first, *zoom and filter*, and then *details-on-demand* (which has been called the Information Seeking Mantra [32]). First, the data analyst needs to get an overview of the data. In the overview, the data analyst identifies interesting patterns or groups in the data and focuses on one or more of them. For analyzing the patterns, the data analyst needs to drill down and access details of the data. Visualization technology may be used for all three steps of the data exploration process. Visualization techniques are useful for showing an overview of the data, allowing the data analyst to identify interesting subsets. In this step, it is important to keep the overview visualization while focusing on the subset using another visualization technique. An alternative is to distort the overview visualization in order to focus on the interesting subsets. This can be performed by dedicating a larger percentage of the display to the interesting subsets while decreasing screen utilization for uninteresting data. To further explore the interesting subsets, the data analyst needs a drill-down capability in order to observe the details about the data. Note that visualization technology not only provides the base visualization techniques for all three steps but also bridges the gaps between the steps. Visual data mining can be seen as a hypothesis-generation process; the visualizations of the data allow the data analyst to gain insight into the data and come up with new hypotheses. The verification of the hypotheses can also be done via data visualization, but it may also be accomplished by automatic techniques from statistics, pattern recognition, or machine learning. As a result, visual data mining usually allows faster data exploration and often provides better results, especially in cases in which automatic data-mining algorithms fail. In addition, visual data exploration techniques provide a much higher degree of user satisfaction and confidence in the findings of the exploration. This fact leads to a high demand for visual exploration techniques and makes them indispensable in conjunction with automatic exploration techniques.

Visual data mining is based on an automatic part, the data-mining algorithm, and an interactive part, the visualization technique. There are three common approaches to integrate the human in the data exploration process to realize different kinds of approaches to visual data mining (Fig. 43.3):

- **Preceding Visualization (PV)**: Data is visualized in some visual form before running a data-mining algorithm. By interaction with the raw data, the data analyst has full control over the analysis in the search space. Interesting patterns are discovered by exploring the data.

- **Subsequent Visualization (SV)**: An automatic data-mining algorithm performs the data-mining task by extracting patterns from a given dataset. These patterns are visualized to make them interpretable for the data analyst. Subsequent visualizations enable the data analyst to specify feedbacks. Based on the visualization, the data analyst may want to return to the data-mining algorithm and use different input parameters to obtain better results.

- **Tightly Integrated Visualization (TIV)**: An automatic data-mining algorithm performs an analysis of the data but does not produce the final results. A visualization technique is used to present the intermediate results of the data exploration process. The combination of some automatic data-mining algorithms and visualization techniques enables specified user feedback for the next data-mining run. Then, the data analyst identifies the interesting patterns in the visualization of the intermediate results based on his domain knowledge. A motivation of this approach is to achieve independence of the data-mining algorithms from the application. A given automatic data-mining algorithm can be very useful in one domain but may have drawbacks in some other domain. Since there is no automatic data-mining algorithm (with one parameter setting) suitable for all application domains, tightly integrated

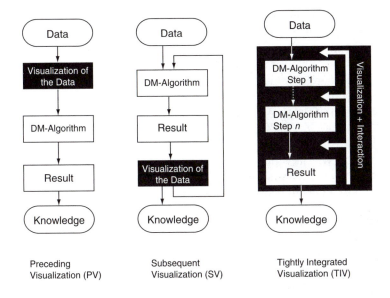

Figure 43.3   Overview of different approaches of human involvement.

visualization leads to a better understanding of the data and the extracted patterns.

In addition to the direct involvement of the human, the main advantages of visual data exploration over automatic data-mining techniques are the following:

- Visual data exploration can easily deal with highly nonhomogeneous and noisy data.

- Visual data exploration is intuitive and requires no understanding of complex mathematical or statistical algorithms or parameters.

- Visualization can provide a qualitative overview of the data, allowing data phenomena to be isolated for further quantitative analysis.

Visual data-mining techniques have proven to be of high value in exploratory data analysis and have a high potential for exploring large databases. Visual data exploration is especially useful when little is known about the data and the exploration goals are vague. Since the data analyst is directly involved in the exploration process, shifting and adjusting the explor-

ation goals is automatically done if necessary. In the next sections, we show that the integration of the human in the data-mining process and applying human perceptual abilities to the analysis of large datasets can help to provide more effective results in important data-mining application domains, such as in the mining for association rules, clustering, classification, and text retrieval.

43.3   Association Rules

The goal of association rule generation is to find interesting patterns and trends in transaction databases. Association rules are statistical relations between two or more items in the dataset. In a supermarket basket application, associations express the relations between items that are bought together. It is, for example, interesting if we find out that in 70% of the cases when people buy bread, they also buy milk. Association rules tell us that the presence of some items in a transaction imply the presence of other items in the same transaction with a certain probability, called confidence. A second important parameter is the support of an association rule, which is defined as the percentage of transactions in which the items co-occur.

Let $I = \{i_1, \ldots i_n\}$ be a set of items and let D be a set of transactions, where each transaction T is a set of items such that $T \subseteq I$. An association rule is an implication of the form $X \Rightarrow Y$, where $X \in I$, $Y \in I$, and $X, Y \neq \emptyset$. The confidence c is defined as the percentage of transactions that contain Y, given X. The support is the percentage of transactions that contain both X and Y. For given support and confidence levels, there are efficient algorithms to determine all association rules [1]. A problem, however, is that the resulting set of association rules is usually very large, especially for low support and confidence levels. Using higher support and confidence levels may not be effective, since useful rules may then be overlooked.

Visualization techniques have been used to overcome this problem and to allow an interactive selection of good support and confidence levels. Fig. 43.4 shows SGI MineSet's *Association Rule Visualizer* [17], which maps the left- and right-hand sides of the rules to the x- and y-axes of the plot and shows the confidence as the height of the bars and the support as the height of the discs. The color of the bars shows the interestingness of the rule. Using the visualization, the user is able to see groups of related rules and the impact of different confidence and support levels. The number of rules that can be visualized, however, is limited, and the visualization does not support combinations of items on the left- or right-hand side of the association rules. Fig. 43.5 shows two alternative visualizations called mosaic and double-decker plots [15]. The basic idea is to partition a rectangle on the y-axis according to one attribute and make the regions proportional to the sum of the corresponding data values. Compared to bar charts, mosaic plots use the height of the bars instead of the width to show the parameter value. Then each resulting area is split in the same way according to a second attribute. The coloring reflects the percentage of data items that fulfill a third attribute. The visualization shows the support and confidence values of all rules of the form $X_1 X_2 \Rightarrow Y$. Mosaic plots are restricted to two attributes on the left side of the association rule. Double-decker plots can be used to show more than two attributes on the left side. The idea is to show a hierarchy of attributes on the bottom (Heineken, Coke, chicken, in the example shown in Fig. 43.5) corresponding to the left-hand side of the asso-

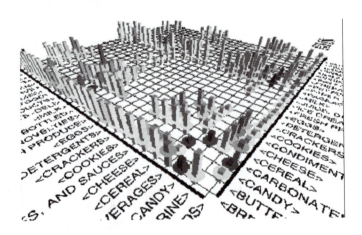

Figure 43.4. MineSet's Association Rule Visualizer [17] maps the left- and right-hand sides of the rules to the x- and y-axes of the plot and shows the confidence as the height of the bars and the support as the height of the discs; color of the bars shows the interestingness of the rule (example visualization shows market basket data for customer buying patterns). © SGI. (See also color insert.)

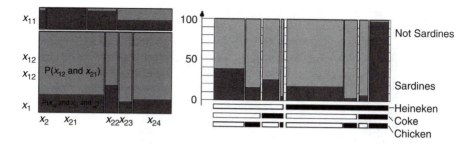

Figure 43.5   Association rule visualization [15] partitions a rectangle on the y-axis according to one attribute and makes the regions proportional to the sum of the corresponding data values. © ACM (a) Mosaic plot: 2D mosaic plot of attributes Ax_1 and Ax_2; highlighting shows up in the mosaic plot as a third dimension (b) Double-decker plot: example visualization shows a hierarchy of supermarket basket items: Heineken, Coke, chicken, and sardines.

ciation rules; the bars on the top correspond to the number of items in the corresponding subset of the database and therefore visualize the support of the rule. The colored areas in the bars correspond to the percentage of data transactions that contain an additional item (sardines, in Fig. 43.5) and therefore correspond to the support. Other approaches to association rule visualization include graphs with nodes corresponding to items and arrows corresponding to implications as used in DBMiner [16] and association matrix visualizations to cluster-related rules [12].

43.4   Classification

Classification is the process of developing a classification model based on a training dataset with known class labels. To construct the classification model, the attributes of the training dataset are analyzed and an accurate description or model of the classes based on the attributes available in the dataset is developed. The class descriptions are used then to classify data for which the class labels are unknown. Classification is sometimes also called *supervised learning* because the training set is used to teach the system how to classify the data. There are many algorithms for solving classification talks. The most popular approaches are algorithms that inductively construct decision trees. Examples are ID3 [25], CART [7], ID5 [34,35], C4.5 [26], SLIQ [22], and SPRINT [30]. In addition, there

are approaches that use neural networks, genetic algorithms, or Bayesian networks to solve the classification problem. Since most algorithms work as black-box approaches it is often difficult to understand and optimize the decision model. Problems such as over-fitting or tree pruning are difficult to tackle.

Visualization techniques can help to overcome these problems. The decision tree visualizer in SGI's MineSet system [17] shows an overview of the decision tree together with important parameters such as the attribute value distributions. The system allows an interactive selection of the attributes shown and helps the user understand the decision tree. A more sophisticated approach that also helps in decision tree construction is visual classification, as proposed by Ankerst et al. [3]. The basic idea is to show each attribute value by a colored pixel and arrange them in bars. The pixels of each attribute bar are sorted separately, and the attribute with the purest value distribution is selected as the split attribute of the decision tree. The procedure is repeated until all leaves correspond to pure classes. An example of the decision tree resulting from this process is shown in Fig. 43.7. Compared to a standard visualization of a decision tree, additional information is provided that is helpful for explaining and analyzing the decision tree, namely the following:

- Size of the nodes (number of training records corresponding to the node).

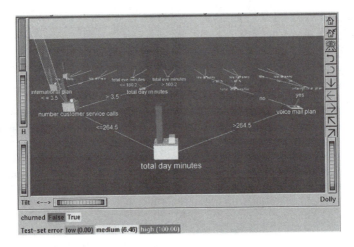

Figure 43.6   MineSet's Decision Tree Visualizer [17] displays decision trees as 3D landscapes; each node contains bars whose height, color, and disk correspond to important parameters. © SGI. (See also color insert.)

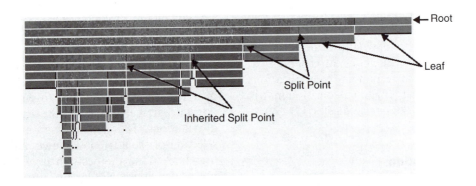

Figure 43.7   Visual Classification [3] shows attribute values by colored pixels arranged in bars (here we see a decision tree for DNA segment data with 19 attributes). © ACM. (See also color insert.)

- Quality of the split (purity of the resulting partitions).
- Class distribution (frequency and location of the training instances of all classes).

Some of this information might also be provided by annotating the standard visualization of a decision tree (e.g., annotating the nodes with the number of records or the gini-index), but this approach clearly fails for more complex information such as the class distribution. In general, visualizations can help us to better understand the classification models and to easily interact with the classification algorithms in order to optimize the model generation and classification process.

43.5   Clustering

Clustering is the process of finding a partitioning of the dataset into homogeneous subsets called clusters. Unlike classification, clustering is *unsupervised learning*. This means that the classes are unknown and no training set with class labels is available. A wide range of clustering

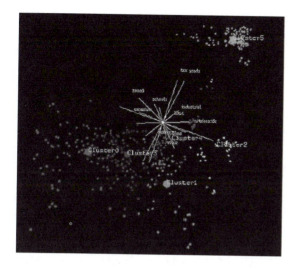

Figure 43.8 Visualization based on a projection into 3D space [39]: 3D cluster-guided projection, where the 3D subspace is determined by centroids of 4 clusters 0, 1, 3, 5. © ACM. (See also color insert.)

(e.g., x-y plots), but in higher-dimensional space the impact is much more difficult to understand. Some higher-dimensional techniques try to determine 2D or 3D projections of the data that retain the properties of the high-dimensional clusters as much as possible [39]. Fig. 43.8 shows a 3D projection of a dataset consisting of five clusters.

While this approach works well with low- to medium-dimensional datasets, it is difficult to apply to large high-dimensional datasets, especially if the clusters are not clearly separated and the dataset also contains noise (data that does not belong to any cluster). In this case, more sophisticated visualization techniques are needed to guide the clustering process, select the right clustering model, and adjust the parameter values appropriately. An example of a system that uses visualization techniques to help in high-dimensional clustering is OPTICS [2]. The idea of OPTICS (*Ordering Points To Identify the Clustering Structure*) is to create a 1D ordering of the database representing its density-based clustering structure. Fig. 43.9 shows a 2D example dataset together with its reachability distance plot. Intuitively, points within a cluster are close in the generated 1D ordering and their reachability distance shown in Fig. 43.9 is similar. Jumping to another cluster results in higher reachability distances. The idea works for data of arbitrary dimension. The reachability plot provides a visualization of the inherent clustering structure and is therefore valuable for

algorithms have been proposed in the literature, including density-based methods such as kernel density estimation [29] and linkage-based methods [6]. Most algorithms use assumptions about the properties of the clusters that are either used as defaults or have to be given as input parameters. Depending on the parameter values, the user gets differing clustering results. In 2D or 3D space, the impact of different algorithms and parameter settings can easily be explored using simple visualizations of the resulting clusters

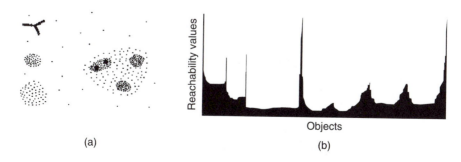

(a)　　　　　(b)

Figure 43.9 OPTICS Visual Clustering [2]. (a) Example dataset; (b) reachability plot. © ACM.

understanding the clustering and guiding the clustering process.

Another interesting approach is the *HD-Eye* system [14]. The *HD-Eye* system considers the clustering problem a partitioning problem and supports a tight integration of advanced clustering algorithms and state-of-the-art visualization techniques, allowing the user to directly interact in the crucial steps of the clustering process. The crucial steps are the selection of dimensions to be considered, the selection of the clustering paradigm, and the partitioning of the dataset. Novel visualization techniques are employed to help the user identify the most interesting projections and subsets as well as the best separators for partitioning the data. Fig. 43.10 shows an example screenshot of the *HD-Eye* system with its basic visual components for cluster separation. The separator tree represents the clustering model produced so far in the clustering process. The *abstract iconic displays* (top-right

and bottom-middle in Fig. 43.10) visualize the partitioning potential of a large number of projections. The properties are based on histogram information of the point density in the projected space. The number of data points belonging to the maximum corresponds to the color of the icon. The color follows a given color table ranging from dark colors for large maxima to bright colors for small maxima. The measure of how well a maximum is separated from the others corresponds to the shape of the icon, and the degree of separation varies from sharp spikes for well separated maxima to blunt spikes for badly separated maxima. The *color- and curve-based point density displays* present the density of the data and allow a better understanding of the data distribution, which is crucial for an effective partitioning of the data. The visualizations are used to decide which dimensions are used for the partitioning. In addition, the partitioning can be specified interactively

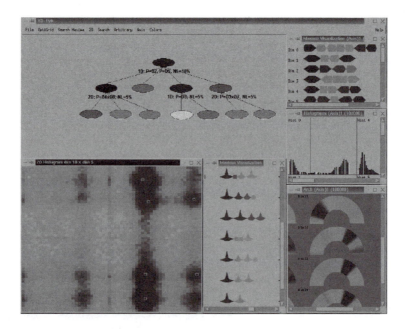

Figure 43.10 *HD-Eye* screenshot [14] showing different visualizations of projections and the separator tree. Clockwise from the top: separator tree, iconic representation of 1D projections, 1D projection histogram, 1D color-based density plots, iconic representation of multidimensional projections, and color-based 2D density plot (example visualization shows a large molecular-biology dataset). © IEEE. (See also color insert.)

directly within the visualizations, allowing the user to define nonlinear partitionings.

43.6 Text

With the growing importance of electronic media for storing and exchanging text documents, there is also a growing interest in tools that can help us find and sort information included in the text documents. Text documents are semistructured data, in that they are neither completely unstructured nor completely structured. For example, a document may contain some structured fields, such as title, authors, publication date, length, and category, as well as largely unstructured text components, such as abstract and content. Text mining is a process of finding patterns in text databases and may be defined as the process of analyzing text to extract information from it. Complete understanding of natural-language text is not immediately attainable, and therefore text mining focuses on extracting a small amount of information with high reliability. The goals of the

text-mining process are automatic document clusterization/categorization, assignment of keywords to text documents, topic identification and tracking in ordered (time) sequences of text documents, searching documents based on the content categories and not only keywords, generation and analysis of user profiles based on the usage of text databases, and other related problems. A wide range of automatic text-mining algorithms have been proposed in the literature over the last few decades [10,11].

An interesting visual data-mining approach is ThemeRiver [13]. ThemeRiver visualizes thematic variations over time within a large collection of documents. The thematic changes are shown in the context of a timeline and corresponding external events. The timeline within the document collection, selected thematic content, and thematic strength are directly indicated by the directed flow, composition, and changing width of the visualized river. The directed flow from left to right is interpreted as movement through time. At any point in time, the vertical distance, or width, of the river indicates the collective strength of the selected

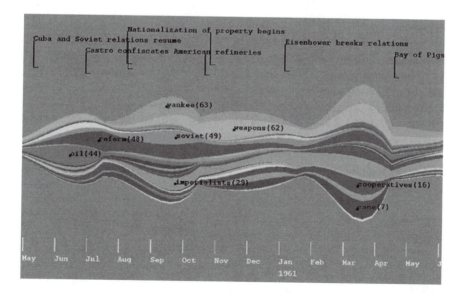

Figure 43.11 ThemeRiver [13]: visualization of thematic changes in documents (example visualization shows Castro data from November 1959 through June 1961). © IEEE. (See also color insert.)

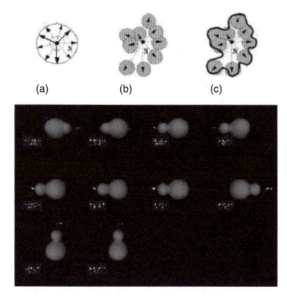

(a)   (b)   (c)

Figure 43.12 Shape-based Visual Interface for Text Retrieval [27]: shape-based visualization of query results for the key words lion, sheep, mouse, and wolf. © ACM. (See also color insert.)

themes. Colored "currents" flowing within the river represent individual themes. Vertical width of the river segments indicates decreasing or increasing strength of the themes.

Another interesting approach is the shape-based Visual Interface for Text Retrieval [27]. This exploration system uses procedurally generated shapes coupled with an underlying text-retrieval engine. Traditional text-based queries and summarization are enhanced with a visual interface based on 3D shapes (glyphs). The interface allows visualization of multidimensional relationships among documents and perception of more information than with conventional text-based interfaces.

43.7   Conclusion

The exploration of large datasets is an important but difficult problem. Information visualization techniques can be useful in solving this problem. Visual data exploration has a high potential, and many applications such as fraud detection and data mining can use information visualization technology for improved data analysis.

Avenues for future work include the tight integration of visualization techniques with traditional techniques from such disciplines as statistics, machine learning, operations research, and simulation. Integration of visualization techniques and these more established methods would combine fast automatic data-mining algorithms with the intuitive power of the human mind, improving the quality and speed of the data-mining process. Visual data-mining techniques also need to be tightly integrated with the systems used to manage the vast amounts of relational and semistructured information, including database management and data warehouse systems. The ultimate goal is to bring the power of visualization technology to every desktop to allow a better, faster, and more intuitive exploration of very large data resources. This will not only be valuable in an economic sense but will also stimulate and delight the user.

References

1. D. Agarwal, H. Mannila, R. Srikant, H. Toivonen, and A. Verkamo. Fast discovery of association rules. *Advances in Knowledge Discovery and Data Mining*, pages 307–328, 1996.

2. M. Ankerst, M. Breunig, H. Kriegel, and J. Sander. OPTICS: Ordering points to identify the clustering structure. *Proc. ACM SIGMOD '99, Int. Conf on Management of Data*, pages 49–60, 1999.

3. M. Ankerst, M. Ester, and H. Kriegel. Towards an effective cooperation of the computer and the user for classification. *SIGKDD Int. Conf. On Knowledge Discovery & Data Mining (KDD 2000)*, pages 179–188, 2000.

4. M. Ankerst, D. A. Keim, and H.-P. Kriegel. Recursive pattern: A technique for visualizing very large amounts of data. In *Proc. Visualization '95*, pages 279–286, 1995.

5. M. Ankerst, D. A. Keim, and H.-P. Kriegel. Circle segments: A technique for visually exploring large multidimensional data sets. In *Visualization '96, Hot Topic Session*, San Francisco, CA, 1996.

6. H. H. Bock. *Automatic Classification*. Vandenhoeck and Ruprecht, Göttingen, 1974.

7. L. Breiman, J. Friedman, R. Olshen, and C. Stone. *Classification and Regression Trees*. Wadsworth and Brooks, Monterey, CA, 1984.

8. S. Card, J. Mackinlay, and B. Shneiderman. *Readings in Information Visualization*. Morgan Kaufmann, 1999.

9. H. Chernoff. The use of faces to represent points in k-dimensional space graphically. *Journal Amer. Statistical Association*, 68:361–368, 1973.

10. J. Han and M. Kamber. *Data Mining: Concepts and Techniques*. Morgan Kaufmann Publishers, 2001.

11. D. J. Hand, H. Mannila, and P. Smyth. *Principles of Data Mining*. MIT Press, 2001.

12. M. Hao, M. Hsu, U. Dayal, S. F. Wei, T. Sprenger, and T. Holenstein. Market basket analysis visualization on a spherical surface. *Visual Data Exploration and Analysis Conference*, 2001.

13. S. Havre, B. Hetzler, L. Nowell, and P. Whitney. Themeriver: visualizing thematic changes in large document collections. *Transactions on Visualization and Computer Graphics*, 2001.

14. A. Hinneburg, D. Keim, and M. Wawryniuk. HD-Eye: visual mining of high-dimensional data. *IEEE Computer Graphics and Applications*, 19(5), 1999.

15. H. Hofmann, A. Siebes, and A. Wilhelm. Visualizing association rules with interactive mosaic plots. *SIGKDD Int. Conf. On Knowledge Discovery & Data Mining (KDD 2000)*, Boston, MA, 2000.

16. D. T. Inc. Dbminer, http://www.dbminer.com, 2001.

17. S. G. Inc. Mineset, http://www.sgi.com/software/mineset, 2001.

18. A. Inselberg and B. Dimsdale. Parallel coordinates: A tool for visualizing multi-dimensional geometry. In *Proc. Visualization 90, San Francisco, CA*, pages 361–370, 1990.

19. B. Johnson and B. Shneiderman. Treemaps: A space-filling approach to the visualization of hierarchical information. In *Proc. Visualization '91 Conf*, pages 284–291, 1991.

20. D. Keim. Visual exploration of large databases. *Communications of the ACM*, 44(8):38–44, 2001.

21. D. Keim and M. Ward. Visual data mining techniques. In *Intelligent Data Analysis: An Introduction* (D. Hand and M. Berthold, Eds.). Berlin, Springer Verlag, 2002.

22. M. Mehta, R. Agrawal, and J. Rissanen. SLIQ: A fast scalable classifier for data mining. *Conf. on Extending Database Technology (EDBT), Avignon, France*, 1996.

23. R. M. Pickett. *Visual Analyses of Texture in the Detection and Recognition of Objects*. Academic Press, New York, 1970.

24. R. M. Pickett and G. G. Grinstein. Iconographic displays for visualizing multidimensional data. In *Proc. IEEE Conf. on Systems, Man and Cybernetics*, pages 514–519, 1988.

25. J. R. Quinlan. Induction of decision trees. *Machine Learning*, pages 81–106, 1986.

26. J. R. Quinlan. *C4.5: Programs for Machine Learning*. Morgan Kaufmann, Los Altos, CA, 1993.

27. R. M. Rohrer, J. L. Sibert, and D. S. Ebert. A shape-based visual interface for text retrieval. *IEEE Computer Graphics and Applications*, 19(5):40–47, 1999.

28. H. Schumann and W. Müller. *Visualisierung: Grundlagen und Allgemeine Methoden*. Berlin, Springer, 2000.

29. D. W. Scott. *Multivariate Density Estimation*. Wiley and Sons, 1992.

30. J. Shafer, R. Agrawal, and M. Mehta. SPRINT: A scalable parallel classifier for data mining. *Conf. on Very Large Databases*, 1996.

31. B. Shneiderman. Tree visualization with treemaps: A 2D space-filling approach. *ACM Transactions on Graphics*, 11(1):92–99, 1992.

32. B. Shneiderman. The eye have it: A task by data type taxonomy for information visualizations. In *Visual Languages*, 1996.

33. B. Spence. *Information Visualization*. Pearson Education Higher Education publishers, UK, 2000.

34. P. E. Utgoff. Incremental induction of decision trees. *Machine Learning*, 4:161–186, 1989.

35. P. E. Utgoff, N. C. Berkman, and J. A. Clouse. Decision tree induction based on efficient tree restructuring. *Machine Learning*, 29:5–44, 1997.

36. J. Walter and H. Ritter. On interactive visualization of high-dimensional data using the hyperbolic plane. In *Proc. ACM SIGKDD International Conference on Knowledge Discovery and Data Mining*, pages 123–131, 2002.

37. M. O. Ward. Xmdvtool: Integrating multiple methods for visualizing multivariate data. In *Proc. Visualization 94, Washington, DC*, pages 326–336, 1994.

38. C. Ware. *Information Visualization: Perception for Design*. Morgan Kaufmann, 2000.

39. L. Yan. Interactive exploration of very large relational data sets through 3d dynamic projections. *SIGKDD Int. Conf. On Knowledge Discovery & Data Mining (KDD 2000)*, Boston, MA, 2000.

44   Visualization in Weather and Climate Research

DON MIDDLETON and TIM SCHEITLIN
National Center for Atmospheric Research

BOB WILHELMSON
National Center for Supercomputing Applications
University of Illinois

44.1   A Brief History

Early in the 20th century, Lewis Fry Richardson, a British mathematician, formulated numerical approximations for the general circulation of the atmosphere and in 1922 published *Weather Prediction by Numerical Process*. At a time long before the advent of computers, he tried to imagine how it might be possible to solve those complex equations in order to make a weather prediction. He imagined a stadium filled with people, each equipped with a slide rule and tasked with calculating a small part of an overwhelmingly large problem. With a human conductor guiding the process, somehow such a symphony of parallel calculation might be able to produce solutions to the problem. Dr. Richardson was truly a visionary because, of course, that's essentially how we simulate weather and climate today: on parallel computers, and building on his formative work.

In the mid-1950s the availability of digital computers offered enhanced opportunities for effectively solving Richardson's equations, and pioneering work began in developing the first global general circulation models (GCMs) for the atmosphere. At about the same time, and into the early 1960s, a series of the world's first meterological satellites—the TIROS (Television Infrared Observation Satellite) series—was launched and scientists were afforded an entirely new view of our planet and its weather patterns. As research progressed in the area of simulation and observation, analysis and visualization capabilities became increasingly important and the development of early visualization software began. One of the prime examples of this was NCAR Graphics, a suite of FORTRAN libraries that started out as a basic 1D/2D package for plotting graphs and contours of global and regional scientific data.

In the mid-1960s, Warren Washington and Akira Kasahara of the National Center for Atmospheric Research (NCAR) began developing a numerical GCM of the atmosphere, still building largely upon Dr. Richardson's early work. The numbers were starting to pour out, and interest in developing better insight into the results grew as well. These pioneers set out on an aggressive effort to visualize their simulation results and, in an incredibly laborious process, used their early visualization tools to produce remarkably sophisticated 3D visualizations (projections of the globe) of their data. Using early digital film recorders, they recorded 3D animations of simulated time-evolving weather using some clever techniques, such as defocusing the camera on an image of precipitable water so that the end result would resemble clouds. The image shown in Fig. 44.1 is a single frame from one of these early movies that has recently been restored from archives. The first paper on this was published in 1967

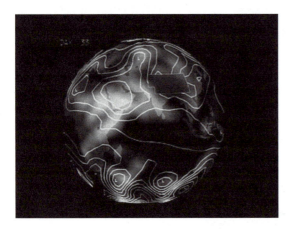

Figure 44.1 Early GCM. (See also color insert.)

in *The Monthly Weather Review*, entitled "NCAR Global General Circulation Model of the Atmosphere." In 1975, Richard Anthes developed one of the first multilayer numerical hurricane models. In contrast to the early climate modelers, he created an animation by producing a large sequence of printouts, each page representing a single time-step from his model. By hand, he registered each printout on a table and photographed it, ultimately producing a 16 mm film animation of his simulation.

In 1987, visualization was identified in a seminal NSF-sponsored report [17] as a critical capability for dealing with "fire hoses of data," and this triggered new work on visualization at universities and the then-new National Science Foundation (NSF) supercomputing centers. One of the best-known efforts was undertaken by the National Center for Supercomputing Applications (NCSA) in the area of weather, where researchers developed a beautiful visualization of a long-lasting supercell thunderstorm, the type of storm that produces the largest tornadoes. One frame of this movie is shown in Fig. 44.2. Various rendering techniques were used to illustrate the water and ice structure of a storm, how air moves and rotates in and around the storm, and how different physical processes influence storm rotation near the ground. Techniques used included weightless tracer particles repre-

sented by balls and ribbons, color-filled slice planes, and solid and transparent isosurfaces. The animation was a large team effort, as reported by Wilhelmson et al. [34]. Storyboards and mock-ups were created and a variety of visualization techniques explored. Over the 11-month period leading to the release of the storm video, four scientific animators, using Wavefront software, worked on the project for approximately 1 person-year along with scriptwriters, artistic consultants, and postproduction personnel. It won a number of awards, appeared in and on the cover of many science and visualization articles and books, and was considered for the National Academy's Short Film—Animation Category Award. It remains a classic of scientific visualization, appearing on the cover and in the discussion of Tufte's 1997 book on *Visual Explanations* [31].

Development of the interactive atmospheric science visualization package known as Vis5D [32] began in 1988, and it remains today an important tool for visualizing weather and climate data on the scientific desktop. A number of other more general-purpose interactive visualization tools appeared on the scene throughout the 1990s (AVS, IBM DX, Iris Explorer, IDL), and their widespread use was propelled by the availability of powerful graphics workstations and ultimately a new generation of visualization-capable PCs. Now, in 2004, the atmospheric science community uses a broad array of old and new visualization tools. Through continued development, NCAR graphics, Vis5D, and IDL have stood the test of time and are widely used in the earth science community. New software developments are being directed at utilizing new hardware capabilities, such as large display walls, and at creating collaborative visualization environments.

This chapter provides a high-level overview of visualization in the areas of weather and climate research. The approach is to provide a general sense of the nature of the research problems, the models (this term is used interchangeably with "simulations") used to conduct numerical experiments, commonly used visualization tools

2:20:00

Figure 44.2   A transparent isosurface is combined with components of other earlier animation segments and includes moving balls representing air motion, a colored horizontal slice through the storm showing the region of precipitation, and a twisting ribbon rising up through the storm. The animators referred to this as the party scene. The rate of twist of the ribbon is related to the magnitude of the streamwise vorticity present in air rising through the storm. Such vorticity is analogous to the rate of rotation of a spiraling football as it moves from the quarterback to the receiver. To fully understand the behavior illustrated, it is important to look at the animation. For example, the colored balls released at regular intervals in a horizontal plane are colored blue when they are sinking and orange when they are rising. In the early portion of the animation, the alternating regions of blue and orange move away from the storm, revealing the presence of wave motion. The various visualization idioms used here are still used today. (See also color insert.)

and techniques, and the underlying data that must be processed, analyzed, visualized, compared, and ultimately understood. When appropriate, some of the real-world challenges and problems that one must surmount while trying to visualize complex geoscientific data in real scientific applications are noted. A number of examples are presented—miniature case studies—that highlight some specific research thrusts, the visualization approaches taken, and what these efforts represent in terms of future requirements and needs. The examples are intended to be illustrative of current practice as well as challenges in weather and climate visualization. Those shown here represent undertakings by close-knit teams of visualization specialists and domain researchers. The chapter will be closed by projecting a few years into the future and speculating about future visualization needs and possible avenues for future visualization research and software development.

44.2   Weather and Climate Research

Understanding and prediction of the weather plays an important role in our day-to-day lives. The weather forecasts that we read in the newspaper or access on the web are the results of years of research and the regular execution of forecast simulations running on supercomputers. Climate change research is central to understanding the natural variability of our planet's rhythms and the potential impact that society at large may be having on our climate system, such as global warming. Put another way, weather is the dynamic, time-evolving behavior of the atmosphere, while climate is the average behavior of the weather or, more precisely, a statistical description of it. Colloquially, climate is what, on average, you expect and weather is what you actually get. A weather model will produce detailed characteristics of temperature, precipitation, humidity, and so forth for a week or two over a spatial domain that ranges from quite small (e.g., a county) to global. A climate model produces similar data on the global and multiple-year scale, but the results are generally processed so that they are expressed in terms of monthly, seasonal, annual, or even decadal averages along with extremes and temporal variability (i.e., a long-term, statistical description).

A variety of weather and climate models are used throughout the world. For example, in the study of weather, this includes the University of Oklahoma's Advanced Regional Prediction System (ARPS) model [37]. Other community models include the Mesoscale Model, Version 5 (MMS5) [19], a long-term joint effort of Penn State University and NCAR; and the Weather Research and Forecasting (WRF) model [36], a new parallel and scalable model under development by NCAR, government agencies, and the university community. These "community models" are developed as community efforts and are freely available for use and modification by the weather research community. ARPS, MM5, and WRF run on most Unix systems and have successfully been deployed on fairly large parallel computational platforms. The datasets that result from even a single simulation can be quite large, and visualization of them presents new challenges, as discussed at the end of the chapter.

Models for climate research are quite complex since they are composed of a broad range of complex interacting physical, chemical, dynamical, and biological processes that span very small to very large temporal and spatial domains. Furthermore, climate research spans observation and simulation of the atmosphere, weather, climate, oceans, cryosphere, chemical processes, ecosystems, space weather, and solar processes. Building a global climate model is a huge undertaking in terms of both science and software engineering. As a result, there are only a handful of large-scale comprehensive models worldwide. One example is the Community Climate System Model (CCSM), which is a multiagency effort in the US. Like its cousins developed at other centers, the CCSM is a fully

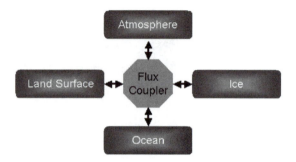

Figure 44.3 Components of the Community Climate System Model (CCSM).

coupled model, meaning that it includes several interacting components (Fig. 44.3).

Models such as this are used to conduct numerical experiments for past, present-day, and future climate. Together with observational data, the volume of data from multiple simulations and multiple models is very large and requires careful analysis and visualization to gain insight into factors that influence our climate.

44.3 Data and Grids

Weather and climate research involves the analysis and visualization of a staggering amount and variety of data, including observational (e.g., satellite, radar, airplane, etc.), simulation, and hybrid forms. While the visualization of observational data is an interesting and important topic, much of the advanced and most challenging visualization work underway is focused on the 3D simulations that are used to conduct numerical experiments and predictions. Thus, the focus herein will primarily be upon the visualization of model data.

Discussing visualization is, to a great degree, a story about correctly processing and visually representing some fairly complex data, and the discussion that follows is aimed at providing some information that should be useful to anyone working in this area. In general, weather and climate models produce georeferenced, 3D, time-evolving data of substantial volume in various formats. Dealing correctly with commonly used computational grid topologies, different data formats or interfaces, and map projections is often an early stumbling block for the tool or algorithm developer/user.

In terms of data formats and interfaces, there has been steady progress towards the adoption of the network Common Data Form (netCDF) [22] standard. NetCDF defines a general model for scientific data and provides a standard interface (API) for access along with an underlying format that is system-independent and network transparent. Even though netCDF itself provides a certain level of standarization for data access, it is still possible to store the data in an almost infinite number of different ways. In order to construct visualization and analysis capabilities that enable the user to easily ingest and explore the resulting data, it is necessary that standards be adopted for the *metadata* such that variable names, units, and coordinate systems may be expressed using an agreed-upon vocabulary. Large communities have come together and developed *conventions*, which do precisely this. Building on a couple of earlier efforts, the netCDF Climate and Forecast Metadata Convention [23] has recently emerged. The "CF Convention," as it is commonly known, is already in use by some efforts (discussed later) and appears to be poised for some level of international acceptance.

Another primary source of data that is of interest to the climate community is satellite observation data, and the Hierarchical Data Format (HDF4) has been used in this area for several years. During this decade, NASA's Earth Observing System (EOS) program and others like it will launch quite a number of new satellite platforms, and it appears likely that data production will be in the newer HDF5 [12] format. HDF5 is a significant advance over HDF4 and offers provisions for more elaborate data models along with features that provide the potential for much greater performance and volume-handling capacity. As such, it is also drawing interest from the modeling community, which is very demanding relative to HDF5's capabilities. It is also expected

that in the near future HDF5 capabilities will be integrated with netCDF. A third important data category is that of "reanalysis," which is essentially a hybrid class of climate data: observational data is fed into a special model that reconstructs a historical record. The World Meteorological Organization (WMO) has developed a standard for gridded data called GRIB (Gridded Binary), and it is used extensively worldwide for reanalysis and weather-data distribution. GRIB data tends to be somewhat less portable and self-describing than netCDF.

The data that is produced by the MM5 model has its own unique format and a suite of processing capabilities that accompany the model. MM5 produces time-evolving data and, like most weather models, does not operate on a strictly Cartesian grid. The horizontal dimension is typically referenced to one of several map projections, while the vertical dimension uses a "sigma" coordinate system in which pressure levels "follow" the terrain. The WRF model produces very similar data to MM5, but it uses the netCDF interface for data output, and exploratory work is underway to utilize HDF5. WRF also uses a terrain-following coordinate system, but instead of pressure, it currently offers a choice of vertical levels referenced to either mass or height, with mass being the preferred formulation. Weather models typically provide "nested grid" capability, where a given simulation will be run across multiple progressively finer grids, with the finest grid covering a region of particular interest (e.g., a hurricane). At present, none of the weather models described above adhere to the CF conventions.

As discussed earlier, climate models consist of a number of different, interacting components. The CCSM model has adopted the netCDF interface for data generation and has also recently begun using the CF convention. Many of the current formulations for the atmospheric component of climate models use the spectral method [33], and this is true for CCSM as well. The spatial resolution of these models is expressed as a series truncation number. For example, the CCSM has been run in production mode for several years at "T42" resolution for the atmosphere, where "T42" specifies a resolution in the spectral domain that translates to a 128×64 Gaussian grid over the globe (examples of exploratory experiments at higher resolutions are presented later). The other model components have their own unique grids, and these can sometimes be challenging to deal with. For example, the current CCSM ocean model uses a curvilinear grid that has one or two poles displaced onto land, an approach that provides computational advantages. The sea ice model also uses a curvilinear grid.

In general, one of the challenging aspects of visualizing weather and climate data is certainly all of the different grid topologies that one must deal with. Data may be resampled onto Cartesian grids, but this can introduce undesirable errors and artifacts. Developing visualization and analysis tools that deal correctly with these data and grid issues will be a growing challenge, as climate and weather researchers adopt new grid topologies for computational and scientific reasons. In the near future, geodesic grids, new formulations of pole-displaced grids, and various terrain-following coordinate systems can be expected. Looking further out, generalized multiscale and adaptive-grid models will further challenge the visualization community.

44.4 Visualization and Analysis Tools

While visualization practitioners can and do produce elegant 3D renderings of climate simulation and observed data, in practice both weather and climate researchers generally employ highly quantitative analysis, and most use 1D and 2D visual representations. The tools of the trade in this realm feature data ingest, data processing, analysis functions, and a variety of traditional visualization capabilities. Virtually all of the work requires georeferenced display and flexibility in portraying

various map projections. In terms of visualization techniques, they are the traditional line graphs, contour diagrams, streamlines, velocity vectors, maps, and so forth. In fact, there are suites of tools that are commonly used by both climate and weather researchers for regular work.

For example, in the weather community, an application known as RIP (Read/Interpolate/Plot—see MM5 reference) will ingest MM5 data and produce a variety of 2D visualizations that are commonly used by weather researchers. RIP can also process, regrid, and format MM5 data so that it can be ingested by the popular Vis5D application, and this capability is frequently used by practicing researchers who want to gain deeper insight into the 3D structure and time-evolving characteristics of their simulation results.

The climate research community generally uses one or more of a collection of visualization and analysis tools that have emerged over the last decade or so. All of these provide a fourth-generation scripting language interface with some amount of interactive display and control capabilities. Ferret [9], developed by NOAA's Pacific Marine Environmental Laboratories (PMEL), is used for climate research in general and is particularly popular with the oceanographic community. IDL, a commercial product by Research Systems, Inc., is another widely used package with an extensive array of processing and statistical functions and a suite of 3D rendering capability as well. The Grid Analysis and Display System (GrADS) [11] was developed at the Center for Ocean, Land, Atmosphere Studies (COLA) and has a long-time following in the climate community. The National Center for Atmospheric Research in Boulder, CO has developed NCAR Command Language (NCL) [20], which serves as the primary analysis and visualization package for the NCAR CCSM. The Program for Climate Model Diagnosis and Intercomparison (PCMDI) at Lawrence Livermore National Laboratories distributes the Climate Data Analysis Tools (CDAT) package [2]. CDAT is based on the Python language and provides an interactive interface for certain functions.

There is an enormous amount of overlap in the functionality of all of these tools, and each one has its strengths and its followings. In general, they all deal with essentially the same types of scientific data, offer the same sorts of visualization techniques, and have an emphasis on maps and georeferenced data. The actual languages vary from application to application and are generally custom and unique, with the exception of CDAT with its Python binding (a Python binding for NCL functionality is underway as well). None of them offer very good capabilities for 3D rendering, but they are often used in a complementary manner with other community and commercial tools to create the more advanced visualizations. It is not uncommon to find a researcher who uses several of these, because no single application delivers all of the required capabilities.

These tools have relevance for the visualization specialist, whether they are developing 2D, 3D, or even virtual-reality environments for the exploration of climate data. Specifically, a substantial amount of data processing and analysis is required in advance of such efforts, and the tools enumerated above can be effectively used to ingest various types of data, subset them, and process and output them into forms that can then be used for other pursuits while maintaining data integrity and correctness.

Visualization specialists use a wide variety of 3D capabilities to visualize weather and climate research data, and many of them are mentioned in the examples that follow. The Visualization Tookit (VTK) has also been used to good effect for one-off applications but has not yet been used to construct generalized applications for this domain. In terms of practicing scientists, it is somewhat rare to find climate researchers using 3D visualization much at all, but it is fairly common in the area of weather research. This is due mostly to simple differences between the two areas of research: climate is focused more on long-term trends on a global scale, whereas weather research deals with dynamic phenomena on much smaller scales. For 3D visualization, Vis5D is probably the most popular application and has arguably done more to promote the use of advanced scientific visualization than any other single application. NCAR has enhanced the Vis5D tool so that it is capable of dealing with fairly large datasets and may be used in stereo or 3D display environments. There is also a variation called CAVE5D, which was developed for use in the immersive CAVE environment. A newcomer to the visualization arena is the VisAD framework [13], which is discussed in Chapter 34. VisAD is a very interesting approach for a number of reasons but especially for its elegant data model, which shows potential for addressing many of the problems associated with the visualization of multiple disparate data types (data fusion). There is also a new VisAD-based application under development called the Integrated Data Viewer (IDV) [15], which, like Vis5D, is targeted at meteorological applications. VisAD and IDV are developed in Java, and their suitability for some of the larger problems described here hinges on how well Java can perform for large data and computational applications.

44.5 Visualization Case Studies in Weather and Climate Research

Most of us are very familiar with visualization as it is most commonly used in weather. We see examples of this on the evening news, in the newspaper, and on the web. 2D forecast maps and real-time observation of satellite and radar data are the primary visual representations in present-day weather forecasting, along with a long tradition of using glyphs, including those on weather reports used to indicate observed sky, precipitation, and temperature forecasts. It is also common to create contour-filled images that show temperature or pressure or some other field such as radar reflectivity. These 2D images often have country or state maps, rivers, or roads that act as references. Television forecasters typically use very readable images with a small amount of information while researchers are typically interested in a higher density of information and often use multiple techniques for different variables in a single image (e.g., temperature contours and shaded pressure distribution). Thus, in this section, the focus will be mostly on model visualization approaches that are used extensively in the research (as opposed to the forecasting) community.

A recent example of 2D visual comparison between model and observed behavior is shown in Fig. 44.4. A series of tornadic storms moved through the Fort Worth, TX metro area on March 29, 2000. The top row in the figure shows an hourly sequence of reflectivity images from the Fort Worth WSR-88D (NEXRAD) radar. Shown in the lower three panels is the equivalent radar reflectivity from a 3-kilometer grid forecast made using the University of Oklahoma Advanced Regional Prediction System (ARPS), initialized at 2300 UTC with Fort Worth NEXRAD radar and other data. The degree of agreement between observations and

forecast, even out to 4 hours, is remarkably good. However, without radar data assimilation in the model's initial conditions, the tornadic storms to the north of Fort Worth are completely absent (not shown)—thus highlighting the value of radar data in storm-scale numerical weather prediction.

44.5.1   Visualizing Severe Weather: Storms and Hurricanes

Advancing predictive capabilities for severe weather such as strong storm systems, hurricanes, typhoons, and cyclones is a major focus of the weather research community. 3D visualization is widely used by researchers to understand the resulting data, the size and complexity of which typically provide fertile ground for visualization work.

A good example of an MM5-based research effort is a recent study of Hurricane Opal [30]. At landfall near Pensacola Beach, FL on October 4, 1995, Opal was a Category 3 storm with 115 mph sustained winds and gusts over land as

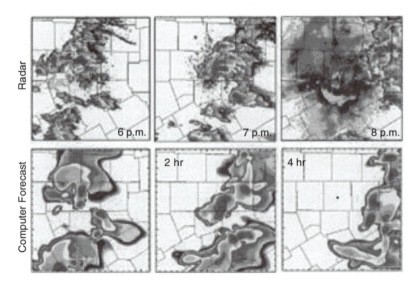

Figure 44.4   The top row is the reflectivity from the Fort Worth Doppler radar, and the bottom row is from a numerical prediction model. The times at the top of the figure are local Fort Worth time, and the times in the forecast are the length of time from the start of the simulation. Fort Worth is located at the star (original figure in color).

high as 144 mph. Along with waves reaching 18 feet above sea level on the Gulf Coast, Opal caused $2 billion worth of damage, nine fatalities, and up to 10 inches of rain over parts of Florida, Alabama, and Georgia. Further, 22 tornadoes were attributed to the hurricane. Multiday numerical simulations were conducted with MM5 using two-way nesting with grid resolutions of 90, 30, 10, 3.3, and 1.1 km. The inner grid was $460 \times 460 \times 35$ grid-points and over 100 GB of data were produced during the 90 hr time period from when the hurricane formed over the Gulf of Mexico and subsequently made landfall. The aim of the study was to investigate the behavior of resolved severe convection within a hurricane simulation. The resulting simulation of Hurricane Opal [29] reached Category 5 (slightly more intense than the observed Opal, which only reached strong Category 4 intensity). Atmospheric conditions in the model were discovered that would support mini-supercells that could produce tornadoes, although the resolution was not sufficient to capture the tornadoes themselves.

Fig. 44.5 shows a volume-rendered visualization of one of the Opal simulations as it approached land using Pixar's Renderman [28]. Rainbands are evident to the east and southeast of the hurricane center. All five nested grids were incorporated in the visualization using the finest grid resolution available at any location (typically, researchers only visualize one grid at a time). Using all five grids helped the researchers at the location to uncover a problem that occurred at the boundaries between different grids, which was subsequently corrected. This problem was particularly evident when performing the volume integration from above the hurricane, and its discovery illustrates the value of creating visualizations across all grids used in adaptive or nested grid simulations, as well as the general importance of visualization as a model-debugging capability.

Another freely available volume visualization tool, Hierarchical Volume Renderer (HVR) [14], was used to reveal fine-scale banding features within the simulated hurricane core region (not shown) as also observed. HVR, which was recently developed in the Laboratory for

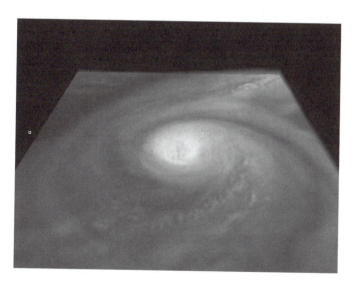

Figure 44.5 A volume visualization of a Hurricane Opal simulation was created using Renderman by NCSA's David Bock with land terrain generated by NCSA's Rob Stein. The volume rendered view is based on the model water vapor field. (See also color insert.)

Computational Science and Engineering at the University of Minnesota, was intended for creating animations from very large datasets.

The numerical model study of Hurricane Diana [6,7], which had landfall off the coast of North Carolina in 1984, recently pushed both computational and visualization capabilities by utilizing a data grid four times larger than that used in Hurricane Opal. Using a massively parallel IBM supercomputer, researchers ran an MM5 simulation of Diana for a period of two simulated days on 1060 × 1060 grid points at 1.2 km horizontal resolution with 37 levels in the vertical. The model used 552 processors for 2 days and produced 100 GB of output data. A top-down view of the simulated cloud water is shown in Fig. 44.6.

Diana is an interesting case study from several standpoints. For one, it constituted the highest resolution over the largest spatial domain that any MM5 researcher has tried to date. Postcomputational analysis and visualization of such a large model can be a very demanding process. In the example here, a Compaq 4100 cluster (8 nodes) was employed for MM5 preprocessing, two 8-processor SGI Origin 2000 systems were used for data postprocessing, and an 8-processor SGI Onyx system was used for 3D visualization. When the model runs, it saves data at periodic intervals, which in this case ended up being hourly. For the first time in this particular area of weather modeling, each snapshot of the model's state surpassed the 2GB level (32-bit addressing limit), and this outstripped the capabilities of a suite of software that had been used for a decade or more, including the venerable Vis5D application. Even after applications were adapted, the model output was generally manually subsetted so that analysis and visualization tools could perform at reasonably interactive rates.

The model produced a wealth of detail and, in retrospect, researchers would have liked to save output every 5 minutes in order to be able to study the detailed time evolution of the hurricane's eye formation. This would have resulted

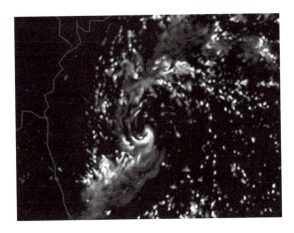

Figure 44.6 Grey-scale rendering of Diana simulation showing convective clouds from above the hurricane and to the east of the North Carolina coast. The eye is detectible near the middle of the image.

in 1–2 *terabytes* of data for the single simulation. One of the researchers involved with this study ventured that the overall analysis process was so difficult that rapid progress towards running weather models at this resolution could be hindered unless there are major advances in the visualization and analysis tools that the scientists have access to. This highlights the fact that scientific progress is not just about 3D visualization, but about overall workflow: data ingest, processing, regridding, analysis, 2D visualization, and, ultimately, 3D exploration and analysis. Tools of practice would benefit greatly from multiresolution techniques and parallel-rendering approaches that operate on the native model grid topologies, and effectively dealing with multiterabyte datasets will almost certainly require remote visualization capabilities.

44.5.2 Using Visualization to Intercompare Model and Observed Data: Typhoon Herb

As shown earlier in this section, one way to qualitatively judge the accuracy of a computer simulation is to visually compare data generated by a computer model with data collected by satellite, ground observation systems, or radar. In this example, observed typhoon data recorded at a WSR-88D radar installation in Northern Taiwan is compared with data generated by a computer model using 3D visualization techniques. A side-by-side comparison helped demonstrate how closely computer models are able to track storm systems and how well they can match the circulation and precipitation characteristics of severe weather events.

On July 31, 1996, Typhoon Herb roared across Northern Taiwan. High winds, floods, and landslides resulted in the loss of many lives and extensive damage to property, including heavy damage to a WSR-88D radar facility located 30 km east-northeast of Taipei. The facility, operated by Taiwan's Central Weather Bureau, collected reflectivity and velocity data as the typhoon approached, but it was forced to shut down as the eye of the storm drew closer and the threat of water damage to the radar electronics increased. High winds eventually

ripped the fiberglass panels from the radar instrumentation, but not before several hours of data were collected from the approaching storm.

Researchers later used MM5 to simulate this storm system and to compare the observed and model data. One frame of a side-by-side stereo 3D visualization is shown in Fig. 44.7. In this example, observed typhoon data recorded at a WSR-88D radar installation in Northern Taiwan are compared with data generated by a computer model. A combination of tools was used in the visualization process, including Vis5D, various data resampling packages, and Alias|Wavefront's Maya application for animation control and final rendering.

There were several challenges in producing the comparison due to differences in spatial resolution, differences in data quality, and differences in variable sets. For example, the observed data had a much higher resolution (1.5 km) than the model data (6.6 km). Several iterations of a low-pass filter were applied along each coordinate axis of the observed data to

Figure 44.7 Isosurface visualization of Typhoon Herb and radar return signal, with simulated data on the left and observed data on the right. The vertical axis is the height of the domain, and the domain is roughly over Taiwan. (See also color insert.)

Figure 44.8 A second isosurface visualization of Typhoon Herb and radar return signal, with simulated data on the left and observed data on the right.

smooth it and eliminate noise and artifacts. Also, as a result of how the radar beams project out in a conical pattern, the observed data contained areas of missing data, especially in the upper levels of the domain and directly above the radar site. The missing data resulted in a visible hole above the WSR-88D radar location and an absence of structure in the upper domain. Computer models, on the other hand, produce no missing data, and therefore there were no holes or gaps in the structure of the visualization produced from the MM5 data. Finally, there was also visual disparity because each dataset was composed of different variables. The visualization of the observed data depicted radar reflectivity as measured by Doppler radar, but the visualization of the model data depicted the snow, graupel (sleet), and rainwater as produced by MM5. Despite this, the results were qualitatively similar, but there were noticeable differences in the shape of the isosurface contours.

Even though the differences in spatial resolution, data quality, and variable sets resulted in significant disparity between the visualizations, when the datasets were animated in time and compared side by side, striking similarities were revealed. The circulation and the storm tracks were almost identical, and the microphysics and precipitation patterns compared very well. Both the observed data and the model data produced

an elliptical-shaped eye at the typhoon's center, and the rotation rates matched very closely. Also, in both cases, the visualizations revealed strong upslope conditions on the west side of the island, a condition that produced heavy rainfall, mudslides, and flooding.

This comparison of observed and model data reveals some of the practical challenges that must be addressed when comparing datasets from different sources. It also shows that even when two datasets differ in many respects, useful comparisons can still be made by employing 3D visualization and animation techniques to compare structure and motion characteristics.

44.5.3 Tornado Revealed

Advances in computational capability have enabled researchers not only to simulate severe storms with increased accuracy but also to begin simulating smaller scale features associated with these storms, such as tornadoes. The most damaging tornados are produced by supercell storms, and it has been the goal of some modelers to simulate the process of tornadogenesis within these storms. In order to do this most accurately, the models need to be run with grid spacings on 10–20 m, rather than the 1000 m used in the simulation of Hurricane Opal (Fig. 44.9). Such simulations have not yet been

Figure 44.9 Visualization of a developing tornado revealed through the use of thousands of circulating particles. (See also color insert.)

carried out. However, resolutions of 50–200 m have been achieved through nesting or horizontal grid stretching. One such storm simulation was carried out using the COMMAS nested grid nonhydrostatic numerical model by Wicker and Wilhelmson [35]. Horizontal resolution in the finest grid encompassing the tornado was 200 m, while the coarsest resolution was 1800 m. An initial storm evolved into a supercell, and shortly thereafter several tornadic events occurred. A special animation was created for use in OMNIMAX theaters for the film Stormchasers [24] that focused on the most spectacular of the simulated tornados, which lasted approximately 10 minutes, had ground relative wind speeds of more than 60 m/s (134 mph) near the ground, and had a 40 millibar pressure drop.

Isosurfaces of the growing thunderstorm reveal common shapes associated with supercell storms (dome, vault, and anvil). The viewer sees the early development of the precipitation region of the storm (the region typically seen by radar, which is somewhat different than that seen by an observer watching a real storm develop) from above the ground and to the east of the storm. An abstract prairie landscape with a blue horizon is used to anchor the audience with familiar real-world cues. The main source of lighting came from the west in a tone evocative of afternoon sunlight. This environment was designed to create the time and place in which a tornado can occur.

As the storm develops, the viewer begins to descend and move around the southern part of the storm in order to gain a different perspective and prepare to move closer to the tornado as it develops. The tornado is revealed through the use of thousands of particles after the lower part of the precipitation isosurface is removed (see Fig. 44.9). This is necessary in order to "see" the tornado, which is embedded in the precipitation region but near its edge. Other images from the complete animation can be found on the NCSA web site [21].

The images in the animation were computed from the approximately 40 GB of simulation data saved from the numerical simulation. The storm isosurfaces were rendered using the Wavefront Technology's Advanced Visualizer. There are actually two isosurfaces used in every image; they define surfaces very close to one another. This was done to soften the surface and to give a perception of depth. The particle representation and shading techniques were developed especially for this project at NCSA. The final frames were composed of up to four separate images: the prairie landscape, the shadow, the cloud, and the particles. Each image was produced at a resolution of 2048 × 1536 pixels.

A weightless particle moving with the flow field was represented as a view-aligned, texture-mapped polygon. The polygon started at the location of the particle for a given time and then stretched back for some specified duration to previous locations of the particle. The shading of a particle was based on a simple reflection/transmission model for a sphere. Each particle was shadowed by all the other particles between it and the light source. The goal of the shading technique was to make the volume of the tornado more prevalent than the individual particles.

44.5.4 Tornadic Storm Fest

Forecasting of tornadoes through direct numerical simulation is unlikely in the next decade because the models will not run faster than the actual weather when the billion or more gridpoints needed are employed. However, through the use of dynamically adjustable nested grids, mesoscale models like the ones discussed earlier will, in the future, run at 1–3 km horizontal resolution in predictive mode. In preparation for this, high-resolution research simulations are being conducted to study how well models can predict storm behavior. One example was presented in Fig. 44.4. Fig. 44.10 shows another example from a simulation carried out to study the development of storms on April 19, 1996 that produced 36 tornadoes in Illinois. The storm cells formed over eastern Missouri and moved into Illinois with some becoming supercells. The frame is from an animation done with IRIS Explorer, and it shows a series of storms revealed through the use of isosurfaces enclosing regions of precipitation, some of them supercells. At the surface, regions of high humidity are represented by green and those of low humidity by blue. Contours are also used to represent surface moisture in more detail. This image again illustrates the need to composite various visualization techniques to get a better integrated, quantitative understanding of storm development.

44.5.5 Seeing the Unseen: Visualizing Clear Air Turbulence

Most of us have read about clear air turbulence (CAT) or heard about it on the news: an aircraft is traveling along when suddenly it encounters a violent disturbance in the atmospheric flow. Results range from frightened passengers all the way to severely damaged airframes and sometimes crashes and loss of life. Better understanding of turbulence (in all of its many forms, not just atmospheric) is an enormous scientific challenge and an extremely lucrative area for advanced visualization because of the need to qualitatively study extremely complex flow behavior and structure. In the case of CAT, we have strong and obvious motivations for understanding the behavior of turbulence: the hope of being able to detect it better and ultimately predict and avoid it. The example shown in Fig. 44.11 depicts a simulation of a CAT scenario associated with a 1992 incident where a DC-8

Figure 44.10   Model results from the simulation of the April 19, 1996 tornado outbreak in Illinois. (See also color insert.)

cargo plane encountered severe clear air turbulence and lost an engine and about 6 meters of wing (the crew managed to land the plane safely). Scientists at NCAR and NOAA worked together to develop a better understanding of the event, beginning with an initial analysis of Doppler radar data that revealed horizontally aligned vortex tubes (HVTs), which are somewhat like tornadoes parallel to the ground. In order to try to understand the potential origins of turbulence, the Clark-Hall model was configured with a five-level nested grid where the largest domain roughly covered the United States and the finest domain covered a 24 km square region over Boulder, Colorado.

The Clark-Hall model uses a "zeta" coordinate system, where the vertical dimension is specified as a percentage of distance from the surface of terrain to the top of the domain in meters. Model data for the innermost domain (wind, temperature, vorticity, etc.) were resampled onto a Cartesian grid with an irregular vertical interval for use with Vis5D. Landsat-7 Thematic Mapper imagery was georeferenced to the model domain and texture-mapped onto the model's own terrain field. Vis5D was used to explore and study the data (with researchers, flight safety experts, and pilots) to identify good depictions of the interesting phenomena and to produce geometric

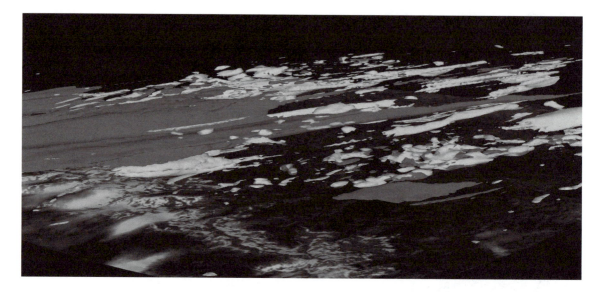

Figure 44.11 Simulation of clear air turbulence with jet stream in blue, enstrophy in red, and fast moving air parcels in yellow. The visualization was created using a combination of IDL, Vis5D, and the Persistence of Vision (POV) raytracer run on a 128-processor SGI Origin system. (See also color insert.)

scene descriptions that were ultimately rendered in parallel using ray-tracing techniques on a 128-processor SGI system.

There are several observations to make about this case study. The model results were interactively explored and animated for viewing with stereo/3D glasses. While this is almost always fun and interesting, in this case it proved to be invaluable for studying some of the complex structures that appeared in the data. There were some relationships that were extremely difficult to discern without them. The incorporation of remotely sensed data (i.e., the Landsat information) was awkward and accomplished manually, which points to the need for integrating Geographic Information System (GIS) capabilities and data interfaces into visualization tools. Interestingly, the researchers found that volume visualization for this study was not as interesting as more traditional isosurface approaches; they wanted to be able to communicate precisely what a given visual representation meant, and isosurfaces provided a more quantitative means

of accomplishing this. Like many of the case studies presented earlier, tools need to deal with the model data in its native coordinate system, and the sheer volume of the data underscores earlier comments about the need for multiresolution techniques and better rendering capabilities.

The turbulent processes that arose in this simulation were very complex, and the model results showed that the observed CAT derives from interactions between wavelike jet-stream disturbances and mountain-forced internal gravity waves. The work ultimately resulted in several publications [4,5] on a new form of turbulence, and the lead researcher indicated that insight into the complex problem would not have been possible without advanced visualization.

44.5.6 Visualizing Specialized Mesoscale Models: Wildfires

Wildfires are unpredictable, dangerous events that often lead to extensive property damage, severe ecosystem destruction, and loss of life. A number of research efforts are currently underway to improve the scientific understanding of wildfires, and these efforts typically combine observational programs (monitoring how real fires behave) and modeling efforts. In the example presented here, a specialized version [3] of the Clark–Hall model, mentioned earlier, was used to simulate wildfires by coupling a nested mesoscale simulation with a small inner domain where combustion occurs. The "combustion domain" included fuels, terrain, burnt areas, wind, smoke, and turbulent features along with a grid resolution of 60 m. Fig. 44.12 shows a grey-scale rendition of a volume and flow visualization developed as part of a study on how forest fires propagate by dispersing burning "brands." Fig. 44.13 shows another view that includes combustible ground cover, the enstrophy (a measure of rotation), and a volume rendering of temperature. This study used a combination of IDL, Vis5D, and ray-tracing software.

Most of the challenges discussed earlier relative to weather models hold here as well. This particular effort did not present a particularly large volume of data, but future research goals aim to refine the grid resolution down into the 1–10 m range, which will boost the data volumes substantially. This is also another domain where GIS data (e.g., ground cover, terrain) will need to be visualized along with model data. Visualizing fires is an area that could benefit from improved rendering capabilities for phenomena such as smoke and mixed volumetric, geometric, and flow visualization. Lastly, this is an example where visualization serves well to broaden public awareness of research: an image from this study was published in *Wired* magazine [18].

44.5.7 Weather Visualization for the General Public

Different approaches for visualization are relevant for different audiences: what a scientist needs in order to understand a phenomenon is often very different from what the general public, or a student, needs to understand a phenomenon. Weather is, of course, something that just about everyone is interested in, and one can observe progress even in the visualizations provided on the evening news. Fig. 44.14 shows a particularly nice example of this. The visualization approach was developed as part of a joint project between the German weather service (DWD) and a private company (ASK Innovative Visuals) with a target audience of television viewers. DWD runs a weather forecast model and supplies the data to ASK, which produces a time-evolving animation of the forecast. The idea here is to make it very realistic and understandable by the average viewer. Fig. 44.14 is a grey-scale rendition, so it's a little difficult to make everything out, but the scene includes European topography, satellite imagery, cities (shown as icons), rain, snow, and even snow accumulation on the ground. While this is shown on television to a general audience, it begins to hint at the possibility of constructing virtual environments where this level of realism is brought to bear in the rendering of clouds, precipitation, snow, ground cover, and man-made features.

Figure 44.12 Volume visualization of a forest fire and "burning brand" lofting (particles). Image courtesy NCAR. (See also color insert.)

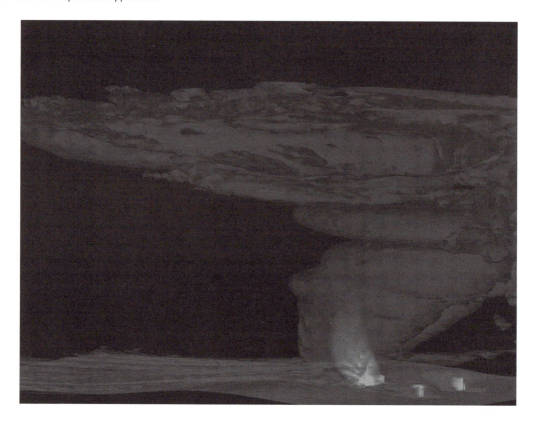

Figure 44.13   Visualization of turbulence and heat fields evolving in a fire model. Image courtesy NCAR. (See also color insert.)

Figure 44.14   Visualization of a weather forecast for the general public, showing snow, rain (streaks), and snow accumulation on the ground (light shaded areas). Image courtesy of DWD, German weather center, ASK, Inc.

44.5.8 An Experimental High-Resolution Global Atmospheric Simulation

Scientifically and societally, there is great interest in regional climate trends, and therefore, high-resolution model experiments are needed. In the US, production models are beginning to use T85 resolution ($\sim$ 140 kilometers horizontally), and recently US and Japanese researchers conducted experiments at T170 resolution ($\sim$ 70 km horizontally on a 512 $\times$ 256 Gaussian grid) using the CCM-3 (Community Climate Model Version 3) model [16], and 2D and 3D visualizations were subsequently produced. The experiment not only had a high spatial resolution but also was sampled hourly rather than on a monthly or seasonal average basis, which is more typical of climate work. A climate model simulates weather, which is averaged in order to evaluate the simulated climate. A 2D visualization approach was developed to reveal the weather patterns that the climate model integrates over the course of time. The image in Fig. 44.15 is a snapshot from a one-year-long animation sequence of data from the CCM3 model, sampled hourly. The main goal in producing this sequence was to develop a realistic-looking visualization of Earth with water vapor rendered to look like clouds (clouds are not water vapor). The rendering was created entirely in the data-processing and imaging space, but the technique was essentially the same as using volume rendering with a parallel projection and no lighting model. Interesting variations on this would include employing fractal models driven by underlying cloud data and more advanced rendering techniques for the clouds themselves.

The visualization was created using a suite of custom-developed data ingest and field-to-image conversion tools. Paralleling some of the challenges associated with developing effective volume visualizations, very careful tuning of the color and transparency transfer functions was required to highlight specific features of the model results. In the images shown here and in the color plates, detail in the tropics was emphasized. Other renditions were created to reveal detail and structure in the higher latitudes.

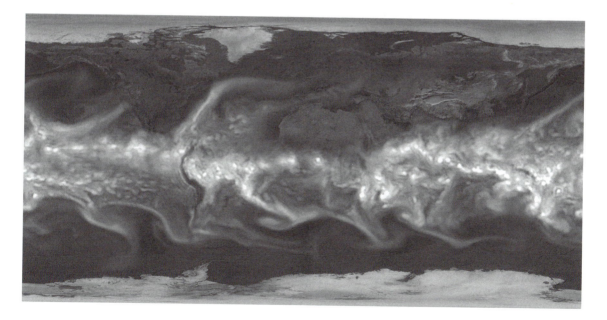

Figure 44.15 Visualization of water vapor from an experimental version of the Community Climate System Model (CCM3) at T170 resolution. Water vapor is rendered in white. Image courtesy NCAR. (See also color insert.)

44.5.9   A Glimpse into the Future of Climate Modeling

Some of the world's fastest computers are used to simulate weather and climate. Perhaps the premiere example of this is Japan's Earth Simulator Center, where a dedicated facility has been constructed and equipped with an enormous NEC supercomputer. It is currently being used to explore the frontiers of simulating various components of the Earth System, such as weather, climate, chemistry, and even earthquakes. Fig. 44.16 depicts a global climate model run at T1279 resolution [25]. The detail produced is stunning, and at roughly 10 km resolution, it is comparable to the resolution at which some weather models are being run today.

44.5.10   Towards Global Earth System Modeling

Relentless growth in computational power offers many opportunities for improving Earth System models, their predictive capability, and the overall understanding of our planet. As described above, increasing the spatial resolution of the model components is one important direction, especially with regard to the atmosphere and oceans. The ability to conduct "ensemble runs," where a small number of closely related experiments are conducted with minor changes, is another. Continued improvements in the formulation of all the various underlying processes are also important. As we look towards the future, however, perhaps the most exciting advances will be in the movement towards much more comprehensive models, where the spatial domain ultimately goes from the surface of the Earth through the stratosphere and beyond, and many more processes are coupled into the simulation framework. This is a movement into the realm of Earth System modeling, where the model-component diagram in Fig. 44.3 expands to include biogeochemical processes, the carbon cycle, atmospheric chemistry, and the interactions among these various processes and life on Earth. Ultimately these Earth System models will incorporate terrestrial and oceanic biological components such as forests and coral populations. Fig. 44.17 shows a frame from an animated visualization of a simulated hypothetical coral reef and its growth and evolution since

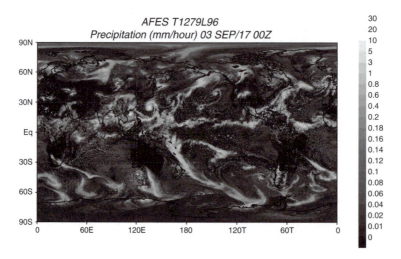

Figure 44.16   Precipitation from a T1279 global atmospheric model. Image courtesy of the Japanese Earth Simulator Center.

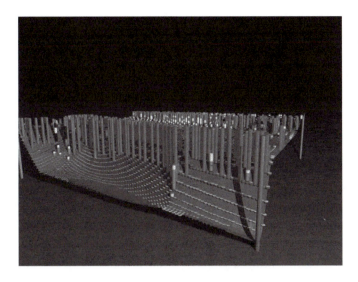

Figure 44.17   A snapshot of coral reef growth since the last glacial maximum. The height of the columns represents growth, and shading (coloring) indicates specific coral varieties. Image courtesy NCAR.

the last ice age. The visualization was created using custom data-processing tools and the POV ray-tracer. Some day such a biological component could be included as part of an overall Earth System model, where the biological system would respond and interact with simulated changes in the ocean temperatures, salinity, and chemistry.

By the end of the decade we can expect to see the atmospheric part of global Earth System models running at resolutions that will rival those of our mesoscale (weather) models of today. While the study of climate will continue to focus, by definition, on the statistical behavior of the climate system, the complex interplay among the various components (atmosphere, oceans, biological, etc.—many of these simulated at very high resolution) will provide many opportunities and challenges for new visualization work. Examples include the interactions between the oceans and atmospheric convection, the evolution of turbulence in oceans, the appearance and behavior of cloud systems, and the interplay of chemical and biological processes.

44.6   Visualization Challenges and Futures

Our ability to generate data through simulation and observation is growing very rapidly and, in many respects, this growth is outpacing our ability to explore, analyze, and ultimately understand the data. There are, of course, a lot of positive developments. Inexpensive PCs with very powerful graphics and large storage arrays make it possible to engage in advanced visualization on the scientific desktop. There are encouraging shifts towards standardized data formats and conventions, a trend that makes it possible to develop analysis and visualization tools that can operate on native data in a useful way. Fast wide-area networking provides dramatic improvements in accessing remote data. There is also a lot of exciting work underway in new display environments, collaborative capabilities, and web-based portals for data analysis and visualization. Unfortunately, most researchers who used advanced visualization to understand their data still use Vis5D, IDL, and a collection of other applications that are, in general, not nearly up to dealing with the volume and complexity of the simulation results commonly encountered today. In the remaining paragraphs, some of the future needs and interesting new developments are briefly covered.

44.6.1   Emerging Visualization Displays and Environments

As datasets become larger and more complex, new display technologies and visualization environments become increasingly interesting. Rapid progress in the development of new and less expensive display technologies opens up a number of interesting possibilities for enhanced display of scientific data. It was mentioned earlier that scientists in both weather and climate often run ensembles, or collections of related models. This turns out to be an area where researchers are faced with a formidable volume of data, and they may want to rapidly sift through it in order to prioritize particular instances for deeper quantitative study. Over the last several years, there has been a lot of activity aimed at developing display systems that offer much greater display real estate than the traditional monitor that most people have on their desk.

Tiled display walls provide a large-format environment for presenting high-resolution visualizations by coupling together the output from a collection of projectors. The NCSA has used its walls for viewing and studying very large images, such as those from GOES or MISR satellite instruments. Currently this includes 1 km visible GOES images at approximately 14,000 × 12,000 pixels per image and single pass, 250 m MISR images at approximately 80,000 × 9,000 pixels. The wall is located in a room where these images can be viewed in a classroom setting or by a small- to moderate-sized group studying them. Viewers can zoom and pan through the image much as they do with imaging software with single computer screen display.

Alternatively, the wall can be used to display a set of images or animations such as those from multiple simulations within an ensemble modeling study or of different fields from one simulation. Lock-step animation and rotation of these images with standard video display controls, along with collocation capabilities across all images, enhances viewing and understanding. For example, the results from simulations carried out using the WRF are displayed in 44.18. In this figure, 20 of the 40 tiles available

Figure 44.18   Tiled wall display of ensemble WRF model simulation. Image courtesy NCSA. (See also color insert.)

were used to display several fields from a small set of simulations in both 2D and 3D. The display software allows the individual windows to be sized and moved around the wall as shown. In this way, the researcher can order the images based on visual inspection. This type of display is important for MEAD, an NCSA Alliance Expedition that is focused on developing and adapting cyberinfrastructure for retrospective study of a variety of weather events including hurricanes, severe storms, and mesoscale ocean phenomena. The goal of MEAD is to integrate model computation (WRF atmospheric and ocean models), grid workflow management, data management, model coupling, data analysis/mining, and visualization so that the user can launch and study tens to hundreds of simulations distributed across the Grid and TeraGrid.

Virtual environments and interactive stereo have been fairly common in research environments for quite some time now, but they have not found significant penetration into academic science departments because of cost and com-plexity. The GeoWall Consortium [10] is changing this by developing a specification for an inexpensive but powerful stereo/3D visualization environment that can be easily and affordably deployed. There are now several dozen GeoWalls in various departments around the world, with active development underway to create imaging and visualization applications that work within. This nice development is made possible by advances in commodity display technology, inexpensive Linux-based PCs, and graphics cards that are very fast and offer an OpenGL interface such that existing applications may easily be ported into the environment. The stereo/3D version of Vis5D is being used as one of the primary data visualization tools for the GeoWall. Fig. 44.19 shows a GeoWall in use in a geology classroom.

NOAA's Forecast Systems Lab has recently developed their Science on a Sphere (SoS) environment, which uses commodity projectors to display a "tiled" image on a large, plastic sphere. This has been a very popular develop-

Figure 44.19 GeoWall in a geology labroom. Image courtesy Paul Morin, University of Minnesota. (See also color insert.)

ment and has been demonstrated at recent conferences and increasingly to schools and educational groups, as shown in Fig. 44.20. Just about any data that can be projected onto a map can be shown to good effect on the SOS. The SOS has been instrumented as a 3D display as well, using passively polarized projection.

44.6.2 Environments for Collaborative Knowledge Development

Visualization is one important piece in a large scientific workflow that encompasses theory, observation, simulation, an enormous amount of data processing, analysis (both visual and statistical, qualitative and quantitative), and, increasingly, collaboration with other scientists. The Access Grid (AG) [1] is an important new development in collaboration technology and is fundamentally changing how scientific research and many other endeavors are conducted. In its early incarnations, the AG provided an environment where people could see and talk to each other and share PowerPoint presentations. There is a desire and an opportunity to develop a much richer environment that encompasses visualization and analysis applications so that understanding data can be accomplished in a collaborative mode. Early work to incorporate

interactive 3D visualization into the AG has already begun [26] and is a focal point of new efforts in the realm of electronically mediated collaborative environments.

Scientific portals that combine visualization and collaboration are becoming increasingly interesting. Fig. 44.21 shows the AG collaborative environment hosting people along with a collaboration portal (CHEF, from the University of Michigan), which in turn hosts a web-based portal for scientific data. In this case, the portal is providing shared data access and visualization of vegetation and ecosystem data. Weaving useful analysis and visualization capabilities into collaborative frameworks appears to be an attractive opportunity. In many cases, broad multidisciplinary teams will be needed to make progress on big problems, like global climate change, and full-featured collaborative environments can facilitate scientific progress here.

44.6.3 Visualization Frameworks and Applications

Terascale computing—and eventually petascale computing—will offer myriad opportunities for new scientific discovery in weather and climate. Researchers need visualization tools that can

Figure 44.20   A classroom visualizing weather data on NOAA's Science on a Sphere (SoS) display. Image courtesy NOAA. (See also color insert.)

Figure 44.21 The AccessGrid, hosting the CHEF portal, hosting the NOAA Live Access Server (LAS), hosting data and visualization for the Vegetation and Ecosystem Mapping Project (VEMAP). (See also color insert.)

keep pace with advances in computational capabilities and models, but there is a growing consensus that they are falling behind. At present, there is an array of community and commercial software applications, each of which provides some subset of processing capabilities with varying levels of 2D and/or 3D visualization. There is marginal interoperability, little utilization of parallel computing capabilities, and an overall brittle handling of different data formats and models. Much of the available software is near the end of life relative to the terascale problems that are emerging in present-day research activities. Without substantial advances in our visualization tools, many Earth System researchers will encounter barriers to understanding their data. In some areas this is already occurring, and as more and more scientists make the transition to large-scale parallel models, the problem will escalate.

As powerful as new desktop systems are and will be, the flood of data that is a growing reality across the Earth System sciences will still dwarf them. The personal computing systems of researchers will need to be coupled in a distributed, collaborative mode to large-scale, parallel systems that effectively support the analysis of extremely large datasets. The research community needs applications that enable large-scale data processing and at the same time couple

quantitative analysis with exploratory visualization. These next-generation tools will need to employ much more ingenious approaches, including more sophisticated data models, multiresolution techniques, level-of-detail (LOD) views, hierarchical data representation, parallel processing and rendering algorithms, region-of-interest rendering, and effective distributed operation. Visualization software developers need a robust framework upon which to build viable applications and to integrate the most promising results of the computational science research advances. The underlying data framework not only needs to be very high performance, it needs to be exceptionally versatile with regard to geoscientific data.

In climate and weather research, one promising trend is a growing emphasis upon the sustainability of developing the large and incredibly complex models that the community uses. As a result, there is enhanced attention focused on the software engineering of these systems. In 2001, the European Union started the PRISM project—the Program for Integrated Earth System Modeling [27]. At about the same time in the US, NASA funded a new project called the Earth System Modeling Framework (ESMF) [8]. PRISM currently appears to be primarily focused on climate research, while ESMF aims to address core software infrastructure for global coupled climate models as well as weather models. Similar thrusts are required for the analysis and visualization area, and there are some obvious advantages in developing direct linkage between visualization applications/frameworks and the model frameworks themselves.

44.7 Summary

The weather and climate research community has been a major driver of the development of advanced visualization capabilities for several decades now. There are many exciting new developments in visualization technology and methods that have the potential to play a significant role in advancing scientific progress and discovery in an age of terascale (and petascale) computing during the next decade. However, new frameworks and applications are needed that can cope with the sheer volume and growing complexity of the data and provide well integrated quantitative and qualitative capabilities. Domain researchers and computer scientists have a fine opportunity to work together to realize this.

References

1. AccessGrid, http://wwwfp.mcs.anl.gov/fl/access-grid/
2. CDAT, http://esg.llnl.gov/cdat/
3. T. L. Clark, M. A. Jenkins, J. Coen, and D. Packham. A coupled atmospheric-fire model: convective Froude number and dynamic fingering. *Intl. Journal of Wildland Fire* 6:177–190, 1996.
4. T. L. Clark, W. D. Hall, R. M. Kerr, and D. E. Middleton. Observations and simulations of clear air turbulence: case study of coherent structures during the December 9, 1992 front range windstorm. *J. Atmospheric Science*, 1999.
5. T. L. Clark, W. D. Hall, R. M. Kerr, D. E. Middleton, L. Radke, F. M. Ralph, and P. J. Newman. On the origins of aircraft-damaging clear-air turbulence during the 9 December 1992 Colorado downslope windstorm: numerical simulations and comparisons with observations. *J. Atmospheric Science* 57:1105-1131, 2000.
6. C. A. Davis and L. F. Bosart. Numerical simulations of the genesis of hurricane Diana (1984). Part I: Control Simulation. *Monthly Weather Review* 129(8):1859–1881, 2001.
7. C. A. Davis and L. F. Bosart. Numerical simulations of the genesis of hurricane Diana (1984). Part II: Sensitivity of Track and Intensity Prediction. *Monthly Weather Review* 130(5):1100–1124, 2002.
8. Earth System Modeling Framework, http://www.esmf.ucar.edu
9. FERRET, http://ferret.pmel.noaa.gov/Ferret/
10. GeoWall, http://www.geowall.org
11. GrADS, http://grads.iges.org/grads/
12. HDF5, http://hdf.ncsa.uiuc.edu/ HDF5/
13. W. Hibbard, C. Rueden, S. Emmerson, T. Rink, D. Glowacki, D. Murrat, T. Whittaker, D. Fulker, and J. Anderson. Java distributed objects for numerical visualization in VisAD. *Communications of the ACM* 45(4):160-170, 2001.
14. Hierarchical Volume Renderer, http://www.lcse.umn.edu/hvr/hvr.html.
15. IDV, http://my.unidata.ucar.edu/content/software/metapps/index.html
16. T. J. Kiehl, J. J. Hack, G. B. Bonan, B. A. Boville, D. L. Williamson, and P. J. Rasch. The national center for atmospheric research community climate model: CCM3. *J. Climate* 11:1131–1149, 1998.
17. B. H. McCormick, T. A. DeFanti, and M. D. Brown. Visualization in scientific computing. *Computer Graphics* 21(6), 1987.
18. D. Middleton, T. L. Clark, and J. Coen. Forest fire on a small hill. *Wired*, June 2000.
19. MM5, http://www.mmm.ucar.edu/mm5/mm5-home.html
20. NCL, http://ngwww.ucar.edu/ncl/index.html
21. NCSA Stormchasers Imagery, http://redrock.ncsa.uiuc.edu/AOS/imax.html
22. netCDF, http://www.unidata.ucar.edu/packages/netcdf/
23. netCDF Climate and Forecast Metadata Convention, http://www.cgd.ucar.edu/cms/eaton/cf-metadata/index.html
24. NOVA/WGBH. *Stormchasers*. M.F. Films, 1995.
25. W. Ohfuchi, S. Shingu, H. Fuchigami, and M. Yamada. Dependence of the parallel performance of the atmospheric general circulation model for the earth simulator on problem size. *NEC Research and Development: Special Issue on High Performance Computing*, 44(1):99–103, 2003.

26. R. Olson and M. E. Papka. Remote visualization with Vic/Vtk. *Visualization 2000 Hot Topics*. Salt Lake City, Utah, 2000.

27. Program for Integrated Earth System Modeling (PRISM), http://prism.enes.org/main.html

28. Renderman, https://renderman.pixar.com

29. G. Romine. A high resolution numerical simulation of the landfall of Hurricane Opal. Unpublished paper, Department of Atmospheric Sciences, University of Illinois at Urbana-Champaign, 1995.

30. Simulation of Hurricane Opal, http://pampa.ncsa.uiuc.edu/~romine/opal.html

31. E. R. Tufte. *Visual Explanations: Images and Quantities, Evidence and Narrative*. Cheshire, CT, Graphics Press, 1997.

32. VIS5D, http://www.ssec.wisc.edu/_billh/vis5d.html

33. W. M. Washington and C. L. Parkinson. *An Introduction to Three-Dimensional Climate Modeling*. Mill Valley, CA, University Science Books, 1986.

34. R. B. Wilhelmson, B. Jewett, C. Shaw, L. Wicker, M. Arrott, M. Bajuk, C. Bushell, J. Thingvold, and J. Yost. A study of the evolution of a numerically modeled severe storm. *International Journal of Supercomputer Applications* 4(2):20-36, 1990.

35. L. J. Wicker and R. B. Wilhelmson. Simulation and analysis of tornado development and decay within a three-dimensional supercell thunderstorm. *Journal of Atmospheric Science* 52:2675-2703.

36. WRF, http://wrf-model.org

37. M. Xue, D.-H. Wang, J.-D. Gao, K. Brewster, and K. K. Droegemeier. The advanced regional prediction system (ARPS): storm-scale numerical weather prediction and data assimilation. *Meteorological and Atmospheric Physics* 82:139-170, 2003.

45   Painting and Visualization

ROBERT M. KIRBY
Scientific Computing and Imaging Institute
University of Utah

DANIEL F. KEEFE and DAVID H. LAIDLAW
Department of Computer Science
Brown University

45.1   Introduction

Art, in particular painting, has had clear impacts on the style, techniques, and processes of scientific visualization. Artists strive to create visual forms and ideas that are evocative and convey meaning or tell a story. Over time, painters and other artists have developed sophisticated techniques, as well as a finely tuned aesthetic sense, to help accomplish their goals. As visualization researchers, we can learn from this body of work to improve our own visual representations. We can study artistic examples to learn what art works and what does not, we can study the visual design process to learn how to design better visualization artifacts, and we can study the pedagogy for training new designers and artists so we can better train visualization experts and better evaluate visualizations. The synergy between art and scientific visualization, whether manifested in collaborative teams, new painting-inspired visualization techniques, or new visualization methodologies, holds great potential for the advancement of scientific visualization and discovery.

Scientific visualization applications can be loosely divided into two categories: expository and exploratory. In this chapter, we will focus on exploratory applications. Exploratory applications typically represent complicated scientific data as fully as possible so that a scientific user can interactively explore it. Per the scientific method, a scientist gathers data to test a hypothesis, but the binary answer to that test is usually just a beginning (Fig. 45.1). From the data come ideas for the next hypothesis, insights about the scientific area of study, and predictive models upon which further scientific advances can be made. Exploration of increasingly complicated and interrelated data become a means to that end.

One of the most complicated types of data that scientists wish to explore and understand comes in the form of multivalued, multidimensional fields. There are a number of visualization application areas that work with this type of data, including fluid dynamics and medicine. These data are difficult to understand because so many variables, or values, are of interest to the scientists. The challenge comes in understanding the correlations and dependencies between all of the values. For example, 2D fluid-flow simulations produce a 2D vector field that is sometimes time-varying. From this field, additional scalar, vector, and tensor fields are often derived, each relating to the others and providing a different view of the whole. Displaying such multivalued data all together is difficult, even in 2D. It requires showing six to 10 different values within a single image. For 3D fluid flow, the data exist within a volume. Representing a 3D vector field alone is a challenge; representing such a vector field together with derived scalar, vector, and tensor fields is an extremely difficult problem in visual representation.

We will begin with a narrative of some of our work in the area of representing multivalued

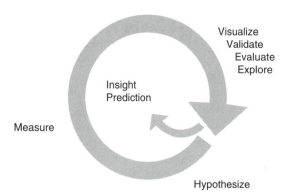

Figure 45.1 Exploratory scientific visualization is a specific instance of the scientific method. It begins with a hypothesis about some physical phenomenon. It continues with the collection of data that is expected to validate the model. Visualization of the data then helps in the validation of the hypothesis and in generating new hypotheses and insights, often iteratively.

data, illustrating more specifically some of the ways in which art can be brought to bear on scientific visualization. We will then give a broader survey of scientific visualization work that has been influenced by art, followed by a discussion of some of the open issues in this area, which will tie back to studying art, design, and art education.

45.2 Mimicking Artists: Strokes, Design, Critiques, and Sketching

Perhaps the most compelling reasons for visualization researchers to look toward oil painting, and to art in general, are the visual richness and visual effectiveness of the art that we see in our everyday lives. Paintings and reproductions are accessible in museums, in posters, in calendars, and on the web because there is a demand for them—they are broadly appealing and often convey a meaning or narrative to which we can relate.

Besides their obvious visual appeal, we can learn from art, artists, and art teachers what is visually compelling, what works for specific visual goals, how to tell if something is working, the process of visual design, and the process of learning visual design. Over the last several years we have been exploring each of these areas, and we will try to illustrate some of what we have learned with examples from those efforts.

45.2.1 Strokes

Some of our earliest attempts to borrow ideas from the art world began with trips to museums to view paintings and loosely emulate the techniques that we saw there. We were expertly accompanied by artist davidkremers, who guided us through the collections, showing us what he felt would be most relevant to our scientific visualization process. We absorbed ideas, transformed them to our digital medium, and generated a series of visual representations of multivalued data.

This stage was motivated by Meier's work to create painterly animations [25]. Her haystack image (Fig. 45.2) illustrates how brushstrokes can be layered to build up a compelling visual image. This same layering process is common in oil painting, although deconstructing it is more difficult.

In our early examples we used software that created data-driven visualization by layering "strokes" onto a 2D "canvas." Many visual characteristics of the strokes were set directly from the data, with the mapping under the control of the user. The images are data driven but are not guided by a particular scientific problem; they are based on experimentation with a new medium. Some of our experiments involved varying stroke shape, texture, color, and size; changing relationships among layers; and modifying the placement of strokes. Fig. 45.3 shows some examples.

In one example of a technique we worked to mimic, a painter uses a lightly loaded brush to paint over a dry, but previously painted, region. The texture of the underlying dry paint catches wet paint off the brush, leaving small textured bits of paint. Our version, shown in Fig. 45.4, used small strokes in a layer atop much larger ones, placed in only a small portion of the image, and in a contrasting color.

Figure 45.2   Meier layered strokes to build up computer paintings, much as painters layer their strokes to build up an oil painting. The stroke layers are shown here as they accrete. Here, the layers are organized around form and lighting, but other organizing principles can work in other contexts. (See also color insert.)

From this work we sensed potential. Some of the images created are visually compelling, and the sources of inspiration seem only touched upon. We were also excited by the potential to incorporate time into visualization design. By mapping some parts of data to quickly seen visual cues and others to visual cues that are seen less quickly, the order in which data is seen in a visualization may be controlled.

This early work also reminded us that there was no evidence that these images would have scientific value. While they were data driven in the sense that data values controlled many of the visual attributes in the images, they were not

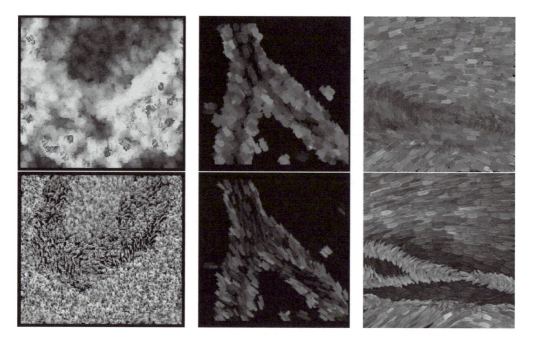

Figure 45.3 Several early "painterly" visualizations. We experimented with varying the visual representation of underlying data by changing stroke shapes, texture, color, size, and placement. The top and bottom image in each pair use the same underlying data. (See also color insert.)

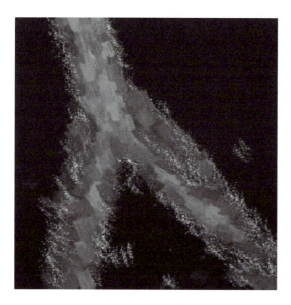

Figure 45.4 Scumbling, or lightly painting over an already-painted region, is an example of a painting technique we mimicked. We used small strokes to emulate the small bits of paint left behind. (See also color insert.)

targeted at solving a specific scientific problem. Indeed, measuring the effectiveness of visualization methods is a controversial and difficult problem.

It was also pointed out to us that design decisions sometimes have unintended consequences. For example, some of the painterly experiments had a sense of depth from regions that were lighter or darker. Qualities like this can be difficult for an untrained eye to notice but can dramatically affect our perception of a scene or data.

45.2.2 Designing Scientific Visualizations

As a follow-up to our early experimentation, we created a set of visualizations addressing three scientific applications using multivalued 2D imaging data: sections of 3D tensor-valued MR images [19], 2D fluid flow (and derived quantities) [16], and six-valued multiecho MR images [22]. We will discuss in this section the painting-related motivation behind the 2D flow application and also try to provide some insight into the issues with which we grappled.

In [16] we examined the scientific problem of understanding fluid flowing past a cylinder. The primary focus of the study was to visualize multivalued data. Within the study of fluid mechanics, many mathematical constructs are used to enhance our understanding of physical phenomena. Visualization techniques are often used as tools for developing physical intuition of these quantities. One important question, however, is: what do we visualize? To maximize their potential to cross-correlate information, scientists usually want to maximize the amount of comprehensible data presented in one visualization. For example, scientists often choose to examine derived quantities, such as vorticity, along with standard quantities, such as velocity and pressure, in an effort to fully understand the underlying process of fluid flow.

We illustrate the complexity of this issue by displaying velocity and vorticity simultaneously (Fig. 45.5). Vorticity is a classic example of a mathematical construct that provides information not immediately apparent in the velocity field. When examining only the velocity field, it is difficult to see that there is a rotational component of the flow in the far wake region of the cylinder (to the right). But when vorticity is combined with the velocity field, the underlying dynamics of vortex generation and advection are more apparent.

Although vorticity cannot be measured directly, its relevance to fluid flow was recognized as early as 1858 with Helmholtz's pioneering work. Vorticity as a physical concept is not intuitive to all, yet visualizations of experiments demonstrate its usefulness and hence account for its popularity. Vorticity is derived from velocity (and vice versa, under certain constraints) [27]. A function and its derivative are similarly related. Hence, vorticity does not provide any new information that is not already available from the velocity field, but it does emphasize the rotational component of the flow. The latter is clearly demonstrated in Fig. 45.5, where the rotational component is not apparent when one merely views the velocity.

Other derived quantities, such as the rate of strain tensor, the turbulent charge, and the turbulent current, can be of value in the same way as vorticity. Since examination of the rate of strain tensor, the turbulent charge, and the turbulent current within the fluids community is relatively new, few people have ever seen visualizations of these quantities in well known fluid mechanics problems. Simultaneous display of the velocity and the quantities derived from it is done both to allow the fluids researcher to examine these new quantities against the canvas of previously examined and understood quantities and also to allow the fluids researcher to accelerate her or his understanding of these new quantities by visually correlating them with well known fluid phenomena.

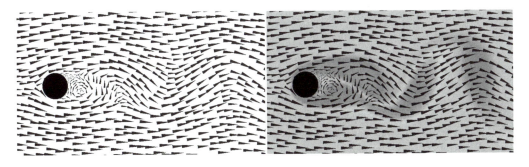

Figure 45.5   Typical visualization methods for 2D flow past a cylinder at Reynolds number 100. On the left, we show only the velocity field. On the right, we simultaneously show velocity and vorticity. Vorticity represents the rotational component of the flow. Clockwise vorticity is blue; counterclockwise is yellow. (See also color insert.)

In our painting-inspired visualizations of fluid flow, we sought representations inspired by the brushstrokes artists apply in layers to create an oil painting. We copied the idea of using a primed canvas or underpainting that shows through the layers of strokes. Rules borrowed from art guided our choice of colors, texture, visual elements, composition, and focus to represent data components. These ideas are discussed in more depth by Laidlaw et al. [18,19].

In one of our visual designs, shown in Fig. 45.6 (left), we wanted the viewer to read first velocity from the visualization, and then vorticity and its relationship to velocity. Because of the complexity of the second-order rate of strain tensor, we want it to be read last. We describe the layers here from the bottom up, beginning with a primed canvas, adding an underpainting, representing the tensor values transparently over that, and finishing with a very dark, high-contrast representation of the velocity vectors.

- **Primer**: The bottom layer of the visualization is light gray, selected because it would show through the transparent layers to be placed on top.

- **Underpainting**: The next layer encodes the scalar vorticity value in semitransparent color. Since the vorticity is an important part of fluid behavior, we emphasized it by mapping it onto three visual cues: color, el-

lipse opacity, and ellipse texture contrast (see the next category). Clockwise vorticity is blue, and counterclockwise vorticity is yellow. The layer is almost transparent where the vorticity is zero, but it reaches 75% opacity for the largest magnitudes, emphasizing regions where the vorticity is nonzero.

- **Ellipse layer**: This layer shows the rate of strain tensor and also gives additional emphasis to the vorticity. The logarithms of the rates of strain in each direction scale the radii of a circular brush shape to match the shape that a small circular region would have after being deformed. The principal deformation direction is mapped to the direction of the stroke to orient the ellipse. The strokes are placed to cover the image densely, but with minimal overlap. The color and transparency of the ellipses are taken from the underpainting, so they blend well and are visible primarily where the vorticity magnitude is large. Finally, a texture whose contrast is weighted by the vorticity magnitude gives the ellipses a visual impression of spinning where the vorticity is larger.

- **Arrow layer**: The arrow layer represents the velocity field measurements: the arrows point in the direction of the velocity, and the brush area is proportional to the speed. We chose a dark blue that contrasts with the

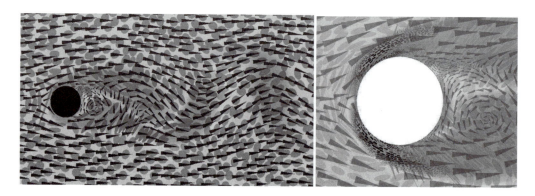

Figure 45.6 (Left) Visualization of 2D flow. Velocity, vorticity, and rate of strain (including divergence and shear) are all encoded in image layers. (Right) Additional values for turbulent charge and turbulent current for Reynolds number 100 flow are added to the visualization. A total of nine values are simultaneously displayed. (See also color insert.)

light underpainting and ellipses so that the velocities would be read first. The arrows are spaced so that strokes overlap end-to-end but are well separated side-to-side. This draws the eye along the flow.

- **Mask layer**: The final layer is a white mask covering the image where the cylinder was located.

In a second visual design, shown in Fig. 45.6 (right), we added two additional derived flow quantities: turbulent current, a vector, and turbulent charge, a scalar. The layers from the first design were changed to make the ellipses and arrows less contrasting and an additional layer added atop them.

- **Turbulent sources layer**: In this layer we encode both the turbulent charge and the turbulent current. The current is encoded in the size and orientation of the vector value just as the velocity in the arrow layer. The charge is mapped to the color of the strokes. Green strokes represent negative charge and red strokes represent positive. The magnitude of the charge is mapped to opacity. Where the charge is large, we get dark, opaque, high-contrast strokes that strongly assert their presence. Where the charge is small, the strokes disappear and do not clutter the image. For these quantities that tend to lie near surfaces, this representation makes very efficient use of visual bandwidth. The strokes in this layer are much smaller than the strokes in the arrow layer. This allows for finer detail to be represented for the turbulent sources, which tend to be more localized. It also helps the turbulent sources layer to be more easily distinguished from the arrows layer than in the previous visualization, where the stroke sizes were closer and, therefore, harder to disambiguate visually.

The use of these painting and design concepts helped us create a visual representation for the data that encoded all of the data for a more holistic understanding. The images in this 2D flow example, and in the other application areas

described elsewhere, simultaneously display six to nine data values while qualitatively representing the underlying phenomena, emphasizing different data values to different degrees, and displaying different portions of the data from different viewing distances. These qualities lead a viewer through the temporal cognitive process of understanding interrelationships in the data, much as a painting can lead a viewer through the visual narrative designed by the painter.

We were left with several observations and questions from this work. First, the images became more iconic than our early experiments as they were targeted at specific scientific applications. They have a less painterly look, as a result.

Also, once again, the question arises, how can we evaluate visualizations? User studies are a stock visualization answer, but we also wondered if we could borrow from art and art education in evaluating visualizations.

45.2.3 Art Education

Perhaps the most important educational tool to the art instructor is the critique, or crit for short. Art critiques can take many different forms, but in a typical classroom (a group-critique setting) they often involve displaying the work of all the students and then moving from piece to piece discussing and dissecting the visual decisions and techniques employed. The instructor running the critique usually has very specific goals in mind for the process and leads the discussion and criticism in a direction that culminates in the transmission of some design concept or theory to the students.

Critiques are a checkpoint along a path to creating visually refined imagery. They are almost always a part of a larger, iterative process. The lessons learned in a critique should carry on to future work, either in the form of a refinement of an initial design based on feedback, or as a lesson applied to a completely new design in the future. A critique that does not lead to new thought or work by the student is a failure.

Our initial experience applying the concept of critiques to visualization problems is encouraging. The critique framework, especially when expert artistic illustrators, designers, and instructors are involved, may offer an excellent alternative or complimentary approach to the traditional user studies used to evaluate visualizations.

Some of our experience with this framework came in the form of a class we taught in conjunction with Fritz Drury, head of the Illustration department at the Rhode Island School of Design (RISD). The class was composed half of RISD students and half of Brown University students. Our focus for a semester was to learn how to visually represent time-varying 3D fluid-flow data generated computationally. We started our exploration of visual representation with 2D fluid-flow problems and eventually

created visualizations of 3D flow that run in a CAVE virtual reality display. Throughout the process, students worked on weekly design assignments, and each week these were expertly critiqued to teach the class how to create successful designs from both visual and scientific standpoints. The importance of enabling a scientist to perform a specific task, such as locating areas of high vorticity within a flow, was a new constraint for RISD design students. The depth of understanding reached by the class on the effects of color, texture, form, and iconic representation upon human perception, particularly in virtual reality, was new territory for all the students.

Some results from a 2D flow visualization design assignment are shown in Fig. 45.7. Input

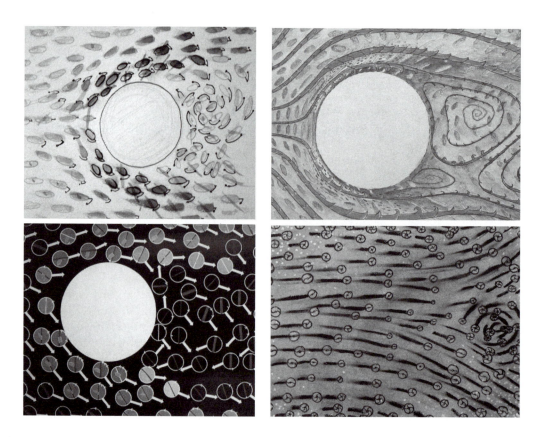

Figure 45.7 Students in a joint computer science/art scientific visualization class generated creative multivalued 2D flow visualizations. (See also color insert.)

from the critique of these works helped shape the students' future assignments as well as the final class projects in virtual reality. Based on feedback in weekly critiques, most designs in the class were eventually refined to the point that they were perceptually sound, were useful for scientific inquiry, and maintained a pleasing aesthetic.

One conclusion from this class experience is that, particularly in complicated, multivariate visualization problems, the design process is extremely important. When approaching these difficult visualization problems, it is rare for an initial visualization design to be visually coherent enough for scientists to use successfully. Iterating upon a visualization design takes time. Critiques can certainly help in this process.

Quickly sketching out design ideas and refining them again and again, each time evaluating them from the standpoint of the target audience's scientific goals, is one of the best ways to refine a design. For 2D visualization problems, this is often easily accomplished with traditional artistic tools. In fact, Fig. 45.8 shows some of the designs that attendees of the SIGGRAPH 2001 course entitled "Non-Photorealistic Rendering in Scientific Visualization" [10] were able to create in an afternoon. These were quick sketches made with paint, markers, etc. They represent experimentation and thinking outside the box. This type of effort is needed for complex visualization problems, the type to which art-based visualization methods are perhaps most suited. When we move to 3D visualization problems, quick sketches and visualization prototypes become much more difficult to make and critique.

45.2.4   Sketching and Prototyping for Virtual Reality

Currently, it can take a long time to advance from an initial art-based visualization idea sketched out on a piece of paper to a useful

Figure 45.8   Examples of 2D flow visualizations developed by students in a SIGGRAPH 2001 course. (See also color insert.)

visualization. One of the most time-consuming parts of this process is refining and iterating on the design; iteration is an essential part of the design process.

For some media, it is important to do much of the refinement step within the final medium itself. For 2D visualizations, this is less of a concern because traditional 2D media can do a fairly complete job of mimicking what can be seen on a computer screen. Thus, visualization designers can sketch out ideas, critique them, and revise them, all without the time-consuming step of implementing the design on the computer. However, for virtual reality (VR) and other 3D computer mediums, it is difficult to mock up and accurately critique a visualization without actually going through the trouble of programming it and experiencing it. Prototyping designs with traditional 2D and 3D artistic media is still beneficial for VR-based visualizations, but the insight that can be gained from critiquing these prototypes is limited because so many of our physical and perceptual cues change when we enter a virtual environment. Dimension, scale, colors, composition, interaction, and sense of presence all change as we move from a 2D representation of the idea to a complete virtual world.

Recently, we have started to take a new approach to prototyping and design in 3D that mimics a traditional 2D artistic process. The cornerstone of this approach is the CAVE-based VR system, CavePainting [15]. CavePainting uses a prop and gesture-based interface derived from a traditional oil-painting process to allow an artist to paint 3D forms directly in VR using a six degree-of-freedom tracker. While the interaction is based on painting techniques with which the artist is already familiar, the resulting "paintings" are a form of zero-gravity sculpture that bears little resemblance to a flat oil painting. Nevertheless, the quick, loose, stroke-based style of CavePainting makes it an excellent candidate for testing the feasibility of extending painting inspired visualization techniques to 3D problems and prototyping 3D visualization designs.

Through use of this tool, designers have been able to refine 3D visualization techniques quickly from within VR. The immediate advantage of this approach is that designers can visually critique a Cave-based visualization during the early stages of design. At this point in the process, even dramatic changes to the approach are easy to make. In our experience, design changes are often discussed and sketched out in 3D *during* a critique. Our vision for this approach to visualization design is that the ability to more quickly produce and iterate on designs within VR will decrease the time that it takes us to converge on scientifically useful visualizations.

This vision has played out in some of our initial work with the visualization class described above. As we continue to develop this prototyping tool and achieve a tighter coupling with scientific needs, we anticipate that prototyping designs in VR will allow us to spend much more time iteratively designing for VR visualizations and less time implementing complex visualization approaches that eventually prove to be less perceptually sound and scientifically useful than originally planned.

We further explore some of the issues raised in this section after providing historical perspective in the next section.

45.3 Historical Perspective: The Connections Between Art and Science

We now present a historical perspective on the connections between art and science, with particular emphasis on the efforts that have been made over the last 10 years to unite scientific visualization with other visual science disciplines. This section is by no means comprehensive; our goal is to provide a broad overview of the current stream of momentum from which painterly methods have derived over the past 20 years or so.

We partition this section into two subsections, a conceptual history and then practical connections between art and science. The former traces the steady infusion of artistic ideas and concepts into the scientific visualization community, whereas the latter presents current applications, both explicit and tacit, of painterly concepts in the development of visualization methodologies.

45.3.1 History of Art-Related Scientific Visualization

For at least the last six centuries, artists have striven to develop methods for distilling complex scene information into oil-painting representations. Some of this work was even directed at scientific topics, including astronomy and fluid flow. Within the last 20 years, there has been a renewed recognition that concepts from art and visual disciplines are not orthogonal to the goals of scientific visualization. Victoria Interrante succinctly presents the similarities and differences between visualization and art [10]. She states, "Visualization can be viewed as the art of creating a pictorial representation that eloquently conveys the layered complexity of the information in a complicated dataset." In the same article, however, she also emphasizes how visualization and art are different:

> "Visualization differs from art in that its ultimate goal is not to please the eye or to stir the senses but, far more mundanely, to communicate information—to portray a set of data in a pictorial form that facilitates its understanding. As such, the ultimate success of a visualization can be objectively measured in terms of the extent to which it proves useful in practice. But to take the narrow view that aesthetics don't matter is to overlook the complexity of visual understanding."

Early pioneers in this field, such as Donna Cox, who holds positions in both the School of Art and Design and the National Center for Supercomputing Applications at the University of Illinois, Urbana-Champaign, understood the potential of bringing scientists and visual design artists together. In 1987 Cox developed the concept of "Renaissance Teams," a team of domain experts and visualization experts whose goal was to determine visual representations that both appropriately and instructively presented domain specific scientific data.

In her 1995 essay "Art, Science" Vibeke Sorensen, professor and founding chair of the Division of Animation and Digital Arts in the School of Cinema–Television at the University of Southern California, alludes to the necessity of such "Renaissance Teams" to effectively counter the divisional chasm between artistic and scientific disciplines that has been caused by specialization. She argues that in the mind of most scholars, the ideal of the artist–scientist as an integrated, educated individual culminated in Leonardo da Vinci, who represents the union of artist and scientist. Although considered by some to be the epitome of the artist–scientist combination, the da Vinci ideal was soon lost to specialization. As our quest for knowledge produced a plethora of different subfields of science, the communication between different disciplines disintegrated, and in particular the ties between art and science were severed in the name of scientific objectiveness. Sorensen, however, asserts in her published articles on art and science her strong conviction that artists have an important role to play in the further development of science and technology. In particular, the means of restoring the ideal artist–scientist is through interdisciplinary research collaborations in which there is a synergy of many different disciplines, scientific and artistic.

There have been several attempts to foster this cultural crossover through panels and workshops. For instance, in 1998, David Laidlaw organized a panel at IEEE Visualization 1998 entitled "Art and Visualization: Oil and Water?" [23] whose purpose was to explore such questions as "How can artistic experience benefit visualization?" and "What artistic disciplines have the most to offer?" J. Edward Swan organized a panel at IEEE Visualization 1999 entitled "Visualization Needs More Visual Design!" [30], the purpose of which was to argue two main points: that utilizing visual design may be difficult but is important for visualization, and that, in general, the scientific community needs to work harder to tap into the centuries' worth of design knowledge that exists in fields such as art, music, theater, cartography, and architecture. In 2001, Theresa-Marie Rhyne organized a panel at IEEE Visualization 2001 entitled "Realism, Expressionism, and Abstraction: Applying Art Techniques to Visualization" [26], which explored the artistic transition between realism, expressionism, and abstraction and attempted to examine if such a progression also exists within the field of scientific visualization. One conclusion of that panel, articulated by Chris Healey, is that "the appropriate use of perceptual cues can significantly enhance a viewer's ability to explore, analyze, validate and discover." In that same year, two SIGGRAPH 2001 courses were dedicated to artistic topics. Sara Diamond organized a class entitled "Visualization, Semantics, and Aesthetics" and Chris Healey organized a class entitled "Nonphotorealistic Rendering in Scientific Visualization," both of which further explored the connection between scientific visualization and artistic sciences. At a different forum, Felice Frankel, a research scientist in the School of Science at the Massachusetts Institute of Technology (MIT), organized what has been referred to as a groundbreaking conference at MIT entitled "Image and Meaning: Envisioning and Communicating Science and Technology," which was an initiative to promote new collaborations among scientists, image experts, and science writers. Her new book captures some of the excitement of the conferences [7]. The following year, at SIGGRAPH 2002, Kwan-Liu Ma organized a course entitled "Recent Advances in Non-Photorealistic Rendering for Art and Visualization" whose expressed purpose was to give a concise introduction to nonphotorealistic rendering in the context of generation of artistic imagery and perceptually effective scientific visualization. Along the same lines, Non-Photorealistic Animation and Rendering (NPAR) in 2002 had a section specifically devoted to painterly rendering. Though the section was not limited to scientific visualization, its focus was on the exploration of interjecting painterly ideas into the visualization process.

Interest in collaboration between the arts and science has not remained confined to conferences and workshops; it has also spilled over into the archival publication realm. Laidlaw published [18] an article entitled "Loose, Artistic 'Textures' for Visualization" in which he encouraged the scientific community to search beyond what perceptual psychologists understand about visual perception into the fundamental lessons that can be learned from art and art history. Herman and Duke, in their article entitled "Minimal Graphics" [5], explored what can be learned from artistic traditions with respect to representing only salient features in a visualization. Taylor, in his article entitled "Visualizing Multiple Scalar Fields on the Same Surface" [31], reviewed and augmented with his own work ideas for visualizing multivalued data fields built upon artistic ideas. This small sampling is meant not to be all-inclusive, but rather to show that mainstream publishing venues are also seeing the wave of the collaborative mixing of art and science.

In summary, over the past 20 years there have been many efforts to, as Sorensen describes, resurrect the artist–scientist combination found in da Vinci. In our modern times, the process of scientific investigation often requires extensive specialization into the nuances of one particular field of discovery, making a da Vinci-like combination of the artist–scientist in a single personage an extremely difficult, yet worthwhile, goal [36]. In today's world, the synergistic interdependence of "Renaissance Teams," in which experts from many different disciplines combine their efforts, offers the most likely means for achieving a productive fusion of art and science. Slowly but surely this message is being disseminated through conference panels, workshops, and publications.

45.3.2 Practical Connections Between Art and Science

We now present three areas in which, whether explicitly or tacitly, ideas from painting have been applied to scientific visualization. We categorize these areas as multivalued data visualization, flow visualization, and computer graphics painting. Again, our purpose is not necessarily to provide a comprehensive listing of all scientific visualization efforts that could be classified as exhibiting painterly themes, but rather to illustrate the point that scientific visualization as a discipline has been attempting to answer some of the same questions as other visual art disciplines, namely, how to effectively present information in a form that is comprehensive, yet uncluttered.

45.3.2.1 Multivalued Data Visualization

Hesselink et al. [11] give an overview of research issues in visualization of vector and tensor fields. While they describe several methods that apply to specific problems, primarily for vector fields, the underlying data are still difficult to comprehend; this is particularly true for tensor fields. "Feature-based" methods, i.e., those that visually represent only important data values, are promising.

Statistical methods such as principal component analysis (PCA) [14] and eigenimage filtering [37] can be used to reduce the number of relevant values in multivalued data; this is often a worthwhile tradeoff. In reducing the dimensionality, these methods inevitably lose information from the data. The approach taken in the fluid-flow example presented earlier complements these data-reduction methods by increasing the number of data values that can be visually represented.

Different visual attributes of icons can be used to represent each value of a multivalued dataset. Haber and McNable [8] mapped temperature, pressure, and velocity of injected plastic to geometric prisms that sparsely cover the volume of a mold. Similarly, Chernoff [3] mapped data values to icons of faces; features like the curve of the mouth or size of the eyes encoded different values. In both cases, the icons capture many values simultaneously but can obscure the continuous nature of fields. A more continuous representation using small line

segment-based icons shows multiple values more continuously [6].

Layering has been used in scientific visualization to show multiple items; Interrante et al. show [12,13] surfaces with transparent stroked textures without completely obscuring what is behind them. The layering we presented earlier in the fluid-flow example is more in the spirit of oil painting, where layers are used more broadly, often as an organizing principle.

45.3.2.2 Flow Visualization

A number of flow-visualization methods display multivalued data. The examples by Max et al. [24] and Crawfis et al. [4] combine surface geometries representing cloudiness with volume rendering of arrows representing wind velocity. In some cases, renderings are also placed on top of an image of the ground. Unlike our 2D examples, however, the phenomena are 3D and the layering represents this third spatial dimension. Similarly, van Wijk [34] uses surface particles, or small facets, to visualize 3D flow: the particles are spatially isolated and are again rendered as 3D objects.

A "probe" or parameterized icon can display detailed information for one location within a 3D flow [35]; it faithfully captures velocity and its derivatives at that location, but it does not display them globally.

Spot noise [33] and line integral convolution [2] methods generate texture with structure derived from 2D flow data; the textures show the velocity data but do not directly represent any additional information, e.g., divergence or shear. Van Wijk [33] mentions that spot noise can be described as a weighted superposition of many "brushstrokes," but he does not explore the concept. The method presented in the previous fluid-flow example takes the placement of the strokes to a more carefully structured level. Of course, placement can be optimized in a more sophisticated manner, as demonstrated by Turk and Banks [32].

45.3.2.3 Computer Graphics Painting

Haeberli [9] was the first to experiment with painterly effects in computer graphics. Meier

[25] extended the approach for animation and further refined the use of layers and brushstrokes characteristic for creating effective imagery. Both of these efforts were aimed toward creating art, however, and not toward scientific visualization. Along similar lines, other researchers [28,38,39] used software to create pen-and-ink illustrations for artistic purposes. The pen-and-ink approach has successfully been applied to 2D tensor visualization [29]. In Laidlaw et al. [20], painterly concepts were presented for visualizing diffusion tensor images of the mouse spinal cord.

45.4 Open Issues

The previous sections suggest some open issues, which we will discuss in more detail here.

45.4.1 Evaluation

One of the most difficult aspects of developing new visualization methods is evaluating their success, and this is certainly true for methods that are motivated by painting and art. For many exploratory applications, the best measure of success is the acceleration of scientific discovery and insight in other disciplines, but that is virtually impossible to measure quantitatively, even with a crystal ball. Scientific advances are dependent on many factors, and visualization tools are only one. Even a significant increase could be lost in the variance caused by the others.

We must revert to less direct measures. These may be judgments about an algorithm's elegance, simplicity, or speed. They may be about the accuracy or speed of a group of users in performing specific, well defined tasks. Or they may be about a visualization's aesthetics, ability to display certain features in data, or appeal to domain scientists.

The first type of algorithmic measure is well understood in computer science. We know elegance and simplicity when we see it, and we can easily measure speed and talk about how it scales with problem size. While these are im-

portant, their connection to how well a tool will advance scientific discovery is tenuous, at best. There has been many an algorithm that has scaled nicely with problem size and yet provided no new insight into the scientific problem that was being visualized.

The second type of measure, which is results from performance-based user studies, is appealing because such results are both quantitative and objective [17]. For example, for six methods of visualizing 2D fluid-flow data, we measured user accuracy and performance in locating critical points in 2D flow, identifying their types, and visually creating integral lines [21]. With the results, we compared the six methods and drew some conclusions about which features of each may have accounted for good performance on these specific tasks. On the other hand, a leap of faith is required to generalize these results more broadly to other visualization methods, particularly exploratory ones, or even to other tasks. Finding features faster and more accurately could speed the advance of science, but we cannot know for certain. One clear contribution of these kinds of measures is the very explicit set of visualization goals that must be defined in order to perform tests.

The third type of measure is more subjective. Here we might ask domain scientists whether they like a method, or appeal to reviewers to judge whether a certain feature is adequately represented visually and whether that is important. This tends to be faster to evaluate than more formal performance-based user studies and can often evaluate larger conceptual advances, but at the cost of some quantization and objectivity, and often with implicit assumptions. For example, domain scientists may understandably be biased against unfamiliar methods, even if the unfamiliar methods will be more effective after a learning period. This kind of measure may come the closest to addressing our original question about advancing science.

Each type of measure has its place. What relates the second and third types is the choices that must be made about the important visualization goals to target and the specific population to evaluate them. With explicit design goals, the third type of measure may be particularly valuable. In fact, this kind of evaluation is very similar to art critiques and has the potential to advance our field more quickly. They can provide measures of new methodology. They can help educate both visualization researchers and designers. They can also help clarify visualization goals. They should be used more broadly and incorporated into what we teach our visualization students.

45.4.2 Visualization Goals

An essential step in critiquing or evaluating visualization methods is defining explicit visualization goals. Too often visual appeal, or even glitz, is confused with effectiveness. Only explicit goals can be effectively evaluated.

Defining visualization goals is an iterative process and should be driven by the underlying scientific applications [1]. As our understanding of a scientific problem moves forward, so will our design goals for visualization methods to address that problem. Our understanding of visualization will also help us to bring effective methods from one scientific domain to bear on others.

It is important to understand that different scientific questions will imply different visualization goals, sometimes contradictory goals. No one visualization method is right. Some people claim that more is better. This is likely to be true for some kinds of exploration, but for expository visualizations, "less is more" is more likely true.

45.4.3 Design, Engineering, and Science Collaborations

Designers, engineers, and scientists are brought together because their skills and their disciplines can benefit from collaborations. For scientists, the benefit of collaboration is the potential for increased scientific understanding that can result from clearer, more perceptually sound visualizations. Artists hold one key to making

these visualizations a reality. For artists, the win in scientific visualization collaboration comes in many forms. First, working with scientific visualization opens the door to working with a variety of new media. Virtual reality, volume rendering, and other advanced computer graphics techniques are just beginning to migrate out of the graphics research community. Through visualization research, artists have the opportunity to be at the forefront of learning, working with, and even influencing recently created computer media. As illustrators, artists are also drawn to visualization problems because of the complexity of the situations that they represent. These types of problems are exciting because they push theories of visual representation to their limits. In addition to these factors, art educational institutions are beginning to become interested in scientific visualization collaborations because of the potential job opportunities that may be available for their students in the future. As the embrace of artistic insight continues to grow within scientific fields, we will develop a need for a new generation of artists that are adept at understanding and interacting with scientists and that specialize in illustrating the new scientific phenomenon that our technology helps us to explore.

While there is often some overlap in critical knowledge and techniques within design, art, engineering, and science, the terminology, goals, and methods of each are often as different as they are advanced. In scientific visualization, collaborative efforts require insight, communication, and education from all those involved.

45.4.3.1 Designer Education

The first area for designers to master when applying their skills to visualization problems is the new media that they may be using. Computer graphics in some form is now common at most design schools. In our experience, most potential design or illustration collaborators are familiar with programs such as Adobe Photoshop and occasionally a 3D modeling package. However, many of the visualization approaches in which designers can be most helpful to scientists today utilize more recent computer graphics techniques, such as volume rendering or virtual reality environments. Many basic design principles transcend the differences between various media, but clearly some time is needed for designers to experiment and eventually become proficient within a new medium.

Prototyping systems, such as the CavePainting-based virtual reality system described in Section 45.2.4, offer a transitional tool for designers. Designers are given an intuitive interface for creating VR worlds that can be targeted towards an artistic purpose or a scientific design. This allows for experimentation and gives designers a chance to learn the properties and limitations of a medium that they might not have without becoming proficient graphics programmers. There is much room for experimentation here in creating tools for quickly iterating on complicated interactive 3D visualizations.

In addition to learning how to use new media, designers must also learn the language and goals of their collaborators' disciplines. Understanding the scientific goals behind a visualization is the most important element for designers to grasp. It is nearly impossible to create a good visualization when you do not know what you are trying to show. This does not mean that the designer needs to be an expert in the scientific field. This is an unrealistic goal, but designers must be prepared to work with scientists to understand their goals and needs. This can be a difficult process, as the languages of the two disciplines are often quite different. For example, to a scientist looking at a point in a visualization, "value" means 10 m/s, a measurement of an experimental quantity. To an artist, "value" means the lightness or darkness of the region. Even simple conversations can become exercises in creating a common language of communication.

Cross-discipline initiatives, such as the Brown University and RISD cross-registered course, "Interdisciplinary Scientific Visualization," and RISD's newly created program in digital media,

will help to tighten the threads connecting the art world and the visualization community. These ventures, and similar ones at other institutions, will help to develop a language for collaboration and teach scientists, engineers, programmers, and artists to understand each others' goals and work together, as in Donna Cox's renaissance teams, to realize their designs.

45.4.3.2 Engineering and Scientific Education

As for designers, it is important for scientists, engineers, and programmers to not only master the new media that computers provide but also understand the scientific goals behind the visualization. The mastery of computer media should cover potential uses of current hardware and software solutions. It is also important for the computer experts in a collaboration to provide tools to other collaborators that they can use. This may be as simple as providing digital or physical printouts of imagery. It may be as complex as a virtual reality prototyping system. It is imperative that engineers and programmers find the means for including scientists and designers in the design loop. Technological barriers often make this difficult. However, any visualization collaboration will be enhanced by quickly establishing a means for overcoming the obstacles to communication and design input presented by differences in computing facilities and experience.

Finally, it is critically important for scientists to appreciate design and the aesthetic sense that designers have developed through their training and experience. This leads to a recognition of the potential that design has for furthering scientific discovery, a necessary ingredient for a successful collaboration. Often, this appreciation is best accomplished through experience in artistic projects and classes.

45.4.3.3 Education and the Renaissance Person

Most of the scientific visualization approaches we have discussed up to this point involve significant interdisciplinary collaboration by multiple people. It is interesting to note that what this approach strives to create through collaboration is the equivalent of a Leonardo da Vinci: a scientist and artist acting as one. Artistic insight feeds into and illustrates scientific discovery, while scientific discovery pushes the limits of artistic representation and understanding. In a sense, there is a continuum between science and art, and each individual spans some portion of that continuum. The more that one learns about the other's field, the more of the continuum one covers. As scientists learn more about design and art through collaborations, classes, and experience, they break down the barriers between the two disciplines, develop a new visual language and understanding, and make it easier for the collaborative processes to succeed. The same is true for artists and designers. As they come to understand science and its goals, they become, more and more, renaissance people, spanning the entire continuum. Perhaps only a very few will reach da Vinci status, but the future collaborations of all who strive to understand their collaborators' fields will be enhanced by their increased knowledge.

As interdisciplinary initiatives continue to grow in universities and research settings worldwide, we are beginning to see a change in the way science and art are taught. There is a tighter bond between the two and a greater appreciation for how the two disciplines can work together to help achieve the goals of each. By structuring our teaching to embrace this principle, we have the ability to foster a new generation of renaissance people and skilled collaborators.

45.5 Summary

In this chapter we have narrated some of our own experiments with merging concepts from art and design into the scientific visualization process, particularly for exploratory applications that work with multivalued data. We have also surveyed related work to give some context for others aiming to continue explorations into the synergy between these two disciplines. It is clear to us that there remains much visualization knowledge to mine from the world of painting, art, and design. Some of this knowledge is about visual representations, but there are design and pedagogical components as well that will play a role in educating visualization researchers and in evaluating visualization methods. Collaboration in the form of renaissance teams and the development of renaissance scholars will advance our field, and tools that amplify the output of designers by better leveraging their design capabilities without taxing their stamina and patience will be critical to this advancement.

Acknowledgments

We would like to acknowledge Prof. George Em Karniadakis and the CRUNCH group for support with fluid computations. All fluid computations were performed using the spectral/hp element code Nektar.

This work was supported by NSF (CCR-96-19649, CCR-9996209, CCR-0086065) and NSF (ASC-89-20219) as part of the NSF STC for Computer Graphics and Scientific Visualization; and the Human Brain Project with contributions from the National Institute on Drug Abuse, the National Institute of Mental Health, and the National Institute on Biomedical Imaging and Bioengineering.

References

1. F. P. Brooks, Jr. The computer scientist as a toolsmith II. *Communications of the ACM*, 39(3): 61–68, 1996.
2. B. Cabral and L. C. Leedom. Imaging vector fields using line integral convolution. *Computer Graphics (SIGGRAPH '93 Proceedings)*, 27:263–272, 1993.
3. H. Chernoff. The use of faces to represent points in k-dimensional space graphically. *Journal of the American Statistical Association*, 68(342):361–368, 1973.

4. R. Crawfis, N. Max, and B. Becker. Vector field visualization. *IEEE Computer Graphics and Applications*, 14(5):50–56, 1994.

5. D. Duke and I. Herman. Minimal graphics. *Computer Graphics and Applications*, 21(6), 2001.

6. R. F. Erbacher, G. Grinstein, J. P. Lee, H. Levkowitz, L. Masterman, R. Pickett, and S. Smith. Exploratory visualization research at the University of Massachusetts at Lowell. *Computers and Graphics*, 19(1):131–139, 1995.

7. F. Frankel. *Envisioning Science: The Design and Craft of the Science Image*. MIT Press, 2002.

8. R. B. Haber and D. A. McNabb. Visualization idioms: A conceptual model for scientific visualization systems. *Visualization in scientific computing*, pages 74–93, 1990.

9. P. E. Haeberli. Paint by numbers: Abstract image representations. *Computer Graphics (SIGGRAPH '90 Proceedings)*, 24:207–214, 1990.

10. C. G. Healey, V. Interrante, davidkremers, D. H. Laidlaw, and P. Rheingans. Nonphotorealistic rendering in scientific visualization. *Course Notes of SIGGRAPH 2001*, Course 32, 2001.

11. L. Hesselink, F. H. Post, and J. J. van Wijk. Research issues in vector and tensor field visualization. *IEEE Computer Graphics and Applications*, 14(2):76–79, 1994.

12. V. Interrante, H. Fuchs, and S. M. Pizer. Conveying the 3D shape of smoothly curving transparent surfaces via texture. *IEEE Transactions on Visualization and Computer Graphics*, ISSN. 3(2):1077–2626, 1997.

13. V. L. Interrante. Illustrating surface shape in volume data via principal direction-driven 3D line integral convolution. *SIGGRAPH 97 Conference Proceedings*, pages 109–116, 1997.

14. K. Jain. *Fundamentals of Digital Image Processing*. Prentice Hall, 1989.

15. D. Keefe, D. Acevedo, T. Moscovich, D. H. Laidlaw, and J. LaViola. Cavepainting: A fully immersive 3D artistic medium and interactive experience. *Proceedings of ACM Symposium on Interactive 3D Graphics 2001*, pages 85–93, 2001.

16. M. Kirby, H. Marmanis, and D. H. Laidlaw. Visualizing multivalued data from 2D incompressible flows using concepts from painting. *Proceedings of IEEE Visualization 1999*, pages 333–340, 1999.

17. R. Kosara, C. G. Healey, V. Interrante, D. H. Laidlaw, and C. Ware. Thoughts on user studies: Why, how, and when. *Computer Graphics and Applications*, 2003.

18. D. H. Laidlaw. Loose, artistic "textures" for visualization. *IEEE Computer Graphics and Applications*, 21(2):6–9, 2001.

19. D. H. Laidlaw, E. T. Ahrens, D. Kremers, M. J. Avalos, C. Readhead, and R. E. Jacobs. Visualizing diffusion tensor images of the mouse spinal cord. *Proceedings of IEEE Visualization 1998*, pages 127–134, 1998.

20. D. H. Laidlaw, E. T. Ahrens, D. Kremers, M. J. Avalos, C. Readhead, and R. E. Jacobs. Visualizing diffusion tensor images of the mouse spinal cord. *Proceedings of Visualization '98*. IEEE Computer Society Press, 1998.

21. D. H. Laidlaw, M. Kirby, J. S. Davidson, T. Miller, M. DaSilva, W. H. Warren, and M. Tarr. Quantitative comparative evaluation of 2D vector field visualization methods. *Proceedings of IEEE Visualization 2001*, pages 143–150, 2001.

22. D. H. Laidlaw, D. Kremers, E. T. Ahrens, and M. J. Avalos. Visually representing multivalued scientific data using concepts from oil painting. *SIGGRAPH '98 Visual Proceedings*, page 249, 1998.

23. D. H. Laidlaw, D. Kremers, F. Frankel, V. Interrante, and T. F. Banchoff. Art and visualization: Oil and water? In *Visualization '98 Proceedings*, pages 507–509, 1998.

24. N. Max, R. Crawfis, and D. Williams. Visualization for climate modeling. *IEEE Computer Graphics and Applications*, 13(4):34–40, 1993.

25. B. J. Meier. Painterly rendering for animation. *SIGGRAPH 96 Conference Proceedings*, pages 477–484, 1996.

26. T.-M. Rhyne, D. H. Laidlaw, C. G. Healey, V. Interrante, and D. Duke. Realism, expressionism, and abstraction: Applying art techniques to visualization. *Visualization '01 Proceedings*, pages 523–526, 2001.

27. P. Saffman. *Vortex Dynamics*. Cambridge University Press, Cambridge, UK, 1992.

28. M. P. Salisbury, S. E. Anderson, R. Barzel, and D. H. Salesin. Interactive pen-and-ink illustration. *Proceedings of SIGGRAPH '94*, pages 101–108, 1994.

29. M. P. Salisbury, M. T. Wong, J. F. Hughes, and D. H. Salesin. Orientable textures for image-based pen-and-ink illustration. *SIGGRAPH 97 Conference Proceedings*, pages 401–406, 1997.

30. J. E. Swan, V. Interrante, D. H. Laidlaw, T.-M. Rhyne, and T. Munzner. Visualization needs more visual design! Sensory design issues as a driving problem for visualization research. *Visualization '99 Proceedings*, pages 485–490, 1999.

31. R. M. Taylor II. Visualizing multiple scalar fields on the same surface. *IEEE Computer Graphics and Applications*, 22(2):6–10, 2002.

32. G. Turk and D. Banks. Image-guided streamline placement. *SIGGRAPH 96 Conference Proceedings*, pages 453–460, 1996.

33. J. J. van Wijk. Spot noise-texture synthesis for data visualization. *Computer Graphics (SIGGRAPH '91 Proceedings)*, 25:309–318, 1991.

34. J. J. van Wijk. Flow visualization with surface particles. *IEEE Computer Graphics and Applications*, 13(4):18–24, 1993.

35. J. J. van Wijk, A. S. Hin, W. C. de Deeuw, and F. H. Post. Three ways to show 3d fluid flow. *IEEE Computer Graphics and Applications*, 14(5):33–39, 1994.

36. K. Walker. Virtual da Vinci. *Shift Magazine*, 10(2), 2002.

37. J. P. Windham, M. A. Abd-Allah, D. A. Reimann, J. W. Froelich, and A. M. Haggar. Eigenimage filtering in MR imaging. *Journal of Computer Assisted Tomography*, 12(1):1–9, 1988.

38. G. Winkenbach and D. H. Salesin. Computer-generated pen-and-ink illustration. *Proceedings of SIGGRAPH '94*, pages 91–100, 1994.

39. G. Winkenbach and D. H. Salesin. Rendering parametric surfaces in pen and ink. *SIGGRAPH 96 Conference Proceedings*, pages 469–476, 1996.

46   Visualization and Natural Control Systems for Microscopy

RUSSELL M. TAYLOR II, DAVID BORLAND, FREDERICK P. BROOKS, JR., MIKE FALVO,
KEVIN JEFFAY, GAIL JONES, DAVID MARSHBURN, STERGIOS J. PAPADAKIS,
LU-CHANG QIN, ADAM SEEGER, F. DONELSON SMITH,
DIANNE SONNENWALD, RICHARD SUPERFINE, SEAN WASHBURN,
CHRIS WEIGLE, MARY WHITTON, and LEANDRA VICCI
University of North Carolina at Chapel Hill

MARTIN GUTHOLD
Wake Forest University

TOM HUDSON
University of North Carolina at Wilmington

PHILLIP WILLIAMS
NASA Langley Research Center

WARREN ROBINETT
http://www.warrenrobinett.com

46.1   Introduction

The world now stands at the threshold of the age of nanotechnology, which biologists have been exploring for years. Imagination has already leapt ahead to the day when it will be possible to touch proteins within living cells, to tug on DNA as it is transcribed, and to manipulate molecules one atom at a time. To reach these goals, scientists need *instruments* and *interfaces* that extend their eyes and hands into this new nanoscale world. This chapter is about the construction of such interfaces.

For 10 years, the growing Nanoscale Science Research Group (NSRG) at the University of North Carolina at Chapel Hill has been building visualization systems that intuitively map the additional sensing made available by various microscopes into the human senses and intuitively control systems that project human actions directly into this world. The NSRG is composed of teams of computer scientists, physicists, materials scientists, information scientists, and educators. Three systems have been developed to the point where they have been used in physical science experiments:

- The *nanoManipulator* (nM) provides an interactive 3D graphics and force-feedback (haptic) interface to atomic force microscopes (AFMs) to enable scientists to naturally control experiments as if they could directly see, touch, and manipulate nanometer-scale objects on surfaces. Begun in 1991, this system has been used to perform a wide variety of experiments on viruses [12], carbon nanotubes [11,14,15,45,46], fibrin (the fiber that forms blood clots) [28], and DNA [25].

- The *Nanometer Imaging and Manipulation System* (NIMS) augments the nM with a scanning electron microscope (SEM), using projective texture mapping and manual align-

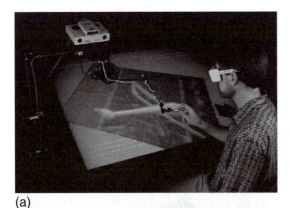

(a)

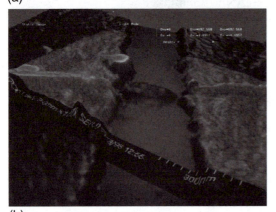

(b)

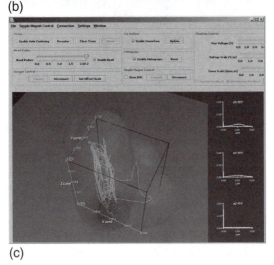

(c)

Figure 46.1   (a) UNC nanoManipulator being used to manipulate carbon nanotubes. (b) UNC NIMS overlays SEM image on AFM surface display. (c) UNC's 3D Force Microscope views and manipulates biological samples. (See also color insert.)

ment of SEM and AFM datasets to enable viewing during direct manipulation of samples inside the SEM. The goal is to use visualization hardware and software to combine the two microscopes into one virtual microscope that includes the capabilities of each and mitigates their limitations. Begun in 1998, this system has been used to perform experiments on carbon nanotubes and their use in actuating devices (from MEMS to NEMS) [67,68,69].

- The *3-Dimensional Force Microscope* (3DFM) provides an interactive 3D graphics and haptic interface to a custom 3D optically tracked, magnetically driven force microscope that can track and control submicron beads on and near living cells. A recently completed prototype of this system is being used to study viscosity and force in lung cell cultures to investigate the causes and mechanisms of cystic fibrosis.

The NSRG has also begun to design and develop interfaces for a new microscopy system:

- The Keck *Atomic Imaging and Manipulation System* (AIMS) will add atomic-scale manipulation capabilities to a transmission electron microscope (TEM) that is capable of near-atomic-resolution imaging of carbon nanotubes and other small structures. This system will be used to study the details of atomic lattice deformations for nanotube structures under stress.

This chapter presents these microscope systems, along with brief descriptions of the science experiments driving the development of each system. Beginning with a discussion of the philosophy that has driven the NSRG and the methods used, it describes the lessons learned during system development, including both useful directions and blind alleys. It also describes techniques to enable telemicroscopy in the context of remote experiments and outreach.

46.1.1   NSRG Philosophy and Methods

The NSRG aims to provide tools that are, like a lens, transparent and easy to use, yet as powerful and versatile as contemporary computing technology can make them—tools that enable direct viewing of, and interaction with, real and simulated molecules, viruses, and cells. Virtual filters enable the transformation and overlay of multiple datasets in order to map them from the raw instrument data formats onto more natural and useful views. Haptic (force-feedback) display coupled to the microscope's probe enables real-time exploration of the properties of real objects, touching and moving them to feel how they respond. The goal is to enable the scientist to pay great attention to the experiment and little attention to the tools, rapidly and easily chasing down "what if" scenarios as they present themselves.

Fred Brooks put forward the two major philosophies that have guided this research [7,8]. The first is the "driving problem" method of doing computer science research. This posits that excellent computer science research arises from tackling a real-world problem and addressing it on its own terms, as a total system problem, and aiming to satisfy not just computer scientists but professional practitioners in the problem domain. This requires developers to face all aspects of a problem, not merely the tractable or publishable aspects.

The second major research philosophy is that human–machine systems can address more difficult problems than can machines alone, an idea that can be cast as the saying, "Intelligence

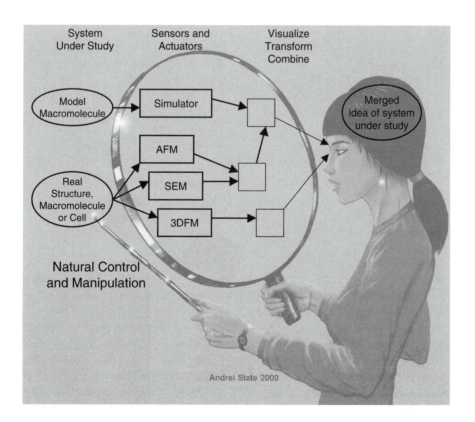

Figure 46.2 Pictorial representation of the NSRG's goal of providing tools that are versatile, powerful, and transparent to use; this will enable the scientist to concentrate on the experiment rather than the tools.

Amplification is better than Artificial Intelligence." This posits that at any given level of technological advancement a person plus a machine can beat a machine-only system. This suggests building human–computer shared-work systems, where the human provides creativity, pattern matching, and decision making in the presence of incomplete information and the computer provides precise recollection from large databases and performs the tedious transformations from instrumentation space to the 3D world.

46.2   nanoManipulator

How do you examine an unfamiliar object? You look at it. When possible, you pick it up, hold it at arm's length, and turn it around. You may squeeze or prod it to determine its stiffness; a fingernail feels for grooves or surface texture. If the object is on a surface, you may use a fingertip or pen to roll it around.

The nanoManipulator System (nM) provides a scientist with the ability to perform these actions on objects as small as single molecules while at the same time quantitatively measuring both the surface shape and forces applied. The nM uses the ultra-sharp tip of an atomic-force

microscope (AFM) as a tool both to scan and to modify samples. It uses advanced computer graphics to display the scanned surface to the user. A force-feedback device (like a robot arm, but used to present forces to the user) enables the user to feel and modify the surface [10,17,26,57,58]. It is basically a teleoperation system that operates at a scale difference of about 100,000 to 1.

The diagram in Fig. 46.4 shows the basic operation of the AFM, which uses a fine tip at the end of a cantilever to scan and push objects on a surface. The cantilever bends when it comes into contact with the surface or objects on it, causing deflection of a laser beam that bounces off the cantilever. The deflection is detected by a four-quadrant photodiode, which is able to measure both the normal and the lateral force applied by the tip to the surface. The cantilever is very sensitive: sub-nanoNewton forces can be measured. For imaging, the tip is scanned across the surface in a raster pattern. Feedback moves the sample up and down to maintain a constant (very small) applied force. The resulting trajectory yields the topography of the surface. The user can also direct the tip with the robot arm, feeling around on the surface. To modify the surface, the force applied by the tip is increased. The user's hand motions are

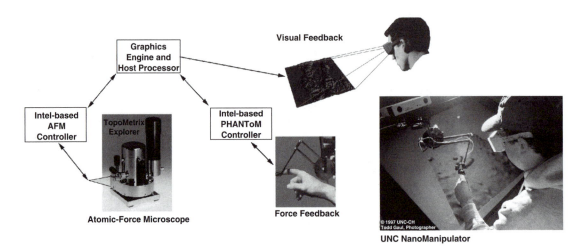

Figure 46.3   nanoManipulator system diagram. Separate threads control data acquisition, visual rendering, and force display. (See also color insert.)

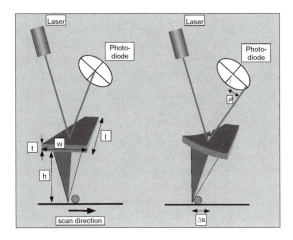

Figure 46.4 An atomic-force microscope (AFM) detects forces by observing deflections in a laser that bounces off of a flexible cantilever.

scaled down by a factor of up to a million, enabling sub-nanometer control over the position of the AFM tip.

46.2.1   Driving Problems

46.2.1.1   Virus Particles

One of the earliest biological applications was the study of tobacco-mosaic virus (TMV). The nM was used to probe the mechanical properties of the virus. Fig. 46.5a shows an AFM image of a TMV that has been dragged across a graphite substrate with the AFM tip. The resulting bent shape is the balance between the bending rigidity of the virus and the friction between the TMV and the substrate. In the image, brightness corresponds to surface height. The darker gray line drawn along the central axis of the TMV was found by the medial-axis location software developed by the UNC MIDAG group [18].

Fig. 46.5b shows where the mechanical equations for beam bending for a beam under uniformly distributed force are fit to the shape of the TMV found with the medial-axis software. This fit yields the ratio of the distributed frictional force to the bending rigidity of the TMV.

Another series of investigations has explored the physical properties and surface interactions of adenovirus. Adenovirus is an icosahedral virus that is being used by the UNC Gene Therapy Center as a vector for gene therapy. The elastic properties of the virus in air and in liquid were studied by placing the AFM probe on top of an individual virus using haptic feedback. A semiautomatic position vs. force measurement tool was then used to map the response of the virus to increasing force [43]. The nM is also being used to push adenovirus across different surfaces to investigate the adhesion between the virus and each surface and to determine whether the viruses slide or roll. Fig. 46.6 shows two adenovirus particles, one of which has been dimpled at the top using the nM.

46.2.1.2   Carbon Nanotubes

Carbon nanotubes are of interest both because of their mechanical properties (they are the strongest known material) and because of their electrical properties (they are insulating or conducting depending on the details of their construction). They are also interesting because they are atomically precise constructions with atomic spacing exactly matching that of graphite. The nM system has been used to probe all of these characteristics. It was used to probe the bending and buckling behavior of the tubes by manip-

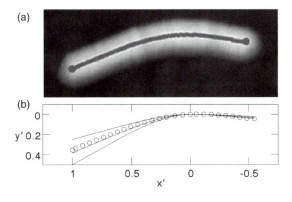

Figure 46.5 The center of a bent tobacco-mosaic virus is (a) estimated and (b) used to derive a one-parameter fit.

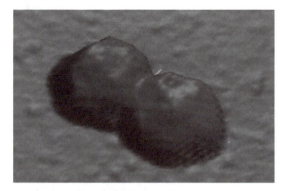

Figure 46.6 Two icosahedral adenovirus capsids. The nanoManipulator was used to press a dimple into the one on the right. (See also color insert.)

ulating them by hand on various surfaces [11,13]. It revealed that the atoms in the tube lock in like gear teeth at the appropriate orientation on graphite, causing them to roll or slide depending on orientation [9,15,16]. Adding electrical measurements to the manipulation ability revealed that the electrical resistance through the tubes to a graphite surface is an order of magnitude smaller when they are in alignment than when they are not [46]. The effects of strain on conductance were also explored [45]. Each of these experiments required carefully controlled manipulation of tubes. Fig. 46.7 shows a configuration where one tube was pushed end-on to slide over another pair of tubes; the AFM probe was then carefully inserted by feel between the first two tubes to peel them apart—leaving the third tube suspended between them.

46.2.1.3 *Blood Clotting Disorders*

Blood clots are composed of blood cells trapped in a matrix of fibrin fibers. While the fibrin from normal, "wild type" clots prevents excess bleeding, mutant versions found in clotting disorders form clots either too easily (leading to stroke) or too poorly (causing excess bleeding). Although the bulk properties of the clots found in these disorders have been measured and the specific gene defects for many of these variants are known, the mechanisms by which the defects

Figure 46.7 The nanoManipulator was used to push a carbon nanotube on top of two others; haptic feedback enabled the scientist to peel apart the lower tubes. (See also color insert.)

cause large-scale change to the properties are not. The strength, thickness, and stickiness of fibrin variants are being studied to help bridge this knowledge gap. Wild-type and mutant fibers are formed and deposited on a surface. Their height is measured using an AFM, which is then employed to cut the fibers. During cutting, measured force profiles reveal the strength of adhesion between the fiber and surface, the springiness of the fibers, and their rupture strength. Imaging after manipulation reveals whether the fibers undergo elastic or plastic deformation. Fig. 46.8 shows a fibrin fiber after being cut with an AFM (which tore out the portion seen on the right side of the image) [28].

46.2.2 System Description

Achieving a working system meant overcoming the following computer-science challenges:

- Real-time rendering of a large and dynamically updated surface model.
- Integration of haptics, teleoperation, and a virtual-reality system.
- Real-time, low-latency, distributed heterogeneous computing.
- Network-aware real-time AFM control.

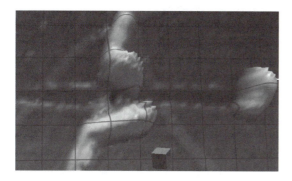

Figure 46.8 Here, the nanoManipulator system is being used to measure the rupture strength of a fibrin fiber. (See also color insert.)

The **dynamic, large-model rendering challenge** was initially tackled using the world's fastest graphics computer (Pixel-Planes 5, developed at UNC under DARPA funding) [19]. This machine was fast enough to render the number of triangles needed for the scanned surfaces, but its graphics pipeline was unable to handle the required dynamic update of the surface as the microscope scanned. This was addressed by reprogramming Pixel-Planes' parallel array of geometry processors and developing a more efficient protocol for sending updates from the host processor to the graphics subsystem [58]. The rendering challenge has been solved over time by advances in rendering technology, to the point where the nM now runs effectively on a laptop with an Nvidia GeForce2Go graphics processor.

The **haptics subsystem** was initially implemented using the Argonne III Remote Manipulator (ARM) developed for use by Ming Ouh-Young in an earlier UNC project—his Docker program, which simulated the docking of drugs within protein receptor sites [6]. This used *Armlib*, a UNC-developed network-aware ARM server. When the higher-performance Phantom haptic display from SensAble Devices became available, the nM was ported to it. Because there was no haptic-control software library available for the device, the Armlib software was modified to also control the Phantom [42]. The nM currently runs

on top of the manufacturer-supplied *GHOST* software toolkit, using the UNC-developed public-domain *Virtual Reality Peripheral Network* (VRPN) library as the network layer that enables remote graphics machines to control the device [60,61]. The physical separation of microscope and haptic display pushed the development of novel *intermediate representations* to enable both the microscope feedback and haptic display feedback to proceed at their required rates of 100 kHz and 1 kHz while updates between them occurred over the network at only 30 Hz. The resulting techniques were published by Mark et al. [42].

The **real-time, low-latency, distributed heterogeneous computing** challenges were first addressed by adopting the techniques that had been developed to enable the Docker program to do its parallel computation on a MasPar computing array connected to a Vax, its haptic display on a Sparcstation connected to the Argonne ARM, and its graphics display on an SGI or Evans & Sutherland display console. This required careful control over network link settings (using TCP_NODELAY for one-way streams and limiting the attempted packet transmission rate to that sustainable by the network); common numerical encodings (using the *htonl()* functions for integer conversion, and hand-tuned binary conversions between IEEE and Vax floating point because binary-to-ASCII routines were too slow in this extreme-low-latency environment); careful management of relative loop rates, buffering, and pipelining to enable data to be ready when needed without introducing unacceptable delays; and the development of robust, operating system–independent remote process creation/destruction to enable rapid startup and to avoid leaving behind cycle-burning processes that occupied resources when the system was shut down. These techniques have been formalized and documented over time, and they make up the core of the public-domain VRPN library and are described by Taylor et al. [61].

The **network-aware real-time AFM control system** was developed in three steps.

- The first implementation controlled a user-built scanning tunneling microscope (STM) sent from Stan William's group at UCLA; it ran on Microsoft's DOS operating system and coordinated network activity using Sun's *PC-NFS* network stack, a finite-state machine to implement the control interface, digital-to-analog control using Data Translation boards, and external Hewlett-Packard pulse generators for tip voltage bias and modification by voltage pulses. A nonlinear analog feedback control system for the STM was designed and built to improve the instrument's performance [57,58].

- The second implementation was for a Digital Instruments (DI) brand Atomic Force Microscope. DI kept its interfaces proprietary, requiring extreme measures to achieve external program control over the instrument. This made it necessary to use a second, auxiliary control computer that drove the DI's signals by replacing analog multiplexer chips on the control system board and effecting changes in the digital control by transmitting characters through a serial port to the keyboard buffer on the main control computer.

- The third implementation was done in software on top of the TopoMetrix (since then, the company has merged into ThermoMicroscopes and now Veeco) control software under Microsoft *Windows*. TopoMetrix provided the source code to their control system as part of an equipment donation to the project. The current version uses a custom VRPN object type [56] as the network transport layer.

46.2.3 Lessons Learned

Several system features have proven very useful by enabling new types of experiments or revealing previously unseen phenomena. These features are listed in detail by Taylor et al. [59], along with the particular insights revealed by each. A summary of the highlights for each area follows:

- **Graphics.** Augmenting the standard real-time 2D view with user-controlled, real-time, publication-quality 3D views enables the scientist to gain insight during experimentation that otherwise would be missed. Subtleties of shape and interactions between 3D objects become clearer when viewed in their natural 3D context and from changing viewpoints ("The map is not the terrain.")

- **Haptics.** Touch enables the scientist to find the correct location to measure or start an experiment, even in the presence of drift and positioner nonlinearities. During manipulation, the probe is busy, so no new scanned images can be produced: the user works blind. However, forces are continually measured. When these forces are displayed, this feedback during manipulation enables understanding and control of the path of delicate modification ("pushing bags of Jell-O across a table in the dark with a screwdriver without breaking them"). Slow, deliberate feeling can find the location of objects that scanning would knock aside.

- **Virtual tips.** Switching between an oscillating mode for imaging and contact mode for modification enables imaging of fragile specimens, which are then modified with known force. A sewing-machine mode enables finer lines to be formed in thin films without tearing. A virtual whiskbroom enables extended structures (such as TMV) to be moved as units.

- **Replay.** Storing the entire experiment and enabling replay lets the scientist see things missed the first time around, as well as enabling the application of new analysis techniques to old experiments. (Replay can be thought of as a "flight data recorder.") Data is exported in a variety of formats to existing image-analysis tools (Kaleidagraph, SPIP, ThermoMicroscopes) and standard file formats (TIFF, PPM, ASCII).

The lessons learned, in full detail, are as follows:

Design in machine independence and replay. Because the microscope control computer and the graphics computer for the nM system were of different architectures and communicated across a network, the system had to be designed to use a machine-independent wire protocol to communicate between them. This has borne much fruit: it simplified both the storage and replay of experiments and the porting of parts of the system to different architectures with time, and it enabled remote access. One specific design decision that enables replay is to make the application rely only on responses from the microscope to determine the system state (whether it is in modify mode or touch mode, the actual size of a region that is selected, etc.). This also made the system robust to instrument limitations: if asked to do something beyond its capabilities, a particular microscope would either not respond or reply with a clipped version of the request. This required a few specialized state-indicating messages to be added to the network interface: "Tell me I think I'm in modify mode now," and "Tell me I think I'm in touch mode now." These messages preserve the user's intent, which cannot be inferred from other nM system parameters.

Design for more than one instrument. When scientists perform an experiment, it is often the case that they use more than one instrument. For some nanotube experiments, a computer-controlled voltage source and current monitor were used to explore changes in conductance while the AFM was used to apply strain to the tubes. The control and measurement for all instruments needed to be time-aligned within the tolerances of the experiment design and ideally be controlled from within a single framework. This is true both during the experiment and during replay. This becomes even clearer in some of the later systems that explicitly combine two or more instruments.

Give them the data. Whenever scientists think that they have obtained a new insight into a problem, they seek to verify the insight by comparing its predictions with the data at hand. This requires access to the data from the experiments, not just derived visualizations of the data. It is tempting to design a new user interface for display and analysis of raw data values, a process that may take weeks or months. Often, the data in a very raw ASCII format that can be imported into spreadsheets or other analysis tools is really what is needed. Providing the data in this format makes it available to the scientist sooner and involves spending less time designing tools that mimic those that already exist.

There was no one "best" immersive interface for this system. Different people preferred different interfaces. Washburn preferred wearing a head-mounted display (HMD) when doing experiments because of its stereo display and the ease of navigating using motions of head and hand. Falvo preferred operating using a non-head-tracked stereo projection display because of its higher resolution compared to the HMD. Stan Williams preferred to direct by watching a non-stereo-projected display while Taylor was driving the experiment from inside the HMD interface. One trend has emerged as experiment lengths stretched to hours: for this application, scientists found the additional headwear and eyestrain required for stereo head-tracking not to be worth the benefit of using the technologies available.

Exact calibration may not be worth the effort. The goal of constructing the nanoWorkbench system displayed in the nM system diagram was to align the graphics and haptics display within 1 mm (1 display pixel) of the virtual model to enable direct and precise interaction with the model. After years spent chasing this goal, the scientists asked that the force display be offset from the visual display by several centimeters so that the end-effector would not visually obscure the point of contact. A task-level analysis should have been done before pursuing the technology-driven goal of exact alignment. Alignment was both more difficult and less useful than expected.

46.2.4 nM Summary

The nM system has been in continuous use exploring and modifying surfaces since 1993 and biological samples since 1995. It is the longest-running of the systems described here, and it is in the most advanced state of development. The nM system currently has 101 hierarchically grouped functions, each asked for by a user to address a particular challenge in an experiment. It has been ported from its initial configuration using a custom-built haptic display device, a custom-built SPM controller, and a custom-built graphics supercomputer to a configuration running entirely on commercially available equipment (PC-based graphics, the Phantom haptic display, and TopoMetrix-derived AFMs). It was developed into the commercial *NanoManipulator DP-100* by Aron Helser at 3rdTech and has been sold to NASA and a number of university departments [30].

The nM brings advanced visualization, analysis, and modification tools to bear *as experiments are happening*, providing immediate feedback that can be used to select the most promising step forward at each stage of an experiment. It forms the base software for expansion into the NIMS. It also forms the base for the telemicroscopy efforts. Despite the system's long tenure, new capabilities are continually being added to support new experimental needs.

46.3   NIMS: nM + SEM

Although the nM enables both precise manipulation and imaging, they are separate functions. Because there is only one probe in the AFM, at one time it can be used for either imaging or manipulation. The scientist can feel the force on the probe at the point of contact but cannot watch the probe and sample deformation during manipulation. The end configuration of objects on the surface is imaged after manipulations by a subsequent raster scan of the sample. The *Nanoscale Imaging and Manipulation System* (NIMS) incorporates the AFM into a Hitachi S4700 scanning electron microscope (SEM), a 1.5 nm resolution instrument that enables electron microscope imaging during AFM manipu-

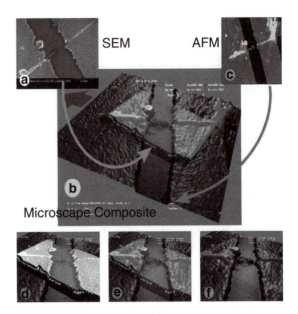

Figure 46.9 The NIMS system combines the display of SEM and AFM data from the same specimen. (See also color insert.)

lation. This effectively "turns the lights on" for the user while he or she is manipulating samples.

Fig. 46.9 shows the overview of AFM and SEM scans of the same area, a carbon nanotube that was draped between two raised electrodes and then broken. The simplest combination of the two datasets is shown here: the datasets were aligned by hand and the SEM was laid over the underlying AFM topography using projective texture mapping. The user can adjust the relative mixture of the two datasets.

46.3.1   Driving Problems

Thermally Actuated Mobile Structures (TAMS): Bimetallic multiarmed structures with smallest dimensions under a micron are being formed with the intention of driving their motion using differential heating. An eight-legged version in Fig. 46.10 is one of the prototype designs. The inset shows the AFM probe in the

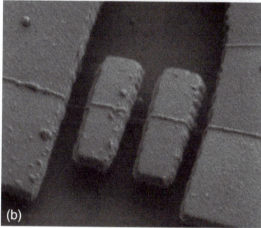

Figure 46.10 (a) Bimetallic "spider" structure that can be thermally actuated (the inset shows an AFM probe measuring the stiffness). (b) Torsion oscillator composed of two paddles suspended on a carbon nanotube.

NIMS pressing on one from above to measure its stiffness and see its response.

Carbon Nanotubes (CNT): Building on the fabrication and analysis described in the nM section, NSRG scientists are designing and characterizing nanoscale mechanisms such as the torsion oscillator seen in Fig 46.10. The structure is composed of two metallic paddles fabricated on a suspended CNT between two metallic leads [69].

Problems: How to measure shape deformation from one or more SEM views; how to determine Z position of AFM tip within the SEM projection view to enable rapid and safe manipulation.

46.3.2 System Description

New interaction modes were added to those of the stand-alone nM to support experiments on fragile structures. One enabled control of the AFM probe in three dimensions, rather than keeping it always touching the surface. Another provided the ability to drop down onto the surface from above and measure the position offset of an object as the force was uniformly increased and then decreased.

A method of calibration between SEM and AFM images has been developed that enables the system to show the AFM probe in its proper 3D location compared to the AFM scan. The method uses manually selected corresponding points in the AFM and SEM images to solve for the transformation between the images. This has been extended to include calibration between the AFM probe position, the SEM image, and a geometric model of the surface being studied to enable manipulation experiments on fragile samples without requiring a complete AFM scan of the sample.

Projective texture mapping is used to display the AFM scan, SEM image, surface model, and AFM probe positions within the same image to provide an optimal understanding of the sample, to enable planning of intricate manipulations and electron beam lithography.

46.3.3 Results

Figs. 46.11 and 46.12 show two steps in an experiment that shows the NIMS being used as a combined tool, employing the capabilities of the SEM and the AFM together. The top image shows a carbon nanotube draped over a gap between the tip of an AFM probe (upper right corner) and one-half of a MEMS test structure. Direct 3D control of the AFM probe is being used to touch one end of the tube to the surface at the correct location. Once there, the AFM probe is locked into place and the electron beam is switched from scanning mode to focusing its energy at the point of contact between the tube and the surface. This causes carbon atoms

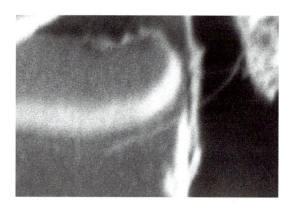

Figure 46.11 A carbon nanotube suspended from the AFM tip onto a test structure before being welded in place using the SEM's electron beam.

in the "vacuum" to accrete, effectively welding one end of the tube down.

Fig. 46.12 shows the case after each end of the tube has been "welded" down on opposite sides of the test structure. The SEM beam has returned to scanning, and the AFM is being used to test the tube's strength (the tip is blurred because it is in motion). In the final portion of the experiment (not shown here), the AFM probe was used to move one arm of the test structure. This caused flexion of the tube and

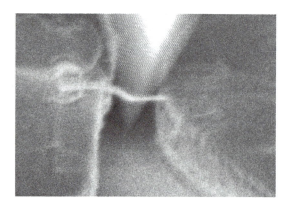

Figure 46.12 A carbon nanotube welded to both sides of the test structure being probed using the AFM.

then broke the connection between the tube and one end of the structure—the weld failed before the tube ruptured.

46.3.4 Lessons Learned

Look to add-ons when the manufacturer doesn't supply a programmable interface. The software interface provided by a standard SEM add-on controller from EDAX was used to provide scanning and directed-beam modification control within the SEM. This enabled the NIMS system to control beam parameters and scanning, without which the combined instrument would have been impossible to build. The EDAX control is performed through attachments that have become standard in the SEM industry. It exports a library of functions intended for scripting that was used in the NIMS system to integrate the SEM with the rest of the system.

Advancing science and computer science. Sometimes, the most acceleration of an experiment comes from applying pedestrian computer science to the most time-consuming part of an experiment. Spending time developing tools of this type helps cement the usefulness of the computer science and can make colleagues willing to spend time in system development. It can also result in science publications coauthored by computer science students.

46.4 3DFM

The AFM has two major drawbacks for biological imaging. First, the measuring probe is attached to a cantilever for position control and force sensing. It cannot probe beneath objects, only the tops of surface-bound objects. Second, it cannot go inside living cells because the cantilever would have to penetrate the cell membrane.

Freeing the tip of the probe from the cantilever alleviates both of these problems. This requires new methods for tracking the tip and for applying forces to it. This has been done

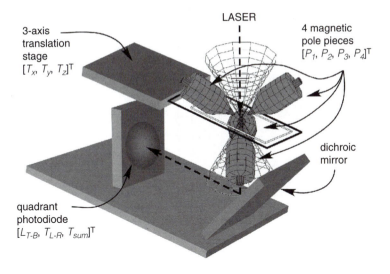

Figure 46.13 A 3DFM system diagram. The position of a bead is measured using laser scatter onto a quadrant photodiode. Forces are applied using magnetic pole pieces. The bead is kept centered in the laser using a 3-axis translation state.

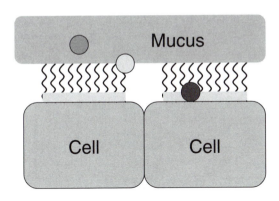

Figure 46.14 Viscosity and force measurements of interest to cystic fibrosis researchers are indicated by the three beads.

using an optical beam in a laser tweezers configuration [21], where a focused laser beam is used both to apply forces and to track the particle position. Whereas this technique has made possible experiments in single molecule dynamics [55], the optical beam can generate only relatively small forces, normally up to several tens of picoNewtons [44]. This is insufficient to break covalent bonds, or to measure the full mechanical properties of biological fibers such as microtubules. Also, the method of applying the force is nonspecific, causing the beam to accumulate extraneous material.

The NSRG physical science team has invented a 3D free-particle force microscope (3DFM) that uses magnetic beads to apply forces using techniques similar to those used by Bausch et al. [4,5]. The particle is tracked using optical light scattering, as in laser, tweezers [2,22,23]. Fig. 46.13 shows the components of the integrated system, including a 3-axis translation stage that is used to move the sample so that the bead remains centered in the laser beam.

46.4.1 Driving Problem

Cystic Fibrosis: The UNC Cystic Fibrosis (CF) Center is investigating the mechanisms by which CF affects its victims. As seen in Fig. 46.14, scientists there hope to place sub-micron fluorescent beads in the viscous mucus layer to study its viscosity and motion, attached to the cilia that beat to move the mucus, and attached to cell surfaces. The beads will be viewed using either 2D widefield optical or 3D confocal microscopy. Tracking bead diffusion in the mucus enables calculation of viscosity of different fluid layers. Applying forces to the beads on the cilia will help determine system reactions to force and stall force of cilia. Applying forces to cell surfaces will enable determination of mechanical deformations and system responses.

Problems: How to control the positions and forces to enable application of forces using the beads and measuring the system response to those forces; how to display volumetric viscosity information, lines and volumes of bead travel, and surfaces of cells without confusing the user.

46.4.2   System Description

The tracking and position-control portions of the system are not described here because they function essentially as black boxes from the visualization and user-interface points of view. The magnet control system, on the other hand, has a feature that is exposed to the user interface designer. Because the four magnetic poles in the system (Fig. 46.15) can only generate forces towards each pole, forces in arbitrary directions must be broken down into time-sequential tugs towards each of the poles. This design enables

Figure 46.15 Photograph of four magnetic poles in a 3DFM surrounding a specimen mounted between two cover slips. (See also color insert.)

the poles to be very close to the sample (thus applying more force), but it requires that the system map the force commands from the user to sequences of forces to be applied by the magnets.

A prototype user interface for the 3DFM built using the visualization toolkit (VTK) [65] and Java Swing is shown in Fig. 46.16. It includes a 2D section for control over visualization and microscope parameters, as well as connections to the bead tracker, video stream, magnet controls, and haptic device.

The 3D section (shown here in a monoscopic view but also displayable in stereo) displays the current location of a tracked bead as a wireframe sphere. This sphere is centered in the live video display (the gray plane with the dark spot surrounding the sphere in the image). This plane of video moves with the bead through the volume. The sphere leaves a yellow line as a trail, showing where the bead has moved during an experiment. A transparent shell can also be drawn around the volume that has been "carved out" by the bead as it has moved along the trace; it shows the boundary of the explored region. To the right are three histograms of the bead's motion in X, Y, and Z.

46.4.3   Results

The 3DFM has been used to estimate the viscosity of corn-syrup test samples by tracking the Brownian motion of included beads, and to apply force to estimate the viscosity by observing bead velocity. It has also been used to track the motion of a bead attached to a group of cilia on a lung cell culture and to apply forces to the cilia through the bead.

46.4.4   Lessons Learned

Displaying intent while recording details. An attempt to provide the most faithful force representation to the users by driving the force display with the same alternating force sequence used to drive the magnet cores resulted in force display that was uninformative and difficult to

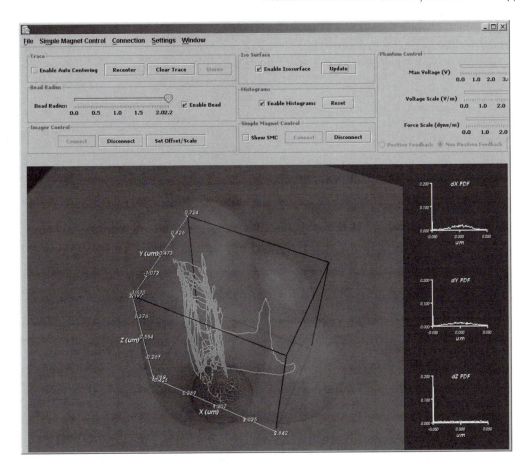

Figure 46.16 3DFM user interface prototype. The current position of the bead is indicated by the green wire-frame sphere. Its trajectory is shown by the yellow trace. The brownish translucent surface shows the volume carved out during travel. The grey-scale plane shows live video from an optical microscope centered on the bead. (See also color insert.)

control. Displaying the average force to the user was more satisfactory; the actual sequence of forces is recorded to the experiment log so that analysis can be done using the detailed force information.

Build in annotation support. The experiment-data logging system for the 3DFM is being augmented with a mechanism to enable scientists to record text comments that are time-aligned with the experiment data. These comments can be added either during the experiment or during replay of a previous experiment. This is being added by request of the scientists, so that they can record significant events as well as interesting locations in the experiment.

46.5 AIMS: TEM + MEMS

The NIMS provides a resolution of about 2 nm, which is too coarse to resolve the individual layers in a carbon nanotube or the fine details of other molecular systems. The Atomic Imaging and Manipulation System (AIMS) will address this limitation by combining a 200 kV field emission transmission electron microscope (TEM) with MEMS-based manipulation and electrical measurement. The TEM provides better than 0.2 nm resolution for imaging the atomic sidewalls of carbon structures and the atomic positions within nanoparticles. AIMS is being developed as a unified scientific exploration system that will be capable of simultaneous manipulation, measurement, and atomic-scale imaging.

46.5.1 Driving Problems

There are two broad classes of experiments driving AIMS development: mechanical contact and electrical transport in nanoscale junctions.

Nanoscale Mechanical Contact: NSRG researchers seek to explore the configuration of the atoms in the contact region between carbon nanotubes. This study will include the distortion of atomic arrangements and the rebonding of atoms across the interface, energy loss, and electron transport. It is predicted that distortion of the contacting surfaces (Fig. 46.17) occurs because of the strong attractive forces that bind materials together, but no one has imaged these interfaces in contact for moving nanoscale devices. It has been predicted that the local distortion can dramatically change the properties of the interface, increasing energy loss during motion and enhancing electron transport.

While a TEM by itself can image contact regions, manipulation capabilities are essential for creating particular arrangements of interest (tee junctions, sliding rails, etc.), and for creating motion. AIMS will be used to explore atomic-scale distortion during motion, interfacial wear at the atomic scale during the sliding of lattices, and the atomic origins of friction and energy flow.

Nanoscale Electrical Junctions: The AIMS will also be used to move nanomaterials into atomic contact. Experiments target both nanotube–nanotube junctions and nanotube–nanoparticle junctions (Fig. 46.18). For the former, new carbon structures with positive and negative curvature have been proposed to form integrated tee junctions. In this case, the nanotubes are not simply lying on top of one another, they are

Rajeev Dassani

Figure 46.17 Artistic conception of constrained electron flows between carbon nanotube and surface. Charge is injected from the AFM probe above the nanotube. (See also color insert.)

Figure 46.18 Crossed-nanotube junction, with each tube connected to electrical leads for measurement.

intimately connected like the tee junction of a water pipe. Fig. 46.18 is an SEM image of a crossed-nanotube device created in the NSRG laboratory (the nanotubes, about 2 nm in diameter, are the very thin crossing connections).

Problems: How to provide an atomic-scale view of these devices and enable creation of the proposed integrated tee junctions; how to move nanotubes into atomic contact and then induce rebonding of the carbon lattices through heating or electron bombardment; how to provide *in situ* electronic characterization to monitor the atomic bonding.

46.5.2   System Planning

The AIMS project presents difficult challenges in the integration of atomic-scale motion control, force sensing, and *in situ* electrical characterization within the extremely tight confines of the sample stage of a TEM. This integration within the tight confines of the TEM sample volume will initially be implemented using microelectromechanical systems (MEMS). MEMS technology applies processing techniques common to silicon electronic device fabrication to create actuating and sensing systems integrated onto silicon chips.

The user interface paradigms employed to build the NIMS system will be reused in the AIMS system. Although there are no AFM scans in AIMS, the models and projective texture alignment techniques will be similar. The first step has been preparing prototype applications using different visualization display libraries to select the most effective existing toolkit for this application.

46.5.3   Lessons Learned

Include software interface criteria in the instrument purchase decision. During the TEM selection process, availability of real-time digital access to the control and imaging systems of the TEM was a requirement. This disqualified one manufacturer whose image quality was slightly better than that of the system that was purchased.

Talk with others who have interfaced to each instrument. Talking with experts in other groups who have attempted to digitally control a par-

ticular instrument can reveal pitfalls and suggest required system components: sending a student to Mark Ellisman's NCRR helped determine which camera to use.

46.6   TeleMicroscopy

Remote use of AFM, SEM, and TEM systems to view samples is becoming widespread: the nM software has been used by the IN-VSEE group at Arizona State as a base to provide remote web-based access to AFMs [3,53]. Mark Ellisman's NIH Resource at UCSD routinely uses a TEM from Japan remotely [29,47]. The MAGIC group at CSU Hayward has remote access tools for SEM, SPM, and confocal microscopes [49]. Oak Ridge National Laboratory has a web-based interface to its electron microscopes and is developing remote manipulation techniques [20]. The Bugscope project at the Beckman Institute has a complete educational system built around remote access to an SEM [48]. CERN is developing an *OpenLab for Nanotechnology* that will interface with live instruments on the Grid [24]. There are other groups as well. The manufacturer JEOL has even provided a web-based interface to its SEM systems [70].

46.6.1   Low-Latency Remote Microscope Control

Remote control of microscope viewing parameters and viewing of the resulting images requires high-bandwidth connections to support interactive use. Manipulation experiments impose the additional challenge of providing remote haptic interaction for touching and manipulating the sample.

Effective deployment of such networked virtual-environment systems requires paying special attention to network latency, jitter, and loss [32,36]. Graphical and VE applications have particularly stringent latency requirements because their interfaces are interactive: users directly manipulate parameters controlling the images they see, using continuous input devices such as mice. The usability of interactive interfaces degrades significantly when visual feedback is not essentially immediate [33].

Providing stable and accurate force-feedback control during remote experiments is even more challenging and falls into the domain of remote teleoperation. When force feedback is being used simultaneously, or user input is driving a control loop, response time becomes even more critical: 50 ms of latency in flight simulators reduces performance, and just a little more causes system instability [66].

The nM system operates over a network by default, so it might seem that operating it over wide-area networks would be straightforward. Indeed, there have been several instances of successful operation over distance: Internet2 network engineers provided a dedicated, low-latency link from an AFM at UNC to a graphics and haptics user interface located in Washington D.C. for the 1999 Internet2 conference [35]. A similar Internet2/T1 connection from Ohio was used during the BioMEMS and Biomedical Nanotechnology World 2000 conference [31]. (There is a video showing this in operation [27].) For Orange County High School, McDougle Middle School, and Stanback Middle School (all near UNC), the round-trip network latency is acceptable to enable remote experiments.

There have also been unsuccessful attempts: An Internet2-based link to Microsoft Research in Washington was created in 2001, over which the latency was too high to provide reliable manipulation in the absence of application adaptations. Internet-based connections through a network reflector at Louisiana State University had unacceptably high loss.

Hudson has developed network-level and application-specific adaptations to minimize, hide, or enable the user to deal with higher latency and jitter [33,34,36]. Networking adaptations within the transport protocol are used to reduce jitter and latency, and application-level adaptations deal with the latency and jitter that remain. Providing the appropriate intermediate representation [1] was critical to achieving stable and responsive haptic display in this virtual-environment application, especially when operating over a wide-area network [33,42].

46.6.2 Remote Microscope-Based Distributed Collaboration

A *collaboratory* was defined by Bill Wulf in 1989 as "a center without walls, in which researchers can perform their research without regard to physical location—interacting with colleagues, accessing instrumentation, sharing data and computational resources, and accessing information in digital libraries." As a step towards this goal, a collaborative version of the nM system has been developed with which two users who are remote from each other and the AFM can share microscope control and visualization [50,51]. As seen in Fig. 46.19, each user has his or her own nM display and control interface (which can be either shared or private) as well as a second computer to support shared work. This shared-work computer runs Microsoft Netmeeting to provide video conferencing between the two ends and enables sharing of word processor and analysis packages. Two video cameras (only one or the other sending data at any time) are located at each end, one stationary and providing a head and shoulders view of the user for conversations, and one on a gooseneck that can be positioned as desired to share views of hand drawings and other things in the room. Hands-free telephones were used for audio.

The collaboratory system allows scientists to dynamically switch between working together in shared mode and working independently in private mode (Fig. 46.20). In shared mode, remote (i.e., non-collocated) collaborators view and analyze the same (scientific) data. Mutual awareness is supported via multiple pointers, each showing the focus of attention and interaction state for one collaborator. Optimistic concurrency techniques are used in shared mode [52], eliminating explicit floor control and enabling collaborators to perform visualization operations synchronously. Because of the risk of damage to an AFM, control of the microscope tip is explicitly passed between collaborators.

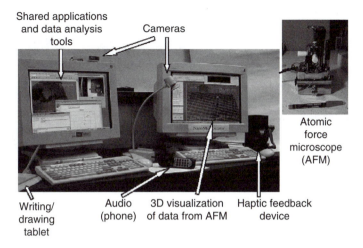

Shared applications and data analysis tools

Cameras

Atomic force microscope (AFM)

Writing/ drawing tablet

Audio (phone)

3D visualization of data from AFM

Haptic feedback device

Figure 46.19 The distributed collaborative nanoManipulator (nM) system provides each collaborator with a custom nM linked to their peers' system plus standard televideo and application-sharing tools.

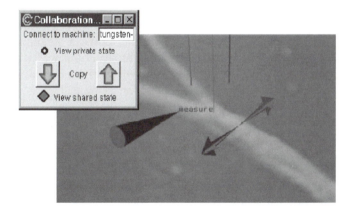

Figure 46.20 Both local actions and collaborator actions are indicated to each user of the distributed collaborative nanoManipulator. Private state can be copied to and from shared state to facilitate easy transitions between individual and shared work.

In private mode, each collaborator can independently analyze the same or different data from stored experiments generated previously or from a live microscope. When switching back to private from shared mode, collaborators return to the exact data and setting they were using.

A report on a repeated-measures, controlled experiment evaluating the collaborative nano-Manipulator was done by Sonnenwald et al. [52]. Twenty pairs of upper-level undergraduate science majors participated in two lab sessions, one session face-to-face using the standard nanoManipulator and the other session using the Collaborative nanoManipulator. As expected, participants reported disadvantages to collaborating remotely. When working remotely, interaction was less personal, individuals received fewer cues from their partners, and some tasks, such as sharing math formulas, were more difficult. However, participants also reported that some of these disadvantages are not significant in scientific work contexts, and that coping strategies, or work-arounds, can reduce the impact of other disadvantages.

Participants reported that remote collaboration also provided several advantages compared with face-to-face collaboration, including the ability to more easily explore the system and their ideas independently, and increased productivity with the ability to work simultaneously on the data visualization. While the statistical analysis of graded lab reports produced a null result, i.e., the scores in the collaborative condition were not significantly lower than those in the face-to-face condition, considering both the quantitative and qualitative data, the NSRG collaboration team concludes that there is positive promise for effective remote scientific collaboration.

The study participants were asked what they liked and didn't like about the system in post-experiment interviews. Many participants reported that they liked the ability to simultaneously adjust visualization-model parameters when in shared mode and found that using the explicit floor control in shared applications running in NetMeeting hindered their work.

46.6.3 Remote Microscopy for K–12 Science Classes: Visualization for Education

For 6 years, a team of educators, physicists, material scientists, and computer scientists has taken "reverse field trips" to local middle and high school classes to enable students to participate in multidisciplinary science (Fig. 46.21). Students were able to control the microscope and experiment with viruses using the nM interface over the Internet to control an AFM at UNC. Studies were done on the educational impacts of designed learning experiences on students' knowledge of viruses, nanoscale science, scale, and the nature of science [37].

Figure 46.21 The distributed nanoManipulator system has been taken to local high school and middle school science classes to let students directly view and manipulate viruses and nanotubes.

This remote microscopy enabled students to experience the interdisciplinary nature of cutting-edge science firsthand [38–41,54,64]. Student response to the visits was overwhelmingly positive. Formal questionnaires showed strong positive shifts in attitudes towards science and the process of science, particularly for girls [39]. Students' written evaluations of the project were also very compelling:

"In the course of a week, I have learned so much. Coming into this experiment we knew so little about viruses, and now we can describe their size, some of their characteristics, and how viruses infect you and make you sick. The visiting scientists have inspired me and so many others to join a field in science. They have lit a flame that cannot be put out."
(Female high-school student, 1999)

Data showed that students were highly motivated and interested in learning about nanoscience and that they learned more about viruses, scale, science processes, and scientists [38,39].

These reverse field trips continue to press the limits of the available network. Although the schools each have fractional T1 lines to points of presence very near UNC's, providing responsive control required that Internet use in other classrooms be curtailed during the remote microscope operation. Firewalls and network address translation prevent the use of UDP, thus imposing TCP congestion control on all streams to and from the AFM. As the field trips continue to schools with networks that are further (in terms of packet hops) from UNC, or with more loss, new techniques will be needed to maintain responsive control.

46.6.4 Lessons Learned

Network latency is the critical parameter for remote haptics. Providing haptic feedback requires different networking characteristics than remote control of viewing parameters. Whereas high bandwidth is required to send microscope images and video across the network, latency and jitter are the critical parameters for remote haptic display.

Shared and private spaces. Shared and private spaces are important to enable each scientist to explore visualizations and hypotheses independently.

Design primarily for collaborative science rather than social interaction. For this system, designing to support collaborative science enabled remote use that was as effective and as satisfying as sharing a single local system. Some preferred the remote system. Another local researcher (David Stotts) has also found that sometimes people prefer working remotely for pair programming because it reduces the amount of time spent socializing during the work.

Optimistic concurrency control was needed to make collaboration work with acceptable latency in this application. Compared to Netmeeting, which was operating over the same network using explicit token passing, the latency was much lower.

Asynchronous remote procedure calls (RPC), where the calling process does not wait for the return before continuing to process other events, enabled the decoupling of system responsiveness from network latency; the user interface continues with the latest available data, and callbacks are used to update the display as requested data arrives.

Providing the appropriate intermediate representation [1] is critical to achieving stable and responsive haptic display in this virtual-environment application, especially when operating over a wide-area network [33,42].

Use telephone for remote audio. Compared to the audio included in remote conferencing systems, telephone audio had lower latency and better quality and was easier to use.

46.7 Conclusions

Several lessons learned during the development of these microscope sessions are listed here. These are more general lessons not specific to one system.

Begin software development at least as soon as hardware development. It was possible to obtain the software interface before the TEM was delivered. This has enabled software development to commence before system integration. The development of the other microscope systems showed us repeatedly that software development should be begun as early as possible. Some manufacturers provide simulators for their instruments; these can be very useful for debugging.

Partner with experts in required technologies. The MEMS designs for the AIMS system were advanced through collaboration with Mike Sinclair at Microsoft Research, who has done dozens of preliminary designs. Design and manufacturing are also proceeding with the help of Shuo-Hung Chang's nanotechnology center at the National Taiwan University.

Stefan Seelecke's group at North Carolina State University is working with the NSRG on the development of shape-memory alloy actuators and advanced control systems for magnetic and piezo-ceramic actuators for several of the microscopes described here. These systems serve as driving problems for Seelecke's own research.

Jean-Marc Brequet and his students within the nanorobotics group at the EPFL in Lausanne, Switzerland have completed two design iterations on compact piezoceramic-based 2-axis translators for use in the 3DFM system. The second design is also being incorporated into the NIMS system to enable larger range on the AFM.

Partnering with outside experts in required technology can enable a group to concentrate on its strengths and leave other research parts of the system to others.

Build on existing visualization toolkits. Whereas the nM interface was based on custom rendering and interaction codes [10,17,57,59], the 3DFM prototype is based on the Visualization Toolkit [62,63,65]. This has enabled rapid implementation and testing of different visualization techniques during interface development. There have been cases during development where a bug deep inside the various toolkits caused several weeks of searching. These weeks-long chases also happened during the development of the nM system, but when they were fixed in the 3DFM the rest of the toolkits were still available for application development.

Effective before cost-effective. The NSRG attempts to use the best available computer technology to develop *effective* systems for use by the physical science team, which then become *cost-effective* and can be deployed on widely available hardware as technology marches on.

Acknowledgments

Warren Robinett and R. Stanley Williams had the initial idea to hook up a scanning tunneling microscope to a virtual-reality interface; the nanoManipulator was born, which led to all of the work described here. The following individuals participated in the development of one or more of the above systems: Benito Valencia Avila, Ron Bergquist, Gary Bishop, Alexandra Bokinsky, David Borland, Brian Boyd, Don Brenner, Fred Brooks, Stephen Brumback, Lisa Cameron, Anthony Canning, Chun-Fa Chang, Jun Chen, Zhi (James) Chen, Jason Clark, Michelle Clark, Greg Clary, Nathan Conrad, Jeremy Cribb, Lin Cui, Jeremy Cummings, Elan Dassani, Rajeev Dassani, C. William Davis, Kalpit Desai, Chris Dwyer, Stephen Ehmann, Dorothy Erie, Mike Falvo, Mark Finch, Jay Fisher, Joseph Fletcher III, Mark Foskey, Darlene Freedman, Charlampos (Haris) Fretzagias, Yoni Fridman, Jacob Furst, Ashes Ganguly, Jai Glasgow, David Glick, Brian Grant, Xiaohu Guan, Martin Guthold, Adam Hall, Jonathan Halper, Gregory Hamamgian, Jing Hao, David Harrison, Bil Hays, Chris Healey, Aron Helser, Amy Henderson, Youn-Joo Heo, Andrea Hilchey, Mark Hollins, Phil Holman, Mave Houston, Tom Hudson, David Jack, Kevin Jeffay, Ja-Yeon Jeong, Gail Jones, Jeff Juliano, Kurtis Keller, Jake Kitchener, Sang-Uok Kum, Eileen Kupstas-Soo, Tom Lassanske, Bin Li, Alena Lieto, Ming Lin, Qiang Liu, Shawn Liu, Noel Llopis-Artime, Tanner Lovelace, Claudia Low, Kelly Maglaughlin, Jim Mahaney, Renee Maheshwari, David Marshburn, Chuck Mason, Garrett Matthews, Roberto Melo, Jameson Miller, Paul Morris, Atsuko Negishi, Shoji Okimoto, Timothy O'Brien, Shayne O'Neill, Lu-Chang Qin, Sherry Palmer, Stergios Papadakis, Ramkumar Parameswaran, Kimberly Passarella-Jones, Aarish Patel, Scott Paulson, Steve Pizer, Daniel Plaisted, Leila Plummer, Latita Pratt, Sharif Razzaque, Daniel Rohrer, Kent Rosenkoetter, Stefan Sain, Adam Seeger, Stefan Seelecke, Tatsuhura Segi, Woojin Seok, Deborah Sill, Kwan Skinner, Don Smith, Neal Snider, Eric Snyder, Diane Sonnenwald, Michael Stadermann, Anthony Steed, Josh Steele, Onejae Sul, Rich Superfine, Russ Taylor, Pichet Thiansathaporn, John Thomas, Kelly van Busum, Gokul Varadhan, Leandra Vicci, Frederic Voegele, Hong Wang, Sean Washburn, Chris Weigle, Mary Whitton, Ben Wilde, Phillip Williams, Bill Wright, Dongxiang Wu, Joe Yandle, and Christine Yao.

Due to the number of years and people included in this list and the frailty of human memory, it is almost certain that someone important to the development of at least one of the projects was not included. The authors apologize for any such oversight.

Section 46.2

Current Project Personnel: *Computer Science Toolbuilders:* Russ Taylor, Mary Whitton, Leandra Vicci, Steve Pizer, Paul Morris, David Marshburn, Aron Helser, Tom Hudson, Yonatan Fridman, Jameson Miller. *Physical Scientist Collaborators:* Richard Superfine, Sean Washburn, Mike Falvo, Stergios Papadakis, Garrett Matthews, Michael Stadermann, Adam Hall, Rohit Prakash, Dorothy Erie. *Information Science Collaborators:* Diane Sonnenwald, Kelly Maglaughlin. *Education Collaborators:* Gail Jones, Dennis Kubasko, Michele Kloda, Tom Trettor, Atsuko Negishi.

Project Funding and Support: NIH NCRR program, grant number 5-P41-RR02170, has supported the development and application to biology throughout. NSF's HPCC program supported advanced visualization, scalability, and network access through grant number ASC-9527192. NSF's ARI program, grant number DMR-9512431, supported installation of the first nanoManipulator in the scientists' laboratory. Initial support for bringing the system from UCLA to UNC was provided by NSF's SGER program, grant number IIS-9202424.

Section 46.3

Current Project Personnel: *Computer Science Toolbuilders:* Russ Taylor, Leandra Vicci, Steve Pizer, Paul Morris, David Marshburn, Adam Seeger, David Borland, Yonatan Fridman. *Physical Science Collaborators:* Richard Superfine, Sean Washburn, Mike Falvo, Stefan Seelecke, Stergios Papadakis, Michael Stadermann, Onejae Sul, Hakan Deniz, Adam Hall, Aarish Patel, Rohit Prakash.

Project Funding and Support: ARO funded the equipment purchased for this system through two successive DURIP awards in 1998 and 1999. The research and tool development for use in carbon nanotubes has been supported by ONR through the MURI program.

Section 46.4

Current Project Personnel: *Computer Science Toolbuilders:* Russ Taylor, Mary Whitton, Leandra Vicci, Gary Bishop, Greg Welch, Prasun Dewan, Paul Morris, David Marshburn, Kurtis Keller, Chris Weigle, Haris Fretzagias, Jonathan Robbins, Tatsuhiro Segi, Ben Wilde, Rajeev Dassani. *Physical Science Toolbuilders:* Richard Superfine, Tim O'Brien, Stefan Seelecke (NCSU), Kalpit Desai, Jay Fisher, Jeremy Cribb, Debbie Sill. *Physical Science Collaborators:* Garrett Matthews, C. William Davis, Lisa Cameron.

Project Funding and Support: NIH NCRR program, grant number 5-P41-RR02170 has supported the development throughout. NIH NIBIB has provided 5 years of support for development of a beyond-prototype system including a confocal microscope. The Cystic Fibrosis Center at UNC has provided support for equipment and personnel.

Section 46.5

Current Project Personnel: *Computer Science Toolbuilders:* Russ Taylor, Leandra Vicci, Steve Pizer, Paul Morris, Kurtis Keller, David Borland, Yonatan Fridman. *Physical Science Toolbuilders and Collaborators:* Richard Superfine, Sean Washburn, Mike Falvo, Lu-Chang Qin, Stefan Seelecke (NCSU), Stergios Papadakis.

Project Funding and Support: The W. M. Keck foundation provided the TEM. UNC Chapel Hill has provided engineering and graduate-student support.

Section 46.6

Current Project Personnel: *Computer Science Toolbuilders:* Mary Whitton, David Marshburn, Tom Hudson, Jameson Miller, Kent Rosenkoetter. *Information Science Toolbuilders:* Diane Sonnenwald, Kelly Maglaughlin. *Physical Science Collaborators:* Martin Guthold (Wake Forrest), Roger Cubicciotti (NanoMedica). *Education Toolbuilders and Collaborators:* Gail Jones, Dennis Kubasko, Michele Kloda, Tom Trettor, Atsuko Negishi.

Project Funding and Support: A supplement to the NIH NCRR program, grant number 5-P41-RR02170 has supported the development throughout. NSF's HPCC program supported network access through grant number ASC-9527192. NSF's ROLE program EDU-0087389 has supported studying the educational impact of bringing the system to K-12 schools. The

UNC Chapel Hill Chancellor's Office provided seed funding for the K–12 outreach.

References

1. Y. Adachi, T. Kumano, and K. Ogino. Intermediate Representation for Stiff Virtual Objects. In *Proc. IEEE Virtual Reality Annual International Symposium (VRAIS'95)*, pages 203–210, 1995.

2. M. W. Allersma, F. Gittes, M. J. deCastro, R. J. Stewart, and C. F. Schmidt. 2D particle tracking of NCD motility by back focal plane interferometry. *Biophys. J.* 74, pages 1074–1085, 1998.

3. N. K. B. Anshuman Razdan, B. L. Ashish Amresh, B. L. Ramakrishna, Ed Ong, and Junyi Sun. Remote control and visualization of scanning probe microscopes via the web. *Webnet Journal*, pages 20–26, 2001.

4. A. Bausch, F. Ziemann, A. A. Boulbitch, K. Jacobson, and E. Sackmann. Local measurements of viscoelastic parameters of adherent cell surfaces by magnetic bead microrheometry. *Biohpys. J.*, 75, pages 2038–2049, 1998.

5. A. R. Bausch, W. Möller, and E. Sackmann. Measurement of local viscosity and forces in living cells by magnetic tweezers. *Biophys. J.*, 76, pages 573–579, 1999.

6. F. P. Brooks, Jr., M. Ouh-Young, J. J. Batter, and P. J. Kilpatrick. Project GROPE—haptic displays for scientific visualization. In *Computer Graphics: Proceedings of SIGGRAPH '90*, pages 177–185, 1990.

7. F. P. Brooks, Jr. The computer 'scientist' as toolsmith: studies in interactive computer graphics. *Proc. International Federation of Information Processing Congress '77*, pages 625–634, 1977.

8. F. P. Brooks, Jr. The computer scientist as toolsmith II. *SIGGRAPH '94: Computer Graphics*, pages 281–287, 1996.

9. M. R. Falvo. Nanometer scale tribology of carbon nanotubes. *Centennial Meeting of the American Physical Society*, Atlanta, GA, 1999.

10. M. R. Falvo, M. Finch, V. Chi, S. Washburn, R. M. Taylor II, F. P. Brooks, Jr., and R. Superfine. The nanoManipulator: a teleoperator for manipulating materials at the nanometer scale. *Proceedings of the 5th International Symposium on the Science and Engineering of Atomically Engineered Materials*, pages 579–586, 1996.

11. M. R. Falvo, G. J. Clary, R. M. I. Taylor, V. Chi, F. P. Brooks, Jr., S. Washburn, and R. Superfine. Bending and buckling of carbon nanotubes under large strain. *Nature* 389(9):582–584, 1997.

12. M. R. Falvo, S. Washburn, R. Superfine, M. Finch, F. P. Brooks, Jr., V. Chi, and R. M. I. Taylor. Manipulation of individual viruses: friction and mechanical properties. *Biophys. J.* 72:1396–1403, 1997.

13. M. R. Falvo, G. Clary, A. Helser, S. Paulson, R. M. Taylor II, V. Chi, F. P. Brooks, Jr., S. Washburn, and R. Superfine. Nanomanipulation experiments exploring frictional and mechanical properties of carbon nanotubes. *Microscopy and Microanalysis* 4:504–512, 1999.

14. M. R. Falvo, R. M. Taylor, A. Helser, V. Chi, F. P. Brooks, Jr., S. Washburn, and R. Superfine. Nanometre-scale rolling and sliding of carbon nanotubes. *Nature* 397(6716):236–239, 1999.

15. M. R. Falvo, J. Steele, R. M. Taylor II, and R. Superfine. Gearlike rolling motion mediated by commensurate contact: carbon nanotubes on HOPG. *Phys. Rev. B* 62:10665–10667, 2000.

16. M. R. Falvo, J. Steele, R. M. Taylor II, and R. Superfine. Evidence of commensurate contact and rolling motion: AFM manipulation studies of carbon nanotubes on HOPG. *Tribology Letters* 9:73–76, 2000.

17. M. Finch, V. Chi, R. M. Taylor II, M. Falvo, S. Washburn, and R. Superfine. Surface modification tools in a virtual environment interface to a scanning probe microscope. *Computer Graphics: Proceedings of the ACM Symposium on Interactive 3D Graphics*, pages 13–18, 1995.

18. D. S. Fritsch, D. Eberly, S. M. Pizer, and M. J. McAuliffe. Stimulated cores and their applications in medical imaging. *Information Processing in Medical Imaging*, pages 365–368, 1995.

19. H. Fuchs, J. Poulton, J. Eyles, T. Greer, J. Goldfeather, D. Ellsworth, S. Molnar, G. Turk, B. Tebbs, and L. Israel. Pixel-planes 5: a heterogeneous multiprocessor graphics system using processor-enhanced memories. *SIGGRAPH '89*. ACM SIGGRAPH, 1989.

20. A. Geist. Remote control of scientific instruments: electron microscope project, http://wikihip.cern.ch/twiki/bin/view/Openlab/

21. L. P. Ghislain and W. W. Webb. Scanning force microscope based on an optical trap. *Opt. Lett.* 18:1678–1680, 1993.

22. F. Gittes and C. F. Schmidt. Thermal noise limitations on micromechanical experiments. *Eur. Biophys. J.* 27:75–81, 1998.

23. F. Gittes and C. F. Schmidt. Interference model for back focal plane displacement detection in optical tweezers. *Optics Lett.*, 1998.

24. F. Grey. OpenLab for Nanotechnology, http://wikihip.cern.ch/twiki/bin/view/Openlab/OpenLab

25. M. Guthold, M. Falvo, W. G. Matthews, S. Paulson, A. Negishi, S. Washburn, R. Superfine, F. P. Brooks, Jr., and R. M. Taylor. Investigation and modification of molecular structures with the nanomanipulator. *Journal of Molecular Graphics and Modeling* 17(3):187–197, 1999.

26. M. Guthold, M. R. Falvo, W. G. Matthews, S. Paulson, S. Washburn, D. Erie, R. Superfine, F. P. Brooks, Jr., and R. M. Taylor. Controlled manipulation of molecular samples with the nanoManipulator. *IEEE/ASME Transactions on Mechatronics*, 5(2):189–198, 2000.

27. M. Guthold and A. Helser. Remote AFM manipulation of fibrin fibers over internet2, http://www.cs.unc.edu/Research/nano/document archive/demonstrations/2000_Guthold_Bio-MEMS_movie.mov.

28. M. Guthold, J. Mullin, S. Lord, D. Erie, R. Superfine, and R. Taylor. Controlled manipulation of individual fibrin molecules. *Biophys. J.* 78:A53, 2000.

29. M. Hadida, Y. Kadobayashi, S. Lamont, H. W. Braun, B. Fink, T. Hutton, A. Kamrath, H. Mori, M. H. Ellisman. Advanced networking for telemicroscopy. *Proceedings of the 10th Annual Internet Society Conference (INET2000)*. Yokohama, Japan, 2000.

30. A. Helser. NanoManipulator, http://www.nanomanipulator.com/NanoManipulator.htm.

31. A. Helser and M. Guthold. Remote AFM manipulation of fibrin fibers over internet2. *BioMEMs & Biomedical Nanotechnology WORLD 2000*. Columbus, OH, 2000.

32. T. Hudson, D. Sonnenwald, K. Maglaughlin, M. Whitton, and R. Bergquist. Enabling distributed collaborative science: the collaborative nanoManipulator. In *Video Proceedings of ACM Conference on Computer Supported Collaborative Work 2000*, 2000.

33. T. Hudson, A. Helser, D. H. Sonnenwald, and M. C. Whitton. Managing collaboration in the distributed nanoManipulator. *IEEE Virtual Reality 2003*. Los Angeles, CA, 2003.

34. T. C. Hudson, M. C. Weigle, K. Jeffay, and R. M. Taylor II. Experiments in best-effort multimedia networking for a distributed virtual environment. *Proceedings of Multimedia Computing and Networking*, pages 88–98, 2001.

35. K. Jeffay and R. Taylor. Network support for distributed virtual environments: the telenanoManipulator. *Internet2 Conference*. Washington, DC, 1999.

36. K. Jeffay, T. Hudson, and M. Parris. Beyond audio and video: multimedia networking support for distributed, immersive virtual environment. *27th EUROMICRO Conference*, pages 300–307, 2001.

37. G. Jones. Nanoscale science education, http://www.cs.unc.edu/Research/nano/ed/

38. G. M. Jones, R. Superfine, and R. M. Taylor II. Virtual viruses. *Science Teacher* 66(7):48–50, 1999.

39. M. Jones, T. Andre, R. Superfine, R. Taylor. Learning at the nanoscale: the impact of students' use of remote microscopy on concepts of viruses, scale, and microscopy. *Journal of Research in Science Teaching* 40(3), 2003.

40. M. Jones, A. Bokinsky, T. Andre, D. Kubasko, A. Negishi, R. Taylor, R. Superfine. NanoManipulator applications in education: the impact of haptic experiences on students' attitudes and concepts. *IEEE Computer Science Haptics 2002 Symposium*, pages 295–298, 2002.

41. M. Jones, A. Bokinsky, T. Tretter, A. Negishi, D. Kubasko, R. Superfine, R. Taylor. Atomic force microscopy with touch: educational applications, *Science, Technology and Education of Microscopy*, A. Mendez-Vilas, Editor. Madrid, Formatex, 2002.

42. W. Mark, S. Randolph, M. Finch, J. V. Verth, and R. M. Taylor II. Adding force feedback to graphics systems: issues and solutions. *Computer Graphics: Proceedings of SIGGRAPH '96*, pages 447–452, 1996.

43. W. G. Matthews, A. Negishi, A. Seeger, R. Taylor, D. M. McCarty, R. J. Samulski, and R. Superfine. Elasticity and binding of adenovirus in air and in liquid. *Biophys. J.*, page A27, 1999.

44. A. Mehta, J. T. Finer, and J. A. Spudich. Reflections of a lucid dreamer: optical trap design considerations. *Methods in Cell Biology*, pages 47–69, 1998.

45. S. Paulson, M. R. Falvo, N. Snider, A. Helser, T. Hudson, A. Seeger, R. M. Taylor, R. Superfine, and S. Washburn. *In situ* resistance measurements of strained carbon nanotubes. *Applied Physics Letters*, 75(19):2936–2938, 1999.

46. S. Paulson, A. Helser, M. B. Nardelli, R. M. Taylor. II, M. Falvo, R. Superfine, and S. Washburn. Tunable resistance of a carbon nanotube–graphite interface. *Science* 290:1742–1744, 2000.

47. S. Peltier, M. Hadida, D. Levy, J. Crum, M. Wong, S. Lamont, M. H. Ellisman. Web-based telemicroscopy with the JEM-4000EX. *Jeol-*

News, Special Edition for Microscopy and Microanalysis, 2000.

48. C. S. Potter, B. Carragher, M. Ceperley, C. Conway, B. Grosser, J. Hanlon, C. Hoyer, N. Kisseberth, S. Robinson, J. Sapp, P. Soskin, D. Stone, U. Thakkar, D. Weber. Bugscope: A sustainable web-based telemicroscopy project for K-12 classrooms. *Proceedings of Microscopy and Microanalysis 99*:514–515, 1999.

49. N. Smith. Microscope and graphic imaging center: remote control of instrumentation, http://www.csuhayward.edu/SCI/sem/remote.html.

50. D. H. Sonnenwald, E. Kupstas Soo, and R. Superfine. A multi-dimensional evaluation of the nanomanipulator, a scientific collaboration system. *ACM SIGGROUP Bulletin* 20(2):46–50, 1999.

51. D. H. Sonnenwald, R. E. Bergquist, K. L. Maglaughlin, E. Kupstas-Soo, and M. C. Whitton. Designing to support collaborative scientific research across distances: the nanomanipulator environment. In *Collaborative Virtual Environments*, E. Churchill, D. Snowdon, and A. Munro (Eds.). London, Springer Verlag, 2001.

52. D. H. Sonnenwald, M. Whitton, and K. Maglaughlin. Evaluating a scientific collaboratory: results of a controlled experiment. *ACM Transactions on Computer Human Interaction*, 2003.

53. J. Sun and A. Razdan. Remote control and visualization of scanning probe microscope via web. *IEEE Second Workshop on Multimedia Tools and Applications*, pages 209–214, 1999.

54. R. Superfine, M. G. Jones, and R. Taylor. Touching viruses in a networked microscopy outreach project. In *K-12 Outreach from University Science Departments*. Raleigh, NC, North Carolina State University, pages 151–153, 2000.

55. K. Svoboda and S. M. Block. Biological applications of optical forces. *Annu. Rev. Biophys. Biomol. Struct.* 23:247–295, 1994.

56. Taylor. VRPN: A device-independent, network-transparent VR peripheral system. *Virtual Environment Research Talk Series*, 2001.

57. R. M. Taylor II, W. Robinett, V. L. Chi, F. P. Brooks, Jr., W. V. Wright, R. S. Williams, and E. J. Snyder. The nanoManipulator: A virtual-reality interface for a scanning tunneling microscope. *SIGGRAPH 93*, pages 127–134, 1993.

58. R. M. Taylor II. The nanoManipulator: a virtual-reality interface to a scanning tunneling microscope. *Computer Science*, page 139, 1994.

59. R. M. Taylor II, J. Chen, S. Okimoto, N. Llopis-Artime, V. L. Chi, F. P. Brooks, Jr., M. Falvo, S. Paulson, P. Thiansathaporn, D. Glick, S. Washburn, and R. Superfine. Pearls found on the way to the ideal interface for scanned-probe microscopes. *Proceedings of IEEE Visualization '97*, pages 467–470, 1997.

60. R. M. Taylor II. Network access to a PHANToM through VRPN. *PHANToM User's Group Workshop*, page 4. Dedham, MA, 1998.

61. R. M. Taylor II, T. C. Hudson, A. Seeger, H. Weber, J. Juliano, and A. T. Helser. VRPN: A device-independent, network-transparent VR peripheral system. *ACM Symposium on Virtual Reality Software & Technology 2001*. Banff Centre, Canada, 2001.

62. R. M. Taylor II. A 3D force microscope: manipulation/measurement in cellular systems. *CISMM External Advisory Board Meeting* Chapel Hill, NC, 2002.

63. R. M. Taylor. Nanoscale computer science. NC, *CSIT Visualization Lecture Series*. Tallahassee, FL, 2002.

64. T. Tretter and M. G. Jones. Different worlds: The importance of size. *Science Teacher*, in press.

65. VTK home page, www.vtk.org.

66. C. D. Wickens and P. Baker. Cognitive issues in virtual reality. In *Virtual Environments and Advanced Interface Design*, W. Barfield and I. Furness (Eds.). Oxford University Press: New York, 1995.

67. P. A. Williams, S. J. Papadakis, M. R. Falvo, A. M. Patel, A. Sinclair, A. Seeger, A. Helser, R. M. Taylor II, S. Washburn, and R. Superfine. Controlled placement of an individual carbon nanotube onto a microelectromechanical structure. *Applied Physics Letters* 80(14):2574–2576, 2002.

68. P. A. Williams, S. J. Papadakis, N. E. Snider, H. Deniz, M. R. Falvo, S. Washburn, R. Superfine, and R. M. Taylor II. Progress on field emission studies of individual, cantilevered multi-walled carbon nanotubes. *American Physical Society March Meeting 2002*. Indianapolis, in 2002.

69. P. A. Williams, S. J. Papadakis, A. M. Patel, M. R. Falvo, S. Washburn, and R. Superfine. Torsional response and stiffening of individual multiwalled carbon nanotubes. *Physical Review Letters*, 89(25):25502-1–25502-4, 2002.

70. A. Yamada. The remote control scanning microscope with web operation interface (Web SEM) http://www.jeoleuro.com/news/jeolnews/NEWSHOME/News%20home/25/.

47 Visualization for Computational Accelerator Physics

KWAN-LIU MA, GREG SCHUSSMAN, and BRETT WILSON
University of California at Davis

47.1 Introduction

High-energy physics is about the study of the smallest elementary particles, the building blocks of the universe. New discoveries in high-energy physics often lead to fundamental advances in other disciplines such as astronomy, biology, environmental science, materials science, and medicine. Particle accelerators are used in the laboratory by high-energy physicists to study the properties of these particles, how they are created, and how they interact under controlled conditions. Further study of fundamental particle properties requires the design and construction of new accelerators to provide higher-energy particle collisions. The design, construction, and operation of particle accelerators are very expensive and involve large-scale effort by teams of scientists and engineers from various disciplines.

Computer simulations are used in the design of next-generation particle accelerators for modeling—for example, for the acceleration and steering of particle beams. To meet design requirements and to reduce cost and technological risk in the later stages of accelerator design, construction, and operations, very high-resolution modeling is essential. The scale and complexity of these computer simulations requires the use of powerful high-performance computing platforms using software and algorithms targeted to parallel and distributed environments, as well as advanced data analysis and visualization tools that make it possible to understand the resulting terabytes of simulation data.

In this chapter, we illustrate the visualization challenges introduced by the most advanced particle accelerator simulations and describe visualization solutions derived to address these challenges. Two primary visualization problems are considered: first, the problem of visualizing very dense point data (i.e., particles), and second, visualizing very dense line data (e.g., electric and magnetic field lines). While these two problems are specific to accelerator physics data, the techniques we describe here are also suited to any other applications concerned with the visualization of particle and field line data.

47.2 Visualizing Beam Dynamics Simulations

A powerful accelerator directs beams of particles to create a large number of head-on collisions to produce new particles. The first type of particle accelerator simulations that we consider models a large number of charged particles as they move through the accelerator and respond to various forces [7]. The resulting datasets consist of hundreds of millions to billions of particles for each time-step, making it impossible to render in real time or even to fit into the memory of most PCs. One approach is to convert the particles to volumetric data representing point density and use texture-mapping hardware to render to the screen [5]. However, the size of volumes that can be efficiently visualized in this manner is limited by the amount of available texture memory, as well as the fill rate of the available hardware. In addition, high-resolution representations present challenges with regard to the available network bandwidth, disk space, and time required to process the data. Even though this approach does give good interactivity and compact data size, many fine details can be lost, especially in the very low-density zone in which scientists are most interested. A more effective approach is based on a hybrid data storage and rendering method [3], which allows scientists to visualize and explore the data at interactive rates while maintaining much of the important detail of the original data.

47.2.1   A Hybrid Visualization Technique

The hybrid technique leverages the speed of texture-based hardware volume rendering to represent large features and the flexibility of point-based rendering to represent fine details. The foundation of the hybrid method is the use of low-resolution volume rendering in the areas of low interest/detail, and the use of point-based methods to enhance or replace areas of high interest/detail. Thus, storage, transfer, and rendering resources are put to more efficient use than with volumetric or particle rendering alone.

The interactivity offered by the hybrid method makes choosing viewing parameters and transfer functions for subsequent higher-quality rendering an easy job, and the storage savings mean that the data can be more efficiently transferred from the computer that generated it to a remote computer on a scientist's desk thousands of miles away.

This approach has been tested on data obtained from several large-scale beam dynamics simulations. Each particle in these simulations consists of spatial coordinates (x, y, z) and momenta (p_x, p_y, p_z) in double-precision. The space of coordinates and momenta is called phase space. Plots of higher-dimensional projections of phase space data offer the possibility of providing improved insight into complex beam dynamics phenomena.

The primary simulation, consisting of 100 million particles, requires 5GB of storage per time-step. An additional dataset, the initial time-step of a billion-point simulation, requires 48 GB of storage. These sizes make data impractical to move and impossible for most computers to handle.

Fig. 47.1 shows a comparison of a standard volumetric rendering to a mixed (point and volumetric) rendering of the same object. The mixed rendering is able to more clearly resolve the horizontal stratifications in the right arm, and it also reveals thin horizontal stratifications in the left arm not visible in the volume rendering from this angle. Note that the bands near the edges are part of the data, not rendering artifacts.

Images for four different distributions, including (x, y, z), (x, px, y), (x, px, z), and (px, py, pz), of the data at time-step 180 are displayed in Fig. 47.2. The simulation corresponds to an intense beam propagating in a magnetic quadrupole channel. The beam propagates in the z-direction, with focusing provided in the transverse (x and y) directions.

47.2.1.1   *Point Selection Criteria*

In order to construct a hybrid representation, we must decide how to classify points (i.e., particles) as being rendered directly or simulated via volume rendering. For this dataset, the most detailed and important area to visualize is the very low-density beam halo [6]. This area poses additional problems for volumetric representation because it is thousands of times less dense than the beam core, meaning there will be difficulty computing the precise density for a given region and assigning that density one of a limited number of palette values.

Therefore, we choose points in areas of low density to render directly, and the remaining areas of high density are rendered using fast low-resolution volume rendering. This allows the fine detail of the beam halo to be accurately represented at the full data resolution while maintaining interactivity by reducing the amount of data transferred and rendered.

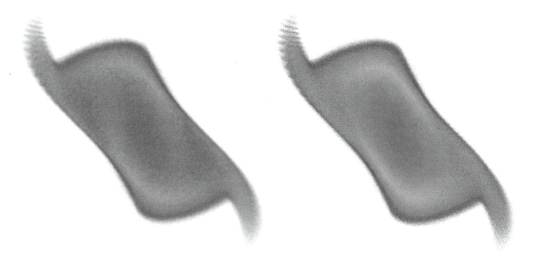

Figure 47.1   Comparison of (left) a volume rendering and (right) a mixed volume/point rendering of the phase plot (x, P_x, y) of frame 170. The volume rendering has a resolution of 256^3. The mixed rendering, with a volumetric resolution of 64^3 and 2 million points, provides more detail than the volume rendering while displaying at a much higher frame rate. (See also color insert.)

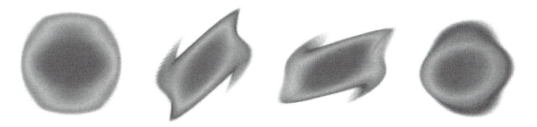

Figure 47.2   Selected distributions for time-step 180. From left to right: (x, y, z), (x, Px, y), (x, Px, z), and (Px, Py, Pz). (See also color insert.)

47.2.1.2   *Preprocessing*

The hybrid representation of the data is computed on the same parallel supercomputer at NERSC/LBL that generated the original simulation: an IBM SP RS/6000 with 2,944 processors. Preprocessing consists of two steps: partitioning and extraction. Partitioning is a one-time process that adds structure to the originally unstructured particle data. Extraction is a fast process that quickly extracts a hybrid representation with given parameters from the partitioned data.

The partitioning program organizes the unstructured point data into an octree. It is provided a time-step number, a plot type (since there are six parameters per point, various 3D plots can be generated), and a maximal subdivision level. It then reads in all the points and inserts them into an octree. The levels of subdivision of the octree are limited by the maximal subdivision level, which prevents the octree from becoming impractically large. This octree is written out to disk in two parts: one part contains all the particles of the simulation, and the other contains the octree nodes themselves. In the particle files, particles in the same octree node are grouped together, and the groups are sorted in order of increasing density. Each node in the octree then contains an offset into the particle file and the the number of particles in its group.

The partitioning program takes about 7 minutes per time-step for the 100-million-particle simulation. Since it is primarily input/output (I/O)-bound, processing time-scales linearly as the number of points increases. If the amount of data exceeds the amount of memory available on one node of the supercomputer, it can also be run on multiple nodes; in this case, the volume is divided up between nodes and particles are assigned to the corresponding nodes once they are read from disk. Since the partitioned representation contains all the data present in the original representation, it is possible (although it has not yet been implemented) to discard the original data and convert between different plot type partitionings.

The extraction program converts the partitioned data into the hybrid representation. It is given a partitioned frame and a threshold density. Particles in octree nodes below the threshold density are stored in the hybrid representation. All other points (those in the higher-density regions) are discarded (Fig. 47.3). To accomplish this, the extraction program reads in the octree and determines which nodes should contain stored points. Since the particle file is sorted in order of increasing density, all particles required for any hybrid representation are in a contiguous block at the beginning of the file. This portion of the particle data is just copied to the output; no computation is necessary for the particles, and discarded particles are never read from disk.

The threshold density parameter provided to the extraction program allows the user to balance file size and visual accuracy for a given application. A high threshold value will yield large file sizes, but larger areas of the rendering can be drawn using the more accurate point-rendering method. A low threshold value will yield smaller file sizes appropriate for viewing multiple frames simultaneously or quickly transferring over a network at the expense of having a thinner halo region representable by points. Because the extraction process is fast, different hybrid representations can be created and discarded as needed.

47.2.1.3 Viewing

A separate view program with an interactive transfer function editor is used on a desktop PC to visualize the partitioned data generated by the parallel computer. The volume transfer function (Fig. 47.3b) maps point density to color and opacity for the volume-rendered portion of the image. Typically, a step function is used to map low-density regions to 0 (fully transparent) and higher-density regions to some

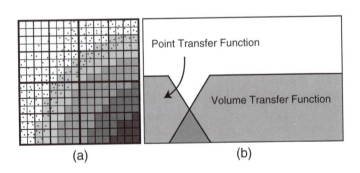

(a)   (b)

Figure 47.3 A hybrid data representation. (a) An image is created by classifying each octree node as belonging to a volume-rendered region or a point-rendered region, depending upon the transfer function for each region (the regions can overlap, as in this example). The combination of the two regions defines the output image. (b) The relationship between the two transfer functions. The two transfer functions can be edited together or separately.

low constant so that one can see inside the volume. The program also allows a ramp to transition between the high and low values, so the artificial boundary of the volume-rendered region is less visible.

The point transfer function (Fig. 47.3b) maps density to number of points rendered for the point-rendered portion of the image. Below a certain threshold density, the data is rendered as points; above that threshold, no points are drawn. Intermediate values are mapped to the fraction of points drawn. When the transfer function's value is at 0.75 for some density, for example, it means that three out of every four points are drawn for areas of that density. This allows the user to see fewer points if too many points are obscuring important features, or to make rendering faster. It also allows a smooth transition between point-rendered portions of the image and non-point-rendered portions. Fig. 47.4 displays parts of a hybrid rendering.

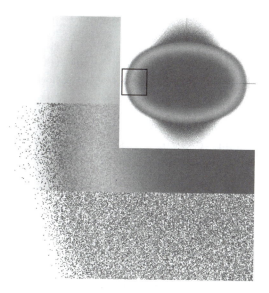

Figure 47.4 Portions of a hybrid rendering on a sphere-like (x, y, z) distribution showing (top) the volume-rendered portion, (middle) the combined hybrid rendering, and (bottom) the point-rendered portion alone. The front half of the sphere has been clipped; the points obscured by the volume rendering are on the far side. The points shown here are completely opaque so that they are more visible. (See also color insert.)

By nature, the two transfer functions are inverses of each other. Changing one results in an equal and opposite change in the other. This way, the user can change the boundary between the volume- and point-rendered regions of the image (up until the boundary specified during preprocessing, beyond which no points are available).

47.2.1.4 Results

The hybrid beam-rendering method is effective for a variety of simulation configurations and visualization requirements. The user can tailor the hybrid output to range from large, very accurate representations to small, less accurate representations (which still preserve as much interesting data as possible).

The hybrid method can produce very compact representations, allowing multiple time-steps to fit into memory. Reasonably high-quality pictures can be made with hybrid data smaller than 100MB, so a high-end PC is capable of holding around 10 time-steps in memory at once. The previewing program allows the user to step through frames using the keyboard. If a frame is already in memory, it can be displayed instantaneously; the volume texture and display lists are already loaded into video memory, or they can be quickly swapped in by the display driver. If a frame is not in memory, it is loaded from disk, a process that takes around 10 seconds for a 100MB time-step. This allows very efficient exploration of the beam's evolution over time; if the step size is small enough, individual particles can be seen moving between frames.

Fig. 47.5 displays selected frames from a simulation over 350 time-steps for the (x, y, z) distribution of the data. All frames use the same view looking down z, the beam's axis. The quadrupole magnets are alternately focusing and defocusing in the x and y directions, resulting in the fourfold symmetry seen in the figure. At this resolution, each time-step is about 500MB, allowing only two to fit into memory at once. However, hybrid frames are often smaller; these use a conservative point density threshold.

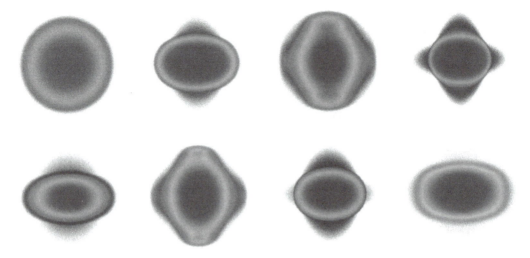

Figure 47.5 Selected time-steps from a simulation over 350 time-steps for the x-y-z distribution of the data. (Top) frames 1, 50, 100, 150. (Bottom) frames 200, 250, 300, 350. (See also color insert.)

Sometimes scientists intentionally misalign the particle beam to study how to correct beam alignment and size. Fig. 47.6 displays selected frames from an animation of a misaligned beam that could not achieve a proper focus.

In addition to scaling in the time dimension, the hybrid algorithm also scales well in terms of simulation sizes. Because the output data size does not necessarily depend on the input data size, large simulations approaching 1 billion particles can be reduced to the same size of hybrid representation as the smaller simulations. The large simulation's point-based halo region will be thinner than the smaller simulation, but that has little effect on the quality of the resulting image: regardless of the simulation size, points at the high-density halo cutoff region are typically so dense that they visually merge into a volume anyway.

One important effect that occurs in larger simulations is that the octree must be subdivided more finely where there is a high gradient. This occurs both in very large simulations and in smaller simulations with very focused beams. If a higher level of subdivision is not used, the outline of the lowest-level octree nodes will be visible at the boundary of the halo region. For low gradients, a shallower depth of octree subdivision can be used without introducing significant artifacts, saving valuable space.

For visualizing the particle beam data, volume rendering lacks the spatial resolution and the dynamic range to resolve regions with very low density, areas that may be of significant interest to researchers. Point-based rendering alone lacks the interactive speed and the ability to run on a desktop workstation that the hybrid approach provides. Furthermore, point-based rendering for low-density areas provides more room for feature enhancements. Because points are drawn dynamically, they could be drawn (in terms of color or opacity) based on some dynamically calculated property that the scientist is interested in, such as temperature or emittance. Volume-based rendering, because it is limited to precalculated data, cannot allow dynamic changes like these.

47.2.2 Summary

Large-scale beam dynamics simulations are used to understand the physics of intense beams, including the important phenomena of

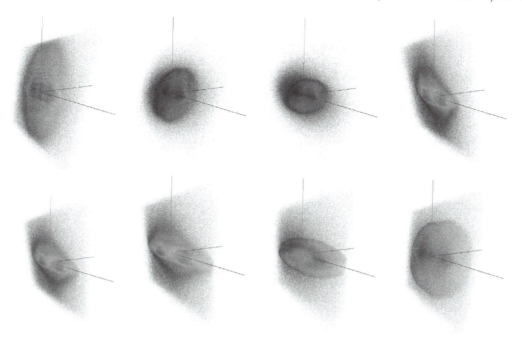

Figure 47.6 Selected frames from the visualization of a misoriented particle beam using the hybrid rendering technique. (See also color insert.)

halo formation. The simulation codes currently run on parallel computers operated at NERSC at the Lawrence Berkeley National Laboratory. Simulations have been performed with up to 500 million particles [6,7], which is approaching the number that is required for the next-generation accelerator designs. The large dynamic range of beam density involved (roughly six orders of magnitude) represents a challenge to scientific visualization and has led to the development of a new hybrid visualization algorithm. This new algorithm enables the core and the low-density halo to be visualized simultaneously with an efficiency that allows interactive data analysis.

Many visualization challenges still remain in the field of particle accelerator simulation. We need the capability to track down the origin of halo particles, which could be done by tagging particles and representing their trajectories as streamlines in animations. For the simulation of gas jets used in advanced accelerator applications, visualization is needed to view the geom-

etry of the gas nozzle and to understand how the shape of the nozzle affects the gas flow. Finally, in the design of laser- or plasma-based accelerators, visualization techniques are needed to study and optimize the structure of the fields used to accelerate the particles.

47.3   Visualizing Electromagnetic Field Data

The other type of simulation we consider is the time-domain evolution of electromagnetic fields in 3D accelerator structures, and the influence of these fields on particle behavior. For the Next Linear Collider (NLC), the design parameters are strongly affected by the need to suppress wake fields, in order to reduce or eliminate destructive particle effects such as dark current capture and breakdown. In order to produce accurate and meaningful simulation results, in terms of both prediction and verification, it is necessary to model accelerator cavities to high accuracy. One representative simulation is based on a parallel time-domain electromagnetic field solver running on unstructured hexahedral meshes. This code models the reflection and transmission properties of open structures in an accelerator design [11]. Fig. 47.7 shows a cutaway view of a 30-cell particle accelerator structure. To achieve the needed accuracy, the simulations must not proceed faster than electromagnetic information can physically flow through mesh elements. In order to satisfy this *Courant Condition* on that mesh, a very small time-step is required; simulating 100 nanoseconds in the real world requires millions of time-steps. The parallel simulation code is scalable in terms of both the the number of mesh elements and the number of particles. Fig. 47.8 shows the domain decomposition of the structure for a parallel simulation.

Each run of the simulation, for example, on a 32-node PC cluster can produce terabytes of data. The fields produced by this code can be used with another parallel code to simulate particle tracking, which can also produce very large particle path data.

A scalable solution is required for visualizing such large and complex electromagnetic fields and particle paths. The main challenge is concerned with interactively displaying a dense collection of intertwined lines in a way that shows clear spatial relationships between them, with unambiguous global or local detail. In this case, a compact representation for the field lines, combined with hardware-assisted perceptually effective rendering techniques, results in the interactivity that is key to insightful visualization.

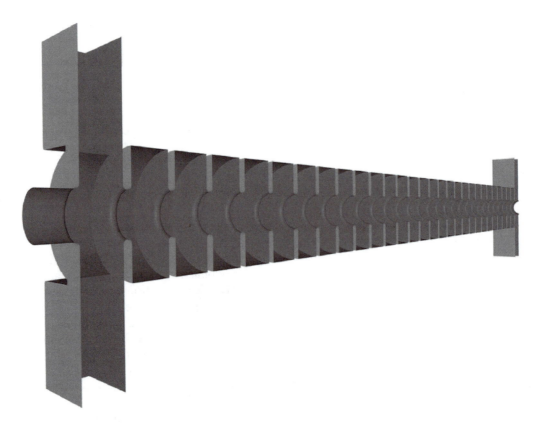

Figure 47.7   Cutaway surface visualization of a 30-cell structure.

Figure 47.8 Visualization of the domain decomposition for a parallel simulation.

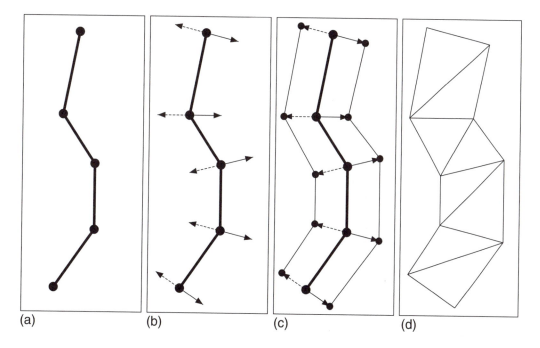

(a)   (b)   (c)   (d)

Figure 47.9 A sequence of points (a) is converted to a triangle strip by the addition of positive and negative sideways offset vectors (b) to produce new points (c), which become vertices for a triangle strip (d).

47.3.1 A Compact Representation for Field Lines

The problem of drawing lines to show the structure of vector fields has been studied extensively. Work has also been done to use alternative representation of lines like tubes and ribbons to improve perception of their structures or additional physical properties of the data. We have developed a flexible and scalable representation that we call self-orienting surfaces (SOS) for illustrating field lines [8]. This representation uses hardware texturing to efficiently display field properties using ribbons that appear essentially identical to tubes.

Each self-orienting surface is a triangle strip (or quad strip) that is constructed from a sequence of points along a curve, an associated sequence of tangent vectors, and a viewing position. Fig. 47.9 shows the process of generating such triangle strips. The triangle strip always orients toward the observer, which makes aligning a texture to the strip easy. For example, the tube-like appearance is made possible by use of hardware-accelerated bump mapping. Compared to polygonal tubes, self-orienting triangle strips are much more compact, resulting in significant savings in storage and rendering. These savings are summarized in Table 47.1. Timing was performed on a 1.4 GHz Pentium 4 PC running RedHat Linux 7.3 and using OpenGL on an nVidia GeForce 3 Ti500.

Self-orienting surfaces provide very convenient geometry for texturing. Because the strips orient themselves in a view-dependent way, the texture coordinates for moving across the strip become view-independent. Difficulties that can occur with polygonal tubes are avoided (e.g., orientation of textured glyphs, or pinched sections of the tube due to axial twisting.) Figs. 47.10a, 47.10b, and 47.10c show conventional line drawing, illuminated field lines, and streamtubes, respectively, for illustrating both the electric field and the magnetic field inside a 3-cell linear accelerator structure. As shown in Fig. 47.10d, the self-orienting surfaces rendered with hardware bump-mapping provide a nearly identical visual effect while using only a very small number of triangles, about five to six times fewer than a typical streamtube representation would require.

47.3.2 Seeding Strategy and Incremental Visualization

A key task in field line visualization is the selection of seed points for streamline integration. Much work has been done [2,4,10] for providing aesthetically pleasing streamlines through careful selection of seed points. The emphasis is generally on producing a visually uniform density of streamlines in the final image. Our approach is to select seeds so that the local density anywhere in the final distribution of field lines is approximately proportional to the local magnitude of the underlying field. When this approach is applied to electromagnetic fields, the resulting image is intuitive for physicists, because the densities of electric and magnetic flux lines are proportional to the corresponding field strength.

The implementation of our seeding strategy consists of computing a desired average number of field lines to pass through each element of the mesh. This is the average field intensity at the element's vertices multiplied by the volume of the element. These numbers are then scaled so that the sum over all elements is equal to the total maximum number of field lines to preintegrate. The algorithm consists of selecting the element that most needs an additional field line, picking a random seed point within that element, and integrating the field line from there. During integration, as each new element is visited, that element's desired number of field lines is decremented by one. This selection and integration process repeats until the total desired number of field lines for the entire mesh has been obtained. By keeping track of how many field lines already pass through the elements, disproportionately high densities of field lines are

Table 47.1 Timing results are given for different line representations: polygonal tube in immediate mode or in a display list, and hardware bump-mapped SOS. Hardware bump-mapped SOS runs an average of 1.4 times faster than polygonal tubes in a display list, and an average of 24.4 times faster than polygonal tubes computed on the fly. All times are in seconds obtained on a 1.4 GHz Pentium 4 PC with an nVidia GeForce 3 Ti500.

	150 lines	800 lines	10k lines
Polygonal tube (immediate mode)	0.445	3.001	32.8
Polygonal tube (display list)	0.027	0.173	1.82
Hardware SOS	0.019	0.124	1.28

avoided. By always choosing the element that most needs an additional field line, the images that result from rendering the first n field lines are always nearly correct in showing field line density proportional to the magnitude of the underlying field.

This incremental approach addresses the challenge in presenting extremely dense collections of field lines. Although transparency effects also help address the challenge, they are only useful up to moderate field line density. At extreme densities, transparency effects result in images qualitatively similar to those produced by direct volume rendering. Simple direct volume rendering suffers from ambiguity resulting from several factors. The perspective depth cues and lighting cues are absent, and different combinations of thickness, opacity, and coloration assigned to the data can composite to produce the same color and intensity in a final image. Furthermore, for large datasets, limited precision of the alpha channel can permit significant accumulation of quantization error when many lines of near-minimal transparency are composited. An interactive animation of our incremental approach avoids these ambiguities. By sweeping from a minimum to a maximum number of field lines, one gets a compelling sense of the structure and magnitude of the fields being built up. It is clear where the strong regions are, because sparse lines appear first, and these lines have good perspective and lighting depth cues. As more field lines are added, the strong regions become more dense and the weaker regions start to fill in. One meaningful order is to first load the lines corresponding to the highest-magnitude field regions. From there, progressively weaker field lines are loaded in. In each image, the density of field lines is approximately proportional to the magnitude of the underlying field. In this way, each image attempts to be the most accurate representation of field magnitude possible, given the number of field lines used. The set of field lines in each image in the sequence is a superset of those field lines in the preceding image.

47.3.3 Perceptually Effective Visualization

In order to better understand a large number of intertwined field lines, we cannot neglect perceptual issues. Proper use of illumination, haloing, transparency, and other visual cues can help clarify spatial relationships and reveal hidden information. In this section, we describe how to incorporate perceptually effective enhancement methods into the self-orienting surface representation to increase the information level and clarity of the picture.

47.3.3.1 Illumination

While the illuminated field lines technique [9] can help determine the shading of a field line, this technique is less effective for enabling accurate interpretation of the spatial relationships between similarly oriented adjacent or overlapping lines, as pointed out by Interrante and Grosch [1]. In particular, thin lines could look artificial because the texture does not vary sideways across the width of the lines. Our illuminated triangle strips offer not only improved visual clarity, comparable to the volume LIC approach [1], but also the critical interactivity needed for efficient data exploration. Fig. 47.10e demonstrates the effect of enhanced lighting. The enhanced lighting is hardware accelerated and carries no significant performance penalty over a single light source.

47.3.3.2 Haloing

Adding halos can clarify the spatial relationships between overlapping lines. Our self-orienting surfaces representation is superior to the illuminated field lines with halos. The illuminated-lines images do not provide a perspective depth cue, whereas the self-orienting surfaces do. At medium depth, a cross-section of the haloed lines appears as one or two black pixels on either side of a few illuminated pixels. There is an abrupt transition from the black region to the illuminated region. This can be thought of as an approximation for Phong illumination of a tube with a headlight. The diffuse and specular components remain at the middle of the cross section because that is where the surface normal vector is most closely parallel to the viewing and light vectors. Assuming a small or nonexistent ambient contribution, the cross-section edges are dark because the surface normal is orthogonal to the viewing and lighting vectors. Our self-orienting surfaces use texture to effectively capture the same surface normal vectors that a polygonal tube would have, so for self-orienting surfaces the lighting appears exact.

At first glance, comparison of the two techniques at medium depth shows little difference. However, at near depth, self-orienting surfaces look better. The perspective widening of the self-orienting surfaces provides a significant depth cue. If the widths of the haloed lines are scaled up to match, the sharp transition from black halo to illuminated region becomes very apparent. What was a reasonable approximation at several pixels wide becomes noticeably incorrect when scaled up. In contrast, self-orienting surfaces show even more clearly the Phong illumination model at work, providing a smooth and very convincing cross-section.

47.3.3.3 Transparency

For very dense line data, as displayed in Fig. 47.10f, it can be difficult to unambiguously perceive the details in a region in the interior of the 3D field. When sufficiently dense, surrounding lines can occlude the interior structures. One approach is to "cut away" the data that is not in the region of interest. While effective, as shown in Fig. 47.10g, in other cases this could take away the global context for the current region of interest. The other approach is to leave the region of interest opaque while using transparency to deemphasize the remaining data. As a result, as shown in Fig. 47.10h, the interior structures can remain clear, and the global context is not lost. Transparency in

complex scenes requires back-to-front compositing for a correct image. Depth sorting is not practical for very large data. Our approach can be coupled with the order-independent transparency technique supported by graphics hardware but would require disabling bump mapping and finer tessellation of self-orienting surfaces.

47.3.4 Results

Fig. 47.11 shows images of four selected time-steps from the simulation of the three-cell structure. The ability to animate field lines in the temporal domain is particularly valuable. For example, from these four images, scientists can examine and verify the propagation of the radio frequency (RF) waves. Storing the precomputed field lines rather than the raw data can significantly cut down the data storage and transfer requirements, making interactive interrogation of the time-varying electromagnetic field-lines data possible. The typical savings is about a factor of 25, which would allow many time-steps of electromagnetic field lines to reside in memory for interactive viewing.

Note that simulation of a 12-cell accelerator structure reaches steady state at about 40 nanoseconds, which corresponds to 326,700 time-steps. Since it would take about 80 MB of storage space to save one time-step of the electric and magnetic fields together, over 26 terabytes of storage space would be needed for the overall dataset. Storing the preintegrated field lines instead and using the seeding strategy described make it possible for us to visualize the data. For a large dataset, it is desirable to parallelize the field line calculations on a PC cluster to speed up this preprocessing task.

The sequence of images in Fig. 47.12 shows incremental loading of field lines with line transparency and color assigned according to the field strength. The key is that the scientist is allowed to interactively change these visualization and viewing parameters and then see the resulting visualization immediately.

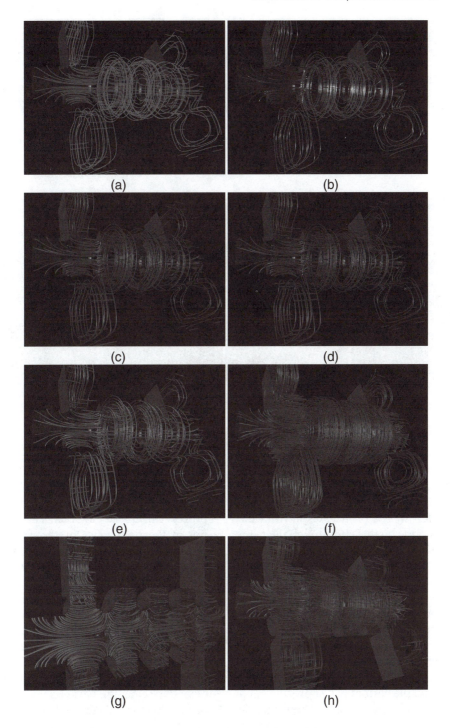

Figure 47.10   Visualization of an electromagnetic field corresponding to a section of an accelerator structure. (a) Conventional line drawing; (b) illuminated streamline technique; (c) conventional streamtube technique; (d) self-orienting surface technique; (e) with enhanced lighting; (f) dense lines; (g) cutting away; (h) use of transparency. (See also color insert.)

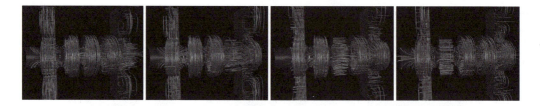

Figure 47.11   Selected time-steps that show RF waves propagating in through the input ports (left) and out through the output ports (right). (See also color insert.)

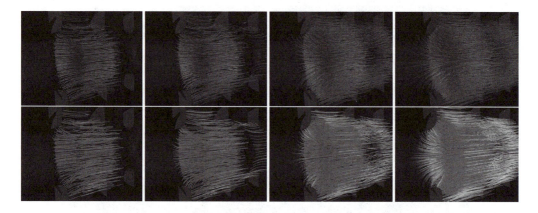

Figure 47.12   From left to right, incremental loading and rendering of electric field lines. (Top) with the line opacity proportional to local field strength. (Bottom) with enhancement using both color and transparency. (See also color insert.)

Fig. 47.13 shows a cutaway view of the inside of a three-cell particle accelerator structure to display the result of particle tracking. Power, in the form of RF waves, propagates in from the vertical input ports on the left, through the main structure, and then out through the vertical output ports on the right. Particles to be accelerated would enter from the left and exit to the right. The surface is pseudo-colored according to the rainbow, where blue represents lowest magnitude and red represents highest magnitude. The structure's surface in the top half of the image indicates the maximum magnetic field strength at any time during steady-state operation. Similarly, the bottom half shows the maximum electric field. The red paths show particle trajectories for a 50MeV field gradient, and the green paths show the trajectories for a 100MeV field gradient.

Under the influence of strong electric fields, undesirable electrons can be pulled free from the metal walls of the accelerator structure in a process called field emission. These electrons are not in the right place at the right time to be accelerated down to the end of the accelerator. However, they are still influenced by the fields within the accelerator structure. They can reach relativistic speeds, and they tend to hit the walls of the accelerator structure. This, in turn, can release more undesirable particles in a process called secondary emission. These particles can hit the surface again, releasing even more particles in a chain reaction. The field-emission and secondary-emission particles under the influence of the electromagnetic fields within the accelerator result in a phenomenon called dark current capture. If enough of these particles hit the same region on the inner surface of the accelerator

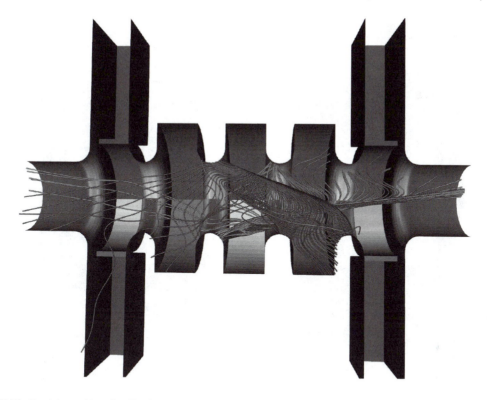

Figure 47.13 Particle tracking visualization. (See also color insert.)

structure, they can cause physical damage. In even greater numbers, these particles can produce a catastrophic event called breakdown.

The simulation that produced the data for this image released particles from a specific point on the surface repeatedly at even intervals. By symmetry, the red and green path starting points are the same. This image shows that the strength of the field gradient affects the number of particles that move from one accelerator structure into the next. It also shows that these particles, depending on when they are released, can move forward or backward along the length of the accelerator structure. Notice that the tube-like appearance effectively serves the purpose of haloing, clarifying depth relationships between the different paths, even when they are of the same color. The paths also remain distinct from the pseudo-colored background, even when the colors are similar.

Finally, Fig. 47.14 shows frames from an animation sequence, another capability made possible by efficient memory and graphics hardware utilization. Primary particles are released at evenly spaced time intervals. Under the influence of the time-varying electromagnetic fields, these particles follow the red paths. When a primary particle hits the wall of the accelerator structure, additional secondary particles can be released. The secondary particles are also influenced by the electromagnetic fields, and they follow the green paths. Not only does the rendering technique permit smooth animation, the view can be interactively changed while the animation takes place.

47.3.5 Summary

Interactive visualization is vital for understanding complex phenomena that involve particle

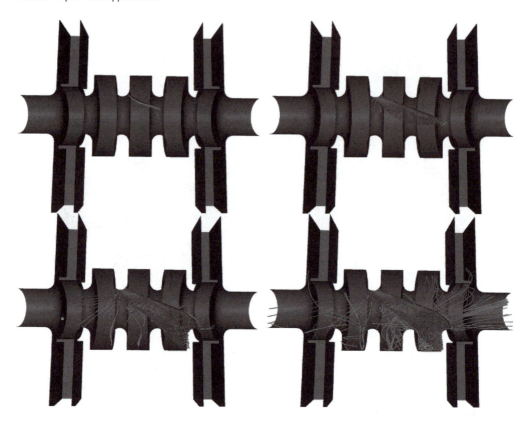

Figure 47.14 Primary particles released at evenly spaced time intervals follow the paths shown in red. When these particles hit the walls of the accelerator structure, they can liberate secondary particles, which follow the paths shown in green. (See also color insert.)

dynamics in complicated 3D geometries and the intricate interactions between particles and the structure surface. The latter process includes both field-emitted and secondary particles, and it produces plasma and x-rays. To capture all these effects in a full-scale simulation on an actual structure under realistic operating conditions presents a monumental challenge in data management, storage, manipulation (I/O), and, most importantly, visualization. The end-to-end modeling will produce terabytes of unstructured, time-varying data consisting of multiple field and particle species that have to be visualized individually and simultaneously on both local and global scales. The development of effective visualization tools to meet this challenge is of the highest priority because they are crucial to the discovery and understanding of the physics involved.

Compact graphics representations like the self-orienting surfaces can effectively cut down both storage and computational requirements without degrading image quality to enable interactive field-line and particle path visualization on a commodity PC. Further savings and interactivity can be obtained by using a wider version of the self-orienting surfaces to give the impression of the field density while rendering only a small number of self-orienting surfaces, with line density textured according to local field strength. The reduction in the number of lines that must be traced and plotted can help maintain a desirable level of interactivity.

However, for extremely dense line data, the SOS-projected line thickness can become less than about 3 pixels, which would be too wide to produce meaningful images because a mass of lines close to the observer can completely occlude any lines behind her or him. If an overview of an extremely large and dense dataset is desired (as is the case for dark current particle paths), the occlusion that helped for the relatively sparse (though still dense) datasets becomes a hindrance. What we need is a technique that allows us to scale down the widths of lines by many orders of magnitude, thereby further reducing mutual occlusion of lines within the dataset. A viable approach is to sample extremely dense line data into a fixed-size and -resolution anisotropic voxel representation that enables visualization of very dense regions of extremely thin lines that would otherwise be individually discarded by other techniques due to quantization error in graphics hardware [12]. Such a voxel representation allows arbitrarily large line datasets to be compressed enough to provide meaningful global overview of the simulation data on a single PC. SOS is then more appropriate for a close-up view of the data, while a direct rendering of the voxel representation gives an overview.

Acknowledgments

The authors are grateful for the funding support provided by the National Science Foundation and the Department of Energy, and the computing resources provided by the National Energy Research Scientific Computing Center at the Lawrence Berkeley National Laboratory. The authors would especially like to thank Pat McCormick at the Los Alamos National Laboratory, the DOE SciDAC particle accelerator project team, and the Visualization Group at the Lawrence Berkeley National Laboratory for the valuable discussions and assistance they have provided to us.

References

1. V. Interrante and C. Grosch. Strategies for effectively visualizing 3D flow with volume LIC. In *Proceedings of Visualization '97*, pages 421–424, 1997.
2. B. Jobard and W. Lefer. Creating evenly-spaced streamlines of arbitrary density. In W. Lefer and M. Grave (Eds.), *Visualization in Scientific Computing '97*. New York, Springer Verlag, pages 43–56, 1997.
3. K.-L. Ma, G. Schussman, B. Wilson, K. Kwok, J. Qiang, and R. Ryne. Advanced visualization technology for terascale particle accelerator simulations. *Proceedings of Supercomputing 2002 Conference*, 2002.
4. X. Mao, Y. Hatanaka, H. H., and A. Imamiya. Image-guided streamline placement on curvilinear grid surfaces. *Proceedings of Visualization '98 Conference*, pages 135–142, 1998.
5. P. McCormick, J. Qiang, and R. Ryne. Visualizing high-resolution accelerator physics. *IEEE Computer Graphics and Applications* 19(5): 11–13, 1999.
6. J. Qiang and R. Ryne. Beam halo studies using a 3D particle-core model. *Physical Review Special Topics—Accelerators and Beams*, 3(064201), 2000.
7. J. Qiang, R. Ryne, S. Habib, and V. Decyk. An object-oriented parallel particle-in-cell code for beam dynamics simulation in linear accelerators. *Journal of Computational Physics* 163(434), 2000.
8. G. Schussman and K.-L. Ma. Scalable self-orienting surfaces: A compact, texture-enhanced representation for interactive visualization of 3d vector fields. In *Pacific Graphics 2002*, pages 356–363, 2002.
9. D. Stalling, M. Zockler, and H.-C. Hege. Fast display of illuminated field lines. *IEEE Transactions on Visualization and Computer Graphics* 3(2):118–128, 1997.
10. G. Turk and D. Banks. Image-guided streamline placement. *Computer Graphics* 30:453–460, 1996.
11. M. Wolf, A. Guetz, and C.-K. Ng. Modeling large accelerator structures with the parallel field solver tau3p, submitted to ACES (Applied Computational Electromagnetic Society) Conference, http://scidac.nersc.gov/accelerator/pdf/wolf_aces.pdf
12. G. Schussman and K.-L. Ma. Anisotropic volume rendering for extremely dense, thin lines data. In *Proceedings of the IEEE Visualization 2004 Conference* (in press).

Index